MERCURY

Outb...
2001-09 REPAIR MANUAL
ALL 2-STROKE ENGINES

SELOC

Managing Partners	Dean F. Morgantini
	Barry L. Beck
Executive Editor	Kevin M. G. Maher, A.S.E.
Production Managers	Melinda Possinger
	Ronald Webb

Manufactured in USA
© 2009 Seloc Publishing
104 Willowbrook Lane
West Chester, PA 19382
ISBN-13: 978-089330-067-8
ISBN-10: 0-89330-067-5
7890123456 0987654321

www.selocmarine.com
1-866-SELOC55

CONTENTS

CONTENTS

SAFETY NOTICE

Proper service and repair procedures are vital to the safe, reliable operation of all marine engines, as well as the personal safety of those performing repairs. This manual outlines procedures for servicing and repairing engines and drive systems using safe, effective methods. The procedures contain many NOTES, CAUTIONS and WARNINGS which should be followed, along with standard procedures, to minimize the possibility of personal injury or improper service which could damage the vehicle or compromise its safety.

It is important to note that repair procedures and techniques, tools and parts for servicing these engines, as well as the skill and experience of the individual performing the work, vary widely. It is not possible to anticipate all of the conceivable ways or conditions under which the engine may be serviced, or to provide cautions as to all possible hazards that may result. Standard and accepted safety precautions and equipment should be used during cutting, grinding, chiseling, prying, or any other process that can cause material removal or projectiles.

Some procedures require the use of tools specially designed for a specific task. Before substituting another tool or procedure, you must be completely satisfied that neither your personal safety, nor the performance of the vessel, will be endangered. All procedures covered in this manual requiring the use of special tools will be noted at the beginning of the procedure by means of an **OEM symbol**

Additionally, any procedure requiring the use of an electronic tester or scan tool will be noted at the beginning of the procedure by means of a **DVOM symbol**

Although information in this manual is based on industry sources and is complete as possible at the time of publication, the possibility exists that some manufacturers made later changes which could not be included here. While striving for total accuracy, Seloc Publishing cannot assume responsibility for any errors, changes or omissions that may occur in the compilation of this data. We must therefore warn you to follow instructions carefully, using common sense. If you are uncertain of a procedure, seek help by inquiring with someone in your area who is familiar with these motors before proceeding.

PART NUMBERS

Part numbers listed in this reference are not recommendations by Seloc Publishing for any particular product brand name, simply iterations of the manufacturer's suggestions. They are also references that can be used with interchange manuals and aftermarket supplier catalogs to locate each brand supplier's discrete part number.

SPECIAL TOOLS

Special tools are recommended by the manufacturers to perform a specific job. Use has been kept to a minimum, but, where absolutely necessary, they are referred to in the text by the part number of the manufacturer if at all possible; and also noted at the beginning of each procedure with one of the following symbols: **OEM** or **DVOM.**

The **OEM** symbol usually denotes the need for a unique tool purposely designed to accomplish a specific task, it will also be used, less frequently, to notify the reader of the need for a tool that is not commonly found in the average tool box.

The **DVOM** symbol is used to denote the need for an electronic test tool like an ohmmeter, multi-meter or, on certain later engines, a scan tool.

These tools can be purchased, under the appropriate part number, from your local dealer or regional distributor, or an equivalent tool can be purchased locally from a tool supplier or parts outlet. Before substituting any tool for the one recommended, read the SAFETY NOTICE at the top of this page.

Providing the correct mix of service and repair procedures is an endless battle for any publisher of "How-To" information. Users range from first time do-it yourselfers to professionally trained marine technicians, and information important to one is frequently irrelevant to the other. The editors at Seloc Publishing strive to provide accurate and articulate information on all facets of marine engine repair, from the simplest procedure to the most complex. In doing this, we understand that certain procedures may be outside the capabilities of the average DIYer. Conversely we are aware that many procedures are unnecessary for a trained technician.

SKILL LEVELS

In order to provide all of our users, particularly the DIYers, with a feeling for the scope of a given procedure or task before tackling it we have included a rating system denoting the suggested skill level needed when performing a particular procedure. One of the following icons will be included at the beginning of most procedures:

① *EASY*

EASY. These procedures are aimed primarily at the DIYer and can be classified, for the most part, as basic maintenance procedures; battery, fluids, filters, plugs, etc. Although certainly valuable to any experience level, they will generally be of little importance to a technician.

② *MODERATE*

MODERATE. These procedures are suited for a DIYer with experience and a working knowledge of mechanical procedures. Even an advanced DIYer or professional technician will occasionally refer to these procedures. They will generally consist of component repair and service procedures, adjustments and minor rebuilds.

③ *DIFFICULT*

DIFFICULT. These procedures are aimed at the advanced DIYer and professional technician. They will deal with diagnostics, rebuilds and internal engine/drive components and will frequently require special tools.

④ *SKILLED*

SKILLED. These procedures are aimed at highly skilled technicians and should not be attempted without previous experience. They will usually consist of machine work, internal engine work and gear case rebuilds.

Please remember one thing when considering the above ratings—they are a guide for judging the complexity of a given procedure and are subjective in nature. Only you will know what your experience level is, and only you will know when a procedure may be outside the realm of your capability. First time DIYer, or life-long marine technician, we all approach repair and service differently so an easy procedure for one person may be a difficult procedure for another, regardless of experience level. All skill level ratings are meant to be used as a guide only! Use them to help make a judgement before undertaking a particular procedure, but by all means read through the procedure first and make your own decision—after all, our mission at Seloc is to make boat maintenance and repair easier for everyone whether you are changing the oil or rebuilding an engine. Enjoy boating!

ALL RIGHTS RESERVED

ACKNOWLEDGMENTS

Seloc Publishing expresses appreciation to the following companies who supported the production of this book:
- Marine Mechanics Institute—Orlando, FL
- Belks Marine—Holmes, PA

Thanks to John Hartung and Judy Belk of Belk's Marine for for their assistance, guidance, patience and access to some of the motors photographed for this manual.

Seloc Publishing would like to express thanks to the fine companies who participate in the production of all our books:
- Hand tools supplied by Craftsman are used during all phases of our vehicle teardown and photography.
- Many of the fine specialty tools used in our procedures were provided courtesy of Lisle Corporation.
- Much of our shop's electronic testing equipment was supplied by Universal Enterprises Inc. (UEI).

1

GENERAL INFORMATION, SAFETY AND TOOLS

HOW TO USE THIS MANUAL

This manual is designed to be a handy reference guide to maintaining and repairing your Mercury/Mariner Outboard. We strongly believe that regardless of how many or how few year's experience you may have, there is something new waiting here for you. Throughout this manual we'll refer to models by their CORPORATE PARENT'S name of Mercury, rather than constantly repeat Mercury/Mariner, but know that the information pertains to Mariner outboards of the same cubic inch and HP designation.

This manual covers the topics that a factory service manual (designed for factory trained mechanics) and a manufacturer owner's manual (designed more by lawyers than boat owners these days) covers. It will take you through the basics of maintaining and repairing your outboard, step-by-step, to help you understand what the factory trained mechanics already know by heart. By using the information in this manual, any boat owner should be able to make better informed decisions about what they need to do to maintain and enjoy their outboard.

Even if you never plan on touching a wrench (and if so, we hope that we can change your mind), this manual will still help you understand what a mechanic needs to do in order to maintain your engine.

Can You Do It?

If you are not the type who is prone to taking a wrench to something, NEVER FEAR. The procedures provided here cover topics at a level virtually anyone will be able to handle. And just the fact that you purchased this manual shows your interest in better understanding your outboard.

You may even find that maintaining your outboard yourself is preferable in most cases. From a monetary standpoint, it could also be beneficial. The money spent on hauling your boat to a marina and paying a tech to service the engine could buy you fuel for a whole weekend of boating. And, if you are really that unsure of your own mechanical abilities, at the very least you should fully understand what a marine mechanic does to your boat. You may decide that anything other than maintenance and adjustments should be performed by a mechanic (and that's your call), but if so you should know that every time you board your boat, you are placing faith in the mechanic's work and trusting him or her with your well-being, and maybe your life.

It should also be noted that in most areas a factory-trained mechanic will command a hefty hourly rate for off site service. If the tech comes to you this hourly rate is often charged from the time they leave their shop to the time that they return home. When service is performed at a boat yard, the clock usually starts when they go out to get the boat and bring it into the shop and doesn't end until it is tested and put back in the yard. The cost savings in doing the job yourself might be readily apparent at this point.

Of course, if even you're already a seasoned Do-It-Yourselfer or a Professional Technician, you'll find the procedures, specifications, special tips as well as the schematics and illustrations helpful when tackling a new job on a motor.

■ To help you decide if a task is within your skill level, procedures will often be rated using a wrench symbol in the text. When present, the number of wrenches designates how difficult we feel the procedure to be on a 1-4 scale. For more details on the wrench icon rating system, please refer to the information under Skill Levels at the beginning of this manual.

Where to Begin

Before spending any money on parts, and before removing any nuts or bolts, read through the entire procedure or topic. This will give you the overall view of what tools and supplies will be required to perform the procedure or what questions need to be answered before purchasing parts. So read ahead and plan ahead. Each operation should be approached logically and all procedures thoroughly understood before attempting any work.

Avoiding Trouble

Some procedures in this manual may require you to "label and disconnect . . . " a group of lines, hoses or wires. Don't be lulled into thinking you can remember where everything goes - you won't. If you reconnect or install a part incorrectly, the motor may operate poorly, if at all. If you hook up electrical wiring incorrectly, you may instantly learn a very expensive lesson.

A piece of masking tape, for example, placed on a hose and another on its fitting will allow you to assign your own label such as the letter "A", or a short name. As long as you remember your own code, you can reconnect the lines by matching letters or names. Do remember that tape will dissolve when saturated in some fluids (especially cleaning solvents). If a component is to be washed or cleaned, use another method of identification. A permanent felt-tipped marker can be very handy for marking metal parts; but remember that some solvents will remove permanent marker. A scribe can be used to carefully etch a small mark in some metal parts, but be sure NOT to do that on a gasket-making surface.

SAFETY is the most important thing to remember when performing maintenance or repairs. Be sure to read the information on safety in this manual.

Maintenance or Repair?

Proper maintenance is the key to long and trouble-free engine life, and the work can yield its own rewards. A properly maintained engine performs better than one that is neglected. As a conscientious boat owner, set aside a Saturday morning, at least once a month, to perform a thorough check of items that could cause problems. Keep your own personal log to jot down which services you performed, how much the parts cost you, the date, and the amount of hours on the engine at the time. Keep all receipts for parts purchased, so that they may be referred to in case of related problems or to determine operating expenses. As a do-it-yourselfer, these receipts are the only proof you have that the required maintenance was performed. In the event of a warranty problem (on new motors), these receipts can be invaluable.

It's necessary to mention the difference between maintenance and repair. Maintenance includes routine inspections, adjustments, and replacement of parts that show signs of normal wear. Maintenance compensates for wear and deterioration. Repair implies that something has broken or is not working. A need for repair is often caused by lack of maintenance.

For example: draining and refilling the gearcase oil is maintenance recommended by all manufacturers at specific intervals. Failure to do this can allow internal corrosion or damage and impair the operation of the motor, requiring expensive repairs. While no maintenance program can prevent items from breaking or wearing out, a general rule can be stated: MAINTENANCE IS CHEAPER THAN REPAIR.

Directions and Locations

◆ See Figure 1

Two basic rules should be mentioned here. First, whenever the Port side of the engine (or boat) is referred to, it is meant to specify the left side of the engine when you are sitting at the helm. Conversely, the Starboard means your right side. The Bow is the front of the boat and the Stern or Aft is the rear.

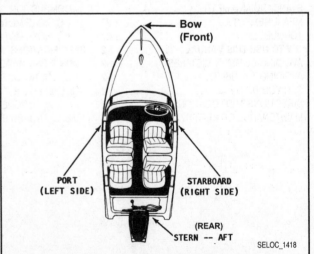

SELOC_1418

Fig. 1 Common terminology used for reference designation on boats of all size. These terms are used through out the text

Most screws and bolts are removed by turning counterclockwise, and tightened by turning clockwise. An easy way to remember this is: righty-tighty; lefty-loosey. Corny, but effective. And if you are really dense (and we have all been so at one time or another), buy a ratchet that is marked ON and OFF (like Snap-on® ratchets), or mark your own. This can be especially helpful when you are bent over backwards, upside down or otherwise turned around when working on a boat-mounted component.

Professional Help

Occasionally, there are some things when working on an outboard that are beyond the capabilities or tools of the average Do-It-Yourselfer (DIYer). This shouldn't include most of the topics of this manual, but you will have to be the judge. Some engines require special tools or a selection of special parts, even for some basic maintenance tasks.

Talk to other boaters who use the same model of engine and speak with a trusted marina to find if there is a particular system or component on your engine that is difficult to maintain.

You will have to decide for yourself where basic maintenance ends and where professional service should begin. Take your time and do your research first (starting with the information contained within) and then make your own decision. If you really don't feel comfortable with attempting a procedure, DON'T DO IT. If you've gotten into something that may be over your head, don't panic. Tuck your tail between your legs and call a marine mechanic. Marinas and independent shops will be able to finish a job for you. Your ego may be damaged, but your boat will be properly restored to its full running order. So, as long as you approach jobs slowly and carefully, you really have nothing to lose and everything to gain by doing it yourself.

On the other hand, even the most complicated repair is within the ability of a person who takes their time and follows the steps of a procedure. A rock climber doesn't run up the side of a cliff, he/she takes it one step at a time and in the end, what looked difficult or impossible was conquerable. Worry about one step at a time.

Purchasing Parts

◆ See Figures 2 and 3

When purchasing parts there are two things to consider. The first is quality and the second is to be sure to get the correct part for your engine. To get quality parts, always deal directly with a reputable retailer. To get the proper parts always refer to the model number from the information tag on your engine prior to calling the parts counter. An incorrect part can adversely affect your engine performance and fuel economy, and will cost you more money and aggravation in the end.

Just remember a tow back to shore will cost plenty. That charge is per hour from the time the towboat leaves their home port, to the time they return to their home port. Get the picture. . .$$$?

So whom should you call for parts? Well, there are many sources for the parts you will need. Where you shop for parts will be determined by what kind of parts you need, how much you want to pay, and the types of stores in your neighborhood.

Your marina can supply you with many of the common parts you require. Using a marina as your parts supplier may be handy because of location (just walk right down the dock) or because the marina specializes in your particular brand of engine. In addition, it is always a good idea to get to know the marina staff (especially the marine mechanic).

The marine parts jobber, who is usually listed in the yellow pages or whose name can be obtained from the marina, is another excellent source for parts. In addition to supplying local marinas, they also do a sizeable business in over-the-counter parts sales for the do-it-yourselfer.

Almost every boating community has one or more convenient marine chain stores. These stores often offer the best retail prices and the convenience of one-stop shopping for all your needs. Since they cater to the do-it-yourselfer, these stores are almost always open weeknights, Saturdays, and Sundays, when the jobbers are usually closed.

The lowest prices for parts are most often found in discount stores or the auto department of mass merchandisers. Parts sold here are name and private brand parts bought in huge quantities, so they can offer a competitive price. Private brand parts are made by major manufacturers and sold to large chains under a store label. And, of course, more and more large automotive parts retailers are stocking basic marine supplies.

Avoiding the Most Common Mistakes

There are 3 common mistakes in mechanical work:

1. Following the incorrect order of assembly, disassembly or adjustment. When taking something apart or putting it together, performing steps in the wrong order usually just costs you extra time; however, it CAN break something. Read the entire procedure before beginning disassembly. Perform everything in the order in which the instructions say you should, even if you can't immediately see a reason for it. When you're taking apart something that is very intricate, you might want to draw a picture of how it looks when assembled at one point in order to make sure you get everything back in its proper position. When making adjustments, perform them in the proper order; often, one adjustment affects another, and you cannot expect satisfactory results unless each adjustment is made only when it cannot be changed by subsequent adjustments.

■ **Digital cameras are handy. If you've got access to one, take pictures of intricate assemblies during the disassembly process and refer to them during assembly for tips on part orientation.**

2. Over-torquing (or under-torquing). While it is more common for over-torquing to cause damage, under-torquing may allow a fastener to vibrate loose causing serious damage. Especially when dealing with plastic and aluminum parts, pay attention to torque specifications and utilize a torque wrench in assembly. If a torque figure is not available, remember that if you are using the right tool to perform the job, you will probably not have to strain yourself to get a fastener tight enough. The pitch of most threads is so slight that the tension you put on the wrench will be multiplied many times in actual force on what you are tightening.

Fig. 2 By far the most important asset in purchasing parts is a knowledgeable and enthusiastic parts person

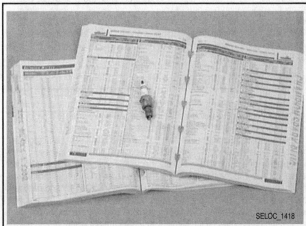

Fig. 3 Parts catalogs, giving application and part number information, are provided by manufacturers for most replacement parts

3. Cross-threading. This occurs when a part such as a bolt is screwed into a nut or casting at the wrong angle and forced. Cross-threading is more likely to occur if access is difficult. It helps to clean and lubricate fasteners, then to start threading with the part to be installed positioned straight inward. Always start a fastener, etc. with your fingers. If you encounter resistance, unscrew the part and start over again at a different angle until it can be inserted and turned several times without much effort. Keep in mind that some parts may have tapered threads, so that gentle turning will automatically bring the part you're threading to the proper angle, but only if you don't force it or resist a change in angle. Don't put a wrench on the part until it has been tightened a couple of turns by hand. If you suddenly encounter resistance, and the part has not seated fully, don't force it. Pull it back out to make sure it's clean and threading properly.

BOATING SAFETY

In 1971 Congress ordered the U.S. Coast Guard to improve recreational boating safety. In response, the Coast Guard drew up a set of regulations.

Aside from these federal regulations, there are state and local laws you must follow. These sometimes exceed the Coast Guard requirements. This section discusses only the federal laws. State and local laws are available from your local Coast Guard. As with other laws, "Ignorance of the boating laws is no excuse." The rules fall into two groups: regulations for your boat and required safety equipment on your boat.

Regulations For Your Boat

Most boats on waters within Federal jurisdiction must be registered or documented. These waters are those that provide a means of transportation between two or more states or to the sea. They also include the territorial waters of the United States.

DOCUMENTING OF VESSELS

A vessel of five or more net tons may be documented as a yacht. In this process, papers are issued by the U.S. Coast Guard as they are for large ships. Documentation is a form of national registration. The boat must be used solely for pleasure. Its owner must be a citizen of the U.S., a partnership of U.S. citizens, or a corporation controlled by U.S. citizens. The captain and other officers must also be U.S. citizens. The crew need not be.

If you document your yacht, you have the legal authority to fly the yacht ensign. You also may record bills of sale, mortgages, and other papers of title with federal authorities. Doing so gives legal notice that such instruments exist. Documentation also permits preferred status for mortgages. This gives you additional security, and it aids in financing and transfer of title. You must carry the original documentation papers aboard your vessel. Copies will not suffice.

REGISTRATION OF BOATS

If your boat is not documented, registration in the state of its principal use is probably required. If you use it mainly on an ocean, a gulf, or other similar water, register it in the state where you moor it.

If you use your boat solely for racing, it may be exempt from the requirement in your state. Some states may also exclude dinghies, while others require registration of documented vessels and non-power driven boats.

All states, except Alaska, register boats. In Alaska, the U.S. Coast Guard issues the registration numbers. If you move your vessel to a new state of principal use, a valid registration certificate is good for 60 days. You must have the registration certificate (certificate of number) aboard your vessel when it is in use. A copy will not suffice. You may be cited if you do not have the original on board.

NUMBERING OF VESSELS

A registration number is on your registration certificate. You must paint or permanently attach this number to both sides of the forward half of your boat. Do not display any other number there.

The registration number must be clearly visible. It must not be placed on the obscured underside of a flared bow. If you can't place the number on the bow, place it on the forward half of the hull. If that doesn't work, put it on the superstructure. Put the number for an inflatable boat on a bracket or fixture. Then, firmly attach it to the forward half of the boat. The letters and numbers must be plain block characters and must read from left to right. Use a space or a hyphen to separate the prefix and suffix letters from the numerals. The color of the characters must contrast with that of the background, and they must be at least three inches high.

In some states your registration is good for only one year. In others, it is good for as long as three years. Renew your registration before it expires. At that time you will receive a new decal or decals. Place them as required by state law. You should remove old decals before putting on the new ones. Some states require that you show only the current decal or decals. If your vessel is moored, it must have a current decal even if it is not in use.

If your vessel is lost, destroyed, abandoned, stolen, or transferred, you must inform the issuing authority. If you lose your certificate of number or your address changes, notify the issuing authority as soon as possible.

SALES AND TRANSFERS

Your registration number is not transferable to another boat. The number stays with the boat unless its state of principal use is changed.

HULL IDENTIFICATION NUMBER

A Hull Identification Number (HIN) is like the Vehicle Identification Number (VIN) on your car. Boats built between November 1, 1972 and July 31, 1984 have old format HINs. Since August 1, 1984 a new format has been used.

Your boat's HIN must appear in two places. If it has a transom, the primary number is on its starboard side within two inches of its top. If it does not have a transom or if it was not practical to use the transom, the number is on the starboard side. In this case, it must be within one foot of the stern and within two inches of the top of the hull side. On pontoon boats, it is on the aft crossbeam within one foot of the starboard hull attachment. Your boat also has a duplicate number in an unexposed location. This is on the boat's interior or under a fitting or item of hardware.

LENGTH OF BOATS

For some purposes, boats are classed by length. Required equipment, for example, differs with boat size. Manufacturers may measure a boat's length in several ways. Officially, though, your boat is measured along a straight line from its bow to its stern. This line is parallel to its keel.

The length does not include bowsprits, boomkins, or pulpits. Nor does it include rudders, brackets, outboard motors, outdrives, diving platforms, or other attachments.

CAPACITY INFORMATION

◆ **See Figure 4**

Manufacturers must put capacity plates on most recreational boats less than 20 feet long. Sailboats, canoes, kayaks, and inflatable boats are usually exempt. Outboard boats must display the maximum permitted horsepower of their engines. The plates must also show the allowable maximum weights of the people on board. And they must show the allowable maximum combined weights of people, engine(s), and gear. Inboards and stern drives need not show the weight of their engines on their capacity plates. The capacity plate must appear where it is clearly visible to the operator when underway. This information serves to remind you of the capacity of your boat under normal circumstances. You should ask yourself, "Is my boat loaded above its recommended capacity" and, "Is my boat overloaded for the present sea and wind conditions?" If you are stopped by a legal authority, you may be cited if you are overloaded.

CERTIFICATE OF COMPLIANCE

◆ **See Figure 4**

Manufacturers are required to put compliance plates on motorboats greater than 20 feet in length. The plates must say, "This boat," or "This equipment complies with the U. S. Coast Guard Safety Standards in effect on the date of certification." Letters and numbers can be no less than one-eighth of an inch high. At the manufacturer's option, the capacity and compliance plates may be combined.

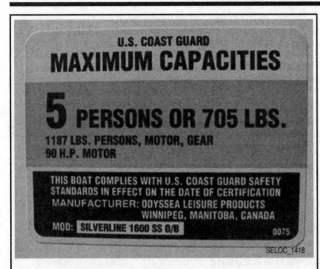

Fig. 4 A U.S. Coast Guard certification plate indicates the amount of occupants and gear appropriate for safe operation of the vessel (and allowable engine size for outboard boats)

VENTILATION

A cup of gasoline spilled in the bilge has the potential explosive power of 15 sticks of dynamite. This statement, commonly quoted over 20 years ago, may be an exaggeration; however, it illustrates a fact. Gasoline fumes in the bilge of a boat are highly explosive and a serious danger. They are heavier than air and will stay in the bilge until they are vented out.

Because of this danger, Coast Guard regulations require ventilation on many powerboats. There are several ways to supply fresh air to engine and gasoline tank compartments and to remove dangerous vapors. Whatever the choice, it must meet Coast Guard standards.

■ **The following is not intended to be a complete discussion of the regulations. It is limited to the majority of recreational vessels. Contact your local Coast Guard office for further information.**

General Precautions

Ventilation systems will not remove raw gasoline that leaks from tanks or fuel lines. If you smell gasoline fumes, you need immediate repairs. The best device for sensing gasoline fumes is your nose. Use it! If you smell gasoline in a bilge, engine compartment, or elsewhere, don't start your engine. The smaller the compartment, the less gasoline it takes to make an explosive mixture.

Ventilation for Open Boats

In open boats, gasoline vapors are dispersed by the air that moves through them. So they are exempt from ventilation requirements.

To be "open," a boat must meet certain conditions. Engine and fuel tank compartments and long narrow compartments that join them must be open to the atmosphere." This means they must have at least 15 square inches of open area for each cubic foot of net compartment volume. The open area must be in direct contact with the atmosphere. There must also be no long, unventilated spaces open to engine and fuel tank compartments into which flames could extend.

Ventilation for All Other Boats

Powered and natural ventilation are required in an enclosed compartment with a permanently installed gasoline engine that has a cranking motor. A compartment is exempt if its engine is open to the atmosphere. Diesel powered boats are also exempt.

VENTILATION SYSTEMS

There are two types of ventilation systems. One is "natural ventilation." In it, air circulates through closed spaces due to the boat's motion. The other type is "powered ventilation." In it, air is circulated by a motor-driven fan or fans.

Natural Ventilation System Requirements

A natural ventilation system has an air supply from outside the boat. The air supply may also be from a ventilated compartment or a compartment open to the atmosphere. Intake openings are required. In addition, intake ducts may be required to direct the air to appropriate compartments.

The system must also have an exhaust duct that starts in the lower third of the compartment. The exhaust opening must be into another ventilated compartment or into the atmosphere. Each supply opening and supply duct, if there is one, must be above the usual level of water in the bilge. Exhaust openings and ducts must also be above the bilge water. Openings and ducts must be at least three square inches in area or two inches in diameter. Openings should be placed so exhaust gasses do not enter the fresh air intake. Exhaust fumes must not enter cabins or other enclosed, non-ventilated spaces. The carbon monoxide gas in them is deadly.

Intake and exhaust openings must be covered by cowls or similar devices. These registers keep out rain water and water from breaking seas. Most often, intake registers face forward and exhaust openings aft. This aids the flow of air when the boat is moving or at anchor since most boats face into the wind when properly anchored.

Power Ventilation System Requirements

◆ **See Figure 6**

Powered ventilation systems must meet the standards of a natural system, but in addition, they must also have one or more exhaust blowers. The blower duct can serve as the exhaust duct for natural ventilation if fan blades do not obstruct the air flow when not powered. Openings in engine compartment, for carburetion are in addition to ventilation system requirements.

Required Safety Equipment

Coast Guard regulations require that your boat have certain equipment aboard. These requirements are minimums. Exceed them whenever you can.

TYPES OF FIRES

There are four common classes of fires:
• Class A - fires are of ordinary combustible materials such as paper or wood.
• Class B - fires involve gasoline, oil and grease.
• Class C - fires are electrical.
• Class D - fires involve ferrous metals

One of the greatest risks to boaters is fire. This is why it is so important to carry the correct number and type of extinguishers onboard.

The best fire extinguisher for most boats is a Class B extinguisher. Never use water on Class B or Class C fires, as water spreads these types of fires. Additionally, you should never use water on a Class C fire as it may cause you to be electrocuted.

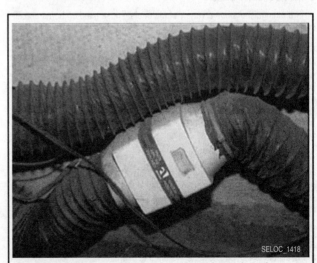

Fig. 5 Typical blower and duct system to vent fumes from the engine compartment

FIRE EXTINGUISHERS

◆ See Figure 6

If your boat meets one or more of the following conditions, you must have at least one fire extinguisher aboard. The conditions are:
• Inboard or stern drive engines
• Closed compartments under seats where portable fuel tanks can be stored
• Double bottoms not sealed together or not completely filled with flotation materials
• Closed living spaces
• Closed stowage compartments in which combustible or flammable materials are stored
• Permanently installed fuel tanks
• Boat is 26 feet or more in length.

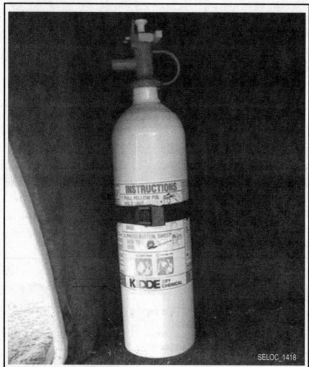

Fig. 6 An approved fire extinguisher should be mounted close to the operator for emergency use

Contents of Extinguishers

Fire extinguishers use a variety of materials. Those used on boats usually contain dry chemicals, Halon, or Carbon Dioxide (CO2). Dry chemical extinguishers contain chemical powders such as Sodium Bicarbonate - baking soda.

Carbon dioxide is a colorless and odorless gas when released from an extinguisher. It is not poisonous but caution must be used in entering compartments filled with it. It will not support life and keeps oxygen from reaching your lungs. A fire-killing concentration of Carbon Dioxide can be lethal. If you are in a compartment with a high concentration of CO2, you will have no difficulty breathing. But the air does not contain enough oxygen to support life. Unconsciousness or death can result.

Halon Extinguishers

Some fire extinguishers and "built-in" or "fixed" automatic fire extinguishing systems contain a gas called Halon. Like carbon dioxide it is colorless and odorless and will not support life. Some Halons may be toxic if inhaled.

To be accepted by the Coast Guard, a fixed Halon system must have an indicator light at the vessel's helm. A green light shows the system is ready. Red means it is being discharged or has been discharged. Warning horns are available to let you know the system has been activated. If your fixed Halon system discharges, ventilate the space thoroughly before you enter it. There are no residues from Halon but it will not support life.

Although Halon has excellent fire fighting properties; it is thought to deplete the earth's ozone layer and has not been manufactured since January 1, 1994. Halon extinguishers can be refilled from existing stocks of the gas until they are used up, but high federal excise taxes are being charged for the service. If you discontinue using your Halon extinguisher, take it to a recovery station rather than releasing the gas into the atmosphere. Compounds such as FE 241, designed to replace Halon, are now available.

Fire Extinguisher Approval

Fire extinguishers must be Coast Guard approved. Look for the approval number on the nameplate. Approved extinguishers have the following on their labels: "Marine Type USCG Approved, Size. . ., Type. . ., 162.208/," etc. In addition, to be acceptable by the Coast Guard, an extinguisher must be in serviceable condition and mounted in its bracket. An extinguisher not properly mounted in its bracket will not be considered serviceable during a Coast Guard inspection.

Care and Treatment

Make certain your extinguishers are in their stowage brackets and are not damaged. Replace cracked or broken hoses. Nozzles should be free of obstructions. Sometimes, wasps and other insects nest inside nozzles and make them inoperable. Check your extinguishers frequently. If they have pressure gauges, is the pressure within acceptable limits? Do the locking pins and sealing wires show they have not been used since recharging?

Don't try an extinguisher to test it. Its valves will not reseat properly and the remaining gas will leak out. When this happens, the extinguisher is useless.

Weigh and tag carbon dioxide and Halon extinguishers twice a year. If their weight loss exceeds 10 percent of the weight of the charge, recharge them. Check to see that they have not been used. They should have been inspected by a qualified person within the past six months, and they should have tags showing all inspection and service dates. The problem is that they can be partially discharged while appearing to be fully charged.

Some Halon extinguishers have pressure gauges the same as dry chemical extinguishers. Don't rely too heavily on the gauge. The extinguisher can be partially discharged and still show a good gauge reading. Weighing a Halon extinguisher is the only accurate way to assess its contents.

If your dry chemical extinguisher has a pressure indicator, check it frequently. Check the nozzle to see if there is powder in it. If there is, recharge it. Occasionally invert your dry chemical extinguisher and hit the base with the palm of your hand. The chemical in these extinguishers packs and cakes due to the boat's vibration and pounding. There is a difference of opinion about whether hitting the base helps, but it can't hurt. It is known that caking of the chemical powder is a major cause of failure of dry chemical extinguishers. Carry spares in excess of the minimum requirement. If you have guests aboard, make certain they know where the extinguishers are and how to use them.

Using a Fire Extinguisher

A fire extinguisher usually has a device to keep it from being discharged accidentally. This is a metal or plastic pin or loop. If you need to use your extinguisher, take it from its bracket. Remove the pin or the loop and point the nozzle at the base of the flames. Now, squeeze the handle, and discharge the extinguisher's contents while sweeping from side to side. Recharge a used extinguisher as soon as possible.

If you are using a Halon or carbon dioxide extinguisher, keep your hands away from the discharge. The rapidly expanding gas will freeze them. If your fire extinguisher has a horn, hold it by its handle.

Legal Requirements for Extinguishers

You must carry fire extinguishers as defined by Coast Guard regulations. They must be firmly mounted in their brackets and immediately accessible.

A motorboat less than 26 feet long must have at least one approved hand-portable, Type B-1 extinguisher. If the boat has an approved fixed fire extinguishing system, you are not required to have the Type B-1 extinguisher. Also, if your boat is less than 26 feet long, is propelled by an outboard motor, or motors, and does not have any of the first six conditions described at the beginning of this section, it is not required to have an extinguisher. Even so, it's a good idea to have one, especially if a nearby boat catches fire, or if a fire occurs at a fuel dock.

A motorboat 26 feet to less than 40 feet long, must have at least two Type B-1 approved hand-portable extinguishers. It can, instead, have at least one Coast Guard approved Type B-2. If you have an approved fire extinguishing system, only one Type B-1 is required.

A motorboat 40 to 65 feet long must have at least three Type B-1 approved portable extinguishers. It may have, instead, at least one Type B-1 plus a Type B-2. If there is an approved fixed fire extinguishing system, two Type B-1 or one Type B-2 is required.

WARNING SYSTEM

Various devices are available to alert you to danger. These include fire, smoke, gasoline fumes, and carbon monoxide detectors. If your boat has a galley, it should have a smoke detector. Where possible, use wired detectors. Household batteries often corrode rapidly on a boat.

There are many ways in which carbon monoxide (a by-product of the combustion that occurs in an engine) can enter your boat. You can't see, smell, or taste carbon monoxide gas, but it is lethal. As little as one part in 10,000 parts of air can bring on a headache. The symptoms of carbon monoxide poisoning - headaches, dizziness, and nausea - are like seasickness. By the time you realize what is happening to you, it may be too late to take action. If you have enclosed living spaces on your boat, protect yourself with a detector.

PERSONAL FLOTATION DEVICES

Personal Flotation Devices (PFDs) are commonly called life preservers or life jackets. You can get them in a variety of types and sizes. They vary with their intended uses. To be acceptable, PFDs must be Coast Guard approved.

Type I PFDs

A Type I life jacket is also called an offshore life jacket. Type I life jackets will turn most unconscious people from facedown to a vertical or slightly backward position. The adult size gives a minimum of 22 pounds of buoyancy. The child size has at least 11 pounds. Type I jackets provide more protection to their wearers than any other type of life jacket. Type I life jackets are bulkier and less comfortable than other types. Furthermore, there are only two sizes, one for children and one for adults.

Type I life jackets will keep their wearers afloat for extended periods in rough water. They are recommended for offshore cruising where a delayed rescue is probable.

Type II PFDs

A Type II life jacket is also called a near-shore buoyant vest. It is an approved, wearable device. Type II life jackets will turn some unconscious people from facedown to vertical or slightly backward positions. The adult size gives at least 15.5 pounds of buoyancy. The medium child size has a minimum of 11 pounds. And the small child and infant sizes give seven pounds. A Type II life jacket is more comfortable than a Type I but it does not have as much buoyancy. It is not recommended for long hours in rough water. Because of this, Type IIs are recommended for inshore and inland cruising on calm water. Use them only where there is a good chance of fast rescue.

Type III PFDs

◆ See Figure 7

Type III life jackets or marine buoyant devices are also known as flotation aids. Like Type IIs, they are designed for calm inland or close offshore water where there is a good chance of fast rescue. Their minimum buoyancy is 15.5 pounds. They will **NOT** turn their wearers face up.

Type III devices are usually worn where freedom of movement is necessary. Thus, they are used for water skiing, small boat sailing, and fishing among other activities. They are available as vests and flotation coats. Flotation coats are useful in cold weather. Type IIIs come in many sizes from small child through large adult.

Life jackets come in a variety of colors and patterns - red, blue, green, camouflage, and cartoon characters. From purely a safety standpoint, the best color is bright orange. It is easier to see in the water, especially if the water is rough.

Type IV PFDs

◆ See Figures 8 and 9

Type IV ring life buoys, buoyant cushions and horseshoe buoys are Coast Guard approved devices called throwables. They are made to be thrown to people in the water, and should not be worn. Type IV cushions are often used as seat cushions. But, keep in mind that cushions are hard to hold onto in the water, thus, they do not afford as much protection as wearable life jackets.

The straps on buoyant cushions are for you to hold onto either in the water or when throwing them, they are **NOT** for your arms. A cushion should never be worn on your back, as it will turn you face down in the water.

Type IV throwables are not designed as personal flotation devices for unconscious people, non-swimmers, or children. Use them only in emergencies. They should not be used for, long periods in rough water.

Ring life buoys come in 18, 20, 24, and 30 in. diameter sizes. They usually have grab lines, but you will need to attach about 60 feet of polypropylene line to the grab rope to aid in retrieving someone in the water. If you throw a ring, be careful not to hit the person. Ring buoys can knock people unconscious

Type V PFDs

Type V PFDs are of two kinds, special use devices and hybrids. Special use devices include boardsailing vests, deck suits, work vests, and others. They are approved only for the special uses or conditions indicated on their labels. Each is designed and intended for the particular application shown on its label. They do not meet legal requirements for general use aboard recreational boats.

Hybrid life jackets are inflatable devices with some built-in buoyancy provided by plastic foam or kapok. They can be inflated orally or by cylinders of compressed gas to give additional buoyancy. In some hybrids the gas is released manually. In others it is released automatically when the life jacket is immersed in water.

The inherent buoyancy of a hybrid may be insufficient to float a person unless it is inflated. The only way to find this out is for the user to try it in the water. Because of its limited buoyancy when deflated, a hybrid is recommended for use by a non-swimmer only if it is worn with enough inflation to float the wearer.

If they are to count against the legal requirement for the number of life jackets you must carry, hybrids manufactured before February 8, 1995 must be worn whenever a boat is underway and the wearer must not go below decks or in an enclosed space. To find out if your Type V hybrid must be worn to satisfy the legal requirement, read its label. If its use is restricted it will say, "REQUIRED TO BE WORN" in capital letters.

Hybrids cost more than other life jackets, but this factor must be weighed against the fact that they are more comfortable than Types I, II or III life jackets. Because of their greater comfort, their owners are more likely to wear them than are the owners of Type I, II or III life jackets.

Fig. 7 Type III PFDs are recommended for inshore/inland use on calm water (where there is a good chance of fast rescue)

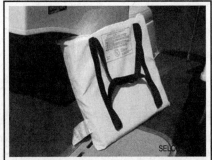

Fig. 8 Type IV buoyant cushions are thrown to people in the water. If you can squeeze air out of the cushion, it should be replaced

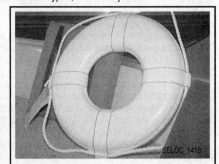

Fig. 9 Type IV throwables, such as this ring life buoy, are not designed for unconscious people, non-swimmers, or children

The Coast Guard has determined that improved, less costly hybrids can save lives since they will be bought and used more frequently. For these reasons, a new federal regulation was adopted effective February 8, 1995. The regulation increases both the deflated and inflated buoyancy's of hybrids, makes them available in a greater variety of sizes and types, and reduces their costs by reducing production costs.

Even though it may not be required, the wearing of a hybrid or a life jacket is encouraged whenever a vessel is underway. Like life jackets, hybrids are now available in three types. To meet legal requirements, a Type I hybrid can be substituted for a Type I life jacket. Similarly Type II and III hybrids can be substituted for Type II and Type III life jackets. A Type I hybrid, when inflated, will turn most unconscious people from facedown to vertical or slightly backward positions just like a Type I life jacket. Type II and III hybrids function like Type II and III life jackets. If you purchase a new hybrid, it should have an owner's manual attached that describes its life jacket type and its deflated and inflated buoyancys. It warns you that it may have to be inflated to float you. The manual also tells you how to don the life jacket and how to inflate it. It also tells you how to change its inflation mechanism, recommended testing exercises, and inspection or maintenance procedures. The manual also tells you why you need a life jacket and why you should wear it. A new hybrid must be packaged with at least three gas cartridges. One of these may already be loaded into the inflation mechanism. Likewise, if it has an automatic inflation mechanism, it must be packaged with at least three of these water sensitive elements. One of these elements may be installed.

Legal Requirements

A Coast Guard approved life jacket must show the manufacturer's name and approval number. Most are marked as Type I, II, III, IV or V. All of the newer hybrids are marked for type.

You are required to carry at least one wearable life jacket or hybrid for each person on board your recreational vessel. If your vessel is 16 feet or more in length and is not a canoe or a kayak, you must also have at least one Type IV on board. These requirements apply to all recreational vessels that are propelled or controlled by machinery, sails, oars, paddles, poles, or another vessel. Sailboards are not required to carry life jackets.

You can substitute an older Type V hybrid for any required Type I, II or III life jacket provided:

1. Its approval label shows it is approved for the activity the vessel is engaged in

2. It's approved as a substitute for a life jacket of the type required on the vessel

3. It's used as required on the labels

and

4. It's used in accordance with any requirements in its owner's manual (if the approval label makes reference to such a manual.)

A water skier being towed is considered to be on board the vessel when judging compliance with legal requirements.

You are required to keep your Type I, II or III life jackets or equivalent hybrids readily accessible, which means you must be able to reach out and get them when needed. All life jackets must be in good, serviceable condition.

General Considerations

The proper use of a life jacket requires the wearer to know how it will perform. You can gain this knowledge only through experience. Each person on your boat should be assigned a life jacket. Next, it should be fitted to the person who will wear it. Only then can you be sure that it will be ready for use in an emergency. This advice is good even if the water is calm, and you intend to boat near shore.

Boats can sink fast. There may be no time to look around for a life jacket. Fitting one on you in the water is almost impossible. Most drownings occur in inland waters within a few feet of safety. Most victims had life jackets, but they weren't wearing them.

Keeping life jackets in the plastic covers they came wrapped in, and in a cabin, assure that they will stay clean and unfaded. But this is no way to keep them when you are on the water. When you need a life jacket it must be readily accessible and adjusted to fit you. You can't spend time hunting for it or learning how to fit it.

There is no substitute for the experience of entering the water while wearing a life jacket. Children, especially, need practice. If possible, give your guests this experience. Tell them they should keep their arms to their sides when jumping in to keep the life jacket from riding up. Let them jump in and see how the life jacket responds. Is it adjusted so it does not ride up? Is it the proper size? Are all straps snug? Are children's life jackets the right

sizes for them? Are they adjusted properly? If a child's life jacket fits correctly, you can lift the child by the jacket's shoulder straps and the child's chin and ears will not slip through. Non-swimmers, children, handicapped persons, elderly persons and even pets should always wear life jackets when they are aboard. Many states require that everyone aboard wear them in hazardous waters.

Inspect your lifesaving equipment from time to time. Leave any questionable or unsatisfactory equipment on shore. An emergency is no time for you to conduct an inspection.

Indelibly mark your life jackets with your vessel's name, number, and calling port. This can be important in a search and rescue effort. It could help concentrate effort where it will do the most good.

Care of Life Jackets

Given reasonable care, life jackets last many years. Thoroughly dry them before putting them away. Stow them in dry, well-ventilated places. Avoid the bottoms of lockers and deck storage boxes where moisture may collect. Air and dry them frequently.

Life jackets should not be tossed about or used as fenders or cushions. Many contain kapok or fibrous glass material enclosed in plastic bags. The bags can rupture and are then unserviceable. Squeeze your life jacket gently. Does air leak out? If so, water can leak in and it will no longer be safe to use. Cut it up so no one will use it, and throw it away. The covers of some life jackets are made of nylon or polyester. These materials are plastics. Like many plastics, they break down after extended exposure to the ultraviolet light in sunlight. This process may be more rapid when the materials are dyed with bright dyes such as "neon" shades.

Ripped and badly faded fabrics are clues that the covering of your life jacket is deteriorating. A simple test is to pinch the fabric between your thumbs and forefingers. Now try to tear the fabric. If it can be torn, it should definitely be destroyed and discarded. Compare the colors in protected places to those exposed to the sun. If the colors have faded, the materials have been weakened. A life jacket covered in fabric should ordinarily last several boating seasons with normal use. A life jacket used every day in direct sunlight should probably be replaced more often.

SOUND PRODUCING DEVICES

All boats are required to carry some means of making an efficient sound signal. Devices for making the whistle or horn noises required by the Navigation Rules must be capable of a four-second blast. The blast should be audible for at least one-half mile. Athletic whistles are not acceptable on boats 12 meters or longer. Use caution with athletic whistles. When wet, some of them come apart and loose their "pea." When this happens, they are useless.

If your vessel is 12 meters long and less than 20 meters, you must have a power whistle (or power horn) and a bell on board. The bell must be in operating condition and have a minimum diameter of at least 200mm (7.9 in.) at its mouth.

VISUAL DISTRESS SIGNALS

◆ See Figure 10

Visual Distress Signals (VDS) attract attention to your vessel if you need help. They also help to guide searchers in search and rescue situations. Be sure you have the right types, and learn how to use them properly.

It is illegal to fire flares improperly. In addition, they cost the Coast Guard and its Auxiliary many wasted hours in fruitless searches. If you signal a distress with flares and then someone helps you, please let the Coast Guard or the appropriate Search And Rescue (SAR) Agency know so the distress report will be canceled.

Recreational boats less than 16 feet long must carry visual distress signals on coastal waters at night. Coastal waters are:

• The ocean (territorial sea)
• The Great Lakes
• Bays or sounds that empty into oceans
• Rivers over two miles across at their mouths upstream to where they narrow to two miles.

Recreational boats 16 feet or longer must carry VDS at all times on coastal waters. The same requirement applies to boats carrying six or fewer passengers for hire. Open sailboats less than 26 feet long without engines are exempt in the daytime as are manually propelled boats. Also exempt are boats in organized races, regattas, parades, etc. Boats owned in the United States and operating on the high seas must be equipped with VDS.

A wide variety of signaling devices meet Coast Guard regulations. For pyrotechnic devices, a minimum of three must be carried. Any combination can be carried as long as it adds up to at least three signals for day use and at least three signals for night use. Three day/night signals meet both requirements. If possible, carry more than the legal requirement.

■ **The American flag flying upside down is a commonly recognized distress signal. It is not recognized in the Coast Guard regulations, though. In an emergency, your efforts would probably be better used in more effective signaling methods.**

Types of VDS

VDS are divided into two groups; daytime and nighttime use. Each of these groups is subdivided into pyrotechnic and non-pyrotechnic devices.

Daytime Non-Pyrotechnic Signals

A bright orange flag with a black square over a black circle is the simplest VDS. It is usable, of course, only in daylight. It has the advantage of being a continuous signal. A mirror can be used to good advantage on sunny days. It can attract the attention of other boaters and of aircraft from great distances. Mirrors are available with holes in their centers to aid in "aiming." In the absence of a mirror, any shiny object can be used. When another boat is in sight, an effective VDS is to extend your arms from your sides and move them up and down. Do it slowly. If you do it too fast the other people may think you are just being friendly. This simple gesture is seldom misunderstood, and requires no equipment.

Daytime Pyrotechnic Devices

Orange smoke is a useful daytime signal. Hand-held or floating smoke flares are very effective in attracting attention from aircraft. Smoke flares don't last long, and are not very effective in high wind or poor visibility. As with other pyrotechnic devices, use them only when you know there is a possibility that someone will see the display.

To be usable, smoke flares must be kept dry. Keep them in airtight containers and store them in dry places. If the "striker" is damp, dry it out before trying to ignite the device. Some pyrotechnic devices require a forceful "strike" to ignite them.

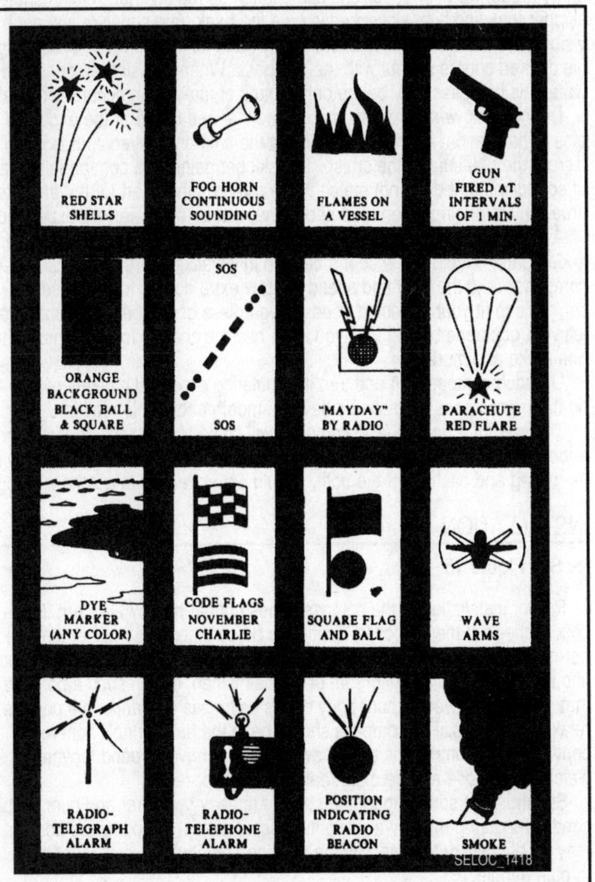

All hand-held pyrotechnic devices may produce hot ashes or slag when burning. Hold them over the side of your boat in such a way that they do not burn your hand or drip into your boat.

Nighttime Non-Pyrotechnic Signals

An electric distress light is available. This light automatically flashes the international Morse code SOS distress signal (••• --- •••). Flashed four to six times a minute, it is an unmistakable distress signal. It must show that it is approved by the Coast Guard. Be sure the batteries are fresh. Dated batteries give assurance that they are current.

Under the Inland Navigation Rules, a high intensity white light flashing 50-70 times per minute is a distress signal. Therefore, use strobe lights on inland waters only for distress signals.

Nighttime Pyrotechnic Devices

◆ See Figure 11

Aerial and hand-held flares can be used at night or in the daytime. Obviously, they are more effective at night.

Currently, the serviceable life of a pyrotechnic device is rated at 42 months from its date of manufacture. Pyrotechnic devices are expensive. Look at their dates before you buy them. Buy them with as much time remaining as possible.

Like smoke flares, aerial and hand-held flares may fail to work if they have been damaged or abused. They will not function if they are or have been wet. Store them in dry, airtight containers in dry places. But store them where they are readily accessible.

Aerial VDSs, depending on their type and the conditions they are used in, may not go very high. Again, use them only when there is a good chance they will be seen.

A serious disadvantage of aerial flares is that they burn for only a short time; most burn for less than 10 seconds. Most parachute flares burn for less than 45 seconds. If you use a VDS in an emergency, do so carefully. Hold hand-held flares over the side of the boat when in use. Never use a road hazard flare on a boat; it can easily start a fire. Marine type flares are specifically designed to lessen risk, but they still must be used carefully.

Aerial flares should be given the same respect as firearms since they are firearms! Never point them at another person. Don't allow children to play with them or around them. When you fire one, face away from the wind. Aim it downwind and upward at an angle of about 60 degrees to the horizon. If there is a strong wind, aim it somewhat more vertically. Never fire it straight up. Before you discharge a flare pistol, check for overhead obstructions that might be damaged by the flare. An obstruction might deflect the flare to where it will cause injury or damage.

RED STAR SHELLS	FOG HORN CONTINUOUS SOUNDING	FLAMES ON A VESSEL	GUN FIRED AT INTERVALS OF 1 MIN.
ORANGE BACKGROUND BLACK BALL & SQUARE	SOS	"MAYDAY" BY RADIO	PARACHUTE RED FLARE
DYE MARKER (ANY COLOR)	CODE FLAGS NOVEMBER CHARLIE	SQUARE FLAG AND BALL	WAVE ARMS
RADIO-TELEGRAPH ALARM	RADIO-TELEPHONE ALARM	POSITION INDICATING RADIO BEACON	SMOKE

Fig. 10 Internationally accepted distress signals

Fig. 11 Moisture-protected flares should be carried onboard any vessel for use as a distress signal

Disposal of VDS

Keep outdated flares when you get new ones. They do not meet legal requirements, but you might need them sometime, and they may work. It is illegal to fire a VDS on federal navigable waters unless an emergency exists. Many states have similar laws.

Emergency Position Indicating Radio Beacon (EPIRB)

There is no requirement for recreational boats to have EPIRBs. Some commercial and fishing vessels, though, must have them if they operate beyond the three-mile limit. Vessels carrying six or fewer passengers for hire must have EPIRBs under some circumstances when operating beyond the three-mile limit. If you boat in a remote area or offshore, you should have an EPIRB. An EPIRB is a small (about 6 to 20 in. high), battery-powered, radio transmitting buoy-like device. It is a radio transmitter and requires a license or an endorsement on your radio station license by the Federal Communications Commission (FCC). EPIRBs are either automatically activated by being immersed in water or manually by a switch.

Courtesy Marine Examinations

One of the roles of the Coast Guard Auxiliary is to promote recreational boating safety. This is why they conduct thousands of Courtesy Marine Examinations each year. The auxiliarists who do these examinations are well-trained and knowledgeable in the field.

These examinations are free and done only at the consent of boat owners. To pass the examination, a vessel must satisfy federal equipment requirements and certain additional requirements of the coast guard auxiliary. If your vessel does not pass the Courtesy Marine Examination, no report of the failure is made. Instead, you will be told what you need to correct the deficiencies. The examiner will return at your convenience to redo the examination.

If your vessel qualifies, you will be awarded a safety decal. The decal does not carry any special privileges, it simply attests to your interest in safe boating.

BOATING EQUIPMENT (NOT REQUIRED BUT RECOMMENDED)

Although not required by law, there are other pieces of equipment that are good to have onboard.

Oar/Paddle (Second Means of Propulsion)

All boats less than 16 feet long should carry a second means of propulsion. A paddle or oar can come in handy at times. For most small boats, a spare trolling or outboard motor is an excellent idea. If you carry a spare motor, it should have its own fuel tank and starting power. If you use an electric trolling motor, it should have its own battery.

Bailing Devices

All boats should carry at least one effective manual bailing device in addition to any installed electric bilge pump. This can be a bucket, can, scoop, hand-operated pump, etc. If your battery "goes dead" it will not operate your electric pump.

First Aid Kit

◆ See Figure 12

All boats should carry a first aid kit. It should contain adhesive bandages, gauze, adhesive tape, antiseptic, aspirin, etc. Check your first aid kit from time to time. Replace anything that is outdated. It is to your advantage to know how to use your first aid kit. Another good idea would be to take a Red Cross first aid course.

Anchors

◆ See Figure 13

All boats should have anchors. Choose one of suitable size for your boat. Better still, have two anchors of different sizes. Use the smaller one in calm water or when anchoring for a short time to fish or eat. Use the larger one when the water is rougher or for overnight anchoring.

Carry enough anchor line, of suitable size, for your boat and the waters in which you will operate. If your engine fails you, the first thing you usually should do is lower your anchor. This is good advice in shallow water where you may be driven aground by the wind or water. It is also good advice in windy weather or rough water, as the anchor, when properly affixed, will usually hold your bow into the waves.

VHF-FM Radio

Your best means of summoning help in an emergency or in case of a breakdown is a VHF-FM radio. You can use it to get advice or assistance from the Coast Guard. In the event of a serious illness or injury aboard your boat, the Coast Guard can have emergency medical equipment meet you ashore.

■ **Although the VHF radio is the best way to get help, in this day and age, cell phones are a good backup source, especially for boaters on inland waters. You probably already know where you get a signal when boating, keep the phone charged, handy and off (so it doesn't bother you when boating right?). Keep phone numbers for a local dockmaster, coast guard, tow service or maritime police unit handy on board or stored in your phone directory.**

Compass

SELECTION

◆ See Figure 14

The safety of the boat and her crew may depend on her compass. In many areas, weather conditions can change so rapidly that, within minutes, a skipper may find himself socked in by a fog bank, rain squall or just poor visibility. Under these conditions, he may have no other means of keeping to his desired course except with the compass. When crossing an open body of water, his compass may be the only means of making an accurate landfall.

During thick weather when you can neither see nor hear the expected aids to navigation, attempting to run out the time on a given course can disrupt the pleasure of the cruise. The skipper gains little comfort in a chain of soundings that does not match those given on the chart for the expected area. Any stranding, even for a short time, can be an unnerving experience.

A pilot will not knowingly accept a cheap parachute. By the same token, a good boater should not accept a bargain in lifejackets, fire extinguishers, or compass. Take the time and spend the few extra dollars to purchase a compass to fit your expected needs. Regardless of what the salesman may tell you, postpone buying until you have had the chance to check more than one make and model.

Lift each compass, tilt and turn it, simulating expected motions of the boat. The compass card should have a smooth and stable reaction.

The card of a good quality compass will come to rest without oscillations about the lubber's line. Reasonable movement in your hand, comparable to the rolling and pitching of the boat, should not materially affect the reading.

INSTALLATION

◆ See Figure 15

Proper installation of the compass does not happen by accident. Make a critical check of the proposed location to be sure compass placement will permit the helmsman to use it with comfort and accuracy. First, the compass should be placed directly in front of the helmsman, and in such a position that it can be viewed without body stress as he sits or stands in a posture of relaxed alertness. The compass should be in the helmsman's zone of comfort. If the compass is too far away, he may have to bend forward to watch it; too close and he must rear backward for relief.

Second, give some thought to comfort in heavy weather and poor visibility conditions during the day and night. In some cases, the compass position may be partially determined by the location of the wheel, shift lever and throttle handle.

Fig. 12 Always carry an adequately stocked first aid kit on board for the safety of the crew and guests

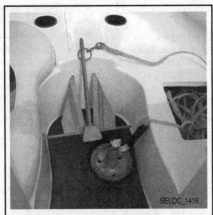

Fig. 13 Choose an anchor of sufficient weight to secure the boat without dragging

Fig. 14 Don't hesitate to spend a few extra dollars for a reliable compass

Third, inspect the compass site to be sure the instrument will be at least two feet from any engine indicators, bilge vapor detectors, magnetic instruments, or any steel or iron objects. If the compass cannot be placed at least two feet (six feet would be better but on a small craft, let's get real two feet is usually pushing it) from any one of these influences, then either the compass or the other object must be moved, if first order accuracy is to be expected.

Once the compass location appears to be satisfactory, give the compass a test before installation. Hidden influences may be concealed under the cabin top, forward of the cabin aft bulkhead, within the cockpit ceiling, or in a wood-covered stanchion.

Move the compass around in the area of the proposed location. Keep an eye on the card. A magnetic influence is the only thing that will make the card turn. You can quickly find any such influence with the compass. If the influence cannot be moved away or replaced by one of non-magnetic material, test to determine whether it is merely magnetic, a small piece of iron or steel, or some magnetized steel. Bring the north pole of the compass near the object, then shift and bring the south pole near it. Both the north and south poles will be attracted if the compass is demagnetized. If the object attracts one pole and repels the other, then the compass is

magnetized. If your compass needs to be demagnetized, take it to a shop equipped to do the job PROPERLY.

After you have moved the compass around in the proposed mounting area, hold it down or tape it in position. Test everything you feel might affect the compass and cause a deviation from a true reading. Rotate the wheel from hard over-to-hard over. Switch on and off all the lights, radios, radio direction finder, radio telephone, depth finder and, if installed, the shipboard intercom. Sound the electric whistle, turn on the windshield wipers, start the engine (with water circulating through the engine), work the throttle, and move the gear shift lever. If the boat has an auxiliary generator, start it.

If the card moves during any one of these tests, the compass should be relocated. Naturally, if something like the windshield wipers causes a slight deviation, it may be necessary for you to make a different deviation table to use only when certain pieces of equipment are operating. Bear in mind, following a course that is off only a degree or two for several hours can make considerable difference at the end, putting you on a reef, rock or shoal.

Check to be sure the intended compass site is solid. Vibration will increase pivot wear.

Now, you are ready to mount the compass. To prevent an error on all courses, the line through the lubber line and the compass card pivot must be exactly parallel to the keel of the boat. You can establish the fore-and-aft line of the boat with a stout cord or string. Use care to transfer this line to the compass site. If necessary, shim the base of the compass until the stile-type lubber line (the one affixed to the case and not gimbaled) is vertical when the boat is on an even keel. Drill the holes and mount the compass.

COMPASS PRECAUTIONS

◆ **See Figures 16, 17 and 18**

Many times an owner will install an expensive stereo system in the cabin of his boat. It is not uncommon for the speakers to be mounted on the aft bulkhead up against the overhead (ceiling). In almost every case, this position places one of the speakers in very close proximity to the compass, mounted above the ceiling.

You probably already know that a magnet is used in the operation of the speaker. Therefore, it is very likely that the speaker, mounted almost under the compass in the cabin will have a very pronounced effect on the compass accuracy.

Consider the following test and the accompanying photographs as proof:

First, the compass was read as 190 degrees while the boat was secure in her slip.

Next, a full can of soda in an aluminum can was placed on one side and the compass read as 204 degrees, a good 14 degrees off.

Next, the full can was moved to the opposite side of the compass and again a reading was observed, this time as 189 degrees, 11 degrees off from the original reading.

Finally, the contents of the can were consumed, the can placed on both sides of the compass with NO effect on the compass reading.

Two very important conclusions can be drawn from these tests.

• Something must have been in the contents of the can to affect the compass so drastically.

Fig. 15 The compass is a delicate instrument which should be mounted securely in a position where it can be easily observed by the helmsman

• Keep even innocent things clear of the compass to avoid any possible error in the boat's heading.

■ **Remember, a boat moving through the water at 10 knots on a compass error of just 5 degrees will be almost 1.5 miles off course in only ONE hour. At night, or in thick weather, this could very possibly put the boat on a reef, rock or shoal with disastrous results.**

Tools and Spare Parts

◆ **See Figures 19 and 20**

Carry a few tools and some spare parts, and learn how to make minor repairs. Many search and rescue cases are caused by minor breakdowns that boat operators could have repaired. Carry spare parts such as propellers, fuses or basic ignition components (like spark plugs, wires or even ignition coils) and the tools necessary to install them.

Fig. 16 This compass is giving an accurate reading, right?

Fig. 17 . . .well think again, as seemingly innocent objects may cause serious problems. . .

Fig. 18 . . .a compass reading off by just a few degrees could lead to disaster

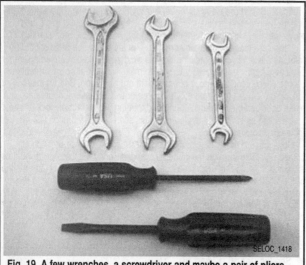

Fig. 19 A few wrenches, a screwdriver and maybe a pair of pliers can be very helpful to make emergency repairs

Fig. 20 A flashlight with a fresh set of batteries is handy when repairs are needed at night. It can also double as a signaling device

SAFETY IN SERVICE

It is virtually impossible to anticipate all of the hazards involved with maintenance and service, but care and common sense will prevent most accidents.

The rules of safety for mechanics range from "don't smoke around gasoline," to "use the proper tool(s) for the job." The trick to avoiding injuries is to develop safe work habits and to take every possible precaution. Whenever you are working on your boat, pay attention to what you are doing. The more you pay attention to details and what is going on around you, the less likely you will be to hurt yourself or damage your boat.

Do's

• Do keep a fire extinguisher and first aid kit handy.
• Do wear safety glasses or goggles when cutting, drilling, grinding or prying, even if you have 20-20 vision. If you wear glasses for the sake of vision, wear safety goggles over your regular glasses.
• Do shield your eyes whenever you work around the battery. Batteries contain sulfuric acid. In case of contact with the eyes or skin, flush the area with water or a mixture of water and baking soda; then seek immediate medical attention.

• Do use adequate ventilation when working with any chemicals or hazardous materials.
• Do disconnect the negative battery cable when working on the electrical system. The secondary ignition system contains EXTREMELY HIGH VOLTAGE. In some cases it can even exceed 50,000 volts. Furthermore, an accidental attempt to start the engine could cause the propeller or other components to rotate suddenly causing a potentially dangerous situation.
• Do follow manufacturer's directions whenever working with potentially hazardous materials. Most chemicals and fluids are poisonous if taken internally.
• Do properly maintain your tools. Loose hammerheads, mushroomed punches and chisels, frayed or poorly grounded electrical cords, excessively worn screwdrivers, spread wrenches (open end), cracked sockets, or slipping ratchets can cause accidents.
• Likewise, keep your tools clean; a greasy wrench can slip off a bolt head, ruining the bolt and often harming your knuckles in the process.
• Do use the proper size and type of tool for the job at hand. Do select a wrench or socket that fits the nut or bolt. The wrench or socket should sit straight, not cocked.

• Do, when possible, pull on a wrench handle rather than push on it, and adjust your stance to prevent a fall.

• Do be sure that adjustable wrenches are tightly closed on the nut or bolt and pulled so that the force is on the side of the fixed jaw. Better yet, avoid the use of an adjustable if you have a fixed wrench that will fit.

• Do strike squarely with a hammer; avoid glancing blows.

• Do use common sense whenever you work on your boat or motor. If a situation arises that doesn't seem right, sit back and have a second look. It may save an embarrassing moment or potential damage to your beloved boat.

Don'ts

• Don't run the engine in an enclosed area or anywhere else without proper ventilation - EVER! Carbon monoxide is poisonous; it takes a long time to leave the human body and you can build up a deadly supply of it in your system by simply breathing in a little every day. You may not realize you are slowly poisoning yourself.

• Don't work around moving parts while wearing loose clothing. Short sleeves are much safer than long, loose sleeves. Hard-toed shoes with neoprene soles protect your toes and give a better grip on slippery surfaces. Jewelry, watches, large belt buckles, or body adornment of any kind is not safe working around any craft or vehicle. Long hair should be tied back under a hat.

• Don't use pockets for toolboxes. A fall or bump can drive a screwdriver deep into your body. Even a rag hanging from your back pocket can wrap around a spinning shaft.

• Don't smoke when working around gasoline, cleaning solvent or other flammable material.

• Don't smoke when working around the battery. When the battery is being charged, it gives off explosive hydrogen gas. Actually, you shouldn't smoke anyway, it's bad for you. Instead, save the cigarette money and put it into your boat!

• Don't use gasoline to wash your hands; there are excellent soaps available. Gasoline contains dangerous additives that can enter the body through a cut or through your pores. Gasoline also removes all the natural oils from the skin so that bone dry hands will suck up oil and grease.

• Don't use screwdrivers for anything other than driving screws! A screwdriver used as a prying tool can snap when you least expect it, causing injuries. At the very least, you'll ruin a good screwdriver.

TROUBLESHOOTING

Troubleshooting can be defined as a methodical process during which one discovers what is causing a problem with engine operation. Although it is often a feared process to the uninitiated, there is no reason to believe that you cannot figure out what is wrong with a motor, as long as you follow a few basic rules.

To begin with, troubleshooting must be systematic. Haphazardly testing one component, then another, **might** uncover the problem, but it will more likely waste a lot of time. True troubleshooting starts by defining the problem and performing systematic tests to eliminate the largest and most likely causes first.

Start all troubleshooting by eliminating the most basic possible causes. Begin with a visual inspection of the boat and motor. If the engine won't crank, make sure that the kill switch or safety lanyard is in the proper position. Make sure there is fuel in the tank and the fuel system is primed before condemning the carburetor or fuel injection system. On electric start motors, make sure there are no blown fuses, the battery is fully charged, and the cable connections (at both ends) are clean and tight before suspecting a bad starter, solenoid or switch.

The majority of problems that occur suddenly can be fixed by simply identifying the one small item that brought them on. A loose wire, a clogged passage or a broken component can cause a lot of trouble and are often the cause of a sudden performance problem.

The next most basic step in troubleshooting is to test systems before components. For example, if the engine doesn't crank on an electric start motor, determine if the battery is in good condition (fully charged and properly connected) before testing the starting system. If the engine cranks, but doesn't start, you know already know the starting system and battery (if it cranks fast enough) are in good condition, now it is time to look at the ignition or fuel systems. Once you've isolated the problem to a particular system, follow the troubleshooting/testing procedures in the section for that system to test either subsystems (if applicable, for example: the starter circuit) or components (starter solenoid).

Basic Operating Principles

◆ **See Figures 21 and 22**

Before attempting to troubleshoot a problem with your motor, it is important that you understand how it operates. Once normal engine or system operation is understood, it will be easier to determine what might be causing the trouble or irregular operation in the first place. System descriptions are found throughout this manual, but the basic mechanical operating principles for both 2-stroke engines (like most of the outboards covered here) and 4-stroke engines (like some outboards and like your car) are given here. A basic understanding of both types of engines is useful not only in understanding and troubleshooting your outboard, but also for dealing with other motors in your life.

All motors covered by this manual (and probably MOST of the motors you own) operate according to the Otto cycle principle of engine operation. This means that all motors follow the stages of intake, compression, power and exhaust. But, the difference between a 2- and 4-stroke motor is in how many times the piston moves up and down within the cylinder to accomplish this. On 2-stroke motors (as the name suggests) the four cycles take place in 2 movements (one up and one down) of the piston. Again, as the name suggests, the cycles take place in 4 movements of the piston for 4-stroke motors.

2-STROKE MOTORS

The 2-stroke engine differs in several ways from a conventional four-stroke (automobile or marine) engine.

1. The intake/exhaust method by which the fuel-air mixture is delivered to the combustion chamber.
2. The complete lubrication system.
3. The frequency of the power stroke.

Let's discuss these differences briefly (and compare 2-stroke engine operation with 4-stroke engine operation.)

Intake/Exhaust

◆ **See Figures 23, 24 and 25**

Two-stroke engines utilize an arrangement of port openings to admit fuel to the combustion chamber and to purge the exhaust gases after burning has been completed. The ports are located in a precise pattern in order for them to be open and closed off at an exact moment by the piston as it moves up and down in the cylinder. The exhaust port is located slightly higher than the fuel intake port. This arrangement opens the exhaust port first as the piston starts downward and therefore, the exhaust phase begins a fraction of a second before the intake phase.

Actually, the intake and exhaust ports are spaced so closely together that both open almost simultaneously. For this reason, 2-stroke engines from some manufacturers (though few if any Mercury/Mariners) utilize deflector-type pistons. This design of the piston top serves two purposes very effectively.

First, it creates turbulence when the incoming charge of fuel enters the combustion chamber. This turbulence results in a more complete burning of the fuel than if the piston top were flat. The second effect of the deflector-type piston crown is to force the exhaust gases from the cylinder more rapidly. Although this configuration is used in many older outboards, it is generally not found only many Mercury/Mariner motors. Instead many of the Mercury/Mariner motors could be referred to as Loop charged.

Loop charged motors, or as they are commonly called "loopers", differ in how the air/fuel charge is introduced to the combustion chamber. Instead of the charge flowing across the top of the piston from one side of the cylinder to the other the use a looping action on top of the piston as the charge is forced through irregular shaped openings cut in the piston's skirt. In a looper motor, the charge is forced out from the crankcase by the downward motion of the piston, through the irregular shaped openings and transferred upward by long, deep grooves in the cylinder wall. The charge completes its looping action by entering the combustion chamber, just above the piston, where the

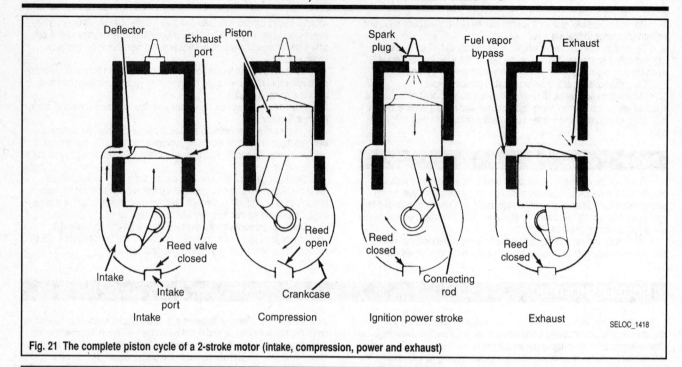

Fig. 21 The complete piston cycle of a 2-stroke motor (intake, compression, power and exhaust)

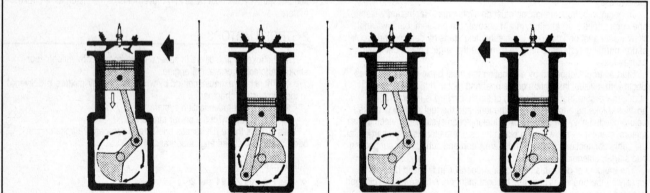

Fig. 22 The complete piston cycle of a 4-stroke motor (intake, compression, power and exhaust

upward motion of the piston traps it in the chamber and compresses it for optimum ignition power.

Unlike the knife-edged deflector top pistons used in non-looper (or cross flow) motors, the piston domes on Loop motors are relatively flat.

These systems of intake and exhaust are in marked contrast to individual intake and exhaust valve arrangement employed on four-stroke engines (and the mechanical methods of opening and closing these valves).

■ It should be noted here that there are some 2-stroke engines that utilize a mechanical valve train, though it is very different from the valve train employed by most 4-stroke motors. Rotary 2-stroke engines use a circular valve or rotating disc that contains a port opening around part of one edge of the disc. As the engine (and disc) turns, the opening aligns with the intake port at and for a predetermined amount of time, closing off the port again as the opening passes by and the solid portion of the disc covers the port.

Lubrication

A 2-stroke engine is lubricated by mixing oil with the fuel. Therefore, various parts are lubricated as the fuel mixture passes through the crankcase and the cylinder. In contrast, four-stroke engines have a crankcase containing oil. This oil is pumped through a circulating system and returned to the crankcase to begin the routing again.

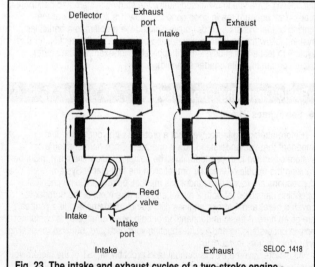

Fig. 23 The intake and exhaust cycles of a two-stroke engine - Cross flow (CV) design shown

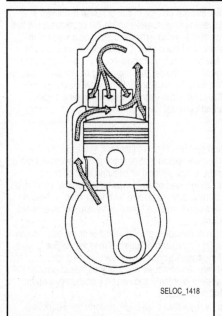

Fig. 24 Cross-sectional view of a typical loop-charged cylinder, showing charge flow while piston is moving downward

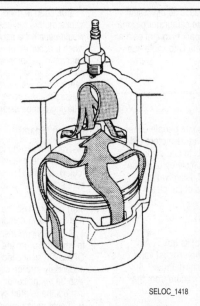

Fig. 25 Cutaway view of a typical loop-charged cylinder, depicting exhaust leaving the cylinder as the charge enters through 3 ports in the piston

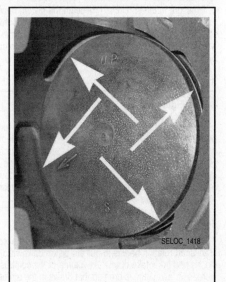

Fig. 26 The combustion chamber of a typical looper, notice the piston is far enough down the cylinder bore to reveal intake and exhaust ports

Power Stroke

The combustion cycle of a 2-stroke engine has four distinct phases.
1. Intake
2. Compression
3. Power
4. Exhaust

The four phases of the cycle are accomplished with each up and down stroke of the piston, and the power stroke occurs with each complete revolution of the crankshaft. Compare this system with a four-stroke engine. A separate stroke of the piston is required to accomplish each phase of the cycle and the power stroke occurs only every other revolution of the crankshaft. Stated another way, two revolutions of the four-stroke engine crankshaft are required to complete one full cycle, the four phases.

Physical Laws
◆ See Figure 27

The 2-stroke engine is able to function because of two very simple physical laws.

One: Gases will flow from an area of high pressure to an area of lower pressure. A tire blowout is an example of this principle. The high-pressure air escapes rapidly if the tube is punctured.

Two: If a gas is compressed into a smaller area, the pressure increases, and if a gas expands into a larger area, the pressure is decreased.

If these two laws are kept in mind, the operation of the 2-stroke engine will be easier understood.

Actual Operation
◆ See Figure 21

■ **The engine described here is of a carbureted type. EFI and DFI/Optimax motors operate similarly for intake of the air charge and for exhaust of the unburned gasses. Obviously though, the very nature of fuel injection changes the actual delivery of the fuel/oil charge.**

Beginning with the piston approaching top dead center on the compression stroke: the intake and exhaust ports are physically closed (blocked) by the piston. During this stroke, the reed valve is open (because as the piston moves upward, the crankcase volume increases, which reduces the crankcase pressure to less than the outside atmosphere (creates a vacuum under the piston). The spark plug fires; the compressed fuel-air mixture is ignited; and the power stroke begins.

As the piston moves downward on the power stroke, the combustion chamber is filled with burning gases. As the exhaust port is uncovered, the

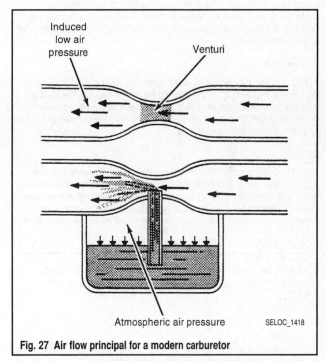

Fig. 27 Air flow principal for a modern carburetor

gases, which are under great pressure, escape rapidly through the exhaust ports. The piston continues its downward movement. Pressure within the crankcase (again, under the piston) increases, closing the reed valves against their seats. The crankcase then becomes a sealed chamber so the air-fuel mixture becomes compressed (pressurized) and ready for delivery to the combustion chamber. As the piston continues to move downward, the intake port is uncovered. The fresh fuel mixture rushes through the intake port into the combustion chamber striking the top of the piston where it is deflected along the cylinder wall. The reed valve remains closed until the piston moves upward again.

When the piston begins to move upward on the compression stroke, the reed valve opens because the crankcase volume has been increased, reducing crankcase pressure to less than the outside atmosphere. The intake and exhaust ports are closed and the fresh fuel charge is compressed inside the combustion chamber.

Pressure in the crankcase (beneath the piston) decreases as the piston moves upward and a fresh charge of air flows through the carburetor picking up fuel. As the piston approaches top dead center, the spark plug ignites the air-fuel mixture, the power stroke begins and one complete Otto cycle has been completed.

4-STROKE MOTORS

◆ **See Figure 22**

The 4-stroke motor may be easier to understand for some people either because of its prevalence in automobile and street motorcycle motors today or perhaps because each of the four strokes corresponds to one distinct phase of the Otto cycle. Essentially, a 4-stroke engine completes one Otto cycle of intake, compression, ignition/power and exhaust using two full revolutions of the crankshaft and four distinct movements of the piston (down, up, down and up).

Intake

The intake stroke begins with the piston near the top of its travel. As crankshaft rotation begins to pull the piston downward, the exhaust valve closes and the intake opens. As volume of the combustion chamber increases, a vacuum is created that draws in the air/fuel mixture from the intake manifold.

Compression

Once the piston reaches the bottom of its travel, crankshaft rotation will begin to force it upward. At this point the intake valve closes. As the piston rises in the bore, the volume of the sealed combustion chamber (both intake and exhaust valves are closed) decreases and the air/fuel mixture is compressed. This raises the temperature and pressure of the mixture and increases the amount of force generated by the expanding gases during the Ignition/Power stroke.

Ignition/Power

As the piston approaches top dead center (the highest point of travel in the bore), the spark plug will fire, igniting the air/fuel mixture. The resulting combustion of the air/fuel mixture forces the piston downward, rotating the crankshaft (causing other pistons to move in other phases/strokes of the Otto cycle on multi-cylinder motors).

Exhaust

As the piston approaches the bottom of the Ignition/Power stroke, the exhaust valve opens. When the piston begins its upward path of travel once again, any remaining unburned gasses are forced out through the exhaust valve. This completes one Otto cycle, which begins again as the piston passes top dead center, the intake valve opens and the Intake stroke starts.

COMBUSTION

Whether we are talking about a 2- or 4-stroke engine, all Otto cycle, internal combustion engines require three basic conditions to operate properly,

1. Compression
2. Ignition (Spark)
3. Fuel

A lack of any one of these conditions will prevent the engine from operating. A problem with any one of these will manifest itself in hard-starting or poor performance.

Compression

An engine that has insufficient compression will not draw an adequate supply of air/fuel mixture into the combustion chamber and, subsequently, will not make sufficient power on the power stroke. A lack of compression in just one cylinder of a multi-cylinder motor will cause the motor to stumble or run irregularly.

But, keep in mind that a sudden change in compression is unlikely in 2-stroke motors (unless something major breaks inside the crankcase, but that would usually be accompanied by other symptoms such as a loud noise when it occurred or noises during operation). On 4-stroke motors, a sudden change in compression is also unlikely, but could occur if the timing belt or chain was to suddenly break. Remember that the timing belt/chain is used to synchronize the valve train with the crankshaft. If the valve train suddenly ceases to turn, some intake and some exhaust valves will remain open, relieving compression in that cylinder.

Ignition (Spark)

Traditionally, the ignition system is the weakest link in the chain of conditions necessary for engine operation. Spark plugs may become worn or fouled, wires will deteriorate allowing arcing or misfiring, and poor connections can place an undue load on coils leading to weak spark or even a failed coil. The most common question asked by a technician under a no-start condition is: "do I have spark and fuel" (as they've already determined that they have compression).

A quick visual inspection of the spark plug(s) will answer the question as to whether or not the plug(s) is/are worn or fouled. While the engine is shut **OFF** a physical check of the connections could show a loose primary or secondary ignition circuit wire. An obviously physically damaged wire may also be an indication of system problems and certainly encourages one to inspect the related system more closely.

If nothing is turned up by the visual inspection, perform the Spark Test provided in the Ignition System section to determine if the problem is a lack of or a weak spark. If the problem is not compression or spark, it's time to look at the fuel system.

Fuel

If compression and spark is present (and within spec), but the engine won't start or won't run properly, the only remaining condition to fulfill is fuel. As usual, start with the basics. Is the fuel tank full? Is the fuel stale? If the engine has not been run in some time (a matter of months, not weeks) there is a good chance that the fuel is stale and should be properly disposed of and replaced.

■ **Depending on how stale or contaminated (with moisture) the fuel is, it may be burned in an automobile or in yard equipment, though it would be wise to mix it well with a much larger supply of fresh gasoline to prevent moving your driveability problems to that motor. But it is better to get the lawn tractor stuck on stale gasoline than it would be to have your boat motor quit in the middle of the bay or lake.**

For hard starting motors, is the choke or primer system operating properly. Remember that the choke/prime should only be used for **cold** starts. A true cold start is really only the first start of the day, but it may be applicable to subsequent starts on cooler days, if the engine sat for more than a few hours and completely cooled off since the last use. Applying the primer to the motor for a hot start may flood the engine, preventing it from starting properly. One method to clear a flood is to crank the motor while the engine is at wide-open throttle (allowing the maximum amount of air into the motor to compensate for the excess fuel). But, keep in mind that the throttle should be returned to idle immediately upon engine start-up to prevent damage from over-revving.

Fuel delivery and pressure should be checked before delving into the carburetor(s) or fuel injection system. Make sure there are no clogs in the fuel line or vacuum leaks that would starve the motor of fuel.

Make sure that all other possible problems have been eliminated before touching the carburetor. It is rare that a carburetor will suddenly require an adjustment in order for the motor to run properly. It is much more likely that an improperly stored motor (one stored with untreated fuel in the carburetor) would suffer from one or more clogged carburetor passages sometime after shortly returning to service. Fuel will evaporate over time, leaving behind gummy deposits. If untreated fuel is left in the carburetor for some time (again typically months more than weeks), the varnish left behind by evaporating fuel will likely clog the small passages of the carburetor and cause problems with engine performance. If you suspect this, remove and disassemble the carburetor following procedures under Fuel System.

SHOP EQUIPMENT

Safety Tools

WORK GLOVES

◆ See Figure 28

Unless you think scars on your hands are cool, enjoy pain and like wearing bandages, get a good pair of work gloves. Canvas or leather gloves are the best. And yes, we realize that there are some jobs involving small parts that can't be done while wearing work gloves. These jobs are not the ones usually associated with hand injuries.

A good pair of rubber gloves (such as those usually associated with dish washing) or vinyl gloves is also a great idea. There are some liquids such as solvents and penetrants that don't belong on your skin. Avoid burns and rashes. Wear these gloves.

And lastly, an option. If you're tired of being greasy and dirty all the time, go to the drug store and buy a box of disposable latex gloves like medical professionals wear. You can handle greasy parts, perform small tasks, wash parts, etc. all without getting dirty! These gloves take a surprising amount of abuse without tearing and aren't expensive. Note however, that some people are allergic to the latex or the powder used inside some gloves, so pay attention to what you buy.

EYE AND EAR PROTECTION

◆ See Figures 29 and 30

Don't begin any job without a good pair of work goggles or impact resistant glasses! When doing any kind of work, it's all too easy to avoid eye injury through this simple precaution. And don't just buy eye protection and leave it on the shelf. Wear it all the time! Things have a habit of breaking, chipping, splashing, spraying, splintering and flying around. And, for some reason, your eye is always in the way!

If you wear vision-correcting glasses as a matter of routine, get a pair made with polycarbonate lenses. These lenses are impact resistant and are available at any optometrist.

Often overlooked is hearing protection. Engines and power tools are noisy! Loud noises damage your ears. It's as simple as that! The simplest and cheapest form of ear protection is a pair of noise-reducing ear plugs. Cheap insurance for your ears! And, they may even come with their own, cute little carrying case.

More substantial, more protection and more money is a good pair of noise reducing earmuffs. They protect from all but the loudest sounds. Hopefully those are sounds that you'll never encounter since they're usually associated with disasters.

WORK CLOTHES

Everyone has "work clothes." Usually these consist of old jeans and a shirt that has seen better days. That's fine. In addition, a denim work apron is a nice accessory. It's rugged, can hold some spare bolts, and you don't feel bad wiping your hands or tools on it. That's what it's for.

When working in cold weather, a one-piece, thermal work outfit is invaluable. Most are rated to below freezing temperatures and are ruggedly constructed. Just look at what local marine mechanics are wearing and that should give you a clue as to what type of clothing is good.

Chemicals

There is a whole range of chemicals that you'll find handy for maintenance and repair work. The most common types are: lubricants, penetrants and sealers. Keep these handy. There are also many chemicals that are used for detailing or cleaning.

When a particular chemical is not being used, keep it capped, upright and in a safe place. These substances may be flammable, may be irritants or might even be caustic and should always be stored properly, used properly and handled with care. Always read and follow all label directions and be sure to wear hand and eye protection!

LUBRICANTS & PENETRANTS

◆ See Figure 31

Anti-seize is used to coat certain fasteners prior to installation. This can be especially helpful when two dissimilar metals are in contact (to help prevent corrosion that might lock the fastener in place). This is a good practice on a lot of different fasteners, BUT, NOT on any fastener that might vibrate loose causing a problem. If anti-seize is used on a fastener, it should be checked periodically for proper tightness.

Lithium grease, chassis lube, silicone grease or a synthetic brake caliper grease can all be used pretty much interchangeably. All can be used for coating rust-prone fasteners and for facilitating the assembly of parts that are a tight fit. Silicone and synthetic greases are the most versatile.

■ **Silicone dielectric grease is a non-conductor that is often used to coat the terminals of wiring connectors before fastening them. It may sound odd to coat metal portions of a terminal with something that won't conduct electricity, but here is it how it works. When the connector is fastened the metal-to-metal contact between the terminals will displace the grease (allowing the circuit to be completed). The grease that is displaced will then coat the non-contacted surface and the cavity around the terminals, SEALING them from atmospheric moisture that could cause corrosion.**

Silicone spray is a good lubricant for hard-to-reach places and parts that shouldn't be gooped up with grease.

Penetrating oil may turn out to be one of your best friends when taking something apart that has corroded fasteners. Not only can they make a job easier, they can really help to avoid broken and stripped fasteners. The most familiar penetrating oils are Liquid Wrench® and WD-40®. A newer penetrant, PB Blaster® works very well (and has become a mainstay in our shops). These products have hundreds of uses. For your purposes, they are vital!

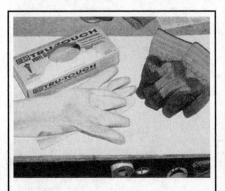

Fig. 28 Three different types of work gloves. The box contains latex gloves

Fig. 29 Don't begin major repairs without a pair of goggles for your eyes and earmuffs to protect your hearing

Fig. 30 Things have a habit of, splashing, spraying, splintering and flying around during repairs

Before disassembling any part, check the fasteners. If any appear rusted, soak them thoroughly with the penetrant and let them stand while you do something else (for particularly rusted or frozen parts you may need to soak them a few days in advance). This simple act can save you hours of tedious work trying to extract a broken bolt or stud.

SEALANTS

◆ See Figures 32 and 33

Sealants are an indispensable part for certain tasks, especially if you are trying to avoid leaks. The purpose of sealants is to establish a leak-proof bond between or around assembled parts. Most sealers are used in conjunction with gaskets, but some are used instead of conventional gasket material.

The most common sealers are the non-hardening types such as Permatex® No.2 or its equivalents. These sealers are applied to the mating surfaces of each part to be joined, then a gasket is put in place and the parts are assembled.

■ **A sometimes overlooked use for sealants like RTV is on the threads of vibration prone fasteners.**

One very helpful type of non-hardening sealer is the "high tack" type. This type is a very sticky material that holds the gasket in place while the parts are being assembled. This stuff is really a good idea when you don't have enough hands or fingers to keep everything where it should be.

The stand-alone sealers are the Room Temperature Vulcanizing (RTV) silicone gasket makers. On some engines, this material is used instead of a gasket. In those instances, a gasket may not be available or, because of the shape of the mating surfaces, a gasket shouldn't be used. This stuff, when used in conjunction with a conventional gasket, produces the surest bonds.

RTV does have its limitations though. When using this material, you will have a time limit. It starts to set-up within 15 minutes or so, so you have to assemble the parts without delay. In addition, when squeezing the material out of the tube, don't drop any glops into the engine. The stuff will form and set and travel around a cooling passage, possibly blocking it. Also, most types are not fuel-proof. Check the tube for all cautions.

CLEANERS

◆ See Figures 34 and 35

There are two basic types of cleaners on the market today: parts cleaners and hand cleaners. The parts cleaners are for the parts; the hand cleaners are for you. They are **NOT** interchangeable.

There are many good, non-flammable, biodegradable parts cleaners on the market. These cleaning agents are safe for you, the parts and the environment. Therefore, there is no reason to use flammable, caustic or toxic substances to clean your parts or tools.

As far as hand cleaners go; the waterless types are the best. They have always been efficient at cleaning, but they used to all leave a pretty smelly odor. Recently though, most of them have eliminated the odor and added stuff that actually smells good. Make sure that you pick one that contains lanolin or some other moisture-replenishing additive. Cleaners not only remove grease and oil but also skin oil.

■ **Most women already know to use a hand lotion when you're all cleaned up. It's okay. Real men DO use hand lotion too! Believe it or not, using hand lotion BEFORE your hands are dirty will actually make them easier to clean when you're finished with a dirty job. Lotion seals your hands, and keeps dirt and grease from sticking to your skin.**

Fig. 31 Keep a supply of anti-seize, penetrating oil, lithium grease, electronic cleaner4 and silicone spray

Fig. 32 Sealants are essential for preventing leaks

Fig. 33 On some engines, RTV is used instead of gasket material to seal components

Fig. 34 Citrus hand cleaners not only work well, but they smell pretty good too. Choose one with pumice for added cleaning powe

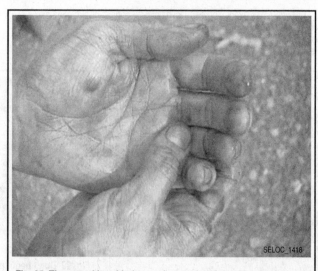

Fig. 35 The use of hand lotion seals your hands and keeps dirt and grease from sticking to your skin

TOOLS

◆ See Figure 36

Tools; this subject could fill a completely separate manual. The first thing you will need to ask yourself, is just how involved do you plan to get. If you are serious about maintenance and repair you will want to gather a quality set of tools to make the job easier, and more enjoyable. BESIDES, TOOLS ARE FUN!!!

Almost every do-it-yourselfer loves to accumulate tools. Though most find a way to perform jobs with only a few common tools, they tend to buy more over time, as money allows. So gathering the tools necessary for maintenance or repair does not have to be an expensive, overnight proposition.

When buying tools, the saying "You get what you pay for ..." is absolutely true! Don't go cheap! Any hand tool that you buy should be drop forged and/or chrome vanadium. These two qualities tell you that the tool is strong enough for the job. With any tool, go with a name that you've heard of before, or, that is recommended buy your local professional retailer. Let's go over a list of tools that you'll need.

Most of the world uses the metric system. However, some American-built engines and aftermarket accessories use standard fasteners. So, accumulate your tools accordingly. Any good DIYer should have a decent set of both U.S. and metric measure tools.

■ Don't be confused by terminology. Most advertising refers to "SAE and metric", or "standard and metric." Both are misnomers. The Society of Automotive Engineers (SAE) did not invent the English system of measurement; the English did. The SAE likes metrics just fine. Both English (U.S.) and metric measurements are SAE approved. Also, the current "standard" measurement IS metric. So, if it's not metric, it's U.S. measurement.

Hand-Tools

SOCKET SETS

◆ See Figures 37 thru 43

Socket sets are the most basic hand tools necessary for repair and maintenance work. For our purposes, socket sets come in three drive sizes: 1/4 inch, 3/8 inch and 1/2 inch. Drive size refers to the size of the drive lug on the ratchet, breaker bar or speed handle.

A 3/8 inch set is probably the most versatile set in any mechanic's toolbox. It allows you to get into tight places that the larger drive ratchets can't and gives you a range of larger sockets that are still strong enough for heavy-duty work. The socket set that you'll need should range in sizes from 1/4 inch through 1 inch for standard fasteners, and a 6mm through 19mm for metric fasteners.

You'll need a good 1/2 inch set since this size drive lug assures that you won't break a ratchet or socket on large or heavy fasteners. Also, torque wrenches with a torque scale high enough for larger fasteners are usually 1/2 inch drive.

Plus, 1/4 inch drive sets can be very handy in tight places. Though they usually duplicate functions of the 3/8 in. set, 1/4 in. drive sets are easier to use for smaller bolts and nuts.

As for the sockets themselves, they come in shallow (standard) and deep lengths as well as 6 or 12 point. The 6 and 12 points designation refers to how many sides are in the socket itself. Each has advantages. The 6 point socket is stronger and less prone to slipping which would strip a bolt head or nut. 12 point sockets are more common, usually less expensive and can operate better in tight places where the ratchet handle can't swing far.

Standard length sockets are good for just about all jobs, however, some stud-head bolts, hard-to-reach bolts, nuts on long studs, etc., require the deep sockets.

Most marine manufacturers use recessed hex-head fasteners to retain many of the engine parts. These fasteners require a socket with a hex shaped driver or a large sturdy hex key. To help prevent torn knuckles, we would recommend that you stick to the sockets on any tight fastener and leave the hex keys for lighter applications. Hex driver sockets are available individually or in sets just like conventional sockets.

More and more, manufacturers are using Torx® head fasteners, which were once known as tamper resistant fasteners (because many people did not have tools with the necessary odd driver shape). Since Torx® fasteners have become commonplace in many DIYer tool boxes, manufacturers designed newer tamper resistant fasteners that are essentially Torx® head bolts that contain a small protrusion in the center (requiring the driver to contain a small hole to slide over the protrusion. Tamper resistant fasteners are often used where the manufacturer would prefer only knowledgeable mechanics or advanced Do-It-Yourselfers (DIYers) work.

TORQUE WRENCHES

◆ See Figure 44

In most applications, a torque wrench can be used to ensure proper installation of a fastener. Torque wrenches come in various designs and most stores will carry a variety to suit your needs. A torque wrench should be used any time you have a specific torque value for a fastener. Keep in mind that because there is no worldwide standardization of fasteners, so charts or figure found in each repair section refer to the manufacturer's fasteners. Any general guideline charts that you might come across based on fastener size (they are sometimes included in a repair manual or with torque wrench packaging) should be used with caution. Just keep in mind that if you are using the right tool for the job, you should not have to strain to tighten a fastener.

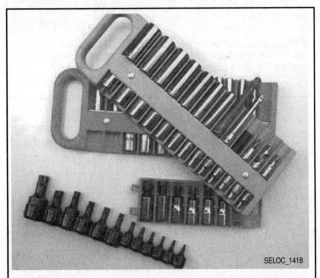

SELOC_1418

Fig. 36 Socket holders, especially the magnetic type, are handy items to keep tools in order

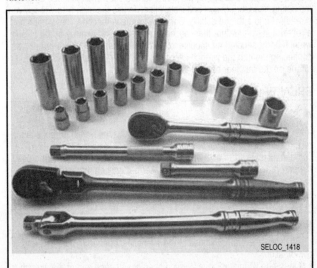

SELOC_1418

Fig. 37 A 3/8 in. socket set is probably the most versatile tool in any mechanic's tool box

Fig. 38 A swivel (U-joint) adapter (left), a wobble-head adapter (center) and a 1/2 in.-to-3/8 in. adapter (right)

Fig. 39 Ratchets come in all sizes and configurations from rigid to swivel-headed

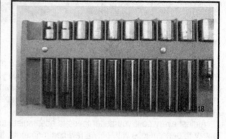

Fig. 40 Shallow sockets (top) are good for most jobs. But, some bolts require deep sockets (bottom)

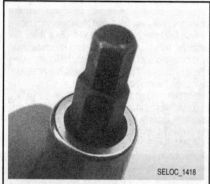

Fig. 41 Hex-head fasteners require a socket with a hex shaped driver

Fig. 42 Torx® drivers . . .

Fig. 43 . . . and tamper resistant drivers are required to remove special fasteners

Beam Type

◆ See Figures 45 and 46

The beam type torque wrench is one of the most popular styles in use. If used properly, it can be the most accurate also. It consists of a pointer attached to the head that runs the length of the flexible beam (shaft) to a scale located near the handle. As the wrench is pulled, the beam bends and the pointer indicates the torque using the scale.

Click (Breakaway) Type

◆ See Figures 47 and 48

Another popular torque wrench design is the click type. The clicking mechanism makes achieving the proper torque easy and most use a ratcheting head for ease of bolt installation. To use the click type wrench you pre-adjust it to a torque setting. Once the torque is reached, the wrench has a reflex signaling feature that causes a momentary breakaway of the torque wrench body, sending an impulse to the operator's hand. But be careful, as continuing the turn the wrench after the momentary release will increase torque on the fastener beyond the specified setting.

BREAKER BARS

◆ See Figure 49

Breaker bars are long handles with a drive lug. Their main purpose is to provide extra turning force when breaking loose tight bolts or nuts. They come in all drive sizes and lengths. Always take extra precautions and use the proper technique when using a breaker bar (pull on the bar, don't push, to prevent skinned knuckles).

WRENCHES

◆ See Figures 50 thru 54

Basically, there are 3 kinds of fixed wrenches: open end, box end, and combination.

Open-end wrenches have 2-jawed openings at each end of the wrench. These wrenches are able to fit onto just about any nut or bolt. They are extremely versatile but have one major drawback. They can slip on a worn or rounded bolt head or nut, causing bleeding knuckles and a useless fastener.

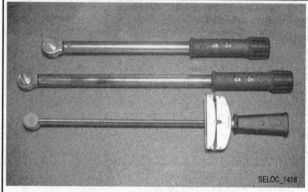

Fig. 44 Three types of torque wrenches. Top to bottom: a 3/8 in. drive beam type that reads in inch lbs., a 1/2 in. drive clicker type and a 1/2 in. drive beam type

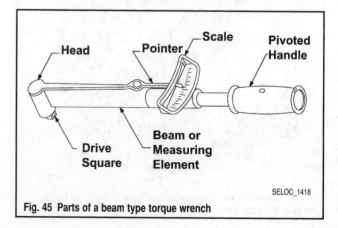

Fig. 45 Parts of a beam type torque wrench

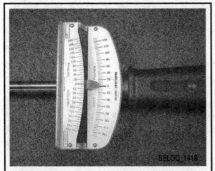

Fig. 46 A beam type torque wrench consists of a pointer attached to the head that runs the length of the flexible beam (shaft) to a scale located near the handle

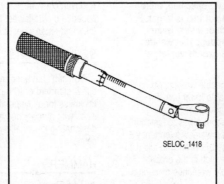

Fig. 47 A click type or breakaway torque wrench - note this one has a pivoting head

Fig. 48 Setting the torque on a click type wrench involves turning the handle until the specification appears on the dial

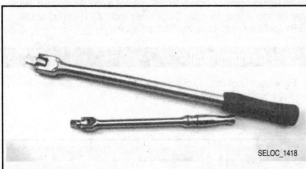

Fig. 49 Breaker bars are great for loosening large or stuck fasteners

■ Line wrenches are a special type of open-end wrench designed to fit onto more of the fastener than standard open-end wrenches, thus reducing the chance of rounding the corners of the fastener.

Box-end wrenches have a 360° circular jaw at each end of the wrench. They come in both 6 and 12 point versions just like sockets and each type has some of the same advantages and disadvantages as sockets.

Combination wrenches have the best of both. They have a 2-jawed open end and a box end. These wrenches are probably the most versatile.

As for sizes, you'll probably need a range similar to that of the sockets, about 1/4 in. through 1 in. for standard fasteners, or 6mm through 19mm for metric fasteners. As for numbers, you'll need 2 of each size, since, in many instances, one wrench holds the nut while the other turns the bolt. On most fasteners, the nut and bolt are the same size so having two wrenches of the same size comes in handy.

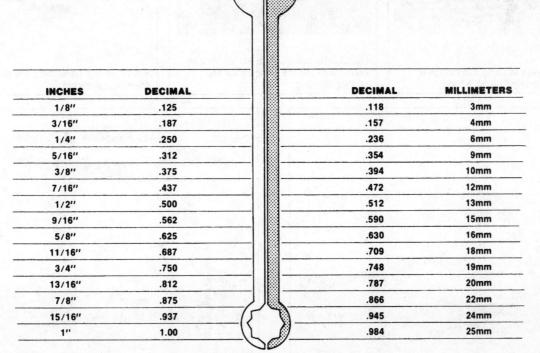

INCHES	DECIMAL	DECIMAL	MILLIMETERS
1/8"	.125	.118	3mm
3/16"	.187	.157	4mm
1/4"	.250	.236	6mm
5/16"	.312	.354	9mm
3/8"	.375	.394	10mm
7/16"	.437	.472	12mm
1/2"	.500	.512	13mm
9/16"	.562	.590	15mm
5/8"	.625	.630	16mm
11/16"	.687	.709	18mm
3/4"	.750	.748	19mm
13/16"	.812	.787	20mm
7/8"	.875	.866	22mm
15/16"	.937	.945	24mm
1"	1.00	.984	25mm

Fig. 50 Comparison of U.S. measure and metric wrench sizes

■ Although you will typically just need the sizes we specified, there are some exceptions. Occasionally you will find a nut that is larger. For these, you will need to buy ONE expensive wrench or a very large adjustable. Or you can always just convince the spouse that we are talking about SAFETY here and buy a whole (read expensive) large wrench set.

One extremely valuable type of wrench is the adjustable wrench. An adjustable wrench has a fixed upper jaw and a moveable lower jaw. The lower jaw is moved by turning a threaded drum. The advantage of an adjustable wrench is its ability to be adjusted to just about any size fastener.

The main drawback of an adjustable wrench is the lower jaw's tendency to move slightly under heavy pressure. This can cause the wrench to slip if it is not facing the right way. Pulling on an adjustable wrench in the proper direction will cause the jaws to lock in place. Adjustable wrenches come in a large range of sizes, measured by the wrench length.

PLIERS

◆ See Figure 55

Pliers are simply mechanical fingers. They are, more than anything, an extension of your hand. At least 3 pairs of pliers are an absolute necessity - standard, needle nose and slip joint.

In addition to standard pliers there are the slip-joint, multi-position pliers such as ChannelLock® pliers and locking pliers, such as Vise Grips®.

Slip joint pliers are extremely valuable in grasping oddly sized parts and fasteners. Just make sure that you don't use them instead of a wrench too often since they can easily round off a bolt head or nut.

Locking pliers are usually used for gripping bolts or studs that can't be removed conventionally. You can get locking pliers in square jawed, needle-nosed and pipe-jawed. Locking pliers can rank right up behind duct tape as the handy-man's best friend.

SCREWDRIVERS

You can't have too many screwdrivers. They come in 2 basic flavors, either standard or Phillips. Standard blades come in various sizes and thickness for all types of slotted fasteners. Phillips screwdrivers come in sizes with number designations from 1 on up, with the lower number designating the smaller size. Screwdrivers can be purchased separately or in sets.

HAMMERS

◆ See Figure 56

You need a hammer for just about any kind of work. You need a ball-peen hammer for most metal work when using drivers and other like tools. A plastic hammer comes in handy for hitting things safely. A soft-faced dead-blow hammer is used for hitting things safely and hard. Hammers are also VERY useful with non air-powered impact drivers.

Other Common Tools

There are a lot of other tools that every DIYer will eventually need (though not all for basic maintenance). They include:
- Funnels
- Chisels

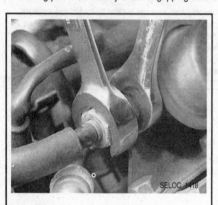

Fig. 51 Always use a backup wrench to prevent rounding flare nut fittings

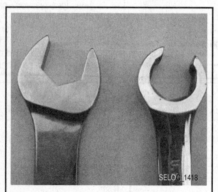

Fig. 52 Note how the flare wrench jaws are extended to grip the fitting tighter and prevent rounding

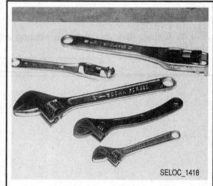

Fig. 53 Several types and sizes of adjustable wrenches

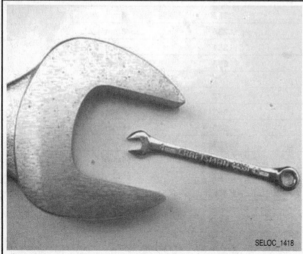

Fig. 54 You may find a nut that requires a particularly large or small wrench (that is usually available at your local tool store)

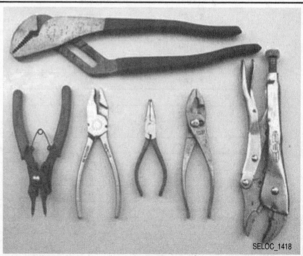

Fig. 55 Pliers come in many shapes and sizes. You should have an assortment on hand

Fig. 56 Three types of hammers. Top to bottom: ball peen, rubber dead-blow, and plastic

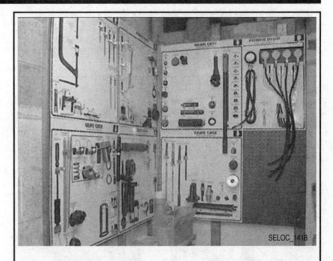

Fig. 57 Almost every marine engine around today requires at least one special tool to perform a certain task

- Punches
- Files
- Hacksaw
- Portable Bench Vise
- Tap and Die Set
- Flashlight
- Magnetic Bolt Retriever
- Gasket scraper
- Putty Knife
- Screw/Bolt Extractors
- Prybars

Hacksaws have just one use - cutting things off. You may wonder why you'd need one for something as simple as maintenance or repair, but you never know. Among other things, guide studs to ease parts installation can be made from old bolts with their heads cut off.

A tap and die set might be something you've never needed, but you will eventually. It's a good rule, when everything is apart, to clean-up all threads, on bolts, screws or threaded holes. Also, you'll likely run across a situation in which you will encounter stripped threads. The tap and die set will handle that for you.

Gasket scrapers are just what you'd think, tools made for scraping old gasket material off of parts. You don't absolutely need one. Old gasket material can be removed with a putty knife or single edge razor blade. However, putty knives may not be sharp enough for some really stubborn gaskets and razor blades have a knack of breaking just when you don't want them to, inevitably slicing the nearest body part! As the old saying goes, "always use the proper tool for the job". If you're going to use a razor to scrape a gasket, be sure to always use a blade holder.

Putty knives really do have a use in a repair shop. Just because you remove all the bolts from a component sealed with a gasket doesn't mean it's going to come off. Most of the time, the gasket and sealer will hold it tightly. Lightly inserting a putty knife at various points between the two parts will break the seal without damage to the parts.

A small - 8-10 in. (20-25cm) long - prybar is extremely useful for removing stuck parts.

■ **Never use a screwdriver as a prybar! Screwdrivers are not meant for prying. Screwdrivers, used for prying, can break, sending the broken shaft flying!**

Screw/bolt extractors are used for removing broken bolts or studs that have broken off flush with the surface of the part.

Special Tools

◆ **See Figure 57**

Almost every marine engine around today requires at least one special tool to perform a certain task. In most cases, these tools are specially designed to overcome some unique problem or to fit on some oddly sized component.

When manufacturers go through the trouble of making a special tool, it is usually necessary to use it to ensure that the job will be done right. A special tool might be designed to make a job easier, or it might be used to keep you from damaging or breaking a part.

Don't worry, MOST maintenance procedures can either be performed without any special tools OR, because the tools must be used for such basic things, they are commonly available for a reasonable price. It is usually just the low production, highly specialized tools (like a super thin 7-point star-shaped socket capable of 150 ft. lbs. (203 Nm) of torque that is used only on the crankshaft nut of the limited production what-dya-callit engine) that tend to be outrageously expensive and hard to find. Hopefully, you will probably never need such a tool.

Special tools can be as inexpensive and simple as an adjustable strap wrench or as complicated as an ignition tester. A few common specialty tools are listed here, but check with your dealer or with other boaters for help in determining if there are any special tools for YOUR particular engine. There is an added advantage in seeking advice from others, chances are they may have already found the special tool you will need, and know how to get it cheaper (or even let you borrow it).

Electronic Tools

BATTERY TESTERS

The best way to test a non-sealed battery is using a hydrometer to check the specific gravity of the acid. Luckily, these are usually inexpensive and are available at most parts stores. Just be careful because the larger testers are usually designed for larger batteries and may require more acid than you will be able to draw from the battery cell. Smaller testers (usually a short, squeeze bulb type) will require less acid and should work on most batteries.

Electronic testers are available and are often necessary to tell if a sealed battery is usable. Luckily, many parts stores have them on hand and are willing to test your battery for you.

BATTERY CHARGERS

◆ **See Figure 58**

If you are a weekend boater and take your boat out every week, then you will most likely want to buy a battery charger to keep your battery fresh. There are many types available, from low amperage trickle chargers to electronically controlled battery maintenance tools that monitor the battery voltage to prevent over or undercharging. This last type is especially useful if you store your boat for any length of time (such as during the severe winter months found in many Northern climates).

Even if you use your boat on a regular basis, you will eventually need a battery charger. The charger should be used anytime the boat is going to be in storage for more than a few weeks or so. Never leave the dock or loading ramp without a battery that is fully charged.

Also, some smaller batteries are shipped dry and in a partial charged state. Before placing a new battery of this type into service it must be filled and properly charged. Failure to properly charge a battery (which was shipped dry) before it is put into service will prevent it from ever reaching a fully charged state.

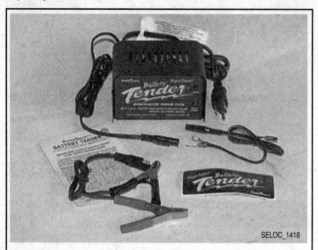

Fig. 58 The Battery Tender® is more than just a battery charger, when left connected, it keeps your battery fully charged

MULTI-METERS (DVOMS)

◆ See Figure 59

Multi-meters or Digital Volt Ohmmeter (DVOMs) are an extremely useful tool for troubleshooting electrical problems. They can be purchased in either analog or digital form and have a price range to suit any budget. A multi-meter is a voltmeter, ammeter and ohmmeter (along with other features) combined into one instrument. It is often used when testing solid state circuits because of its high input impedance (usually 10 mega ohms or more). A brief description of the multi-meter main test functions follows:

• Voltmeter - the voltmeter is used to measure voltage at any point in a circuit or to measure the voltage drop across any part of a circuit. Voltmeters usually have various scales and a selector switch to allow the reading of different voltage ranges. The voltmeter has a positive and a negative lead. To avoid the possibility of damage to the meter, whenever possible, connect the negative lead to the negative (-) side of the circuit (to ground or nearest the ground side of the circuit) and connect the positive lead to the positive (+) side of the circuit (to the power source or the nearest power source). Luckily, most quality DVOMs can adjust their own polarity internally and will

indicate (without damage) if the leads are reversed. Note that the negative voltmeter lead will always be black and that the positive voltmeter will always be some color other than black (usually red).

• Ohmmeter - the ohmmeter is designed to read resistance (measured in ohms) in a circuit or component. Most ohmmeters will have a selector switch which permits the measurement of different ranges of resistance (usually the selector switch allows the multiplication of the meter reading by 10, 100, 1,000 and 10,000). Some ohmmeters are "auto-ranging" which means the meter itself will determine which scale to use. Since the meters are powered by an internal battery, the ohmmeter can be used like a self-powered test light. When the ohmmeter is connected, current from the ohmmeter flows through the circuit or component being tested. Since the ohmmeter's internal resistance and voltage are known values, the amount of current flow through the meter depends on the resistance of the circuit or component being tested. The ohmmeter can also be used to perform a continuity test for suspected open circuits. In using the meter for making continuity checks, do not be concerned with the actual resistance readings. Zero resistance, or any ohm reading, indicates continuity in the circuit. Infinite resistance indicates an opening in the circuit. A high resistance reading where there should be little or none indicates a problem in the circuit. Checks for short circuits are made in the same manner as checks for open circuits, except that the circuit must be isolated from both power and normal ground. Infinite resistance indicates no continuity, while zero resistance indicates a dead short.

✳✳ WARNING

Never use an ohmmeter to check the resistance of a component or wire while there is voltage applied to the circuit.

• Ammeter - an ammeter measures the amount of current flowing through a circuit in units called amperes or amps. At normal operating voltage, most circuits have a characteristic amount of amperes, called "current draw" which can be measured using an ammeter. By referring to a specified current draw rating, then measuring the amperes and comparing the two values; one can determine what is happening within the circuit to aid in diagnosis. An open circuit, for example, will not allow any current to flow, so the ammeter reading will be zero. A damaged component or circuit will have an increased current draw, so the reading will be high. The ammeter is always connected in series with the circuit being tested. All of the current that normally flows through the circuit must also flow through the ammeter; if there is any other path for the current to follow, the ammeter reading will not be accurate. The ammeter itself has very little resistance to current flow and, therefore, will not affect the circuit, but, it will measure current draw only when the circuit is closed and electricity is flowing. Excessive current draw can blow fuses and drain the battery, while a reduced current draw can cause motors to run slowly, lights to dim and other components to not operate properly.

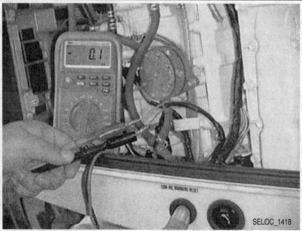

Fig. 59 Multi-meters, such as this one from UEI, are an extremely useful tool for troubleshooting electrical problems

Fig. 60 Cylinder compression test results are extremely valuable indicators of internal engine condition

GAUGES

Compression Gauge

◆ See Figure 60

An important element in checking the overall condition of your engine is to check compression. This becomes increasingly more important on outboards with high hours. Compression gauges are available as screw-in types and hold-in types. The screw-in type is slower to use, but eliminates the possibility of a faulty reading due to pressure escaping by the seal. A compression reading will uncover many problems that can cause rough running. Normally, these are not the sort of problems that can be cured by a tune-up.

Vacuum/Pressure Gauge/Pump

◆ See Figures 61, 62 and 63

Vacuum gauges are handy for discovering air leaks, late ignition or valve timing, and a number of other problems. A hand-held vacuum/pressure pump can be purchased at many automotive or marine parts stores and can be used for multiple purposes. The gauge can be used to measure vacuum or pressure in a line, while the pump can be used to manually apply vacuum or pressure to a solenoid or fitting. The hand-held pump can also be used to power-bleed trailer and tow vehicle brakes.

Measuring Tools

Eventually, you are going to have to measure something. To do this, you will need at least a few precision tools.

MICROMETERS & CALIPERS

Micrometers and calipers are devices used to make extremely precise measurements. The simple truth is that you really won't have the need for many of these items just for routine maintenance. But, measuring tools, such as an outside caliper can be handy during repairs. And, if you decide to tackle a major overhaul, a micrometer will absolutely be necessary.

Should you decide on becoming more involved in boat engine mechanics, such as repair or rebuilding, then these tools will become very important. The success of any rebuild is dependent, to a great extent on the ability to check the size and fit of components as specified by the manufacturer. These measurements are often made in thousandths and ten-thousandths of an inch.

Micrometers

◆ See Figures 64 and 65

A micrometer is an instrument made up of a precisely machined spindle that is rotated in a fixed nut, opening and closing the distance between the

Fig. 61 Vacuum gauges are useful for troubleshooting including testing some fuel pumps

Fig. 62 You can also use the gauge on a hand-operated vacuum pump for tests

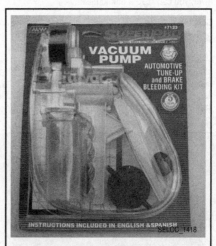

Fig. 63 Hand-held vacuum/pressure pumps are available at most parts stores

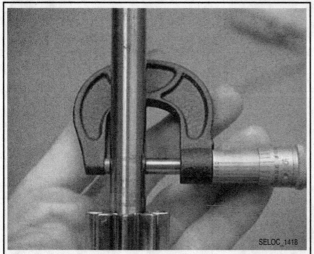

Fig. 64 Outside micrometers measure thickness, like shims or a shaft diameter

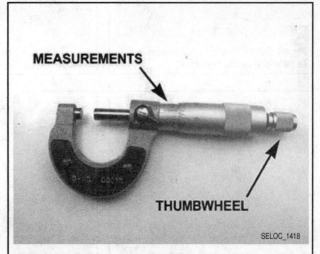

Fig. 65 Be careful not to over-tighten the micrometers always use the thumbwheel

end of the spindle and a fixed anvil. When measuring using a micrometer, don't over-tighten the tool on the part as either the component or tool may be damaged, and either way, an incorrect reading will result. Most micrometers are equipped with some form of thumbwheel on the spindle that is designed to freewheel over a certain light touch (automatically adjusting the spindle and preventing it from over-tightening).

Outside micrometers can be used to check the thickness of parts such as shims or the outside diameter of components like the crankshaft journals. They are also used during many rebuild and repair procedures to measure the diameter of components such as the pistons. The most common type of micrometer reads in 1/1000 of an inch. Micrometers that use a vernier scale can estimate to 1/10 of an inch.

Inside micrometers are used to measure the distance between two parallel surfaces. For example, in powerhead rebuilding work, the "inside mike" measures cylinder bore wear and taper. Inside mikes are graduated the same way as outside mikes and are read the same way as well.

Remember that an inside mike must be absolutely perpendicular to the work being measured. When you measure with an inside mike, rock the mike gently from side to side and tip it back and forth slightly so that you span the widest part of the bore. Just to be on the safe side, take several readings. It takes a certain amount of experience to work any mike with confidence.

Metric micrometers are read in the same way as inch micrometers, except that the measurements are in millimeters. Each line on the main scale equals 1mm. Each fifth line is stamped 5, 10, 15 and so on. Each line on the thimble scale equals 0.01 mm. It will take a little practice, but if you can read an inch mike, you can read a metric mike.

Calipers

◆ See Figures 66, 67 and 68

Inside and outside calipers are useful devices to have if you need to measure something quickly and absolute precise measurement is not necessary. Simply take the reading and then hold the calipers on an accurate steel rule. Calipers, like micrometers, will often contain a thumbwheel to help ensure accurate measurement.

DIAL INDICATORS

◆ See Figure 69

A dial indicator is a gauge that utilizes a dial face and a needle to register measurements. There is a movable contact arm on the dial indicator. When the arm moves, the needle rotates on the dial. Dial indicators are calibrated to show readings in thousandths of an inch and typically, are used to measure end-play and runout on various shafts and other components.

Dial indicators are quite easy to use, although they are relatively expensive. A variety of mounting devices are available so that the indicator can be used in a number of situations. Make certain that the contact arm is always parallel to the movement of the work being measured.

TELESCOPING GAUGES

◆ See Figure 70

A telescope gauge is really only used during rebuilding procedures (NOT during basic maintenance or routine repairs) to measure the inside of bores. It can take the place of an inside mike for some of these jobs. Simply insert the gauge in the hole to be measured and lock the plungers after they have contacted the walls. Remove the tool and measure across the plungers with an outside micrometer.

DEPTH GAUGES

◆ See Figure 71

A depth gauge can be inserted into a bore or other small hole to determine exactly how deep it is. One common use for a depth gauge is measuring the distance the piston sits below the deck of the block at top dead center. Some outside calipers contain a built-in depth gauge so you can save money and buy just one tool.

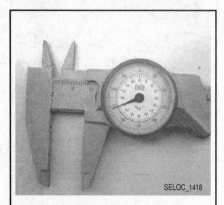

Fig. 66 Calipers are the fast and easy way to make precise measurements

Fig. 67 Calipers can also be used to measure depth . . .

Fig. 68 . . . and inside diameter measurements, to 0.001 in. accuracy

Fig. 69 This dial indicator is measuring the end-play of a crankshaft during a powerhead rebuild

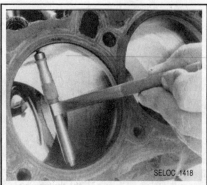

Fig. 70 Telescoping gauges are used during powerhead rebuilding procedures to measure the inside diameter of bores

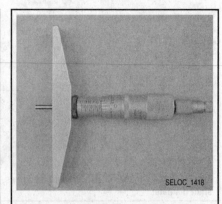

Fig. 71 Depth gauges are used to measure the depth of bore or other small holes

FASTENERS, MEASUREMENTS AND CONVERSIONS

Bolts, Nuts and Other Threaded Retainers

◆ See Figures 72 and 73

Although there are a great variety of fasteners found in the modern boat engine, the most commonly used retainer is the threaded fastener (nuts, bolts, screws, studs, etc). Most threaded retainers may be reused, provided that they are not damaged in use or during the repair.

■ Some retainers (such as stretch bolts or torque prevailing nuts) are designed to deform when tightened or in use and should not be reused.

Whenever possible, we will note any special retainers which should be replaced during a procedure. But you should always inspect the condition of a retainer when it is removed and you should replace any that show signs of damage. Check all threads for rust or corrosion that can increase the torque necessary to achieve the desired clamp load for which that fastener was originally selected. Additionally, be sure that the driver surface itself (on the fastener) is not compromised from rounding or other damage. In some cases a driver surface may become only partially rounded, allowing the driver to catch in only one direction. In many of these occurrences, a fastener may be installed and tightened, but the driver would not be able to grip and loosen the fastener again. (This could lead to frustration down the line should that component ever need to be disassembled again).

If you must replace a fastener, whether due to design or damage, you must always be sure to use the proper replacement. In all cases, a retainer of the same design, material and strength should be used. Markings on the heads of most bolts will help determine the proper strength of the fastener. The same material, thread and pitch must be selected to assure proper installation and safe operation of the motor afterwards.

Thread gauges are available to help measure a bolt or stud's thread. Most part or hardware stores keep gauges available to help you select the proper size. In a pinch, you can use another nut or bolt for a thread gauge. If the bolt you are replacing is not too badly damaged, you can select a match by finding another bolt that will thread in its place. If you find a nut that will thread properly onto the damaged bolt, then use that nut as a gauge to help select the replacement bolt. If however, the bolt you are replacing is so badly damaged (broken or drilled out) that its threads cannot be used as a gauge, you might start by looking for another bolt (from the same assembly or a similar location) which will thread into the damaged bolt's mounting. If so, the other bolt can be used to select a nut; the nut can then be used to select the replacement bolt.

In all cases, be absolutely sure you have selected the proper replacement. Don't be shy, you can always ask the store clerk for help.

✳✳ WARNING

Be aware that when you find a bolt with damaged threads, you may also find the nut or tapped bore into which it was threaded has also been damaged. If this is the case, you may have to drill and tap the hole, replace the nut or otherwise repair the threads. Never try to force a replacement bolt to fit into the damaged threads.

Torque

Torque is defined as the measurement of resistance to turning or rotating. It tends to twist a body about an axis of rotation. A common example of this would be tightening a threaded retainer such as a nut, bolt or screw. Measuring torque is one of the most common ways to help assure that a threaded retainer has been properly fastened.

When tightening a threaded fastener, torque is applied in three distinct areas, the head, the bearing surface and the clamp load. About 50 percent of the measured torque is used in overcoming bearing friction. This is the friction between the bearing surface of the bolt head, screw head or nut face and the base material or washer (the surface on which the fastener is rotating). Approximately 40 percent of the applied torque is used in overcoming thread friction. This leaves only about 10 percent of the applied torque to develop a useful clamp load (the force that holds a joint together). This means that friction can account for as much as 90 percent of the applied torque on a fastener.

Standard and Metric Measurements

Specifications are often used to help you determine the condition of various components, or to assist you in their installation. Some of the most common measurements include length (in. or cm/mm), torque (ft. lbs., inch lbs. or Nm) and pressure (psi, in. Hg, kPa or mm Hg).

In some cases, that value may not be conveniently measured with what is available in your toolbox. Luckily, many of the measuring devices that are available today will have two scales so U.S. or Metric measurements may easily be taken. If any of the various measuring tools that are available to you do not contain the same scale as listed in your specifications, use the conversion factors that are provided in the Specifications section to determine the proper value.

The conversion factor chart is used by taking the given specification and multiplying it by the necessary conversion factor. For instance, looking at the first line, if you have a measurement in inches such as "free-play should be 2 in." but your ruler reads only in millimeters, multiply 2 in. by the conversion factor of 25.4 to get the metric equivalent of 50.8mm. Likewise, if a specification was given only in a Metric measurement, for example in Newton Meters (Nm), then look at the center column first. If the measurement is 100 Nm, multiply it by the conversion factor of 0.738 to get 73.8 ft. lbs.

A - Length
B - Diameter (major diameter)
C - Threads per inch or mm
D - Thread length
E - Size of the wrench required
F - Root diameter (minor diameter)

SELOC_1418

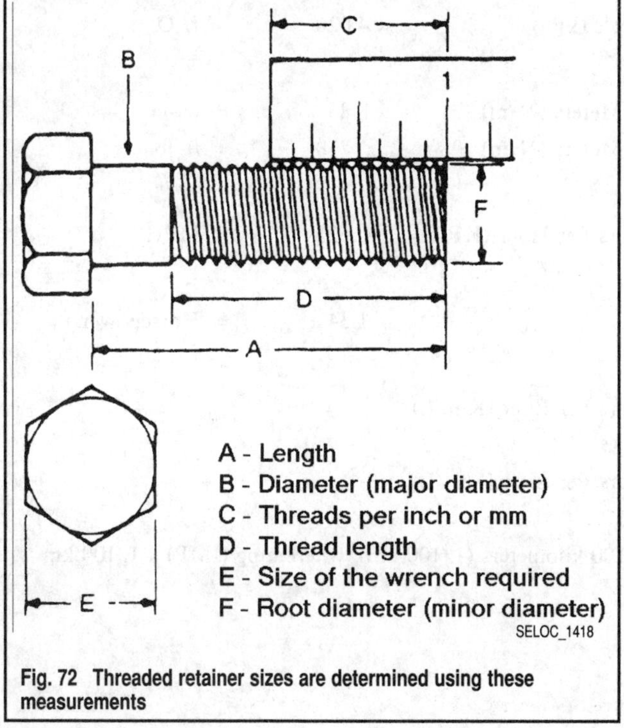

Fig. 72 Threaded retainer sizes are determined using these measurements

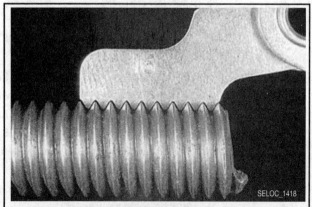

SELOC_1418

Fig. 73 Thread gauges measure the threads-per-inch and the pitch of a bolt or stud's threads

SPECIFICATIONS

CONVERSION FACTORS

LENGTH–DISTANCE

Inches (in.)	x 25.4	= Millimeters (mm)	x .0394	= Inches
Feet (ft.)	x .305	= Meters (m)	x 3.281	= Feet
Miles	x 1.609	= Kilometers (km)	x .0621	= Miles

VOLUME

Cubic Inches (in3)	x 16.387	= Cubic Centimeters	x .061	= in3
IMP Pints (IMP pt.)	x .568	= Liters (L)	x 1.76	= IMP pt.
IMP Quarts (IMP qt.)	x 1.137	= Liters (L)	x .88	= IMP qt.
IMP Gallons (IMP gal.)	x 4.546	= Liters (L)	x .22	= IMP gal.
IMP Quarts (IMP qt.)	x 1.201	= US Quarts (US qt.)	x .833	= IMP qt.
IMP Gallons (IMP gal.)	x 1.201	= US Gallons (US gal.)	x .833	= IMP gal.
Fl. Ounces	x 29.573	= Milliliters	x .034	= Ounces
US Pints (US pt.)	x .473	= Liters (L)	x 2.113	= Pints
US Quarts (US qt.)	x .946	= Liters (L)	x 1.057	= Quarts
US Gallons (US gal.)	x 3.785	= Liters (L)	x .264	= Gallons

MASS–WEIGHT

Ounces (oz.)	x 28.35	= Grams (g)	x .035	= Ounces
Pounds (lb.)	x .454	= Kilograms (kg)	x 2.205	= Pounds

PRESSURE

Pounds Per Sq. In. (psi)	x 6.895	= Kilopascals (kPa)	x .145	= psi
Inches of Mercury (Hg)	x .4912	= psi	x 2.036	= Hg
Inches of Mercury (Hg)	x 3.377	= Kilopascals (kPa)	x .2961	= Hg
Inches of Water (H_2O)	x .07355	= Inches of Mercury	x 13.783	= H_2O
Inches of Water (H_2O)	x .03613	= psi	x 27.684	= H_2O
Inches of Water (H_2O)	x .248	= Kilopascals (kPa)	x 4.026	= H_2O

TORQUE

Pounds–Force Inches (in–lb)	x .113	= Newton Meters (N·m)	x 8.85	= in–lb
Pounds–Force Feet (ft–lb)	x 1.356	= Newton Meters (N·m)	x .738	= ft–lb

VELOCITY

Miles Per Hour (MPH)	x 1.609	= Kilometers Per Hour (KPH)	x .621	= MPH

POWER

Horsepower (Hp)	x .745	= Kilowatts	x 1.34	= Horsepower

FUEL CONSUMPTION*

Miles Per Gallon IMP (MPG)	x .354	= Kilometers Per Liter (Km/L)
Kilometers Per Liter (Km/L)	x 2.352	= IMP MPG
Miles Per Gallon US (MPG)	x .425	= Kilometers Per Liter (Km/L)
Kilometers Per Liter (Km/L)	x 2.352	= US MPG

*It is common to covert from miles per gallon (mpg) to liters/100 kilometers (1/100 km), where mpg (IMP) x 1/100 km = 282 and mpg (US) x 1/100 km = 235.

TEMPERATURE

Degree Fahrenheit (°F)	= (°C x 1.8) + 32
Degree Celsius (°C)	= (°F – 32) x .56

SELOC_1418

Metric Bolts						
Relative Strength Marking	4.6, 4.8			8.8		
Bolt Markings						
Usage	Frequent			Infrequent		
Bolt Size	Maximum Torque			Maximum Torque		
Thread Size x Pitch (mm)	Ft-Lb	Kgm	Nm	Ft-Lb	Kgm	Nm
6 x 1.0	2–3	.2–.4	3–4	3–6	.4–.8	5–8
8 x 1.25	6–8	.8–1	8–12	9–14	1.2–1.9	13–19
10 x 1.25	12–17	1.5–2.3	16–23	20–29	2.7–4.0	27–39
12 x 1.25	21–32	2.9–4.4	29–43	35–53	4.8–7.3	47–72
14 x 1.5	35–52	4.8–7.1	48–70	57–85	7.8–11.7	77–110
16 x 1.5	51–77	7.0–10.6	67–100	90–120	12.4–16.5	130–160
18 x 1.5	74–110	10.2–15.1	100–150	130–170	17.9–23.4	180–230
20 x 1.5	110–140	15.1–19.3	150–190	190–240	26.2–46.9	160–320
22 x 1.5	150–190	22.0–26.2	200–260	250–320	34.5–44.1	340–430
24 x 1.5	190–240	26.2–46.9	260–320	310–410	42.7–56.5	420–550

SELOC_1418

Typical Torque Values - Metric Bolts

SAE Bolts

SAE Grade Number	1 or 2			5			6 or 7		
Bolt Markings Manufacturers' marks may vary—number of lines always two less than the grade number.							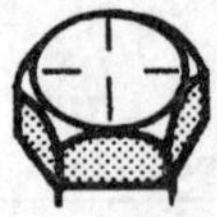		
Usage	Frequent			Frequent			Infrequent		
Bolt Size (inches)—(Thread)	Maximum Torque			Maximum Torque			Maximum Torque		
	Ft-Lb	kgm	Nm	Ft-Lb	kgm	Nm	Ft-Lb	kgm	Nm
¹⁄₄—20	5	0.7	6.8	8	1.1	10.8	10	1.4	13.5
—28	6	0.8	8.1	10	1.4	13.6			
⁵⁄₁₆—18	11	1.5	14.9	17	2.3	23.0	19	2.6	25.8
—24	13	1.8	17.6	19	2.6	25.7			
³⁄₈—16	18	2.5	24.4	31	4.3	42.0	34	4.7	46.0
—24	20	2.75	27.1	35	4.8	47.5			
⁷⁄₁₆—14	28	3.8	37.0	49	6.8	66.4	55	7.6	74.5
—20	30	4.2	40.7	55	7.6	74.5			
¹⁄₂—13	39	5.4	52.8	75	10.4	101.7	85	11.75	115.2
—20	41	5.7	55.6	85	11.7	115.2			
⁹⁄₁₆—12	51	7.0	69.2	110	15.2	149.1	120	16.6	162.7
—18	55	7.6	74.5	120	16.6	162.7			
⁵⁄₈—11	83	11.5	112.5	150	20.7	203.3	167	23.0	226.5
—18	95	13.1	128.8	170	23.5	230.5			
³⁄₄—10	105	14.5	142.3	270	37.3	366.0	280	38.7	379.6
—16	115	15.9	155.9	295	40.8	400.0			
⁷⁄₈— 9	160	22.1	216.9	395	54.6	535.5	440	60.9	596.5
—14	175	24.2	237.2	435	60.1	589.7			
1— 8	236	32.5	318.6	590	81.6	799.9	660	91.3	894.8
—14	250	34.6	338.9	660	91.3	849.8			

SELOC_1418

Typical Torque Values - U.S. Standard Bolts

2

MAINTENANCE & TUNE-UP

GENERAL INFORMATION (WHAT EVERYONE SHOULD KNOW ABOUT MAINTENANCE)

At Seloc, we estimate that 75% of engine repair work can be directly or indirectly attributed to lack of proper care for the engine. This is especially true of care during the off-season period. There is no way on this green earth for a mechanical engine, particularly an outboard motor, to be left sitting idle for an extended period of time, say for 6 months, and then be ready for instant satisfactory service.

Imagine, if you will, leaving your car or truck for 6 months, and then expecting to turn the key, having it roar to life, and being able to drive off in the same manner as a daily occurrence.

Therefore it is critical for an outboard engine to either be run (at least once a month), preferably, in the water and properly maintained between uses or for it to be specifically prepared for storage and serviced again immediately before the start of the season.

Only through a regular maintenance program can the owner expect to receive long life and satisfactory performance at minimum cost.

Many times, if an outboard is not performing properly, the owner will "nurse" it through the season with good intentions of working on the unit once it is no longer being used. As with many New Year's resolutions, the good intentions are not completed and the outboard may lie for many months before the work is begun or the unit is taken to the marine shop for repair.

Imagine, if you will, the cause of the problem being a blown head gasket. And let us assume water has found its way into a cylinder. This water, allowed to remain over a long period of time, will do considerably more damage than it would have if the unit had been disassembled and the repair work performed immediately. Therefore, if an outboard is not functioning properly, do not stow it away with promises to get at it when you get time, because the work and expense will only get worse the longer corrective action is postponed. In the example of the blown head gasket, a relatively simple and inexpensive repair job could very well develop into major overhaul and rebuild work.

Maintenance Equals Safety

OK, perhaps no one thing that we do as boaters will protect us from risks involved with enjoying the wind and the water on a powerboat. But, each time we perform maintenance on our boat or motor, we increase the likelihood that we will find a potential hazard before it becomes a problem. Each time we inspect our boat and motor, we decrease the possibility that it could leave us stranded on the water.

In this way, performing boat and engine service is one of the most important ways that we, as boaters, can help protect ourselves, our boats, and the friends and family that we bring aboard.

Outboards On Sail Boats

Owners of sailboats pride themselves in their ability to use the wind to clear a harbor or for movement from Port A to Port B, or maybe just for a day sail on a lake. For some, the outboard is carried only as a last resort - in case the wind fails completely, or in an emergency situation or for ease of docking.

Therefore, in some cases, the outboard is stowed below, usually in a very poorly ventilated area, and subjected to moisture and stale air - in short, an excellent environment for "sweating" and corrosion.

If the owner could just take the time at least once every month, to pull out the outboard, clean it up, and give it a short run, not only would he/she have "peace of mind" knowing it will start in an emergency, but also maintenance costs will be drastically reduced.

Maintenance Coverage In This Manual

At Seloc, we strongly feel that every boat owner should pay close attention to this section. We also know that it is one of the most frequently used portions of our manuals. The material in this section is divided into sections to help simplify the process of maintenance. Be sure to read and thoroughly understand the various tasks that are necessary to keep your outboard in tip-top shape.

Topics covered in this section include:

1. General Information (What Everyone Should Know About Maintenance) - an introduction to the benefits and need for proper maintenance. A guide to tasks that should be performed before and after each use.

2. Lubrication Service - after the basic inspections that you should perform each time the motor is used, the most frequent form of periodic maintenance you will conduct will be the Lubrication Service. This section takes you through each of the various steps you must take to keep corrosion from slowly destroying your motor before your very eyes.

3. Engine Maintenance - the various procedures that must be performed on a regular basis in order to keep the motor and all of its various systems operating properly.

4. Boat Maintenance - the various procedures that must be performed on a regular basis in order to keep the boat hull and its accessories looking and working like new.

5. Tune-Up - also known as the pre-season tune-up, but don't let the name fool you. A complete tune-up is the best way to determine the condition of your outboard while also preparing it for hours and hours of hopefully trouble-free enjoyment. And if you use your boat enough during a single season, a 2nd or even 3rd tune-up could be required.

6. Winter Storage and Spring Commissioning Checklists - use these sections to guide you through the various parts of boat and motor maintenance that protect your valued boat through periods of storage and return it to operating condition when it is time to use it again.

7. Specification Charts - located at the end of the section are quick-reference, easy to read charts that provide you with critical information such as General Engine Specifications, Maintenance Intervals and Capacities.

Engine & Serial Number Identification

◆ See Figures 1 thru 5

For many years the engine serial numbers were Mercury's key to engine changes. These numbers identify the year of manufacture, the horsepower

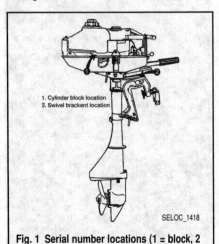

1. Cylinder block location
2. Swivel brackent location

SELOC_1418

Fig. 1 Serial number locations (1 = block, 2 = swivel bracket) - 2.5/3.3 hp outboards

SELOC_1418

Fig. 2 A serial number tag is found on top of the swivel bracket for 2.5/3.3 hp outboards

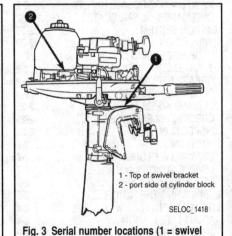

1 - Top of swivel bracket
2 - port side of cylinder block

SELOC_1418

Fig. 3 Serial number locations (1 = swivel bracket, 2 = block) - 4/5 hp outboards

rating and the parts book identification. If any correspondence or parts are required, it is still a good idea to use the engine serial number to make SURE you get the right part.

However, we're happy to report that more and more Mercury has come in line with other engine manufacturers and is sticking closer and closer to model years than ever before. At least 1 of the serial number tags used by Mercury on their newer 2-strokes includes a specific spot which displays the MODEL YEAR of the motor (along with the serial number, hp rating, weight rating, and year of manufacture).

■ **Also remember that the serial number tag establishes the model year and the year in which the engine was produced which is (or are) not necessarily the year of first installation.**

Two serial numbers tags were placed on each of these motors. One is attached to the powerhead itself, usually on the lower starboard side of the powerhead, but on some 75 hp and larger motors it may be instead affixed to the TOP of the powerhead.

All motors are also equipped with a serial number ID tag on the swivel bracket. For the smallest motors (2.5/3.3 hp models) this tag may be on the top center of the swivel bracket, but for most Mercury motors it is found on the starboard SIDE of the swivel bracket.

For more information on engine identification and specifications, please refer to the General Engine Specifications charts later in this section.

Before/After Each Use

As stated earlier, the best means of extending engine life and helping to protect yourself while on the water is to pay close attention to boat/engine maintenance. This starts with an inspection of systems and components before and after each time you use your boat.

A list of checks, inspections or required maintenance can be found in the Maintenance Intervals Chart at the end of this section. Some of these inspections or tasks are performed before the boat is launched, some only after it is retrieved and the rest, both times.

VISUALLY INSPECTING THE BOAT & MOTOR

◆ **See Figures 6 and 7**

Both before each launch and immediately after each retrieval, visually inspect the boat and motor as follows:

1. **Check the fuel and oil levels** according to the procedures in this manual. Do NOT launch a boat without properly topped off fuel and (if applicable) oil tanks. It is not worth the risk of getting stranded or of damage to the motor. Likewise, upon retrieval, check the oil and fuel levels while it is still fresh in your mind. This is a good way to track fuel consumption (one indication of engine performance). For oil injected 2-stroke motors, compare the fuel consumption to the oil consumption (a dramatic change in proportional use may be an early sign of trouble).

2. **Check for signs of fuel or oil leakage.** Probably as important as making sure enough fuel and oil is onboard, is the need to make sure that no dangerous conditions might arise due to leaks. Thoroughly check all hoses, fittings and tanks for signs of leakage. Oil leaks may cause the boat to become stranded, or worse, could destroy the motor if undetected for a significant amount of time. Fuel leaks can cause a fire hazard, or worse, an explosive condition. This check is not only about properly maintaining your boat and motor, but about helping to protect your life.

✳✳ SELOC CAUTION

On fuel injected motors (ESPECIALLY EFI or Optimax), fuel is pumped at high pressure through various lines under the motor cowl. The smallest leak will allow for fuel to spray in a fine, atomized and highly combustible stream from the damaged hose/fitting. It is critical that you remove the cowling and turn the key to the ON position (to energize the fuel pump and begin building system pressure) for a quick check before starting the motor (to ensure that no leaks are present). Even so, leaks may not show until the motor is operating so it is a good idea to either leave the cover off until the motor is running or to remove it again later in the day to double-check that you are leak free. Of course, if you DO remove the cover with the engine running take GREAT care to prevent contact with any moving parts.

3. **Inspect the boat hull and engine cases** for signs of corrosion or damage. Don't launch a damaged boat or motor. And don't surprise yourself dockside or at the launch ramp by discovering damage that went unnoticed last time the boat was retrieved. Repair any hull or case damage now.

4. **Check the battery** connections to make sure they are clean and tight. A loose or corroded connection will cause charging problems (damaging the system or preventing charging). There's only one thing worse than a dead battery dockside/launch ramp and that's a dead battery in the middle of a bay, river or worse, the ocean. Whenever possible, make a quick visual check of battery electrolyte levels (keeping an eye on the level will give some warning of overcharging problems). This is especially true if the engine is operated at high speeds for extended periods of time.

5. **Check the propeller (impeller on jet drives) and gearcase.** Make sure the propeller shows no signs of damage. A broken or bent propeller may allow the engine to over-rev and it will certainly waste fuel. The gearcase should be checked before and after each use for signs of leakage. Check the gearcase oil for signs of contamination if any leakage is noted. Also, visually check behind the propeller for signs of entangled rope or fishing lines that could cut through the lower gearcase propeller shaft seal. This is a common cause of gearcase lubricant leakage, and eventually, water

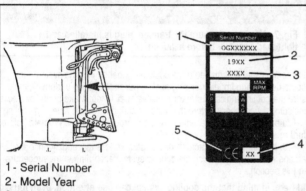

1 - Serial Number
2 - Model Year
3 - Model Description
4 - Year Manufactured
5 - Certified Europe Insignia

SELOC_1418

Fig. 4 All 6-250 hp motors should have a serial number tag like this on the starboard side of the swivel bracket

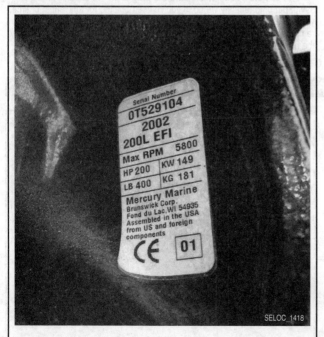

SELOC_1418

Fig. 5 A typical serial number tag found on Mercury swivel brackets

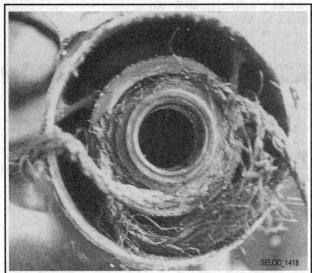

Fig. 6 Rope or fishing line entangled behind the propeller can cut through the seal, allowing water in or lubricant out

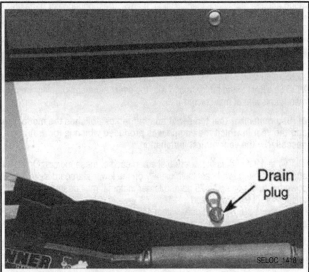

Fig. 7 Always make sure the transom plug is installed and tightened securely before a launch

contamination that can lead to gearcase failure. Even if no gearcase leakage is noted when the boat is first retrieved, check again next time before launching. A nicked seal might not seep fluid right away when still swollen from heat immediately after use, but might begin seeping over the next day, week or month as it sat, cooled and dried out.

6. **Check all accessible fasteners for tightness.** Make sure all easily accessible fasteners appear to be tight. This is especially true for the propeller nut, any anode retaining bolts, all steering or throttle linkage fasteners and the engine clamps or mounting bolts. Don't risk losing control or becoming stranded due to loose fasteners. Perform these checks before heading out, and immediately after you return (so you'll know if anything needs to be serviced before you want to launch again.)

7. **Check operation of all controls including the throttle/shifter, steering and emergency stop/start switch and/or safety lanyard.** Before launching, make sure that all linkage and steering components operate properly and move smoothly through their range of motion. All electrical switches (such as power trim/tilt) and especially the emergency stop system(s) must be in proper working order. While underway, watch for signs that a system is not working or has become damaged. With the steering, shifter or throttle, keep a watchful eye out for a change in resistance or the start of jerky/notchy movement.

8. **Check the water pump intake grate and water indicator.** The water pump intake grate should be clean and undamaged before setting out.

Remember that a damaged grate could allow debris into the system that could destroy the impeller or clog cooling passages. Once underway, make sure the cooling indicator stream is visible at all times. Make periodic checks, including 1 final check before the motor is shut down each time. If a cooling indicator stream is not present at any point, troubleshoot the problem before further engine operation.

9. **If used in salt, brackish or polluted waters** thoroughly rinse the engine (and hull), then flush the cooling system according to the procedure in this section.

■ Keep in mind that the cooling system can use attention, even if used in fresh waters. Sand, silt or other deposits can help clog passages, chemicals or pollutants can speed corrosion. It's a good idea to flush your motor after every use, regardless of where you use it.

10. **Visually inspect all anodes** after each use for signs of wear, damage or to make sure they just plain didn't fall off (especially if you weren't careful about checking all the accessible fasteners the last time you launched).

11. **For Pete's sake, make sure the plug is in!** We shouldn't have to say it, but unfortunately we do. If you've been boating for any length of time, you've seen or heard of someone whose backed a trailer down a launch ramp, forgetting to check the transom drain plug before submerging (literally) the boat. Always make sure the transom plug is installed and tight before a launch.

LUBRICATION

About Lubrication

◆ See Figure 8

An outboard motor's greatest enemy is corrosion. Face it, oil and water just don't mix and, as anyone who has visited a junkyard knows, metal and water aren't the greatest of friends either. To expose an engine to a harsh marine environment of water and wind is to expect that these elements will take their toll over time. But, there is a way to fight back and help prevent the natural process of corrosion that will destroy your beloved boat motor.

Various marine grade lubricants are available that serve two important functions in preserving your motor. Lubricants reduce friction on metal-to-metal contact surfaces and, they also displace air and moisture, therefore slowing or preventing corrosion damage. Periodic lubrication services are your best method of preserving an outboard motor. Marine lubricants are designed for the harsh environment to which outboards are exposed and are designed to stay in place, even when submerged in water (and we've got some bathing suits with marine grease stains to attest to this).

Lubrication takes place through various forms. For all engines, internal moving parts are lubricated by engine oil, in the case of 2-stroke motors it is through oil contained in the fuel/oil mixture. Also, on all conventional motors

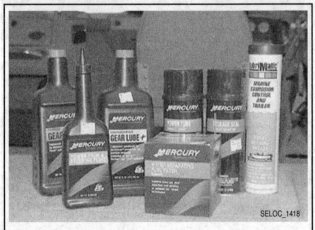

Fig. 8 Mercury/Quicksilver products will do much to keep the outboard unit running right

(as opposed to jet drives) the gearcase is filled with gear oil that lubricates the driveshaft, propshaft, gears and other internal gearcase components. The gear oil should be periodically checked and replaced following the appropriate Engine Maintenance procedures. Perform these services based on time or engine use, as outlined in the Maintenance Intervals chart at the end of this section.

For motors equipped with power trim/tilt, the fluid level and condition in the reservoir should be checked periodically to ensure proper operation. Proper fluid level not only ensures that the system will function properly, but also helps lubricate and protect the internal system components from corrosion.

Most other forms of lubrication occur through the application of some form of marine grade (usually either Quicksilver Anti-Corrosion Grease or Quicksilver 2-4-C with Teflon), either applied by hand (an old toothbrush can be helpful in preventing a mess) or using a grease gun to pump the lubricant into grease fittings (also known as zerk fittings). When using a grease gun, do not pump excessive amounts of grease into the fitting. Unless otherwise directed, pump until either the rubber seal (if used) begins to expand or until the grease just begins to seep from the joints of the component being lubricated (if no seal is used).

To ensure your motor is getting the protection it needs, perform a visual inspection of the various lubrication points at least once a week during regular seasonal operation (this assumes that the motor is being used at least once a week). Follow the recommendations given and perform the various lubricating services at least every 60 days when the boat is operated in fresh water or every 30 days when the boat is operated in salt, brackish or polluted waters. We said **at least** meaning you should perform these services more often, if a need is discovered by your weekly inspections.

■ Jet drive models require one form of lubrication EVERY time that they are used. The jet drive bearing should be greased, following the procedure given in this section, after every day of boating. But don't worry, it only takes a minute once you've done it before.

Fig. 9 Various lubrication points on the powerhead should be maintained regularly to ensure a long service life including all rotating. . .

Lubricating the Motor

◆ See Figures 9 thru 15

The first thing you should do upon purchasing a new or "new to you" motor is to remove the engine top cover and look for signs of grease. Note all components that have been freshly greased (or if the motor has been neglected that shows signs of wear or dirt/contamination that has collected on the remnants of old applied grease). If the motor shows signs of dirt, corrosion or wear, clean those components thoroughly and apply a fresh coat of grease.

Thereafter, lubricate those surfaces at LEAST every 60 days (more often if used in salt or corrosive environments) and grease all necessary surfaces regularly to keep them clean and well lubricated. As a general rule of thumb any point where two metallic mechanical parts connect and push, pull, turn, slide, pivot on each other should be greased. For most motors this will include shift and throttle cables and/or linkage, steering and swiveling points and items such as the cowl clamp bolts (on smaller motors) and top cover or cowling clamp levers.

■ For more information on greasing and lubrication points, check your owner's manual. Most Owner's manuals will provide 1 or more illustrations to help you properly identify all necessary greasing points.

Points such as the swivel bracket and/or the tilt tube will normally be equipped with grease (zerk) fittings. For these, use a grease gun to carefully pump small amounts of grease into the fittings, displacing some of the older grease and lubricating the internal surfaces of the swivel and tilt tubes. Some engine cowl levers require a dab of lubricant be applied manually over sliding surfaces, but some may also equipped with grease fittings for lubrication using a grease gun.

✳✳ SELOC CAUTION

When lubricating the steering linkage grease fitting the cable core must be completely retracted into the housing BEFORE applying the lubricant to the fitting. Failure to do this may cause hydraulic lock of the cable.

Items without a grease fitting, such as the steering ram, cable ends, shifter and carburetor or throttle body linkage all must normally be greased by hand using a small dab of lubricant. Be sure not to over apply grease as it is just going to get over everything and exposed grease will tend to attract and hold dirt or other particles of general crud. For this reason it is always a good idea to wipe away the old grease before applying fresh lubricant to these surfaces.

■ The automatic belt tensioner assembly on some motors, like the mid-range EFI models, may contain a zerk fitting for greasing. When present, it's usually pretty obvious, but check the assembly on fuel injected models just to be sure.

Lubrication Inside the Boat

The following points inside the boat will also usually benefit from lubrication with an all purpose marine grade lubricant:

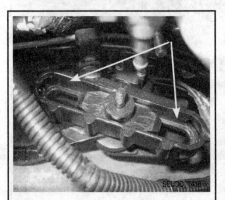

Fig. 10 . . .or sliding linkage points and cable ends

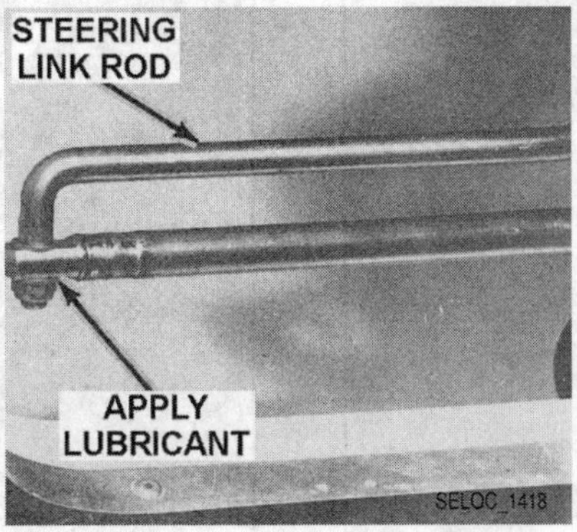

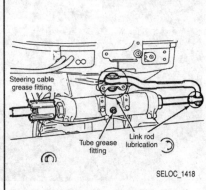

Fig. 12 Steering cable grease fittings and pivot point lubrication points

Fig. 11 Apply a marine grade grease to the steering linkage

Fig. 13 The swivel bracket usually contains grease fittings for the steering linkage. . .

Fig. 14 . . . and for the vertical portion of the swivel housing

Fig. 15 The propeller should be removed periodically to clean and re-grease the prop shaft splines

• Remote control cable ends next to the hand nut. DO NOT over-lubricate the cable.
• Steering arm pivot socket.
• Exposed shaft of the cable passing through the cable guide tube.
• Steering link rod to steering cable.

Lower Unit

◆ See Figures 16 and 17

Regular maintenance and inspection of the lower unit is critical for proper operation and reliability. A lower unit can quickly fail if it becomes heavily contaminated with water or excessively low on oil. The most common cause of a lower unit failure is water contamination.

Water in the lower unit is usually caused by fishing line or other foreign material, becoming entangled around the propeller shaft and damaging the seal. If the line is not removed, it will eventually cut the propeller shaft seal and allow water to enter the lower unit. Fishing line has also been known to cut a groove in the propeller shaft if left neglected over time. This area should be checked frequently.

■ Though the gearcase drain plugs are found on the side of the housing for most models, on most Pro/XS/Sport model gearcases the plugs are found at the bottom center of the propeller shaft bearing carrier, just behind the propeller (which must be removed for access). One exception though - on 15-spline shaft Pro XS models, the plug is at the port side at the tip of the nose cone.

OIL RECOMMENDATIONS

◆ See Figure 18

Use only Quicksilver Gear Lube or and equivalent high quality SAE 85-90 weight hypoid gear oil. These oils are proprietary lubricants designed to ensure optimal performance and to minimize corrosion in the lower unit.

■ Remember, it is this lower unit lubricant that prevents corrosion and lubricates the internal parts of the drive gears. Lack of lubrication due to water contamination or the improper type of oil can cause catastrophic lower unit failure.

Fig. 16 This lower unit was destroyed because the bearing carrier froze due to lack of lubrication.

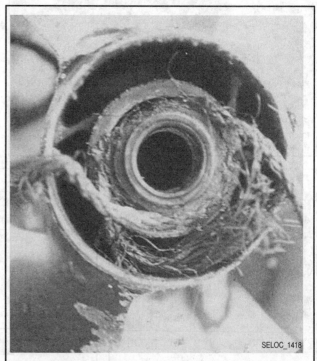

Fig. 17 Fishing line entangled behind the prop can actually cut through the seal

CHECKING GEARCASE OIL LEVEL & CONDITION

◆ See Figures 19 and 20

Visually inspect the gearcase before and after each use for signs of leakage. At least monthly, or as needed, remove the gearcase level plug in order to check the lubricant level and condition as follows:

1. Position the engine in the upright position with the motor shut off for at least 1 hour. Whenever possible, checking the level overnight cold will give a true indication of the level without having to account for heat expansion.

2. Disconnect the negative battery cable or remove the propeller for safety.

✳✳ SELOC CAUTION

Always observe extreme care when working anywhere near the propeller. Take steps to ensure that no accidental attempt to start the engine occurs while work is being performed or remove the propeller completely to be safe.

3. Position a small drain pan under the gearcase, then unthread the drain/filler plug at the bottom of the housing (behind the propeller on most Pro/XS/Sport models) and allow a small sample (a teaspoon or less) to drain from the gearcase. Quickly install the drain/filler plug and tighten securely.

4. Examine the gear oil as follows:

a. Visually check the oil for obvious signs of water. A small amount of moisture may be present from condensation, especially if a motor has been stored for some time, but a milky appearance indicates that either the fluid has not been changed in ages or the gearcase allowing some water to intrude. If significant water contamination is present, the first suspect is the propeller shaft seal.

b. Dip an otherwise clean finger into the oil, then rub a small amount of the fluid between your finger and your thumb to check for the presence of debris. The lubricant should feel smooth. A **very** small amount of metallic shavings may be present, but should not really be felt. Large amounts of grit or metallic particles indicate the need to overhaul the gearcase looking for damaged/worn gears, shafts, bearings or thrust surfaces.

■ **If a large amount of lubricant escapes when the level/vent plug is removed in the next step, either the gearcase was seriously overfilled on the last service, the crankcase is still too hot from running the motor in gear (and the fluid is expanded) or a large amount of water has entered the gearcase. If the later is true, some water should escape before the oil and/or the oil will be a milky white in appearance (showing the moisture contamination).**

5. Next, remove the level/vent plug from the top of the gearcase and ensure the lubricant level is up to the bottom of the level/vent plug opening. A very small amount of fluid may be added through the level plug, but larger

amounts of fluid should be added through the drain/filler plug opening to make certain that the case is properly filled. If necessary, add gear oil until fluid flows from the level/vent opening. If much more than 1 oz. (29 ml) is required to fill the gearcase, check the case carefully for leaks. Install the drain/filler plugs and/or the level/vent plug, then tighten both securely.

■ **One trick that makes adding gearcase oil less messy is to install the level/vent plug BEFORE removing the pump from the drain/filler opening and threading the drain/filler plug back into position.**

6. Once fluid is pumped into the gearcase, let the unit sit in a shaded area for at least 1 hour for the fluid to settle. Recheck the fluid level and, if necessary, add more lubricant.

7. Install the propeller and/or connect the negative battery cable, as applicable.

DRAINING & FILLING

◆ See Figures 21 and 22

✳✳ SELOC CAUTION

The EPA warns that prolonged contact with used engine oil may cause a number of skin disorders, including cancer! You should make every effort to minimize your exposure to used engine oil. Protective gloves should be worn when changing the oil. Wash your hands and any other exposed skin areas as soon as possible after exposure to used engine oil. Soap and water or waterless hand cleaner should be used.

1. Place a suitable container under the lower unit. It is usually a good idea to place the outboard in the tilted position so the drain plug is at the lowest position on the gearcase, this will help ensure the oil drains fully.

2. Loosen the oil level/vent plug (or plugs if they are separate) on the lower unit. This step is important! If the oil level/vent plug(s) cannot be loosened or removed, you cannot refill the gearcase with fluid and purge it of air.

■ **Never remove the vent or filler plugs when the lower unit is hot. Expanded lubricant will be released through the hole.**

3. Remove the oil level/vent plug(s) from the top of the gearcase, followed by the drain/filler plug from the lower end of the gearcase.

■ **Though the gearcase drain plugs are found on the side of the housing for most models, on most Pro/XS/Sport model gearcases the plugs are found at the bottom center of the propeller shaft bearing carrier, just behind the propeller (which must be removed for access). One exception though, on 15-spline shaft Pro XS models, the plug is at the port side at the tip of the nose cone.**

4. Allow the lubricant to completely drain from the lower unit.

5. If applicable, check the magnet end of the drain screw for metal

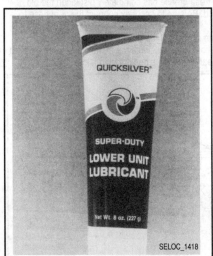

Fig. 18 Use only a suitable gear lube, like Quicksilver Gear Lube or Quicksilver High Performance/Super Duty gear lube

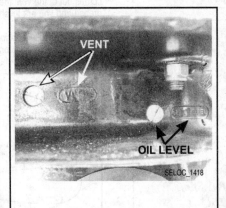

Fig. 19 Mercury often labels the oil vent/level plug (or plugs) on their gearcases

Fig. 20 Typical oil vent/level and drain plug locations on most Mercury gearcases

particles. Some amount of metal is considered normal wear is to be expected but if there are signs of metal chips or excessive metal particles, the gearcase needs to be disassembled and inspected.

6. Inspect the lubricant for the presence of a milky white substance, water or metallic particles. If any of these conditions are present, the lower unit should be serviced immediately.

7. Place the outboard in the proper position for filling the lower unit (straight up and down). The lower unit should not list to either port or starboard and should be completely vertical.

8. Insert the lubricant tube into the oil drain hole at the bottom of the lower unit and inject lubricant until the excess begins to come out the oil level hole.

■ **The lubricant must be filled from the bottom to prevent air from being trapped in the lower unit. Air displaces lubricant and can cause a lack of lubrication or a false lubricant level in the lower unit.**

9. Oil should be squeezed in using a tube or with the larger quantities, by using a pump kit to fill the gearcase through the drain plug.

■ **One trick that makes adding gearcase oil less messy is to install the level/vent plug(s) BEFORE removing the pump from the drain/filler opening and threading the drain/filler plug back into position.**

10. Using new gaskets/washers (if equipped) install the oil level/vent plug(s) first, then install the oil fill plug.

11. Wipe the excess oil from the lower unit and inspect the unit for leaks.

12. Place the used lubricant in a suitable container for transportation to an authorized recycling facility.

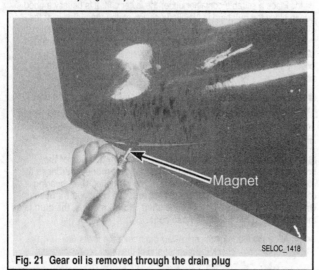

Fig. 21 Gear oil is removed through the drain plug

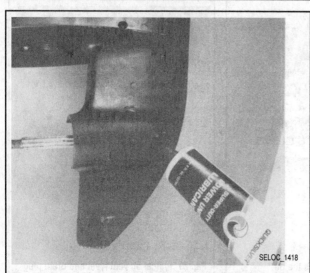

Fig. 22 The lower unit is also refilled through the drain plug

◆ **See Figure 23**

Jet drive models require special attention to ensure that the driveshaft bearing remains properly lubricated.

Mercury recommends that you lubricate the jet drive bearing using a grease gun after EACH day's use. However, at an absolute minimum, use the grease fitting every 10 hours (in fresh water) or 5 hours (in salt water). Also, after about every 30 hours of operation, the drive bearing grease must be replaced. Follow the appropriate procedure:

RECOMMENDED LUBRICANT

Use Quicksilver 2-4-C with Teflon or an equivalent water-resistant NLGI No. 1 lubricant.

DAILY BEARING LUBRICATION

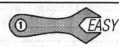

◆ **See Figures 24 and 25**

A grease fitting is located under a vent hose on the lower port side of the jet drive. Disconnect the hose from the fitting, then use a grease gun to apply enough grease to the fitting to **just** fill the vent hose. Basically, grease is pumped into the fitting until the old grease just starts to come out from the passages through the hose coupling, then reconnect the hose to the fitting.

■ **Do not attempt to just grasp the vent hose and pull, as it is a tight fit and when it does come off, you'll probably go flying if you didn't prepare for it. The easier method of removing the vent hose from the fitting is to deflect the hose to one side and snap it free from the fitting.**

GREASE REPLACEMENT

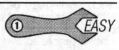

◆ **See Figures 24 and 26**

A grease fitting is located under a vent hose on the lower port side of the jet drive. This grease fitting is utilized at the end of each day's use to add fresh grease to the jet drive bearing. But, every 30 hours and/or 30 days (depending on the amount of use), the grease should be completely replaced. This is very similar to the daily greasing, except that a lot more grease it used. Disconnect the hose from the fitting (by deflecting it to the side until it snaps free from the fitting), then use a grease gun to apply enough grease to the fitting until grease exiting the assembly fills the vent hose. Then, continue to pump grease into the fitting to force out all of the old grease (you can tell this has been accomplished when fresh grease starts to come out of the vent instead of old grease, which will be slightly darker due to minor contamination from normal use). When nothing but fresh grease comes out of the vent the fresh grease has completely displaced the old grease and you are finished. Be sure to securely connect the vent hose to the fitting.

Each time this is performed, inspect the grease for signs of moisture contamination or discoloration. A gradual increase in moisture content over a few services is a sign of seal wear that is beginning to allow some seepage.

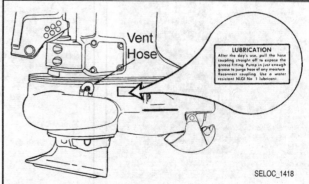

Fig. 23 Jet drive models require lubrication of the bearing after each day of use, sometimes they are equipped with a label on the housing to remind the owner

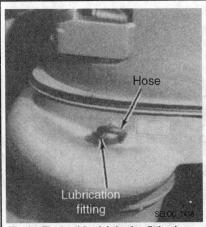

Fig. 24 The jet drive lubrication fitting is found under the vent hose

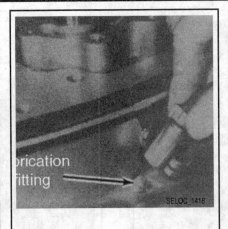

Fig. 25 Attach a grease gun to the fitting for lubrication

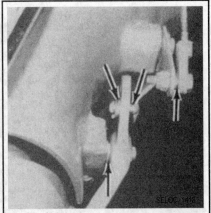

Fig. 26 Also, coat the pivot points of the jet linkage with grease periodically

Very dark or dirty grease may indicate a worn seal (inspect and/or replace the seal, as necessary to prevent severe engine damage should the seal fail completely).

■ **Keep in mind that some discoloration of the grease is expected when a new seal is broken-in. The discoloration should go away gradually after 1 or 2 additional grease replacement services.**

Whenever the jet drive bearing grease is replaced, take a few minutes to apply some of that same water-resistant marine grease to the pivot points of the jet linkage.

Power Trim/Tilt Reservoir

RECOMMENDED LUBRICANT

Quicksilver Power Trim and Steering Fluid is a highly refined hydraulic fluid with a high detergent content and additives to keep seals pliable. A high grade automatic transmission fluid such as Dexron® II or Type F/FA, may also be used if the recommended fluid is not available.

CHECKING FLUID LEVEL/CONDITION

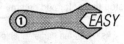

◆ **See Figure 27**

The fluid in the power trim/tilt reservoir should be checked periodically to ensure it is full and is not contaminated. To check the fluid, tilt the motor upward to the full tilt position, then manually engage the tilt support for safety and to prevent damage. Loosen and remove the filler plug using a suitable socket, wrench or even large screwdriver (as many of the Mercury PTT reservoir caps are also slotted) and make a visual inspection of the fluid. It should seem clear and not milky. The level is proper if, with the motor at full tilt, the level is even with the bottom of the filler plug hole.

Fig. 27 Typical Mercury PTT reservoir and cap

ENGINE MAINTENANCE

Engine Covers (Top Cover & Cowling)

REMOVAL & INSTALLATION

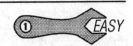

◆ **See Figures 28 and 29**

Removal of the top cover is necessary for the most basic of maintenance and inspection procedures. The cover should come off before and after each use in order to perform these basic safety checks.

In addition, the top cover seals to the engine lower cowling which comes in 1 of 2 styles, either 1-piece or 2-piece set-up. Typically, the 1-piece lowers are designed to sit low enough in relation to the powerhead that they do

not interfere with most maintenance or repair items. However, most SPLIT, 2-piece lowers sit higher and may be unbolted and removed for access during service. Removal of the split 2-piece lower covers is normally pretty straight forward in that most fasteners are exposed and obvious. However, some may be threaded along the inside lip and care must be taken to make sure all fasteners are released. NEVER force a plastic cover during removal, always double-check yourself and look for what you might have missed.

Whenever the top cover is removed, check the cover seal (a rubber insulator normally used between the cover and cowling) for wear or damage. Take the opportunity to perform a quick visual inspection for leaking hoses, chaffing wires etc. Get to know where things are placed under the cover so you'll recognize instantly if something is amiss. When installing the cover, care must be taken to ensure wiring and hoses are in their original positions to prevent damage. Once the cover is installed, make sure the latches grab it securely to prevent it from coming loose and flying off while underway.

Fig. 28 All Mercury motors have a top cover. . .

Fig. 29 . . . and a 1 or 2 piece lower cowling

Cooling System

FLUSHING THE COOLING SYSTEM

OEM ① EASY

◆ See Figures 30 thru 34

The most important service that you can perform on your motor's cooling system is to flush it periodically using fresh, clean water. This should be done immediately following any use in salt, brackish or polluted waters in order to prevent mineral deposits or corrosion from clogging cooling passages. Even if you do not always boat in salt or polluted waters, get used to the flushing procedure and perform it often to ensure no silt or debris clogs your cooling system over time.

■ Flush the cooling system after any use in which the motor was operated through suspended/churned-up silt, debris or sand.

Although the flushing procedure should take place right away (dockside or on the trailer), be sure to protect the motor from damage due to possible thermal shock. If the engine has just been run under high load or at continued high speeds, allow time for it to cool to the point where the powerhead can be touched. Do not pump very cold water through a very hot engine, or you are just asking for trouble. If you trailer your boat short distances, the flushing procedure can probably wait until you arrive home or wherever the boat is stored, but ideally it should occur within an hour of use in salt water. Remember that the corrosion process begins as soon as the motor is removed from the water and exposed to air.

The flushing procedure is not used only for cooling system maintenance, but it is also a tool with which a technician can provide a source of cooling water to protect the engine (and water pump impeller) from damage anytime the motor needs to be run out of the water. **Never** start or run the engine out of the water, even for a few seconds, for any reason. Water pump impeller damage can occur instantly and damage to the engine from overheating can follow shortly thereafter. If the engine must be run out of the water for tuning or testing, always connect an appropriate flushing device **before** the engine is started and leave it turned on until **after** the engine is shut off.

✳✳ SELOC WARNING

ANYTIME the engine is run, the first thing you should do is check the cooling stream or water indicator. All Mercury's are equipped with some form of a cooling stream indicator towards the aft portion of the lower engine cover. Anytime the engine is operating, a steady stream of water should come from the indicator, showing that the pump is supplying water to the engine for cooling. If the stream is ever absent, stop the motor and determine the cause before restarting.

As we stated earlier, flushing the cooling system consists of supplying fresh, clean water to the system in order to clean deposits from the internal passages. If the engine is running, the water does not normally have to be pressurized, as it is delivered through the normal water intake passages and the water pump (the system can self flush if supplied with clean water). Smaller, portable engines can be flushed by mounted them in a test tank (a sturdy, metallic 30 gallon drum or garbage pail filled with clean water).

On almost all Mercury motors, a generic flush adapter is available that will allow you to attach a garden hose to supply water through the normal water intake grate(s) on the lower unit. A few of the smallest motors (from 2.5-15 hp) utilize a water intake grate which is mounted to the underside of the anti-cavitation plate, so the adapter would have to have ONE suction cup that fits over the grate and a metal frame which would hold it securely to the plate. The propeller must normally be removed not only for safety on these models, but for the necessary clearance to install the adapter.

All of the 20 hp and larger Mercury models use a lower unit where the water intake grate(s) is(are) mounted on the side of the lower unit. On these models a generic (ear-muff type) flush adapter may be used (or for most models another version that uses both suction cups but no metal frame, instead a ratcheting pull strip attaches them together THROUGH the water intake grate. Either way these generic adapters fit over the engine water intakes on the gearcase and, when connected to a garden hose, will provide a constant supply of water for the water pump impeller to pump through the motor when running.

■ When running the engine on a flushing adapter using a garden hose, it is best to make sure the hose delivers about 20-40 psi (140-300 kPa) of pressure.

Some models, like the 1.5L OptiMax motors or any of the V6 motors, are usually equipped with a built-in flushing adapter. On 1.5L OptiMax motors the adapter consists of a short hose at the front, starboard side of the motor, just under the front lip of the cowling. The hose contains a plug that either pops off so a small hose adapter can be threaded into place or a threaded cover which is unscrewed so a garden hose can be directly attached to the flushing hose. On V6 models the adapter is found toward the center of the split-line at the aft end of the lower cowling. The adapter itself consists of a plastic plug, threaded into a housing attached to the powerheads cooling passages. The plug consists of the same threads at the male end of a standard garden hose, therefore you can remove the plug and thread the garden hose directly to the motor.

In either case, the built in flushing adapter can be used with the engine stopped OR running at idle, whichever is preferred. It is important that when using this fitting the water supply does NOT exceed 45 psi (310 kPa), but this is usually accomplished simply by not leaving the hose valve fully opened (though delivery pressures will vary, so you might want to double check the supply in question).

■ **Jet drive models are equipped with a flushing port mounted under a flat head screw directly above the jet drive bearing grease fitting. For more information, please refer to Flushing Jet Drives, later in this section.**

For safety, the propeller should be removed ANYTIME the motor is run on the trailer or on an engine stand. We realize that this is not always practical when flushing the engine on the trailer, but cannot emphasize enough how much caution must be exercised to prevent injury to you or someone else. Either take the time to remove the propeller or take the time to make sure no-one or nothing comes close enough to it to become injured. Serious personal injury or death could result from contact with the spinning propeller. And it only take a moment of inattention for someone to accidentally shift the gearcase into gear when someone is near the propeller to cause a tragedy.

1. Check the engine top case and, if necessary remove it to check the powerhead and ensure it is cooled enough to flush without causing thermal shock.

2. Prepare the engine for flushing depending on the method you are using as follows:

a. If using a test tank, make sure the tank is made of sturdy material, then securely mount the motor to the tank. If necessary, position a wooden plank between the tank and engine clamp bracket for thickness. Fill the tank so the water level is at least 4 in. (10cm) above the anti-ventilation plate (above the water inlet).

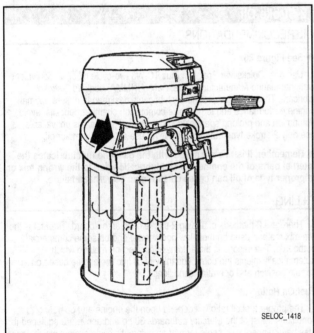

Fig. 30 All models may be flushed in a test tank. Smaller ones, in a garbage pail

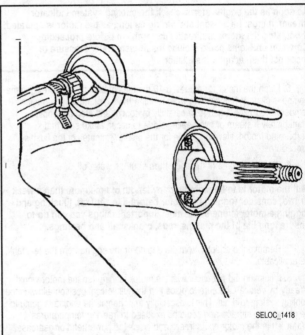

Fig. 31 The easiest way to flush most models is using a clamp-type adapter

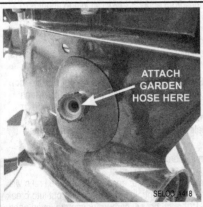

Fig. 32 This adapter attaches through the water intake grate

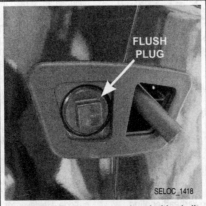

Fig. 33 All V6 motors are equipped with a built in flush adapter on the back of cowling. . .

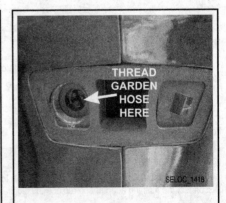

Fig. 34 . . .remove the plastic plug and attach a standard garden hose

b. If using a flushing adapter of either the generic clamp-type or specific port-type for your model attach the water hose to the flush test adapter and connect the adapter to the motor following the instructions that came with the adapter. If the motor is to be run (for flushing or testing) it is normally a good idea to position the outboard vertically (especially on oil injected models) and remove the propeller, for safety. Also, be sure to position the water hose so it will not contact with moving parts (tie the hose out of the way with mechanic's wire or wire ties, as necessary).

■ When using a clamp-type adapter, position the suction cup(s) over water intake grate(s) in such a way that they form tight seals. A little pressure seepage should not be a problem, but look to the water stream indicator once the motor is running to be sure that sufficient water is reaching the powerhead.

3. Unless using a test tank, turn the water on, making sure that pressure does not exceed 45 psi (310 kPa).

4. If using a test tank or if the motor must be run for testing/tuning procedures, start the engine and run in neutral until the motor reaches operating temperature. The motor will continue to run at fast idle until warmed.

✳✳ SELOC WARNING

As soon as the engine starts, check the cooling system indicator stream. It must be present and strong as long as the motor is operated. If not, stop the motor and rectify the problem before proceeding. Common problems could include insufficient water pressure or incorrect flush adapter installation.

5. Flush the motor for at least 5-10 minutes or until the water exiting the engine is clear. When flushing while running the motor, check the engine temperature (using a gauge or carefully by touch) and stop the engine immediately if steam or overheating starts to occur. Make sure that carbureted motors slow to low idle for the last few minutes of the flushing procedure.

6. Stop the engine (if running), **then** shut the water off.

■ If the motor is to be stored for any length of time more than a week or two, consider fogging the motor before it is shut off. Flushing and fogging a motor is one of the most important things you can do to ensure long life of the pistons, rods, crankshaft and bearings.

7. Remove the adapter from the engine or the engine from the test tank, as applicable.

8. If flushing did not occur with the motor running (so the motor would already by vertical), be sure to place it in the full vertical position allowing the cooling system to drain. This is especially important if the engine is going to be placed into storage and could be exposed to freezing temperatures. Water left in the motor could freeze and crack the powerhead or gearcase.

FLUSHING JET DRIVES

◆ See Figure 35

Regular flushing of the jet drive will prolong the life of the powerhead, by clearing the cooling system of possible obstructions.

Jet drives are equipped with a plug on the port side just above the lubrication hose. You'll need to install a suitable hose coupling (like Quicksilver # 24789A1 or equivalent) to a garden hose in order to attach the hose to the jet drive assembly.

1. Remove the plug and gasket, install the flush adapter, connect the garden hose and turn on the water supply.

2. Start and operate the powerhead at a fast idle for about 5-10 minutes. Disconnect the flushing adapter and replace the plug.

■ The procedure just described will only flush the powerhead cooling system, not the jet drive and drive impeller. To flush the jet drive unit, direct a stream of high pressure water through the intake grille.

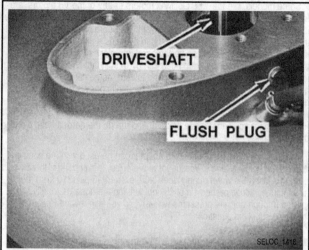

Fig. 35 Jet drives are equipped with a flushing plug on the side of the unit

2-Stroke Oil

OIL RECOMMENDATIONS

◆ See Figure 36

Use only Quicksilver Premium Plus TC-W3 two-cycle oil or an equivalent National Marine Manufacturers Association (NMMA) certified 2-stroke lubricant. These oils are proprietary lubricants designed to ensure optimal engine performance and to minimize combustion chamber deposits, avoid detonation and prolong spark plug life. If certified lubricant is unavailable, use only 2-stroke type outboard oil. Never use automotive motor oil.

■ Remember, it is this oil, mixed with the gasoline that lubricates the internal parts of the engine. Lack of lubrication due to the wrong mix or improper type of oil can cause catastrophic powerhead failure.

FILLING

There are 2 methods of adding 2-stroke oil to an outboard. The first is the pre-mix method used on most low horsepower and on some commercial outboards. The second is the automatic oil injection method which automatically injects the correct quantity of oil into the engine based on throttle position and operating conditions.

Fuel:Oil Ratio

The proper fuel:oil ratio will depend upon the engine and operating conditions. Most of the Mercury outboards 30 hp and above are equipped (or may be equipped) with some form of an automatic oiling system which varies from a purely mechanical single point variable ratio system to electronic single (EFI) and multi-point (Optimax) systems.

Generally speaking the single-point systems are usually designed to maintain an oil ratio that varies from about 80-120:1 at idle to about 50:1 at WOT without adding anything to the fuel tank. The sophisticated multi-point system found on Optimax motors can vary oil delivery GREATLY from lows of 300-400:1 at idle and highs of 40-60:1 at WOT.

When an oil injection system is not used, all Mercury carbureted motors covered here require a pre-mix of 50:1 fuel/oil ratio.

■ Check the Capacities Specifications chart at the end of this section for more details on your motor. But realize that these are the specs listed in Mercury service literature. Because your engine may differ slightly from service manual specification, refer to your owner's manual or a reputable dealer to be certain that you use the proper mixture for your motor.

In many cases, the normal operational fuel:oil ratio should be increased (doubled for carbureted motors) during break-in of a new or rebuilt powerhead. On Optimax and EFI motors that enrichment of the oil ratio will occur automatically by the computer control module once it is put into break-in mode. However for MOST (though not all) carbureted motors, even ones with an oil injection system, you'll have to ensure the proper oil ratio is followed for the first 4 hours of operation (or longer if the first tank full outlasts that period of time). This means that a carbureted motor that NORMALLY runs a 50:1 ratio should run a 25:1 ratio during this break-in period.

■ **Most motors equipped with the carbureted motor's mechanical single point variable ratio oiling system should run a 50:1 ratio of pre-mix in the boat or portable fuel tank during break-in. This, added to the normal output of the oiling system will produce the proper 25:1 or 50:1 break-in ratio required to ensure proper seating of the pistons and rings during break-in. Pre-mix should be discontinued on oil injected motors as soon as possible after the 4 hours break-in (basically as soon as the last tank of pre-mix is used up).**

■ **A FEW Mercury models DO NOT use a higher fuel:oil ratio during break-in, specifically the 2.5-5 hp models, as well as the 20/20 Jet/25 hp models. But again, we refer you to the fuel/oil ratio information found in the Capacities chart.**

Pre-Mix

◆ See Figure 37

Mixing the engine lubricant with gasoline before pouring it into the tank is by far the simplest method of lubrication for 2-stroke outboards. However, this method is the messiest and causes the most amount of harm to our environment.

The most important part of filling a pre-mix system is to determine the proper fuel/oil ratio. Most manufacturers use a 50:1 ratio (that is 50 parts of fuel to 1 part of oil) or a 100:1 ratio (and as we've already stated, Mercury seems to stick with 50:1 for these motors, but consult your Owner's manual if you are uncertain what the appropriate ratio should be for your engine).

The procedure itself is uncomplicated, but you've got a couple options depending on how the fuel tank is set-up for your boat. To fill an empty portable tank, add the appropriate amount of oil to the tank, then add gasoline and close the cap. Rock the tank from side-to-side to gently agitate the mixture, thereby allowing for a thorough mixture of gasoline and oil. When just topping off built-in or larger portable tanks, it is best to use a separate 3 or 6 gallon (11.4 or 22.7 L) mixing tank in the same manner as the portable tank noted earlier. In this way a more exact measurement of fuel can occur in 3 or 6 gallon increments (rather than just directly adding fuel to the tank and realizing that you've just added 2.67 gallons of gas and need to ad, uh, a little less than 8 oz of oil for a 50:1 ratio, but exactly how many ounces would that be?) Use of a mixture tank will prevent the need for such mathematical equations. Of course, the use of a mixing tank may be inconvenient or impossible under certain circumstances, so the next best method for topping off is to take a good guess (but be a little conservative to prevent an excessively rich oil ratio). Either add the oil and gasoline at the same time, or add the oil first, then add the gasoline to ensure proper mixing. For measurement purposes, it would obviously be more exact to add the gasoline first, then add a suitable amount of oil to match it. The problem with adding gasoline first is that unless the tank could be thoroughly agitated afterward (and that would be **really** difficult on built-in tanks), the oil might not mix properly with the gasoline. Don't take that unnecessary risk.

To determine the proper amount of oil to add to achieve the desired fuel:oil ratio, refer to the Fuel:Oil Ratio chart at the end of this section.

Oil Injection

◆ See Figure 38

Most 30 hp and larger Mercury motors are equipped with 1 of 3 possible oil injection systems; a mechanical single point variable ratio system (carbureted motors), electronic single point variable ratio (EFI motors) or electronic multi-point variable ratio (Optimax motors).

What they all have in common is a powerhead mounted oil reservoir which is used to feed oil to the system. Although all systems use some form of Low-Oil warning alarm, the level should be checked before and after EVERY outing to make sure the system is continuing to operate properly. It's not a bad idea to use a piece of tape to mark the level before an outing (especially after repairs or a tune-up) so that you can watch the level slowly drop through operation.

The procedure for filling these systems is simple. Unthread the oil reservoir cap and fill the oil tank to the proper capacity. If for some reason air is allowed to enter the system (the level is dropped WAY too low or a line has been disconnected) a oil pump/system priming procedure must be followed to purge any air from the line(s), for details or for more information on these systems, please refer to the Lubrication & Cooling System section.

■ **It is highly advisable to carry several spare bottles of 2-stroke oil with you onboard.**

Fuel Filter/Fuel Water Separator

◆ See Figure 39

A fuel filter is designed to keep particles of dirt and debris from entering the carburetor(s) or fuel injectors clogging the tiny internal passages. A small speck of dirt or sand can drastically affect the ability of the fuel system to deliver the proper amount of air and fuel to the engine. If a filter becomes clogged, it will quickly impede the flow of gasoline. This could cause lean fuel mixtures, hesitation and stumbling and idle problems in carburetors. Although a clogged fuel passage in a fuel injected engine could also cause lean symptoms and idle problems, dirt can also prevent a fuel injector from closing properly. A fuel injector that is stuck partially open by debris will cause the engine to run rich due to the unregulated fuel constantly spraying from the pressurized injector.

Regular cleaning or replacement of the fuel filter (depending on the type or types used) will decrease the risk of blocking the flow of fuel to the engine, which could leave you stranded on the water. It will also decrease the risk of damage to the small passages of a carburetor or fuel injector that could require more extensive and expensive replacement. Keep in mind that fuel filters are usually inexpensive and replacement is a simple task. Service your fuel filter on a regular basis to avoid fuel delivery problems.

We tried to find a pattern to tell us what type of filter will be installed on a powerhead, based on the year or model of the engine. However, Mercury did not seem to have a whole lot of rhyme or reason when it comes to choosing a fuel filter for their outboards (or if they did, we couldn't figure it out).

Fig. 36 This scuffed piston is an example of damage caused by improper 2-stroke oil/fuel mixture

Fig. 37 With portable tanks, either add the oil and gasoline at the same time, or add the oil first, then add the gasoline to ensure proper mixing

Fig. 38 Oil injected motors normally have an oil reservoir tank mounted to the powerhead

Fig. 39 Most boats with integral fuel tanks are also rigged with some form of in-line fuel filter/water separator

Loosely, we'll say that the smallest of Mercury outboards (2.5-5 hp) use disposable, inline filters (or fuel tank strainers on the smallest of outboards), while many (but not all) of the larger carbureted motors (6 hp and up) TEND to use a serviceable filter element that can be removed from a housing/cup assembly for cleaning or replacement. However there are exceptions to this and many larger motors (65 hp and up, including some commercial or work models) may be equipped with a non-serviceable inline filter as well. As far as we can tell, ALL of the EFI or Optimax models are equipped with a disposable water-separating fuel filter on or near the vapor separator assembly, though they could also be equipped with a disposable or serviceable filter before the lift pump.

Keep in mind that the type of fuel filter used on your boat/engine will vary not only with the year and model, but also with the accessories and rigging. Because of the number of possible variations it is impossible to accurately give instructions based on model. Instead, we will provide instructions for the different types of filters the manufacturer used on various families of motors or systems with which they are equipped. To determine what filter(s) are utilized by your boat and motor rigging, trace the fuel line from the tank to the fuel pump and then from the pump to the carburetor(s) or EFI fuel vapor separator tank. Most outboards place a filter (disposable or serviceable) found inline just before the fuel pump. Non-serviceable inline filters are replaced by simply removing the clamps, disconnecting the hoses and installing a new filter. When installing a new disposable inline filter, make sure the arrow on the filter points in the direction of fuel flow.

Some motors have a fuel filter mounted in the fuel tank itself. On many small Mercury motors with portable fuel tanks a fuel filter element is normally found on the fuel tank pickup which must be removed from the tank in order to service it. For larger motors, in addition to the fuel filter mounted on the engine, a filter is usually found inside or near the fuel tank. Because of the large variety of differences in both portable and fixed fuel tanks, it is impossible to give a detailed procedure for removal and installation. However, keep in mind that most in-tank filters are simply a screen on the pickup line inside the fuel tank. Filters of this type rarely require service or attention, but if the tank is removed for cleaning the filter will usually only need to be cleaned and returned to service (assuming they are not torn or otherwise damaged).

■ **Most EFI motors are equipped with a replaceable fuel filter/water separator elements (many of them almost look like automotive oil filters) on or near the fuel vapor separator tank.**

Depending upon the boat rigging a fuel filter/water separator may also be found inline between a boat mounted fuel tank and the motor. Boat mounted fuel filter/water separators are normally of the spin-on filter type (resembling automotive oil filters) but will vary greatly with boat rigging. Replacement of filters/separators is the same as a typical automotive oil filter replacement, just make sure to have a small drain basin handy to catch escaping fuel and make sure to coat the rubber gasket with a small dab of engine oil during installation.

RELEASING FUEL SYSTEM PRESSURE (EFI/OPTIMAX MODELS)

◆ **See Figures 40 thru 43**

On fuel injected engines, always relieve system pressure prior to disconnecting any fuel system component, fitting or fuel line.

✳✳ SELOC CAUTION

Exercise extreme caution whenever relieving fuel system pressure to avoid fuel spray and potential serious bodily injury. Please be advised that fuel under pressure may actually penetrate the skin or any part of the body it contacts.

To avoid the possibility of fire and personal injury, always disconnect the negative battery cable (this will not only prevent sparks but prevent fuel from possibly spraying out of open fittings should you or someone else accidentally turn the ignition key to on while you are working).

Keep in mind that EFI motors have a single high-pressure fuel circuit and therefore only 1 Schrader valve fitting from which to relieve system pressure, but Optimax motors have both a low-pressure AND high-pressure electric fuel pump circuit, each with their own Schrader valve fittings from which you should relieve pressure (unless you're certain you'll only be opening up access to one or the other). For the most part, the Schrader valve fittings for EFI motors (and low-pressure fitting for Optimax motors) are found on or near the vapor separator tank assembly. The high-pressure fitting for Optimax motors is normally found ON the air/fuel rail assembly.

✳✳ SELOC WARNING

Although Mercury has at time suggested it, and we know it is FAIRLY common practice, we cannot in good conscience recommend relieving system pressure by simply wrapping the Schrader valve with a shop rag to absorb the spray, then using a screwdriver or other tool to depress/open the valve. The fuel spray will be so well atomized and so easily ignited as to make it a very dangerous situation.

Always place a shop towel or cloth around the fitting or connection prior to loosening to absorb any excess fuel due to spillage. Ensure that all fuel spillage is removed from engine surfaces. Ensure that all fuel soaked clothes or towels in suitable waste container.

1. Remove the plastic cap over the pressure port located either on/near the vapor separator tank or on the combination air/fuel rail, as applicable. For more details, please refer to the information on Fuel Injection in the Fuel System section.
2. Attach a fuel pressure gauge to the pressure test port Schrader valve and direct the bleed valve hose into a suitable container. Carefully open the bleed valve allowing fuel to drain into the container until system pressure is released.

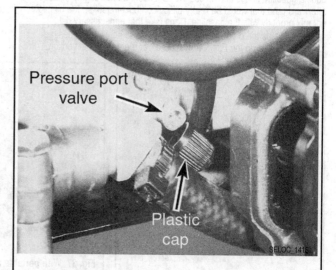

Fig. 40 On EFI and Optimax motors, use the pressure port valve(s) to release system pressure

Fig. 41 Typical EFI vapor separator tank with pressure port

Fig. 42 Optimax motors may also have pressure ports along the middle. . .

Fig. 43 . . . or at the top of the air/fuel rail assembly

FUEL FILTER SERVICE

Disposable Inline Filters

◆ See Figure 44

As noted earlier, a number of outboards are equipped with a disposable inline filter. Generally speaking the smallest of Mercury outboards (2.5-5 hp) use disposable, inline filters (or fuel tank strainers on the smallest of outboards), while MANY (but not all) of the larger carbureted motors (6 hp and up) TEND to use a serviceable filter element that can be removed from a housing/cup assembly for cleaning or replacement. However there are exceptions to this and many larger motors (65 hp and up, including some commercial or work models) may be equipped with a non-serviceable inline filter as well. As far as we can tell ALL of the EFI or Optimax models are equipped with a disposable water-separating fuel filter on or near the vapor separator assembly, though they could also be equipped with a disposable or serviceable filter before the lift pump.

When equipped with a disposable inline filter it is usually found inline just before the mechanical fuel pump HOWEVER, we've seen them plumbed AFTER the fuel pump and before the carburetors on a lot of Mercury models.

This type of filter is a sealed canister type (usually plastic) and cannot be cleaned, so service is normally limited to replacement. Because of the relative ease and relatively low expense of a filter (when compared with the time and hassle of a carburetor overhaul or fuel injection repair) we encourage you to replace the filter at least annually.

When replacing the filter, release the hose clamps (they may be secured by spring-type clamps that are released by squeezing the tabs using a pair of pliers or by threaded clamps or even wire ties that must be cut) and slide them back on the hose, past the raised portion of the filter inlet/outlet nipples.

Once a clamp is released, position a small drain pan or a shop towel under the filter and carefully pull the hose from the nipple. Allow any fuel remaining in the filter and fuel line to drain into the drain pan or catch fuel with the shop towel. Repeat on the other side, noting which fuel line connects to which portion of the filter (for assembly purposes). Inline filters are usually marked with an arrow indicating fuel flow. The arrow should point towards the fuel line that runs to the motor (not the fuel tank).

Before installation of the new filter, make sure the hoses are in good condition and not brittle, cracking and otherwise in need of replacement. During installation, be sure to fully seat the hoses, then place the clamps over the raised portions of the nipples to secure them. Spring clamps will weaken over time, so replace them if they've lost their tension. If wire ties or adjustable clamps were used, be careful not to over-tighten the clamp. If the clamp cuts into the hose, it's too tight; loosen the clamp or cut the wire tie (as applicable) and start again.

Serviceable Canister-Type Inline Filters

◆ See Figures 45 and 46

It's probably safe to say that about half of these motors (including most of the 6-25 hp models, as well as many of the non-oil injected versions of the

Fig. 44 Typical Mercury disposable inline filter

Fig. 45 Typical Mercury serviceable canister-type filter

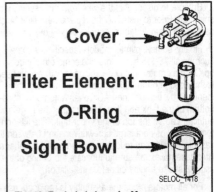

Fig. 46 Exploded view of a Mercury serviceable filter (with threaded bolt instead of a lockring)

65 hp and larger models) are equipped with a serviceable canister-type inline filter. These can be identified by their design and shape, which varies from the typical inline filter. A typical, disposable inline filter will have a simple round canister to which the fuel lines attach at either end. The serviceable inline filters used by these motors usually have 2 fuel inlets at the top of a filter housing and the serviceable element is located in a bowl or cap that is threaded up from underneath the housing (or is secured by a lockring which is threaded up from underneath the housing at the lip of the bowl).

When servicing these filters it may be possible to access and unthread the bowl without disconnecting any hoses, but if so, you'll have to hold the housing steady to prevent stressing and damaging the fuel lines. However, on some motors access to the bowl is impossible without first disconnecting 1 or more fuel hoses and repositioning the assembly.

■ **On some models the bowl/canister itself threads into the housing, but on others a separate knurled lockring is threaded over the bowl/canister and onto the housing to secure the assembly. If present, turn the lockring to loosen the housing, not the bowl itself.**

Serviceable filter elements should be cleaned carefully using solvent and inspected for clogging, tears or damage. If problems are found, normally just the element itself will require replacement. You really should always replace the O-ring, it's cheap insurance, but you CAN reuse it in a pinch, as long as it is not torn, cut or deformed. While you're servicing the filter always inspect and replace any damaged hose or clamp as you would with any other fuel filter.

Disposable Inline Fuel/Water Separators

◆ **See Figures 47 and 48**

As stated earlier, any model with a fuel tank integrated into the boat MAY be rigged with a boat mounted fuel/water separator. When equipped this type of fuel filter and water separator assembly normally resembles a spin-on automotive oil filter and it is basically serviced in the same way, loosened and spun off, the replacement spun on and tightened by hand. It is normally a good idea to use a small dab of 2-stroke engine oil on the sealing ring to make sure it both seals and releases easily once it comes time for replacement.

✳✳ SELOC WARNING

DO NOT remove the engine-mounted fuel filter on EFI or Optimax motors without first Releasing Fuel System Pressure as detailed earlier in this section.

In addition, all EFI and Optimax models are equipped with some form of a water-separating fuel filter. On MOST models, it also takes the form of a spin-on/spin-off filter (again, just like an automotive oil filter) with the difference that MOST models also contain a wire (usually connected to the bottom of the filter) for the water sensor. This spin-on filter is usually found either threaded upward to the bottom of the vapor separator tank assembly OR it may be threaded to a separate filter mounting boss, usually located in the same general area as the vapor separator tank.

Be sure to disconnect the wire from the assembly before attempting to spin it off. Prior to installing the replacement filter be sure to lubricate the seal using a small dab of 2-stroke engine oil.

Finally, some Optimax models (specifically some 2003 or later models such as all 75-115 hp motors and some other models like the 3.0L 200-250 hp Optimax motors) may be equipped with a water-separating fuel filter that is threaded into the TOP of the vapor separator tank. This type is easily identified by, first off, the LACK of an automotive style filter hanging from the bottom of the tank and then secondly, by the large, round, castellated nut/filter cap on the top of the vapor tank. The filter on these models is removed by using a large screwdriver shaft between the lugs of the castellated filter cap to loosen and unthread the filter assembly. As with other filter assemblies, be sure to lubricate the O-ring of the replacement filter using 2-stroke engine oil before installation.

For all motors, once the assembly is replaced, use the primer bulb to pressurize the system and visually check for leaks. We also think it is a good idea to cycle the ignition key or crank the motor to pressurize the high-pressure portion of the system and check again.

Fig. 47 Typical Mercury spin-on filter/water-separator found on many EFI and Optimax motors

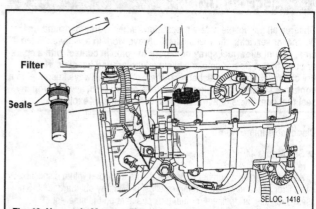

Fig. 48 New-style Mercury filter/water-separator found on some 2003 and later Optimax motors

Propeller

As you know, the propeller is actually what moves the boat through the water. The propeller operates in water in much the same manner as a wood screw or auger passing through wood. The propeller "bites" into the water as it rotates. Water passes between the blades and out to the rear in the shape of a cone. This "biting" through the water is what propels the boat.

Most Mercury outboards are equipped, from the factory, with a through-the-propeller exhaust, meaning that exhaust gas is routed out through the propeller.

GENERAL PROPELLER INFORMATION

Diameter and Pitch

◆ **See Figures 49 and 50**

Only 2 dimensions of the propeller are of real interest to the boat owner: diameter and pitch. These 2 dimensions are stamped on the propeller hub and always appear in the same order, the diameter first and then the pitch.

The diameter is the measured distance from the tip of 1 blade to the tip of the other.

The pitch of a propeller is the angle at which the blades are attached to the hub. This figure is expressed in inches of water travel for each revolution of the propeller. In our example of a 9 7/8 in. x 10 1/2 in., the propeller should travel 10 1/2 inches through the water each time it revolves. If the propeller action was perfect and there was no slippage, then the pitch multiplied by the propeller rpm would be the boat speed.

Most outboard manufacturers equip their units with a standard propeller, having a diameter and pitch they consider to be best suited to the engine and boat. Such a propeller allows the engine to run as near to the rated rpm and horsepower (at full throttle) as possible for the boat design.

The blade area of the propeller determines its load-carrying capacity. A 2-blade propeller is used for high-speed running under very light loads.

A 4-blade propeller is installed in boats intended to operate at low speeds under very heavy loads such as tugs, barges or large houseboats. The 3-blade propeller is the happy medium covering the wide range between high performance units and load carrying workhorses.

Propeller Selection

There is no one propeller that will do the proper job in all cases. The list of sizes and weights of boats is almost endless. This fact, coupled with the many boat-engine combinations, makes the propeller selection for a specific purpose a difficult task. Actually, in many cases the propeller may be changed after a few test runs. Proper selection is aided through the use of charts set up for various engines and boats. These charts should be studied and understood when buying a propeller. However, bear in mind that the charts are based on average boats with average loads; therefore, it may be necessary to make a change in size or pitch, in order to obtain the desired results for the hull design or load condition.

Propellers are available with a wide range of pitch. Remember, a low pitch design takes a smaller bite of water than a high pitch propeller. This means the low pitch propeller will travel less distance through the water per revolution. However, the low pitch will require less horsepower and will allow the engine to run faster.

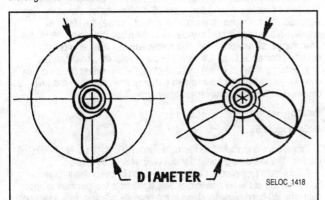

Fig. 49 Diameter and pitch are the 2 basic dimensions of a propeller. Diameter is measured across the circumference of a circle scribed by the propeller blades

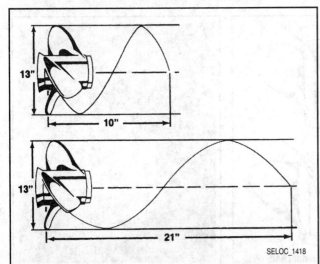

Fig. 50 This diagram illustrates the pitch dimension of a propeller. The pitch is the theoretical distance a propeller would travel through water if there were no friction

All engine manufacturers design their units to operate with full throttle at, or in the upper end of the specified operating rpm. If the powerhead is operated at the rated rpm, several positive advantages will be gained.

- Spark plug life will be increased.
- Better fuel economy will be realized.
- Steering effort is often reduced.
- The boat and power unit will provide best performance.

Therefore, take time to make the proper propeller selection for the rated rpm of the engine at full throttle with what might be considered an average load. The boat will then be correctly balanced between engine and propeller throughout the entire speed range.

A reliable tachometer must be used to measure powerhead speed at full throttle, to ensure that the engine achieves full horsepower and operates efficiently and safely. To test for the correct propeller, make a test run in a body of smooth water with the lower unit in forward gear at full throttle. If the reading is above the manufacturer's recommended operating range, try propellers of greater pitch, until one is found allowing the powerhead to operate continually within the recommended full throttle range.

If the engine is unable to deliver top performance and the powerhead is properly tuned, then the propeller may not be to blame. Operating conditions have a marked effect on performance. For instance, an engine will lose rpm when run in very cold water. It will also lose rpm when run in salt water, as compared with fresh water. A hot, low-barometer day will also cause the engine to lose power.

Cavitation

◆ See Figure 51

Cavitation is the forming of voids in the water just ahead of the propeller blades. Marine propulsion designers are constantly fighting the battle against the formation of these voids, due to excessive blade tip speed and engine wear. The voids may be filled with air or water vapor, or they may actually be a partial vacuum. Cavitation may be caused by installing a piece of equipment too close to the lower unit, such as the knot indicator pickup, depth sounder or bait tank pickup.

Vibration

The propeller should be checked regularly to ensure that all blades are in good condition. If any of the blades become bent or nicked, this condition will set up vibrations in the drive unit and motor. If the vibration becomes very serious, it will cause a loss of power, efficiency, and boat performance. If the vibration is allowed to continue over a period of time, it can have a damaging effect on many of the operating parts.

Vibration in boats can never be completely eliminated, but it can be reduced by keeping all parts in good working condition and through proper maintenance and lubrication. Vibration can also be reduced in some cases by increasing the number of blades. For this reason, many racers use two-blade propellers, while luxury cruisers have four- and five-blade propellers installed.

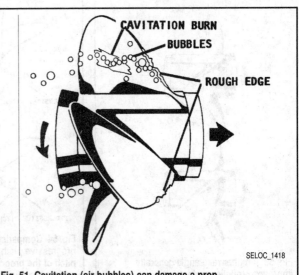

Fig. 51 Cavitation (air bubbles) can damage a prop

Shock Absorbers

◆ See Figure 52

The shock absorber in the propeller plays a very important role in protecting the shafting, gears and engine against the shock of a blow, should the propeller strike an underwater object. The shock absorber allows the propeller to stop rotating at the instant of impact, while the power train continues turning.

How much impact the propeller is able to withstand, before causing the shock absorber to slip, is calculated to be more than the force needed to propel the boat, but less than the amount that could damage any part of the power train. Under normal propulsion loads of moving the boat through the water, the hub will not slip. However, it will slip if the propeller strikes an object with a force that would be great enough to stop any part of the power train.

If the power train was to absorb an impact great enough to stop rotation, even for an instant, something would have to give, resulting in severe damage. If a propeller is subjected to repeated striking of underwater objects, it will eventually slip on its clutch hub under normal loads. If the propeller should start to slip, a new shock absorber/cushion hub will have to be installed by a propeller repair shop.

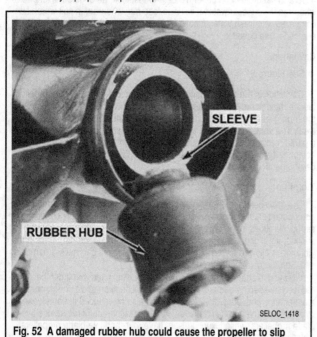

Fig. 52 A damaged rubber hub could cause the propeller to slip

Propeller Rake

◆ See Figure 53

If a propeller blade is examined on a cut extending directly through the center of the hub, and if the blade is set vertical to the propeller hub, the propeller is said to have a zero degree (0°) rake. As the blade slants back, the rake increases. Standard propellers have a rake angle from 0° to 15°.

A higher rake angle generally improves propeller performance in a cavitating or ventilating situation. On lighter, faster boats, a higher rake often will increase performance by holding the bow of the boat higher.

Progressive Pitch

◆ See Figure 54

Progressive pitch is a blade design innovation that improves performance when forward and rotational speed is high and/or the propeller breaks the surface of the water.

Progressive pitch starts low at the leading edge and progressively increases to the trailing edge. The average pitch over the entire blade is the number assigned to that propeller. In the illustration of the progressive pitch, the average pitch assigned to the propeller would be 21.

Cupping

◆ See Figure 55

If the propeller is cast with an edge curl inward on the trailing edge, the blade is said to have a cup. In most cases, cupped blades improve performance. The cup helps the blades to HOLD and not break loose, when operating in a cavitating or ventilating situation.

A cup has the effect of adding to the propeller pitch. Cupping usually will reduce full-throttle engine speed about 150-300 rpm below that of the engine equipped with the same pitch propeller without a cup to the blade. A propeller repair shop is able to increase or decrease the cup on the blades. This change, as explained, will alter powerhead rpm to meet specific operating demands. Cups are rapidly becoming standard on propellers.

In order for a cup to be the most effective, the cup should be completely concave (hollowed) and finished with a sharp corner. If the cup has any convex rounding, the effectiveness of the cup will be reduced.

Rotation

◆ See Figure 56

Propellers are manufactured as right-hand (RH) rotation or left-hand (LH) rotation. The standard propeller for outboard units is RH rotation.

A right-hand propeller can easily be identified by observing it. Observe how the blade of the right-hand propeller slants from the lower left to upper right. The left-hand propeller slants in the opposite direction, from lower right to upper left.

When the RH propeller is observed rotating from astern the boat, it will be rotating clockwise when the outboard unit is in forward gear. The left-hand propeller will rotate counterclockwise.

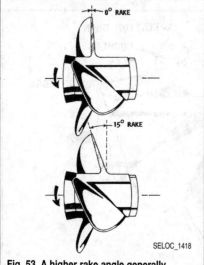

Fig. 53 A higher rake angle generally improves propeller performance

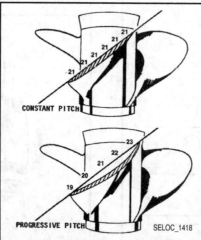

Fig. 54 Comparison of a constant and progressive pitch propeller. Notice how the pitch of the progressive propeller (right) changes to give the blade more thrust

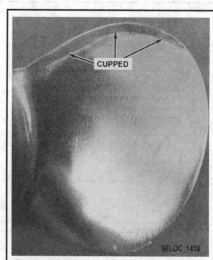

Fig. 55 Propeller with a cupped leading edge. Cupping gives the propeller a better hold in the water

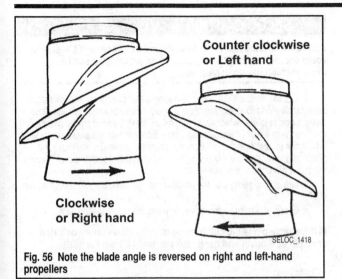

Fig. 56 Note the blade angle is reversed on right and left-hand propellers

Counter clockwise or Left hand

Clockwise or Right hand

INSPECTION

◆ See Figures 57, 58 and 59

The propeller should be inspected before and after each use to be sure the blades are in good condition. If any of the blades become bent or nicked, this condition will set up vibrations in the motor. Remove and inspect the propeller. Use a file to trim nicks and burrs. Take care not to remove any more material than is absolutely necessary.

✳✳ SELOC CAUTION

Never run the engine with serious propeller damage, as it can allow for excessive engine speed and/or vibration that can damage the motor. Also, a damaged propeller will cause a reduction in boat performance and handling.

Also, check the rubber and splines inside the propeller hub for damage. If there is damage to either of these, take the propeller to your local marine dealer or a "prop shop". They can evaluate the damaged propeller and determine if it can be saved by re-hubbing.

Additionally, the propeller should be removed AT LEAST every 100 hours of operation or at the end of each season, whichever comes first for cleaning, greasing and inspection. Whenever the propeller is removed, apply a fresh coating of an all-purpose water-resistant marine grade grease to the propeller shaft and the inner diameter of the propeller hub. This is necessary to prevent possible propeller seizure onto the shaft that could lead to costly or troublesome repairs. Also, whenever the propeller is removed, any material entangled behind the propeller should be removed before any damage to the shaft and seals can occur. This may seem like a waste of time at first, but the small amount of time involved in removing the propeller

is returned many times by reduced maintenance and repair, including the replacement of expensive parts.

■ Propeller shaft greasing and debris inspection should occur more often depending upon motor usage. Frequent use in salt, brackish or polluted waters would make it advisable to perform greasing more often. Similarly, frequent use in areas with heavy marine vegetation, debris or potential fishing line would necessitate more frequent removal of the propeller to ensure the gearcase seals are not in danger of becoming cut.

REMOVAL & INSTALLATION

The propeller is secured to the gearcase propshaft either by a drive pin on only the smallest of the 2-stroke portable motors, or by a castellated hex nut on all other Mercury outboards. It is pretty easy to tell the difference, just take a look at the propeller, if you see a nut, that's how it is secured. If however, you're working on a 2.5/3.3 hp motor, you should see a cotter pin going through the nose cone of the propeller itself, telling you that it is a shear pin prop.

For models secured by a hex nut, the propeller is driven by a splined connection to the shaft and the rubber drive hub found inside the propeller. The rubber hub provides a cushioning that allows softer shifts, but more importantly, it provides some measure of protection for the gearcase components in the event of an impact.

On 2.5/3.3 hp motors, where the propeller is retained by a drive pin, impact protection is provided by the drive pin itself. The pin is designed to break or shear when a specific amount of force is applied because the propeller hits something.

In both cases (rubber hubs or shear pins) the amount of force necessary to break the hub or shear the pin is supposed to be just less than the amount of force necessary to cause gearcase component damage. In this way, the hope is that the propeller and hub or shear pin will be sacrificed in the event of a collision, but the more expensive gearcase components will survive unharmed. Although these systems do supply a measure of protection, this, unfortunately, is not always the case and gearcase component damage will still occur with the right impact or with a sufficient amount of force.

✳✳ SELOC WARNING

Do not use excessive force when removing the propeller from the hub as excessive force can result in damage to the propeller, shaft and, even other gearcase components. If the propeller cannot be removed by normal means, consider having a reputable marine shop remove it. The use of heat or impacts to free the propeller will likely lead to damage.

■ Clean and lubricate the propeller and shaft splines using a high-quality, water-resistant, marine grease every time the propeller is removed from the shaft. This will help keep the hub from seizing to the shaft due to corrosion (which would require special tools to remove without damage to the shaft or gearcase.)

Fig. 57 This propeller is long overdue for repair or replacement

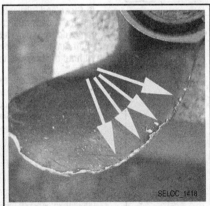

Fig. 58 Although minor damage can be dressed with a file. . .

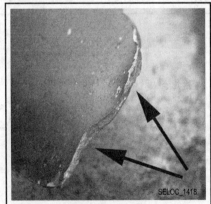

Fig. 59 . . . a propeller specialist should repair large nicks or damage

Mercury has varied their recommendations of what lubricant to apply to the propeller shaft over the years. And the fact that they often recommend more than one possible lubricant even for a given motor makes us think that it is not terribly important WHAT lubricant to use, as long as whatever it is was made for the marine environment. Mercury has advised using the following lubricants on the propeller shafts for motors covered here

- Perfect Seal
- Special Lubricant 101
- 2-4-C Marine Lubricant with Teflon
- Quicksilver Anti-Corrosion Grease

Actually, the first 3 are specifically recommended for 2.5/3.3 hp motors, the final 3 are specifically recommended for 30/40 hp (2-cylinder models) and the last 2 are specifically recommended for everything else. However, as we've said, we feel any marine grade grease or anti-seize should be fine.

Many outboards are equipped with aftermarket propellers. Because of this, the attaching hardware may differ slightly from what is shown. Contact a reputable propeller shop or marine dealership for parts and information on other brands of propellers.

Shear Pin Props (2.5/3.3 Hp Motors)

◆ See Figures 60 and 61

The propeller on the smallest of the portable Mercury 2-stroke outboards (2.5/3.3 hp) is secured to the propshaft using a drive pin. Be sure to always keep a spare drive pin handy when you are onboard the boat. Remember that a sheared drive pin will leave you stranded on the water. A damaged shear pin can also contribute to motor damage, exposing it to over-revving while trying to produce thrust. The pin itself is locked in position by the propeller/cone that is in turn fastened held in position over the drive pin by a cotter pin. ALWAYS replace the cotter pin once it has been removed. Remember that should the cotter pin fail, you could be diving to recover your propeller.

1. Disconnect the spark plug lead from the plug for safety.

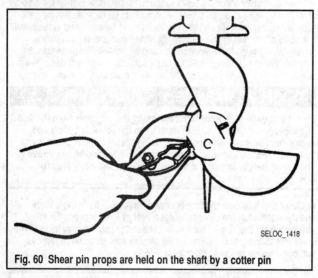

SELOC_1418

Fig. 60 Shear pin props are held on the shaft by a cotter pin

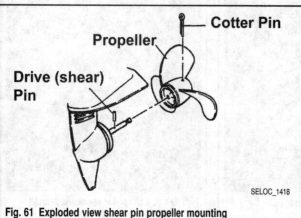

SELOC_1418

Fig. 61 Exploded view shear pin propeller mounting

2. Cut the ends off the cotter pin using a pair of wire cutters (as that is easier than trying to straighten them in most cases) or straighten the ends using a pair of pliers (whichever you prefer). Next, free the pin by grabbing the head with a pair of needlenose pliers. Either tap on the pliers gently with a hammer to help free the pin from the propeller cap or carefully use the pliers as a lever by carefully prying back against the propeller cone. Discard the cotter pin once it is removed.

3. Grasp and gently pull the propeller off the drive pin and the propeller shaft.

4. Grasp and remove the drive pin using the needlenose pliers.

■ **If the drive pin is difficult to remove, use a small punch or a new drive pin as a driver and gently tap the pin free from the shaft.**

To Install:

5. Clean the propeller hub and shaft splines, then apply a fresh coating of a water-resistant, marine grease.

6. Insert the drive pin into the propeller shaft.

7. Align the propeller, then carefully slide it over the shaft.

8. Install a new cotter pin, then spread the pin ends in order to form tension and secure them.

9. Reconnect the spark plug lead.

Castellated Nut Props

◆ See Figures 62 thru 72

The propeller is fastened to the gearcase propshaft by a nut which is USUALLY secured by either a cotter pin or a lockwasher. The propeller itself is driven by a splined connection to the shaft and the rubber drive hub found inside the propeller. The rubber hub provides a cushioning that allows softer shifts, but more importantly, it provides some measure of protection for the gearcase components in the event of an impact. In this way, the hope is that the propeller and hub will be sacrificed in the event of a collision, but the more expensive gearcase components will survive unharmed. Although these systems do supply a measure of protection, this, unfortunately, is not always the case and gearcase component damage will still occur with the right impact or with a sufficient amount of force.

■ **Clean and lubricate the propeller and shaft splines using a high-quality, water-resistant, marine grease every time the propeller is removed from the shaft. This will help keep the hub from seizing to the shaft due to corrosion (which would require special tools to remove without damage to the shaft or gearcase.)**

Many outboards are equipped with aftermarket propellers. Because of this, the attaching hardware may differ slightly from what is shown. Contact a reputable propeller shop or marine dealership for parts and information on other brands of propellers.

On almost all Mercury outboards the propeller is held in place over the shaft splines by a large castellated nut. The nut is so named because, when viewed from the side, it appears similar to the upper walls or tower of a castle.

On most Mercury outboards, the propeller is held in place over the shaft splines by a nut and some form of tabbed nut retainer/lock-washer. The most popular versions of these propellers are the FloTorque I and FloTorque II drive hubs, which come in various sizes and contain slightly different combinations of thrust hubs, retainers, drive sleeves etc.

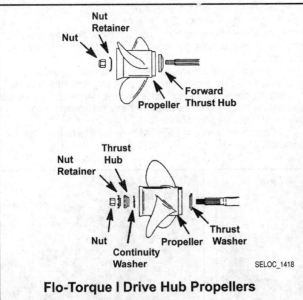

Flo-Torque I Drive Hub Propellers

Fig. 62 Exploded view of common Mercury/Mariner propeller mounting - FloTorque I Drive Hub models

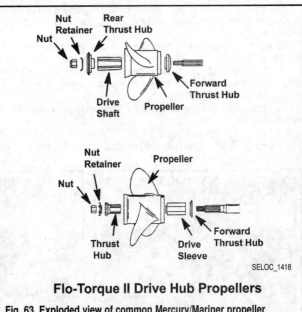

Flo-Torque II Drive Hub Propellers

Fig. 63 Exploded view of common Mercury/Mariner propeller mounting - FloTorque II Drive Hub models

■ **When servicing all propellers, ESPECIALLY ones mounted on Mercurys, take note of the order of assembly and the direction each propeller retaining component faces as they are removed. Be sure to assemble each component in the same order and facing the same direction as noted during removal.**

For safety, the nut usually is locked in place by a cotter pin or by tabs of a retainer that keep it from loosening while the motor is running. When equipped with a cotter pin, the pin passes through a hole in the propeller shaft, as well as through the notches in the sides of the castellated nut. When equipped with a tabbed retainer, the tabs (usually at least 3) are bend into position against the nut, keeping it from turning.

■ **Be sure to install a new cotter pin anytime the propeller is removed and, perhaps more importantly, make sure the cotter pin is of the correct size and is made of materials designed for marine use. The tabbed retainer used by most Mercs can usually be reused, but inspect the tabs for weakening and replace it when one or more of the tabs become questionable.**

Whenever working around the propeller, check for the presence of black rubber material in the drive hub (don't confuse bits of black soot/carbon deposits from the exhaust as rubber hub material) and spline grease. Presence of this material normally indicates that the hub has turned inside the propeller bore (have the propeller checked by a propeller repair shop). Keep in mind that a spun hub will not allow proper torque transfer from the motor to the propeller and will allow the engine to over-rev in attempting to produce thrust. If the propeller has spun on the hub it has been weakened and is more likely to fail completely in use.

1. For safety, disconnect the negative cable (if so equipped) and/or disconnect the spark plug leads from the plugs (ground the leads to prevent possible ignition damage should the motor be cranked at some point before the leads are reconnected to the spark plugs).

✳✳ SELOC CAUTION

Don't ever take the risk of working around the propeller if the engine could accidentally be started. Always take precautions such as disconnecting the spark plug leads and, if equipped, the negative battery cable.

2. If equipped with a cotter pin, cut the ends off the pin using wire cutters (as that is usually easier than trying to straighten them in most cases) or straighten the ends of the pin using a pair of pliers, whichever you prefer. Next, free the pin by grabbing the head with a pair of needle-nose pliers. Either tap on the pliers gently with a hammer to help free the pin from the nut or carefully use the pliers as a lever by prying back against the castellated nut. Discard the cotter pin once it is removed.

3. If equipped with a tabbed lock-washer/retainer, use a small punch to carefully bend the tabs away from the nut.

4. Place a block of wood between the propeller and the anti-ventilation housing to lock the propeller and shaft from turning, then loosen and remove the castellated nut. Note the orientation, then remove the washer and/or splined spacer (on most Mercs this would mean a nut retainer, followed by a drive sleeve on FloTorque II models), from the propeller shaft.

■ **Pro/XS/Sport models usually use 3 washers between the nut and propeller, a large washer closest to the propeller, a belleville washer in the center and another plain washer against the nut.**

5. Slide the propeller from the shaft. If the prop is stuck, use a block of wood to prevent damage and carefully drive the propeller from the shaft.

■ **If the propeller is completely seized on the shaft, have a reputable marine or propeller shop free it. Don't risk damage to the propeller or gearcase by applying excessive force.**

6. Note the direction in which the inner or forward thrust washer is facing (since some motors or aftermarket props may use a thrust washer equipped with a fishing line trap that must face the proper direction if it is to protect the gearcase seal). Remove the thrust washer from the propshaft (if the washer appears stuck, tap lightly to free it from the propeller shaft).

7. Clean the thrust washer(s), propeller, drive sleeve and shaft splines (as equipped) of any old grease. Small amounts of corrosion can be removed carefully using steel wool or fine grit sandpaper.

8. Inspect the shaft for signs of damage including twisted splines or excessively worn surfaces. Rotate the shaft while looking for any deflection. Replace the propeller shaft if these conditions are found. Inspect the thrust washer for signs of excessive wear or cracks and replace, if found.

To Install:

9. Apply a fresh coating of a water-resistant grease to all surfaces of the propeller shaft and to the splines inside the propeller hub.

■ **Some people prefer to use Anti-Seize on the hub splines, which is acceptable in most applications, but if used you should double-check after the first 10-20 hours of service to make sure the anti-seize is holding up to operating conditions.**

10. Position the inner (forward) thrust washer over the propshaft in the direction noted during removal. (Generally speaking on Mercurys, the flat shoulder should face forward toward the gearcase while the bevel/shoulder faces rearward toward the propeller, but this varies).

11. On Mercury motors which contain the larger Flo-Torque II Drive Hub Propeller there is normally a drive sleeve which is installed at this point. On the smaller models the sleeve usually installed from the other side of the propeller.

12. Carefully slide the propeller onto the propshaft, rotating the propeller to align the splines. Push the propeller forward until it seats against the thrust washer.

13. Install the splined and/or plain spacer (or drive sleeve and nut retainer, as applicable) onto the propeller shaft, as equipped.

14. Place a block of wood between the propeller and housing to hold the prop from turning, then thread the castellated nut onto the shaft with the cotter pin grooves facing outward.

15. Tighten the nut to specification using a suitable torque wrench as follows:.
- 4/5 hp motors: 150 inch lbs./12.5 ft. lbs. (17 Nm)
- 6/8 hp and 9.9/10/15 hp motors: 70 inch lbs./6 ft. lbs. (7.9 Nm)

- 20/25 hp motors: 16.7 ft. lbs. (22.6 Nm)
- 30 hp and larger motors: 55 ft. lbs. (74.6 Nm)

16. On tabbed retainer models, bend the tabs up against the flats on the propeller. On most of the larger models you should be able to bend 3 tabs into position.

17. On cotter pin retained models, install a new cotter pin through the grooves in the nut that align with the hole in the propshaft. If the cotter pin hole and the grooves do not align, tighten or loosen the nut very slightly, just enough to align them. Once the cotter pin is inserted, spread the ends sufficiently to lock the pin in place.

18. Connect the spark plug leads and/or the negative battery cable, as applicable.

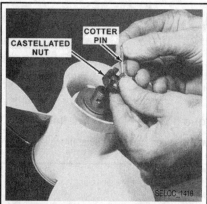
Fig. 64 Most Mercury props are fastened by a castellated nut (which is usually secured by a cotter pin or lockwasher)

Fig. 65 In all cases, to loosen the nut use a block of wood to keep the prop from turning

Fig. 66 A stuck propeller can be freed with heat. . .

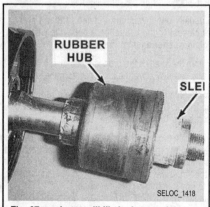
Fig. 67 . . . but you'll likely destroy the rubber hub

Fig. 68 If so, use a chisel to careful remove the frozen sleeve

Fig. 69 For installation position the inner spacer or thrust hub (as equipped) first. . .

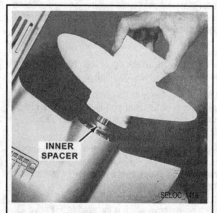
Fig. 70 . . . on most models with the shoulder facing the prop

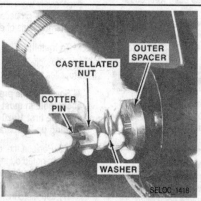
Fig. 71 Then install the outer components such as outer spacer, washer (if used), nut and cotter pin (as applicable)

Fig. 72 Assembled and ready to go!

Jet Drive Impeller

◆ See Figure 73

A jet drive motor uses an impeller enclosed in a jet drive housing instead of the propeller used by traditional gearcases. Outboard jet drives are designed to permit boating in areas prohibited to a boat equipped with a conventional propeller outboard drive system. The housing of the jet drive barely extends below the hull of the boat allowing passage in ankle deep water, white water rapids, and over sand bars or in shoal water which would foul a propeller drive.

The outboard jet drive provides reliable propulsion with a minimum of moving parts. It operates, simply stated, as water is drawn into the unit through an intake grille by an impeller. The impeller is driven by the driveshaft off the powerhead's crankshaft. Thrust is produced by the water that is expelled under pressure through an outlet nozzle that is directed away from the stern of the boat.

As the speed of the boat increases and reaches planing speed, only the very bottom of the jet drive where the intake grille is mounted facing downward remains in contact with the water.

The jet drive is provided with a reverse-gate arrangement and linkage to permit the boat to be operated in reverse. When the gate is moved downward over the exhaust nozzle, the pressure stream is deflected (reversed) by the gate and the boat moves sternward.

Conventional controls are used for powerhead speed, movement of the boat, shifting and power trim and tilt.

INSPECTION

◆ See Figure 74

The jet impeller is a precisely machined and dynamically balanced aluminum spiral. Close observation will reveal drilled recesses at exact locations used to achieve this delicate balancing. Excessive vibration of the jet drive may be attributed to an out-of-balance condition caused by the jet impeller being struck excessively by rocks, gravel or from damage caused by cavitation "burn".

The term cavitation "burn" is a common expression used throughout the world among people working with pumps, impeller blades, and forceful water movement. These "burns" occur on the jet impeller blades from cavitation air bubbles exploding with considerable force against the impeller blades. The edges of the blades may develop small dime-size areas resembling a porous sponge, as the aluminum is actually "eaten" by the condition just described.

Excessive rounding of the jet impeller edges will reduce efficiency and performance. Therefore, the impeller and intake grate (that protects it from debris) should be inspected at regular intervals.

Before and after each use, make a quick visual inspection of the intake grate and impeller, looking for obvious signs of damage. Always clear any debris such as plastic bags, vegetation or other items that sometimes become entangled in the water intake grate before starting the motor. If the intake grate is damaged, do not operate the motor, or you will risk destroying the impeller if rocks or other debris are drawn upward by the jet drive. If possible, replace a damaged grate before the next launch. This makes inspection after use all that much more important. Imagine the disappointment if you only learn of a damaged grate while inspecting the motor immediately prior to the next launch.

An obviously damaged impeller should be removed and either repaired or replaced depending on the extent of the damage. If rounding is detected, the impeller can be placed on a work bench and the edges restored to as sharp a condition as possible, using a file. Draw the file in only one direction. A back-and-forth motion will not produce a smooth edge. Take care not to nick the smooth surface of the jet impeller. Excessive nicking or pitting will create water turbulence and slow the flow of water through the pump. For more details on impeller replacement or service, please refer to the information on Jet Drives in the Gearcase section of this manual.

Fig. 73 Instead of a traditional gearcase, jet drives use an impeller (it's sort of an enclosed propeller) mounted in a jet drive housing that just barely extends below the boat's hull

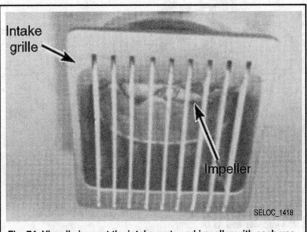

Fig. 74 Visually inspect the intake grate and impeller with each use

CHECKING IMPELLER CLEARANCE

◆ See Figures 75 and 76

Proper operation of the jet drive depends upon the ability to create maximum thrust. In order for this to occur the clearance between the outer edge of the jet drive impeller and the water intake housing cone wall should be maintained at approximately 1/32 in. (0.79mm). This distance can be checked visually by shining a flashlight up through the intake grille and estimating the distance between the impeller and the casing cone, as indicated in the accompanying illustrations. But, it is not humanly possible to accurately measure this clearance by eye. Close observation between outings is fine to maintain a general idea of impeller condition, but, at least annually, the clearance must be measured using a set of long feeler gauges.

✳✳ SELOC CAUTION

Whenever working around the impeller, ALWAYS disconnect the negative battery cable and/or disconnect the spark plug leads to make sure the engine cannot be accidentally started during service. Failure to heed this caution could result in serious personal injury or death in the event that the engine is started.

When checking clearance, a feeler gauge larger than the clearance specification should not fit between the tips of the impeller and the housing.

A gauge within specification should fit, but with a slight drag. A smaller gauge should fit without any interference whatsoever. Check using the feeler gauge at various points around the housing, while slowly rotating the impeller by hand.

After continued normal use, the clearance will eventually increase. In anticipation of this the manufacturer mounts the impeller deep in a tapered housing, and positions spacers beneath the impeller to hold it in position. The spacers are used to position the impeller along the driveshaft with the desired clearance between the jet impeller and the housing wall. When clearance has increased, spacers are removed from underneath the impeller and repositioned behind it, dropping the impeller slightly in the housing and thereby decreasing the clearance again.

If adjustment is necessary, refer to the Jet Drive procedures under Gearcase in this manual for impeller removal, shimming and installation procedures. Follow the appropriate parts of the Removal & Disassembly, as well as Assembly procedures for impeller service.

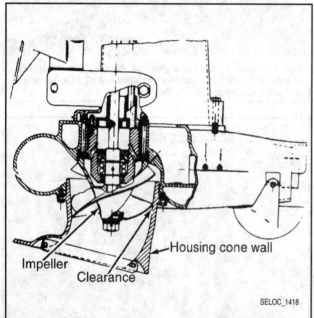

Fig. 75 Jet drive impeller clearance is the gap between the edges of the impeller and its housing

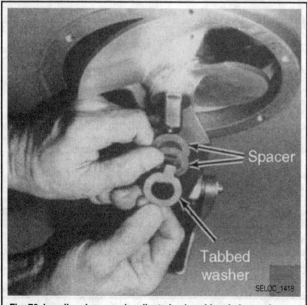

Fig. 76 Impeller clearance is adjusted using shims below and above the impeller

Anodes (Zincs)

◆ See Figures 77 and 78

The idea behind anodes (also known as sacrificial anodes) is simple: When dissimilar metals are submerged in water and a small electrical current is leaked between or amongst them, the less-noble metal (galvanically speaking) is sacrificed (corrodes).

The zinc alloy of which most anodes are made is designed to be less noble than the aluminum alloy of which your outboard is constructed. If there's any electrolysis, and there almost always is, the inexpensive zinc anodes are consumed in lieu of the expensive outboard motor.

■ We say the zinc alloy of which MOST anodes are made because the anodes recommended for fresh water applications may also be made of a different material. Some manufacturers recommend that you use magnesium anodes for motors used solely in freshwater applications. Magnesium offers a better protection against corrosion for aluminum motors, however the ability to offer this protection generally makes it too active a material for use in salt water applications. Because zinc is the more common material many people (including us) will use the word zinc interchangeably with the word anode, don't let this confuse you as we are referring equally to all anodes, whether they are made of zinc or magnesium.

These zincs require a little attention in order to make sure they are capable of performing their function. Anodes must be solidly attached to a clean mounting site. Also, they must not be covered with any kind of paint, wax or marine growth.

Fig. 77 Anodes come in various shapes and sizes

Fig. 78 Extensive corrosion of an anode suggests a problem or a complete disregard for maintenance

INSPECTION

◆ **See Figures 77 and 78**

Visually inspect the anodes, especially gearcase mounted ones, before and after each use. You'll want to know right away if it has become loose or fallen off in service. Periodically inspect them closely to make sure they haven't eroded too much. At a certain point in the erosion process, the mounting holes start to enlarge, which is when the zinc might fall off. Obviously, once that happens your engine no longer has any protection. Generally, a zinc anode is considered worn if it has shrunken to 2/3 or less than the original size. To help judge this, buy a spare and keep it handy (in the boat or tow vehicle for comparison).

If you use your outboard in salt water or brackish water, and your zincs never seem to wear, inspect them carefully. Paint, wax or marine growth on zincs will insulate them and prevent them from performing their function properly. They must be left bare and must be installed onto bare metal of the motor. If the zincs are installed properly and not painted or waxed, inspect around them for sings of corrosion. If corrosion is found, strip it off immediately and repaint with a rust inhibiting paint. If in doubt, replace the zincs.

On the other hand, if your zinc seems to erode in no time at all, this may be a symptom of the zincs themselves. Each manufacturer uses a specific blend of metals in their zincs. If you are using zincs with the wrong blend of metals, they may erode more quickly or leave you with diminished protection.

At least annually or whenever an anode has been removed or replaced, check the mounting for proper electrical contact using a multi-meter. Set the multi-meter to check resistance (ohms), then connect 1 meter lead to the anode and the other to a good, unpainted or un-corroded ground on the motor. Resistance should be very low or zero. If resistance is high or infinite, the anode is insulated and cannot perform its function properly.

SERVICING

◆ **See Figures 79, 80 and 81**

Most Mercury outboards contain 1 or 2 anodes on the gearcase (though there MAY be more mounted to the powerhead on some models). Generally speaking the smallest models (2.5-15 hp) have a single anode mounted to the gearcase, either right under the anti-cavitation plate/just above the propeller (all but 4/5 hp motors) or on the starboard side of the gearcase just a little rearward of the oil level/vent plug (4/5 hp motors). Most mid-range motors (20-125 hp) utilize a trim tab anode (except for the Optimax motors which use the same set-up as large hp models). The larger models (135 hp and up) motors should all be equipped with a single gearcase anode mounted in a bore right above the anti-cavitation plate. And, most models 30 hp and up should be equipped with an additional anode mounted to the bottom of the transom bracket. Regardless of their locations, there are some fundamental rules to follow that will give your boat's sacrificial anodes the ability to do the best job protecting your boat's underwater hardware that they can.

The first thing to remember is that zincs are electrical components and like all electrical components, they require good clean connections. So after

you've undone the mounting hardware and removed last year's zincs, you want to get the zinc mounting sites clean and shiny.

Get a piece of coarse emery cloth or some 80-grit sandpaper. Thoroughly rough up the areas where the zincs attach (there's often a bit of corrosion residue in these spots). Make sure to remove every trace of corrosion.

Zincs are attached with stainless steel machine screws that thread into the mounting for the zincs. Over the course of a season, this mounting hardware is inclined to loosen. Mount the zincs and tighten the mounting hardware securely. Tap the zincs with a hammer hitting the mounting screws squarely. This process tightens the zincs and allows the mounting hardware to become a bit loose ill the process. Now, do the final tightening. This will insure your zincs stay put for the entire season.

Drive Belt (Compressor and/or Alternator)

Most V6 models, including all carbureted motors and most EFI motors (except the 2001 150-200 hp 2.5L EFI models) and all Optimax motors are equipped with a drive belt for the compressor (Optimax only) and/or alternator. On most of these models the belt is of a multi-ribbed serpentine type, such as the type most commonly found on cars and trucks. However, it is possible that some of the EFI models, especially those which did not use an automatic tensioner, may be equipped with a simple v-belt.

Either way the belts should be inspected periodically to ensure they are fit for service (that they are not deteriorating or stretching). On MOST models, Mercury recommends inspecting the belts every year or 100 hours, whichever comes first. However literature for the standard 3.0L Optimax models (not the Pro/XS/Sport version, but we would apply to them as well) specifically suggests REPLACING the belt pro-actively every 3 years or 300 hours, whichever comes first. And honestly, considering the fact that the motor CANNOT run without this belt on Optimax models (or cannot run for long without it on EFI models) we would really tend to concur that this is the smarter course of action to take on ALL motors equipped with such a drive belt. If you decide NOT to do this, then AT LEAST carry a spare to get you home should it fail.

INSPECTION

◆ **See Figures 82 thru 85**

On MOST of these models (except in this case the 2001 200-250 hp 3.0L EFI motors), the belt tension is automatically adjusted through the use of a spring loaded belt-tensioner (again, a kudos to the same prevalent automotive technology). However, for the 2001 200-250 hp 3.0L EFI motors belt tension is manually adjusted by moving the alternator mounting position through a slotted range on one of the mounting brackets. Inspection will vary slightly based on the belt type and the method with which it is tensioned.

■ **On most models you'll have to remove the flywheel cover for access. On some models, like on V6 Optimax models, you'll need to disconnect the vent hose from the fitting on the flywheel cover.**

Both V-belts and serpentine belts should be inspected on a regular basis for signs of glazing or cracking. A glazed belt will be perfectly smooth from

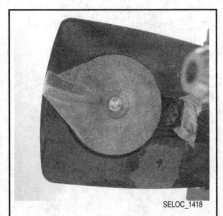

Fig. 79 Many mid-sized Mercury outboards use a trim tab anode

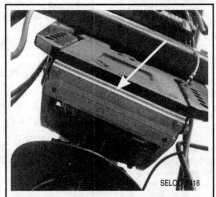

Fig. 80 And most mid-size or larger also have one on the bottom of the transom bracket

Fig. 81 The larger Mercurys use an anode mounted just above the anti-cavitation plate

Fig. 82 Most V6 models use a belt driven alternator. . .

Fig. 83 . . .and the same belt drives the compressor on Optimax models

Fig. 84 Some belt tensioners are equipped with a grease fitting for lubrication

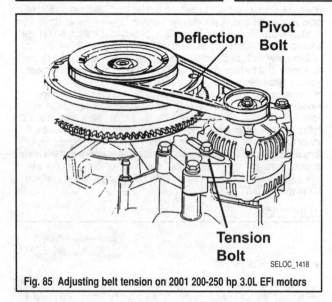
Fig. 85 Adjusting belt tension on 2001 200-250 hp 3.0L EFI motors

slippage, while a good belt will have a slight texture of fabric visible. Cracks will usually start at the inner edge of the belt and run outward. A worn or damaged drive belts should be replaced immediately.

■ Serpentine belts will commonly show cracks across parts of their sections. This in and of itself is not enough to condemn one, however watch for large cracks spanning multiple ribs or large pieces of missing rib sections, this is usually a sign that the belt is wearing and should be replaced to prevent an untimely failure.

To check tension on the manually adjusted belt found on 2001 200-250 hp 3.0L models measure at a point halfway between the pulleys by pressing on the belt with moderate thumb pressure. The belt should deflect about 1/2 in. (12.7mm) in response to a 22 lbs. (30 N) side load. If belt tension is incorrect, loosen the 2 alternator mounting bolts, then pivot the alternator to achieve the proper load. REMEMBER YOU DON'T WANT IT TOO TIGHT EITHER, as that will quickly ruin the alternator bearings. Once adjusted, tighten the alternator adjustment and pivot bolts to 40 ft. lbs. (54 Nm).

■ On 2001 200-250 hp 3.0L models with manual tension adjustment, Mercury recommends using a standard belt tension adjustment tool to measure and properly set belt tension. When using a generic tool for this the belt tension is set to 80 lbs. (356 Nm), this should achieve the proper side load tension we outlined checking earlier.

On models with a tensioner pulley, the belt tension is automatically regulated by a spring loaded mechanism in the tensioner pulley. Mercury does not provide any specification to determine when the tensioner is worn out, but suffice it to say that if a belt slips, replace the belt. If a NEW belt continues to slip (and you're sure that you've used the correct sized replacement), then you should replace the tensioner. Either way, periodically check the belt for glazing or other signs of slippage. Besides, as we said earlier, replacing the belt periodically to prevent a potential failure in use is

probably a good idea anyway (and it is a good way to come up with a spare belt to keep on board to limp you home in the event of a failure).

■ The automatic belt tensioner assembly on some motors, like the mid-range EFI models, may contain a zerk fitting for greasing. When present, it's usually pretty obvious, but check the assembly on fuel injected models just to be sure. When present, be sure to grease it lightly with each engine lubrication service.

REMOVAL & INSTALLATION

◆ See Figures 85 thru 89

■ When replacing belts, we recommend cleaning the inside of the belt pulleys to extend the service life of the belts. Never use automotive belts, marine belts used on your engine are heavy duty and not interchangeable.

On models with an automatic tensioner pulley, removal and installation of the belt is a relatively simple matter of using a ratchet or breaker bar in the square fitting on the tensioner assembly to rotate the tensioner pulley back off of the belt. This removes the tension sufficiently that you can slip it over one or more of the pulleys and remove it from the motor.

■ On most models you'll have to remove the flywheel cover for access. On some models, like on V6 Optimax models, you'll need to disconnect the vent hose from the fitting on the flywheel cover.

A couple of points to keep in mind. FIRST, pay attention to belt routing, YES, it's pretty simple on these motors, but take a quick digital pic or make a

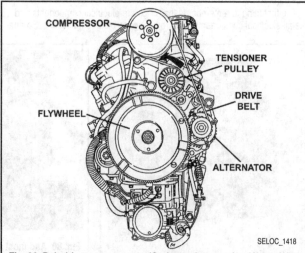
Fig. 86 Belt driven components (Optimax shown, other V6 models similar but no compressor)

Fig. 87 Typical belt routing - carb or EFI motor (note models will vary)

Fig. 88 Typical belt routing - Optimax motor (note models will vary)

Fig. 89 Belt tensioners are designed to receive a socket or breaker bar drive

mental note just to be sure, it will save you a couple of head scratching minutes when you're ready to install it again. More importantly, on a few EFI models the belt may actually be routed to a pulley BELOW the flywheel instead of on top. This doesn't pose a huge problem, but sometimes another component, like the crankshaft position sensor must be repositioned to sneak the belt out from under the flywheel. Lastly, and MOST importantly, be gentle with the tensioner assembly. Don't allow it to snap back into position (whether or not the belt is there) as this can damage the assembly. Always release and reset tension gradually and gently.

As for the few models that require manual tension adjustment (specifically the 2001 150-200 hp 2.5L models), belt replacement is a simple matter of loosening the alternator mounting bolts and pivoting it inward to release tension, then slipping the belt off the pulleys. Installation is the reverse, but be sure to follow the belt tension adjustment guidelines found in this section under INSPECTION.

Air Filter (Optimax Air Compressor)

◆ See Figures 90 thru 93

REMOVAL & INSTALLATION

Although the shape and placement of the filter varies slightly from model-to-model (and sometimes year-to-year) most Optimax motors (except Pro/XS/Sport models) are equipped with an air filter for the air compressor assembly. This filter element should be replaced approximately every 100 hours or once a season, whichever comes first.

✳✳ SELOC WARNING

NEVER, and we mean NEVER, operate the motor without this air filter in place. Contaminants would not only potentially damage the compressor cylinder bore, but air from the cylinder is pressurized and pumped to the air/fuel mixture rail and the direct injectors. All of these components could potentially be damaged.

Basically, the filter is mounted in the engine flywheel cover or side cover/air intake silencer assembly. For access to the filter itself you MAY need to remove the flywheel cover and/or the side cover itself, as access to the filter assembly is normally on the back or underside (powerhead side) of the side cover/flywheel cover. However, on some models, such as the 2003 and later 200-250 3.0L motors, access MAY be possible with the cover installed. For more details, access the filter as follows, depending upon the model:

• On 75-115 hp (3-cyl) models, the filter is behind an access cover on the top corner of the side cover. The access cover is secured by 3 retaining screws. Back out the screws and remove the access cover, then pull the small square filter out of the side cover.

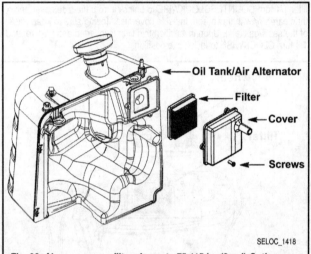

Fig. 90 Air compressor filter element - 75-115 hp (3-cyl) Optimax Models

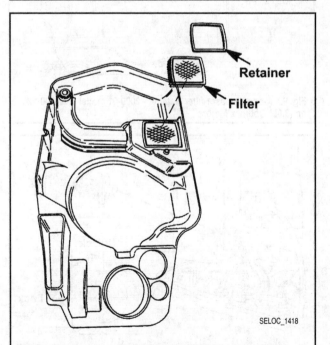

Fig. 91 Air compressor filter element - 135-200 hp (2.5L) Optimax Models

✳✳ SELOC CAUTION

Use care when removing the flywheel cover on Optimax models. Some are retained strictly by pushpins and rubber grommets and the covers themselves can be fragile and break if mis-handled. Make sure there are no bolts or screws securing the cover (or remove them if there are), then CAREFULLY support the cover at the connection points by reaching underneath and pushing upward (instead of just grabbing the cover flange and pulling upward).

• On 135-200 hp (2.5L) models, the filter is secured to the corner of the flywheel cover by a retainer that is snapped into place. To remove the filter carefully unsnap the retainer and pull the element from the cover.

• On 200-250 hp (3.0L) motors, there are two very different designs. On 2001-02 models, the filter is mounted to the top end of the flywheel cover, under an access cover that is secured by 4 screws. During installation be sure to coat the threads of the 4 access cover retaining screws using Loctite® 271 or an equivalent threadlocking compound. However, starting in 2003 the filter is attached to a hose on the lower corner of the side cover. On these late-models, the filter (with hose attached) is removed by rotating the filter 1/4 turn COUNTERCLOCKWISE to unlock it from the side cover and then carefully withdrawn from the side cover itself (being sure to keep track of the seals/gaskets). Upon installation this filter is inserted and then rotated 1/4 turn CLOCKWISE to lock it into position.

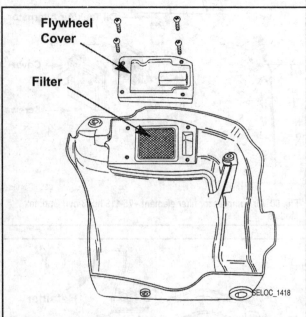

Fig. 92 Air compressor filter element - 2000-01 model year 200-225 hp (3.0L) Optimax Motors

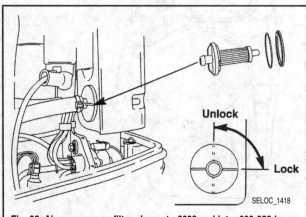

Fig. 93 Air compressor filter element - 2003 and later 200-250 hp (3.0L) Optimax Motors

Oil Filter (Inline Filter for Injection System)

REMOVAL & INSTALLATION

◆ **See Figures 94 and 95**

All Optimax and a handful of EFI motors are equipped with one or more inline oil filters for the engines oil fuel supply. On MOST motors, the filter is normally found inline between the remote oil tank and the engine mounted reservoir. On V6 Optimax motors this line is quickly identified by the blue stripe. On 3-cylinder Optimax motors the filter is usually mounted between the engine oil reservoir and the oil pump.

Now, ONLY the service literature for the 1.5L and 3.0L Optimax motors mentions a periodic replacement of this filter. However, again, we feel that it is a cheap enough part to justify replacing it periodically in order to prevent clogs. Remember a clog would allow the oil level to drop and potentially starve (if only for a short time) the powerhead of needed oil flow.

In all cases, when replacement is necessary be sure to pay close attention to the direction of flow marked on the old filter BEFORE removal. The size/shape and flanges of the new filter may vary, but in the end, you MUST install the new filter with the arrow pointing in the same direction. This means TOWARDS the reservoir if it was mounted between the remote tank and engine reservoir OR AWAY from the reservoir if it was mounted between the reservoir and oil pump.

Be sure to have a rag handy to catch any oil which will escape from the lines when you are doing this. Also, once completed, be sure to properly prime the oiling system before the motor is operated. For more details about priming the oil system and bleeding the air from it, please refer to the Lubrication & Cooling System section.

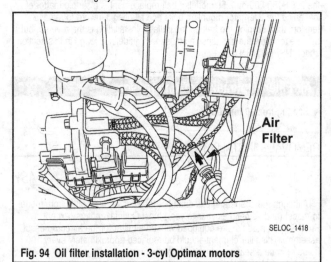

Fig. 94 Oil filter installation - 3-cyl Optimax motors

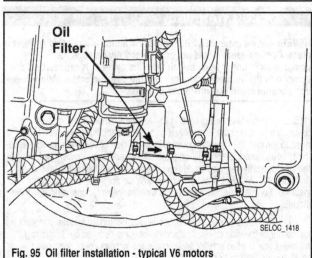

Fig. 95 Oil filter installation - typical V6 motors

BOAT MAINTENANCE

Batteries

◆ **See Figures 96 and 97**

Batteries require periodic servicing, so a definite maintenance program will help ensure extended life. A failure to maintain the battery in good order can prevent it from properly charging or properly performing its job even when fully charged. Low levels of electrolyte in the cells, loose or dirty cable connections at the battery terminals or possibly an excessively dirty battery top can all contribute to an improperly functioning battery. So battery maintenance, first and foremost, involves keeping the battery full of electrolyte, properly charged and keeping the casing/connections clean of corrosion or debris.

If a battery charges and tests satisfactorily but still fails to perform properly in service, 1 of 3 problems could be the cause.

1. An accessory left on overnight or for a long period of time can discharge a battery.

2. Using more electrical power than the stator assembly or lighting coil can replace would slowly drain the battery during motor operation, resulting in an undercharged condition.

3. A defect in the charging system. A faulty stator assembly or lighting coil, defective regulator or rectifier or high resistance somewhere in the system could cause the battery to become undercharged.

■ **For more information on marine batteries, please refer to Battery in the Ignition & Electrical Systems section.**

MAINTENANCE

◆ **See Figures 97 thru 100**

Electrolyte Level

The most common and important procedure in battery maintenance is checking the electrolyte level. On most batteries, this is accomplished by removing the cell caps and visually observing the level in the cells. The bottom of each cell normally is equipped with a split vent which will cause the surface of the electrolyte to appear distorted when it makes contact. When the distortion first appears at the bottom of the split vent, the electrolyte level is correct. Smaller marine batteries are sometimes equipped with translucent cases that are printed or embossed with high and low level markings on the side. On some of these, shining a flashlight through the battery case will help make it easier to determine the electrolyte level.

During hot weather and periods of heavy use, the electrolyte level should be checked more often than during normal operation. Add distilled water to bring the level of electrolyte in each cell to the proper level. Take care not to overfill, because adding an excessive amount of water will cause loss of electrolyte and any loss will result in poor performance, short battery life and will contribute quickly to corrosion.

■ **Never add electrolyte from another battery. Use only distilled water. Even tap water may contain minerals or additives that will promote corrosion on the battery plates, so distilled water is always the best solution.**

Fig. 96 Explosive hydrogen gas is released from the batteries in a discharged state. This one exploded when something ignited the gas. Explosions can be caused by a spark from the battery terminals or jumper cables

Fig. 97 Ignoring a battery (and corrosion) to this extent is asking for it to fail

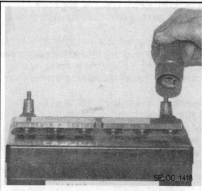

Fig. 98 Place a battery terminal tool over posts, then rotate back and forth . . .

Fig. 99 . . . until the internal brushes expose a fresh, clean surface on the post

Fig. 100 Clean the insides of cable ring terminals using the tool's wire brush

Although less common in marine applications than other uses today, sealed maintenance-free batteries also require electrolyte level checks, through the window built into the tops of the cases. The problem for marine applications is the tendency for deep cycle use to cause electrolyte evaporation and electrolyte cannot be replenished in a sealed battery. Although, more and more companies are producing a maintenance-free batteries for marine applications and their success should be noted.

The second most important procedure in battery maintenance is periodically cleaning the battery terminals and case.

Cleaning

Dirt and corrosion should be cleaned from the battery as soon as it is discovered. Any accumulation of acid film or dirt will permit a small amount of current to flow between the terminals. Such a current flow will drain the battery over a period of time.

Clean the exterior of the battery with a solution of diluted ammonia or a paste made from baking soda and water. This is a base solution that will neutralize any acid that may be present. Flush the cleaning solution off with plenty of clean water.

✳✳ SELOC WARNING

Take care to prevent any of the neutralizing solution from entering the cells as it will quickly neutralize the electrolyte (ruining the battery).

Poor contact at the terminals will add resistance to the charging circuit. This resistance may cause the voltage regulator to register a fully charged battery and thus cut down on the stator assembly or lighting coil output adding to the low battery charge problem.

At least once a season, the battery terminals and cable clamps should be cleaned. Loosen the clamps and remove the cables, negative cable first. On batteries with top mounted posts, if the terminals appear stuck, use a puller specially made for this purpose to ensure the battery casing is not damaged. NEVER pry a terminal off a battery post. Battery terminal pullers are inexpensive and available in most parts stores.

Clean the cable clamps and the battery terminal with a wire brush until all corrosion, grease, etc., is removed and the metal is shiny. It is especially important to clean the inside of the clamp thoroughly (a wire brush or brush part of a battery post cleaning tool is useful here), since a small deposit of foreign material or oxidation there will prevent a sound electrical connection and inhibit either starting or charging. It is also a good idea to apply some dielectric grease to the terminal, as this will aid in the prevention of corrosion.

After the clamps and terminals are clean, reinstall the cables, negative cable last, do not hammer the clamps onto battery posts. Tighten the clamps securely but do not distort them. To help slow or prevent corrosion, give the clamps and terminals a thin external coating of grease after installation.

Check the cables at the same time that the terminals are cleaned. If the insulation is cracked or broken or if its end is frayed, that cable should be replaced with a new one of the same length and gauge.

TESTING

◆ **See Figure 101**

A quick check of the battery is to place a voltmeter across the terminals. Although this is by no means a clear indication, it gives you a starting point when trying to troubleshoot an electrical problem that could be battery related. Most marine batteries will be of the 12 volt DC variety. They are constructed of 6 cells, each of which is capable of producing slightly more than 2 volts, wired in series so that total voltage is 12 and a fraction. A fully charged battery will normally show more than 12 and slightly less than 13 volts across its terminals. But keep in mind that just because a battery reads 12.6 or 12.7 volts does NOT mean it is fully charged. It is possible for it to have only a surface charge with very little amperage behind it to maintain that voltage rating for long under load. A discharged battery will read some value less than 12 volts, but can normally be brought back to 12 volts through recharging. Of course a battery with one or more shorted or

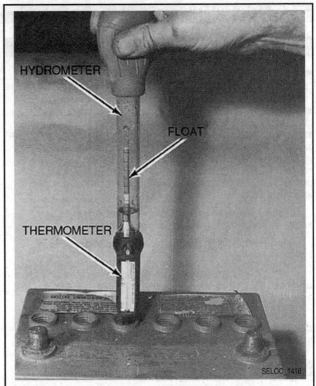

Fig. 101 A hydrometer is the best method for checking battery condition

un-chargeable cells will also read less than 12, but it cannot be brought back to 12+ volts after charging. For this reason, the best method to check battery condition on most marine batteries is through a specific gravity check or a load test.

A hydrometer is a device that measures the density of a liquid when compared to water (specific gravity). Hydrometers are used to test batteries by measuring the percentage of sulfuric acid in the battery electrolyte in terms of specific gravity. When the condition of the battery drops from fully charged to discharged, the acid is converted to water as electrons leave the solution and enter the plates, causing the specific gravity of the electrolyte to drop.

It may not be common knowledge but hydrometer floats are calibrated for use at 80°F (27°C). If the hydrometer is used at any other temperature, hotter or colder, a correction factor must be applied.

■ **Remember, a liquid will expand if it is heated and will contract if cooled. Such expansion and contraction will cause a definite change in the specific gravity of the liquid, in this case the electrolyte.**

A quality hydrometer will have a thermometer/temperature correction table in the lower portion, as illustrated in the accompanying illustration. By measuring the air temperature around the battery and from the table, a correction factor may be applied to the specific gravity reading of the hydrometer float. In this manner, an accurate determination may be made as to the condition of the battery.

When using a hydrometer, pay careful attention to the following points:

1. Never attempt to take a reading immediately after adding water to the battery. Allow at least 1/4 hour of charging at a high rate to thoroughly mix the electrolyte with the new water. This time will also allow for the necessary gases to be created.

2. Always be sure the hydrometer is clean inside and out as a precaution against contaminating the electrolyte.

3. If a thermometer is an integral part of the hydrometer, draw liquid into it several times to ensure the correct temperature before taking a reading.

4. Be sure to hold the hydrometer vertically and suck up liquid only until the float is free and floating.

5. Always hold the hydrometer at eye level and take the reading at the surface of the liquid with the float free and floating.

6. Disregard the slight curvature appearing where the liquid rises against the float stem. This phenomenon is due to surface tension.

7. Do not drop any of the battery fluid on the boat or on your clothing, because it is extremely caustic. Use water and baking soda to neutralize any battery liquid that does accidentally drop.

8. After drawing electrolyte from the battery cell until the float is barely free, note the level of the liquid inside the hydrometer. If the level is within the charged (usually green) band range for all cells, the condition of the battery is satisfactory. If the level is within the discharged (usually white) band for all cells, the battery is in fair condition.

9. If the level is within the green or white band for all cells except one, which registers in the red, the cell is shorted internally. No amount of charging will bring the battery back to satisfactory condition.

10. If the level in all cells is about the same, even if it falls in the red band, the battery may be recharged and returned to service. If the level fails to rise above the red band after charging, the only solution is to replace the battery.

■ **An alternate way of testing a battery is to perform a load test using a special Carbon-Pile Load Tester. These days most automotive and many marine parts stores contain a tester and will perform the check for free (hoping that your battery will fail and they can sell you another). Essentially a load test involves placing a specified load (current drain/draw) on a fully-charged battery and checking to see how it performs/recovers. This is the only way to test the condition of a sealed maintenance-free battery.**

STORAGE

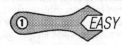

If the boat is to be laid up (placed into storage) for the winter or anytime it is not going to be used for more than a few weeks, special attention must be given to the battery. This is necessary to prevent complete discharge and/or possible damage to the terminals and wiring. Before putting the boat in storage, disconnect and remove the batteries. Clean them thoroughly of any dirt or corrosion and then charge them to full specific gravity readings. After they are fully charged, store them in a clean cool dry place where they will not be damaged or knocked over, preferably on a couple blocks of wood. Storing the battery up off the deck, will permit air to circulate freely around and under the battery and will help to prevent condensation.

Never store the battery with anything on top of it or cover the battery in such a manner as to prevent air from circulating around the filler caps. All batteries, both new and old, will discharge during periods of storage, more so if they are hot than if they remain cool. Therefore, the electrolyte level and the specific gravity should be checked at regular intervals. A drop in the specific gravity reading is cause to charge them back to a full reading.

In cold climates, care should be exercised in selecting the battery storage area. A fully-charged battery will freeze at about 60°F below zero. The electrolyte of a discharged battery, almost dead, will begin forming ice at about 19°F above zero.

■ **For more information on batteries and the engine electrical systems, please refer to the Ignition & Electrical section of this manual.**

INSPECTION & CARE

◆ **See Figures 102, 103 and 104**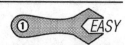

Fiberglass reinforced plastic hulls are tough, durable and highly resistant to impact. However, like any other material they can be damaged. One of the advantages of this type of construction is the relative ease with which it may be repaired.

A fiberglass hull has almost no internal stresses. Therefore, when the hull is broken or stove-in, it retains its true form. It will not dent to take an out-of-shape set. When the hull sustains a severe blow, the impact will be either absorbed by deflection of the laminated panel or the blow will result in a definite, localized break. In addition to hull damage, bulkheads, stringers and other stiffening structures attached to the hull may also be affected and therefore, should be checked. Repairs are usually confined to the general area of the rupture.

■ **The best way to care for a fiberglass hull is to wash it thoroughly, immediately after hauling the boat while the hull is still wet. The next best way to care for your hull is to give it a waxing a couple of times per season. Your local marina or boat supply store should be able to help you find some high quality boat soaps and waxes.**

A foul bottom can seriously affect boat performance. This is one reason why racers, large and small, both powerboat and sail, are constantly giving attention to the condition of the hull below the waterline.

In areas where marine growth is prevalent, a coating of vinyl, anti-fouling bottom paint should be applied if the boat is going to be left in the water for extended periods of time such as all or a large part of the season. If growth has developed on the bottom, it can be removed with a diluted solution of muriatic acid applied with a brush or swab and then rinsed with clear water. Always use rubber gloves when working with Muriatic acid and take extra care to keep it away from your face and hands. The fumes are toxic. Therefore, work in a well-ventilated area or if outside, keep your face on the windward side of the work.

■ **If marine growth is not too severe you may avoid the unpleasantness of working with muriatic acid by trying a power washer instead. Most marine vegetation can be removed by pressurized water and a little bit of scrubbing using a rough sponge (don't use anything that will scratch or damage the surface).**

Barnacles have a nasty habit of making their home on the bottom of boats that have not been treated with anti-fouling paint. Actually they will not harm the fiberglass hull but can develop into a major nuisance.

If barnacles or other crustaceans have attached themselves to the hull, extra work will be required to bring the bottom back to a satisfactory condition. First, if practical, put the boat into a body of fresh water and allow it to remain for a few days. A large percentage of the growth can be removed in this manner. If this remedy is not possible, wash the bottom thoroughly with a high-pressure fresh water source and use a scraper. Small particles of hard shell may still hold fast. These can be removed with sandpaper.

Fig. 102 The best way to care for a fiberglass hull is to wash it thoroughly

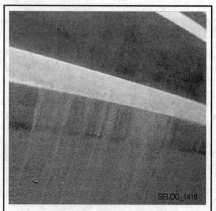

Fig. 103 If marine growth is a problem, apply a coating of anti-foul bottom paint

Fig. 104 Fiberglass, vinyl and rubber care products, like those from Maguire's protect your boat

Interior

INSPECTION & CARE

No one wants to walk around in bare feet on a boat whose deck or carpet is covered in fish guts right? It's not just a safety hazard, it's kind of nasty. Taking time to wash down and clean your boat's interior is just as important to the long term value of your boat as it is to your enjoyment. So take time, after every outing to make sure your baby is clean on the inside too.

Always try to find gentle cleaners for your vinyl and plastic seats. Harsh chemicals and abrasives will do more harm than good. Take care with guests aboard, as more than one brand of sun-tan lotion has been known to cause stains. Some people get carried away, forbidding things like cheesy coated chips/snacks or mustards on board. Don't let keeping your boat clean so much of an obsession that you forget to enjoy it, just keep a bottle of cleaner handy for quick spill clean-ups. And keeping it handy will ensure you'll be more likely to wipe things down after a fun-filled outing.

Be sure to always test a cleaner on a hidden or unexposed area of your carpet or vinyl before soaking things down with it. If it does not harm or damage the color of your finish, you're good to go.

When we trailer our boats, we sometimes find it more convenient to hit a spray-it-yourself car wash on the way home. This gives us a chance to spray down the boat hull, trailer and tow vehicle before we get home and turn our attention to engine flushing and wiping down/cleaning the interior.

If you're lucky enough to have snap out marine carpet, remove it and give it a good wash down once in a while. This allows you to spray down the deck as well. Hang the carpet to dry and reinstall once it is ready. If you've got permanently installed marine carpet, you can spray it down too, just make sure you can give it a chance to dry before putting the cover back on.

■ **For permanently installed marine carpet, try renting a rug steam-cleaner at least once a season and give it a good deep cleaning. We like to do it at the beginning and the end of each season!**

TUNE-UP

Introduction to Tune-Ups

A proper tune-up is the key to long and trouble-free outboard life and the work can yield its own rewards. Studies have shown that a properly tuned and maintained outboard can achieve better fuel economy than an out-of-tune engine. As a conscientious boater, set aside a Saturday morning, say once a month, to check or replace items which could cause major problems later. Keep your own personal log to jot down which services you performed, how much the parts cost you, the date and the number of hours on the engine at the time. Keep all receipts for such items as oil and filters, so that they may be referred to in case of related problems or to determine operating expenses. These receipts are the only proof you have that the required maintenance was performed. In the event of a warranty problem on newer engines, these receipts will be invaluable.

The efficiency, reliability, fuel economy and enjoyment available from boating are all directly dependent on having your outboard tuned properly. The importance of performing service work in the proper sequence cannot be over emphasized. Before making any adjustments, check the specifications. Never rely on memory when making critical adjustments.

Before tuning any outboard, insure it has satisfactory compression. An outboard with worn or broken piston rings, burned pistons or scored cylinder walls, will not perform properly no matter how much time and expense is spent on the tune-up. Poor compression must be corrected or the tune-up will not give the desired results.

The extent of the engine tune-up is usually dependent on the time lapse since the last service. In this section, a logical sequence of tune-up steps will be presented in general terms. If additional information or detailed service work is required, refer to the section of this manual containing the appropriate instructions.

Tune-Up Sequence

A tune-up can be defined as pre-determined series of procedures (adjustments, tests and worn component replacements) that are performed to bring the engine operating parameters back to original condition. The series of steps are important, as the later procedures (especially adjustments) are often dependent upon the earlier procedures. In other words, a procedure is performed only when subsequent steps would not change the result of that procedure (this is mostly for adjustments or settings that would be incorrect after changing another part or setting). For instance, fouled or excessively worn spark plugs may affect engine idle. If adjustments were made to the idle speed or mixture of a carbureted engine **before** these plugs were cleaned or replaced, the idle speed or mixture might be wrong after replacing the plugs. The possibilities of such an effect become much greater when dealing with multiple adjustments such as timing, idle speed and/or idle mixture. Therefore, be sure to follow each of the steps given here. Since many of the steps listed here are full procedures in themselves, refer to the procedures of the same name in this section for details.

■ **Computer controlled ignition and fuel components on more and more modern outboards have lessened the amount of steps necessary for a pre-season tune-up, but not completely eliminated the need for replacing worn components. EFI and Optimax motors may not allow for many (or often even ANY) timing or mixture adjustments, however they still have mechanical and electrical components that wear making compression checks, spark plug/wire replacement, and component inspection steps all that much more important.**

A complete pre-season tune-up should be performed at the beginning of each season or when the motor is removed from storage. Operating conditions, amount of use and the frequency of maintenance required by your motor may make one or more additional tune-ups necessary during the season. Perform additional tune-ups as use dictates.

1. Before starting, inspect the motor thoroughly for signs of obvious leaks, damage and loose or missing components. Make repairs, as necessary.

✳✳ SELOC CAUTION

We can't emphasize enough how important is this first step, ESPECIALLY on Optimax motors. The high-pressure air/fuel system of these motors makes fuel system/line integrity a very important safety issue.

2. Check all accessible bolts and fasteners and tighten any that are loose.

3. Perform a compression check to make sure the motor is mechanically ready for a tune-up. An engine with low compression on one or more cylinder should be overhauled, not tuned. A tune-up will not be successful without sufficient engine compression. Refer to the Compression Testing in this section.

■ **If this tune-up is occurring immediately after removing an engine from storage, be sure to start and run the motor using the old plugs first (if possible) while burning off the fogging oil. Then install the new spark plugs once the compression test is completed!**

4. Since the spark plugs must be removed for the compression check, take the opportunity to inspect them thoroughly for signs of oil fouling, carbon fouling, damage due to detonation, etc. Clean and gap the plugs or, better yet, install new plugs as no amount of cleaning will precisely match the performance and life of new plugs. Refer to Spark Plugs, in this section. Also, this is a good time to check the spark plug wires as well. Please refer to Spark Plug Wires, in this section.

■ **We don't care how little you use your motor, there is usually no excuse for not installing new plugs at the beginning of each season. If only because when storing a motor the fogging oil will go a long way to fouling even a decent set of spark plugs. Remember, the secondary ignition circuit is the most likely performance problem that occurs on an outboard.**

5. Visually inspect all ignition system components for signs of obvious defects. Look for signs of burnt, cracked or broken insulation. Replace wires

or components with obvious defects. If spark plug condition suggests weak or no spark on one or more cylinders, perform ignition system testing to eliminate possible worn or defective components. Refer to the Ignition System Inspection procedures in this section and the Ignition and Electrical System section.

6. Visually inspect all engine wiring and, if equipped, the battery and starter motor. A quick starter motor draw test can tell you a lot about the condition of your electrical starting system.

7. Remove and clean (on serviceable filters) or replace the inline filter and/or fuel pump filter, as equipped. Refer to the Fuel Filter procedures in this section. Perform a thorough inspection of the fuel system, hoses and components. Replace any cracked or deteriorating hoses. If carburetor adjustment or overhaul is necessary, perform these procedures before proceeding.

8. Pressurize the fuel system according to the procedures found in the Fuel System section, then check carefully for leaks. Again, this is important on all motors, but even more so on EFI motors and critical on Optimax engines!

9. Perform engine Timing & Synchronization adjustments as described in this section.

■ Although many of the motors covered here allow for certain ignition timing and, if applicable, carburetor adjustment procedures, none of them require the level of tuning attention that was once the norm. Many of the motors are equipped with electronic ignition systems that limit or eliminate timing adjustments. Carburetors used on many of these ALL of these models (at least sold in the US) are U.S. EPA regulated and contain few mixture adjustments. The air/fuel mixture is completely computer controlled on fuel injected motors and allows for no adjustment.

10. Except for jet drive models, remove the propeller in order to thoroughly check for leaks at the shaft seal. Inspect the propeller or impeller condition, look for nicks, cracks or other signs of damage and repair or replace, as necessary. If available, install a test wheel to run the motor in a test tank after completion of the tune-up. If no test wheel is available, lubricate the shaft/splines, then install the propeller or rotor. Refer to the procedures for Propeller or Jet Drive Impeller in this section, as applicable.

11. Change the gearcase oil as directed under the Gearcase Oil procedures in this section. If you are conducting a pre-season tune-up and the oil was changed immediately prior to storage this is not necessary. But, be sure to check the oil level and condition. Drain the oil anyway if significant contamination is present.

■ Anytime large amounts of water or debris is present in the gearcase oil, be sure to troubleshoot and repair the problem before returning the gearcase to service. The presence of water may indicate problems with the seals, while debris could a sign that overhaul is required.

12. Perform a test run of the engine to verify proper operation of the starting, fuel, oil and cooling systems. Although this can be performed using a flush/test adapter or even on the boat itself (if operating with a normal load/passengers), the preferred method is the use of a test tank. Keep in mind that proper operation without load and at low speed (on a flush/test adapter) doesn't really tell you how well the motor will run under load. If possible, run the engine, in a test tank using the appropriate test wheel. Monitor the cooling system indicator stream to ensure the water pump is working properly. Once the engine is fully warmed, slowly advance the engine to wide-open throttle, then note and record the maximum engine speed. Refer to the Tune-Up Specifications chart to compare engine speeds with the test propeller minimum rpm specifications. If engine speeds are below specifications, yet engine compression was sufficient at the beginning of this procedure, recheck the fuel and ignition system adjustments (as well as the condition of the prop if you're not using a test wheel).

Compression Testing

The quickest way to gauge the condition of an internal combustion engine is through a compression check. In order for an internal combustion engine to work properly, it must be able to generate sufficient compression in the combustion chamber to take advantage of the explosive force generated by the expanding gases after ignition.

If the combustion chambers or reed valves are worn or damaged in some fashion as to allow pressure to escape, the engine cannot develop sufficient horsepower. Under these circumstances, combustion will not occur properly meaning that the air/fuel mixture cannot be set to maximize power and

minimize emissions. Obviously, it is useless to try to tune an engine with extremely low or erratic compression readings, since a simple tune-up will not cure the problem. An engine with poor compression on one or more cylinders should be overhauled.

The pressure created in the combustion chamber may be measured with a gauge that remains at the highest reading it measures during the action of a one-way valve. This gauge is inserted into the spark plug hole and held or threaded in position while the motor is cranked. A compression test will uncover many mechanical problems that can cause rough running or poor performance.

If the powerhead shows any indication of overheating, such as discolored or scorched paint, inspect the cylinders visually through the spark plug hole or transfer ports for possible scoring. It is possible for a cylinder with satisfactory compression to be scored slightly. Also, check the water pump, since a faulty water pump can cause an overheating condition.

A secondary and more involved leak test can be performed, using a partially stripped powerhead who's intake and exhaust ports are blocked off by temporary plates, allowing the powerhead to be pressurized and checked for leaks.

TUNE-UP COMPRESSION CHECK

◆ See Figure 105

■ A compression check requires a compression gauge and a spark plug port adapter that matches the plug threads of your motor.

When analyzing the results of a compression check, generally the actual amount of pressure measured during a compression check is not quite as important as the variation from cylinder-to-cylinder on the same motor. On multi-cylinder powerheads the rule of thumb is that variations in pressure up to about 15 psi (103 kPa) or 15% (although on these motors since most compression should be above 100 psi/103 KPa, 15% would be MORE of a variance than 15 psi/103 kPa) may be considered normal (when comparing the lowest compression reading to the highest during a given test). Since there are no other cylinders to compare with on single cylinder powerheads you will generally compare the reading to the one that you took and recorded when it was new (you did do a compression test on it when it was new, didn't you?). Again, generally speaking, a drop of about 15 psi (103 kPa) from the normal compression pressure you established when it was new is cause for concern.

Ok, for the point of arguments sake let's say you bought the engine used or never checked compression the first season or so, assuming it wasn't something you needed to worry about. You're not alone, many of us have done that. But now that you're reading this it is your chance to take the data and note it for future.

Mercury does not publish compression specifications for some of their motors, but we've listed the normal compression specs for as many as we could locate here. As a rule of thumb on MOST Mercury motors, 90-120 psi (621-856 kPa) is at the lower limit of acceptable compression (at least for

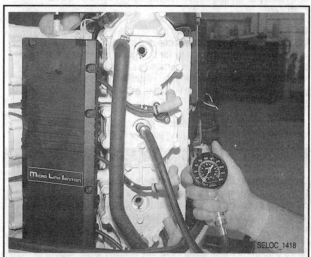

Fig. 105 Typical two-stroke powerhead secondary compression test

mid-range and larger motors). Published compression specifications for these models are as follows:

- 4/5 hp motors Minimum: 90 psi (621 kPa)
- 6/8 and 9.9/10/15 hp motors Normal 115-120 psi (787-856), Minimum: 100 psi (685 kPa)
- 20/20 Jet/25 hp motors Minimum: about 120 psi (856 kPa)
- 75/65 Jet/90 hp motors Minimum: about 120 psi (856 kPa)
- 80 Jet/115/125 hp motors Minimum: about 120 psi (856 kPa)
- 75/90/115 hp Optimax motors Normal 90-110 psi (621-758 kPa)
- 135-200 hp (2.5L) motors Normal 110-130 psi (759-896 kPa)
- 200-250 hp (3.0L) motors Normal 90-110 psi (621-758 kPa), except for Pro/XS/Sport models which should be 100-110 psi (689-758 kPa)

You might be interested to note that, a quick check of the specifications from other manufacturers shows that a range of about 78 psi (556 kPa) to 153 psi (1079 kPa) is not surprising for many 2-stroke motors (again generally the smaller hp motors having lower ranges and larger motors having higher ranges).

If a specification is not available for your motor, put the most weight on a comparison of the readings from the other cylinders on the same motor (or readings when the motor was new or even from the last tune-up).

When taking readings during the compression check, repeat the procedure a few times for each cylinder, recording the highest reading for that cylinder. Then, for all multi-cylinder motors, the compression reading on the lowest cylinder should be within about 15 psi (103 kPa) of the highest reading. If not, consider 15% to be the absolute limit. If the reading in the lowest cylinder is less than 85% of the reading in the highest cylinder, it's time to find and remedy the cause.

■ **If the powerhead has been in storage for an extended period, the piston rings may have relaxed. This will often lead to initially low and misleading readings. Always run an engine to normal operating temperature to ensure that the readings are accurate.**

■ **If you've never removed the spark plugs from this cylinder head before, break each one loose and retighten them before starting the motor in order to make sure they will not seize in the head once it is warmed. Better yet, remove each one and coat the threads very lightly with some fresh anti-seize compound.**

Prepare the engine for a compression test as follows:

1. Run the engine until it reaches operating temperature. The engine is at operating temperature a few minutes after the powerhead becomes warm to the touch and the stream of water exiting the cooling indicator becomes warm. If the test is performed on a cold engine, the readings will be considerably lower than normal, even if the engine is in perfect mechanical condition.

2. Label and disconnect the spark plug wires. Always grasp the molded cap and pull it loose with a twisting motion to prevent damage to the connection.

3. Clean all dirt and foreign material from around the spark plugs, and then remove all the plugs. Keep them in order by cylinder for later evaluation.

■ **On many Mercury motors you can disable the ignition system by leaving the safety lanyard disconnected but still use the starter motor to turn the motor. This is handy for things like compression tests or distributing fogging oil. To be certain use a spark plug gap tester on 1 lead and crank the motor using the keyswitch. If no spark is present, you're good to go, if not, you'll have to ground the spark plug leads to the cylinder head.**

4. Ground the spark plug leads to the engine to render the ignition system inoperative while performing the compression check.

✳✳ SELOC CAUTION

Grounding the spark plug leads not only protects the ignition system from potential damage that may be caused by the excessive load placed on operating the system with the wires disconnected, but more importantly, protects you from the dangers of arcing. The ignition system operates at extremely high voltage and could cause serious shocks. Also, keep in mind that you're cranking an engine with open spark plug ports which could allow any remaining fuel vapors to escape become ignited by arcing current.

5. Insert a compression gauge into the No. 1, top, spark plug opening.

6. Move the throttle to the wide open position in order to make sure the throttle plates are not restricting air flow. If necessary you may have to spin the propeller shaft slowly by hand while advancing the throttle in order to get the shifter into gear.

7. Crank the engine with the starter through at least 4-5 complete strokes with the throttle at the wide-open position, to obtain the highest possible reading. Record the reading.

■ **On electric start motors, it is very important to use a freshly charged cranking battery as a weakened battery will cause a slower than normal cranking speed, reducing the compression reading.**

8. Repeat the test and record the compression for each cylinder.

9. A variation between cylinders is far more important than the actual readings. A variation of more than about 15 psi (103 kPa), between cylinders indicates the lower compression cylinder is defective. Not all engines will exhibit the same compression readings. In fact, two identical engines may not have the same compression. Generally, the rule of thumb is that the lowest cylinder should be within 15% of the highest (difference between the 2 readings).

10. If compression is low in one or more cylinders, the problem may be worn, broken, or sticking piston rings, scored pistons or worn cylinders.

LOW COMPRESSION

Compression readings that are generally low indicate worn, broken, or sticking piston rings, scored pistons or worn cylinders, and usually indicate an engine that has a lot of hours on it. Low compression in 2 adjacent cylinders (with normal compression in the other cylinders) indicates a blown head gasket between the low-reading cylinders. Other problems are possible (broken ring, hole burned in a piston), but a blown head gasket is most likely.

■ **Use of an engine cleaner, available at any automotive parts house, will help to free stuck rings and to dissolve accumulated carbon. Follow the directions on the container.**

To test a cylinder or motor further, add a few drops of engine oil to the cylinder and recheck.

- If compression is higher with oil added to the cylinder, suspect a worn or damaged piston.
- If compression is the same as without oil, suspect one or more defective rings or a defective piston, but you can also suspect oil seals or reed valves.
- If compression is well above specification, suspect the carbon deposits are on the cylinder head and/or piston crown. Significant carbon deposits will lead to pre-ignition and other performance problems and should be cleaned (either using an additive or by removing the cylinder head for manual cleaning).

You're final testing option is to acquire or fabricate some plates to block off intake and exhaust ports, then pressurize the crankcase and check for leaks. For more details, please refer to Leak Testing.

LEAK TESTING

Because the 2-stroke powerhead is a pump, the crankcase must be sealed against pressure created on the down stroke of the piston and vacuum created when the piston moves toward top dead center. If there are air leaks into the crankcase, insufficient fuel (or additional air on Optimax motors) will be brought into the crankcase and into the cylinder for normal combustion.

■ **If it is a very small leak, the powerhead will run poorly, because the fuel mixture will be lean and cylinder temperatures will be hotter than normal.**

Air leaks are possible around any seal, O-ring, cylinder block mating surface or gasket. Always replace O-rings, gaskets and seals when service work is performed.

If the powerhead is running, soapy water can be sprayed onto the suspected sealing areas. If bubbles develop, there is a leak at that point. Oil around sealing points and on ignition parts under the flywheel indicates a crankcase leak.

The base of the powerhead and lower crankshaft seal is impossible to check on an installed powerhead. When every test and system have been checked out and the bottom cylinder seems to be effecting performance, then the lower seal should be tested.

In some cases adapter plates available from tool manufacturers to seal the inlet, exhaust and base of the powerhead. Adapter plates can also be manufactured by cutting metal block off plates from pieces of plate steel or aluminum. A pattern made from the gaskets can be used for an accurate shape. Seal these plates using rubber or silicone gasket making compound.

1. Install adapter plates over the intake ports and the exhaust ports to completely seal the powerhead.

■ **When installing the adapter plates, make sure to leave the water jacket holes open.**

2. Into one adapter, place an air fitting which will accept a hand air pump.

3. Using the hand pump (or another regulated air source), pressurize the crankcase to 5 pounds of pressure.

4. Spray soapy water around the lower seal area and other sealed areas watching for bubbles which indicate a leaking point.

5. Turn the powerhead upside down and fill the water jacket with water. If bubbles show up in the in the water when a positive pressure is applied to the crankcase, there may be cracks or corrosion holes in the cooling system passages. These holes can cause a loss of cooling system effectiveness and lead to overheating.

6. After the pressure test is completed, pull a vacuum to stress the seals in the opposite direction and watch for a pressure drop.

7. Note the leaking areas and replace the seals or gaskets.

Spark Plugs

◆ **See Figure 106**

The spark plug performs four main functions:
• First and foremost, it provides spark for the combustion process to occur.
• It also removes heat from the combustion chamber.
• Its removal provides access to the combustion chamber (for inspection or testing) through a hole in the cylinder head.
• It acts as a dielectric insulator for the ignition system.

It is important to remember that spark plugs do not create heat, they help remove it. Anything that prevents a spark plug from removing the proper amount of heat can lead to pre-ignition, detonation, premature spark plug failure and even internal engine damage, especially in 2-stroke engines.

In the simplest of terms, the spark plug acts as the thermometer of the engine. Much like a doctor examining a patient, this "thermometer" can be used to effectively diagnose the amount of heat present in each combustion chamber.

Spark plugs are valuable tuning tools, when interpreted correctly. They will show symptoms of other problems and can reveal a great deal about the engine's overall condition. Evaluating the appearance of the spark plug's firing tip, gives visual cues to determine the engine's overall operating condition, get a feel for air/fuel ratios and even diagnose driveability problems.

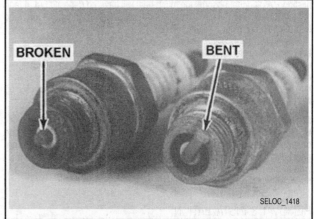

Fig. 106 Damaged spark plugs. Notice the broken electrode on the left plug. The electrode must be found and retrieved prior to returning the powerhead to service

As spark plugs grow older, they lose their sharp edges as material from the center and ground electrodes slowly erodes away. As the gap between these 2 points grows, the voltage required to bridge this gap increases proportionately. The ignition system must work harder to compensate for this higher voltage requirement and hence there is a greater rate of misfires or incomplete combustion cycles. Each misfire means lost horsepower, reduced fuel economy and higher emissions. Replacing worn out spark plugs with new ones (with sharp new edges) effectively restores the ignition system's efficiency and reduces the percentage of misfires, restoring power, economy and reducing emissions.

■ **Although spark plugs can typically be cleaned and gapped if they are not excessively worn, no amount of cleaning or gapping will return most spark plugs to original condition and it is usually best to just go ahead and replace them.**

How long spark plugs last will depend on a variety of factors, including engine compression, fuel used, gap, center/ground electrode material and the conditions in which the outboard is operated.

SPARK PLUG HEAT RANGE

◆ **See Figures 107, 108 and 109**

Spark plug heat range is the ability of the plug to dissipate heat from the combustion chamber. The longer the insulator (or the farther it extends into the engine), the hotter the plug will operate; the shorter the insulator (the closer the electrode is to the engine's cooling passages) the cooler it will operate.

Selecting a spark plug with the proper heat range will ensure that the tip maintains a temperature high enough to prevent fouling, yet cool enough to prevent pre-ignition. A plug that absorbs little heat and remains too cool will quickly accumulate deposits of oil (for 2-strokes) and carbon since it won't be able to burn them off. This leads to plug fouling and consequently to

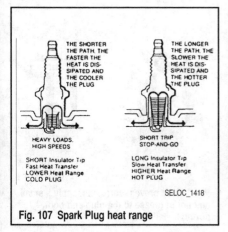

Fig. 107 Spark Plug heat range

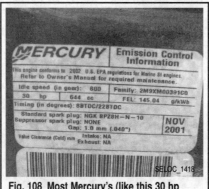

Fig. 108 Most Mercury's (like this 30 hp motor) have a label which lists spark plug type and gap...

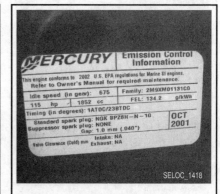

Fig. 109 ...it's an Emission Control Information label like on this 115 hp model

misfiring. A plug that absorbs too much heat will have no deposits but, due to the excessive heat, the electrodes will burn away quickly and might also lead to pre-ignition or other ignition problems.

Pre-ignition takes place when plug tips get so hot that they glow sufficiently to ignite the air/fuel mixture before the actual spark occurs. This early ignition will usually cause a pinging during heavy loads and if not corrected, will result in severe engine damage. While there are many other things that can cause pre-ignition, selecting the proper heat range spark plug will ensure that the spark plug itself is not a hot-spot source.

■ **The manufacturer recommended spark plugs are listed in the Tune-Up Specifications chart. Recommendations may also be present on one or more labels affixed to your motor. When the label disagrees with the chart, we'd normally defer to the label as it may reflect a change that was made mid-production and not reflected in the manufacturer's service literature.**

REMOVAL & INSTALLATION

◆ **See Figures 110 thru 115**

■ **New technologies in spark plug and ignition system design have greatly extended spark plug life over the years. But, spark plug life will still vary greatly with engine tuning, condition and usage. In general, 2-stroke motors are a little tougher on plugs, especially if great care is not taken to maintain proper oil/fuel mixtures on pre-mix motors.**

Typically spark plugs will require replacement once a season. The electrode on a new spark plug has a sharp edge but with use, this edge becomes rounded by wear, causing the plug gap to increase. As the gap increases, the plug's voltage requirement also increases. It requires a greater voltage to jump the wider gap and about 2-3 times as much voltage to fire a plug at high speeds than at idle.

■ **Fouled plugs can cause hard-starting, engine mis-firing or other problems. You don't want that happening on the water. Take time, at least once a month to remove and inspect the spark plugs. Early signs of other tuning or mechanical problems may be found on the plugs that could save you from becoming stranded or even allow you to address a problem before it ruins the motor.**

Tools needed for spark plug replacement include: a ratchet, short extension, spark plug socket (there are 2 types; either 13/16 inch or 5/8 inch, depending upon the type of plug), a combination spark plug gauge and gapping tool and a can of anti-seize type compound.

1. When removing spark plugs from multi-cylinder motors, work on one at a time. Don't start by removing the plug wires all at once, because unless you number them, they may become mixed up. Take a minute before you begin and number the wires with tape.

2. For safety, disconnect the negative battery cable or turn the battery switch **OFF**.

3. If the engine has been run recently, allow the engine to thoroughly cool (unless performing a compression check and then you should have already broken them loose once when cold and retightened them before warming the motor, so they should have less of a tendency to stick). Attempting to remove plugs from a hot cylinder head could cause the plugs to seize and damage the threads in the cylinder head, especially on aluminum heads!

■ **To ensure an accurate reading during a compression check, the spark plugs must be removed from a hot engine. But, DO NOT force a plug if it feels like it is seized. Instead, wait until the engine has cooled, remove the plug and coat the threads lightly with anti-seize then reinstall and tighten the plug, then back off the tightened position a little less than 1/4 turn. With the plug(s) installed in this manner, re-warm the engine and conduct the compression check.**

4. Carefully twist the spark plug wire boot to loosen it, then pull the boot using a twisting motion to remove it from the plug. Be sure to pull on the boot

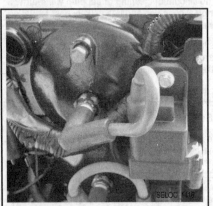

Fig. 110 The spark plugs are threaded into the center of the cylinder cover

Fig. 111 Gently grasp the boot, then twist and pull the wire from the plug

Fig. 112 Next unthread the plug using a ratchet and socket

Fig. 113 ALWAYS thread plugs by hand to prevent cross-threading...

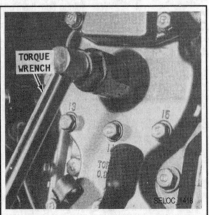

Fig. 114 ... then use a torque wrench to tighten the plug to spec

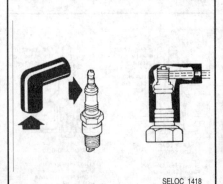

Fig. 115 To prevent corrosion, apply a small amount of grease to the plug and boot during installation

and not on the wire, otherwise the connector located inside the boot may become separated from the high-tension wire.

■ **If removal is difficult (or on motors where the spark plug boot is hard to grip because of access) a spark plug wire removal tool is recommended as it will make removal easier and help prevent damage to the boot and wire assembly. Most tools have a wire loom that fits under the plug boot so the force of pulling upward is transmitted directly to the bottom of the boot.**

5. Using compressed air (and safety glasses), blow debris from the spark plug area to assure that no harmful contaminants are allowed to enter the combustion chamber when the spark plug is removed. If compressed air is not available, use a rag or a brush to clean the area. Compressed air is available from both an air compressor or from compressed air in cans available at photography stores. In a pinch, blow up a balloon and use the escaping air to blow debris from the spark plug port(s).

■ **Remove the spark plugs when the engine is cold, if possible, to prevent damage to the threads. If plug removal is difficult, apply a few drops of penetrating oil to the area around the base of the plug and allow it a few minutes to work.**

6. Using a spark plug socket that is equipped with a rubber insert to properly hold the plug, turn the spark plug counterclockwise to loosen and remove the spark plug from the bore.

✳✳ SELOC WARNING

Avoid the use of a flexible extension on the socket. Use of a flexible extension may allow a shear force to be applied to the plug. A shear force could break the plug off in the cylinder head, leading to costly and/or frustrating repairs. In addition, be sure to support the ratchet with your other hand - this will also help prevent the socket from damaging the plug.

7. Evaluate each cylinder's performance by comparing the spark condition. Check each spark plug to be sure they are from the same plug manufacturer and have the same heat range rating. Inspect the threads in the spark plug opening of the block and clean the threads before installing the plug.

8. When purchasing new spark plugs, always ask the dealer if there has been a spark plug change for the engine being serviced. Sometimes manufacturers will update the type of spark plug used in an engine to offer better efficiency or performance.

9. Always use a new gasket (if applicable). The gasket must be fully compressed on clean seats to complete the heat transfer process and to provide a gas tight seal in the cylinder.

10. Inspect the spark plug boot for tears or damage. If a damaged boot is found, the spark plug boot and possibly the entire wire will need replacement.

11. Check the spark plug gap prior to installing the plug. Most spark plugs do not come gapped to the proper specification.

12. Apply a thin coating of anti-seize on the thread of the plug. This is extremely important on aluminum head engines to prevent corrosion and heat from seizing the plug in the threads (which could lead to a damaged cylinder head upon removal).

13. Carefully thread the plug into the bore by hand. If resistance is felt before the plug completely bottoms, back the plug out and begin threading again.

✳✳ SELOC WARNING

Do not use the spark plug socket to thread the plugs. Always carefully thread the plug by hand or using an old plug wire/boot to prevent the possibility of cross-threading and damaging the cylinder head bore. An old plug wire/boot can be used to thread the plug if you turn the wire by hand. Should the plug begin to cross-thread the wire will twist before the cylinder head would be damaged. This trick is useful when accessories or a deep cylinder head design prevents you from easily keeping fingers on the plug while it is threaded by hand.

14. Carefully tighten the spark plug to specification using a torque wrench, as follows:
- 2.5/3.3 hp motors: 240 inch lbs./20 ft. lbs. (27.1 Nm)
- 4/5 hp motors: 168 inch lbs./14 ft. lbs. (19.7 Nm)
- 6-250 hp motors: 90 inch lbs./20 ft. lbs. (27.1 Nm) or finger tight PLUS 1/4 turn.

■ **Whenever possible, spark plugs should be tightened to the factory torque specification. If a torque wrench is not available, and the plug you are installing is equipped with a crush washer, tighten the plug until the washer seats, then tighten it an additional 1/4 turn to crush the washer.**

15. Apply a small amount of a silicone dielectric grease to the ribbed, ceramic portion of the spark plug lead and inside the spark plug boot to prevent sticking, then install the boot to the spark plug and push until it clicks into place. The click may be felt or heard. Gently pull back on the boot to assure proper contact.

16. If applicable, connect the negative battery cable or turn the battery switch **ON**.

17. Test run the outboard (using a test tank or flush fitting) and ensure proper operation.

READING SPARK PLUGS

◆ **See Figures 116 thru 121**

Reading spark plugs can be a valuable tuning aid. By examining the insulator firing nose color, you can determine much about the engine's overall operating condition.

In general, a light tan/gray color tells you that the spark plug is at the optimum temperature and that the engine is in good operating condition.

Dark coloring, such as heavy black wet or dry deposits usually indicate a fouling problem. Heavy, dry deposits can indicate an overly rich condition, too cold a heat range spark plug, possible vacuum leak, low compression, overly retarded timing or too large a plug gap.

If the deposits are wet, it can be an indication of a breached head gasket, oil control from ring problems (on 4-stroke engines) or an extremely rich condition, depending on what liquid is present at the firing tip.

Also look for signs of detonation, such as silver specs, black specs or melting or breakage at the firing tip.

Compare your plugs to the illustrations shown to identify the most common plug conditions.

Fouled Spark Plugs

A spark plug is "fouled" when the insulator nose at the firing tip becomes coated with a foreign substance, such as fuel, oil or carbon. This coating makes it easier for the voltage to follow along the insulator nose and leach back down into the metal shell, grounding out, rather than bridging the gap normally.

Fuel, oil and carbon fouling can all be caused by different things but in any case, once a spark plug is fouled, it will not provide voltage to the firing tip and that cylinder will not fire properly. In many cases, the spark plug cannot be cleaned sufficiently to restore normal operation. It is therefore recommended that fouled plugs be replaced.

Signs of fouling or excessive heat must be traced quickly to prevent further deterioration of performance and to prevent possible engine damage.

Overheated Spark Plugs

When a spark plug tip shows signs of melting or is broken, it usually means that excessive heat and/or detonation was present in that particular combustion chamber or that the spark plug was suffering from thermal shock.

Since spark plugs do not create heat by themselves, one must use this visual clue to track down the root cause of the problem. In any case, damaged firing tips most often indicate that cylinder pressures or temperatures were too high. Left unresolved, this condition usually results in more serious engine damage.

Detonation refers to a type of abnormal combustion that is usually preceded by pre-ignition. It is most often caused by a hot spot formed in the combustion chamber.

As air and fuel is drawn into the combustion chamber during the intake stroke, this hot spot will "pre-ignite" the air fuel mixture without any spark from the spark plugs.

Detonation

Detonation exerts a great deal of downward force on the pistons as they are being forced upward by the mechanical action of the connecting rods. When this occurs, the resulting concussion, shock waves and heat can be severe. Spark plug tips can be broken or melted and other internal engine components such as the pistons or connecting rods themselves can be damaged.

Left unresolved, engine damage is almost certain to occur, with the spark plug usually suffering the first signs of damage.

■ **When signs of detonation or pre-ignition are observed, they are symptom of another problem. You must determine and correct the situation that caused the hot spot to form in the first place.**

INSPECTION & GAPPING

◆ **See Figures 122 and 123**

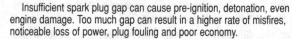

A particular spark plug might fit hundreds of powerheads and although the factory will typically set the gap to a pre-selected setting, this gap may not be the right one for your particular powerhead.

Insufficient spark plug gap can cause pre-ignition, detonation, even engine damage. Too much gap can result in a higher rate of misfires, noticeable loss of power, plug fouling and poor economy.

■ **Refer to the Tune-Up Specifications chart for spark plug gaps.**

Check spark plug gap before installation. The ground electrode (the L-shaped one connected to the body of the plug) must be parallel to the center electrode and the specified size wire gauge must pass between the electrodes with a slight drag.

Do not use a flat feeler gauge when measuring the gap on a used plug, because the reading may be inaccurate. A round-wire type gapping tool is the best way to check the gap. The correct gauge should pass through the electrode gap with a slight drag. If you're in doubt, try a wire that is one size smaller and one larger. The smaller gauge should go through easily, while the larger one shouldn't go through at all.

Fig. 116 A normally worn spark plug should have light tan or gray deposits on the firing tip (electrode)

Fig. 117 A carbon-fouled plug, identified by soft, sooty black deposits, may indicate an improperly tuned powerhead

Fig. 118 This spark plug has been left in the powerhead too long, as evidenced by the extreme gap. Plugs with such an extreme gap can cause misfiring and stumbling accompanied by a noticeable lack of power

Fig. 119 An oil-fouled spark plug indicates a powerhead with worn piston rings or a malfunctioning oil injection system that allows excessive oil to enter the combustion chamber

Fig. 120 A physically damaged spark plug may be evidence of severe detonation in that cylinder. Watch the cylinder carefully between services, as a continued detonation will not only damage the plug but will most likely damage the powerhead

Fig. 121 A bridged or almost bridged spark plug, identified by the build-up between the electrodes caused by excessive carbon or oil build up on the plug

Wire gapping tools usually have a bending tool attached. **USE IT!** This tool greatly reduces the chance of breaking off the electrode and is much more accurate. Never attempt to bend or move the center electrode. Also, be careful not to bend the side electrode too far or too often as it may weaken and break off within the engine, requiring removal of the cylinder head to retrieve it.

Spark Plug Wires

TESTING

Each time you remove the engine cover, visually inspect the spark plug wires for burns, cuts or breaks in the insulation. Check the boots on the coil and at the spark plug end. Replace any wire that is damaged.

Once a year, usually when you change your spark plugs, check the resistance of the spark plug wires with an ohmmeter and note it for future reference. Wires with excessive resistance will cause misfiring and may make the engine difficult to start. In addition worn wires will allow arcing and misfiring in humid conditions.

Remove the spark plug wire from the engine. Test the wires by connecting 1 lead of an ohmmeter to the coil end of the wire and the other lead to the spark plug end of the wire. Mercury doesn't provide specifications for most models, but generally you can expect wires to measure from 0-7000 ohms per foot of wire. One exception for Mercury are the 75-115 hp Optimax motors, for which Mercury states you should get a reading of about 1170-1330 ohms. If a spark plug wire is found to have excessive (high) resistance, the entire set should be replaced.

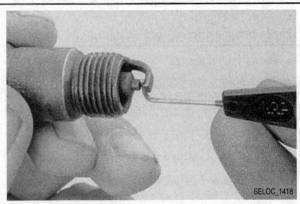

Fig. 122 Use a wire-type spark plug gapping tool to check the distance between center and ground electrodes

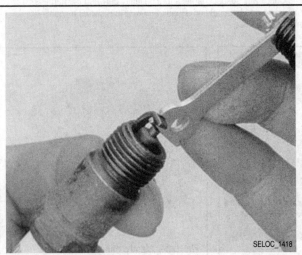

Fig. 123 Most plug gapping tools have an adjusting fitting used to bend the ground electrode

■ Keep in mind that just because a spark plug wire passes a resistance test doesn't mean that it is in good shape. Cracked or deteriorated insulation will allow the circuit to misfire under load, especially when wet. Always visually check wires to cuts, cracks or breaks in the insulation. If found, run the engine in a test tank or on a flush device either at night (looking for a bluish glow from the wires that would indicate arcing) or while spraying water on them while listening for an engine stumble.

Regardless of resistance tests and visual checks, it is never a bad idea to replace spark plug leads at least every couple of years, and to keep the old ones around for spares. Think of spark plug wires as a relatively low cost item that whose replacement can also be considered maintenance.

REMOVAL & INSTALLATION

When installing a new set of spark plug wires, replace the wires one at a time so there will be no confusion. Coat the inside of the boots with a dielectric grease to prevent sticking and corrosion. Install the boot firmly over the spark plug until it clicks into place. The click may be felt or heard. Gently pull back on the boot to assure proper contact. Repeat the process for each wire.

■ It is vitally important to route the new spark plug wire the same as the original and install it in a similar manner on the powerhead. Improper routing of spark plug wires may cause powerhead performance problems.

Ignition System

INSPECTION

◆ See Figure 124

Modern electronic ignition systems have become one of the most reliable components on an outboard. There is very little maintenance involved in the operation of these ignition systems and even less to repair if they fail. Most systems are sealed and there is no option other than to replace failed components.

Just as a tune-up is pointless on an engine with no compression, installing new spark plugs will not do much for an engine with a damaged ignition system. At each tune-up, visually inspect all ignition system components for signs of obvious defects. Look for signs of burnt, cracked or

Fig. 124 The CDI modules comprise a major portion of the typical Mercury ignition system

broken insulation. Replace wires or components with obvious defects. If spark plug condition suggests weak or no spark on one or more cylinders, perform ignition system testing to eliminate possible worn or defective components.

If trouble is suspected, it is very important to narrow down the problem to the ignition system and replace the correct components rather than just replace parts hoping to solve the problem. Electronic components can be very expensive and are usually not returnable.

Refer to the Ignition & Electrical Systems section for more information on troubleshooting and repairing ignition systems.

These days, all Mercury outboards are equipped with some form of a Capacitor Discharge Ignition (CDI) System (CDI, CDI Modular or a PCM/ECM controlled CDI system). The only possible adjustment on most CDI systems would be timing (Idle, Carb Pickup and/or WOT) and that varies by engine/model. For more details, please refer to the Timing & Synchronization procedure for your motor.

Various engines are equipped with a PCM/ECM controlled version of the CDI system (specifically EFI and Optimax motors). This is essentially a standard CDI type ignition with computer controls. There are normally no adjustable components in these systems, but for more details on a particular engine, please refer to Timing & Synchronization in this section.

Electrical System Checks

CHECKING THE BATTERY

Difficulty in starting accounts for almost half of the service required on boats each year. Some years ago, a survey by Champion Spark Plug Company indicated that roughly one third of all boat owners experienced a "won't start" condition in a given year. When an engine won't start, most people blame the battery when, in fact, it may be that the battery has run down in a futile attempt to start an engine with other problems.

Maintaining your battery in peak condition may be thought of as either tune-up or maintenance material. Most wise boaters will consider it to be both. A complete check up of the electrical system in your boat at the beginning of the boating season is a wise move. Continued regular maintenance of the battery will ensure trouble free starting on the water.

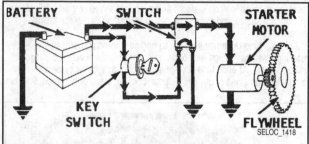

Fig. 125 Functional diagram of a typical cranking circuit

Details on battery service procedures are included under Batteries in Boat Maintenance. The following is a list of basic electrical system service procedures that should be performed as part of any tune-up.
- Check the battery for solid cable connections
- Check the battery and cables for signs of corrosion damage
- Check the battery case for damage or electrolyte leakage
- Check the electrolyte level in each cell
- Check to be sure the battery is fastened securely in position
- Check the battery's state of charge and charge as necessary
- Check battery voltage while cranking the starter. Voltage should remain above 9.5 volts
- Clean the battery, terminals and cables
- Coat the battery terminals with dielectric grease or terminal protector

CHECKING THE STARTER MOTOR

◆ See Figure 125

The starter motor system generally includes the battery, starter motor, solenoid, ignition switch and in most cases, a relay.

The frequency of starts governs how often the motor should be removed and reconditioned. Actually, when the starter starts to act up really governs that.

When checking the starter motor circuit during a tune-up, ensure the battery has the proper rating and is fully charged. Many starter motors are needlessly overhauled, when the battery is actually the culprit.

Connect 1 lead of a voltmeter to the positive terminal of the starter motor. Connect the other meter lead to a good ground on the engine. Check the battery voltage under load by turning the ignition switch to the **START** position and observing the voltmeter reading. If the reading is 9.5 volts or greater, and the starter motor fails to operate, repair or replace the starter motor.

CHECKING THE INTERNAL WIRING HARNESS

◆ See Figures 126, 127 and 128

Corrosion is probably a boater's worst enemy. It is especially harmful to wiring harnesses and connectors. Small amounts of corrosion can cause havoc in an electrical system and make it appear as if major problems are present.

The following are a list of checks that should be performed as part of any tune-up.
- Perform a through visual check of all wiring harnesses and connectors on the vessel
- Check for frayed or chafed insulation, loose or corroded connections between wires and terminals
- Unplug all suspect connectors and check terminal pins to be sure they are not bent or broken, then lubricate and protect all terminal pins with dielectric grease to provide a water tight seal
- Check any suspect harness for continuity between the harness connection and terminal end. Repair any wire that shows no continuity (infinite resistance)

Fig. 126 Any time electrical gremlins are present, start by visually checking for loose or corroded connections. . .

Fig. 127 . . .then check to make sure harnesses are tight and secure. . .

Fig. 128 . . .lastly, check the harness connectors for pins which are bent, broken or corroded

Fuel System Checks

FUEL INSPECTION

Most often, tune-ups are performed at the beginning of the boating season. If the fuel system in your boat was properly winterized, there should be no problem with starting the outboard for the first time in the spring. If problems exist, perform the following checks.

1. If the condition of the fuel is in doubt, drain, clean, and fill the tank with fresh fuel.

2. Visually check all fuel lines for kinks, leaks, deterioration or other damage.

** SELOC CAUTION

We cannot over-emphasize how important it is to visually check the fuel lines and fittings for any sign of leakage. This is even more important on the high-pressure fuel circuits of the EFI and especially the Optimax motors. It is a good idea to prime the all fuel systems using the primer bulb and, on EFI/Optimax systems, check the lines/fittings of the high-pressure fuel circuit after the motor has been operated (meaning the system has been under normal operating pressures).

3. Disconnect the fuel lines and blow them out with compressed air to dislodge any contamination or other foreign material.

4. Check the line between the fuel pump and the carburetor (or the vapor separator tank on fuel injected motors) while the powerhead is operating and the line between the fuel tank and the pump when the powerhead is not operating. A leak between the tank and the pump many times will not appear when the powerhead is operating, because the suction created by the pump drawing fuel will not allow the fuel to leak. Once the powerhead is shut down and the suction no longer exists, fuel may begin to leak.

** SELOC CAUTION

DO NOT do anything around the fuel system without first reviewing the warnings and safety precautions in the Fuel System section. Remember that fuel is highly combustible and all potential sources of ignition from cigarettes and open flame to sparks must be kept far away from the work area.

Timing & Synchronization

◆ **See Figures 129, 130 and 131**

Timing and synchronization on an outboard engine is extremely important to obtain maximum efficiency. The powerhead cannot perform properly and produce its designed horsepower output if the fuel and ignition systems have not been precisely adjusted.

Outboards equipped with a mechanical advance type Capacitor Discharge Ignition (CDI) system use a series of link rods between the carburetor and the ignition base plate assembly. At the time the throttle is opened, the ignition base plate assembly is rotated by means of the link rod, thus advancing the timing.

On electronically controlled ignition models, a microcomputer decides when to advance or retard the timing based on input from various sensors. Therefore, there is no link rod between the magneto control lever and the stator assembly.

Many models have timing marks on the flywheel and CDI base. A timing light is normally used to check the ignition timing with the powerhead operating (dynamically). An alternate method is to check the timing with the powerhead not operating (static timing). This second method requires the use of a dial indicator gauge.

Various models have unique methods of checking ignition timing. As appropriate, these differences will be explained in detail in the text.

In simple terms, synchronization is timing the fuel system to the ignition. As the throttle is advanced to increase powerhead rpm, the fuel and the ignition systems are both advanced equally and at the same rate.

Any time the fuel system or the ignition system on a powerhead is serviced to replace a faulty part or any adjustments are made for any reason, powerhead timing and synchronization must be carefully checked and verified.

■ **Before making any adjustments to the ignition timing or synchronizing the ignition to the fuel system, both systems should be verified to be in good working order.**

Timing and synchronizing the ignition and fuel systems on an outboard motor are critical adjustments. Certain pieces of the following equipment are essential and are called out repeatedly in this section. This equipment must be used as described, unless otherwise instructed by the equipment manufacturer. Naturally, the equipment is removed following completion of the adjustments.

Some manufacturers recommend the use of a test wheel instead of a normal propeller in order to put a load on the engine and propeller shaft. The use of the test wheel prevents the engine from excessive rpm.

• Dial Indicator - Top dead center (TDC) of the No. 1 (top) piston must be precisely known before the timing adjustment can be made. TDC can only be determined through installation of a dial indicator into the No. 1 spark plug opening

• Timing Light - During many procedures in this section, the timing mark on the flywheel must be aligned with a stationary timing mark on the engine while the powerhead is being cranked or is running. Only through use of a timing light connected to the No. 1 spark plug lead, can the timing mark on the flywheel be observed while the engine is operating

• Tachometer - A tachometer connected to the powerhead must be used to accurately determine engine speed during idle and high-speed adjustment. Engine speed readings range from 0-6,000 rpm in increments of 100 rpm. Choose a tachometer with solid state electronic circuits which eliminates the need for relays or batteries and contribute to their accuracy. For maximum performance, the idle rpm should be adjusted under actual operating conditions. Under such conditions it might be necessary to attach a tachometer closer to the powerhead than the one installed on the control panel.

• Flywheel Rotation - The instructions may call for rotating the flywheel until certain marks are aligned with the timing pointer. When the flywheel must be rotated, always move the flywheel in the indicated direction. If the

Fig. 129 Typical Mercury timing pointer (built into flywheel cover)...

Fig. 130 ...and timing marks

Fig. 131 Sometimes you need to paint the relevant marks white to make them more visible

flywheel should be rotated in the opposite direction, the water pump impeller vanes would be twisted. Should the powerhead be started with the pump tangs bent back in the wrong direction, the tangs may not have time to bend in the correct direction before they are damaged. The least amount of damage to the water pump will affect cooling of the powerhead

• Test Tank - Since the engine must be operated at various times and engine speeds during some procedures, a test tank or moving the boat into a body of water, is necessary. If installing the engine in a test tank, outfit the engine with an appropriate test propeller

✳✳ SELOC CAUTION

Water must circulate through the lower unit to the powerhead anytime the powerhead is operating to prevent damage to the water pump in the lower unit. Just 5 seconds without water will damage the water pump impeller.

■ Remember the powerhead will not start without the emergency tether in place behind the kill switch knob.

✳✳ SELOC CAUTION

Never operate the powerhead above a fast idle with a flush attachment connected to the lower unit. Operating the powerhead at a high rpm with no load on the propeller shaft could cause the powerhead to runaway causing extensive damage to the unit.

2.5/3.3 Hp Models

IGNITION TIMING

The ignition advance and timing of these models is not adjustable.

THROTTLE JET NEEDLE

◆ See Figures 132, 133 and 134

Air/fuel mixtures on this carburetor are controlled by a tapered main throttle jet needle which passes through the throttle body from the mixing chamber assembly into a main nozzle. The combination of needle taper and needle position (along with the main jet) make for an automatically adjusting fuel delivery orifice which varies with engine speed/throttle position. The needle itself is held in position in the mixing chamber by an E-clip which may be repositioned upward or downward for adjustment. HOWEVER, once the clip is set to the proper position for the motor NO PERIODIC adjustments should be necessary (unless the motor is moved to or from high altitude conditions).

If adjustment is necessary the top of the carburetor must be partially disassembled for access to the main Jet. Since the jet is a tapered needle which is thicker on top and thinner on the bottom, moving the E-clip to a different position will change what portion of the taper is in the main jet at any given throttle position. Moving the clip UPWARD on the needle will allow the needle to sit lower (i.e. placing a fatter part of the needle in the jet), making the orifice smaller or LEANING out the mixture. Moving the clip DOWNWARD on the needle will allow it to sit higher (i.e. placing a thinner part of the needle in the jet), making the orifice bigger or ENRICHENING the mixture.

Generally speaking it is safer for the motor to be adjusted a little too rich, rather than a little too lean.

1. Remove the knobs from the throttle and choke levers and lift away the front cover from the outboard for access.

2. Remove the throttle valve assembly and bracket from the top of the carburetor. With a pair of pliers, unscrew the throttle valve assembly (mixing chamber cover) to allow the throttle cable end to clear the recess in the base of the throttle valve and to slide down the slot.

3. Disassemble the throttle valve, consisting of the throttle valve, spring, jet needle, (with E-clip normally on the second groove), jet retainer and throttle cable end. The tapered jet needle varies the fuel flow as the throttle is opened.

■ Relocating the E-clip closer to the end (top) of the jet needle will lean the air/fuel mixture; lowering the E-clip will enrich the mixture. It is preferable to operate the powerhead with a mixture that is a little too rich rather than too lean. A too lean air/fuel mixture will cause the powerhead to overheat. A powerhead operating too hot will cause pre-ignition. If the condition is not corrected, the powerhead could be severely damaged.

■ Here's a TIP. If the air/fuel mixture is too rich, the rpm will decrease as the throttle is opened beyond the 2/3 to full throttle range.

4. If the correct mixture is not obtainable by changing the position of the E-clip on the throttle needle, the main jet located inside the fuel bowl must be changed from the standard jet size of No. 92 on the CDI type ignition system.

IDLE SPEED

◆ See Figure 135

1. Connect a tachometer to the powerhead. Connect a flushing attachment to the lower unit if the outboard is not in a test tank or body of water.

✳✳ SELOC CAUTION

Water must circulate through the lower unit to the powerhead ant time the powerhead is operating to prevent damage to the water pump in the lower unit. Just 5 seconds without water will damage the water pump.

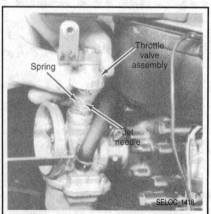

Fig. 132 Unscrew the mixing chamber cover and lift out the throttle valve assembly

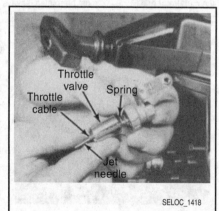

Fig. 133 Disassemble the throttle valve components for access to the needle

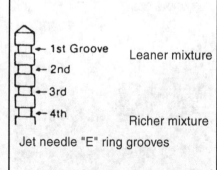

Fig. 134 Raising or lowering the needle clip will adjust the air/fuel mixture

Fig. 135 The idle speed screw is located on the port side of the carburetor - 2.5/3.3 hp motors

2. Start the engine and allow it to warm up to operating temperature.

3. Adjust the throttle lever to the lowest speed and then adjust the idle speed screw on the side of the carburetor until the powerhead idles at the speed specified in the Tune-Up Specifications chart (which should be about 900-1000 rpm).

4/5 Hp Models

IGNITION TIMING

The timing on these models is factory set and cannot be adjusted. However, the following procedures can be used to check timing mark alignment and then check timing, thereby confirming that the ignition system is operating properly.

If the inspection indicates the timing to be incorrect, the fault may either be mechanical or electrical.

A mechanical fault can only be directly related to the flywheel. Either the flywheel has been installed without the Woodruff key of the key has been sheared off.

An electrical problem may be caused by a fault in the ignition system. The ignition system uses an electronic spark advance. Detailed testing procedures for the capacitor charging coil, trigger and ignition coil are covered in the Ignition & Electrical section.

Static Timing (Checking Pointer Alignment)

◆ See Figure 136

1. Begin by removing the cowling.

2. Remove the spark plug and install a dial indicator in the spark plug hole. Now, slowly rotate the flywheel clockwise and determine TDC for the cylinder using the dial indicator.

3. After TDC has been determined, observe the 2 vertical lines embossed on the port side of the flywheel. The aft line should align with split line of the cylinder block and the crankcase cover.

• If the aft line is centered as explained, the flywheel has been installed correctly and the Woodruff key is in place.

• If the aft line is not centered, the hand rewind starter and flywheel must be removed and the Woodruff key and keyway checked for damage and if it is installed correctly.

Dynamic Timing

◆ See Figures 137 and 138

1. Remove the dial indicator, install the spark plug and connect a timing light to the powerhead.

2. Mount the outboard in a test tank or on a boat in a body of water with the boat well secured to the dock or slip.

3. Obtain and connect a timing light to the high tension spark plug lead as per the manufacturer's instructions.

4. Start the powerhead and allow it to reach the correct operating temperature.

5. Allow it run at idle speed in **FORWARD**. Aim the timing light at the port side of the powerhead.

6. The split line between the cylinder block and the crankcase cover should be misaligned by approximately 1/4 in. (6.4mm) toward the forward side of the aft line embossed on the flywheel. This position corresponds to 5°BTDC advance at the flywheel.

7. If the timing is not as indicated and the powerhead operates roughly or misfires, the problem is most likely electrical. Detailed procedures to test the capacitor charging coil, the trigger and the ignition coil are outlined in "Ignition and Electrical" section.

8. Twist the throttle grip to the **FAST** position with the outboard still in **FORWARD**. Aim the timing light at the port side of the powerhead.

9. The forward line embossed on the flywheel should align with the split line between the cylinder block and the crankcase cover. This position corresponds to 30°BTDC advance at the flywheel.

10. If the timing is not as indicated and the powerhead operates roughly or misfires, the problem is most likely electrical. Detailed procedures for testing the capacitor charging coil, trigger and ignition coil are covered in the Ignition & Electrical section.

Fig. 136 Timing mark alignment is correct if the piston is at TDC when the flywheel TDC mark aligns with the crankcase split line

Fig. 137 First check for proper idle timing (about 5° BTDC)

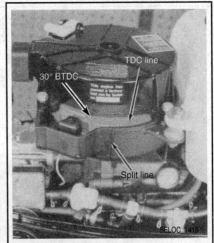

Fig. 138 Then watch for the timing to advance toward WOT timing as the throttle is opened

IDLE SPEED & MIXTURE ADJUSTMENT

◆ See Figure 139

More than likely the motor should NOT require periodic adjustments to the idle mixture screw. However, IF the carburetor is replaced or rebuilt OR if a proper idle speed cannot be set and maintained using only the idle speed screw, then the idle mixture should be adjusted.

1. If the carburetor was replaced or rebuilt, set the initial screw positions as follows:

a. Rotate the throttle twist grip to the idle position (on tiller handle models), or place the remote control handle in the neutral (idle) position (on remote control equipped models).

b. Loosen the set screw to allow the throttle arm to move freely on the wire.

c. Back the idle screw off the throttle arm.

d. Turn the idle speed screw inward (clockwise) until if just touches the throttle arm, then inward 2 additional turns to slightly open the throttle plate in the carburetor.

e. With the throttle arm of the carburetor against the idle speed screw, pull the slack from the throttle wire and secure the wire into the retainer by tightening the set-screw.

f. Turn the low speed mixture screw 1 1/2 turns outward (counterclockwise) from a lightly seated position.

✳✳ SELOC CAUTION

AND WE MEAN LIGHTLY seated position. If you crank down on that screw you'll ruin the taper on the mixture screw needle.

Once the initial idle speed settings have been established, the idle speed can be set using the following procedures:

2. Start the engine and allow it to warm up to normal operating temperature.

3. Throttle the engine back to the idle position and shift into forward gear.

4. Adjust the idle speed screw to attain an idle speed of 850 rpm.

5. If the idle mixture adjustment requires further adjustment:

a. Slowly turn the low speed mixture (pilot) screw counterclockwise until the engine starts to "load up" or fire unevenly because of an over-rich mixture.

b. Slowly turn the low speed mixture (pilot) screw clockwise until the engine picks-up speed and fires evenly.

✳✳ SELOC CAUTION

Do not adjust leaner than necessary for reasonably smooth idling. It is preferable to set it a little rich than too lean.

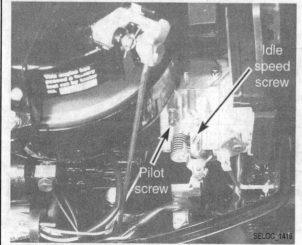

Fig. 139 Idle speed and mixture (pilot) screw locations - 4/5 hp motors

c. If the engine hesitates during acceleration after adjusting the low speed mixture, the mixture is too lean. Richen the low speed mixture slightly until the engine accelerates smoothly.

d. Return the engine to idle and recheck the idle rpm (850, in gear). Readjust the idle speed screw if necessary.

6. Shift the engine to neutral and shut down the engine.

6/8 & 9.9/10/15 Hp Models

The timing and synchronization procedures for these units require that the powerhead be operated at idle rpm and at wide-open throttle. Therefore, the outboard unit must be placed in a test tank or a body of water with the boat well secured to the dock.

Remove the upper cowling to gain full access to the adjustment screws and timing marks. Check the condition and security of the tiller handle or remove the throttle cables. Check the cable movement for idle and full throttle position in both forward and reverse gears. Check the jam-nuts for proper travel and to eliminate all slack.

■ **Idle speed and mixture adjustments are NOT necessary UNLESS components have been repaired/replaced (such as the carburetor) or unless the motor is not idling properly.**

IDLE SPEED

◆ See Figures 140 and 141

1. Begin by shifting the outboard into the **NEUTRAL** position. On tiller models, move the twist grip to the **SLOW** position.

2. Push the primer/fast idle knob completely inward and then rotate the knob to the full clockwise position.

3. Rotate the idle speed screw counterclockwise until the screw is off the cam follower. Now, rotate the idle speed screw inward (clockwise) until it just makes contact with the cam follower.

4. Once contact is made, rotate the idle speed screw 1/2 turn more further inward. This action will just barely open the throttle plate.

■ **There are models which are marketed WITHOUT an idle speed screw. When applicable there is a throttle cam locking screw which should be loosened so the cam follower can be pushed into contact with the cam (then the screw is re-tightened). On models without an idle speed screw the carburetor is set at the factory to maintain idle speeds of about 575-725 rpm.**

5. Start and warm the powerhead to normal operating temperature, then adjust the isle speed screw so as to achieve the proper idle speed specification (as listed in the Tune-Up Specifications chart).

■ **If the idle mixture or idle timing requires adjustment you will likely need to come back and reset the idle speed again after the timing and synchronization procedures are complete.**

6. If necessary proceed to Low Speed Mixture or Ignition Timing adjustments.

7. Shut the powerhead down and install the access plug, if it was removed or install the upper powerhead cowling.

LOW SPEED MIXTURE

Static Adjustment (Preliminary Set-Up)

◆ See Figure 142

1. If the model being serviced has an air intake cover, remove the low speed mixture screw access plug. On other models, the low speed mixture screw is exposed when the upper cowling is removed.

2. Rotate the low speed screw slowly clockwise (inward) until it JUST BARELY seats and then back it out counterclockwise about 1 1/2 turns (for details, please refer to the Carburetor Set-Up Specifications chart from the Fuel System section).

Fig. 140 Push the primer/fast idle knob completely inward and then rotate as directed

Fig. 141 Turn the idle speed screw counterclockwise until it is off the cam follower

Fig. 142 If equipped with an air intake cover, remove the low speed mixture screw access plug

✳✳ SELOC WARNING

Moving the low speed screw inward after it seats will damage the needle and seat.

3. If any further adjustments are to be made, do not install the access plug or the upper cowling at this time.

Dynamic Adjustment (Running Adjustment)

◆ See Figure 143

■ Make an effort to prevent the powerhead from shutting down during the following adjustments.

✳✳ SELOC CAUTION

Water must circulate through the lower unit to the powerhead ant time the powerhead is operating to prevent damage to the water pump in the lower unit. Just 5 seconds without water will damage the water pump.

1. Start the powerhead and allow it to reach operating temperature.
2. Move the throttle to the idle speed position and allow the rpm to stabilize.
3. Push the primer/fast idle knob inward and then rotate it to the full counterclockwise position.
4. Shift the unit into **FORWARD**.
5. With the powerhead operating at idle rpm, rotate the low speed mixture screw counterclockwise (to enrichen the fuel mixture) until the powerhead begins to misfire because of an overly-rich fuel mixture.

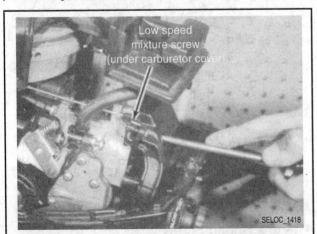

Fig. 143 Making a dynamic, low-speed mixture screw adjustment (powerhead operating). Screwdriver is placed through the carburetor cover opening, the low speed mixture screw is adjusted for proper air/fuel ratio at idle speed

6. Now, slowly rotate the low speed mixture screw clockwise and count the turns until the powerhead is operating evenly and the rpm increases. Continue rotating the mixture screw clockwise until the powerhead begins to slow down and eventually begins to misfire because of a too lean air/fuel mixture.
7. Count and move the low speed mixture screw back to the halfway point between the too rich mixture and the too lean mixture.

■ Do not adjust the mixture ant leaner than necessary to obtain a smooth idle. It is better for the mixture to be on the "too rich" side rather than the "too lean" side.

IGNITION TIMING

Maximum (WOT) Advance

◆ See Figures 144 and 145

1. Mount the outboard in a test tank or body of water with the boat well secured to the dock or slip.
2. Remove the powerhead cowling.
3. Connect a fuel line to a fuel source.
4. Install a tachometer to the powerhead following the manufacturer instructions.
5. Connect a timing light to the No.1 spark plug lead.

✳✳ SELOC CAUTION

Water must circulate through the lower unit to the powerhead any time the powerhead is operating to prevent damage to the water pump in the lower unit. Just 5 seconds without water will damage the water pump.

6. Start the engine and allow it reach operating temperature.
7. Move the shift lever to the **FORWARD** position.
8. Advance the throttle to the wide open throttle position.
9. Check the maximum timing mark on the flywheel. The 36° Before Top Dead Center (BTDC) mark should be aligned with the pointer.
10. Adjust the timing by loosening the jam-nut (if equipped) on the maximum spark advance screw. Rotate the advance spark screw until the maximum spark timing mark on the flywheel is aligned with the pointer.
11. If equipped with a jam nut, once the timing marks are aligned, tighten the jam-nut to hold the adjustment.
12. Return the throttle to the idle position, leave the unit in the **FORWARD** position and perform the idle timing adjustment.

Idle Timing

◆ See Figures 140 and 146

1. Check and/or adjust the Maximum Advance (WOT) timing.
2. Push the fast idle knob inward and at the same time rotate the knob counterclockwise as far as possible.

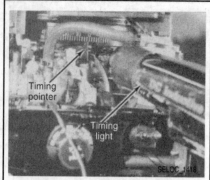

Fig. 144 Check the WOT timing using a timing light. .

Fig. 145 ...and adjust as necessary using the maximum spark advance screw

Fig. 146 Adjusting the idle timing

■ It may be necessary to adjust the idle speed screw in order to obtain a satisfactory idle rpm.

3. Reduce powerhead rpm to idle speed.

4. Aim the timing light at the pointer, then adjust the idle timing screw until the pointer aligns somewhere in the 7-9° BTDC range (the specified idle timing as indicated in the Tune-Up Specifications chart).

5. After the timing is set, either adjust the Idle Wire or shut down the powerhead, then disconnect the tachometer and timing light.

IDLE WIRE

◆ See Figure 147

1. Push the fast idle knob inward and then turn the knob counterclockwise as far as possible.

2. Shift the unit into **NEUTRAL**.

3. Adjust the upper screw on the ratchet to remove any clearance between the idle wire and the trigger. The wire should barely make contact with the trigger.

4. Check the fast idle speed by turning the knob to the fully clockwise position. Fast idle should be about 1500-2000 rpm.

5. Shut down the powerhead, then disconnect the tachometer.

20/20 Jet/25 Hp Models

The timing and synchronization procedures for these units require that the powerhead be operated at idle rpm and at wide-open throttle. Therefore, outboard unit must be placed in a test tank or a body of water with the boat well secured to the dock or in a slip. Never attempt to make the following adjustments with a flush attachment connected to the lower unit. The powerhead operating at high rpm with such a device attached, would cause a runaway condition from a lack of load on the propeller, causing extensive damage.

Mount the outboard unit in a test tank or on a boat well secured. Connect a fuel line from a fuel source to the powerhead. Remove the powerhead cowling to gain access to adjustment locations.

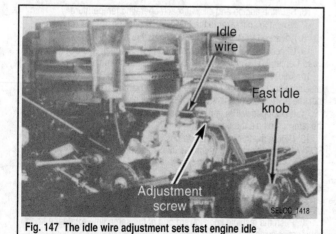

Fig. 147 The idle wire adjustment sets fast engine idle

On electric start models, check to be sure the shift and throttle cables and wiring harness are all properly installed and connected and on the manual start models; check the cables from the steering handle and remove any slack through the adjusting jam-nuts.

CARBURETOR & LINKAGE (PRELIMINARY ADJUSTMENTS)

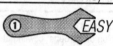

For proper timing and synchronization you must start with the appropriate carburetor and shift linkage adjustments.

Tiller Handle & Side Shift Models

◆ See Figure 148

1. Check all electrical connections (including the battery, if equipped) to make sure they are clean and tight.

2. Shift the outboard to **NEUTRAL**. On side shift models, place the throttle twist grip in the **SLOW** position.

3. Push the choke/primer knob all the way in. On models equipped with a primer, turn the knob to the full counterclockwise position.

4. Check the tiller handle cables for proper adjustment. If necessary, adjust the cables and remove any slack.

5. With the outboard in **NEUTRAL**, back the idle speed screw off the cam follower.

6. Now, turn the idle speed screw in until the cam follower just makes contact with the throttle cam, then turn it inward 1 additional turn to slightly open the throttle plate.

7. Place the throttle twist grip to the **FULL THROTTLE** position. Verify that the throttle plate can achieve wide open throttle. If adjustment is necessary, adjust the tiller cables or throttle link rod as follows:

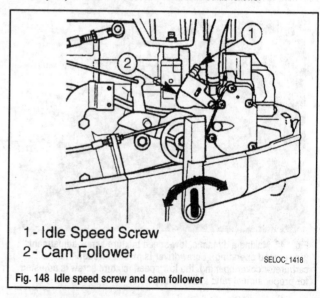

1 - Idle Speed Screw
2 - Cam Follower

Fig. 148 Idle speed screw and cam follower

■ The throttle return spring on the cam follower plate should just contact the fuel pump housing at wide open throttle position. Do not allow the throttle spring to act as a throttle stop.

a. On tiller shift models, the throttle return spring on the cam follower plate should JUST contact the fuel pump housing at WOT. If not, adjust the throttle link rod, but DO NOT allow the throttle spring to act as a throttle stop.

b. On side shift models, adjust the throttle cable jam nuts so the throttle plate opens fully and the throttle return spring on the cam follower plate JUST contact the fuel pump housing. Make sure all slack is removed from the throttle cables.

Remote Control Models

◆ See Figures 149, 150 and 151

This procedure assumes that the shift and throttle cables are disconnected from the powerhead. The procedure can still be performed even with the cables in place, however we recommend making sure that the cables do not bind movement of the throttle during adjustment. If necessary, shift the remote at the same time you manually move the throttle lever during the procedure.

1. Check all electrical connections to ensure that they are tight and secure (including battery connections on electric start models)

2. Push the primer enrichener to the full in position and turn the primer knob to full counterclockwise position.

3. Manually pull the throttle lever to the full aft position until the stop tab on the throttle arm contacts the stop tab on the shift platform. (This step assumes the throttle and shift cables are disconnected, if necessary, shift the remote in order to allow the throttle lever to move properly).

4. Back the idle speed screw off the cam follower.

5. Now, turn the idle speed screw inward until the cam follower just makes contact with the throttle cam, then turn it in again, 1 additional turn, to slightly open the throttle plate.

■ The throttle return spring on the cam follower plate should just contact the fuel pump housing at wide open throttle position. If not, adjust the throttle link rod. Do not allow the throttle spring to act as a throttle stop.

6. Manually push the throttle lever forward until it contacts thy throttle stop on the shift platform (again, shifting the remote also, if necessary). Verify that the throttle plate can achieve wide open throttle and that no pre-load exists on the throttle return spring behind the cam follower plate. Adjust the throttle link rod if necessary.

■ If the shift and/or throttle cables are currently disconnected from the powerhead, connect and adjust them at this time. Keep in mind that the shift cable is the FIRST cable to move when the remote control handle is moved into gear. Always install the shift cable first.

7. If the shift cable is currently disconnected, install it to the powerhead as follows:

a. Move the remote control handle to the full REVERSE position, then slowly rotate the propeller by hand and manually move the shift lever on the powerhead to reverse (move it toward the rear). Once the propeller is locked in position, you'll know the gearcase is shifted into Reverse.

b. Open the cable retainer cover on the lower engine cowling and remove the barrel holder and front rubber grommet.

c. Connect the shift cable onto the shift lever pin, then lock it in place with the retainer latch.

d. Adjust the shift cable barrel so it will fit into the bottom hole of the barrel holder in the lower engine cowling. Make sure the barrel holder will slide freely into the retaining pocket WITHOUT pre-loading the cable itself.

e. Check shift adjustment by making sure the remote shifts the gearcase into forward, neutral and reverse (you'll likely have to slowly spin the propeller by hand while attempting to make those shifts).

8. If the throttle cable is currently disconnected, install it to the powerhead (after the shift cable is already installed) as follows:

a. Position the remote control handle fully forward to the Wide Open Throttle (WOT) position.

b. Install the throttle cable to the throttle pin, then lock it in to place with the retainer latch.

c. Move the throttle lever until the tab contacts the throttle stop, then adjust the barrel on the cable so the barrel fits into the barrel holder.

d. Slip the barrel into the holder, then place the holder into the retaining pocket.

e. Check the throttle cable adjustment by making sure the throttle fully opens and closes when moving the remote from NEUTRAL to WOT and back again. Double-check that the tab continues to contact the throttle stop.

f. Position the rubber seal with holes facing toward the front onto the control cables, then install the cables, barrel holder and rubber seal into the cable holder. Lock the barrel holder in position with the cable retainer latch.

Idle Mixture Preliminary Adjustment (After Repair Or Rebuild)

◆ See Figure 152

Don't needless mess with a proper running carburetor. This adjustment should REALLY only be necessary if the carburetor has been rebuilt or replaced (or to compensate for some other major mechanical/physical change).

1. Remove the access plug from the carburetor intake cover.

2. Turn the idle mixture screw in (clockwise) until lightly seated, then back out to the initial setting noted in the Carburetor Set-Up Specifications chart from the Fuel System section.

IGNITION TIMING

◆ See Figure 153

✲✲ SELOC WARNING

Reminder! The outboard MUST be mounted in a test tank or on a launched boat, securely affixed to the dock or trailer before proceeding with this adjustment.

1. Obtain and connect a timing light to the No.1 spark plug lead.
2. Start the powerhead and then shift the unit into FORWARD.
3. Allow the powerhead to warm to normal operating temperature.

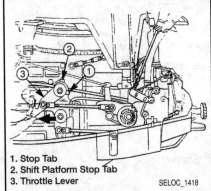

1. Stop Tab
2. Shift Platform Stop Tab
3. Throttle Lever

SELOC_1418

Fig. 149 Pull the throttle lever to the full aft position (so the throttle arm stop tab contacts the shift platform stop tab)

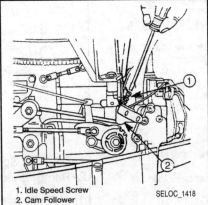

1. Idle Speed Screw
2. Cam Follower

SELOC_1418

Fig. 150 Back the idle speed screw off the cam follower

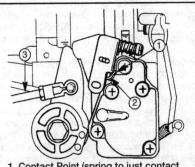

1. Contact Point (spring to just contact fuel pump housing)
2. Throttle Return Spring
3. Throttle Link Rod

SELOC_1418

Fig. 151 Idle speed screw and throttle cam

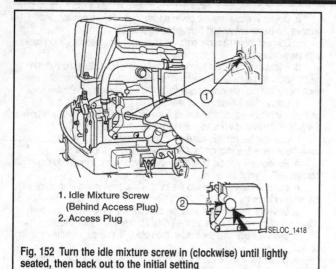

1. Idle Mixture Screw
   (Behind Access Plug)
2. Access Plug

SELOC_1418

Fig. 152 Turn the idle mixture screw in (clockwise) until lightly seated, then back out to the initial setting

4. Advance the throttle to the wide open throttle position. Direct the timing light to the flywheel and pointer mark on the side of the powerhead.

5. If the maximum (WOT) timing requires adjustment, loosen the maximum timing jam nut and turn the maximum timing adjustment screw in/out to obtain the specified timing. Turning the screw OUT will ADVANCE the timing and turning the screw IN will RETARD the timing.

6. Next, reduce the engine speed to idle. If necessary, skip ahead to the Idle Speed & Mixture Adjustment procedure in order to maintain a proper idle speed, then come back and check/adjust idle timing.

7. If the idle timing requires adjustment, loosen the idle timing jam nut and turn the idle adjustment screw in/out to obtain the specified timing. Turning the screw IN will ADVANCE the timing and turning the screw OUT will RETARD the timing.

IDLE SPEED & MIXTURE

MODERATE

◆ See Figures 152, 154 and 155

■ A FEW models MAY NOT be equipped with an idle speed screw. On these models the carburetors are adjusted at the factory to maintain proper idle speed if the idle fuel mixture is correct.

The idle speed itself is controlled by the "idle speed" screw (a spring loaded screw that is used to physically hold the carburetor throttle plates open slightly when the throttle linkage is in neutral/minimum position). However, since the idle mixture can affect the idle speed as well both adjustments are part of the same procedure.

We caution you though that periodic adjustment of the idle mixture should NOT be necessary. Generally speaking the only time you should touch the mixture screw is if something major has changed with the motor and fuel system (carburetor rebuild, carburetor replacement, MAJOR powerhead repair or if the motor was moved to/from high altitude operation).

1. If not done already for Ignition Timing adjustment, supply a source of cooling water to the motor, then connect a tachometer.

2. If it will be necessary to adjust the idle mixture (and it was not done already during the Carburetor & Linkage Preliminary Adjustments), then remove the rubber button on the face of the air silencer to gain access to the low speed adjustment screw. (Again, if not done already, rotate the low speed adjustment screw clockwise until it barely seats and then back it out counterclockwise the proper number of turns as specified in the Carburetor Set-Up Specifications chart from the Fuel System section.

3. Start the outboard. After the outboard starts, push the primer knob to the "Full In" position and immediately check for water at the tell tale and idle relief holes. After allowing the engine to warm up, turn the primer knob to "Full Counterclockwise" position.

4. Run the engine at 3500 rpm and check the stop button function and lanyard stop switch function.

5. Adjust the idle speed screw in the **FORWARD** position.

6. Using the idle speed screw, adjust the motor to run at approximately the specified Idle Speed, as listed in the Tune-Up Specifications chart.

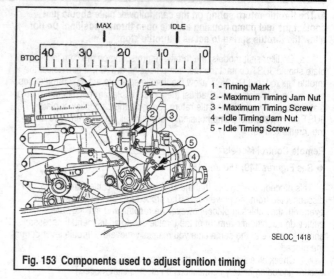

1 - Timing Mark
2 - Maximum Timing Jam Nut
3 - Maximum Timing Screw
4 - Idle Timing Jam Nut
5 - Idle Timing Screw

SELOC_1418

Fig. 153 Components used to adjust ignition timing

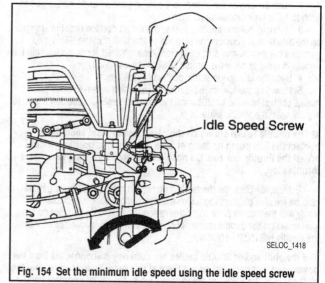

Idle Speed Screw

SELOC_1418

Fig. 154 Set the minimum idle speed using the idle speed screw

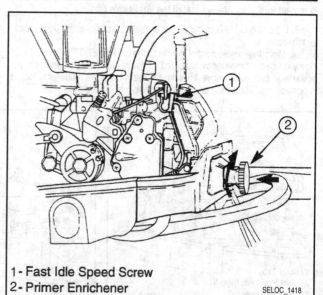

1 - Fast Idle Speed Screw
2 - Primer Enrichener

SELOC_1418

Fig. 155 With the engine in neutral and the primer enrichener pushed to the full in and full clockwise position, adjust the fast idle speed screw to specification

■ When setting idle mixture, do not adjust leaner than necessary to attain reasonably smooth idling. When in doubt, set the idle to the slightly richer side of the highest rpm.

7. Adjust the carburetor for the best performance, after clearing the engine. With the engine running at idle speed in **FORWARD** gear, turn the mixture screw in (clockwise) until the engine begins to lose rpm, fire unevenly and or misfires. Back out the screw 1/4 turn or more.

8. Check for a too lean mixture on acceleration (engine will hesitate or stall on acceleration). Readjust the mixture if necessary. Install the access plug.

9. With the engine in Neutral and the primer enrichener pushed to the **Full In** and **Full Clockwise** position, adjust the fast idle speed screw to specification.

10. Check the primer enrichener circuit by pulling the primer enrichener knob to the full out position. The engine speed should drop a minimum of 300 rpm and remain running for approximately five (5) seconds before stalling.

■ Some late-models may be equipped with an electronic choke/enrichener whose position is adjustable if it is disturbed. On these models, make sure that with the engine OFF the gap between the electric choke plunger and carburetor lever is 0.025 in. (0.635mm). If adjustment is necessary, loosen the 2 choke cover plate screws and move the plate to obtain the cap.

30/40 Hp 2-Cylinder Models

The timing and synchronization procedures for these units require that the powerhead be operated at idle rpm and at wide-open throttle. Therefore, outboard unit must be placed in a test tank or a body of water with the boat well secured to the dock or in a slip. Never attempt to make the following adjustments with a flush attachment connected to the lower unit. The powerhead operating at high rpm with such a device attached, would cause

a runaway condition from a lack of load on the propeller, causing extensive damage.

Mount the outboard unit in a test tank or on a boat well secured. Connect a fuel line from a fuel source to the powerhead. Remove the powerhead cowling to gain access to adjustment locations.

These procedures can be performed both on Tiller and Remote control models, however for the later either make sure the remote cables are disconnected during the preliminary adjustments OR make sure that they are not preloading and interfering with those adjustments.

IDLE SPEED & MIXTURE ADJUSTMENTS

◆ See Figures 156 thru 159

1. Check all electrical connections (including the battery, if equipped) to make sure they are clean and tight.

2. With the engine off and the gear shift in the Neutral position, loosen the cam follower screw.

3. Back off the idle speed screw until the throttle shutter positioner does not touch the taper of the idle speed screw. (Throttle plate closed).

4. Loosen the throttle cable jam nuts (located right below the throttle cam assembly).

5. With the throttle at the idle position, place the cam follower roller against the throttle cam. Center the roller with raised mark on the throttle cam by adjusting the position of the throttle cable sleeves in the mounting bracket.

■ When positioning throttle cables, a minimum of 1/16 in. (1.59 mm) to a maximum of 1/8 in. (3.18 mm) slack must be allowed to prevent throttle cables from binding. (Rock throttle cam side to side and measure the amount of throttle cam travel at link rod ball).

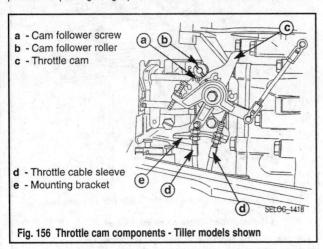

a - Cam follower screw
b - Cam follower roller
c - Throttle cam

d - Throttle cable sleeve
e - Mounting bracket

SELOC_1418

Fig. 156 Throttle cam components - Tiller models shown

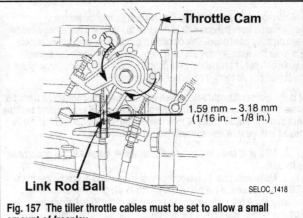

Fig. 157 The tiller throttle cables must be set to allow a small amount of freeplay

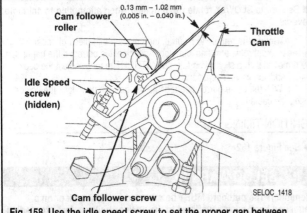

Fig. 158 Use the idle speed screw to set the proper gap between the cam and follower at idle

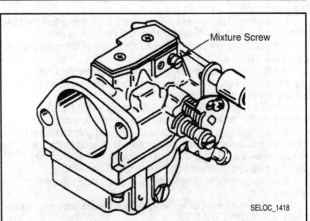

Fig. 159 The idle mixture screw should only be adjusted after major changes (such as component rebuild or replacement)

6. Tighten the throttle cable jam nuts.

7. With the cam follower resting on the throttle cam, tighten the cam follower screw.

8. With the throttle at idle position, turn the idle speed screw clockwise in until a gap of 0.005-0.040 in. (0.13-1.02mm) is achieved between the throttle cam and cam follower.

9. If the carburetor was replaced or rebuilt (or the powerhead received a major overhaul or was moved to/from high altitude conditions) place the idle mixture screw in the proper initial position by turning it inward (clockwise) until it is LIGHTLY seated, then back out to an initial setting of 1 1/2 turns.

10. With the engine in a test tank or otherwise with the drive unit in the water, start and warm the motor to normal operating temperature.

11. Once the motor is fully warmed use the idle speed screw on the throttle linkage to adjust the speed to specification (as listed in the Tune-Up Specifications chart).

12. If the motor will not maintain a proper idle (or if idle mixture adjustment is otherwise necessary for reasons listed earlier), adjust the idle mixture as follows:

a. With the engine running at idle speed and the gear box in the **FORWARD** position, turn the mixture screw in (clockwise) until the engine starts to bog down and misfire. Back out the screw 1/4 turn or more.

■ When setting the idle mixture, do not adjust leaner than necessary to attain reasonably smooth idling. When in doubt, stay to the slightly rich side of the adjustment.

13. Check for too lean of mixture on acceleration. (engine will "bog" on acceleration). Readjust as necessary.

14. Readjust idle speed screw so the engine idles at the specified rpm.

15. Check and adjust the ignition timing (first Idle and then WOT).

REMOTE CONTROL CABLE INSTALLATION

◆ See Figures 160 and 161

This procedure assumes that the shift and throttle cables are disconnected from the powerhead, but that the throttle linkage has already been properly positioned. This procedure can also be used to disconnect and reset the remote cables in case there is a problem with the current adjustment, but IT should not be a periodic adjustment. Once properly set, the cables should require no further adjustment.

■ If the shift and/or throttle cables are currently disconnected from the powerhead, connect and adjust them at this time. Keep in mind that the shift cable is FIRST cable to move when the remote control handle is moved into gear. Always install the shift cable first.

1. If the shift cable is currently disconnected, install it to the powerhead as follows:

a. Make sure both the remote control handle and gearcase are in the **NEUTRAL** position. If necessary slowly rotate the propeller by hand and manually move the shift lever on the powerhead to Neutral. Once the propeller is free, you'll know the gearcase is shifted into Neutral.

b. Measure the distance between the mounting pin and the dead center of the barrel holder.

c. Push inward on the cable until resistance is felt, then adjust the cable barrel until it is the proper distance measured in the previous step.

d. Place the barrel into the bottom hole in the barrel holder, then fasten the cable to the mounting pin with a retainer.

e. Check shift adjustment by making sure the remote shifts the gearcase into forward, neutral and reverse (you'll likely have to slowly spin the propeller by hand while attempting to make those shifts).

2. If the throttle cable is currently disconnected, install it to the powerhead (after the shift cable is already installed) as follows:

a. Make sure both the remote control handle and gearcase are in the **NEUTRAL** position. If necessary slowly rotate the propeller by hand and manually move the shift lever on the powerhead to Neutral. Once the propeller is free, you'll know the gearcase is shifted into Neutral.

b. Connect the cable to the throttle lever, then secure with the latch retainer.

c. Adjust the cable barrel so that the installed throttle cable will hold the throttle arm AGAINST THE STOP.

d. Check the throttle adjustment by shifting the outboard into gear and out again a few times, activating the throttle linkage.

a - Move cable barrel to attain the measured distance
b - Cable barrel

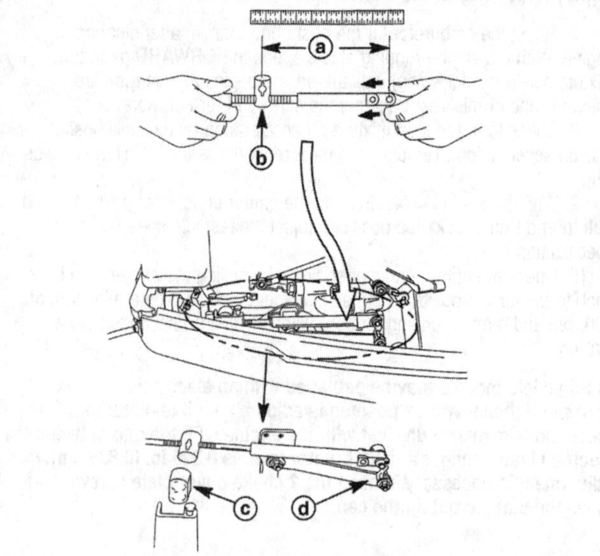

c - Barrel holder– place barrel into bottom hole
d - Retainer

SELOC_1418

Fig. 160 Shift cable installation and adjustment

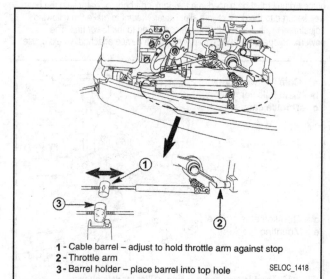

1 - Cable barrel – adjust to hold throttle arm against stop
2 - Throttle arm
3 - Barrel holder – place barrel into top hole

SELOC_1418

Fig. 161 Throttle cable installation and adjustment

■ Be sure to SLOWLY rotate the propeller when attempting to shift into reverse.

e. Return the remote to Neutral, then place a thin piece of paper between the throttle arm and idle stop. Adjustment is correct if the paper can be removed with a slight drag felt, but WITHOUT tearing. Adjust the cable barrel additionally as necessary.

f. With the cable barrel inserted into the barrel holder, lock the holder using the cable latch.

IGNITION TIMING

◆ See Figure 162

✳✳ SELOC WARNING

Reminder! The outboard MUST be mounted in a test tank or on a launched boat, securely affixed to the dock or trailer before proceeding with this adjustment.

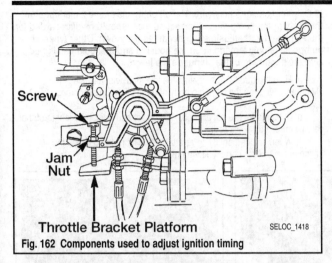

Screw

Jam Nut

Throttle Bracket Platform

SELOC_1418

Fig. 162 Components used to adjust ignition timing

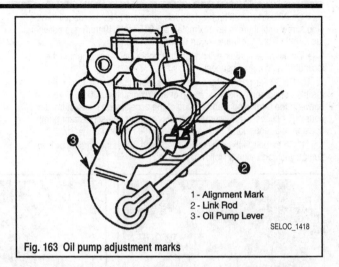

1 - Alignment Mark
2 - Link Rod
3 - Oil Pump Lever

SELOC_1418

Fig. 163 Oil pump adjustment marks

The ignition trigger is mounted on a rotating assembly that will physically change position relative to the powerhead (and therefore relative to the spinning flywheel magnets). By rotating the link forward or backward in relation to the motion of the flywheel you will change the ignition timing (the time at which the trigger sends the signal to fire the spark plugs).

Idle timing is adjusted by changing the length of the throttle cam-to-trigger link rod.

Wide open throttle (WOT) or maximum timing is adjusted using a screw and locknut on the throttle cam which works to mechanically control/limit the amount of motion allowed by the throttle cam (and therefore that same cam-to-trigger link rod, when the cam is in the WIDE OPEN THROTTLE position).

1. Obtain and connect a timing light to the No.1 spark plug lead.
2. Start the powerhead and then shift the unit into **FORWARD**.
3. Allow the powerhead to warm to normal operating temperature.
4. Check and set the idle timing first. With the engine running in gear at the properly adjusted idle speed timing should be 7-9° Before Top Dead Center (BTDC). If adjustment is necessary shut the powerhead down, unsnap the trigger link rod socket off the ball stud at one end and proceed as follows:
   • Shortening the rod length will RETARD ignition timing.
   • Extending the rod length will ENHANCE ignition timing.
5. Snap the link rod back onto the ball stud and restart the powerhead. Allow it to warm up slightly again, then cycle the throttle open and closed again 3 times. Allow the idle to stabilize, then recheck. If necessary, adjust the idle speed screw again to achieve the proper idle specification.
6. Next, check the WOT timing. If adjustment is necessary, proceed as follows:
   a. Turn the engine Off.
   b. Position the twist grip or remote control (if the cables are connected) to the WOT setting.
   c. Loosen the Jam nut on the WOT screw, then turn the screw COUNTERCLOCKWISE until it no longer makes contact with the throttle bracket platform.
   d. Next, turn the screw inward (CLOCKWISE) until it JUST makes contact with the throttle bracket platform, THEN give the screw 1 additional turn in (CLOCKWISE).
   e. Hold the screw from turning and tighten the jam nut.
7. Restart the motor, then double-check the Idle and WOT settings.

OIL PUMP

◆ See Figure 163

While holding the throttle arm at idle position, adjust the length of the pump link rod so the stamped mark of the oil pump body aligns with the stamped mark of the oil pump lever.

40-125 Hp Carbureted 3 & 4-Cylinder Models

The adjustment procedures for all 3 and 4-cylinder models are very similar. Therefore these procedures apply to all carbureted versions of 40/40 Jet/50/55/60 hp and 75/65 Jet/90 hp 3-cylinder models as well as 80 Jet/115/125 hp 4-cylinder models.

CARBURETOR SYNCHRONIZATION

◆ See Figures 164 thru 172

On these models carburetor synchronization is a relatively straightforward process where the linkage screws are loosened, then the throttle shutters held closed while the linkage screws are tightened again to hold that position. Afterwards the throttle linkage is adjusted to a proper starting point.

As such, once adjusted the throttle linkage should maintain this relationship between the carburetors and not require any periodic attention. However, should the carburetors or linkage be removed, replaced or otherwise serviced, or should the linkage come loose for any reason, start your timing and synchronization procedures by making sure the carbs are mechanically synchronized as follows:

1. Remove the attaching screws securing the sound box (attenuator) cover to the powerhead and then remove the cover for access and so you can observe the throttle butterflies.
2. Check that the throttle cable is not holding the throttle open. If it is, disconnect the cable barrel from the barrel retainer or disconnect the cable completely to make sure it does not interfere with throttle position during this adjustment.
3. Loosen but do not remove the cam follower adjustment screw (the Phillips screw located just under the roller for the cam follower).
4. Now, loosen but do not remove the 2 (3-cyl) or 3 (4-cyl) synchronizing screws (the Phillips screws on the carburetor throttles where they connect to the tie bar).
5. Peer into the throttle bores to make sure the shutters (butterflies) are completely closed. While holding the shutters closed (as necessary), tighten the 2 or 3 synchronizing screws (as applicable). Double-check the throttle shutters and readjust as necessary.
6. Loosen the locknut on the idle stop screw (the horizontally threaded screw toward the bottom of the throttle lever). Hold the throttle lever aft until the idle stop screw rests against the stop on the outer exhaust cover.
7. Position the throttle roller against the throttle cam. Adjust the idle stop screw until the raised mark on the cam is aligned with the center of the throttle roller. Tighten the locknut on the idle stop screw.
8. Pull back and hold the throttle lever against the idle stop. Check the gap between the throttle cam and the roller on the cam follower - it should be 0.005-0.020 in. (0.13-0.51 mm). Position the cam follower to obtain the correct gap, then tighten the Philips screw loosened at the beginning of this procedure.
9. Loosen the locknut on the full throttle stop screw. Adjust the full throttle stop screw and at the same time observe all the throttle shutter valves. The shutter valves must be in the full open position. Check to be sure that the valves themselves do not act as a throttle stop. This is accomplished slightly differently depending upon the model as follows:
   • On 40/40 Jet/50/55/60 hp 3-cyl models, continue to adjust the full throttle stop screw until there is approximately 0.015 in. (0.40 mm) free-play exists in the throttle linkage at wide open throttle.
   • On 75/65 Jet/90 hp 3-cyl models after the full throttle stop screw is adjusted to allow the throttle shutters to open fully, turn the screw in (CLOCKWISE) 1/2 of an additional turn.

• On 80 Jet/115/125 hp 4-cyl models, continue to adjust the full throttle stop screw until there is approximately a 0.015 in. (0.40 mm) gap between the throttle lever and the roller on the cam follower.

• On all models, after setting the full throttle stop screw, tighten the locknut to hold the adjustment.

10. On 80 Jet/115/125 hp 4-cyl models, hold the throttle cam in the full throttle position and check for a gap of approximately 0.030 in. (0.76 mm) between the throttle cam and the top of the acceleration pump aluminum housing. If necessary, loosen the screws mounting the accelerator pump housing and reposition it slightly to achieve the proper gap.

11. On models with oil injection, check and adjust the carburetor-to-oil pump synchronization as follows:

■ Oil pump designs may vary. The oil pump may have 2 markings on the body of the pump. If so, the second mark should be disregarded for the alignment and adjustment procedures as shown. Other types of pumps have only a marking on the housing and it should be aligned with the beginning of the cutout on the lever.

a. Hold the throttle arm against the idle stop. Visually check the alignment marks on the oil pump body and the oil pump synchronization lever.

b. If the marks are not aligned, disconnect and adjust the length of the carburetor/oil pump link rod.

c. When finished adjusting re-attach the link rod and check the alignment marks for correct alignment.

Fig. 164 View of the throttle cam and cam follower roller

Fig. 165 View of the synchronizing and cam follower screws

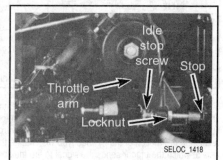
Fig. 166 Once the throttle shutters are set, use the idle stop screw to align the mark on the throttle cam with the roller. . .

Fig. 167 Alternative location for Idle Stop Screw and Idle Timing Screw on late-model motor

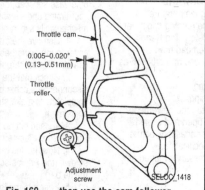
Fig. 168 . . . then use the cam follower screw to set the proper gap between the roller and cam

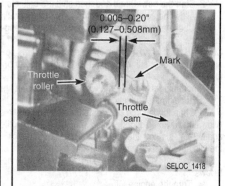
Fig. 169 Reposition the cam follower as necessary so the gap is correct at the roller, then tighten the follower screw

Fig. 170 Next adjust the full throttle stop screw to make sure the shutters don't act at throttle stops (40/40 Jet/50/55/60 hp shown, others similar)

Fig. 171 On models with oil injection, be sure to align the mark on the lever with the mark on the pump housing

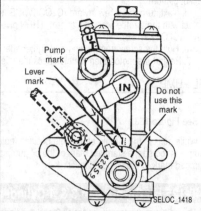
Fig. 172 Detailed view of the dual-marked oil pump (note the one mark to the right is not used)

IGNITION TIMING

MODERATE

◆ See Figure 173

Timing adjustments are made at cranking speed on these motors. In order to prepare the motor for checking and adjustment you must first make sure that the battery is in good condition and fully charged. Likewise, you must make sure that all connections are clean and tight and the wiring is not damaged.

Before beginning be sure to remove all but the No. 1 spark plug wires and ground them to the cylinder head. This will both prevent the motor from starting AND will prevent damage to the ignition system and danger from spark. If you have a spark tester (basically a clip-on, threaded adjustable spark plug) disconnect the No. 1 spark plug wire from the plug and connect it to the spark plug testing tool.

Even though you're only cranking the motor, we still recommend a source of water for the lower unit to prevent potential damage to the water pump if it is repeatedly cranked dry. Besides, after checking/adjusting timing you'll need to run the engine to check WOT timing and to check/adjust idle speed you'll need a source of cooling water anyway. Although keep in mind that in order to run the engine at WOT you REALLY need a test tank or you need to launch and secure the boat.

Also, be sure not to crank the motor for extended periods of time and to allow the starter motor to cool off for AT LEAST 30 seconds to a few minutes between cranking to prevent overheating and damage.

Idle Timing

◆ See Figures 167, 173 and 174

1. Disconnect the spark plug leads from the plugs and ground each lead to the powerhead (one great method for this is to use spark testing tools).
2. If easy enough, disconnect the fuel supply line to the powerhead. (Although advisable for extended cranking/testing, this step is not critical).
3. Disconnect the remote throttle cable barrel nut at the powerhead barrel retainer.
4. Connect a timing light to the No. 1 (Top) cylinder high tension lead.

■ On these models the throttle lever is rotated COUNTERCLOCKWISE toward the idle stop (this is because the idle stop screw is located on the lower portion of the lever and the stop is to your right when facing the lever).

5. Make sure the throttle lever on the side of the powerhead is being held fully against the idle stopper (an assistant can be helpful here).
6. Crank the powerhead with the starter motor and at the same time, aim the timing light at the flywheel cover timing pointer and compare with the idle timing specifications in the Tune-Up Specifications chart.
7. If adjustment is necessary, loosen the locknut on the idle timing screw (threaded diagonally downward on the top/center of the throttle lever). Adjust the idle timing screw until the timing pointer aligns (while cranking the motor) with specified timing mark. Tighten the locknut to hold this position.
8. Double-check the timing after the locknut has been tightened, then proceed to WOT (Maximum Timing) adjustment.

Maximum Timing

◆ See Figures 173 and 175

1. Check and adjust the Idle Timing as detailed earlier in this section. Use the same conditions (spark plug wires grounded, cooling water for cranking motor etc) to check the wide open throttle (WOT) or maximum timing using a timing light while cranking the motor with the starter.

■ On these models the throttle lever is rotated CLOCKWISE toward the WOT stop (this is because the maximum stop screw is located on the upper portion of the lever and the stop is to your right when facing the lever).

2. Make sure the throttle lever on the side of the powerhead is being held fully against the maximum advance screw stopper (an assistant can be helpful here).
3. Crank the powerhead with the starter motor and at the same time, aim the timing light at the flywheel cover timing pointer and compare with the WOT timing specifications in the Tune-Up Specifications chart.

■ The Tune-Up Specifications chart correctly lists the WOT timing specification as the target timing for when the engine is RUNNING above 3000 rpm. BUT the spark advance characteristics of the ignition system on these models will automatically reduce the WOT timing 2° BTDC when the motor is running above 3000 rpm from wherever it was set during cranking. Therefore add 2° BTDC to the specification when setting while cranking. For example, the specification for 4-cylinder motors is 23° BTDC, so you should set it at 25° BTDC with the motor cranking. In all cases, run the motor at or above 3000 rpm after adjustment to verify proper adjustment.

4. If adjustment is necessary, loosen the locknut on the idle timing screw (threaded horizontally from left to right when looking at the lever, upward

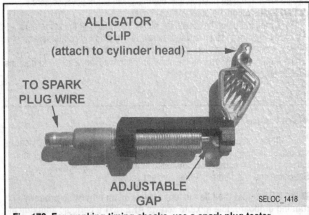

Fig. 173 For cranking timing checks, use a spark plug tester attached to the No. 1 plug wire

Fig. 174 Adjust the idle timing screw until the timing pointer is correctly aligned (and later so that proper idle speed is maintained)

Fig. 175 Use the maximum (WOT stop) timing screw to set the full timing advance

toward the top of the throttle lever). Adjust the WOT timing screw until the timing pointer aligns (while cranking the motor) with specified the timing mark (plus 2° BTDC as noted). Tighten the locknut to hold this position.

5. Reconnect all of the spark plug leads and, if disconnected, the fuel line.

6. With the motor either in a test tank or with the launched boat secured to the trailer or dock, start and allow the engine to run to normal operating temperature.

7. With the motor fully warmed, place it in gear and run the motor at or above 3000 rpm, then double-check the WOT timing adjustment. Shut the motor down and readjust, as necessary.

8. Once you are finished remove the timing light.

9. Loosen the locknut on the maximum spark advance screw. Hold the throttle lever in the aft position until the maximum spark advance screw contacts the stop on the outer exhaust cover.

10. Crank the powerhead with the starter motor and at the same time adjust the maximum spark advance screw until the timing pointer is aligned in the correct position as found in the specifications chart.

11. Tighten the locknut on the maximum spark advance screw to hold the adjustment. Remove the timing light.

IDLE SPEED

◆ **See Figures 174 and 176**

1. If the carburetors where (or the powerhead was) rebuilt or replaced set each idle mixture screw to the initial setting by LIGHTLY seating it then backing it out the number of turns listed in the Carburetor Set-Up Specifications chart from the Fuel System section.

■ **DO NOT adjust the idle mixture screws on an otherwise properly running outboard. Idle mixture screws ARE NOT a periodic adjustment and should only be used after a major repair/replacement or if the motor cannot otherwise be properly tuned.**

2. With a source of cooling water still supplied as from the previous adjustments, start and allow the engine to warm to normal operating temperature. Allow the engine to run in Forward gear at idle speed.

3. If the idle mixture screws are being reset, turn each screw INWARD (clockwise) until the engine starts to bog down and misfire, then back it out 1/4 turn or more. Remember, it is better to set the motor a little too rich than a little too lean (so back it out a little more when in doubt). After each adjustment advance the engine speed above idle for a few seconds, then allow it to return to idle and stabilize before proceeding with further adjustment of that or another idle mixture screw.

4. If (or once) the idle mixture is properly set, check the idle speed using a tachometer while manually holding the throttle lever against the throttle stop. If the speed is not within specification use the Idle Timing screw to reset the idle speed within spec.

5. Once adjustment is complete tighten the locknut and shut the powerhead down.

6. If removed to prevent interference with adjustments, reconnect the remote throttle cable barrel to the barrel holder or throttle arm.

Fig. 176 Idle mixture screw

75/90/115 Hp Optimax Models

The ignition and fuel systems on these models are controlled by the Engine Control Module (ECM or sometimes referred to a PCM, Powerhead Control Module) which performs the following functions:

• Calculates the precise fuel and ignition timing requirements based on engine speed, throttle position, manifold pressure and coolant temperature.

• Controls the fuel injectors for each cylinder, direct injectors for each cylinder and ignition for each cylinder.

• Controls and monitors the Oil Injection system

• Monitors the air compressor system.

• Controls all alarm horn and warning lamp functions.

• Supplies the tachometer signal to the gauge.

• Controls the rpm limiter function.

• Records engine running information.

Because of the vast extent of electronic monitoring and control MOST functions, including Ignition Timing and Idle Speed are fully under the control of the ECM and are not adjustable. Idle rpm can be spot-checked using a tachometer, but any operation out of specification must be troubleshot using the information on the Optimax system from the Fuel System section (unless a mechanical condition such as throttle cable binding can be found and attributed to the problem).

Unlike some fuel systems both the throttle position sensor (TPS) and crankshaft position sensor (CPS), two vital inputs to the Optimax system, are NOT adjustable.

This means that the only potential adjustment for Timing and Synchronization on these motors is the Throttle Linkage. And that said, it should NOT be necessary to adjust the throttle linkage periodically only to check and adjust it after some component of the throttle system (such as the cables themselves) have been replaced. However, if the powerhead is not operating properly, a simple check of the throttle linkage adjustment is a good way to eliminate one potential mechanical cause for trouble.

THROTTLE LINKAGE ADJUSTMENT

Models Through Serial Number 1B226999

◆ **See Figures 177 thru 180**

As stated earlier, this adjustment should NOT be necessary on a periodic basis, but should only be required after some portion of the throttle system has been repositioned or replaced.

1. Remove the engine top cover and the air intake silencer for access and so you can observe the butterfly in the throat of the throttle body.

2. Verify that the throttle cam ends of the throttle cables (the ends of the cables on the lower side of the adjustment bracket, closest to the throttle body) are properly set before starting. The throttle cable sheaves should extend through the bracket about 0.40-0.48 in. (9.1-12.1mm) below the bracket. This means the measurement from the underside of the bracket to the end of the threads (which includes the thickness of the jam nut on the underside of the bracket) should be within that range. If necessary, loosen the jam nuts then adjust the cable housing as necessary and retighten the jam nuts before proceeding.

3. With the throttle arm against the idle stop, adjust the LOWER throttle cable to remove any excess slack.

4. While holding the throttle arm against the idle stop, adjust the UPPER throttle cable to remove and excess cable slack.

✱✱ SELOC CAUTION

Excessive tension on the upper throttle cable will prevent the throttle arm from returning freely to the idle stop. MAKE SURE not to *just* remove the slack, but not to adjust the throttle cables too tight.

5. With the remote set in the Idle/Neutral position peer into the throttle body and observe the position of the throttle butterfly. Use a feeler gauge to verify that there is approximately a 0.016 in. (0.40mm) gap between the throttle shutter and the throttle bore. If out of spec (and the throttle cable is not holding it open), use the throttle shutter stop screw (threaded vertically downward through a locknut and bracket to connect an arm of the throttle lever) to achieve the proper gap.

6. Hold the throttle arm against the idle stop and verify that the throttle roller is in the throttle cam pocket AND that the roller rotates freely (does not bind against the cam). If the roller is not positioned correctly, use the roller

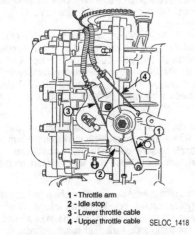

Fig. 177 Remove excess slack from the throttle cables, but do NOT over tension them

1 - Throttle arm
2 - Idle stop
3 - Lower throttle cable
4 - Upper throttle cable

SELOC_1418

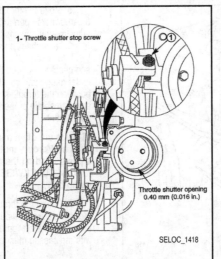

1 - Throttle shutter stop screw

Throttle shutter opening 0.40 mm (0.016 in.)

SELOC_1418

Fig. 178 The throttle shutter closed gap is adjusted using the shutter screw

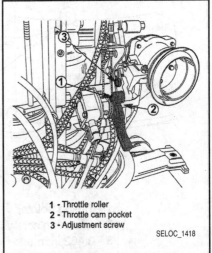

1 - Throttle roller
2 - Throttle cam pocket
3 - Adjustment screw

SELOC_1418

Fig. 179 The roller arm screw is used to ensure there is no binding against the cam

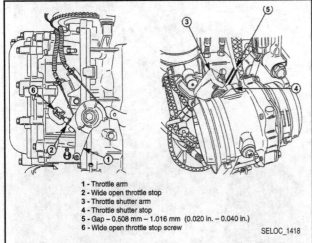

1 - Throttle arm
2 - Wide open throttle stop
3 - Throttle shutter arm
4 - Throttle shutter stop
5 - Gap – 0.508 mm – 1.016 mm (0.020 in. – 0.040 in.)
6 - Wide open throttle stop screw

SELOC_1418

Fig. 180 The WOT screw is used to ensure a proper gap between the shutter arm and stop

arm adjustment screw (on the side of the arm assembly) to obtain the correct alignment.

7. Hold the throttle arm against the Wide Open Throttle (WOT) stop and check the gap between the throttle shutter arm and the throttle shutter stop, it should be 0.020-0.060 in. (0.508-1.016mm). If necessary, use the WOT stop screw (threaded diagonally downward into the throttle bracket) to adjust the maximum throttle position to obtain the proper gap.

8. Once the throttle cable and linkage is properly adjusted, install the air intake silencer and verify proper motor operation. If adjustments are in spec and idle speed/throttle performance is off, eliminate all potential mechanical problems before proceeding with Optimax system troubleshooting.

Models w/Serial Number 1B227000 & Above

◆ See Figures 177 and 180

As stated earlier, this adjustment should NOT be necessary on a periodic basis, but should only be required after some portion of the throttle system has been repositioned or replaced.

1. Remove the engine top cover and the air intake silencer for access and so you can observe the butterfly in the throat of the throttle body.

2. Verify that the throttle cam ends of the throttle cables (the ends of the cables on the lower side of the adjustment bracket, closest to the throttle body) are properly set before starting. The threaded throttle cable sheaves should extend through the bracket about 0.54-0.62 in. (13.63-15.83mm) below the bracket. This means the measurement from the underside of the

bracket to the end of the threads (which includes the thickness of the jam nut on the underside of the bracket) should be within that range. If necessary, loosen the jam nuts then adjust the cable housing as necessary and retighten the jam nuts before proceeding.

3. Next, verify that the cable ends are properly inserted into the throttle cam.

4. Move to the other end of the cables (the linkage on the port side near the Vapor Separator Tank assembly). Loosen the WOT (upper) cable nuts, and remove the cable from the bracket, then loosen the idle cable nuts (leaving the cable in the bracket).

5. Make sure the throttle lever is contacting the lever stop, then pull on the idle throttle cable (lower cable) to remove any slack and thread the top nut so it touches the cable bracket. Next thread the bottom nut to secure the idle cable to the bracket and tighten the nut securely.

6. Now pull on the WOT (upper) cable to remove any clack and thread the top nut so it touches the cable bracket. Next thread the bottom nut to secure the WOT cable to the cable bracket and tighten the nut securely.

7. Just in front of the throttle lever (still on the port side) locate and loosen the WOT stop adjustment screw.

8. Pull the throttle lever to the WOT position and check the throttle shutter arm and throttle shutter stop at the throttle body. Adjust the WOT stop adjustment screw so the shutter arm at the throttle body has 0.02-0.04 in. (0.508-1.016mm) of clearance. Once this clearance has been obtained, tighten the nut on the WOT securely to hold this position.

150-200 Hp (2.5L) V6 Carbureted Models

IGNITION TIMING POINTER ADJUSTMENT

◆ See Figure 181

Before the ignition timing can be checked or adjusted you need to verify that the timing pointer is properly aligned and has not been moved or has not shifted slightly during service. In order to do this you need a dial gauge to perform precision measurements on piston position.

1. Remove all spark plugs to relieve engine compression allowing you to very slowly and gently rotate the flywheel.

2. Install a dial indicator into the No. 1 cylinder spark plug opening (top cylinder, starboard bank).

3. Slowly rotate the flywheel clockwise (when viewed from above) until the No. 1 piston is at Top Dead Center (TDC), which is the very top of the piston travel. Set the dial indicator to "0".

4. Slowly rotate the flywheel COUNTERCLOCKWISE until the dial indicator needle is about 1/4 turn beyond the 0.462 in. (11.73mm) mark.

5. Next, slowly rotate the flywheel back CLOCKWISE again until the dial indicator reads exactly 0.462 in. (11.73 mm).

6. NOW, observe the timing pointer on the flywheel cover and the 0.462 in (11.73mm) mark on the flywheel. If the flywheel pointer is not exactly on

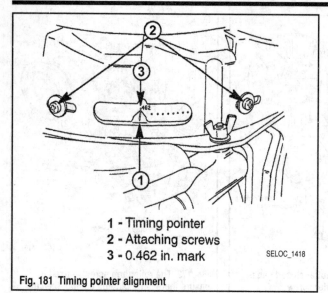

1 - Timing pointer
2 - Attaching screws
3 - 0.462 in. mark

SELOC_1418

Fig. 181 Timing pointer alignment

the mark, loosen the pointer adjustment screws (on either side of the timing pointer window) and align the pointer with the 0.462 in. (11.73mm) mark. Tighten the pointer adjustment screws.

■ **On some models with a 1 piece flywheel, a 4/5 (45°) mark on the flywheel is equal to the 0.462 mark on flexplate type flywheels.**

7. Remove the dial indicator from the No. 1 spark plug opening.

8. Proceed with carburetor synchronization (IF the carburetor linkage has been disturbed) OR with Ignition Timing checks and adjustments, as applicable.

CARBURETOR SYNCHRONIZATION

 MODERATE

◆ **See Figures 182 thru 185**

1. Remove the air box cover from the powerhead both for access and so you can peer into the throttle bores to check shutter positions.

2. Loosen the 3 carburetor synchronization screws (there is on 1 the linkage for each carburetor) to allow the shutter plates to fully close.

3. Position the throttle lever so the idle stop screw is against the idle stop, then move the roller arm until the roller itself LIGHTLY touches the throttle cam. In this position, use the idle stop screw to adjust the position of the throttle lever so that the mark on the throttle cam aligns with the center of the roller.

4. Keeping the roller aligned with the mark on the throttle lever, retighten the 3 carburetor synchronization screws.

5. Move the throttle lever and check to be sure the shutter plates on all 3 carburetors open and close simultaneously. Repeat this portion of the adjustment if any shutter plate lags.

6. Next check and adjust the WOT position. Move the throttle lever to WOT. Make sure the carburetor shutters are fully open at wide open throttle and check to be sure the throttle shutters do not act as throttle stops. Use a feeler gauge and check for 0.010-0.015 in. (0.25mm-0.38mm), clearance between the throttle shaft arm and the stop. If necessary, use the WOT stop screw to obtain proper clearance, then tighten the locknut to hold the adjustment.

■ **There are often multiple marks on the link arms for the oil pumps used by Mercury motors. Where there are multiple marks on the arm of a pump used on these models be sure to align the mark positioned furthest CLOCKWISE away from the link ball stud with the mark on the pump body.**

7. Return the throttle linkage to the idle position, then check and adjust the carburetor-to-oil pump synchronization. With the throttle linkage in the idle position the alignment mark on the oil pump link arm should be aligned with the casting mark on the oil pump body. If the mark is not aligned,

Fig. 182 The synchronization screws control the position of each carb's shutter in relation to the throttle linkage

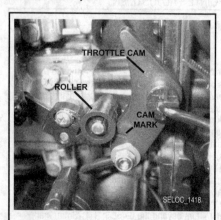

Fig. 183 Align the mark on the throttle cam with the roller, then tighten the carb sync screws

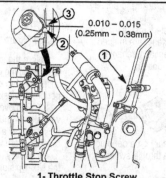

0.010 – 0.015
(0.25mm – 0.38mm)

1- Throttle Stop Screw
2- Throttle Shaft Arm
3- Stop

SELOC_1418

Fig. 184 At WOT set the proper gap between the throttle shaft arm and the throttle stop

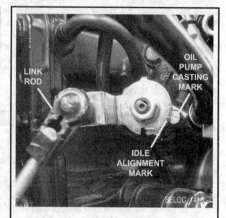

Fig. 185 Carburetor-to-oil pump synchronization at idle

disconnect the linkage rod from the oil pump lever. Screw the rod end (in or out, as necessary) to align the mark on the oil pump body with the mark on the lever. Connect the rod end back onto the oil pump lever.

IGNITION TIMING

◆ **See Figures 173 and 186 thru 189**

Timing adjustments are made at cranking speed on these motors. In order to prepare the motor for checking and adjustment you must first make sure that the battery is in good condition and fully charged. Likewise, you must make sure that all connections are clean and tight and the wiring is not damaged.

Before beginning be sure to remove all but the No. 1 spark plug wires and ground them to the cylinder head. This will both prevent the motor from starting AND will prevent damage to the ignition system and danger from spark. If you have a spark tester (basically a clip-on, threaded adjustable spark plug) disconnect the No. 1 spark plug wire from the plug and connect it to the spark plug testing tool.

Even though you're only cranking the motor, we still recommend a source of water for the lower unit to prevent potential damage to the water pump if it is repeatedly cranked dry. Besides, after checking/adjusting timing you'll need to run the engine to both VERIFY the cranking adjustments (as the nature of the ignition system may allow for differences between timing at cranking and at running speeds) and to check/adjust idle speed, so you'll need a source of cooling water for that anyway.

Also, during timing setting procedures, be sure not to crank the motor for extended periods of time and to allow the starter motor to cool off for AT

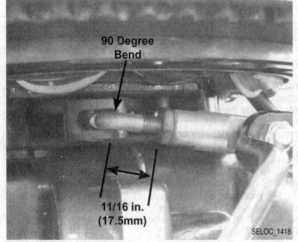

Fig. 186 Before adjusting timing, verify the proper length of the trigger rod

LEAST 30 seconds to a few minutes between cranking to prevent overheating and damage.

1. If not already done to verify the position of the timing pointer, disconnect the spark plug leads from the plugs and ground each lead to the powerhead (one great method for this is to use spark testing tools).

■ **Mercury usually recommends removing all of the spark plugs during the timing checking and adjustment. This is probably to relieve engine compression and allow the starter motor to more easily and freely spin the motor.**

2. Disconnect the fuel supply line and the remote wiring harness from the powerhead.

3. Disconnect the remote throttle cable barrel at the powerhead barrel retainer.

4. If the ignition trigger link rod (the rod which goes from the top of the throttle arm to the trigger base assembly under the flywheel) was removed at any point, check to verify it is still 11/16 in. (17.5mm) in length from the portion where the threads protrude out of the locknut to the end of the rod where it bends 90°. If necessary, remove the rod and reset it to the correct length before proceeding.

5. Connect a timing light to the No. 1 (Top) cylinder high tension lead.

6. Place the remote shift lever in the **NEUTRAL** position.

7. Hold the throttle lever against the WOT (maximum advance) stop screw and have an assistant crank the powerhead with the starter motor while you aim the timing light at the timing decal on the flywheel.

■ **Ignition may retard as engine speed approaches maximum RPM.**

8. Compare the current WOT timing setting with the specs listed in the Tune-Up Specifications chart in this section. If necessary, loosen the locknut and adjust the WOT (maximum throttle advance) stop screw until the correct WOT timing setting is reached, then tighten the locknut hold the adjustment.

9. With the outboard still in Neutral, now set the Idle & Pick-Up timing. Hold the throttle lever against the idle stop screw and have an assistant crank the powerhead with the starter motor while you continue to aim the timing light at the timing decal on the flywheel.

■ **The Idle/Primary Pick-Up timing will also determine powerhead speed at idle. Refer to the Idle Speed Adjustment procedures to check and verify this setting.**

10. Compare the current Idle/Pick-up timing setting with the specs listed in the Tune-Up Specifications chart in this section. If necessary, loosen the locknut and adjust the primary pick-up stop screw until the correct Idle/Pick-Up timing setting is reached, then tighten the locknut hold the adjustment.

■ **The remainder of the procedures must be performed with the outboard in a test tank or the boat and outboard in a body of water (launched or secured to a dock/trailer).**

✳✳ SELOC WARNING

DO not connect a flush device to the lower unit for this adjustment, because the outboard unit must be under load with the lower unit in Forward gear.

Fig. 187 WOT (maximum advance) stop screw

Fig. 188 Primary Pick-Up stop screw

Fig. 189 Idle stop screw

11. Remove the timing light from the powerhead No. 1 spark plug lead.

12. If removed to relieve engine compression, install the spark plugs and tighten them to 20 ft. lbs. (27 Nm).

13. Remove all the ground leads from the spark plug wire boots, then reconnect the spark plug leads to the plugs.

14. Connect a tachometer to the powerhead.

15. Reconnect the fuel line to the fuel tank and prime the fuel system.

16. Reconnect the remote harness assembly.

17. Start the powerhead and allow it to warm to operating temperature.

18. Once the powerhead is fully warmed recheck the WOT timing setting with the engine running. Note the different specifications while running listed in the Tune-Up Specifications chart. If adjustments must be made, shut the powerhead down for safety, tweak the WOT screw, then restart and recheck the timing.

■ Since idle speed is determined by the Idle/Pick-Up timing, check and verify that in the next procedure, Idle Speed Adjustment.

IDLE SPEED ADJUSTMENT

◆ See Figures 187 and 190

■ Like the final check of the Ignition Timing procedure, the Idle Speed check and adjustment must be performed with the outboard in a test tank or the boat and outboard in a body of water (launched or secured to a dock/trailer).

The idle speed adjustment procedures can only be performed with the outboard running with the lower unit in Forward gear and the propeller under an actual load condition.

✳✳ SELOC WARNING

DO not connect a flush device to the lower unit for this adjustment, because the outboard unit must be under load with the lower unit in Forward gear.

1. Check and adjust the Ignition Timing as detailed in this section.

2. IF them carburetors where (or the powerhead was) rebuilt or replaced set each idle mixture screw to the initial setting by LIGHTLY seating it then backing it out the number of turns listed in the Carburetor Set-Up Specifications chart from the Fuel System section.

■ DO NOT adjust the idle mixture screws on an otherwise properly running outboard. Idle mixture screws ARE NOT a periodic adjustment and should only be used after a major repair/replacement or if the motor cannot otherwise be properly tuned.

3. Start and allow the engine to warm to normal operating temperature. Allow the engine to run in Forward gear at idle speed.

4. If the idle mixture screws are being reset, turn each screw INWARD (clockwise) until the engine starts to bog down and misfire, then back it out 1/4 turn or more. Remember, it is better to set the motor a little too rich than a little too lean (so back it out a little more when in doubt). After each adjustment advance the engine speed above idle for a few seconds, then allow it to return to idle and stabilize before proceeding with further adjustment of that or another idle mixture screw.

5. If (or once) the idle mixture is properly set, check the Idle Speed using a tachometer while manually holding the throttle lever against the throttle stop. If the speed is not within specification use the primary pick-up screw to reset the idle speed within spec.

6. Double-check the Idle/Pick-Up timing adjustment as this time and tweak the primary pick-up adjustment screw additionally as necessary to get the speed and timing within spec (or as close as possible).

7. Once adjustment is complete tighten the locknut and shut the powerhead down.

8. If removed to prevent interference with adjustments, reconnect the remote throttle cable barrel to the barrel holder or throttle arm.

9. Check and adjust the throttle cable pre-load as follows:

a. With the end of the throttle cable connected to the throttle lever anchor pin, hold the throttle lever against the idle stop. Adjust the throttle cable barrel until it will slip into the barrel retainer on the cable anchor bracket. If necessary, adjust the throttle cable barrel until there is a very light preload on the throttle lever against the idle stop. Lock the barrel in place.

Fig. 190 Idle mixture screws

■ An excessive preload on the throttle cable will cause difficulty when shifting from Forward gear to Neutral.

b. Check the preload by placing a piece of paper between the idle stop screw and the idle stop and then withdrawing it. If the paper does not tear but drag can be felt, the preload is correct. Adjust the cable barrel, if necessary, to obtain the proper preload just described. Install the powerhead cowling.

c. If sufficient throttle cable barrel adjustment is not available, a check must be made for correct installation of the link rod (located between the throttle lever and the throttle cam). The throttle end of the link rod must be threaded into its plastic barrel until it bottoms against the throttle lever casting and then backed out from this position only far enough to obtain correct orientation of the link rod. The link rod must be backed out less than 1 full turn. All timing adjustments must be reset after this procedure has been completed.

10. After completion of the timing and idle adjustments, disconnect the timing light and tachometer, then reinstall the powerhead cowling.

11. If the outboard is not to be used in the very near future, place the throttle in neutral and disconnect the fuel line. Allow the powerhead to operate until most/all fuel is used in the carburetors. Set the key switch to the **OFF**.

150-200 Hp (2.5L) V6 EFI - 2001 Models Only

The EFI models built through serial #OT408999 (for the 2001 or earlier model years) have a radically different throttle body/air induction assembly (and therefore radically different linkage) than the EFI models for 2002 and later and serial #OT409000 and above. As a result the Timing & Synchronization procedures are very different on these otherwise similar outboards.

TIMER POINTER ADJUSTMENT

◆ See Figure 191

Before the ignition timing can be checked or adjusted you need to verify that the timing pointer is properly aligned and has not been moved or has not shifted slightly during service. In order to do this you need a dial gauge to perform precision measurements on piston position.

1. Remove all spark plugs.

2. Install a dial indicator into the No. 1 (top cylinder starboard bank), spark plug opening.

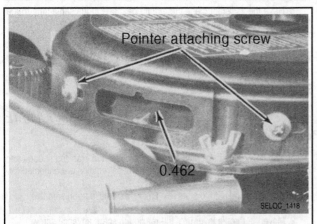

Fig. 191 Slowly rotate the flywheel CLOCKWISE until the dial indicator reads exactly 0.462 in. (11.73 mm), then check the timing pointer to make sure it is aligned with the 0.462 in. (11.73 mm) mark on the flywheel

3. Slowly rotate the flywheel CLOCKWISE until the No. 1 piston is at top dead center (TDC). Set the dial indicator to "0".

4. Slowly rotate the flywheel COUNTERCLOCKWISE until the dial indicator needle is 1/4 turn beyond the 0.462 in. (11.73mm) mark.

5. Slowly rotate the flywheel back - CLOCKWISE - until the dial indicator reads exactly 0.462 in. (11.73mm). Observe the timing pointer on the flywheel cover and the 0.462 in. (11.73 mm) mark on the flywheel.

6. If the flywheel pointer is not exactly on the mark, loosen the pointer adjustment screws and align the pointer with the 0.462 in. (11.73mm) mark.

7. Tighten the pointer adjustment screws and remove the dial indicator from the No. 1 spark plug opening.

TRIGGER LINK ROD & PRELIMINARY ADJUSTMENTS

◆ See Figure 192

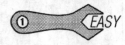

If the link rod (which connects the throttle cam to the ignition timing base assembly in order to mechanically advance timing as engine speed is increased) was disassembled at any point, verify the proper trigger link rod length before checking or adjusting timing. Also, there are few quick steps which must be followed to prepare the motor for throttle and timing adjustments as follows:

1. Measure the length of the trigger link rod from the center line of the 90° bend to the edge of the locknut. A dimension of 11/16 in. (17.5mm) is required for proper timing adjustment.

2. If necessary loosen the locknut and adjust the rod length (on most models it must first be removed from the linkage before it can be adjusted).

3. Once the link rod is adjusted to the proper length, secure the locknut.

4. If not done already to check the timing pointer, tag and disconnect all of the spark plug wires (except the No. 1 cylinder, top-starboard bank), then remove all of the spark plugs (again, except the No. 1 cylinder). Make sure the spark plug wires are all properly grounded to prevent injury or damage.

Fig. 192 Check and adjust the trigger link rod before checking ignition timing

5. Disconnect the remote fuel line from the motor.

6. If removed, connect the remote control electrical harness to the engine harness.

7. Remove the throttle cable barrel from the barrel retainer.

8. Disconnect the control unit harness connectors for the ignition timing procedures.

9. Proceed to Throttle Cam adjustment.

THROTTLE CAM

◆ See Figures 193, 194 and 195

1. Loosen the cam follower screw (the Philips head screw located under the center of the vapor separator tank), so the follower can move freely.

2. Loosen the locknut on the idle stop screw.

3. Let the roller rest on the throttle cam, then adjust the idle stop screw so that the mark on the throttle cam is centered on the roller, then retighten the stop screw locknut.

4. Lastly, hold the throttle arm against the idle stop while you tighten the cam follower screw so the roller is lightly touching the cam.

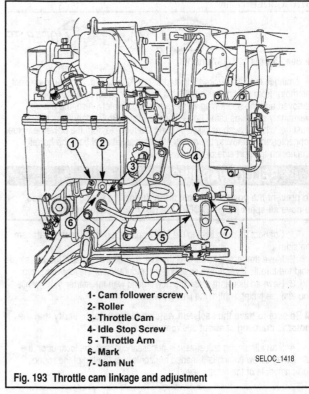

1- Cam follower screw
2- Roller
3- Throttle Cam
4- Idle Stop Screw
5 - Throttle Arm
6- Mark
7- Jam Nut

Fig. 193 Throttle cam linkage and adjustment

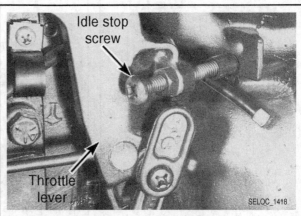

Fig. 194 Loosen the locknut on the idle stop screw

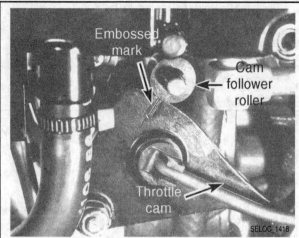

Fig. 195 Make sure the mark on the throttle cam is centered on the cam follower roller

STATIC IDLE TIMING

◆ See Figures 196, 197 and 198

Timing is initially checked/adjusted at cranking speed. In order to prevent the motor from starting, all but the No. 1 cylinder spark plug must be removed and the spark plug wires grounded for safety. Also, the other preliminary steps, as listed under Throttle Link & Preliminary Adjustments must be followed such as disconnecting the fuel line from the powerhead fuel connector and removing the remote throttle cable barrel from the barrel retainer on the port side of the powerhead.

✱✱ SELOC WARNING

To prevent the powerhead from firing during the timing procedures, remove all spark plugs EXCEPT the No. 1 (top, starboard bank).

1. Connect a timing light to the No. 1 spark plug lead (on the starboard top side).
2. Move the throttle lever until the idle stop screw contacts the idle stop. Hold the throttle lever in this position for the duration of the following step.
3. Have an assistant crank the powerhead with the starter motor while you aim the timing light at the timing pointer.

■ Be sure to have the assistant watch a tachometer to verify that the motor is cranking at about 300 rpm.

4. If an idle timing adjustment is necessary, loosen the locknut on the idle timing screw (which is threaded horizontally at the top of the round portion/middle of the throttle lever).

5. Have the assistant crank the motor again and at the same time rotate the idle timing screw until the marks on the timing decal align with the timing pointer as listed in the Tune-Up Specifications chart in this section.

6. After adjustment, tighten the locknut on the idle timing screw to hold this adjustment. Release the throttle lever.

STATIC MAXIMUM TIMING

◆ See Figure 199

Like idle timing, the maximum timing is initially set at cranking speeds and under the same circumstances. The ECM harness remains disconnected for this check/adjustment.

1. Check and adjust the state idle timing, as detailed earlier in this section.
2. Move the throttle lever until the maximum spark advance screw (threaded horizontally through a tab toward the top of the throttle lever, just above the idle timing screw) contacts the stop. Hold the throttle lever in this position for the duration of this step.
3. Have an assistant crank the powerhead with the starter motor while you aim the timing light at the timing pointer.
4. If a maximum timing adjustment is necessary, loosen the locknut on the maximum spark advance screw. Have an assistant crank the powerhead with the starter motor and at the same time rotate the maximum spark advance screw until the marks on the timing decal align with the timing pointer as listed in the Tune-Up Specifications chart in this section.
5. After adjustment, tighten the locknut on the maximum spark advance screw.
6. Set the ignition switch to Off and install the remaining 5 spark plugs.

MAXIMUM THROTTLE ADJUSTMENT

◆ See Figure 200

1. Move the throttle lever forward until the lever contacts the full throttle stop screw. While holding the lever in this position, check that the throttle valves are fully opened while there is still clearance between the throttle shaft arm and the stop on the induction box.
2. If adjustment is necessary, loosen the locknut on the full throttle stop screw (the horizontally mounted screw which is positioned to contact the lower end of the throttle lever) and rotate the screw until all the throttle shutters are fully open (again, making sure a slight clearance is maintained between the throttle shaft arm and the stop on the manifold.).
3. Check for free-play at the cam follower roller and the throttle cam. This indicates there is no binding of the throttle linkage. Readjust the full throttle stop screw, if required.
4. Reconnect the ECM harness which was disconnected for the Static Ignition Timing procedures. Also, if not already done, install the spark plugs and reconnect the fuel line.

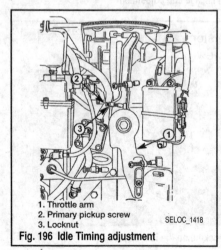

1. Throttle arm
2. Primary pickup screw
3. Locknut

Fig. 196 Idle Timing adjustment

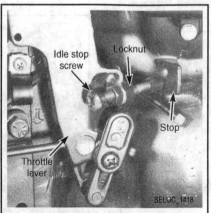

Fig. 197 Move the throttle until the idle stop screw just contacts the idle stop

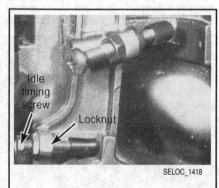

Fig. 198 Location of the idle timing screw

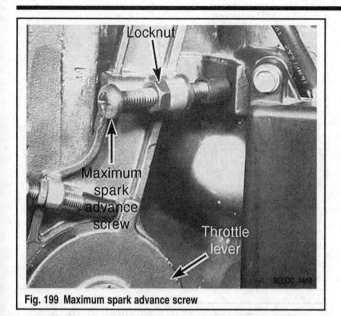

Fig. 199 Maximum spark advance screw

THROTTLE POSITION SENSOR (TPS)

◆ See Figures 201 and 202

The throttle position sensor (TPS) is used for low speed and mid-range power settings. The TPS transmits throttle position information to the ECU controller in the form of a low voltage signal. This low voltage signal will range from 0.20 VDC at the idle position to 7.46 VDC at wide open throttle setting. A lower voltage setting will cause the powerhead to operate on a leaner mixture and a higher voltage will richen the fuel mixture.

Alignment and adjustment of the TPS is a critical step in the timing and synchronizing of the powerhead (IF the switch has been disturbed or you suspect out of range operation). This procedure requires the use of a digital meter and special in-line harness connector (or a home-made inline jumper harness) to monitor the TPS voltage output during throttle operation.

■ Mercury notes that although the Mercury DDT (Digital Diagnostic Terminal) can be used to determine if the sensor is currently within proper operating range of 0.200-0.300 volts at idle, they do NOT recommend using the DDT to adjust the sensor. The reason is that because of the circuitry characteristics of the ECM the DDT may not always accurately display small TPS movements.

1. Locate the tan/black cylinder head temperature sensor leads for the Port cylinder head, then disconnect the 2 lead bullet connectors for the duration of TPS adjustment.

2. Locate the Throttle Position Sensor (a small, cylindrical sensor mounted on the side of the powerhead in a position where it can be moved by the throttle linkage), then trace the wiring and disconnect it from the EFI harness.

3. Install the TPS Test Lead Assembly #91-859199 or an equivalent jumper harness which allows probing a completed circuit for the TPS in between the TPS connector and the EFI harness connector.

4. Connect a digital volt ohmmeter (DVOM) to the test harness with the voltmeter set to the 2 volts DC scale.

5. Disconnect the remote control cable from the throttle lever.

6. Turn the ignition keyswitch to the **ON** position, but don't start the motor.

7. Check the DVOM for voltage readings with the throttle both closed to the idle position and open to the wide open throttle (WOT) position. It should read 0.200-0.300 at idle and approximately 7.46 volts at WOT.

8. If the DVOM is not showing readings within specifications, first loosen the 2 mounting screws slightly, then rotate the sensor fully clockwise (while holding the throttle shaft in the closed position). Then reposition the TPS as necessary so that readings are 0.24-0.26 in this position. Hold the TPS in this position and retighten the mounting screws.

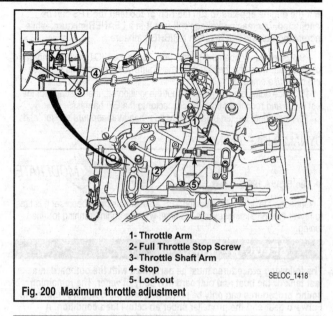

1- Throttle Arm
2- Full Throttle Stop Screw
3- Throttle Shaft Arm
4- Stop
5- Lockout

Fig. 200 Maximum throttle adjustment

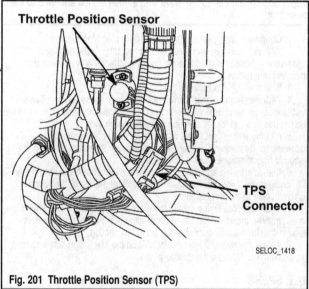

Fig. 201 Throttle Position Sensor (TPS)

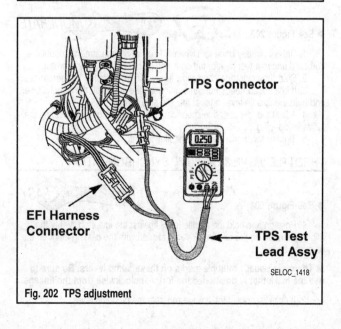

Fig. 202 TPS adjustment

■ If the engine appears to run too rich or too lean the TPS can be readjusted. Decreasing voltage will result in a LEANER mixture, while increasing voltage will result in a RICHER mixture.

9. Again, slowly open the throttle from idle to the WOT position and back to idle again, watching the DVOM to make sure voltage readings increase and decrease smoothly.

10. Once adjustment is finished shut the ignition off, then disconnect the test leads and reconnect the TPS connector to the EFI harness.

11. Reconnect the tan/black port cylinder head temperature sensor leads.

DYNAMIC IDLE TIMING

◆ **See Figure 196**

Once the Maximum Throttle and TPS Adjustments have been set it is time to revisit Ignition Timing, except this time, with the engine running for fine tuning.

✳✳ SELOC WARNING

The following procedures must be performed with the outboard in a test tank or the boat and outboard in a body of water. The remaining timing procedures can only be performed with the outboard running in Forward gear and the propeller under an actual load condition. A proper adjustment cannot be made with a flush device attached to the lower unit.

1. Connect a timing light to the No. 1 spark plug high-tension lead.
2. With the outboard unit mounted in a test tank or the craft and outboard in a body of water, start the powerhead and allow it to warm to operating temperature.
3. Shift the unit into Forward gear.
4. With the throttle cable "barrel" removed from the retainer, move the throttle lever by hand until the idle stop screw contacts the idle stop. Hold the lever in this position for the duration of the following step.
5. Aim the timing light at the timing decal and notate the timing according to the pointer. Compare the reading with the specs listed in the Tune-Up Specifications chart in this section.
6. If an idle timing adjustment is necessary, loosen the locknut on the idle timing screw (which is threaded horizontally at the top of the round portion/middle of the throttle lever).
7. Again, aim the timing light at the pointer and rotate the idle timing screw until the marks on the timing decal align with the timing pointer as listed in the Tune-Up Specifications chart in this section.
8. After adjustment, tighten the locknut on the idle timing screw to hold this adjustment. Release the throttle lever.

IDLE SPEED

◆ **See Figure 203**

1. Unless already done for Dynamic Idle Timing adjustment, mount the outboard unit in a test tank or the craft and outboard in a body of water.
2. Start the powerhead and allow it to warm to operating temperature.
3. If idle speed needs adjustment, loosen the screw on the cam follower and hold the throttle lever against the idle stop screw.
4. Adjust the idle speed to specification by increasing or decreasing the air valve opening.
5. Tighten the screw on the cam follower.

THROTTLE VALVE & OIL PUMP SYNCHRONIZATION

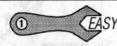

◆ **See Figure 204**

1. Position and hold the throttle lever against the idle stop. Verify the alignment mark on the oil pump lever is aligned with the casting mark on the oil pump body.

■ **There are usually multiple marks on these pump levers. Be sure to use the mark that is positioned the furthest clockwise from the linkage.**

2. If not, disconnect the linkage rod from the oil pump lever.

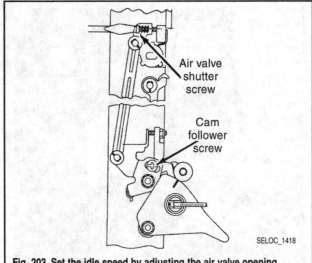

Fig. 203 Set the idle speed by adjusting the air valve opening

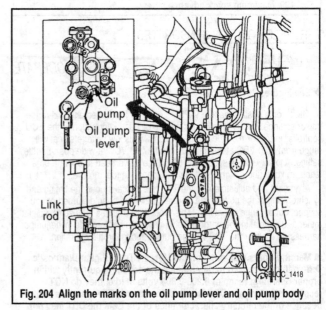

Fig. 204 Align the marks on the oil pump lever and oil pump body

3. Screw the rod end, in or out, to align the oil pump lever mark with the mark on the body casting.
4. Connect the rod end onto the oil pump lever ball.

THROTTLE CABLE PRELOAD

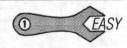

◆ **See Figure 205**

1. Connect the end of the throttle cable to the throttle lever.
2. Move the throttle lever against the idle stop screw and hold in this position. Adjust the throttle cable barrel until it will slip into the barrel recess in the control cable anchor bracket with only a light preload against the idle stop.
3. Lock the throttle cable barrel into the barrel recess on the control cable anchor bracket.

■ **An excessive preload on the throttle cable will cause difficulty when shifting from Forward gear into Neutral.**

4. The preload may be easily checked by placing a piece of paper between the idle stop screw and the idle stop and then withdrawing it by pulling on the end of the paper. If the paper does not tear but drag can be felt, the preload is correct.
5. If necessary, adjust the cable barrel, if necessary, to obtain the proper preload just described.
6. Install the powerhead cowling.

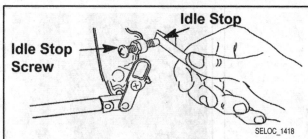

Fig. 205 Checking throttle cable preload using a piece of paper (early-model linkage shown but the principle is the same on all models)

DETONATION CONTROL SYSTEM

■ **This system is used on the 200 hp model only.**

1. If removed after earlier tests, connect a timing light to the No. 1 cylinder spark plug lead.
2. If not done already, mount the outboard unit in a test tank or the craft and outboard in a body of water.

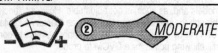

❄❄ SELOC WARNING

To prevent damage or injury from a runaway powerhead, DO NOT attempt this procedure on a flushing device!

3. Start the powerhead and allow it to warm to operating temperature.
4. Shift the unit into Forward gear. Move the throttle lever forward until the powerhead indicates 3500 rpm on the tachometer.
5. Verify the timing has been electronically advanced to 24° BTDC. The advancement indicates the detonation (knock), control circuit is functioning properly.
6. Return the throttle lever to the idle position and shift the outboard into Neutral.
7. NOW check and set the Dynamic Maximum Timing adjustment, as detailed later in this section.

DYNAMIC MAXIMUM TIMING

◆ **See Figure 206**

1. If not done already for earlier checks/adjustments, connect a timing light to the No. 1 cylinder spark plug wire.
2. Again, if not already done, mount the outboard unit in a test tank or the craft and outboard in a body of water.

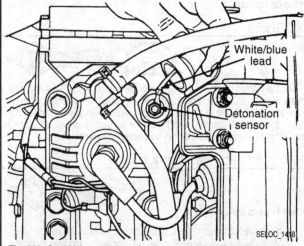

Fig. 206 On 200 hp models only, disconnect the White/Blue lead from the detonation sensor on the port side cylinder head

3. On 200 hp models, disconnect White/Blue lead from the detonation sensor on the port side cylinder head.
4. Start the powerhead and allow it to warm to operating temperature.
5. Shift the unit into Forward gear.
6. Move the throttle lever forward until the maximum spark adjustment screw contacts the stop. This contact should occur when the powerhead approaches 2500 rpm.
7. Check and, if necessary, adjust the maximum spark adjustment screw to align the timing mark on the decal with the timing pointer according to specification (as listed in the Tune-Up Specifications chart), then tighten the locknut to hold the adjustment.
8. Return the throttle lever to the idle position and shift the unit into Neutral.
9. Shut down the powerhead; disconnect the timing light; and, if applicable, reconnect the White/Blue lead to the detonation sensor.

150-200 Hp (2.5L) V6 EFI - 2002 & Later Models

The EFI models for 2002 and later, and serial #OT409000 and above, have a radically different throttle body/air induction assembly (and therefore radically different linkage) than the models produced for 2001 and earlier models years (serial #OT408999 and lower). As a result the Timing & Synchronization procedures are very different on these otherwise similar outboards.

All throttle linkage adjustments are set at the factory and no periodic adjustments should be necessary, unless a component has been replaced.

THROTTLE CAM ADJUSTMENT

◆ **See Figures 207 thru 210**

If the throttle body or linkage has been replaced the throttle cam must be properly adjusted so there is the proper gap between the roller and throttle cam while the linkage is in the idle position.

Because the cam follower screw is a tamper-resistant Torx screw (an internal lobular star shaped screw with a pin in the middle) you'll need a special Torx bit to loosen it. Although you can obtain these special Torx bits from many sources, the Snap-On part number is TTXR25E (meaning it's a T25 Torx bit with a hole drilled in the center to receive the pin).

The throttle linkage is all located on the port side of the motor. The throttle cam and cam follower are at the front port side, just above the vapor separator tank assembly, attached to the side of the throttle body. The throttle arm/lever is about halfway back on the port side, directly below the alternator. The idle stop screw is about 3/4ths of way down the throttle lever, threaded horizontally with the head facing forward and the thread facing aft.

While holding the bottom of the throttle arm aftward against the idle stop, check the gap between the throttle cam and the roller on the cam follower. It should be 0.005-0.020 in. (0.13-0.51mm). If not, adjust it as follows:

1. Using a suitable tamper-resistant Torx bit loosen the cam follower roller screw so that the follower moves freely.
2. Locate the idle stop screw (found about 3/4ths of the way down the throttle lever), then loosen the locknut.
3. Rest the roller on the throttle cam, then turn the idle stop screw until the curved front end of the cam aligns with the center of the roller. Then retighten the idle stop screw jam nut.
4. Hold the throttle arm against the idle stop, then reposition the cam follower to achieve the proper 0.005-0.020 in. (0.13-0.51mm) gap between the roller and throttle cam. Holding the follower in this position, retighten the T25 tamper resistant Torx screw securing the follower.

IGNITION TIMING CHECK

All ignition timing control and adjustment is handled automatically by the engine control module (ECM). Because of this reason there are normally no timing marks or pointer on the motor, so the only method available to check timing is to use a scan tool such as Mercury's digital diagnostic terminal (DDT).

Timing will vary at idle depending upon whether the motor is in gear or in neutral. However, generally speaking the motor will use ignition timing to run at about 700 rpm in neutral and 625 rpm running in forward gear. Though

Fig. 207 Throttle cam and follower (EFI shown, Optimax very similar)

Fig. 208 Snap-on TTXR25E (T25 Tamper Resistant Torx Fastener Bit)

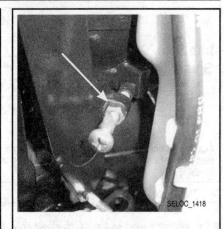

Fig. 209 Idle stop screw

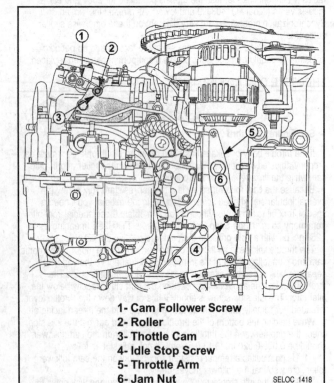

1- Cam Follower Screw
2- Roller
3- Thottle Cam
4- Idle Stop Screw
5- Throttle Arm
6- Jam Nut

Fig. 210 Throttle cam adjustment (EFI shown, Optimax very similar)

keep in mind that timing will be more advanced during warm-up or cold motor operation. Once the motor reaches normal operating temperatures the idle timing will usually vary from 1° to 7° After Top Dead Center (ATDC).

WOT timing will vary from about 18° to 20° Before Top Dead Center (BTDC).

MAXIMUM THROTTLE ADJUSTMENT

◆ See Figures 211, 212 and 213

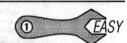

The throttle linkage is all located on the port side of the motor. The throttle cam and cam follower are at the front port side, just above the vapor separator tank assembly, attached to the side of the throttle body. The cam follower is attached to the throttle arm coming out of the throttle body.

The throttle arm/lever is about halfway back on the port side, directly below the alternator. The full throttle stop screw is threaded horizontally in a boss on the side of the powerhead, in such a position that the bottom of the bolts threads will contact the throttle lever on the opposite side of the lever

from where the idle stop screw threads (and a tiny bit lower on the lever).

While holding the bottom of the throttle arm forward against the full throttle stop screw, check the gap between the arm on the throttle shaft and the stop on the throttle body assembly. It should be about 0.020 in. (0.51mm). If not, adjust it as follows:

1. Loosen the jam nut on the full throttle stop screw.
2. Move the throttle arm so that it is resting against the end of the full throttle stop screw, then adjust the screw so that the proper 0.020 in. (0.51mm) gap is measured up at the throttle body, between the throttle arm and the stop.
3. Retighten the locknut on the full throttle stop screw, then check to make sure there is still a slightly free-play between the roller and cam at WOT to prevent the linkage from binding in this position. Readjust as necessary to prevent that potentially dangerous condition of linkage binding at WOT.

THROTTLE POSITION SENSOR

The throttle position sensor is mounted in a fixed position on these models and, as such, is not adjustable. However, function can be monitored through the use of a scan tool such as Mercury's Digital Diagnostic Terminal (DDT) or by using a home-made jumper harness to complete the circuit while still allowing points for a DVOM to probe.

Voltage at idle should be about 0.19-1.0 VDC, while at WOT it should be about 3.45-4.63 VDC. Also, as the throttle is slowly opened from idle to WOT or closed, the meter should show smooth and progressive changes in the voltage.

If voltage is out of range and there is no problem with the wire harness OR the meter shows erratic changes through the throttle range, the sensor should be replaced.

IDLE SPEED

Like the ignition timing, idle speed automatically controlled by the engine control module (ECM). Generally speaking the motor will use ignition timing to run at about 700 rpm in neutral and 625 rpm running in forward gear. If the engine does not idle properly, start by looking for mechanical causes such as worn plugs or a vacuum leak, before suspecting the electrical system.

THROTTLE CABLE INSTALLATION

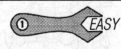

◆ See Figures 205 and 214

1. If removed, connect the end of the throttle cable to the throttle lever.
2. Move the throttle lever against the idle stop screw and hold in this position. Adjust the throttle cable barrel until it will slip into the barrel recess in the control cable anchor bracket with only a light preload against the idle stop.

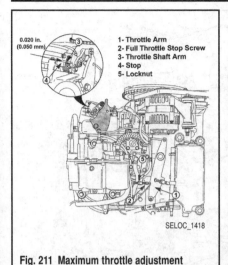

Fig. 211 Maximum throttle adjustment

1- Throttle Arm
2- Full Throttle Stop Screw
3- Throttle Shaft Arm
4- Stop
5- Locknut

Fig. 212 Check for proper gap with no binding at WOT (EFI shown, Optimax similar)

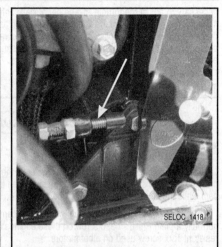

Fig. 213 Full throttle stop screw

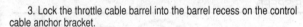

Fig. 214 Throttle and shift cable barrel retainer

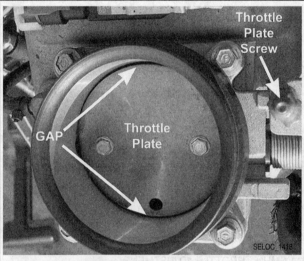

Fig. 215 Checking/adjusting throttle plate gap

3. Lock the throttle cable barrel into the barrel recess on the control cable anchor bracket.

■ An excessive preload on the throttle cable will cause difficulty when shifting from Forward gear into Neutral.

4. The preload may be easily checked by placing a piece of paper between the idle stop screw and the idle stop and then withdrawing it by pulling on the end of the paper. If the paper does not tear but drag can be felt, the preload is correct.

5. If necessary, adjust the cable barrel, if necessary, to obtain the proper preload just described.

6. Install the powerhead cowling.

THROTTLE PLATE SCREW

◆ See Figure 215

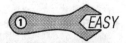

✳✳ SELOC CAUTION

This should almost NEVER be touched, unless for some reason the throttle plate screw has been disturbed from the factory setting.

Should the throttle plate require adjustment, use the screw on the throttle arm to set the plate position so there is a TOTAL throttle plate clearance of 0.040 in. (1.0mm), which translates to a gap of about 0.020 in. (0.50mm) on either side of the plate, as illustrated.

135-175 Hp·(2.5L) V6 OPTIMAX Models

The ignition and fuel systems on these models are controlled by the engine control module (ECM) which performs the following functions:

• Calculates the precise fuel and ignition timing requirements based on engine speed, throttle position, manifold pressure and coolant temperature.

• Controls the fuel injectors for each cylinder, direct injectors for each cylinder and ignition for each cylinder.

• Controls and monitors the Oil Injection system

• Monitors the air compressor system.

• Controls all alarm horn and warning lamp functions.

• Supplies the tachometer signal to the gauge.

• Controls the rpm limiter function.

• Records engine running information.

Because of the vast extent of electronic monitoring and control MOST functions, including Ignition Timing and Idle Speed are fully under the control of the ECM and are not adjustable. Idle rpm can be spot-checked using a tachometer, but any operation out of specification must be troubleshot using the information on the Optimax system from the Fuel System section (unless a mechanical condition such as throttle cable binding can be found and attributed to the problem).

Unlike some fuel systems, both the throttle position sensor (TPS) and crankshaft position sensor (CPS), two vital inputs to the Optimax system, are NOT adjustable.

This means that the only potential adjustment for Timing & Synchronization on these motors is the throttle linkage (throttle cam, maximum throttle setting and throttle plate). And that said, it should NOT be necessary to adjust the throttle linkage periodically only to check and adjust it after some component of the throttle system (such as the cables themselves) have been replaced. However, if the powerhead is not operating properly, a simple check of the throttle linkage adjustment is a good way to eliminate one potential mechanical cause for trouble.

THROTTLE CAM ADJUSTMENT

◆ **See Figures 207, 209 and 210**

If the throttle body or linkage has been replaced the throttle cam must be properly adjusted so there is the proper gap between the roller and throttle cam while the linkage is in the idle position.

Unlike the EFI versions of the same motors the cam follower screw is normally a plain Philip's head screw on these engines and not the tamper-resistant Torx screw used on other motors.

The throttle linkage is all located on the port side of the motor. The throttle cam and cam follower are at the front port side, just above the vapor separator tank assembly, attached to the side of the throttle body. The throttle arm/lever is about halfway back on the port side, directly below the alternator. The idle stop screw is about 3/4ths of way down the throttle lever, threaded horizontally with the head facing forward and the thread facing aft.

While holding the bottom of the throttle arm aftward against the idle stop, check the gap between the throttle cam and the roller on the cam follower. It should be 0.00-0.10 in. (0.00-0.25mm). If not, adjust it as follows:

1. Loosen the cam follower roller screw so that the follower moves freely.

2. Locate the idle stop screw (found about 3/4ths of the way down the throttle lever), then loosen the locknut.

3. Rest the roller on the throttle cam, then turn the idle stop screw until the curved front end of the cam aligns with the center of the roller. Then retighten the idle stop screw jam nut.

4. Hold the throttle arm against the idle stop, then reposition the cam follower to achieve the proper 0.00-0.10 in. (0.00-0.25mm) gap between the roller and throttle cam. Holding the follower in this position, retighten the screw securing the follower.

MAXIMUM THROTTLE ADJUSTMENT

◆ **See Figures 212 and 216**

The throttle linkage is all located on the port side of the motor. The throttle cam and cam follower are at the front port side, just above the vapor separator tank assembly, attached to the side of the throttle body. The cam follower is attached to the throttle arm coming out of the throttle body.

The throttle arm/lever is about halfway back on the port side, directly below the alternator. The full throttle stop screw is threaded horizontally in a boss on the side of the powerhead, in such a position that it is almost fully obscured by the external electric fuel pump mounted to the end of the vapor separator tank assembly.

While holding the bottom of the throttle arm forward against the full throttle stop screw, check the gap between the arm on the throttle shaft and the stop on the attenuator box. It should be about 0.020 in. (0.51mm). If not, adjust it as follows:

1. Loosen the jam nut on the full throttle stop screw.

2. Move the throttle arm so that it is resting against the end of the full throttle stop screw, then adjust the screw so that the proper 0.020 in. (0.51mm) gap is measured up at the throttle body, between the throttle arm and the stop.

3. Retighten the locknut on the full throttle stop screw, then check to make sure there is still a slightly free-play between the roller and cam at WOT to prevent the linkage from binding in this position. Readjust as necessary to prevent that potentially dangerous condition of linkage binding at WOT.

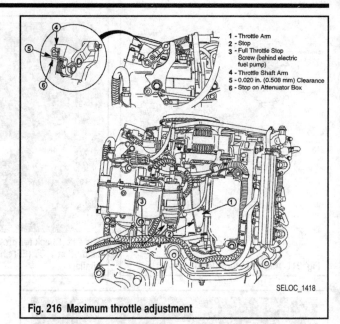

1 - Throttle Arm
2 - Stop
3 - Full Throttle Stop Screw (behind electric fuel pump)
4 - Throttle Shaft Arm
5 - 0.020 in. (0.508 mm) Clearance
6 - Stop on Attenuator Box

SELOC_1418

Fig. 216 Maximum throttle adjustment

THROTTLE POSITION SENSOR

The throttle position sensor is mounted in a fixed position on these models and, as such, is not adjustable. However, function can be monitored through the use of a scan tool such as Mercury's Digital Diagnostic Terminal (DDT) or by using a home-made jumper harness to complete the circuit while still allowing points for a DVOM to probe.

Voltage at idle should be about 3.20-4.90 VDC, while at WOT it should be about 0.10-1.50 VDC. Also, as the throttle is slowly opened from idle to WOT or closed, the meter should show smooth and progressive changes in the voltage.

If voltage is out of range and there is no problem with the wire harness OR the meter shows erratic changes through the throttle range, the sensor should be replaced.

IDLE SPEED

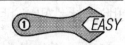

Like the ignition timing, idle speed automatically controlled by the engine control module (ECM). Generally speaking the motor will use ignition timing to run at about 550 rpm. If the engine does not idle properly, start by looking for mechanical causes such as worn plugs or a vacuum leak, before suspecting the electrical system.

THROTTLE PLATE SCREW

◆ **See Figures 217 and 218**

✳✳ **SELOC CAUTION**

This should almost NEVER be touched, unless for some reason the throttle plate screw has been disturbed from the factory setting.

Should the throttle plate require adjustment, use the screw on the throttle arm and a #68 drill bit as a spacer to set the plate position so there is a gap (clearance) of about 0.031 in. (0.7937mm).

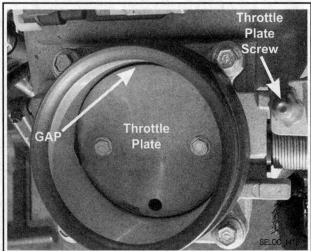

Fig. 217 On these models set the gap on one side of the plate using a #68 drill bit

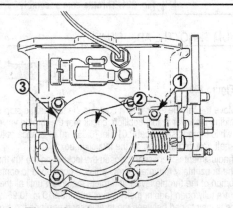

1 - Throttle plate stop screw
2 - Throttle plate clearance
3 - Grommet

SELOC_1418

Fig. 218 On 135 hp models a grommet must be installed for the motor to run properly

225-250 Hp (3.0L) V6 EFI - 2001 Models Only

The EFI models built through serial #OT408999 (for the 2001 or earlier model years) have a radically different throttle body/air induction assembly (and therefore radically different linkage) than the EFI models for 2002 and later and serial #OT409000 or higher. As a result the Timing & Synchronization procedures are very different on these otherwise similar outboards.

TIMER POINTER ADJUSTMENT

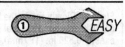

OEM ② MODERATE

◆ See Figure 219

Before the ignition timing can be checked (it is automatically controlled by the ECM and it cannot be adjusted) you need to verify that the timing pointer is properly aligned and has not been moved or has not shifted slightly during service. In order to do this you need a dial gauge to perform precision measurements on piston position.

1. For access, remove the flywheel/alternator cover by lifting the rear of the cover, then sliding the cover back and off the attenuator.

2. Remove all spark plugs.

3. Install a dial indicator into the No. 1 (top cylinder starboard bank), spark plug opening.

4. Slowly rotate the flywheel CLOCKWISE until the No. 1 piston is at top dead center (TDC). Set the dial indicator to "0".

5. Slowly rotate the flywheel COUNTERCLOCKWISE until the dial indicator needle is 1/4 turn beyond the 0.526 in. (11.7mm) mark.

6. Slowly rotate the flywheel back - CLOCKWISE - until the dial indicator reads exactly 0.526 in. (11.7mm). Observe the timing pointer on the flywheel cover and the 0.526 in. (11.7mm) mark on the flywheel.

7. If the flywheel pointer is not exactly on the mark, loosen the pointer adjustment screws and align the pointer with the 0.526 in. (11.7mm) mark.

8. Tighten the pointer adjustment screws and remove the dial indicator from the No. 1 spark plug opening.

9. During the Idle Speed check and adjustment later in this section, also connect a timing light to the No. 1 spark plug lead in order to check the idle and WOT ignition timing. Refer to the Tune-Up specifications chart in this section to see if the motor is operating within the appropriate parameters. If not, look for mechanical causes before troubleshooting the EFI system.

CRANKSHAFT POSITION SENSOR ADJUSTMENT

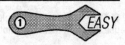

① EASY

◆ See Figure 220

1. If not already done to check/reposition the ignition timer pointer, remove the flywheel/alternator cover by lifting the rear of the cover, then sliding the cover back and off the attenuator.

2. Rotate the flywheel slowly clockwise by hand and align 1 of the flywheel sensor teeth directly in line with (parallel with the end of) the crankshaft position sensor. Stated another way the protrusion of the sensor body must be perpendicular to the flywheel tooth.

3. Using a feeler gauge, measure the gap between the sensor and flywheel tooth. The allowable gap measurement is 0.020-0.060 in. (0.51-1.53mm). If the gap is incorrect, loosen the 2 sensor bracket bolts.

4. Set the gap to the correct measurement and tighten the 2 sensor bracket bolts to 105 in. lbs. (11.8 Nm).

5. Install the flywheel cover.

THROTTLE CAM ADJUSTMENT

① EASY

◆ See Figure 221

If the throttle body or linkage has been replaced the throttle cam must be properly adjusted so there is the proper gap between the roller and throttle cam while the linkage is in the idle position.

Unlike some later EFI versions of the same motors the cam follower screw is normally a plain Philip's head screw on these engines and not the tamper-resistant Torx screw used on other motors.

The throttle linkage is all located on the port side of the motor. The throttle cam and cam follower are at the front port side, just below the vapor separator tank assembly, attached to the side of the throttle body. The throttle arm/lever is about halfway back on the port side. The idle stop screw is about 3/4ths of way down the throttle lever, threaded horizontally with the head facing forward and the thread facing aft.

While holding the bottom of the throttle arm aftward against the idle stop, check the alignment of and gap between the throttle cam and the roller on the cam follower. The mark on the cam should be aligned with the center of the roller and there should be a 0.00-0.10 in. (0.00-0.26mm) gap. If not, adjust it as follows:

1. Loosen the cam follower roller screw so that the follower moves freely.

2. Locate the idle stop screw (found about 3/4ths of the way down the throttle lever), then loosen the locknut.

3. Rest the roller on the throttle cam, then turn the idle stop screw until the mark on the throttle cam is aligned with the center of the roller on the cam follower. Then retighten the idle stop screw jam nut.

4. Hold the throttle arm against the idle stop, then reposition the cam follower to achieve the proper 0.00-0.10 in. (0.00-0.26mm) gap between the roller and throttle cam. Holding the follower in this position, retighten the screw securing the follower.

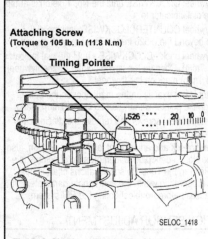

Attaching Screw (Torque to 105 lb. in (11.8 N.m))

Timing Pointer

SELOC_1418

Fig. 219 Slowly rotate the flywheel CLOCKWISE until the dial indicator reads exactly 0.526 in. (11.7mm), then check the timing pointer to make sure it is aligned with the 0.526 in. (11.7 mm) mark on the flywheel

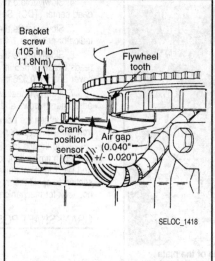

Bracket screw (105 in lb 11.8Nm)

Flywheel tooth

Crank position sensor

Air gap (0.040" +/- 0.020")

SELOC_1418

Fig. 220 After removing the flywheel cover it is possible to check the gap on the crankshaft position sensor

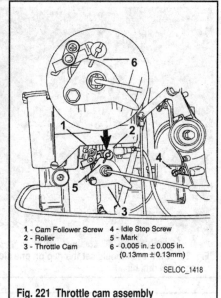

1 - Cam Follower Screw
2 - Roller
3 - Throttle Cam
4 - Idle Stop Screw
5 - Mark
6 - 0.005 in. ± 0.005 in. (0.13mm ± 0.13mm)

SELOC_1418

Fig. 221 Throttle cam assembly

THROTTLE VALVE & OIL PUMP SYNCHRONIZATION

◆ See Figure 222

This adjustment is necessary for models equipped with a mechanical oil pump assembly.

1. Position and hold the throttle lever against the idle stop. Verify the alignment mark on the oil pump lever is aligned with the casting mark on the oil pump body.

■ There are usually multiple marks on these pump levers. Be sure to use the mark that is positioned the furthest clockwise from the linkage.

2. If not, disconnect the linkage rod from the oil pump lever.

3. Screw the rod end, in or out, to align the oil pump lever mark with the mark on the body casting.

4. Connect the rod end onto the oil pump lever ball.

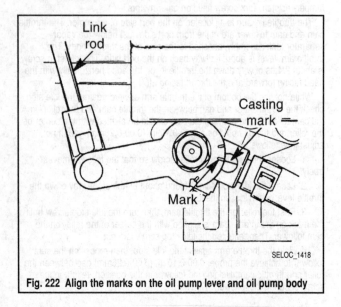

Link rod

Casting mark

Mark

SELOC_1418

Fig. 222 Align the marks on the oil pump lever and oil pump body

MAXIMUM THROTTLE ADJUSTMENT

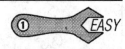

◆ See Figure 223

1. Move the throttle lever until it contacts the full throttle stop screw. While holding the lever in this position, check that the throttle valves are fully opened while there is still about 0.010 in. (0.25mm) clearance between the throttle shaft arm and the stop on the induction box.

2. If adjustment is necessary, loosen the locknut on the full throttle stop screw (the horizontally mounted screw which is positioned to contact the upper portion of the throttle lever) and rotate the screw until all the throttle shutters are fully open (again, making sure about 0.010 in. (0.25mm) of clearance is maintained between the throttle shaft arm and the stop on the manifold.).

3. Check for free-play at the cam follower roller and the throttle cam. This indicates there is no binding of the throttle linkage. Readjust the full throttle stop screw, if required.

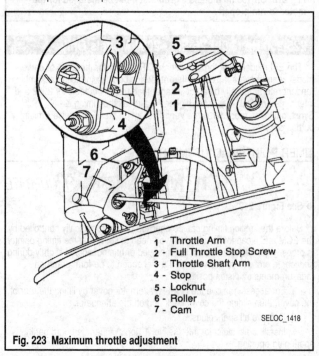

1 - Throttle Arm
2 - Full Throttle Stop Screw
3 - Throttle Shaft Arm
4 - Stop
5 - Locknut
6 - Roller
7 - Cam

SELOC_1418

Fig. 223 Maximum throttle adjustment

THROTTLE POSITION SENSOR (TPS) ADJUSTMENT

◆ **See Figure 224**

Alignment and adjustment of the TPS is a critical step in the timing and synchronizing of the powerhead (IF the switch has been disturbed or you suspect out of range operation). This procedure requires the use of a meter (digital or analog, though the later can be VERY difficult to read the low-voltage output of the sensor at certain throttle positions) and special in-line harness connector (or a home-made inline jumper harness) to monitor the TPS voltage output during throttle operation.

1. The throttle position sensor (TPS) is used for low speed and mid-range power settings. The TPI transmits throttle position information to the ECU controller in the form of a low voltage signal. This low voltage signal will range from about 0.90 VDC at the idle position to 4.05 VDC at wide open throttle setting. A lower voltage setting will cause the powerhead timing to retard and a higher voltage will advance the timing. Typically, a 0.050 volt change in the TPI setting, results in a 1° change in the powerhead timing.

■ **The TPS is a small cylindrical-bodied sensor mounted to the side of the motor in such a position where it can be directly affected by positioning of the throttle shaft. It is normally secured by 2 Phillip's head screws.**

2. Locate the throttle position sensor (a small, cylindrical sensor mounted on the side of the powerhead in a position where it can be moved by the throttle linkage), then trace the wiring and disconnect it from the EFI harness.

3. Install the TPS Test Lead Assembly #84-825207A1, or an equivalent jumper harness which allows probing a completed circuit for the TPS, in between the TPS connector and the EFI harness connector.

4. Connect a voltmeter to the test harness with the meter set to the 20 volts DC scale.

5. Turn the ignition keyswitch to the **ON** position, but don't start the motor.

6. Check the DVOM for voltage readings with the throttle both closed to the idle position and open to the Wide Open Throttle (WOT) position. It should read 0.90-1.00 at idle and approximately 3.55-4.05 volts at WOT.

7. If the DVOM is not showing readings within specifications, first loosen the 2 mounting screws slightly, then rotate the sensor fully clockwise (while holding the throttle shaft in the closed position). Then reposition the TPS as necessary so that readings are within specification and retighten the mounting screws.

■ **If the engine appears to run too rich or too lean the TPS can be readjusted. Decreasing voltage will result in a LEANER mixture, while increasing voltage will result in a RICHER mixture.**

8. Again, slowly open the throttle from idle to the WOT position and back to idle again, watching the DVOM to make sure voltage readings increase and decrease smoothly.

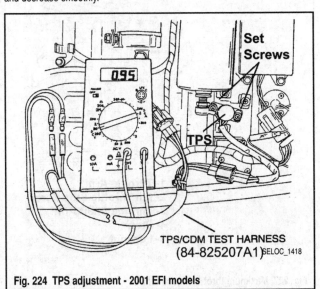

Fig. 224 TPS adjustment - 2001 EFI models

9. Once adjustment is finished shut the ignition "OFF", then disconnect the test leads and reconnect the TPS connector to the EFI harness.

IDLE SPEED

◆ **See Figure 225**

■ **The following procedures must be performed with the outboard in a test tank or the boat and outboard in a body of water. The idle speed adjustment procedures can only be performed with the outboard running with the lower unit in Forward gear and the propeller under an actual load condition.**

✳✳ SELOC WARNING

DO not connect a flush device to the lower unit for this adjustment, because the outboard unit must be under load with the lower unit in Forward gear.

1. If timing is also going to be checked, connect a timing light to the powerhead No. 1 spark plug lead.

2. If removed to set the timing pointer, install and tighten the spark plugs, then connect the plug leads.

3. Connect a tachometer to the powerhead.

4. Connect the fuel line from the fuel tank to the powerhead and prime the fuel system.

5. Start the powerhead and allow it to warm to operating temperature. Place the outboard into Forward gear and monitor the powerhead rpm. It should about 600-700 rpm in gear.

6. If the powerhead rpm is not correct adjust the idle air flow screw at the lower throttle shaft on the PORT side of the induction manifold, then secure the screw using the jam nut.

7. After adjusting idle speed, recheck and if necessary reset, the Throttle Position Sensor, as detailed earlier in this section.

8. Place the throttle in Neutral and disconnect the fuel line to the outboard unit. If the outboard is not to be used for some time, allow the outboard to operate until all fuel is used in the carburetors. Set the key switch to the **OFF** position. Remove the test equipment.

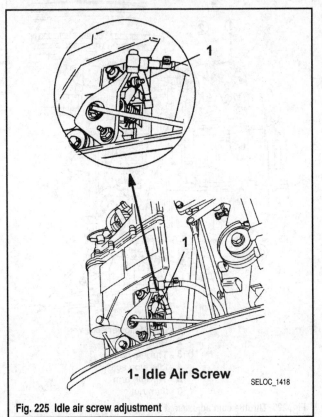

Fig. 225 Idle air screw adjustment

200-250 Hp (3.0L) V6 EFI - 2002 & Later Models

The EFI models for 2002 and later and serial #OT409000 or higher have a radically different throttle body/air induction assembly (and therefore radically different linkage) than the models produced for 2001 and earlier models years (serial #OT408999 and lower). As a result the Timing & Synchronization procedures are very different on these otherwise similar outboards.

All throttle linkage adjustments are set at the factory and no periodic adjustments should be necessary, unless a component has been replaced.

THROTTLE CAM ADJUSTMENT

◆ See Figures 207, 208 and 226

If the throttle body or linkage has been replaced the throttle cam must be properly adjusted so there is the proper gap between the roller and throttle cam while the linkage is in the idle position.

Because the cam follower screw is a tamper-resistant Torx screw (an internal lobular star shaped screw with a pin in the middle) you'll need a special Torx bit to loosen it. Although you can obtain these special Torx bits from many sources, the Snap-On part number is TTXR25E (meaning it's a T25 Torx bit with a hole drilled in the center to receive the pin).

The throttle linkage is all located on the port side of the motor. The throttle cam and cam follower are at the front port side, just above the vapor separator tank assembly, attached to the side of the throttle body. The throttle arm/lever is about halfway back on the port side, directly below the alternator. The idle stop screw is about 3/4ths of way down the throttle lever, threaded pretty much horizontally (maybe tilted back-end up slightly) with the head facing forward and the threads facing aft.

While holding the bottom of the throttle arm aftward against the idle stop, check the gap between the throttle cam and the roller on the cam follower. It should be 0.005-0.020 in. (0.13-0.51mm). If not, adjust it as follows:

1. Using a suitable tamper-resistant Torx bit loosen the cam follower roller screw so that the follower moves freely.

2. Locate the idle stop screw (found about 3/4ths of the way down the throttle lever), then loosen the locknut.

3. Rest the roller on the throttle cam, then turn the idle stop screw until the curved front end of the cam aligns with the center of the roller. Then retighten the idle stop screw jam nut.

4. Hold the throttle arm against the idle stop, then reposition the cam follower to achieve the proper 0.005-0.020 in. (0.13-0.51mm) gap between the roller and throttle cam. Holding the follower in this position, retighten the T25 tamper resistant Torx screw securing the follower.

IGNITION TIMING CHECK

All ignition timing control and adjustment is handled automatically by the engine control module (ECM). Because of this reason there are normally no timing marks or pointer on the motor, so the only method available to check timing is to use a scan tool such as Mercury's Digital Diagnostic Terminal (DDT).

Timing will vary at idle depending upon whether the motor is in gear or in neutral. However, generally speaking the motor will use ignition timing to run at about 700 rpm in neutral and 625 rpm running in forward gear. Though keep in mind that timing will be more advanced during warm-up or cold motor operation. Once the motor reaches normal operating temperatures the idle timing will usually vary from 0° to 6° Before Top Dead Center (BTDC).

WOT timing will be approximately 22° Before Top Dead Center (BTDC).

MAXIMUM THROTTLE ADJUSTMENT

◆ See Figures 212 and 227

The throttle linkage is all located on the port side of the motor. The throttle cam and cam follower are at the front port side, just above the vapor separator tank assembly, attached to the side of the throttle body. The cam follower is attached to the throttle arm coming out of the throttle body.

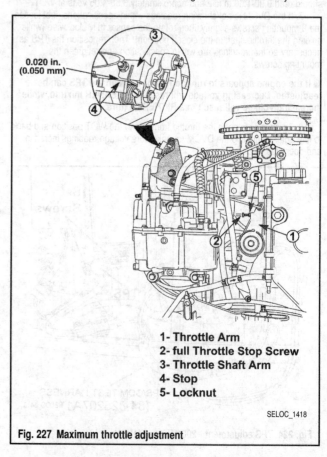

1 - Throttle Arm
2 - full Throttle Stop Screw
3 - Throttle Shaft Arm
4 - Stop
5 - Locknut

SELOC_1418

Fig. 227 Maximum throttle adjustment

1 - Cam follower screw
2 - Roller
3 - Throttle cam
4 - Idle stop screw
5 - Throttle arm
6 - Jam nut

SELOC_1418

Fig. 226 Throttle cam adjustment

The throttle arm/lever is about halfway back on the port side, directly below the alternator. The full throttle stop screw is threaded horizontally in a boss on the upper half of the throttle lever, in such a position that the bottom of the bolts threads will contact the powerhead, limiting throttle lever movement when the top of the lever is moved aftward.

While holding the top of the throttle lever aftward against the full throttle stop screw, check the gap between the arm on the throttle shaft and the stop on the throttle body assembly. It should be about 0.020 in. (0.50mm). If not, adjust it as follows:

1. Loosen the jam nut on the full throttle stop screw.

2. Move the throttle arm so that it is resting against the end of the full throttle stop screw, then adjust the screw so that the proper 0.020 in. (0.50mm) gap is measured up at the throttle body, between the throttle arm and the stop.

3. Retighten the locknut on the full throttle stop screw, then check to make sure there is still a slightly free-play between the roller and cam at WOT to prevent the linkage from binding in this position. Readjust as necessary to prevent that potentially dangerous condition of linkage binding at WOT.

THROTTLE POSITION SENSOR

The throttle position sensor is mounted in a fixed position on these models and, as such, is not adjustable. However, function can be monitored through the use of a scan tool such as Mercury's Digital Diagnostic Terminal (DDT) or by using a home-made jumper harness to complete the circuit while still allowing points for a DVOM to probe.

Voltage at idle should be about 0.19-1.0 VDC, while at WOT it should be about 3.45-4.63 VDC. Also, as the throttle is slowly opened from idle to WOT or closed, the meter should show smooth and progressive changes in the voltage.

If voltage is out of range and there is no problem with the wire harness OR the meter shows erratic changes through the throttle range, the sensor should be replaced.

IDLE SPEED

Like the ignition timing, idle speed automatically controlled by the engine control module (ECM). Generally speaking the motor will use ignition timing to run at about 700 rpm in neutral and 625 rpm running in forward gear. If the engine does not idle properly, start by looking for mechanical causes such as worn plugs or a vacuum leak, before suspecting the electrical system.

THROTTLE CABLE INSTALLATION

◆ **See Figures 205 and 214**

1. If removed, connect the end of the throttle cable to the throttle lever.

2. Move the throttle lever against the idle stop screw and hold in this position. Adjust the throttle cable barrel until it will slip into the barrel recess in the control cable anchor bracket with only a light preload against the idle stop.

3. Lock the throttle cable barrel into the barrel recess on the control cable anchor bracket.

■ **An excessive preload on the throttle cable will cause difficulty when shifting from Forward gear into Neutral.**

4. The preload may be easily checked by placing a piece of paper between the idle stop screw and the idle stop and then withdrawing it by pulling on the end of the paper. If the paper does not tear but drag can be felt, the preload is correct.

5. If necessary, adjust the cable barrel, if necessary, to obtain the proper preload just described.

6. Install the powerhead cowling.

THROTTLE PLATE SCREW

◆ **See Figures 215**

Should the throttle plate require adjustment, use the screw on the throttle arm to set the plate position so there is a TOTAL throttle plate clearance of 0.040 in. (1.0mm), which translates to a gap of about 0.020 in. (0.50mm) on either side of the plate, as illustrated.

200-250 Hp (3.0L) V6 OPTIMAX Models

The ignition and fuel systems on these models are controlled by the engine control module (ECM) which performs the following functions:

• Calculates the precise fuel and ignition timing requirements based on engine speed, throttle position, manifold pressure and coolant temperature.

• Controls the fuel injectors for each cylinder, direct injectors for each cylinder and ignition for each cylinder.

• Controls and monitors the Oil Injection system

• Monitors the air compressor system.

• Controls all alarm horn and warning lamp functions.

• Supplies the tachometer signal to the gauge.

• Controls the rpm limiter function.

• Records engine running information.

Because of the vast extent of electronic monitoring and control, MOST functions, including ignition timing and idle speed, are fully under the control of the ECM and are not adjustable. Idle rpm can be spot-checked using a tachometer or a scan tool, but any operation out of specification must be troubleshot using the information on the Optimax system from the Fuel System section (unless a mechanical condition such as throttle cable binding can be found and attributed to the problem).

Unlike some fuel systems the throttle position sensor (TPS) is mounted in a fixed position and is NOT adjustable. However, unlike some other Optimax systems, the crankshaft position sensor (CPS), another of the vital inputs to the Optimax system IS adjustable, so it is the first thing that you should check before beginning the Timing and Synchronization procedures.

Because so many items (like timing and idle speed are computer controlled) the only potential adjustment for Timing & Synchronization on these motors (other than the CPS) is the throttle linkage (throttle cam, maximum throttle setting and throttle plate). And that said, it should NOT be necessary to adjust the throttle linkage periodically only to check and adjust it after some component of the throttle system (such as the cables themselves) have been replaced. However, if the powerhead is not operating properly, a simple check of the throttle linkage adjustment is a good way to eliminate one potential mechanical cause for trouble.

CRANKSHAFT POSITION SENSOR ADJUSTMENT

◆ **See Figures 228, 229 and 230**

■ **On some models, such as the Pro/XS/Sport motors, the CPS is mounted in a fixed position and cannot be adjusted.**

1. Remove the flywheel cover for access.

2. Rotate the flywheel slowly clockwise by hand and align 1 of the flywheel sensor teeth directly in line with (parallel with the end of) the crankshaft position sensor. Stated another way the protrusion of the sensor body must be perpendicular to the flywheel tooth.

3. Using a feeler gauge, measure the gap between the sensor and flywheel tooth. The allowable gap measurement is 0.025-0.040 in. (0.64-1.02 mm). If the gap is incorrect, loosen the 2 sensor bracket bolts.

4. Set the gap to the correct measurement and tighten the 2 sensor bracket bolts to 105 in. lbs. (11.8 Nm).

5. Install the flywheel cover.

Fig. 228 Early-model CPS

Fig. 229 Late-model CPS

Fig. 230 Crankshaft Position Sensor Gap

THROTTLE CAM ADJUSTMENT

◆ See Figures 208 and 231 thru 234

If the throttle body or linkage has been replaced the throttle cam must be properly adjusted so there is the proper gap between the roller and throttle cam while the linkage is in the idle position.

Although the Mercury factory information does not mention the use of any tamper resistant fasteners (and on the contrary even shows picture of normal Philip's head screws for the cam follower screw), we've found that some of our test engines used a tamper-resistant Torx screw (an internal lobular star shaped screw with a pin in the middle). If your model is so equipped, you'll need a special Torx bit to loosen it. Although you can obtain these special Torx bits from many sources, the Snap-On part number is TTXR25E (meaning it's a T25 Torx bit with a hole drilled in the center to receive the pin).

The throttle linkage is all located on the port side of the motor. The throttle cam and cam follower are at the front port side, just above the vapor separator tank assembly on models through 2002 or just in front of and above the separator tank on 2003 or later models (but in all cases, attached to the side of the throttle body).

The throttle arm/lever is about halfway back on the port side, either right behind the vapor separator tank and fuel pump on models through 2002 or directly BEHIND the vapor separator tank on 2003 and later models. In all cases, the idle stop screw is about 3/4ths of way down the throttle lever, threaded horizontally with the head facing forward and the thread facing aft.

While holding the bottom of the throttle arm aftward against the idle stop, check the gap between the throttle cam and the roller on the cam follower. It should be 0.00-0.10 in. (0.00-0.25mm). If not, adjust it as follows:

1. Loosen the cam follower roller screw so that the follower moves freely.

2. Locate the idle stop screw (found about 3/4ths of the way down the throttle lever), then loosen the locknut.

3. Rest the roller on the throttle cam, then turn the idle stop screw until the curved front end of the cam aligns with the center of the roller. Then retighten the idle stop screw jam nut.

4. Hold the throttle arm against the idle stop, then reposition the cam follower to achieve the proper 0.00-0.10 in. (0.00-0.25mm) gap between the roller and throttle cam. Holding the follower in this position, retighten the screw securing the follower.

MAXIMUM THROTTLE ADJUSTMENT

◆ See Figures 235 thru 238

The throttle linkage is all located on the port side of the motor. The throttle cam and cam follower are at the front port side, just above the vapor separator tank assembly on models through 2002 or just in front of and above the separator tank on 2003 or later models (but in all cases, attached to the side of the throttle body).

The throttle arm/lever is about halfway back on the port side, either right behind the vapor separator tank and fuel pump on models through 2002 or directly BEHIND the vapor separator tank on 2003 and later models.

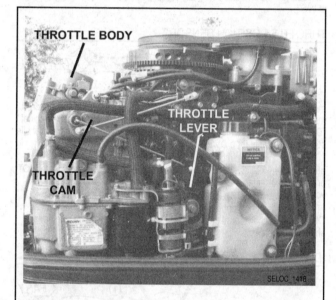

Fig. 231 Throttle cam linkage (early model through 2002 shown)

Fig. 232 Check the gap between the throttle cam and the roller on the follower

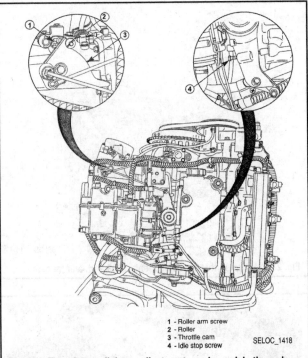

1 - Roller arm screw
2 - Roller
3 - Throttle cam
4 - Idle stop screw

SELOC_1418

Fig. 233 Throttle cam linkage adjustment - early models through 2002

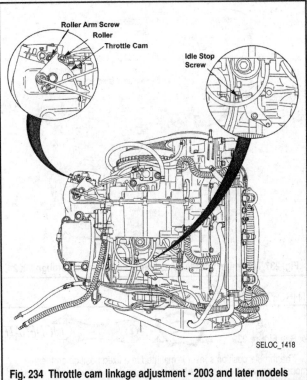

SELOC_1418

Fig. 234 Throttle cam linkage adjustment - 2003 and later models

On early models the full throttle stop screw is located at the top of the throttle lever, just below the linkage, with the head facing forward and the thread facing aft. In this position the bottom of the screw itself will move into contact with a stopper on the powerhead when the top of the lever is moved afterward (opening the throttle).

On some late model motors it appears that the screw may have been relocated to the side of the powerhead, just forward of the bottom of the lever, so that the lever itself will contact the screw when the top of the lever moves aftward and bottom of the lever moves forward (again, opening the throttle). This relocation was necessary because of the difference in vapor separator tank positioning (as it would now be right in front of the screw on early model motors).

While holding the top of the throttle arm rearward or the bottom of the throttle arm forward, as applicable, against the full throttle stop screw, check

the gap between the arm on the throttle shaft and the stop on the attenuator box. It should be about 0.010 in. (0.25mm). If not, adjust it as follows:

1. Loosen the jam nut on the full throttle stop screw.

2. Move the throttle arm so that it is resting against the end of the full throttle stop screw, then adjust the screw so that the proper 0.010 in. (0.25mm) gap is measured up at the throttle body, between the throttle arm and the stop.

3. Retighten the locknut on the full throttle stop screw, then check to make sure there is still a slightly free-play between the roller and cam (the roller will lift from cam) at WOT to prevent the linkage from binding in this position. Readjust as necessary to prevent that potentially dangerous condition of linkage binding at WOT.

SELOC_1418

Fig. 235 Full throttle stop screw on an early model (through 2002) motor

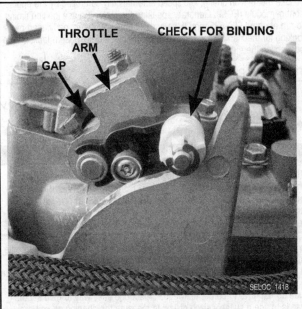

SELOC_1418

Fig. 236 Check for proper gap with no binding at WOT

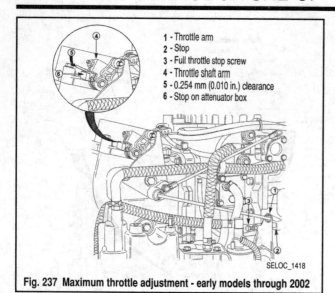

1 - Throttle arm
2 - Stop
3 - Full throttle stop screw
4 - Throttle shaft arm
5 - 0.254 mm (0.010 in.) clearance
6 - Stop on attenuator box

SELOC_1418

Fig. 237 Maximum throttle adjustment - early models through 2002

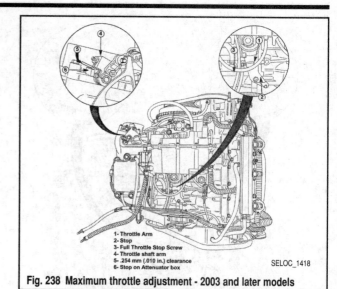

1- Throttle Arm
2- Stop
3- Full Throttle Stop Screw
4- Throttle shaft arm
5- .254 mm (.010 in.) clearance
6- Stop on Attenuator box

SELOC_1418

Fig. 238 Maximum throttle adjustment - 2003 and later models

THROTTLE POSITION SENSOR

◆ See Figure 239

 MODERATE

The throttle position sensor is mounted in a fixed position on these models and, as such, is not adjustable. However, function can be monitored through the use of a scan tool such as Mercury's Digital Diagnostic Terminal (DDT) or by using a home-made jumper harness to complete the circuit while still allowing points for a DVOM to probe.

Voltage at idle should be about 0.40-1.30 VDC, while at WOT it should be about 4.00-4.70 VDC. Also, as the throttle is slowly opened from idle to WOT or closed, the meter should show smooth and progressive changes in the voltage.

If voltage is out of range and there is no problem with the wire harness OR the meter shows erratic changes through the throttle range, the sensor should be replaced.

IDLE SPEED

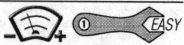 EASY

Like the ignition timing, idle speed automatically controlled by the engine control module (ECM). Generally speaking the motor will use ignition timing to run at about 550-575 rpm. If the engine does not idle properly, start by looking for mechanical causes such as worn plugs or a vacuum leak, before suspecting the electrical system.

THROTTLE PLATE SCREW

 EASY

◆ See Figure 240

✳✳ SELOC CAUTION

This should almost NEVER be touched, unless for some reason the throttle plate screw has been disturbed from the factory setting. As a matter of fact Mercury doesn't even provide specifications for this adjustment on their Pro/XS/Sport models and even say in their own technical literature to "call Mercury Racing Service for specifications."

Should the throttle plate require adjustment, use the screw on the throttle arm to set the plate position so there is a TOTAL throttle plate clearance of 0.149 in. (3.78mm), which translates to a gap of about 0.075 in. (1.89mm) on either side of the plate, as illustrated. Keep in mind that the gaps on either side do not have to be perfectly equal. Also, one easy way to help set the gap is to use a suitably sized drill bit to measure the opening as you are adjusting.

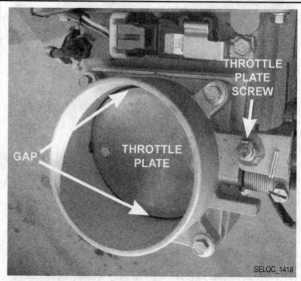

SELOC_1418

Fig. 239 Typical Mercury Optimax Throttle Position Sensor (TPS)

THROTTLE PLATE SCREW

GAP

THROTTLE PLATE

SELOC_1418

Fig. 240 Checking/adjusting throttle plate gap

STORAGE (WHAT TO DO BEFORE AND AFTER)

Winterization

◆ See Figure 238

Taking extra time to store the boat and motor properly at the end of each season or before any extended period of storage will greatly increase the chances of satisfactory service at the next season. Remember, that next to hard use on the water, the time spent in storage can be the greatest enemy of an outboard motor. Ideally, outboards should be used regularly. If weather in your area allows it, don't store the motor, enjoy it. Use it, at least on a monthly basis. It's best to enjoy and service the boat's steering and shifting mechanism several times each month. If a small amount of time is spent in such maintenance, the reward will be satisfactory performance, increased longevity and greatly reduced maintenance expenses.

But, in many cases, weather or other factors will interfere with time for enjoying a boat and motor. If you must place them in storage, take time to properly winterize the boat and outboard. This will be your best shot at making time stand still for them.

For many years there was a widespread belief simply shutting off the fuel at the tank and then running the powerhead until it stops constituted prepping the motor for storage. Right? Well, WRONG! (Even though this is still the method specifically recommended by Mercury for carbureted outboards).

First, it is not possible to remove all fuel in the carburetor or fuel injection system by operating the powerhead until it stops. Considerable fuel will remain trapped in the carburetor float chamber (or vapor separator tank of fuel injected motors) and other passages, especially in the lines leading to carburetors or injectors. The only guaranteed method of removing all fuel from a carbureted motor is to take the physically drain the carburetors from the float bowls. And, though you should drain the fuel from the vapor separator tank on most fuel injected motors, you still will not be able to remove ALL of it from the sealed high-pressure lines.

✳✳ SELOC CAUTION

On most late-model outboards Mercury warns that in areas gasoline containing alcohol (ethanol or methanol) is used, the alcohol content of the gasoline may cause formation of acid during storage. If the outboard is going to be stored for 2 months or more, they recommend draining as much of the remaining fuel as possible from the fuel tank, remote fuel line and engine system.

Depending upon the length of storage you can also use fuel stabilizer as opposed to draining the fuel system, but if the motor is going to be stored for more than a couple of months at a time, draining the system is really the better option.

Fig. 241 Add fuel stabilizer to the system anytime it will be stored without complete draining

■ Up here in the northeastern U.S., we always start adding fuel stabilizer to the fuel tank with every fuel fill up starting sometime in September. That helps to make sure that we'll be at least partially protected if the weather takes a sudden turn and we haven't had a chance to complete winterization yet.

Proper storage involves adequate protection of the unit from physical damage, rust, corrosion and dirt. The following steps provide an adequate maintenance program for storing the unit at the end of a season.

■ **If your outboard requires one or more repairs, PERFORM THEM NOW or during the off-season. Don't wait until the sun is shining, the weather is great and you want to be back on the water. That's not the time to realize you have to pull of the gearcase and replace the seals. Don't put a motor that requires a repair into storage unless you plan on making the repair during the off-season. It's too easy to let it get away from you and it will cost you in down time next season.**

PREPPING FOR STORAGE

Where to Store Your Boat and Motor

Ok, a well lit, locked, heated garage and work area is the best place to store you precious boat and motor, right? Well, we're probably not the only ones who wish we had access to a place like that, but if you're like most of us, we place our boat and motor wherever we can.

Of course, no matter what storage limitations are placed by where you live or how much space you have available, there are ways to maximize the storage site.

If possible, select an area that is dry. Covered is great, even if it is under a carport or sturdy portable structure designed for off-season storage. Many people utilize canvas and metal frame structures for such purposes. If you've got room in a garage or shed, that's even better. If you've got a heated garage, God bless you, when can we come over? If you do have a garage or shed that's not heated, an insulated area will help minimize the more extreme temperature variations and an attached garage is usually better than a detached for this reason. Just take extra care to make sure you've properly inspected the fuel system before leaving your boat in an attached garage for any amount of time.

If a storage area contains large windows, mask them to keep sunlight off the boat and motor otherwise, use a high-quality, canvas cover over the boat, motor and if possible, the trailer too. A breathable cover is best to avoid the possible build-up of mold or mildew, but a heavy duty, non-breathable cover will work too. If using a non-breathable cover, place wooden blocks or length's of 2 x 4 under various reinforced spots in the cover to hold it up off the boat's surface. This should provide enough room for air to circulate under the cover, allowing for moisture to evaporate and escape.

Whenever possible, avoid storing your boat in industrial buildings or parks areas where corrosive emissions may be present. The same goes for storing your boat too close to large bodies of saltwater. Hey, on the other hand, if you live in the Florida Keys, we're jealous again, just enjoy it and service the boat often to prevent corrosion from causing damage.

Finally, when picking a place to store your motor, consider the risk or damage from fire, vandalism or even theft. Check with your insurance agent regarding coverage while the boat and motor is stored.

Storage Checklist (Preparing the Boat & Motor)

◆ See Figures 242 thru 246

The amount of time spent and number of steps followed in the storage procedure will vary with factors such as the length of planed storage time and the conditions under which boat and motor are to be stored, as well as the particular model you are servicing (or more often it's fuel system Carb, EFI or Optimax) and your personal decisions regarding storage.

But, even considering the variables, plans can change, so be careful if you decide to perform only the minimal amount of preparation. A boat and motor that has been thoroughly prepared for storage can remain so with minimum adverse affects for as short or long a time as is reasonably necessary. The same cannot be said for a boat or motor on which important winterization steps were skipped.

Fig. 242 Be sure to fog the motor through the spark plug ports. . .

Fig. 243 . . .and drain the carb float bowls before storage

Fig. 244 Multiple carburetors will mean multiple float bowls to drain

Fig. 245 Most EFI and Optimax motor have a drain plug on the vapor separator tank, just like carburetor float bowls, they should be drained

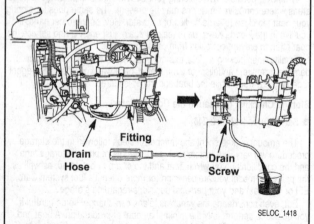

Fig. 246 Draining the water and fuel from the filter side of the vapor separator tank on 1.5L Optimax motors and 2003 or later 3.0L Optimax motors

Storage involves a number of things, but most importantly it means:
- Making sure the motor is properly drained of water which could freeze
- Making sure the motor is either drained of fuel or the fuel is treated.
- Making sure the motor is lubricated both inside and out.

When it comes to internal engine lubrication, Mercury has 2 different recommendations. For carbureted motors they recommend that you should use Storage Seal Rust Inhibitor. Mercury claims that Storage Seal is a blend of corrosion inhibiting additives designed to coat engine components during engine storage to reduce the formation of surface rust.

HOWEVER, it is important to note that Mercury DOES NOT recommend the use of Storage seal on EFI and Optimax motors. This is because Storage Seal uses a thick base designed to protect the internal engine surfaces. Mercury has noted that in some applications they've found a build-up of the thick base inside the injectors, causing the injectors to stick or operate incorrectly. For this reason, on EFI and Optimax motors Mercury recommends Quickleen and 2-cycle engine oil as a replacement for Storage Seal Rust Inhibitor. The Quickleen is used to clean the injectors, and the engine oil lubricates, as well as prevents rust from forming.

✳✳ SELOC CAUTION

If possible always store your Mercury vertically on the boat or on a suitable engine stand. Also, no motor should be placed on its side until after ALL water has drained, otherwise water may enter a cylinder through an exhaust port causing corrosion (or worse, may become trapped in a passage and freeze causing cracks in the powerhead or gearcase!). Check your owner's manual for more details if you need to store a motor on its side. Motors stored outside should not be tilted upward as water may collect in the propeller/exhaust outlet and freeze damaging the gearcase.

1. Thoroughly wash the boat motor and hull. Be sure to remove all traces of dirt, debris or marine life. Check the water stream fitting, water inlet(s) and, on jet models, the impeller grate for debris. If equipped, inspect the speedometer opening at the leading edge of the gearcase or any other gearcase drains for debris (clean debris with low-pressure compressed air or a piece of thin wire). Mercury also recommends disconnecting the speedometer pickup tube temporarily to make sure it drains completely, then reconnecting it before storage.

■ As the season winds down and you are approaching your last outing, start treating the fuel system then to make sure it thoroughly mixes with all the fuel in the tank. By the last outing of the season you should already have a protected fuel system.

✳✳ SELOC WARNING

Besides treating the fuel system to prevent evaporation or clogging from deposits left behind, coating all bearing surfaces in the motor with FRESH, clean oil is the most important step you can take to protect the engine from damage during storage. Mercury recommends varying methods for this, depending upon engine and fuel type. Details are included in the following steps.

2. Stabilize the engine's fuel supply using a high quality fuel stabilizer (of course the manufacturer recommends using Quicksilver Gasoline Stabilizer) and take this opportunity to thoroughly flush the engine cooling system at the same time as follows:

a. Add an appropriate amount of fuel stabilizer to the fuel tank and, if the tank is not going to be drained, top it off to minimize the formation of moisture through condensation in the fuel tank.

■ **When using a portable fuel tank, add the conditioner, then agitate the tank to ensure proper mixing with the fuel. When using a permanent mounted tank, mix the amount of conditioner necessary for the whole tank with 1 gallon of gasoline in a portable container, agitate the container, then pour this mixture into the tank.**

b. On 1.5L Optimax motors and 2003 and later 3.0L Optimax motors, drain all water and fuel from the vapor separator tank so that it can be filled with the storage mixture later in this procedure. The tank assembly on these models (which includes a water separating fuel filter) contains a drain hose attached to the bottom of the tank. The hose is attached to a drain screw/valve for the filter area on the left end of the assembly (when looking at the tank), and a drain screw/valve fitting for the main tank toward the right end (but not quite all the way to the end of the housing). First, disconnect the hose from the valve/fitting at the right end first and direct the hose into a suitable container, then open the drain screw on the face of the left end of the tank. Drain all water from the assembly, then tighten the drain screw and reconnect the hose to the valve/fitting toward the right end of the assembly. Next, disconnect the hose from the valve on the left end of the tank (again directing it into a suitable container) and open the screw for the right end (to drain the main tank of fuel). Once the fuel has drained, close the screw and reconnect the hose again.

c. On all EFI and Optimax motors, remove the water separating fuel filter.

d. On 1.5L and 3.0L Optimax motors, pre-mix 2 teaspoons of QuickKleen Lubricant and 2 teaspoons of Fuel Stabilizer in a separate container. Remove the fuel/water stabilizer and pour the mixture directly into the tank (except 3.0L models through 2002, which use a different type of water separating fuel filter). On 3.0L models through 2002, you'll have to pour the mixture directly into the replacement water separating fuel filter before installation.

■ **We don't know why Mercury does not recommend the previous step for 2.5L Optimax motors, but we could find no specific instructions to treat the fuel system/vapor separator tank in the same manner for 2.5L models.**

e. On all EFI and Optimax motors, install the new water separating fuel filter and prime the fuel system.

f. Attach a flushing attachment as a cooling water/flushing source. For details, please refer to the information on Flushing the Cooling System, in this section.

g. Start and run the engine at fast idle approximately 10-15 minutes for most models (please note that recommendations for Optimax engine running times vary from 5 minutes for 1.5L motors or on Pro/XS/Sport models to 10 minutes for standard 3.0L motors and 25 minutes for 2.5L motors). This will ensure the entire fuel supply system contains the appropriate storage mixtures.

■ **On EFI motors or Optimax motors where you are planning on draining the vapor separator tank completely the run time becomes less of a critical issue.**

h. On 60 hp and smaller or 150 hp and lager carbureted motors, after the motor has run for a few minutes, disconnect the fuel supply line (if a quick-connect if used, if not shut the motor down and disconnect it manually, then restart the motor). Once the engine begins to stall for lack of fuel begin spraying Quicksilver Storage Seal or an equivalent fogging oil into the carburetor throat (or carburetor throats). Continue to spray until the engine stalls from lack of fuel.

i. Stop the engine and remove the flushing source, keeping the outboard perfectly vertical. Allow the cooling system to drain completely, especially if the outboard might be exposed to freezing temperatures during storage.

✳✳ SELOC WARNING

NEVER keep the outboard tilted when storing in below-freezing temperatures as water could remain trapped in the cooling system. Any water left in cooling passages might freeze and could cause severe engine damage by cracking the powerhead or gearcase.

3. Finish fogging the motor manually through the spark plug ports as follows (note that this is the ONLY method recommended for fogging all EFI and Optimax motors, as well as the 65 Jet-90 hp 1386cc carbureted models):

a. On multi-cylinder motors, tag and disconnect the spark plug leads.

b. Disconnect the spark plug lead(s), then remove the spark plug(s) as described under Spark Plugs.

c. Proceed as follows depending upon motor:
• On all carbureted motors EXCEPT 65 Jet-90 hp (1386cc) models, spray a 5 second blast of Quicksilver Storage Seal around the inside of each cylinder.
• On 65 Jet-90 hp (1386cc) models, inject a small amount of 2-stroke engine oil into each cylinder.
• On EFI and Optimax models, inject about 1 oz (30ml) of 2-stroke engine oil into each cylinder.

d. Turn the flywheel slowly by hand (clockwise, in the normal direction of rotation) to distribute the oil evenly across the cylinder walls. If necessary, re-spray or inject more oil into each cylinder when that cylinder's piston reaches the bottom of its travel. Reinstall and tighten the spark plugs, but leave the leads disconnected (and GROUNDED) to prevent further attempts at starting until the motor is ready for re-commissioning.

■ **On motors equipped with a rope start handle, the rope can be used to turn the motor slowly and carefully using the rope starter. On other models, turn the flywheel by hand or using a suitable tool, but be sure to ALWAYS turn the engine in the normal direction of rotation (normally clockwise on these motors).**

4. Drain and refill the engine gearcase while the oil is still warm (for details, refer to the Gearcase Oil procedures in this section). Take the opportunity to inspect for problems now, as storage time should allow you the opportunity to replace damaged or defective seals. More importantly, remove the old, contaminated gear oil now and place the motor into storage with fresh oil to help prevent internal corrosion.

5. On carbureted motors, if the motor is to be stored for any length of time more than one off-season you really MUST drain the carburetor float bowls. Honestly, it is a pretty easy task and we'd recommend doing that for all motors, even if they are only going to be stored for a few months. To drain the float bowls locate the drain screw on the bottom of each bowl, place a small container under the bowl and remove the screw. Repeat for the remaining float bowls on multiple carburetor motors.

6. On EFI or Optimax motors, it is recommended that you drain the vapor separator tank. On some models there MAY be a drain screw, but it is more likely that you'll have to remove the tank from the powerhead and drain it manually.

7. On models equipped with portable fuel tanks, disconnect and relocate them to a safe, well-ventilated, storage area, away from the motor. Drain any fuel lines that remain attached to the tank. It's a tough call whether or not to drain a portable tank. Plastic tanks, drain em' and burn the fuel in something else. Metal tanks, well, draining them will expose them to moisture and possible corrosion, while topping them off will help prevent this, so it probably makes more sense to top them off with treated fuel.

8. Remove the battery or batteries from the boat and store in a cool dry place. If possible, place the battery on a smart charger or Battery Tender®, otherwise, trickle charge the battery once a month to maintain proper charge.

✳✳ SELOC WARNING

Remember that the electrolyte in a discharged battery has a much lower freezing point and is more likely to freeze (cracking/destroying the battery case) when stored for long periods in areas exposed to freezing temperatures. Although keeping the battery charged offers one level or protection against freezing; the other is to store the battery in a heated or protected storage area.

9. On models equipped with a boat mounted fuel filter or filter/water canister, clean or replace the boat mounted fuel filter at this time. If the fuel system was treated, the engine mounted fuel filters should be left intact, so the sealed system remains filled with treated fuel during the storage period.

10. On motors with external oil tanks, if possible, leave the oil supply line connected to the motor. This is the best way to seal moisture out of the system. If the line must be disconnected for any reason (such as to remove the motor or oil tank from the boat), seal the line by sliding a snug fitting cap over the end.

11. Perform a complete lubrication service following the procedures in this section.

12. Except for Jet Drive models, remove the propeller and check thoroughly for damage. Clean the propeller shaft and apply a protective coating of grease.

13. On Jet models, thoroughly inspect the impeller and check the impeller clearance. Refer to the procedures in this section.

14. Check the motor for loose, broken or missing fasteners. Tighten fasteners and, again, use the storage time to make any necessary repairs.

15. Inspect and repair all electrical wiring and connections at this time. Make sure nothing was damaged during the season's use. Repair any loose connectors or any wires with broken, cracked or otherwise damaged insulation.

16. Clean all components under the engine cover and apply a corrosion preventative spray.

17. Too many people forget the boat and trailer, don't be one of them.

  a. Coat the boat and outside painted surfaces of the motor with a fresh coating of wax, then cover it with a breathable cover

  b. If possible place the trailer on stands or blocks so the wheels are supported off the ground.

  c. Check the air pressure in the trailer tires. If it hasn't been done in a while, remove the wheels to clean and repack the wheel bearings.

18. Sleep well, since you know that your baby will be ready for you come next season.

Re-Commissioning

REMOVAL FROM STORAGE

The amount of service required when re-commissioning the boat and motor after storage depends on the length of non-use, the thoroughness of the storage procedures and the storage conditions.

At minimum, a thorough spring or pre-season tune-up and a full lubrication service is essential to getting the most out of your engine. If the engine has been properly winterized, it is usually no problem to get it in top running condition again in the springtime. If the engine has just been put in the garage and forgotten for the winter, then it is doubly important to perform a complete tune-up before putting the engine back into service. If you have ever been stranded on the water because your engine has died and you had to suffer the embarrassment of having to be towed back to the marina you know how it can be a miserable experience. Now is the time to prevent that from occurring.

CLEARING A SUBMERGED MOTOR

Unfortunately, because an outboard is mounted on the exposed transom of a boat, and many of the outboards covered here are portable units that are mounted and removed on a regular basis, an outboard can fall overboard. Ok, it's relatively rare, but it happens often enough to warrant some coverage here. The best way to deal with such a situation is to prevent it, by keeping a watchful eye on the engine mounting hardware (bolts and/or clamps). But, should it occur, here's how to salvage, service and enjoy the motor again.

In order to prevent severe damage, be sure to recover an engine that is dropped overboard or otherwise completely submerged as soon as possible. It is really best to recover it immediately. But, keep in mind that once a

■ Although you should normally replace your spark plugs at the beginning of each season, we like to start and run the engine (for the spring compression check) using the old spark plugs. Why? Well, on that first start-up you're going to be burning off a lot of fogging oil, why expose the new plugs to it? Besides you have to remove the plugs in order to perform the tune-up compression check anyway so you can install the new ones at that time.

Take the opportunity to perform any annual maintenance procedures that were not conducted immediately prior to placing the motor into storage. If the motor was stored for more than one off-season, pay special attention to inspection procedures, especially those regarding hoses and fittings. Check the engine gear oil for excessive moisture contamination. The same goes for engine crankcase oil on 4-strokes or oil tanks on 2-strokes, so equipped. If necessary, change the gearcase or engine oil to be certain no bad or contaminated fluids are used.

■ Although not absolutely necessary, it is a good idea to ensure optimum cooling system operation by replacing the water pump impeller at this time. Impeller replacement was once an annual ritual, we now hear that most impellers seem to do well for 2-3 seasons, and some more. You'll have to make your own risk assessment here, but keep an eye on your cooling indicator stream get to know how strong the spray looks (and how warm it feels) to help you make your decision.

Other items that require attention include:
1. Install the battery (or batteries) if so equipped.
2. Inspect all wiring and electrical connections. Rodents have a knack for feasting on wiring harness insulation over the winter. If any signs of rodent life are found, check the wiring carefully for damage, do not start the motor until damaged wiring has been fixed or replaced.
3. On motors with a remote oil tank, if the line was disconnected, remove the cover and reconnect the line, then make sure the system is primed to ensure proper operation once the motor is started. For more details, refer to the Lubrication & Cooling section.
4. If not done when placing the motor into storage clean and/or replace the fuel filters at this time.

■ Portable fuel tanks should be emptied and cleaned using solvent. Take the opportunity to thoroughly inspect the condition of the tank. For more details on fuel tanks, please refer to the Fuel System section.

5. If the fuel tank was emptied, or if it must be emptied because the fuel is stale fill the tank with fresh fuel. Keep in mind that even fuel that was treated with stabilizer will eventually become stale, especially if the tank is stored for more than one off-season. Pump the primer bulb and check for fuel leakage or flooding at the carburetor.
6. Attach a flush device or place the outboard in a test tank and start the engine. Run the engine at idle speed and warm it to normal operating temperature. Check for proper operation of the cooling, electrical and warning systems.

✱✱ SELOC CAUTION

Before putting the boat in the water, take time to verify the drain plug is installed. Countless number of spring boating excursions have had a very sad beginning because the boat was eased into the water only to have the boat begin to fill with it.

submerged motor is recovered exposure to the atmosphere will allow corrosion to begin etching highly polished bearing surfaces of the crankshaft, connecting rods and bearings. For this reason, not only do you have to recover it right away, but you should service it right away too. Make sure the motor is serviced within about 2-3 hours of initial submersion.

OK, maybe now you're saying "2-3 hours, it will take me that long to get it to a shop or to my own garage!" Well, if the engine cannot be serviced immediately (or sufficiently serviced so it can be started), re-submerge it in a tank of fresh water to minimize exposure to the atmosphere and slow the corrosion process. Even if you do this, do not delay any more than absolutely necessary, service the engine as soon as possible. This is especially important if the engine was submerged in salt, brackish or polluted water as even submersion in fresh water will not preserve the engine indefinitely. Service the engine, at the **MOST** within a few days of protective submersion.

After the engine is recovered, vigorously wash all debris from the engine using pressurized freshwater.

■ **If the engine was submerged while still running, there is a good chance of internal damage (such as a bent connecting rod). Under these circumstances, don't start the motor, follow this procedure, ESPECIALLY the part about removing the spark plugs and draining the motor of any water which might be inside and then turning it over slowly by hand while feeling for mechanical problems. If necessary, refer to Powerhead Overhaul for complete disassembly and repair instructions.**

✳✳ SELOC WARNING

NEVER try to start a recovered motor until at least the steps dealing with draining the motor and checking to see it if is hydro-locked or damaged are performed. Keep in mind that attempting to start a hydro-locked motor could cause major damage to the powerhead, including bending or breaking a connecting rod.

If the motor was submerged for any length of time it should be thoroughly disassembled and cleaned. Of course, this depends on whether water intruded into the motor or not. To determine this check remove the spark plugs and check for water in the cylinders and check the gearcase oil (on all motors) for signs of contamination.

The extent of cleaning and disassembly that must take place depends also on the type of water in which the engine was submerged. Engines totally submerged, for even a short length of time, in salt, brackish or polluted water will require more thorough servicing than ones submerged in fresh water for the same length of time. But, as the total length of submerged time or time before service increases, even engines submerged in fresh water will require more attention. Complete powerhead disassembly and inspection is required when sand, silt or other gritty material is found inside the engine cover.

Many engine components suffer the corrosive effects of submersion in salt, brackish or polluted water. The symptoms may not occur for some time after the event. Salt crystals will form in areas of the engine and promote significant corrosion.

Electrical components should be dried and cleaned or replaced, as necessary. If the motor was submerged in salt water, the wire harness and connections are usually affected in a shorter amount of time. Since it is difficult (or nearly impossible) to remove the salt crystals from the wiring connectors, it is best to replace the wire harness and clean all electrical component connections. The starter motor, relays and switches on the engine usually fail if not thoroughly cleaned or replaced.

To ensure a through cleaning and inspection:

1. Remove the engine cover and wash all material from the engine using pressurized freshwater. If sand, silt or gritty material is present inside the engine cover, completely disassemble and inspect the powerhead.

2. Dry all electrical components with compressed air.

3. On Optimax motors, start draining the water as follows:

a. On 2.5L motors, remove the 4 bolts securing the throttle plate assembly, then remove the plate.

b. Tilt the outboard so that water will drain from the crankcase area into the air plenum. Use a sponge, hose or air pump to remove the water.

4. Tag (except on single cylinder motors) and disconnect the spark plugs leads. Be sure to grasp the spark plug cap and not the wire, then twist the cap while pulling upward to free it from the plug. Remove the spark plugs. For more details, refer to the Spark Plug procedure in this section.

■ **If water is found in the cylinders, MOST of the water can be removed at this stage, simply by placing the outboard in a horizontal position with the spark plug ports facing downward and slowly attempting to rotate the flywheel by hand. This will start to blow any large amounts of water in the cylinder.**

✳✳ SELOC CAUTION

When attempting to turn the flywheel for the first time after the submersion, be sure to turn it SLOWLY, feeling for sticking or binding that could indicate internal damage from hydro-lock. This is a concern, especially if the engine was cranked before the spark plug(s) were removed to drain water or if the engine was submerged while still running.

5. On carbureted and EFI motors, proceed as follows:

a. Place the engine horizontally with the spark plug ports facing downward, then pour alcohol into the carburetor throat(s) (or through the throttle body on EFI motors) while slowly turning the flywheel by hand (keep in mind that alcohol is hygroscopic, meaning it will absorb water so it is very helpful when trying to purge moisture from the combustion chambers and engine internals).

b. Turn the outboard over so the plug ports are now facing upward and pour alcohol directly into the spark plug openings, again turning the flywheel slowly.

c. Turn the outboard over yet AGAIN so the ports are downward and pour engine oil directly into the carburetor throat(s) (or throttle body on EFI motors) while rotating the flywheel to distribute oil throughout the crankcase.

6. Disconnect the fuel supply line from the engine, then drain and clean all fuel lines. Depending on the circumstances surrounding the submersion, inspect the fuel tank for contamination and drain, if necessary.

7. On EFI motors, drain the fuel system as follows:

a. Remove the drain plug from the vapor separator tank and drain the tank of any water and/or fuel, then reinstall the plug.

b. Remove the water separating fuel filter and drain the contents, then reinstall or replace the filter assembly.

8. On Optimax motors, drain the fuel system as follows:

a. Drain the fuel (and water) from the vapor separator tank. Either there will be a drain plug at the front of the housing OR on 1.5L models and 2003 or later 3.0L models, there is a drain screw/valve for the water separator chamber and usually a second for the fuel portion of the chamber. After draining install the drain plug or tighten the drain screws/valves, as applicable.

b. Remove the fuel hose from the bottom port side of the fuel rail, then drain the fuel/water from the assembly and reconnect the hose.

c. On models that utilize a water separating fuel filter that is NOT integral with the vapor separator tank (which would have been drained 2 steps ago), remove the filter and drain the contents, then reinstall or replace the filter assembly.

9. On Optimax motors, drain water from the air compressor system as follows:

a. If equipped (normally Pro/XS/Sport models do not utilize one), either replace or carefully air dry the compressor air filter.

b. Disconnect the air outlet hose from the air compressor and drain water from the compressor hose, then reconnect the hose.

c. On 2.5L and 3.0L motors, disconnect the air hose from the bottom port side fuel rail and drain the water, then reconnect the hose.

10. On ALL models, place the outboard horizontally with the spark plug ports facing upward, now pour about 1 oz (30ml) of engine oil directly into EACH spark plug hole. Rotate the flywheel slowly by hand several time to distribute the oil in the cylinders.

11. Install the spark plug(s) and reconnect the wiring.

12. Support the engine in the normal upright position.

13. On carbureted motors, remove the carburetors and drain or disassemble as necessary.

14. On motors with a mechanical fuel pump, remove the pump and drain or disassemble and clean as necessary.

15. Reinstall the carburetor(s) and/or fuel pump, as applicable.

16. On models equipped with an oil injection system, drain water from it as follows:

a. If equipped with a remote oil tank, remove the hose from the fitting on top of the oil reservoir or, for most models, disconnect the black hose without the blue stripe from the pulse fitting on the starboard side of the motor. Drain any water in the hose, then reconnect.

b. If you found water in the hose, check the remote oil tank for water, then drain the tank if any is found.

c. If the motor is equipped with an engine mounted reservoir, remove the reservoir from the engine and drain all oil and water.

17. If equipped with an electric starter, disassemble the starter motor and dry the brush contacts, armature and other corrodible parts. Reassemble the starter and install it back on the motor.

18. Remove all other external electrical components for disassembly and cleaning. Spray all connectors with electrical contact cleaner, then apply a small amount of dielectric grease prior to reconnection to help prevent corrosion. On electric start models, remove, disassemble and clean the starter components. For details on the electrical system components, refer to the Ignition & Electrical section.

19. On Optimax motors, prime the oil injection pump as follows:

a. Fill the engine fuel system with fuel, then connect the fuel hose and squeeze the primer bulb until it feels firm.

b. Turn the ignition key to the **ON** position (without trying to start), then within the first 10 seconds of turning the key, move the remote control harness from Neutral into Forward Gear and back to Neutral again 3-5 times, this will automatically start the priming process. Listen for an audible click at the oil pump which will tell you it is priming, but be patient as it takes a few minutes for the pump to complete the process.

20. Check the engine gearcase oil for contamination. Refer to the procedures for Gearcase Oil in this section. The gearcase is sealed and, if the seals are in good condition, should have survived the submersion without contamination. But, if contamination is found, look for possible leaks in the seals, then drain the gearcase and make the necessary repairs before refilling it. For more details, refer to the section on Gearcases.

21. Reassemble the motor and mount the engine or place it in a test tank. Start and run the engine for 1 hour. If the engine won't start, remove the spark plugs again and check for signs of moisture on the tips. If necessary, use compressed air to clean moisture from the electrodes or replace the plugs.

22. Stop the engine.

23. Perform all other lubrication services.

24. Try not to let it get away from you (or anyone else) again!

SPECIFICATION CHARTS

General Engine Specifications

Model (HP)	Engine Type	Year	Displacement cu.in. (cc)	Bore and Stroke in. (mm)	Appx Weight lb. (kg) *	Oil Injection System	Ignition System	Starting System	Cooling System
2.5	1-cyl	2001-05	4.6 (74.6)	1.85 x 1.69 (47 x 43)	27.5-28.5 (12.5-13.0)	Pre-Mix	CDI	Manual	WC
3.3	1-cyl	2001-05	4.6 (74.6)	1.85 x 1.69 (47 x 43)	28.5-30 (12.9-14.0)	Pre-Mix	CDI	Manual	WC
4	1-cyl	2001-05	6.2 (102)	2.16 x 1.69 (55 x 43)	44-45 (20)	Pre-Mix	CDI	Manual	WC
5	1-cyl	2001-05	6.2 (102)	2.16 x 1.69 (55 x 43)	44-45 (20)	Pre-Mix	CDI	Manual	WC
6	2-cyl	2001-03	13 (210)	2.13 x 1.77 (54 x 45)	73-80 (31-36)	Pre-Mix	CDI	Manual (electric opt.)	WC w/ Tstat (opt.)
8	2-cyl	2001-05	13 (210)	2.13 x 1.77 (54 x 45)	73-80 (31-36)	Pre-Mix	CDI	Manual (electric opt.)	WC w/ Tstat (opt.)
9.9	2-cyl	2001-05	16 (262)	2.38 x 1.80 (60 x 46)	74-84 (34-38)	Pre-Mix	CDI	Manual or Electric	WC w/ Tstat (opt.)
10	2-cyl	2001	16 (262)	2.38 x 1.80 (60 x 46)	75 (32)	Pre-Mix	CDI	Manual (electric opt.)	WC w/ Tstat
15	2-cyl	2001-05	16 (262)	2.38 x 1.80 (60 x 46)	74-83 (34-37)	Pre-Mix	CDI	Manual or Electric	WC w/ Tstat
20	2-cyl	2001-03	24 (400)	2.56 x 2.36 (65 x 60)	112-117 (51-53)	Pre-Mix	CDI	Manual or Electric	WC w/ Tstat
20Jet (25)	2-cyl	2001-05	24 (400)	2.56 x 2.36 (65 x 60)	124 (56)	Pre-Mix	CDI	Manual	WC w/ Tstat
25	2-cyl	2001-05	24 (400)	2.56 x 2.36 (65 x 60)	112-117 (51-53)	Pre-Mix	CDI	Manual or Electric	WC w/ Tstat
30	2-cyl	2001-02	39 (644)	2.99 x 2.80 (76 x 71)	156-166 (71-75)	SPVR	Modular CDI	Manual or Electric	WC w/ Tstat & PC
40	2-cyl	2001-03	39 (644)	2.99 x 2.80 (76 x 71)	173-176 (78-80)	SPVR	Modular CDI	Manual or Electric	WC w/ Tstat & PC
40	3-cyl	2001-05	59 (967)	2.99 x 2.80 (76 x 71)	199-205 (90-93)	SPVR	Modular CDI	Manual or Electric	WC w/ Tstat & some w/ PC
40Jet (60)	3-cyl	2001-05	59 (967)	2.99 x 2.80 (76 x 71)	233 (106)	SPVR	Modular CDI	Electric	WC w/ Tstat & some w/ PC
50	3-cyl	2001-09	59 (967)	2.99 x 2.80 (76 x 71)	199-205 (90-93)	SPVR	Modular CDI	Manual or Electric	WC w/ Tstat & some w/ PC
55	3-cyl	2001	59 (967)	2.99 x 2.80 (76 x 71)	220 (100)	SPVR	Modular CDI	Manual or Electric	WC w/ Tstat & some w/ PC
60	3-cyl	2001-05	59 (967)	2.99 x 2.80 (76 x 71)	219-242 (99-110)	SPVR	Modular CDI	Electric	WC w/ Tstat & some w/ PC
75	3-cyl	2001-05	85 (1386)	3.50 x 2.93 (89 x 75)	303-305 (137-139)	SPVR	Modular CDI	Electric	WC w/ Tstat & PC
65 Jet (90)	3-cyl	2001-09	85 (1386)	3.50 x 2.93 (89 x 75)	315 (143)	SPVR	Modular CDI	Electric	WC w/ Tstat & PC
90	3-cyl	2001-09	85 (1386)	3.50 x 2.93 (89 x 75)	303-323 (137-146)	SPVR	Modular CDI	Electric	WC w/ Tstat & PC
75 Optimax	3-cyl	2004-09	93 (1526)	3.63 x 3.00 (92 x 76)	360-372 (163-169)	EMP	PCM 038	Electric (turnkey)	WC w/ Tstat & PC
80 Jet (115) Opti	3-cyl	2009	93 (1526)	3.63 x 3.00 (92 x 76)	360-375 (163-170)	EMP	PCM 038	Electric (turnkey)	WC w/ Tstat & PC
90 Optimax	3-cyl	2004-09	93 (1526)	3.63 x 3.00 (92 x 76)	360-372 (163-169)	EMP	PCM 038	Electric (turnkey)	WC w/ Tstat & PC
115 Optimax	3-cyl	2004-09	93 (1526)	3.63 x 3.00 (92 x 76)	360-375 (163-170)	EMP	PCM 038	Electric (turnkey)	WC w/ Tstat & PC
80 Jet (115)	4-cyl	2001-05	113 (1848)	3.50 x 2.93 (89 x 75)	357 (162)	SPVR	Modular CDI	Electric	WC w/ Tstat & PC
115	4-cyl	2001-05	113 (1848)	3.50 x 2.93 (89 x 75)	347-348 (157-158)	SPVR	Modular CDI	Electric	WC w/ Tstat & PC
125	4-cyl	2001-05	113 (1848)	3.50 x 2.93 (89 x 75)	347-348 (157-158)	SPVR	Modular CDI	Electric	WC w/ Tstat & PC

SELOC_1418

General Engine Specifications

Model (HP)	Engine Type	Year	Displacement cu.in. (cc)	Bore and Stroke in. (mm)	Appx Weight lb. (kg)*	Oil Injection System	Ignition System	Starting System	Cooling System
110 Jet (150) Opti	V-6 (60°)	2009	153 (2507)	3.50 x 2.65 (89 x 67)	431-462 (195-210)	EMP	PCM 038	Electric (turnkey)	WC w/ Tstat & PC
135 Optimax	V-6 (60°)	2001-03	153 (2507)	3.50 x 2.65 (89 x 67)	431-462 (195-210)	EMP	Digital Inductive (PCM 555)	Electric (turnkey)	WC w/ Tstat & PC
	V-6 (60°)	2004-09	153 (2507)	3.50 x 2.65 (89 x 67)	431-462 (195-210)	EMP	PCM 038	Electric (turnkey)	WC w/ Tstat & PC
150 (XR6,Classic)	V-6 (60°)	2001-05	153 (2507)	3.50 x 2.65 (89 x 67)	406-416 (184-189)	SPVR	Modular CDI	Electric	WC w/ Tstat & PC
150 EFI	V-6 (60°)	2001	153 (2507)	3.50 x 2.65 (89 x 67)	416-434 (189-197)	ESP	Modular CDI w/ elect control (ECM)	Electric (turnkey)	WC w/ Tstat & PC
	V-6 (60°)	2001-09	153 (2507)	3.50 x 2.65 (89 x 67)	425-434 (193-197)	ESP	PCM 038	Electric (turnkey)	WC w/ Tstat & PC
150 Optimax	V-6 (60°)	2001-03	153 (2507)	3.50 x 2.65 (89 x 67)	431-462 (195-210)	EMP	Digital Inductive (PCM 555)	Electric (turnkey)	WC w/ Tstat & PC
	V-6 (60°)	2004-09	153 (2507)	3.50 x 2.65 (89 x 67)	431-462 (195-210)	EMP	PCM 038	Electric (turnkey)	WC w/ Tstat & PC
175 EFI	V-6 (60°)	2001	153 (2507)	3.50 x 2.65 (89 x 67)	416-434 (189-197)	ESP	Modular CDI w/ elect control (ECM)	Electric (turnkey)	WC w/ Tstat & PC
	V-6 (60°)	2002	153 (2507)	3.50 x 2.65 (89 x 67)	425-434 (193-197)	ESP	PCM 038	Electric (turnkey)	WC w/ Tstat & PC
175 Optimax	V-6 (60°)	2001-03	153 (2507)	3.50 x 2.65 (89 x 67)	431-462 (195-210)	EMP	Digital Inductive (PCM 555)	Electric (turnkey)	WC w/ Tstat & PC
	V-6 (60°)	2004-09	153 (2507)	3.50 x 2.65 (89 x 67)	431-462 (195-210)	EMP	PCM 038	Electric (turnkey)	WC w/ Tstat & PC
200	V-6 (60°)	2001-05	153 (2507)	3.50 x 2.65 (89 x 67)	406-414 (184-188)	SPVR	Modular CDI	Electric	WC w/ Tstat & PC
200 EFI (2.5L)	V-6 (60°)	2001	153 (2507)	3.50 x 2.65 (89 x 67)	416-434 (189-197)	ESP	Modular CDI w/ elect control (ECM)	Electric (turnkey)	WC w/ Tstat & PC
	V-6 (60°)	2002-05	153 (2507)	3.50 x 2.65 (89 x 67)	425-434 (193-197)	ESP	PCM 038	Electric (turnkey)	WC w/ Tstat & PC
200 EFI (3.0L)	V-6 (60°)	2004-05	185 (3032)	3.63 x 3.00 (92 x 76)	458-479 (208-217)	ESP	PCM 038	Electric (turnkey)	WC w/ Tstat & PC
200 Optimax	V-6 (60°)	2001-03	185 (3032)	3.63 x 3.00 (92 x 76)	497-544 (225-247)	EMP	Digital Inductive (PCM 555)	Electric (turnkey)	WC w/ Tstat & PC
	V-6 (60°)	2004-09	185 (3032)	3.63 x 3.00 (92 x 76)	497-544 (225-247)	EMP	PCM 038	Electric (turnkey)	WC w/ Tstat & PC
225 EFI	V-6 (60°)	2001	185 (3032)	3.63 x 3.00 (92 x 76)	445-493 (202-224)	ESP	Modular CDI w/ elect control (ECM)	Electric (turnkey)	WC w/ Tstat & PC
	V-6 (60°)	2002-05	185 (3032)	3.63 x 3.00 (92 x 76)	479-493 (218-224)	ESP	PCM 038	Electric (turnkey)	WC w/ Tstat & PC
225 Optimax	V-6 (60°)	2001-03	185 (3032)	3.63 x 3.00 (92 x 76)	497-544 (225-247)	ESP	Digital Inductive (PCM 555)	Electric (turnkey)	WC w/ Tstat & PC
	V-6 (60°)	2004-09	185 (3032)	3.63 x 3.00 (92 x 76)	497-544 (225-247)	EMP	PCM 038	Electric (turnkey)	WC w/ Tstat & PC
250 EFI	V-6 (60°)	2001	185 (3032)	3.63 x 3.00 (92 x 76)	458-497 (208-225)	ESP or SPVR	Modular CDI w/ elect control (ECM)	Electric (turnkey)	WC w/ Tstat & PC
	V-6 (60°)	2002-05	185 (3032)	3.63 x 3.00 (92 x 76)	458-479 (208-218)	ESP or SPVR	PCM 038	Electric (turnkey)	WC w/ Tstat & PC
250 Optimax	V-6 (60°)	2006-09	185 (3032)	3.63 x 3.00 (92 x 76)	505-544 (229-247)	EMP	PCM 038	Electric (turnkey)	WC w/ Tstat & PC

CDI - Capacitor Discharge Ignition   EMP/ESP - Electronic Multipoint/Single Point   SPVR - Single Point Variable Ratio   PC - Pressure Controlled   Tstat - Thermostat   WC - Water Cooled

* NOTE: Measurements are approximate and usually, minus fluids and propeller. Generally speaking, pre-mix carbureted versions are lightest and BIGFOOT, EFI, or OPTIMAX models are heaviest, but there are some exceptions. See owner's manual

SELOC_1418

Maintenance Intervals Chart

Component	Each Use or As Needed	Initial 10-Hour Check (Break-In)	Every 12mths/100hrs ①	Off Season or Multiple Year/hrs
Alternator drive belt (Some EFI)		I	I	I
Anode(s)	I	I	I	I
Battery condition and connections (if equipped)*	I (condition/connections), T		I	I
Battery charge / fluid level (if equipped)*	I (at least monthly)		I	
Boat hull*	I		I	I
Bolts and nuts (all accessible fasteners)*	I	I, T	I, T	I
Carburetor (adjust - if necessary) ②		I, A	I, A	I, A
Case finish and condition (inspect & wash/wax)	C, salt/brackish/polluted water	I	I, C	I, C
Compressor & alternator drive belt (Optimax)		I	I	I, R 3/300 ⑧
Compressor air intake filter (Optimax)			R	R
Decarbon Pistons ③			P	P
Driveshaft splines (and shift shaft splines, if app) ④			L ④	L ④
Electricical wiring and connectors*		I	I	I
Emergency stop switch, clip &/or lanyard*	I	I	I	I
Engine mounting bolts	I	I	I	I
Flush cooling system	salt/brackish/polluted water	P	P	P
Fuel filter and/or water separator (C or R, as app)		P	P	P
Fuel hose and system components*	I	I	I	I
Fuel tank (integral or portable)	I		C (portable p-up), I	C (portable p-up), I
Gear oil	I (for signs of leakage)	R	R	R
Idle speed	I (look/listen for changes)		A	A
Ignition timing (carbureted models only)		I, A	I, A	I, A
Impeller clearance/intake grate (jet models)	I (visually inspect impeller)	I	I	I
Jet drive bearing lubrication ⑤	L, fill vent hose after each day	L	L	L
Lubrication points	I	⑥	⑥	⑥
Oil filter, inline (oil injection system Opti/EFI)				R, 3/300
Power trim and tilt (if equipped)	I (check fluid level monthly)	I	I	I
Propeller	I	I	I	I
Propeller shaft / prop fastener (nut and/or pin)	I	I	I, L / T	I, L / T
Carbon fiber reeds (Pro/XS/Sport models)			I (look for chips/cracks)	
Remote control* (check for looseness or binding)	I	I	I, A	I, A
Spark plugs	R (as needed)	I	I	I
Steering cable* (also check for looseness or bind)	L (as needed) / I	L / I	L / I	L / I
Steering link/rod fasteners (check for looseness)*	I, T	I, T	I, T	I, T
Thermostat and Pressure Control Valve			I	I
Throttle (tiller models)*		Check, A		Check, A
Tune-up	A (as needed)		I (annually)	Pre-season tune-up
Water pump impeller				varies ⑦
Water pump intake grate and indicator	I			

A-Adjust   I-Inspect and Clean, Adjust, Lubricate or Replace, as necessary   R-Replace   P - Perform

C-Clean   L-Lubricate   T-Tighten

* Denotes possible safety item (although, all maintenance inspections/service can be considered safety related when it means not being stranded on the water should a component fail.)

① Many items are listed for both every 100 hours and off season. Since many boaters use their crafts less than 100 hours a year, these items should at least be performed annually. If you find yourself right around 100 hours per season, try to time the service so it occurs immediately prior to placing the motor in storage. Also, half these service intervals for all Pro/XS/Sport models (see your owner's manual for further details)

② Recommendations vary regarding the need to adjust the carburetor (as opposed to only checking how it is operating). The most recent advice we've received from Merc not to tamper with a carb that is running correctly

③ Mercury manuals for 6 hp and larger motors (except Optimax and most EFI motors) recommend decarboning of the pistons using Power Tune Engine Cleaner annually or every 100 hours. However, note that the 3.0L EFI manual through 2001 also recommends decarboning.

④ Mercury manuals list lubrication of the driveshaft splines every year, but we recommend waiting for impeller replacement and doing both at the same time

⑤ Lubricate the jet drive bearing at the end of EVERY day's use and every 10 hours. Also, every 50 hours replace the grease using the same lubrication fitting, but adding enough to purge the old grease

⑥ Varies with use, generally every 30 days when used in salt, brackish or polluted water and every 60 days when used in fresh water

⑦ Water pump impeller inspection recommendations vary from replace every 1 year/100 hours 200/225 3.0L EFI motors through 2001 and all 135-200 2.5L motors, or to replace every 3 years/300 hours for ALL other models (it is likely that earlier models would be included too after first impeller replacement)

⑧ Most Merc manuals recommend inspection every 3 years/300 hours, but the 3.0L Optimax manual recommends REPLACEMENT at that time

TWO-STROKE MOTOR FUEL:OIL RATIO CHART

Desired Fuel:Oil Ratio	Amount of oil needed when mixed with:				
	3 G (11.4 L) of Gas	6 G (22.7 L) of Gas	18 G (68.1 L) of Gas	30 G (114 L) of Gas	45 G (171 L) of Gas
100:1 (1% oil)	4 fl. oz. (118 mL)	8 fl. oz. (236 mL)	24 fl. oz. (708 mL)	40 fl. oz. (1180 mL)	60 fl. oz. (1770 mL)
50:1 (2% oil)	8 fl. oz. (236 mL)	16 fl. oz. (473 mL)	48 fl. oz. (1419 mL)	80 fl. oz. (2360 mL)	120 fl. oz. (3.54 L)
25:1 (4% oil)	16 fl. oz. (473 mL)	32 fl. oz. (946 mL)	96 fl. oz. (2838 mL)	160 fl. oz. (4.73 L)	240 fl. oz. (7.1 L)

NOTE: Fuel:Oil ratios listed here are for calcuation purposes. Refer to the fuel:oil recommendations for your engine before mixing. Remember that a pre-mix system designed to produce a 50:1 ratio will produce a 25:1 ratio if a 50:1 ratio is already in the fuel tank feeding the motor.

SELOC_1418

Capacities

Model (HP)	No of Cyl	Year	Displace cu.in. (cc)	Model	Injection Oil Qt (L)	Lower Unit Oz. (ml)	Fuel Tank Gal. (L)	Min Fuel Octane	Fuel/Oil Ratio Break-in / Normal
2.5/3.3	1	2001-05	4.6 (74.6)	all	PreMix	3.0 (90)	0.375 (1.4)	87	50:1 / 50:1
4/5	1	2001-05	6.2 (102)	all	PreMix	6.6 (195)	0.66 (2.5)	87	50:1 / 50:1
6/8	2	2001-05	13 (210)	all	PreMix	6.8 (201)	Remote	87	25:1 / 50:1
9.9	2	2001-05	16 (262)	All	PreMix	6.8 (201)	Remote	87	25:1 / 50:1
10/15	2	2001-05	16 (262)	All	PreMix	6.8 (201)	Remote	87	25:1 / 50:1
20/20 Jet/25	2	2001-05	24 (400)	All	PreMix	8.8 (260)	Remote	87	50:1 / 50:1
30/40	2	2001-03	39 (644)	All	1.6 (1.5)	14.9 (441)	Remote	87	25:1 / 50:1
40/40 Jet/50/55	3	2001-09	59 (967)	All	3.0 (2.8)	14.9 (441)	Remote	87	25:1 / 50:1
60	3	2001-09	59 (967)	Standard	3.0 (2.8)	11.5 (340)	Remote	87	25:1 / 50:1
	3	2001-09	59 (967)	Bigfoot	3.0 (2.8)	22.5 (665.4)	Remote	87	25:1 / 50:1
65 Jet/75/90	3	2001-09	85 (1386)	All	3.2 (3.0)	22.5 (665.4)	Remote	87	25:1 / 50:1
75/80 Jet/90/115 Optimax	3	2004-09	93 (1526)	All	5.0 (4.7)	22.5 (665.4)	Remote	87	ECM controlled
80 Jet/115/125	4	2001-05	113 (1848)	All	5.13 (4.9)	22.5 (665.4)	Remote	87	25:1 / 50:1
135/110 Jet/150/175 Optimax	V6	2001-09	153 (2507)	All	①	22.5 (665.4) ③	Remote	87	ECM controlled
150 (XR6,Classic) and 200	V6	2001-05	153 (2507)	All	12 (11.4)	22.5 (665.4)	Remote	87	25:1 / 50:1
150/175/200 (2.5L) EFI	V6	2001-09	153 (2507)	All	12 (11.4)	22.5 (665.4)	Remote	87	ECM controlled
200/225/250 (3.0L) EFI	V6	2001-05	185 (3032)	All	12 (11.4)	28 (828)	Remote	87	25:1/50:1-ECM controlled
200/225/250 Optimax	V6	2001-09	185 (3032)	All	① ②	27 (800) ④	Remote	87	ECM controlled

n/a - not available

① Oil tank capacity not listed for most Optimax motors, however they likely use the same 3 gallon (11.4 liter) tank

② Most models equipped with a 0.88 Qt. (0.84L) engine mounted transfer tank

③ For 175 pro/XS models, check owners manual capacity may be higher

④ Specification is regular Optimax models, for Pro/XS/Sport models capacity is 28 Oz. (828 ml)

TUNE-UP SPECIFICATIONS

Model (HP)	No of Cyl	Year	Displace cu.in. (cc)	Spark Plug NGK	Spark Plug Champion	Spark Plug Gap Inch(mm)	Idle Timing °BTDC	Pick-Up	Max BTDC Cranking	Running WOT Timing °BTDC	Idle Speed RPM (In Gear)	WOT RPM
2.5	1	2001-05	4.6 (74.6)	BPR6HS-10	RL87YC	0.040 (1.0)	-		-	-	900-1000	4000-5000
3.3	1	2001-05	4.6 (74.6)	BPR6HS-10	RL87YC	0.040 (1.0)	-		-	-	900-1000	4500-5500
4	1	2001-05	6.2 (102)	BP7HS-10	L82YC	0.040 (1.0)	5° BTDC			28°-32° BTDC	800-900	4500-5500
5	1	2001-05	6.2 (102)	BP7HS-10	L82YC	0.040 (1.0)	5° BTDC			28°-32° BTDC	800-900	4500-5500
6	2	2001-03	13 (210)	BP8HN-10[1]	-	0.040 (1.0)	7°-9° BTDC			36° BTDC	575-725	4000-5000
8	2	2001-05	13 (210)	BP8HN-10[1]	-	0.040 (1.0)	7°-9° BTDC			36° BTDC	675-775	4500-5500
9.9	2	2001-05	16 (262)	BP8HS-15[1]	-	0.060 (1.5)	7°-9° BTDC			36° BTDC	675-775	5000-6000
10	2	2001	16 (262)	BP8HS-15[1]	-	0.060 (1.5)	7°-9° BTDC			36° BTDC	800-900	5000-6000
15	2	2001-05	16 (262)	BP8HS-15[1]	-	0.060 (1.5)	7°-9° BTDC			36° BTDC	675-775	5000-6000
20	2	2001-03	24 (400)	BP8HN-10[1]	-	0.040 (1.0)	5°-7° BTDC			24°-26° BTDC	700-800	4500-5500
20 Jet	2	2001-05	24 (400)	BP8HN-10[1]	-	0.040 (1.0)	5°-7° BTDC			24°-26° BTDC	700-800	4500-5500
25	2	2001-05	24 (400)	BP8HN-10[1]	-	0.040 (1.0)	5°-7° BTDC			24°-26° BTDC	700-800	5000-6000
30	2	2001-02	39 (644)	BP8HN-10[1]	-	0.040 (1.0)	7°-9° BTDC			22°-28° BTDC [2]	700-800	4500-5500
40 Standard	2	2001-03	39 (644)	BP8HN-10[1]		0.040 (1.0)	7°-9° BTDC			22°-28° BTDC [2]	700-800	5000-5500
40 Seapro-Marathon	2	2001-03	39 (644)	BP8HN-10[1]		0.040 (1.0)	7°-9° BTDC			22°-28° BTDC [2]	700-800	4500-5500
40	3	2001-05	59 (967)	BP8HN-10[1]	-	0.040 (1.0)	2° ATDC - 2° BTDC		24° BTDC	22° BTDC	650-700	5000-5500
40 Jet Standard	3	2001-05	59 (967)	BP8HN-10[1]	-	0.040 (1.0)	6° BTDC - 7° BTDC		24° BTDC	22° BTDC	725-825	5000-5500
50	3	2001-09	59 (967)	BP8HN-10[1]	-	0.040 (1.0)	2° ATDC - 2° BTDC		24° BTDC	22° BTDC	650-700	5000-5500
55	3	2001	59 (967)	BP8HN-10[1]	-	0.040 (1.0)	2° ATDC - 2° BTDC		18° BTDC	16° BTDC	650-700	5000-5500
60 Standard	3	2001-09	59 (967)	BP8HN-10[1]	-	0.040 (1.0)	2° ATDC - 2° BTDC		24° BTDC	22° BTDC	650-700	5000-5500
60 Seapro-Marathon	3	2001-09	59 (967)	BP8HN-10[1]	-	0.040 (1.0)	2° ATDC - 2° BTDC		18° BTDC	16° BTDC	650-700	5000-5500
65 Jet	3	2001-09	85 (1386)	BUHW-2		Surface Gap/(No spec)	2° ATDC - 6° BTDC		22° BTDC	20° BTDC	650-700	5000-5500
75	3	2001-05	85 (1386)	BUHW-2		Surface Gap/(No spec)	2° ATDC - 6° BTDC		20° BTDC	18° BTDC	650-700	4750-5250
90	3	2001-09	85 (1386)	BUHW-2		Surface Gap/(No spec)	2° ATDC - 6° BTDC		22° BTDC	20° BTDC	650-700	5000-5500
75 Optimax	3	2004-09	93 (1526)	PZFR6H[8]	-	0.030 (0.8) [4]	Timing controlled by the ECM and is not adjustable				625-675	5000-5750
90 Optimax	3	2004-09	93 (1526)	PZFR6H[8]	-	0.030 (0.8) [4]	Timing controlled by the ECM and is not adjustable				625-675	5000-5750
80 Jet/115 Optimax	3	2004-09	93 (1526)	PZFR6H[8]	-	0.030 (0.8) [4]	Timing controlled by the ECM and is not adjustable				625-675	5000-5750
80 Jet	4	2001-05	113 (1848)	BP8HN-10[1]	-	0.040 (1.0)	4° ATDC - 2° BTDC		25° BTDC	23° BTDC	650-700	4750-5250
115	4	2001-05	113 (1848)	BP8HN-10[1]	-	0.040 (1.0)	4° ATDC - 2° BTDC		25° BTDC	23° BTDC	650-700	4750-5250
125	4	2001-05	113 (1848)	BP8HN-10[1]	-	0.040 (1.0)	4° ATDC - 2° BTDC		25° BTDC	23° BTDC	650-700	4750-5250

SELOC_1418

TUNE-UP SPECIFICATIONS

Model (HP)	No of Cyl	Year	Displace cu.in. (cc)	Spark Plug NGK	Spark Plug Champion	Spark Plug Gap Inch(mm)	Idle Timing °BTDC	Pick-Up	Max BTDC Cranking	Running WOT Timing °BTDC	Idle Speed RPM (In Gear)	WOT RPM
135 Optimax	V6	2001-09	153 (2507)	PZFR5F-11	QC12GMC	0.040 (1.0)	Timing controlled by the ECM and is not adjustable				525-575	5000-5500
150 (XR6,Classic)	V6	2001-05	153 (2507)	BP28HS-10	-	0.040 (1.0)	0°-9° ATDC	0°-9° ATDC	26° BTDC	20° BTDC	625-725	5000-5500
150 EFI	V6	2001	153 (2507)	BP28HS-10	-	0.040 (1.0)	0°-9° ATDC	0°-9° ATDC	22° BTDC	16° BTDC	600-700	5000-5600
150 EFI	V6	2002-09	153 (2507)	BP28HS-10	-	0.040 (1.0)	1°-7° ATDC	-	-	18° BTDC	600-700	5250-5750
110 Jet/150 Optimax	V6	2001-09	153 (2507)	PZFR5F-11	QC12GMC	0.040 (1.0)	Timing controlled by the ECM and is not adjustable				525-575	5250-5750
175 EFI	V6	2001	153 (2507)	BP28HS-10	-	0.040 (1.0)	0°-9° ATDC	0°-9° ATDC	26° BTDC	20° BTDC	600-700	5000-5600
175 EFI	V6	2002	153 (2507)	BP28HS-10	-	0.040 (1.0)	1°-7° ATDC	-	-	18° BTDC	600-700	5250-5750
175 Optimax	V6	2001-09	153 (2507)	PZFR5F-11 ⑨	QC12GMC	0.040 (1.0)	Timing controlled by the ECM and is not adjustable				525-575	5250-5750 ⑩
200	V6	2001-05	153 (2507)	BP28HS-10	-	0.040 (1.0)	0°-9° ATDC	0°-9° ATDC	24° BTDC	18° BTDC	600-700	5000-5500
200 EFI (2.5L)	V6	2001	153 (2507)	BP28HS-10	-	0.040 (1.0)	0°-9° ATDC	0°-9° ATDC	24° BTDC	18° BTDC	600-700	5000-5800
200 EFI (2.5L)	V6	2002-05	153 (2507)	BP28HS-10	-	0.040 (1.0)	1°-7° ATDC	-	-	20° BTDC	600-700	5250-5750
200 EFI (3.0L)	V6	2004-05	185 (3032)	1ZFR5G ⑦	QL77CC	0.040 (1.0)	0°-6° BTDC ⑤	-	-	22° BTDC ⑤	575-675	⑥
200 Optimax	V6	2001-09	185 (3032)	1ZFR5G ⑦	-	0.031 (0.8)	Timing controlled by the ECM and is not adjustable				550-575	5000-5750
225 EFI	V6	2001	185 (3032)	-	QL77CC	0.035 (0.89)	4°-8° ATDC ⑤	-	-	24°BTDC @ 5000 rpm ⑤	600-700	5000-5800
225 EFI	V6	2002-05	185 (3032)	-	QL77CC	0.040 (1.0)	0°-6° BTDC ⑤	-	-	24°BTDC @ 5800 rpm ⑤	575-675	5000-5500
225 Optimax	V6	2001-09	185 (3032)	1ZFR5G ⑦	-	0.031 (0.8)	Timing controlled by the ECM and is not adjustable				550-575	5000-5750 ⑩
250 EFI	V6	2001	185 (3032)	-	QL77CC	0.035 (0.89)	4°-8° ATDC ⑤	-	-	24°BTDC @ 5000 rpm ⑤	600-700	5000-5800
250 EFI	V6	2002-05	185 (3032)	-	QL77CC	0.040 (1.0)	0°-6° BTDC ⑤	-	-	28°BTDC @ 5800 rpm ⑤	575-675	5000-5500
250 Optimax	V6	2006-09	185 (3032)	1ZFR5G ⑦	-	0.031 (0.8)	Timing controlled by the ECM and is not adjustable				550-575	5500-6000 ⑩

① Can also use resistor spark plug BP28H-N-10
② Spec is for reference only, timing is set by adjusting 1 turn CLOCKWISE after contacting the throttle plate
③ Spec is for all EXCEPT WME-38 Carb on 40MLL models, for which the spec is 1 1/2 - 2
④ Sources conflict. Some say spec listed, others say 0.043 in. (1.1mm), if possible check emissions label
⑤ All specs are for reference or checking only, timing is ECM controlled and NOT adjustable on these models
⑥ Spec varies, for 3.0L Work motors should be 4500-5500, but standard 200 EFI are 5000-5700
⑦ Standard Models can also utilize NGK plug PZFR5F-11 with gap of 0.043 in. (1.1mm), For XS models use IZFR6J with gap of 0.032 (0.81)
⑧ Can also utilize NGK plug IZFR5J
⑨ For Pro/XS/Sport Models, check emissions label for plug and gap
⑩ Spec is for standard models, Pro/XS/Sport models are 5500-6000 rpm

SELOC_1418

SPARK PLUG DIAGNOSIS

Tracking Arc
High voltage arcs between a fouling deposit on the insulator tip and spark plug shell. This ignites the fuel/air mixture at some point along the insulator tip, retarding the ignition timing which causes a power and fuel loss.

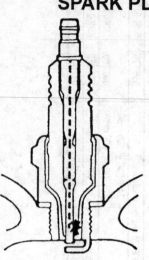

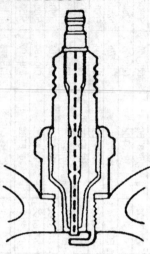

Wide Gap
Spark plug electrodes are worn so that the high voltage charge cannot arc across the electrodes. Improper gapping of electrodes on new or "cleaned" spark plugs could cause a similar condition. Fuel remains unburned and a power loss results.

Flashover
A damaged spark plug boot, along with dirt and moisture, could permit the high voltage charge to short over the insulator to the spark plug shell or the engine. A buttress insulator design helps prevent high voltage flashover.

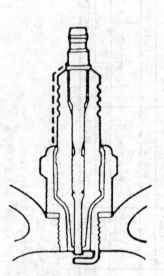

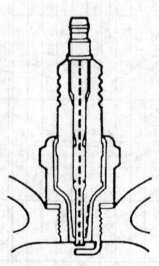

Fouled Spark Plug
Deposits that have formed on the insulator tip may become conductive and provide a "shunt" path to the shell. This prevents the high voltage from arcing between the electrodes. A power and fuel loss is the result.

Bridged Electrodes
Fouling deposits between the electrodes "ground out" the high voltage needed to fire the spark plug. The arc between the electrodes does not occur and the fuel air mixture is not ignited. This causes a power loss and exhausting of raw fuel.

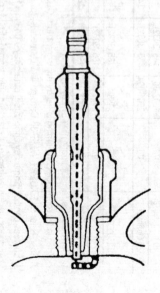

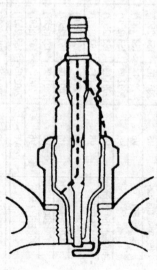

Cracked Insulator
A crack in the spark plug insulator could cause the high voltage charge to "ground out." Here, the spark does not jump the electrode gap and the fuel air mixture is not ignited. This causes a power loss and raw fuel is exhausted.

SELOC_1418

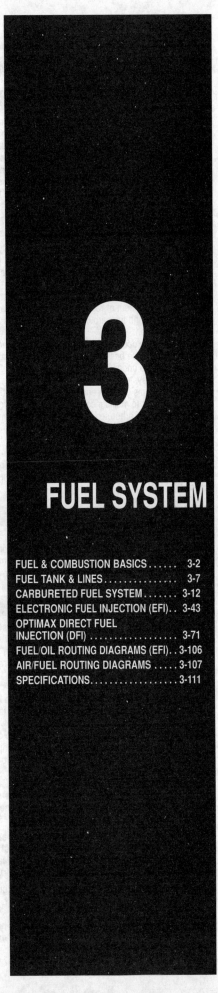

3

FUEL SYSTEM

FUEL & COMBUSTION BASICS

✳✳ SELOC CAUTION

If equipped, disconnect the negative battery cable anytime work is performed on the engine, especially when working on the fuel system. This will help prevent the possibility of sparks during service (from accidentally grounding a hot lead or powered component). Sparks could ignite vapors or exposed fuel. Disconnecting the cable on electric start motors will also help prevent the possibility fuel spillage if an attempt is made to crank the engine while the fuel system is open.

✳✳ SELOC CAUTION

Fuel leaking from a loose, damaged, or incorrectly installed hose or fitting may cause a fire or an explosion. ALWAYS pressurize the fuel system and run the motor while inspecting for leaks after servicing any component of the fuel system.

To tune, troubleshoot or repair the fuel system of an outboard motor you must first understand the carburetion or fuel injection, and ignition principles of engine operation.

If you have any doubts concerning your understanding of engine operation, it would be best to study The Basic Operating Principles of an engine as detailed under Troubleshooting in Section 1, before tackling any work on the fuel system.

The fuel systems used on engines covered by this guide range from single carburetors to multiple carburetors or various forms of electronic fuel injection systems. Carbureted motors utilize various means of enriching fuel mixture for cold starts, including a manual choke (2.4-5 hp models), manual primer (most 6 hp and larger manual start models) or electric primer solenoid (most 6 hp and larger electric start models). Also, some larger models may use a thermal air valve which restricts intake air at idle along with an electric primer assembly.

Fuel System Service Cautions

There is no way around it. Working with gasoline can provide for many different safety hazards and requires that extra caution is used during all steps of service. To protect yourself and others, you must take all necessary precautions against igniting the fuel or vapors (which will cause a fire at best or an explosion at worst).

✳✳ SELOC CAUTION

Take extreme care when working with the fuel system. NEVER smoke (it's bad for you anyhow, but smoking during fuel system service could kill you much faster!) or allow flames or sparks in the work area. Flames or sparks can ignite fuel, especially vapors, resulting in a fire at best or an explosion at worst.

For starters, disconnect the negative battery cable **EVERY** time a fuel system hose or fitting is going to be disconnected. It takes only 1 moment of forgetfulness for someone to crank the motor, possibly causing a dangerous spray of fuel from the opening. This is especially true on EFI or Optimax systems that contain an electric fuel pump, where simply turning the key on (without cranking) will energize the pump and result in high-pressure fuel spray in certain fuel system lines/fittings.

Gasoline contains harmful additives and is quickly absorbed by exposed skin. As an additional precaution, always wear gloves and some form of eye protection (regular glasses help, but only safety glasses can really protect your eyes).

■ **Throughout service, pay attention to ensure that all components, hoses and fittings are installed them in the correct location and orientation to prevent the possibility of leakage. Matchmark components before they are removed as necessary.**

Because of the dangerous conditions that result from working with gasoline and fuel vapors always take extra care and be sure to follow these guidelines for safety:
- Keep a Coast Guard-approved fire extinguisher handy when working.
- Allow the engine to cool completely before opening a fuel fitting. Don't allow gasoline to drip on a hot engine.

- The first thing you must do after removing the engine cover is to check for the presence of gasoline fumes. If strong fumes are present, look for leaking or damage hoses, fittings or other fuel system components and repair.
- Do not repair the motor or any fuel system component near any sources of ignition; including sparks, open flames, or anyone smoking.
- Clean up spilled gasoline right away using clean rags. Keep all fuel soaked rags in a metal container until they can be properly disposed of or cleaned. NEVER leave solvent, gasoline or oil soaked rags in the hull.
- Don't use electric powered tools in the hull or near the boat during fuel system service or after service, until the system is pressurized and checked for leaks.
- Fuel leaking from a loose, damaged or incorrectly installed hose or fitting may cause a fire or an explosion. ALWAYS pressurize the fuel system and run the motor while inspecting for leaks after servicing any component of the fuel system.

Fuel

◆ See Figure 1

Fuel recommendations have become more complex as the chemistry of modern gasoline changes. The major driving force behind the many of the changes in gasoline chemistry was the search for additives to replace lead as an octane booster and lubricant. These additives are governed by the types of emissions they produce in the combustion process. Also, the replacement additives do not always provide the same level of combustion stability, making a fuel's octane rating less meaningful.

In the 1960's and 1970's, leaded fuel was common. The lead served two functions. First, it served as an octane booster (combustion stabilizer) and second, in 4-stroke engines, it served as a valve seat lubricant. For 2-stroke engines, the primary benefit of lead was to serve as a combustion stabilizer. Lead served very well for this purpose, even in high heat applications.

For decades now, all lead has been removed from the refining process. This means that the benefit of lead as an octane booster has been eliminated. Several substitute octane boosters have been introduced in the place of lead. While many are adequate in automobile engines, most do not perform nearly as well as lead did, even though the octane rating of the fuel is the same.

OCTANE RATING

◆ See Figure 1

■ **These Mercury outboards may all run on a minimum of U.S. pump fuel labeled 87 Octane.**

A fuel's octane rating is a measurement of how stable the fuel is when heat is introduced. Octane rating is a major consideration when deciding whether a fuel is suitable for a particular application. For example, in an engine, we want the fuel to ignite when the spark plug fires and not before, even under high pressure and temperatures. Once the fuel is ignited, it must burn slowly and smoothly, even though heat and pressure are building up while the burn occurs. The unburned fuel should be ignited by the traveling flame front, not by some other source of ignition, such as carbon deposits or the heat from the expanding gasses. A fuel's octane rating is known as a measurement of the fuel's anti-knock properties (ability to burn without exploding). Essentially, the octane rating is a measure of a fuel's stability.

Usually a fuel with a higher octane rating can be subjected to a more severe combustion environment before spontaneous or abnormal combustion occurs. To understand how 2 gasoline samples can be different, even though they have the same octane rating, we need to know how octane rating is determined.

The American Society of Testing and Materials (ASTM) has developed a universal method of determining the octane rating of a fuel sample. The octane rating you see on the pump at a gasoline station is known as the pump octane number. Look at the small print on the pump. The rating has a formula. The rating is determined by the R+M/2 method. This number is the average of the research octane reading and the motor octane rating.

- The Research Octane Rating is a measure of a fuel's anti-knock properties under a light load or part throttle conditions. During this test, combustion heat is easily dissipated.

Fig. 1 Damaged piston, possibly caused by; using too-low an octane fuel or by insufficient oil

• The Motor Octane Rating is a measure of a fuel's anti-knock properties under a heavy load or full throttle conditions, when heat buildup is at maximum.

Because a 2-stroke engine has a power stroke every revolution, with heat buildup every revolution, it tends to respond more to the motor octane rating of the fuel than the research octane rating. Therefore, in an outboard motor, the motor octane rating of the fuel is the best indication of how it will perform.

VAPOR PRESSURE

Fuel vapor pressure is a measure of how easily a fuel sample evaporates. Many additives used in gasoline contain aromatics. Aromatics are light hydrocarbons distilled off the top of a crude oil sample. They are effective at increasing the research octane of a fuel sample but can cause vapor lock (bubbles in the fuel line) on a very hot day. If you have an inconsistent running engine and you suspect vapor lock, use a piece of clear fuel line to look for bubbles, indicating that the fuel is vaporizing.

One negative side effect of aromatics is that they create additional combustion products such as carbon and varnish. If your engine requires high octane fuel to prevent detonation, de-carbon the engine more frequently with an internal engine cleaner to prevent ring sticking due to excessive varnish buildup.

ALCOHOL-BLENDED FUELS

When the Environmental Protection Agency mandated a phase-out of the leaded fuels in January of 1986, fuel suppliers needed an additive to improve the octane rating of their fuels. Although there are multiple methods currently employed, the addition of alcohol to gasoline seems to be favored because of its favorable results and low cost. Two types of alcohol are used in fuel today as octane boosters, methanol (wood alcohol) or ethanol (grain alcohol).

When used as a fuel additive, alcohol tends to raise the research octane of the fuel, so these additives will have limited benefit in an outboard motor. There are, however, some special considerations due to the effects of alcohol in fuel.

• Since alcohol contains oxygen, it replaces gasoline without oxygen content and tends to cause the air/fuel mixture to become leaner.

• On older outboards, the leaching affect of alcohol will, in time, cause fuel lines and plastic components to become brittle to the point of cracking. Unless replaced, these cracked lines could leak fuel, increasing the potential for hazardous situations. However, later model outboards (such as the models covered here) utilize fuel lines which are much more resistant to alcohol in the fuel mixture.

■ **Modern outboard fuel lines and plastic fuel system components have been specially formulated to resist alcohol leaching effects.**

• When alcohol blended fuels become contaminated with water, the water combines with the alcohol then settles to the bottom of the tank. This leaves the gasoline (and the oil for models using premix) on a top layer.

■ **Modern outboard fuel lines and plastic fuel system components have been specially formulated to resist alcohol leaching effects.**

RECOMMENDATIONS - THE BOTTOM LINE WITH FUELS

If we could buy fuel of the correct octane rating, free of alcohol and aromatics, this would be our first choice.

Mercury continues to recommend unleaded fuel. This is almost a redundant recommendation due to the near universal unavailability of any other type fuel.

According to the fuel recommendations that come with your outboard, there is no engine in the product line that requires more than 87. An 87 octane rating generally means bottom grade unleaded (at least in the U.S.). Premium unleaded is more stable under severe conditions, but also produces more combustion products. Therefore, when using premium unleaded, more frequent de-carboning is necessary.

Regardless of the fuel octane rating you choose, try to stay with a name brand fuel. You never know for sure what kinds of additives or how much is in off brand fuel.

HIGH ALTITUDE OPERATION

At elevated altitudes there is less oxygen in the atmosphere than at sea level. Less oxygen means lower combustion efficiency and less power output. As a general rule, power output is reduced 3% for every thousand feet above sea level. At ten thousand feet, power is reduced 30% percent from that available at sea level.

Re-jetting carburetors for high altitude does not restore this lost power. It simply corrects the air-fuel ratio for the reduced air density, and makes the most of the remaining available power. If you re-jet an engine, you are locked into the higher elevation. You cannot operate at sea level until you re-jet for sea level. Understand that going below the elevation jetted for your motor will damage the engine with overly lean fuel:air ratios. As a general rule, jet for the lowest elevation anticipated. Spark plug insulator tip color is the best guide for high altitude jetting.

If you are in an area of known poor fuel quality, you may want to use fuel additives. Today's additives are mostly alcohol and aromatics, and their effectiveness may be limited. It is difficult to find additives without ethanol, methanol, or aromatics. If you use octane boosters frequent de-carboning may be necessary. If possible, the best policy is to use name brand pump fuel with no additional additives except those sold by Mercury such as Quicksilver Fuel Stabilizer.

CHECKING FOR STALE/CONTAMINATED FUEL

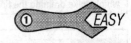

◆ **See Figure 2 and 3**

Outboard motors often sit weeks at a time making them the perfect candidate for fuel problems. Gasoline has a short life, as combustibles begin evaporating almost immediately. Even when stored properly, fuel starts to deteriorate within a few months, leaving behind a stale fuel mixture that can cause hard-starting, poor engine performance and even lead to possible engine damage.

Furthermore, as gasoline evaporates it leaves behind gum deposits that can clog filters, lines and small passages. Although the sealed high-pressure fuel system of an EFI or Optimax motor is less susceptible to fuel evaporation, the low-pressure fuel systems of all engines can suffer the effects. However carburetors, due to their tiny passages and naturally vented designs are the most susceptible components on non-fuel injected motors.

As mentioned under Alcohol-Blended fuels, modern fuels contain alcohol, which is hygroscopic (meaning it absorbs water). And, over time, fuel stored in a partially filled tank or a tank that is vented to the atmosphere will absorb water. The water/alcohol settles to the bottom of the tank, promoting rust (in metal tanks) and leaving a non-combustible mixture at the bottom of a tank that could leave a boater stranded.

One of the first steps to fuel system troubleshooting is to make sure the fuel source is not at fault for engine performance problems. Check the fuel if the engine will not start and there is no ignition problem. Stale or contaminated fuels will often exhibit an unusual or even unpleasant unusual odor.

■ The best method of disposing stale fuel is through a local waste pick-up service, automotive repair facility or marine dealership. But, this can be a hassle. If fuel is not too stale or too badly contaminated, it may be mixed with greater amounts of fresh fuel and used to power lawn/yard equipment or (as long as we're not talking about disposing pre-mix here) even an automobile (if greatly diluted so as to prevent misfiring, unstable idle or damage to the automotive engine). But we feel that it is much less of a risk to have a lawn mower stop running because of the fuel problem than it is to have your boat motor quit or refuse to start.

Mercury carburetors are normally equipped with a float bowl drain screw that can be used to drain fuel from the carburetor for storage or for inspection. The vapor separator tank on fuel injected models is also usually equipped with either a drain screw or drain valve (the screws are not unlike the float bowl screws on carburetor motors). The drain screws on these motors are usually pretty easy to access, but on a few motors it may be easier to drain a fuel sample from the hoses leading to or from the low-pressure fuel filter or fuel pump. Removal and installation instructions for the fuel filters are provided in the Maintenance Section, while fuel pump procedures are found in this section. To check for stale or contaminated fuel:

Fig. 2 Mercury float bowls are usually equipped with drain screws which make draining the float bowls or taking a fuel sample a snap

Fig. 3 Commercial additives, such as Sta-bil or Mercury Stabilizer may be used to help prevent "souring"

1. Disconnect the negative battery cable for safety. Secure it or place tape over the end so that it cannot accidentally contact the terminal and complete the circuit.

2. On carbureted motors, remove the float bowl drain screw (and orifice plug, if equipped), then allow a small amount of fuel to drain into a glass container.

3. On fuel injected motors, remove the float bowl drain screw from the bottom of the vapor separator tank or use the drain valve located in the same position (depending on the year and model). You can also disconnect the fuel supply hose from the pump or low-pressure fuel filter (as desired).

4. If there is no fuel present in the float bowl, vapor tank or fuel line (as applicable) squeeze the fuel primer bulb to force fuel through the lines and to the opening. If necessary, you can always disconnect the fuel line between the low-pressure filter and pump or, on models with serviceable filter elements simply unscrew the housing to get a sample.

■ If a sample cannot be obtained from the float bowls/separator tank, fuel filter or pump supply hose, there is a problem with the fuel tank-to-motor fuel circuit. Check the tank, primer bulb, fuel hose, fuel pump, fitting or carburetor float inlet needle (as necessary).

5. Check the appearance and odor of the fuel. An unusual smell, signs of visible debris or a cloudy appearance (or even the obvious presence of water) points to a fuel that should be replaced.

6. If contaminated fuel is found, drain the fuel system and dispose of the fuel in a responsible manner, then clean the entire fuel system.

■ If debris is found in the fuel system, clean and/or replace all fuel filters.

7. When finished, reconnect the negative battery cable, then properly pressurize the fuel system and check for leaks.

Fuel System Pressurization

When it comes to safety and outboards, the condition of the fuel system is of the utmost importance. The system must be checked for signs of damage or leakage with every use and checked, especially carefully when portions of the system have been opened for service.

The best method to check the fuel system is to visually inspect the lines, hoses and fittings once the system has been properly pressurized.

All engines, carbureted or fuel injected utilize a low-pressure fuel circuit which delivers fuel from a portable or boat mounted tank to the powerhead via fuel hoses and the action of a diaphragm-displacement fuel pump.

Fuel injected engines utilize a multi-stage fuel system, the low-pressure circuit (fuel tank-to-pump and pump-to-vapor separator), as well as a high-pressure circuit (vapor separator-to-fuel injectors). Though keep in mind that there is a difference between the high-pressure circuit of EFI motors and that of Optimax motors.

On EFI motors, a high-pressure fuel pump submerged in the vapor separator tank provides a constant supply of fuel to the fuel rails at an operating pressure somewhere in the 36-39 psi (248-269 kPa) range depending upon the year and model. Pressure is maintained by a regulator assembly which directs excess fuel/pressure back to the separator tank.

Optimax motors, being direct injection units which combine pressurized fuel with pressurized air in a combination air/fuel rail assembly, operate the fuel side of the circuit in a slightly different manner. Fuel is still delivered to the water separator assembly or water separator portion of the vapor separator tank in the same basic manner, but from there a low-pressure electric fuel pump (or boost pump as it is often called by Mercury) increases fuel pressure slightly as the fuel is fed to the fuel chamber of the vapor separator tank. Depending upon the year and model this boost pump might be located externally on the end of the vapor separator tank or may be submerged inside the assembly. In the fuel chamber of all Optimax motors a high-pressure fuel pump takes this fuel and delivers it to the air/fuel rail assembly at an operating pressure somewhere in the neighborhood of 87-110 psi (600-758 kPa) or higher, again, depending upon the year and model.

RELEASING FUEL SYSTEM PRESSURE (EFI/OPTIMAX MODELS)

◆ See Figures 4, 5 and 6

On fuel injected engines, always relieve system pressure prior to disconnecting any fuel system component, fitting or fuel line.

✳✳ SELOC CAUTION

Exercise extreme caution whenever relieving fuel system pressure to avoid fuel spray and potential serious bodily injury. Please be advised that fuel under pressure may actually penetrate the skin or any part of the body it contacts.

To avoid the possibility of fire and personal injury, always disconnect the negative battery cable (this will not only prevent sparks but prevent fuel from possibly spraying out of open fittings should you or someone else accidentally turn the ignition key to on while you are working).

Keep in mind that EFI motors have a single high-pressure fuel circuit and therefore only 1 Schrader valve fitting from which to relieve system pressure, but Optimax motors have both a low-pressure and high-pressure electric fuel pump circuit, each with their own Schrader valve fittings from which you should relieve pressure (unless you're certain you'll only be opening up access to one or the other). For the most part the Schrader valve fittings for EFI motors (and low-pressure fitting for Optimax motors) are found on or near the vapor separator tank assembly. The high-pressure fitting for Optimax motors is normally found on the air/fuel rail assembly.

✳✳ SELOC WARNING

Although Mercury has at times suggested it, and we know it is FAIRLY common practice, we cannot in good conscience recommend relieving system pressure by simply wrapping the Schrader valve with a shop rag to absorb the spray, then using a screwdriver or other tool to depress/open the valve. The fuel spray will be so well atomized and so easily ignited as to make it a very dangerous situation.

Always place a shop towel or cloth around the fitting or connection prior to loosening to absorb any excess fuel due to spillage. Ensure that all fuel spillage is removed from engine surfaces. Ensure that all fuel soaked clothes or towels in suitable waste container.

1. Remove the plastic cap over the pressure port located either on/near the vapor separator tank or on the combination air/fuel rail, as applicable. For more details, please refer to the information on Fuel Injection in the Fuel System section.

2. Attach a fuel pressure gauge to the pressure test port Schrader valve and direct the bleed valve hose into a suitable container. Carefully open the bleed valve allowing fuel to drain into the container until system pressure is released.

PRESSURIZING THE FUEL SYSTEM (CHECKING FOR LEAKS)

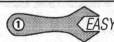 EASY

Carbureted Motors

Carbureted engines are equipped with a low-pressure fuel system, making pressure release before service a relative non-issue. However, if the primer bulb is firm when you go to disconnect a fuel line, there will be some fuel leakage once the hose is disconnected, so always have a rag or handy during service. But, even a low-pressure fuel system should be checked following repairs to make sure that no leaks are present. Only by checking a fuel system under normal operating pressures can you be sure of the system's integrity.

✳✳ SELOC CAUTION

Fuel leaking from a loose, damaged or incorrectly installed hose or fitting may cause a fire or an explosion. Always pressurize the fuel system and run the motor while inspecting for leaks after servicing any component of the fuel system.

Most carbureted engines (except some integral tank models with gravity feed) utilize a fuel primer bulb mounted inline between the fuel tank and engine. On models so equipped, the bulb can be used to pressurize that portion of the fuel system. Squeeze the bulb until it and the fuel lines feel firm with gasoline. At this point check all fittings between the tank and motor for signs of leakage and correct, as necessary.

Once fuel reaches the engine it is the job of the fuel pump(s) to distribute it to the carburetor(s). No matter what system you are inspecting, start and run the motor with the engine top case removed, then check each of the system hoses, fittings and gasket-sealed components to be sure there is no leakage after service.

Fuel Injected Motors

The low-pressure fuel delivery circuit of an EFI or Optimax motor is pressurized and checked using the primer bulb, in the same manner as the low-pressure circuit of a carbureted motor. For more details, refer to the section on Carbureted Motors.

However, the vapor separator tank contains an electric high-pressure pump which is used on the high-pressure circuit to build system pressure (and a low-pressure boost pump on Optimax motors). Normally this pump is energized for a few seconds when the ignition switch is turned on (even before the engine is started). If so, you can usually hear the pump run for a few seconds when the switch is turned on. For these models, turn the ignition switch on, wait a few seconds, and then turn the switch off again. Repeat a few times until system pressure starts to build (you can monitor this using a pressure gauge attached to the fuel pressure test port).

■ The best way to make sure no leaks occur during service is to visually inspect the motor while it is running at idle (that way all fuel circuits are pressurized and operating as normal).

Fig. 4 Typical EFI vapor separator tank with pressure port

Fig. 5 Optimax motors may also have pressure ports along the middle. . .

Fig. 6 . . . or at the top of the air/fuel rail assembly

⁂ SELOC CAUTION

As usual, use extreme caution when working around an operating engine. Keep hands, fingers, hair, loose clothing, etc as far away as possible from moving components. Severe injury or death could occur from getting caught in rotating components.

Combustion

A 2 cycle engine has a power stroke every revolution of the crankshaft. A 4 cycle engine has a power stroke every other revolution of the crankshaft. Therefore, the 2 cycle engine has twice as many power strokes for any given RPM. If the displacement of the 2 types of engines is identical, then the 2 cycle engine has to dissipate twice as much heat as the 4 cycle engine. In such a high heat environment, the fuel must be very stable to avoid detonation. If any parameters affecting combustion change suddenly (the engine runs lean for example), uncontrolled heat buildup occurs very rapidly in a 2 cycle engine.

ABNORMAL COMBUSTION

There are 2 types of abnormal combustion:
• Pre-ignition - Occurs when the air-fuel mixture is ignited by some other incandescent source other than the correctly timed spark from the spark plug.
• Detonation - Occurs when excessive heat and or pressure ignites the air/fuel mixture rather than the spark plug. The burn becomes explosive.

In both cases, pistons still on their way up towards TDC meet the force of early ignited expanding air/fuel mixture, which rattles the pistons back and forth against the cylinder walls. This causes a metallic rattling sound, similar to castanets and can damage from rapid wear at best to blowing holes in the tops of the pistons at worst.

FACTORS AFFECTING COMBUSTION

The combustion process is affected by several interrelated factors. This means that when one factor is changed, the other factors also must be changed to maintain the same controlled burn and level of combustion stability.

Compression

Determines the level of heat buildup in the cylinder when the air-fuel mixture is compressed. As compression increases, so does the potential for heat buildup. Over time, the build-up of carbon deposits can increase compression (by decreasing the available size of the combustion chamber). Similarly, machining the cylinder head (to correct for warpage) can increase compression. Increases of compression often bring with them the need to increase fuel octane to prevent abnormal combustion.

Ignition Timing

Determines when the ignited gasses will start to expand in relation to the motion of the piston. If the gasses begin to expand too soon, such as they would during pre-ignition or in an overly advanced ignition timing, the motion of the piston opposes the expansion of the gasses, resulting in extremely high combustion chamber pressures and heat.

As ignition timing is retarded, the burn occurs later in relation to piston position. This means that the piston has less distance to travel under power to the bottom of the cylinder, resulting in less usable power.

Fuel Mixture

Determines how efficient the burn will be. A rich mixture burns cooler and slower than a lean one. However, if the mixture is too lean, it can't become explosive. The slower the burn, the cooler the combustion chamber, because pressure buildup is gradual.

Fuel Quality (Octane Rating)

Determines how much heat is necessary to ignite the mixture. Once the burn is in progress, heat is on the rise. The unburned poor quality fuel is ignited all at once by the rising heat instead of burning gradually as a flame front of the burn passing by. This action results in detonation (pinging). Think of octane rating as the measure of how stable and resistant to radical combustion a fuel is. The higher the octane rating, the more stable/resistant is the fuel.

Other Factors

In general, anything that can cause abnormal heat buildup can be enough to push an engine over the edge to abnormal combustion, if any of the 4 basic factors previously discussed are already near the danger point, for example, excessive carbon buildup raises the compression and retains heat as glowing embers.

COMBUSTION RELATED PISTON FAILURES

◆ See Figure 7

When an engine has a piston failure due to abnormal combustion, fixing the mechanical portion of the engine is the easiest part of the equation. The hard part is determining what caused the problem in order to prevent a repeat failure. Think back to the 4 basic areas that affect combustion to find the cause of the failure.

Since you probably removed the cylinder head, inspect the failed piston and look for excessive deposit buildup that could raise compression or retain heat in the combustion chamber. Statically check the wide open throttle timing. Be sure that the timing is not over advanced. It is a good idea to seal these adjustments with paint or liquid electrical tape to detect tampering.

Look for a fuel restriction that could cause the engine to run lean. Don't forget to check the fuel pump, fuel tank and lines, especially if a built in tank is used. Be sure to check the anti-siphon valve on built in tanks. If everything else looks good, the final possibility is poor quality fuel.

SELOC_1418

Fig. 7 This burned piston is typical of a combustion related failure. Combustion temperature was so hot it melted the top of the piston

FUEL TANK & LINES

✳✳ SELOC CAUTION

If equipped, disconnect the negative battery cable anytime work is performed on the engine, especially when working on the fuel system. This will help prevent the possibility of sparks during service (from accidentally grounding a hot lead or powered component). Sparks could ignite vapors or exposed fuel. Disconnecting the cable on electric start motors will also help prevent the possibility fuel spillage if an attempt is made to crank the engine while the fuel system is open.

✳✳ SELOC CAUTION

Fuel leaking from a loose, damaged or incorrectly installed hose or fitting may cause a fire or an explosion. Always pressurize the fuel system and run the motor while inspecting for leaks after servicing any component of the fuel system.

If a problem is suspected in the fuel supply, tank and/or lines, by far the easiest test to eliminate these components as possible culprits is to substitute a known good fuel supply. This is known as running a motor on a test tank (as opposed to running a motor *IN* a test tank, which is an entirely different concept). If possible, borrow a portable tank, fill it with fresh gasoline and connect it to the motor.

■ When using a test fuel tank, make sure the inside diameter of the fuel hose and fuel fittings is at least equal to the size of the hose used on the usual supply line (otherwise you could cause fuel starvation problems or other symptoms which will make troubleshooting nearly impossible.

Fuel Tank

◆ See Figures 8 and 9

There are 3 different types of fuel tanks that might be used along with these motors. The very smallest motors covered by this guide may be equipped with an integral fuel tank mounted to the powerhead. But, most motors, even some of the smaller ones, may be rigged using either a portable fuel tank or a boat mounted tank. In both cases, a tank that is not mounted to the engine itself is commonly called a remote tank.

■ Although many dealers rig boats using Mercury fuel tanks, there are many other tank manufacturers and tank designs may vary greatly. Your outboard might be equipped with a tank from the engine manufacturer, the boat manufacturer or even another tank manufacturer. Although components used, as well as the techniques for cleaning and repairing tanks are similar for almost all fuel tanks be sure to use caution and common sense. Since design vary greatly we cannot provide step-by-step service procedures for the fuel tanks you might encounter. Instead, we've provided information about tanks and how you can decide when service is necessary. Refer to a reputable marine repair shop or marine dealership when parts are needed for aftermarket fuel tanks.

Whether or not your boat is equipped with a boat mounted, built-in tank depends mostly on the boat builder and partially on the initial engine installer. Boat mounted tanks can be hard to access (sometimes even a little hard to find if parts of the deck must be removed). When dealing with boat mounted tanks, look for access panels (as most manufacturers are smart or kind enough to install them for tough to reach tanks). At the very least, all manufacturers must provide access to fuel line fittings and, usually, the fuel level sender assembly.

No matter what type of tank is used, all must be equipped with a vent (either a manual vent or an automatic 1-way check valve) which allows air in (but the later check valve should prevent vapors from escaping). An inoperable vent (one that is blocked in some fashion) would allow the formation of a vacuum that could prevent the fuel pump from drawing fuel from the tank. A blocked vent could cause fuel starvation problems. Whenever filling the tank, check to make sure air does not rush into the tank each time the cap is loosened (which could be an early warning sign of a blocked vent).

Fig. 8 Typical aftermarket portable fuel tank

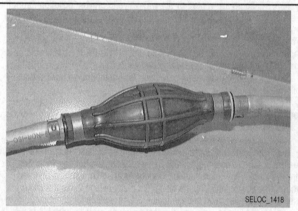

Fig. 9 Both portable and built-in tanks are connected to the motor using a fuel line equipped with a primer bulb

If fuel delivery problems are encountered, first try running the motor with the fuel tank cap removed to ensure that no vacuum lock will occur in the tank or lines due to vent problems. If the motor runs properly with the cap removed but stalls, hesitates or misses with the cap installed, you know the problem is with the tank vent system.

SERVICE

Integral Fuel Tanks

◆ See Figure 10

On small Mercury powerheads equipped with an integral fuel tank above the powerhead, instead of an in-tank filter there is usually a filler neck screen and/or a an inline filter between the tank and carburetor.

Regardless, the best way to check fuel flow is to disconnect the fuel line at the carburetor and fuel should flow from the line if the shut-off valve is open. If fuel is not present, the usual cause is a likely a plugged line or filter.

Would you believe, many times a lack of fuel at the carburetor is caused because the vent on the fuel tank was not opened or has become clogged/stuck? As mentioned under Fuel Tank in this section, it is important to make sure the vent is open at all times during engine operation to prevent formation of a vacuum in the tank that could prevent fuel from reaching the fuel pump or carburetor.

To remove the tank, proceed as follows:

1. Remove the engine cover(s) as detailed in Engine Covers (Top Cover & Cowling) in the Engine Maintenance section.

2. If equipped, make sure the fuel tank valve is shut off.

■ **On models not equipped with a fuel tank mounted shut off valve, have a shop rag, large funnel and gas can handy to drain the fuel into when the hose is removed from the carburetor.**

3. Disconnect the fuel supply line from the fuel tank shut-off valve or trace the line down past the filter to the pump or carburetor and disconnect it there, whichever your preference.

4. Remove the fasteners (usually nuts and lockwashers, but they may vary), flat washers fastening the tank to the powerhead, then make sure no other wires, components or brackets will interfere with tank removal.

5. Carefully lift the fuel tank from the powerhead.

6. Check the tank, filler cap, gasket and, if equipped, filler screen, for wear or damage. Replace any components as necessary to correct any leakage (if found.)

7. If cleaning is necessary, flush the tank with a small amount of solvent or gasoline, then drain and dispose of the flammable liquid properly.

8. If equipped, check the tank cushions for excessive deterioration, wear or damage and replace, as necessary.

9. When installing the tank, be sure all fuel lines are connected properly and that the retaining screws are securely bolted in place.

10. Refill the tank and pressure test the system by opening the fuel valve, then starting and running the engine.

Portable Fuel Tanks

Modern fuel tanks are vented to prevent vapor-lock of the fuel supply system, but some are vented by a 1-way valve to prevent pollution through the evaporation of vapors. A squeeze bulb is used to prime the system until the powerhead is operating. Once the engine starts, the fuel pump, mounted on the powerhead pull fuel from the tank and feeds the carburetor(s) or fuel vapor separator tank (EFI). The pick-up unit in the tank is usually sold as a complete unit, but without the gauge and float.

To disassemble and inspect or replace tank components, proceed as follows:

1. For safety, remove the filler cap and drain the tank into a suitable container.

2. Disconnect the fuel supply line from the tank fitting.

3. Fuel pick-up units are usually 1 of 2 possible types:

a. Threaded fittings are mounted to the tops of tanks using a Teflon sealant or tape to seal the threads. They may contain right or left-hand threads, depending upon the tank manufacturer so care should be used when removing or installing the fitting. On many plastic tanks the pick-up itself cannot be removed and the tank must be replaced if damaged.

b. Bolted fittings are used, predominantly on metallic portable tanks. To replace the pick-up unit, first remove the screws (normally 4) securing the unit in the tank. Next, lift the pick-up unit up out of the tank. On models with an integrated fuel gauge on the pick-up, there are normally screws securing the gauge to the bottom of the pick-up unit.

■ **If the pick-up unit is not being replaced, clean and check the screen for damage. It is possible to bend a new piece of screen material around the pick-up and solder it in place without purchasing a complete new unit.**

4. If equipped with a level gauge assembly, check for smooth, and non-binding movement of the float arm and replace if binding is found. Check the float itself for physical damage or saturation and replace, if found.

5. Check the fuel tank for dirt or moisture contamination. If any is found use a small amount of gasoline or solvent to clean the tank. Pour the solvent in and slosh it around to loosen and wash away deposits, then pour out the solvent and recheck. Allow the tank to air dry, or help it along with the use of an air hose from a compressor.

※※ **SELOC CAUTION**

Use extreme care when working with solvents or fuel. Remember that both are even more dangerous when their vapors are concentrated in a small area. No source of ignition from flames to sparks can be allowed in the workplace for even an instant.

To Install:

6. On models with bolted fuel fittings:

a. Attach the fuel gauge to the new pick-up unit and secure it in place with the screws.

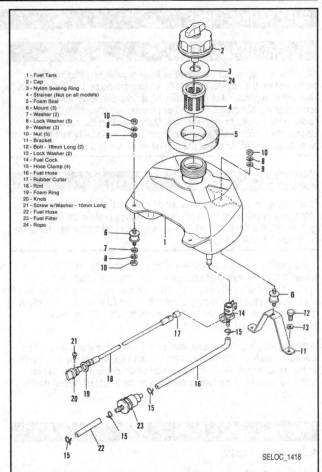

1 - Fuel Tank
2 - Cap
3 - Nylon Sealing Ring
4 - Strainer (Not on all models)
5 - Foam Seal
6 - Mount (3)
7 - Washer (2)
8 - Lock Washer (5)
9 - Washer (3)
10 - Nut (5)
11 - Bracket
12 - Bolt - 16mm Long (2)
13 - Lock Washer (2)
14 - Fuel Cock
15 - Hose Clamp (4)
16 - Fuel Hose
17 - Rubber Collar
18 - Rod
19 - Foam Ring
20 - Knob
21 - Screw w/Washer - 10mm Long
22 - Fuel Hose
23 - Fuel Filter
24 - Rope

SELOC_1418

Fig. 10 Exploded view of a typical Mercury integral tank (4 hp shown)

b. Clean the old gasket material from fuel tank and, if being used, the old pick-up unit. Position a new gasket/seal, then work the float arm down through the fuel tank opening, and at the same time the fuel pick-up tube into the tank. It will probably be necessary to exert a little force on the float arm in order to feed it all into the hole. The fuel pick-up arm should spring into place once it is through the hole.

c. Secure the pick-up and float unit in place with the attaching screws.

7. On models with threaded fuel fittings, carefully clean the threads on the fitting and the tank, removing any traces of old sealant or tape. Apply a light coating of Teflon plumbers tape or sealant to the threads, then carefully thread the fitting into position until lightly seated.

8. If removed, connect the fuel line to the tank, then pressurize the fuel system and check for leaks.

Boat-Mounted Fuel Tanks

The other type of remote fuel tank sometimes used on these models (usually only on the larger models) is a boat mounted built-in tank. Depending on the boat manufacturer, built-in tanks may vary greatly in actual shape/design and access. All should be of a 1-way vented valve type to prevent a vacuum lock, but capped to prevent evaporation design.

Most boat manufacturers are kind enough to incorporate some means of access to the tank should fuel lines, fuel pick-up or floats require servicing. But, the means of access will vary greatly from boat-to-boat. Some might contain simple access panels, while others might require the removal of one or more minor or even major components for access. If you encounter difficulty, seek the advice of a local dealer for that boat builder. The dealer or his/her techs should be able to set you in the right direction.

※※ **SELOC CAUTION**

Observe all fuel system cautions, especially when working in recessed portions of a hull. Fuel vapors tend to gather in enclosed areas causing an even more dangerous possibility of explosion.

Fuel Lines & Fittings

◆ See Figure 11

In order for an engine to run properly it must receive an uninterrupted and unrestricted flow of fuel. This cannot occur if improper fuel lines are used or if any of the lines/fittings are damaged. Too small a fuel line could cause hesitation or missing at higher engine rpm. Worn or damaged lines or fittings could cause similar problems (also including stalling, poor/rough idle) as air might be drawn into the system instead of fuel. Similarly, a clogged fuel line, fuel filter or dirty fuel pick-up or vacuum lock (from a clogged tank vent as mentioned under Fuel Tank) could cause these symptoms by starving the motor for fuel.

If fuel delivery problems are suspected, check the tank first to make sure it is properly vented, then turn your attention to the fuel lines. First check the lines and valves for obvious signs of leakage, then check for collapsed hoses that could cause restrictions.

■ If there is a restriction between the primer bulb and the fuel tank, vacuum from the fuel pump may cause the primer bulb to collapse. Watch for this sign when troubleshooting fuel delivery problems.

✳✳ SELOC CAUTION

Only use the proper fuel lines containing suitable Coast Guard ratings on a boat. Failure to do so may cause an extremely dangerous condition should fuel lines fail during adverse operating conditions.

TESTING

◆ See Figure 12

A lack of adequate fuel supply will cause the powerhead to run lean or loose rpm. If a fuel tank other than the one supplied or recommended by Mercury is being used you should make sure the fuel cap has an adequate air vent. Also, verify the size of the fuel line from the tank to be sure it is of sufficient size to accommodate the powerhead demands.

An adequate size line would be anywhere from 5/16-3/8 in. (7.94-9.52mm) Inner Diameter (ID).

■ Mercury typically lists 5/16 in. (7.94mm) as the minimum ID for fuel hoses on their larger outboards.

Check the fuel strainer on the end of the pick-up in the fuel tank to be sure it is not tool small or clogged. Check the fuel pick-up tube to make sure it is large enough and is not kinked or clogged. Be sure to check the inline filter to make sure sufficient quantities of fuel can pass through it.

If necessary, perform a Fuel Line Quick Check to see if the problem is fuel line or delivery system related.

Fuel Line Quick Check

◆ See Figure 11

Stalling, hesitation, rough idle, misses at high rpm are all possible results of problems with the fuel lines. A quick visual check of the lines for leaks, kinked or collapsed lengths or other obvious damage may uncover the problem. If no obvious cause is found, the problem may be due to a restriction in the line or a problem with the fuel pump.

If a fuel delivery problem due to a restriction or lack of proper fuel flow is suspected, operate the engine while attempting to duplicate the miss or hesitation. While the condition is present, squeeze the primer bulb rapidly to manually pump fuel from the tank to (and through) the fuel pump to the carburetors (or the vapor separator tank on fuel injected motors). If the engine then runs properly while under these conditions, suspect a problem with a clogged restricted fuel line, a clogged fuel filter or a problem with the fuel pump.

Checking Fuel Flow at Motor

◆ See Figures 11 and 13 thru 16

To perform a more thorough check of the fuel lines and isolate or eliminate the possibility of a restriction, proceed as follows:

1. For safety, disconnect the spark plug leads, then ground each of them to the powerhead to prevent sparks and to protect the ignition system.

2. Disconnect the fuel line from the engine (quick-connector on many models). On other models you may have to pick a fitting from which to disconnect the fuel line. You might choose the fuel pump, the filter or even the carburetor(s), but keep in mind that the further down the system you check, the more possible restrictions you're testing. If fuel flow is not adequate, you'll have to repeat the test further upstream (closer to the tank).

3. Place a suitable container over the end of the fuel line to catch the fuel discharged. If equipped with a quick-connector, insert a small screwdriver into the end of the line to hold the valve open.

4. Squeeze the primer bulb and observe if there is satisfactory fuel flow from the line. If there is no fuel discharged from the line, the check valve in the squeeze bulb may be defective, or there may be a break or obstruction in the fuel line.

5. If there is a good fuel flow, reconnect the tank-to-motor fuel supply line and disconnect the fuel line from the carburetor(s) or fuel injection vapor separator tank or even low-pressure electric boost pump (as applicable), directing that line into a suitable container. Crank the powerhead. If the fuel pump is operating properly, a healthy stream of fuel should pulse out of the line. If sufficient fuel does not pulse from the line, compare flow at either side of the inline fuel filter (if equipped) or check the fuel pump.

6. Continue cranking the powerhead and catching the fuel for about 15 pulses to determine if the amount of fuel decreases with each pulse or maintains a constant amount. A decrease in the discharge indicates a restriction in the line. If the fuel line is plugged, the fuel stream may stop. If there is fuel in the fuel tank but no fuel flows out the fuel line while the powerhead is being cranked, the problem may be in one of several areas:

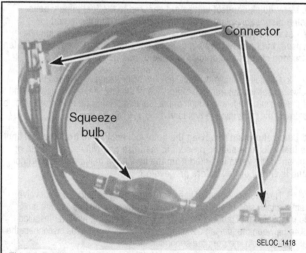

Fig. 11 Problems with the fuel lines or fittings can manifest themselves in engine performance problems, especially at WOT

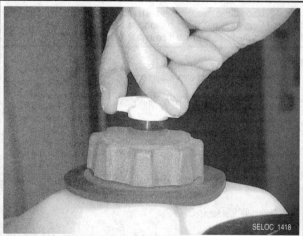

Fig. 12 One of the first things to check when troubleshooting a fuel delivery problem is the fuel tank vent (make sure it is open and not stuck closed)

Fig. 13 Disconnect the fuel supply from fuel pump or engine quick-connect to check fuel flow. . .

Fig. 14 . . .you could even disconnect it at a fuel filter. But in all cases, squeeze the primer bulb and watch for unrestricted fuel flow

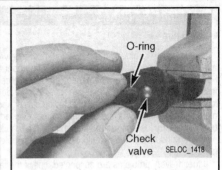

Fig. 15 Typical fuel quick-connector with O-ring and check valve visible (you'll have to depress the check valve in order to see fuel flow with the connector removed from the engine

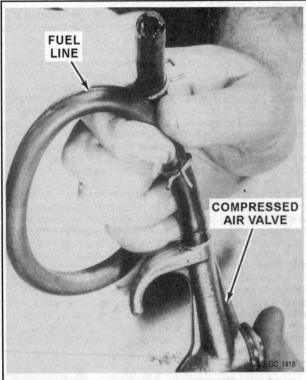

Fig. 16 Many times, restrictions such as foreign matter may be cleared from fuel lines using compressed air. But be careful to make sure the hose is pointed away from yourself or others

    a. Plugged fuel line from the fuel pump to the carburetor(s) or vapor separator tank (EFI)/boost pump (Optimax).

    b. Defective O-ring in fuel line connector into the fuel tank.

    c. Defective O-ring in fuel line connector into the engine.

    d. Defective fuel pump.

    e. The line from the fuel tank to the fuel pump may be plugged; the line may be leaking air; or the squeeze bulb may be defective.

    f. Defective fuel tank.

    7. If the engine does not start even though there is adequate fuel flow from the fuel line, the float inlet needle valve and the seat may be gummed together and prevent adequate fuel flow into the float bowl or fuel injection vapor separator tank/boost pump.

Checking the Primer Bulb

◆ See Figures 11 and 17

    The way most outboards are rigged, fuel will evaporate from the system during periods of non-use. Also, anytime quick-connect fittings on portable tanks are removed, there is a chance that small amounts of fuel will escape and some air will make it into the fuel lines. For this reason, outboards are normally rigged with some method of priming the fuel system through a hand-operated pump (primer bulb).

    When squeezed, the bulb forces fuel from inside the bulb, through the 1-way check valve faced toward the motor, thereby filling the carburetor float bowl(s) or fuel injection vapor separator tank with the fuel necessary to start the motor. When the bulb is released, the 1-way check valve on the opposite end (tank side of the bulb) opens under vacuum to draw fuel from the tank and refill the bulb.

    When using the bulb, squeeze it gently as repetitive or forceful pumping may flood the carburetor(s) or potentially, though not likely, overfill the vapor separator tank on fuel injected motors. The bulb is operating normally if a few squeezes will cause it to become firm, meaning the float bowl/tank is full, and the float valve is closed. If the bulb collapses and does not regain its shape, the bulb must be replaced (or there is a restriction between the bulb and the tank).

    For the bulb to operate properly, both check valves must operate properly and the fuel lines from the check valves back to the tank or forward to the motor must be in good condition (properly sealed). To check the bulb and check valves use hand operated vacuum/pressure pump (available from most marine or automotive parts stores):

    1. Remove the fuel hose from the tank and the motor, then remove the clamps for the quick-connect valves at the ends of the hose.

■ Many quick-connect valves are secured to the fuel supply hose using disposable plastic ties that must be cut and discarded for removal. If equipped, spring-type or threaded metal clamps may be reused, but be sure they are in good condition first. Do not over-tighten threaded clamps and crack the valve or cut the hose.

    2. Carefully remove the fuel line connector (fitting or quick-connect valve, as applicable) from the motor side of the fuel line, then place the end of the line into the filler opening of the fuel tank. Gently pump the primer bulb to empty the hose into the fuel tank.

■ Be careful when removing the quick-connect valve from the fuel line as fuel will likely still be present in the hose and will escape (drain or splash) if the valve is jerked from the line. Also, make sure the primer bulb is empty of fuel before proceeding.

    3. Next, remove the fuel line connector (fitting or quick-connect valve) from the tank side of the fuel line, draining any residual fuel into the tank. Boat mounted fuel tanks often have a water separating fuel filter installed inline between the fuel tank and the motor, on these systems disconnect the tank/primer bulb line from the separator.

■ For proper orientation during testing or installation, the primer bulb is marked with an arrow that faces the engine side check valve.

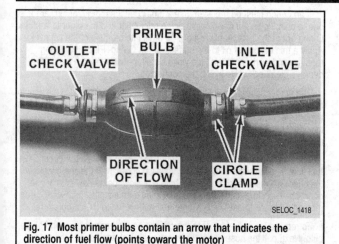

Fig. 17 Most primer bulbs contain an arrow that indicates the direction of fuel flow (points toward the motor)

4. Securely connect the pressure pump to the hose on the tank side of the primer bulb. Using the pump, slowly apply pressure while listening for air escaping from the end of the hose that connects to the motor. If air escapes, both 1-way check valves on the tank side and motor side of the prime bulb are opening.

5. If air escapes prior to the motor end of the hose, hold the bulb, check valve and hose connections under water (in a small bucket or tank). Apply additional air pressure using the pump and watch for escaping bubbles to determine what component or fitting is at fault. Repair the fitting or replace the defective hose/bulb component.

6. If no air escapes, attempt to draw a vacuum from the tank side of the primer bulb. The pump should draw and hold a vacuum without collapsing the primer bulb, indicating that the tank side check valve remained closed.

7. Securely connect the pressure pump to the hose on the motor side of the primer bulb. Using the pump, slowly apply pressure while listening for air escaping from the end of the hose that connects to the tank. This time, the check valve on the tank side of the primer bulb should remain closed, preventing air from escaping or from pressurizing the bulb. If the bulb pressurizes, the motor side check valve is allowing pressure back into the bulb, but the tank side valve is operating properly.

8. Replace the bulb and/or check valves if they operate improperly.

SERVICE

◆ See Figures 13, 18 and 19

■ When replacing fuel lines, make sure the inside diameter of the fuel hose and fitting is at least the same diameter as the hose it is replacing. Also, be certain to use only marine fuel line the meets or exceeds United States Coast Guard (USCG) A1 or B1 guidelines.

Whenever work is performed on the fuel system, check all hoses for wear or damage. Replace hoses that are soft and spongy or ones that are hard and brittle. Fuel hoses should be smooth and free of surface cracks, and they should definitely not have split ends (there's a bad hair joke in there, but we won't sink that low). Do not cut the split ends of a hose and attempt to reuse it, whatever caused the split (most likely time and deterioration) will cause the new end to follow soon. Fuel hoses are safety items, don't scrimp on them, instead, replace them when necessary. If one hose is too old, check the rest, as they are likely also in need of replacement.

When replacing fuel lines only use replacement hoses or other marine fuel supply lines that meet United States Coast Guard (USCG) requirements A1 or B1 for marine applications. All lines must be of the same inner diameter as the original to prevent leakage and maintain the proper seal that is necessary for fuel system operation.

■ Using a smaller fuel hose than originally specified/equipped could cause fuel starvation problems leading to misfiring, hesitation, rough idling and possibly even engine damage.

The USCG ratings for fuel supply lines have to do with whether or not the lines have been testing regarding length of time it might take for them to succumb to flame (burn through) in an emergency situation. A line is "A" rated if it passes specific requirements regarding burn-through times, while "B" rated lines are not tested in this fashion. The A1 and B1 lines (normally recommended on these applications) are capable of containing liquid fuel at all times. The A2 and B2 rated lines are designed to contain fuel vapor, but not liquid.

✳✳ SELOC CAUTION

To help prevent the possibility of significant personal injury or death, do not substitute "B" rated lines when "A" rated lines are required. Similarly, *do not* use "A2" or "B2" lines when "A1" or "A2" lines are specified.

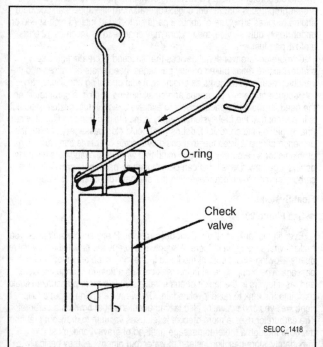

Fig. 18 Use 2 picks, punches or other small tool to replace quick-connect O-rings. One to push the valve and the other work the O-ring free

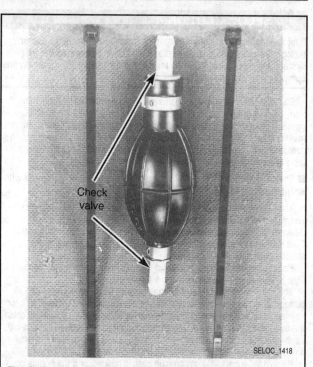

Fig. 19 A squeeze bulb kit usually includes the bulb, 2 check valves, and 2 tie straps (or other replacement clamps)

Various styles of fuel line clamps may be found on these motors. Many applications will simply secure lines with plastic wire ties or special plastic locking clamps. Although some of the plastic locking clamps may be released and reconnected, it is usually a good idea to replace them. Obviously wire ties are cut for removal, which requires that they be replaced.

Some applications use metal spring-type clamps, that contain tabs which are squeezed allowing the clamp to slid up the hose and over the end of the fitting so the hose can be pulled from the fitting. Threaded metal clamps are nice since they are very secure and can be reused, but do not over tighten threaded clamps as they will start to cut into the hose and they can even damage some fittings underneath the hose. Metal clamps should be replaced anytime they've lost tension (spring type clamps), are corroded, bent or otherwise damaged.

■ **The best way to ensure proper fuel fitting connection is to use the same size and style clamp that was originally installed (unless of course the "original" clamp never worked correctly, but in those cases, someone probably replaced it with the wrong type before you ever saw it).**

To avoid leaks, replace all displaced or disturbed gaskets, O-rings or seals whenever a fuel system component is removed.

On many installations, the fuel line is provided with quick-disconnect fittings at the tank and at the powerhead. If there is reason to believe the problem is at the quick-disconnects, the hose ends can be replaced as an assembly, or new O-rings may be installed. A supply of new O-rings should be carried on board for use in isolated areas where a marine store is not available (like dockside, or worse, should you need one while on the water). For a small additional expense, the entire fuel line can be replaced and eliminate this entire area as a problem source for many future seasons. (If the fuel line is replaced, keep the old one around as a spare, just in case).

If an O-ring must be replaced, use 2 small punches, picks or similar tools, one to push down the check valve of the connector and the other to work the O-ring out of the hole. Apply just a drop of oil into the hole of the connector. Apply a thin coating of oil to the surface of the O-ring. Pinch the O-ring together and work it into the hole while simultaneously using a punch to depress the check valve inside the connector.

The primer squeeze bulb can be replaced in a short time. A squeeze bulb assembly kit, complete with the check valves installed, may be obtained from your dealer or marine parts retailer. The replacement kit will usually also include 2 tie straps (or some other type of clamp) to secure the bulb properly in the line.

An arrow is clearly visible on the squeeze bulb to indicate the direction of fuel flow. The squeeze bulb must be installed correctly in the line because the check valves in each end of the bulb will allow fuel to flow in only one direction. Therefore, if the squeeze bulb should be installed backwards, in a moment of haste to get the job done, fuel will not reach the carburetor(s).

To replace the bulb, first undo the clamps on the hose at each end of the bulb. Next, pull the hose out of the check valves at each end of the bulb. New clamps are included with a new squeeze bulb.

If the fuel line has been exposed to considerable sunlight, it may have become hardened, causing difficulty in working it over the check valve. To remedy this situation, simply immerse the ends of the hose in boiling water for a few minutes to soften the rubber. The hose will then slip onto the check valve without further problems. After the lines on both sides have been installed, snap the clamps in place to secure the line. Check a second time to be sure the arrow is pointing in the fuel flow direction, towards the powerhead.

CARBURETED FUEL SYSTEM

✳✳ SELOC CAUTION

If equipped, disconnect the negative battery cable anytime work is performed on the engine, especially when working on the fuel system. This will help prevent the possibility of sparks during service (from accidentally grounding a hot lead or powered component). Sparks could ignite vapors or exposed fuel. Disconnecting the cable on electric start motors will also help prevent the possibility fuel spillage if an attempt is made to crank the engine while the fuel system is open.

✳✳ SELOC CAUTION

Fuel leaking from a loose, damaged or incorrectly installed hose or fitting may cause a fire or an explosion. Always pressurize the fuel system and run the motor while inspecting for leaks after servicing any component of the fuel system.

Carbureted motors covered in this section are equipped with anything from 1 single barrel carburetor to 3 dual-barrel carbs (one barrel per cylinder). Although on initial inspection some of the larger carbureted motors may look somewhat complicated, they're actually very basic, especially when compared with other modern fuel systems (such as an automotive or marine fuel injection system).

■ **The carburetors used on 4-25 hp motors are normally equipped with an integral fuel pump assembly.**

The entire system essentially consists of a fuel tank, a fuel supply line, and a mechanical fuel pump assembly mounted to the powerhead all designed to feed the carburetor with the fuel necessary to power the motor. Cold starting is normally enhanced by the use of a choke plate or an enrichment circuit.

For information on fuels, tanks and lines please refer to the sections on Fuel System Basics and Fuel Tanks & Lines.

The most important fuel system maintenance that a boat owner can perform is to stabilize fuel supplies before allowing the system to sit idle for any length of time more than a few weeks (or better yet, to completely drain the carburetors.) The next most important item is to provide the system with fresh gasoline if the system has stood idle for any length of time, especially if it was without fuel system stabilizer during that time.

If a sudden increase in gas consumption is noticed, or if the engine does not perform properly, a carburetor overhaul, including cleaning or replacement of the fuel pump may be required.

Description & Operation

◆ See Figures 20 and 21

GENERAL INFORMATION

The Role of a Carburetor

◆ See Figures 20 and 21

The carburetor is merely a metering device for mixing fuel and air in the proper proportions for efficient engine operation. At idle speed, an outboard engine requires a mixture of about 8 parts air to 1 part fuel. At high speed or under heavy duty service, the mixture may change to as much as 12 parts air to 1 part fuel.

Carburetors are wonderful devices that succeed in creating precise air/fuel mixture ratios based on tiny passages, needle jets or orifices and the variable vacuum that occurs as engine rpm and operating conditions vary.

Because of the tiny passages and small moving parts in a carburetor (and the need for them to work precisely to achieve exact air/fuel mixture ratios) it is important that the fuel system integrity is maintained. Introduction of water (that might lead to corrosion), debris (that could clog passages) or even the presence of unstabilized fuel that could evaporate over time can cause big problems for a carburetor. Keep in mind that when fuel evaporates it leaves behind a gummy deposit that can clog those tiny passages, preventing the carburetor (and therefore preventing the engine) from operating properly.

Float Systems

◆ See Figure 20

Ever lift the tank lid off the back of your toilet. Pretty simple stuff once you realize what's going on in there. A supply line keeps the tank full until a valve opens allowing all or some of the liquid in the tank to be drawn out through a passage. The dropping level in the tank causes a float to change position, and, as it lowers in the tank it opens a valve allowing more pressurized liquid back into the tank to raise levels again. OK, we were talking about a toilet right, well yes and no, we're also talking about the float bowl on a carburetor. The carburetor uses a more precise level, uses vacuum to draw out fuel from the bowl through a metered passage (instead of gravity) and, most importantly, store gasoline instead of water, but otherwise, they basically work in the same way.

A small chamber in the carburetor serves as a fuel reservoir. A float valve admits fuel into the reservoir to replace the fuel consumed by the engine. If

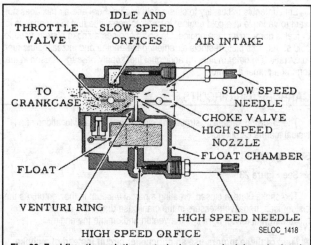

Fig. 20 Fuel flow through the venturi, showing principle and related parts controlling intake and outflow (carburetor with manual choke circuit shown)

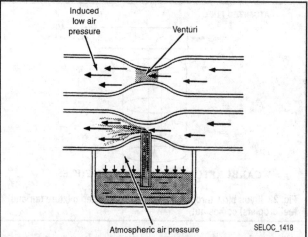

Fig. 21 Air flow principle of a modern carburetor, demonstrates how the low pressure induced behind the venturi draws fuel through the high speed nozzle

the carburetor has more than 1 reservoir, the fuel level in each reservoir (chamber) is controlled by identical float systems.

Fuel level in each chamber is extremely critical and must be maintained accurately. Accuracy is obtained through proper adjustment of the float. This adjustment will provide a balanced metering of fuel to each cylinder at all speeds. Improper levels will lead to engine operating problems. Too high a level can promote rich running and spark plug fouling, while excessively low float bowl fuel levels can cause lean conditions, possibly leading to engine damage.

Following the fuel through its course, from the fuel tank to the combustion chamber of the cylinder, will provide an appreciation of exactly what is taking place. In order to start the engine, the fuel must be moved from the tank to the carburetor by a squeeze bulb installed in the fuel line. This action is necessary because the fuel pump does not have sufficient pressure to draw fuel from the tank during cranking before the engine starts.

The fuel for some small horsepower units is gravity fed from a tank mounted at the rear of the powerhead. Even with the gravity feed method, a small fuel pump may be an integral part of the carburetor for some manufacturers.

After the engine starts, the fuel passes through the pump to the carburetor. All systems have some type of filter installed somewhere in the line between the tank and the carburetor. Over the years many carburetors have utilized a filter as an integral part of the carburetor.

At the carburetor, the fuel passes through the inlet passage to the needle and seat, and then into the float chamber (reservoir). A float in the chamber rides up and down on the surface of the fuel. After fuel enters the chamber and the level rises to a predetermined point, a tang on the float closes the inlet needle and the flow entering the chamber is cut off. When fuel leaves the chamber as the engine operates, the fuel level drops and the float tang allows the inlet needle to move off its seat and fuel once again enters the chamber. In this manner, a constant reservoir of fuel is maintained in the chamber to satisfy the demands of the engine at all speeds.

A fuel chamber vent hole is located near the top of the carburetor body to permit atmospheric pressure to act against the fuel in each chamber. This pressure assures an adequate fuel supply to the various operating systems of the powerhead.

Air/Fuel Mixture
◆ See Figure 21

A suction effect is created each time the piston moves upward in the cylinder. This suction draws air through the throat of the carburetor. A restriction in the throat, called a venturi, controls air velocity and has the effect of reducing air pressure at this point.

The difference in air pressures at the throat and in the fuel chamber, causes the fuel to be pushed out of metering jets extending down into the fuel chamber. When the fuel leaves the jets, it mixes with the air passing through the venturi. This fuel/air mixture should then be in the proper proportion for burning in the cylinders for maximum engine performance.

In order to obtain the proper air/fuel mixture for all engine speeds, high- and low-speed orifices or needle valves are installed. On most modern

powerheads the high-speed needle valve has been replaced with a fixed high-speed orifice (to more discourage tampering and to help maintain proper emissions under load). There is no adjustment with the orifice type. The needle valves are used to compensate for changing atmospheric conditions. The low-speed needles, on the other hand, are still provided so that air/fuel mixture can be precisely adjusted for idle conditions other than what occurs at atmospheric sea-level. Although the low-speed needle should not normally require periodic adjustment, it can be adjusted to compensate for high-altitude (river/lake) operation or to adjust for component wear within the fuel system.

A throttle valve controls the flow of air/fuel mixture drawn into the combustion chambers. A cold powerhead requires a richer fuel mixture to start and during the brief period it is warming to normal operating temperature. A choke valve is often placed ahead of the metering jets and venturi. As this valve begins to close, the volume of air intake is reduced, thus enriching the mixture entering the cylinders. When this choke valve is fully closed, a very rich fuel mixture is drawn into the cylinders.

■ **Some carburetors do not use a choke valve, but instead use an enrichening circuit which mechanically provides more fuel to the carburetor or intake during cold start-up.**

The throat of the carburetor is usually referred to as the barrel. Carburetors with single, double, or 4 barrels have individual metering jets, needle valves, throttle and, if applicable, choke plates for each barrel. Single and 2 barrel carburetors are fed by a single float and chamber.

CARBURETOR BASIC FUNCTIONS

◆ See Figure 22

The carburetor systems on these engines require careful cleaning and adjustment if problems occur. The carburetors themselves may seem a little complicated but they are not too complex to understand. All carburetors operate on the same principles.

Traditional carburetor theory often involves a number of laws and principles. To troubleshoot carburetors learn the basic principles, watch how the carburetor comes apart, trace the circuits, see what they do and make sure they are clean. These are the basic steps for troubleshooting and successful repair.

The diagram illustrates several carburetor basics. If you blow through the horizontal straw an atomized mixture (air and fuel droplets) comes out. When you blow through the straw a pressure drop is created in the straw column inserted in the liquid. In a carburetor this is mostly air and a little fuel. The actual ratio of air to fuel differs with engine conditions but is usually from 15 parts air to one part fuel at optimum cruise to as little as 7 parts air to one part fuel at full choke.

If the top of the container is covered and sealed around the straw what will happen? No flow. This is typical of a clogged carburetor bowl vent.

If the base of the straw is clogged or restricted what will happen? No flow or low flow. This represents a clogged main jet.

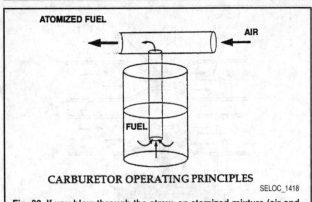

CARBURETOR OPERATING PRINCIPLES

SELOC_1418

Fig. 22 If you blow through the straw, an atomized mixture (air and fuel droplets) comes out

If the liquid in the glass is lowered and you blow through the straw with the same force what will happen? Not as much fuel will flow. A lean condition occurs.

If the fuel level is raised and you blow again at the same velocity what happens? The result is a richer mixture.

Many Mercury carburetors control air flow semi-independently of engine speed (rpm). This is done with a throttle plate. The throttle plate works in conjunction with other systems or circuits to deliver correct mixtures within certain rpm bands. The idle circuit pilot outlet controls from 0-1/8 throttle. The series of small holes in the carburetor throat called transition holes control the 1/8-3/8 throttle range. At wide open throttle the main jet handles most of the fuel metering chores, but the low and mid-range circuits continue to supply part of the fuel.

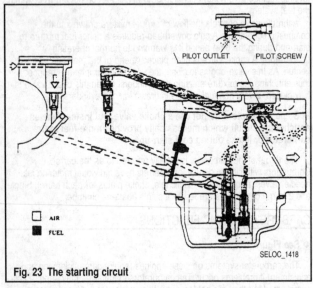

SELOC_1418

Fig. 23 The starting circuit

Enrichment is necessary to start a cold engine. Fuel and air mix does not want to vaporize in a cold engine. In order to get a little fuel to vaporize, a lot of fuel is dumped into the engine. On some engines a choke plate is used for cold starts. This plate restricts air entering the engine and increases the fuel to air ratio. On other motors an additional fuel supply passage is used in the form of a primer assembly.

CARBURETOR CIRCUITS

The following section illustrates the circuit functions and locations of a typical marine carburetor.

Starting Circuit

◆ See Figure 23

The choke plate is closed, creating a partial vacuum in the venturi. As the piston rises, negative pressure in the crankcase draws the rich air-fuel mixture from the float bowl into the venturi and on into the engine.

■ **Most of these motors use an enrichener circuit instead of a choke plate. An enrichener allows the full amount of air to pass through the carburetor during cold-start up, but it fattens the air/fuel mixture by providing additional fuel through a dedicated (usually electronically controlled) passage. Same result, different method. Ironically, throughout the industry and other types of small motors an enrichener circuit is often referred to as a "choke" even though it works in the exact opposite manner of a traditional "choke plate."**

Low Speed Circuit

◆ See Figure 24

0-1/8th throttle, when the pressure in the crankcase is lowered, the air-fuel mixture is discharged into the venturi through the pilot outlet because the throttle plate is closed. No other outlets are exposed to low venturi pressure. The fuel is metered by the pilot jet. The air is metered by the pilot air jet. The combined air-fuel mixture is regulated by the pilot air screw.

Mid-Range Circuit

◆ See Figure 25

1/8-3/8ths throttle, as the throttle plate continues to open, the air-fuel mixture is discharged into the venturi through the bypass holes. As the throttle plate uncovers more bypass holes, increased fuel flow results because of the low pressure in the venturi. Depending on the model, there could be 2, 3 or 4 bypass holes.

High Speed Circuit

◆ See Figure 26

3/8ths to wide-open throttle, as the throttle plate moves toward wide open, we have maximum air flow and very low pressure. The fuel is metered through the main jet, and is drawn into the main discharge nozzle. Air is metered by the main air jet and enters the discharge nozzle, where it combines with fuel. The mixture atomizes, enters the venturi, and is drawn into the engine.

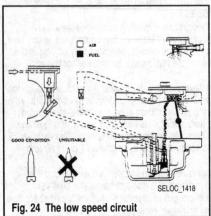

SELOC_1418

Fig. 24 The low speed circuit

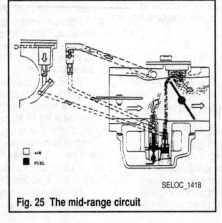

SELOC_1418

Fig. 25 The mid-range circuit

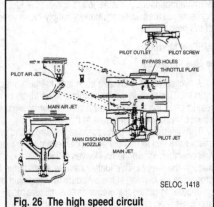

SELOC_1418

Fig. 26 The high speed circuit

FUEL PUMP BASIC FUNCTIONS

◆ **See Figures 27, 28 and 29**

A fuel pump is a basic mechanical device that utilizes crankcase positive and negative pressures (alternating pressure and vacuum) to pump fuel from the fuel tank to the carburetors.

This device contains a flexible diaphragm and 2 check valves (flappers or fingers) that control flow. As the piston goes up, crankcase pressure drops (negative pressure or vacuum) and the inlet valve opens, pulling fuel from the tank. As the piston nears TDC, pressure in the pump area is neutral (atmospheric pressure). At this point both valves are closed. As the piston comes down, pressure goes up (positive pressure) and the fuel is pushed toward the carburetor bowl by the diaphragm through the now open outlet valve.

■ **This system works because the check valves only allow fuel to flow in one direction. Should one of the valves fail, the pump will not operate correctly.**

This is a reliable method to move fuel but can have several problems. Sometimes an engine backfire can rupture the diaphragm (though this is rare on pumps like the Mercury designs which are not mounted directly to the powerhead as a vacuum source, but instead connect to the powerhead vacuum source through a hose). The diaphragm and valves are moving parts subject to wear. The flexibility of the diaphragm material can go away, reducing or stopping flow. Rust or dirt can hang a valve open and reduce or stop fuel flow.

Most pumps consist of a diaphragm, 2 similar spring loaded disc valves, one for inlet (suction) and the other for outlet (discharge) and a small opening leading directly into the crankcase bypass. The suction and compression created, as the piston travels up and down in the cylinder, causes the diaphragm to flex.

As the piston moves upward, the diaphragm will flex inward displacing volume on its opposite side to create suction. This suction will draw liquid fuel in through the inlet disc valve.

When the piston moves downward, compression is created in the crankcase. This compression causes the diaphragm to flex in the opposite direction. This action causes the discharge valve disc to lift off its seat. Fuel is then forced through the discharge valve into the carburetor.

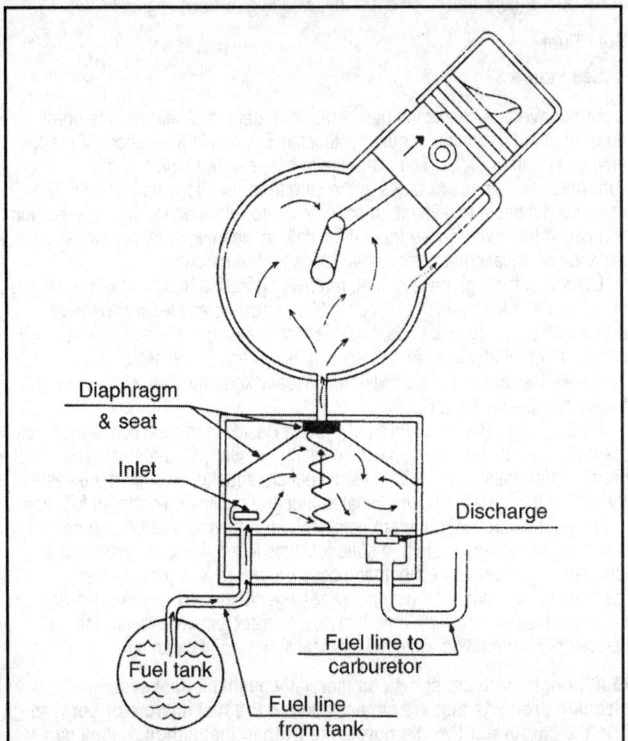

Generally speaking this type of pump has the capacity to lift fuel 2 feet and deliver approximately 5 gallons per hour at about 4 psi pressure.

Problems with the fuel pump are limited to possible leaks in the flexible neoprene suction lines, a punctured diaphragm, air leaks between sections of the pump assembly or possibly from the disc valves not seating properly.

The pump is activated by 1 cylinder. If this cylinder indicates a wet fouled condition, as evidenced by a wet fouled spark plug, be sure to check the fuel pump diaphragm for possible puncture or leakage.

Troubleshooting the Fuel System

◆ **See Figures 30 and 31**

Troubleshooting fuel systems requires the same techniques used in other areas. A thorough, systematic approach to troubleshooting will pay big rewards. Build your troubleshooting checklist, with the most likely offenders at the top. Use your experience to adjust your list for local conditions. At one time or another nearly everyone has been tempted to jump into the carburetor based on nothing more than a vague hunch. Pause a moment and review the facts when this urge occurs.

■ **Keep in mind that the number one cause of Fuel System problems is the Ignition System - meaning that many ignition system problems are mis-diagnosed as fuel system problems.**

In order to accurately troubleshoot a carburetor or fuel system problem, you must first verify that the problem is fuel related. Many symptoms can have several different possible causes. Be sure to eliminate mechanical and electrical systems as the potential fault. Carburetion is a common cause of most engine problems, but there are many other possibilities.

One of the toughest tasks with a fuel system is the actual troubleshooting. Several tools are at your disposal for making this process very simple. A timing light works well for observing carburetor spray patterns. Look for the proper amount of fuel and for proper atomization in the 2 fuel outlet areas (main nozzle and bypass holes). The strobe effect of the lights helps you see in detail the fuel being drawn through the throat of the carburetor. On multiple carburetor engines, always attach the timing light to the cylinder you are observing so the strobe doesn't change the appearance of the patterns. If you need to compare 2 cylinders, change the timing light hookup each time you observe a different cylinder.

Fig. 27 Simplified drawing of the fuel pump with the powerhead piston on the upward stroke. The springs to preload the discs are not shown for clarity)

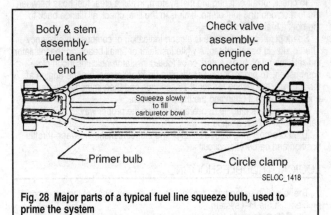

Fig. 28 Major parts of a typical fuel line squeeze bulb, used to prime the system

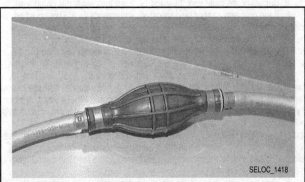

Fig. 29 A squeeze bulb is used to prime the fuel pump prior to initial start

Pressure testing fuel pump output can determine whether sufficient fuel is being supplied for the fuel spray and if the fuel pump diaphragms are functioning correctly. A pressure gauge placed between the fuel pump(s) and the carburetor(s) will test the entire fuel delivery system. Normally a fuel system problem will show up at high speed where the fuel demand is the greatest. A common symptom of a fuel pump output problem is surging at wide open throttle, while still operating normally at slower speeds. To check the fuel pump output (on models where the pump *isn't* integral with the carburetor), install the pressure gauge and accelerate the engine to wide open throttle. Observe the pressure gauge needle. It should always swing up to some value above 2 psi (14 kPa), usually something in the 5-6 psi (34-41 kPa) area and remain steady. This reading would indicate a system that is functioning properly.

■ **More specific vacuum testing procedures can be found for models with non-carburetor integral fuel pumps under Fuel Pump, in this section.**

If the needle gradually swings down toward zero, fuel demand is greater than the fuel system can supply. This reading isolates the problem to the fuel delivery system (fuel tank or line). To confirm this, an auxiliary tank should be installed and the engine retested. Be aware that a bad anti-siphon valve on a built-in tank can create enough restriction to cause a lean condition and serious engine damage.

If the needle movement becomes erratic, suspect a ruptured diaphragm in the fuel pump.

A quick way to check for a ruptured fuel pump diaphragm is, while the engine is at idle speed, to squeeze the primer bulb and hold steady firm pressure on it. If the diaphragm is ruptured, this will cause a rough running condition because of the extra fuel passing through the diaphragm into the crankcase. After performing this test you should check the spark plugs. If the spark plugs are OK, but the fuel pumps are still suspected, you should remove the fuel pumps and completely disassemble them. Rebuild or replace the pumps as needed.

To check the boat's fuel system for a restriction, install a vacuum gauge in the line before the fuel pump. Run the engine under load at wide open throttle to get a reading. Vacuum should read no more than 4.5 in. Hg (15.2 kPa) for engines up to and including 200 hp and should not exceed 6.0 in. Hg (20.2 kPa) for engines greater than 200 hp.

To check for air entering the fuel system, install a clear fuel hose between the fuel screen and fuel pump. If air is in the line, check all fittings back to the boat's fuel tank.

Spark plug tip appearance is a good indication of combustion efficiency. The tip should be a light tan. A white insulator or small beads on the insulator indicate too much heat. A dark or oil fouled insulator indicates incomplete combustion. To properly read spark plug tip appearance, run the engine at the rpm you are testing for about 15 seconds and then immediately turn the engine Off without changing the throttle position.

Reading spark plug tip appearance is also the proper way to test jet verifications in high altitude.

The accompanying illustration explains the relationship between throttle position and carburetion circuits.

LOGICAL TROUBLESHOOTING

The following paragraphs provide an orderly sequence of tests to pinpoint problems in the fuel system.
1. Gather as much information as you can.
2. Duplicate the condition. Take the boat out and verify the complaint.
3. If the problem cannot be duplicated, you cannot fix it. This could be a product operation problem.
4. Once the problem has been duplicated, you can begin troubleshooting. Give the entire unit a careful visual inspection. You can tell a lot about the engine from the care and condition of the entire rig. What's the condition of the propeller and the lower unit? Remove the hood and look for any visible signs of failure. Are there any signs of head gasket leakage? Is the engine paint discolored from high temperature or are there any holes or cracks in the engine block? Perform a compression (or better yet, if you have the equipment, a leak down test). While cranking the engine during the compression test, listen for any abnormal sounds. If the engine passes these simple tests we can assume that the mechanical condition of the engine is good. All other engine mechanical inspection would be too time consuming at this point.
5. Your next step is to isolate the fuel system into 2 sub-systems. Separate the fuel delivery components from the carburetors. To do this, substitute a known good fuel supply for the regular boat's fuel supply. Use a

6 gallon portable tank and fuel line. Connect the portable fuel supply directly to the engine fuel pump, bypassing the boat fuel delivery system. Now test the engine. If the problem is no longer present, you know where to look. If the problem is still present, further troubleshooting is required.
6. When testing the engine, observe the throttle position when the problem occurs. This will help you pinpoint the circuit that is malfunctioning. Carburetor troubleshooting and repair is very demanding. You must pay close attention to the location, position and sometimes the numbering on each part removed. The ability to identify a circuit by the operating rpm it affects is important. Often your best troubleshooting tool is a can of cleaner. This can be used to trace those mystery circuits and find that last speck of dirt. Be careful and wear safety glasses when using this method.

■ **The last step of fuel system troubleshooting is to adjust or rebuild and then adjust the carburetor. We say it is the last step, because it is the most involved repair procedures on the fuel system and should only be performed after all other possible causes of fuel system trouble have been eliminated.**

COMMON PROBLEMS

Fuel Delivery

◆ **See Figures 30 and 31**

Many times fuel system troubles are caused by a plugged fuel filter, a defective fuel pump, or by a leak in the line from the fuel tank to the fuel pump. Aged fuel left in the carburetor and the formation of varnish could cause the needle to stick in its seat and prevent fuel flow into the bowl. A defective choke may also cause problems. Would you believe, a majority of starting troubles, which are traced to the fuel system, are the result of an empty fuel tank or aged fuel.

If fuel delivery problems are suspected, refer to the testing procedures in Fuel Tank and Lines to make sure the tank vent is working properly and that there are no leaks or restrictions that would prevent fuel from getting to the pump and/or carburetor(s).

A blocked low-pressure fuel filter causes hard starting, stalling, misfire or poor performance. Typically the engine malfunction worsens with increased engine speed. This filter prevents contaminants from reaching the low-pressure fuel pump. Refer to the Fuel Filter in the section on Maintenance & Tune-Up for more details on checking, cleaning or replacing fuel filters.

Sour Fuel

◆ **See Figure 32**

Under average conditions (temperate climates), fuel will begin to break down in about 4 months. A gummy substance forms in the bottom of the fuel tank and in other areas. The filter screen between the tank and the carburetor and small passages in the carburetor will become clogged. The gasoline will begin to give off an odor similar to rotten eggs. Such a condition can cause the owner much frustration, time in cleaning components, and the expense of replacement or overhaul parts for the carburetor.

Even with the high price of fuel, removing gasoline that has been standing unused over a long period of time is still the easiest and least expensive preventative maintenance possible. In most cases, this old gas can be used without harmful effects in an automobile using regular gasoline.

A gasoline preservative additive will usually keep the fuel fresh for up to twelve months (or longer).

Refer to the information on Fuel System Basics in this section, specifically the procedure under Fuel entitled Checking For Stale/Contaminated Fuel will provide information on how to determine if stale fuel is present in the system. If draining the system of contaminated fuel and refilling it with fresh fuel does not make a difference in the problem, look for restrictions or other problems with the fuel delivery system. If stale fuel was left in the tank/system for a long period of time and evaporation occurred, there is a good chance that the carburetor is gummed (tiny passages are clogged by deposits left behind when the fuel evaporated). If no fuel delivery problems are found, the carburetor(s) should be removed for disassembly and cleaning.

■ **Although there are some commercially available fuel system cleaning products that are either added to the fuel mixture or sprayed into the carburetor throttle bores, the truth is that although they can provide some measure of improvement, there is no substitute for a thorough disassembly and cleaning. The more fuel which was allowed to evaporate, the more gum or varnish may have been left behind and the more likely that only a disassembly will be able to restore proper performance.**

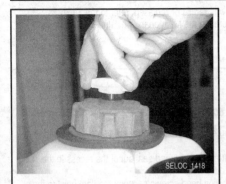

Fig. 30 One of the first things to check on the fuel system is to make sure the tank vent is open. . .

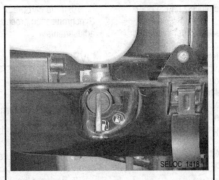

Fig. 31 . . .and that the fuel petcock (if equipped) is also in the open or run position

Fig. 32 The use of an approved fuel additive will prevent fuel from souring for up to twelve months

Choke Problems

◆ See Figure 33

When the engine is hot, the fuel system can cause starting problems. After a hot engine is shut down, the temperature inside the fuel bowl may rise to 200° F (94° C) and cause the fuel to actually boil. All carburetors are vented to allow this pressure to escape to the atmosphere. However, some of the fuel may percolate over the high-speed nozzle.

If the choke should stick in the open position, the engine will be hard to start. If the choke should stick in the closed position, the engine will flood, making it very difficult to start.

In order for this raw fuel to vaporize enough to burn, considerable air must be added to lean out the mixture. Therefore, the best remedy for a flooded motor is to remove the spark plugs, ground the leads, crank the powerhead through about ten revolutions, clean the plugs, reinstall the plugs, and start the engine.

■ Another common solution to a flooded motor is to open the throttle all the way while cranking (using the additional air to help clear/burn the excess fuel in the combustion chambers), but the problem with marine engines is that might put the gearcase in Forward (unless there is a neutral lockout on the shifter) which will prevent the starter from working on most models.

If the needle valve and seat assembly is leaking, an excessive amount of fuel may enter the intake tract. After the powerhead is shut down, the pressure left in the fuel line will force fuel past the leaking needle valve. This extra fuel will raise the level in the fuel bowl and cause fuel to overflow into the intake.

A continuous overflow of fuel into the intake tract may be due to a sticking inlet needle or to a defective float, which would cause an extra high level of fuel in the bowl and overflow.

Rough Engine Idle

■ As with all troubleshooting procedures, start with the easiest items to check/fix and work towards the more complicated ones.

If an engine does not idle smoothly, the most reasonable approach to the problem is to perform a tune-up to eliminate such areas as:
- Faulty spark plugs
- Timing and synchronization out of adjustment

Other problems that can prevent an engine from running smoothly include:
- An air leak in the intake manifold
- Uneven compression between the cylinders
- Sticky or broken reeds

Of course any problem in the carburetor affecting the air/fuel mixture will also prevent the engine from operating smoothly at idle speed. These problems usually include:
- Too high a fuel level in the bowl
- A heavy float
- Leaking needle valve and seat
- Defective automatic choke
- Improper adjustments for idle mixture or idle speed

"Sour" fuel (fuel left in a tank without a preservative additive) will cause an engine to run rough and idle with great difficulty.

Excessive Fuel Consumption

◆ See Figures 34 and 35

Excessive fuel consumption can result from 1 of 4 conditions, or some combination of the four.
1. Inefficient engine operation.
2. Damaged condition of the hull, outdrive or propeller, including excessive marine growth.
3. Poor boating habits of the operator.
4. Leaking or out of tune carburetor.

If the fuel consumption suddenly increases over what could be considered normal, then the cause can probably be attributed to the engine or boat and not the operator (unless he/she just drastically changed the manner in which the boat is operated).

Marine growth on the hull can have a very marked effect on boat performance. This is why sail boats always try to have a haul-out as close to race time as possible. While you are checking the bottom take note of the

Fig. 33 Fouled spark plug, possibly caused by over-choking or a malfunctioning enrichment circuit

Fig. 34 Marine growth on the lower unit will create "drag" and seriously hamper boat performance

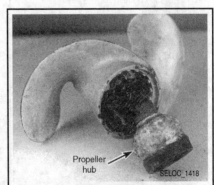

Fig. 35 A corroded hub on a small engine propeller. Hub and propeller damage will also cause poor performance

propeller condition. A bent blade or other damage will definitely cause poor boat performance.

If the hull and propeller are in good shape, then check the fuel system for possible leaks. Check the line between the fuel pump and the carburetor while the engine is running and the line between the fuel tank and the pump when the engine is not running. A leak between the tank and the pump many times will not appear when the engine is operating, because the suction created by the pump drawing fuel will not allow the fuel to leak. Once the engine is turned off and the suction no longer exists, fuel may begin to leak.

If a minor tune-up has been performed and the spark plugs and engine timing/synchronization are properly adjusted, then the problem most likely is in the carburetor, indicating an overhaul is in order. Check for leaks at the needle valve and seat. Use extra care when making any adjustments affecting the fuel consumption, such as the float level or automatic choke.

Engine Surge

If the engine operates as if the load on the boat is being constantly increased and decreased, even though an attempt is being made to hold a constant engine speed, the problem can most likely be attributed to the fuel pump (or a restriction between the tank and powerhead). Refer to Fuel Tank & Lines in this section for information on checking the lines for restrictions and checking fuel flow. Also, refer to Fuel Pump for more information on fuel pump testing, operation and repair.

2.5/3.3 Hp Models

REMOVAL & INSTALLATION

◆ **See Figures 36, 37 and 38**

1. Remove the screws securing the knobs at the end of the choke and throttle levers.
2. Remove the 2 screws securing the rectangular intake cover to the carburetor body. Lift the cover free.

■ **It is not necessary to disconnect the wires attached to the cover. Simply move the cover to one side out of the way.**

3. Close the fuel valve between the fuel tank and the carburetor. Squeeze the wire-type hose clamp on the fuel line enough for the hose to slip free of the inlet fuel fitting.
4. Loosen the clamp screw and remove the carburetor from the powerhead.

To Install:

5. Position a new carburetor-to-power-head seal in the recess of the carburetor throat.
6. Slide the carburetor onto the crank-case cover and secure it in place with the clamp screw.
7. Connect the fuel line to the fuel inlet fitting.
8. Open the fuel valve between the fuel tank and the carburetor.
9. Install the front cover and the knobs onto the choke and throttle levers.

10. If the carburetor was rebuilt or replaced, refer to the Timing & Synchronization procedures in the Maintenance & Tune-Up section for adjustments.

OVERHAUL

◆ **See Figures 39 thru 50**

1. Remove the carburetor from the powerhead, as detailed earlier in this section.
2. Work the carburetor-to-powerhead seal out of the recess in the carburetor throat. Discard the seal.
3. Back out the 2 Phillips head screws securing the float bowl to the carburetor.
4. Remove the float from the carburetor.
5. Slide the hinge pin free and then remove the float hinge and the inlet needle.
6. Lift off and discard the float bowl gasket.
7. Unscrew the main jet from the center of the main nozzle.
8. Use the proper size wrench and remove the main nozzle.
9. Rotate the idle speed screw clockwise and count the number of turns necessary to seat it lightly. After the number of turns has been noted and recorded somewhere, back out the idle speed screw. Slide the spring free of the screw.
10. Loosen the retainer nut with the proper size wrench. Rotate the nut several turns counterclockwise but do not remove the nut.
11. Unscrew the mixing chamber cover with a pair of pliers. Once the cover is free, lift off the throttle valve assembly. Disconnect the throttle lever from the bracket.
12. Compress the spring in the throttle valve assembly to allow the throttle cable end to clear the recess in the base of the throttle valve and to slide down the slot.
13. Disassemble the throttle valve consisting of the throttle valve, spring, jet needle with an E-clip on the second groove, jet retainer and throttle cable end.

■ **It is not necessary to remove the E-clip from the jet needle, unless replacement is required or if the powerhead is to be operated at a significantly different elevation.**

To Assemble:

■ **The E-clip must be installed into the same groove from which it was removed. If the clip is lowered, the carburetor will allow the powerhead to operate "rich". Raising the E-clip will cause the powerhead to operate "lean".**

14. Assemble the throttle valve components in the following order: Insert the E-clip end of the jet needle into the throttle valve (the end with the recess for the throttle cable end). Place the needle retainer into the throttle valve over the E-clip and align the retainer slot with the slot in the throttle valve.

Fig. 36 Remove the screws securing the knobs on the choke and throttle levers

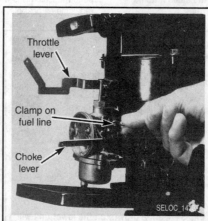

Fig. 37 Disconnect the fuel line from the carburetor

Fig. 38 Loosen the clamp screw to free the carburetor from the motor

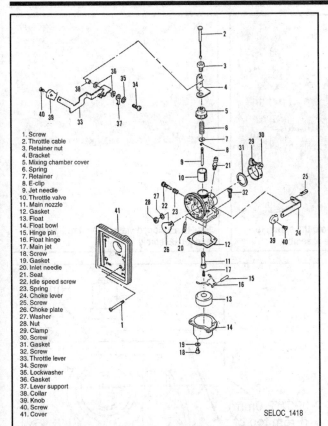

1. Screw
2. Throttle cable
3. Retainer nut
4. Bracket
5. Mixing chamber cover
6. Spring
7. Retainer
8. E-clip
9. Jet needle
10. Throttle valve
11. Main nozzle
12. Gasket
13. Float
14. Float bowl
15. Hinge pin
16. Float hinge
17. Main jet
18. Screw
19. Gasket
20. Inlet needle
21. Seat
22. Idle speed screw
23. Spring
24. Choke lever
25. Screw
26. Choke plate
27. Washer
28. Nut
29. Clamp
30. Screw
31. Gasket
32. Screw
33. Throttle lever
34. Screw
35. Lockwasher
36. Gasket
37. Lever support
38. Collar
39. Knob
40. Screw
41. Cover

SELOC_1418

Fig. 39 Exploded view with major carburetor parts identified - 2.5/3.3 hp models

15. Thread the spring over the end of the throttle cable and insert the cable into the retainer end of the throttle valve. Compress the spring and at the same time guide the cable end through the slot until the end locks into place in the recess. Position the assembled throttle valve in such a manner to permit the slot to slide over the alignment pin while the throttle valve is lowered into the carburetor. Attach the throttle lever to the bracket on top of the throttle valve assembly.

16. Carefully tighten the mixing chamber cover with a pair of pliers.

17. Position the bracket to allow the throttle to just clear the front of the carburetor and then tighten the retainer nut.

18. Slide the spring onto the idle speed screw and then start to thread the screw into the carburetor body. Slowly rotate the screw until it seats lightly. From this position, back out the screw the same number of complete turns as noted during removal. If the count was lost, back the screw out 2 turns as a preliminary adjustment. A fine idle adjustment will be made later.

19. Thread the main nozzle into the carburetor body and tighten it just "snug" with the proper size wrench. do not over-tighten the nozzle.

20. Install the main jet into the main nozzle and tighten it just "snug" with a screwdriver.

21. On CDI ignition models, use a No. 92 main jet and for magneto ignition models use a No. 94 main jet.

22. Position a new float bowl gasket in place and then install the inlet needle.

23. Hold the carburetor body in a perfect upright position on a firm surface. Set the hinge in position and then slide the hinge pin into place through the hinge.

24. Check the float hinge adjustment. The vertical distance between the top of the hinge and the top of the gasket should be 0.090 in. (2.0mm). Carefully bend the hinge, as necessary, to achieve the required measurement. Install the float.

25. Place the float bowl in position on the carburetor body and then secure it with the 2 Phillips head screws.

Fig. 40 Loosen and remove the 2 float bowl screws to remove the bowl

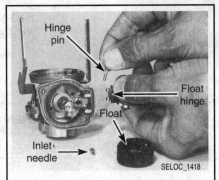

Fig. 41 Remove the hinge pin securing the float assembly

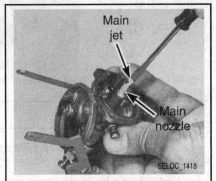

Fig. 42 Unscrew the main jet from the nozzle. . .

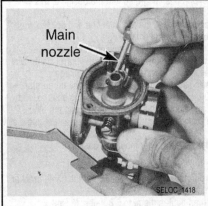

Fig. 43 . . .then unscrew the main nozzle

Fig. 44 Note the position of the idle screw before removal (by counting the turns necessary to seat it)

Fig. 45 Loosen the retainer nut. . .

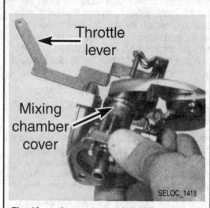

Fig. 46 . . .then unscrew the mixing chamber cover

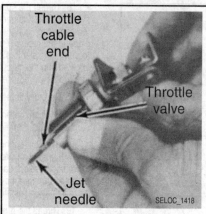

Fig. 47 Compress and hold the spring to remove the throttle valve assembly

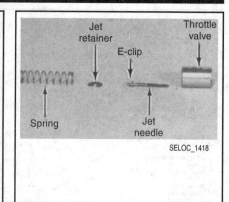

Fig. 48 Exploded view of the throttle valve and jet assembly

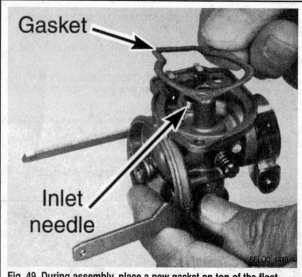

Fig. 49 During assembly, place a new gasket on top of the float bowl chamber. . .

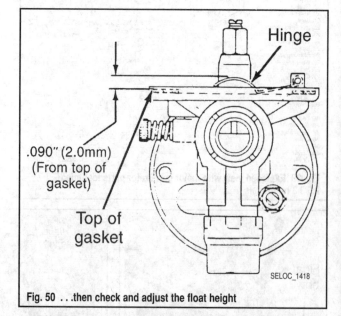
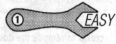

Fig. 50 . . .then check and adjust the float height

CLEANING & INSPECTION

◆ See Figure 39

MODERATE

Remove and disassemble the carburetor as detailed in this section under Removal & Installation and under Overhaul.

Never dip rubber parts, plastic parts, diaphragms or pump plungers in carburetor cleaner, because they tend to absorb liquid and expand. These parts should be cleaned only in solvent and then blown dry with compressed air immediately.

Place all of the metal parts in a screen-type tray and dip them in carburetor cleaner until they appear completely clean - free of gum and varnish which accumulates from stale fuel. Blow the parts dry with compressed air.

Blow out all of the passageways in the carburetor with compressed air. Check all of parts and passageways to be sure they are clear and not clogged with any deposits.

Never use a piece of wire or any type of pointed instrument to clean drilled passages or calibrated holes in a carburetor.

Carefully inspect the casting for cracks, stripped threads or plugs for any sign of leakage. Inspect the float hinge in the hinge pin area for wear and the float for any sign of leakage.

Examine the inlet needle for wear. If there is any evidence of wear, the needle must be replaced.

Always replace any worn or damaged parts. A carburetor service kit is available at modest cost from the local Mercury/Mariner dealer. The kit will contain all necessary parts to perform the usual carburetor overhaul.

4/5 Hp Models

REMOVAL & INSTALLATION

◆ See Figures 51 thru 55

EASY

The 4/5 hp models use a round bowl single float carburetor with "Keihin" integral fuel pump. The fuel pump is integrated into the design of this carburetor.

1. Squeeze the wire clamp around the fuel inlet line and pull the line from the pump. Plug the fuel line to prevent leakage.
2. Loosen but do not remove, the screw retaining the throttle wire in the barrel clamp. Pull the throttle wire downward, to clear the barrel clamp. Pry the choke link from the plastic retainer on the carburetor linkage.
3. Remove the 2 screws securing the baffle cover to the carburetor.
4. Loosen the 2 carburetor retaining bolts. These 2 long bolts extend through the baffle bracket, the body of the carburetor and into the block.
5. Lift away the baffle bracket and carburetor with integral fuel pump still attached.
6. Remove and discard the carburetor mounting gasket.

To Install:

7. Slide the 2 long carburetor retaining bolts through the baffle bracket, carburetor body and mounting gasket.
8. Start the bolts and then tighten them, to 5.8 ft. lbs. (8 Nm).
9. Secure the baffle cover to the baffle bracket using 2 screws and washers. Tighten the screws securely.

Fig. 51 Release the clamp and carefully remove the fuel line

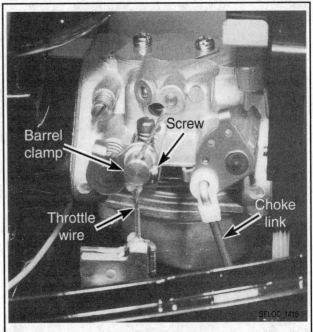

Fig. 52 Disconnect the throttle wire and the choke link

Fig. 53 Unbolt the air intake (baffle cover) from the carb

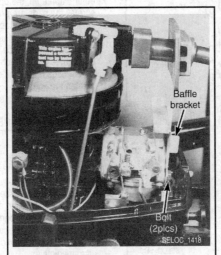

Fig. 54 Loosen the 2 carburetor retaining bolts and remove the carb

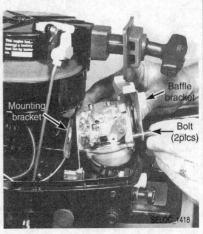

Fig. 55 During installation, position the mounting bracket, carb and baffle bracket, then thread the screws

10. Snap the choke link into the plastic fitting on the starboard side of the carburetor.

11. Check the action of the choke knob to be sure there is no evidence of binding.

12. Slide the throttle cable up into the barrel clamp from the bottom up but do not secure the wire until the following adjustment has been made. Back off the vertical idle speed screw until it no longer contacts the throttle arm. Rotate the same screw clockwise until it barely makes contact with the throttle arm, then continue to rotate the screw 2 more turns to slightly open the throttle plate. Grasp the cable end and lightly pull up to take up the slack. Tighten the screw on the clamp to secure the cable in position.

13. Slide the inlet fuel hose onto the fitting on the carburetor. Squeeze the 2 ends of the wire clamp and bring the clamp up over the inlet fitting.

14. Mount the outboard unit in a test tank, on the boat in a body of water or connect a flush attachment and hose to the lower unit. Connect a tachometer to the powerhead.

15. If the carburetor was rebuilt or replaced, refer to the Timing & Synchronization procedures in the Maintenance & Tune-Up section for adjustments.

OVERHAUL

◆ See Figures 56 thru 64

The 4/5 hp models use a round bowl single float carburetor with "Keihin" integral fuel pump. The fuel pump is integrated into the design of this carburetor.

1. Remove the carburetor from the powerhead, as detailed earlier in this section.

2. Remove the pilot (low speed mixture) screw and spring. The number of turns out from a lightly seated position will be given during assembly, but it is never a bad idea to lightly seat it first, counting and noting the number of turns just to be sure.

3. Remove the 2 screws securing the top cover and lift off the cover. Gently pry the 2 rubber plugs out with an awl. Remove the oval air jet cover and the round bypass cover.

4. Remove the 4 screws securing the fuel pump to the carburetor body. Disassemble the fuel pump components in the following order: the pump

cover, the outer gasket, the outer diaphragm, the pump body, the inner diaphragm and finally the inner gasket.

5. Remove the screws from both sides of the fuel pump body. Remove the check valves.

6. Remove the 4 screws securing the float bowl cover in place. Lift off the float bowl. Remove and discard the rubber sealing ring.

7. Remove the small Phillips screw securing the float hinge to the mounting posts. Lift out the float, the hinge pin and the needle valve. Slide the hinge pin free of the float.

8. Slide the wire attaching the needle valve to the float free of the tab.

9. Use the proper size slotted screwdriver and remove the main jet, then unscrew the main nozzle from beneath the main jet. Remove the plug and then unscrew the pilot jet located beneath the plug.

To Assemble:

10. Invert the carburetor and allow the float to rest on the needle valve. Measure the distance between the top of the float and the mixing chamber housing. This distance should be 1/2 in. (13mm). This dimension, with the carburetor inverted, places the lower surface of the float parallel to the carburetor body. If the dimension is not within the limits listed, the needle valve must be replaced.

11. carefully bend the float arm, as required, to obtain a satisfactory measurement. Install the float.

12. Slide the wire attached to the needle valve onto the float tab.

13. Guide the hinge pin through the float hinge. Lower the float and needle assembly down into the float chamber and guide the needle valve into the needle seat. Check to be sure the hinge pin indexes into the mounting posts. Secure the pin in place with the small Phillips screw.

14. Insert the rubber sealing ring into the groove in the float bowl. Install the float bowl and secure it in place with the 4 Phillips screws.

15. Place the check valves, one at a time, in position on both sides of the fuel pump body. Secure each valve with the attaching screw.

16. Assemble the fuel pump components onto the carburetor body in the following order: the inner gasket, the inner diaphragm, the pump body, the outer diaphragm, the outer gasket and finally the pump cover. Check to be sure all the parts are properly aligned with the mounting holes. Secure it all in place with the 4 attaching screws.

17. Install the oval air jet cover and the round bypass cover in their proper recesses. Place the top cover over them, no gasket is used. Install and tighten the 2 attaching screws.

18. Slide the spring over the pilot screw and then install the screw. Tighten the screw until it barely seats. Back the pilot screw out 1 1/2 turns.

CLEANING & INSPECTION

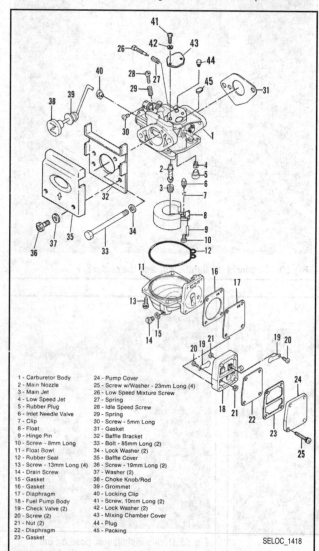

1 - Carburetor Body	24 - Pump Cover
2 - Main Nozzle	25 - Screw w/Washer - 23mm Long (4)
3 - Main Jet	26 - Low Speed Mixture Screw
4 - Low Speed Jet	27 - Spring
5 - Rubber Plug	28 - Idle Speed Screw
6 - Inlet Needle Valve	29 - Spring
7 - Clip	30 - Screw - 5mm Long
8 - Float	31 - Gasket
9 - Hinge Pin	32 - Baffle Bracket
10 - Screw - 8mm Long	33 - Bolt - 85mm Long (2)
11 - Float Bowl	34 - Lock Washer (2)
12 - Rubber Seal	35 - Baffle Cover
13 - Screw - 13mm Long (4)	36 - Screw - 19mm Long (2)
14 - Drain Screw	37 - Washer (2)
15 - Gasket	38 - Choke Knob/Rod
16 - Gasket	39 - Grommet
17 - Diaphragm	40 - Locking Clip
18 - Fuel Pump Body	41 - Screw, 10mm Long (2)
19 - Check Valve (2)	42 - Lock Washer (2)
20 - Screw (2)	43 - Mixing Chamber Cover
21 - Nut (2)	44 - Plug
22 - Diaphragm	45 - Packing
23 - Gasket	

SELOC_1418

Fig. 56 Exploded drawing of the round bowl, single float carburetor, with integrated Keihin fuel pump. Major parts are identified

♦ **See Figure 56**

Remove and disassemble the carburetor as detailed in this section under Removal & Installation and under Overhaul.

Never dip rubber parts, plastic parts, diaphragms or pump plungers in carburetor cleaner, because they tend to absorb liquid and expand. These parts should be cleaned only in solvent and then blown dry with compressed air immediately.

Place all of the metal parts in a screen-type tray and dip them in carburetor cleaner until they appear completely clean - free of gum and varnish which accumulates from stale fuel. Blow the parts dry with compressed air.

② MODERATE

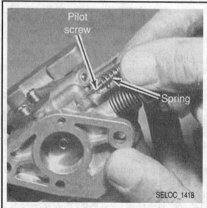

Fig. 57 Remove the pilot screw and spring

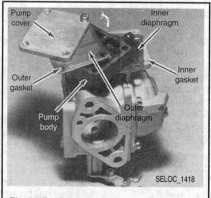

Fig. 58 Partially exploded view of the built-in fuel pump used on these carbs

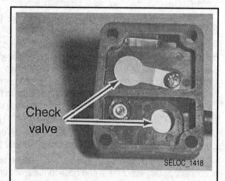

Fig. 59 View of the check valve used by the fuel pump assembly

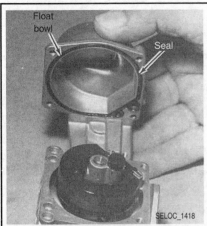

Fig. 60 Remove the float bowl and discard the old rubber seal

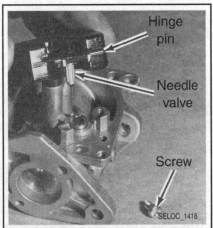

Fig. 61 Remove the screw securing the float hinge, then remove the float assembly

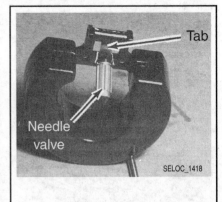

Fig. 62 Remove the needle valve and wire from the float

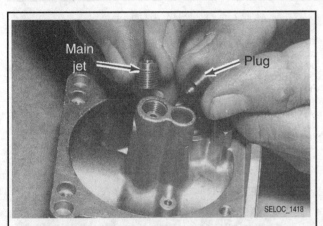

Fig. 63 Remove the main jet and nozzle from 1 bore, then remove the plug and pilot jet from the other bore

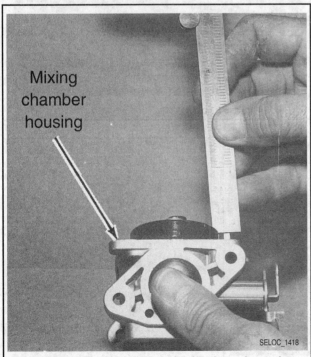

Fig. 64 Measure float height from the bowl mating surface to the top of the float

Blow out all of the passageways in the carburetor with compressed air. Check all of parts and passageways to be sure they are clear and not clogged with any deposits.

Never use a piece of wire or any type of pointed instrument to clean drilled passages or calibrated holes in a carburetor.

Carefully inspect the casting for cracks, stripped threads or plugs for any sign of leakage. Inspect the float hinge in the hinge pin area for wear and the float for any sign of leakage.

Examine the inlet needle for wear. If there is any evidence of wear, the needle must be replaced.

Always replace any worn or damaged parts. A carburetor service kit is available at modest cost from the local Mercury/Mariner dealer. The kit will contain all necessary parts to perform the usual carburetor overhaul.

6/8, 9.9/10/15 & 20/20 Jet/25 Hp Models

REMOVAL & INSTALLATION

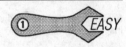

◆ See Figures 65, 66 and 67

The 6/8, 9.9/10/15 and 20/20 Jet/25 hp models are normally equipped some version with a Walbro WM or WMC carburetor with integral fuel pump. The WM and WMC carburetors can be identified by the embossed letters on the mounting flange. Some of the 20 hp and larger models are equipped with an electric actuated choke mounted on top of the carburetor (to the mixing chamber) instead of the manual knob actuated version used on most smaller models. However, the balance of the carburetor assembly is the same.

Don't let the moniker "choke" knob fool you, it is really a primer knob and serves to supply extra fuel to the motor, not to restrict (choke) the air supply.

1. If equipped, slightly inboard of the lower engine cowling slightly to the right of the choke/primer knob, locate the primer cam retaining screw, then loosen and remove the screw.

2. Also if equipped, use a pair of pliers, remove the circlip securing the choke knob in place immediately inboard of the engine cowling. Remove the choke knob by pulling it straight out.

3. Unsnap the idle wire from the primer bracket/fast idle lever.

4. Using a pair of snips, cut the Sta-Strap from the fuel hose to the carburetor. Pull the fuel line free of the fitting.

5. Remove the carburetor mounting nuts (usually 2) and lift the carburetor off the powerhead. Remove the bleed line from the fitting in the bottom of the carburetor.

To Install:

6. If removed (and when applicable) install the primer assembly components and/or air intake cover to the carburetor.

7. Reconnect the bleed hose to the fitting on the carburetor and position a new carburetor gasket onto the powerhead.

8. Install the carburetor to the powerhead and secure it in place with the 2 nuts.

9. Connect the fuel hose and secure it with a new Sta-Strap.

10. Reconnect the idle wire to the fast idle lever.

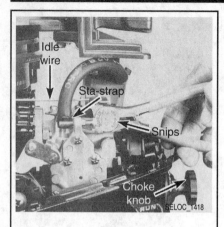

Fig. 65 For removal, cut the Sta-Strap and disconnect the fuel line from the carburetor

Fig. 66 During installation, position the choke/primer knob (if equipped) and install the circlip. . .

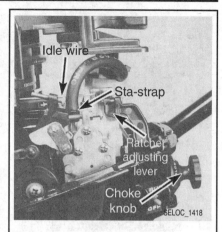

Fig. 67 . . .and be sure to use a new Sta-Strap to secure the fuel line

11. If equipped, push down on the primer arm, then insert the choke/primer knob, bezel and slide block into the primer assembly. Install the screw to secure the slide block in place.

12. Also if equipped, align the notch in the back side of the primer/choke bezel with the tab on the bottom cowl, then secure the bezel in place using the retaining clip.

13. Mount the outboard unit in a test tank, on a boat in a body of water or connect a flush attachment to the lower unit.

14. If the carburetor was rebuilt or replaced, refer to the Timing & Synchronization procedures in the Maintenance & Tune-Up section for adjustments.

OVERHAUL

② MODERATE

◆ See Figures 68 thru 78, 81 and 82

■ For carburetor exploded view(s), please refer to Cleaning & Inspection, in this section.

1. Remove the carburetor from the powerhead, as detailed earlier in this section.

2. Remove the high speed jet/float bowl retainer from the bottom of the carburetor. Keep track of the retainer, jet and gasket.

3. Lift the fuel bowl off the carburetor. The primer system is attached to the bowl and therefore, will come with it.

4. Lift out the fuel float for access.

5. Back out the screw securing the float hinge pin in place. Pull the hinge pin and then lift the hinge out of the carburetor body.

6. Remove the float bowl inlet needle from its position under the hinge.

7. Remove the carburetor bowl gasket from its recess.

8. Remove the screws (usually 2) securing the mixing chamber cover or the electric choke (on some 20 hp and larger models) to the top of the carburetor body. Remove the cover or electric choke assembly and then the gasket.

9. Remove the screws (usually 5) securing the fuel pump cover in place. Remove in order, the cover, gasket and then the fuel pump check valve diaphragm assembly. These components vary slightly by model, so pay attention to them as they are removed. Most late-models include a fuel pump body (to which the check valves and diaphragm assembly is mounted) and a second gasket.

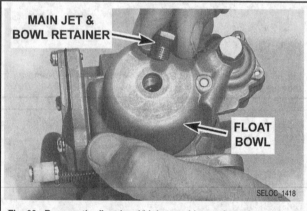

Fig. 68 Remove the float bowl/high speed jet retainer plug

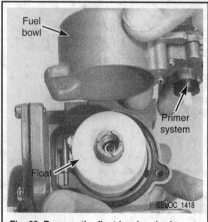

Fig. 69 Remove the float bowl and primer system assembly

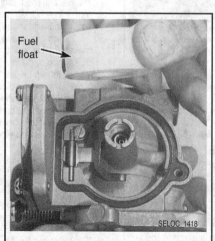

Fig. 70 Remove the float. . .

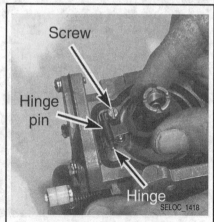

Fig. 71 . . .along with the screw, hinge pin and hinge

■ Make sure you lightly seat the low speed mixture needle, counting the number of turns it takes, as an aid during assembling to making a rough bench adjustment.

10. Lightly seat the low speed mixture needle while counting the number of turns it was out from a seated position and record that for reference during assembly. Then, carefully unthread the low speed mixture screw. The spring will come with the screw.

11. Remove the screws (usually 4) securing the primer system cover to the float bowl flange. Gently lift the cover and the primer diaphragm, gasket and spring will then be free. A seal is installed in the cover.

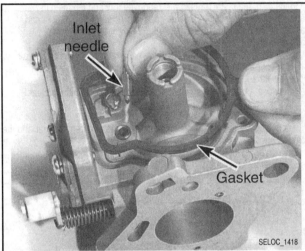

Fig. 72 Don't forget the fuel inlet needle or the float bowl gasket

12. Remove the base plug from the underside of the float bowl flange (primer system housing). Take care not to lose the spring from the shaft of the plug or the tiny check ball(s). Remove the washer.

To Assemble:

13. Drop the tiny check ball(s) and then the little spring into the cavity of the primer system housing. Set a new washer in place on the housing, then thread the plug into place and tighten it securely.

14. Turn the carburetor body over and set the spring in place in the primer housing.

15. Position a new gasket in place on the housing. Set the primer diaphragm on top of the spring with the holes in the diaphragm for the mounting screws roughly aligned with the holes in the primer housing.

16. Slip the seal over the lip on the cover and compress the spring with the primer system cover allowing the shaft of the primer diaphragm to pass up through the hole in the seal and the holes in the diaphragm to align perfectly with the holes in the housing.

17. Secure the cover in place with the attaching screws and lockwashers. Tighten the screws securely.

18. Slide the spring onto the shaft of the low speed mixture screw. Thread the screw into the carburetor housing and seat it very lightly.

■ Never tighten the screw, because such action would damage the tip.

19. From the lightly seated position, back the screw out the same number of turns recorded during disassembly. If the number of turns was lost, back the screw out as follows, depending upon the model:
- 6 hp - 1-1 1/2 turns
- 8-15 hp - 1 1/4-1 3/4 turns
- 20 hp - 3/4-1 1/4 turns
- 20 Jet - 1-2 turns
- 25 hp - 1-1 1/2 turns

■ This initial adjustment should allow the powerhead to be started and adjusted after assembly.

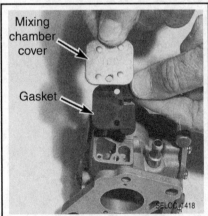

Fig. 73 If necessary, remove the mixing chamber cover and gasket (or electric choke assembly)

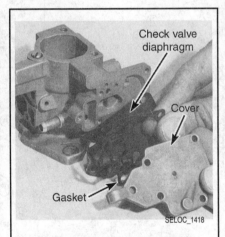

Fig. 74 Remove the fuel pump assembly (varies by model)

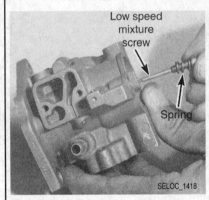

Fig. 75 Lightly seat the low speed mixture screw counting the number of turns for reference, then remove it from the carburetor

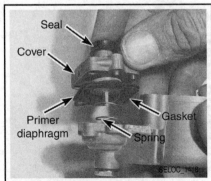

Fig. 76 If necessary, remove the primer assembly components. . .

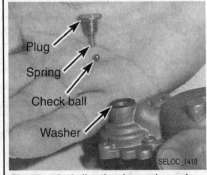

Fig. 77 . . .including the plug, spring and check ball(s)

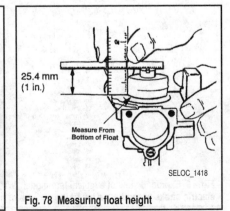

Fig. 78 Measuring float height

20. Assemble the fuel pump components (valve body, check valve diaphragm etc) with the holes in the gasket(s) for the mounting screws aligned with the holes in the body. Position a new gasket on top of the diaphragm with the mounting holes aligned.

21. Place the fuel pump cover onto the carburetor body and secure it with the attaching screws and lockwashers. Tighten the screws securely.

22. Position a new gasket in place on the mixing chamber. Install the mixing chamber cover onto the carburetor and secure it with the screws. Tighten the screws securely.

23. Position a new carburetor bowl gasket into place on the carburetor body. Slide the inlet needle into its hole in the body.

24. Slide the hinge pin through the hinge. Next, install the hinge and secure it in place with the screw. Tighten the screw snugly.

■ **Check to be sure the hinge will move without binding.**

25. Slide the float down over the shaft and onto the hinge.

26. Check the float level. With the carburetor inverted and the float in position on the hinge assembly. Measure from a line even with the carburetor body (as protruding through the center of the float) to the very bottom of the float (well, bottom in this inverted position, but normally it would be the top). The distance should be about 1 in. (25.4mm). If necessary, adjust the float level by ever so carefully bending the hinge *slightly*. Just a "whisker" of change in the hinge will move the float.

27. Cover the float with the float bowl and primer system housing.

28. Install the high speed jet along with the jet/float bowl retainer, using a new gasket. Tighten the retainer securely.

CLEANING & INSPECTION

◆ **See Figures 79 thru 82**

Remove and disassemble the carburetor as detailed in this section under Removal & Installation and under Overhaul.

Never dip rubber parts, plastic parts, diaphragms or pump plungers in carburetor cleaner, because they tend to absorb liquid and expand. These parts should be cleaned only in solvent and then blown dry with compressed air immediately.

Place all of the metal parts in a screen-type tray and dip them in carburetor cleaner until they appear completely clean - free of gum and varnish which accumulates from stale fuel. Blow the parts dry with compressed air.

Blow out all of the passageways in the carburetor with compressed air. Check all of parts and passageways to be sure they are clear and not clogged with any deposits.

Never use a piece of wire or any type of pointed instrument to clean drilled passages or calibrated holes in a carburetor.

Carefully inspect the casting for cracks, stripped threads or plugs for any sign of leakage. Inspect the float hinge in the hinge pin area for wear and the float for any sign of leakage.

Examine the inlet needle for wear. If there is any evidence of wear, the needle must be replaced.

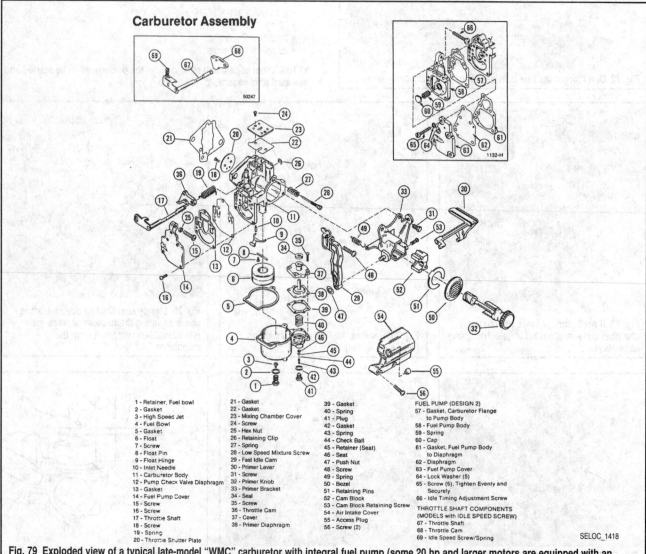

Carburetor Assembly

1 - Retainer, Fuel bowl	21 - Gasket	39 - Gasket	**FUEL PUMP (DESIGN 2)**
2 - Gasket	22 - Gasket	40 - Spring	57 - Gasket, Carburetor Flange
3 - High Speed Jet	23 - Mixing Chamber Cover	41 - Plug	to Pump Body
4 - Fuel Bowl	24 - Screw	42 - Gasket	58 - Fuel Pump Body
5 - Gasket	25 - Hex Nut	43 - Spring	59 - Spring
6 - Float	26 - Retaining Clip	44 - Check Ball	60 - Cap
7 - Screw	27 - Spring	45 - Retainer (Seat)	61 - Gasket, Fuel Pump Body
8 - Float Pin	28 - Low Speed Mixture Screw	46 - Seat	to Diaphragm
9 - Float Hinge	29 - Fast Idle Cam	47 - Push Nut	62 - Diaphragm
10 - Inlet Needle	30 - Primer Lever	48 - Screw	63 - Fuel Pump Cover
11 - Carburetor Body	31 - Screw	49 - Spring	64 - Lock Washer (5)
12 - Pump Check Valve Diaphragm	32 - Primer Knob	50 - Bezel	65 - Screw (5), Tighten Evenly and
13 - Gasket	33 - Primer Bracket	51 - Retaining Pins	Securely
14 - Fuel Pump Cover	34 - Seal	52 - Cam Block	66 - Idle Timing Adjustment Screw
15 - Screw	35 - Screw	53 - Cam Block Retaining Screw	
16 - Screw	36 - Throttle Cam	54 - Air Intake Cover	**THROTTLE SHAFT COMPONENTS**
17 - Throttle Shaft	37 - Cover	55 - Access Plug	**(MODELS with IDLE SPEED SCREW)**
18 - Screw	38 - Primer Diaphragm	56 - Screw (2)	67 - Throttle Shaft
19 - Spring			68 - Throttle Cam
20 - Throttle Shutter Plate			69 - Idle Speed Screw/Spring

SELOC_1418

Fig. 79 Exploded view of a typical late-model "WMC" carburetor with integral fuel pump (some 20 hp and larger motors are equipped with an electric choke assembly on top of the carburetor instead of the primer knob)

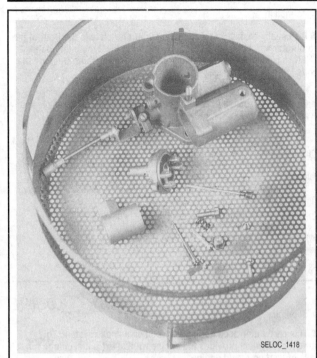

Fig. 80 Only metallic parts should be submerged in carburetor cleaner

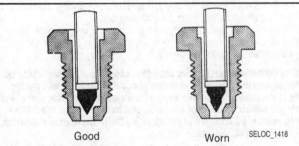

Fig. 81 Comparison of a worn and new needle and seat arrangement. The worn needle would have to be replaced for full carburetor efficiency

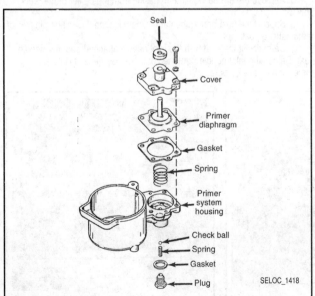

Fig. 82 Exploded drawing of the primer system with major parts identified

Always replace any worn or damaged parts. A carburetor service kit is available at modest cost from the local Mercury/Mariner dealer. The kit will contain all necessary parts to perform the usual carburetor overhaul.

30/40 Hp 2-Cylinder Models

Several different versions of the Walbro WME carburetor are used. The units are almost identical except for the main jet size, though some fittings or linkage may vary slightly in shape from the illustrations.

REMOVAL & INSTALLATION

◆ See Figures 83, 84 and 85

1. On electric start models, disconnect the negative battery cable for safety.
2. On oil injected models remove the bolts for the oil tank top cover, then reposition the cover. Tag and disconnect the wiring and oil line for the oil tank (a hose pincher or plug for the line will be handy if the tank is not to be completely drained), then remove the oil tank from the powerhead for access to the carburetor behind it.
3. Loosen the 2 retaining screws then remove the carburetor cover/air intake silencer assembly.
4. Disconnect the fuel line from the carburetor.
5. Disconnect the throttle linkage.
6. Remove the 2 hex bolts threaded through the carburetor plate and carburetor into the intake manifold. Support the carburetor as the bolts are unthreaded, then carefully remove the carburetor from the powerhead.

Fig. 83 On oil injected motors you must remove the oil tank top cover. . .

Fig. 84 . . .and oil tank itself for access to the carburetor

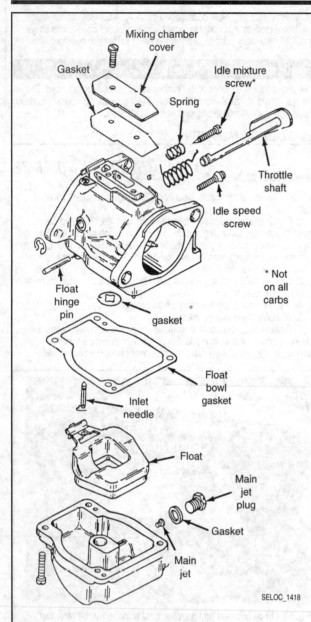

Fig. 85 Exploded view of the main components of a typical Walbro WME carburetor (note the Idle Speed Screw is only found in that location on 2-cylinder models)

To Install:

■ It's usually easiest to position the carburetor plate to the front of the carburetor and hold it in position by inserting the 2 mounting bolts, THEN to place the gasket on the back side of the carburetor using the bolts to hold it there.

7. Position the gasket, carburetor and mounting plate to the intake manifold and thread the 2 mounting bolts. Tighten the bolts alternately and evenly to 100 inch lbs. (11.3 Nm).

8. Connect the fuel line and the throttle linkage.

9. Pressurize the fuel system using the fuel line squeeze bulb and check the carburetor and fuel lines for leaks.

10. Install the air intake silencer/carburetor cover and tighten the retaining screws.

11. On oil injected models, install the oil tank and tank cover assembly. Tighten the bolts securely.

12. On electric start models, reconnect the negative battery cable.

13. If the carburetor was rebuilt or replaced, refer to the Timing & Synchronization procedures in the Maintenance & Tune-Up section for adjustments.

OVERHAUL

◆ See Figures 85 thru 94   ② ◁ MODERATE

■ This carburetor does not have a traditional choke system. This enrichener system provides the engine with an extra fuel charge for ease of starting a cold engine.

1. Remove the carburetor from the powerhead, as detailed earlier in this section.

2. Remove the screws (usually 2), then remove the mixing chamber cover and gasket from the top of the carburetor body.

3. Remove the float bowl plug/main jet retainer, then remove the gasket and, if necessary the main jet. Drain as much of the fuel present in the float bowl as possible through the opening.

■ The float bowl/main jet plug provides access for a screwdriver to be inserted into the opening and allows removal of the main jet from the side of the center turret. This design facilitates changing the size of the main jet, without removal of the carburetor from the powerhead. Main jet sizes must be changed when operating the powerhead at elevations higher than 2,500 feet above sea level.

4. Invert the carburetor for access to the float bowl retaining screws. Loosen the screws (usually 4), then lift off the fuel bowl. Remove and discard the gasket.

5. Remove the nozzle well (stem) gasket from the center of the fuel bowl.

6. Support the float and at the same time push the float hinge pin free of the mounting posts.

7. Carefully lift the float with the inlet needle attached from the needle seat. Detach the inlet needle from the float. The needle seat is not removable on this carburetor.

Fig. 86 Float bowl plug and idle mixture screw locations

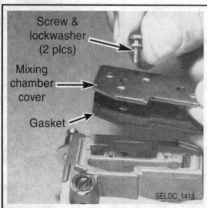

Fig. 87 Remove the mixing chamber cover and gasket

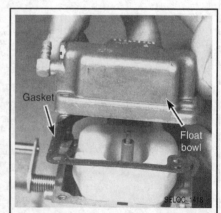

Fig. 88 After draining, remove the float bowl and gasket

8. If necessary to remove the idle mixture screw, first turn the screw inward while counting the turns necessary to lightly seat it, then unthread and remove the screw and spring.

■ **Further disassembly of the carburetor is not necessary in order to clean it properly.**

To Assemble:

9. If removed, install the idle mixture screw and spring. Thread the screw inward slowly and gently until it is lightly seated, then unthread the screw either the same number of turns noted during removal or if that number was not recorded, 1 1/4-1 3/4 turns (for all except 40 hp motors with the WME-38 carburetor, for which the initial setting should be 1 1/2-2 turns). This setting should allow the powerhead to be started for adjustment.

10. Hook the inlet needle spring over the float tab and lower the needle into its seat with the float hinge between the mounting posts.

11. Slide the float hinge pin through the posts to secure the float. Center the pin between the posts. Place the stem gasket over the center turret.

12. Hold the carburetor body in the inverted position - the same position it has been in since the start of the assembling procedures - with the float resting on the inlet needle. Measure the distance between the float bowl gasket surface (no gasket in place) and a point on the float directly opposite the hinge. Notice the surface of the float curves downward. Therefore, the measurement point is the lowest on the horizontal surface. The distance should be 9/16 in. (14.29mm).

13. If the distance is not as specified, then remove the float hinge pin to free the float. The float height adjustment may be made by bending the tab on which the inlet needle hangs.

14. Repeat the adjustments until the specified distance between the float bowl gasket surface and the lowest edge of the float is obtained.

15. If removed, install the main jet into the center turret of the fuel bowl. Tighten the jet securely. Install the gasket and main jet plug in the exterior wall of the float bowl and tighten the plug securely.

16. With the carburetor still inverted, position a new gasket onto the body. Place the fuel bowl in position and secure it in place with the 4 screws. Tighten the screws securely.

17. Position a new gasket over the mixing chamber. Install the cover and secure it in place with the attaching hardware.

CLEANING & INSPECTION

◆ **See Figures 85, 95 and 96**

Remove and disassemble the carburetor as detailed in this section under Removal & Installation and under Overhaul.

Never dip rubber parts, plastic parts, diaphragms or pump plungers in carburetor cleaner, because they tend to absorb liquid and expand. These parts should be cleaned only in solvent and then blown dry with compressed air immediately.

Place all of the metal parts in a screen-type tray and dip them in carburetor cleaner until they appear completely clean - free of gum and varnish which accumulates from stale fuel. Blow the parts dry with compressed air.

Blow out all of the passageways in the carburetor with compressed air. Check all of parts and passageways to be sure they are clear and not clogged with any deposits.

Never use a piece of wire or any type of pointed instrument to clean drilled passages or calibrated holes in a carburetor.

Carefully inspect the casting for cracks, stripped threads or plugs for any sign of leakage. Inspect the float hinge in the hinge pin area for wear and the float for any sign of leakage.

Examine the inlet needle for wear. If there is any evidence of wear, the needle must be replaced.

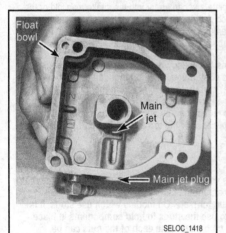

Fig. 89 See how the main jet is accessible through the float bowl plug

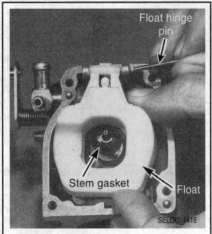

Fig. 90 Remove the hinge pin in order to free the float. . .

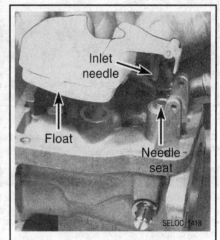

Fig. 91 . . .then carefully remove the float and inlet needle

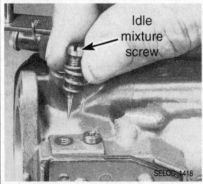

Fig. 92 If necessary, remove the idle mixture screw and spring

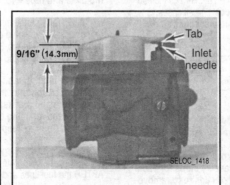

Fig. 93 During assembly measure the float height

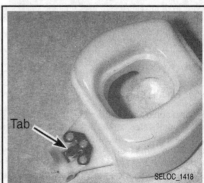

Fig. 94 If necessary, adjust the float height by bending the tab

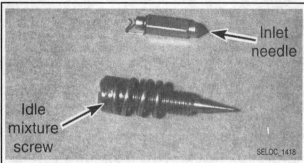

Fig. 95 Inspect the taper of the inlet needle and the idle mixture screw for evidence of a worn groove

Fig. 96 Never use a piece of wire or any pointed tool to clean drilled passages or calibrated holes in a carburetor

Always replace any worn or damaged parts. A carburetor service kit is available at modest cost from the local Mercury/Mariner dealer. The kit will contain all necessary parts to perform the usual carburetor overhaul.

40-125 Hp 3 & 4-Cylinder Models

These models are all equipped with versions of the Walbro WMA carburetor, similar to the type used on 30/40 hp 2-cylinder models, except that on those models only 1 carburetor is installed and on these there are multiple carburetors (one per cylinder).

Though technically several different versions of the Walbro WME carburetor are used, the units are almost identical except for the main jet size and the fact that idle speed is centrally controlled on the linkage instead of directly at the single carb used on the 2-cylinder models. In all cases, identification letters are stamped on the side of the carburetor throttle body.

REMOVAL & INSTALLATION

◆ See Figures 97 thru 101

■ Unlike some motors the multiple carburetor assemblies on these Mercury outboards can be awkward to remove as an assembly. This is because the outer carburetor plates usually share the same mounting bolts at the carbs themselves. For this reason the procedure is written for removal of a single carburetor. Repeat the necessary steps for each subsequent carburetor, as necessary. However, if you desire to remove the carburetors all together as an assembly, don't disconnect the enrichener/fuel lines or the throttle linkage which run from carb-to-carb and this will help keep them together as a unit.

1. On electric start models, disconnect the negative battery cable for safety.

2. Remove the engine cowling for access.

3. Loosen the retaining screw(s), then remove the carburetor cover/air intake silencer assembly. The number of screws used and the design of the cover will vary from as few as 1 screw at the top of the cover or 4 toward the center to 6 or 8 screws around the perimeter of the assembly.

4. On 75 hp and larger motors equipped with oil injection, remove the screws from the oil tank support (where they connect to the top of the air box backing plate/carburetor plate). In some cases carburetor removal and installation will be easier if you also disconnect the oil lines and completely remove the oil tank.

5. On 75 hp and larger motors it is difficult to access the carburetor fuel lines or linkage with the air box backing plate/carburetor plate still installed. Remove the fasteners securing the plate (and carburetors), then remove the plate assembly. On 3-cylinder motors there are usually 4 nuts and 2 bolts while on 4-cylinder motors there are usually 6 nuts and 2 bolts.

■ **Reminder, if you want to remove multiple carburetors as an assembly leave the throttle linkage and fuel lines connected between them, just disconnect them from the powerhead linkage and fuel line.**

6. Disconnect the fuel line from the carburetor (or carburetors) being removed.

7. Disconnect the throttle linkage from the carburetor (or carburetors) being removed.

■ **When removing multiple carburetors from the manifold at the same time, take a moment to identify each carburetor to ensure each will be installed back in its original position, because on some applications each carburetor is slightly different.**

8. On 60 hp and smaller motors, remove either the 2 hex bolts threaded through the carburetor plate and carburetor into the intake manifold or the nuts from the studs which protrude through the plate, carb and intake manifold. Support the carburetor as the fasteners are unthreaded, then carefully remove the carburetor from the powerhead.

■ **On most applications the linkage on each carburetor will vary slightly with the carb's positioning. Also, on most 40-60 hp models the float bowl on the upper carburetor will vary from the others, as it is used to gravity feed the enrichener circuit. On larger models the enrichener circuit was redesigned by about 1996 to feed through a T-fitting on the fuel supply line for the top and middle carburetors.**

To Install:

■ **Installation will vary very slightly depending upon whether the motor uses studs or bolts to mount the carburetor and, of course, the size/shape of the carburetor plate. On models which use studs, it is convenient and easy to use the studs to hold components in place (gaskets, carburetors, carb plate) until each of the nuts can be installed. However, on models which use bolts you may have to temporarily install each of one at a time (without the mounting plates), then go back and remove the bolts from one side of the carbs to install the plate and repeat for the other side.**

9. On 60 hp and smaller motors, position the gasket, carburetor (in the proper position as noted during removal) and mounting plate to the intake manifold and thread the 2 mounting bolts or nuts (as applicable). Tighten the fasteners alternately and evenly until secure.

■ **Keep in mind that on models with bolts instead of studs, you can cheat and temporarily install a set of studs to hold the gaskets, carburetors and plate(s) in position, then remove the studs one-by-one installing the retaining bolts in their place, but allowing the rest of the studs and bolts to keep the components in this position as you work.**

10. On 75 hp and larger motors, position the gasket(s) over the studs, followed by the carburetor(s). You'll install the carburetor plate and fasteners AFTER the fuel lines and throttle linkage.

11. Connect the fuel line(s) and the throttle linkage.

12. Pressurize the fuel system using the fuel line squeeze bulb and check the carburetors and fuel lines for leaks.

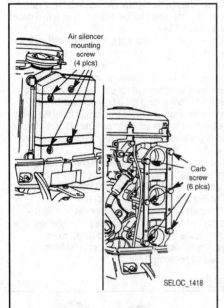

Fig. 97 Typical air silencer cover and carburetor mounting plates

Fig. 98 Carburetor mounting plate from a large 4-cylinder motor (note the 8 carb/plate retainers)

Fig. 99 Disconnect the fuel hoses and linkage. . .

Fig. 100 . . .then loosen the fasteners and remove the carburetor

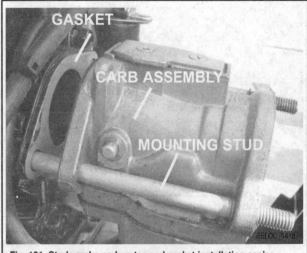

Fig. 101 Studs make carburetor and gasket installation easier

13. On 75 hp and larger motors install the air box backing plate/carburetor plate and secure using the fasteners (nuts and bolts). Install the oil tank retainers (or reinstall the tank with lines, if the tank was completely removed for access). Tighten all fasteners securely.

14. Install the air intake silencer/carburetor cover and tighten the retaining screws.

15. On electric start models, reconnect the negative battery cable.

16. If the carburetors were disconnected from each other, rebuilt or replaced, refer to the Timing & Synchronization procedures in the Maintenance & Tune-Up section for adjustments.

17. Test run the motor checking again for leaks, then install the engine cowling.

OVERHAUL

◆ See Figures 85 and 102

MODERATE

Overhaul, as well as Cleaning & Inspection of the WMA carburetor is identical (except for initial pilot screw settings) to the procedures given earlier in this section for the WMA carburetor used on 30/40 hp 2-cylinder models. For details, please refer to those procedures.

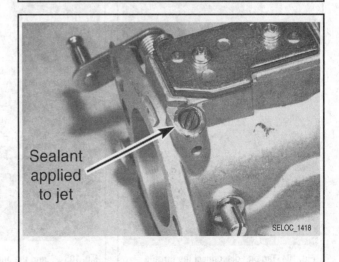

Sealant applied to jet

Fig. 102 This passage should remain un-tampered and sealed

■ The WMA Carburetors used on 3- and 4-cylinder models may be equipped with a sealed adjustment point (jet) in the top rear of the carburetor body (immediately next to both the intake manifold mounting point and the mixing chamber cover). The only information Mercury publishes about this point is that it is set at the factory, sealed, and SHOULD NOT be touched in the field. They go on to explain that the type of sealant used should not be affected by conventional carburetor cleaners.

If the idle mixture screw (pilot screw) was disturbed and you did not record the previous setting, the initial adjustment should be 1 to 1 1/2 turns out from a LIGHTLY seated position for all models except the 75 hp motor, for which the initial spec is 7/8 - 1 3/8 turns.

150/200 Hp (2.5L) V6 Models

The 2000 and later 150/200 hp carbureted (2.5L) V6 models are equipped with a Walbro WMV carburetor. These newer carburetors are provided with 2 off idle jets, 2 main (high speed) jets, one common back draft jet and 2 adjustable idle mixture screws. They can be identified by having 2 barrels and only 1 float used to feet 2 cylinders each on these V6 motors (meaning they are installed in a grouping of 3 carburetors).

The main jets, located on each side of the carburetor fuel bowl, control the high speed air/fuel mixture. Fuel is drawn from the fuel bowl, through the main jets. A set of discharge tubes guide the fuel into the carburetor venturi. A jet with a small orifice will provide a lean air/fuel mixture. A jet with a larger orifice will provide a rich air/fuel mixture.

The idle circuit consists of 2 externally adjustable fuel mixture screws and 2 idle air bleed jets. The idle circuit operates independently of the progression and main jet circuits.

The 2 fuel mixture screws provide a maintenance capability of lean or richen the idle fuel mixture for a smoother idle speed. The screws have a limiter cap installed, from the factory, to limit the adjustments to 1/2 turn. When adjusting the fuel mixture screws all screws must be turned the same amount and in the same direction for the engine to operate efficiently at idle speeds. Turning the screws clockwise will lean the fuel mixture and turning them counterclockwise will richen the fuel mixture.

Generally speaking you should not remove the limiter caps to further richen or lean the fuel mixture. If the limiter caps and fuel mixture screws are removed for carburetor cleaning, count the number of turns clockwise (in) until the screw is lightly seated and then remove the screws.

The back draft circuit consists of a vent jet which supplies less than atmospheric pressure to the fuel bowl during mid-range power settings. The lower atmospheric pressure in the fuel bowl during mid-range power operation causes the fuel flow to become slightly lean and results in improved performance and fuel economy.

REMOVAL & INSTALLATION

◆ See Figures 103 thru 106

Because the jetting may vary from carburetor-to-carburetor on a given motor (meaning that the carbs feeding different cylinders will vary slightly in jetting), the carburetors must remain in the exact position from where they are removed. Be sure to identify each carburetor position prior to removal.

■ Some people will leave the fuel/enrichment lines and throttle linkage attached from carburetor-to-carburetor in order to remove the 3 carbs as an assembly. Frankly, it can work, but it can also be awkward. But the choice is yours.

1. Disconnect the negative battery cable for safety during the procedure.
2. Unlatch and remove the cowling from the powerhead.
3. Remove the bolts (usually 6) securing the air box and heat shield assembly to the carburetors, then remove the air box assembly for access.
4. Disconnect the throttle linkage either at the powerhead connection or from each carburetor throttle lever (depending upon whether you are removing each carburetor individually or as an assembly) by inserting a flat blade screwdriver between the throttle lever and the nylon linkage arm. Pry the nylon linkage arm off the end of the throttle lever.

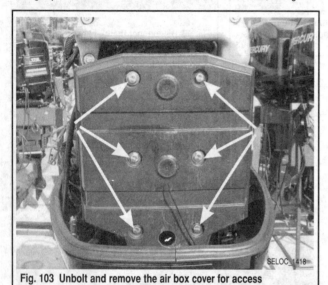

Fig. 103 Unbolt and remove the air box cover for access

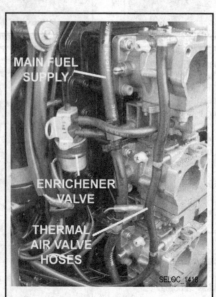

Fig. 104 Tag and disconnect the throttle linkage...

Fig. 105 ...and the necessary fuel/enrichment/air valve lines

Fig. 106 Each carburetor is secured by 2 nuts and 2 Allen head bolts

5. Disconnect the oil pump link rod from the main throttle lever (usually right behind the center carburetor).

6. Loosen the clamps or cut the tie-straps and disconnect the fuel supply hoses, fuel enrichment hoses and the thermal air valve hoses from the carburetor fittings. Again, if you are trying to remove all carbs as an assembly, leave the hoses which run from carb-to-carb connected. If you are separating the carbs, BE SURE TO mark each of the lines for positioning (Top, Center and Bottom) before removal.

7. Remove the 2 nuts and 2 Allen head bolts securing each carburetor to the intake manifold and lift the carburetor(s) free of the powerhead. Discard the carburetor gasket(s).

To Install:

8. Position a new carburetor gasket onto the intake manifold studs for each carburetor that was removed.

9. Install each carburetor in the same position from which it was removed - Top, Center and Bottom - as identified during the removal procedures.

10. Secure each carburetor to the powerhead with the 2 nuts and 2 Allen head screws.
Connect the fuel enrichment hoses, fuel lines and the thermal air valve hoses to the carburetor fittings as tagged or disconnected during removal. Secure the lines and hoses with clamps or tie-straps.

11. Use the fuel primer bulb to pressurize the system and check for leaks.

12. Connect the throttle linkage to each carburetor throttle lever (or the main throttle linkage to the carburetor assembly) by pushing the nylon linkage arm onto the end of the throttle lever.

13. Connect the oil pump link rod to the throttle lever assembly.

14. Install the air box over the carburetors, aligning the holes in the cover with the carburetors. Secure the air box to the carburetors with 6 bolts and tighten the bolts securely.

15. Connect the negative battery lead.

■ When operating the outboard for the first time following a carburetor overhaul, be sure to check all fuel line connections and the carburetor float bowl seams for any sign of fuel leakage. With the powerhead operating, check for fuel dribbling from the carburetor vent tube. Fuel leaks from the vent tube and float bowl seams indicate the float level is not set properly or a needle valve is sticking.

16. If the carburetors were disconnected from each other, rebuilt or replaced, refer to the Timing & Synchronization procedures in the Maintenance & Tune-Up section for adjustments.

17. After the powerhead has been operated and verified no fuel leaks exist, install the cowling over the powerhead and latch it closed.

OVERHAUL

◆ See Figures 107 thru 120

■ Proper identification of the carburetor by model number is critical when purchasing replacement parts. Without these numbers the dealer has no way of ordering the correct parts. The model WMV carburetor is identified by stampings on the top or side (usually starboard side) of the main body. The models most often found on these motors are the WMV-16 (150 hp motors) and the WMV - 18 (200 hp motors).

1. Remove the carburetor from the powerhead, as detailed earlier in this section.

2. Remove the screws (usually 2) securing each of the mixing chamber covers to the top of the carburetor body. Remove the covers, then remove and discard the gaskets.

3. Loosen and remove the main jet access plug on one side of the float bowl. Discard the gasket on the plug.

Fig. 107 You can service some components with the carbs installed. . .

Fig. 108 . . .like main jets (accessed behind the this plug). . .

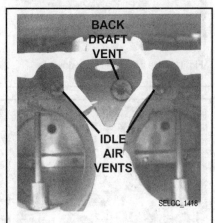

Fig. 109 . . .and idle or back draft vents

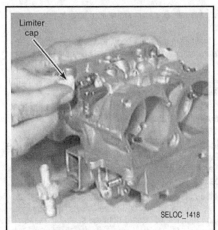

Fig. 110 During a complete overhaul, remove the limiter cap. . .

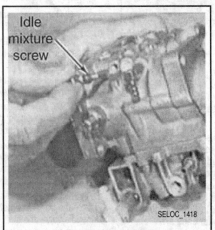

Fig. 111 . . .then unthread the idle screw (after seating it and counting the turns)

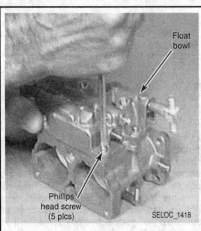

Fig. 112 Loosen the float bowl retaining screws. . .

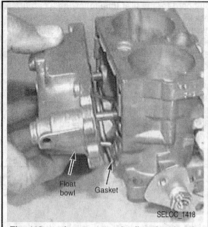

Fig. 113 . . .then remove the float bowl and gasket

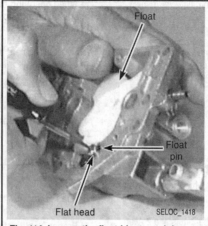

Fig. 114 Loosen the float hinge retaining screw. . .

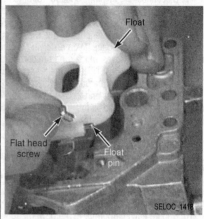

Fig. 115 . . .then remove the screw, hinge pin and float

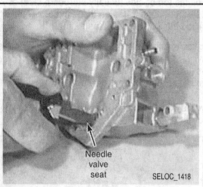

Fig. 116 Use a flat head screwdriver to unthread. . .

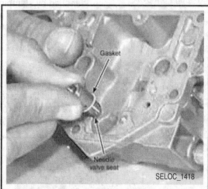

Fig.117 . . .and remove the inlet valve seat (and gasket)

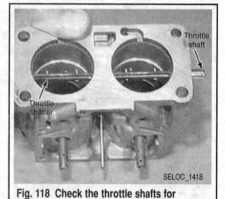

Fig. 118 Check the throttle shafts for sticking or binding

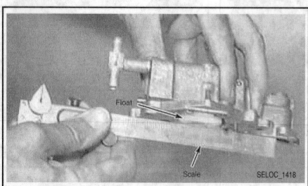

Fig. 119 Before installation make sure the float is even with the edge of the float bowl

4. Insert a long flat blade screwdriver down inside the counterbore and unscrew the main jet. Withdraw the main jet from the bore. Set the jet to one side.

5. Repeat steps to remove the other main jet access plug and main jet from the other side of the carburetor body.

■ **Limiter caps are installed on the idle air mixture screw at the factory. The caps are necessary to prevent an over rich or too lean a fuel mixture but do allow some movement for idle speed fuel mixture settings. If it is necessary to remove the idle mixture screws from the carburetor for proper cleaning, remove the limiter cap and turn the screw gently clockwise, while counting the number of turns, until it is lightly seated. The mixture screw must be set back to this number of turns from the lightly seated position when the limiter caps are installed. If this spec is not recorded or if this does not provide a good starting point for mixture adjustment you can also use the factory build spec of 1 1/8-1 3/8 turns out from seated.**

6. Slip the nylon limiter cap off the end of the idle mixture screw. Using a flat blade screwdriver slowly and gently turn the mixture screw clockwise and count the number of turns until it is lightly seated. Record this number for assembly purposes.

7. Unthread the idle mixture screw fully from the bore. Do not lose the spring over the screw shank. Repeat for the other idle mixture screw on the other side of the carburetor.

8. The 2 idle air vent jets and the back draft vent jet are all installed in the front of the carburetor body. If the carburetor is being disassembled only for general cleaning and these jets are not to be replaced, it is recommended these jets not be disturbed and a good quality carburetor spray cleaner be applied. If the jets are being replaced, remove each jet using a flat blade screw driver.

9. Remove the screws (usually 6) securing the float bowl to the carburetor main body.

10. Separate the float bowl from the main body and slide the bowl assembly free of the carburetor. Discard the gasket between the float bowl and main body.

11. Hold the float bowl carefully and remove the flat head screw securing the float pin to the bowl assembly.

12. Lift out the screw, float and pin from the float bowl. Examine the float assembly for signs of cracking or saturation of fuel. Replace the float assembly if found to be defective.

13. Lift the fuel needle valve from the needle seat assembly. Examine the needle valve for signs of wear and deterioration. If the needle valve is worn, replace the needle valve and seat as an assembly. Discard the needle valve if a new assembly is to be installed.

14. Use a "wide" flat blade screwdriver and remove the needle valve seat from the bowl assembly. Discard the gasket on the valve seat. If a new needle valve and seat is to be installed, discard both the seat and gasket.

15. Push and pull on the throttle shaft, if there is sign of excessive "play" in the shaft and/or the shutters are gouging the throttle bore heavily. The throttle shaft and bushings are defective or worn and must be replaced.

■ Further disassembly of the carburetor is not necessary in order to clean it properly. Normally it is not necessary to disassemble the throttle shutters unless there is excessive "play" or the shutters are binding. Before disassembling the shutters and throttle shaft, first check with the local dealer for parts availability. The shutters on some carburetor models cannot be serviced as individual parts. In some cases the throttle body must be replaced.

16. Operate the throttle shutters open and closed several times checking for any sticking or binding action. If the throttle shutters show any sign of binding or sticking they must be replaced.

To Assemble:

17. Slip a new gasket over the threaded end of the fuel inlet valve seat. Install the seat into the float bowl assembly. Use a wide-bladed screwdriver to tighten the valve seat.

18. Insert the needle valve into the seat and then install the float and float pin. Be sure to center the float pin into the recess of the float bowl. Install the flat head screw to secure the float.

19. Pick up the float bowl and invert the assembly to allow the float to hang free. Place a machinist scale across the flat surface of the float bowl directly under the float. The correct float height is the float just making contact with the scale but the float does not drop below the horizontal surface of the float bowl.

20. If the float height is not correct, carefully bend the small metal tab on the tip of the float to adjust the float height.

21. Place a new gasket over the float bowl. Guide the float bowl onto the carburetor body. Secure the float bowl to the carburetor body with 5 Phillips head screws.

22. Thread the idle mixture screw and spring into the bore. Slowly and gently turn the mixture screw all the way down until it is lightly seated. Back out the mixture screw the exact number of turns recorded during disassembly. If the count was lost or not remembered, back it out 1 1/8-1 3/8 turns as a "rough" bench adjustment at this time. Fit the limiter cap over the end of the screw at approximately the 12 o'clock position. Repeat this step for the other idle mixture screw.

23. Install the main jet into the bore using a long narrow screwdriver. Be careful not to cross thread the jet. Place a new gasket over the end of the access plug and then install the plug and tighten it securely. Repeat this step for the other main jet and access plug.

24. Install the idle air jets and back draft vent jet, if any were removed during the disassembly procedures. Be sure to install the proper size jet for the carburetor model (same size as was removed or see a Mercury parts dealer to determine the appropriate replacement size).

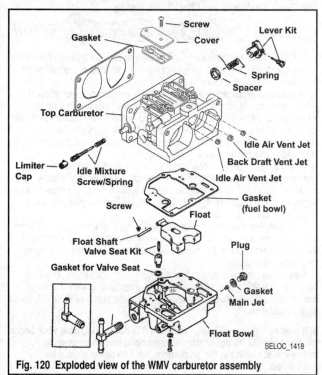

Fig. 120 Exploded view of the WMV carburetor assembly

Screw, Gasket, Cover, Lever Kit, Spring, Spacer, Top Carburetor, Limiter Cap, Idle Mixture Screw/Spring, Screw, Float Shaft Valve Seat Kit, Gasket for Valve Seat, Idle Air Vent Jet, Back Draft Vent Jet, Idle Air Vent Jet, Gasket (fuel bowl), Float, Plug, Gasket Main Jet, Float Bowl

SELOC_1418

CLEANING & INSPECTION

◆ See Figure 120

Remove and disassemble the carburetor as detailed in this section under Removal & Installation and under Overhaul.

Never dip rubber parts, plastic parts, diaphragms or pump plungers in carburetor cleaner, because they tend to absorb liquid and expand. These parts should be cleaned only in solvent and then blown dry with compressed air immediately.

Place all of the metal parts in a screen-type tray and dip them in carburetor cleaner until they appear completely clean - free of gum and varnish which accumulates from stale fuel. Blow the parts dry with compressed air.

Blow out all of the passageways in the carburetor with compressed air. Check all of parts and passageways to be sure they are clear and not clogged with any deposits.

Never use a piece of wire or any type of pointed instrument to clean drilled passages or calibrated holes in a carburetor.

Carefully inspect the casting for cracks, stripped threads or plugs for any sign of leakage. Inspect the float hinge in the hinge pin area for wear and the float for any sign of leakage.

Examine the inlet needle for wear. If there is any evidence of wear, the needle must be replaced.

Always replace any worn or damaged parts. A carburetor service kit is available at modest cost from the local Mercury/Mariner dealer. The kit will contain all necessary parts to perform the usual carburetor overhaul.

Fuel Pump

TESTING

Lack of an adequate fuel supply will cause the engine to run lean, lose rpm or cause piston scoring due to a lack of lubricant that is carried in the fuel. If an integral fuel pump carburetor is installed, the fuel pressure cannot be checked (as fuel delivery passage is internal to the carb so there is no way to place a gauge between the pump and carb). On integral carb models, other methods must be used to check pump integrity.

Generally speaking, with a multiple carburetor installation, fuel pressure at the top carburetor should be checked whenever insufficient fuel is suspected because it is the farthest from the fuel pump.

✳✳ SELOC CAUTION

Observe all applicable safety precautions when working around fuel. Whenever servicing the fuel system, always work in a well ventilated area. Do not allow fuel spray or vapors to come in contact with a spark or open flame. Keep a dry chemical fire extinguisher near the work area. Always keep fuel in a container specifically designed for fuel storage, also, always properly seal fuel containers to avoid the possibility of fire or explosion.

The problem most often seen with fuel pumps is fuel starvation, hesitation or missing due to inadequate fuel pressure/delivery. In extreme cases, this might lead to a no start condition, but that is pretty rare as the primer bulb should at least allow the operator to fill the float bowl or vapor separator tank. More likely, pump failures are not total, and the motor will start and run fine at idle, only to miss, hesitate or stall at speed when pump performance falls short of the greater demand for fuel at high rpm.

Before replacing a suspect fuel pump, be absolutely certain the problem is the pump and NOT with fuel tank, lines or filter. A plugged tank vent could create vacuum in the tank that will overpower the pump's ability to create vacuum and draw fuel through the lines. An obstructed line or fuel filter could also keep fuel from reaching the pump. Any of these conditions could partially restrict fuel flow, allowing the pump to deliver fuel, but at a lower pressure/rate. A pump delivery or pressure test under these circumstances would give a low reading that might be mistaken for a faulty pump. Before testing the fuel pump, refer to the testing procedures found under Fuel Lines & Fittings to ensure there are no problems with the tank, lines or filter.

A quick check of fuel pump operation is to gently squeeze the primer bulb with the motor running. If a seemingly rough or lean running condition (especially at speed) goes away when the bulb is squeezed, the fuel pump is suspect.

If inadequate fuel delivery is suspected and no problems are found with the tank, lines or filters, a conduct a quick-check to see how the pump affects performance. Use the primer bulb to supplement fuel pump. This is done by operating the motor under load and otherwise under normal operating conditions to recreate the problem. Once the motor begins to hesitate, stumble or stall, pump the primer bulb quickly and repeatedly while listening for motor response. Pumping the bulb by hand like this will force fuel through the lines to the vapor separator tank, regardless of the fuel pump's ability to draw and deliver fuel. If the engine performance problem goes away while pumping the bulb, and returns when you stop, there is a good chance you've isolated the low pressure fuel pump as the culprit. Depending upon the model you may be able to perform a pressure or vacuum check or you'll have to disassemble the pump to physically inspect the check valves and diaphragms (your only option on carburetor-integrated pumps, but it's not that difficult and can be done on ALL pumps).

✳✳ SELOC WARNING

Never run a motor without cooling water. Use a test tank, a flush/test device or launch the craft. Also, never run a motor at speed without load, so for tests running over idle speed, make sure the motor is either in a test tank with a test wheel or on a launched craft with the normal propeller installed.

Checking Fuel Pump Lift (Vacuum)

◆ See Figures 121

Fuel system vacuum testing is an excellent way to pinpoint air leaks, restricted fuel lines and fittings or other fuel supply related performance problems.

The standard square fuel pump used on non-carburetor integrated models (generally speaking that is 30 hp or larger Mercury motors) is designed to lift fuel vertically about 60 in. (1524mm). But those capabilities are dependent upon there being no other restrictions in the system and a fuel hose that is at least 5/16 in. (7.9mm) inner diameter (I.D.). Each restriction that is added to the system (such as a fuel filter, valve or fitting) will reduce the amount of lift available to the system.

Using a gauge, a T-fitting and a length of clear hose you can check the amount of vacuum in the system, as well as visually inspect for air bubbles (which would indicate a leak).

1. Disconnect the fuel inlet line (tank side) from the fuel pump and connect a clear piece of fuel hose to the pump inlet nipple.

2. Next, connect a T-fitting to the end of the clear line and connect the fuel inlet line to the other side of the T-fitting.

3. Lastly, connect the vacuum gauge to the T-fitting.

4. Supply the motor with a source of cooling water.

5. First check to be sure the pump itself is capable of producing enough vacuum by pinching off the fuel supply line on the tank side of the T-fitting, then starting and running the engine while watching the gauge. If the pump produces 2.5 in. Hg (6.35 cm Hg/8.44 kPa) or MORE of vacuum remove the hose pincher from the fuel supply line and continue the test. If the pump reading is below 2.5 in. Hg (6.35 cm Hg/8.44 kPa), then check the following:

• Pump check valves and/or diaphragm may be defective

• Pump may contain an air leak (requiring a new gasket, rebuild or replacement, be sure to check the fittings as well)

• There may be a problem with low crankcase pressure (check for plugged pump vacuum passageways)

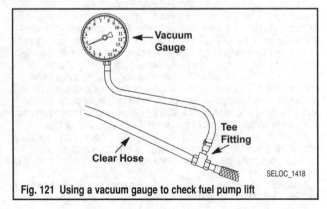

Fig. 121 Using a vacuum gauge to check fuel pump lift

6. With the pincher removed from the tube, watch the clear fuel line for signs of bubbles. If none are found, continue with the test. However, if bubbles are present, check the following for sources of the air leak:

• Pick-up tube in fuel tank
• Outlet fitting in fuel tank
• Fuel inlet hose may be improperly clamped at fitting
• Fuel tank valve may be leaking
• If equipped, fuel line from kicker engine into main fuel line may be leaking

7. If there are no air leaks, operate the engine until warmed to normal operating temperature. Run the motor at both idle and a little above idle (not too much above idle if on a flush fitting instead of a test tank or launched craft). As engine speed increases there should be a slight increase in vacuum however the reading should be at or below 2.5 in. Hg (6.35 cm Hg/8.44 kPa). If the reading is above 2.5 in. Hg (6.35 cm Hg/8.44 kPa), then check the following:

• Is the anti-siphon valve restricted?
• Is there a restriction in the primer bulb assembly?
• Is there a kinked or collapsed fuel hose?
• Is there a plugged water separating fuel filter rigged to the craft?
• Is there a restriction in the fuel line thru-hull fitting?
• If equipped, is there a restriction in the fuel tank switching valve(s)?
• Is there a plugged fuel tank pick-up screen.

8. Once you are finished, shut down the motor, then remove the tubing, T-fitting and gauge. Reconnect the fuel inlet line.

Checking Pump Pressure

A basic low-pressure fuel pump pressure check was mentioned earlier in this section under Troubleshooting the Fuel System. Unfortunately, Mercury does not publish pump pressure specifications for their carbureted 2-stroke motors, so the information should only be used to decide whether or not further diagnosis seems necessary on those motors.

■ **Mercury DOES however publish some pump pressure information on their 2-stroke EFI motors which are of the same basic design as the pump used on carbureted models. Furthermore, comparison of the vacuum lift tests for the EFI pumps vs. the carbureted motor pumps show the specifications and procedures to be the same so it is likely that the pressure specifications would be valid, at least as a guide, on carbureted motors. For more details, please refer to the Fuel Pump, Testing procedures in the EFI section.**

Fuel Pump Diaphragm and Check Valve Testing

For non-carburetor integrated fuel pumps you can use a simple hand-held vacuum/pressure pump and gauge to perform basic tests of the fuel pump diaphragm and check valve conditions.

Unfortunately Mercury does not publish any specifications for this test either, so we do not recommend that you necessarily condemn a fuel pump for this reason, However, you can use this as one possible method to decide if you should at least disassemble the pump for further inspection or not. However, conversely, if the pump does pass this test, we feel it is likely that the pump is not your problem.

Attach a hand-held vacuum/pressure pump to the fuel pump inlet fitting (the fitting to which the filter/tank line connects). Using the pump apply 7 psi (50 kPa) of pressure while manually restricting the outlet fitting(s) using your finger. If the diaphragm is in good condition it will hold the pressure for at least 10 seconds.

Next, switch to the vacuum fitting on the pump and draw 4.3 psi (30 kPa) of negative pressure (vacuum) on the fuel pump inlet fitting. This checks if the 1-way check valve in the pump remains closed. It should hold vacuum for at least 10 seconds.

Now, move the pressure pump to the fuel pump outlet fitting. This time cover the inlet fitting (and other outlet fitting, if applicable) with your finger and apply the same amount of pressure. Again, the pressure must hold for at least 10 seconds.

■ **If pressure does not hold, verify that it is not leaking past your finger (on the opposite fitting) or from a pump/hose test connection. If leakage is occurring in the diaphragm, the fuel pump should be overhauled.**

Visually Inspecting the Pump Components

◆ See Figure 122

The only way to inspect most diaphragm-displacement fuel pumps (including all carburetor-integrated pumps) is through disassembly and visual inspection. Remove and/or disassemble the pump according to the procedures found either in this section and any additional information which may be found under Carburetor Service (specifically Overhaul, for carburetor-integrated models).

Wash all metal parts thoroughly in solvent, and then blow them dry with compressed air. Use care when using compressed air on the check valves. Do not hold the nozzle too close because the check valve can be damaged from an excessive blast of air.

Inspect each part for wear and damage. Visually check the pump body/cover assembly for signs of cracks or other damage. Verify that the valve seats provide a flat contact area for the valve. Tighten all check valve connections firmly as they are replaced.

Check the diaphragms for pin holes by holding it up to the light. If pin holes are detected or if the diaphragm is not pliable, it must be replaced.

■ If you've come this far and are uncertain about pump condition, replace the diaphragms and check valves, you've already got to replace any gaskets or O-rings which were removed. Once the pump is rebuilt, you can remove it from your list of potential worries for quite some time.

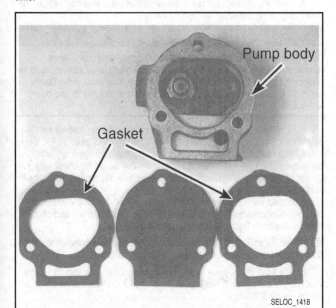

Fig. 122 Major components of a typical fuel pump assembly

REMOVAL & INSTALLATION

4/5, 6/8, 9.9/10/15 and 20/20 Jet/25 Hp Models

The fuel pumps on these models are integral with the carburetors. The mounting bolts are the same screws which hold the assembly together, so removal is also essentially disassembly. For this reason, details are covered under Overhaul in this section.

30-200 HP

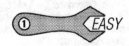

◆ See Figures 123, 124 and 125

1. On 30/40 hp 2-cylinder motors equipped with Power Trim Tilt (PTT), remove the screw securing the top of each trim relay to the side of the powerhead (just above the fuel pump), then reposition the relays for access to the fuel pump.

2. On 40/40 Jet/50/55/60 hp 3-cylinder models, remove the oil reservoir for access.

3. Cut away the Sta-Straps from the 3 hoses at the fuel pump. Tag each of the hoses to ensure proper installation. The 2 hoses at top and bottom of the pump are fuel lines (one in from tank and one out to carburetor) and the one on the pump cover is a pulse hose from the powerhead which supplies the vacuum necessary for the pump to operate properly.

4. Disconnect the top and bottom fuel lines. Use a golf tee or a stubby pencil to plug the end of each disconnected hose to prevent the loss of fuel.

5. Disconnect the pulse hose from the front surface (cover) of the pump.

6. Observe the 4 bolts threaded through the pump cover. The bolt closest to the pulse nipple on the cover and the bolt diagonally across from it (normally both M6 bolts) are used to fasten the pump to the block. The other 2 bolts (normally M5) are used hold the pump components together. Remove only the 2 bolts securing the pump to the powerhead, then carefully lift the pump clear of the block.

7. If equipped, remove and discard the mounting gasket.

To Install:

8. Place the fuel pump on the powerhead base using a new mounting gasket (if equipped).

■ When equipped with a gasket, it is usually easiest to insert the 2 mounting screws through the pump before positioning it to the powerhead so it can hold the gasket in place as you align the pump.

9. Tighten the 2 pump retaining bolts to alternately to 55 inch lbs. (6 Nm).

10. Reconnect the fuel lines as tagged during removal.

11. Reconnect the vacuum (pulse) line to the cover of the pump assembly.

12. Secure the lines using new wire-ties.

13. On 40/40 Jet/50/55/60 hp 3-cylinder models, install the oil reservoir.

14. On 30/40 hp 2-cylinder motors equipped with power trim/tilt (PTT), reposition trim relays and secure using the bolt at the top of each bracket.

Fig. 123 Unbolt and reposition the relays for access on 30/40 hp 3-cyl motors

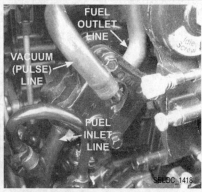

Fig. 124 Tag and disconnect the 3 hoses attached to the pump (80 Jet/115/125 hp motors shown)

Fig. 125 Only remove the 2 bolts holding the pump to the powerhead (150/200 hp 2.5L motors shown)

OVERHAUL

4/5 Hp Models

♦ **See Figure 56**

The fuel pump can be removed for service without removing the carburetor.

1. Disconnect the fuel hose from the fuel pump inlet.
2. Remove the 4 fuel pump screws, then and separate the pump cover, body and gasket from the carburetor. The components should come off in the following order:
- Pump cover
- Cover gasket
- Outer diaphragm
- Pump body
- Inner diaphragm
- Carburetor mounting flange gasket

3. If necessary, loosen the nut and bolt which secures each check valve to the pump body and remove the check valve assemblies.
4. Wash all parts thoroughly in solvent and then blow them dry with compressed air. Use care when using compressed air on the check valves. Do not hold the nozzle too close because the check valve can be damaged from an excessive blast of air.
5. Inspect each part for wear and damage. Verify that the valve seats provide a flat contact area for the valve disc. Tighten all elbows and check valve connections firmly as they are replaced.
6. Test each check valve by blowing through it with your mouth. In one direction the valve should allow air to pass through. In the other direction, air should not pass through.
7. Check the diaphragm for pin holes by holding it up to the light. If pin holes are detected or if the diaphragm is not pliable, it must be replaced.

To Assemble:

8. If removed, center the reed valve (for each check valve assembly) over the valve seat, then secure using the screw and nut.
9. Install a new gasket between the fuel pump and the carburetor.
10. Install the fuel pump assembly onto the carburetor. Make sure all the components are aligned properly. Sometimes it is easier to pre-assemble the cover gasket and outer diaphragm between the pump cover and body, then to place the inner diaphragm on the back of the assembly, holding it all in alignment as you position it on the carburetor. In this later case it is often easier to insert one or more mounting screws through the assembly to maintain alignment as it is installed.

■ **Make sure the gaskets and diaphragms are properly aligned. Do not use gasket sealing compound on any of the pump components.**

11. Install the 4 mounting screws and tighten securely.

6/8, 9.9/10/15 and 20/20 Jet/25 Hp Models

♦ **See Figure 81**

1. If equipped, turn the fuel shut-off valve to the **OFF** position.
2. Remove the 5 fuel pump screws, then and separate the pump cover, body and gasket from the carburetor. The components should come off in the following order:
- Pump cover
- Diaphragm (note for inspection, the diaphragm MUST be flat)
- Cover and diaphragm gasket
- Cap and spring assembly
- Pump body
- Carburetor mounting flange gasket

3. Discard the used gaskets.
4. Wash all parts thoroughly in solvent and then blow them dry with compressed air. Use care when using compressed air on the check valves. Do not hold the nozzle too close because the check valve can be damaged from an excessive blast of air.
5. Inspect each part for wear and damage. Verify that the valve seats provide a flat contact area for the valve disc. Tighten all elbows and check valve connections firmly as they are replaced.
6. Check the diaphragm for pin holes by holding it up to the light. Also, make sure the diaphragm is FLAT. If pin holes are detected or if the diaphragm is not flat or not pliable, it must be replaced.

To Assemble:

7. Install a new gasket between the fuel pump and the carburetor.
8. Assemble the pump starting with the body, then adding the spring, cap, cover and diaphragm gasket, followed by the cover and gasket. Hold these components together with some finger pressure, and perhaps insert one or more of the mounting bolts through the cover and body to maintain alignment.
9. Place the carburetor mounting flange gasket on the back of the carburetor body, then carefully (while still holding the components together in a sandwich), position the assembly to the carburetor. Slowly thread the mounting bolts.
10. Tighten the 5 mounting screws securely.

30-200 HP

♦ **See Figures 126 and 127**

1. Remove the pump from the powerhead. For details, refer to the procedure located earlier in this section.
2. Lay the pump on a suitable work surface
3. Remove the screws holding the fuel pump assembly together.

■ **Keep close track of the order in which the pump components are assembled in the pump housing. Compare the locations of the components to the accompanying diagram and note differences (if any) for installation purposes.**

4. Carefully separate the pump cover, gaskets and diaphragms from the pump body. Always discard the used gaskets!

✳✳ SELOC CAUTION

Do not remove the check valves unless they are defective. Once removed, the valves cannot be used again. If the check valves are to be replaced, take time to Observe and remember how each valve faces, because it must be installed in exactly the same manner or the pump will not function.

5. To remove a check valve for replacement, grasp the retainer with a pair of needlenose pliers and pull the valve from the valve seat.
6. Wash all parts thoroughly in solvent and then blow them dry with compressed air. Use care when using compressed air on the check valves. Do not hold the nozzle too close because the check valve can be damaged from an excessive blast of air.
7. Inspect each part for wear and damage. Verify that the valve seats provide a flat contact area for the valve disc. Tighten all elbows and check valve connections firmly as they are replaced.

■ **On models equipped with the black rubber check valves, check each disc to make sure the black coating is not coming off.**

8. Test each check valve by blowing through it with your mouth. In one direction the valve should allow air to pass through. In the other direction, air should not pass through.

■ **On some models (specifically on 75/65 Jet/90 hp and 80 Jet/115/125 hp models), the manufacturer says you can also test the check valves using a low pressure air source. Apply 1-3 psi (6.8-20.5 kPa) of air pressure and observe to make sure the check ball is moving.**

9. Check the diaphragm for pin holes by holding it up to the light. If pin holes are detected or if the diaphragm is not pliable, it must be replaced.

To Assemble:

The fuel pump rebuild kit will contain new gaskets, diaphragms and check valve components. Each check valve consists of a large rubber disc, a smaller plastic disc and a valve retainer OR a large plastic disc and a retainer, depending on the design and kit.

■ **The plastic discs are more resistant to damage from fuel additives than the rubber discs. For this reason most new overhaul kits should contain the plastic discs.**

10. To install the check valves and retainers, proceed as follows:
   a. Insert the fingers of the retainer into the smaller plastic disc and then the larger rubber disc (or the single plastic disc, as applicable).
   b. Install the disc and retainer assembly onto the fuel pump body. Push in the retainer until the collar and both discs are tightly pressed against the pump body.

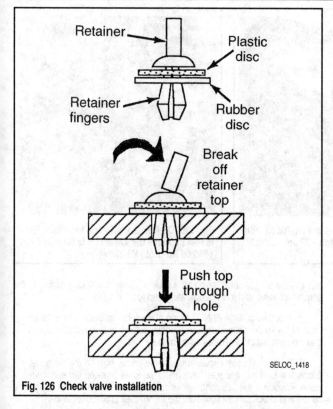

Fig. 126 Check valve installation

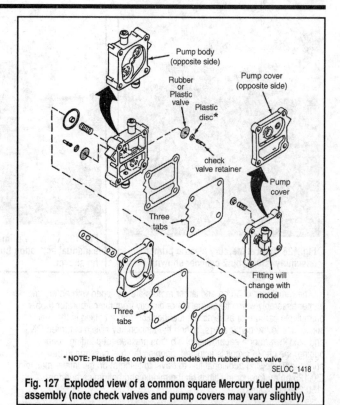

* NOTE: Plastic disc only used on models with rubber check valve

SELOC_1418

Fig. 127 Exploded view of a common square Mercury fuel pump assembly (note check valves and pump covers may vary slightly)

c. Bend the end of the retainer from side to side until it breaks away from the collar.

d. Install the broken off end through the hole in the collar and through the disc(s). Use a hammer and tap the retainer end down into place. As this piece is forced down, it will spread the fingers of the retainer and secure the check valve within the pump body.

e. Place the larger of the 2 caps and boost springs into place on the back side of the pump body.

■ **All the layered components of the fuel pump have notches which must be aligned during assembling. With the check valves in place in the pump body, you can begin final assembly of the fuel pump. Assembly will be much easier if you use two 3 in. (76mm) long 1/4 in. screws or dowels inserted through the large pump mounting bolt holes to align the pump components.**

11. Insert two 1/4 in. dowels through the pump mounting holes in the pump cover, then place the cover with the interior side facing upward on the worksurface.

12. Position the coil spring and cap, then carefully place the diaphragm in position over the dowels. Make sure the 3-tabs on the diaphragm are aligned in the same direction and the single tab on the pump cover.

13. Position the gasket over the dowels, with the 2-tabs facing the same direction as the tabs on the diaphragm and cover.

14. Install the fuel pump body down over the assembly, again with the tab in proper alignment.

15. Now install the coil spring and cap in the pump body.

16. Install the gasket with the tabs aligned in the proper direction.

17. Position the diaphragm over the dowels (aligning the tabs, you should be used to that by now).

18. Lastly, position the fuel pump base (rear cover) down over the assembly.

19. Grasp the assembly firmly (holding it together against diaphragm/spring pressure), then invert the assembly so you can insert the 2 pump assembly screws.

20. Recheck the alignment of all pump components, then insert and thread the 2 pump assembly screws. Tighten the screws to 55 inch lbs. (6 Nm).

21. Install the pump assembly, as detailed earlier in this section.

Enrichener Systems

DESCRIPTION & OPERATION

◆ **See Figures 128, 129 and 130**

Because a cold motor requires a fatter (more rich) fuel mixture for starting all motors are equipped with some method of providing either less air (a choke plate) or more fuel (a primer or enrichener circuit) for cold starts.

The smallest of motors covered here, the 2.5-5 hp motors are normally equipped with a mechanical choke plate. Actuation is rather straightforward, as the operator must manually move the linkage through the use of a choke knob. Potential problems with this system are limited to binding or broken linkage and a mechanical inspection can quickly verify or eliminate this concern. For more details on these motors, please refer to the Overhaul procedures for their carburetors.

Other small motors 6-25 hp utilize a carburetor which is equipped with a primer circuit. Like the smallest Mercury motors actuation is still manual, whereby the operator must pull a knob. However, rather than blocking air flow using a choke plate, the knob and related linkage work by providing additional fuel through the carburetor primer circuit. The primer assembly consists of a diaphragm, spring and check ball assembly installed in a dedicated portion of the carburetor body, attached to the float bowl. If the outboard is hard to start possible primer system problems could include a cut diaphragm, check ball stuck closed or a plugged primer passage. Should the motor smoke excessively at idle the primer check ball may not be fully seating due to varnish, debris or a damaged/weak spring. In all cases, inspection must be performed through disassembly, therefore for more details, please refer to the Overhaul procedures for these carburetors.

30 hp and larger carbureted Mercury motors are all equipped with either a manual or electric primer assembly. The manual primers (usually restricted to manual start 2-cyl and 3-cyl motors) consist of a primer bulb which allows the operator to manually pump fuel from the carburetor float bowl to fitting(s) on the intake manifold. Like the mechanical systems noted for smaller models this system is easily checked visually. If necessary you can disconnect the line(s) at the manifold and check to make sure fuel is expelled when pumped and/or check to make sure the passages into the manifold are not blocked.

Fig. 128 Typical Mercury electric primer assembly - 30/40 hp 2-cylinder shown

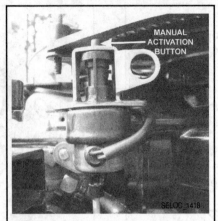

Fig. 129 Mercury primers are equipped with a manual activation button - 80 Jet/115/125 hp shown

Fig. 130 On Mercury primers, the upper line is fuel IN, while the lower line is fuel OUT - 150/200 hp (2.5L) V6 shown

The remainder of 30 hp and larger motors are equipped with an electric primer assembly which takes fuel either directly from the carburetor (upper on multiple carburetor assemblies) fuel bowl or from a t-fitting in the fuel supply line to the carburetor(s). When the ignition key switch is turned ON and held inward or a separate choke button pressed (depending upon rigging) the choke circuit is actuated, opening the fuel passage in the enrichener allowing additional fuel to travel to fitting(s) on the intake manifold (30-125 hp motors) or to fittings on the middle and bottom carburetors (150/200 hp motors).

Mercury electric primers are also equipped with a manual activation button on one end (top or bottom) of the assembly. The primers are equipped with a manual activation button to ensure ease of cold start should the electronic circuit fail while the boat is on the water. Should it become necessary to use the manual button, squeeze the main fuel line primer bulb until firm, then depress and hold the button on the primer assembly for about 5 seconds, then immediately crank/start the motor. Repeat, if necessary, to start the motor.

TESTING THE ELECTRIC PRIMER

◆ See Figures 128, 129 and 130

There are multiple tests that can be used to determine if the electric primer circuit is working. With the motor running at idle speed and already at normal operating temperature there are 2 quick checks which can tell you if the electrical and mechanical parts of the circuit are working. With the engine idling as noted, push inward on the Key or the Choke Button (as applicable) - if the electric and mechanical parts of the enrichener circuit are working, additional fuel will enter the fuel system causing the motor to either stumble or to increase idle speed noticeably. If nothing occurs, locate the button on the end of the enrichener valve, then press and hold the button. If only the mechanical portion of the system is working then the engine should stumble or there should be a change in idle as long as the button is depressed.

If you wish to test the system thoroughly and isolate any problems, proceed as follows:

1. Push inward on the key switch or choke button while listening at the valve for a click. If there is a click the electrical portion of the circuit is working, proceed with the next step to check the fuel circuit. If no click is heard, proceed as follows to check the electrical circuit:

   a. If no click is heard, use a test light or Digital Volt Ohmmeter (DVOM) to check for battery voltage on the Yellow/Black wire at the terminal block (normally located on the side of the powerhead) on the engine each time the key is held inward or the choke button is depressed.

   b. If battery voltage IS found, check for a loose or corroded connection between the Yellow/Black wire and Black wire (both under the same screw at the terminal block). Make sure the Black wire at the terminal block is properly grounded. Repair any connections which are not clean and tight, then

recheck the enrichener listening for a click. If power is definitely getting to the enrichener valve and it is not operating, replace the valve.

   c. If battery voltage IS NOT found at the terminal block, check for an open between the Yellow/Black wire between the key switch/choke button and the terminal block. Repair as necessary.

2. To check the fuel circuit, squeeze the primer bulb on the main fuel supply line until firm and then disconnect the lower hose on the enrichener valve. Position a suitable container under the valve, then push the key/choke button and check for fuel flow from the fitting, then proceed as follows:

   a. If NO fuel flows from the valve, remove the upper hose from the fitting on the enrichener valve. If fuel flows from the upper hose, but NOT from the lower valve fitting; the valve is clogged or otherwise defective and must be replaced. If no fuel flows from the upper hose, check the fuel supply line to the enrichener.

   b. If fuel flows from the lower hose on the enrichener, the valve is good. If there are still problems with the system, make sure the hose(s) and fitting(s) which are fed by the valve are clean and free of obstructions.

FUEL PRIMER HOSE ROUTING DIAGRAMS

◆ See Figures 131 thru 136

The following diagrams provide the most typical hose routing diagrams for the electric fuel primer assemblies. When applicable, manual primer bulb diagrams are usually similar, though they may not always contain all components, such as the oil injection system.

THERMAL AIR VALVE CIRCUIT

◆ See Figure 137

In addition to an electric fuel primer circuit, the 150/200 hp 2.5L V6 motors also utilize a Thermal Air Valve Circuit to help with cold motor operation. The thermal air valve operates as an idle air restrictor based on engine operating temperature. As long as the valve (which is threaded into the side of the starboard cylinder head, underneath the No. 3 spark plug) is below 100°F (38°C) the internal passages of the valve remain closed, limiting the amount of air provided to the carburetor idle circuits and therefore causing idle air fuel mixture to be richer. Once the engine warms, the thermal valve will open, allowing additional idle air and balancing the mixture.

Failure of the valve in a closed position will cause the powerhead to run rich when hot. Failure of the valve in an open position will cause the powerhead to run lean when cold. Testing is a simple matter of making sure the valve passages (you can manually try to blow air through them) remain closed when below 100°F (38°C) and open when above that temperature.

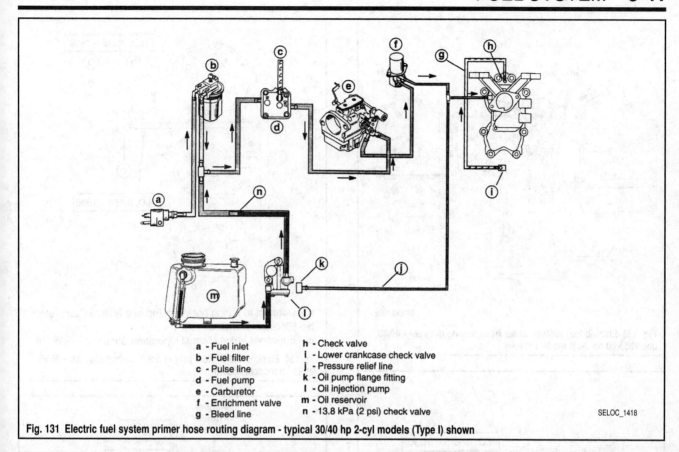

a - Fuel inlet
b - Fuel filter
c - Pulse line
d - Fuel pump
e - Carburetor
f - Enrichment valve
g - Bleed line

h - Check valve
i - Lower crankcase check valve
j - Pressure relief line
k - Oil pump flange fitting
l - Oil injection pump
m - Oil reservoir
n - 13.8 kPa (2 psi) check valve

SELOC_1418

Fig. 131 Electric fuel system primer hose routing diagram - typical 30/40 hp 2-cyl models (Type I) shown

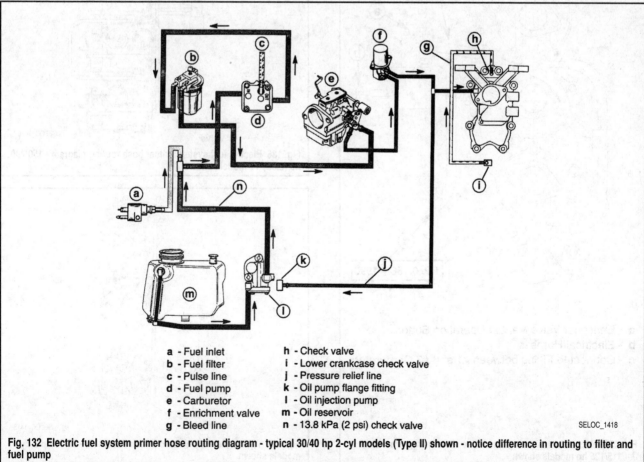

a - Fuel inlet
b - Fuel filter
c - Pulse line
d - Fuel pump
e - Carburetor
f - Enrichment valve
g - Bleed line

h - Check valve
i - Lower crankcase check valve
j - Pressure relief line
k - Oil pump flange fitting
l - Oil injection pump
m - Oil reservoir
n - 13.8 kPa (2 psi) check valve

SELOC_1418

Fig. 132 Electric fuel system primer hose routing diagram - typical 30/40 hp 2-cyl models (Type II) shown - notice difference in routing to filter and fuel pump

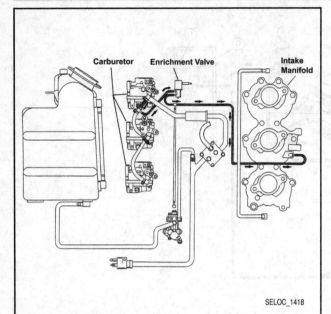

SELOC_1418

Fig. 133 Electric fuel system primer hose routing diagram - 40/40 Jet/50/55/60 hp 3-cyl models shown

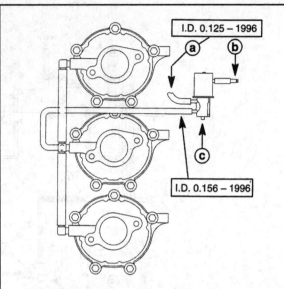

a - Connect to Fitting between Top and Middle Carburetors
b - Electrical Harness
c - Enrichner Valve Manual Operation Button

SELOC_1418

Fig. 134 Electric fuel system primer hose routing diagram - 75/65 Jet/90 hp models shown

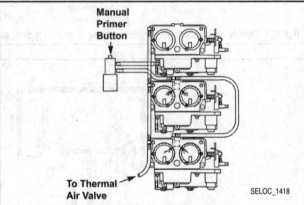

SELOC_1418

Fig. 136 Electric fuel system primer hose routing diagram - 150/200 hp (2.5L) V6 models shown

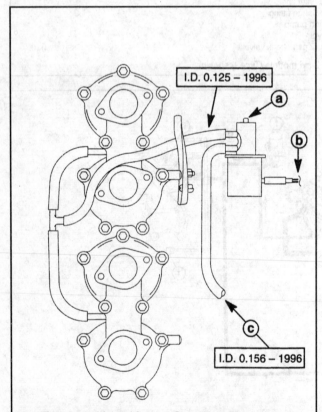

a - Enrichner Valve Manual Operation Button
b - Electrical Harness
c - Connect to Fitting between #1 and #2 Carburetors

SELOC_1418

Fig. 135 Electric fuel system primer hose routing diagram - 80 Jet/115/125 hp models shown

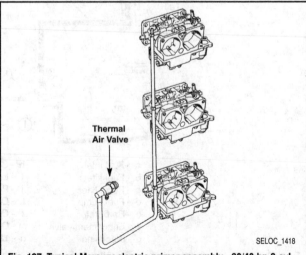

SELOC_1418

Fig. 137 Typical Mercury electric primer assembly - 30/40 hp 2-cyl models shown

ELECTRIC FUEL INJECTION (EFI)

Description & Operation

FUEL INJECTION BASICS

Fuel injection is not a new invention. Even as early as the 1950s, various automobile manufacturers experimented with mechanical-type injection systems. There was even a vacuum tube equipped control unit offered for one system! This might have been the first "electronic fuel injection system."

Early problems with fuel injection revolved around the control components. The electronics were not very smart or reliable. These systems have steadily improved since. Today's fuel injection technology, responding to the need for better economy and emission control, has become amazingly reliable and efficient. Computerized engine management, the brain of fuel injection, continues to get more reliable and more precise.

Components needed for a basic computer-controlled system are as follows:

• A computer-controlled engine manager, which is normally called the Electronic Control Unit (or ECU, known as an Electronic or Engine Control Module a.k.a ECM on most Mercury motors), with a set of internal maps to follow. Changes to fuel and timing are based on input information matched to the map programs.

• A set of sensors or input devices to inform the ECM of engine performance parameters.

• A set of output devices, each controlled by the ECM, to modify fuel delivery and timing.

This list gets a little more complicated when you start to look at specific components. Some fuel injection systems may have twenty or more input devices. On many systems, output control can extend beyond fuel and timing. Most modern systems provide more than just the basic functions but are still straight forward in their layout.

There are several fuel injection delivery methods. Throttle body injection is relatively inexpensive and is used widely in stern drive applications. This is usually a low pressure system running at 15 psi or less. Often an engine with a single carburetor was selected for throttle body injection. The carburetor was recast to hold a single injector and the original manifold was retained. Throttle body injection is not as precise or efficient as port injection.

Multi-port fuel injection is defined as one or more electrically activated fuel injectors for each cylinder. Multi-port injection generally operates at higher pressures than throttle body systems (equal to or in excess of 35.5 psi).

Port injectors can be triggered 2 ways. One system uses simultaneous injection. All injectors are triggered at once. The fuel "hangs around" until the pressure drop in the cylinder pulls the fuel into the combustion chamber.

The second type is more precise and follows the firing order of the engine. Each cylinder gets a squirt of fuel precisely when needed.

MERCURY ELECTRONIC FUEL INJECTION

◆ See Figures 138 and 139

The type fuel injection used on Mercury EFI models is known as an Indirect Multi-Port Fuel Injection system, because the fuel is injected into the intake manifold before entering the combustion chamber.

By design, the method of injection is also sometimes referred to as a Port Tuned Injection system, as each port has minimum and equal restriction to ensure all ports pass the same amount of air into the crankcase. Pairs of injectors are pulsed sequentially and timed to the induction of air into the crankcase.

A microprocessor housed in the electronic control module (ECM) accepts data from a number of sensors and computes the new ideal air/fuel ratio and the fuel injectors deliver the correct amount of fuel. Based on the information received, the ECM signals each fuel injector to inject a precise and correct amount of fuel. The system provides the correct air/fuel ratio for all powerhead loads, rpm and temperature conditions. System input and output devices are as follows:

■ **The Mercury EFI system evolved for the 2002 model year, which included a change in shape and location (and often function/specification) for many of its components. Throughout this section we'll refer to early-model EFI meaning systems through 2001 and late-model EFI meaning 2002 and later systems.**

Fig. 138 View of a typical Mercury EFI motors with some major parts identified

Major components that are common to the various versions of the Mercury EFI systems include:
• Electronic/Engine Control Module (ECM)
• Engine Temperature Sensor (ETS)
• Air Temperature Sensor (ATS)
• Manifold Absolute Pressure (MAP) Sensor
• Throttle Position Sensor (TPS)
• Fuel Injectors
• Mechanical Fuel Pump
• Electric Fuel Pump
• Vapor Separator
• Fuel Pressure Regulator
• Detonation/Knock Sensor and Module (not used on all models)

In addition the system is designed to incorporate the functions of the Ignition and the Warning system.

Most of these components are covered in this section. Where applicable, with sub-sections for Description & Operation, Testing and/or Removal & Installation.

Troubleshooting Electronic Fuel Injection

BASIC DIAGNOSIS & PRELIMINARY TESTS

Diagnosis of electronic fuel injection is generally based on symptom diagnosis and stored fault codes. Diagnosis of a running problem or as it is commonly called "driveability problem" requires attention to detail and following the diagnostic procedures in the correct order. Resist the temptation to begin extensive testing before completing the preliminary diagnostic steps. A preliminary or visual inspection must be completed in detail before diagnosis begins. In many cases this will shorten diagnostic time and often cure the problem without the need for involved electronic testing.

There are 2 basic ways to check your fuel system for problems. They are by symptom diagnosis and by the on-board computer self-diagnostic system. The first place to start is always the preliminary inspection. Intermittent problems are the most difficult to locate. If the problem is not present at the time you are testing you may not be able to locate the fault.

The visual inspection of all components is possibly the most critical step of diagnosis. A detailed examination of connectors, wiring and vacuum hoses can often lead to a repair without further diagnosis.

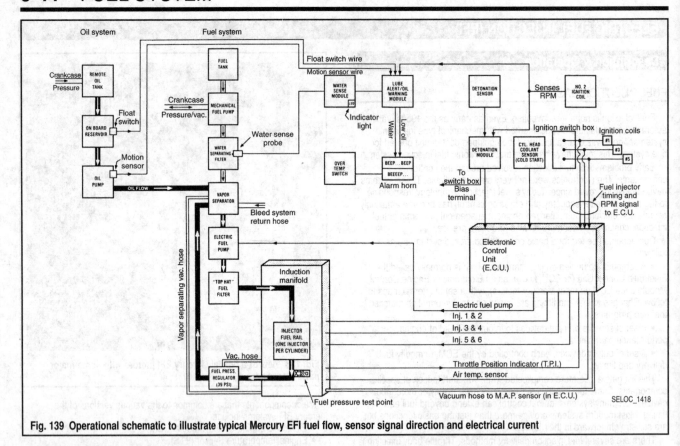

Fig. 139 Operational schematic to illustrate typical Mercury EFI fuel flow, sensor signal direction and electrical current

Also, take into consideration if the powerhead has been serviced recently. Sometimes things get reconnected in the wrong place or not at all. A careful inspector will check the undersides of hoses as well as the integrity of hard-to-reach hoses blocked by the air silencer or other components. Wiring should be checked carefully for any sign of strain, burning, crimping or terminals pulled from their connectors.

Checking connectors at components or in harnesses is required, usually, pushing them together will reveal a loose fit. Also, check electrical connectors for corroded, bent, damaged, improperly seated pins and bad wire crimps to terminals. Pay particular attention to ground circuits, making sure they are not loose or corroded. Any component or wiring in the vicinity of a fluid leak or spillage should be given extra attention during inspection. Remember how corrosive an environment it is that an outboard engine operates in. Salt spray is especially corrosive to electrical connectors.

Additionally, inspect maintenance items such as belt condition and tension and the battery charge and condition. Any of these very simple items may affect the system enough to set a fault code.

When diagnosing by symptom, the first step is to find out if the problem really exists. This may sound like a waste of time but you must be able to recreate the problem before you begin testing. This is called an "operational check". Each operational check will give either a positive or negative answer (symptom). A positive answer is found when the check gives a positive result (the instruments function when you turn the key). A negative answer is found when the check gives a negative result (the instruments do not function when you turn the key). After performing several operational checks, a pattern may develop. This pattern is used in the next step of diagnosis to determine related symptoms.

In order to determine related symptoms, perform operational checks on circuits related to the problem circuit (the instruments and the tilt and trim system do not work). These checks can be made without the use of any test equipment. Simply follow the wires in the wiring harness or, if available, obtain a copy of your engine's specific wiring diagram. If you see that the instruments and the tilt and trim system are on the same circuit, first check the instruments to see if they work. Then check the tilt and trim system. If the neither instruments or tilt and trim system work, this tells you that there is a problem in that circuit. Perform additional operational checks on that circuit and compile a list of symptoms.

When analyzing your answers, a defect will always lie between a check which gave a positive answer and one which gave a negative answer. Look

at your list of symptoms and try to determine probable areas to test. If you get negative answers on related circuits, then maybe the problem is at the common junction. After you have determined what the symptoms are and where you are going to look for defects, develop a plan for isolating the trouble. Ask a knowledgeable mechanic which components frequently fail on your vehicle. Also notice which parts or components are easiest to reach and how can you accomplish the most by doing the least amount of checks.

A common way of diagnosis is to use the split-in-half technique. Each test that is made essentially splits the trouble area in half. By performing this technique several times the area where a problem is located becomes smaller and smaller until the problem can be isolated in a single wire or component. This area is most commonly between the 2 closest checkpoints that produced a negative answer and a positive answer.

After the problem is located, perform the repair procedure. This may involve replacing a component, repairing a component or damaged wire or making an adjustment.

■ **Never assume a component is defective until you have thoroughly tested it.**

The final step is to make sure the complaint is corrected. Remember that the symptoms that you uncover may lead to several problems that require separate repairs. Repeat the diagnosis and test procedures repeatedly until all negative symptoms are corrected.

Two simple tests, pressure and volume, can easily verify the mechanical integrity of the powerhead's fuel supply and return systems. Problems within the fuel supply or return system can cause many driveability problems because the control system can't compensate for a mechanical problem. Nor can the system adjust the air/fuel mixture when the system is in open loop or back-up mode. Moreover, a control system's parameters are programmed based on a set flow rate and system pressure. Therefore, even if the computer were adjusting the air/fuel mixture, the computer's calculations would be wrong whenever the system pressure or volume was incorrect.

■ **When testing fuel system pressure, always use a pressure gauge that is capable of handling the system pressure. Old style vacuum/pressure gauges will not work with most fuel injection systems**

The fuel system's pressure can be thoroughly verified by performing both static and dynamic tests. Dynamic tests are used to determine the system's overall operating condition while the engine is running. For example, if the pressure is low, the problem is most likely to be in the supply or high side of the system. Likewise, if the fuel pressure is high, the problem is going to be on the return or low side of the fuel system. All pressure tests should be taken on the high or supply side of the fuel system.

If a dynamic test reveals that the pressure is below manufacturer specifications, the problem is caused by one or more of the following items: weak pumps, clogged or restricted fuel filter or lines, external leaks, bad check valves or a faulty fuel pressure regulator. But if a test reveals pressures above the desired specification, then the problem is caused by one or more of these items: restricted or clogged return lines, vacuum leak at the pressure regulator or a defective fuel pressure regulator.

However, if the pressure is high on a returnless fuel system, the cause, most likely, is a defective pressure regulator or clogged/faulty injectors. The causes are slightly different on returnless systems because the regulator is located in or near the vapor tank and there is no separate fuel return line used on these systems. However, IF the system uses a remotely mounted fuel pressure regulator, they will require a short return line. Thus, there is no excess fuel beyond the injectors and the measured pressure is already the regulated pressure.

Previously we mentioned that a static pressure test should also be performed. This test can be performed before or after a dynamic test with the engine off. These tests are useful for determining the sealing capacity of the system. If the fuel system leaks down after the engine is shut off, the vehicle will experience extended crank times while the system builds up enough fuel pressure to start. If the system is prone to losing residual pressure too quickly, it also usually suffers from low fuel pressure during the dynamic test. In addition to verifying that the system is holding residual pressure, a static test can also verify that the system is receiving the initial priming pulse at startup.

However, don't assume that the correct amount of fuel is being delivered just because the system pressure is within specifications. The only way to measure fuel volume is to open the fuel system into a graduated container and measure the fuel output while cranking the engine. Some manufacturers specify that a certain amount of fuel be dispersed over a short period of time but generally the pump should be able to supply at least 1.5 oz. (45 ml.) for approximately 2 seconds. If the fuel output is below the recommended specification, there is a restriction in the fuel supply or the fuel pump is weak.

In summary, pressure and volume tests provide information about the overall condition of the fuel system. The information obtained from these tests will set the direction and path of your diagnosis.

When troubleshooting the fuel injection system, always remember to make absolutely sure that the problem is with the fuel injection system and not another component. This particular system has very sophisticated electronic controls, which in most cases, can only be diagnosed and repaired by an authorized dealer who has the expensive diagnostic equipment.

There are some basic troubleshooting procedures that can be done before delving into the electronics:

Many times fuel system troubles are caused by a plugged fuel filter, a defective fuel pump or by a leak in the line from the fuel tank to the fuel pump. A defective choke may also cause problems. Would you believe a majority of starting troubles which are traced to the fuel system are the result of an empty fuel tank or aged sour fuel? Believe it.

Under average conditions (temperate climates), fuel will begin to break down in about 4 months. A gummy substance forms in the bottom of the fuel tank and in other areas. The filter screen between the tank and the carburetor and small passages in the carburetor will become clogged. The gasoline will begin to give off an odor similar to rotten eggs. Such a condition can cause the owner much frustration, time in cleaning components and the expense of replacement or overhaul parts for the carburetor.

Even with the high price of fuel, removing gasoline that has been standing unused over a long period of time is still the easiest and least expensive preventative maintenance possible. In most cases, this old gas can be used without harmful effects.

The gasoline preservative will keep the fuel fresh for up to twelve months. If this particular product is not available in your area, other similar additives are produced under various trade names.

Fuel injected engines don't actually have a choke circuit in the fuel system. These engines rely on the ECM to control the air/fuel mixture and injector function according to the electronic signals sent by a host of sensors and monitors. If a fuel injected engine suffers a driveability problem and it can be traced to the fuel injection system, then the sensors and computer-controlled output components must be tested and repaired or replaced.

SELF-DIAGNOSTIC SYSTEM

Mercury recommends the use of the Quicksilver Digital Diagnostic Terminal (DDT) for use with most EFI testing and troubleshooting. The DDT is a scan tool designed specifically to interact with the Mercury EFI and Optimax ECMs. As such it is capable of functions such as:
- Data Monitoring (real time feedback from engine sensors)
- Fault Status (reporting of ECM controlled or monitored circuits whose values have gone beyond the expected operating parameters)
- System Info (an archival of expected data values from ECM monitored and controlled systems)
- History (the ability to read or clear stored fault data in the ECMs memory)
- Special Functions (a number of tests or functions that the ECM can perform on demand when told by the DDT to do so, such as Oil Pump Prime or Resetting Break-In Oil delivery or Cylinder Misfire Test).

In some cases, aftermarket software may be available to mimic some or all of the functions of the DDT by using your laptop or PDA. However, at this time we have no source to recommend for such software.

COMPONENT TESTING

Once a problem has been narrowed to one or more components (either through deduction from basic system testing or with the help of the ECM Self-Diagnostic System), proceed to the individual component testing procedures found in this section in order to verify the problem component before replacement. In some cases, no specific tests are available, like with the ECM itself, and you can only deduce it is the bad component by eliminating everything else first. Be as certain as you can before replacing a component, especially since most are non-returnable or worse, might even become damaged by the still existing fault after installation if you are incorrect.

Fuel Injectors, Fuel Rail & Intake Manifold
DESCRIPTION & OPERATION

◆ See Figures 140 and 141

EFI motors are equipped with one separate fuel injector per cylinder. The 6 fuel injectors force fuel under pressure into the intake manifold. Each injector is mounted on a fuel rail extending the length of the intake manifold.

Fuel enters the rail at one end. A fuel pressure test point is provided at the other end of the rail on some motors (predominantly early-models through 2001) for troubleshooting purposes. Fuel is evenly distributed from the rail to each injector. The injectors are held in place with fragile wire clips.

The injectors used in this type fuel injection system are solenoid operated. Each consists of a valve body, a needle valve and valve seat. A small voltage is sent from the ECM to each injector. When this voltage is applied to the windings of the solenoid, a magnetic field is induced around the needle valve. The valve lifts off its seat and fuel is allowed to pass between the needle valve and the needle seat. Because the fuel is pressurized, a spray emerges from the injector nozzle. The nozzle spray angle of each injector remains constant and is the same for all 6 injectors. A small return spring seats the needle valve back onto the seat, the instant the voltage is removed.

The time interval for the injector to be open and emitting fuel is called the "pulse width". The actual "pulse width" for the injector is controlled by the ECM and must be measured in microseconds.

Two O-rings are used to secure each injector in the fuel rail. One O-ring provides a seal between the injector nozzle and the intake manifold. The other O-ring provides a seal between the injector and the fuel inlet connection on the fuel rail. Both O-rings prevent excessive injector vibration. These O-rings are replaceable and are included in an injector overhaul kit.

As stated the fuel injectors and rail assembly as installed to the intake manifold, the design of which varies slightly between early-model (through 2001) and late-model (2002 or later) EFI systems.

On early-model EFI systems the intake manifold (also called an induction manifold by Mercury) contains either 2 or 4 small bore throttle butterflies (shutter valves) facing horizontally outward (which are used to control engine speed). The whole assembly is bolted to the reed valve block on the end of the powerhead (in the same position as the carburetors would be mounted on a carb version of the same motors).

Late-model EFI systems use an intake manifold that is shaped somewhat differently. For starters they are equipped with a single, large bore throttle shutter assembly which is mounted facing vertically upward on top of the intake manifold cover. The fuel rail and injectors themselves are mounted directly to the reed valve adaptor plate.

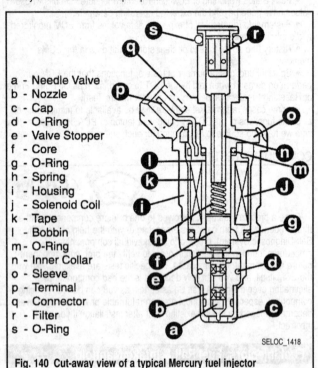

a - Needle Valve
b - Nozzle
c - Cap
d - O-Ring
e - Valve Stopper
f - Core
g - O-Ring
h - Spring
i - Housing
j - Solenoid Coil
k - Tape
l - Bobbin
m - O-Ring
n - Inner Collar
o - Sleeve
p - Terminal
q - Connector
r - Filter
s - O-Ring

SELOC_1418

Fig. 140 Cut-away view of a typical Mercury fuel injector

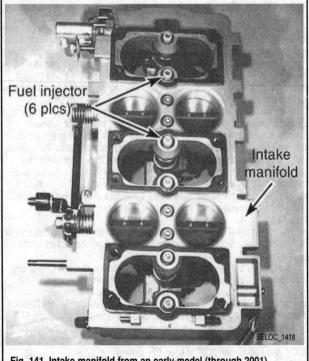

Fuel injector
(6 plcs)

Intake manifold

SELOC_1418

Fig. 141 Intake manifold from an early-model (through 2001) powerhead

TESTING THE INJECTORS

◆ **See Figures 142, 143 and 144**

There are multiple potential ways to test the fuel injectors. Remember that because an injector is an electrically operated solenoid valve which opens and closes against spring pressure, an injector will make a soft ticking noise during operation.

The easiest and most common way to quick-test and injector is to physically touch it using your hand, a long screwdriver or a stethoscope in order to hear or feel this clicking during operation. The problem is that unlike many engine designs, the Mercury EFI system covers the injectors under part of the intake manifold assembly, so you have to at least partially remove the intake manifold (the outer cover) for access to the injectors before this can be done. This could make starting or running the motor impractical or impossible, especially on late-model (2002 or later models) EFI motors where the throttle valve assembly is mounted to the intake manifold cover and would be removed at this point.

You still could remove the cover and crank the motor to listen or feel for injector clicking, but at that point the test is not as quick and easy as most technicians would prefer if they are trying to quickly eliminate the fuel injectors from their list of possible problems.

■ **Keep in mind that the Mercury DDT includes the function of performing an Injector Operational test, so the use of a scan tool is one quick and easy way to at least actuate and check the injectors whether or not they are accessible.**

Because the injectors are not out in the open Mercury recommends an initial test through the wiring harness with the use of a DVOM to check circuit and injector winding resistance. Though resistance checks are normally not 100% reliable, this method tends to give accurate results the vast majority of the time on injectors. To test the injectors and electrical harness, proceed as follows:

1. On 2.5L EFI motors through 2001 Mercury allows for a running dead-cylinder test. Although this test may be valid for other years and models, since Mercury does not recommend it we cannot be certain it is safe for the electronics on other motors. In order to determine if rough running or other suspected problems are due to 1 or more dead cylinders, proceed as follows:

a. With the outboard in the water or a test tank start and run the motor, allowing it to warm to normal operating temperature.

b. Once fully-warmed, run the engine at 2000-2500 rpm.

c. With the motor running at speed disconnect 1 spark plug lead at a time, while listening for rpm change. If 1 or more leads come off WITHOUT any change what-so-ever, you've located a dead cylinder. At that point, determine if the lead was creating spark (you can use a timing light or spark tester for this, for more details please refer to the Ignition System) and IF it was creating spark it is time to suspect the fuel injector for that cylinder.

d. Shut the motor down and check the injectors as necessary.

Harness
Connector

Fuel
injector

SELOC_1418

Fig. 142 Fuel injector assembly

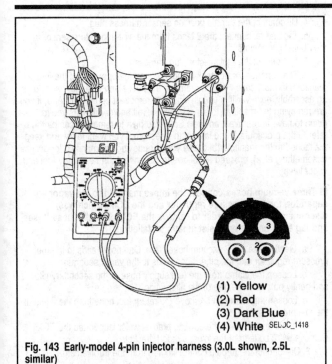

(1) Yellow
(2) Red
(3) Dark Blue
(4) White   SELOC_1418

Fig. 143 Early-model 4-pin injector harness (3.0L shown, 2.5L similar)

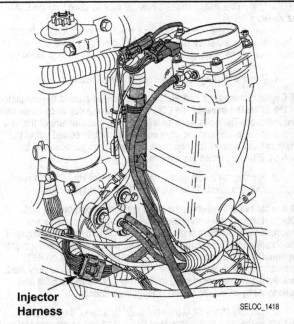

Injector Harness   SELOC_1418

Fig. 144 Late-model multi-pin injector harness (2.5L shown, 3.0L similar)

2. Disconnect the appropriate wiring harness for access to the injector wiring as follows:

• On early-models (through 2001), disconnect the 4-pin harness connector located right under the vapor separator tank assembly. On 3.0L motors the harness should be near and a little below the air temperature sensor. The wiring harness should contain White (cyl. #'s 1 and 2), Dark Blue (cyl. #'s 3 and 4) and Yellow leads (cyl. #'s 5 and 6), plus a Red (positive) lead.

• On late-models (2002 and later), disconnect the large multi-pin injector harness connector located on the front, starboard side of the motor, toward the bottom of the intake manifold assembly. The connector should contain 6 Red wires plus 6 wires of other colors one for each injector: Brown (Cyl # 1), White (Cyl # 2), Orange (Cyl # 3), Yellow (Cyl # 4), Lt. Blue (Cyl # 5) and Purple (Cyl # 1).

3. Set the DVOM to read resistance on low ohm scale, then check resistance of each injector harness and injector wiring by probing across the Red lead to the appropriate color lead as noted in the last step.

• On early-model 2.5L motors (through 2001), resistance should be 0.9-1.3 ohms for each circuit. If readings are 2.0-2.4 ohms for a given circuit ONE of the injectors of the pair that you are testing does NOT have a complete circuit.

• On early-model 3.0L motors (through 2001), resistance should be 5-6 ohms for each circuit. If readings are 11-12 ohms for a given circuit, one of the injectors of the pair that you are testing does NOT have a complete circuit.

• On late-models (2002 and later), resistance should be 11.8-12.8 ohms per injector circuit.

4. If the injectors test good, but a fuel problem is still suspected, check fuel pressure next.

REMOVAL & INSTALLATION

Early-Models (Through 2001)

DIFFICULT

2.5L Motors

◆ See Figures 145 thru 162

1. For safety, disconnect the negative battery cable.
2. Release the fuel system pressure and, if necessary, drain the Vapor Separator Tank.
3. Remove the ECM for access, as follows:

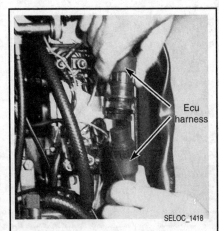

Ecu harness   SELOC_1418

Fig. 145 Disconnect the main ECM wiring harness

Water sensing module
Bolt
Screw (2 pcs)   SELOC_1418

Fig. 146 Remove the water sensing module and lower ECM mounting bolt. . .

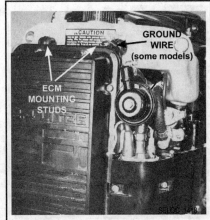

GROUND WIRE (some models)
ECM MOUNTING STUDS   SELOC_1418

Fig. 147 . . .then remove the upper ECM mounting studs. . .

a. Unplug the ECM harness connector located along side the ECM (starboard side).

b. Remove the 2 screws securing the water sensing module to the powerhead. The left screw has 2 small black grounding leads attached. The module will still be connected to the powerhead by other leads. Move the module aside.

c. Remove the large bolt in the center at the base of the ECM.

■ Do not disturb the small Phillips head screws around the perimeter of the ECM. The cover of the ECM will not come away once these bolts are removed. The bolts were installed during manufacture of the unit and then the cover and the case were hermetically sealed using the Phillips head screws for alignment. Therefore, no purpose will be served, if these screws are removed.

d. Remove the 2 nuts on the studs at the top of the ECM. On some models, the right stud may have a small Black ground wire over it.

■ It is far easier to disconnect leads and leave components still attached to the intake manifold, because some components are mounted behind others and should not be disturbed. For example: if the throttle position sensor is misaligned during installation, a digital type multimeter is required to correctly reset the sensor on its mounting bracket. Voltage in the 1/10 range must be accurately read. A misaligned sensor could send misleading signals to the ECM and consequently affect the EFI and the ignition timing.

e. Carefully lift the ECM from the studs for access, then tag and disconnect the vacuum line from the ECM at the fitting on the manifold next to the pressure port. Remove the ECM completely from the powerhead.

4. Disconnect the throttle position sensor harness plug.

5. Tag and disconnect the 2 leads from the air temperature sensor at their bullet connectors.

6. Disconnect the fuel injector plug harness.

7. The vapor separator tank is mounted to the upper side of the intake manifold. In most cases you can remove the tank along with the manifold as an assembly, however it is usually easier to remove it completely. And, if you are removing the manifold to service the fuel rail and/or injectors you're going to have to remove it anyway. So, if you plan to remove it completely, refer to the procedure found later in this section. If you want to try and keep it attached to the manifold (because you're removing the manifold for access to something else), proceed as follows to disconnect the necessary fuel and vapor lines:

■ Three vacuum hoses connect the intake manifold to the vapor separator, the fuel pressure regulator and the ECM (which was disconnected earlier in order to remove the ECM). Identify these hoses and tag each one as an assist in later installation.

a. Gently ease the 2 vacuum hoses free. One hose leads to the fuel pressure regulator and the other hose leads to the vapor separator.

b. Loosen the clamp from the fuel supply hose at the secondary filter and gently pull the hose free of the filter fitting.

c. Loosen the clamp and remove the return fuel hose from the fitting at the fuel pressure regulator.

d. If necessary, lift off the cowling bracket which supported the ECM. Remove the 2 bolts from the lower support bracket.

8. Disconnect the throttle control link rod from the throttle cam.

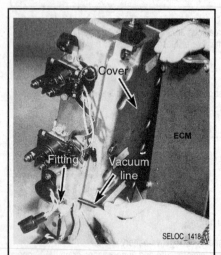

Fig. 148 . . .and carefully remove the ECM from the powerhead

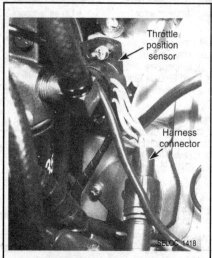

Fig. 149 Disconnect the TPS harness. . .

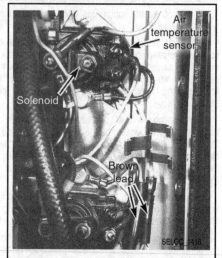

Fig. 150 . . .and the ATS bullet connectors

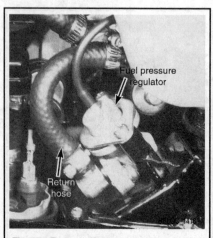

Fig. 151 Tag and disconnect the fuel and vacuum lines from the assembly

Fig. 152 Disconnect the oil control and throttle link rods

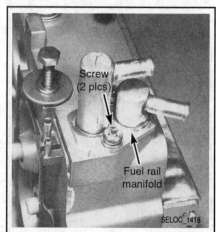

Fig. 153 Loosen the screws securing the fuel rail manifold. . .

9. Pry the oil pump control link rod from the ball joint on the intake manifold, leaving the rod on the oil pump.

10. Note the location of the grounding leads (there are usually 2, but sometimes 3 on older models) on the front face of the intake manifold cover, as an aid during installation.

11. Remove the twelve bolts securing the manifold and cover to the powerhead. On some applications all bolts may not be the same length. If so make a note as to length and location.

12. Tag and disconnect the bleed hoses from the manifold fittings (there are normally at least 2 hoses towards the base of the manifold. It is usually easier to pull the manifold away from the powerhead for access.

13. Lift the intake manifold from the powerhead. Remove and discard the gasket. Remove the front cover and gasket from the manifold.

14. If necessary, tag and remove the vacuum and bleed hoses from the intake manifold.

15. If not done already, remove the vapor separator assembly from the intake manifold.

16. Remove the 2 Phillips head screws with captive lockwashers securing the fuel rail to the intake manifold, then gently pull the fuel rail manifold from the intake manifold.

■ Two tubes are installed under the fuel rail manifold which will either come away with the fuel rail manifold or stay inside the intake manifold.

17. Remove the 2 tubes which connect the fuel passages of the fuel rail manifold to the fuel rail. Both tubes are identical and have an O-ring at each end. The O-rings need not be removed unless they are no longer fit for service. Remove the 2 O-rings which seal the fuel rail-to-intake manifolds.

18. To remove the fuel rail from the manifold, remove the 4 long screws (usually Allen head type screws) from the injector side of the manifold. Support the fuel rail and turn the manifold over.

19. Separate the fuel rail from the fuel injectors and the manifold.

20. Remove the 2 ferrules (fuel rail guides) from the manifold. These 2 ferrules may have remained either with the manifold or in the fuel rail.

21. Remove the 2 Phillips head screws with captive lockwashers from the fuel rail support tube. Lift the support tube from the rail. Lift out the 2 tiny plugs and a total of 4 O-rings from the rail.

22. Tag the injector harness connectors for identification, then disconnect them from each of the injectors. Using a small slotted screwdriver, pry the retaining clip from the injector harness connector, then separate the connector from the injector.

23. Slowly pull each injector straight up and free of the manifold. With the fuel rail out of the way, the injectors are held in the manifold by rubber sealing rings only. These rings may harden or become gummy, depending on use, making removal of the injectors difficult.

■ Each injector contains a removable/replaceable filter in the top where it connects to the fuel rail, along with an O-ring and small seal on that end. A larger removable/replaceable sealing ring is used on the lower end, where it connects to the manifold. Although you don't have to replace these components, it's usually a good idea as long as you're here.

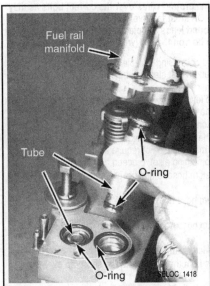

Fig. 154 . . .then separate the manifold and connecting tubes from the fuel rail

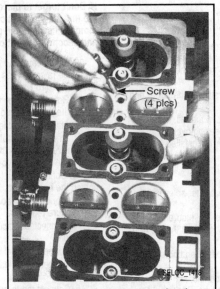

Fig. 155 Remove the screws securing the fuel rail

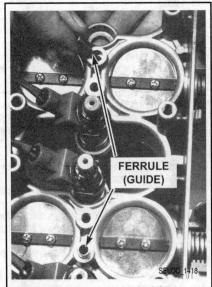

Fig. 156 Be sure to keep track of the 2 ferrules (fuel rail guides)

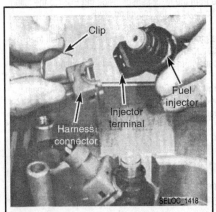

Fig. 157 Disengage the fuel injector connectors. . .

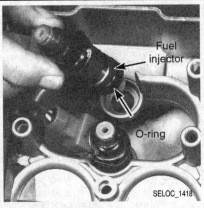

Fig. 158 . . .then remove the injectors (keeping track of seals)

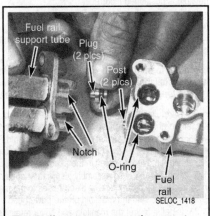

Fig. 159 If necessary remove the support tube from the fuel rail

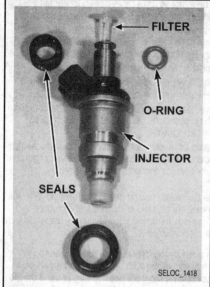

Fig. 160 Exploded view of the fuel injector components

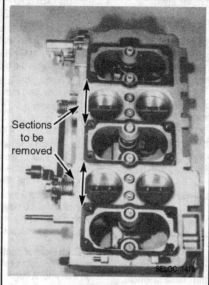

Fig. 161 The intake manifold gasket is modified after installation

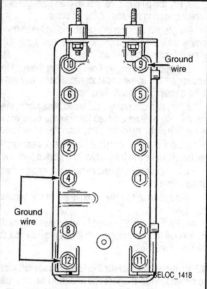

Fig. 162 Intake manifold/cover torque sequence

To Install:

24. Install each injector assembly to the manifold making sure the seals, O-ring and filter are all in position.

25. Position the retaining clip around the injector wire harness. Push the injector harness connector firmly over the injector terminal. The harness can only be connected one way. Slide the clip into place to secure the harness connector to the injector.

26. If removed, install the 2 tiny plugs into the fuel rail, with the larger end of the plug going into the fuel rail first.

27. Install the 2 sealing O-rings over the 2 fuel ports in the rail. Install the fuel rail support tube over the rail with the 2 notches indexing over the 2 posts on the rail end. This ensures the support can only be installed one way over the rail.

28. Install and tighten the 2 Phillips head screws with captive lockwashers securing the fuel rail support tube to the rail itself.

29. Position the 2 ferrules (fuel rail guides) over the 2 outer holes in the intake manifold.

30. Lower the fuel rail over the ferrules and injectors. Make sure the rail is seated properly over each injector and none are askew. Support the rail with one hand and turn the manifold over.

31. Install the 4 long Allen head fuel rail retaining bolts, then tighten them alternately and evenly to 35 inch lbs. (4 Nm).

32. Insert the 2 identical tubes, with an O-ring at each end, into the intake manifold. These tubes must seat in the fuel rail support tube. Position 2 O-rings over the ports. Install the fuel rail manifold over the tubes. The manifold can only be installed properly one way. The bolt holes will not align if the manifold is positioned incorrectly.

33. Install and tighten the 2 Phillips head screws securing the fuel rail manifold to the intake manifold.

34. If removed, connect the engine bleed and vacuum hoses to the manifold.

■ In order to help hold the intake manifold gasket (and cover) in place during installation and to help with gasket alignment use four M8 x 1.25 x 38mm long bolts (one at each corner of the manifold assembly).

35. Carefully install the cover seal to the manifold, then install the cover and temporarily secure using the 4 M8 bolts referred to earlier.

36. Position a new gasket on the intake manifold. This gasket is purchased and installed in one piece. After installation, 4 sections of the gasket are removed as noted in the accompanying illustration.

37. Hold the manifold assembly just in front of its mounting point on the powerhead, then reconnect the bleed hoses to the manifold fittings. With the hoses in place, seat the manifold on the powerhead.

38. Install a few of the manifold and cover retaining screws to hold the assembly in place, then remove the 4 M8 bolts and finish threading the rest of the manifold bolts.

■ Don't forget to place the ground wires over the correct bolts as they are threaded in position. On most later versions of this early-model motor/EFI system the ground leads are placed on the top left cover bolt and second from bottom bolt on the left (when looking at the cover).

39. Tighten the manifold bolts using multiple passes of the proper torque sequence (a criss-cross pattern that starts at the center of works outward) to 90 inch lbs. (10 Nm).

40. Rotate the oil pump lever clockwise and snap the oil pump control link rod on the ball joint onto the intake manifold throttle linkage.

41. Attach the throttle control link rod to the throttle cam.

42. If removed, install the vapor separator tank assembly. If the tank was not removed (perhaps the manifold did not require service but you just removed it for access to something else), proceed as follows:

a. Connect the fuel return hose to the fuel pressure regulator and tighten the hose clamp.

b. Connect the vacuum hose from the intake manifold to the fitting on the pressure regulator.

c. Connect the vacuum hose from the intake manifold to the vapor separator.

43. Connect the fuel injector harness plug.

44. Connect the 2 leads from the air temperature sensor to the 2 bullet connectors.

45. Connect the throttle position sensor harness plug.

46. Install the ECM as follows:

a. Hold the ECM in position just in front of the intake manifold and connect the vacuum hose from the MAP sensor inside the ECM to the fitting on the intake, then seat the ECM on the mounting studs.

b. Secure the ECM to the mounting studs using the nuts (remember, some models may have a ground one on the right stud), and install the ECM lower mounting bolt. Tighten the nuts and bolt to 45 inch lbs. (5 Nm).

c. Install the water sensing module and tighten the retaining screws to 25 inch lbs. (3 Nm).

d. Reconnect the large, round, ECM harness connector to the main powerhead harness connector. The connection can only be made one way. Secure the harness into the bracket with the connecting ring either above or below the bracket arms.

3.0L Motors

◆ See Figures 153 thru 158, 160, 161, 163 and 164

1. For safety, disconnect the negative battery cable.

2. Release the fuel system pressure and, if necessary, drain the vapor separator tank.

3. Disconnect the throttle position sensor harness plug (the large plug almost directly under the starter motor).

4. Tag and disconnect the 2 leads from the air temperature sensor at their bullet connectors (to locate the connectors, follow the wiring back from the sensor which is mounted in the side of the manifold itself).

5. Disconnect the fuel injector plug harness (the round multi-pin connector found just below the air temperature sensor bullets).

6. Remove the 2 bolts securing the water separating fuel filter bracket to the manifold assembly (the bolts are located in the bracket at the top of the filter assembly), then reposition the assembly aside with the hoses and filter still attached.

■ The vapor separator tank is mounted to the upper side of the intake manifold. In most cases on these motors you can simply unbolt it and position it aside with most of the connections still attached. However, you may decide it is easier or more convenient (especially for other related work) to remove it completely from the motor. If so, refer to the Vapor Separator procedure, later in this section.

7. Remove the 2 screws securing the fuel inlet/outlet pipes (from the top of the vapor separator tank) to the side of the manifold.

■ Two tubes are installed under the pressure regulator joint which will either come away with the fuel rail manifold or stay inside the intake manifold.

8. Remove the 2 tubes which connect the fuel passages of the pressure regulator joint to the fuel rail. Both tubes are identical and have an O-ring at each end. The O-rings need not be removed unless they are no longer fit for service. Remove the 2 O-rings which seal the fuel rail-to-intake manifolds.

9. Tag and disconnect the MAP sensor hose (found at the end of the vapor separator tank, on the top, powerhead side of the tank).

10. Tag and disconnect the vapor separator tank hose (found toward the middle, top of the tank on a fitting which faces the powerhead).

11. Tag and disconnect the fuel pressure regulator hose to the top manifold fitting.

12. Tag and disconnect the engine bleed hose (with the white filter) to the fitting under the MAP sensor.

13. Carefully pry the throttle control link from the throttle cam using a small, flat-bladed screwdriver or prytool.

14. Remove the 3 bolts securing the vapor separator tank assembly to the side of the manifold (2 on top and 1 on the bottom), then reposition the tank off to one side with the remaining connections still intact.

15. Remove the twelve bolts securing the manifold and cover to the powerhead.

Because the screws that secure the manifold cover to the powerhead are ALSO the screws that secure the manifold itself, be sure to support the assembly as the last screws are removed, in order to prevent the assembly from falling and becoming damaged.

16. Lift the intake manifold from the powerhead. Remove and discard the gasket. Remove the front cover and gasket from the manifold.

17. To remove the fuel rail from the manifold, remove the 4 long screws (usually Allen head type screws) and the single Philip's head screw from the injector side of the manifold. Support the fuel rail and turn the manifold over.

18. Separate the fuel rail from the fuel injectors and the manifold.

■ The injectors are secured to the fuel rail and manifold simply by rubber seals (and physically by the clamp load of the rail when it is bolted over top of them to the manifold). As the rail is pulled away from the manifold the injectors usually pull out of the rail first, but they may pull out of the manifold instead, so take your time to prevent damaging them should they fall away from the assembly as the rail is separated.

19. Remove the 2 ferrules (fuel rail guides) from the manifold. These 2 ferrules may have remained either with the manifold or in the fuel rail.

20. If desired, disassemble the fuel rail components as follows:

a. Remove the 4 screws securing the inlet/outlet elbow to the fuel rail, then remove the elbow.

b. Check the O-rings on the elbow for cuts or abrasions and replace, as necessary (for the record, we ALWAYS suggest replacing O-rings that have been in service for any length of time).

c. Remove the transfer tubes from the external fuel pressure regulator joint. Again, inspect the O-rings on the transfer tubes and the regulator joint for cuts or abrasions and replace, as necessary.

21. To remove the injectors, tag the injector harness connectors for identification, then disconnect them from each of the injectors. Using a small slotted screwdriver, pry the retaining clip from the injector harness connector, then separate the connector from the injector.

22. Slowly pull each injector straight up and free of the manifold. With the fuel rail out of the way, the injectors are held in the manifold by rubber sealing rings only. These rings may harden or become gummy, depending on use, making removal of the injectors difficult.

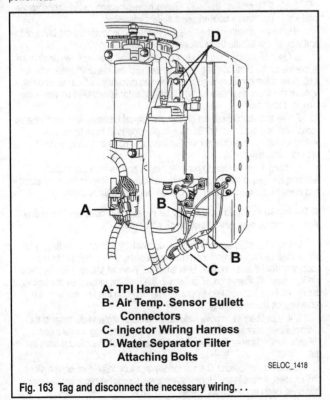

A- TPI Harness
B- Air Temp. Sensor Bullett
   Connectors
C- Injector Wiring Harness
D- Water Separator Filter
   Attaching Bolts

SELOC_1418

Fig. 163 Tag and disconnect the necessary wiring. . .

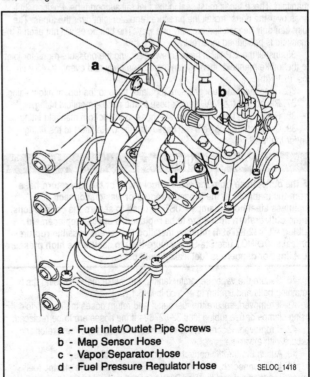

a - Fuel Inlet/Outlet Pipe Screws
b - Map Sensor Hose
c - Vapor Separator Hose
d - Fuel Pressure Regulator Hose

SELOC_1418

Fig. 164 . . .and hoses in order to remove the intake assembly

■ Each injector contains a removable/replaceable filter in the top where it connects to the fuel rail, along with an O-ring and small seal on that end. A larger removable/replaceable sealing ring is used on the lower end, where it connects to the manifold. Although you don't have to replace these components, it's usually a good idea as long as you're here.

To Install:

23. Install each injector assembly to the manifold making sure the seals, O-ring and filter are all in position.

24. Position the retaining clip around the injector wire harness. Push the injector harness connector firmly over the injector terminal. The harness can only be connected one way. Slide the clip into place to secure the harness connector to the injector.

25. If disassembled, prepare the fuel rail for installation to the manifold as follows:

a. Replace any O-rings which were identified as worn, damaged or suspect in any way.

b. Install the inlet/outlet elbow to the fuel rail and secure by tightening the screws to 18 inch lbs. (2 Nm).

26. Position the 2 ferrules (fuel rail guides) over the 2 outer holes in the intake manifold.

27. Lower the fuel rail over the ferrules and injectors. Make sure the rail is seated properly over each injector and none are askew. Support the rail with 1 hand and turn the manifold over.

28. Install the 4 long Allen head fuel rail retaining bolts and the Philips screw, then tighten them alternately and evenly to a torque value of 45 inch lbs. (5.1 Nm).

29. Thoroughly inspect the intake manifold cover seal and replace if necessary. If removed, carefully install the cover seal to the manifold, then install the cover over the manifold.

30. Position a new gasket on the intake manifold (you may wish to use a couple of the cover/manifold bolts to help hold it in position during alignment.

31. Hold the cover, manifold and gasket assembly just in front of the powerhead, then align it and begin threaded in the cover/manifold bolts.

32. Finish threading the rest of the manifold bolts, then tighten them to 19 ft. lbs. (26 Nm) using multiple passes of a criss-cross pattern.

33. Reposition the vapor separator tank in position near the side of the intake manifold, however don't thread the retaining bolts yet as it may be easier to reconnect some hoses with the assembly still loose.

34. Insert the 2 identical tubes, with an O-ring at each end, into the intake manifold. These tubes must seat in the fuel rail support tube. Position 2 O-rings over the ports. Install the pressure regulator joint over the tubes. The joint can only be installed properly one way. The bolt holes will not align if the manifold is positioned incorrectly.

35. Install the 2 Phillips head screws securing the pressure regulator joint to the intake manifold, then tighten them alternately and evenly to 45 inch lbs. (5.1 Nm).

36. Reconnect the fuel pressure regulator hose to the top manifold fitting.

37. Reconnect the MAP sensor hose to the middle manifold fitting.

38. Reconnect the vapor separator hose to the bottom manifold fitting.

39. Connect the engine bleed hose (with the white filter) to the fitting under the MAP sensor.

✳✳ SELOC WARNING

IF the outlet hose from the electric fuel pump or the fuel return hose from the manifold to the pressure regulator was disconnected, stainless steel hose clamps MUST be used to secure the connections. If the outlet/return hoses are to be replaced, a special replacement tubing kit (#32-827694) must be installed to prevent possible ruptures or leaks. DO NOT use Sta-Straps (wire-ties) to secure the high pressure fuel lines or dangerous fuel leakage will occur.

40. Secure the vapor separator tank to the intake manifold using the 3 retaining bolts and tighten to 45 inch lbs. (5.1 Nm).

41. If removed, secure the fuel outlet and return hoses to the manifold using clamps or use tubing kit # 32-827694 if the hoses are to be replaced.

42. If disconnected, reconnect the oil inlet hose to the vapor separator and secure with a wire tie.

43. Attach the throttle control link rod to the throttle cam.

44. If disconnected, attach the Red (positive) lead to the electric fuel pump Port terminal and the Red/Purple (negative) lead to the Starboard terminal of the pump.

45. Reposition the water separating filter and bracket assembly, then secure using the 2 retaining bolts.

46. Connect the fuel injector harness plug.

47. Connect the 2 leads from the air temperature sensor to the 2 bullet connectors.

48. Connect the throttle position sensor harness plug.

49. Reconnect the negative battery cable, then pressurize the fuel system and check for leaks.

Late-Models (2002 and Later)

◆ See Figures 165 thru 170

Unlike the earlier EFI models where an intake manifold cover mounted over a combined throttle body/intake manifold assembly to which the fuel rail was attached, the 2002 and later models utilize a throttle body mounted on top of an intake manifold (air plenum assembly). On these later models the combined throttle body/intake manifold attach directly the reed block, to which the fuel rail is bolted.

The vapor separator tank assembly and intake manifold/throttle body assembly must be removed for access to the fuel rail and injectors.

The fuel injectors themselves are retained to the fuel rail using Torx® head screws. In most cases they are the tamper proof screws that not only require a hex lobular head, but also one that is drilled in the center as there is a pin in the center of the fastener which prevents the use a standard Torx® bit. Mercury recommends the use of the Tamper Proof Torx Screw Set (#91-881828) to remove the screws.

1. For safety, disconnect the negative battery cable.

2. Release the fuel system pressure and, if necessary, drain the vapor separator tank.

3. Remove the vapor separator tank assembly, as detailed later in this section.

4. Disconnect the throttle cam link rod and the throttle position sensor link rod.

5. Tag and disconnect the following wiring:

a. The air temperature sensor (ATS) on top of the throttle body/intake manifold assembly.

b. The manifold absolute pressure (MAP) sensor, also on top of the throttle body/intake manifold assembly.

c. On 2.5L motors, the main engine harness (large round connector on the side of the throttle body/intake manifold assembly.

d. The large multi-pin fuel injector harness connector (at the base of the port side of the throttle body/intake manifold assembly).

6. On 2.5L motors, tag and disconnect the fuel pressure regulator hose at the top of the port side of the throttle body/intake manifold assembly (on 3.0L motors the pressure regulator hose was disconnected earlier in order to remove the vapor separator tank, as the regulator is mounted on top of the tank for those models).

7. Tag and disconnect the oil pump electrical harness, along with the oil pump inlet and outlet hoses. Be sure to plug the inlet hose to prevent leakage (and it is a good idea to plug the outlet hose as well to prevent system contamination).

8. Using a criss-crossing pattern, loosen and remove the 12 bolts retaining the intake manifold (and throttle body assembly) to the reed block and powerhead. Remove the intake manifold assembly and gasket.

■ The intake manifold gasket may be reused IF it does not show any signs of tears, cuts, abrasions or oil saturation.

9. The fuel rail is secured to the reed block by 4 bolts. If necessary to service the fuel rail or injectors, remove the bolts, then carefully life the assembly from the reed block. Unlike the early-model Mercury EFI systems the injectors will always come away with the fuel rail, since they are bolted to the rail and not just secured by positioning between the fuel rail and rail mounting on the powerhead.

10. If necessary to remove individual injectors, tag and disconnect the injector wiring harness. Then, loosen the injector retaining screw using a suitable Torx® Tamper Resistant bit and carefully pull the injector from the rail.

11. Inspect the injector O-rings for any signs of wear, cuts or abrasions and replace, if necessary. Honestly, we always suggest replacing O-rings that have been in service for any length of time.

Fig. 165 Remove the vapor separator tank assembly

Fig. 166 Disconnect the throttle cam and TPS link rods. . .

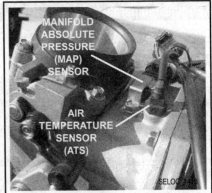

Fig. 167 . . .tag and disconnect the necessary wiring (like the ATS and MAP sensor)

Fig. 168 On 2.5L motors, unplug the main engine harness connector

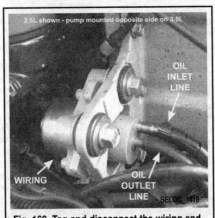

Fig. 169 Tag and disconnect the wiring and lines from the oil pump

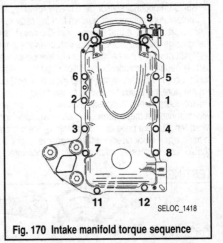

Fig. 170 Intake manifold torque sequence

To Install:

12. Make sure the O-rings are in place on the injectors. Apply a light coating of 2-stroke oil to the O-rings and O-ring contact surfaces on the reed block assembly plate.

13. If removed, install the injectors to the fuel rail and tighten the retaining bolts to 27 inch lbs. Reconnect the injector wiring as tagged during removal.

14. If removed, install the fuel inlet pipe toward the center of the fuel rail and tighten the retaining screw to 65 inch lbs. (7 Nm).

15. With the O-rings lightly oiled as noted earlier, carefully align the fuel rail and injectors to the reed block and seat the assembly to the block.

16. Install the fuel rail retaining bolts and tighten to 65 inch lbs. (7 Nm).

17. Position the intake manifold gasket with the sealing bead facing toward the intake manifold, then install the intake manifold and carefully thread the 12 retaining screws. Tighten the screws to 125 inch lbs. (14 Nm) using multiple passes of counterclockwise spiraling pattern that starts at the center and works its way outwards.

■ Mercury publications conflict as to the proper torque spec for the intake manifold/air plenum. In one place they give the torque sequence and value of 125 inch lbs. (14 Nm), while in another place they give no torque sequence and a value of 15 ft. lbs. (20 Nm).

18. Reconnect the oil pump electrical harness, along with the oil pump inlet and outlet hoses. Connect the hoses as tagged during removal, however keep in mind that the same basic pump is mounted on opposite sides of the powerhead depending upon whether you are working on a 2.5L or a 3.0L motors. On a 2.5L motors, the inlet hose is slightly higher and more inboard then the fitting for the outlet hose (and the wiring is on the bottom). However, on a 3.0L motors although the inlet hose is still more inboard than the outlet hose, it's slightly lower (and the wiring is on the top).

19. On 2.5L motors, reconnect the fuel pressure regulator hose to the top of the port side of the throttle body/intake manifold assembly.

20. Reconnect the following wiring, as tagged during removal:

a. The air temperature sensor (ATS) on top of the throttle body/intake manifold assembly.

b. The manifold absolute pressure (MAP) sensor, also on top of the throttle body/intake manifold assembly.

c. On 2.5L motors, the main engine harness (large round connector on the side of the throttle body/intake manifold assembly.

d. The large multi-pin fuel injector harness connector (at the base of the port side of the throttle body/intake manifold assembly).

21. Reconnect the throttle cam link rod and the throttle position sensor link rod.

22. Install the vapor separator tank assembly, as detailed later in this section.

23. Reconnect the negative battery cable, then pressurize the fuel system and check for leaks.

CLEANING & INSPECTION

The intake manifold, with sensors attached, may be carefully cleaned using a parts brush and solvent. Try not to get the solvent on the sensor or the wire harnesses.

Never dip rubber parts or plastic parts in carburetor cleaner. Blow the manifold dry with compressed air.

Move each throttle shaft back-and-forth to check for wear. If any shaft appears to be too loose, replace the manifold. Individual replacement parts are not available.

Check action of the throttle return springs. Remove and replace as necessary.

Inspect the main body of the intake manifold and cover gasket surfaces for cracks and burrs which might cause a leak.

If the injectors are removed, check the O-rings or seals carefully for signs of cuts, abrasion wear and replace, as necessary. As we've said many times before, anytime a component which is sealed with an O-ring is removed, we recommend replacing the O-ring if it has been in service for any length of time.

Inspect the sensors as far as possible without disturbing them. Check the wire harnesses for signs of chafing, cracks or corroded connections.

Engine Temperature Sensors (ETS)

DESCRIPTION & OPERATION

◆ See Figures 171, 172 and 173

Each EFI engine uses 1 or more engine temperature sensors (ETS) in order to determine powerhead operating conditions. Signals from the sensor(s) are used to determine air/fuel mixture (whether the powerhead is cold and therefore requires a more rich mixture) and are used by the warning system to let the operator know of an impending dangerous powerhead overheat condition.

Early-model EFI systems (through 2001) utilized just a single sensor mounted at the center of the to mid-to-upper portion of the port cylinder head.

Late-model (2002 or later) systems are normally equipped with that same sensor (called an analog temperature sensor on these models) mounted in the same location, PLUS an additional digital temperature sensor mounted toward the top of each cylinder head. Indications are that on late-model EFI systems, only the digital sensors feed signals to the ECM, while the analog sensor is an optional sensor used chiefly for an optional temperature gauge.

Temperature sensors for modern fuel injection systems are normally thermistors, meaning that they are variable resistors or electrical components that change their resistance value with changes in temperature. From the test data provided by the manufacturer it appears that Mercury sensors are usually negative temperature coefficient (NTC) thermistors. Whereas the resistance of most thermistors (and most electrical circuits) increases with temperature increases (or lowers as the temperature goes down), an NTC sensor operates in an opposite manner. The resistance of an NTC thermistor (like that used in these sensors) goes down as temperature rises (or goes up when temperature goes down). Therefore, the most important thing to keep in mind when quick checking these sensors is that their resistance readings should have an inverse relationship to temperature.

TESTING

◆ See Figures 171, 172 and 173

Temperature sensors are among the easiest components of the EFI system to check for proper operation. That is because the operation of an NTC thermistor is basically straightforward. In general terms, raise the temperature of the sensor and resistance should go down. Lower the temperature of the sensor and resistance should go up. The only real concern during testing is to make clean test connections with the probe and to use accurate (high quality) testing devices including a DVOM and a relatively accurate thermometer or thermosensor.

A quick check of the circuit and/or sensor can be made by disconnecting the sensor wiring and checking resistance (comparing specifications to the ambient temperature of the motor and sensor at the time of the test). Keep in mind that this test can be misleading as it could mask a sensor that reads incorrectly at other temperatures. Of course, a cold engine can be warmed and checked again in this manner.

More detailed testing involves removing the sensor and suspending it in a container of liquid (Mercury recommends water), then slowly heating or cooling the liquid while watching sensor resistance changes on a DVOM. This method allows you to check for problems in the sensor as it heats across its entire operating range.

Since the sensor specifications vary by model and sensor, check the Test Specifications for your motor.

■ In addition to the tests detailed here, the DDT or an equivalent scan tool can be used to check the sensor circuits through the ECM. But realize that an out of spec reading this way could be the fault of the wiring and not just the sensor.

Sensor Test Specifications

◆ See Figures 174 thru 177

■ Test readings may vary from specification as much as 10% before there is a cause for concern.

Quick Test

◆ See Figures 174 thru 177

A quick check of a temperature sensor can be made using a DVOM set to the resistance scale and applied across the sensor terminals. The DVOM can be connected to directly to the sensor, or to the sensor pigtail, as the wiring varies by model. Use a thermometer or a thermo-sensor to determine ambient engine/sensor temperature before checking resistance.

On the digital and analog temperature sensors used by 2002 and later models, there should be no continuity between either of the colored sensor leads and ground.

On the analog style engine temperature sensor used through 2001, there is a black ground lead which connects to the clamp retaining bolt. Resistance checks should be between the connectors for the colored sensor wires and NOT this ground lead. Furthermore on some models, like the 3.0L, one of the tan/black leads is grounded THROUGH the engine harness, however when testing the sensor itself neither tan/black lead should be grounded otherwise the sensor is shorted internally and grounded through the sensor case.

Even if the sensor tests ok cold, the sensor might read incorrectly hot (or anywhere in between). If trouble is suspected, reconnect the circuit, then start and run the engine to normal operating temperature. After the engine is fully warmed, shut the engine **OFF** and recheck the sensor hot. If the sensor checks within specification hot, it is still possible that another temperature point in between cold and fully-warmed specifications could be causing a problem, but not likely.

■ Test readings may vary from specification as much as 10% before there is a cause for concern.

The sensor can be removed and checked using the Comprehensive Test in this section or other causes for the symptoms can be checked. If the sensor was checked directly and looks good, but there are still problems with

Fig. 171 Early-model port cylinder head temp sensor

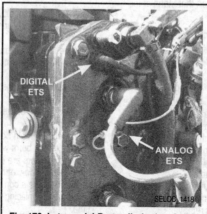

Fig. 172 Late-model Port cylinder head with both analog and digital temp sensors

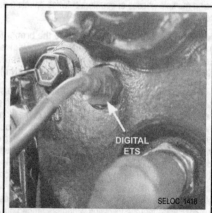

Fig. 173 Late-model starboard cylinder head and digital temp sensor

the circuit, be sure to check the wiring harness between the ECM and the sensor for continuity. Excessive resistance due to loose connections or damage in the wiring harness can cause the sensor signals to read out of range. Remember, never take resistance readings on the ECM harness without first disconnecting the harness from the ECM.

Comprehensive Test

◆ See Figures 174 thru 177

It is important to EFI operation that the temperature sensors provides accurate signals across the entire operating range and not just when fully hot or fully cold. For this reason, when resistance specifications are available, it is best to test the sensor by watching resistance constantly as the sensor is heated from a cold temperature to the upper end of the engine's operating range (or cooled from a hot temperature). The most accurate way to do this is to suspend the sensor in a container of water, connect a DVOM and slowly heat (or cool) the liquid while watching resistance on the meter.

To perform this check, you will need a high quality (accurate) DVOM, a thermometer (or thermo-sensor, some multi-meters are available with thermo-sensor adapters), a length of wire, a metal or laboratory grade glass container and a heat source (such as a hot plate or camp stove). A DVOM with alligator clip style probes will make this test a lot easier, otherwise alligator clip adapters can be used, but check before testing to make sure they do not add significant additional resistance to the circuit. This check is performed by connecting the 2 alligator clips together and checking for a very low or 0 resistance reading. If readings are higher than 0, record the value to subtract from the sensor resistance readings that are taken with the clips in order to compensate for the use of the alligator clips.

1. Remove the temperature switch as detailed in this section.
2. Suspend the sensor and the thermometer or thermo-sensor probe in a container of cool water or 4-stroke engine oil.

■ To ensure accurate readings make sure the temperature sensor and the thermometer are suspended in the liquid and are not touching the bottom or sides of the container (as the temperature of the container may vary somewhat from the liquid contained within and sensor or thermometer held in suspension).

3. Set the DVOM to the resistance scale, then attach the probes to the sensor or sensor pigtail terminals, as applicable.
4. Allow the temperature of the sensor and thermometer to stabilize, then note the temperature and the resistance reading. If a resistance specification is provided for low-temperatures, you may wish to add ice to the water in order to cool it down and start the test well below ambient temperatures.
5. Use the hot plate or camp stove to slowly raise the temperature. Watch the display on the DVOM closely (in case there are any sudden dips or spikes in the reading which could indicate a problem). Continue to note resistance readings as the temperature rises to 68°F (20°C) and 212°F (100°C). Of course by 212 degrees, your water should be boiling, so it is time to stop the test before you make a mess or get scalded. Up to that point, the meter should show a steady decrease in resistance that is proportional to the rate at which the liquid is heated. Extreme peaks or

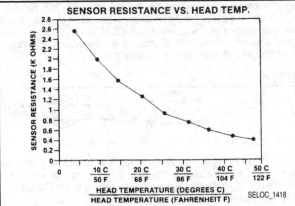

Fig. 174 Engine temperature sensor specifications - early-model 2.5L & 3.0L motors (through 2001)

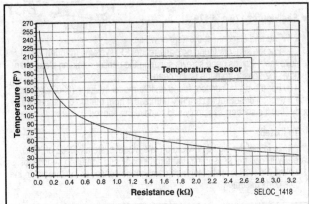

Fig. 175 Analog Temperature sensor specifications - late-model 2.5L motors (2002 and later)

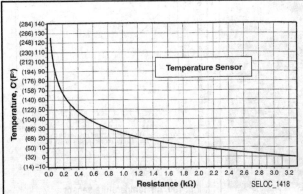

Fig. 176 Analog Temperature sensor specifications - late-model 3.0L motors (2002 and later)

Temperature Sensor Specifications		
Fahrenheit	Centigrade	OHMS
257	125	340
248	120	390
239	115	450
230	110	517
221	105	592
212	100	680
203	95	787
194	90	915
185	85	1070
176	80	1255
167	75	1480
158	70	1752
149	65	2083
140	60	2488
131	55	2986
122	50	3603
113	45	4370
104	40	5327
95	35	6530
86	30	8056
77	25	10000
68	20	12493
59	15	15714
50	10	19903
41	5	25396
32	0	32654
14	−10	55319
5	−15	72940

SELOC_1418

Fig. 177 Digital Temperature sensor specifications - late-model 2.5L & 3.0L motors (2002 and later)

valleys in the sensor signal should be rechecked to see if they are results of sudden temperature increases or a possible problem with the sensor.

6. Compare the readings to the Sensor Test Specifications.

■ **Test readings may vary from specification as much as 10% before there is a cause for concern.**

REMOVAL & INSTALLATION

◆ **See Figures 171, 172 and 173**

1. To remove the sensors found on early-model EFI systems (through 2001) or the analog sensors on late-model systems (2002 or later), proceed as follows:

   a. Remove the retaining bolt and retainer plate.
   b. Disconnect the wiring and remove the sensor.

2. To remove the digital sensors on late-model systems (2002 or later), proceed as follows:

   a. Disconnect the sensor wiring.
   b. Using a wrench or socket, carefully unthread and remove the sensor.

To Install:

3. To install the digital sensors on late-model systems (2002 or later), proceed as follows:

   a. Carefully thread the sensor into the cylinder head and tighten to 14 inch lbs. (1.6 Nm). Do NOT over-tighten and damage the sensor or the threads in the cylinder head.
   b. Reconnect the sensor wiring.

4. To install the sensors found on early-model EFI systems (through 2001) or the analog sensors on late-model systems (2002 or later), proceed as follows:

5. Connect the sensor wiring and install the sensor into the cylinder head.

6. Install the retainer plate and screw. Tighten the screw to 16.5 ft. lbs./200 inch lbs. (22.4 Nm).

Air Temperature Sensor (ATS)

DESCRIPTION & OPERATION

◆ **See Figures 178 and 179**

On early-model EFI systems (through 2001) the air temperature sensor is located on the starboard side of the intake manifold or the air intake silencer assembly. On these motors the sensor contains a round flange which is secured by 3 Philips head screws.

On late-model EFI systems (2002 and later) the digital air temperature sensor has been relocated to the top of the throttle body (on top of the forward end of the powerhead, just behind the throttle valve). On these models the sensor contains a hex and is usually threaded into position.

■ On late-model EFI systems (2002 and later) the ATS is physically the same component as the digital ETS mounted to the cylinder heads. Testing and service is identical to that for the ETS.

In all cases, the sensor measures the ambient air temperature and conducts this information in the form of an electrical signal to the ECM. As the air temperature changes, the amount of oxygen per cubic foot also changes. The quantity of available oxygen has an effect on combustion and therefore must be taken into account when computing the ideal air/fuel ratio.

Temperature sensors for modern fuel injection systems are normally thermistors, meaning that they are variable resistors or electrical components that change their resistance value with changes in temperature. From the test data provided by the manufacturer it appears that Mercury sensors are usually negative temperature coefficient (NTC) thermistors. Whereas the resistance of most thermistors (and most electrical circuits) increases with temperature increases (or lowers as the temperature goes down), an NTC sensor operates in an opposite manner. The resistance of an NTC thermistor (like that used in these sensors) goes down as temperature rises (or goes up when temperature goes down). Therefore, the most important thing to keep in mind when quick checking these sensors is that their resistance readings should have an inverse relationship to temperature.

TESTING

◆ **See Figures 177 thru 180**

The ATS can be tested in the identical manner as the ETS. As a matter of fact, for 2002 and later models the specifications are identical as well. For more information, please refer to the testing procedures for the ETS found earlier in this section.

REMOVAL & INSTALLATION

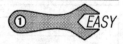

◆ **See Figures 178 and 179**

1. Tag and disconnect the sensor wiring.

2. For the sensors found on early-model EFI systems (through 2001), remove the 3 Philips head screws securing the sensor retaining plate, then remove the sensor from the powerhead.

3. For the digital sensors found on late-model systems (2002 or later), use a wrench or socket to carefully unthread it from the top of the throttle body (throttle shutter) assembly and remove the sensor.

To Install:

4. To install the digital sensors on late-model systems (2002 or later), carefully thread the sensor into the throttle body and tighten to 14 inch lbs. (1.6 Nm). Do NOT over-tighten and damage the sensor or the threads in the throttle body assembly.

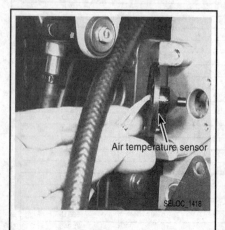

Fig. 178 Early-model (through 2001) air temperature sensor (ATS)

Fig. 179 Late-model (2002 and later) air temperature sensor (ATS)

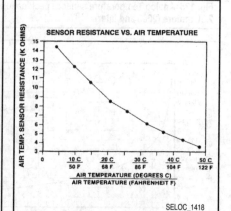

Fig. 180 Air Temperature sensor specifications - early-model 2.5L & 3.0L motors (through 2001)

5. To install the sensors found on early-model EFI systems (through 2001), insert the sensor into position, then use the retaining screws to secure it.

6. Reconnect the sensor wiring.

Manifold Absolute Pressure (MAP) Sensor

DESCRIPTION & OPERATION

◆ See Figures 181 and 182

Like many EFI systems Mercury's EFI delivers fuel based on mapping which requires inputs as to how much air the engine is ingesting. The most important inputs come from the air temperatures sensor, throttle position sensor and finally, the MAP sensor. If you think about it, the amount of air which is coming into the motor is dependent not only on the throttle position (physical size of the air entrance) but the temperature and density of that air. The density portion of the equation is supplied by the MAP sensor.

The location of the sensor varies somewhat by model, as follows:
• On early-model (through 2001) 2.5L EFI motors, the sensor is actually located inside the ECM.

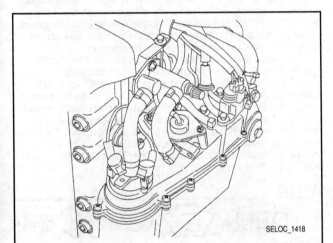

Fig. 181 MAP sensor location - early-model (through 2001) 3.0L EFI motors

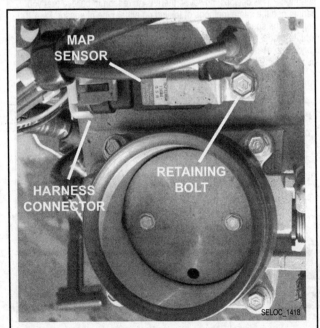

Fig. 182 MAP sensor location - late-model (2002 and later) EFI motors

• On early-model (through 2001) 3.0L EFI motors, the sensor is bolted to the top of the vapor separator tank.
• On late-model (2002 and later) 2.5L & 3.0L EFI motors, the sensor is mounted directly to the top of the intake manifold/throttle body assembly.

On early-model EFI motors because the sensor is mounted remotely from the manifold, a vacuum hose is necessary to connect the manifold to the sensor (and provide intake manifold pressure).

However, regardless of location, the sensor operates in the same basic manner. The sensor is a flexible type resistor. As the pressure changes, the resistor flexes and its resistance and the voltage applied across the sensor changes. This change is registered at the ECM.

Two conditions can affect the pressure in the manifold. The first and most common condition is a reduction in manifold pressure when load on the powerhead is increased. When the operator places a power demand on the powerhead by advancing the throttle, intake manifold pressure is reduced. Conversely, a reduction in intake manifold pressure, indicates an additional load has been placed on the powerhead.

The second condition which may affect the manifold pressure is operation of the powerhead at high altitudes.

TESTING

◆ See Figures 181, 182 and 183

MAP sensor testing will vary slightly with the EFI system. In all cases, the DDT or an equivalent scan tool can be used to monitor sensor output and can be used to determine if the sensor (or circuit) is defective. When using a scan tool, the numerical value displayed on the tool should increase as the throttle position is advanced (or decrease as the throttle is closed).

However, if a scan tool is not available, there are alternate ways of testing the sensor, again, depending upon the type of EFI system used.

On early-model systems (through 2001), the most simple way to test the MAP sensor is using a hand-vacuum pump on the sensor vacuum line. With the motor running at normal operating temperature, disconnect the MAP sensor vacuum hose from the intake manifold and connect it to a vacuum pump. Draw a vacuum on the line and observe/listen for a change in engine operation as the ECM responds to the change in vacuum. If the ECM does not respond to any changes in vacuum, the MAP sensor is defective and should be replaced.

On late-model EFI systems (2002 and later), a vacuum pump CANNOT be used (at least as easily) as the sensor is mounted directly to the manifold (there is no vacuum hose to detach and connect to the sensor). However, Mercury provides data in order to perform resistance tests with the sensor removed from the motor or installed to the motor, but engine not operating. Use the accompanying illustration for test connections and specifications.

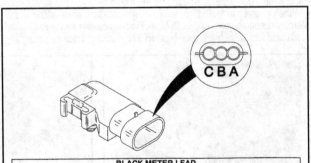

RED METER LEAD		BLACK METER LEAD		
		BLK/ORG PIN A	YEL PIN B	PPL/YEL PIN C
	BLK/ORG PIN A	X	95000 - 105000 ohm	3900 - 4100 ohm
	YEL PIN B	95000 - 105000 ohm	X	95000 - 105000 ohm
	PPL/YEL PIN C	3900 - 4100 ohm	95000 - 105000 ohm	X

Fig. 183 MAP sensor resistance testing - late-model (2002 and later) EFI motors

REMOVAL & INSTALLATION

◆ **See Figures 181 and 182**

Removal and installation of the sensor is pretty straight forward on these motors, but since the location varies by model, so does the procedure, as follows:

• On early-model (through 2001) 2.5L EFI motors, the sensor is actually located inside the ECM. Unfortunately the ECM is a permanently sealed unit, so there is no way to replace the JUST the sensor itself.

• On early-model (through 2001) 3.0L EFI motors, the sensor is bolted to the top of the vapor separator tank, at the opposite end from the electric fuel pump. The sensor is normally secured to the top of the tank by 2 bolts, threaded downward from the top. In order to remove the sensor, disconnect the vacuum hose and the wiring, then loosen the bolts and remove the sensor. Installation is obviously the reverse, but in some cases, it may be easier to disconnect/connect the vacuum hose while the sensor is already removed from the vapor separator tank.

• On late-model (2002 and later) 2.5L & 3.0L EFI motors, the sensor is mounted directly to the top of the intake manifold/throttle body assembly. Removal is a simple matter of disconnecting the sensor harness, loosening the single mounting bolt and pull the sensor up from the top of the manifold. During installation, tighten the retaining bolt to 80 inch lbs. (9 Nm).

Crankshaft Position Sensor (CPS)

The ECM needs to know exactly where each piston is in its 2-stroke cycle in order to accomplish both ignition timing and fuel injection functions. Where carbureted and the 2001 or earlier 2.5L V6 (150-200 hp) EFI motors accomplish this task using a trigger coil, all other EFI motors and all OptiMax motors utilize a crankshaft position sensor (CPS) to provide this data to the ECM.

Because the ignition systems are operated in much the same manner, most testing and service information on the CPS can be found in the Ignition and Electrical System section.

The sensor works in basically the same fashion as a trigger coil, by picking up a magnetic field created by magnets attached to the spinning flywheel, therefore the sensor itself must be positioned a certain distance away from the flywheel.

For most models the CPS is not adjustable, so simply installing it and tightening the bolts will position it the proper distance from the flywheel. However, on 3.0L EFI motors through 2001 and some Optimax motors the sensor mounting bracket allows for adjustment. On these models you must set a specific air gap when installing the sensor in order to ensure it will operate correctly. Specifications for the CPS air gap are available in the Fuel Injection Component Adjustment Specifications chart in this section, however they are normally around 0.025-0.040 in. (0.635-1.010mm) for OptiMax motors and 0.020-0.060 (0.51-1.53) for EFI motors.

When it comes to testing, no resistance specifications are available in the Mercury documentation for 2.5L Optimax motors, however the sensor (calibrated for use with the same gap) on 3.0L motors is listed as having a resistance spec of 300-340 ohms. More details on testing this sensor can be found in the Ignition and Electrical System section. Refer to the Ignition System Component Testing specifications chart in that section for data on sensor testing.

Throttle Position Sensor (TPS)

DESCRIPTION & OPERATION

◆ **See Figures 184, 185 and 186**

On early-model (through 2001) EFI systems, the throttle position sensor (TPS) is mounted on the starboard side of the powerhead directly onto the bottom throttle shaft. The holes in the sensor's mounting flange are slotted to allow for minute adjustments. As such the sensor must be checked and adjusted anytime it is removed from the powerhead.

On late-model (2002 and later) EFI systems, the TPS is mounted on the port side of the powerhead, and NOT directly on the throttle shaft. For these models the TPS was relocated a little further back on the powerhead, just behind the Vapor Separator Tank and is actuated by a throttle link. The sensor on these models is mounted in a fixed position and no adjustment is necessary or possible.

In all cases, the sensor is an encased potentiometer that sends a signal to the ECM indicating how far open or closed the throttle shaft is at any given moment. The ECM processes this throttle opening as an indication of powerhead load for a specific rpm and uses it in helping to determine the amount of fuel which should be delivered by the injectors.

In both cases, the body of the sensor is stationary with a small shaft emerging from the center (early-models) or from the upper edge (late-models) of the sensor. The shaft is connected to the throttle valve (early-model) or throttle link (late-model). As the throttle is advanced, movement is transferred to the sensor and the internal resistance of the sensor changes.

The ECM sends a reference voltage signal to the sensor through the harness and then measures the amount of voltage which returns. So, as the throttle is opened or closed and the internal resistance of the sensor changes, the variable voltage signal sent to the ECM is changed, proportionally to the changes in throttle position.

TESTING

◆ **See Figures 184, 185 and 186**

Mercury's recommended testing differs slightly between early-model and late-model TPS designs, however, there is no reason to believe that you cannot use the same methods of testing on either type of sensor. Plus, if some of the special tools are not available (such as the scan tool necessary for late-model tests) there may be some alternate tests which you can conduct to at least give you some idea of the sensors condition. Either way, use the following tests as you see fit:

1. If you have access to a scan tool like the DDT, proceed as follows:
   a. Hook up the scan tool.

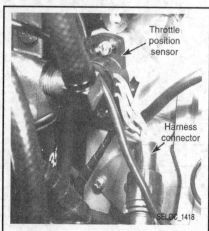

Fig. 184 TPS and wiring harness - early-model motors

Fig. 185 TPS and link rod location - late-model motors

Fig. 186 TPS and wiring harness - late-model motors

b. Provide a suitable source of cooling, then start and warm the engine to normal operating temperature.

c. Slowly open and close the throttle while reading the TPS value on the scan tool. The numbers should increase or decrease linearly in proportion to throttle position. On late-model EFI systems (2002 or later) the numbers should INCREASE as the throttle is opened and DECREASE as the throttle is closed.

2. If you have access to a TPS test harness such as #91-859199 for early-model (through 2001) 2.5L motors or #84-825207A1 for early-model (through 2001) 3.0L motors proceed as follows:

■ The TPS test harness is basically a set of jumper wires which will connect inline with the TPS-to-ECM wiring, along with a T which allows you to connect a DVOM to monitor the unbroken circuit during operation. If you do not have access to the factory kit you can fashion your own jumper wires either using the same type of connectors as the factor harness or using some other type of clips/connectors which will attach to the connector terminals of the factory harness, but beware, do NOT allow wires to short. Although Mercury does not seem to offer a test harness for late-model (2002 and later) EFI systems, there is no reason you could not fashion your own jumper in the same manner.

a. Disengage the TPS harness and install a suitable test harness which allows you to probe the unbroken TPS circuit using a DVOM.

b. On early-model (through 2001) 2.5L EFI motors, locate the tan/black cylinder head temperature sensor leads for the Port cylinder head, then disconnect the 2 lead bullet connectors for the duration of TPS adjustment.

c. Turn the ignition keyswitch to the **ON** position, but don't start the motor.

d. Check the DVOM for voltage readings with the throttle both closed to the idle position and open to the wide open throttle (WOT) position. Compare what you get to the Throttle Position Sensor Output Voltage specifications found in the Fuel Injection Component Adjustment Specification chart found in this section.

e. Again, slowly open the throttle from idle to the WOT position and back to idle again, watching the DVOM to make sure voltage readings increase and decrease smoothly.

f. On early-model (through 2001) 2.5L EFI motors, once testing is complete, be sure to remember to reconnect the tan/black port cylinder head temperature sensor leads.

3. Although not a test specifically recommended by Mercury, If you have no access to jumper wires to test the completed circuit, you can use a DVOM to check resistance of the sensor itself to make sure resistance changes smoothly between idle and WOT positions. For this test, you can disconnect the TPS wiring and probe the sensor harness directly (NOT THE ECM side of the harness!). Slowly open and close the throttle, watching the ohm readings on the meter for flat spots or spikes.

4. Lastly, Mercury has noted that some TPS faults are heat related and as such they recommend one final quick-check (at least they recommend it on late-model, 2002 or later EFI models, though we can't see why it's not worth a try on early-model EFI models too). With the motor running on a suitable cooling source, use a heat gun to carefully apply heat to the TPS near the electrical connection, until it is warm to the touch. While heat is applied listen for any of the following signs of a possible problem with the TPS:

• Engine rpm change

• Illumination of the Check Engine light
• A momentary activation of the warning horn signal
• TPS voltage value change (1/2 volt) as seen on the scan tool or DVOM with test harness and no accompanying throttle change.

✱✱ SELOC CAUTION

Keep in mind that excessive heat will damage the TPS, so use caution and don't heat it up too much.

REMOVAL & INSTALLATION

② *MODERATE*

◆ See Figures 184, 185 and 186

The big difference between the TPS used on early-model (through 2001) EFI systems and the one used on late-model (2002 or later) systems is that the late-model sensors are mounted in a fixed position and cannot be adjusted. It makes removal & installation a whole lot quicker and easier.

1. Tag and disconnect the wiring for the sensor.
2. On late-model motors, disconnect the sensor link rod
3. On early-model motors, matchmark the original location of the throttle position sensor before removal (this will give you the best starting position for adjustment after installation).
4. Remove the 2 sensor mounting screws (early-model) or the 3 sensor mounting bolts (late-model), as applicable, then lift the sensor from the powerhead.

To Install:

5. Position the TPS on the powerhead. On early-model motors, check to be sure the throttle position sensor shaft engages the slot on the throttle shaft.
6. Secure the sensor with the attaching hardware. On early-model motors, be sure to align the matchmarks made earlier, then temporarily tighten the retaining screws (as you'll have to loosen them again in order to adjust the sensor anyway). On late-model motors tighten the retaining bolts to 70 inch lbs. (8 Nm).
7. On early-model motors, properly adjust the TPS. For details, please refer to the Timing & Synchronization procedures found in the Maintenance & Tune-Up section.
8. Reconnect the sensor wiring.
9. On late-model motors, connect the sensor link rod.

Detonation (Knock) Sensor and Module

DESCRIPTION & OPERATION

◆ See Figures 187, 188 and 189

Some models are equipped with a detonation (knock) sensor threaded into one or both of the cylinder heads. On early-model EFI systems so equipped (usually only some 2.5L motors through 2001) the sensor is threaded into the top of the portside cylinder head. On late-model EFI systems so equipped (usually 2.5L or 3.0L 200 hp or 200 HP Work Models

Fig. 187 Detonation sensor threaded into port head - early-model EFI

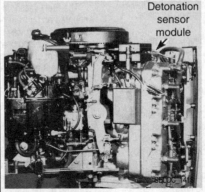

Fig. 188 Detonation sensor control module - early-model EFI

Fig. 189 Detonation sensor - late-model EFI

only) there is 1 sensor threaded into the center of each cylinder head, JUST below the middle spark plug.

In all cases, the sensor is able to detect the frequency of vibrations associated with pre-ignition and detonation, approximately, 8,000 Hertz. If either of these conditions is due to fuel of an insufficient octane rating (less than 87) or a sudden change in loading of the powerhead, the sensor will be activated.

An electronic signal is sent to the detonation module (early-models only) and the ECM (all models). The result of these signals is ignition timing being retarded by as much as 8° to help stop knocking and prevent potential damage to the motor that can be caused by knocking.

TESTING

◆ **See Figures 187, 188 and 189**

The recommended test procedures for the detonation sensor vary with system. The early-model systems can be tested with a DVOM, while Mercury claims the late-model systems can only be tested using the DDT or an equivalent scan tool. Some technicians have been known to test knock sensors by running the motor in a test tank at mid-to-full throttle and watching or listening for changes in engine timing when the powerhead is tapped with a large brass mallet to simulate knocking. When done correctly a properly operating knock sensor system will retard engine timing momentarily in response to the vibrations caused by the mallet.

On late-model systems Mercury states the only way to check the system is to run the motor (under load in the water or a test tank) while using a scan tool such as the DDT to check function. Using the Special Functions feature of the scan tool run the Knock Output Load Test and watch the tool. As the throttle is advanced and the engine run under load (in gear), the numerical value displayed by the scan tool should increase. As the throttle is then retarded the value should decrease. If so the knock sensor circuit is functioning. If the values do NOT change as the throttle setting and load varies, the knock sensor or circuit in the ECM is defective.

■ **Never operate the engine at high speed with a flush device attached. The engine, operating at high speed with such a device attached, could run-away from lack of a load on the propeller, causing extensive damage.**

On early-model systems use the following procedure and a DVOM to check the overall system operation, circuit, module and sensor respectively.

1. For starters perform an operational check to see if the knock system is operating properly. This check will eliminate the possibility of improper fuel delivery due to a failure in the detonation control system.

   a. Place the motor in a test tank or the boat in the water secured to sturdy dock so the motor can be run under load.

   b. Connect a timing light or a scan tool to monitor ignition timing.

   c. Start and warm the engine to normal operating temperature.

   d. With the motor fully warmed, advance the throttle to 3000-3500 rpm in gear while watching the ignition timing. If timing advanced from 18 to 24° BTDC, the system is functioning properly. However if the timing advances to 18-24° BUT retards while maintaining a steady throttle position the system is functioning and detonation may be occurring (to be sure you can check the sensor and module next). If no timing advancement occurs check the ignition system to be sure, but denotation could be occurring preventing advance.

2. To check the circuit, proceed as:

   a. Remove the single white/blue lead from the detonation controller to the detonation sensor at the screw terminal of the sensor. Now connect one end of a jumper wire directly to the module where the white/blue lead normally connects.

   b. Make contact with the other end of the jumper wire to a suitable ground on the powerhead.

   c. Turn the ignition key to the **RUN** position (without starting the motor). Using a scan tool like the DDT check the knock voltage output, it should be 6 volts or more.

   d. Start the engine and allow it to run at idle, rechecking the scan tool for knock voltage. The output should be less than 1 volt. If voltage does NOT drop below 1 volt (and no detonation is occurring) the module is defective.

3. To check the detonation sensor itself, proceed as follows:

   a. Disconnect the single white/blue lead from the detonation sensor itself.

b. Connect the red lead from a DVOM to the terminal on the end of the sensor.

c. Connect the black DVOM lead to the sensor housing.

d. Turn the DVOM to the 200 MV scale, AC.

e. With the outboard in the water or a test tank start and operate the engine at idle speed.

f. Watch the meter, you should see an power reading of 0.075-0.120 volts AC. Slowly raise the engine speed and watch power output, voltage should increase with rpm. If voltage is within range, but problems are suspected, check the detonation module. If voltage is NOT within range, replace the sensor (tighten the sensor to 144 inch lbs./16 Nm).

4. To check the detonation module, proceed as follows:

   a. Insert a probe (like a paper clip) into the back of the bullet connector on the gray/white lead for the detonation module.

   b. With the meter set on the 20V DC scale, make contact with the red meter lead to the probe/clip from the module and make contact with the black meter lead to a suitable ground on the powerhead.

   c. Start the powerhead and allow it to reach operating temperature at idle speed.

   d. With the motor warm and running at idle, the DVOM should register 0.30-0.70 volts DC. If the voltage does not fall within this range at idle the control module MAY be bad.

   e. Increase powerhead rpm with the motor in gear to 3000-4000 rpm. The meter should register near zero volts (approximately 0.01 volts DC). If the voltage is within range the control unit is functioning and there is no detonation occurring. However, if the voltage reading is above 0.01 volts DC and between 1.0-6.6 volts at 3000-4000 rpm then detonation is occurring. A constant voltage of 6.6 volts DC usually indicates a bad detonation control module.

Water Sensor

DESCRIPTION, OPERATION & TESTING

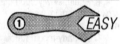

EFI motors are equipped with a simple, one wire water sensor usually mounted to the bottom of the water separating fuel filter. The sensor is designed to provide a ground for the warning circuit when water is present, thus enabling the operator to remedy the situation before damage can occur to the powerhead.

Testing is a simple manner of removing the wiring connector (the wire is normally Tan or Tan with a colored tracer) from the sensor probe then grounding the wiring to the powerhead either with the motor running or, in some cases, just with the ignition key in the **RUN** position. If the warning horn sounds in either of those conditions the circuit can be considered operational. The sensor itself can be tested by exposing it to water in those same conditions, either by putting water in the filter (not the best method) or by removing the probe from the filter assembly and dipping it into water while the ignition key is in the **RUN** position.

Electronic/Engine Control Module (ECM)

DESCRIPTION & OPERATION

The fuel injection system is controlled by the ECM, an onboard computer mounted as far from heat and vibration as possible. The computer is sealed unit and is in no way serviceable. The ECM receives signals from numerous sensors on the powerhead. From the signals received, the ECM determines the amount of fuel to be injected. This computer also determines the timing of the spark at the spark plugs.

The amount of fuel injected is determined by how long each injector nozzle remains open. This is commonly referred to as the "pulse width". The nozzle opens and closes in response to signals from the ECM.

The ECM receives 3 types of input signals - Analog, Digital and Pressure Differential.

• Analog signals change with changing conditions. For example: the coolant temperature sensor will have more electrical resistance when the powerhead is cold, than when the powerhead is hot.

• Digital signals are a series of on and off pulses. These pulses are counted by the computer to determine a condition. For example: the signal sent from the No. 2 ignition coil will provide the computer with information on powerhead rpm.

• Pressure differential signals received through vacuum lines connecting the intake manifold to the ECM indicate powerhead loading to sensors within the computer.

Sensors located on the powerhead (or inside the ECM in some cases) provide information to the computer on powerhead load, rpm, temperature and other conditions affecting operation.

The computer is programmed or provided with instructions, to produce correct air/fuel mixtures and throttle openings for varying conditions.

■ **The early-model (through 2001) 3.0L motors were equipped with 2 ECMs, one for the Ignition System and the other for the Fuel System. Other models integrate all functions into a single control unit.**

There are no tests available for the ECM. If problems are suspected with the Fuel Injection or Ignition systems, you'll have to test the rest of the system thoroughly, eliminating all other possible causes before condemning the ECM. Remember, they are not inexpensive components, and they are normally NOT RETURNABLE!

REMOVAL & INSTALLATION

Early-Model 2.5L (Through 2001)

◆ **See Figures 145 thru 148**

1. For safety and to protect the ECM itself, disconnect the negative battery cable.
2. Unplug the ECM harness connector located along side the ECM (starboard side).
3. Remove the 2 screws securing the water sensing module to the powerhead. The left screw has 2 small black grounding leads attached. The module will still be connected to the powerhead by other leads. Move the module aside.
4. Remove the large bolt in the center at the base of the ECM.

■ **Do not disturb the small Phillips head screws around the perimeter of the ECM. The cover of the ECM will not come away once these bolts are removed. The bolts were installed during manufacture of the unit and then the cover and the case were hermetically sealed using the Phillips head screws for alignment. Therefore, no purpose will be served, if these screws are removed.**

5. Remove the 2 nuts on the studs at the top of the ECM. On some models, the right stud may have a small Black ground wire over it.
6. Carefully lift the ECM from the studs for access, then tag and disconnect the vacuum line from the ECM at the fitting on the manifold next to the pressure port. Remove the ECM completely from the powerhead.

To Install:

7. Hold the ECM in position just in front of the intake manifold and connect the vacuum hose from the MAP sensor inside the ECM to the fitting on the intake, then seat the ECM on the mounting studs.
8. Secure the ECM to the mounting studs using the nuts (remember, some models MAY have a ground one on the right stud). And install the ECM lower mounting bolt. Tighten the nuts and bolt to 45 inch lbs. (5 Nm).
9. Install the water sensing module and tighten the retaining screws to 25 inch lbs. (3 Nm).
10. Reconnect the large, round, ECM harness connector to the main powerhead harness connector. The connection can only be made one way. Secure the harness into the bracket with the connecting ring either above or below the bracket arms.
11. Reconnect the negative battery cable.

Except Early-Model 2.5L (Through 2001)

◆ **See Figure 190**

For most Mercury EFI systems removal or installation of the ECM is pretty straight forward, meaning tag and disconnect the connectors, then unbolt and remove the unit. However there are a couple of points to keep in mind.

• Use care when handling the ECM or any electronic components. It's always a good idea to touch the powerhead immediately each time before touching the ECM. This will allow you to release any static electricity which can actually harm some sensitive circuits on the ECM.
• On early-model (through 2001) 3.0L motors there are 2 ECMs. The upper ECM is for the Ignition System, the lower ECM is for the FUEL system.

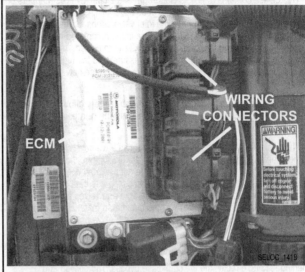

Fig. 190 Typical late-model Mercury ECM

• On some models one or more of the bolts may also contain a ground strap. This is usually true on the lower right bolt of the fuel injection ECM for the early-model 3.0L models, but may be found elsewhere as well. Pay attention to any ground straps, when present, so that you can return them to their original positions.
• Upon installation, don't over-tighten the bolts and possibly crack the housing or mounting. On early-models the bolts should be tightened to 80 inch lbs. (9 Nm). On late-models (2002 or later) the bolts should be tightened to 70 inch lbs. (8 Nm).

Mechanical Fuel Pump

DESCRIPTION & OPERATION

◆ **See Figure 191**

The fuel delivery system on Mercury EFI motors includes both a low-pressure and high-pressure circuit. The low-pressure circuit is fed by a mechanical fuel pump which draws fuel from the fuel tank, pressurizes it and pushes it toward the Vapor Separator Tank assembly (which acts like a large float bowl/fuel reservoir from which the electric high-pressure fuel pump can draw fuel). Generally speaking, fuel from the tank passes through the primary fuel filter, a water separating filter and then on to the vapor separator.

The fuel pump is a basic mechanical device that utilizes crankcase positive and negative pressures (alternating pressure and vacuum) to pump fuel through the supply lines.

Fig. 191 Mechanical low-pressure pump from a Mercury EFI motor

This device contains a flexible diaphragm and 2 check valves (flappers or fingers) that control flow. As the piston goes up, crankcase pressure drops (negative pressure or vacuum) and the inlet valve opens, pulling fuel from the tank. As the piston nears TDC, pressure in the pump area is neutral (atmospheric pressure). At this point both valves are closed. As the piston comes down, pressure goes up (positive pressure) and the fuel is pushed toward the carburetor bowl by the diaphragm through the now open outlet valve.

■ **This system works because the check valves only allow fuel to flow in one direction. Should one of the valves fail, the pump will not operate correctly.**

This is a reliable method to move fuel but can have several problems. Sometimes an engine backfire can rupture the diaphragm (though this is rare on pumps like the Mercury design which are not mounted directly to the powerhead as a vacuum source, but instead connect to the powerhead vacuum source through a hose). The diaphragm and valves are moving parts subject to wear. The flexibility of the diaphragm material can go away, reducing or stopping flow. Rust or dirt can hang a valve open and reduce or stop fuel flow.

Most pumps consist of a diaphragm, 2 similar spring loaded disc valves, one for inlet (suction) and the other for outlet (discharge) and a small opening leading directly into the crankcase bypass. The suction and compression created, as the piston travels up and down in the cylinder, causes the diaphragm to flex.

As the piston moves upward, the diaphragm will flex inward displacing volume on its opposite side to create suction. This suction will draw liquid fuel in through the inlet disc valve.

When the piston moves downward, compression is created in the crankcase. This compression causes the diaphragm to flex in the opposite direction. This action causes the discharge valve disc to lift off its seat. Fuel is then forced through the discharge valve into the carburetor.

Problems with the fuel pump are limited to possible leaks in the flexible neoprene suction lines, a punctured diaphragm, air leaks between sections of the pump assembly or possibly from the disc valves not seating properly.

The pump is usually activated by 1 cylinder. If this cylinder indicates a wet fouled condition, as evidenced by a wet fouled spark plug, be sure to check the fuel pump diaphragm for possible puncture or leakage.

TESTING

◆ **See Figure 191**

Lack of an adequate fuel supply can cause the engine to run lean, lose rpm or cause piston scoring due to a lack of lubricant that is carried in the fuel.

Generally speaking, checking the low-pressure delivery circuit can be a good way to eliminate lack of fuel as a potential problem with an EFI system.

✳✳ SELOC CAUTION

Observe all applicable safety precautions when working around fuel. Whenever servicing the fuel system, always work in a well ventilated area. Do not allow fuel spray or vapors to come in contact with a spark or open flame. Keep a dry chemical fire extinguisher near the work area. Always keep fuel in a container specifically designed for fuel storage, also, always properly seal fuel containers to avoid the possibility of fire or explosion.

The problem most often seen with fuel pumps is fuel starvation, hesitation or missing due to inadequate fuel pressure/delivery. In extreme cases, this might lead to a no start condition, but that is pretty rare as the primer bulb should at least allow the operator to fill the vapor separator tank. More likely, pump failures are not total, and the motor will start and run fine at idle, only to miss, hesitate or stall at speed when pump performance falls short of the greater demand for fuel at high rpm.

Before replacing a suspect fuel pump, be absolutely certain the problem is the pump and NOT with the fuel tank, lines or filter. A plugged tank vent could create vacuum in the tank that will overpower the pump's ability to create vacuum and draw fuel through the lines. An obstructed line or fuel filter could also keep fuel from reaching the pump. Any of these conditions could partially restrict fuel flow, allowing the pump to deliver fuel, but at a lower pressure/rate. A pump delivery or pressure test under these circumstances would give a low reading that might be mistaken for a faulty pump. Before testing the fuel pump, refer to the testing procedures found under Fuel Lines & Fittings to ensure there are no problems with the tank, lines or filter.

A quick check of fuel pump operation is to gently squeeze the primer bulb with the motor running. If a seemingly rough or lean running condition (especially at speed) goes away when the bulb is squeezed, the fuel pump is suspect. Another quick check is to shut the motor down when it starts to miss (or leave it off if it stalls) and then drain the vapor separator tank. If little or no fuel comes out, the mechanical pump and/or low-pressure circuit are suspect.

If inadequate fuel delivery is suspected and no problems are found with the tank, lines or filters, a conduct a quick-check to see how the pump affects performance. Use the primer bulb to supplement fuel pump. This is done by operating the motor under load and otherwise under normal operating conditions to recreate the problem. Once the motor begins to hesitate, stumble or stall, pump the primer bulb quickly and repeatedly while listening for motor response. Pumping the bulb by hand like this will force fuel through the lines to the vapor separator tank, regardless of the fuel pump's ability to draw and deliver fuel. If the engine performance problem goes away while pumping the bulb, and returns when you stop, there is a good chance you've isolated the low pressure fuel pump as the culprit. On most EFI models, you should be able to perform a pressure or vacuum check, but of course, you can always disassemble the pump to physically inspect the check valves and diaphragms.

✳✳ SELOC WARNING

Never run a motor without cooling water. Use a test tank, a flush/test device or launch the craft. Also, never run a motor at speed without load, so for tests running over idle speed, make sure the motor is either in a test tank with a test wheel or on a launched craft with the normal propeller installed.

Checking Fuel Pump Lift (Vacuum)

◆ **See Figure 192**

Fuel system vacuum testing is an excellent way to pinpoint air leaks, restricted fuel lines and fittings or other fuel supply related performance problems.

The standard square fuel pump used on carbureted and EFI Mercury motors is designed to lift fuel vertically about 60 in. (1524mm). But those capabilities are dependent upon there being no other restrictions in the system and a fuel hose that is at least 5/16 in. (7.9mm) Inner Diameter (I.D.). Each restriction that is added to the system (such as a fuel filter, valve or fitting) will reduce the amount of lift available to the system.

Using a gauge, a T-fitting and a length of clear hose you can check the amount of vacuum in the system, as well as visually inspect for air bubbles (which would indicate a leak).

1. Disconnect the fuel inlet line (tank side) from the fuel pump and connect a clear piece of fuel hose to the pump inlet nipple.

2. Next, connect a T-fitting to the end of the clear line and connect the fuel inlet line to the other side of the T-fitting.

3. Lastly, connect the vacuum gauge to the T-fitting.

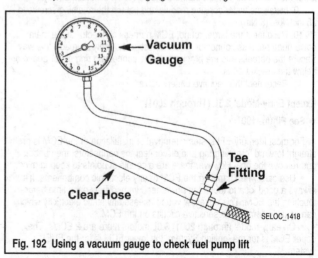

Fig. 192 Using a vacuum gauge to check fuel pump lift

4. Supply the motor with a source of cooling water.

5. First check to be sure the pump itself is capable of producing enough vacuum by pinching off the fuel supply line on the tank side of the T-fitting, then starting and running the engine while watching the gauge. If the pump produces 2.5 in. Hg (6.35 cm Hg/8.44 kPa) or MORE of vacuum remove the hose pincher from the fuel supply line and continue the test. If the pump reading is below 2.5 in. Hg (6.35 cm Hg/8.44 kPa), then check the following:
- Pump check valves and/or diaphragm may be defective
- Pump may contain an air leak (requiring a new gasket, rebuild or replacement, be sure to check the fittings as well)
- There may be a problem with low crankcase pressure (check for plugged pump vacuum passageways)

6. With the pincher removed from the tube, watch the clear fuel line for signs of bubbles. If none are found, continue with the test. However, if bubbles are present, check the following for sources of the air leak:
- Pick-up tube in fuel tank
- Outlet fitting in fuel tank
- Fuel inlet hose may be improperly clamped at fitting
- Fuel tank valve may be leaking
- If equipped, fuel line from kicker engine into main fuel line may be leaking

7. If there are no air leaks, operate the engine until warmed to normal operating temperature. Run the motor at both idle and a little above idle (not too much above idle if on a flush fitting instead of a test tank or launched craft). As engine speed increases there should be a slight increase in vacuum however the reading should be at or below 2.5 in. Hg (6.35 cm Hg/8.44 kPa). If the reading is above 2.5 in. Hg (6.35 cm Hg/8.44 kPa), then check the following:
- Is the anti-siphon valve restricted?
- Is there a restriction in the primer bulb assembly?
- Is there a kinked or collapsed fuel hose?
- Is there a plugged water separating fuel filter rigged to the craft?
- Is there a restriction in the fuel line thru-hull fitting?
- If equipped, is there a restriction in the fuel tank switching valve(s)?
- Is there a plugged fuel tank pick-up screen.

8. Once you are finished, shut down the motor, then remove the tubing, T-fitting and gauge. Reconnect the fuel inlet line.

Checking Pump Pressure

◆ See Figure 191

Fuel system pressure testing is a great way to check the condition of the pump itself, especially if questions concerning its condition are raised by a pump lift test.

The specifications for all EFI motor low-pressure pumps are nearly identical, and though the differences are probably not significant, we'll publish them here just to be sure we've given the big picture.

The following are the operating pressures listed for the mechanical fuel pump used on early-model (through 2001) EFI systems:
- Operating pressure at idle: 2-3 psi (13.8-20.7 kPa)
- Minimum pressure at idle: 1 psi (6.9 kPa)
- Operating pressure at WOT: 6-8 psi (41.4-55.2 kPa)
- Minimum pressure at WOT: 4 psi (27.6 kPa)

The following are the operating pressures listed for the mechanical fuel pump used on late-model (2002 and later) EFI systems:
- Operating pressure at idle: 2-3 psi (13.7-20.5 kPa)
- Minimum pressure at idle: 1 psi (6.8 kPa)
- Operating pressure at WOT: 8-10 psi (55-69 kPa)
- Minimum pressure at WOT: 3 psi (20.5 kPa)
- Maximum pressure at WOT: 10 psi (69 kPa)

Using a pressure gauge, short length of hose and a T-fitting you can check the amount of pressure being developed by the pump under idle and/or WOT conditions as follows:

1. Disconnect the fuel outlet line (powerhead side) from the fuel pump and connect the short length of hose to the pump outlet nipple.

2. Next, connect a T-fitting to the end of the line and connect the fuel outlet line to the other side of the T-fitting.

3. Lastly, connect the pressure gauge to the T-fitting.

4. Supply the motor with a source of cooling water.

5. Start and run the motor to normal operating temperature, then operate it at both idle and/or WOT and not the operating pressures, as compared with the specifications listed at the beginning of this procedure. If the pump is not making proper pressure and there are no restrictions in the

fuel supply line (and no mechanical problems with the powerhead producing vacuum pulses), then the pump should be overhauled or replaced.

6. Once you are finished, shut down the motor, then remove the tubing, T-fitting and gauge. Reconnect the fuel inlet line.

Fuel Pump Diaphragm and Check Valve Testing

◆ See Figure 191

With the use of a hand-held vacuum/pressure pump you can check the integrity of the pump diaphragm and the check valves.

Attach a hand-held vacuum/pressure pump to the fuel pump inlet fitting (the fitting to which the filter/tank line connects). Using the pump apply 7 psi (50 kPa) of pressure while manually restricting the outlet fitting using your finger. If the diaphragm is in good condition it will hold the pressure for at least 10 seconds.

Next, switch to the vacuum fitting on the pump and draw 4.3 psi (30 kPa) of negative pressure (vacuum) on the fuel pump inlet fitting. This checks if the 1-way check valve in the pump remains closed. It should hold vacuum for at least 10 seconds.

Now, move the pressure pump to the fuel pump outlet fitting. This time cover the inlet fitting (and other outlet fitting, if applicable) with your finger and apply the same amount of pressure. Again, the pressure must hold for at least 10 seconds.

■ **If pressure does not hold, verify that it is not leaking past your finger (on the opposite fitting) or from a pump/hose test connection. If leakage is occurring in the diaphragm, the fuel pump should be overhauled.**

Visually Inspecting the Pump Components

◆ See Figure 193

The final way to inspect any diaphragm-displacement fuel pump is through disassembly and visual inspection. Remove and/or disassemble the pump according to the procedures found either in this section.

Wash all metal parts thoroughly in solvent, and then blow them dry with compressed air. Use care when using compressed air on the check valves. Do not hold the nozzle too close because the check valve can be damaged from an excessive blast of air.

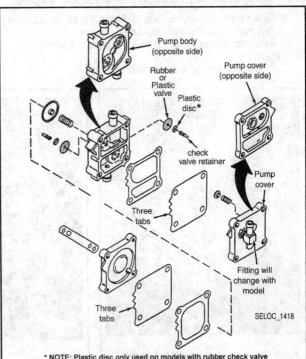

SELOC_1418

* NOTE: Plastic disc only used on models with rubber check valve

Fig. 193 Exploded view of a common square Mercury fuel pump assembly (note check valves and pump covers may vary slightly)

Inspect each part for wear and damage. Visually check the pump body/cover assembly for signs of cracks or other damage. Verify that the valve seats provide a flat contact area for the valve. Tighten all check valve connections firmly as they are replaced.

Check the diaphragms for pin holes by holding it up to the light. If pin holes are detected or if the diaphragm is not pliable, it MUST be replaced.

■ If you've come this far and are uncertain about pump condition, replace the diaphragms and check valves, you've already got to replace any gaskets or O-rings which were removed. Once the pump is rebuilt, you can remove it from your list of potential worries for quite some time.

REMOVAL & INSTALLATION

◆ See Figures 191, 194 and 195

1. Disconnect the negative battery cable for safety.
2. Cut away the Sta-Straps from the 3 hoses at the fuel pump. Tag each of the hoses to ensure proper installation. The 2 hoses at top and bottom of the pump are fuel lines (one in from tank and one out to the powerhead) and the one on the pump cover is a pulse hose from the powerhead which supplies the vacuum necessary for the pump to operate properly.
3. Disconnect the top and bottom fuel lines. Use a golf tee or a stubby pencil to plug the end of each disconnected hose to prevent the loss of fuel.
4. Disconnect the pulse hose from the front surface (cover) of the pump.
5. Observe the 4 bolts threaded through the pump cover. The bolt closest to the pulse nipple on the cover and the bolt diagonally across from it (normally both M6 bolts) are used to fasten the pump to the block. The other 2 bolts (normally M5) are used hold the pump components together. Remove only the 2 bolts securing the pump to the powerhead, then carefully lift the pump clear of the block.
6. If equipped, remove and discard the mounting gasket.

To Install:

7. Place the fuel pump on the powerhead base using a new mounting gasket (if equipped).

■ When equipped with a gasket, it is usually easiest to insert the 2 mounting screws through the pump before positioning it to the powerhead so it can hold the gasket in place as you align the pump.

8. Tighten the 2 pump retaining bolts to alternately to 55 inch lbs. (6 Nm).
9. Reconnect the fuel lines as tagged during removal.
10. Reconnect the vacuum (pulse) line to the cover of the pump assembly.
11. Reconnect the negative battery cable.

OVERHAUL

◆ See Figures 193 and 196

1. Remove the pump from the powerhead. For details, refer to the procedure located earlier in this section.
2. Lay the pump on a suitable work surface
3. Remove the screws holding the fuel pump assembly together.

■ Keep close track of the order in which the pump components are assembled in the pump housing. Compare the locations of the components to the accompanying diagram and note differences (if any) for installation purposes.

4. Carefully separate the pump cover, gaskets and diaphragms from the pump body. Always discard the used gaskets!

✳✳ SELOC CAUTION

Do not remove the check valves unless they are defective. Once removed, the valves cannot be used again. If the check valves are to be replaced, take time to Observe and remember how each valve faces, because it must be installed in exactly the same manner or the pump will not function.

5. To remove a check valve for replacement, grasp the retainer with a pair of needlenose pliers and pull the valve from the valve seat.
6. Wash all parts thoroughly in solvent and then blow them dry with compressed air. Use care when using compressed air on the check valves. Do not hold the nozzle too close because the check valve can be damaged from an excessive blast of air.
7. Inspect each part for wear and damage. Verify that the valve seats provide a flat contact area for the valve disc. Tighten all elbows and check valve connections firmly as they are replaced.

■ On models equipped with the black rubber check valves, check each disc to make sure the black coating is not coming off.

8. Test each check valve by blowing through it with your mouth. In one direction the valve should allow air to pass through. In the other direction, air should not pass through.
9. Check the diaphragm for pin holes by holding it up to the light. If pin holes are detected or if the diaphragm is not pliable, it must be replaced.

To Assemble:

The fuel pump rebuild kit will contain new gaskets, diaphragms and check valve components. Each check valve consists of a large rubber disc, a smaller plastic disc and a valve retainer or a large plastic disc and a retainer, depending on the design and kit.

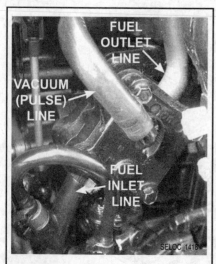

Fig. 194 Tag and disconnect the 3 hoses attached to the pump (typical Mercury pump shown)

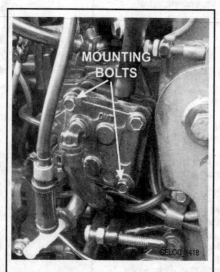

Fig. 195 Only remove the 2 bolts holding the pump to the powerhead

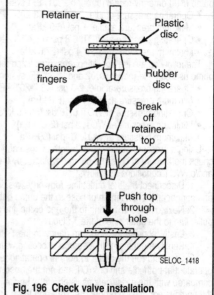

Fig. 196 Check valve installation

■ The plastic discs are more resistant to damage from fuel additives than the rubber discs. For this reason most new overhaul kits should contain the plastic discs.

10. To install the check valves and retainers, proceed as follows:

a. Insert the fingers of the retainer into the smaller plastic disc and then the larger rubber disc (or the single plastic disc, as applicable).

b. Install the disc and retainer assembly onto the fuel pump body. Push in the retainer until the collar and both discs are tightly pressed against the pump body.

c. Bend the end of the retainer from side to side until it breaks away from the collar.

d. Install the broken off end through the hole in the collar and through the disc(s). Use a hammer and tap the retainer end down into place. As this piece is forced down, it will spread the fingers of the retainer and secure the check valve within the pump body.

e. Place the larger of the 2 caps and boost springs into place on the back side of the pump body.

■ All the layered components of the fuel pump have notches which must be aligned during assembling. With the check valves in place in the pump body, you can begin final assembly of the fuel pump. Assembly will be much easier if you use two 3 in. (76mm) long 1/4 in. screws or dowels inserted through the large pump mounting bolt holes to align the pump components.

11. Insert two 1/4 in. dowels through the pump mounting holes in the pump cover, then place the cover with the interior side facing upward on the worksurface.

12. Position the coil spring and cap, then carefully place the diaphragm in position over the dowels. Make sure the 3 tabs on the diaphragm are aligned in the same direction and the single tab on the pump cover.

13. Position the gasket over the dowels, with the 2 tabs facing the same direction as the tabs on the diaphragm and cover.

14. Install the fuel pump body down over the assembly, again with the tab in proper alignment.

15. Now install the coil spring and cap in the pump body.

16. Install the gasket with the tabs aligned in the proper direction.

17. Position the diaphragm over the dowels (aligning the tabs, you should be used to that by now).

18. Lastly, position the fuel pump base (rear cover) down over the assembly.

19. Grasp the assembly firmly (holding it together against diaphragm/spring pressure), then invert the assembly so you can insert the 2 pump assembly screws.

20. Recheck the alignment of all pump components, then insert and thread the 2 pump assembly screws. Tighten the screws to 55 inch lbs. (6 Nm).

21. Install the pump assembly, as detailed earlier in this section.

Electric Fuel Pump

DESCRIPTION & OPERATION

◆ See Figure 197

On all EFI motors, an electric fuel pump is mounted inside the vapor separator tank (VST) assembly on the front, port side of the powerhead. The purpose of the electric fuel pump is to take fuel from the separator tank and to deliver it at high-pressure (much higher than the vacuum boost/lift pump could achieve). Although operating pressure will vary slightly by model, it will be somewhere in the 34-45 psi (234-310 kPa) range. The high-pressure fuel is necessary for proper injector operation, as fuel must be delivered at a constant pressure in order for the ECM to successfully meter it by varying injector on time.

The electric (booster) fuel pump is a roller-cell type and is NOT serviceable. Once opened, the airtight seal is lost and the integrity of the seal cannot be regained. The pump must be replaced.

This booster pump is unique because the pump and electric motor are housed together in one permanently sealed case, constantly surrounded with fuel. Yes, the fuel actually flows past the electric motor brushes. If the case remains absolutely airtight, there is no danger of explosion. However, if the case develops a crack or is deformed in any way, there is the distinct possibility air may enter the case and together with the fuel form an explosive mixture.

Fig. 197 Many VST mounted components can be serviced with the VST still on the powerhead

■ Fuel by itself will not ignite while exposed to a spark from the electric motor brushes, because there is no oxygen present for combustion. A mixture of 0.7 parts air to 1.3 parts fuel becomes explosive.

Pump action will commence the moment the ignition key switch is rotated to the **ON** position. If the key switch is held in this position for more than 30 continuous seconds, the ECM will cut off electrical power to the pump and of course the pumping action will cease.

✳✳ SELOC WARNING

The powerhead must never be cranked without an adequate supply of fuel to the booster pump. If the electric pump draws air into the pump case, an explosive mixture may then form. On the other hand, a fuel injection system cannot be flooded. The fuel pressure regulator will activate the return circuit and excess fuel will not flood the crankcase, as in a conventional carbureted powerhead. As explained earlier, the excess fuel will be returned to the vapor separator.

On early-model EFI systems fuel travels both directions to the fuel rail and excess fuel returns to the tank. However, on late-model EFI systems fuel only travels one direction toward the rail, while excess pressure is redirected back into the tank before it leaves the tank/pump assembly.

TESTING

◆ See Figures 197 and 198

The EFI electric fuel pump will energize to prime the fuel system any time the ignition key is turned to **ON** or **RUN** (as opposed to **START**). If the engine is not cranked and started the pump will cease to run, but the fact that it will run briefly should be enough to both test the delivery pressure of the pump as well as verify that the circuit is working.

■ If the pump does not energize, grab a test light or a DVOM and troubleshoot the pump circuit. Start at the pump connector and work your way back. First make sure the connector is clean and tight, if that doesn't work, use the test light or meter to check for power.

The amount of time the pump will energize when the key is first turned to the **ON/RUN** position will vary slightly. It seems from Mercury literature that the early-model systems (through 2001) may energize for as long as 30 seconds, however on late-model systems (2002 or later) the pump should only energize for 2-4 seconds.

If you listen each time the key is turned to **ON/RUN**, you will actually hear the pump run. If you hear it, but suspect a fuel delivery problem the next step

Fig. 198 Attach a pressure gauge to the pressure test port - late-model EFI shown

is to connect a fuel pressure gauge and check the pump delivery pressure to make sure it is within spec as follows:

1. Obtain a pressure gauge capable of registering accurately from at least 30-50 psi (207-345 kPa).

2. Locate the fuel pressure test port, usually found on or near the top of the vapor separator tank assembly. Remove the plastic cap from the pressure port.

3. Connect the gauge to the pressure port.

4. Rotate the ignition key to the **ON/RUN** position without cranking the motor and allow the system to pressurize. If necessary (especially on late-model EFI systems) cycle the ignition key back Off for a few seconds, then On again allowing the pump to run for a second pressurization cycle. Note the reading on the gauge. Normal powerhead operation requires pressure of:

- On early-model (through 2001) EFI systems - 34-36 psi (234-248 kPa).
- On late-model (2002 and later) EFI systems - 41-45 psi (283-310 kPa).

5. If the reading is low, the cause may be a restriction in a fuel line, a problem with fuel delivery to the vapor tank, a fuel pump failure, possibly even a clogged fuel filter.

6. If fuel pressure is above normal, check the pressure regulator.

7. If there is no reading, the pump is defective and must be replaced.

On late-model systems, if pressure is low and no causes can be found Mercury recommends performing an amperage draw test on the electric fuel pump. Using an ammeter check the pump draw, it should be 3.5-4.5 amps for 2.5L motors or 5-6 amps for 3.0L motors. If amperage draw is within spec, no restrictions or low-pressure fuel delivery problems can be found, but pressure is still too low, replace the regulator.

Also on late-model systems, Mercury publishes a pump motor resistance specification. The pump should show 0.4-1.0 ohms resistance across when checked with an ohmmeter.

REMOVAL & INSTALLATION

◆ See Figure 197

The electric fuel pump is mounted to the underside of the vapor separator tank (VST) Assembly cover. Although it may be possible to unbolt and remove the cover with the tank assembly still installed, you'll still have to disconnect almost all of the same wires and hoses (since most are attached to cover mounted components in the first place), so it is probably easier just to remove the whole vapor separator tank assembly. Either way, for details, please refer to Vapor Separator Tank procedures in this section.

Vapor Separator Tank (VST) Assembly

DESCRIPTION & OPERATION

◆ See Figure 197

The vapor separator tank (VST) assembly is a major component of the EFI system. The tank itself serves multiple purposes, though its primary

purpose is to provide a fuel reservoir from which the electric high-pressure fuel pump can draw an un-interrupted supply of fuel to feed the injectors.

The tank also serves as a mounting point for multiple EFI system components including:
- The electric fuel pump
- The fuel pressure regulator
- The fuel pressure test port (all except early-model 3.0L motors through 2001)
- The MAP Sensor (early-model 3.0L motors through 2001)

On late-model EFI systems the tank also serves as a water separating point.

REMOVAL & INSTALLATION

◆ See Figures 197 and 199 thru 202

Most VST mounted components CAN be serviced with the tank assembly still installed on the powerhead. Components like the fuel pressure regulator or, on early-model 3.0L motors, the MAP sensor can easily be unbolted and removed. Components like the electric fuel pump assembly are a little more involved and would require some of the hoses and wiring to be disconnected from the VST cover so the cover can be unbolted and lifted up sufficiently for access to the pump in the tank. At that point, it is probably just easier to remove the whole VST assembly, but the choice remains yours.

This procedure includes the removal and installation of the complete of the VST assembly. If you wish to just remove one of the components, follow only the necessary portions of the procedure, along with the appropriate part of the overhaul procedure found later in this section.

✷✷ SELOC CAUTION

Before the electric fuel pump, vapor separator tank or the fuel pressure regulator is removed, the EFI system must be depressurized.

1. Release the fuel system pressure, as detailed earlier in this section.

2. Disconnect the negative battery cable for safety.

3. Remove the plug from the bottom of the VST assembly and drain the fuel into a suitable container.

4. On early-model EFI systems (through 2001) tag and disconnect the following components from VST assembly, as follows:

a. If necessary for additional clearance, remove the oil reservoir from the powerhead.

b. On 3.0L motors disconnect the MAP sensor wiring harness and vacuum hose towards the rear top of the VST.

c. Disconnect the VST bleed hose (usually only on 3.0L models) and the VST vacuum hose both toward the center, top of the VST.

d. Disconnect the vacuum hose from the top center of the pressure regulator.

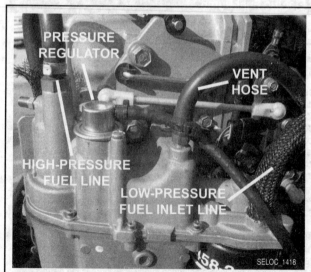

Fig. 199 Tag and disconnect hoses and wiring from the VST - late-model shown

e. Disconnect the main fuel supply line from the rear of the VST assembly. It's a good idea to plug this line to prevent system contamination.

f. Disconnect the oil inlet hose from the VST. Definitely plug this line to prevent an oily mess and to prevent system contamination.

g. Tag and disconnect the fuel pump wiring from the terminals at the front top of the VST.

h. Although you CAN cut the clamps and disconnect the high-pressure inlet/outlet fuel lines from the VST, you'll need a special crimping tool to seal them upon installation. The easier and more recommended method is to

Fig. 200 Don't forget the water sensor connector on late-model units

remove the screws (usually 2) securing the inlet/outlet fuel manifold adapter plate assembly to the side of the intake manifold, this way the manifold and O-ring sealed tubes can simply be pulled out of the fuel joint with the hoses still attached. The O-rings can be inspected and easily replaced, if necessary, upon installation.

■ On late-model EFI systems there is no way around the need to disconnect some of the high pressure fuel lines, therefore it will be necessary for you to obtain a suitable crimp tool in order to put new steel crimped clamps on the fuel lines (well, at least on the fuel output hose).

5. On late-model EFI systems (2002 and later) tag and disconnect the following components from VST assembly, as follows:

a. Disconnect the VST vent hose from the vertical fitting at the top center of the VST cover.

b. Disconnect the pressure regulator vacuum hose from the horizontal fitting on the regulator itself.

c. Disconnect the VST fuel output hose from the top front of the cover, centered over the electric fuel pump.

d. Disconnect the oil input hose which is routed up the front of the assembly to the powerhead side of the cover.

e. Unplug the wiring harness for the electric fuel pump, positioned vertically along the front of the VST.

f. Unplug the water sensor bullet connector from the rear bottom of the VST.

g. Disconnect the fuel supply hose from the top rear of the VST.

The vapor separator and fuel regulator assembly is now disconnected from the system ready for removal from the powerhead.

6. Remove the 3 attaching bolts (2 at the top and 1 at the bottom for early-model EFI or 2 at the bottom and 1 at the top for late-model EFI) and then remove the assembly from the powerhead.

7. If necessary, refer to Overhaul in this section for details on VST disassembly and assembly.

To Install:

8. If removed for clearance, install the oil reservoir to the powerhead.

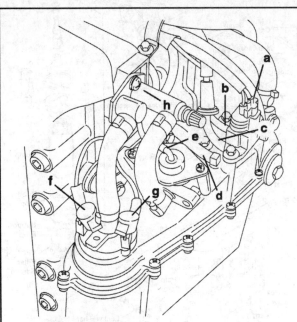

a - MAP Sensor Harness
b - MAP Sensor Manifold Hose
c - Separator Bleed Hose
d - Separator Vacuum Hose
e - Pressure Regulator Manifold Hose
f - POSITIVE (+) Lead
g - NEGATIVE (–) Lead
h - 2 Screws (1 hidden)

SELOC_1418

Fig. 201 Major VST hose and wiring connections - early-model EFI units (3.0L shown, 2.5L similar but no MAP sensor on VST)

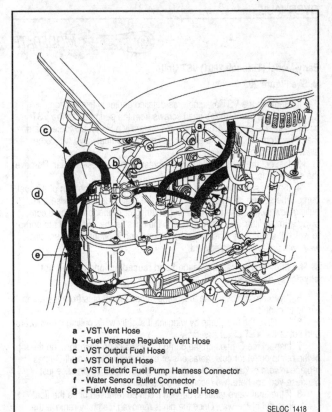

a - VST Vent Hose
b - Fuel Pressure Regulator Vent Hose
c - VST Output Fuel Hose
d - VST Oil Input Hose
e - VST Electric Fuel Pump Harness Connector
f - Water Sensor Bullet Connector
g - Fuel/Water Separator Input Fuel Hose

SELOC_1418

Fig. 202 Major VST hose and wiring connections - late-model EFI units

9. On early-models inspect the O-rings on the ends of the fuel joint-to-fuel elbow tubes and the O-rings on the intake manifold adaptor plate and replace, as necessary. Install the tubes into place in the joint/intake manifold assembly.

10. Position the assembled vapor separator tank assembly against the powerhead, then loosely install the VST retaining bolts.

11. Again, for early-models, position the inlet/outlet fuel manifold adapter plate to the tubes and to the mating surface on the intake manifold. Thread the 2 retaining screws and tighten to 45 inch lbs. (5 Nm).

12. Tighten the VST assembly retaining bolts to 45 inch lbs. (5 Nm) for early-models or to 140 inch lbs. (16 Nm) for late-models.

13. Reconnect the wiring and hoses as tagged during removal. If a high-pressure fuel line was disconnected, be sure to secure it using a new metal crimp-style clamp using an appropriate crimping tool.

✳✳ SELOC WARNING

Remember, the ONE place you really don't want a fuel leak is on the high-pressure fuel circuit. A leak could cause an explosive spray of fuel/fuel vapors under the hood of the outboard.

14. Reconnect the negative battery cable, then prime the low-pressure and high pressure fuel systems in order to check for leaks. The low-pressure system is easily primed manually using the primer bulb. Once the primer bulb is firm the VST assembly should be full and you can cycle the ignition On for a few seconds, the Off and then On again in order to energize the fuel pump and allow the high-pressure system to build pressure.

15. Provide a cooling source of water, then start and run the engine to check the repair and again inspect for leaks.

✳✳ SELOC WARNING

Watch for leaks around all fuel hose connections which were disturbed. Because the system is pressurized, a leak will not appear as drips. Instead fuel will spray all over the powerhead. Should this occur, shut down the powerhead immediately. Do not forget to depressurize the system at the pressure port again before making necessary repairs.

OVERHAUL

 ②　MODERATE

Early-Model (Through 2001) VST Units

◆ See Figure 203

1. Remove the VST assembly, as detailed earlier in this section.

2. Remove the 7 Philips head screws from the perimeter of the VST cover. Once the screws are all removed, carefully pull upward to separate the cover from the tank. The electric fuel pump and float valve will come away attached to the cover.

3. Inspect the cover seal for cuts, abrasions or excessive wear. Remove the seal if replacement is necessary.

4. If necessary, remove the final filter screen from the bottom of the fuel pump. The screen is removed by rotating the filter counterclockwise and pulling downward. Take note if the support ring or rubber pad should come away with the filter. The pad aligns with the boss on the bottom of the pump, to which the filter connects, while the support ring is oriented in a specific way, with the tabs facing upward.

■ The filter screen can be cleaned and reused as long as it is not physically damaged.

5. If you are removing the fuel pump remove the 2 nuts from the pump wiring terminals.

6. Remove the fuel pump by wiggling it slightly while pulling it downward and out of the VST cover assembly.

7. Inspect the O-rings on the pump terminals (2 on each) and on the top of the pump outlet for cuts, abrasions or excessive wear. Like all O-rings, replace if worn or damaged, or replace for good workshop practice just because you are here (we recommend the later).

8. If the float valve requires service it can be removed once the float pivot pin is out of the way. Once the pin is removed, carefully remove the float and needle for inspection or replacement. If the needle seat is damaged the entire separator cover must be replaced. Make sure the float itself has not absorbed any fuel and is not deteriorating.

✳✳ SELOC CAUTION

DO NOT make any attempts to adjust the float height or the float limit bracket. The float height is set at the factory and no adjustment is necessary.

To Assemble:

9. If removed, attach the float needle to the arm, then guide the float and needle assembly into position under the float drop limit bracket, placing the needle into the seat.

10. Secure the pivot pin through the float arms, then check to make sure the float moves freely through the pivot range.

11. With the O-rings in place, carefully insert the pump into the underside of the VST cover. Because the wire terminals are different diameters it should only fit one way. Tighten the nuts to secure the pump to the cover. The Positive terminal nut should be tightened to 6 inch lbs. (0.7 Nm), while the Negative terminal nut should be tightened to 16 inch lbs.(1.8 Nm) for 2.5L motors or 8 inch lbs.(0.9 Nm) for 3.0L motors.

12. If removed install the rubber pad to the bottom of the pump, so it fits flush on the bottom of the pump, with the cutout positioned over the lower pump boss for the pick-up filter screen.

13. If removed, install the pump support ring, which also fits only one way, with the tabs facing upward.

14. If removed, install the final filter by pushing upward and then rotating the filter clockwise to lock it onto the pump.

15. Make sure the VST cover-to-tank seal is in position. Sometimes after use the O-ring material will swell duel to fuel and air exposure and will not fit back into the tank groove. If so, replace the seal.

✳✳ SELOC WARNING

NEVER, EVER, EVER cut the seal in an effort to make it fit better. Dangerous fuel leaks could be caused by such an action.

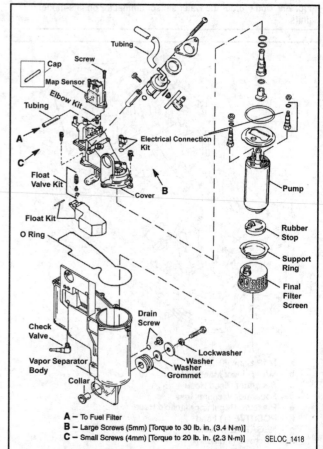

Fig. 203 Exploded view of the early-model VST assembly - 3.0L shown (2.5L very similar)

7. Installation is essentially the reverse of removal, making sure to securely tighten the retaining screws.

8. Reconnect the negative battery cable, then prime the low-pressure and high pressure fuel systems in order to check for leaks. The low-pressure system is easily primed manually using the primer bulb. Once the primer bulb is firm the VST assembly should be full and you can cycle the ignition On for a few seconds, the Off and then On again in order to energize the fuel pump and allow the high-pressure system to build pressure.

9. Provide a cooling source of water, then start and run the engine to check the repair and again inspect for leaks.

✳✳ SELOC WARNING

Watch for leaks around all fuel hose connections which were disturbed. Because the system is pressurized, a leak will not appear as drips. Instead fuel will spray all over the powerhead. Should this occur, shut down the powerhead immediately. Do not forget to depressurize the system at the pressure port again before making necessary repairs.

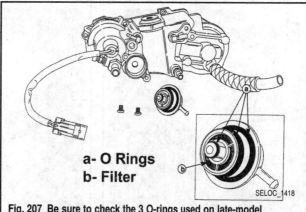

**a- O Rings
b- Filter**

SELOC_1418

Fig. 207 Be sure to check the 3 O-rings used on late-model pressure regulators

OPTIMAX DIRECT FUEL INJECTION (DFI)

Description & Operation

◆ See Figures 208 thru 212

Combustion air enters the cowl through holes located in the top aft end of the cowl. The cowl liner on most models then directs air to the bottom of the powerhead. This limits the exposure of salt air to the components inside the engine cowl.

■ **On 2.5L models the intake air routing in the cowl includes an area where excess moisture is dispersed out of the rear of the cowling without even entering the engine compartment to further limit exposure of the powerhead to salt.**

Once inside the cowl, the air enters the plenum through the throttle shutter which is located in the plenum assembly. The air then continues through the reed valves and into the crankcase. The throttle shutter is actuated by the throttle shaft. Mounted on a separate shaft is the throttle position sensor (TPS). This sensor tells the engine control module (ECM) the position of the throttle.

■ **On all 1.5L motors and some V6 models (specifically some 2003 and later models that are equipped with digital throttle and shift) are equipped with a computer control unit which is called the propulsion control module (PCM) instead of ECM. Function of the PCM is identical to that of an ECM, except that it also controls the digital throttle and shift functions. For this reason, throughout this repair guide any reference to an ECM is meant to mean PCM as well, unless otherwise noted.**

On most models, if the TPS should fail, the dash mounted check engine light will flash and the warning horn will sound. Generally the engine speed will be Reduced by 20-25% and the ECM will reference the MAP sensor for fuel calibration.

Air from inside the engine cowl is drawn into the compressor through the flywheel cover (3.0L models through 2002 or all 2.5L models) or through an air filter in the air attenuator cover (1.5L models and 2003 or later 3.0L models). In both cases, the cover/attenuator assembly acts like a muffler to quiet compressor noise and contains a filter to prevent the ingestion of debris into the compressor.

The compressor is driven by a serpentine belt from a pulley mounted on the flywheel and is automatically self adjusting using a single idler pulley. This air compressor is a single cylinder unit containing a connecting rod, piston, rings, bearings, reed valves and a crankshaft. The compressor is water cooled to lower the temperature of the air charge and is lubricated by oil from the engine oil pump assembly.

As the compressor piston moves downward inside the cylinder, air is pulled through the filter, reed valves and into the cylinder. After the compressor piston change direction, the intake reeds close and the exhaust reeds open allowing the compressed air into the hose leading to the air/fuel rails.

The air fuel rails contain 2 passages: the first is for fuel, the second is for air. The air passage is common between all the cylinders included in the rail. A hose connects the rail air passage (starboard on V6 motors) to the air compressor. On V6 motors, another hose connects the starboard air rail passage to the port air rail passage.

An air pressure regulator will limit the amount of pressure developed inside the air passages to some level below the pressure of the fuel inside the fuel passages, although the exact pressures vary by model as follows:

• On 75-115 hp 1.5L motors the regulator maintains air pressure at about 14 psi (97 kPa) below the pressure of the fuel meaning about 94 psi (648 kPa) air vs. 108 psi (745 kPa) fuel.

• On 135-175 hp 2.5L motors the regulator maintains air pressure at about 10 psi (69 kPa) below the pressure of the fuel meaning about 80 psi (552 kPa) air vs. 90 psi (621 kPa) fuel.

• On conventional 200-250 hp 3.0L motors the regulator maintains air pressure at about 10.5 psi (72 kPa) below the pressure of the fuel meaning about 80 psi (552 kPa) air vs. 90.5 psi (624 kPa) fuel.

• On Pro/XS/Sport 225/250 hp 3.0L motors the regulator maintains air pressure to 15 psi (103 kPa) below the pressure of the fuel meaning about 91-95 psi (627-655 kPa) air vs. 106-110 psi (627-655 kPa) fuel.

Air exiting the pressure regulator is returned into a muffler located in the cowling for 75-115 hp 1.5L motors, the exhaust adaptor and exits through the propeller on the 135-175 hp 2.5L motors, or returns to the air plenum on the 200-250 hp 3.0L motors.

SELOC_1418

Fig. 208 Typical OptiMax fuel rail assembly

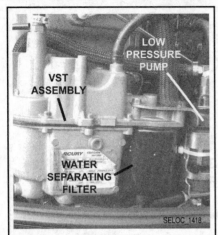

Fig. 209 Typical Mercury vapor separator tank (VST) assembly (with water separating filter and, this one also has an external low-pressure fuel pump)

Fig. 210 Engine mounted oil reservoir on the 115 hp engine . . .

Fig. 211 . . . and the remote oil tank mounted beneath the transom

Fig. 212 Electrically operated OptiMax solenoid design oil pump assembly

Fuel for the engine is stored in a typical tank. A primer bulb is installed into the fuel line to allow the priming of the fuel system. A crankcase (most models) or a vapor separator tank (2003 or later 1.5L and 3.0L models) mounted pulse driven diaphragm fuel pump draws fuel through the fuel line, primer bulb, fuel pump assembly and then pushes the fuel through a water separating fuel filter. This filter removes any contaminates and water before fuel reaches the vapor separator.

Fuel vapors are vented through a hose into the air compressor inlet in the front of the flywheel cover for most models, except for 1.5L motors on which the vent is run through the bottom cowling to the atmosphere and on 2003 or later 3.0L models where the vent runs to the adaptor plate.

The electric fuel pump is different than the fuel pump that is utilized on the standard EFI engine (non DFI) as it must be capable of much higher fuel pressure. The DFI electric pumps are capable of developing fuel pressures in excess of the 90-110 psi (621-758 kPa) pressures necessary to maintain fuel pressure above that of air pressure inside the air rail, again the exact specifications varying with model as noted earlier.

Fuel from the vapor separator is supplied to the fuel rail(s) as follows:
• On 75-115 hp 1.5L motors, the fuel goes to the top of the fuel rail.
• On 135-175 hp 2.5L motors, the fuel is supplied first to the bottom of the Starboard fuel rail.
• On 200-250 hp 3.0L motors, the fuel is supplied first to the bottom of the Port fuel rail.

On V6 models, a fuel line connects the bottom of the first rail to the opposite fuel rail. On all motors, fuel is stored inside the rail until an injector opens.

A fuel pressure regulator controls pressure in the fuel rails and allows excess fuel to return into the vapor separator. The fuel regulator not only controls fuel pressure but maintains it at approximately 10-15 psi (69-103 kPa) higher than whatever the air rail pressure is (depending upon the model). The fuel regulator diaphragm is held closed with a spring that requires that amount of force (from 10-15 psi/69-103 kPa) in order to move the diaphragm off the diaphragm seat. The back side of the diaphragm is exposed to air rail pressure. As the air rail pressure increases, the fuel pressure needed to open the regulator will equally increase.

For example, looking at a typical 2.5L motor (which requires 10 psi more fuel pressure than air) if there is 50 psi of air pressure on the rail side of the diaphragm, 60 psi of the pressure will be required to open the regulator. The math isn't as neat for the other models, like the 1.5L motor (which requires 14 psi fuel pressure than air), but the concept is the same. On the 1.5L, if there is 50 psi of air pressure on the rail side of the diaphragm, 64 psi of the pressure will be required to open the regulator.

On 1.5L motors there is only 1 fuel rail, and it is water cooled. On V6 motors, the port fuel rail is water cooled. Also on V6 motors, in order to equalize the pulses generated by the pumps (both air and fuel) a tracker diaphragm is installed in the starboard rail. The tracker diaphragm is positioned between the fuel and air passages. The tracker diaphragm is a rubber diaphragm which expands and retracts depending upon which side of the diaphragm senses the pressure increase (pulse).

Oil in these motors is not mixed with fuel before entering the combustion chamber. On 1.5L motors oil is stored inside an engine mounted reservoir, however a remote reservoir option is available for these motors.

On all V6s and on 1.5L motors so equipped, oil is stored in a remote oil reservoir located somewhere near the transom of the boat. Crankcase pressure will force oil from the remote oil tank into the oil reservoir on the side (V6) or front (1.5L motors) of the powerhead. Oil will flow from the oil reservoir into the oil pump.

The oil pump is a solenoid design. It is activated by the ECM and includes 7 pistons with corresponding discharge ports. The oil pump is mounted directly onto the powerhead. Each cylinder is lubricated by 1 of the discharge ports. The oil is discharged into the crankcase. The seventh passage connects to the hose that leads to the air compressor for lubrication. Excess oil from the compressor returns into the cylinder bores on 1.5L motors, is returned to the main bearings on 2.5L motors or returned to the plenum and is ingested through the crankcase on 3.0L motors.

The ECM will change the discharge rate of the oil pump, depending upon engine demand. Therefore the oil ratio varies with engine rpm and load. The ECM will also pulse the pump on initial start up to fill the oil passages eliminating the manually need to bleed the oil system. The ECM provides additional oil for break in, as determined by its internal clock, but the break-in clock may be reset in case the powerhead is overhauled or replaced.

Generally speaking, the electrical system consists of:
• Engine Control Module (ECM)

- Air Temperature Sensor (ATS)
- Crankshaft Position Sensor (CPS, which monitors flywheel speed & crankshaft position)
- Throttle Position Sensor (TPS)
- Manifold Absolute Pressure (MAP) sensor
- Engine Temperature Sensor (ETS)
- Water Temperature Sensor (WTS)
- Compressor Temperature Sensor (CTS)
- Tracker Valve (V6 models)
- Fuel Pressure Regulator
- Air Pressure Regulator
- Air Compressor assembly
- Ignition Coils
- Fuel Injectors (fuel & direct)
- Shift Interrupt Switch

The engine requires a battery to start (the ignition and injection will not occur if the battery is dead). However, the system will run off the alternator.

The operation of the system happens in milliseconds (ms) exact timing is critical for engine performance. As the crankshaft rotates, air is drawn into the crankcase through the throttle shutter, into the plenum and through the reed valves.

As the piston nears bottom dead center (BDC), air from the crankcase is forced through the transfer system into the cylinder. As the crankshaft continues to rotate, the exhaust and intake ports close. With these ports closed, fuel can be injected into the cylinder. The ECM will receive a signal from the throttle position sensor, engine temperature sensor and the crank position sensor (flywheel speed and position sensor). With this information the ECM refers to its internal fuel calibration (maps) to determine when to activate (open and close) the injectors and fire the ignition coils (therefore precisely controlling the amount of fuel and the ignition timing).

With the piston in the correct position, the ECM opens the fuel injector, 90-110 psi (621-758 kPa) of fuel (depending upon the model) is discharged into the machined cavity inside the air chamber of the air/fuel rail. This mixes the fuel with the air charge. Next, the direct injector will open, discharging the air/fuel mixture into the combustion chamber. The direct injector directs the mixture at the bowl located in the top of the piston. The pistons bowl directs the air/fuel mixture into the center of the combustion chamber. This air/fuel mixture is then ignited by the spark plug.

To aid in starting when the air rail pressure is low and before the compressor has time to build pressure, some direct injectors are held open by the ECM. This allows the compression from inside the cylinders to pressurize the air rail faster (1 or 2 strokes or 60°of crankshaft rotation).

Idle quality is controlled by fuel volume and fuel timing. The throttle shutter will be open at idle speed. On V6 motors the shift cut-out switch will interrupt the fuel to 3 of the cylinders to assist in shifting. On 1.5L motors the shift switch, mounted on the exhaust adaptor plate is activated by a notch in the upper shift shaft. The switch has a closed circuit in gear and open when in neutral. The PCM uses this signal to change fuel delivery amounts and ignition timing to limit powerhead rpm.

The TPS signals the ECM to change the fuel and spark without movement of the throttle shutters. The throttle cam is manufactured to allow the TPS shaft to move before opening the throttle shutter.

Troubleshooting the OptiMax Fuel Injection System

BASIC DIAGNOSIS & PRELIMINARY TESTS

Diagnosis of electronic fuel injection is generally based on symptom diagnosis and stored fault codes. Diagnosis of a running problem or as it is commonly called "driveability problem" requires attention to detail and following the diagnostic procedures in the correct order. Resist the temptation to begin extensive testing before completing the preliminary diagnostic steps. A preliminary or visual inspection must be completed in detail before diagnosis begins. In many cases this will shorten diagnostic time and often cure the problem without the need for involved electronic testing.

■ **The problem with checking for trouble-codes on these systems is that a scan tool, such as the Digital Diagnostic Terminal (DDT) is necessary in order to communicate with the ECM.**

There are 2 basic ways to check your fuel system for problems. They are by symptom diagnosis and by the on-board computer self-diagnostic system. The first place to start is always the preliminary inspection. Intermittent problems are the most difficult to locate. If the problem is not present at the time you are testing you may not be able to locate the fault.

The visual inspection of all components is possibly the most critical step of diagnosis. A detailed examination of connectors, wiring and vacuum hoses can often lead to a repair without further diagnosis.

Also, take into consideration if the powerhead has been serviced recently. Sometimes things get reconnected in the wrong place or not at all. A careful inspector will check the undersides of hoses as well as the integrity of hard-to-reach hoses blocked by the air silencer or other components. Wiring should be checked carefully for any sign of strain, burning, crimping or terminals pulled from their connectors.

Checking connectors at components or in harnesses is required, usually, pushing them together will reveal a loose fit. Also, check electrical connectors for corroded, bent, damaged, improperly seated pins and bad wire crimps to terminals. Pay particular attention to ground circuits, making sure they are not loose or corroded. Any component or wiring in the vicinity of a fluid leak or spillage should be given extra attention during inspection. Remember how corrosive an environment it is that an outboard engine operates in. Salt spray is especially corrosive to electrical connectors.

Additionally, inspect maintenance items such as belt condition and tension and the battery charge and condition. Any of these very simple items may affect the system enough to set a fault code.

When diagnosing by symptom, the first step is to find out if the problem really exists. This may sound like a waste of time but you must be able to recreate the problem before you begin testing. This is called an "operational check". Each operational check will give either a positive or negative answer (symptom). A positive answer is found when the check gives a positive result (the instruments function when you turn the key). A negative answer is found when the check gives a negative result (the instruments do not function when you turn the key). After performing several operational checks, a pattern may develop. This pattern is used in the next step of diagnosis to determine related symptoms.

In order to determine related symptoms, perform operational checks on circuits related to the problem circuit (the instruments and the tilt and trim system do not work). These checks can be made without the use of any test equipment. Simply follow the wires in the wiring harness or, if available, obtain a copy of your engine's specific wiring diagram. If you see that the instruments and the tilt and trim system are on the same circuit, first check the instruments to see if they work. Then check the tilt and trim system. If the neither the instruments or tilt and trim system work, this tells you that there is a problem in that circuit. Perform additional operational checks on that circuit and compile a list of symptoms.

When analyzing your answers, a defect will always lie between a check which gave a positive answer and one which gave a negative answer. Look at your list of symptoms and try to determine probable areas to test. If you get negative answers on related circuits, then maybe the problem is at the common junction. After you have determined what the symptoms are and where you are going to look for defects, develop a plan for isolating the trouble. Ask a knowledgeable mechanic which components frequently fail on your vehicle. Also notice which parts or components are easiest to reach and how can you accomplish the most by doing the least amount of checks.

A common way of diagnosis is to use the split-in-half technique. Each test that is made essentially splits the trouble area in half. By performing this technique several times the area where a problem is located becomes smaller and smaller until the problem can be isolated in a single wire or component. This area is most commonly between the 2 closest checkpoints that produced a negative answer and a positive answer.

After the problem is located, perform the repair procedure. This may involve replacing a component, repairing a component or damaged wire or making an adjustment.

■ **Never assume a component is defective until you have thoroughly tested it.**

The final step is to make sure the complaint is corrected. Remember that the symptoms that you uncover may lead to several problems that require separate repairs. Repeat the diagnosis and test procedures repeatedly until all negative symptoms are corrected.

Two simple tests, pressure and volume, can easily verify the mechanical integrity of the powerheads fuel supply and return systems. Problems within the fuel supply or return system can cause many driveability problems because the control system can't compensate for a mechanical problem. Nor can the system adjust the air/fuel mixture when the system is in open loop or back-up mode. Moreover, a control system's parameters are programmed based on a set flow rate and system pressure. Therefore, even if the computer were adjusting the air/fuel mixture, the computer's calculations would be wrong whenever the system pressure or volume was incorrect.

■ **When testing fuel system pressure, always use a pressure gauge that is capable of handling the system pressure. Old style vacuum/pressure gauges will not work with most fuel injection systems**

The fuel system's pressure can be thoroughly verified by performing both static and dynamic tests. Dynamic tests are used to determine the system's overall operating condition while the engine is running. For example, if the pressure is low, the problem is most likely to be in the supply or high side of the system. Likewise, if the fuel pressure is high, the problem is going to be on the return or low side of the fuel system. All pressure tests should be taken on the high or supply side of the fuel system.

A static pressure test should also be performed. This test can be performed before or after a dynamic test with the engine off. These tests are useful for determining the sealing capacity of the system. If the fuel system leaks down after the engine is shut off, the vehicle will experience extended crank times while the system builds up enough fuel pressure to start. If the system is prone to losing residual pressure too quickly, it also usually suffers from low fuel pressure during the dynamic test. In addition to verifying that the system is holding residual pressure, a static test can also verify that the system is receiving the initial priming pulse at startup.

However, don't assume that the correct amount of fuel is being delivered just because the system pressure is within specifications. The only way to measure fuel volume is to open the fuel system into a graduated container and measure the fuel output while cranking the engine. Some manufacturers specify that a certain amount of fuel be dispersed over a short period of time but generally the pump should be able to supply at least 1.5 oz. (45 ml.) for approximately 2 seconds. If the fuel output is below the recommended specification, there is a restriction in the fuel supply or the fuel pump is weak.

In summary, pressure and volume tests provide information about the overall condition of the fuel system. The information obtained from these tests will set the direction and path of your diagnosis.

When troubleshooting the fuel injection system, always remember to make absolutely sure that the problem is with the fuel injection system and not another component. This particular system has very sophisticated electronic controls, which in most cases, can only be diagnosed and repaired by an authorized dealer who has the expensive diagnostic equipment.

There are some basic troubleshooting procedures that can be done before delving into the electronics:

Many times fuel system troubles are caused by a plugged fuel filter, a defective fuel pump or by a leak in the line from the fuel tank to the fuel pump. A defective choke may also cause problems. Would you believe, a majority of starting troubles which are traced to the fuel system are the result of an empty fuel tank or aged sour fuel.

Under average conditions (temperate climates), fuel will begin to break down in about 4 months. A gummy substance forms in the bottom of the fuel tank and in other areas. The filter screen between the tank and the carburetor and small passages in the carburetor will become clogged. The gasoline will begin to give off an odor similar to rotten eggs. Such a condition can cause the owner much frustration, time in cleaning components and the expense of replacement or overhaul parts for the carburetor.

Even with the high price of fuel, removing gasoline that has been standing unused over a long period of time is still the easiest and least expensive preventative maintenance possible. In most cases, this old gas can be used without harmful effects.

The gasoline preservative will keep the fuel fresh for up to twelve months. If this particular product is not available in your area, other similar additives are produced under various trade names.

Fuel injected engines don't actually have a choke circuit in the fuel system. These engines rely on the ECM to control the air/fuel mixture and injector function according to the electronic signals sent by a host of sensors and monitors. If a fuel injected engine suffers a driveability problem and it can be traced to the fuel injection system, then the sensors and computer-controlled output components must be tested and repaired or replaced.

TESTING W/O THE DIGITAL DIAGNOSTIC TERMINAL (DDT)

Troubleshooting without the Digital Diagnostic Terminal (DDT) or equivalent tester is limited in many cases to checking resistance on some of the sensors. Typical failures usually do not involve the ECM. Connectors, set-up and mechanical wear are usually most likely to fail. The OptiMax models can be much more difficult to check without a scan tool such as the DDT.

1. Make sure that all spark plug wires are securely installed all the way on the coil towers.

2. The engine may not run or may not run above idle if the wrong spark plugs are installed.

3. Swap ignition coils to see if the problem follows that particular coil or if the problem stays with original cylinder.

■ **The ECM is capable of performing a cylinder misfire test to isolate problem cylinders. Once a suspect cylinder is located, an output load test on the ignition coil, fuel injector and direct injector may be initiated through the use of the DDT.**

4. Any sensor or connection can be disconnected and reconnected while the engine is running without damaging the ECM. Disconnecting the crank position sensor will stop the engine.

■ **Any sensor that is disconnected while the engine is running will be recorded as a Fault in the ECM Fault History. Use the DDT to view and clear the fault history when the repair or troubleshooting is finished.**

5. If all cylinders exhibit similar symptoms, the problem is with a sensor or harness input to the ECM.

6. If the problem is speed related or intermittent, it is probably connector or contact related. Inspect the connectors for corrosion, loose wires or loose pins. Secure the connector seating by using dielectric compound.

7. Inspect the harness for obvious damage, such as, pinched wires and chaffing.

8. Secure grounds and all connections involving ring terminals. Use Liquid Neoprene to seal the terminals.

9. Check fuel pump connections fuel pump pressure.

10. Check air compressor pressure.

TESTING WITH THE DIGITAL DIAGNOSTIC TERMINAL (DDT)

◆ **See Figures 213 and 214**

Locate the DDT interface connector, normally found on the side of the powerhead, near the ECM. If necessary, refer to the wiring diagrams and use the wire colors to help identify the connector.

Attach the diagnostic cable to the ECM diagnostic connector and plug in the software cartridge. You will then be able to monitor the sensors and ECM data values including status switches.

The ECM program can help diagnose intermittent engine problems. It will record the state of the engine sensors and switches for a period of time and then can be played back to review the recorded information.

Various versions of the DDT software are available and the tool menus differ slightly with each version. However all versions should allow both an Auto Self Test and a Manual Test function. During the Manual Test there should be an option for Fault Status or Status Switches which will allow the DDT to illuminate one or more of the 8 built-in test lights should a fault be present.

The DDT fault/status lights each have a specific meaning. Numbering them from 1-8, arranged top-to-bottom, left side of the tool first, then right side, the meanings are as follows:

- 1 - Ignition
- 2 - Injector
- 3 - Pump

Fig. 213 All OptiMax models have a diagnostic connector to interface with the DDT (though many late-model connectors are smaller)

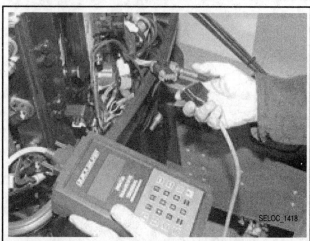

Fig. 214 Connect the DDT to the onboard ECM diagnostic cable to troubleshoot the system

- 4 - Sensors
- 5 - Switches
- 6 - Miscellaneous
- 7 - RPM Limit
- 8 - Break-In

SELF-DIAGNOSTIC SYSTEM

Mercury recommends the use of the Quicksilver Digital Diagnostic Terminal (DDT) for use with most EFI testing and troubleshooting. The DDT is a scan tool designed specifically to interact with the Mercury EFI and Optimax ECMs. As such it is capable of functions such as:

- Data Monitoring (real time feedback from engine sensors)
- Fault Status (reporting of ECM controlled or monitored circuits whose values have gone beyond the expected operating parameters)
- System Info (an archival of expected data values from ECM monitored and controlled systems)
- History (the ability to read or clear stored fault data in the ECMs memory)
- Special Functions (a number of tests or functions that the ECM can perform on demand when told by the DDT to do so, such as Oil Pump Prime or Resetting Break-In Oil delivery or Cylinder Misfire Test).

In some cases, aftermarket software may be available to mimic some or all of the functions of the DDT by using your laptop or PDA. However, at this time we have no source to recommend for such software.

COMPONENT TESTING

Once a problem has been narrowed to one or more components (either through deduction from basic system testing or with the help of the ECM Self-Diagnostic System), proceed to the individual component testing procedures found in this section in order to verify the problem component before replacement. In some cases, no specific tests are available, like with the ECM itself, and you can only deduce it is the bad component by eliminating EVERYTHING else first. Be as certain as you can before replacing a component, especially since most are non-returnable or worse, might even become damaged by the still existing fault after installation if you are incorrect.

Air Compressor

DESCRIPTION & OPERATION

◆ See Figure 215

Air from inside the engine cowl is drawn into the compressor through the flywheel cover (3.0L models thru 2002 or all 2.5L models) or through an air filter in the air attenuator cover (1.5L models and 2003 or later 3.0L models).

Fig. 215 Typical OptiMax air compressor (early-model 3.0L shown)

In both cases, the cover/attenuator assembly acts like a muffler to quiet compressor noise and contains a filter to prevent the ingestion of debris into the compressor.

The compressor is driven by a serpentine belt from a pulley mounted on the flywheel and is automatically self adjusting using a single idler pulley. This air compressor is a single cylinder unit containing a connecting rod, piston, rings, bearings, reed valves and a crankshaft. The compressor is water cooled to lower the temperature of the air charge and is lubricated by oil from the engine oil pump assembly.

As the compressor piston moves downward inside the cylinder, air is pulled through the filter, reed valves and into the cylinder. After the compressor piston change direction, the intake reeds close and the exhaust reeds open allowing the compressed air into the hose leading to the air/fuel rails.

AIR/FUEL PRESSURE TESTING

◆ See Figures 215 thru 225

For this test you'll need a set of pressure gauges capable of reading at least 100-110 psi (689-758 kPa) of pressure (preferably more if you want a cushion to protect the gauges in case of an over-pressurized situation), depending upon the model and the highest expected test specifications listed later in this procedure. Of course, it's also a good idea to use a gauge capable of accurately reading pressures a good 10-20% higher than the highest expected pressure.

1. Install a suitable pressure gauge assembly to the pressure test valves for air and fuel. On conventional 2.5L and 3.0L motors Mercury recommends #91-852087A1/A2/A3 (actually A1/A2 is all that should be necessary for 2.5L models). On 1.5L models, and 3.0L Pro/XS/Sport models, Mercury recommends #91-881834A1. Connect the gauges as follows:

- 75/90/115 Hp (1.5L) Optimax: an air pressure test valve is located on the lower half of the fuel rail, somewhere toward the middle of the rail assembly, while the fuel pressure test valve is either located on the BOTTOM of the rail or on the top half of the fuel rail assembly, also toward the middle of the rail.
- 135/150/175 Hp (2.5L) Optimax V6: an air pressure test valve is located toward the middle of the starboard fuel rail, while the fuel pressure valve is located on TOP of the starboard rail.
- 200/225/250 Hp (3.0L) Optimax V6: an air pressure test valve is located toward the middle of the port side fuel rail, while the fuel pressure valve is located on TOP of the starboard rail.

2. With the gauges connected and the motor equipped with a source of cooling for safety (to protect the impeller), crank the motor for 15 seconds using the starter motor, then check the gauges, readings should be as follows depending upon the model:

- 75/90/115 Hp (1.5L) Optimax: the air pressure gauge should read 92-96 psi (634-662 kPa) and the fuel gauge should read 106-110 psi (731-758 kPa). Although it is preferable that the pressures are both within the given

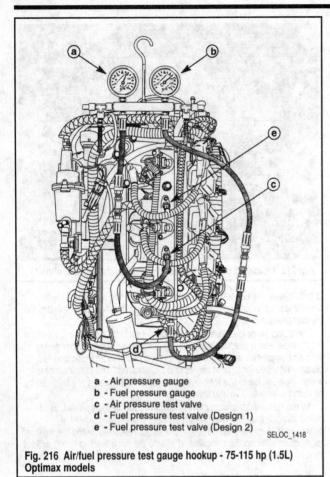

a - Air pressure gauge
b - Fuel pressure gauge
c - Air pressure test valve
d - Fuel pressure test valve (Design 1)
e - Fuel pressure test valve (Design 2)

SELOC_1418

Fig. 216 Air/fuel pressure test gauge hookup - 75-115 hp (1.5L) Optimax models

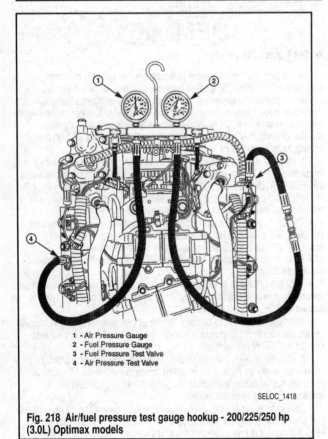

1 - Air Pressure Gauge
2 - Fuel Pressure Gauge
3 - Fuel Pressure Test Valve
4 - Air Pressure Test Valve

SELOC_1418

Fig. 218 Air/fuel pressure test gauge hookup - 200/225/250 hp (3.0L) Optimax models

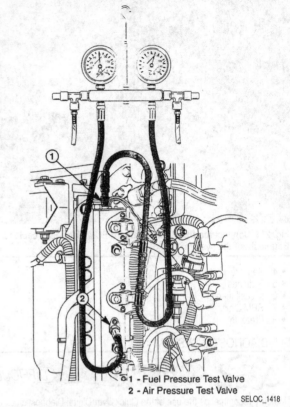

1 - Fuel Pressure Test Valve
2 - Air Pressure Test Valve

SELOC_1418

Fig. 217 Air/fuel pressure test gauge hookup - 135-175 hp (2.5L) Optimax models

spec, it is even more important that the difference between the pressures is only 13-15 psi (89.6-103.4 kPa).

• 135/150/175 Hp (2.5L) Optimax V6: the air pressure gauge should read 77-81 psi (530-558 kPa) and the fuel gauge should read 87-91 psi (600-627 kPa). Although it is preferable that the pressures are both within the given spec, it is even more important that the difference between the pressures is only 9-11 psi (62-76 kPa).

• Conventional 200/225/250 Hp (3.0L) Optimax V6: the air pressure gauge should read 78-82 psi (538-565 kPa) and the fuel gauge should read 88-92 psi (607-634 kPa). Although it is preferable that the pressures are both within the given spec, it is even more important that the difference between the pressures is only 9-11 psi (62-76 kPa).

• Pro/XS/Sport 225/250 Hp (3.0L) Optimax V6: the air pressure gauge should read 91-95 psi (627-655 kPa) and the fuel gauge should read 106-110 psi (731-758 kPa). Although it is preferable that the pressures are both within the given spec, it is even more important that the difference between the pressures is only about 15 psi (103 kPa).

■ **Keep in mind that sufficient pressure does not guarantee proper fuel/air volume.**

3. To confirm sufficient air volume, use a pair of needlenose pliers to momentarily pinch off the air discharge hose on the port side of the powerhead while an assistant cranks the motor. Air pressure should quickly rise to about 120 psi (827 kPa) or more. If the air pressure does not rise quickly as noted check for the following:
• Dirty air compressor filter
• Leaking air hoses
• Broken air compressor reed valves
• Scuffed air compressor piston
• Defective direct injector

4. To confirm sufficient fuel volume, use a pair of needlenose pliers to momentarily pinch off the fuel return hose on the port side of the powerhead while an assistant cranks the motor. Fuel pressure should quickly rise to about 120 psi (827 kPa) or more. If the fuel pressure does not rise quickly as noted check for the following:
• Improper fuel level in the vapor separator tank (VST)
• Improper output from the low pressure fuel pump

PROBLEM	CORRECTIVE ACTION
Fuel Pressure and Air Pressure are Both Low	1. Inspect air compressor air intake (air filter in flywheel cover) for blockage.
	2. Remove air compressor cylinder head and inspect for scuffing of cylinder wall. Inspect for broken reeds and/or reed stops.
	3. Tracker Valve – Remove and inspect diaphragm for cuts or tears and seat damage on diaphragm and rail.
	4. Air Regulator – Remove and inspect diaphragm for cuts or tears on diaphragm and rail.
Fuel Pressure Low or Fuel Pressure Drops while Running (Air Pressure Remains Normal)	1. Each time key is turned to the RUN position, both electric pumps should operate for 2 seconds. If it they do not run, check 20 ampere fuse and wire connections.
	2. If pumps run but have no fuel output, check vapor separator (remove drain plug) for fuel.
	3. If no fuel present in vapor separator, check fuel/water separator for debris. Check crankcase mounted fuel pump for output.
	4. Check high pressure pump amperage draw. Normal draw is 6 – 9 amperes; if draw is below 2 amperes, check fuel pump filter (base of pump) for debris. If filter is clean, replace pump. If amperage is above 9 amperes, pump is defective – replace pump. Check low pressure output – 6–9 psi. Check low pressure electric fuel pump amperage draw. Normal draw is 1 – 2 amperes; if draw is below 1 ampere, check for blockage between pump inlet fitting and vapor separator tank. If ampere draw is above 2 amperes, replace pump.
	5. Fuel Regulator – Remove and inspect diaphragm for cuts or tears.
Fuel Pressure High and Air Pressure is Normal	1. Stuck check valve in fuel return hose.
	2. Debris blocking fuel regulator hole.
	3. Faulty pressure gauge
Fuel and Air Pressure Higher than Normal	1. Debris blocking air regulator passage.
	2. Air dump hose (rail to air plenum) blocked/plugged.

SELOC_1418

Fig. 219 Air and fuel pressure troubleshooting chart - 135-175 hp (2.5L) Optimax models

PROBLEM	CORRECTIVE ACTION
Fuel pressure and air pressure are both low	1. Inspect air compressor air intake for blockage.
	2. Inspect air compressor for scuffing of cylinder wall. Inspect for broken reeds and/or reed stops.
	3. Tracker Valve – inspect diaphragm for cuts or tears and seat damage on diaphragm and rail.
	4. Air Regulator – inspect diaphragm for cuts or tears on diaphragm and rail.
Fuel pressure low or fuel pressure drops while running (air pressure remains normal)	1. Each time key is turned to the RUN position, both electric pumps should operate for 2 seconds. If they do not run, check 20 ampere fuse and wire connections.
	2. If pumps run but have no fuel output, check vapor separator (remove drain plug) for fuel.
	3. If no fuel present in vapor separator, check fuel/water separator for debris. Check crankcase mounted fuel pump for output.
	4. **Model 2000/2001/2002** – Check high pressure pump amperage draw. Normal draw is 6 - 9 amperes; if draw is below 2 amperes, check fuel pump filter (base of pump) for debris. If filter is clean, replace pump. If amperage is above 9 amperes, pump is defective – replace pump. Check low pressure output – 41-62 kPa (6-9) psi. Check low pressure electric fuel pump amperage draw. Normal draw is 1 - 2 amperes; if draw is below 1 ampere, check for blockage between pump inlet fitting and vapor separator tank. If ampere draw is above 2 amperes, replace pump. **Model 2003 and Newer** – High and low pressure pumps are on the same electrical circuit. Amperage draw for both combined should be 10 – 14 amperes. If amperage is low, check for insufficient fuel to pumps or an open circuit in one or both pumps. If amperage is high, check low pressure pump output. It should be 137.9 – 206.8 kPa (20 – 30 psi). If pressure is low, replace low pressure pump. If pressure is correct, replace high pressure pump.
	5. Fuel Regulator – inspect diaphragm for cuts or tears.
Fuel pressure high and air pressure is normal	1. Stuck check valve in fuel return hose.
	2. Debris blocking fuel regulator hole.
	3. Faulty pressure gauge
Fuel and air pressure higher than normal	1. Debris blocking air regulator passage.
	2. Air dump hose (rail to air plenum) blocked/plugged.

SELOC_1418

Fig. 220 Air and fuel pressure troubleshooting chart - 75-115 hp (1.5L) and conventional 200-250 Hp (3.0L) Optimax models (note Tracker Valve not used on 1.5L models)

Component or Problem	Possible Causes	Corrective Action
Fuel pressure and air pressure are *both* low	Air compressor	Provide 80 psi of shop air to the air rail test (Schrader) valve. Inspect the air compressor for leaks. Leaking air through the compressor indicated bad reeds.
		Inspect for broken reeds.
		Remove the air compressor cylinder head and inspect the cylinder wall and piston for scuffing.
		Replace components as necessary.
	Air compressor inlet	Inspect the compressor air inlet for blockage, kinks, or tears.
		Clean, correct, or replace components as necessary.
	Starboard rail air inlet hose	Inspect the air inlet hose to the starboard rail for blockage, kinks, or tears.
		Clean, correct, or replace components as necessary.

SELOC_1418

Fig. 221 Air and fuel pressure troubleshooting chart - 225/250 hp (3.0L) Pro/XS/Sport Optimax models (fuel and air pressure low)

Component or Problem	Possible Causes	Corrective Action
Fuel pressure and air pressure are *both* low	Starboard to port rail air transfer hose	Inspect the air transfer hose from the starboard to port rails for blockage, kinks, or tears.
		Clean, correct, or replace components as necessary.
	Tracker valve	Remove/inspect the tracker valve diaphragm for cuts, tears, or abrasions.
		Inspect the diaphragm seat and rail for damage. Replace components as necessary.
	Direct injector	Injector may be damaged or stuck open. Perform a resistance check or leak test to determine cause.
		Check o-ring for nicks or cuts.
Fuel pressure and air pressure are *both* low	Air regulator	Provide 80 psi of shop air to the air rail test (Schrader) valve.
	(This should be the last component to check/remove)	Visually inspect the air regulator cover. If air is escaping through the small hole, the regulator is faulty. Replace the port fuel rail.
		If necessary, remove/inspect the air regulator diaphragm for cuts, tears, or abrasions.
		Inspect the diaphragm seat and rail for damage. If the air regulator is faulty, replace the port fuel rail
	Air rail test (Schrader) valve	Damaged valve. Replace the valve with an Air Valve Kit.
Fuel pressure is low or fuel pressure drops while running	Tracker valve	Remove/inspect the tracker valve diaphragm for cuts, tears, or abrasions.
Air pressure remains normal		Inspect the diaphragm seat and rail for damage. Replace components as necessary.
		Check the air system for fuel contamination.
	Fuel hoses	Inspect the fuel hoses for blockage, kinks, or tears.
		Clean, correct, or replace components as necessary.
	Fuel regulator	Remove/inspect the fuel regulator diaphragm for cuts, tears, or abrasions.
		Inspect the diaphragm seat and rail for damage. If the fuel regulator is faulty, replace the port fuel rail.
		Check the air system for fuel contamination.

SELOC_1418

Fig. 222 Air and fuel pressure troubleshooting chart - 225/250 hp (3.0L) Pro/XS/Sport Optimax models (fuel and air pressure low - cont'd / fuel pressure low or drops while running)

Component or Problem	Possible Causes	Corrective Action
Fuel pressure is low or fuel pressure drops while running Air pressure remains normal	Electric fuel pump(s), VST, or filters	Ensure that the lanyard switch is in the correct position.
		Turn the ignition switch to the "RUN" position. Both electric pumps should operate for 2 seconds. If they do not run, check the 20 amp fuse and wire connections.
		Replace components as necessary.
	Electric fuel pump(s), VST, or filters	If the pumps run, but have low fuel pressure, remove the drain plug from the vapor separator tank (VST) and inspect for fuel.
		If the VST does not contain fuel: Prime the system (primer bulb) and check the water separator for debris. Clean/replace as necessary.
		If the bulb primes the fuel system, perform a vacuum test on the pulse pump.
		If there is still no fuel in the VST, check all components prior to the VST for restrictions.
		If no restrictions are found, inspect/replace the water separating fuel filter.
		If there is still no fuel, inspect the VST filter and needle/seat for contamination or obstruction. Replace as necessary.

SELOC_1418

Fig. 223 Air and fuel pressure troubleshooting chart - 225/250 hp (3.0L) Pro/XS/Sport Optimax models (fuel pressure low or drops while running - cont'd)

Component or Problem	Possible Causes	Corrective Action
Fuel pressure is low or fuel pressure drops while running Air pressure remains normal	Electric fuel pump(s), VST, or filters	If the VST contains fuel:
		Pinch the fuel return line to the high-pressure fuel pump/VST. The pressure at the fuel test valve should exceed 110 psi.
		If necessary, check the high pressure pump amperage draw (in series).
		Normal draw is 6 - 9 amps. If the draw is below 2 amps, check the fuel pump filter at the base of the pump for debris.
		If filter is clean and the draw is below 2 amps, replace the pump.
		If the draw is above 9 amps, replace the pump.
	Pulse pump	Check the crankcase-mounted pulse fuel pump for output. Normal output is 2-3 psi at idle, 1 psi minimum.
		Service/replace if necessary.
	Pulse Pump anti-siphon valve	Check for restricted fuel flow.
	Fuel primer bulb	Ensure that the primer bulb is firm. Inspect hose for leaks or blockage.
	Primer bulb check valve	Ensure that the check valve is operational.
	Fuel test (Schrader) valve	Damaged valve. Replace the valve with a Fuel Valve Kit.
Fuel pressure is high and air pressure is normal	Fuel return hose	Inspect the fuel return hose (to VST) for blockage or stuck check valve.
	Fuel regulator	Inspect the fuel regulator hole for blockage. Clean or correct as necessary.
		Inspect the diaphragm seat and rail for damage. If the fuel regulator is faulty or incorrectly calibrated, replace the port fuel rail.
		Inspect the diaphragm seat and the rail for damage. Replace the rail as necessary.

SELOC_1418

Fig. 224 Air and fuel pressure troubleshooting chart - 225/250 hp (3.0L) Pro/XS/Sport Optimax models (fuel pressure low or drops while running - cont'd / fuel pressure high w/air pressure normal)

Component or Problem	Possible Causes	Corrective Action
Fuel and air pressure are *both* high	Air vent hose	Remove the air vent hose at the rail and check air pressure at idle. If the pressures return to normal, the air vent hose is blocked or otherwise restricted.
		Inspect the air vent hose to the air plenum for blockage or kinks. Clear or correct as necessary.
	Air regulator	Check the air pressure regulator for blockage or kinks.
		Clean or correct as necessary. If the air regulator is faulty or incorrectly calibrated, replace the port rail.
Air and fuel pressure differential out of range. Normal differential is 103 kPa (15 psi).	Tracker valve	Remove/inspect the tracker valve diaphragm for cuts, tears, or abrasions.
		Check the diaphragm seat and rail for damage. Replace components as necessary.
	Fuel regulator	Remove/inspect the fuel regulator diaphragm for cuts, tears, or abrasions.
		Inspect the fuel regulator hole for blockage. Clean or correct as necessary.
		Inspect the diaphragm seat and rail for damage. If the fuel regulator is faulty, replace the port fuel rail.
		Check the air system for fuel contamination.
	Failed fuel injector	Inspect the injector o-ring for nicks, cuts, or abrasions.
		Replace the injector as necessary.
	Injector stuck open	Perform a fuel injector leak test.
		Replace components as necessary.
	Fuel rail damaged	Check for fuel in the air system.
		Inspect the offending rail valves and regulators. If all components check out, replace the rail.

SELOC_1418

Fig. 225 Air and fuel pressure troubleshooting chart - 225/250 hp (3.0L) Pro/XS/Sport Optimax models (fuel and air pressure both high / air and fuel pressure differential out of range)

REMOVAL & INSTALLATION

◆ See Figures 215 and 226 thru 229

■ There are a lot of hoses connected to the air compressor, be sure to tag them all during disassembly to ensure proper reconnection.

1. For safety, disconnect then negative battery cable.
2. Remove the top cowling.

■ For V6 motors, prior to removing the flywheel cover, remove the vent hose from the fitting on the flywheel cover.

3. Remove the flywheel cover.
4. Use a 3/8 in. drive on belt tensioner arm to relieve belt tension (use the drive in the hole on the automatic tensioner arm to pivot the tensioner pulley out of contact with the belt). Then remove the belt.

✳✳ SELOC CAUTION

If the engine has been run recently, the air pressure outlet hose fittings may be extremely hot. Allow the components to cool off before beginning disassembly.

5. Remove 2 screws securing retainer plate in order to disconnect the air pressure outlet hose. Inspect the O-rings on the air pressure hose fitting for cuts and abrasions. Replace the O-rings as necessary.
6. Disconnect the compressor water inlet and outlet hoses as follows:
• 1.5L Optimax: the water outlet hose runs from the compressor to the top of the rail, while the water outlet hose (to the fuel rail) connects to the middle of the compressor cover.
• 2.5L Optimax: the water inlet hose runs from the compressor to the top of the rail, while the water outlet-to-tell tale (cooling stream indicator) hose connects to the bottom of the compressor.

• 3.0L Optimax: the water inlet hose runs from the compressor to the top of the rail, while the water outlet-to-tell tale (cooling stream indicator) hose connects to the side of the compressor cover.

7. On 1.5L models, disconnect the wiring for the air compressor temperature sensor (follow the wiring back from the sensor mounted in the top of the compressor cover). If necessary, also remove the sensor. Also for these models, disconnect the compressor air inlet hose. Lastly, on these models the pulley blocks access to the compressor retaining bolts so loosen the 6 bolts securing the pulley and remove it for access.

8. On 3.0L models, disconnect the wiring bullet connectors for the air compressor temperature sensor (follow the wiring back from the sensor mounted in the compressor cover down to the bullets found just below the compressor, just to the right of the water inlet hose).

9. Disconnect the air compressor oil inlet hose, as well as the excess oil return hoses.

10. Remove the fasteners securing the compressor to the powerhead, then carefully remove the compressor, as follows:

• 1.5L Optimax: there are 3 bolts threaded downward from the top (pulley side) of the compressor into the powerhead.

• 2.5L Optimax: there are 2 bolts and 1 nut securing the compressor to the powerhead.

• 3.0L Optimax: there are 4 bolts securing the compressor to the powerhead.

11. If necessary, disassemble the compressor for service, as detailed under Overhaul, in this section.

To Install:

12. If overhauled, assemble the compressor once again, as detailed under Overhaul, in this section.

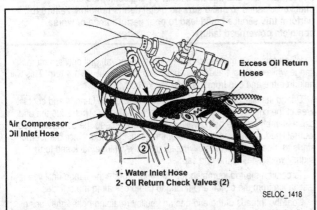

Fig. 226 Example of some of the oil, water and air lines used by the compressor - 1.5L models shown

1- Water Inlet Hose
2- Oil Return Check Valves (2)

SELOC_1418

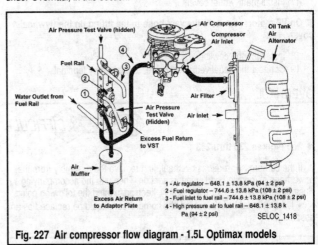

Fig. 227 Air compressor flow diagram - 1.5L Optimax models

1 - Air regulator – 648.1 ± 13.8 kPa (94 ± 2 psi)
2 - Fuel regulator – 744.6 ± 13.8 kPa (108 ± 2 psi)
3 - Fuel inlet to fuel rail – 744.6 ± 13.8 kPa (108 ± 2 psi)
4 - High pressure air to fuel rail – 648.1 ± 13.8 kPa (94 ± 2 psi)

SELOC_1418

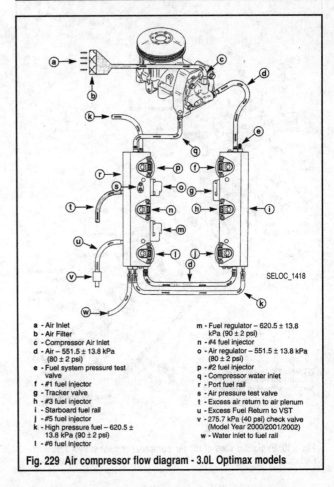

Fig. 228 Air compressor flow diagram - 2.5L Optimax models

a - Air Inlet
b - Air Filter
c - Compressor Oil Inlet
d - Oil to Upper Main Bearing
e - Compressor Air Inlet
f - Compressor Water Inlet
g - Fuel System Pressure Test Valve
h - #2 Fuel Injector
i - Port Fuel Rail
j - Excess Fuel Return to VST
k - #4 Fuel Injector
l - 40 psi Check Valve
m - Air Pressure Test Valve
n - Excess Air Return to Exhaust Adaptor Plate
o - #6 Fuel Injector

p - Water Inlet to Fuel Rail
q - High Pressure Fuel [89 ± 2 psi (613.5 ± 13.8 kPa)]
r - Air [79 ± 2 psi (544.0 ± 13.8 kPa)]
s - Air Regulator [79 ± 2 psi (544.0 ± 13.8 kPa)]
t - Fuel Regulator [89 ± 2 psi (613.5 ± 13.8 kPa)]
u - Water Outlet (tell-tale)
v - #1 Fuel Injector
w - Starboard Fuel Rail
x - #3 Fuel Injector
y - #5 Fuel Injector
z - Tracker Valve
aa - Low Pressure (Air)
ab - High Pressure (Air)
ac - Oil to Lower Crankshaft Bearing

SELOC_1418

Fig. 229 Air compressor flow diagram - 3.0L Optimax models

a - Air Inlet
b - Air Filter
c - Compressor Air Inlet
d - Air – 551.5 ± 13.8 kPa (80 ± 2 psi)
e - Fuel system pressure test valve
f - #1 fuel injector
g - Tracker valve
h - #3 fuel injector
i - Starboard fuel rail
j - #5 fuel injector
k - High pressure fuel – 620.5 ± 13.8 kPa (90 ± 2 psi)
l - #6 fuel injector

m - Fuel regulator – 620.5 ± 13.8 kPa (90 ± 2 psi)
n - #4 fuel injector
o - Air regulator – 551.5 ± 13.8 kPa (80 ± 2 psi)
p - #2 fuel injector
q - Compressor water inlet
r - Port fuel rail
s - Air pressure test valve
t - Excess air return to air plenum
u - Excess Fuel Return to VST
v - 275.7 kPa (40 psi) check valve (Model Year 2000/2001/2002)
w - Water inlet to fuel rail

SELOC_1418

13. Install the compressor to the powerhead and tighten the fasteners, as follows:

• 1.5L Optimax: tighten the bolts to 18 ft. lbs. (24.4 Nm). Also on these models install the pulley, then apply Loctite® 271 or equivalent threadlocker to the bolts and tighten them to 170 inch lbs. (19.2 Nm).

• 2.5L Optimax: tighten the bolts to 41.5 ft. lbs. (56 Nm) and the nut to 25 ft. lbs. (34 Nm).

• Conventional 3.0L Optimax: tighten the bolts to 20 ft. lbs. (27 Nm).

• Pro/XS/Sport 3.0L Optimax: tighten the bolts to 30 ft. lbs. (41 Nm).

14. Reconnect the oil, water and air hoses, as tagged during removal.

15. On models so equipped, reconnect the wiring for the compressor temperature sensor.

16. Install the drive belt and then use a 3/8 in. (9.5mm) drive on belt tensioner arm to adjust the belt tension.

17. Install the flywheel cover.

■ On V6 motors, connect the vent hose to the fitting on the flywheel cover.

18. Install the top cowling.

19. Connect the negative battery cable, to the battery terminal.

OVERHAUL

 DIFFICULT

◆ **See Figures 230 thru 233**

If the compressor requires overhaul, first remove the assembly from the powerhead as detailed earlier in this section. Then use the accompanying illustrations as a guide for disassembly, keeping in mind the following points.

• If the cylinder bore is scored, the compressor should be replaced as an assembly.

• The piston and rings are NOT sold separately and must be replaced as an assembly if either is worn or damaged. Similarly, the connecting rod and bearings should be replaced as an assembly as well.

• During disassembly, inspect all O-rings for cuts, abrasions or excessive wear and replace, as necessary.

• Check the end-cap bearing assembly for any signs of binding or roughness.

• Check the reed plate for broken or chipped reeds or reed stops. Check the reed stand (open), it should be a maximum of 0.010 in. (0.254mm).

✳✳ SELOC CAUTION

IF the compressor is being overhauled due to an internal failure (such as a broken reed, scuffed piston, bearing failure or so on) ALL air hoses, fuel rails and injectors must be disassembled and checked/cleaned to make sure no metal debris remains. Failure to perform this service could lead to poor performance or worse, complete powerhead failure.

When removing the end cap (after pulley removal) pull outward on the flange while rotating it alternately back and forth about 1/8-1/4 turn. This will help free the end cap from the compressor body.

During installation, be sure to lubricate the end cap O-rings and contact area in the compressor body using 2-stroke engine oil. Then slide the end cap assembly into the body, while making sure to keep the connecting rod journal lined up with the open end of the connecting rod. Once the journal starts to enter the rod, carefully rotate the pulley flange back and forth slightly while seating the end cap.

To confirm the rod journal-to-rod alignment rotate the flange until you feel resistance from the piston attempting to compress air in the cylinder.

Coat the threads of the end-cap and pulley retaining bolts lightly using Loctite® 271 or an equivalent threadlocking compound.

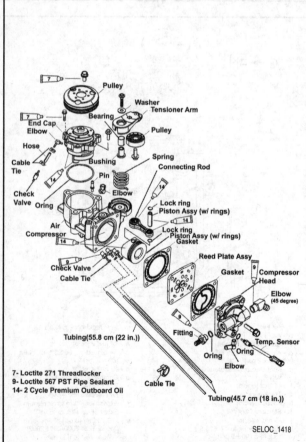

Fig. 230 Exploded view of the air compressor assembly - 1.5L Optimax models

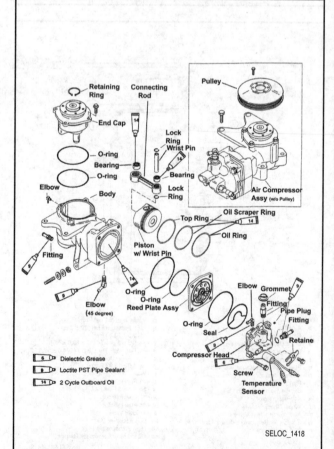

Fig. 231 Exploded view of the air compressor assembly - 2.5L Optimax models

Air Pressure Regulator

DESCRIPTION & OPERATION

◆ **See Figure 234**

The air pressure regulator is located on the fuel rail (on the port side fuel rail for V6 motors) and is designed to limit the air pressure inside the rails to approximately 80 psi (552 kPa) at idle on all except 1.5L models, where it limits air pressure to about 94 psi (648 kPa) at idle.

The air regulator uses a spring (pressure) to control the air pressure. This spring holds the diaphragm against the diaphragm seat. The contact area blocks (closes) the air inlet passage from the excess air, return passage.

As the air pressure rises (below the diaphragm), it must reach a pressure equal to or greater than the spring pressure holding the diaphragm closed. Once this pressure is achieved, the spring collapses, allowing the diaphragm to move. The diaphragm moves away from the diaphragm seat, allowing air to exit through the diaphragm seat, into the excess air passage leading to the air plenum (1.5L and 3.0L motors) or exhaust adaptor plate (2.5L motors).

TESTING

OEM ② **MODERATE**

◆ **See Figure 234**

The air pressure regulator is tested in the same basic manner as the Air Compressor air/fuel pressure test noted earlier in this section. You'll need to hook up a pressure gauge to the air pressure valve, as detailed in that procedure, then run the motor at idle, in Neutral in order to observe air pressures.

If the pressure is within spec (92-96 psi/634-662 kPa for 1.5L motors, 77-82 psi/530-565 kPa for 2.5L and conventional 3.0L motors or 91-95 psi/627-655 kPa for Pro/XS/Sport 3.0L motors), the regulator is ok.

Proceed as follows, depending upon the model

1. On 1.5L motors, momentarily pinch off the air discharge hose and watch the pressure:

a. If pressure was in spec, then rapidly rises, approaching 108 psi (745 kPa) or more, the regulator and compressor are fine.

b. If pressure was in spec, but does NOT rise to 108 psi (745 kPa) or more, then the air regulator is ok, but compressor output is low.

c. If pressure was NOT in spec, but rises once the output line is pinched, then air compressor output and volume are ok, but the air regulator is faulty. Check the regulator diaphragm, spring and cup.

d. If the pressure was NOT in spec and pinching the output line did not help, then there is an obstruction in the compressor inlet/outlet lines or the compressor is faulty.

■ **On 2.5L and 3.0L motors, if the pressure was within spec, no further testing is recommended.**

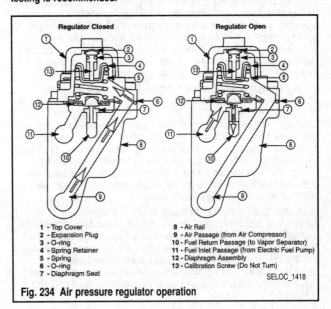

Fig. 234 Air pressure regulator operation

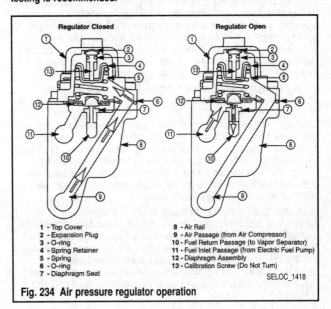

Fig. 232 Exploded view of the air compressor assembly - 3.0L Optimax models (2003 or later shown, early-models very similar)

OptiMax Air Compressor Mechanical Specifications

Component	Standard (in.)	Metric (mm)
Compressor Output		
1.5L motors	94 psi @ idle	648 kPa @ idle
2.5L motors	80 psi @ idle / 110 psi @ WOT	552 kPa @ idle / 758 kPa @ WOT
3.0L motors	80 psi @ idle / 110 psi @ WOT	552 kPa @ idle / 758 kPa @ WOT
Cylinder block displacement		
1.5L motors	3.66 cu. In.	60cc
2.5L and 3.0L motors	7.07 cu. In.	116cc
Cylinder bore		
1.5L motors	2.0472	52
2.5L and 3.0L motors	2.5591	65
Cylinder bore type		Cast Iron
Cylinder out of round (max)	0.001	0.025
Cylinder taper (max)	0.001	0.025
Cylinder wear (max)	0.001	0.025
Piston diameter		
1.5L motors ①	2.0449-2.0457	51.94-51.96
2.5L and 3.0L motors ②	2.5574-2.5582	64.96-64.98
Ring Gap		
1.5L motors		
Top and Middle	0.0059-0.0098 ③	0.15-0.25 ③
Bottom	0.0039-0.0140	0.10-0.35
2.5L and 3.0L motors		
Top and Middle	0.0059-0.0098	0.15-0.25
Bottom	0.0039-0.0140	0.10-0.35
Reed stand open	0.010	0.25

① Measured 90 degrees to piston pin and 0.30 in. (7.62mm) from bottom edge of skirt
② Measured 90 degrees to piston pin and 0.50 in. (12.7mm) from bottom edge of skirt
③ 75/90/115 hp motors may have a gap up to 0.0118 in. (0.30mm)

SELOC_1418

Fig. 233 Air Compressor Mechanical Specifications

2. For 2.5L and 3.0L motors, if the pressure was low, proceed as follows:

a. Remove the regulator and clamp, then inspect the O-ring.

b. If the O-ring was OK, but pressure was still low, check for air exiting the air bypass. If so, the regulator is faulty. If not, check for an obstruction in the compressor inlet/outlet (and if none is found, suspect a faulty compressor).

3. For 2.5L and 3.0L motors, if the pressure was high, proceed as follows:

a. Remove the air bypass hose from the exhaust adaptor and re-measure air pressure (running in neutral, at idle speed). If air pressure is STILL high, suspect a failed air regulator. If pressure if not normal, suspect a plugged bypass circuit in the adapter plate.

REMOVAL & INSTALLATION

◆ See Figures 235 and 236

The air pressure regulator can usually be removed from the air/fuel rail assembly with the assembly still installed on the powerhead. However, because it is mounted on the side of the rail (facing inward towards the other rail on most V6 motors) it can be a lot easier with the assembly removed (especially if it is of the 4 bolt design, described later). HOWEVER, regardless, for safety, if the rail is still installed you must first release the air and fuel pressure using the test valves on the rail.

There are 2 types of air pressure regulators used on these powerheads. One type uses only 2 screws to secure a flange retainer over the regulator, holding the regulator to the fuel rail. For these designs, removal and installation is a simple matter of loosening the bolts, removing the retainer and then removing the regulator assembly from the fuel rail. Installation is the reverse of the removal although it is recommended that you coat the O-rings with engine oil and the threads of the retaining bolts on these models lightly with anti-seize or using Quicksilver 2-4-C w/Teflon before installation and then tighten them to 70 inch lbs. (8 Nm).

The other type of pressure regulator used on various models (including most/all 1.5L motors as well as some early-model 2.5L motors and some/most 3.0L motors) is an assembled regulator consisting of a housing, cup, diaphragm and spring. On these models, a round housing with a square flange is secured to the fuel rail by 4 bolts. You cannot just unbolt and remove the housing, as spring pressure could force the regulator off the rail with great force as the last of the bolts are removed. Therefore on these models Mercury suggests the use of the Air Regulator Installation Tool (#91-889431) for safety and to ease removal or installation of the assembly.

On these assembled regulators, remove 2 of the 4 bolts (pick one on each side, diagonally across from each other), then install the tool over top of the regulator, using those bolt holes to secure the tool to the fuel rail. The tool contains a bridge and a threaded central post (not wholly unlike a puller assembly, and you may be able to use a puller as a substitute, if you can position it properly. With the tool in position and the center post in contact with the regulator, gradually loosen and remove the remaining 2 bolts, then slowly loosen the center post of the tool allowing the spring pressure to slowly release as the regulator is pushed away from the fuel rail.

Once spring pressure is relieved, remove the tool and the regulator housing from the fuel rail, but keeping track of the regulator components. In order coming off the fuel rail they are the housing, cup, spring, and diaphragm. During assembly of this style regulator you should coat the diaphragm mating surface lightly using Quicksilver 2-4-C w/Teflon, plus, like the other style regulator you should apply a light coating of 2-4-C or an anti-seize compound to the threads of the regulator housing retaining bolts. During installation you can the tool to center and compress the assembly over the spring and fuel rail, or you can simply obtain slightly longer bolts which will make the housing easier to install.

By using a pair of 5mm x 25mm long retaining bolts (with a set of flat washers to protect the housing), position regulator assembly and thread the 2 longer bolts into place (at opposite corners of the housing). Tighten the bolts alternately and evenly, slowly drawing the housing down onto the rail, until 2 of the shorter normal housing bolts can be threaded and used to completely seat the housing. With the housing seated, remove the longer bolts and thread the final 2 housing bolts, then tighten them all to 70 inch lbs. (8 Nm).

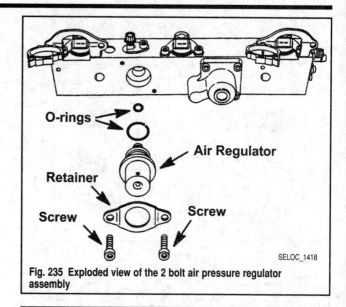

Fig. 235 Exploded view of the 2 bolt air pressure regulator assembly

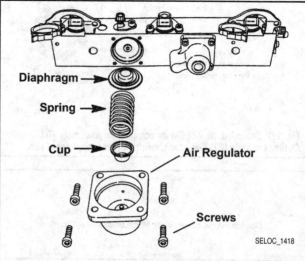

Fig. 236 Exploded view of the 4 bolt assembled pressure regulator

Temperature Sensors (ATS, CTS and ETS)

DESCRIPTION & OPERATION

◆ See Figures 237 and 238

Each Optimax motor is enabled with a number of temperature sensors which are used either to determine correct operating parameters (whether the motor is cold or up to temperature for air/fuel mixtures, or even temperature of the air, to help determine how dense it is/how much oxygen might be present) or to monitor operating conditions of the motor itself (such as to prevent damage from overheating).

Potential sensors used on these motors include:
- Air Temperature Sensor (ATS)
- Compressor Temperature Sensor (CTS)
- Engine Temperature Sensor (ETS)

In all cases, the sensors themselves operate on the same basic principles.

Temperature sensors for modern fuel injection systems are normally thermistors, meaning that they are variable resistors or electrical components that change their resistance value with changes in temperature. From the test data provided by the manufacturer it appears that Mercury sensors are usually negative temperature coefficient (NTC) thermistors. Whereas the resistance of most thermistors (and most electrical circuits) increases with temperature increases (or lowers as the temperature goes down), an NTC sensor operates in an opposite manner. The resistance of an NTC thermistor (like that used in these sensors) goes down as temperature rises (or goes up when temperature goes down). Therefore, the most important thing to keep in mind when quick checking these sensors is that their resistance readings should have an inverse relationship to temperature.

Sensors can be found in the following locations, depending upon the sensor and model:

• Air Temperature Sensor (ATS) - is normally located on the air intake plenum assembly. At the lower starboard side of the plenum on 1.5L motors or on the top of the plenum, just behind the throttle bore (near the MAP sensor) on 2.5L and 3.0L motors. The ATS on most models is a small hexagon sensor, threaded into position.

• Compressor Temperature Sensor (CTS) - is, not surprisingly, mounted on the side of the air compressor for all models. It is the same type of sensor as the ATS on 1.5L motors (a small hexagon sensor, normally threaded into position) but on 2.5L and 3.0L motors it is usually retained by a bracket and retaining bolt.

• Engine Temperature Sensor (ETS) - there is normally an ETS mounted toward the top of each cylinder head. On 1.5L and 3.0L motors, it is usually the small hexagonal, threaded sensor, like that used for the ATS. On 2.5L motors it is normally a small, hexagonal threaded sensor for the starboard cylinder head while the port cylinder head uses the same small, cylindrical sensor which is secured by a retainer bracket and mounting bolt, like that used by the CTS on V6 motors as well.

Fig. 237 Example of a cylindrical, retainer secured temperature sensor

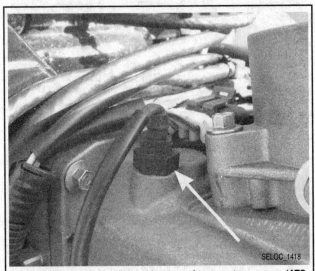

Fig. 238 Example of a threaded hexagonal temperature sensor (ATS for 2.5L and 3.0L shown)

TESTING

◆ **See Figures 237 and 238**

Temperature sensors are among the easiest components of the OptiMax system to check for proper operation. That is because the operation of an NTC thermistor is basically straightforward. In general terms, raise the temperature of the sensor and resistance should go down. Lower the temperature of the sensor and resistance should go up. The only real concern during testing is to make clean test connections with the probe and to use accurate (high quality) testing devices including a DVOM and a relatively accurate thermometer or thermo-sensor.

A quick check of the circuit and/or sensor can be made by disconnecting the sensor wiring and checking resistance (comparing specifications to the ambient temperature of the motor and sensor at the time of the test). Keep in mind that this test can be misleading as it could mask a sensor that reads incorrectly at other temperatures. Of course, a cold engine can be warmed and checked again in this manner.

More detailed testing involves removing the sensor and suspending it in a container of liquid (Mercury recommends water), then slowly heating or cooling the liquid while watching sensor resistance changes on a DVOM. This method allows you to check for problems in the sensor as it heats across its entire operating range.

Since the sensor specifications vary by model and sensor, check the Test Specifications for your motor.

■ **In addition to the tests detailed here, the DDT or an equivalent scan tool can be used to check the sensor circuits through the ECM. But realize that an out of spec reading this way could be the fault of the wiring and not just the sensor.**

Sensor Test Specifications
◆ **See Figure 239**

■ **Test readings may vary from specification as much as 10% before there is a cause for concern.**

Regarding the accompanying sensor specifications, here's the deal. Mercury only specifies that the accompanying chart of temperature sensor specs applies to the engine temperature sensor (ETS) on their OptiMax models. Now, it's the SAME specifications as used for both the ETS and ATS on most late-model EFI powerheads, and the specs do specifically apply to both hexagonal threaded ET sensors and cylindrical retainer/bracket secured

Temperature Sensor Specifications		
Fahrenheit	Centigrade	OHMS
257	125	340
248	120	390
239	115	450
230	110	517
221	105	592
212	100	680
203	95	787
194	90	915
185	85	1070
176	80	1255
167	75	1480
158	70	1752
149	65	2083
140	60	2488
131	55	2986
122	50	3603
113	45	4370
104	40	5327
95	35	6530
86	30	8056
77	25	10000
68	20	12493
59	15	15714
50	10	19903
41	5	25396
32	0	32654
14	−10	55319
5	−15	72940

Fig. 239 Temperature sensor specifications - Optimax Motors (see accompanying data before condemning sensor on these specs)

ET sensors on Optimax Models. Therefore we feel it is quite likely that the specs will be applicable to the compressor (CTS) and air (ATS) sensors on Optimax models as well, but we can't be 100% certain.

So, IF a CTS or ATS tests within these specs, we'd suggest you consider it good. But if a CTS or ATS tests out of these specs, you might want to compare it to the same test on a new sensor just to be sure before replacing it.

As with all linear NTC sensors, you should expect a relatively steady change in resistance as the sensor is heated or chilled. Watch for severe drop offs/spikes or even a reverse in readings, as potential indicators of a bad sensor.

■ On Pro/XS/Sport models Mercury gives test specifications for all of the motor temperature sensors which look a lot like the specifications provided in the accompanying chart, except that that the ohms reading should be divided by 10 (move the decimal point one place to the left) and allow for an error rate of 10% in either direction. This seems strange to us, as if the Mercury info for the Pro/XS/Sport models was drawn up for a meter set to read Kilo-ohms instead of just ohms, but there is no mention of that in their literature, so use caution when interpreting results, as Mercury might really mean 653 ohms at 95°F (35°C) or 1249 at 68°F (20°C) and not 6530 and 12,493 respectively as they do on other motors.

Quick Test

◆ **See Figure 239**

A quick check of a temperature sensor can be made using a DVOM set to the resistance scale and applied across the sensor terminals. The DVOM can be connected to directly to the sensor, or to the sensor pigtail, as the wiring varies by model. Use a thermometer or a thermo-sensor to determine ambient engine/sensor temperature before checking resistance.

On the ETS for V6 motors, there should be no continuity between either of the colored sensor leads and ground. Also, on the 2.5L motors there should be no continuity between the Black/Orange lead and either the Green or Tan/Green wire.

Even if the sensor tests ok cold, the sensor might read incorrectly hot (or anywhere in between). If trouble is suspected, reconnect the circuit, then start and run the engine to normal operating temperature. After the engine is fully warmed, shut the engine **OFF** and recheck the sensor hot. If the sensor checks within specification hot, it is still possible that another temperature point in between cold and fully-warmed specifications could be causing a problem, but not likely.

■ **Test readings may vary from specification as much as 10% before there is a cause for concern.**

The sensor can be removed and checked using the Comprehensive Test in this section or other causes for the symptoms can be checked. If the sensor was checked directly and looks good, but there are still problems with the circuit, be sure to check the wiring harness between the ECM and the sensor for continuity. Excessive resistance due to loose connections or damage in the wiring harness can cause the sensor signals to read out of range. Remember, never take resistance readings on the ECM harness without first disconnecting the harness from the ECM.

Comprehensive Test

◆ **See Figure 239**

It is important to OptiMax operation that the temperature sensors provides accurate signals across the entire operating range and not just when fully hot or fully cold. For this reason, when resistance specifications are available, it is best to test the sensor by watching resistance constantly as the sensor is heated from a cold temperature to the upper end of the engine's operating range (or cooled from a hot temperature). The most accurate way to do this is to suspend the sensor in a container of water, connect a DVOM and slowly heat (or cool) the liquid while watching resistance on the meter.

To perform this check, you will need a high quality (accurate) DVOM, a thermometer (or thermo-sensor, some multi-meters are available with thermo-sensor adapters), a length of wire, a metal or laboratory grade glass container and a heat source (such as a hot plate or camp stove). A DVOM with alligator clip style probes will make this test a lot easier, otherwise

alligator clip adapters can be used, but check before testing to make sure they do not add significant additional resistance to the circuit. This check is performed by connecting the 2 alligator clips together and checking for a very low or 0 resistance reading. If readings are higher than 0, record the value to subtract from the sensor resistance readings that are taken with the clips in order to compensate for the use of the alligator clips.

1. Remove the temperature switch as detailed in this section.
2. Suspend the sensor and the thermometer or thermo-sensor probe in a container of cool water or 4-stroke engine oil.

■ **To ensure accurate readings make sure the temperature sensor and the thermometer are suspended in the liquid and are not touching the bottom or sides of the container (as the temperature of the container may vary somewhat from the liquid contained within and sensor or thermometer held in suspension).**

3. Set the DVOM to the resistance scale, then attach the probes to the sensor or sensor pigtail terminals, as applicable.
4. Allow the temperature of the sensor and thermometer to stabilize, then note the temperature and the resistance reading. If a resistance specification is provided for low-temperatures, you may wish to add ice to the water in order to cool it down and start the test well below ambient temperatures.
5. Use the hot plate or camp stove to slowly raise the temperature. Watch the display on the DVOM closely (in case there are any sudden dips or spikes in the reading which could indicate a problem). Continue to note resistance readings as the temperature rises to 68°F (20°C) and 212°F (100°C). Of course by 212 degrees, your water should be boiling, so it is time to stop the test before you make a mess or get scalded. Up to that point, the meter should show a steady decrease in resistance that is proportional to the rate at which the liquid is heated. Extreme peaks or valleys in the sensor signal should be rechecked to see if they are results of sudden temperature increases or a possible problem with the sensor.
6. Compare the readings to the Sensor Test Specifications.

■ **Test readings may vary from specification as much as 10% before there is a cause for concern.**

REMOVAL & INSTALLATION

◆ **See Figures 237 and 238**

As noted earlier, the sensors can be found in the following locations, depending upon the sensor and model:

• Air Temperature Sensor (ATS) - is normally located on the air intake plenum assembly. At the lower starboard side of the plenum on 1.5L motors or on the top of the plenum, just behind the throttle bore (near the MAP sensor) on 2.5L and 3.0L motors. The ATS on most models is a small hexagon sensor, threaded into position.

• Compressor Temperature Sensor (CTS) - is, not surprisingly, mounted on the side of the air compressor for all models. It is the same type of sensor as the ATS on 1.5L motors (a small hexagon sensor, normally threaded into position) but on 2.5L and 3.0L motors it is usually secured by a bracket and retaining bolt.

• Engine Temperature Sensor (ETS) - there is normally an ETS mounted toward the top of each cylinder head. On 1.5L and 3.0L motors, it is usually the small hexagonal, threaded sensor, like that used for the ATS. On 2.5L motors it is normally a small, hexagonal threaded sensor for the starboard cylinder head while the port cylinder head uses the same small, cylindrical sensor which is secured by a retainer bracket and mounting bolt, like that used by the CTS on V6 motors as well.

1. To remove cylindrical sensors (secured by a retainer and mounting bolt), proceed as follows:
   a. Remove the retaining bolt and retainer plate.
   b. Disconnect the wiring and remove the sensor.
2. To remove hexagonal threaded sensors, proceed as follows:
   a. Disconnect the sensor wiring.
   b. Using a wrench or socket, carefully unthread and remove the sensor.

To Install:

3. To install the threaded hexagonal sensors, proceed as follows:
   a. Carefully thread the sensor into the cylinder head and tighten to 13-14 inch lbs. (1.4-1.6 Nm). Do NOT over-tighten and damage the sensor or the threads in the cylinder head.

b. Reconnect the sensor wiring.

4. To install the cylindrical sensors (secured by a retainer and mounting bolt), proceed as follows:

5. Connect the sensor wiring and install the sensor into the cylinder head.

6. Install the retainer plate and screw. Tighten the screw to 16.5 ft. lbs./200 inch lbs. (22.4 Nm).

Crankshaft Position Sensor (CPS)

◆ See Figure 240

The ECM needs to know exactly where each piston is in its 2-stroke cycle in order to accomplish certain ignition timing and fuel injection functions. 2-stroke EFI and Optimax motors use a crankshaft position sensor (CPS) to provide this data to the ECM.

The sensor works by picking up a magnetic field created by magnets on the flywheel, therefore the sensor itself must be positioned a certain distance away from the flywheel. On some Optimax motors this position is adjustable and so a specific air gap of 0.025-0.040 in. (0.635-1.010mm) for most motors, or 0.20 in. (5mm) for 75-115 hp Optimax motors with an adjustable sensor, must be measured and maintained anytime the sensor is disturbed.

■ **Mercury states that on 75-115 hp Optimax motors Serial #1785000 or higher the CPS is no longer adjustable.**

When it comes to testing, no resistance specifications are available in the Mercury documentation for 2.5L motors, however the sensor (calibrated for use with the same gap) on 3.0L motors is listed as having a resistance spec of 300-340 ohms. Mercury literature revised in 2007 states that the CPS resistance on 1.5L motors is 300-350 ohms.

Because the ignition systems are operated in much the same manner, more information on the CPS can be found in the Ignition & Electrical System section.

Throttle Position Sensor (TPS)

DESCRIPTION & OPERATION

◆ See Figure 241

The throttle position sensor (TPS) is an encased potentiometer that sends a signal to the ECM indicating how far open or closed the throttle shaft is at any given moment. The ECM processes this throttle opening as an indication of powerhead load for a specific rpm and uses it in helping to determine the amount of fuel which should be delivered by the injectors.

On 1.5L motors, the sensor is mounted to the front of the motor where it attaches to and directly reads movement of the throttle body linkage. The sensor housing is found just inboard of the throttle cables.

On 2.5L and 3.0L motors, the sensor is mounted to the port side of the powerhead a little above and behind the vapor separator tank assembly (2.5L, it's a little further above and behind on the 3.0L motor). On all V6

motors, the TPS does not directly mount to the throttle shaft, but instead is actuated by a link rod which connects a moving arm on the TPS housing to the throttle linkage.

This sensor operates much like the rheostat on a light. In all cases, the body of the sensor is stationary with a small shaft emerging from the center of the sensor (protruding straight outward on 1.5L models, or connecting to an arm on 2.5L and 3.0L models). The shaft is connected to the throttle shutter (1.5L model) or throttle link (2.5L and 3.0L models). As the throttle is advanced, movement is transferred to the sensor and the internal resistance of the sensor changes.

The ECM sends a reference voltage signal to the sensor through the harness and then measures the amount of voltage which returns. So, as the throttle is opened or closed and the internal resistance of the sensor changes, the variable voltage signal sent to the ECM is changed, proportionally to the changes in throttle position.

TESTING

◆ See Figure 241

The BEST method of testing a TPS is to use a scan tool, such as the Quicksilver DDT, to monitor sensor circuit voltage during engine operation (or with the ignition ON and the engine not running). On early-model EFI systems, Mercury used to sell jumper connectors which would allow a technician to probe the completed TPS circuit and there is no reason to believe that a home-made set of jumper wires couldn't be used to do the same. However, use caution, as shorting wires together during this test could result in costly ECM damage.

If the TPS signal goes out of the expected range the ECM will normally illuminate the Check Engine light and sound the warning horn. While the signal remains out of range the ECM will reduce/limit engine rpm. The first things to check under these circumstances are:

• Make sure the throttle cables are properly adjusted. The throttle stop screw on the throttle arm MUST be against the throttle stop on the cylinder block with the engine is first started. The cable barrel must be set so it is preloaded slightly, 1-2 barrel cable turns of preload should be sufficient.

• The operator must not be attempting to advance the throttle during start-up.

• Check the throttle cam-to-throttle roller adjustment. If the roller is not down in the pocket (valley) on the cam, then it may have a tendency to ride up/down on the cam, causing the TPS values to change.

• Perform heat and pressure tests on the TPS housing, as detailed in the accompanying procedure (with or without a scan tool).

If you don't have access to a suitable scan tool and you don't want to try to fabricate a jumper harness there are still a few basic tests in the accompanying procedure that you can perform.

To test a TPS, proceed as follows:

Fig. 240 Typical Mercury CPS

Fig. 241 Typical Mercury TPS - V6 models shown (1.5L models similar)

1. If you have access to a scan tool like the DDT, proceed as follows:

a. Hook up the scan tool.

b. Provide a suitable source of cooling, then start and warm the engine to normal operating temperature.

c. Slowly open and close the throttle while reading the TPS value on the scan tool. The numbers should increase or decrease linearly in proportion to throttle position. On 1.5L and 3.0L motors, the voltage should INCREASE as the throttle is opened and DECREASE as the throttle is closed. On 2.5L motors, voltage should DECREASE as the throttle is opened and INCREASE as the throttle is closed. For more details, please refer to the Fuel Injection Component Adjustment Specifications chart in this section.

2. If you fabricate a TPS test harness proceed as follows:

■ The TPS test harness is basically a set of jumper wires which will connect inline with the TPS-to-ECM wiring, along with a T which allows you to connect a DVOM to monitor the unbroken circuit during operation. You can fashion your own jumper wires either using the same type of connectors as the factor harness or using some other type of clips/connectors which will attach to the connector terminals of the factory harness, but beware, do NOT allow wires to short.

a. Disengage the TPS harness and install a suitable test harness which allows you to probe the unbroken TPS circuit using a DVOM.

b. Turn the ignition keyswitch to the **ON** position, but don't start the motor.

c. Check the DVOM for voltage readings with the throttle both closed to the idle position and open to the wide open throttle (WOT) position. Compare what you get to the Throttle Position Sensor Output Voltage specifications found in the Fuel Injection Component Adjustment Specification chart found in this section.

d. Again, slowly open the throttle from idle to the WOT position and back to idle again, watching the DVOM to make sure voltage readings increase and decrease smoothly.

3. Although not a test specifically recommended by Mercury, IF you have no access to jumper wires to test the completed circuit, you can use a DVOM to check resistance of the sensor itself to make sure resistance changes smoothly between idle and WOT positions. For this test, you can disconnect the TPS wiring and probe the sensor harness directly (NOT the ECM side of the harness!). Slowly open and close the throttle, watching the ohm readings on the meter for flat spots or spikes.

4. Mercury has noted that some TPS faults are heat related and as such they recommend a quick-check. With the motor running on a suitable cooling source, use a heat gun to carefully apply heat to the TPS near the electrical connection, until it is warm to the touch. While heat is applied listen for any of the following signs of a possible problem with the TPS:

• Engine rpm change
• Illumination of the Check Engine light
• A momentary activation of the warning horn signal
• TPS voltage value change (1/2 volt) as seen on the scan tool or DVOM with test harness and no accompanying throttle change.

✳✳ SELOC CAUTION

Keep in mind that excessive heat will damage the TPS, so use caution and don't heat it up too much.

5. Mercury has also noted that some TPS faults are pressure (mechanically) related and as such they recommend a quick-check. With the motor running on a suitable cooling source, carefully press on the sensor housing, just inboard of the electrical connection. BE SURE NOT TO ROTATE the sensor or, on V6 motors, the sensor arm. Watch the DDT or ohmmeter for changes in the sensor output, it should change by values less than hundredths or thousands of a volt. If greater changes occur, replace the sensor.

REMOVAL & INSTALLATION

◆ See Figures 241, 242 and 243

Removal or installation of the TPS is a simple matter of locating it (at the front of the motor on 1.5L models, or on the upper port side of the motor on V6 models), then unplugging the wiring and unbolting the sensor. On V6 models, you'll also have to carefully disconnect the sensor arm from the throttle link rod.

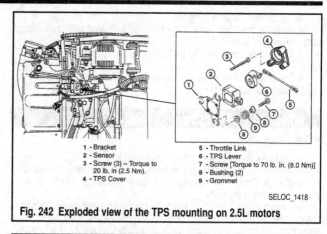

1 - Bracket
2 - Sensor
3 - Screw (3) – Torque to 20 lb. in (2.5 Nm).
4 - TPS Cover
5 - Throttle Link
6 - TPS Lever
7 - Screw [Torque to 70 lb. in. (8.0 Nm)]
8 - Bushing (2)
9 - Grommet

SELOC_1418

Fig. 242 Exploded view of the TPS mounting on 2.5L motors

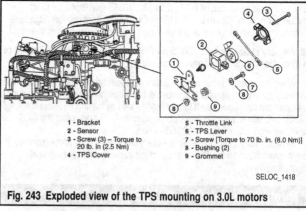

1 - Bracket
2 - Sensor
3 - Screw (3) – Torque to 20 lb. in. (2.5 Nm)
4 - TPS Cover
5 - Throttle Link
6 - TPS Lever
7 - Screw [Torque to 70 lb. in. (8.0 Nm)]
8 - Bushing (2)
9 - Grommet

SELOC_1418

Fig. 243 Exploded view of the TPS mounting on 3.0L motors

On all of these models the sensor is mounted in a fixed position, so during assembly it is only necessary to properly seat the sensor (aligning the throttle shaft on 1.5L motors or connecting the throttle link rod on V6 models).

Be sure to tighten the screws securely, but DO NOT over-tighten and crack the housing. In later service literature Mercury often recommends coating the threads of the TPS retaining screws using Loctite® 271 or equivalent.

Manifold Absolute Pressure (Map) Sensor

DESCRIPTION & OPERATION

◆ See Figure 244

Like many EFI systems Mercury's OptiMax delivers fuel based on mapping which requires inputs as to how much air the engine is ingesting. The most important inputs come from the air temperatures sensor (ATS), throttle position sensor (TPS) and finally, the MAP sensor. If you think about it, the amount of air which is coming into the motor is dependent not only on the throttle position (physical size of the air entrance) but the temperature and density of that air. The density portion of the equation is supplied by the MAP sensor.

On all Optimax models covered here the MAP sensor is located on the top of the air plenum. Unfortunately (from a testing standpoint) on these models the sensor is mounted in such a way as it directly reads manifold vacuum, without the aid of a vacuum hose which could be easily removed and attached to a vacuum gauge for testing purposes.

The sensor is a flexible type resistor. As the pressure changes, the resistor flexes and its resistance and the voltage applied across the sensor changes. This change is registered at the ECM.

Two conditions can affect the pressure in the manifold. The first and most common condition is a reduction in manifold pressure when load on the powerhead is increased. When the operator places a power demand on the powerhead by advancing the throttle, intake manifold pressure is reduced. Conversely, a reduction in intake manifold pressure, indicates an additional load has been placed on the powerhead.

The second condition which may affect the manifold pressure is operation of the powerhead at high altitudes.

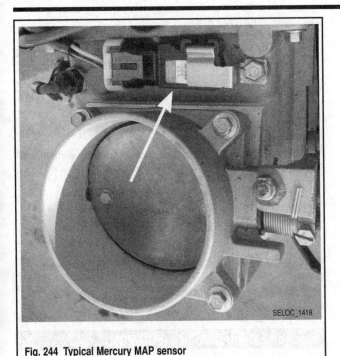

Fig. 244 Typical Mercury MAP sensor

Meter Test Leads		Meter Scale	Reading (Ω)
Red	Black		@ 20 °C (68 °F)
A	B	Auto	4.28 kΩ ± 30%
A	C	Auto	98.6 kΩ ± 30%
B	A	Auto	4.28 kΩ ± 30%
B	C	Auto	102.6 kΩ ± 30%
C	A	Auto	98.6 kΩ ± 30%
C	B	Auto	102.6 kΩ ± 30%

SELOC_1418

Fig. 245 MAP Sensor Testing - 1.5L Optimax models (terminals are A, C, B left to right viewed with sensor probe toward the bottom)

Component or Problem	Possible Causes	Corrective Action
Fuel pressure and air pressure are *both* low	Air compressor	Provide 80 psi of shop air to the air rail test (Schrader) valve. Inspect the air compressor for leaks. Leaking air through the compressor indicated bad reeds.
		Inspect for broken reeds.
		Remove the air compressor cylinder head and inspect the cylinder wall and piston for scuffing.
		Replace components as necessary.
	Air compressor inlet	Inspect the compressor air inlet for blockage, kinks, or tears.
		Clean, correct, or replace components as necessary.
	Starboard rail air inlet hose	Inspect the air inlet hose to the starboard rail for blockage, kinks, or tears.
		Clean, correct, or replace components as necessary.

SELOC_1418

Fig. 246 MAP Sensor Testing - Pro/XS/Sport 3.0L Optimax models

TESTING

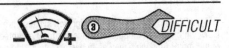

◆ See Figures 244, 245 and 246

The best method of testing a MAP sensor is to monitor the circuit voltage using a scan tool, such as the DTT. The second best method is to monitor the circuit voltage (using jumper wires to keep the circuit complete) or to monitor resistance, as an artificial source of vacuum is applied to the sensor. Unfortunately the second method requires the sensor to be removed and some form of fitting fabricated to hold the vacuum hose from a hand held vacuum pump over the opening in the sensor which normally protrudes into the intake plenum.

To make matters worse, Mercury does not publish resistance figures for this sensor (at least on most models, one exception is for the 1.5L models, and another is for the Pro/XS/Sport versions of the 3.0L model, charts are included here for both of them), so we cannot provide a reference scale. However, keep in mind that resistance should increase or decrease in a linear fashion with changes in vacuum.

A quick check of the sensor is to remove it from the manifold, then start and run the engine. With the engine running at normal operating temperature use a hand-held vacuum pump to apply vacuum to the sensor (where it attaches to the plenum). Listen for a change in engine rpm as vacuum is supplied or released. This would indicate the ECM is responding to changes in the MAP sensor signal.

REMOVAL & INSTALLATION

◆ See Figures 244

Like many sensors, removal and installation of the sensor is pretty straight forward on these motors

1. Disconnect the MAP sensor wiring connector at the harness.
2. Remove the retaining bolt and remove the MAP sensor.

To Install:

3. Connect the MAP sensor wiring harness at the connector.
4. Install the MAP sensor and tighten the retaining bolt.

Shift Interrupt Switch

DESCRIPTION & OPERATION

◆ See Figures 247 thru 250

The shift interrupt switch is designed to reduce the torque load on the gear case components to assist in shifting. The switch is monitored by the ECM which will interrupt the fuel flow momentarily to the cylinders when the engine speed exceeds 600 rpm in neutral.

The switch function can be monitored by the DDT. The DDT will display **ON** when the outboard is in neutral and **OFF** when in gear.

The switch is open (no continuity) when the outboard is in gear and closed (continuity) when the outboard is in neutral.

If shift operation is difficult, the shift interrupt function can be checked by the DDT or an multi-meter - for open or closed operation and for a continuity check of the switch harness for shorts or open wiring.

TESTING

◆ See Figures 247 thru 250

1. Locate the switch as follows:
• 1.5L: the switch is found on the exhaust adapter plate, at the base of the powerhead, right below the ATS).
• 2.5L and 3.0L: the switch is located on the side of the powerhead, right below the throttle control arm, in a position where it can be actuated by movement of the throttle/shift linkage.
2. If necessary for access, remove the bottom cowling.
3. Disconnect the switch wiring.

4. Using a multi-meter, check the continuity of the switch.

5. With the outboard in neutral, the switch is closed (showing continuity) and normally open (no continuity) when in gear.

6. If the switch does not test within specification, it may be faulty and will need to be replaced.

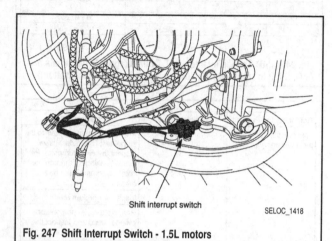

Fig. 247 Shift Interrupt Switch - 1.5L motors

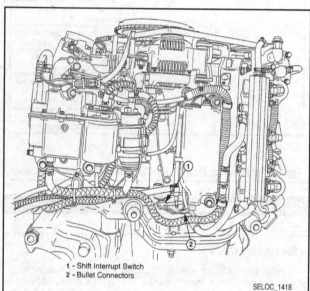

1 - Shift Interrupt Switch
2 - Bullet Connectors

Fig. 248 Shift Interrupt Switch - 2.5L motors (and 3.0L Pro/XS/Sport motors)

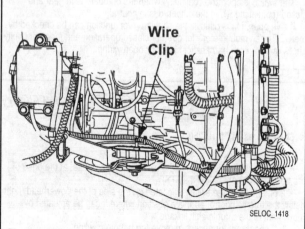

Fig. 249 Shift Interrupt Switch - 3.0L motors thru 2002

REMOVAL & INSTALLATION

◆ **See Figures 247 thru 250**

1. Locate the switch as follows:

• 1.5L: the switch is found on the exhaust adapter plate, at the base of the powerhead, right below the ATS).

• 2.5L and 3.0L: the switch is located on the side of the powerhead, right below the throttle control arm, in a position where it can be actuated by movement of the throttle/shift linkage.

2. If necessary for access, remove the bottom cowling.

3. Disconnect the shift interrupt switch bullet connectors.

4. Remove the retaining screws and remove the switch.

To Install:

5. Install the switch on the powerhead and secure with the retaining screws.

6. Connect the switch bullet connectors.

7. If removed, install the bottom cowling.

Electronic Control Module (ECM/PCM)

DESCRIPTION & OPERATION

◆ **See Figure 251**

The OptiMax fuel injection and ignition systems are controlled by the ECM, an onboard computer which is normally mounted on the starboard side of the motor (in such a way as to best isolate it from heat and vibration). The computer is sealed unit and is in no way serviceable. The ECM receives signals from numerous sensors on the powerhead. From the signals received, the ECM determines the amount of fuel to be injected. This computer also determines the timing of the spark at the spark plugs.

On some models (such as the 1.5L motors, as well as some 2003 and later V6 motors which may be equipped with digital throttle and shift) the ECM is given the additional functions of controlling the digital throttle and shift functions. On these models the additional functions come along with an additional name, and the ECM for these powerheads is usually referred to as a propulsion control module (PCM). Regardless, because the balance of functions are the same, the term ECM is meant to include PCMs throughout this guide, unless specifically stated otherwise.

There are no tests available for the ECM. If problems are suspected with the Fuel Injection or Ignition systems, you'll have to test the rest of the system thoroughly, eliminating all other possible causes before condemning the ECM. Remember, they are not inexpensive components, and they are normally non-returnable!

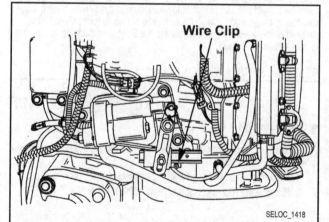

Fig. 250 Shift Interrupt Switch - 2003 and later conventional 3.0L motors

Fig. 251 Typical Mercury ECM mounting (3.0L shown)

REMOVAL & INSTALLATION

◆ See Figures 251 and 252

■ If a replacement PCM is installed to the port motor of a twin outboard boat that is equipped with SmartCraft gauges, Monitor or System View, the replacement PCM unit will have to be programmed AS a port engine. The DDT or an equivalent scan tool must be used to perform this function.

The ECM is mounted to the starboard side of the powerhead. The ECM is normally bolted in position using bushings and grommets which help to isolate it from the heat and vibration of the powerhead.

✳✳ SELOC CAUTION

Use care when handling the ECM or any electronic components. It's always a good idea to touch the powerhead immediately each time before touching the ECM. This will allow you to release any static electricity which can actually harm some sensitive circuits on the ECM.

1. For safety and to protect the ECM itself, disconnect the negative battery cable.
2. Remove the engine top cowling.

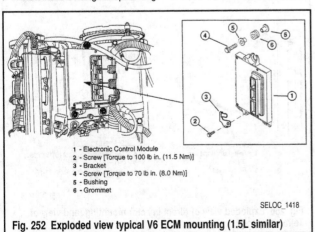

1 - Electronic Control Module
2 - Screw [Torque to 100 lb in. (11.5 Nm)]
3 - Bracket
4 - Screw [Torque to 70 lb in. (8.0 Nm)]
5 - Bushing
6 - Grommet

SELOC_1418

Fig. 252 Exploded view typical V6 ECM mounting (1.5L similar)

3. If necessary for access, remove the flywheel cover.
4. Tag and disengage the ECM wiring harness connectors.
5. Remove the 3 bolts securing the ECM to the powerhead.

■ On some models there are also screws securing metal brackets to the face of the ECM assembly. It is unclear if these screws and brackets must be removed or not, so use caution when removing the ECM.

To Install:

6. Secure the ECM to the powerhead with the 3 retaining bolts and tighten to specification, as noted:
• 1.5L: tighten all of the retaining bolts to torque to 35 inch lbs. (4 Nm).
• Conventional 2.5L and 3.0L: tighten bolts which go through the grommet and bushings to 70 inch lbs. (8 Nm) and bolts which are use to retain a small metal bracket (if removed) to 100 inch lbs. (11.5 Nm). On Pro/XS/Sport models, tighten all of the bolts to 100 inch lbs. (11.5 Nm).
7. Engage the ECM wiring harness connectors as tagged during removal.
8. If removed, install the flywheel cover.
9. Install the top cowling and reconnect the negative battery cable.

Direct Injector

DESCRIPTION & OPERATION

◆ See Figure 253

The direct injectors (1 per cylinder) are used to inject an air/fuel mixture into the cylinders. The direct injector uses a tapered tip and seat. The tapered tip is ground to provide a leak-proof seal. The direct injectors themselves are mounted directly to the cylinder heads and are located between the fuel rails and the heads.

The tapered tip is controlled by the ECM through the use of a solenoid inside the injector body. After the ECM signals the fuel injector to open, high pressure fuel is discharged into the machined cavity inside the air chamber of the air/fuel rail. This mixes the fuel with the air charge. Next, the ECM signals the direct injector to open, discharging the mixed fuel/air mixture into the combustion chamber.

The direct injector sprays the air/fuel mixture at the bowl located in the top of the piston. The piston's bowl directs the air/fuel mixture into the center of the combustion chamber. The air/fuel mixture is then ignited by the spark plug.

Fig. 253 Direct injectors are mounted between the cylinder head and fuel rail

TESTING

Resistance

◆ See Figure 253

1. Disengage the wiring harness connector from the direct injector.
2. Connect the leads of a multi-meter set to read resistance to the terminals of the injector connector.
3. The resistance of the direct injector coil windings should measure 1-1.6 ohms.
4. You can also check for a short to ground using the ohmmeter. Connect 1 meter lead to either of the injector connector pins, then connect the other lead either directly to the cylinder head (injector installed) or to the lower body of the injector where it would mount to the cylinder head (injector removed from the powerhead). After checking with one connector pin, move the first lead to the other pin and recheck. There should be no continuity between either pins of the injector harness connector and ground.
5. If the injector does not test within specification, it may be faulty and will need to be replaced.

Leakage

◆ See Figure 254

1. Remove the direct injector from the cylinder head, as detailed in this section.
2. Attach a Gearcase Leakage Tool (such as #FT-8950, or equivalent) to the discharge side of the injector.
3. Pump up leakage tool to indicate 25-30 psi (172.4-206.8 kPa).
4. Direct injector should not leak down more than 1/2 psi (3.5 kPa) in 1 minute.
5. If injector does not meet the above specifications, the injector must be replaced.

■ **Starting with the Pro/XS/Sport 3.0L model technical literature Mercury started to advise an easier test for all injectors. Remove all of the spark plugs, then using the air test Schrader valve found on the port rail, apply 80 psi (552 kPa) of pressure. Listen at each spark plug port for the sound of air leaking from the injector for that cylinder and replace any injector that leaks.**

REMOVAL & INSTALLATION

◆ See Figures 255 and 256

1. For safety, properly relieve the fuel system pressure and disconnect the negative battery cable.
2. For access, remove the fuel rail and fuel injector assembly from the cylinder head, as detailed later in this section.
3. Remove the harness connectors from the direct injectors.
4. Remove the direct injectors (1 per cylinder) from the cylinder head. On 1.5L motors, removal is much easier with the use of the direct injector removal tool #91-883521 (which basically looks like a spanner with an arced end which fits on the injector body).

■ **If the cylinder head is going to be replaced, be sure to remove the cup washers from each direct injector port (by prying carefully outward with the flat tip of a small prytool). These washers must be reinstalled with retainers into the new cylinder head, as the washers provide tension between the injectors, cylinder head and fuel rails.**

To Install:

5. Inspect the tip of the injector (on the end which protrudes into the combustion chamber) for signs of carbon buildup. Any carbon buildup on the tip of the direct injector my be removed by using a brass wire brush.
6. Inspect the O-rings (on one each side of the injector) for cuts or deformities. Replace any cut, damaged or otherwise excessively worn O-rings.
7. Inspect the injector Teflon sealing ring (white) for sign of combustion blow-by (if present, the Teflon ring will be streaked brownish black). If blow-

by is present, replace the Teflon sealing ring. If blow-by is not present, the sealing ring may be reused.

■ **Later editions of the Mercury service literature, such as for the 1.5L models or for the Pro/XS/Sport 3.0L models say to use #91-851908-2 as the Teflon seal ring sizing tool and #91-851908-3 as the seal ring installation tool.**

8. If the Teflon seal requires replacement, use Teflon ring installation tool (#91-851980) to slide the new seal onto the injector. Following the installation of the Teflon ring use sizing tool (#91-851980-1) to compress the Teflon seal to aid in the installation of the injector into the cylinder head.
9. Carefully slide the fuel rail over the mounting studs and onto the direct injectors.

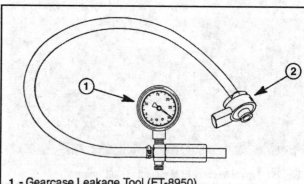

1 - Gearcase Leakage Tool (FT-8950)
2 - Direct Injector

SELOC_1418

Fig. 254 A pressure gauge set is required to check for direct injector leakage

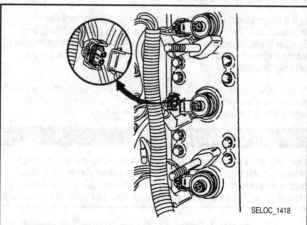

SELOC_1418

Fig. 255 Direct injector mounting and wiring connectors

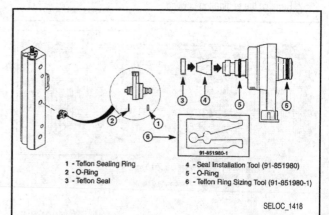

1 - Teflon Sealing Ring
2 - O-Ring
3 - Teflon Seal
4 - Seal Installation Tool (91-851980)
5 - O-Ring
6 - Teflon Ring Sizing Tool (91-851980-1)

SELOC_1418

Fig. 256 Exploded view of direct injector mounting (and view of Teflon ring tool)

10. Secure each fuel rail with 2 nuts. Tighten the nuts to 33 ft. lbs. (45 Nm).

11. Reinstall the direct injector harness connectors.

12. Reconnect the negative battery cable, then pressurize the fuel system and check for leaks.

Fuel Injector

DESCRIPTION & OPERATION

◆ See Figure 257

The fuel injectors (1 per cylinder) are used to provide fuel from the fuel rail to the direct injectors.

The injectors used in this type fuel injection system are solenoid operated. Each consists of a valve body, a needle valve and valve seat. A small voltage is sent from the ECM to each injector. When this voltage is applied to the windings of the solenoid, a magnetic field is induced around the needle valve. The valve lifts off its seat and fuel is allowed to pass between the needle valve and the needle seat. Because the fuel is pressurized, a spray emerges from the injector nozzle. A small return spring seats the needle valve back onto the seat, the instant the voltage is removed.

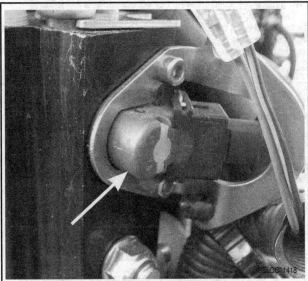

Fig. 257 Optimax fuel injector (1/cyl)

The time interval for the injector to be open and emitting fuel is called the "pulse width". The actual "pulse width" for the injector is controlled by the ECM and must be measured in microseconds.

Two O-rings are used to secure each injector in the fuel rail. Both O-rings prevent excessive injector vibration. These O-rings are replaceable and are included in an injector overhaul kit.

After receiving information from the TPS, engine temperature sensor and crankshaft position sensor, the ECM refers to the fuel calibration maps and determines when to activate the fuel injectors. When the piston is in the correct position, the ECM signals the injector and high pressure fuel is discharged into a machined cavity inside the air chamber of the air/fuel rail, and the resulting high pressure air/fuel mixture is used to feed the direct injectors.

TESTING

◆ See Figure 257

■ **A leaking fuel injector will normally allow fuel to condense in the air side of the air/fuel rail. A quick way to check for a leaking injector is to depressurize the air side (using the Schrader valve) and to watch/smell for fuel vapors. If vapors exist, pull the rail back for visibility and re-pressurize the system WITHOUT allowing the motor to crank (just turn the key to the ON position, don't turn it to START). With the system re-pressurized, check each injector tip for droplets of fuel. Also, keep in mind that multiple static tests will eventually induce fuel into the air side as a natural occurrence. If so the engine will usually surge upon initial restart.**

1. Disengage the wiring harness connector from the fuel injector.

2. Connect the leads of a multi-meter set to read resistance to the terminals of the injector connector.

3. The resistance of the injector coil windings should measure 1.7-1.9 ohms.

4. If the injector does not test within specification, it may be faulty and should be replaced.

REMOVAL & INSTALLATION

◆ See Figures 257 thru 260

1. For safety, properly relieve the fuel system pressure and disconnect the negative battery cable.

■ **Although it is not absolutely necessary to remove the fuel rail from the powerhead, it usually does make the procedure a little easier.**

2. If necessary/desired, remove the Fuel Rail assembly from the powerhead as detailed, later in this section.

3. Remove 2 screws securing injector.

4. Use a cotter pin extractor tool (or other small prytool) in the pry holes to carefully loosen the grip the injector O-rings have on the fuel rail, then remove the injectors.

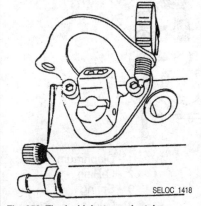

Fig. 258 The fuel injector and retainer bracket are secured by 2 screws

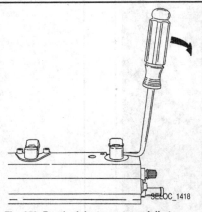

Fig. 259 Pry the injector out carefully to prevent damage

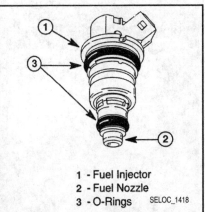

1 - Fuel Injector
2 - Fuel Nozzle
3 - O-Rings

Fig. 260 The injector O-rings must be free of cuts, abrasions or debris

5. Inspect fuel injector orifices for foreign debris, O-rings for cuts or abrasions and plastic components for heat damage. Replace any components as required.

6. An ohm test of the fuel injector may be made by connecting test leads to the injector terminals. The ohm reading should be 1.7-1.9 ohms.

To Install:

■ **Apply anti-seize grease or Quicksilver 2-4-C w/Teflon to the threads of the fuel injector retaining screws.**

7. Insert fuel injector into the fuel rail with connector pins facing (inwards) towards the center of the engine. On most engines, this also means to face the pins towards the air and fuel pressure regulator side of the rail.

■ **Carefully rotate the injector back and forth, just slightly, in order to seat the injector O-rings in the fuel rail.**

8. Secure the injector with the retainer and 2 screws. Torque the screws to 70 inch lbs. (8.0 Nm).

9. If removed, install the fuel rail assembly to the powerhead as detailed, later in this section.

10. Reconnect the negative battery cable, then pressurize the fuel system and check for leaks.

Fuel Pressure Regulator

DESCRIPTION & OPERATION

◆ **See Figure 261**

The fuel pressure regulator is located directly on the fuel rail as follows:
• 1.5L Optimax: it is the top of the 2 regulators located on the side of the rail. Where the air pressure regulator found below it has a completely round housing, the fuel pressure regulator has an oval protrusion coming out of the side of the round body.
• 2.5L Optimax: the fuel pressure regulator is located on the side (top half) of the port fuel rail. Like the 1.5L motor, it is found higher up on the rail than the round-bodied air pressure regulator. Also like the 1.5L motors, the fuel pressure regulator has an oval protrusion coming out of the side of the round body.
• 3.0L Optimax: the fuel pressure regulator is located on the side (bottom half) of the port fuel rail. Unlike the other OptiMax motors, it is found lower down on the rail than the round-bodied air pressure regulator. However, like the other motors, the fuel pressure regulator has an oval protrusion coming out of the side of the round body.

In all cases, the fuel pump is capable of delivering more fuel than the engine can consume. Excess fuel flows through the fuel pressure regulator, interconnecting passages/hoses and back to the vapor separator tank. This constant flow of fuel means that the fuel system is always supplied with fuel which has been cooled, thereby preventing the formation of fuel vapor bubbles and minimizing the chances of vapor lock.

The fuel pressure regulator is calibrated to raise the fuel pressure some 10-15 psi (69-103 kPa) above the air pressure (depending upon the model). This regulator relies on both air and spring pressure to control the fuel pressure. Inside the regulator assembly is a 10-15 lbs. spring, which holds the diaphragm against the diaphragm seat. The contact between the diaphragm and diaphragm seat closes the passage between the incoming fuel (from the electric fuel pump) and the fuel return passage.

When the engine is not running (and there is therefore no air pressure on the spring side of the diaphragm) the fuel pressure required to move the diaphragm is 10-15 psi (69-103 kPa), again, the exact pressure depends upon the model.

With the engine running, air pressure from the air compressor (which is controlled by the air pressure regulator to be 80-95 psi/552-655 kPa, depending upon the model) is routed through the air passages, to the spring side of the fuel pressure regulator diaphragm.

The air pressure (80-95 psi/552-655 kPa) and spring pressure (10-15 psi/69-103 kPa) combine to regulate the fuel system pressure to 90-110 psi (621-758 kPa).

TESTING

For details on fuel pressure testing, please refer to the Air/Fuel Pressure Testing procedure found under Air Compressor earlier in this section.

If fuel pressure is low, out of spec, but the volume is sufficient then either the fuel pump is not creating enough pressure, or there is a problem with the regulator. Since the regulator is a mechanical component, you can disassemble it to inspect the spring and diaphragm with relative ease.

REMOVAL & INSTALLATION

◆ **See Figures 261 and 262**

1. For safety, properly relieve the fuel system pressure and disconnect the negative battery cable.

■ **Although it is not absolutely necessary to remove the fuel rail from the powerhead, it usually does make the procedure a little easier.**

2. If necessary/desired, remove the fuel rail assembly from the powerhead as detailed, later in this section.

Fig. 261 Optimax fuel pressure regulator

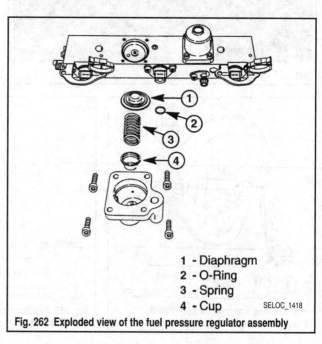

1 - Diaphragm
2 - O-Ring
3 - Spring
4 - Cup

SELOC_1418

Fig. 262 Exploded view of the fuel pressure regulator assembly

3. Alternately and evenly remove the 4 screws securing the regulator to the fuel rail. The spring under the housing will push the regulator away from the rail, so hold it steady as the last of the screws are removed.

4. Remove the regulator, cup, spring, diaphragm and, if necessary, O-ring.

5. Inspect the regulator diaphragm for cuts, nicks or abrasions.

6. Inspect the regulator housing O-ring for cuts and abrasions. If any parts are damaged, replace components as required.

To Install:

■ **The spring used in the fuel pressure regulator is significantly softer than the one used in the air pressure regulator. For this reason, it is really not necessary to use a tool to slowly compress the spring as the housing is installed. However, Mercury has noted that some techs have damaged the diaphragm in their relative haste during installation, so on some publications they recommend using a set of Alignment Pins (# 91-889431) to alternately and evenly compress the spring and seat the housing. An alternative to these pins would be 2 bolts of the same thread as housing mounting bolts, but of longer length so they could be threaded before the spring is compressed. These bolts could be slowly tightened to draw the housing into position so that 2 of the regular mounting bolts could be threaded and the alignment pins/bolts removed.**

7. Apply a light coat of Quicksilver 2-4-C w/Teflon to the diaphragm surface and O-ring to aid in the retention of the diaphragm and O-ring on the fuel rail during reassembly. Also apply a light coating to the threads of the regulator mounting bolts.

8. Position the diaphragm on the fuel rail.

9. Position the O-ring on the fuel rail.

10. Position the spring and cup onto the diaphragm.

11. Place the cover over the spring/cup/diaphragm assembly and secure it with 4 screws. Tighten the screws alternately and evenly until finger tight, then use 2 passes of a criss-crossing torque sequence to tighten the bolts, first to 50 inch lbs. (5.6 Nm) and finally to 70 inch lbs. (8.0 Nm).

12. If removed, install the Fuel Rail assembly to the powerhead as detailed, later in this section.

13. Reconnect the negative battery cable, then pressurize the fuel system and check for leaks.

Mechanical Fuel Pump

DESCRIPTION & OPERATION

◆ See Figure 263

The fuel delivery system on Mercury OptiMax motors includes both a low-pressure and high-pressure circuits. The initial low-pressure circuit is fed by a mechanical fuel pump which draws fuel from the fuel tank, pressurizes it and pushes through a water separating fuel filter and to the vapor separator tank reservoir (which acts like a large float bowl/fuel reservoir from which the electric low-pressure pump can draw fuel to feed the electric high-pressure fuel pump). The low-pressure mechanical fuel pump is a basic mechanical device that utilizes crankcase positive and negative pressures (alternating pressure and vacuum) to pump fuel through the supply lines.

This device contains a flexible diaphragm and 2 check valves that control flow. As the piston goes up, crankcase pressure drops (negative pressure or vacuum) and the inlet valve opens, pulling fuel from the tank. As the piston nears TDC, pressure in the pump area is neutral (atmospheric pressure). At this point both valves are closed. As the piston comes down, pressure goes up (positive pressure) and the fuel is pushed toward the carburetor bowl by the diaphragm through the now open outlet valve.

■ **This system works because the check valves only allow fuel to flow in one direction. Should one of the valves fail, the pump will not operate correctly.**

This is a reliable method to move fuel but can have several problems. Sometimes an engine backfire can rupture the diaphragm (though this is rare on pumps like the Mercury design which are not mounted directly to the powerhead as a vacuum source, but instead connect to the powerhead vacuum source through a hose). The diaphragm and valves are moving parts subject to wear. The flexibility of the diaphragm material can go away, reducing or stopping flow. Rust or dirt can hang a valve open and reduce or stop fuel flow.

Most pumps consist of a diaphragm, 2 similar spring loaded disc valves, one for inlet (suction) and the other for outlet (discharge) and a small opening leading directly into the crankcase bypass. The suction and compression created, as the piston travels up and down in the cylinder, causes the diaphragm to flex.

As the piston moves upward, the diaphragm will flex inward displacing volume on its opposite side to create suction. This suction will draw liquid fuel in through the inlet disc valve.

When the piston moves downward, compression is created in the crankcase. This compression causes the diaphragm to flex in the opposite direction. This action causes the discharge valve disc to lift off its seat. Fuel is then forced through the discharge valve into the carburetor.

Problems with the fuel pump are limited to possible leaks in the flexible neoprene suction lines, a punctured diaphragm, air leaks between sections of the pump assembly or possibly from the disc valves not seating properly.

The pump is usually activated by 1 cylinder. If this cylinder indicates a wet fouled condition, as evidenced by a wet fouled spark plug, be sure to check the fuel pump diaphragm for possible puncture or leakage.

TESTING

◆ See Figure 263

Lack of an adequate fuel supply can cause the engine to run lean, lose rpm or cause piston scoring due to a lack of lubricant that is carried in the fuel.

Generally speaking, checking the mechanical low-pressure delivery circuit can be a good way to eliminate lack of fuel as a potential problem with an OptiMax system.

✳✳ SELOC CAUTION

Observe all applicable safety precautions when working around fuel. Whenever servicing the fuel system, always work in a well ventilated area. Do not allow fuel spray or vapors to come in contact with a spark or open flame. Keep a dry chemical fire extinguisher near the work area. Always keep fuel in a container specifically designed for fuel storage, also, always properly seal fuel containers to avoid the possibility of fire or explosion.

The problem most often seen with fuel pumps is fuel starvation, hesitation or missing due to inadequate fuel pressure/delivery. In extreme cases, this might lead to a no start condition, but that is pretty rare as the primer bulb should at least allow the operator to fill the vapor separator tank. More likely, pump failures are not total, and the motor will start and run fine at idle, only to miss, hesitate or stall at speed when pump performance falls short of the greater demand for fuel at high rpm.

Fig. 263 Mechanical low-pressure pump from a Mercury OptiMax motor

Before replacing a suspect fuel pump, be absolutely certain the problem is the pump and NOT with fuel tank, lines or filter. A plugged tank vent could create vacuum in the tank that will overpower the pump's ability to create vacuum and draw fuel through the lines. An obstructed line or fuel filter could also keep fuel from reaching the pump. Any of these conditions could partially restrict fuel flow, allowing the pump to deliver fuel, but at a lower pressure/rate. A pump delivery or pressure test under these circumstances would give a low reading that might be mistaken for a faulty pump. Before testing the fuel pump, refer to the testing procedures found under Fuel Lines & Fitting to ensure there are no problems with the tank, lines or filter.

A quick check of fuel pump operation is to gently squeeze the primer bulb with the motor running. If a seemingly rough or lean running condition (especially at speed) goes away when the bulb is squeezed, the fuel pump is suspect. Another quick check is to shut the motor down when it starts to miss (or leave it off if it stalls) and then drain the vapor separator tank. If little or no fuel comes out, the mechanical pump low-pressure circuit is suspect.

If inadequate fuel delivery is suspected and no problems are found with the tank, lines or filters, conduct a quick-check to see how the pump affects performance. Use the primer bulb to supplement fuel pump. This is done by operating the motor under load and otherwise under normal operating conditions to recreate the problem. Once the motor begins to hesitate, stumble or stall, pump the primer bulb quickly and repeatedly while listening for motor response. Pumping the bulb by hand like this will force fuel through the lines to the vapor separator tank, regardless of the fuel pump's ability to draw and deliver fuel. If the engine performance problem goes away while pumping the bulb, and returns when you stop, there is a good chance you've isolated the low pressure fuel pump as the culprit. On OptiMax models, you should be able to perform a pressure or vacuum check, but of course, you can always disassemble the pump to physically inspect the check valves and diaphragms.

✹✹ SELOC WARNING

Never run a motor without cooling water. Use a test tank, a flush/test device or launch the craft. Also, never run a motor at speed without load, so for tests running over idle speed, make sure the motor is either in a test tank with a test wheel or on a launched craft with the normal propeller installed.

Checking Fuel Pump Lift (Vacuum)

◆ See Figure 264

Fuel system vacuum testing is an excellent way to pinpoint air leaks, restricted fuel lines and fittings or other fuel supply related performance problems.

The standard square fuel pump used on carbureted and fuel-injected (both EFI and OptiMax) Mercury motors is designed to lift fuel vertically about 60 in. (1524mm). But those capabilities are dependent upon there being no other restrictions in the system and a fuel hose that is at least 5/16

in. (7.9mm) Inner Diameter (I.D.). Each restriction that is added to the system (such as a fuel filter, valve or fitting) will reduce the amount of lift available to the system.

Using a gauge, a T-fitting and a length of clear hose you can check the amount of vacuum in the system, as well as visually inspect for air bubbles (which would indicate a leak).

1. Disconnect the fuel inlet line (tank side) from the fuel pump and connect a clear piece of fuel hose to the pump inlet nipple.
2. Next, connect a T-fitting to the end of the clear line and connect the fuel inlet line to the other side of the T-fitting.
3. Lastly, connect the vacuum gauge to the T-fitting.
4. Supply the motor with a source of cooling water.
5. First check to be sure the pump itself is capable of producing enough vacuum by pinching off the fuel supply line on the tank side of the T-fitting, then starting and running the engine while watching the gauge. If the pump produces 2.5 in. Hg (6.35 cm Hg/8.44 kPa) or MORE of vacuum remove the hose pincher from the fuel supply line and continue the test. If the pump reading is below 2.5 in. Hg (6.35 cm Hg/8.44 kPa), then check the following:
   • Pump check valves and/or diaphragm may be defective
   • Pump may contain an air leak (requiring a new gasket, rebuild or replacement, be sure to check the fittings as well)
   • There may be a problem with low crankcase pressure (check for plugged pump vacuum passageways)
6. With the pincher removed from the tube, watch the clear fuel line for signs of bubbles. If none are found, continue with the test. However, if bubbles are present, check the following for sources of the air leak:
   • Pick-up tube in fuel tank
   • Outlet fitting in fuel tank
   • Fuel inlet hose may be improperly clamped at fitting
   • Fuel tank valve may be leaking
   • If equipped, fuel line from kicker engine into main fuel line may be leaking

■ **There are some other possible causes of bubbles in the fuel line, which are sometimes overlooked. A low fuel level in the tank could allow air to be drawn in the line. Also, winter grade fuel may sometimes be prone to forming bubbles (vaporizing) when used in higher temperatures.**

7. If there are no air leaks, operate the engine until warmed to normal operating temperature. Run the motor at both idle and a little above idle (not too much above idle if on a flush fitting instead of a test tank or launched craft). As engine speed increases there should be a slight increase in vacuum however the reading should be at or below 2.5 in. Hg (6.35 cm Hg/8.44 kPa). If the reading is above 2.5 in. Hg (6.35 cm Hg/8.44 kPa), then check the following:
   • Is the anti-siphon valve restricted?
   • Is there a restriction in the primer bulb assembly?
   • Is there a kinked or collapsed fuel hose?
   • Is there a plugged water separating fuel filter rigged to the craft?
   • Is there a restriction in the fuel line thru-hull fitting?
   • If equipped, is there a restriction in the fuel tank switching valve(s)?
   • Is there a plugged fuel tank pick-up screen.

■ **You can also move the section of clear fuel line to the pump outlet fitting and check for bubbles. If bubbles are present there, but were not in the earlier test, then suspect a problem with the pump (bad gaskets or diaphragm, possibly loose screws)**

8. Once you are finished, shut down the motor, then remove the tubing, T-fitting and gauge. Reconnect the fuel inlet line.

Checking Pump Pressure

◆ See Figure 263

Fuel system pressure testing is a great way to check the condition of the pump itself, especially if questions concerning its condition are raised by a pump lift test.

The specifications for all OptiMax motor pumps are nearly identical, the only difference being in the published minimum pressure at WOT.

The following are the operating pressures listed for the mechanical fuel pumps used on OptiMax systems:
   • Operating pressure at idle: 2-3 psi (13.8-20.7 kPa)
   • Minimum pressure at idle: 1 psi (6.8 kPa)
   • Operating pressure at WOT: 6-8 psi (41.0-54.8 kPa)

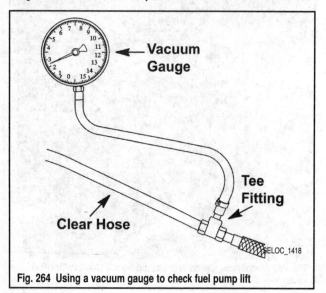

Fig. 264 Using a vacuum gauge to check fuel pump lift

- Maximum pressure at WOT: 10 psi (68.5 kPa)
- Minimum pressure at WOT (1.5L and 3.0L motors): 4 psi (27.4 kPa)
- Minimum pressure at WOT (2.5L motors): 3 psi (20.5 kPa)

Using a pressure gauge, short length of hose and a T-fitting you can check the amount of pressure being developed by the pump under idle and/or WOT conditions as follows:

1. Disconnect the fuel outlet line (powerhead side) from the fuel pump and connect the short length of hose to the pump outlet nipple.

2. Next, connect a T-fitting to the end of the line and connect the fuel outlet line to the other side of the T-fitting.

3. Lastly, connect the pressure gauge to the T-fitting.

4. Supply the motor with a source of cooling water.

5. Start and run the motor to normal operating temperature, then operate it at both idle and/or WOT and not the operating pressures, as compared with the specifications listed at the beginning of this procedure. If the pump is not making proper pressure and there are no restrictions in the fuel supply line (and no mechanical problems with the powerhead producing vacuum pulses), then the pump should be overhauled or replaced.

6. Once you are finished, shut down the motor, then remove the tubing, T-fitting and gauge. Reconnect the fuel inlet line.

Fuel Pump Diaphragm & Check Valve Testing

◆ See Figure 263

With the use of a hand-held vacuum/pressure pump you can check the integrity of the pump diaphragm and the check valves.

Attach a hand-held vacuum/pressure pump to the fuel pump inlet fitting (the fitting to which the filter/tank line connects). Using the pump apply about 7 psi (50 kPa) of pressure while manually restricting the outlet fitting using your finger. If the diaphragm is in good condition it will hold the pressure for at least 10 seconds.

Next, switch to the vacuum fitting on the pump and draw about 4 psi (30 kPa) of negative pressure (vacuum) on the fuel pump inlet fitting. This checks if the 1-way check valve in the pump remains closed. It should hold vacuum for at least 10 seconds.

Now, move the pressure pump to the fuel pump outlet fitting. This time cover the inlet fitting (and other outlet fitting, if applicable) with your finger and apply the same amount of pressure. Again, the pressure must hold for at least 10 seconds.

■ **If pressure does not hold, verify that it is not leaking past your finger (on the opposite fitting or fittings) when applicable or from a pump/hose test connection. If leakage is occurring in the diaphragm, the fuel pump should be overhauled.**

Visually Inspecting the Pump Components

◆ See Figure 265

The final way to inspect any diaphragm-displacement fuel pump is through disassembly and visual inspection. Remove and/or disassemble the pump according to the procedures found either in this section.

Wash all metal parts thoroughly in solvent, and then blow them dry with compressed air. Use care when using compressed air on the check valves. Do not hold the nozzle too close because the check valve can be damaged from an excessive blast of air.

Inspect each part for wear and damage. Visually check the pump body/cover assembly for signs of cracks or other damage. Verify that the valve seats provide a flat contact area for the valve. Tighten all check valve connections firmly as they are replaced.

Check the diaphragms for pin holes by holding it up to the light. If pin holes are detected or if the diaphragm is not pliable, it MUST be replaced.

■ **If you've come this far and are uncertain about pump condition, replace the diaphragms and check valves, you've already got to replace any gaskets or O-rings which were removed. Once the pump is rebuilt, you can remove it from your list of potential worries for quite some time.**

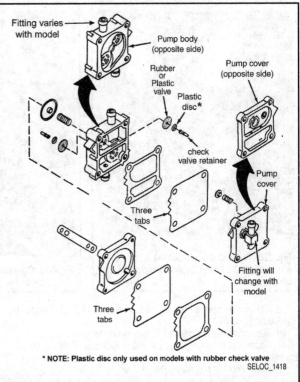

* NOTE: Plastic disc only used on models with rubber check valve

SELOC_1418

Fig. 265 Exploded view of a common square Mercury fuel pump assembly (note check valves and pump covers may vary slightly)

REMOVAL & INSTALLATION

◆ See Figures 263, 266 and 267

1. Disconnect the negative battery cable for safety.

2. Cut away the Sta-Straps from the 2 or 3 hoses at the fuel pump. Tag each of the hoses to ensure proper installation. The 2 hoses at top and bottom of the pump are fuel lines (one in from tank and one out to the VST, usually the lower one is inlet from tank and the upper one is outlet to the VST) and the one on the pump cover is a pulse hose from the powerhead which supplies the vacuum necessary for the pump to operate properly.

■ **The upper fuel fitting on 1.5L motors and on 2003 or later 3.0L motors connects directly to the vapor separator tank (VST) without the use of a hose. Instead the fitting is sealed using 2 O-rings, which can be accessed once the pump housing is unbolted and pulled away from the VST.**

3. Disconnect the top (unless it is a fitting which connects directly to the VST) and bottom fuel lines. Use a golf tee or a stubby pencil to plug the end of each disconnected hose to prevent the loss of fuel.

4. Disconnect the pulse hose from the front surface (cover) of the pump.

5. Observe the 4 bolts threaded through the pump cover. The bolt closest to the pulse nipple on the cover and the bolt diagonally across from it (normally both M6 bolts) are used to fasten the pump to the block (or VST, as applicable). The other 2 bolts (normally M5) are used hold the pump components together. Remove only the 2 bolts securing the pump to the powerhead (or VST), then carefully lift the pump clear.

6. If equipped, remove and discard the mounting gasket.

To Install:

7. On 1.5L motors and 2003 or later 3.0L motors, check the 2 O-rings for the upper fuel pump-to-VST fuel fitting. If the O-rings show any sign of cuts, abrasions or excessive wear, replace them to ensure a good seal.

8. Place the fuel pump on the powerhead base (or VST, as applicable) using a new mounting gasket (if equipped).

■ **When equipped with a gasket, it is usually easiest to insert the 2 mounting screws through the pump before positioning it to the powerhead so it can hold the gasket in place as you align the pump.**

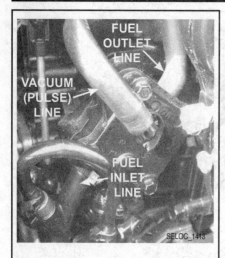

Fig. 266 Tag and disconnect the 3 hoses attached to the pump (typical Mercury pump shown)

Fig. 267 Only remove the 2 bolts holding the pump to the powerhead

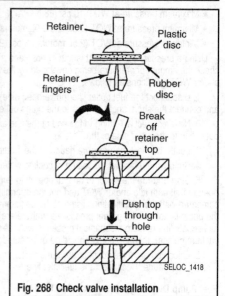

Fig. 268 Check valve installation

9. Tighten the 2 pump retaining bolts to alternately to 55-60 inch lbs. (6.0-6.8 Nm).

10. Reconnect the fuel lines as tagged during removal.

11. Reconnect the vacuum (pulse) line to the cover of the pump assembly.

12. Reconnect the negative battery cable.

OVERHAUL

◆ See Figures 265 and 268

1. Remove the pump from the powerhead. For details, refer to the procedure located earlier in this section.

2. Lay the pump on a suitable work surface

3. Remove the screws holding the fuel pump assembly together.

■ **Keep close track of the order in which the pump components are assembled in the pump housing. Compare the locations of the components to the accompanying diagram and note differences (if any) for installation purposes.**

4. Carefully separate the pump cover, gaskets and diaphragms from the pump body. Always discard the used gaskets!

✳✳ SELOC CAUTION

Do not remove the check valves unless they are defective. Once removed, the valves cannot be used again. If the check valves are to be replaced, take time to Observe and remember how each valve faces, because it must be installed in exactly the same manner or the pump will not function.

5. To remove a check valve for replacement, grasp the retainer with a pair of needlenose pliers and pull the valve from the valve seat.

6. Wash all parts thoroughly in solvent and then blow them dry with compressed air. Use care when using compressed air on the check valves. Do not hold the nozzle too close because the check valve can be damaged from an excessive blast of air.

7. Inspect each part for wear and damage. Verify that the valve seats provide a flat contact area for the valve disc. Tighten all elbows and check valve connections firmly as they are replaced.

■ **On models equipped with the black rubber check valves, check each disc to make sure the black coating is not coming off.**

8. Test each check valve by blowing through it with your mouth. In one direction the valve should allow air to pass through. In the other direction, air should not pass through.

9. Check the diaphragm for pin holes by holding it up to the light. If pin holes are detected or if the diaphragm is not pliable, it must be replaced.

To Assemble:

The fuel pump rebuild kit will contain new gaskets, diaphragms and check valve components. Each check valve consists of a large rubber disc, a smaller plastic disc and a valve retainer or a large plastic disc and a retainer, depending on the design and kit.

■ **The plastic discs are more resistant to damage from fuel additives than the rubber discs. For this reason most new overhaul kits should contain the plastic discs.**

10. To install the check valves and retainers, proceed as follows:

a. Insert the fingers of the retainer into the smaller plastic disc and then the larger rubber disc (or the single plastic disc, as applicable)..

b. Install the disc and retainer assembly onto the fuel pump body. Push in the retainer until the collar and both discs are tightly pressed against the pump body.

c. Bend the end of the retainer from side to side until it breaks away from the collar.

d. Install the broken off end through the hole in the collar and through the disc(s). Use a hammer and tap the retainer end down into place. As this piece is forced down, it will spread the fingers of the retainer and secure the check valve within the pump body.

e. Place the larger of the 2 caps and boost springs into place on the back side of the pump body.

■ **All the layered components of the fuel pump have notches which must be aligned during assembling. With the check valves in place in the pump body, you can begin final assembly of the fuel pump. Assembly will be much easier if you use two 3 in. (76mm) long 1/4 in. screws or dowels inserted through the large pump mounting bolt holes to align the pump components.**

11. Insert two 1/4 dowels through the pump mounting holes in the pump cover, then place the cover with the interior side facing upward on the worksurface.

12. Position the coil spring and cap, then carefully place the diaphragm in position over the dowels. Make sure The 3 tabs on the diaphragm are aligned in the same direction and the single tab on the pump cover.

13. Position the gasket over the dowels, with the 2 tabs facing the same direction as the tabs on the diaphragm and cover.

14. Install the fuel pump body down over the assembly, again with the tab in proper alignment.

15. Now install the coil spring and cap in the pump body.

16. Install the gasket with the tabs aligned in the proper direction.

17. Position the diaphragm over the dowels (aligning the tabs, you should be used to that by now).

18. Lastly, position the fuel pump base (rear cover) down over the assembly.

19. Grasp the assembly firmly (holding it together against diaphragm/spring pressure), then invert the assembly so you can insert the 2 pump assembly screws.

20. Recheck the alignment of all pump components, then insert and thread the 2 pump assembly screws. Tighten the screws to 55 inch lbs. (6 Nm).

21. Install the pump assembly, as detailed earlier in this section.

Electric Fuel Pump(s)

DESCRIPTION & OPERATION

◆ **See Figure 269**

Where carbureted motors use only the mechanical low-pressure fuel pump, and EFI motors use both the mechanical pump and an electrical high-pressure pump, the OptiMax motors use 3, yes THREE separate fuel pumps to provide the necessary high-pressure fuel required by the air/fuel rail assembly and direct injectors.

For all OptiMax motors the mechanical low-pressure fuel pump fills a fuel reservoir in the vapor separator tank (VST) assembly on the port side of the powerhead. A low-pressure electric fuel pump takes fuel from that reservoir and delivers it under pressure to the high-pressure electric fuel pump which then boosts the pressure many times over to the operating pressures described earlier in this section, generally 90-110 psi (621-758 kPa) depending upon the model.

On all 2.5L motors and on 2002 or earlier 3.0L motors, this low-pressure electric pump is mounted externally, on the end of the VST assembly. However, on all 1.5L motors and 2003 or later 3.0L motors the low-pressure pump is mounted inside the VST assembly.

On ALL models a high-pressure electric pump is mounted inside the VST assembly and is used to feed the air/fuel rail with fuel that is already at system operational pressures.

The electric fuel pumps are sealed and are NOT serviceable. Once opened, the airtight seal is lost and the integrity of the seal cannot be regained. The pump must be replaced.

Pump action will commence the moment the ignition key switch is rotated to the **ON** position. If the key switch is held in this position for any length of time, the ECM will cutoff electrical power to the pump and of course the pumping action will cease.

On all OptiMax systems fuel travels in both directions to the fuel rail and excess fuel returns to the VST via the fuel pressure regulator mounted on the fuel rail.

TESTING

Low Pressure Pump Output

The low-pressure pump test will vary depending upon whether the model on which you are working has an external low-pressure electric pump (all 2.5L motors and 2002 or earlier 3.0L motors) or an internal (all 1.5L motors and 2003 or later 3.0L motors) low-pressure electric pump.

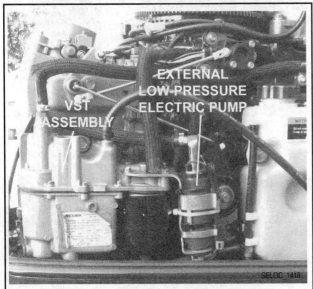

Fig. 269 VST assembly with external low-pressure pump

On models with an external pump you're going to need a T-fitting and a short length of hose in order to connect the fuel pressure gauge inline with the pump outlet line (on top of the pump).

On models with an internal pump Mercury has provided a Schrader valve on the lower corner of the VST to which a fuel gauge assembly can be directly connected.

After completing fuel pressure tests on models with external pumps, reconnect and secure fuel outlet hose to fuel pump with full circle stainless clamps.

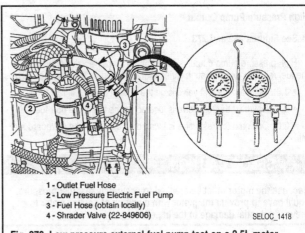

1 - Outlet Fuel Hose
2 - Low Pressure Electric Fuel Pump
3 - Fuel Hose (obtain locally)
4 - Shrader Valve (22-849606)

Fig. 270 Low-pressure external fuel pump test on a 2.5L motor. . .

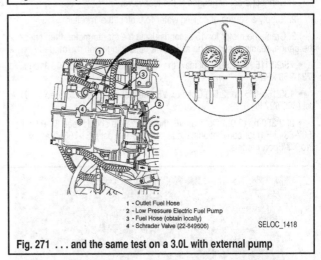

1 - Outlet Fuel Hose
2 - Low Pressure Electric Fuel Pump
3 - Fuel Hose (obtain locally)
4 - Schrader Valve (22-849606)

Fig. 271 . . . and the same test on a 3.0L with external pump

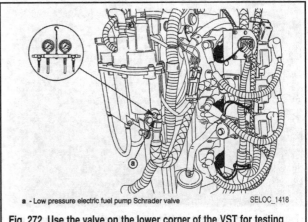

a - Low pressure electric fuel pump Schrader valve

Fig. 272 Use the valve on the lower corner of the VST for testing internal low-pressure pumps

Connect the gauge set and, if you plan to test with the motor running (not just by cycling the ignition key to actuate the pump), then provide the motor with a source of cooling water.

■ **When testing, be sure to use a gauge set with the proper range in order to obtain the correct reading. On models with external pumps, the pressures are low enough that Mercury suggests using the air side of their air/fuel gauge set (# 91-852087).**

Results should be as follows:

• External pumps - fuel pressure should be 6-9 psi (41.37-62.05 kPa).

• Internal pumps - fuel pressure should be 20-30 psi (138-207 kPa).

High Pressure Pump Output

◆ **See Figures 269 and 273**

High-pressure pump output is tested at the pressure valve provided on the fuel rail. The exact location varies slightly, based on the model as follows:

• 1.5L Optimax: the test valve is on the bottom of the fuel rail.

• 2.5L Optimax: the test valve is located on the top of the port fuel rail.

• 3.0L Optimax: the test valve is located on the top of the starboard fuel rail.

✳✳ SELOC CAUTION

Because the motor must be cranked to accurately test fuel pressure, you'll have to provide the motor with a source of cooling water to prevent potential damage to the impeller.

1. Connect a set of gauges to the test valve.

2. Supply a source of cooling water to protect the impeller.

3. Crank the motor for 15 seconds using the starter motor, then check the gauge, readings should be as follows depending upon the model:

• 75/90/115 hp (1.5L) Optimax: the fuel gauge should read 106-110 psi (731-758 kPa).

• 135/150/175 hp (2.5L) Optimax V6: the fuel gauge should read 87-91 psi (600-627 kPa).

• 200-250 hp (3.0L) Optimax V6: the fuel gauge should read 88-92 psi (607-634 kPa) for conventional models or 106-110 psi (731-758 kPa) for Pro/XS/Sport models.

Fig. 273 A test fuel pressure test valve is located on one end of the fuel rail

REMOVAL & INSTALLATION

Low-Pressure Electric Pump

◆ **See Figures 269 and 274 thru 276**

Two types of low-pressure electric pumps are used on OptiMax models, internal and external. On motors with an internal type pump, the low-pressure electric fuel pump is located inside the vapor separator tank (VST) assembly and can be removed once the assembly is removed and disassembled. For details, please refer to the Vapor Separator Tank (VST) assembly Removal & Installation, and Overhaul procedures, later in this section.

■ **Although there is nothing particularly difficult about this procedure, it does involve cutting and removing crimped metal fuel clamps which, for safety reasons, must be replaced with the same type of clamp during installation. In order to ensure a tight seal without leaks Mercury requires the use of a clamp crimping tool (#91-803146T, or the Snap-On equivalent YA3080).**

✳✳ SELOC WARNING

Using a clamp crimping tool other than that specified could result in a crimp that is too loose or too tight. Either could lead to a damaged fuel hose, which could cause a dangerous fuel leak. Similarly, screw type metal clamps may damage the hose and should not be used.

On motors with an external low-pressure electric fuel pump, proceed as follows:

1. For safety, properly relieve the fuel system pressure and disconnect the negative battery cable.

2. Tag and disconnect the wiring for the pump. You may have to cut a wire tie in order to relocate the harness out of the way.

3. Tag and disconnect the fuel lines at the top and bottom of the pump assembly. Carefully cut away the crimped portion of the old metal clamp, then peel away the clamp so that you can loosen and remove the hose.

4. Cut the wire ties used to secure the pump, then pull the pump from the mounting bracket attached to the VST.

To Install:

5. Connect the hoses to the fittings on the top and bottom of the pump assembly, then position the pump in the mounting bracket on the VST. Use new wire ties to secure the pump to the bracket.

Fig. 274 Mercury external low-pressure electric fuel pump

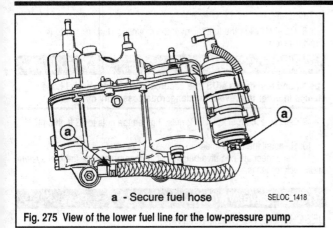

a - Secure fuel hose   SELOC_1418

Fig. 275 View of the lower fuel line for the low-pressure pump

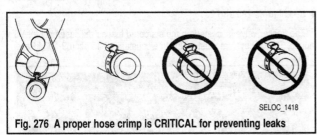

SELOC_1418

Fig. 276 A proper hose crimp is CRITICAL for preventing leaks

6. Using a suitable crimping tool, install new metal clamps on the hose. Make sure the resulting crimp is not too loose (there should not be a large gap between the crimped portion of the clamp), nor too tight (the pieces of the clamp at the center of the crimp should not touch, there should a small gap).

7. Reconnect the pump wiring and position it as noted during removal. If necessary, secure it using a new wire tie.

8. Reconnect the negative battery cable, then pressurize the fuel system and check carefully for leaks.

High-Pressure Electric Pump

◆ See Figure 269

The high-pressure pump is located inside the vapor separator tank (VST) assembly and can be removed once the assembly is removed and disassembled. For details, please refer to the Vapor Separator Tank (VST) Assembly Removal & Installation, and Overhaul procedures, later in this section.

Fuel (and Air) Rail

DESCRIPTION & OPERATION

◆ See Figure 277

The air fuel rails contain 2 passages: the first is for fuel, the second is for air. The air passage is common between all the cylinders included in the rail. A hose connects the rail air passage (starboard on V6 motors) to the air compressor. On V6 motors, another hose connects the starboard air rail passage to the port air rail passage.

Also on V6 models, a fuel line connects the bottom of the first rail to the opposite fuel rail. For all motors, fuel is stored inside the rail until an injector opens.

A fuel pressure regulator controls pressure in the fuel rails and allows excess fuel to return into the vapor separator. The fuel regulator not only controls fuel pressure but maintains it at approximately 10-15 psi (69-103 kPa) higher than whatever the air rail pressure is (depending upon the model). The fuel regulator diaphragm is held closed with a spring that requires that amount of force (from 10-15 psi/69-103 kPa) in order to move the diaphragm off the diaphragm seat. The back side of the diaphragm is exposed to air rail pressure. As the air rail pressure increases, the fuel pressure needed to open the regulator will equally increase.

For example, looking at a typical 2.5L motor (which requires 10 psi more fuel pressure than air) if there is 50 psi of air pressure on the rail side of the diaphragm, 60 psi of the pressure will be required to open the regulator. The

SELOC_1418

Fig. 277 An air/fuel rail assembly is mounted directly over the cylinder head

math isn't as neat for the other models, like the 1.5L motor (which requires 14 psi fuel pressure than air), but the concept is the same. For the 1.5L, if there is 50 psi of air pressure on the rail side of the diaphragm, 64 psi of the pressure will be required to open the regulator.

On 1.5L motors, there is only 1 fuel rail, and it is water cooled. On V6 motors, the port fuel rail is water cooled. Also on V6 motors, in order to equalize the pulses generated by the pumps (both air and fuel) a tracker diaphragm is installed in the starboard rail. The tracker diaphragm is positioned between the fuel and air passages. The tracker diaphragm is a rubber diaphragm which expands and retracts depending upon which side of the diaphragm senses the pressure increase (pulse).

REMOVAL & INSTALLATION

 ②   MODERATE

◆ See Figures 277 thru 281

1. For safety, properly relieve the air and fuel system pressures and disconnect the negative battery cable.

■ **Air rail pressure is released in the same manner as fuel system pressure (using the appropriate Schrader valve). To help locate the right valve, refer to the Air/Fuel Pressure Testing procedures found under Air Compressor earlier in this section.**

2. Remove the fuel injector harness from each injector by compressing the spring clip with a flat tip screw driver while pulling on the connector.

■ **Always remove the fuel/air hose and fitting together by removing fitting retainer rather than cutting the clamps. One of the reasons for this (besides the pain in the butt of crimping clamps to reinstall the hoses) is that IF the air hose is removed from the barbed fitting, the interior of the hose will be damaged requiring replacement.**

3. Loosen the Allen head screws at the ends of the fuel rail assembly (which secure the fuel, air and, if applicable, water hoses) to the fuel rail, then carefully disconnect the assemblies from the rail(s).

■ **It is recommended that direct injectors remain in the cylinder head (if they are not to be replaced) while removing the fuel rail. The direct injectors have a Teflon seal which may expand if the injector is removed from the head. This expansion may cause reinstallation difficulties or require the replacement of the seal.**

4. Remove the 2 nuts securing the fuel rail.

5. As the fuel rail is removed, use a flat tip screwdriver to hold the direct injectors in the cylinder head.

■ On V6 motors, the starboard fuel rail contains 3 fuel injectors and a tracker valve. The port fuel rail contains 3 fuel injectors, 1 fuel regulator and 1 air regulator. On 4-cyl motors, the fuel rail contains 4 fuel injectors, 1 fuel regulator and 1 air regulator.

6. After all fuel injectors, air regulator, tracker valve (if applicable), fuel regulator, inlet hoses and outlet hoses have been removed, the fuel rails may be flushed out with a suitable parts cleaning solvent. Use compressed air to remove any remaining solvent.

7. Each fuel/air inlet or outlet hose adaptor has at least 1 O-ring seal (though usually 2 per adaptor). Theses O-rings should be inspected for cuts, deformities or abrasions and replaced as necessary when the fuel rail is disassembled for inspection and cleaning.

✳✳ SELOC CAUTION

If a mechanical failure occurred in the air compressor the fuel rail and attached hoses MUST be removed and flushed with a parts cleaning solution in order to remove any potential metal debris. Failure to do this could blow the powerhead.

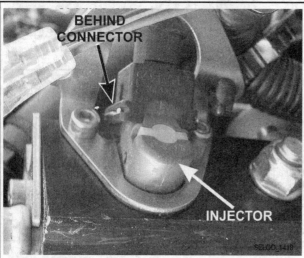

Fig. 278 Unplug the injector connectors (spring clip is on other side)

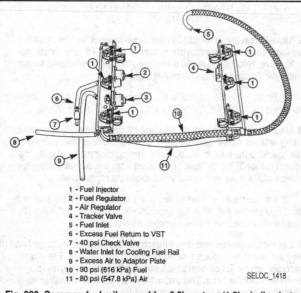

1 - Fuel Injector
2 - Fuel Regulator
3 - Air Regulator
4 - Tracker Valve
5 - Fuel Inlet
6 - Excess Fuel Return to VST
7 - 40 psi Check Valve
8 - Water Inlet for Cooling Fuel Rail
9 - Excess Air to Adaptor Plate
10 - 90 psi (616 kPa) Fuel
11 - 80 psi (547.8 kPa) Air

SELOC_1418

Fig. 280 Common fuel rail assembly - 2.5L motors (1.5L similar, but all in one rail)

To Install:

8. Carefully slide the fuel rail over the mounting studs and onto the direct injectors.

✳✳ SELOC WARNING

The air and fuel hoses MUST be secured with stainless steel hose clamps in order to prevent the dangerous possibility of leaks.

9. Secure each fuel rail with 2 nuts. Tighten the nuts to 33 ft. lbs. (45 Nm).

10. Reinstall the direct injector harness connectors.

11. Reconnect the negative battery cable, then pressurize the fuel system and check for leaks.

Tracker Valve (V6 Only)

DESCRIPTION & OPERATION

◆ See Figures 280 and 281

The tracker valve is located on the starboard fuel/air rail assembly for V6 motors. The DFI system must maintain a constant 10-15 psi (69-103 kPa)

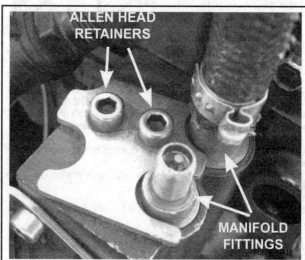

Fig. 279 Remove the Allen head bolts securing hose fittings to the rails

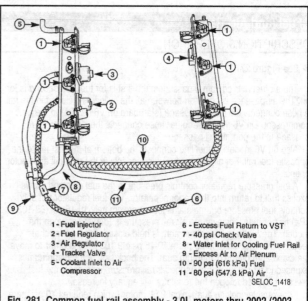

1 - Fuel Injector
2 - Fuel Regulator
3 - Air Regulator
4 - Tracker Valve
5 - Coolant Inlet to Air Compressor
6 - Excess Fuel Return to VST
7 - 40 psi Check Valve
8 - Water Inlet for Cooling Fuel Rail
9 - Excess Air to Air Plenum
10 - 90 psi (616 kPa) Fuel
11 - 80 psi (547.8 kPa) Air

SELOC_1418

Fig. 281 Common fuel rail assembly - 3.0L motors thru 2002 (2003 and later similar)

pressure difference between the fuel pressure and air pressure in the rails (depending upon the model), at all times. The tracker is designed to maintain the pressure differential when the air or fuel pressure suddenly raises (i.e. pulses generated by the compressor's piston or by the fuel injectors opening and closing). The tracker contains a spring on the air side of the diaphragm. This spring positions the diaphragm against the diaphragm's seat (when the engine is not running).

After the engine starts and the fuel and air pressure reach normal operating range, the fuel pressure will compress the spring and the diaphragm will move slightly away from the seat (to a neutral position). At this point the pressure on both sides of the tracker diaphragm is equal (spring pressure + air pressure).

Any air or fuel pressure "spikes" on one side of the diaphragm will transfer this pressure rise to the other system (air or fuel) on the other side of the diaphragm. Both systems will have a momentary increase in pressure so that the difference between the air and fuel systems can be maintained.

■ **To prevent excessive wear in the seat, the tracker is calibrated to allow the diaphragm to be slightly away from the seat during normal operation.**

REMOVAL & INSTALLATION

◆ **See Figures 280 and 281**

The tracker valve can usually be removed from the air/fuel rail assembly with the assembly still installed on the powerhead. However, because it is mounted on the side of the rail (facing inward towards the other rail) it can be a lot easier with the assembly removed. However, regardless, for safety, if the rail is still installed you must first release the air and fuel pressure using the test valves on the rail.

1. For safety, properly relieve the air and fuel system pressures, then disconnect the negative battery cable.

■ **Air rail pressure is released in the same manner as fuel system pressure (using the appropriate Schrader valve). To help locate the right valve, refer to the Air/Fuel Pressure Testing procedures found under Air Compressor earlier in this section.**

2. If necessary/desired, remove the fuel rail assembly from the powerhead as detailed, earlier in this section.

■ **Although it is not absolutely necessary to remove the fuel rail from the powerhead, it usually does make the procedure a little easier.**

3. Alternately and evenly remove the 4 screws securing the tracker valve to the fuel rail. The spring under the housing may push the valve away from the rail slightly, but tension should not be too bad.

4. Remove the tracker valve housing, followed by the spring, diaphragm, and, if necessary, the O-ring.

To Install:

5. Apply a light coat of Quicksilver 2-4-C w/Teflon to the tracker diaphragm and cover O-ring to aid in their retention on the fuel rail while reinstalling the tracker valve to the fuel rail.

6. Apply anti-seize grease or Quicksilver 2-4-C to the tracker valve attaching screw threads.

7. Position the diaphragm, spring and O-ring onto the fuel rail.

8. Place the cover over the diaphragm/spring/O-ring assembly and secure it with 4 screws. Tighten the screws to 70 inch lbs. (8.0 Nm).

9. Reconnect the negative battery cable, then pressurize the fuel system and check for leaks.

Vapor Separator

DESCRIPTION & OPERATION

◆ **See Figures 282**

The vapor separator tank (VST) Assembly is a major component of the OptiMax system. It is sometimes referred to by Mercury as the fuel management assembly.

The tank itself serves multiple purposes, though it's primary purpose is to provide a fuel reservoir from which the electric low-pressure and high-pressure fuel pumps can draw an un-interrupted supply of fuel to feed the injectors.

The tank also serves as a mounting point for the fuel water separating fuel filter, found either inside the assembly or attached to the assembly depending upon the year and model.

There are basically 2 types of VST assemblies used on these motors. The first, which utilizes an external low-pressure electric pump and water separating fuel filter, is normally found on all 2.5L motors, as well as 2002 and earlier 3.0L motors. A redesigned unit which uses an internal low-pressure electric pump and water separating fuel filter is normally found on all 1.5L motors, as well as 2003 and later 3.0L motors.

Throughout the procedures we'll differentiate between these 2 designs as Internal or External component VST assemblies.

REMOVAL & INSTALLATION

◆ **See Figures 276, 282, 283 and 284**

■ **There are basically 2 types of VST assemblies used on these motors. The first, which utilizes an external low-pressure electric pump and water separating fuel filter, is normally found on all 2.5L motors, as well as 2002 and earlier 3.0L motors. A redesigned unit which uses an internal low-pressure electric pump and water separating fuel filter is normally found on all 1.5L motors, as well as 2003 and later 3.0L motors. Throughout the procedures we'll differentiate between these 2 designs as Internal or External component VST assemblies.**

Fig. 282 Mercury VST assembly (external component type shown)

Fig. 283 Removal involves disconnecting hoses and wiring. . .

Fig. 284 . . .from the VST assembly (and in this case the external pump and filter)

1. For safety, properly relieve the air and fuel system pressures, then disconnect the negative battery cable.

■ **Air rail pressure is released in the same manner as fuel system pressure (using the appropriate Schrader valve). To help locate the right valve, refer to the Air/Fuel Pressure Testing procedures found under Air Compressor earlier in this section.**

2. On 1.5L models, cut the cable tie which secures the VST vent hose above the top center of the VST assembly.

3. On 3.0L motors, tag and disconnect the VST vent hose (at the top center of the VST assembly), then remove the bolt securing the air intake silencer assembly (located on the front of the powerhead, toward the port side).

4. Drain the VST, depending upon the model, as follows

• On internal component VST assemblies, position a suitable container under the center of the VST, then locate the small rubber hose which connects the 2 drain fittings. Disconnect the hose from one end, and use it to direct the fuel emptying from that fitting with the drain screw open into the container. Reconnect the hose to that end and disconnect it from the other end, then repeat using the drain screw and hose to drain to the tank. Once you are finished, tighten both of the drain screws and reinstall the hose.

• On external component VST assemblies, position a suitable contain under the drain plug at the center front (end) of the VST, then remove the plug to drain the fuel. Reinstall the plug once the tank is empty.

5. Disconnect the water separator sensor lead. On external component VSTs it is found directly under the filter assembly. On external component VSTs there is a multi-pin style connector which will be found either horizontally, centered under the VST (for 1.5L motors) or vertically, just under the rear end of the assembly (3.0L motors).

6. Disconnect the electric fuel pump harness connector(s), the location varies slightly by model, as follows:

• On 1.5L Optimax Models, the connector can be found vertically, along the cylinder head end of the VST assembly.

• On 2.5L Optimax Models, there are 2 connectors, one at positioned horizontally above the VST, just forward of its center point, and the other vertically at the rear of the VST, often wire tied to the external low-pressure pump.

• On 3.0L Optimax Models through 2002 (with external component VST assemblies), there are 2 connectors, one at positioned vertically at the front of the VST, alongside the lower, starboard side of the air plenum, between the plenum and VST. The other is positioned vertically at the rear of the VST, often wire tied to the external low-pressure pump.

• On 3.0L Optimax Models 2003 or later (with internal component VST assemblies), there are is normally just one connector (like the 1.5L motors) and it can normally be found positioned vertically, under the rear end of the VST assembly.

■ **Some fuel lines may be secured by Sta-Straps (wire ties), but the majority of fuel lines used on VST assemblies will be secured by crimped metal straps. For these hoses you MUST cut away the old straps to disconnect them, then during installation replace them with new metal straps of the SAME design. In order to do so you'll need the Mercury crimping tool (# 91-803146T or # 91-803146A2) or the Snap-On equivalent (YA3080). However, in some cases you MAY have the option of unbolting the Allen head retainers and disconnecting the hose and fitting both (such as from the fuel rail) in order to avoid having to cut the clamp).**

7. Tag and disconnect the necessary fuel hoses from the VST assembly, the locations of which will vary slightly by model, as follows:

• On 1.5L Optimax Models, tag and disconnect the inlet fuel hose and the pulse vacuum hose from the pulse pump on the front end of the VST. Also disconnect the second pulse pump vacuum hose located just above the pump on the end of the VST. Disconnect the fuel outlet hose (this hose runs from the top, rear of the VST to the fuel rail, you should be able to unbolt the retainers from the rail and disconnect the hose/fitting as an assembly from the rail). Finally, disconnect the fuel return hose (again either from the rail or from the bottom rear of the VST assembly).

• On 2.5L Optimax models, At the front of the motor, just below and inboard from the VST, cut the wire tie securing the fuel lines together. Disconnect the fuel outlet hose, either from the top of the VST or follow it back to the fuel rail (in order to possibly avoid cutting the crimped clamp). Similarly, disconnect the fuel return hose either from the lower portion, inboard side of the front of the VST, or follow it back to the fuel rail. Also, disconnect the fuel hose between the mechanical fuel pump and the VST assembly.

• On 3.0L Optimax models thru 2002 (with external component VST assemblies), disconnect the fuel supply hose from the mechanical pump (it connects to the top of the VST, right above the fuel filter. Disconnect the fuel outlet hose (coming out of the top of the VST and running to the fuel rail) and the fuel inlet hose (coming out of the fuel rail and running into the front end of the VST, just below the cover assembly). In both cases of the high-pressure fuel lines, you may prefer to disconnect them at the fuel rail (by unbolting and removing the fittings) to avoid having to cut the crimp-type clamps.

• On 2003 and later 3.0L Optimax models (with internal component VST assemblies), tag and disconnect the inlet fuel hose and the pulse vacuum hose from the pulse pump on the front end of the VST. Also disconnect the second pulse pump vacuum hose located just behind the VST assembly. Disconnect the fuel outlet hose (this hose runs from the top, rear of the VST to the fuel rail, you should be able to unbolt the retainers from the rail and disconnect the hose/fitting as an assembly from the rail). Finally, disconnect the fuel return hose (again either from the rail or from the bottom rear of the VST assembly).

8. Remove the 3 bolts (2 on the bottom and one at the top) securing the VST assembly to the powerhead, then remove the vapor separator assembly.

9. If it is necessary to service the fuel pumps, water filter or other internal/external components, please refer to Overhaul in this section.

To Install:

10. Loosely install the VST assembly to the powerhead, then tighten the 3 retaining bolts to specification as follows:

• 1.5L Optimax: tighten the bolts tot 18 ft. lbs. (24.4 Nm)
• 2.5L Optimax: tighten the bolts to 140 inch lbs. (16 Nm)
• 3.0L Optimax thru 2002 (with external component VST assemblies): tighten the bolts to 140 inch lbs. (16 Nm)
• 2003 and later 3.0L Optimax (with internal component VST assemblies): tighten the bolts to 21 ft. lbs. (28.5 Nm)

11. Reconnect the fuel and vent or vacuum lines (as applicable) as tagged or noted during disassembly. If there are any questions, refer to the disassembly procedure to determine proper connections. If a high-pressure fuel line was disconnected from the fitting by cutting and removing the crimped clamp. You'll need to use a proper crimping tool to secure the new clamp during installation.

12. Secure hoses with tie-wraps as mentioned and/or noted during disassembly.

13. Reconnect the electric fuel pump harness connector(s).

14. Reconnect the water separator sensor lead.

15. Reconnect the negative battery cable, then pressurize the fuel system and check for leaks.

OVERHAUL

External Component VST Units

◆ See Figure 285

② ◁MODERATE

The external component VST assembly is the early-model design and was generally found on all 2.5L motors, as well as 2002 and earlier 3.0L motors. It is characterized by an external low-pressure electric pump mounted to the rear of the assembly, and an external water separating fuel filter, mounted between the tank and the external fuel pump.

1. Remove the VST assembly, as detailed earlier in this section.

2. Remove the 7 Philips head screws from the perimeter of the VST cover. Once the screws are all removed, carefully pull upward to separate the cover from the tank. The float and the electric fuel pump should come away attached to the cover.

3. Inspect the fuel pump chamber seal for cuts, abrasions or excessive wear. Remove the seal if replacement is necessary.

4. If necessary, remove the final filter screen from the bottom of the fuel pump for cleaning or replacement. The screen must be carefully pried from the opening in the bottom of the pump.

■ **The filter screen can be cleaned and reused as long as it is not physically damaged.**

5. If you are removing the fuel pump carefully wiggle it slightly side-to-side while pulling downward to remove it from the cover. However DO NOT twist it, as the wiring harness could become damaged. Once the pump is down far enough to access the harness, carefully disengage the harness connector to free the pump.

6. Inspect the seal on top of the pump for cuts, abrasions or excessive wear. Remove the seal if replacement is necessary.

7. If the float valve requires service it can be removed once the float pivot pin is out of the way. Once the pin is removed, carefully remove the float and needle for inspection or replacement. If the needle seat is damaged the entire separator cover must be replaced. Make sure the float itself has not absorbed any fuel and is not deteriorating.

✳✳ SELOC CAUTION

DO NOT make any attempts to adjust the float height or the float limit bracket. The float height is set at the factory and no adjustment is necessary.

8. Remove the phenolic sealing plate (basically a gasket plate with an irregularly shaped neoprene seal on both sides of the plate). Inspect the plate and seal for cuts, abrasions or excessive wear and replace, the plate/seal assembly, as needed.

To Assemble:

9. Install the phenolic sealing plate assembly to the VST cover.

10. If removed, attach the float needle to the arm, then guide the float and needle assembly into position, placing the needle into the seat.

11. Secure the pivot pin through the float arms, then check to make sure the float moves freely through the pivot range. Install and tighten the pivot pin screw to 10 inch lbs. (1.1 Nm).

12. Apply a light coating of Quicksilver 2-4-C w/Teflon or an equivalent lubricant to the lips of the pump seals in both the bottom of the VST tank and the top of the cover.

13. Reconnect the pump wiring, then insert the pump into the underside of the VST cover. Carefully seat the pump into the cover, making sure not to pinch the harness.

14. If removed, install the final filter by pushing upward and then rotating the filter clockwise to lock it onto the pump.

15. Carefully position the cover onto the separator tank, making sure to keep the phenolic sealing plate in alignment. Install the retaining screws and tighten securely. Tighten the screws to 30 inch lbs (3.5 Nm).

16. Install the VST assembly, as detailed earlier in this section.

Internal Component VST Units

◆ See Figure 286

The external component VST assembly represents a redesigned later model VST and was generally found on all 1.5L motors, as well as 2003 and later 3.0L motors. It is characterized by an internal low-pressure electric pump which is positioned horizontally in the bottom of the VST tank, as well as an internal water separating fuel filter which can be unthreaded and removed from the top of the tank assembly.

1. Remove the VST assembly, as detailed earlier in this section.

2. Remove the mechanical low-pressure fuel pump from the VST. As usual, 2 of the 4 visible screws secure the pump to the housing, the 2 slightly larger screws (the one closes to the vacuum pulse port on the cover and the

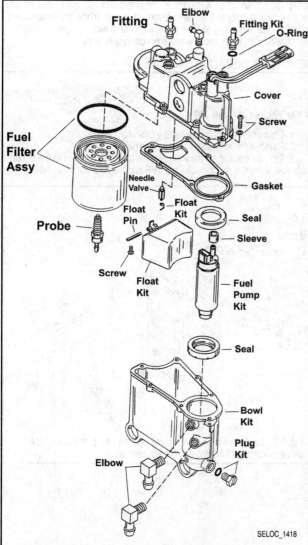

Fig. 285 Exploded view of the external component VST assembly used on 2.5L motors, as well as 2002 and earlier 3.0L motors

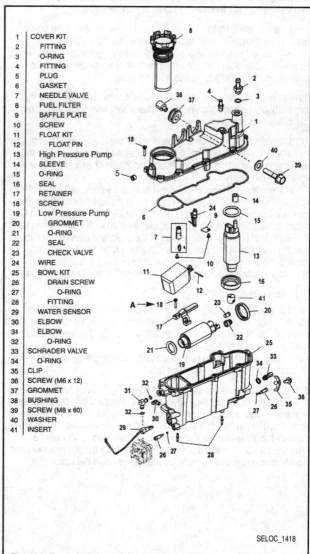

Fig. 286 Exploded view of the late-model VST assembly - 2.5L shown (1.5L and 3.0L very similar)

one diagonally across from it). More details can be found earlier in this section under Mechanical Fuel Pump.

3. Remove the fuel/water separating filter from the top of the VST. The castellated head of the filter assembly can be used to get a grip with a small prytool, then unthread and remove the filter. Once removed, carefully check the 2 O-rings towards the top of the filter assembly for damage and replace, as necessary. Also, inspect and clean or replace the filter element, as necessary. More details can be found under Fuel Filter Service, in the Maintenance & Tune-Up Section.

4. Unthread the water sensor from the lower front portion of the VST. Inspect the sensor for signs of debris or corrosion and clean or replace as required. As usual, be sure to check the sensor O-ring as well.

5. Remove the 9 Philips head screws from the perimeter of the VST cover. Once the screws are all removed, carefully pull upward to separate the cover from the tank. The float valve will come away attached to the cover, however the electric fuel pumps will normally remain behind in the tank.

6. Inspect the cover sealing ring which should remain behind in the grooves on the tank assembly. Check the seal for cuts, abrasions or excessive wear and replace, as necessary.

7. If the float valve requires service it can be removed once the float pivot pin and screw that secures the pin are out of the way. If necessary, loosen the screw, remove the pin and then carefully separate the float and needle valve from the cover assembly. Inspect the float closely for signs of deterioration or signs that it has absorbed fuel and replace, as necessary.

■ Unlike many of the other VST assemblies used on Mercury motors, the float needle AND SEAT are both replaceable on this model tank. Inspect and replace them as an assembly, as required.

8. Loosen the 2 Philips head screws at the bottom rear of the tank, securing the low-pressure electric pump to the bottom of the tank.

9. With the low-pressure pump retaining screws out, both pumps, which are connected by wiring, will be free to remove as an assembly. Lift the low-pressure and high-pressure pumps from the tank.

■ Keep in mind that each pump is also held in position partially by O-ring seals, though the high-pressure pump seal was freed when the VST cover was removed.

10. Inspect the O-rings (one on each end of the low-pressure pump, 2 on the harness connector coming off the low-pressure pump and one on top of the high-pressure pump) for signs of cuts, abrasions or excessive wear and replace, as necessary.

11. There is a fuel director insert and at least 1 seal in the bottom of the VST tank. The high-pressure pump attaches to the fuel director insert. But there are 2 potential designs. One insert fits inside the filter housing on the bottom of the pump and just uses a single round seal. The other insert fits OVER the filter housing on the bottom of the pump and uses both the round seal and a smaller round port seal, which goes on the fuel fitting directly opposite the check valve passage. Most models use the later design. If necessary, remove these components for inspection and/or replacement.

To Assemble:

12. Apply a light coating of oil to the mechanical low-pressure fuel pump elbow O-rings, in order to ease installation onto the VST, then align the pump to the VST and install using a new gasket. Tighten the 2 pump retaining bolts to 60 inch lbs. (6.8 Nm).

13. Apply a light coating of oil to the water sensor O-ring, then carefully thread the sensor into the end of the tank.

■ IF the fuel pump harness is to be replaced it must be routed into the VST BEFORE installing the Red and Black wire pins into the harness connector. Once routed, to ensure correct polarity, the Red wire must be inserted in pin "A" (left side of the connector when looking at it from the lock tab side) and the Black wire must be inserted into pin "B". After the pins are locked into the connector, slide the pin retainer over the harness and lock it in place on the back side of the connector.

14. Apply a light coating of oil to the O-rings on the fuel pumps and pump wiring connector (5 O-rings in total).

15. If removed, attach the electrical harness to each fuel pump. Use a wire tie to secure the high-pressure pump harness at the top of the pump, just below the connector.

16. Install the pumps into the VST, seating the O-rings, then secure the low-pressure pump using the 2 retaining screws. Tighten the screws to 22 inch lbs. (2.5 Nm).

17. If removed, attach the float needle to the arm, then guide the float and needle assembly into position.

18. Insert the pivot pin through the float arms and secure using the retaining screw. Tighten the screw to 10 inch lbs. (1 Nm), then check to make sure the float moves freely through the pivot range.

19. Make sure the VST cover-to-tank seal is in position, then carefully position the cover onto the separator tank, making sure NOT to dislodge the seal. Install the retaining screws and tighten them to 37 inch lbs (4.2 Nm).

20. Apply a light coating of oil to the O-rings on the fuel/water separating filter, then thread the filter into the cover until it seats.

21. Install the VST assembly, as detailed earlier in this section.

Throttle Body, Air Plenum & Reed Block

Though the direct injection system mixes high pressure fuel with high pressure air for fuel atomization, that doesn't provide enough air for the motor to run. The majority of the air necessary for motor operation is still controlled in a traditional manner using a throttle body, air intake plenum and reed assembly. The throttle body contains the butterfly valve which is used to control the amount of air which enters the intake. The plenum routes the air from the throttle body to the valves found in the reed block. The reed valves operate in the same way as they do on carbureted motors, opening and closing with crankcase vacuum and pressure respectively.

REMOVAL & INSTALLATION

◆ See Figures 287 thru 290

1. Disconnect the negative battery cable for safety.
2. Remove the air intake silencer assembly.
3. Tag and disconnect the wiring for the sensors mounted to the throttle body and air intake, this includes the TPS and the MAP for most models,

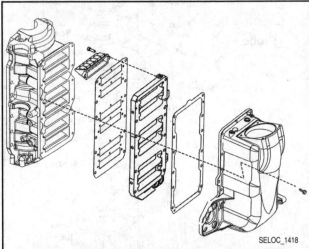

SELOC_1418

Fig. 287 Exploded view of a typical Optimax air plenum and reed valve mounting (3.0L shown, others similar)

and also includes the manifold air temperature (MAT) sensor on 1.5L models.

4. Disconnect the throttle linkage.

5. On 1.5L motors, tag and disconnect the oil line from the top of the reed plate.

6. On V6 models, tag and disconnect the oil lines from the pump. Plug at least the oil inlet line to prevent a mess or contamination. If desired, unbolt and remove the pump from the plenum. For more details, please refer to the procedures in the Cooling & Lubrication section.

7. Using the reverse of the torque sequence (basically a clockwise spiraling pattern that starts at the lower right outer bolt and works its way inward to the center bolts), loosen and remove the air plenum retaining bolts. There are 9 bolts on 1.5L motors and 12 bolts on V6 models.

■ **On V6 models, the bolts that secure the air plenum also secure the reed plate, so use caution when pulling the plenum off as the plate will be loose too. Also for this reason, always be ready to remove and replace the reed plate gasket on these models.**

8. Remove the air intake plenum and throttle body assembly from the motor. Remove and discard the old plenum-to-reed plate gasket. On V6 models, remove the reed plate and discard the old gasket.

9. On 1.5L motors, if necessary to remove the reed plate assembly, loosen and remove the 15 bolts using the reverse of the torque sequence (this results in basically a crossing pattern that works inward from the outside).

10. If necessary for reed valve service, remove the bolts securing each reed valve block to the reed valve plate. Separate the valve blocks for inspection or replacement.

To Install:

■ **Remember to replace all gaskets on components that were disturbed.**

11. If the reed valve blocks were removed from the plate, install them using the retaining screws coated lightly in Loctite® 271 or an equivalent threadlocking compound, then tighten the bolts to 90-100 inch lbs. (10-11 Nm).

12. Install the reed plate to the motor using a new gasket. On 1.5L models install the bolt at the top with the throttle cable retainer clamp, but do not tighten at this point, instead thread the remaining 14 bolts. On these models, tighten all 15 bolts to 18 ft. lbs. (24 Nm) using the proper torque sequence (a crossing pattern that starts at the center and works outward.

13. Install the air intake plenum using a new gasket, loosely installing the retaining bolts (9-12 depending upon the model) to hold the assembly. Tighten the bolts to specification using the proper torque sequence as follows:

• 1.5L models: tighten the 9 bolts to 18 ft. lbs. (24 Nm) using a crossing pattern that starts at the center (middle right) bolts and works outward (ending at the top right bolt).

• All V6 models: use a counterclockwise spiraling pattern that starts at the center bolt on the right and works outward ending at the lower bolt on the right, but the torque spec itself varies. Tighten the bolts to bolts to 175 inch lbs. (20 Nm) for 2.5L models, to 100 inch lbs. (11.5 Nm) on conventional 3.0L models or to 125 inch lbs. (14 Nm) on Pro/XS/Sport 3.0L models.

14. On V6 models, if removed, install the oil pump to the plenum. Either way, reconnect the oil lines to the pump as tagged during removal.

15. On 1.5L motors, reconnect the oil line to the top of the reed plate.

16. Reconnect the throttle linkage.

17. Reconnect the wiring for the sensors as tagged during removal.

18. Install the air intake silencer assembly.

19. Reconnect the negative battery cable.

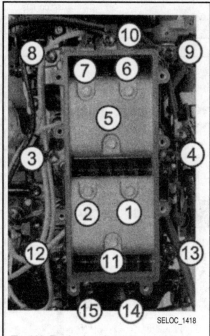

Fig. 288 Reed valve plate torque sequence - 1.5L models

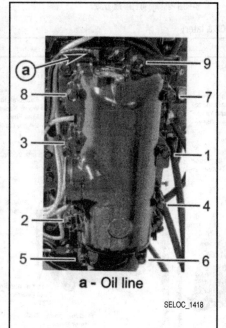

a - Oil line

SELOC_1418

Fig. 289 Air plenum torque sequence - 1.5L models

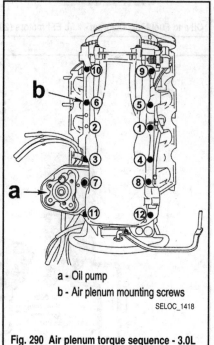

a - Oil pump
b - Air plenum mounting screws

SELOC_1418

Fig. 290 Air plenum torque sequence - 3.0L models

FUEL/OIL ROUTING DIAGRAMS (EFI)

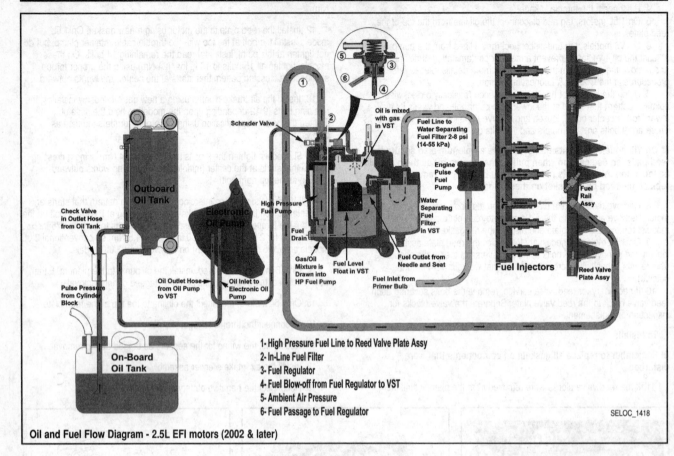

1- High Pressure Fuel Line to Reed Valve Plate Assy
2- In-Line Fuel Filter
3- Fuel Regulator
4- Fuel Blow-off from Fuel Regulator to VST
5- Ambient Air Pressure
6- Fuel Passage to Fuel Regulator

SELOC_1418

Oil and Fuel Flow Diagram - 2.5L EFI motors (2002 & later)

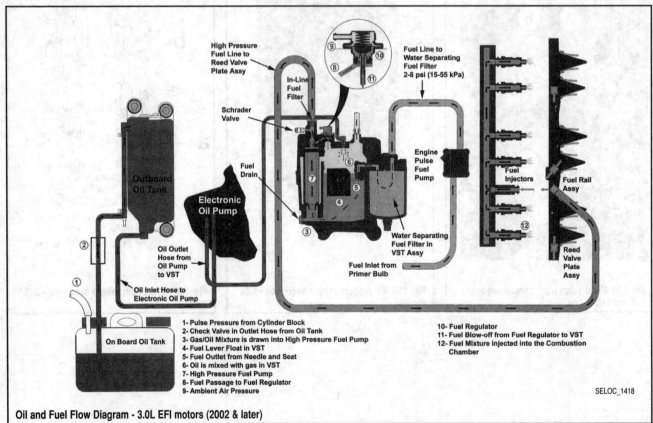

1- Pulse Pressure from Cylinder Block
2- Check Valve in Outlet Hose from Oil Tank
3- Gas/Oil Mixture is drawn into High Pressure Fuel Pump
4- Fuel Lever Float in VST
5- Fuel Outlet from Needle and Seat
6- Oil is mixed with gas in VST
7- High Pressure Fuel Pump
8- Fuel Passage to Fuel Regulator
9- Ambient Air Pressure

10- Fuel Regulator
11- Fuel Blow-off from Fuel Regulator to VST
12- Fuel Mixture injected into the Combustion Chamber

SELOC_1418

Oil and Fuel Flow Diagram - 3.0L EFI motors (2002 & later)

AIR/FUEL ROUTING DIAGRAMS (OPTIMAX)

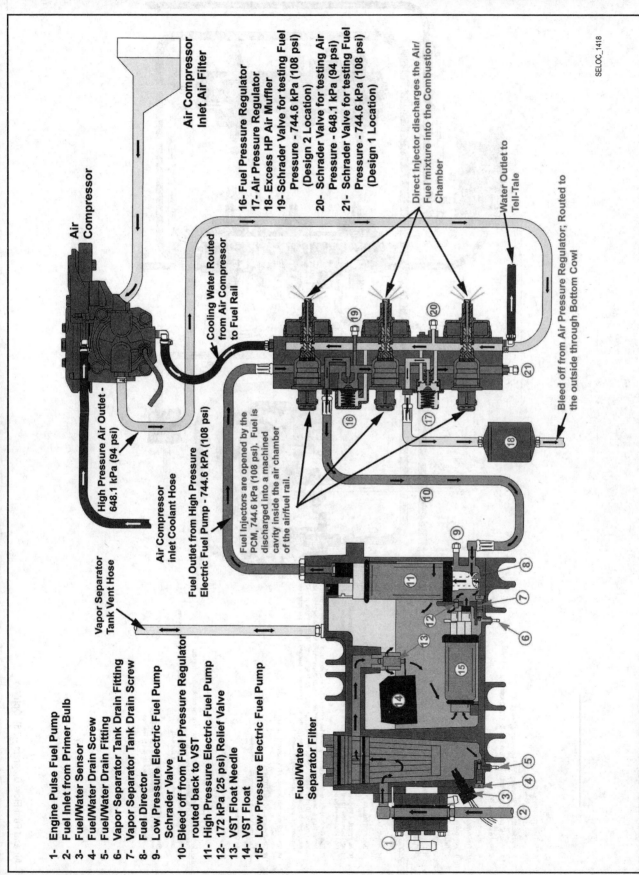

SELOC_1418

1- Engine Pulse Fuel Pump
2- Fuel Inlet from Primer Bulb
3- Fuel/Water Sensor
4- Fuel/Water Drain Screw
5- Fuel/Water Drain Fitting
6- Vapor Separator Tank Drain Fitting
7- Vapor Separator Tank Drain Screw
8- Fuel Director
9- Low Pressure Electric Fuel Pump
10- Bleed off from Fuel Pressure Regulator routed back to VST
11- High Pressure Electric Fuel Pump
12- 172 kPa (25 psi) Relief Valve
13- VST Float Needle
14- VST Float
15- Low Pressure Electric Fuel Pump

16- Fuel Pressure Regulator
17- Air Pressure Regulator
18- Excess HP Air Muffler
19- Schrader Valve for testing Fuel Pressure - 744.6 kPa (108 psi) (Design 2 Location)
20- Schrader Valve for testing Air Pressure - 648.1 kPa (94 psi)
21- Schrader Valve for testing Fuel Pressure - 744.6 kPa (108 psi) (Design 1 Location)

Air Compressor

Air Compressor Inlet Air Filter

High Pressure Air Outlet - 648.1 kPa (94 psi)

Air Compressor inlet Coolant Hose

Cooling Water Routed from Air Compressor to Fuel Rail

Fuel Outlet from High Pressure Electric Fuel Pump - 744.6 kPA (108 psi)

Vapor Separator Tank Vent Hose

Fuel Injectors are opened by the PCM, 744.6 kPa (108 psi). Fuel is discharged into a machined cavity inside the air chamber of the air/fuel rail.

Direct Injector discharges the Air/Fuel mixture into the Combustion Chamber

Water Outlet to Tell-Tale

Bleed off from Air Pressure Regulator; Routed to the outside through Bottom Cowl

Fuel/Water Separator Filter

Air and Fuel Flow Diagram - 1.5L OptiMax

SELOC_1418

1. Fuel inlet from primer bulb
2. Engine Pulse Fuel Pump
3. Fuel line to Water Separating Fuel Filter – 2-8 psi (14-55 kPa)
4. Water Separating Fuel Filter in Vapor Separator Tank (VST) Assembly
5. Fuel outlet from VST
6. Fuel inlet to Low Pressure Electric Fuel Pump
7. Fuel outlet from Low Pressure Electric Fuel Pump – 7-9 psi (48-62 kPa)
8. Fuel inlet to High Pressure Electric Fuel Pump.
9. Relief Passage – Unused fuel returning to VST
10. Air Vent to VST
11. Fuel outlet from High Pressure Electric Fuel Pump – 90 psi (620 kPa)
12. Fuel Return inlet from Fuel Regulator
13. Fuel injector is opened by the ECM, 90 psi (620 kPa) fuel is discharged into a machined cavity inside the air chamber of the air/fuel rail. This mixes the fuel with the air charge.
14. Air inlet to Air Compressor
15. Bleed Off from Air Pressure Regulator, routed to the Exhaust Adaptor and Exits thru the Propeller.
16. Air Pressure Regulator will limit the amount of pressure developed inside the air passages to approximately 10 psi (69 kPa) below the pressure of the fuel inside the fuel passages (i.e. 80 psi [551 kPa] air vs 90 psi [620 kPa] fuel).
17. Fuel Pressure Regulator not only regulates fuel pressure but also regulates it at approx. 10 psi (69 kPa) higher than whatever the air rail pressure is. The fuel regulator diaphragm is held closed with a spring that requires 10 psi (69 kPa) to force the diaphragm off the diaphragm seat. The back side of the diaphragm is exposed to air rail pressure. As the air rail pressure increases, the fuel pressure needed to open the regulator will equally increase.
18. Bleed off from fuel Pressure Regulator, Routed Back to VST.
19. Check Valve - 40 psi (276 kPa)
20. Schrader Valve for Testing Fuel Pressure
21. Air Compressor
22. High Pressure Air Outlet - 80 psi (551 kPa)
23. High Pressure Air inlet to Air/Fuel Rails - 80 psi (551 kPa)
24. Direct Injector discharges the air/fuel mixture into the combustion chamber
25. Schrader Valve for Testing Air Pressure
26. Tracker Valve has a rubber diaphragm which expands and retracts to equalize the pulses developed by the pumps (both air and fuel).
27. Fuel return inlet from Fuel Regulator
28. Water inlet to cool port air/fuel rail and air compressor
29. Cooling water from compressor routed to Tell-Tale

Air and Fuel Flow Diagram - 2.5L OptiMax

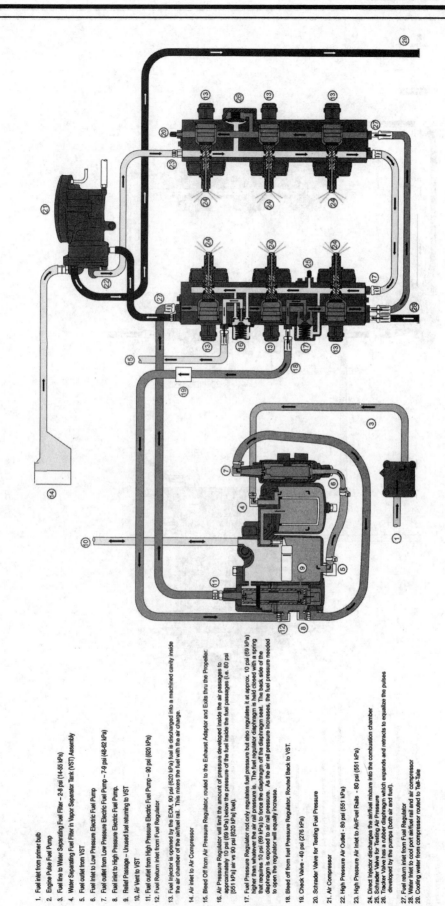

SELOC_1418

1. Fuel inlet from primer bulb
2. Engine Pulse Fuel Pump
3. Fuel line to Water Separating Fuel Filter – 2-8 psi (14-55 kPa)
4. Water Separating Fuel Filter in Vapor Separator Tank (VST) Assembly
5. Fuel outlet from VST
6. Fuel inlet to Low Pressure Electric Fuel Pump
7. Fuel outlet from Low Pressure Electric Fuel Pump – 7-9 psi (48-62 kPa)
8. Fuel inlet to High Pressure Electric Fuel Pump.
9. Relief Passage – Unused fuel returning to VST
10. Air Vent to VST
11. Fuel outlet from High Pressure Electric Fuel Pump – 90 psi (620 kPa)
12. Fuel Return inlet from Fuel Regulator
13. Fuel Injector is opened by the ECM. 90 psi (620 kPa) fuel is discharged into a machined cavity inside the air chamber of the air/fuel rail. This mixes the fuel with the air charge.
14. Air inlet to Air Compressor
15. Bleed Off from Air Pressure Regulator, routed to the Exhaust Adaptor and Exits thru the Propeller.
16. Air Pressure Regulator will limit the amount of pressure developed inside the air passages to approximately 10 psi (69 kPa) below the pressure of the fuel inside the fuel passages (i.e. 80 psi [551 kPa] air vs 90 psi (620 kPa) fuel].
17. Fuel Pressure Regulator not only regulates fuel pressure but also regulates it at approx. 10 psi (69 kPa) higher than whatever the air rail pressure is. The fuel regulator diaphragm is held closed with a spring that requires 10 psi (69 kPa) to force the diaphragm off the diaphragm seat. The back side of the diaphragm is exposed to air rail pressure. As the air rail pressure increases, the fuel pressure needed to open the regulator will equally increase.
18. Bleed off from fuel Pressure Regulator, Routed Back to VST.
19. Check Valve - 40 psi (276 kPa)
20. Schrader Valve for Testing Fuel Pressure
21. Air Compressor
22. High Pressure Air Outlet - 80 psi (551 kPa)
23. High Pressure Air Inlet to Air/Fuel Rails - 80 psi (551 kPa)
24. Direct Injector discharges the air/fuel mixture into the combustion chamber
25. Schrader Valve for Testing Air Pressure
26. Tracker Valve has a rubber diaphragm which expands and retracts to equalize the pulses developed by the pumps (both air and fuel).
27. Fuel return inlet from Fuel Regulator
28. Water Inlet to cool port air/fuel rail and air compressor
29. Cooling water from compressor routed to Tell-Tale

Air and Fuel Flow Diagram - 3.0L OptiMax (thru 2002)

SELOC_1418

1. Engine Pulse Fuel Pump
2. Fuel/Water Separator Filter
3. Vapor Separator Tank Vent Hose
4. Fuel Outlet from High Pressure Electric Fuel Pump – 620 kPa (90 psi)
5. Bleed Off from Air Pressure Regulator; routed to exhaust adaptor and exits through the propeller
6. Air Compressor Inlet Coolant Hose
7. Air Compressor
8. Air Compressor Inlet Air Filter
9. High Pressure Air Outlet – 551 kPa (80 psi)
10. Schrader Valve for testing fuel pressure – 620 kPa (90 psi)
11. Fuel Injector is opened by the ECM/PCU; 620 kPa (90 psi) fuel is discharged into a machined cavity inside the air chamber of the air/fuel rail. This mixes fuel with the air charge.
12. Tracker Valve has a rubber diaphragm which expands and retracts to equalize the pulses developed by the pumps (both air and fuel).
13. Cooling water routed from air compressor to tell-tale
14. Direct Injector discharges the air/fuel mixture into the combustion chamber
15. Schrader Valve for testing air pressure – 551 kPa (80 psi)
16. Water inlet to cool port air/fuel rail and air compressor
17. Fuel Pressure Regulator not only regulates fuel pressure but also regulates it at approximately 69 kPa (10 psi) higher than whatever the air rail pressure is. The fuel regulator diaphragm is held closed with a spring that requires 69 kPa (10 psi) to force the diaphragm off the diaphragm seat. The backside of the diaphragm is exposed to air rail pressure. As the air rail pressure increases, the fuel pressure needed to open the regulator will equally increase.
18. Air Pressure Regulator will limit the amount of pressure developed inside the air passage to approximately 69 kPa (10 psi) below the pressure of the fuel inside the fuel passage.
19. Bleed of from fuel pressure regulator; routed back to vapor separator tank.
20. Low Pressure Electric Fuel Pump Schrader Valve
21. Fuel Director
22. Vapor Separator Tank Drain Screw
23. Vapor Separator Tank Drain Fitting
24. 172 kPa (25 psi) relief valve
25. Low Pressure Electric Fuel Pump
26. Vapor Separator Tank Float Needle
27. Vapor Separator Tank Float
28. Fuel/Water Drain Fitting
29. Fuel/Water Drain Screw
30. Fuel/Water Sensor
31. Fuel Inlet from primer bulb

Air and Fuel Flow Diagram - 3.0L OptiMax (2003 & later)

Carburetor Set-Up Specifications

Model (HP)	No of Cyl	Year	Displace cu.in. (cc)	Pilot Screw Initial Low Speed Setting	Float Height/Drop Setting In. (mm)	Main Jet Sizes In. (mm)
2.5	1	2001-05	4.6 (74.6)	E-clip in second groove	0.090 (2.0) from gasket	#92
3.3	1	2001-05	4.6 (74.6)	E-clip in second groove	0.090 (2.0) from gasket	#94
4	1	2001-05	6.2 (102)	1-1/2	1/2 (13)	0.031 (0.78mm)
5	1	2001-05	6.2 (102)	1-1/2	1/2 (13)	0.032 (0.80mm)
6	2	2001-03	13 (210)	1 to 1-1/2	1.0 (25.4)	varies by altitude
8	2	2001-05	13 (210)	1-1/4 to 1-3/4	1.0 (25.4)	varies by altitude
9.9	2	2001-05	16 (262)	1-1/4 to 1-3/4	1.0 (25.4)	varies by altitude
10	2	2001	16 (262)	1-1/4 to 1-3/4	1.0 (25.4)	varies by altitude
15	2	2001-05	16 (262)	1-1/4 to 1-3/4	1.0 (25.4)	varies by altitude
20	2	2001-03	24 (400)	3/4 to 1-1/4	1.0 (25.4)	0.076 (1.93mm)
20 Jet	2	2001-09	24 (400)	1 to 2	1.0 (25.4)	0.044 (1.12mm)
25	2	2001-09	24 (400)	1 to 1-1/2	1.0 (25.4)	0.076 (1.93mm) ①
30	2	2001-02	39 (644)	1-1/4 to 1-3/4	9/16 (14.3)	0.054 (1.37mm)
40	2	2001-03	39 (644)	1-1/4 to 1-3/4 ②	9/16 (14.3)	0.066 (1.68mm)
40	3	2001-05	59 (967)	1 to 1-1/2	9/16 (14.3)	0.044 (1.12mm)
40 Jet	3	2001-09	59 (967)	1 to 1-1/2	9/16 (14.3)	0.060 (1.52mm)
50	3	2001-09	59 (967)	1 to 1-1/2	9/16 (14.3)	0.052 (1.32mm)
55	3	2001	59 (967)	1 to 1-1/2	9/16 (14.3)	0.058 (1.47mm)
60	3	2001-09	59 (967)	1 to 1-1/2	9/16 (14.3)	0.060 (1.52mm)
65 Jet	3	2001-09	85 (1386)	1 to 1 1/2	9/16 (14.3)	varies by carb and model
75	3	2001-05	85 (1386)	7/8 to 1-3/8	9/16 (14.3)	varies by carb and model
90	3	2001-09	85 (1386)	1 to 1 1/2	9/16 (14.3)	varies by carb and model
80 Jet	4	2001-05	113 (1848)	1 to 1 1/2	9/16 (14.3)	varies by carb and model
115	4	2001-05	113 (1848)	1 to 1 1/2	9/16 (14.3)	varies by carb and model
125	4	2001-05	113 (1848)	1 to 1 1/2	9/16 (14.3)	varies by carb and model
150 (XR6, Classic)	V6	2001-05	153 (2507)	1-1/8 to 1-3/8	even with bowl edge	0.072 (1.83mm)
200	V6	2001-05	153 (2507)	1-1/8 to 1-3/8	even with bowl edge	varies by carb

① Spec is for all models EXCEPT Seapro/Marathon which is 0.080 in. (2.03mm)

② Spec is for all EXCEPT WME-38 Carb on 40MLL models, for which the spec is 1 1/2 - 2

SELOC_1418

Fuel Injection Component Adjustment Specifications

Model (HP)	No of Cyl	Year	Displace cu.in. (cc)	Crankshaft Position Sensor Air Gap in. (mm)	Crankshaft Position Sensor Resistance Ohms	Throttle Position Sensor Output Voltage Voltage @ Idle	Throttle Position Sensor Output Voltage Voltage @ WOT
75 Optimax	3	2004-09	93 (1526)	0.20 (5.0) ①	300-350	0.17-1.05	3.76-4.64
90 Optimax	3	2004-09	93 (1526)	0.20 (5.0) ①	300-350	0.17-1.05	3.76-4.64
110 Jet Optimax	3	2009	93 (1526)	0.20 (5.0) ①	300-350	0.17-1.05	3.76-4.64
115 Optimax	3	2004-09	93 (1526)	0.20 (5.0) ①	300-350	0.17-1.05	3.76-4.64
135 Optimax	V6	2001-09	153 (2507)	0.025-0.040 (0.635-1.010)	not available	3.20-4.90	0.10-1.50
150 EFI	V6	2001	153 (2507)	Uses Trigger Coil see Ignition System Component Testing		0.20-0.30	approx 7.46
	V6	2002-09	153 (2507)	not available	300-350	0.19-1.00	3.45-4.63
110 Jet Optimax	V6	2001-09	153 (2507)	0.025-0.040 (0.635-1.010)	not available	3.20-4.90	0.10-1.50
150 Optimax	V6	2001-09	153 (2507)	0.025-0.040 (0.635-1.010)	not available	3.20-4.90	0.10-1.50
175 EFI	V6	2001	153 (2507)	Uses Trigger Coil see Ignition System Component Testing		0.20-0.30	approx 7.46
	V6	2002	153 (2507)	not available	300-350	0.19-1.00	3.45-4.63
175 Optimax	V6	2001-09	153 (2507)	0.025-0.040 (0.635-1.010) ②	not available	3.20-4.90	0.10-1.50
200 EFI (2.5L)	V6	2001	153 (2507)	Uses Trigger Coil see Ignition System Component Testing		0.20-0.30	approx 7.46
	V6	2002-05	153 (2507)	not available	300-350	0.19-1.00	3.45-4.63
200 EFI (3.0L)	V6	2004-05	185 (3032)	not available	300-350	0.19-1.00	3.45-4.63
200 Optimax	V6	2001-09	185 (3032)	0.025-0.040 (0.635-1.010)	300-340	0.40-1.30	4.00-4.70
225 EFI	V6	2001	185 (3032)	0.020-0.060 (0.51-1.53)	900-1300	0.90-1.00	3.55-4.05
	V6	2002-05	185 (3032)	not available	300-350	0.19-1.00	3.45-4.63
225 Optimax	V6	2001-09	185 (3032)	0.025-0.040 (0.635-1.010) ②	300-340	0.40-1.30	4.00-4.70
250 EFI	V6	2001	185 (3032)	0.020-0.060 (0.51-1.53)	900-1300	0.90-1.00	3.55-4.05
	V6	2002-05	185 (3032)	not available	300-350	0.19-1.00	3.45-4.63
250 Optimax	V6	2006-09	185 (3032)	0.025-0.040 (0.635-1.010) ②	300-340	0.40-1.30	4.00-4.70

① Original Specifications for these motors had a gap of 0.025-0.040 (0.635-1.010), but recommendations were changed when tech literature from Merc was revised in 2007. Also, keep in mind that late-model motors, S/N 1785000 and later utilize sensors that are not adjustable.

② Pro/XS/Sport models are normally NOT ADJUSTABLE and no specification is given for them

SELOC_1418

4

IGNITION & ELECTRICAL SYSTEMS

UNDERSTANDING AND TROUBLESHOOTING ELECTRICAL SYSTEMS

Basic Electrical Theory

◆ See Figure 1

For any 12-volt, negative ground, electrical system to operate, the electricity must travel in a complete circuit. This simply means that current (power) from the positive terminal (+) of the battery must eventually return to the negative terminal (-) of the battery. Along the way, this current will travel through wires, fuses, switches and components. If, for any reason, the flow of current through the circuit is interrupted, the component fed by that circuit would cease to function properly.

Perhaps the easiest way to visualize a circuit is to think of connecting a light bulb (with 2 wires attached to it) to the battery - 1 wire attached to the negative (-) terminal of the battery and the other wire to the positive (+) terminal. With the 2 wires touching the battery terminals, the circuit would be complete and the light bulb would illuminate. Electricity would follow a path from the battery to the bulb and back to the battery. It's easy to see that with wires of sufficient length, our light bulb could be mounted nearly anywhere on the boat. Further, 1 wire could be fitted with a switch inline so that the light could be turned on and off without having to physically remove the wire(s) from the battery.

The normal marine circuit differs from this simple example in two ways. First, instead of having a return wire from each bulb to the battery, the current travels through a single ground wire that handles all the grounds for a specific circuit. Secondly, most marine circuits contain multiple components that receive power from a single circuit. This lessens the overall amount of wire needed to power components.

HOW DOES ELECTRICITY WORK: THE WATER ANALOGY

Electricity is the flow of electrons - the sub-atomic particles that constitute the outer shell of an atom. Electrons spin in an orbit around the center core of an atom. The center core is comprised of protons (positive charge) and neutrons (neutral charge). Electrons have a negative charge and balance out the positive charge of the protons. When an outside force causes the number of electrons to unbalance the charge of the protons, the electrons will split off the atom and look for another atom to balance out. If this imbalance is kept up, electrons will continue to move and an electrical flow will exist.

Many people find electrical theory easier to understand when using an analogy with water. In a comparison with water flowing through a pipe, the electrons would be the water and the wire is the pipe.

The flow of electricity can be measured much like the flow of water through a pipe. The unit of measurement used is amperes, frequently abbreviated as amps (a). You can compare amperage to the volume of water flowing through a pipe (for water that would mean a measurement of mass usually measured in units delivered over a set amount of time such as gallons or liters per minute). When connected to a circuit, an ammeter will measure the actual amount of current flowing through the circuit. When relatively few electrons flow through a circuit, the amperage is low. When many electrons flow, the amperage is high.

Water pressure is measured in units such as pounds per square inch (psi). The electrical pressure is measured in units called volts (v). When a voltmeter is connected to a circuit, it is measuring the electrical pressure.

The actual flow of electricity depends not only on voltage and amperage, but also on the resistance of the circuit. The higher the resistance, the higher the force necessary to push the current through the circuit. The standard unit for measuring resistance is an ohm (<omega>). Resistance in a circuit varies depending on the amount and type of components used in the circuit. The main factors that determine resistance are:

• Material - some materials have more resistance than others. Those with high resistance are said to be insulators. Rubber materials (or rubber-like plastics) are some of the most common insulators used, as they have a very high resistance to electricity. Very low resistance materials are said to be conductors. Copper wire is among the best conductors. Silver is actually a superior conductor to copper and is used in some relay contacts, but its high cost prohibits its use as common wiring. Most marine wiring is made of copper.

• Size - the larger the wire size being used, the less resistance the wire will have (just as a large diameter pipe will allow small amounts of water to just trickle through). This is why components that use large amounts of electricity usually have large wires supplying current to them.

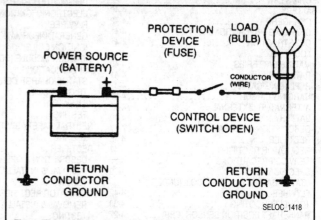

Fig. 1 This example illustrates a simple circuit. When the switch is closed, power from the positive (+) battery terminal flows through the fuse and the switch, and then to the light bulb. The electricity illuminates the bulb and the circuit is completed through the ground wire back to the negative (-) battery terminal

• Length - for a given thickness of wire, the longer the wire, the greater the resistance. The shorter the wire, the less the resistance. When determining the proper wire for a circuit, both size and length must be considered to design a circuit that can handle the current needs of the component.

• Temperature - with many materials, the higher the temperature, the greater the resistance (positive temperature coefficient). Some materials exhibit the opposite trait of lower resistance with higher temperatures (these are said to have a negative temperature coefficient). These principles are used in many engine control sensors (especially those found on microcomputer controlled ignition and fuel injection systems).

OHM'S LAW

There is a direct relationship between current, voltage and resistance. The relationship between current, voltage and resistance can be summed up by a statement known as Ohm's law.

Voltage (E) is equal to amperage (I) times resistance (R): $E = I \times R$

Other forms of the formula are $R = E/I$ and $I = E/R$

In each of these formulas, E is the voltage in volts, I is the current in amps and R is the resistance in ohms. The basic point to remember is that if the voltage of a circuit remains the same, as the resistance of that circuit goes up, the amount of current that flows in the circuit will go down.

The amount of work that electricity can perform is expressed as power. The unit of power is the watt (w). The relationship between power, voltage and current is expressed as:

Power (W) is equal to amperage (I) times voltage (E): $W = I \times E$

This is only true for direct current (DC) circuits; the alternating current formula is a tad different, but since the electrical circuits in most vessels are DC type, we need not get into AC circuit theory.

Electrical Components

POWER SOURCE

◆ See Figure 2

Typically, power is supplied to a vessel by 2 devices: The battery and the stator (or battery charge coil). The stator supplies electrical current anytime the engine is running in order to recharge the battery and in order to operate electrical devices of the vessel. The battery supplies electrical power during starting or during periods when the current demand of the vessel's electrical system exceeds stator output capacity (which includes times when the motor is shut off and stator output is zero).

The Battery

In most modern vessels, the battery is a lead/acid electrochemical device consisting of six 2-volt subsections (cells) connected in series, so that the unit is capable of producing approximately 12 volts of electrical pressure.

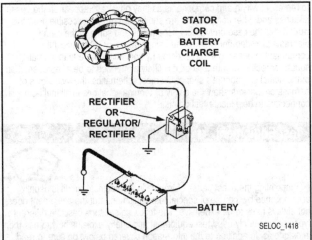

Fig. 2 Functional diagram of a typical charging circuit showing the relationship between the stator (battery charge coil), rectifier (or regulator/rectifier) and battery

Each subsection consists of a series of positive and negative plates held a short distance apart in a solution of sulfuric acid and water.

The 2 types of plates in each battery cell are of dissimilar metals. This sets up a chemical reaction, and it is this reaction which produces current flow from the battery when its positive and negative terminals are connected to an electrical load. Power removed from the battery in use is replaced by current from the stator and restores the battery to its original chemical state.

The Stator

Alternators and generators are devices that consist of coils of wires wound together making big electromagnets. The coil is normally referred to as a stator or battery charge coil. Either, 1 group of coils spins within another set (or a set of permanently charged magnets, usually attached to the flywheel, are spun around a set of coils) and the interaction of the magnetic fields generates an electrical current. This current is then drawn off the coils and fed into the vessel's electrical system.

■ Some vessels utilize a generator instead of an alternator. Although the terms are often misused and interchanged, the main difference is that an alternator supplies alternating current that is changed to direct current for use on the vessel, while a generator produces direct current. Alternators tend to be more efficient and that is why they are used on almost all modern engines.

GROUND

Two types of grounds are used in marine electric circuits. Direct ground components are grounded to the electrically conductive metal through their mounting points. All other components use some sort of ground wire that leads back to the battery. The electrical current runs through the ground wire and returns to the battery through the ground or negative (-) cable; if you look, you'll see that the battery ground cable connects between the battery and a heavy gauge ground wire.

■ A large percentage of electrical problems can be traced to bad grounds.

If you refer back to the basic explanation of a circuit, you'll see that the ground portion of the circuit is just as important as the power feed. The wires delivering power to a component can have perfectly good, clean connections, but the circuit would fail to operate if there was a damaged ground connection. Since many components ground through their mounting or through wires that are connected to an engine surface, contamination from dirt or corrosion can raise resistance in a circuit to a point where it cannot operate.

PROTECTIVE DEVICES

Problems can occur in the electrical system that will cause large surges of current to pass through the electrical system of your vessel. These problems can be the fault of the charging circuit, but more likely would be a problem with the operating electrical components that causes an excessively high

load. An unusually high load can occur in a circuit from problems such as a seized electric motor (like a damaged starter) or the excessive resistance caused by a bad ground (from loose or damaged wires or connections). A short to ground that bypasses the load and allows the battery to quickly discharge through a wire can also cause current surges.

If this surge of current were to reach the load in the circuit, the surge could burn it out or severely damage it. It can also overload the wiring, causing the harness to get hot and melt the insulation. To prevent this, fuses, circuit breakers and/or fusible links are connected into the supply wires of the electrical system. These items are nothing more than a built-in weak spot in the system. When an abnormal amount of current flows through the system, these protective devices work as follows to protect the circuit:

• Fuse - when an excessive electrical current passes through a fuse, the fuse blows (the conductor melts) and opens the circuit, preventing current flow.

• Circuit Breaker - a circuit breaker is basically a self-repairing fuse. It will open the circuit in the same fashion as a fuse, but when the surge subsides, the circuit breaker can be reset and does not need replacement. Most circuit breakers on marine engine applications are self-resetting, but some that operate accessories (such as on larger vessels with a circuit breaker panel) must be reset manually (just like the circuit breaker panels in most homes).

• Fusible Link - a fusible link (fuse link or main link) is a short length of special, high temperature insulated wire that acts as a fuse. When an excessive electrical current passes through a fusible link, the thin gauge wire inside the link melts, creating an intentional open to protect the circuit. To repair the circuit, the link must be replaced. Some newer type fusible links are housed in plug-in modules, which are simply replaced like a fuse, while older type fusible links must be cut and spliced if they melt. Since this link is very early in the electrical path, it's the first place to look if nothing on the vessel works, yet the battery seems to be charged and is otherwise properly connected.

✳✳ SELOC CAUTION

Always replace fuses, circuit breakers and fusible links with identically rated components. Under no circumstances should a component of higher or lower amperage rating be substituted. A lower rated component will disable the circuit sooner than necessary (possibly during normal operation), while a higher rated component can allow dangerous amounts of current that could damage the circuit or component (or even melt insulation causing sparks or a fire).

SWITCHES & RELAYS

◆ See Figure 3

Switches are used in electrical circuits to control the passage of current. The most common use is to open and close circuits between the battery and the various electric devices in the system. Switches are rated according to the amount of amperage they can handle. If a sufficient amperage rated switch is not used in a circuit, the switch could overload and cause damage.

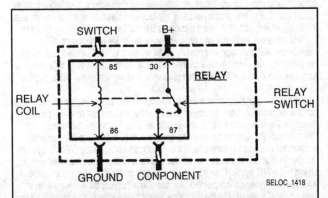

Fig. 3 Relays are composed of a coil and a switch. These 2 components are linked together so that when one is operated it actuates the other. The large wires in the circuit are connected from the battery to one side of the relay switch (B+) and from the opposite side of the relay switch to the load (component). Smaller wires are connected from the relay coil to the control switch for the circuit and from the opposite side of the relay coil to ground

Some electrical components that require a large amount of current to operate use a special switch called a relay. Since these circuits carry a large amount of current, the thickness of the wire in the circuit is also greater. If this large wire were connected from the load to the control switch, the switch would have to carry the high amperage load and the space needed for wiring in the vessel would be twice as big to accommodate the increased size of the wiring harness. A relay is used to prevent these problems.

Think of relays as essentially "remote controlled switches." They allow a smaller current to throw the switch that operates higher amperages devices. Relays are composed of a coil and a set of contacts. When current is passed through the coil, a magnetic field is formed that causes the contacts to move together, closing the circuit. Most relays are normally open, preventing current from passing through the main circuit until power is applied to the coil. But, relays can take various electrical forms depending on the job for which they are intended. Some common circuits that may use relays are horns, lights, starters, electric fuel pumps and other potentially high draw circuits.

LOAD

Every electrical circuit must include a load (something to use the electricity coming from the source). Without this load, the battery would attempt to deliver its entire power supply from one pole to another. This is called a short circuit. All this electricity would take a short cut to ground and cause a great amount of damage to other components in the circuit (including the battery) by developing a tremendous amount of heat. This condition could develop sufficient heat to melt the insulation on all the surrounding wires and reduce a multiple wire cable to a lump of plastic and copper. A short can allow sparks that could ignite fuel vapors or other combustible materials in the vessel, causing an extremely hazardous condition.

WIRING & HARNESSES

The average vessel contains miles of wiring, with hundreds of individual connections. To protect the many wires from damage and to keep them from becoming a confusing tangle, they are organized into bundles, enclosed in plastic or taped together and called wiring harnesses. Different harnesses serve different parts of the vessel. Individual wires are color coded to help trace them through a harness where sections are hidden from view.

Marine wiring or circuit conductors can be either single strand wire, multi-strand wire or printed circuitry. Single strand wire has a solid metal core and is usually used inside such components as stator coil windings, motors, relays and other devices. Multi-strand wire has a core made of many small strands of wire twisted together into a single conductor. Most of the wiring in a marine electrical system is made up of multi-strand wire, either as a single conductor or grouped together in a harness. All wiring is color coded on the insulator, either as a solid color or as a colored wire with an identification stripe. A printed circuit is a thin film of copper or other conductor that is printed on an insulator backing. Occasionally, a printed circuit is sandwiched between 2 sheets of plastic for more protection and flexibility. A complete printed circuit, consisting of conductors, insulating material and connectors is called a printed circuit board. Printed circuitry is used in place of individual wires or harnesses in places where space is limited, such as behind 1-piece instrument clusters.

Since marine electrical systems are very sensitive to changes in resistance, the selection of properly sized wires is critical when systems are repaired. A loose or corroded connection or a replacement wire that is too small for the circuit will add extra resistance and an additional voltage drop to the circuit.

The wire gauge number is an expression of the cross-section area of the conductor. Vessels from countries that use the metric system will typically describe the wire size as its cross-sectional area in square millimeters. In this method, the larger the wire, the greater the number. Another common system for expressing wire size is the American Wire Gauge (AWG) system. As gauge number increases, area decreases and the wire becomes smaller. Using the AWG system, an 18 gauge wire is smaller than a 4 gauge wire. A wire with a higher gauge number will carry less current than a wire with a lower gauge number. Gauge wire size refers to the size of the strands of the conductor, not the size of the complete wire with insulator. It is possible, therefore, to have 2 wires of the same gauge with different diameters because one may have thicker insulation than the other.

It is essential to understand how a circuit works before trying to figure out why it doesn't. An electrical schematic shows the electrical current paths when a circuit is operating properly. Schematics break the entire electrical system down into individual circuits. In most schematics no attempt is made

to represent wiring and components as they physically appear on the vessel; switches and other components are shown as simply as possible. But, this is **not** always the case on Mercury schematics and some of the wiring diagrams provided here. So, when using a Mercury schematic if the component in question is represented by something more than a small square or rectangle with a label, it is likely a intended to be a representation of the actual component's shape. On most schematics, the face views of harness connectors show the cavity or terminal locations in all multi-pin connectors to help locate test points.

Test Equipment

Pinpointing the exact cause of trouble in an electrical circuit is usually accomplished by the use of special test equipment, but the equipment does not always have to be expensive. The following sections describe different types of commonly used test equipment and briefly explains how to use them in diagnosis. In addition to the information covered below, be sure to read and understand the tool manufacturer's instruction manual (provided with most tools) before attempting any test procedures.

JUMPER WIRES

◆ See Figure 4

✳✳ SELOC CAUTION

Never use jumper wires made from a thinner gauge wire than the circuit being tested. If the jumper wire is of too small a gauge, it may overheat and possibly melt. Never use jumpers to bypass high resistance loads in a circuit. Bypassing resistances, in effect, creates a short circuit. This may, in turn, cause damage and fire. Jumper wires should only be used to bypass lengths of wire or to simulate switches.

Jumper wires are simple, yet extremely valuable, pieces of test equipment. They are basically test wires that are used to bypass sections of a circuit. Although jumper wires can be purchased, they are usually fabricated from lengths of standard marine wire and whatever type of connector (alligator clip, spade connector or pin connector) that is required for the particular application being tested. In cramped, hard-to-reach areas, it is advisable to have insulated boots over the jumper wire terminals in order to prevent accidental grounding. It is also advisable to include a standard marine fuse in any jumper wire. This is commonly referred to as a fused jumper. By inserting an in-line fuse holder between a set of test leads, a fused jumper wire is created for bypassing open circuits. Use a 5-amp fuse to provide protection against voltage spikes.

Jumper wires are used primarily to locate open electrical circuits, on either the ground (-) side of the circuit or on the power (+) side. If an electrical component fails to operate, connect the jumper wire between the component and a good ground. If the component operates only with the jumper installed, the ground circuit is open. If the ground circuit is good, but the component does not operate, the circuit between the power feed and component may be open. By moving the jumper wire successively back from the component toward the power source, you can isolate the area of the circuit where the open is located. When the component stops functioning, or the power is cut off, the open is in the segment of wire between the jumper and the point previously tested.

You can sometimes connect the jumper wire directly from the battery to the hot terminal of the component, but first make sure the component uses a full 12 volts in operation. Some electrical components, such as sensors, are designed to operate on smaller voltages like 4 or 5 volts, and running 12 volts directly to these components can damage or destroy them.

TEST LIGHTS

◆ See Figure 5

The test light is used to check circuits and components while electrical current is flowing through them. It is used for voltage and ground tests. To use a 12-volt test light, connect the ground clip to a good ground and probe connectors the pick where you are wondering if voltage is present. The test light will illuminate when voltage is detected. This does not necessarily mean that 12 volts (or any particular amount of voltage) is present; it only means that some voltage is present. It is advisable before using the test light to touch its ground clip and probe across the battery posts or terminals to make sure the light is operating properly and to note how brightly the light glows when 12 volts is present.

Do not use a test light to probe electronic ignition, spark plug or coil wires, as the circuit is much, much higher than 12 volts. Also, never use a pick-type test light to probe wiring on electronically controlled systems unless specifically instructed to do so. Whenever possible, avoid piercing insulation with the test light pick, as you are inviting shorts or corrosion and excessive resistance. But, any wire insulation that is pierced by necessity, must be sealed with silicone and taped after testing.

Like the jumper wire, the 12-volt test light is used to isolate opens in circuits. But, whereas the jumper wire is used to bypass the open to operate the load, the 12-volt test light is used to locate the presence or lack of voltage in a circuit. If the test light illuminates, there is power up to that point in the circuit; if the test light does not illuminate, there is an open circuit (no power). Move the test light in successive steps back toward the power source until the light in the handle illuminates. The open is between the probe and the point that was previously probed.

The self-powered test light is similar in design to the 12-volt test light, but contains a 1.5 volt penlight battery in the handle. It is most often used in place of a multi-meter to check for open or short circuits when power is isolated from the circuit (thereby performing a continuity test).

The battery in a self-powered test light does not provide much current. A weak battery may not provide enough power to illuminate the test light even when a complete circuit is made (especially if there is high resistance in the circuit). Always make sure that the test battery is strong. To check the battery, briefly touch the ground clip to the probe; if the light glows brightly, the battery is strong enough for testing.

■ **A self-powered test light should not be used on any electronically controlled system or component. Even the small amount of electricity transmitted by the test light is enough to damage many electronic components.**

MULTI-METERS

◆ See Figure 6

Multi-meters are extremely useful for troubleshooting electrical problems. They can be purchased in both analog or digital form and have a price range to suit nearly any budget. A multi-meter is a voltmeter, ammeter and ohmmeter (along with other features) combined into one instrument. It is often used when testing solid state circuits because of its high input impedance (usually 10 mega ohms or more). A high-quality digital multi-meter or Digital Volt Ohm Meter (DVOM) helps to ensure the most accurate test results and, although not absolutely necessary for electronic components such as computer controlled ignition systems and charging

systems, is highly recommended. A brief description of the main test functions of a multi-meter follows:

• Voltmeter - the voltmeter is used to measure voltage at any point in a circuit or to measure the voltage drop across any part of a circuit. Voltmeters usually have various scales and a selector switch to allow metering and display of different voltage ranges. The voltmeter has a positive and a negative lead. To avoid damage to the meter, connect the negative lead to the negative (-) side of the circuit (to ground or nearest the ground side of the circuit) and connect the positive lead to the positive (+) side of the circuit (to the power source or the nearest power source). This is mostly a concern on analog meters, as DVOMs are not normally adversely affected (as they are usually designed to take readings even with reverse polarity and display accordingly). Note that the negative voltmeter lead will always be black and that the positive voltmeter will always be some color other than black (usually red).

• Ohmmeter - the ohmmeter is designed to read resistance (measured in ohms) in a circuit or component. Most ohmmeters will have a selector switch which permits the measurement of different ranges of resistance (usually the selector switch allows the multiplication of the meter reading by 10, 100, 1,000 and 10,000). Most modern ohmmeters (especially DVOMs) are auto-ranging which means the meter itself will determine which scale to use. Since ohmmeters are powered by an internal battery, the ohmmeter can be used like a self-powered test light. When the ohmmeter is connected, current from the ohmmeter flows through the circuit or component being tested. Since the ohmmeter's internal resistance and voltage are known values, the amount of current flow through the meter depends on the resistance of the circuit or component being tested. The ohmmeter can also be used to perform a continuity test for suspected open circuits. When using the meter for continuity checks, do not be concerned with the actual resistance readings. Zero resistance, or any ohm reading, indicates continuity in the circuit. Infinite resistance indicates an opening in the circuit. A high resistance reading where there should be none indicates a problem in the circuit. Checks for short circuits are made in the same manner as checks for open circuits, except that the circuit must be isolated from both power and normal ground. Infinite resistance indicates no continuity, while zero resistance indicates a dead short.

Never use an ohmmeter to check the resistance of a component or wire while there is voltage applied to the circuit. Voltage in the circuit can damage or destroy the meter.

• Ammeter - an ammeter measures the amount of current flowing through a circuit in units called amperes or amps. At normal operating voltage, most circuits have a characteristic amount of amperes, called current draw that can be measured using an ammeter. By referring to a specified current draw rating, and then measuring the amperes and comparing the 2 values, you can determine what is happening within the circuit to aid in diagnosis. An open circuit, for example, will not allow any current to flow, so the ammeter reading will be zero. A damaged component or circuit will have an increased current draw, so the reading will be high. The ammeter is always connected in series with the tested circuit. All of the current that normally flows through the circuit must also flow through the ammeter; if there is any other path for the current to follow, the ammeter reading will not be accurate. The ammeter itself has very little resistance to current flow and, therefore, will not affect

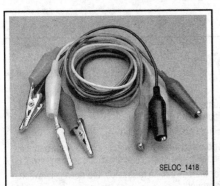

Fig. 4 Jumper wires are simple, but valuable pieces of test equipment

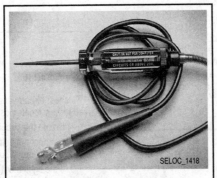

Fig. 5 A 12-volt test light is used to detect the presence of voltage in a circuit

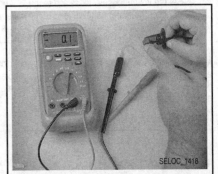

Fig. 6 Multi-meters are probably the most versatile and handy tools for diagnosing faulty electrical components or circuits

the circuit, but it will measure current draw only when the circuit is closed and electricity is flowing. Excessive current draw can blow fuses and drain the battery, while a reduced current draw can cause motors to run slowly, lights to dim and other components to not operate properly.

Troubleshooting Electrical Systems

When diagnosing a specific problem, organized troubleshooting is a must. The complexity of a modern marine vessel demands that you approach any problem in a logical, organized manner. There are certain troubleshooting techniques, however, which are standard:

• **Establish when the problem occurs**. Does the problem appear only under certain conditions? Were there any noises, odors or other unusual symptoms? Isolate the problem area. To do this, make some simple tests and observations, and then eliminate the systems that are working properly. Check for obvious problems, such as broken wires and loose or dirty connections. Always check the obvious before assuming something complicated is the cause.

• **Test for problems systematically to determine the cause once the problem area is isolated**. Are all the components functioning properly? Is there power going to electrical switches and motors? Performing careful, systematic checks will often turn up most causes on the first inspection, without wasting time checking components that have little or no relationship to the problem.

• **Test all repairs after the work is done to make sure that the problem is fixed**. Some causes can be traced to more than 1 component, so a careful verification of repair work is important in order to pick up additional malfunctions that may cause a problem to reappear or a different problem to arise. A blown fuse, for example, is a simple problem that may require more than another fuse to repair. If you don't look for a problem that caused a fuse to blow, a shorted wire (for example) may go undetected and cause the new fuse to blow right away (if the short is still present) or during subsequent operation (as soon as the short returns if it is intermittent).

Experience shows that most problems tend to be the result of a fairly simple and obvious cause, such as loose or corroded connectors, bad grounds or damaged wire insulation that causes a short. This makes careful visual inspection of components during testing essential to quick and accurate troubleshooting.

Electrical Testing

VOLTAGE

◆ See Figure 7

This test determines the voltage available from the battery and should be the first step in any electrical troubleshooting procedure after visual inspection. Many electrical problems, especially on electronically controlled systems, can be caused by a low state of charge in the battery. Many circuits cannot function correctly if the battery voltage drops below normal operating levels.

Loose or corroded battery cable terminals can cause poor contact that will prevent proper charging and full battery current flow.

1. Set the voltmeter selector switch to the 20V position.
2. Connect the meter negative lead to the battery's negative (-) post or terminal and the positive lead to the battery's positive (+) post or terminal.
3. Turn the ignition switch **ON** to provide a small load.
4. A well charged battery should register over 12 volts. If the meter reads below 11.5 volts, the battery power may be insufficient to operate the electrical system properly. Check and charge or replace the battery as detailed under Engine Maintenance before further tests are conducted on the electrical system.

VOLTAGE DROP

◆ See Figure 8

When current flows through a load, the voltage beyond the load drops. This voltage drop is due to the resistance created by the load and also by small resistances created by corrosion at the connectors (or by damaged insulation on the wires). Since all voltage drops are cumulative, the maximum allowable voltage drop under load is critical, especially if there is more than one load in the circuit.

1. Set the voltmeter selector switch to the 20 volts position.
2. Connect the multi-meter negative lead to a good ground.
3. Operate the circuit and check the voltage prior to the first component (load).
4. There should be little or no voltage drop in the circuit prior to the first component. If a voltage drop exists, the wire or connectors in the circuit are suspect.
5. While operating the first component in the circuit, probe the ground side of the component with the positive meter lead and observe the voltage readings. A small voltage drop should be noticed. This voltage drop is caused by the resistance of the component.
6. Repeat the test for each component (load) down the circuit.
7. If an excessively large voltage drop is noticed, the preceding component, wire or connector is suspect.

RESISTANCE

◆ See Figure 9

✳✳ SELOC WARNING

Never use an ohmmeter with power applied to the circuit. The ohmmeter is designed to operate on its own power supply. The normal 12-volt electrical system voltage will damage or destroy many meters!

1. Isolate the circuit from the vessel's power source.
2. Ensure that the ignition key is **OFF** when disconnecting any components or the battery.
3. Where necessary, also isolate at least one side of the circuit to be checked, in order to avoid reading parallel resistances. Parallel circuit resistances will always give a lower reading than the actual resistance of either of the branches.
4. Connect the meter leads to both sides of the circuit (wire or component) and read the actual measured ohms on the meter scale. Make sure the selector switch is set to the proper ohm scale for the circuit being tested, to avoid misreading the ohmmeter test value.

■ The resistance reading of most electrical components will vary with temperature. Unless otherwise noted, specifications given are for testing under ambient conditions of 68°F (20°C). If the component is tested at higher or lower temperatures, expect the readings to vary slightly. When testing engine control sensors or coil windings with smaller resistance specifications (less than 1000 ohms) it is best to use a high quality DVOM and be especially careful of your test results. Whenever possible, double-check your results against a known good part before purchasing the replacement. If necessary, bring the old part to the marine parts dealer and have them compare the readings to prevent possibly replacing a good component.

OPEN CIRCUITS

◆ See Figure 10

This test already assumes the existence of an open in the circuit and it is used to help locate position of the open.

1. Isolate the circuit from power and ground.
2. Connect the self-powered test light or ohmmeter ground clip to the ground side of the circuit and probe sections of the circuit sequentially.
3. If the light is out or there is infinite resistance, the open is between the probe and the circuit ground.
4. If the light is on or the meter shows continuity, the open is between the probe and the end of the circuit toward the power source.

SHORT CIRCUITS

◆ **See Figure 11**

■ **Never use a self-powered test light to perform checks for opens or shorts when power is applied to the circuit under test. The test light can be damaged by outside power.**

1. Isolate the circuit from power and ground.

2. Connect the self-powered test light or ohmmeter ground clip to a good ground and probe any easy-to-reach point in the circuit.

3. If the light comes on or there is continuity, there is a short somewhere in the circuit.

4. To isolate the short, probe a test point at either end of the isolated circuit (the light should be on or the meter should indicate continuity).

5. Leave the test light probe engaged and sequentially open connectors or switches, remove parts, etc. until the light goes out or continuity is broken.

6. When the light goes out, the short is between the last 2 circuit components that were opened.

Wire & Connector Repair

Almost anyone can replace damaged wires, as long as the proper tools and parts are available. Wire and terminals are available to fit almost any need. Even the specialized weatherproof, molded and hard shell connectors used by many marine manufacturers.

Be sure the ends of all the wires are fitted with the proper terminal hardware and connectors. Wrapping a wire around a stud is not a permanent solution and will only cause trouble later. Replace wires one at a time to avoid confusion. Always route wires in the same manner of the manufacturer.

When replacing connections, make absolutely certain that the connectors are certified for marine use. Automotive wire connectors may not meet United States Coast Guard (USCG) specifications.

■ **If connector repair is necessary, only attempt it if you have the proper tools. Weatherproof and hard shell connectors may require special tools to release the pins inside the connector. Attempting to repair these connectors with conventional hand tools will damage them.**

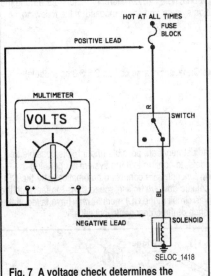

Fig. 7 A voltage check determines the amount of battery voltage available and, as such, should be the first step in any troubleshooting procedure

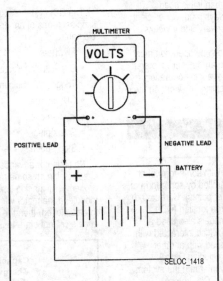

Fig. 8 Voltage drops are due to resistance in the circuit, from the load or from problems with the wiring

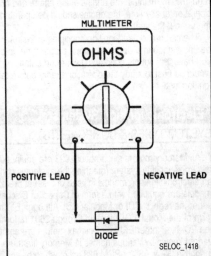

Fig. 9 Resistance tests must be conducted on portions of the circuit, isolated from battery power

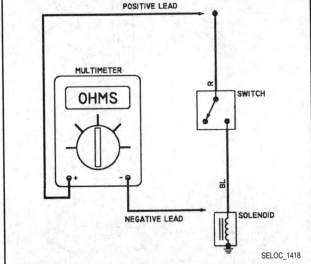

Fig. 10 The easiest way to illustrate an open circuit is to picture a circuit in which the switch is turned OFF (creating an opening in the circuit) that prevents power from reaching the load

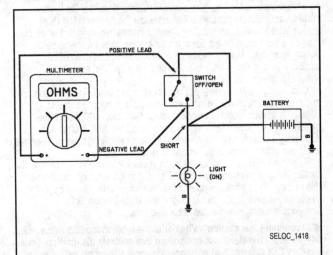

Fig. 11 In this illustration a load (the light) is powered when it should not be (since the switch should be creating an open condition), but a short to power (battery) is powering the circuit. Shorts like this can be caused by chaffed wires with worn or broken insulation

Electrical System Precautions

• Wear safety glasses when working on or near the battery.

• Don't wear a watch with a metal band when servicing the battery or starter. Serious burns can result if the band completes the circuit between the positive battery terminal (or a hot wire) and ground.

• Be absolutely sure of the polarity of a booster battery before making connections. Remember that even momentary connection of a booster battery with the polarity reversed will damage charging system diodes. Connect the cables positive-to-positive, and negative (of the good battery)-to-a good ground on the engine (away from the battery to prevent the possibility of an explosion if hydrogen vapors are present from the electrolyte in the discharged battery). Connect positive cables first (starting with the discharged battery), and then make the last connection to ground on the body of the booster vessel so that arcing cannot ignite hydrogen gas that may have accumulated near the battery.

• Disconnect both vessel battery cables before attempting to charge a battery.

• Never ground the alternator or generator output or battery terminal. Be cautious when using metal tools around a battery to avoid creating a short circuit between the terminals.

• When installing a battery, make sure that the positive and negative cables are not reversed.

• Always disconnect the battery (negative cable first) when charging.

• Never smoke or expose an open flame around the battery. Hydrogen gas is released from battery electrolyte during use and accumulates near the battery. Hydrogen gas is **highly** explosive.

IGNITION SYSTEMS

General Information

The less an outboard engine is operated, the more care it needs. Allowing an outboard engine to remain idle will do more harm than if it is used regularly. To maintain the engine in top shape and always ready for efficient operation at any time, the engine should be operated every 3 to 4 weeks throughout the year.

The carburetion and ignition principles of 2-stroke engine operation must be understood in order to perform a proper tune-up on an outboard motor. If you have any doubts concerning your understanding of engine operation, it would be best to study the operation theory before tackling any work on the ignition system.

Electronic Ignition Systems

SYSTEM TYPES & COMPONENTS

All Mercury motors since about 1993 are equipped with some form of an electronic ignition system (meaning the system is controlled by some form of an electronic ignition module instead of a set of breaker points). Carbureted models are equipped with a form of Capacitor Discharge Ignition (CDI), either the regular CDI or Modular CDI. Although EFI models also used a form of the Modular CDI system through 2001 (a form which interacted with the ECM), all other EFI and Optimax motors are equipped with another variant known by various names in Mercury literature including Digital Inductive, PCM 555 or PCM 038, all of which basically infer that the system is directly controlled by the Propulsion Control Module (PCM, known throughout this guide as ECM).

Probably the largest difference between the variants of the CDI systems and the variants of the PCM systems would be a fundamental difference in what supplies the power for the ignition system. The CDI systems are known as an "alternator driven" system, meaning that the power for the ignition comes from a stator or charge coil found under the flywheel. The PCM systems are referred to by "battery driven" systems, meaning that when the ignition is first turned on the power for the ignition comes from the battery (which ironically is replenished by an ALTERNATOR, and even more ironically, by one that is NOT mounted under the flywheel as with the "alternator driven" systems).

Confused yet? Well, don't be. They are all pretty similar in basic function and pretty straight forward when it comes to troubleshooting them. In all cases, the ignition module (a function integrated into the PCM on those systems or a function often integrated into the CDMs on CDI units so equipped) is used to apply power to the ignition coils, firing them in response to signals from a trigger or pulser coil or multiple coils.

To better understand any one Mercury motor's ignition system, refer to the system descriptions below and to the Description & Operation found later in this section. If you have any doubt as to the components utilized by a given motor, please refer to the Wiring Diagrams and the Ignition System Component testing charts in this section as well.

■ Throughout this section we'll refer to the ignition module when talking about the electronic component that controls the ignition. On Mercury CDI systems that component is most commonly called a control module or switch box. On Mercury PCM systems that function is normally performed BY the PCM/ECM itself. Don't be confused if you're researching a PCM motor and we're calling it a control module, that's just to keep from having to constantly write (and for you to constantly read "control module, switch box, ECM or PCM").

1-Cylinder Ignition

◆ See Figure 12

The CDI systems found on single-cylinder Mercury motors are among the simplest forms of an ignition system and are composed of the following elements:

• Flywheel
• Capacitor charge coil
• Trigger coil (built into Capacitor charge coil on 2.5/3.3 hp models)
• CDI module
• Ignition coil
• Spark plug
• Stop button

No timing or advance adjustments are possible (though it is possible to check timing on some models, to confirm that the system is operating properly). All ignition timing functions are controlled automatically by the CDI module. If the motor will not idle down to trolling speed (runs too fast with the throttle closed) then the bias circuit in the CDI module *may* have failed. If the engine will not start, test each of the components individually.

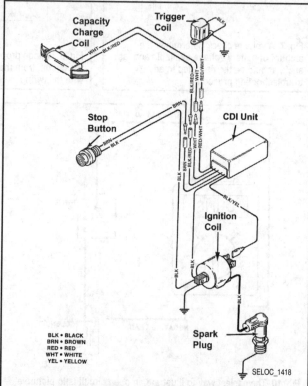

BLK • BLACK
BRN • BROWN
RED • RED
WHT • WHITE
YEL • YELLOW

SELOC_1418

Fig. 12 Typical Mercury 1-cylinder ignition system - 4/5 hp models shown

2-Cylinder Ignition

◆ **See Figure 13**

The CDI systems found on these models, though slightly more complicated than the single-cylinder motors, are still rather straight forward and utilize the following components:
- Flywheel
- Stator (charge coil)
- Trigger coil
- Switch box (ignition module- except on mechanical advance 30/40 hp motors with CDMs)
- 2 Ignition coils (called Capacitor Discharge Modules or CDMs on the 30/40 hp models)
- 2 Spark plugs
- Stop button or ignition keyswitch

Both mechanical advance and electronic advance models are found in these engine families.

On mechanical advance models, ignition timing is controlled (advanced or retarded) by rotating the trigger coil which changes the position of the trigger relative to the magnets spinning with the center hub of the flywheel (thereby altering the point at which the trigger coil gives its signals to fire the ignition coils). The CDMs used on 30/40 hp motors basically act as a combination ignition coil and individual ignition module.

On electronic advance models all timing changes are controlled internally by the ignition module (usually called a switch box on these motors). However, the module is really called a TPM on electronic advance 30/40 hp motors.

The stop circuit works by grounding the stator output.

A timing protection module (TPM) is utilized by electronic advance versions of the 30/40 hp motors and performs the following functions:
- Controls spark timing in relation to engine speed and temperature.
- Quickly advances timing to 25° BTDC under hard acceleration.
- Provides engine overspeed protection, first by retarding timing from 25 to 14° BTDC if engine speed exceeds 5800 rpm, then if rpm continues to increase to 6500 rpm will momentarily shut off the ignition until speed drops below 6500 rpm again.
- Will advance timing, as high as 10° BTDC for cranking and in order to stabilize idle, should idle speed fall below 600 rpm.

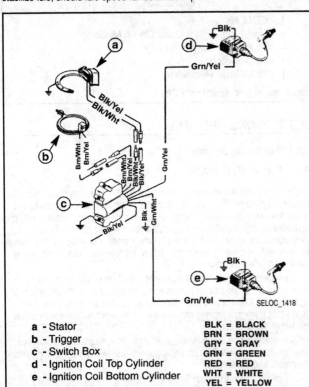

a - Stator
b - Trigger
c - Switch Box
d - Ignition Coil Top Cylinder
e - Ignition Coil Bottom Cylinder

BLK = BLACK
BRN = BROWN
GRY = GRAY
GRN = GREEN
RED = RED
WHT = WHITE
YEL = YELLOW

Fig. 13 Typical Mercury 2-cyl ignition system - common 6/8 and 9.9/10/15 hp models shown

- Provides a warning control for OVERHEAT and LOW-OIL conditions. Should overheating occur, the module will sound the warning horn and intermittently interrupt ignition voltage to the CDMs in order to reduce engine speed to about 2500 rpm. Though engine speed is not limited during LOW-OIL conditions, the warning horn will be activated.

3- and 4-Cyl Ignition Systems (Exc. Optimax Motors)

◆ **See Figure 14**

The CDI systems found on these models are similar to that used by the twins and utilize the following components:
- Flywheel
- Stator (charge coil)
- Trigger coil
- 1 CDM per cylinder
- 1 Spark plug per cylinder
- Stop button or ignition keyswitch (usually the later)

Though earlier versions may have varied, by 2001 all of these motors should be equipped with a modular CDI system that utilizes 1 individual CDM per cylinder. The CDM functions as a combination ignition module/ignition coil.

On these models the stator voltage return path is through the ground wire of another CDM. On most models if the #1 cylinder stator wire is disconnected the engine will stop.

Some models are equipped with an external rev limiter module which functions by using a trigger signal to determine engine speed (rpm) and, if speed exceed as a certain spec the limiter will ground out the CDM capacitor charge to protect the motor.

As with most Mercury ignition systems, the stop circuit works by grounding the stator output.

Carbureted V6 & 2001 EFI Ignition Systems

■ **For ignition system schematics on these models, please refer to the Wiring Diagrams in this section.**

Carbureted V6 motors, as well as the 2001 EFI motors, both utilize a modular CDI system, similar to that used by smaller motors and utilize the following components:
- Flywheel
- Stator (charge coil)
- Trigger coil
- 1 CDM per cylinder
- 1 Spark plug per cylinder
- ECM or control module (to integrate warning and protection functions, as well as EFI control on fuel injected models)
- Ignition keyswitch for stop circuit control
- Shift interrupt switch (some models utilize this switch to ease shifting)

These models are equipped with a modular CDI system that utilizes 1 individual CDM per cylinder. The CDM functions as a combination ignition module/ignition coil.

A control module is utilized to provide over-rev protection (carb models), bias control, shift stabilizer, idle stabilizer, injector timing (EFI models) and low oil warning functions.

Spark timing is achieved either by the movement of the trigger assembly (attached to the throttle/spark arm) for most 2.5L motors or is fully electronic on most 3.0L motors.

As with most Mercury ignition systems, the stop circuit works by grounding the stator output.

OptiMax & 2002 & Later EFI Ignition Systems

■ **For ignition system schematics on these models, please refer to the Wiring Diagrams in this section.**

All OptiMax motors, as well as 2002 or later EFI motors all utilize a battery driven, PCM controlled digital inductive ignition system. The systems are *very* similar to modular CDI, except that the power for ignition comes from the battery instead of directly from charge coil windings. And, at least on the V6 OptiMax motors, the function of the CDMs has been split out again, to separate ignition coils and separate coil drivers. The PCM or Digital Inductive ignitions utilize the following components:
- Battery (fed by a belt driven alternator)
- Flywheel teeth
- Crankshaft Position Sensor (CPS) which works of flywheel teeth
- 1 CDM per cylinder (EFI or 1.5L OptiMax motors) OR 1 coil driver per pair of cylinders with 1 ignition coil per cylinder (V6 OptiMax motors)

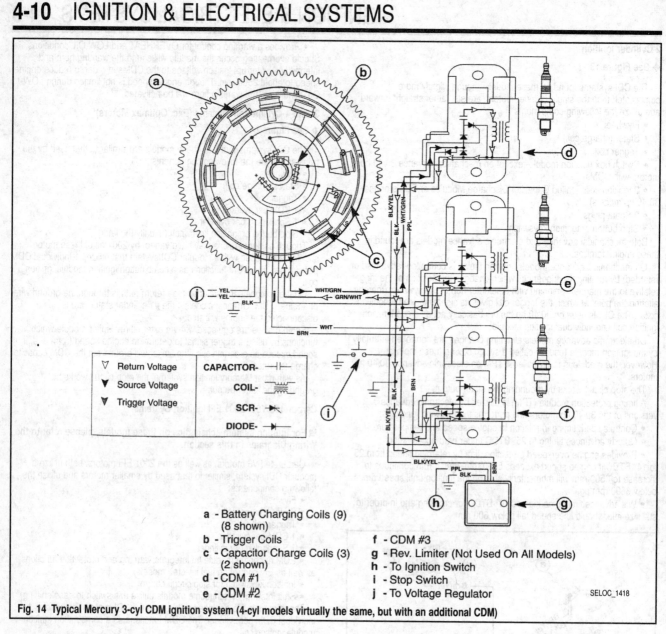

	Return Voltage	CAPACITOR-
▼	Source Voltage	COIL-
▼	Trigger Voltage	SCR-
		DIODE-

a - Battery Charging Coils (9)
    (8 shown)
b - Trigger Coils
c - Capacitor Charge Coils (3)
    (2 shown)
d - CDM #1
e - CDM #2
f - CDM #3
g - Rev. Limiter (Not Used On All Models)
h - To Ignition Switch
i - Stop Switch
j - To Voltage Regulator

SELOC_1418

Fig. 14 Typical Mercury 3-cyl CDM ignition system (4-cyl models virtually the same, but with an additional CDM)

• 1 Spark plug per cylinder
• ECM (usually referred to as the Propulsion Control Module/PCM on these motors)
• Ignition keyswitch for stop circuit control
• Shift interrupt switch (most/all models utilize this switch to ease shifting)

These models are essentially equipped with a PCM run, battery voltage supplied modular CDI system that utilizes either 1 individual CDM per cylinder or 1 coil driver per pair of cylinders along with an individual coil per cylinder.

As with carbureted models, when equipped, the CDM functions as a combination ignition module/ignition coil.

The ECM however, performs a large number of functions on most models, including:

• Controls ignition timing and fuel timing/delivery functions based on sensor inputs including throttle position, manifold pressure and engine temperature.
• Controls the injectors (both regulator fuel injectors and direct injectors on OptiMax models).
• Controls all alarm horn and warning lamp functions.
• Supplies a tachometer signal to gauge
• Controls RPM limit function
• Monitors and reacts to shift interrupt switch
• Records engine running information

DESCRIPTION & OPERATION

CDI & Modular CDI Systems

◆ See Figures 12, 13 and 14

These systems are known as "alternator driven", non-distributor, capacitor discharge ignitions. The major components of the system are generally the flywheel, charge coil (stator), trigger or pulser coil(s), ignition module (switch box), ignition coil(s) and spark plug(s). As you can see most of these components are known by multiple possible names. For instance, the charge coil may also be called a stator; the ignition module may also be called the switch box, CDI Unit, power pack etc.

The engine's flywheel contains magnets carefully positioned to create an electric current as they rotate past specially designed coils of wire (stator and trigger/pulser). Current is created by magnetic induction. Simply put, that means that a magnet moving rapidly near a conductor will induce electrical flow within the conductor.

Conversely, a wire that moves rapidly through a magnetic field will also generate electrical flow. This principle governs the working of electric motors, alternators and generators.

One of the coils under your flywheel is called a charge coil. As the flywheel magnets spin past this coil, they generate in it a fairly high-voltage alternating current that travels to your system's ignition module. This voltage is often in the region of 200 volts AC.

The other ignition system coil(s) found under the flywheel are called the sensor coils, pulser coils or trigger coils. They send an electrical signal to the ignition module to tell it which cylinder to work with at the correct time.

The ignition module unit is the brains of the system and serves several functions. First, it converts the alternating current (AC) from the charge coil into usable direct current (DC). Next it stores the current in a built-in capacitor. The module also interprets the timing signal from the trigger coil - this changes constantly with engine speed and moving the trigger coil's position relative to the flywheel magnets brings about the change. The coil's movement is controlled by a device called a timing plate, to which both the charge coil and trigger coils are mounted. The timing plate moves in response to changes in throttle opening, to which it is mechanically linked. The ignition module also controls the discharge of the capacitor and sends this voltage to the primary winding of the ignition coil for the correct cylinder.

■ **Some motors, generally speaking ones equipped with "modular CDI ignitions" utilize 1 individual capacitor discharge module (CDM) per cylinder. On these motors the CDM functions as a combination ignition module and ignition coil.**

Also (depending on the system) the module may incorporate electronic circuits that limit engine speed and prevent over-revving. Some modules even have a circuit that reduces engine speed if the engine begins to run too hot for any reason. Larger engines often have an automatic ignition advance for initial start-up and for when the engine is running at temperatures less than approximately 100°F.

Manufacturers commonly use 1 power pack or module for each bank of cylinders on carbureted V-type powerheads (unless they use a module CDI system with an individual CDM for each cylinder). One module will control the odd numbered cylinders and the other will service the even numbered cylinders.

The voltage from the module goes to the primary winding of the ignition coil or high-tension coil. You may know this type of coil as a step-up transformer. Here, the voltage is stepped up to between 15,000 and 40,000 volts. That's the 'kind of voltage needed to jump the air gap on the spark plug and ignite the air/fuel mixture in the cylinder. Your high-tension ignition coil has 2 sides, the primary side and the secondary side. It is really 2 coil assemblies combined into one neat, compact case.

The internal construction of a typical ignition coil includes primary and secondary windings. It also uses the principle of magnetic induction with the magnetism generated by the primary (lower-voltage) winding creating a magnetic field around the secondary winding, which has many more windings than the primary coil. The ignition module controls the rapid turning on and off of electrical flow in the primary winding, thereby turning this magnetic field on and off. The rapid movement of this magnetic field past the secondary windings induces electrical current flow. There are a greater the number of turns of wire in the secondary winding, the higher the voltage produced.

As this secondary voltage leaves the center tower of the ignition coil, it travels along the spark-plug wire, which is heavily insulated and designed to carry this high voltage.

If all is well, the high voltage will jump the gap on the spark plug between the center electrode and the ground electrode. On larger engines with surface-gap plugs, the high voltage current will jump from the center electrode to the side of the plug assembly itself, completing a circuit to ground via the engine block.

Last but certainly not least, is the stop control. You need a means to shut your engine off and a good way is to stop the spark plugs from working. Depending on your engine, this may be accomplished by a simple stop button or a key switch on larger engines. This disables the whole ignition system. On most engines, an emergency stop button with an overboard clip and lanyard attachment is standard. This system is wired directly into the ignition module. It functions by creating a momentary short circuit inside the module, grounding the current intended for the high-tension coils and thus shutting off the ignition long enough to stop the engine. Faulty stop circuits are frequently the cause of a no spark condition.

PCM & Digital Inductive Systems

■ **For ignition system schematics on these models, please refer to the Wiring Diagrams in this section.**

The vast majority of EFI and all OptiMax motors are equipped with a battery driven electronic ignition system. On these motors, when the ignition key is turned to **RUN**, battery voltage is applied to the main relay. If the control module (remember it's the PCM/ECM in this case) does not receive a signal from the crankshaft position sensor (CPS) the relay will be turned off again.

When the module receives a CPS signal it will ground the main relay circuit, applying power from the battery (and belt-driven alternator once the motor is running) through a fuse (usually 20 amp) to the positive terminal of the individual ignition coil primary windings (there is normally 1 coil per cylinder).

The negative terminal of the coil primary circuit is connected to engine ground through the module (ECM). Actually, on most EFI models and on 1.5L OptiMax models Mercury states that the ignition coil itself contains an internal driver circuit (meaning it is a CDM unit like those found on the modular CDI ignitions, however wiring diagrams show separate coil drivers on V6 OptiMax models). In all cases, when the circuit is closed a magnetic field will build in the ignition coil. When the circuit is switched off the collapse of the magnetic field will cause the coil to fire on the secondary circuit with a charge as high as 50,000 volts.

A crankshaft position sensor (CPS) is mounted on top of the powerhead, in a position to sense the differences in magnetic field caused by the teeth on the flywheel (54 teeth on most V6 models). This CPS signal is used by the PCM to determine the trigger signals (when to shut off the driver circuit for a given cylinder). When the motor is operated at lower engine RPM and load, most of these systems will repeat the spark plug firing process in quick succession resulting in a multi-strike spark for each combustion event (multiple spark plug firings on the same power stroke for more power and a cleaner burn). Many of these motors require platinum spark plugs (especially OptiMax motors).

Although the name and location of these components appear slightly different than those used on CDI motors, their functions are nearly identical and operation is fundamentally the same.

TROUBLESHOOTING THE IGNITION SYSTEM

Spark Plugs & Checking for Spark

◆ **See Figures 15, 16 and 17**

The absolute first thing that you should check when an ignition problem is suspected is the spark plugs. You want to determine that the plugs (the most common wear item in an electronic system) are in good condition and are producing a good, strong spark. Likewise, a spark check should be conducted using some form of an inexpensive spark checking tool which either installs inline between the secondary ignition wire (spark plug wire) and the spark plug itself, OR substitutes for the spark plug.

1. Check the plug wires to be sure they are properly connected. Check the entire length of the wires from the plug(s) back to the coil(s). If a wire is to be removed from the spark plug, always use a pulling and twisting motion as a precaution against damaging the connection.

2. Attempt to remove the spark plug by hand. This is a rough test to determine if the plug is tightened properly. The attempt to loosen the plug by hand should fail. The plug should be tight and require the proper socket size tool. Remove the spark plug and evaluate its condition. Reinstall the spark plug and tighten to the proper specification.

3. Use a spark tester and check for spark. If a spark tester is not available (and for Pete's sake they're cheap and can be found in almost all auto parts stores) hold the plug wire about 1/4 in. (6.4mm) from the engine (leave the plug in the port for safety). Carefully operate the starter and check for spark. A strong blue spark over a wide gap must be observed when testing in this manner, because under compression a strong spark is necessary in order to ignite the air-fuel mixture in the cylinder. This means it is possible to think a strong spark is present, when in reality the spark will be too weak when the plug is installed. If there is no spark, or if the spark is weak (orange), the trouble is most likely under the flywheel.

Troubleshooting With Minimal Test Equipment

■ **Patterns of the following symptoms and causes conditions have been noted on Mercury motors in the past and should apply to many of the powerheads covered here. Much of these troubleshooting procedures are courtesy of our friends at CDI Electronics (256-772-3829).**

• **Intermittent Firing:** - This problem can be very hard to isolate. A good inductive tachometer can be used to compare the rpm on all cylinders up through wide open throttle. A big difference on 1 or 2 cylinders indicates a problem. Intermittent firing is often caused by a coil (stator, charge or even ignition) which shorts to ground under certain temperature or load conditions.

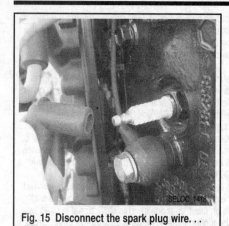

Fig. 15 Disconnect the spark plug wire. . .

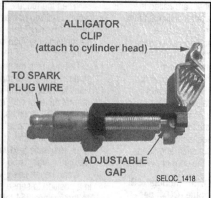

Fig. 16 . . . and check spark using a tester

Fig. 17 Remove and inspect the spark plug

However, it can also be caused by internal problems with the ignition control module or ECM. Troubleshooting should be conducted on components in an attempt to isolate parts of the system while attempting to narrow the problem, but remember that with an intermittent, the problem must be present at the moment of testing for the component to show bad. Otherwise the problem must be replicated. Also, since many intermittent problems only show at a certain temperature or load, static ohm tests on bad components will often still test within spec.

• **One or more cylinders mis-firing:** - Since 2001 or later Mercury motors normally only use either 1 ignition module OR 1 CDM per cylinder look first for problems with the trigger coil. On systems that use multiple CDMs or coil drivers they are normally interchangeable and can be moved from 1 cylinder (or pair) to another to see if the problem follows or remains behind.

• **Engine continuously blows ignition modules:** - When an engine starts blowing modules repeatedly, especially on the same cylinders, replace the ignition coil(s) on those cylinders. The inductive kickback from a bad coil can destroy the packs, even if the coils check good with all known tests. A stator that tests good can also be sending spike voltages to the pack causing them to fail repeatedly.

• **Visually check the stator, trigger and flywheel:** - Cracks, burned marks and bubbling on the stator or trigger indicate a severe problem. If the stator shows bubbling around the battery charge windings, more than likely you will have to replace the rectifier/regulator in addition to the stator. Signs of rubbing on the flywheel usually indicate a bad upper or lower bearing. Check both the outer and trigger magnets for signs of cracking and to be sure that they are not loose.

Troubleshooting Battery Driven Ignitions

DIFFICULT

✷✷ SELOC CAUTION

Some people feel that maintenance free batteries should not be used with these types of ignitions as they tend to overcharge and blow the packs. However, we are not certain that is true on the most modern maintenance free batteries and charging systems.

■ A large portion of the problems with the battery CDMs are caused by low battery voltage, bad ground connections or high battery voltage. Low voltage symptoms are weak fire or weak erratic firing of cylinders. Misfiring after a few minutes of running can be caused by excessive voltage at the pack. Battery reversal will usually destroy battery, CDM units and sometimes the CPS or trigger circuits.

✷✷ SELOC WARNING

Check the voltage on the Red (or Purple) wire at the CDM or coil driver through the rpm range. At no time should the voltage exceed 15.5 Volts DC.

• The first item to check is that all battery and ground connections are solid.
• **Engine Cranks and Fires As Long As The Starter Is Engaged:** This problem usually indicates a bad trigger.

• **Check the Ignition Coil:** An open, cracked or poorly grounded coil can burn out a battery CDM or coil driver.

Troubleshooting Alternator Driven Ignitions

DIFFICULT

■ **Initial peak voltage readings should be taken with everything hooked up.**

• Disconnect the kill wire. Connect a multi-meter between the kill wires and engine ground. Turn the ignition switch **ON** and **OFF** several times. If at any time, you see DC voltage on the kill wires, there is a problem with the harness or ignition switch. Battery voltage on the kill circuit will destroy most CD units.

• **Visually Inspect Stator for Cracks or Varnish Leakage:** If found, replace the stator. Burned marks or discolored areas on the battery charge windings indicate a possible problem with the rectifier.

• **Unit Will Not Fire:** Disconnect kill wire at the pack. Check for broken or bare wires on the unit, stator and trigger. Check the peak voltage of the stator, (for details, please refer to the Ignition System Component Testing chart). Disconnect the rectifier. If the engine fires, replace the rectifier.

• **Engine Will Not Kill:** Check kill circuit in the pack by using a jumper wire connected to the Black/Yellow terminal or wire coming out of the pack and shorting it to ground. If this kills the engine, the kill circuit in the harness or on the boat is bad, possibly the ignition switch.

• **High Speed Miss:** Disconnect the rectifier and retest. If miss is gone, the rectifier is usually at fault. If the miss still exists, check peak voltage of the stator at high speed.

■ **Use caution when doing this and do not exceed the rated voltage range of your meter. The readings should show a smooth climb in voltage. If there is a sudden or fast drop in voltage right before the miss becomes apparent, the stator is usually at fault. If there is no indication of the problem, it could be a small water leak in 1 or 2 cylinders.**

• **Coils Fire with Spark Plugs Out but Not In:** Check for dragging starter or low battery causing slow cranking speed. Test peak voltage on the stator and trigger. Disconnect rectifier, regulator and retest. If the problem goes away, replace the rectifier and/or regulator.

• **Engines Runs Rough on Top or Bottom Two Cylinders (4-Cylinder Engines):** Check peak voltage of the stator between Blue wires and ground. Readings to ground should be fairly equal. If unequal, swap stator leads (such as Blue with Blue/White, Red with Red/White, but see your wiring diagram for details) and see if the problem moves with the stator leads. If it does, replace the stator.

■ **Check trigger resistance between #1 & #2, compare to resistance between #3 and #4. For test purposes only, swap trigger leads 1 & 3, and 2 & 4. If the problem moves, replace the trigger. If it does not move, swap coil primary wires, and replace the pack if the problem remains on the same terminals.**

• **No Fire on One or more pairs of cylinders (V6):** Start diagnosis by checking peak voltage of the stator. On modular systems you should be able to swap coil drivers or CDMs to see if the problem follows.

• **Intermittent Firing On One or More Cylinders:** Check for low voltage from the stator and trigger. Disconnect the rectifier and retest. If the problem disappears, replace the rectifier. On modular systems you should be able to swap CDMs or coil drivers to see if the problem moves.

• **Inline engines with internal exhaust plate:** If engine speeds up when you remove 1 spark plug wire, the internal exhaust plate is more than likely warped.

COMPONENT TESTING

Where possible the components listed in this section will include specific Testing procedures. In some cases those tests will be static (involving resistance readings on a coil winding with the motor not operating) and in others they may be dynamic (involving voltage output with the motor cranking or operating).

Generally speaking dynamic tests are preferred, where available, as a static test can allow an intermittent open or short (which only occurs under heat and/or load) to go undetected.

Some solid state units, such as the ECM or ignition module cannot be tested directly, so testing involves making sure all other components in the system are operating correctly before condemning the solid state unit.

Before delving into specific component tests, there are some basics that must first be checked, as follows:

• Make sure the battery voltage is not too weak for cranking (in case of a no start or hard start problem or weak spark on a battery driven system).

• Check to make sure all connectors are clean and tight. Make sure no wires in the harness are damaged (cut, insulation worn through or crushed which could sever the circuit internally).

• Make sure the spark plugs are in good shape.

• Make sure the powerhead has good compression.

Once the powerhead has passed these tests it is time to dig further into specific component testing.

Resistance Testing Precautions

When performing an electrical resistance test on ignition components the following points should be noted:

• There are no standard specifications for the design of either analog or digital test meters. However, resistance will vary slightly with meter.

• When performing resistance testing, some test meter manufacturers will apply positive battery power to the Red test lead, while others apply the positive voltage to the Black lead. This should not make a difference for most coils or circuits, but *could* be an issue if a resistance check is provided for part of a solid state component.

• Some meters (especially digital meters) will apply a very low voltage to the test leads that may not properly activate some devices.

• Be aware there are ohmmeters that have reverse polarity. If your test results all differ from the chart, swap + and - leads and retest.

■ **Digital meter resistance values are not reliable when testing CDMs, rectifier units, or any device containing semiconductors, transistors, and diodes.**

• Although resistance test specifications can be helpful, they are also very misleading in ignition circuits/components. A component may test within specification during a static resistance check, yet still fail during normal operating conditions. Also, remember that readings will vary both with temperature and by meter.

Compression

Before spending too much time and money attempting to trace a problem to the ignition system, a compression check of the cylinder should be made. If the cylinder does not have adequate compression, troubleshooting and attempted service of the ignition or fuel system will fail to give the desired results of satisfactory engine performance.

For details, please refer to Compression Testing in the Maintenance & Tune-Up section.

IGNITION DIAGNOSTIC CHARTS

Where possible, we've included diagnostic charts for the ignition system to supplement the tests given earlier in this section. Most of the mid-range-to-large models are included.

PROBLEM	CORRECTION
1. No Spark or Weak Spark on Both Cylinders	No Spark – Trigger, Stator, Ignition Switch Box or Bad Ground Connection from Switch Box to Block Weak Spark – Stator
2. No Spark or Weak Spark on 1 Cylinder	Ignition Switch Box or Coil
3. Timing Fluctuates – Note: It is normal for timing to fluctuate 2°-3° @ Idle. – If engine RPM exceeds 5800, switch box will retard timing from 25° BTDC to 15° BTDC – If engine RPM drops below 600, idle stabilizer in switch box will advance timing to as high as 10° BTDC @ cranking speed of 300 RPM.	Shorted Trigger Wire or Ignition Switch Box
4. Timing will not Advance – Note: If timing will not advance on only 1 cylinder, check wiring for shorted trigger wire	Defective Switch Box
5. Engine Misfires @ High RPM	Defective Coil Defective Switch Box
6. Engine Hard to Start when Cold	Defective Trigger Assembly Defective Ignition Switch Box
7. Engine Misfires @ Low RPM but Runs Smooth @ High RPM	Defective Harness or (loose connections) Defective Switch Box Defective Stator
8. Engine Starts Hard when Hot	Defective Switch Box or Trigger
9. Engine Occasionally Misfires	Replace Standard Spark Plug with Inductor Plug Bad Ground Connection from Switch Box to Block

SELOC_1418

Ignition diagnostic charts - 20/20 Jet/25 hp 2-cylinder models

Step	Action	Value	Yes	No	Tools
1	Verify High Tension Leads, Spark Plug and Spark Boots are in good condition. Inspect wires for chafing. **Visual Inspection**	–	Step 2	Replace Failed Component Step 2	High Tension lead pin P/N 84-813706A56
2	Verify 4 Pin Connector Integrity **Visual Inspection**	–	Step 3	Repair/Replace Connector Components Step 3	–
3	Verify Ground from CDM connector to block	0.2 Ohms and below	Step 4	Correct Ground Path Step 4	DMT 2000 Digital Tachometer/ Multi-meter P/N 91-854009A1 & DVA Adaptor P/N 91-89045
4	Test all CDMs at Cranking with Spark Gap Tester **Spark on All CDMs?** Will spark jump a 7/16 in. (11.11 mm) gap?	7/16 in. (11.11 mm) gap	If at least one CDM has spark, continue with Chart #3	Continue with Chart #2	Spark Gap Tester P/N 91-850439

a - Ground
b - Black/Yellow
c - Trigger connection
d - Stator connection

A B C D

SELOC_1418

Ignition diagnostic charts - 30/40 hp 2-cylinder models

CDM diagnostic chart I (CD modules with part #827509) - 3, 4 and 6 cyl models

Step	Action	Value	Yes	No	Tools
1	Verify High Tension Leads, Spark Plug and Spark Boots are in good condition. Inspect wires for chafing. **Visual Inspection**	--	Step 2	Replace Failed Component Step 2	High Tension lead pin P/N 84-813706A56
2	Verify 4 Pin Connector Integrity **Visual Inspection**	--	Step 3	Repair/Replace Connector Components Step 3	--
3	Verify Ground from CDM connector to block	0.2 Ohms and below	Step 4	Correct Ground Path Step 4	DMT 2000 Digital Tachometer/ Multi-meter P/N 91-854009A1 & DVA Adaptor P/N 91-89045
4	Test all CDMs at Cranking with Spark Gap Tester **Spark on All CDMs?** Will spark jump a 7/16 in. (11.11 mm) gap?	7/16 in. (11.11 mm) gap	If at least one CDM has spark, continue with Chart #2	Continue with Chart #3	Spark Gap Tester P/N 91-850439

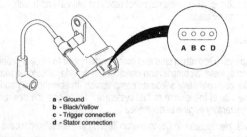

a - Ground
b - Black/Yellow
c - Trigger connection
d - Stator connection

A B C D

SELOC_1418

CDM diagnostic chart II (no spark on any CDM) - 3, 4 and 6 cyl models

Step	Action	Value	Yes	No	Tools
1	With the key switch ON: Verify continuity between disconnected BLK/YEL harness wire and ground. **This Test Checks:** Lanyard Switch Key Switch Rev Limiter (external) Chafed BLK/YEL wire	NO continuity	Step 2	Repair or Replace Component Run Engine Verify Repair Step 9	DMT 2000 Digital Tachometer/ Multi-meter P/N 91-854009A1 & DVA Adaptor P/N 91-89045
2	Un-plug all CDMs and verify continuity between disconnected BLK/YEL harness wire and ground. **This Test Checks:** CDM Harness	NO continuity	Step 3	Repair or Replace Component Run Engine Verify Repair Step 9	DMT 2000 Digital Tachometer/ Multi-meter P/N 91-854009A1 & DVA Adaptor P/N 91-89045
3	Connect CDM one at a time **This Test Checks:** Individual CDM Stop Circuits	Resistance will rise with each CDM connected, full continuity indicates shorted CDM stop circuit.*	Step 4	Repair or Replace Component Run Engine Verify Repair Step 9	DMT 2000 Digital Tachometer/ Multi-meter P/N 91-854009A1 & DVA Adaptor P/N 91-89045 TPI/CDM Test Harness 84-825207A2
4	Check Stator Open circuit voltage at cranking should be no less than 100 Volts on the DVA Resistance between GRN/WHT and WHT/GRN	660-710 Ohms 2, 3 & 4 Cyl. Models 380-430 Ohms 2.0/2.5L 990 - 1210 Ohms 3.0L	Step 5	Replace Stator Run Engine Verify Repair Step 9	DMT 2000 Digital Tachometer/ Multi-meter P/N 91-854009A1 & DVA Adaptor P/N 91-89045

*Diode Readings: Due to the differences in test meters, results other than specified may be obtained. In such a case, reverse meter leads and re-test. If test results then read as specified CDM is O.K. The diode measurements above will be opposite if using a Fluke equivalent multimeter.

SELOC_1418

CDM diagnostic chart II (no spark on any CDM) - 3, 4 and 6 cyl models, cont'd

Step	Action	Value	Yes	No	Tools
5	Check CDM Trigger Input/Crank Shaft Position Sensor Output: Cranking with CDM disconnected. Cranking with CDM connected.	1 Volt and above - CDM disconnected. 0.2 - 5 Volts- CDM connected.	Step 6	2, 3, & 4 cyl Replace Trigger Run Engine Verify Repair Step 9 2.0/2.5L 6 Cyl. - Step 6 3.0L 6 Cyl - Step 7	DMT 2000 Digital Tachometer/ Multi-meter P/N 91-854009A1 & DVA Adaptor P/N 91-89045 TPI/CDM Test Harness 84-825207A2
6	2.0/2.5L V-6 Models Resistance between Trigger wires. Red - White Blue - Purple Brown - Yellow	1100 - 1400 Ohms	Step 8	Replace Control Module Run Engine Verify Repair Step 9	DMT 2000 Digital Tachometer/ Multi-meter P/N 91-854009A1
7	3.0 L V-6 Models Resistance Check Crank Position Sensor	900 - 1300 Ohms	Step 8	Replace Crank Position Sensor Run Engine Verify Repair Step 9	DMT 2000 Digital Tachometer/ Multi-meter P/N 91-854009A1 & DVA Adaptor P/N 91-89045
8	Test all CDMs at Cranking with Spark Gap Tester **Spark on All CDMs?** Will spark jump a 7/16 in. (11.11 mm) gap?	7/16 in. (11.11 mm) gap	Step 9	Verify All Preceding Steps	Spark Gap Tester P/N 91-850439
9	If mis-firing is in a repeatable range: Perform DVA readings on stator and trigger at all running speeds.* 2.0/2.5L V-6 Models Preform Bias circuit tests 200 HP EFI - Disconnect Detonation Module. Check Black/White to ground All other models - Disconnect Black/White to shift switch. Check Black/White to Ground	Stator: 200 Volts and above Trigger: 2 Volts and above -25 to -40 Volts @2500 rpm	Run Engine Verify Repair END	Refer to *Note Below Replace Control Module Run Engine Verify Repair	DMT 2000 Digital Tachometer/ Multi-meter P/N 91-854009A1 & DVA Adaptor P/N 91-89045 TPI/CDM Test Harness 84-825207A2

* Note: Stator tests will only isolate problem down to a charging pair. Further testing is necessary to determine faulty CDM. Disconnecting one CDM of the charging pair is recommended.

SELOC_1418

CDM diagnostic chart III (at least 1 CDM has spark) - 3, 4 and 6 cyl models

Step	Action	Value	Yes	No	Tools
1	Resistance Check ALL CDMs	Refer to chart	Step 3	Replace any CDMs that do not pass specifications even if they fire Step 2	DMT 2000 Digital Tachometer/ Multi-meter P/N 91-854009A1 & DVA Adaptor P/N 91-89045
2	Test all CDMs at Cranking with Spark Gap Tester **Spark on All CDMs** Will spark jump a 7/16 in. (11.11 mm) gap?	7/16 in. (11.11 mm) gap	Run Engine Verify Repair Step 7	Step 3	Spark Gap Tester P/N 91-850439
3	Check CDM Trigger Input: Cranking with CDM disconnected Cranking with CDM connected	1 Volt and above - CDM disconnected. 0.2 - 5 Volts - CDM connected.	Step 6	2, 3, & 4 cyl - Replace Trigger Run Engine Verify Repair Step 7 2.0/2.5L 6 Cyl. - Step 4 3.0L 6 Cyl.- Step 4	DMT 2000 Digital Tachometer/ Multi-meter P/N 91-854009A1 & DVA Adaptor P/N 91-89045 TPI/CDM Test Harness 84-825207A2
4	2.0/2.5L V-6 Models Resistance between Trigger wires Red - White Blue - Purple Brown - Yellow	1100 - 1400 Ohms	Replace Control Module Step 6	Replace Trigger Step 6	DMT 2000 Digital Tachometer/ Multi-meter P/N 91-854009A1
5	3.0 V6 Models Resistance Check Crank Position Sensor	900 - 1300 Ohms	Step 6	Replace Crank Position Sensor Run Engine Verify Repair Step 7	DMT 2000 Digital Tachometer/ Multi-meter P/N 91-854009A1 & DVA Adaptor P/N 91-89045

SELOC_1418

Step	Action	Value	Yes	No	Tools
6	Test all CDMs at Cranking with Spark Gap Tester **Spark on All CDMs?** Will spark jump a 7/16 in. (11.11 mm) gap?	7/16 in. (11.11 mm) gap	Run Engine Verify Repair Step 7	Replace any non-firing CDMs Step 7	Spark Gap Tester P/N 91-850439
7	If mis-firing is in a repeatable range: Perform DVA readings on stator and trigger at all running speeds.* **2.0/2.5L V-6 Models** Preform Bias circuit tests 200 HP EFI - Disconnect Detonation Module. Check Black/White to ground All other models - Disconnect Black/White to shift switch. Check Black/White to Ground	Stator: 200 Volts and above Trigger: 2 Volts and above -25 to -40 Volts @2500 rpm	Run Engine Verify Repair END	Refer to *Note Below Replace Control Module Run Engine Verify Repair	DMT 2000 Digital Tachometer/ Multi-meter P/N 91-854009A1 & DVA Adaptor P/N 91-89045 TPI/CDM Test Harness 84-825207A2 6 Pin Connector

* Note: Stator tests will only isolate problem down to a charging pair. Further testing is necessary to determine faulty CDM. Disconnecting one CDM of the charging pair is recommended.

SELOC_1418

CDM diagnostic chart III (at least 1 CDM has spark) - 3, 4 and 6 cyl models, cont'd

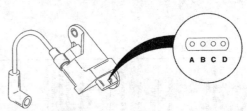

a - Ground
b - Black/Yellow
c - Trigger connection
d - Stator connection

A resistance check is required and can be performed on the CDM as follows:

NOTE: This test can be performed using the test harness (P/N 84-825207A2). Do Not connect the test harness plug to the stator/trigger engine wire harness.

CAPACITOR DISCHARGE MODULE				
Circuit Test	Connect Negative (−) Meter Lead To:	Connect Positive (+) Meter Lead To:	Ohms Scale	Results:
Stop Diode Forward Bias	Green (D) / or Green test harness lead	Black/Yellow (B) / or Black/Yellow test harness lead	R x 100 Diode Reading*	Continuity
Stop Diode Reverse Bias	Black/Yellow (B) / or Black/Yellow test harness lead	Green (D) / or Green test harness lead	R x 100 Diode Reading*	No Continuity
Return Ground Path Diode, Reverse Bias	Green (D) / or Green test harness lead	Ground Pin (A) or Black/Yellow test harness lead	R x 100 Diode Reading*	No Continuity
Return Ground Path Diode, Forward Bias	Ground Pin (A) / or Black test harness lead	Green (D) / or Green test harness lead	R x 100 Diode Reading*	Continuity
CDM Trigger Input Resistance	Ground Pin (A) / or Black test harness lead	White (C) / or White test harness lead	R x 100	1000 - 1250 Ohms
Coil Secondary Impedance	Ground Pin (A) or Black test harness lead	Spark Plug Terminal (At Spark Plug Boot)	R x 100	900 - 1200 Ohms

*Diode Readings: Due to the differences in test meters, results other than specified may be obtained. In such a case, reverse meter leads and re-test. If test results then read as specified CDM is O.K. The diode measurements above will be opposite if using a Fluke equivalent multimeter.

SELOC_1418

CDM testing (modules with part #827509) - 3, 4 and 6 cyl models

2 Cyl.:
CDM #1 gets its charging ground path through CDM #2
CDM #2 gets its charging ground path through CDM #1
A shorted Stop Diode in either CDM would prevent the opposite one from sparking.

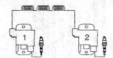

3 Cyl.:
CDM #1 gets its charging ground path through CDM #2 or #3
CDM #2 and #3 get their charging ground path through CDM #1
A shorted Stop Diode in CDM #1 would prevent CDMs #2 and #3 from sparking.
A shorted Stop Diode in CDM #2 or #3 would prevent CDM #1 from sparking.

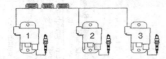

4 Cyl.:
CDM #1 and #2 get their charging ground path through CDM #3 or #4
CDM #3 and #4 get their charging ground path through CDM #1 or #2
A shorted Stop Diode in CDM #1 or #2 would prevent CDMs #3 and #4 from sparking.
A shorted Stop Diode in CDM #3 or #4 would prevent CDM #1 and #2 from sparking.

SELOC_1418

CDM stop diode troubleshooting - 2 thru 4 cyl models

2.0/2.5 Litre 6 Cyl.:
CDM #1, #2 and #3 get their charging ground path through CDM #4, #5 or #6
CDM #4, #5 and #6 get their charging ground path through CDM #1, #2 or #3
A shorted Stop Diode in CDM #1, #2 or #3 would prevent CDMs #4, #5 and #6 from sparking.
A shorted Stop Diode in CDM #4, #5 or #6 would prevent CDMs #1, #2 and #3 from sparking.

3.0 Litre 6 Cyl.:
All CDMs get their charging ground path independently through the stator's white leads.
A shorted Stop Diode in any one CDM will prevent at least 2 other CDMs from sparking.

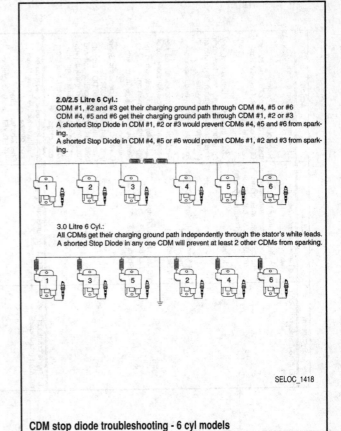

SELOC_1418

CDM stop diode troubleshooting - 6 cyl models

Symptom	Cause	Action
3. Engine idle is rough	3.1 Fouled spark plug	Replace spark plug; if carbon bridges electrode gap or if it is completely black. If it is not firing and is wet with fuel. *Note: If spark plug is grey or completely black with aluminum specs, this indicates a scuffed piston.*
	3.2 Failed fuel injector	Refer to ohm test
	3.3 Bad coil/weak spark	Refer to ohm test
	3.4 Flywheel misaligned during installation	Remove flywheel and inspect.
4. Engine idles fast (rpm >700) or surges	4.1 Fuel leak	Check for fuel entering induction manifold. Fuel pump diaphragm leaking and/or Vapor Separator flooding over.
	4.2 Improper set-up	Check throttle cable & cam roller adjustment.
5. Engine runs rough below 3000 rpm	5.1 Fouled spark plug	See 3.1
	5.2 Throttle misadjusted	Check throttle cam setup on induction manifold. Inspect linkage and roller. If throttle plate stop screws have been tampered with, contact Mercury Marine Service Department for correct adjustment procedures.
	5.3 Bad coil/weak spark	See 3.3
	5.4 TPS malfunction	See 2.2
6. Engine runs rough above 3000 rpm	6.1 Fouled spark plug	See 3.1
	6.2 Speed Reduction	See 7
	6.4 TPS malfunction	See 2.2

SELOC_1418

Ignition troubleshooting - 2002 & later EFI models (6 cyl. w/battery driven ignitions), cont'd

Symptom	Cause	Action
7. Speed Reduction (RPM reduced)	7.1 Low battery voltage ECM requires 8 volts minimum Fuel Pump requires 9 volts	Check battery and/or alternator. Check electrical connections.
	7.2 Overheat condition	Check water pump impeller/cooling system.
	7.3 Oil pump electrical failure	Check electrical connection.
	7.4 TPS failure If TPS fails, rpm is reduced to idle	See 2.2
8. Engine RPM reduced to idle only	8.1 TPS failed	See 2.2
	8.2 Battery voltage below 9.5 volts	Use DDT to monitor system
9. Loss of spark on 1 cylinder	9.1 Loose wire or pin in connectors between ECM and coil primary.	Check connectors.
	9.2 Faulty ignition coil.	Replace coil.
	9.3 Faulty spark plug.	Replace spark plug.
	9.4 Faulty spark plug wire. Note: If spark plug is partially fouled or the plug gap is too small, the DDT may indicate the incorrect cylinder as having an ignition fault. Example: If the DDT indicates an ignition fault on cylinder #4, the problem may be on the prior cylinder in the firing order - I.E. cylinder number #3.	Replace spark plug wire.

SELOC_1418

Ignition troubleshooting - 2002 & later EFI models (6 cyl. w/battery driven ignitions), cont'd

Symptom	Cause	Action
1. Engine cranks but won't start	1.0 Lanyard stop switch in wrong position.	Reset lanyard stop switch.
	1.1 Weak battery or bad starter motor, battery voltage drops below 8 volts while cranking (ECM cuts out below 8volts) (Fuel pump requires 9volts)	Replace/charge battery. Inspect condition of starter motor. Check condition of battery terminals and cables.
	1.2 No fuel	Check that primer bulb is firm. Key-on engine to verify that fuel pump runs for 2 seconds and then turn off. Measure fuel pressure
	1.3 Flywheel misaligned during installation	Remove flywheel and inspect.
	1.4 Blown fuse	Replace fuse. Inspect engine harness and electrical components.
	1.5 Main Power Relay not functioning	Listen for relay to "click" when the key switch is turned on.
	1.6 Spark Plugs	Remove spark plugs from each cylinder. Connect spark plug leads to Spark Gap Tester 91-8302307. Crank engine or use DDT output load test for each ignition coil and observe spark. If no spark is present, replace appropriate ignition coil. If spark is present, replace spark plugs.
1. Engine cranks but will not start (continued)	1.7 ECM not functioning	Check for proper operation by using Inductive Timing Light 91-99379. Check battery voltage (RED/YEL Lead) @ ignition coils. Check for blown fuse (C15). Check battery voltage to fuse from main power relay (PURPLE Lead). Check for shorted stop wire (BLK/YEL). Check crank position sensor setting [0.025 in. – 0.040 in. (0.64 mm – 1.02 mm)] from flywheel or for defective crank position sensor. Defective ECM.
	1.8 Crank Position Sensor not functioning	Power Supply: Clean and inspect remote control male and female harness connectors.
2. Engine cranks, starts and stalls	2.0 Abnormally high friction in engine	– Sensor faulty. – Bad connection – Air gap incorrect
		Check for scuffed piston or other sources of high friction.
	2.1 Air in fuel system/lines	Crank and start engine several times to purge.
	2.2 TPS malfunction	Check motion of throttle arm. Stop nuts should contact block at idle and WOT. Check TPS set-up. Must connect DDT with adapter harness (84-822560A5) to ECM.
	2.3 Remote control to engine harness connection is poor	Clean and inspect male and female connectors.

SELOC_1418

Ignition troubleshooting - 2002 & later EFI models (6 cyl. w/battery driven ignitions)

DFI Troubleshooting Guide

Symptom	Cause	Action
1. Engine cranks but won't start	1.0 Lanyard stop switch in wrong position.	Reset lanyard stop switch.
	1.1 Weak battery or bad starter motor, battery voltage drops below 8 volts while cranking (ECM cuts out below 8 volts) (Fuel pump requires 9 Volts).	Replace/charge battery. Inspect condition of starter motor. Check condition of battery terminals and cables.
	1.2 Low air pressure in rail (less than 70 psi at cranking	Inspect air system for leaks. Inspect air filter for plugging (air pressure measured on port rail). Inspect air compressor reed valves if necessary.
	1.3 No fuel	Check that primer bulb is firm. Key-on engine to verify that fuel pump runs for 2 seconds and then turn off. Measure fuel pressure
	1.4 Low fuel pressure	Check fuel pressure from low pressure electric fuel pump (6–10 psi). Check for fuel leaks. If fuel pressure leaks down faster than air pressure, seals on fuel pump may be leaking. Check air system pressure, see 1.2.
	1.5 Flywheel misaligned during installation	Remove flywheel and inspect.
	1.6 Blown fuse	Replace fuse. Inspect engine harness and electrical components.
	1.7 Main Power Relay not functioning	Listen for relay to click when the key switch is turned on.
	1.8 Spark Plugs*	Remove fuel pump fuse. Unplug all direct injector connectors. Remove spark plugs one at a time from each cylinder. Connect spark plug leads to Spark Gap Tester 91-850439T. Crank engine or use DDT output load test for each ignition coil and observe spark. If no spark is present, replace appropriate ignition coil. If spark is present, replace spark plugs.

*NOTE: * Spark jumping the gap from all cylinders at the same time in the spark gap tool may cause interference in the ECM. The interference may cause the absence of spark on some cylinders and a false diagnosis of a no spark condition. Crank engine over with only one spark plug wire connected to spark gap tool at a time or use DDT to fire one cylinder at a time.*

SELOC_1418

Ignition troubleshooting - OptiMax motors

Symptom	Cause	Action
1. Engine cranks but will not start (continued)	1.9 ECM not functioning	Injection System: Listen for injector ticking when cranking or connect spare injector to each respective harness. Ticking should start after 2 cranking revolutions.

Ignition System:
– Check for proper operation by using Inductive Timing Light 91-99379.
– Check battery voltage (RED/YELLOW Lead) @ ignition coils.
– Check for blown fuse (C15).
– Check battery voltage to fuse from main power relay (PURPLE Lead).
– Check for shorted stop wire (BLACK/YELLOW). |
| | 1.9A Crank Position Sensor not functioning | – Check crank position sensor setting or for defective crank position sensor.

– Defective ECM.

Power Supply: Clean and inspect remote control male and female harness connectors. |
2. Engine cranks, starts and stalls	2.0 Low air pressure in rail	See 1.2
	2.1 Low fuel pressure in rail	See 1.2 and 1.3
	2.2 Abnormally high friction in engine	Check for scuffed piston or other sources of high friction.
	2.3 Air in fuel system/lines	See 1.3 Crank and start engine several times to purge.
	2.4 TPS malfunction	Check motion of throttle arm. Stop nuts should contact block at idle and WOT. Check TPS set-up. Must connect DDT with adapter harness (84-822560A5) to ECM.
	2.5 Remote control to engine harness connection is poor	Clean and inspect male and female connectors.
	2.6 Flywheel misaligned during installation	Remove flywheel and inspect.

SELOC_1418

Ignition troubleshooting - OptiMax motors, cont'd

DFI Troubleshooting Guide (continued)

Symptom	Cause	Action
3. Engine idle is rough	3.1 Low air pressure in rail (less than 79 ± 2 psi while running)	See 1.2
	3.2 Fouled spark plug	Replace spark plug: --If carbon bridges electrode gap or if it is completely black. --If it is not firing and is wet with fuel. Note: If spark plug is gray or completely black with aluminum specs, this indicates a scuffed piston.
	3.3 Failed direct injector	Refer to specifications for ohm test.
	3.4 Failed fuel injector	Refer to specifications for ohm test.
	3.5 Bad coil/weak spark	Refer to specifications for ohm test.
	3.6 Flywheel misaligned during installation	Remove flywheel and inspect.
4. Engine idles fast (rpm >700) or surges	4.1 Broken fuel pressure regulator or tracker diaphragm	Measure fuel pressure. Remove and inspect diaphragms (a special tool is required for assembly).
	4.2 Fuel leak	Check for fuel entering induction manifold or air compressor inlet. Fuel pump diaphragm leaking and/or Vapor Separator flooding over.
	4.3 Tracker Valve spring missing	Inspect tracker valve for proper assembly.
	4.4 Improper set-up	Check throttle cable & cam roller adjustment.
5. Engine runs rough below 3000 rpm	5.1 Fouled spark plug	See 3.2
	5.2 Low air pressure in rail	See 1.2
	5.3 Throttle misadjusted	Check throttle cam setup on induction manifold. Inspect linkage and roller. If throttle plate stop screws have been tampered with, contact Mercury Marine Service Department for correct adjustment procedures.
	5.4 Bad coil/weak spark	See 3.5
	5.5 TPS malfunction	See 2.4
6. Engine runs rough above 3000 rpm	6.1 Fouled spark plug	See 3.2
	6.2 Speed Reduction	See 7
	6.3 Low air pressure in rails	See 1.2
	6.4 TPS malfunction	See 2.4

SELOC_1418

Ignition troubleshooting - OptiMax motors, cont'd

DFI Troubleshooting Guide (continued)

Symptom	Cause	Action
7. Speed Reduction (RPM reduced or limited to 3000)	7.1 Low battery voltage ECM requires 8 volts minimum Fuel Pump requires 9 volts	Check battery and/or alternator. Check electrical connections.
	7.2 Overheat condition (engine and/or air compressor)	Check water pump impeller/cooling system.
	7.3 Oil pump electrical failure	Check electrical connection.
	7.4 TPS failure If TPS fails, rpm is reduced approximately 20% and the ECM refers to the MAP sensor	Check electrical connections.
8. Engine RPM reduced to idle only		Refer to **Section 2D - Guardian System**
9. Loss of spark on 1 cylinder	9.1 Loose wire or pin in connectors between ECM and coil primary.	Check connectors.
	9.2 Faulty ignition coil.	Replace coil.
	9.3 Faulty spark plug.	Replace spark plug.
	9.4 Faulty spark plug wire.	Replace spark plug wire.
	Note: Due to the long spark duration of this ignition system, the DDT may indicate the incorrect cylinder as having an ignition fault. Example: If the DDT indicates an ignition fault on cylinder #4, the problem may be on the prior cylinder in the firing order – I.E. cylinder number #3.	

SELOC_1418

Ignition troubleshooting - OptiMax motors, cont'd

Flywheel

REMOVAL & INSTALLATION

1 & 2-Cylinder Powerheads

◆ See Figures 18 thru 25

1. On electric start models, disconnect the negative battery cable for safety. On manual start models, disconnect and ground the spark plug lead(s) for safety. For twins, be sure to tag the spark plug wires to ensure proper installation.

2. Remove the mounting hardware securing the hand rewind starter assembly or flywheel cover to the powerhead. Remove the hand starter or flywheel cover for access to the flywheel. Most models use 3 or 4 fasteners to secure the starter or cover. On many late-model motors such as the OptiMax models instead of bolts or screws, the flywheel cover is secured by grommets mounted into the cover that are pushed down over top of mounting pins

3. On 2.5/3.3 hp motors the manual starter pulley must be removed in order to install the flywheel holder tool. Loosen the 3 mounting bolts and remove the pulley. On these models use #91-83163M which is like a prybar with a pivoting arm about halfway down its shaft. Both the arm and the end of the shaft have pins that fit into holes in the flywheel (a substitute can usually be fabricated with 2 pieces of flat stock and a number of small bolts.)

4. Some Mercury models use a small plastic cover (usually yellow) over the nut in the center of the flywheel. The purpose of the nut is to seal it from the atmosphere, further reducing the possibility of corrosion. If present, carefully pry and remove the nut cover.

■ Obtain a flywheel holder (#91-52344 works on the ring gear of most 6 hp or larger models) or large strap wrench (for all but the 30/40 hp twins as the flywheel nut torque is just too high). On these motors the flywheel holder is like a prybar with a pivoting arm which is used to lock into the teeth of the ring gear. The strap wrench is used either on the outside circumference of the flywheel itself OR sometimes on the starter pulley, if it is still installed to the flywheel at that point.

5. Secure the flywheel using the holder or strap wrench in order to keep it from turning (remember, you want to avoid turning the flywheel counterclockwise as you could potentially damage the vanes of the water pump impeller). Hold the flywheel from rotating and at the same time loosen and remove the flywheel nut or bolt.

6. On some motors 4/5 hp motors the manual starter pulley must be removed in order to install a puller to the flywheel. On those models, remove the 3 bolts securing the pulley to the flywheel, then remove the pulley.

7. Obtain a threaded flywheel puller tool. On most models a standard steering wheel/gear puller will work, as there are holes in the top of the flywheel that can be used to attach the puller. However, on 30/40 hp models

Fig. 18 Remove the manual starter...

Fig. 19 ...or flywheel cover (as equipped)

Fig. 20 On 2.5/3.3 hp motors you have to remove the starter pulley...

Fig. 21 ...to install the flywheel holder

Fig. 22 Most models use a strap wrench or lock the ring gear

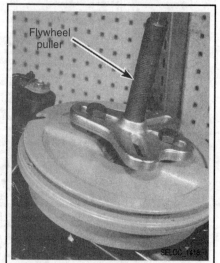

Fig. 23 Most models use a standard gear puller to remove the flywheel...

there are normally no such bolt holes so you'll need to use the crankshaft protector cap (#91-24161) and specialized flywheel puller (#91-849154T1 or #91-73687A1) or equivalents which are designed to work on the flywheel and crankshaft only at the center of the assembly. In all cases, make sure the puller does NOT work on the perimeter of the flywheel.

✳✳ SELOC WARNING

Never attempt to use a jawed puller (which would put stress on the outside edge of the flywheel).

8. Install the puller onto the flywheel, take a strain on the puller with the proper size wrench. Now, continue to tighten on the tool and at the same time, shock the crankshaft with a gentle to moderate tap using a hammer or mallet on the end of the tool. This shock will assist in breaking the flywheel loose from the crankshaft.

9. Lift the flywheel free of the crankshaft. Remove and save the Woodruff key from the recess in the crankshaft.

■ **On most motors the stator coil or the separate charge coil or trigger coil can now be unbolted and removed without further component disassembly.**

To Install:

10. Check the flywheel magnets to ensure they are free of any metal particles. Double check the taper in the flywheel hub and the taper on the crankshaft to verify they are clean and contain no oil.

11. Slide the flywheel down over the crankshaft with the keyway in the flywheel aligned with the Woodruff key in place on the crankshaft. Rotate the

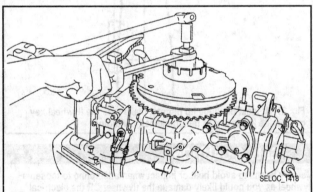

Fig. 24 ...however 30/40 hp motors use a special puller (shown)

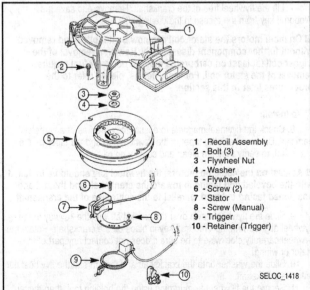

Fig. 25 Exploded view of a typical Mercury flywheel and ignition component assembly - 6/8 and 9.9/10/15 hp motors shown (others similar)

1 - Recoil Assembly
2 - Bolt (3)
3 - Flywheel Nut
4 - Washer
5 - Flywheel
6 - Screw (2)
7 - Stator
8 - Screw (Manual)
9 - Trigger
10 - Retainer (Trigger)

SELOC_1418

flywheel carefully clockwise to be sure it does not contact any part of the stator or wiring.

12. Slide the washer onto the crankshaft, then thread the flywheel nut or bolt onto the crankshaft.

13. If necessary on 4/5 hp models, install the starter pulley to the top of the flywheel, so you can secure the flywheel with a strap wrench while tightening the flywheel nut. Tighten the pulley retaining bolts to 70 inch lbs. (8 Nm).

14. Secure the flywheel from turning using the holding tool, then tighten the flywheel nut to the following torque value for the models listed:
- 2.5/3.3 hp - 30 ft. lbs. (41 Nm)
- 4/5 hp - 40 ft. lbs. (52.2 Nm)
- 6/8 and 9.9/10/15 Hp Models - 50 ft. lbs. (68 Nm)
- 20/20 Jet/25 hp - 50 ft. lbs. (68 Nm)
- 30/40 hp - 95 ft. lbs. (129 Nm)

15. If equipped, install the small plastic nut cover to the center of the flywheel.

16. On 2.5/3.3 hp motors, install the starter pulley and tighten the bolts securely.

17. Install the hand rewind starter or flywheel cover, as applicable.

18. Reconnect the spark plug lead(s) and/or reconnect the negative battery cable, as applicable.

3- and 4-Cylinder Powerheads

◆ **See Figures 26 thru 31**

■ **There is a discrepancy in Mercury technical literature as to the proper torque specification for the flywheel nut on 40/40 Jet/50/55/60 hp 3-cyl models. Some sources show the specification to be 100 ft. lbs. (135.6 Nm), while most show it to be 125 ft. lbs. (169.5 Nm). We suggest that before loosening the nut you use a beam-type torque wrench to check the current torque and help you decide what setting to use during installation. If you only have a clicker type wrench, dial up 90 ft. lbs. first and check to make sure the wrench clicks without nut movement, then dial up another 5 lbs. and retest (repeat until the nut moves and you know about where it was set).**

1. Disconnect the negative battery cable for safety. On manual start models you'll have to disconnect and ground the spark plug leads for safety.

2. Remove the flywheel cover (or hand rewind starter on manual start models).

■ **Obtain a flywheel holder (#91-52344 works on the ring gear). On these motors the flywheel holder is like a prybar with a pivoting arm which is used to lock into the teeth of the ring gear. Whatever flywheel holder is used, it has to be particularly strong, as the flywheel nuts on these models are tightened to about 100 ft. lbs. (136 Nm) or more, depending upon the model.**

3. Some Mercury models use a small plastic cover (usually yellow) over the nut in the center of the flywheel. The purpose of the nut is to seal it from the atmosphere, further reducing the possibility of corrosion. If present, carefully pry and remove the nut cover.

4. Secure the flywheel using the holder in order to keep it from turning (remember, you want to avoid turning the flywheel counterclockwise as you could potentially damage the vanes of the water pump impeller). Hold the flywheel from rotating and at the same time loosen and remove the flywheel nut.

5. Obtain a suitable flywheel puller tool (unlike smaller models a gear type puller is normally not recommended on these motors). You'll need to use the crankshaft protector cap (#91-24161) and specialized flywheel puller (#91-73687A1) or equivalents which are designed to work on the flywheel and crankshaft only at the center of the assembly. If a substitute puller is used, MAKE SURE it does not work on the perimeter of the flywheel.

✳✳ SELOC WARNING

Never attempt to use a jawed puller (which would put stress on the outside edge of the flywheel).

6. Install the puller onto the flywheel, take a strain on the puller with the proper size wrench. Now, continue to slowly tighten the tool until the flywheel breaks loose from the crankshaft.

Fig. 26 Remove the flywheel cover (or manual starter). . .

Fig. 27 . . .for access to the flywheel (note the nut has a cover on this model)

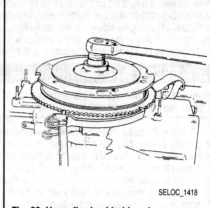

Fig. 28 Use a flywheel holder when loosening or tightening the nut

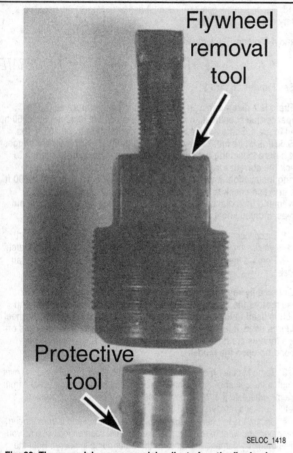

Fig. 29 These models use a special puller to free the flywheel

Fig. 30 Install the puller and slowly free the flywheel

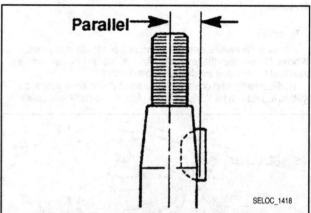

Fig. 31 At least on 40-60 hp motors, make sure the flywheel key outer edge is parallel to the crankshaft centerline

✳✳ SELOC CAUTION

Mercury warns to avoid heat or impact when attempting to loosen a flywheel as you could likely damage the flywheel OR the electrical components which reside underneath.

7. Lift the flywheel free of the crankshaft. Remove and save the Woodruff key from the recess in the crankshaft.

■ On most motors the stator coil can now be unbolted and removed without further component disassembly. However, removal of the trigger coil (at least on carbureted motors) normally first requires removal of the stator coil. For more details, please refer to the procedures later in this section.

To Install:

8. Check the flywheel magnets to ensure they are free of any metal particles. Double check the taper in the flywheel hub and the taper on the crankshaft to verify they are clean and contain no oil.

■ At least on the 40-60 hp motors, the flywheel key should be installed with the beveled end facing in toward the crankshaft and the flat end positioned facing outward, parallel to the centerline of the crankshaft.

9. Slide the flywheel down over the crankshaft with the keyway in the flywheel aligned with the Woodruff key in place on the crankshaft. Rotate the flywheel carefully clockwise to be sure it does not contact any part of the stator or wiring.

10. Slide the washer onto the crankshaft and then thread the flywheel nut onto the crankshaft.

11. Secure the flywheel from turning using the holding tool, then tighten the flywheel nut to the following torque value for the models listed:

• 40/40 Jet/50/55/60 hp 3-cyl - 100 ft. lbs. (135.6 Nm) - but refer to the NOTE at the beginning of this procedure.

• 75/65 Jet/90 hp, 75/90/115 hp Optimax & 80 Jet/115/125 hp -120 ft. lbs. (163 Nm).

12. If equipped, install the small plastic nut cover to the center of the flywheel.

13. Install the flywheel cover (or hand rewind starter on manual start models).

14. Reconnect the spark plug lead(s) and/or reconnect the negative battery cable, as applicable.

6-Cylinder Powerheads

◆ **See Figures 28 thru 30 and 32 thru 35**

1. Disconnect the negative battery cable and/or disconnect and ground the spark plug leads for safety.

■ **Some OptiMax models, such as 2003 & later 3.0L models, have a dedicated air hose to supply the compressor. For others, the cover is part of the air intake system and when installed must be properly seated over the air compressor intake tube.**

2. If equipped, remove the flywheel cover. This varies greatly depending upon the model. Some Carbureted and EFI models have small flywheel covers which are bolted in place. However, most OptiMax and some EFI models have covers which are held in place by rubber grommets which are positioned over small plastic mounting posts. On covers secured by rubber grommets removal is usually just a simple matter of carefully lifting at one or both ends to pull the cover off the grommets. On a few models, such as the 3.0L EFI (thru 2001) the flywheel cover is also attached to the intake air attenuator (the latter of which is secured at its base by 2 latches).

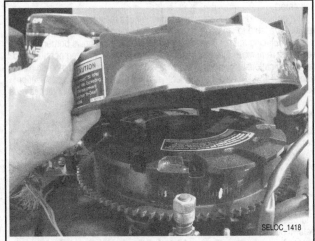

Fig. 32 Remove the flywheel cover (size/shape will vary). . .

Don't FORCE the plastic flywheel cover. If it is not coming off easily, check closely for any fasteners. Also, we've seen some of the OptiMax covers brake easily in the field from rough handling. Especially models where the filter for the air compressor is integrated. On OptiMax models (or any cover secured by rubber grommets) it is a good idea to try and reach under the cover to support it directly where the grommet and pin connect while pulling upward.

3. On all except 2001 2.5L EFI motors, remove the drive belt (compressor and/or alternator belt) from the powerhead. Use a prybar to carefully move the automatic tensioner out of contact with the belt, then slide the belt off the tensioner pulley and slowly allow the tensioner to return to its spring-loaded position. For more details, please refer to the Drive Belt procedure in the Maintenance & Tune-Up section.

4. Check the timing pointer to see if it will interfere with flywheel removal (it will on some models, such as the 2001 3.0L EFI motor). If so, make an alignment mark to ease installation, then loosen the bolt and remove the timing pointer.

■ **Obtain a flywheel holder (#91-52344 works on the ring gear). On these motors the flywheel holder is like a prybar with a pivoting arm which is used to lock into the teeth of the ring gear. Whatever flywheel holder is used, it has to be particularly strong, as the flywheel nuts on these models are tightened to about 120-125 ft. lbs. (163-170 Nm) or so, depending upon the model.**

5. Some Mercury models use a small plastic cover (usually yellow) over the nut in the center of the flywheel. The purpose of the nut is to seal it from the atmosphere, further reducing the possibility of corrosion. If present, carefully pry and remove the nut cover.

6. Secure the flywheel using the holder in order to keep it from turning (remember, you want to avoid turning the flywheel counterclockwise as you could potentially damage the vanes of the water pump impeller). Hold the flywheel from rotating and at the same time loosen and remove the flywheel nut.

7. Obtain a suitable flywheel puller tool (unlike smaller models a gear type puller is normally not recommended on these motors). You'll need to use the crankshaft protector cap (#91-24161) and specialized flywheel puller (#91-73687A1) or equivalents which are designed to work on the flywheel and crankshaft only at the center of the assembly. On Pro/XS/Sport models, Mercury recommends using puller #91-84915T-1 or equivalent. In all cases, if a substitute puller is used, MAKE SURE it does not work on the perimeter of the flywheel.

Never attempt to use a jawed puller (which would put stress on the outside edge of the flywheel).

8. Install the puller onto the flywheel, take a strain on the puller with the proper size wrench. Now, continue to slowly tighten the tool until the flywheel breaks loose from the crankshaft.

Fig. 33 . . .and nut cover. . .

Fig. 34 . . .for access to the flywheel nut

Fig. 35 Remove the drive belt

※※ SELOC CAUTION

Mercury warns to avoid heat or impact when attempting to loosen a flywheel as you could likely damage the flywheel OR the electrical components which reside underneath.

9. Lift the flywheel free of the crankshaft. Remove and save the Woodruff key from the recess in the crankshaft.

To Install:

10. Check the flywheel magnets to ensure they are free of any metal particles. Double check the taper in the flywheel hub and the taper on the crankshaft to verify they are clean and contain no oil.

11. Slide the flywheel down over the crankshaft with the keyway in the flywheel aligned with the Woodruff key in place on the crankshaft. Rotate the flywheel carefully clockwise to be sure it does not contact any part of the stator or wiring.

12. Slide the washer onto the crankshaft and then thread the flywheel nut onto the crankshaft.

13. Secure the flywheel from turning using the holding tool, then tighten the flywheel nut to the following torque value for the models listed:
- 2.5L Models (except OptiMax) - 120 ft. lbs. (163 Nm).
- 2.5L OptiMax Models - 125 ft. lbs. (169.5 Nm).
- 3.0L Models - 125 ft. lbs. (169.5 Nm).

14. If equipped, install the small plastic nut cover to the center of the flywheel.

15. If removed, install the timing pointer, aligning the matchmark made during installation (but be sure to double-check pointer alignment next time your sync and link the motor, meaning next time your perform Timing and Synchronization adjustments).

16. On all except 2001 2.5L EFI motors, install the drive belt.

17. Install the flywheel cover.

18. Reconnect the spark plug lead(s) and/or reconnect the negative battery cable, as applicable.

Stator/Charge Coil

DESCRIPTION & OPERATION

◆ See Figure 36

All carbureted Mercury motors are equipped with some form of a Stator or Charge Coil which is used to generate voltage to power the ignition system. The coil is mounted under the flywheel (centered directly underneath on most models) to interact with 1 or more permanent magnets attached to the underside of the flywheel assembly. As the flywheel rotates, the magnetic force cuts through the coil assembly, generating an electric current that is utilized by the ignition system.

On some smaller motors, the charge coil may be an individual component that can be mounted side-by-side with a battery charge coil and is removed or replaced separately. However, in most instances, the charge coil is part of a round 1-piece stator that, when equipped, also supplies current to the charging system and must be serviced as an assembly.

Refer to the wiring diagrams provided in this section to help determine the system on your engine. Each diagram shows whether a 1-piece stator or an individual charge coil is used.

On all OptiMax motors and most EFI motors (all 2002 or later models) ignition power is supplied by the battery and replenished by the external, belt driven alternator. Those models DO NOT use a stator coil. However, the 2001 EFI models are odd balls, as they are equipped with stators and stator-driven ignition systems (despite the fact that the 3.0L models *also* used an external alternator to charge the battery).

TESTING

 MODERATE

◆ See Figure 36

Stator/charge coil testing typically encompasses making sure the wiring is in good condition, then verifying coil output (using a peak-reading DVOM) while cranking the motor and/or checking resistance (using an ohmmeter) across the coil windings when the motor is not turning. The charge coil and wiring can also be checked for shorts to ground either using the peak-reading DVOM while the motor is cranking or the ohmmeter when the motor is at rest.

Fig. 36 Typical Mercury stator mounted under the flywheel

Problems with the ignition stator/charge coil usually cause a no start condition. But, a partial short of the coil winding can cause hard starting and/or an ignition misfire. The most reliable tests for the charge coil are to dynamically check the output with the engine cranking (or in some cases, running). Of course, dynamic tests require a digital multi-meter capable of reading and displaying peak voltage values (also known as a peak-reading voltmeter).

If a peak-reading voltmeter or adapter is not available, specifications are usually available to statically check the coil winding using an ohmmeter. But, before replacing the coil based only on static test that shows borderline readings, remember that resistance readings will vary with temperature and the specifications provided here are based on a component temperature of about 68°F (20°C). Also, remember that a coil may test within specification statically, but show intermittent faults under engine operating conditions (due to variances in heat and load which are not present during a static test), especially at normal operating temperature.

■ **Both the Resistance Tests and the Voltage Output Tests should be conducted using the information in the Ignition System Component Testing specification chart. All wire connections and values necessary should be listed there.**

Lastly, keep in mind that on models that utilize a 1-piece stator which contains windings BOTH for the ignition system AND for the battery charge system, the tests in this section are isolated to the windings of the IGNITION charge coil, however, if there is a problem with the battery charge coil (as determined using similar tests detailed later in this section under Charging System) then the same Removal & Installation procedure must be followed to replace the 1-piece stator assembly.

Resistance Tests (Static Check)

◆ See Figure 37

1. Using the data from the Ignition System Component Testing specifications chart and, if necessary, the Wiring Diagrams in this section, locate the wiring for the stator/charge coil. Tag and disconnect the wiring.

2. Connect the leads of an ohmmeter (or preferably a DVOM set to read ohms) to the leads as noted in the chart. Red lead to the (+) wire color, and Black lead to the (-) wire color. On some motors the coil circuit should be tested between a given wire color and Gnd (Ground), meaning a good grounding point on the powerhead itself.

■ **Some motors, such as the 2001 3.0L EFI motor, have multiple sets of stator coil windings which much be checked. In the case of the 3.0L EFI motor, there are normally 6 sets of wires to check resistance to ground.**

3. Note the resistance readings and compare with what is listed in the Ignition System Component Testing specifications chart.

■ **DO NOT condemn a coil without testing further. Make sure you've got the right readings by conducting a dynamic Voltage Output Test.**

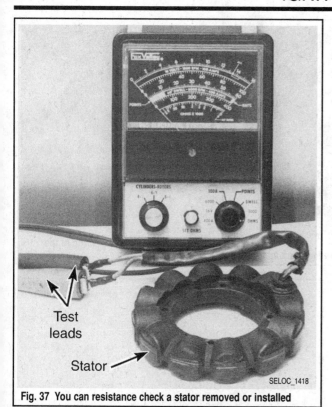

Fig. 37 You can resistance check a stator removed or installed

4. Resistance readings *should* be within the range provided on the chart (which normally gives a 5-10% range). Here's the tricky part. Readings either inside the range or close suggest the possibility of a good coil. However, such static tests really cannot confirm that a component is good. Static tests can hide intermittent shorts or opens that occur during normal operating conditions. For this reason, static tests such as this are really much more accurate when they show a bad component. Readings well out of this range suggest a bad coil winding is very likely.

5. Once the tests are completed, either move to the Voltage Output Test or replace the coil, as necessary.

Voltage Output Test (Dynamic Check)

◆ See Figure 36

1. If not done already for a static Resistance Test, use the data from the Ignition System Component Testing specifications chart and, if necessary, the Wiring Diagrams in this section, to locate the wiring for the stator/charge coil. Tag and disconnect the wiring.

■ **Making the wiring connections can be a challenge for this test. Unless otherwise noted in the chart, all voltage output tests are made under load; meaning with the circuit *complete*. The BEST way to get voltage readings from a complete circuit is to purchase or fabricate a jumper harness with T-connections in it that allow the meter to connect. This is the absolute best method because it will prevent the danger of possibly shorting coil windings to each other during testing (which could damage the stator). Some people will backprobe connectors or worse, pierce the wiring insulation, but both methods CAN lead to trouble down the road so we advise avoiding that, if possible.**

2. Connect the leads of a DVOM set to read peak voltage (for some meters which are not normally capable of reading peak voltage it may be possible to purchase a Peak Reading Voltage Adapter such as CDI Electronics #511-9773) to the leads as noted in the chart. Red lead to the (+) wire color, and Black lead to the (-) wire color. On some motors the coil circuit should be tested between a given wire color and Gnd (Ground), meaning a good grounding point on the powerhead itself.

■ **Some motors, such as the 2001 3.0L EFI motor, have multiple sets of stator coil windings which much be checked. In the case of the 3.0L EFI motor, there are normally 6 sets of wires on which to check voltage output to ground.**

3. Provide the outboard with a good source of cooling water to prevent impeller or powerhead damage.

4. If specifications are provided for cranking output, turn the motor over and check the meter for the appropriate readings. On manual start models it may be better to remove the spark plug(s) and install a spark tester to each spark plug lead, as this will allow the circuit to be complete while relieving engine compression, making it easier to spin for the tests.

5. If specifications are provided for running output, start the motor and allow it to run at various points in the specified test range.

6. Compare the voltage readings with what is listed in the Ignition System Component Testing specifications chart.

7. If the coil tests within specification, reconnect the wiring. If it does not, replace the stator/charge coil as detailed in this section.

REMOVAL & INSTALLATION

◆ **See Figures 36 and 38 thru 42**

1. On electric start models, disconnect the negative battery cable for safety. On manual start models, disconnect and ground the spark plug lead(s) for safety. On multi-cylinder motors, be sure to tag the spark plug wires to ensure proper installation.

2. Remove the flywheel from the powerhead for access as detailed earlier in this section.

✳✳ SELOC CAUTION

The stator/charge coil assembly is mounted in a tight spot under the flywheel. IT IS IMPERATIVE that during removal you pay close attention to how the wiring is routed to ensure proper positioning during installation. Failure to do so may allow the wiring to be positioned in such a manner where it will be contacted by and eventually cut by the flywheel, usually ruining the coil in the process.

3. Remove the stator from the powerhead as follows, depending upon the model:

• **2.5/3.3 hp** - these models use a single combination charge/trigger coil assembly. To remove the coil remove the 2 screws from the coil and the screw and clamp on the wiring opposite the coil and then, disengage the bullet connector underneath the side of the coil mounting plate. It is usually necessary to remove the port side cowling for access.

• **4/5 hp** - these models use a charge coil mounted to the plate under the flywheel (and sometimes a second, almost identical looking lighting coil mounted just opposite it, in the same manner). To remove the coil, remove the 2 screws from the coil and the screw and clamp on the coil wiring. Disconnect the Black/Red and White coil leads from the CDI unit, then remove the coil from the powerhead carefully noting the wire routing. During installation, be sure to apply a light coat of Loctite® "A" to the threads of the mounting screws before installation.

• **6/8, 9.9/10/15 & 20/20 Jet/25 hp** - these motors utilize a charge coil mounted on a semi-circular bracket just above the trigger coil assembly. The stator or charge coil is secured by screws (usually 2 or 3) and connected

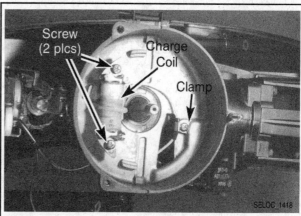

Fig. 38 Individual charge coil used on smaller motors - 2.5/3.3 hp shown

Fig. 39 Loosen and remove the stator retaining bolts

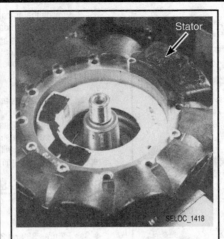

Fig. 40 Note stator and wiring positioning...

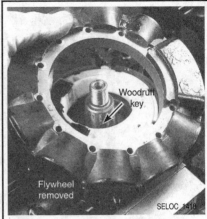

Fig. 41 ...then lift it from the top of the powerhead

through 2 bullet connectors (usually 1 male and 1 female, to ensure proper connections). During installation be sure to coat the threads of the retaining screws using Loctite® 271 or equivalent thread locking material.

• 30/40 hp 2-cyl - these are the first of the models which usually have a 1-piece, round stator (complete with ignition charge coil and lighting/battery charge coil functions all in one). The stator itself is normally secured by 5 screws around the inside hub of the assembly and the wiring should contain multiple bullet connectors. During installation be sure to coat the threads of the retaining screws using Loctite® 222 or equivalent thread locking material.

• 40/40 Jet/50/55/60 hp 3-cyl - on these motors, tag and disconnect the 2 yellow stator leads from the rectifier/regulator as well as the stator leads from the CDM wiring harness. Then remove the 4 screws securing the stator to the trigger bearing cage and remove the stator from the powerhead. During installation be sure to coat the threads of the retaining screws using Loctite® 222 or equivalent thread locking material.

• 75/65 Jet/90 & 80 Jet/115/125 hp - although the set-up on these models is very similar to the 40-60 hp 3-cyl motors, the wire routing may require removal of the starter motor for access. To begin with, remove the 4 screws securing the stator to the powerhead, then trace the wires down by the starter to the switch box. Cut any cable ties as necessary and disconnect the wires from the switch box bullet connectors. During installation be sure to coat the threads of the retaining screws using Loctite® 271 or equivalent thread locking material.

✳✳ SELOC CAUTION

During installation of the stator on 75/65 Jet/90 hp 3-cyl motors, pay particular attention to the stator orientation (as shown in the accompanying illustration) as improper orientation and stator interference with the engine block could cause premature stator failure.

• 150-200 hp (2.5L) carbureted and 2001 EFI V6 - utilize a 1-piece stator assembly under the center of the flywheel. Removal is a simple matter of tagging and disconnecting the wiring (harness connector), then removing the 5 screws securing the stator to the motor upper end cap. During installation be sure to clean the threads of the retaining screws using Loctite® 7649 Primer and then apply a light coating of Loctite® 271 or equivalent threadlocker.

• 200-250 hp (3.0L) V6 - utilize a 1-piece stator assembly under the center of the flywheel. Removal is a simple matter of tagging and disconnecting the wiring (harness connector), then removing the 4 screws securing the stator.

To Install:

4. Position the stator to the powerhead, routing the wiring as noted during removal.

5. Install the stator retaining screws and tighten securely.

6. Install the flywheel to the powerhead.

7. Reconnect the spark plug lead(s) and/or reconnect the negative battery cable, as applicable.

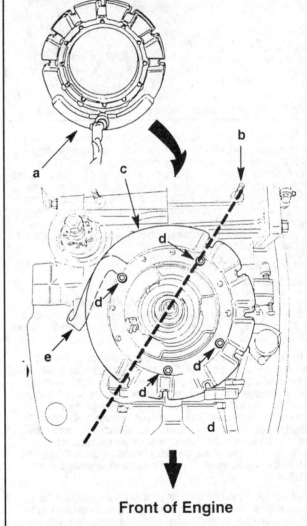

Front of Engine

a - Stator
b - Flywheel Cover Stud
c - High/Low Speed Winding Module of Stator
d - Stator Screws [Apply Loctite 271 to threads] [Torque screws to 60 lb. in. (6.8 N·m)]
e - Stator Harness

Fig. 42 Stator orientation on 75/65 Jet/90 hp 3-cyl motors

Trigger Coils

DESCRIPTION & OPERATION

◆ **See Figures 43 thru 46**

■ **2.5/3.3 hp models utilize a combination stator charge/trigger coil assembly. For details on Operation, Testing or Removal & Installation on this model, please refer to Stator/Charge Coil in this section.**

In years gone by ignition systems worked when a set of points broke the primary circuit for the ignition coil. Breaking that circuit caused the field which was building in the coil to collapse sending voltage through the secondary circuit to the spark plug. This occurred through mechanical means as the rotating crankshaft (or distributor shaft) physically opened and closed the contacts of the points.

Modern electronic ignition systems have replaced mechanical points with an electronically actuated circuit. Trigger coils or crankshaft position sensors (CSP) are used to keep track of where the crankshaft and pistons are at any given moment, though changes in a magnetic field generated by the magnets attached to the spinning flywheel. Therefore the trigger coil (or CPS) must be positioned near the flywheel, much in the same way that the stator coil must be to generate battery and/or ignition voltage.

All carbureted Mercury motors, as well as 2001 2.5L EFI (150-200 hp) motors are equipped with a trigger coil mounted directly under the flywheel (and stator). All other EFI and OptiMax motors are instead equipped with a crankshaft position sensor which is covered later in this section.

Basically, 2 types of trigger coils are found on these motors. On mechanical advance ignition systems the trigger coil is mounted on a moving bearing assembly or plate so that movement from the throttle control linkage can physically reposition it in relation to the flywheel. Moving it in one direction via the linkage will cause the ignition to fire earlier or later in flywheel rotation (i.e. will advance or retard ignition timing). Models with mechanical advance have a link rod extending from the throttle linkage to an arm of the trigger coil assembly sticking out from under the flywheel. This type of trigger coil is almost *always* installed directed under the stator and center of the flywheel.

The second, and probably more common type of trigger coil is the one used on models equipped with electronic timing advance. On these models the trigger coil is installed in a fixed position, either directly under the stator/flywheel (as on mechanical advance models) but is NOT mounted on a moveable plate and contains no link arm, OR it can also be mounted along the outer perimeter of the flywheel. In either case, all engine timing changes are handled electronically by the ignition module or CDMs.

TESTING

General

■ **2.5/3.3 hp models utilize a combination stator charge/trigger coil assembly. For details on Operation, Testing or Removal & Installation on this model, please refer to Stator/Charge Coil in this section.**

Trigger coil testing is normally conducted in the exact same manner as stator/charge coil testing. However, Mercury does not provide the same specifications for all models. For *most* motors, Mercury only provides specs for a Resistance Test OR specs for a Voltage Output Test, though they do provide both for a few motors. In either case, you should start by checking the Ignition System Component Testing specifications chart in this section to decide what data is available for the model on which you are currently working.

Once you have the specifications for your model testing typically encompasses making sure the wiring is in good condition, then verifying coil output (using a peak-reading DVOM) while cranking the motor and/or checking resistance (using an ohmmeter) across the coil windings when the motor is not turning. The trigger coil and wiring can also be checked for shorts to ground either using the peak-reading DVOM while the motor is cranking or the ohmmeter when the motor is at rest.

Problems with the trigger coil will often cause a no start condition. But, a partial short of the coil winding can cause hard starting and/or an ignition misfire. Like the stator/charge coil, the most reliable tests for the trigger coil are to dynamically check the output with the engine cranking (or in some cases, running). Of course, dynamic tests require a digital multi-meter capable of reading and displaying peak voltage values (also known as a peak-reading voltmeter) AND it also assumes that Mercury was kind enough to develop those specs.

If a peak-reading voltmeter or adapter is not available, or if Mercury did not provide any output data, specifications are usually available to statically check the coil winding using an ohmmeter. But, before replacing the coil based only on static test that shows borderline readings, remember that resistance readings will vary with temperature and the specifications provided here are based on a component temperature of about 68°F (20°C). Also, remember that a coil may test within specification statically, but show intermittent faults under engine operating conditions (due to variances in heat and load which are not present during a static test), especially at normal operating temperature.

■ **When possible, both the Resistance Tests and the Voltage Output Tests should be conducted using the information in the Ignition System Component Testing specification chart. All available wire connections and values should be listed there.**

Resistance Tests (Static Check)

◆ **See Figures 43 thru 46**

1. Using the data from the Ignition System Component Testing specifications chart and, if necessary, the Wiring Diagrams in this section, locate the wiring for the trigger coil. Tag and disconnect the wiring.

2. Connect the leads of an ohmmeter (or preferably a DVOM set to read ohms) to the leads as noted in the chart. Red lead to the (+) wire color, and Black lead to the (-) wire color. On a few of the smallest motors the coil circuit should be tested between a given wire color and Gnd (Ground), meaning a good grounding point on the powerhead itself.

■ **Some motors, such as V6 models, have multiple sets of stator coil windings which much be checked. In the case of V6 models, there are normally 3 sets of wires to check, one per pair of cylinders.**

Fig. 43 A mechanical advance trigger coil will be connected to a link rod. . .

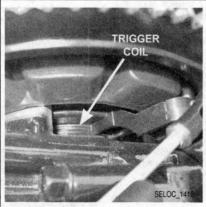

Fig. 44 . . .and tucked just under the flywheel and stator

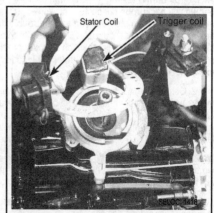

Fig. 45 View of typical trigger coil with the flywheel and stator removed

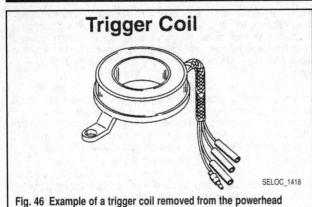

Trigger Coil

SELOC_1418

Fig. 46 Example of a trigger coil removed from the powerhead (note other triggers are also shaped like a small cube)

3. Note the resistance readings and compare with what is listed in the Ignition System Component Testing specifications chart.

■ If it is at all possible, DO NOT condemn a coil without testing further. Make sure you've got the right readings by conducting a dynamic Voltage Output Test (but again, you can only do that when the specs are available).

4. Resistance readings should be within the range provided on the chart (which normally gives a 5-10% range). Here's the tricky part. Readings either inside the range or close suggest the possibility of a good coil. However, such static tests really cannot confirm that a component is good. Static tests can hide intermittent shorts or opens that occur during normal operating conditions. For this reason, static tests such as this are really much more accurate when they show a bad component. Readings well out of this range suggest a bad coil winding is very likely.

5. Once the tests are completed, either move to the Voltage Output Test (if specs are available) or replace the coil, as necessary.

Voltage Output Test (Dynamic Check)

 MODERATE

◆ See Figures 43 thru 46

1. If not done already for a static Resistance Test, use the data from the Ignition System Component Testing specifications chart and, if necessary, the Wiring Diagrams in this section, to locate the wiring for the trigger coil. Tag and disconnect the wiring.

■ Making the wiring connections for the With Load portion of this test can be a challenge. Specifications from the chart under With Load are voltage readings which should be seen under load, meaning with the circuit *complete*. The best way to get voltage readings from a complete circuit is to purchase or fabricate a jumper harness with T-connections in it that allow the meter to connect. This is the absolute best method because it will prevent the danger of possibly shorting coil windings to each other during testing (which could damage the stator). Some people will backprobe connectors or worse, pierce the wiring insulation, but both methods can lead to trouble down the road so we advise avoiding that, if possible. Of course, in this particular case, MOST of the Trigger Coils for which Mercury provided voltage output specs include specs for testing under NO load (circuit disconnected except for the DVOM).

2. Connect the leads of a DVOM set to read peak voltage (for some meters which are not normally capable of reading peak voltage it may be possible to purchase a Peak Reading Voltage Adapter such as CDI Electronics part number 511-9773) to the leads as noted in the chart. Red lead to the (+) wire color, and Black lead to the (-) wire color.

■ Some motors, such as V6 models, have multiple sets of stator coil windings which much be checked. In the case of the V6 motor, there are normally 3 pairs of wires on which to check voltage output.

3. Provide the outboard with a good source of cooling water to prevent impeller or powerhead damage.

4. If specifications are provided for cranking output, turn the motor over and check the meter for the appropriate readings. On manual start models it

may be better to remove the spark plug(s) and install a spark tester to each spark plug lead, as this will allow the circuit to be complete while relieving engine compression, making it easier to spin for the tests.

5. If specifications are provided for running output, start the motor and allow it to run at various points in the specified test range.

6. Compare the voltage readings with what is listed in the Ignition System Component Testing specifications chart.

7. If the coil tests within specification, reconnect the wiring. If it does not, replace the trigger coil.

REMOVAL & INSTALLATION

4/5 Hp Models

 MODERATE

On these models the trigger coil is a small, box shaped sensor which is mounted on the top of the powerhead, *just* outside the perimeter of the flywheel. Because of this it is positioned where you can access it, even, in most cases, with the manual starter and flywheel still installed.

1. Locate and disconnect the Red/White trigger coil wire from the CDI unit (the rectangular box in the rubber brackets near the base of the powerhead).

2. Locate the trigger coil in the small rectangular cutout on the lower flywheel shroud toward the front port side of the powerhead.

3. Loosen and remove the 2 Phillips head screws securing the trigger coil to the lower flywheel shroud. Note the ground wire that is located under the forward of the 2 screws for installation purposes.

4. Carefully remove the trigger coil from the shroud, pulling the coil lead(s) through the rubber grommet.

To Install:

5. Route the coil lead(s) through the rubber grommet, then reconnect the Red/White lead to the CDI unit.

6. Position the trigger coil to the shroud, making sure the ground wire is positioned under coil and secured by the forward of the 2 trigger coil retaining screws.

7. Tighten the 2 Philips head retaining screws securely.

Except 4/5 hp Models

 MODERATE

◆ See Figures 43 thru 46

■ 2.5/3.3 hp models utilize a combination stator charge/trigger coil assembly. For details on Operation, Testing or Removal & Installation on this model, please refer to Stator/Charge Coil in this section.

1. On electric start models, disconnect the negative battery cable for safety. On manual start models, disconnect and ground the spark plug lead(s) for safety. On multi-cylinder motors, be sure to tag the spark plug wires to ensure proper installation.

2. Remove the flywheel from the powerhead for access, as detailed earlier in this section.

3. Remove the stator coil from the powerhead for access as detailed earlier in this section. For some models it will be possible to just unbolt the stator and position it aside with the wiring intact, but for some it may be easier to unplug the wiring and remove it completely.

■ For mechanical advance models, DO NOT change the length of the trigger link rod when disconnecting it from the trigger. If you do, you will change the ignition timing settings and will need to follow the Timing & Synchronization procedures as detailed in the Maintenance & Tune-Up section.

4. Remove the trigger coil from the powerhead as follows, depending upon the model:

• 6/8, 9.9/10/15 and 20/20 Jet/25 hp models - the trigger coil on these models is a cube-shaped component mounted either to a sliding ring (if equipped with mechanical advance) or in a fixed position (electronic advance) mounted under the stator coil. Once the stator is out of the way and the trigger link rod disconnected (for mechanical advance models), removal is a simple matter of unplugging the wiring, releasing the trigger retainer (when equipped) and lifting it free. During installation of mechanical advance models, be sure to apply a light coating of 2-4-C with Teflon or an equivalent marine grease to the sliding surface on the sliding ring.

• 30/40 hp 2-cyl models - the trigger coil on these models is a contained within a round housing, mounted directly beneath the stator assembly. Once the stator is out of the way and the trigger link rod disconnected (for mechanical advance models), removal is a simple matter of unplugging the harness connector and lifting it free.

• 40/40 Jet/50/55/60 hp 3-cyl models - the trigger coil on these models is a contained within a round housing, mounted directly beneath the stator assembly. Once the stator is out of the way and the trigger link rod disconnected (for mechanical advance models), be sure to tag and disconnect the trigger coil wiring (there are usually 4 bullet connectors on these models) from the CDM wiring harness, then lift it free from the bearing cage.

• 75/65 Jet/90 and 80 Jet/115/125 hp models - the trigger coil on these models is a contained within a round housing, mounted directly beneath the stator assembly. Once the stator is out of the way and the trigger link rod disconnected (for mechanical advance models), be sure to tag and disconnect the trigger coil wiring (there are usually 4 bullet connectors on these models) from the CDI switch box, then lift it free from the bearing cage. Just like the stator wiring on these models, the trigger coil wiring MAY be routed/connected in such a way that removal of the starter motor is necessary for access. Be sure to note the location of the wire tie which secures the wires for installation purposes.

• 150-200 hp (2.5L) carbureted or 2001 EFI V6 models - once the stator is unbolted and repositioned (with the wiring intact if desired), the trigger can be accessed, as it is the round housing mounted on top of the powerhead, just below the stator. Remove the link rod from the spark advance lever (for mechanical advance models), then tag and disconnect the trigger coil connector. Remove the trigger assembly from the powerhead, but if the trigger is to be replaced separate the link rod swivel for installation on the new trigger coil. If the link rod lever is disassembled or removed, be sure it is connected so the distance from the end of the elbow to the top of the locknut is 11/16 in. (17.5mm).

To Install:

5. Position the trigger coil to the powerhead, routing the wiring carefully as noted during removal.

6. If equipped, reconnect the trigger coil link rod.

7. Reconnect the trigger coil wiring, as tagged during removal for models with multiple bullet connectors instead of a single harness connector.

8. Install the stator coil to the powerhead, as detailed earlier in this section.

9. Install the flywheel to the powerhead, as detailed earlier in this section.

10. Reconnect the spark plug lead(s) and/or reconnect the negative battery cable, as applicable.

Crankshaft Position Sensor (CPS)

DESCRIPTION & OPERATION

◆ See Figure 47

The ECM needs to know exactly where each piston is in its 2-stroke cycle in order to accomplish both ignition timing and fuel injection functions. Where carbureted and the 2001 2.5L V6 (150-200 hp) EFI motors accomplish this task using a trigger coil, all other EFI motors and all OptiMax motors utilize a crankshaft position sensor (CPS) to provide this data to the ECM.

The sensor works in basically the same fashion as a trigger coil, by picking up a magnetic field created by magnets attached to the spinning flywheel, therefore the sensor itself must be positioned a certain distance away from the flywheel. For most models the CPS is not adjustable, so simply installing it and tightening the bolts will position it the proper distance from the flywheel. However, on 2001 3.0L EFI motors and all 3.0L Optimax motors (along with possibly some early-model 1.5L Optimax motors) the sensor mounting bracket allows for adjustment. On these models you must set a specific air gap when installing the sensor in order to ensure it will operate correctly. Specifications for the CPS air gap are available in the Fuel Injection Component Adjustment Specifications chart in the Fuel System section, however they are normally 0.025-0.040 in. (0.635-1.010mm) for OptiMax motors and 0.020-0.060 (0.51-1.53mm) for EFI motors.

When it comes to testing, no resistance specifications are available in the Mercury documentation for 2.5L motors Optimax motors, however the sensor (calibrated for use with the same gap) on 3.0L motors is listed as having a resistance spec of 300-340 ohms.

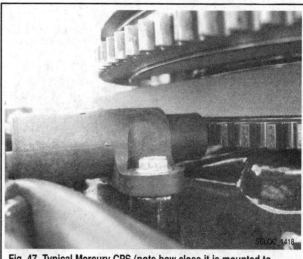

SELOC_1418

Fig. 47 Typical Mercury CPS (note how close it is mounted to flywheel)

Normally on these motors, this is the only sensor which cannot be disconnected while the engine is running. The engine will shut down if it is disconnected or fails.

TESTING

◆ See Figure 47

For most models, specifications are provided either to perform a Resistance Test or a Voltage Output Test, in the same basic manner as the trigger sensor, detailed earlier in this section. However, on a few models (including the 2.5L OptiMax motors) Mercury provides NO test data whatsoever and require all testing be performed using the DDT or an equivalent scan tool. If the sensor fails a Voltage Output Test, check the air gap and retest before replacing it. Also, even though Mercury states that the sensor is not adjustable on some models (like the 2.5L OptiMax motors), they do provide an air gap so we'd suggest checking the gap before condemning a sensor on those motors as well.

REMOVAL & INSTALLATION

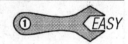

◆ See Figures 48, 49 and 50

There are basically 2 designs for mounting the CPS on these models. The first type, we call early-model, because it was more prevalent on models before 2000. However, in these years spans, as far as we can tell, it seems to be used only on some of the 3.0L motors, specifically the 2001 3.0L EFI models and all of the 3.0L OptiMax models.

The early-model design CPS is bolted by a single screw to a mounting bracket. Then, the mounting bracket and guide bracket assembly is secured by 2 bolts to the top of the powerhead.

The late-model CPS is normally used by all 1.5L and 2.5L EFI and OptiMax motors, as well as 2002 and later 3.0L EFI motors. Mounting for the late-model CPS is a little more simple than the early-model, as the mounting bracket is normally incorporated into the sensor housing, so the 1-piece unit is just secured to the top of the powerhead using to retaining bolts.

■ **Frankly, Mercury tech literature is conflicting on the sensor design for 1.5L Optimax models. In some places it appears they are not adjustable, in others they suggest it IS adjustable (at least on models before Serial #1785000), so expect to find either.**

In all cases, removal and installation is a relatively simple matter of identifying and disconnecting the wiring, then unbolting and removing the sensor. As usual with an electrical component, pay close attention to the wiring harness routing and the use of any wire ties, to ensure proper installation.

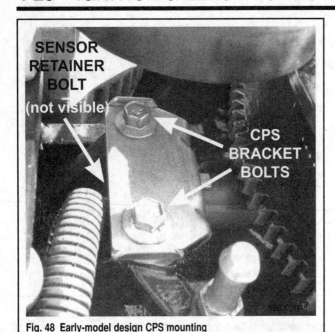

Fig. 48 Early-model design CPS mounting

Fig. 49 Late-model design CPS mounting

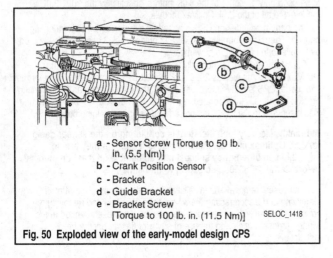

a - Sensor Screw [Torque to 50 lb. in. (5.5 Nm)]
b - Crank Position Sensor
c - Bracket
d - Guide Bracket
e - Bracket Screw [Torque to 100 lb. in. (11.5 Nm)]

Fig. 50 Exploded view of the early-model design CPS

During installation, *some* models (normally only the early-model type sensors) will have slots in the mounting bracket screw holes. When present, the slots are used to adjust the air gap between the sensor and flywheel. If adjustment is possible, use a feeler gauge to properly position the CPS, then tighten the bolts securely and recheck the gap.

Air gap specifications are listed in the Fuel Injection Component Adjustment Specifications chart in the Fuel System section, however they are normally 0.025-0.040 in. (0.635-1.010mm) for most OptiMax motors (except 1.5L models, whose most recent literature suggests that when adjustable they should have a gap of 0.20 in. / 5.0mm) and 0.020-0.060 (0.51-1.53mm) for 3.0L EFI motors.

Ignition Coils

DESCRIPTION & OPERATION

◆ See Figure 51

Besides the spark plugs and wires, the ignition coil is the last major link in the chain that produces spark for ignition. Coils of various size, shape and design are used on Mercury motors. As a matter of fact, a number of motors are equipped with a special type of coil which incorporates part of the functions of the ignition module within it, known as a CDM or capacitor discharge module. Because testing and service of these CDMs varies significantly with standard ignition coil service, we've covered them under Ignition Module/CDM later in this section. However, the theory of operation we cover here is pretty much the same (except that there is no separate ignition module since the CDM combines the function of the coil and module in one unit).

Returning to just stand alone ignition coils, the size, shape and design all depends on how many spark plugs are attached to the coil and whether the motor is designed to fire them individually or simultaneously.

The primary circuit of an ignition coil is connected to the Ignition Module (usually known as the CD unit on these motors). Low voltage power is fed and cut from the ignition coil primary circuit in a manner that induces a high voltage discharge in the coil's secondary winding. When power is cut to the primary circuit, the secondary winding discharges a burst of high voltage through the secondary lead. The voltage then travels through the spark plug and jumps the gap at the plug's tip. The actual voltage jump is the spark referred to when discussing ignition system operation. This spark is what ignites the air/fuel mixture in the combustion chamber and causes the engine to produce power.

TESTING

◆ See Figure 52

Although the best test for an ignition coil is performed using a dynamic ignition coil tester (which will show problems that might occur under load and

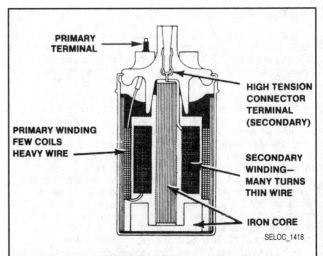

Fig. 51 Conventional ignition coil. All coils work on the same principal: power is transformed from low voltage on the primary side to high voltage on the secondary side

might not be revealed by static test), the simple fact that not everyone has access to a coil tester makes resistance checks useful.

If you do have access to an ignition coil tester, follow the manufacturer's instructions closely in order to prevent damage to the test equipment or the coil assembly.

When checking ignition coils, keep in mind that there are 2 circuits, the primary winding circuit and the secondary winding circuit. Unless coil design prevents it, both need to be checked.

■ **The combined ignition coil/power pack module or CDM used on most 30 hp and larger motors (except V6 OptiMax models) requires slightly more involved test procedures, so is covered separately, later in this section.**

The tester connection procedure for a continuity check will vary slightly depending upon how the coil is constructed. Generally, the primary circuit is the small gauge wire, while the secondary circuit contains the high tension or spark plug lead. The primary circuit is connected to the ignition module (or coil driver on V6 OptiMax models), while the secondary circuit is connected to the spark plug.

Problems with an ignition coil usually cause a no spark condition for the connected spark plug(s). But, a partial internal short or a cracked/damaged coil case that allows voltage leakage could cause an ignition misfire that appears only under certain conditions. If this is the case, the best test for the coil is to use a dynamic ignition tester and try to recreate those conditions.

■ **An ignition coil with an internal or external short will prevent the creation of a strong, blue spark at the plug. If this is suspected, listen closely to the coil during operation for audible clicking noises that may be an indication of a short. External shorts are often visible at night as they cause a blue arc from the coil body to short.**

To ensure the best possible results with resistance testing, remember that specifications will change with temperature and with meters. Whenever possible, use a high quality DVOM for coil resistance testing. The specifications provided are for ignition coils tested when they are 68°F (20°C). That means you can't simply stop a motor that has been operating for 1/2 hour and take a resistance test just because it is 68°F (20°C) outside today. You'd have to wait for the coil to cool off first.

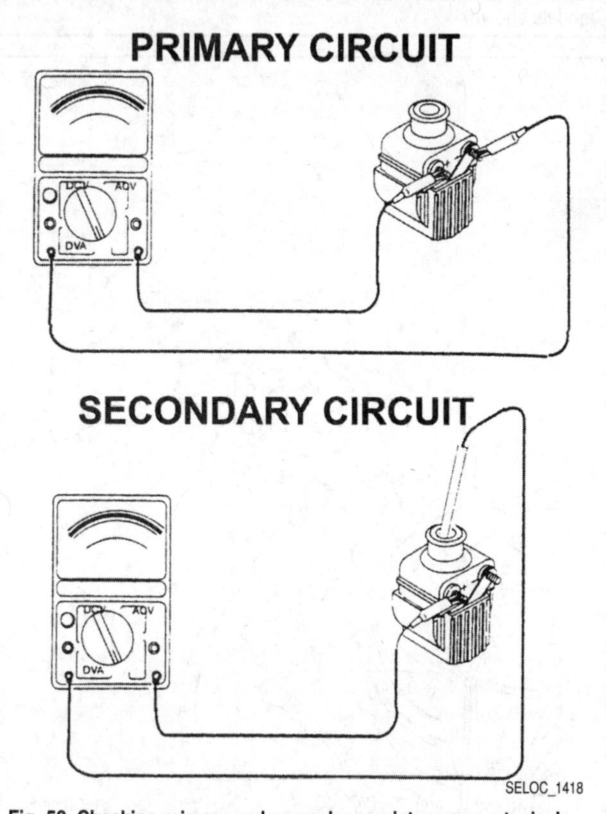

PRIMARY CIRCUIT

SECONDARY CIRCUIT

SELOC_1418

Fig. 52 Checking primary and secondary resistance on a typical Merc coil

In most cases the ignition coil does not have to be removed from the powerhead for testing, as long as the wiring can be disconnected for access to the terminals.

■ **If there is any question as to the location of a coil, find a spark plug in the powerhead and trace the secondary ignition wire (spark plug lead) back to the coil.**

To check the ignition coil(s), proceed as follows (using the data from the Ignition System Component Testing specification chart in this section:

1. If equipped, disconnect the negative battery cable to protect the test equipment and for safety.
2. Note the routing and connection points then disconnect the primary wire or wires from the ignition coil.

■ **When testing multiple ignition coils on the same motor, it is a good idea to only disconnect and test 1 coil at a time, this will help prevent confusion when reconnecting the wiring.**

3. Select the resistance scale on the DVOM.
4. Test the primary and secondary coil windings, depending upon model, as follows:

• 2.5/3.3 hp models - Check primary resistance between the coil Orange lead (which comes in from the CD/Ignition Module) and the Black/While ground wire. The Red (+) meter lead should connect to the Orange coil lead, while the Black (-) meter lead connects to ground. Primary resistance should be less than 1 ohm. Next, leave the Black meter lead on the Black/White ground and move the Red meter lead to the spark plug lead in order to check secondary resistance, the meter should show 3000-4000 ohms.

• 4/5 hp models - Check primary resistance between the coil terminal for the Black/Yellow lead (which comes in from the CD/Ignition Module) and the coil body ground (a tab toward the center of the coil body, located between the B/Y terminal and the secondary spark plug lead). The Red (+) meter lead should connect to the terminal for the B/Y coil lead, while the Black (-) meter lead connects to ground. Primary resistance should be less than 1 ohm, about 0.02-0.38. Next, leave the Red meter lead on the terminal for the B/Y coil lead, and move the Black meter lead to the to the spark plug lead in order to check secondary resistance, the meter should show 3000-4000 ohms.

• 6/8, 9.9/10/15 & 20/20 Jet/25 hp models - Check primary resistance between the positive coil terminal (colored wire, usually G, G/W or G/Y) and the negative coil terminal (to which is normally connected a black ground wire). As you might have guessed, the Red (+) meter lead should connect to the positive terminal (for the colored lead), while the Black (-) meter lead connects to the negative terminal (for the black wire lead). Primary resistance should be less than 1 ohm, about 0.02-0.04. Next, leave the Black meter lead on the negative coil terminal, and move the Red meter lead to the to the spark plug lead in order to check secondary resistance, the meter should show 8000-11,000 ohms.

• 30/40 hp 2-cyl models - are equipped with CDMs, please refer to Ignition Module/CDM later in this section.

• 40/40 Jet/50/55/60 hp 3-cyl models - are equipped with CDMs, please refer to Ignition Module/CDM later in this section.

• 75/65 Jet/90 hp models - are equipped with CDMs, please refer to Ignition Module/CDM later in this section.

• 75/90/115 hp Optimax models - are equipped with CDMs, please refer to Ignition Module/CDM later in this section.

• 80 Jet/115/125 hp models - are equipped with CDMs, please refer to Ignition Module/CDM later in this section.

• 135-200 hp (2.5L) V6 & 200-250 hp (3.0L) V6 models - though MOST are equipped with CDM ignition, the OptiMax models utilize traditional ignition coils with separate coil drivers. For these models, check primary resistance between the colored terminal wire (Green with a stripe) and the Red/Yellow terminal pin. Mercury doesn't specify that you must connect the positive meter lead to the Green/Stripe wire, but it's probably a good idea. Primary resistance should be less than 1 ohm, about 0.38-0.78. Next, check the ignition coil secondary resistance between spark plug lead tower and the ground terminal pin (black wire), the meter should show 8100-8900 ohms.

5. If tests are out of specification, make sure that only the proper wires/connectors were probed. When checking with a ground on the motor, make sure the ground was good (painted and insulated surfaces may prevent you from getting a good ground).
6. Replace the coil(s) if readings are out of specification and no other causes can be located.
7. Reconnect the wiring when testing and/or repairs are complete.
8. If equipped, connect the negative battery cable and verify proper system operation.

REMOVAL & INSTALLATION

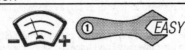

◆ **See Figures 53 thru 58**

On most Mercury motors the ignition coil(s) is(are) found toward the rear of the powerhead (top, side or aft portion) which makes a lot of sense since the coil needs to be in the general vicinity of the spark plug, mounted in the cylinder head, at the rear of the motor. If coil location is not readily apparent, locate a spark plug wire and trace the wire back to the coil. If necessary, refer to the schematics under Wiring Diagrams to help identify the primary wiring for motors equipped with multiple coils.

■ **If the engine has more than 1 spark plug and/or more than 1 primary lead, be sure to tag the wiring before disconnecting it from the ignition coil(s) in order to ensure proper installation and operation.**

Ignition coil removal and installation is normally a relatively straightforward proposition on an outboard. Unplug the wiring, remove the bolts and remove the coil, right, no big deal. And this is true on the single cylinder and inline Mercury motors. Though, it is still important to pay attention to wire routing and connection points. However, on V6 motors the ignition coils are all mounted to a common rubber-mounted electrical bracket (designed to protect the coils from heat and vibration). Unfortunately the coils are mounted to the bracket using bolts and NON-captive locknuts, the latter of which are at the rear of the bracket and cannot be reached with the bracket still on the powerhead. So on V6 motors, you'll have to tag and disconnect all the coil wiring, then remove the bracket assembly for access to the locknuts.

1. Remove the engine top cover for access.
2. On 2.5/3.3 hp models, also remove the Port side lower cowl for access.
3. If equipped, disconnect the negative battery cable for safety.
4. Disconnect the wiring (spark plug leads and primary leads) from the ignition coil(s). On multi-cylinder motors, be sure to tag all of the connections and note wire routing or wire tie points (if the wires must be moved or wire ties cut).
5. Remove the coil(s) from the powerhead, depending upon the model, as follows:

• 2.5/3.3 hp models - On these models the coil is found on the lower, port side of the motor, toward the rear. With the spark plug lead and Orange bullet connector already disconnected, remove the bolt at the top of the coil that secures the coil to the powerhead. Remove the coil.

• 4/5 hp models - With the Black/Yellow lead from the CDI and the spark plug lead disconnected, remove the bolts and washers from either end of the coil assembly. Remove the coil from the powerhead, keeping track of the black ground wire mounting location(s).

■ **On some twin cylinder motors, like most 20/20 Jet/25 hp models, there is an ignition coil cover mounted over the coils. Before you can unbolt and remove them (and sometimes event to access the wiring) you have to remove the retaining bolts (usually 3) and remove the cover.**

• 6/8, 9.9/10/15 and 20/20 Jet/25 hp models - On these models there are 2 ignition coils, each mounted to the top rear of the powerhead, just above the cylinder head. The spark plug lead contains a boot at the coil end which is secured using a wire tie, if the lead or coil must be replaced, carefully cut the wire tie and pull the lead free (making sure to use a new wire tie during installation). Once the nuts are loosened so the positive and negative connections can be removed from the coil terminals, loose and remove the bolt along with the figure 8 shaped retainer and washer which secure the coil to the powerhead.

• 30/40 hp 2-cyl models - are equipped with CDMs, please refer to Ignition Module/CDM later in this section.

• 40/40 Jet/50/55/60 hp 3-cyl models - are equipped with CDMs, please refer to Ignition Module/CDM later in this section.

• 75/65 Jet/90 hp models - are equipped with CDMs, please refer to Ignition Module/CDM later in this section.

• 75/90/115 hp Optimax models - are equipped with CDMs, please refer to Ignition Module/CDM later in this section.

• 80 Jet/115/125 hp models - are equipped with CDMs, please refer to Ignition Module/CDM later in this section.

■ **In some cases, it may be necessary to remove the lower cowling for access to the lower electrical bracket retaining bolts on V6 motors.**

• 135-200 hp (2.5L) V6 & 200-250 hp (3.0L) V6 models - As stated earlier the electrical mounting bracket must be removed for access to the coil mounting locknuts at the back (powerhead) side of the plate. To remove the bracket, loosen and remove the 4 mounting bolts located roughly at the 4 corners of the bracket (keep track of any ground wires, washers, rubber grommets and/or bushing that are used as the bolts are removed). Carefully pull the bracket back, making sure not to lose any grommets or bushings still attached to the bracket. Then loosen the retaining bolt for each coil driver and coil assembly. Remove the coil drivers and coils from the bracket, as necessary.

To install:

6. Carefully clean any traces of dirt, debris or corrosion from the mating surfaces of the coil and powerhead. This is especially important to ensure proper electrical contact for the ground strap.
7. Position the coil to the powerhead (or coils and coil drivers to the electrical component bracket on V6 motors) and tighten the bolts securely.
8. On V6 motors, install the electrical component bracket assembly (with coils and coil drivers installed) to the powerhead making sure any ground

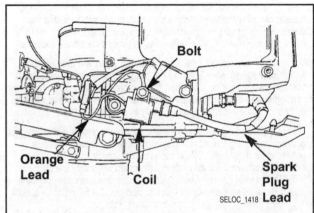

Fig. 53 Ignition coil mounting on a typical 1-cyl Merc (2.5/3.3 hp models shown)

SELOC_1418

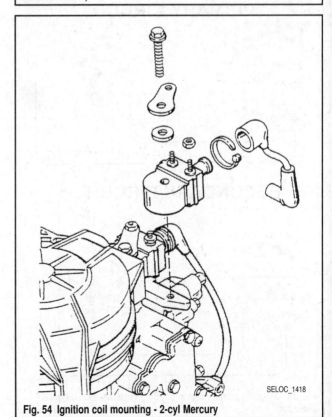

Fig. 54 Ignition coil mounting - 2-cyl Mercury

SELOC_1418

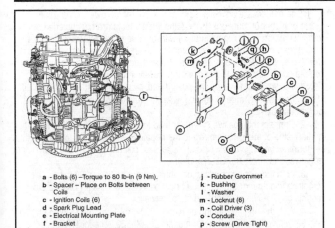

a - Bolts (6) –Torque to 80 lb-in (9 Nm).
b - Spacer – Place on Bolts between Coils
c - Ignition Coils (6)
d - Spark Plug Lead
e - Electrical Mounting Plate
f - Bracket
g - Screw [Torque to 35 lb. in. (4.0 Nm)]
h - Bolts (4) – Torque to 20 lb. ft. (27 N·m)
i - Washer
j - Rubber Grommet
k - Bushing
l - Washer
m - Locknut (6)
n - Coil Driver (3)
o - Conduit
p - Screw (Drive Tight)
q - Ground Strap
r - Model Year 2001

SELOC_1418

Fig. 55 Ignition coil mounting - 150-175 hp OptiMax motors

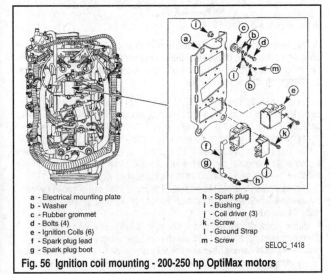

a - Electrical mounting plate
b - Washer
c - Rubber grommet
d - Bolts (4)
e - Ignition Coils (6)
f - Spark plug lead
g - Spark plug boot
h - Spark plug
i - Bushing
j - Coil driver (3)
k - Screw
l - Ground Strap
m - Screw

SELOC_1418

Fig. 56 Ignition coil mounting - 200-250 hp OptiMax motors

wires, washers, rubber grommets and bushings are all in position, then tighten the bracket retaining bolts to 20 ft. lbs. (27 Nm).

9. Reconnect the wiring as tagged during removal. Route and secure the wiring as noted during removal to prevent possible interference with moving components.

10. On 6-25 hp models, once the nuts on the wire terminals are tightened (and be careful NOT to over-tighten and damage the terminals, ruining the coil) apply a light coating of Liquid Neoprene (#92-25711-2) or an equivalent liquid electrical tape/sealant to prevent corrosion.

11. If equipped on twin cylinder motors, install the ignition coil cover.

12. If equipped, reconnect the negative battery cable.

13. If removed, install the lower cowling.

14. Install the engine top cover.

Ignition Module (CD Unit/CDM)

DESCRIPTION & OPERATION

In the old days, breaker points were used to physically open and close the switch responsible for firing the ignition coils. However, with advances in sold state technology, it was eventually found that the job of the points could be controlled more accurately (without the wear, maintenance and potential mechanical problems of points) by using an electronic ignition module. As time went on, manufacturers also quickly realized that they could expand the electronic control to include functions like controlling engine timing (eliminating the need for movable trigger coils and trigger link rods on many models) or even for engine overspeed restriction, engine overheat protection.

On most Mercury motors 30 hp and larger (except V6 OptiMax motors), the ignition module is replaced by the use of capacitor discharge modules (usually 1 per cylinder) which combine the function of and ignition coil and an ignition module, all in one housing. On some carbureted versions of these motors, a timing or warning module, separate from the CDMs is also used to provide things like engine overspeed or overheat protection and ignition timing functions. On EFI or OptiMax motors there will either be an additional ignition control module or the ECM will take over these functions. If in doubt as to the configuration for the module on which you are working, please refer to the Wiring Diagrams in this section.

■ **On V6 OptiMax motors, there is an additional separate component called the Coil Driver, which is mounted with the ignition coils and serves part of the function of the combined CDMs on other models.**

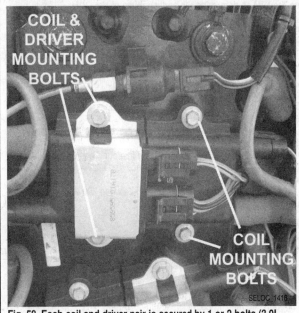

Fig. 57 Coils and coil drivers on V6 motors are mounted to an electrical bracket

Fig. 58 Each coil and driver pair is secured by 1 or 2 bolts (3.0L shown)

TESTING

General

Ignition module or CDM testing varies somewhat from model-to-model depending upon the particular makeup of the system. MOST tests involve resistance checks to determine if the diodes and circuits within the unit are operating correctly (though we've warned in the past that this type of check will sometimes mask an intermittent problem that only occurs under load), while a few checks might dynamic voltage output checks from the unit (to the coil).

Obviously, the functions which are incorporated into an EFI/OptiMax ECM cannot be tested directly, at least without the use of a scan tool, such as the DDT.

2.5/3.3 Hp Models

On these models, Mercury recommends both a Static Diode Check and a Dynamic Voltage Output Test for the CD module. While the static test can be done with nearly any ohmmeter or DVOM, the dynamic test requires a peak reading voltmeter and/or a peak voltage adapter for the meter. For the Dynamic check, you'll have to provide a source of cooling water to prevent possible impeller or powerhead damage.

1. Start by performing the static diode check, as follows:

a. Connect the positive (+) meter lead to the White CDM lead, then connect the negative (-) meter lead to the Orange CDM lead. Continuity should exist.

b. Connect the positive (+) meter lead to the Orange CDM lead and the negative (-) meter lead to ground. Continuity should not exist.

c. If the diodes are not within specification, the CDM may be faulty.

2. Even if the CDM passes the static check it may *still* be bad, to determine for sure, perform a dynamic voltage output test (using a peak reading DVOM or DVOM with a peak reading adapter) as follows:

a. Provide the motor with a source of cooling water to prevent possible impeller or powerhead damage.

b. Connect the positive (+) meter lead to ground and the negative (-) meter lead to the Orange CDM lead (make sure the White lead is reconnected, because that's the power in from the capacitor charge/trigger coil).

c. With the engine cranking or running between 300-2000 rpm, the voltage reading on the meter should be 100-320 volts.

■ **If the engine will not idle down to trolling speed (runs excessively fast the throttle closed), the bias circuit within the CDM may have failed. The CDM will need to be replaced if this is the case.**

3. Once the tests are complete, either replace the faulty CDM or reconnect the wiring and move on to other ignition components.

■ **Also note, if the engine will not start (no spark at the spark plug), use the DVA to determine the serviceability of the CDM.**

4/5 Hp Models

On these motors the only test recommended for the CD module consists of a voltage output test when cranking or running at/near idle speed (300-2000 rpm). This dynamic test requires a peak reading voltmeter and/or a peak voltage adapter for the meter.

1. Provide a source of cooling water to prevent possible impeller or powerhead damage.

2. Connect the positive (+) meter lead to ground and the negative (-) meter lead to the Black/Yellow CDM unit lead (the lead which goes to the ignition coil).

3. With the engine cranking or running between 300-2000 rpm, the reading should be 120-300 volts.

6/8, 9.9/10/15 and 20/20 Jet/25 Hp Models

On these motors the only test recommended for the CD module consists of a voltage output test when cranking or running at/near idle speed (300-1000 rpm) and running above idle (1000-4000 rpm). This dynamic test requires a peak reading voltmeter and/or a peak voltage adapter for the meter.

These models are normally equipped with either a lower output Black stator or higher output Red stator. Although the tests are conducted in an identical manner, CD module output will vary due to the relative difference in output from the stator coils. The larger the motor, the more likely it is to have the Red stator (same goes for the more accessories or features it has, like remote control and/or electric starting). However it is usually pretty easy to determine which stator is on the powerhead, by peering under the flywheel.

1. Provide a source of cooling water to prevent possible impeller or powerhead damage.

2. Connect the positive (+) meter lead to the negative (-) terminal on the primary side of the ignition coil and the negative (-) meter lead to the positive (+) terminal on the primary side of the ignition coil (these are the positive and negative feeds from the CD Module).

3. With the engine cranking or running between 300-1000 rpm (cranking/idle speed) the reading should be 125-260 volts for Black stator models or 125-320 volts for Red stator models.

4. With the engine running between 1000-4000 rpm, the reading should be 200-360 volts for Black stator models or 200-320 volts for Red stator models.

30/40 Hp 2-Cyl Models

◆ **See Figures 59 thru 62**

■ **This test is for CDM's (combination ignition coil/ignition modules) only. Any other type of module, like timing protection, rev limit, ECM, etc. must be tested by a process of elimination, as no direct resistance or voltage checks are available.**

There are 2 slightly different CDMs which may be found on these motors. Since Mercury is not clear on what years or models utilize what CDMs we've provided testing information on both, along with visual and part number identification so you can decide with which tests to proceed.

The first style is CDM part #822779 and it can be visually identified by the smoother, less angular body of the unit along with the fact that it contains an attached harness with 3-pin connector and separate ground strap.

The second style is CDM part #827509, and it is easily identified by the part number which should appear on the face on the unit, just below the built-in 4-pin connector.

Mercury provides for resistance checking of both units, with slightly different results depending upon whether you are using a digital or analog ohmmeter. Tests can be conducted directly on the pins/harness of the unit OR by using a test harness (#84-825207A1 for the first style, and #84-825207A2 for the second style).

Use the accompanying CDM test charts which include both test connection points and expected results in order to perform a resistance check on the CDM units.

40-250 Hp 3-, 4- and 6-Cylinder Models

◆ **See Figures 63 and 64**

This test is for CDM's (combination ignition coil/ignition modules) only. Any other type of module, like timing protection, rev limit, ECM, etc. must be tested by a process of elimination, as no direct resistance or voltage checks are available.

Most 40-250 hp motors are equipped capacitor discharge module units for their ignition system, whether or not they are equipped with an alternator or battery driven ignition. However, there are a few exceptions - specifically, the V6 OptiMax motors are equipped with traditional ignition coils and separate driver modules, therefore these CDM these testing procedures are not applicable to them.

■ **The OptiMax V6 motors do not use CDM assemblies, but instead use 2 ignition coils and 1 coil driver (mounted on top of the 2 coils) for each pair of cylinders. The easiest and most effective test for a coil driver on these motors is simply to swap the position of 2 coil drivers and see if problems move from one pair of cylinders to another. Additionally, on Pro/XS/Sport models you can resistance check the coil drivers directly, details have been provided at the end of this procedure.**

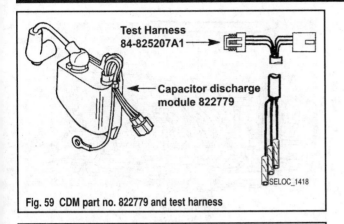

Fig. 59 CDM part no. 822779 and test harness

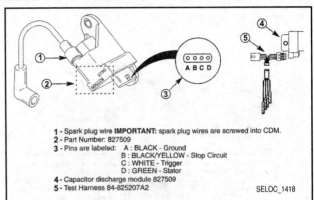

1 - Spark plug wire **IMPORTANT:** spark plug wires are screwed into CDM.
2 - Part Number: 827509
3 - Pins are labeled: A : BLACK - Ground
  B : BLACK/YELLOW - Stop Circuit
  C : WHITE - Trigger
  D : GREEN - Stator
4 - Capacitor discharge module 827509
5 - Test Harness 84-825207A2

SELOC_1418

Fig. 61 CDM part #827509 and test harness

CAPACITOR DISCHARGE MODULE RESISTANCE CHECK-ANALOG METER			
Connect POSITIVE (+) Meter Lead To:	**Connect NEGATIVE (–) Meter Lead To:**	**Ohms Scale**	**Reading**
Ground Lead	WHITE Pin or WHITE Test Harness Lead	R X 1	40 ± 10
GREEN/WHITE Pin or RED Test Harness Lead	Ground Lead	R X 1 Diode Reading	Continuity
Ground Lead	GREEN/WHITE Pin or RED Test Harness Lead	R X 1K Diode Reading	No Continuity
GREEN/WHITE Pin or RED Test Harness Lead	BLACK/YELLOW Pin or BLACK Test Harness Lead	R X 1K Diode Reading	No Continuity
BLACK/YELLOW Pin or BLACK Test Harness Lead	GREEN/WHITE Pin or RED Test Harness Lead	R X 1 Diode Reading	Continuity
Coil Tower	Ground Lead	R X 10	1000 ± 300

NOTE: *Due to the differences in test meters battery polarity, results other than specified may be obtained. In such a case, reverse meter leads and re-test. If test results then read as specified on all tests CDM is O.K. The diode measurements above will be opposite if using a Fluke® equivalent multimeter.*

CAPACITOR DISCHARGE MODULE RESISTANCE CHECK-DIGITAL METER			
Connect Positive (+) Meter Lead To:	**Connect Negative (–) Meter Lead To:**	**Ohms Scale**	**Reading**
Ground Lead	WHITE Pin or WHITE Test Harness Lead	Ω or 200	40 ± 10 Ohms
GREEN/WHITE Pin or RED Test Harness Lead	Ground Lead	▸⊢	OL or OUCH
Ground Lead	GREEN/WHITE Pin or RED Test Harness Lead	▸⊢	0.400 - 0.900
GREEN/WHITE Pin or RED Test Harness Lead	BLACK/YELLOW Pin or BLACK Test Harness Lead	▸⊢	0.400 - 0.900
BLACK/YELLOW Pin or BLACK Test Harness Lead	GREEN/WHITE Pin or RED Test Harness Lead	▸⊢	OL or OUCH or 1.
Coil Tower	Ground Lead	Ω or 2K	0.800 - 1.200 KΩ

SELOC_1418

Fig. 60 CDM part no. 822779 resistance testing - 30/40 hp 2-cyl models

CAPACITOR DISCHARGE MODULE RESISTANCE CHECK - ANALOG METER			
Connect POSITIVE (+) Meter Lead To:	**Connect NEGATIVE (–) Meter Lead To:**	**Ohms Scale**	**Results:**
Ground Pin (A) or BLACK Test Harness Lead	WHITE (C) or WHITE Test Harness Lead	R x 100	1250 ± 300 Ohms
GREEN (D) or GREEN Test Harness Lead	Ground Pin (A) or BLACK Test Harness Lead	R x 100 Diode Reading	Continuity
Ground Pin (A) or BLACK Test Harness Lead	GREEN (D) or GREEN Test Harness Lead	R x 100 Diode Reading	No Continuity
GREEN (D) or GREEN Test Harness Lead	BLACK/YELLOW (B) or BLACK/YELLOW Test Harness Lead	R x 100 Diode Reading	No Continuity
BLACK/YELLOW (B) or BLACK/YELLOW Test Harness Lead	GREEN (D) or GREEN Test Harness Lead	R x 100 Diode Reading	Continuity
Spark Plug Terminal (at spark plug boot)	Ground Pin (A) or BLACK Test Harness Lead	R x 100	1000 ± 300 Ohms

NOTE: *Due to the differences in test meters battery polarity, results other than specified may be obtained. In such a case, reverse meter leads and re-test. If test results then read as specified on all tests CDM is O.K. The diode measurements above will be opposite if using a Fluke equivalent multimeter.*

CAPACITOR DISCHARGE MODULE RESISTANCE CHECK - DIGITAL METER			
Connect Positive (+) Meter Lead To:	**Connect Negative (–) Meter Lead To:**	**Ohms Scale**	**Results:**
Ground Pin (A) or BLACK Test Harness Lead	WHITE (C) or WHITE Test Harness Lead	Ω or 2K	1.125-1.375 KΩ
GREEN (D) or GREEN Test Harness Lead	Ground Pin (A) or BLACK Test Harness Lead	▸⊢	OL or OUCH
Ground Pin (A) or BLACK Test Harness Lead	GREEN (D) or GREEN Test Harness Lead	▸⊢	0.400 - 0.900
GREEN (D) or GREEN Test Harness Lead	BLACK/YELLOW (B) or BLACK/YELLOW Test Harness Lead	▸⊢	0.400 - 0.900
BLACK/YELLOW (B) or BLACK/YELLOW Test Harness Lead	GREEN (D) or GREEN Test Harness Lead	▸⊢	OL or OUCH
Spark Plug Terminal (at spark plug boot)	Ground Pin (A) or BLACK Test Harness Lead	Ω or 2K	0.950 - 1.150 KΩ

SELOC_1418

Fig. 62 CDM part no. 827509 resistance testing - 30/40 hp 2-cyl models

The balance of all 40 hp and larger motors, including all EFI motors, as well as 1.5L OptiMax motors, are equipped with CDMs. The capacitor discharge modules used on these motors takes 1 of 2 forms, either the 4-pin or the 5-pin connector versions.

All carbureted motors, and all 2001 EFI motors are normally equipped with the 4-pin CDMs, while the 1.5L OptiMax motors, as well as all 2002 or later EFI motors are equipped with the 5-pin CDMs.

Mercury provides for resistance checking of both units, with slightly different testing depending upon whether it is a 4-pin or 5-pin unit, due simply to differences in connections and internal circuits. Tests can be conducted directly on the pins/harness of the unit OR on 4-pin CDMs by using a test harness (#84-825207A2).

Use the accompanying CDM test charts which include both test connection points and expected results in order to perform a resistance check on the CDM units.

To test the separate coil drivers on V6 Pro/XS/Sport OptiMax models, proceed as follows:

a. Tag and disconnect the wiring from the coil driver in question. With the driver connector positioned so the locktab is on top use a DVOM to check resistance across the 2 pins to the right (for the Red/Yel and Black wires). Resistance should be 76,500-93,500 ohms.

b. Next use the DVOM to check resistance between the second pin from the right, and in turn, each of the 2 pins to the left (the second pin from the right is for the Black wire, while the other 2 pins toward the left are each Green/Striped wires). In both cases, the Green/Striped wire terminal to the Black wire terminal should read about 9,000-11,000 ohms.

c. Finally, check for continuity between the terminal second from the right on the 4-pin connector (again, that Black wire terminal) and each of the 2 Green/Striped wire terminals on the 2-pin connector. There should be no continuity between the Black wire pin on the 4-pin connector and either of the pins on the 2-pin connector.

REMOVAL & INSTALLATION

2.5/3.3 Hp Models

◆ See Figure 65

 MODERATE

1. Remove the top cowling for access.
2. Remove the Port side lower cowling for access.
3. Disengage the bullet connectors at the CDM on the side of the powerhead. It is a good idea to label the wiring connectors before disassembly. This will aid in assembly.
4. Disconnect the ground lead from the ignition coil retaining bolt.
5. Remove the bolt securing the CDM to the engine block.

To Install:

6. Secure the CDM to the engine block with the retaining bolt.

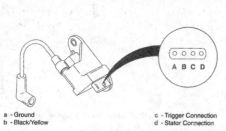

a - Ground
b - Black/Yellow

c - Trigger Connection
d - Stator Connection

A resistance check is required and can be performed on the CDM as follows:

NOTE: *This test can be performed using the test harness (P/N 84-825207A2). DO Not connect the test harness plug to the stator/trigger engine wire harness.*

CAPACITOR DISCHARGE MODULE

Circuit Test	Connect Negative (–) Meter Lead To:	Connect Positive (+) Meter Lead To:	Ohms Scale	Results:
Stop Diode Forward Bias	Green (D)/ or Green test harness lead	Black/Yellow (B)/ or Black/Yellow test harness lead	R x 100 Diode Reading*	Continuity
Stop Diode Reverse Bias	Black/Yellow (B)/ or Black/Yellow test harness lead	Green (D)/ or Green test harness lead	R x 100 Diode Reading*	No Continuity
Return Ground Path Diode, Reverse Bias	Green (D)/ or Green test harness lead	Ground Pin (A) or Black test harness lead	R x 100 Diode Reading*	No Continuity
Return Ground Path Diode, Forward Bias	Ground Pin (A)/ or Black test harness lead	Green (D)/ or Green test harness lead	R x 100 Diode Reading*	Continuity
CDM Trigger Input Resistance	Ground Pin (A)/ or Black test harness lead	White (C)/ or White test harness lead	R x 100	1000 - 1250 Ohms
Coil Secondary Impedance	Ground Pin (A) or Black test harness lead	Spark Plug Terminal (At Spark Plug Boot)	R x 100	900 - 1200 Ohms

*Diode Readings: Due to the differences in test meters, results other than specified may be obtained. In such a case, reverse meter leads and re-test. If test results then read as specified CDM is O.K. The diode measurements above will be opposite if using a Fluke equivalent multimeter.

SELOC_1418

Fig. 63 CDM testing for 4-pin units - 40-250 hp models

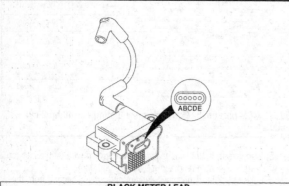

BLACK METER LEAD

		Secondary Tower	EST PIN A	EST Low PIN B	Secondary Low Pin C	Primary Ground Pin D	Battery + Pin E
RED METER LEAD	Secondary Tower	X	No Continuity	No Continuity	850 – 1200 ohm	No Continuity	No Continuity
	EST PIN A	No Continuity	X	8500 – 12000 ohm	No Continuity	29000 – 39000 ohm	11000 – 21000 ohm
	EST Low PIN B	No Continuity	8500 – 12000 ohm	X	No Continuity	39000 – 49000 ohm	21000 – 31000 ohm
	Secondary Low Pin C	850 – 1200 ohm	No Continuity	No Continuity	X	No Continuity	No Continuity
	Primary Ground Pin D	No Continuity	20000 – 30000 ohm	31000 – 41000 ohm	No Continuity	X	13000 – 23000 ohm
	Battery + Pin E	No Continuity	11000 – 21000 ohm	21000 – 31000 ohm	No Continuity	13000 – 23000 ohm	X

EST = Electronic Spark Trigger
EST Low = Return ground path for the trigger signal to the ECM

SELOC_1418

Fig. 64 CDM testing for 5-pin units - 75-250 hp models

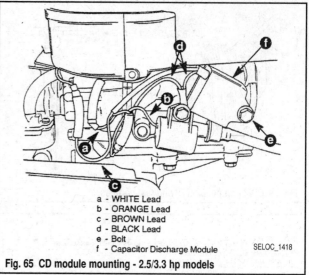

a - WHITE Lead
b - ORANGE Lead
c - BROWN Lead
d - BLACK Lead
e - Bolt
f - Capacitor Discharge Module

SELOC_1418

Fig. 65 CD module mounting - 2.5/3.3 hp models

7. Fasten the ground lead to the block with the ignition coil retaining bolt.
8. Reconnect the bullet connectors as tagged during removal.
9. Reinstall the Port side lower engine cowling.
10. Reinstall the top cowling.

4/5 Hp Models

1. Remove the top cowling for access.
2. Locate the bullet connectors for the CDM, on the side of the powerhead JUST below the ignition coil, then disengage the CDM wiring connectors. It is a good idea to label the wiring connectors before disassembly. This will aid in assembly.
3. Disengage the ground lead from the ignition coil retaining bolt.
4. Slide the CDM out of the rubber brackets and remove it.

To Install:

5. Slide the CDM back into the rubber brackets.
6. Fasten the ground lead to the block with the ignition coil retaining bolt.
7. Reconnect the bullet connectors as tagged during removal.
8. Reinstall the top cowling.

6/8, 9.9/10/15 & 20/20 Jet/25 Hp Models

◆ See Figure 66

1. Remove the top cowling for access.
2. Locate the wiring for the CDM mounted to the side of the powerhead usually the aft Port side).
3. Disengage the bullet connectors at the CDM. It is a good idea to label the wiring connectors before disassembly. This will aid in assembly.
4. Disconnect the CDM ground lead wire(s).
5. Remove the retaining screws (usually 3) and lift off the CDM (also often known as a switch box on these models) from the powerhead.

To Install:

6. Align the CDM (switch box) to the powerhead and install the retaining screws, tightening them firmly.
7. Reconnect the bullet connectors as tagged during removal.
8. Fasten the ground lead wire(s).
9. Reinstall the top cowling.

30/40 Hp 2-Cyl Models

◆ See Figures 67 and 68

These motors are equipped with 1 CDM per cylinder, plus a separate timing protection or rev limiter module. All the modules are secured to an electrical component bracket which is found on the rear starboard side of the powerhead. The CDMs are found on the aft end of the bracket, while the timing/rev limiter module is found at the top, center of the bracket.

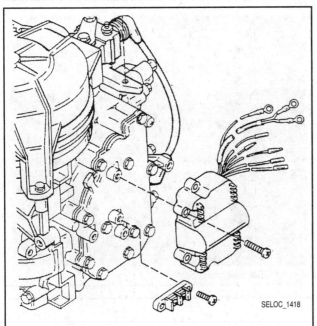

Fig. 66 Typical CDM mounting (shape varies) - assembly 6-25 hp models

Fig. 67 CDM and Timing/Rev limiter locations - 30/40 hp models

Removal of the CDMs is a very straight forward process of tagging and disconnecting the wiring, disconnecting the spark plug wire and then removing the retaining bolts (usually 2). Removal of the timing/rev limiter module is the same, During installation, take care to make sure the all wires are properly routed and connected as noted during removal. Any wire ties which were cut should be replaced to ensure the wiring does not contact any moving components or become pinched from the cowling. Retaining bolts should be tightened securely, but not over-tightened.

40-250 Hp 3-, 4- and 6-Cylinder Models

 MODERATE

◆ See Figures 69 thru 72

■ This procedure applies to *most* Mercury motors 30 hp and larger (except V6 OptiMax motors), where the ignition module is replaced by the use of capacitor discharge modules (usually 1 per cylinder) which combine the function of and ignition coil and an ignition module, all in one housing. On V6 Optimax motors, the function of the ignition module is performed by a combination of the ECM and 1 coil driver per pair of cylinders. The coil driver's are mounted on the Ignition Coils and service procedures are covered under that topic earlier in this section.

Fig. 68 CDM mounting - 30/40 hp 2-cyl models

Fig. 69 CDM mounting - typical inline Merc (4-cylinder shown)

Fig. 70 CDM mounting - carbureted Merc V6

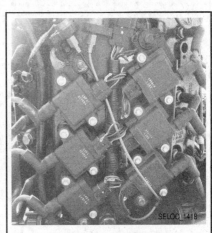

Fig. 71 CDM mounting - EFI Merc V6

These motors are equipped with 1 CDM per cylinder, either located at the rear, side of the powerhead (3-cylinder or 4-cylinder motors) or the center rear of the cylinder head (V6 motors). In all cases they are normally mounted to an electrical component bracket designed to help isolate them from the heat and vibration of the powerhead.

Removal of the CDMs is a very straight forward process of tagging and disconnecting the wiring, disconnecting the spark plug wire and then removing the retaining bolts (usually 2). During installation, take care to make sure the all wires are properly routed and connected as noted during removal. Any wire ties which were cut should be replaced to ensure the wiring does not contact any moving components or become pinched from the cowling. Retaining bolts should be tightened securely, but not over-tightened.

■ Some models may position a ground strap under one or more of the CDM mounting bolts. When present, be sure to re-attach the strap during installation.

Ignition Specifications

◆ See Figure 73

Fig. 72 CDM attachment points

IGNITION SYSTEM COMPONENT TESTING - MERCURY/MARINER 2-STROKE MOTORS

HP	No. of Cyl	Model Year	Displace cu. in. (cc)	Trigger (Pulser/Crankshaft Position) Coils Wire leads (+)	(-)	Resistance Ohms①	Peak Voltage With Load	No Load	RPM	Stator (Charge Coil) Wire leads (+)	(-)	Resistance Ohms①	Peak Voltage With Load While Cranking	While Running	RPM	CDM/Ignition Coil Wire leads or Pins (+)	(-)	Resistance Primary Ohms①	Secondary Ohms①
2.5/3.3	1	2001-05	4.6 (74.6)	W	Gnd	300-400	-	-	-	W	Gnd②	300-400	120-320	120-320	@ 300-2000	O	B/W	< 1 ohm	3,000-4,400
4/5	1	2001-05	6.2 (102)	R/W	Gnd	80-115	-	-	-	B/R	W	93-142	150-325③	150-325③	@ 300-1000	B	Gnd	0.02-0.38	3,000-4,400
6/8 ④	2	2001-05	13 (210)	Br/Y	Br/W	6500-8500	-	-	-	B/W	GND	120-180⑥	10-75	500-300	@ 1000-4000	G/W or G/Y	B	0.02-0.04	8,000-11,000
⑤	2	2001-05	13 (210)	Br/Y	Br/W	6500-8500	-	-	-	G/W	W/G	370-445	150-330⑦	250-330⑦	@ 1000-4000	G or G/Y	B	0.02-0.04	8,000-11,000
9.9/10/15 ④	2	2001-05	16 (262)	Br/Y	Br/W	6500-8500	-	-	-	B/W	GND	120-180⑥	10-75	500-300	@ 1000-4000	G/W or G/Y	B	0.02-0.04	8,000-11,000
⑤	2	2001-05	16 (262)	Br/Y	Br/W	6500-8500	-	-	-	G/W	W/G	370-445	150-330⑦	250-330⑦	@ 1000-4000	G or G/Y	B	0.02-0.04	8,000-11,000
20/20 Jet/25	2	2001-09	24 (400)	Br/Y	Br/W	6500-8500	-	-	-	G/W	W/G	370-445	150-330⑦	250-330⑦	@ 1000-4000	G or G/Y	B	0.02-0.04	8,000-11,000
30/40	2	2001-03	39 (644)	W	Blk	n/a	2-8	2-8	-	W/G	G/W	660-710	190-320	190-320	300-4000	Tower	Gnd	-	800-1200
40/40 Jet/50/55/60	3	2001-09	59 (967)	⑧	Blk	n/a	0.2-5.0	>1	Cranking	W/G	G/W	660-710	100-350	200-350	@ Idle	Pin A	Pin C	1000-1250	900-1200
65 Jet/75/90	3	2001-09	85 (1386)	⑧	Blk	n/a	0.2-5.0	>1	Cranking	W/G	G/W	660-710	100-350	200-350	@ Idle	Pin A	Pin C	1000-1250	900-1200
75/80 Jet/90/115 Optimax	3	2004-09	93 (1526)	R	W	n/a	n/a	n/a	n/a	equipped with 60-amp Alternator						See CDM Testing			850-1200
80 Jet/115/125	4	2001-05	113 (1848)	⑨	Blk	n/a	0.2-5.0	>1	Cranking	W/G	G/W	660-710	100-350	200-350	@ Idle	Pin A	Pin C	1000-1250	900-1200
135/110 Jet/150/175 Opti	V6	2001-09	153 (2507)	R	W	n/a	n/a	n/a	n/a	equipped with 60-amp Alternator						G/stripe	R/Y	0.38-0.78	8100-8900
150 (XR6,Classic) & 200	V6	2001-05	153 (2507)	⑩	⑩	1100-1400	0.2-5.0	>1	Cranking	W/G	G/W	660-710	-	> 200	all	Pin A	Pin C	1000-1250	900-1200
150/175/200 (2.5L) EFI	V6	2001	153 (2507)	⑩	⑩	1100-1400	0.2-5.0	>1	Cranking	W/G	G/W	660-710	-	> 200	all	Pin A	Pin C	1000-1250	900-1200
	V6	2002-09	153 (2507)	R	W	300-350	-	-	-	equipped with 60-amp Alternator						See CDM Testing			850-1200
200/225/250 EFI (3.0L)	V6	2001	185 (3032)	R	W	900-1300	2-10	2-10	Crank & run	⑪	Gnd	990-1210	100-225	250-300	650-3000	Pin A	Pin C	1150-1350	700-1300
	V6	2002-05	185 (3032)	R	W	300-350	-	-	-	equipped with 60-amp Alternator						See CDM Testing			850-1200
200/225/250 Optimax	V6	2001-09	185 (3032)	R	W	300-340	-	-	-	equipped with 60-amp Alternator						G/stripe	R/Y	0.38-0.78	8100-8900

NOTE: Unless stated otherwise, With Load tests are made with circuit connected as normal, while No Load tests are conducted with the component disconnected from the wiring harness

TIP: On models where different stators MAY be installed, use wire colors to help determine which type is installed for the model on which you are working

① Unless noted otherwise all resistance specifications are at an ambient temperature 68 degrees F (20 degrees C). Keep in mind that all resistance readings will vary with temperature and from meter-to-meter

② The functions of the trigger and charge coils are performed by a single Capaciter Charging/Trigger coil on these models

③ Cranking output test is between the B/R lead and Gnd

④ Specs are for 4-amp, BLACK stator models, specs listed are high speed coil (low speed coil is B/Y and GND, 3000-3600 ohms resistance, 150-300 volts @ 300-1000 rpm or 250-360 volts @ 1000-4000 rpm)

⑤ Specs are for 6-amp, RED stator models, check stator voltage ouput between each wire (W/G or G/W) and GND, same output specs for each test

⑥ Also check resistance across both stator leads B/Y and B/W, it should be 3100-3700 ohms.

⑦ Check stator output voltages between each wire and ground (meaning G/W and ground, then W/G and ground), NOT across the two stator wires like the resistance test

⑧ Purple (Cyl No. 1), White (Cyl No. 2) or Brown (Cyl No. 3) for 40-60 hp motors or Purple (Cyl No. 1), Brown (Cyl No. 2) and White (Cyl No. 3) for 65 Jet-90 hp motors

⑨ Purple (Cyl No. 1), White (Cyl No. 2), Brown (Cyl No. 3) or Blue (Cyl No. 4)

⑩ Check trigger between each pair of wires R - W, B - Purple, Br - Y

⑪ Check stator between each G or G w/ tracer (G/R, G/Y, G/B, G/O or G/B) wire and ground

SELOC_1418

Fig. 73 Ignition System Component Testing Specifications

CHARGING CIRCUIT

Description & Operation

◆ See Figures 74, 75 and 76

A charging system is standard on most of the models covered here, with the exception of a few of the smaller rope start models. Even if not standard, a charging system could have been added at the time of engine rigging or even as a later accessory. Regardless, when equipped, the charging system provides current to maintain the battery and to operate other engine/boat mounted electric components.

The charging systems used on these outboards have various outputs (2-amp, 4-amp, 6-amp, 9-amp, 16-amp, 18-amp, 35-amp and so on, up to 60-amps). Output is matched to appropriate load capacities for the motor and loads from accessories due to anticipated boat size. The systems can be broken down into 2 basic types unregulated and regulated (the difference in the later two being that regulated systems protect the battery from overcharging and boiling away electrolyte or possibly even overheating and cracking). In additional, an AC lighting coil may be found on a handful of the smallest rope start models. However, because the voltage produced by this system is Alternating Current (AC), and not Direct Current (DC, which is necessary for 12-volt DC marine batteries), models equipped only with AC lighting coils cannot charge marine batteries.

Regardless of the system used, all types start by generating an AC current in the same basic fashion. Either a stator, battery charge coil or AC lighting coil is either mounted directly under the flywheel (in order to utilize its mechanical spinning motion when the engine is operating) or is built into the belt-driven alternator used on most fuel injected V6 models. Permanent magnets attached to the flywheel (or incorporated into the alternator) generate a spinning magnetic field that cuts through the coil windings producing an alternating current. The coil wiring (yellow and/or yellow with various tracer colors on all under-flywheel models), delivers current to charging system components or to the AC lighting system.

On charging systems (as opposed to AC lighting systems), current is then passed through a series of diodes (electrical components that only pass current in one direction) contained in the rectifier or rectifier/regulator. The diodes convert the current to DC in order to charge the battery and operate DC voltage accessories. On non-regulated systems, charging rates are controlled only by engine speed, as a rectifier only contains circuitry to convert the current, not control it. For this reason, the total output on a non-regulated system is normally lower than a regulated system in order to help prevent the possibility of battery overcharging. However, extended high-speed operation will risk overcharging on some models, so be diligent with battery maintenance (checking and topping-off cell fluid levels) on these models. A non-regulated system can usually be identified by the diamond shaped rectifier which typically includes 3 exposed terminals on the top (as opposed to the box-shaped and sealed regulator/rectifier unit which does not normally have any exposed terminals).

Regulated charging systems utilize a rectifier/regulator that not only converts current to DC, but limits charging system output to about 14.6 volts maximum output to prevent battery overcharging. Visually, the rectifier is identified easily by tracing the wires from the stator coil to the square, sealed regulator assembly.

Servicing charging systems is not difficult if you follow a few basic rules. Always start by verifying the problem. If the complaint is that the battery will not stay charged do not automatically assume that the charging system is at fault. Something as simple as an accessory that draws current with the key off will convince anyone they have a bad charging system. Another culprit is the battery. Remember to clean and service your battery regularly. Battery abuse is the number one charging system problem. The second most common cause of charging system complaints comes from loose or corroded wiring connections, mostly at the battery, but problems can be caused at any charging system harness connection.

On regulated systems, the regulator/rectifier is the brains of the charging system. This assembly controls current flow in the charging system. If battery voltage is low the regulator sends the available current from the rectifier to the battery. If the battery is fully charged the regulator diverts most of the current from the rectifier back to the lighting coil through ground.

Do not expect the regulator/rectifier to send current to a fully charged battery. You may find that you must pull down the battery voltage below 12.5 volts to test charging system output. Running the power trim and tilt will reduce the battery voltage, or running other accessories, especially a spot light or other high-power accessory load will reduce the battery voltage quickly.

When equipped, the regulator/rectifier is the most complex item to troubleshoot. You can avoid troubleshooting the regulator/rectifier by checking around it. Check the battery and charge or replace it as needed. Check the amp output of the stator coil or alternator. If amperage is low check the stator coil for proper resistance and insulation to ground. If all tests and wiring are good, the regulator is likely the culprit, but this can be verified using a variable load tester, if available.

Service Precautions

For safety and to prevent damage to the ECM/PCM on EFI and OptiMax motors or the CD units and, when equipped, regulator/rectifier on carbureted motors the following precautionary measures must be taken when working with the electrical system:

• Wear safety glasses when working on or near the battery.

• Don't wear a watch with a metal band when servicing the battery. Serious burns can result if the band completes the circuit between the positive battery terminal and ground.

• When installing a battery, make sure that the positive and negative cables are not reversed. Making the wrong connections in this fashion, even for a split second, can destroy the charging system diodes in the rectifier or regulator/rectifier.

• Be absolutely sure of the polarity of a booster battery before making connections. Connect the cables positive-to-positive, and negative-to-negative. Connect positive cables first, and then make the last connection to ground on the body of the booster vehicle so that arcing cannot ignite hydrogen gas that may have accumulated near the battery. Even momentary connection of a booster battery with the polarity reversed will damage alternator diodes. This also applies to using a battery charger. Reversed polarity will burn out the alternator and regulator in a matter of seconds.

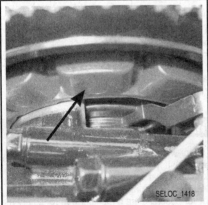

Fig. 74 Power is supplied by a stator coil on carb motors

Fig. 75 Power is supplied by a belt-driven alternator on MOST EFI/OptiMax motors

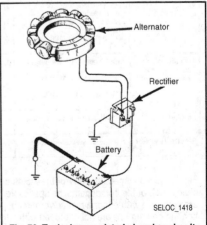

Fig. 76 Typical unregulated charging circuit

• Always disconnect the battery ground cable before disconnecting the alternator lead.

• Always disconnect the battery (negative cable first) when charging it.

• Never ground the stator or alternator output or battery terminal. Be cautious when using metal tools around a battery to avoid creating a short circuit between the terminals. Be just as careful when working around a hot wire (a circuit to which current is currently applied). Because terminal grounding can occur with any hot wire, not just the battery cable, the negative battery cable should be disconnected before servicing any part of the electrical system.

• Never run or crank the engine without the stator coil wiring and battery properly connected. Doing so will damage the diodes or the voltage regulator instantly. For this reason, it is crucial that you check the battery connections periodically, because a terminal that comes loose during service will do more than cause the engine to stop, it may damage the charging system as that occurs.

Troubleshooting the Charging System

GENERAL

◆ See Figure 77

Don't waste your time with haphazard testing. The only way to ensure success (and the only way to avoid the possibility of accidentally replacing a good part) is to perform charging system testing in a logical and systematic manner.

Begin all electronic troubleshooting procedures by ensuring that wiring and connections are in good condition. Check battery condition and connections to make sure there is sufficient voltage to operate the system at start-up.

Before conducting any tests, double-check all wire colors and locations with the Wiring Diagrams provided in this section.

✻✻ SELOC CAUTION

During system testing, ALWAYS follow the steps of our procedures and any tool manufacturer's instructions closely to avoid injury or possible damage to the engine's charging system.

✻✻ SELOC WARNING

If, during testing, the engine is to be run at speeds over 2000 rpm, it must be mounted in a test tank using a suitable test wheel or placed in the water. Running the engine on a flushing device at speeds above 2000 rpm could allow the motor to run overspeed and possibly suffer severe mechanical damage.

Where to Look	Cause	Procedure
Battery	1. Battery defective or worn out 2. Low electrolyte level 3. Terminal connections loose or corroded 4. Excessive electrical load	1. Check condition and charge 2. Add water and recharge 3. Clean and tighten 4. Evaluate accessory loads
Wiring	1. Connections loose or corroded 2. Stator leads shorted or grounded 3. Circuit wiring grounded	1. Clean and tighten 2. Perform ohmmeter tests 3. Perform ohmmeter tests
Coil (Stator, Battery charge, or AC Lighting)	1. Damaged stator windings 2. Weak flywheel magnets 3. Damaged stator leads	1. Perform ohmmeter tests 2. Perform running output tests 3. Perform ohmmeter tests
Rectifier	1. Damaged wiring or diodes	1. Perform rectifier ohmmeter tests
Rectifier/regulator	1. Inoperative rectifier/regulator	1. Perform rectifier/regulator tests

SELOC_1418

Fig. 77 When inspecting and troubleshooting the charging system, check each of the components, as applicable. Charging systems are equipped with either a rectifier or a regulator/rectifier. AC lighting coil models are not equipped with either, nor are they equipped with a battery

Servicing charging systems is not difficult if you follow a few basic rules. Always start by verifying the problem.

• Do not automatically assume that the charging system is at fault.

• A small draw with the key off, a battery with a low electrolyte level, or an overdrawn system can cause the same symptoms.

• It has become common practice on many outboard engines to overload the electrical system with accessories. This places an excessive demand on the charging system. If the system is "overdrawn at the amp bank" then no amount of parts changing will fix it.

The charging system should be inspected if:

• The battery is undercharged (there is insufficient power to crank the starter)

• The battery is overcharged (electrolyte level is low and/or boiling out)

• The voltmeter on the instrument panel (if equipped) indicates improper charging (either high or low) voltage

A thorough, systematic approach to troubleshooting will pay big rewards. Build your troubleshooting check list with the most likely offenders at the top. Do not be tempted to throw parts at a problem without systematically troubleshooting the system first.

The starting point for all charging system problems begins with the inspection of the battery and related wiring. The battery must be in good condition and fully charged before system testing. Perform a visual check of the battery, wiring and fuses. Are there any new additions to the wiring? An excellent clue might be, "Everything was working OK until I added that live well pump." With a comment like this you would know where to check first (is the pump operating properly or did it increase total draw on the system past capacity).

1. Test the battery thoroughly. Check the electrolyte level, the wiring connections and perform a load test to verify condition.

2. Perform a fuse and Red wire (positive battery cable) check with the voltmeter. Verify the ground at the rectifier or regulator/rectifier. Do you have 12 volts and a good fuse? While you are at the Red wire, check alternator output with an ammeter. Be sure the battery is down around 12 volts.

3. Do a draw test if it fits the symptoms. Many times a battery that will not charge overnight or week-to-week has a constant electrical draw applied that is always sapping a small amount of power. Put a test lamp or ammeter in the line with everything off and look for a draw.

4. A similar problem can be a system that is simply overdrawn. The electrical system cannot keep up with the demand. Do a consumption survey. More amps out than the alternator can return will require a different strategy.

5. Next, go to the source. On flywheel systems, check the stator (battery charge) coil for correct resistance and to make sure there are no shorts to ground. For specifications, please refer to the Charting System Testing chart in this section. On Alternator systems there are specific tests which can be conducted stage-by-stage to help isolate the problem. We suggest skipping ahead to Alternator, Testing for more details.

6. Finally, if specs are available for your motor, connect an ammeter and determine the charging system output under idle and full throttle conditions. For specifications, please refer to the Charting System Testing chart in this section.

7. If all these tests fail to pinpoint the problem and you have verified low or no output to the battery then replace the rectifier or regulator/rectifier.

On fully-regulated systems there is really only one cause for undercharging (unless wiring is damaged/corroded or the battery is cannot accept a charge). If other components are in good shape, the only cause for undercharging on a regulated system is that the regulator/rectifier is not working. Refer to the information under regulator/rectifier for more testing hints.

✻✻ SELOC WARNING

Never operate a marine motor without providing a source of cooling water such as a test tank or flush fitting. It takes less than a minute to damage the water pump impeller that could lead to much more expensive powerhead failures later during normal motor usage. Use a suitable test wheel whenever possible, but remember that for safety the propeller should not be installed on motors when running on a flush fitting.

CHARGING SYSTEM TESTS

◆ See Figure 78

✳✳ SELOC WARNING

To protect the motor and test equipment, perform voltage tests with the leads connected and the terminals exposed to accommodate test lead connection (unless specifically directed otherwise by our instructions). All electrical components must be securely grounded to the powerhead any time the engine is cranked or started or certain components could be damaged.

Charging System Quick Test

  MODERATE

◆ See Figure 78

You can use a DVOM and a tachometer to get a quick idea whether or not the charging system is doing its job. For part of this test, the engine is run while a suitable source of water is supplied to the cooling system (such as a test tank or a flush fitting).

Fully regulated charging systems are designed to keep system voltage output around 14.1-14.5 volts in order to charge the battery to a point near 13 volts. Any regulated charging system that is putting out less than 12 volts or more than 15 volts indicates a problem. Undercharging could be a problem with a coil winding/wiring or the regulator/rectifier, while overcharging is usually the fault of the regulator/rectifier, but perform additional component tests as directed in this section to be sure.

■ Remember that charging system problems are often caused by problems with the wiring harness. Loose and corroded connections can cause excessive resistance in the charging circuit. Similarly damaged wiring such as insulation that exposes the wire core because of burning, melting or wear from interference with moving components like the flywheel (on stator/battery charge coil equipped motors) can cause shorts to ground preventing voltage from reaching the rectifier or regulator/rectifier.

✳✳ SELOC WARNING

Never operate a marine motor without providing a source of cooling water such as a test tank or flush fitting. It takes less than a minute to damage the water pump impeller that could lead to much more expensive powerhead failures later during normal motor usage. Use a suitable test wheel whenever possible, but remember that for safety the propeller should not be installed on motors when running on a flush fitting.

1. Make sure the ignition switch and all electrical boat/motor accessories are turned OFF.
2. Connect a shop tachometer to the engine, following the instructions provided from the tool manufacturer.
3. With the engine not running, connect the red DVOM lead to the positive battery terminal and the black lead to a good engine ground. Read and record the voltage. If the battery is fully charged (reads more than 12.6 volts) apply a load to draw the system down slightly for testing purposes. Use boat/motor mounted accessories for a few minutes such as search lights, trolling motors, power trim/tilt to drain the battery slightly, but not below 11.0 volts, then note the reading.
4. Start and run the engine until it warms, then allow it to run at slow idle. Again, check the DVOM and record the voltage produced by the charging system at idle.
5. Advance the throttle until the engine is running at or above 2000 rpm. Check the DVOM and record the voltage produced by the charging system above idle.
6. If the charging system is working at all, the voltage readings taken with the motor running at idle and above 2000 rpm should show an increase of at least 0.3 volt over the static reading of the battery prior to running the motor. If no voltage increase is noted, test stator/battery charge coil and the rectifier or regulator/rectifier as detailed in this section.
7. If the charging system performed properly, but battery the battery is discharging (and the battery tests ok), take additional voltage readings, with the engine running above 2000 rpm, running at idle and engine stopped, but this time, turn on ALL boat and motor accessories. Again, the voltage must be at least 0.3 volt higher with the engine running than stopped or the stator/charge coil and the rectifier or regulator/rectifier should be tested. If

CHARGING SYSTEM TESTING

Model (Hp)	No. of Cyl	Year	Disp cu. in. (cc)	Charg System	Wire Colors	Resistance Ohms ①	Amperage Output @ RPM Minimum	Max (At/Near Peak)
4/5	1	2001-05	6.2 (102)	2-amp charging	Y/R - Y/R	-	-	2 amps by 4500-5000
	1	2001-05	6.2 (102)	4-amp lighting	-	-	-	
6/8	2	2001-05	13 (210)	4-amp (black stator)	Y - Gray	-	0.9 @ 2000 rpm	3.5@4000 / 4.3@6000
	2	2001-05	13 (210)	6-amp (red stator)	Y - Y	0.50	0.75@1000 rpm	4.5@3000 / 6.0@6000
9.9/10/15	2	2001-05	16 (262)	4-amp (black stator)	Y - Gray	-	0.9 @ 2000 rpm	3.5@4000 / 4.3@6000
	2	2001-05	16 (262)	6-amp (red stator)	Y - Y	0.50	0.75@1000 rpm	4.5@3000 / 6.0@6000
20/20 Jet/25	2	2001-09	24 (400)	6-amp	Y - Gray	0.65	0.8 @ 1000 rpm	5.3@3000 / 6.3@6000
30/40	2	2001-03	39 (644)	9-amp Manual	Y - Y	0.50-0.65	0.6 @ 1000 rpm	8 @ 2000 / 9 @ 3000
	2	2001-03	39 (644)	18-amp Electric	Y - Y	0.22-0.29	2.8 @ idle	8@1000 / 18@4000
40/50/55	3	2001-09	59 (967)	9-amp Manual	Y - Y	0.16-0.19	0.6 @ 1000 rpm	8 @ 2000 / 9 @ 3000
	3	2001-09	59 (967)	16-amp	Y - Y	0.16-0.19	2.8 @ idle	9.3@1000 / 16@2000
40 Jet/60	3	2001-09	59 (967)	16-amp	Y - Y	0.16-0.19	2.8 @ idle	9.3@1000 / 16@2000
65 Jet/75/90	3	2001-09	85 (1386)	all	Y - Y	0.16-0.19	2.8 @ idle	9.3@1000 / 17.5@4000
75/80 Jet/90/115 Optimax	3	2004-09	93 (1526)	all	Alternator	-	45 @ 2000	45-60 @ 2000+ rpm
80 Jet/115/125	4	2001-05	113 (1848)	all	Y - Y	0.16-0.19	2.8 @ idle	9.3@1000 / 17.5@4000
110 Jet/135/150/175 Optimax	V6	2001-09	153 (2507)	all	Alternator	-	32-38@2000@ bat	52-60@2000+rpm@ alt
150 (XR6,Classic) & 200	V6	2001-05	153 (2507)	all	Y - Y	0.18-0.45	-	40 @ 5000
150/175/200 (2.5L) EFI	V6	2001	153 (2507)	all	Y - Y	0.18-0.45	-	40 @ 5000
	V6	2002-09	153 (2507)	all	Alternator	-	39@2000@ bat	48@2000@ alternator
200/225/250 EFI (3.0L)	V6	2001	185 (3032)	all	Alternator	-	30 @ 750	60 @ 2000
	V6	2002-05	185 (3032)	all	Alternator	-	39@2000@ bat	48@2000@ alternator
200/225/250 Optimax	V6	2001-09	185 (3032)	all	Alternator	-	-	50-55@2000@alternator

① Unless noted otherwise all resistance specifications are at an ambient temperature 68 degrees F (20 degrees C). Keep in mind that all resistance readings will vary with temperature and from meter-to-meter

SELOC_1418

Fig. 78 Charging System Testing Specifications

component tests show that there is no problem, the system is being overtaxed by accessories (more amps are being withdrawn from the system than are available at the proverbial amp bank, don't use all those accessories at once).

Charging System Running Output Test

◆ See Figure 78

 **DIFFICULT**

Using an ammeter and a tachometer you can determine if the charging system is operating to specification (generating sufficient amperage). This test requires running the engine across a wide portion of its powerband and is really best conducted with the motor in a test tank or attached to a boat that is firmly secured to a dock (or trailer with hefty tie-down straps) or navigated by an assistant. We don't recommend running an engine much above idle on a flush fitting as damage could occur if the engine is run overspeed.

1. Determine proper amperage output for the motor being tested. Refer to the information on Charging System Testing specifications chart in this section.

2. Make sure the ignition switch and all electrical boat/motor accessories are turned **OFF**.

3. Connect a shop tachometer to the engine, following the instructions provided from the tool manufacturer.

4. With the engine not running, connect the red DVOM lead to the positive battery terminal and the black lead to a good engine ground. Read and record the voltage. If the battery is fully charged (reads more than 12.6 volts) apply a load to draw the system down slightly for testing purposes. Use boat/motor mounted accessories for a few minutes such as search lights, trolling motors, power trim/tilt to drain the battery slightly, but not below 11.0 volts, then note the reading.

5. Disconnect the rectifier or rectifier/regulator lead that supplies voltage to the battery.

6. Connect an ammeter with at least 0-40 amp capacity for most motors (0-70 or 80 amps for fuel injected and/or V6 models) in series with the lead (connect the ammeter inline into the harness).

✳✳ SELOC WARNING

Make sure no portion of the wiring (lead, harness or ammeter leads) contacts engine ground while the engine is running or arcing will likely occur.

7. Start and run the engine until it warms, then allow it to run at slow idle. While the engine is running at idle compare the reading on the ammeter with the specified amperage output for that engine rpm in the Charging System Testing specifications chart in this section.

8. Once the engine has warmed fully, slowly advance the throttle noting the ammeter readings at each 500 or 1000 rpm increase in speed. Output should increase linearly according to the Charging System Testing specifications chart in this section. If there is no output or output is incorrect, perform the stator/battery charge coil and rectifier or regulator/rectifier tests in this section, as applicable.

Electric System Current Draw Test

◆ See Figure 78

 **DIFFICULT**

Using an ammeter you can determine if the total load of all electrical accessories on the boat/motor exceed the total capability of the charging system to produce power. The ammeter should be rated for at least 3-5 amps more than the charging system rating, meaning a minimum of 7 amps on the smallest, unregulated charging systems or as much as 65 amps for the most powerful systems (which are typically found on EFI or OptiMax motors).

■ **To help determine the rated output for the charging system of the motor being tested, please refer to the information on Charging System Testing specification chart this section.**

1. Install an ammeter as follows:

a. Unless an inductive ammeter is being used, disconnect the negative cable, followed by the positive cable from the battery. Install the ammeter between the positive battery post and cable, then reconnect the negative battery cable.

b. If using an inductive ammeter, attach the inductive pickup clamp to the positive battery cable following any instructions provided by the tool manufacturer.

2. Turn the ignition keyswitch **ON** to the **RUN** position, but do not start the engine. Turn on all boat/motor accessories (radios, spot lights, trolling motors etc) and note the reading on the ammeter.

3. If the reading exceeds charging system output, turn off accessories one at a time (noting the difference in the reading each time) to determine the individual loads provided by each accessory. This will tell you what accessories can be used together in order to prevent discharging the battery while the motor is running.

Battery

The battery is one of the most important (and vulnerable) components of the electrical system. In addition to providing electrical power to start the engine, it also provides power for operation of the running lights, radio, and electrical accessories.

Because of its job and the potential consequences of a failure (especially in an emergency), the best advice is to purchase a well-known brand, with a sufficient warranty period, from a reputable dealer.

The usual warranty covers a pro-rated replacement policy, which means the purchaser would be entitled to a consideration for the time left on the warranty period if the battery should prove defective before its time.

Do not consider a battery of less rating than the one which originally came with the motor. For Mercury/Mariner motors, most literature specifies a battery of at least 350 Cold Cranking Amps (CCAs), but also adds that a larger capacity unit, one of at least 775 CCAs be used, especially in conditions under 32° F (0° C). But it never hurts to buy a battery of larger capacity. Especially as the boat and motor's electrical system ages, cables crack internally and resistance increases.

■ **Additional information regarding the minimum battery size for your motor should be available in your owner's manual.**

MARINE BATTERIES

◆ See Figure 79

Because marine batteries are required to perform under much more rigorous conditions than automotive batteries, they are constructed differently than those used in automobiles or trucks. Therefore, a marine battery should always be the No. 1 unit for the boat and other types of batteries used only in an emergency.

Fig. 79 A fully charged battery, filled to the proper level with electrolyte, is the heart of the ignition and electrical systems. Engine cranking and efficient performance of electrical items depend on a full rated battery

Marine batteries have a much heavier exterior case to withstand the violent pounding and shocks imposed on it as the boat moves through rough water and in extremely tight turns. The plates are thicker and each plate is securely anchored within the battery case to ensure extended life. The caps are spill proof to prevent acid from spilling into the bilges when the boat heels to one side in a tight turn, or is moving through rough water. Because of these features, the marine battery will recover from a low charge condition and give satisfactory service over a much longer period of time than any type intended for automotive use.

✳✳ SELOC WARNING

Avoid the use an automotive type batteries with an outboard unit. It is not built for the pounding or deep cycling to which this use will subject it and it may be quickly damaged.

BATTERY CONSTRUCTION

◆ **See Figure 80**

A battery consists of a number of positive and negative plates immersed in a solution of diluted sulfuric acid. The plates contain dissimilar active materials and are kept apart by separators. The plates are grouped into elements. Plate straps on top of each element connect all of the positive plates and all of the negative plates into groups.

The battery is divided into cells holding a number of the elements apart from the others. The entire arrangement is contained within a hard plastic case. The top is a 1-piece cover and contains the filler caps for each cell. The terminal posts protrude through the top where the battery connections for the boat are made. Each of the cells is connected to its neighbor in a positive-to-negative manner with a heavy strap called the cell connector.

BATTERY RATINGS

◆ **See Figure 81**

Typically, at least 3 different methods are used to measure and indicate battery electrical capacity:
- Amp/hour rating
- Cold cranking performance
- Reserve capacity

The amp/hour rating of a battery refers to the battery's ability to provide a set amount of amps for a given amount of time under test conditions at a constant temperature. Therefore, if the battery is capable of supplying 4 amps of current for 20 consecutive hours, the battery is rated as an 80 amp/hour battery. The amp/hour rating is useful for some service operations, such as slow charging or battery testing.

Cold cranking performance is measured by cooling a fully charged battery to 0°F (-17°C) and then testing it for 30 seconds to determine the maximum

current flow. In this manner the cold cranking amp rating is the number of amps available to be drawn from the battery before the voltage drops below 7.2 volts.

The illustration depicts the amount of power in watts available from a battery at different temperatures and the amount of power in watts required of the engine at the same temperature. It becomes quite obvious - the colder the climate, the more necessary for the battery to be fully charged.

Reserve capacity of a battery is considered the length of time, in minutes, at 80°F (27°C), a 25 amp current can be maintained before the voltage drops below 10.5 volts. This test is intended to provide an approximation of how long the engine, including electrical accessories, could operate satisfactorily if the stator assembly or lighting coil did not produce sufficient current. A typical rating is 100 minutes.

■ **Mercury often mentions a 4th type of battery output classification, Marine Cranking Amps (MCA). The MCA numbers are higher for a given application than CCAs. Mercury's minimum recommendation for these motors is 465 MCAs, but raises it to 1000 MCAs for operation in ambient temperatures below 32° F (0° C).**

If possible, the new battery should always have a power rating equal to or higher than the original unit.

BATTERY LOCATION

Every battery installed in a boat must be secured in a well protected, ventilated area. If the battery area lacks adequate ventilation, hydrogen gas, which is given off during charging, is very explosive. This is especially true if the gas is concentrated and confined.

Remember, the boat is going to take a pounding at some point and it is critical that the battery is secured so that wire terminals are not subjected to stresses which could break the cables or battery (and could provide for other dangerous conditions, including the potential for shorts, sparks and even an explosion of that hydrogen gas we just mentioned). There are many types of battery tray mounting assemblies available and you will normally find the perfect one for your application on the shelves of boat supply stores. Make sure the tie down bracket or strap is snug, but not over-tightened to the point of distorting or cracking the case.

BATTERY & CHARGING SAFETY PRECAUTIONS

◆ **See Figure 82**

Always follow these safety precautions when charging or handling a battery:
- Wear eye protection when working around batteries. Batteries contain corrosive acid and produce explosive gas as a byproduct of their operation. Acid on the skin should be neutralized with a solution of baking soda and water made into a paste. In case acid contacts the eyes, flush with clear water and seek medical attention immediately.

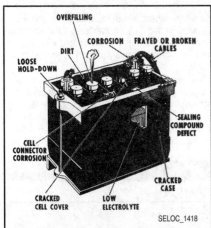

SELOC_1418

Fig. 80 A visual inspection of the battery should be made each time the boat is used. Such a quick check may reveal a potential problem in its early stages. A dead battery in a busy waterway or far from assistance could have serious consequences

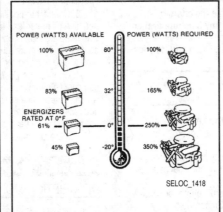

SELOC_1418

Fig. 81 Comparison of battery efficiency and engine demands at various temperatures

SELOC_1418

Fig. 82 Explosive hydrogen gas is released from the batteries in a discharged state. This one exploded when something ignited the gas. Explosions can be caused by a spark from the battery terminals or jumper cables

• Avoid flame or sparks that could ignite the hydrogen gas produced by the battery and cause an explosion. Connection and disconnection of cables to battery terminals is one of the most common potential causes of sparks. Use extreme caution with jumper cables and make sure all switches are **OFF** before making any other connections.

• Always turn a battery charger **OFF**, before connecting or disconnecting the leads. When connecting the leads, connect the positive lead first, then the negative lead, to avoid sparks.

• When lifting a battery, use a battery carrier or lift at opposite corners of the base.

• Ensure there is good ventilation in a room where the battery is being charged.

• Do not attempt to charge or load-test a maintenance-free battery when the charge indicator dot is indicating insufficient electrolyte.

• Disconnect the negative battery cable if the battery is to remain in the boat during the charging process.

• Be sure the ignition switch is **OFF** before connecting or turning the charger **ON**. Sudden power surges can destroy electronic components.

• Use proper adapters to connect charger leads to batteries with non-conventional terminals.

BATTERY CHARGERS

Before using any battery charger, consult the manufacturer's instructions for its use. Battery chargers are electrical devices that change Alternating Current (AC) to a lower voltage of Direct Current (DC) that can be used to charge a marine battery. There are 2 types of battery chargers - manual and automatic.

A manual battery charger must be physically disconnected when the battery has come to a full charge. If not, the battery can be overcharged, and possibly fail. Excess charging current at the end of the charging cycle will heat the electrolyte, resulting in loss of water and active material, substantially reducing battery life.

■ **As a rule, on manual chargers, when the ammeter on the charger registers half the rated amperage of the charger, the battery is fully charged. This can vary, and it is recommended to use a hydrometer to accurately measure state of charge.**

Automatic battery chargers have an important advantage - they can be left connected (for instance, overnight) without the possibility of overcharging the battery. Automatic chargers are equipped with a sensing device to allow the battery charge to taper off to near zero as the battery becomes fully charged. When charging a low or completely discharged battery, the meter will read close to full rated output. If only partially discharged, the initial reading may be less than full rated output, as the charger responds to the condition of the battery. As the battery continues to charge, the sensing device monitors the state of charge and reduces the charging rate. As the rate of charge tapers to zero amps, the charger will continue to supply a few milliamps of current - just enough to maintain a charged condition.

✳✳ SELOC WARNING

Even with automatic storage chargers, don't assume the charger is working properly. When a battery is connected for any time longer than a day, check it frequently (at least daily for the first week and weekly thereafter). Feel the battery case for excessive heat, listen and look for gassing, check the electrolyte levels.

REPLACING BATTERY CABLES

Battery cables don't go bad very often, but like anything else, they can wear out. If the cables on your boat are cracked, frayed or broken, they should be replaced.

When working on any electrical component, it is always a good idea to disconnect the negative (-) battery cable. This will prevent potential damage to many sensitive electrical components.

Always replace the battery cables with one of the same length, or you will increase resistance and possibly cause hard starting. Smear the battery posts with a light film of dielectric grease, or a battery terminal protectant spray once you've installed the new cables. If you replace the cables 1 at a time, you won't mix them up.

■ **Any time you disconnect the battery cables, it is recommended that you disconnect the negative (-) battery cable first. This will prevent you from accidentally grounding the positive (+) terminal when disconnecting it, thereby preventing damage to the electrical system.**

Before you disconnect the cable(s), first turn the ignition to the **OFF** position. This will prevent a draw on the battery which could cause arcing. When the battery cable(s) are reconnected (negative cable last), be sure to check all electrical accessories are all working correctly.

Stator/Battery Charge Coil (Under Flywheel)

DESCRIPTION & OPERATION

All carbureted motors which utilize a charging system, as well as 2001 2.5L EFI motors, are equipped with an under flywheel coil to generate voltage. This coil either takes the form of a battery charge coil (on the smallest of motors) or more likely a 1-piece round stator coil (which also likely provides power for the ignition system on these same motors).

For this reason, service procedures are already covered in the Ignition System section under Stator/Charge Coil. Please refer to Stator/Charge Coil for removal and installation procedures.

However, since the tests found under that heading are for the Ignition Charge Coil windings and NOT the Battery Charge Coil windings, testing is covered separately in this section.

TESTING

◆ **See Figure 37**

The *first* test for stator condition, the Charging System Running Output Test, was listed earlier in this section, under Charging System Tests. If the running output test has led you to test the stator directly, the only remaining option is to check resistance (IF a specification is given) and to make sure it is not shorted to ground. If you've reviewed this section thoroughly, you'll know that resistance checks are not the most reliable way to determine if a coil is good (as they can hide intermittent opens or shorts to ground that only occur under load). They're still better than nothing and, a bad result is usually a good indicator that the coil must be replaced.

The stator coil can be tested installed on the powerhead or removed. Actually, in most cases, it is better to test it while it is installed, in order to check for shorts to ground.

1. Locate and disconnect the Yellow (or Yellow with a tracer color) wiring for the stator (for details, please refer to the Charging System Testing chart, in this section).

■ **The stator wiring runs out from under the flywheel, normally straight to the rectifier or rectifier/regulator.**

2. Connect the positive (+) meter lead to 1 of the stator wires. Make contact with the negative (-) meter lead to the other stator wire. Compare the static resistance readings to the values listed in the Charging System Testing specifications chart.

3. Connect the positive (+) meter lead to 1 of the stator leads and the negative (-) meter lead to a good ground on the powerhead. If the alternator was removed make contact with the metal frame or the ground wire (as equipped). The meter should indicate no continuity (if not, it is shorted to ground and must be replaced).

4. Connect the positive (+) meter lead to the *other* stator lead and the negative (-) meter lead to a good ground on the powerhead. If the alternator was removed make contact with the metal frame or the ground wire (as equipped). The meter should still indicate no continuity (if not, it is shorted to ground and must be replaced).

5. If the alternator fails any one of the above tests, it must be replaced.

Alternator (External)

DESCRIPTION & OPERATION

◆ **See Figure 75**

Unlike Mercury carbureted motors (or the 2001 2.5L EFI motors), all other EFI and OptiMax motors are equipped with an externally mounted, belt-driven alternator, not all that dissimilar to what you find on a car or truck. The relatively high-output alternator is necessary on these motors, almost all of which (excluding the 2001 EFI models) are equipped with a battery-driven ignition system and loads of engine management electronics.

TESTING

Preliminary Diagnosis

Once you've eliminated a problem with the drive belt, the battery itself or that a battery has become discharged by overusing accessories or leaving an accessory on, it's time to systematically check the alternator and charging system. Mercury advises the following tests (in the following order):
- Checking Output Circuit
- Checking Sensing Circuit
- Checking Voltage Output
- Checking Current Output

Checking Output Circuit

◆ See Figure 83

1. Connect the positive (+) voltmeter lead to alternator terminal **B** (output Terminal). Connect the negative (-) meter lead to the case ground on the alternator.

2. Shake the alternator wiring harness. The meter should indicate battery voltage and should not vary. If the proper reading is not obtained, check for loose or dirty connections or damaged wiring.

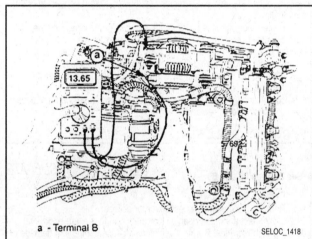

a - Terminal B

SELOC_1418

Fig. 83 Check the output circuit between Terminal B and a good ground (2.5L shown)

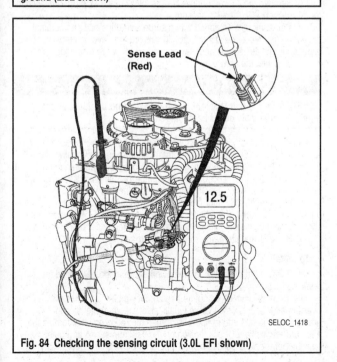

Sense Lead (Red)

12.5

SELOC_1418

Fig. 84 Checking the sensing circuit (3.0L EFI shown)

Checking Sensing Circuit

◆ See Figure 84

1. Locate and unplug the Red and Purple lead connector from the alternator wiring harness.

2. Connect the positive (+) voltmeter lead to the terminal for the Red lead and the negative (-) voltmeter lead to ground.

3. The voltmeter should indicate battery voltage. If the correct voltage is not present, check the sensing circuit (Red lead) for loose or dirty connections or some damaged wiring.

Checking Voltage Output

◆ See Figures 85 and 86

1. Using a 0-20 volt DC voltmeter (or DVOM set to read DC voltage) connect the positive (+) lead of the voltmeter to terminal **B** (output terminal) of the alternator and the negative (-) meter lead to ground.

■ **You can also use a meter with an inductive clamp-on lead. If so, place it over the alternator output lead.**

2. Provide a suitable source of cooling water to the motor, then start the engine allow it to reach operating temperature.

3. Increase the engine rpm from idle to 2000. Normal voltage output should be 13.5-15.1 volts.

4. If the voltage reading is greater than normal the voltage regulator is faulty and will need to be replaced, however the regulator is usually internal to the alternator or bolted to the back of the alternator on these models, so the alternator must be overhauled (if parts are available) or replaced.

5. If the reading is less than normal, you will either need to fabricate a special tool using a piece of stiff wire in order to manually set the alternator to full field output OR you should proceed with Checking Current Output to verify whether the alternator is the problem or not.

■ **Mercury does not specifically mention manually setting the alternator to full field for the following motors - 1.5L OptiMax, 2003 or later 3.0L OptiMax and the 2002 or later EFI motors. Either the alternators on these models lack the terminal F access on the back of the alternator or there is some other unstated concern with manually performing this test on these models.**

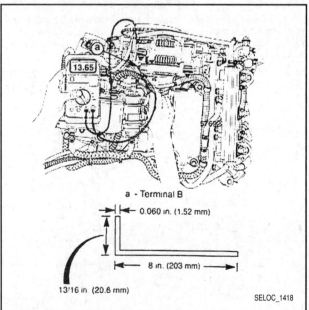

a - Terminal B

0.060 in. (1.52 mm)

8 in. (203 mm)

13/16 in. (20.6 mm)

SELOC_1418

Fig. 85 If voltage output is low, you can fabricate a tool to manually full field the alternator on some models

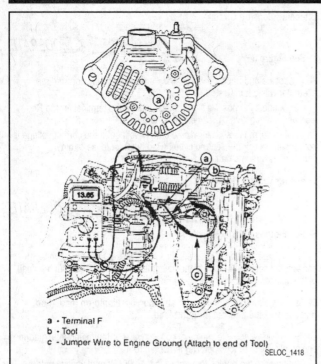

a - Terminal F
b - Tool
c - Jumper Wire to Engine Ground (Attach to end of Tool)

SELOC_1418

Fig. 86 Insert the bent end of the tool through the end cover and ground terminal F

6. On models that you can manually full field the alternator, use a piece of 0.060 in. (1.52mm) thick wire to fabricate a small L-shaped tool which is 8 in. (20.mm) long by 13/16 in. (20.6) on the lower portion of the "L". Insert the bent end of the tool into terminal F at the rear, just about center of the alternator. With Terminal F grounded as directed and the motor running at or above idle, voltage should rise to within normal range (13.5-15.1 volts). If the voltage rises, the regulator is faulty.

7. If the voltage does not rise to within normal specifications with terminal F grounded, you will need to perform the "Checking Current Output" test.

Checking Current Output

 MODERATE

◆ See Figure 87

1. With the engine shut off, install an ammeter (one that is capable of measuring 60-plus amps) either in series between terminal B on the alternator and the positive (+) Terminal of the battery (or using an inductive clamp-on lead on that same wire).

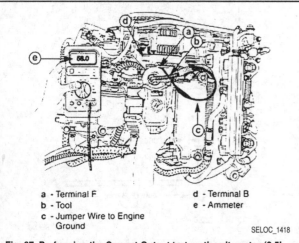

a - Terminal F
b - Tool
c - Jumper Wire to Engine Ground
d - Terminal B
e - Ammeter

SELOC_1418

Fig. 87 Performing the Current Output test on the alternator (2.5L shown)

2. Start the engine and allow it to reach operating temperature.

3. Advance the engine rpm to 2000 rpm.

4. On models where you can access the back of the alternator to manually full field the alternator (for details, please refer to Checking Voltage Output, in this section), insert the fabricated special tool through the end cover and ground terminal **F**.

■ Mercury does not specifically mention manually setting the alternator to full field for the following motors - 1.5L OptiMax, 2003 or later 3.0L OptiMax and the 2002 or later EFI motors. Either the alternators on these models lack the terminal F access on the back of the alternator or there is some other unstated concern with manually performing this test on these models.

5. On models where you CANNOT access the back of the alternator to manually full field the alternator or where it is not recommended by Mercury, supply an external load to the battery (using all possible accessories or some other form of a load. It is possible to use the power trim/tilt to trim the motor fully inward, then continue to hold the trim button for a second or two while taking a reading on the DVOM). It's even better to pre-drain the battery before starting, applying a load like running the trim motor back and forth for at minute or so, then start the motor and immediately start to take readings.

6. Normal output varies by model and RPM. For details, please refer to the Charging System Testing specifications chart in this section. Generally speaking, output should be 40-60 amps @2000-5000 rpm depending upon the particular model.

7. If alternator output is now normal but wasn't earlier, replace the regulator. , However remember, the regulator is usually internal to the alternator or bolted to the back of the alternator on these models, so the alternator must be overhauled (if parts are available) or replaced.

8. If the output is low, make sure the fuse link (if equipped) between terminal **B** and the slave solenoid is good and that the battery cables/connections are in good condition. If no other causes can be found, the alternator should be replaced.

REMOVAL & INSTALLATION

◆ See Figures 88 and 89

 MODERATE

1. Remove the top cowling.
2. Remove the engine flywheel cover for access.
3. Disconnect the battery cables for safety and to prevent accidental grounding of the terminal **B** wiring which could cause component damage or injury.
4. Disconnect the alternator wiring harness.
5. Remove the alternator drive belt as follows (and as further detailed in the Maintenance & Tune-Up section).

• On most models equipped with a serpentine drive belt and automatic tensioner, use a breaker bar to carefully release the drive belt tension, then slip the belt off the alternator pulley and slowly allow the tensioner to return into contact with the belt.

✳✳ SELOC CAUTION

NEVER allow the belt tensioner to snap back into position, as the tensioner assembly could be damaged.

• On model which do NOT use an automatic belt tensioner pulley (such as the 2001 200-250 hp 3.0L EFI motors) belt tension is manually adjusted by moving the alternator mounting position through a slotted range on one of the mounting brackets. Loosen both alternator mounting bolts, then pivot the alternator in the bracket until the belt can be slipped from the pulley.

6. Remove the alternator mounting bolts (usually 2, and usually with loose nuts on the lower end of the bolts). Some models, like the 1.5L motors and some of the 3.0L motors, should have a ground strap under one of the bolts.

7. Remove the alternator from the powerhead.

To Install:

8. Position the alternator to the powerhead and loosely thread the retaining bolts.

9. On models which do not use an automatic belt tensioner, slip the belt over the alternator pulley, then manually pivot the alternator to apply pressure on the belt and temporarily tighten the retaining bolts to hold the alternator in position. Check tension on the belt, measuring at a point halfway between the pulleys by pressing on the belt with moderate thumb

pressure. The belt should deflect about 1/2 in. (12.7mm) in response to a 22 lbs. (30 N) side load. If belt tension is incorrect, loosen the 2 alternator mounting bolts, then pivot the alternator to achieve the proper load. REMEMBER YOU DON'T WANT IT TOO TIGHT EITHER, as that will quickly ruin the alternator bearings. Once adjusted, tighten the alternator adjustment and pivot bolts to 40 ft. lbs. (54 Nm).

■ **On 2001 200-250 hp 3.0L models with manual tension adjustment, Mercury recommends using a standard belt tension adjustment tool to measure and properly set the belt tension. When using a generic tool for this the belt tension is set to 80 lbs. (356 Nm), this should achieve the proper side load tension we outlined checking earlier.**

10. On models that use an automatic belt tensioner, tighten the alternator fasteners to specification (as follows) and then install the drive belt to the pulley using a breaker bar in the tensioner pulley arm:

• 1.5L OptiMax motors: coat the alternator fasteners lightly with Loctite® 271 or equivalent and tighten them to 35 ft. lbs. (47.5 Nm). Also, make sure the ground strap was reinstalled, then cover the strap and bolt with liquid neoprene to prevent corrosion.

• 2.5L motors (EFI & OptiMax), 3.0L OptiMax motors and 2002 or later 3.0L EFI motors: tighten the BOLTS to 40 ft. lbs. (54 Nm) and, for models where the bolts go all the way through the mounting brackets tighten the lower nut(s) to 35 ft. lbs. (47.5 Nm).

11. Reconnect the alternator electrical harness.
12. Reconnect the battery cables.
13. Install the flywheel cover and/or cowling.

Fig. 88 To remove the alternator, remove the bolts (on top). . .

Fig. 89 . . .and the nuts on the bottom of the assembly

Rectifier

DESCRIPTION & OPERATION

The electricity produced by the stator/battery charge coil is AC (Alternating Current). In order to use this current to power most boat accessories or to charge a marine battery it must first be converted to DC (Direct Current). The job of converting the current falls to a rectifier assembly, which is essentially a serried of diodes (one-way electrical valves).

Most smaller Mercury outboards (generally single and twin models) are equipped with an external rectifier. On these models the battery is kept from overcharging by the relatively low output from the motor. However, extended operation (long stretches of time, especially at high speed) can lead to overcharging and excessive gassing at the battery. On these models the rectifier can be visually identified as a small diamond shaped cube with 2 mounting tabs at 2 of it's corners and 3 large, exposed wire terminals on the top of the unit.

Most larger Mercury outboards (including some twins, but mostly 3-cylinder or larger motors) are equipped with a regulator/rectifier assembly (covered later in this section). On these motors the rectifier portion of the assembly performs the same function of converting the current from AC to DC, while the regulator portion controls total current output in order to electronically prevent battery overcharging. The combination regulator/rectifiers tend to be a sealed, black-box type unit with a wiring harness. They also tend to be much larger than the single rectifier units.

Alternator equipped V6 models generally have both a regulator and a rectifier integrated into the alternator assembly (we'll call the internal regulator or internal rectifier to distinguish them from those used on non-alternator models). On some models, the rectifier may be easily removed/replaced simply by unbolting it from the rear of the alternator. On models where the regulator and rectifier are easily serviceable, the regulator tends to be a rectangular finned unit, while the rectifier is a half-mooned shape component.

TESTING

External Rectifiers (Flywheel Stator Models)

◆ **See Figures 90 and 91**

The rectifier can be tested without removal from the powerhead. Disconnect the battery cables from the battery before proceeding with this test.

■ **Depending on the internal polarity of the individual multi-meter used during the following test, the results may be the exact opposite of what is described. Remember the job of the rectifier is to force the current to flow in only one direction. It is done with diodes (one-way electrical valves) so the following test connections, when continuity is allowed, should only indicate continuity in ONE direction.**

Before starting, disconnect the leads from the rectifier terminals.

1. Connect the positive (+) meter lead to the rectifier ground and the negative (-) meter lead alternating between terminals **A** and **C**, then note the results. Swap the positive and negative meter leads and test again (i.e. negative lead to ground this time and positive lead to terminal **A** and then terminal **C**). The meter should show continuity with both terminals, but ONLY under one set of tests (i.e. only while one of the meter leads (either positive or negative) are connected to ground. The meter must show no continuity with the other lead (positive or negative) to ground. If either continuity or no continuity is shown by both sets of tests, replace the rectifier.

■ **Note the lead which was attached to ground the one time the meter showed continuity with terminals A and B in this first step. The results of Steps 2 and 3 will depend upon it.**

2. Connect the negative (-) meter lead to terminal **B** and the positive (+) meter lead alternating between terminals **A** and **C**. Here's where it gets complicated, depending upon the polarity of your meter and the results from the last step:

• If the positive lead to ground yielded continuity in the last step, then this test should show CONTINUITY now.

• If the positive lead to ground yielded no continuity in the last step, then this test should show NO CONTINUITY NOW.

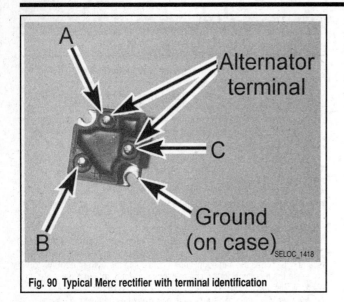

Fig. 90 Typical Merc rectifier with terminal identification

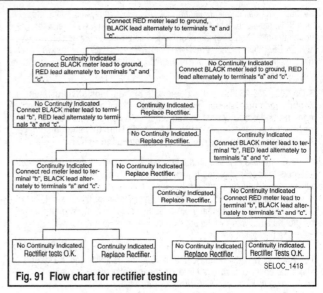

Fig. 91 Flow chart for rectifier testing

If the results are not as expected, replace the rectifier. If this test is correct go to the Step 3 and you SHOULD get the opposite results.

3. Connect the positive (+) meter lead to terminal **B** and the negative (-) meter lead alternating between terminals **A** and **C**. The meter should show the opposite reading of the last step.

• If the FIRST test (Step 1 positive lead to ground and negative lead to terminals **A** and **C**) showed continuity (and the last step also showed continuity) then this test should now show NO CONTINUITY.

• If the FIRST test (Step 1 positive lead to ground and negative lead to terminals **A** and **C**) showed no continuity (and the last step also showed no continuity) then this test should now show CONTINUITY.

If the above readings are not shown, replace the rectifier.

Internal Rectifiers (Alternator Models)

◆ See Figure 92

 MODERATE

On alternator models which use a half-moon shaped rectifier mounted to the rear of the alternator, the rectifier can be removed from the alternator and checked with a DVOM as follows:

1. Connect the positive (+) lead from the meter to the positive stud on the end of the rectifier. Connect the negative (-) lead from the meter in turn to each diode terminal on the rectifier. Each connection should show continuity.

2. Reverse the leads, negative lead to the positive stud and positive lead to each of the diodes. There should now be no continuity for each connection.

3. The rectifier must be replaced if there is either continuity in both directions or no continuity in both directions on any 1 diode.

REMOVAL & INSTALLATION

External Rectifiers (Flywheel Stator Models)

◆ See Figure 90

 MODERATE

1. Disconnect the negative battery cable for safety (so you cannot accidentally ground a hot lead).

2. Locate the rectifier, the diamond shaped electrical component with 3 wire terminals on it's face (it is normally found on the side of the powerhead with other electrical components). Wires from the stator should be running out from under the flywheel toward the rectifier.

3. Tag and disconnect the wiring for the rectifier.

4. Loosen the rectifier fasteners, then remove it from the powerhead.

5. Installation is essentially the reverse of removal. If there were any external ground straps make sure they are properly repositioned as the fasteners are tightened. Be sure that the fasteners and the wire terminal nuts are tightened securely, But DO NOT over-tighten them and crack the housing.

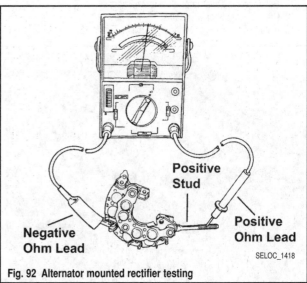

Fig. 92 Alternator mounted rectifier testing

■ After installation it is normally a good idea to coat all terminals with liquid neoprene or a liquid electrical tape, to prevent corrosion on the terminals.

Internal Rectifiers (Alternator Models)

◆ See Figures 93, 94 and 95

 MODERATE

1. Remove the alternator from the powerhead for access.

2. Loosen and remove the 3 screws and the nut which secure the end cover, then remove the end cover and insulator assembly from the back of the alternator for access to the regulator and the rectifier.

3. If necessary, loosen the 4 screws securing the regulator and brush assembly, then remove them from the back of the alternator.

■ Brushes are replaced as an assembly. Inspect the assembly for stuck brushes or excessive brush wear. Normal exposed brush length is 0.158 in. (4.0mm), and a wear limit of 0.059 in. (1.5mm) should be used as a guide for determining their serviceability.

4. Remove the 4 screws securing the rectifier/diode assembly to the alternator, then remove the rectifier assembly.

5. Installation is essentially the reverse of removal. As usual, take care to make sure fasteners are tightened, but not so tight as to crack the components or housing. Both the regulator and rectifier screws may be tightened to 17 inch lbs. (1.9 Nm).

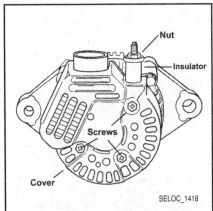

Fig. 93 Remove the end cover and insulator for access

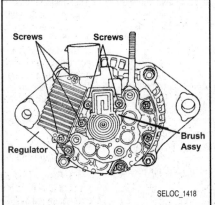

Fig. 94 Loosen the screws and remove the regulator and brush assembly

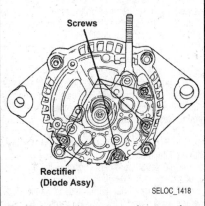

Fig. 95 Loosen the screws and remove the rectifier and diode assembly

Regulator/Rectifier

TESTING

Mercury and Seloc would both really recommend diagnosing a voltage regulator by testing AROUND it, which basically means that you eliminate all other possible causes of a charging system problem, until the Voltage Regulator is the only part left.

That said, where possible, we've provided test specifications which *may* be useful in helping to determine if a regulator is testing properly. However, as we so often warn you, remember that static resistance tests are far from trouble-free.

Also, note that specifications are not available to test in this manner for many models, including all alternator equipped models, and the 75/65 Jet/90 and 80 Jet/115/125 hp motors (though the regulators they use on the inline models are *very* similar in design to that found on 40/40 Jet/50/55/60 hp 3-cylinder models and values MAY be similar to those listed below).

On motors that you are testing around the regulator, remember that excessive voltage is normally a sign of a defective regulator. If voltage isn't sufficient to run accessories, and the stator tests within spec, then you should suspect a bad regulator. One way to test would be to check voltage on the circuit with and without part of the regulator circuit connected (disconnecting 1 of the leads at the regulator terminal and connecting them together using a screw/nut, plus some tape to isolate them). If voltage increases with part of the regulator out of the circuit, and the circuit was undercharging before, the regulator should be considered defective.

To test the regulator on the following models, use an Analog Ohmmeter (Specs for Digital Meters are only provided for some models):
- 25 hp and smaller motors
- 30/40 hp 2-cyl models
- 40/40 Jet/50/55/60 hp 3-cyl models
- 135-200 Hp (2.5L) carbureted, as well as 2001 EFI, V6 Models

1. With an analog ohmmeter set to the R x 10 scale, check the diodes by connecting the positive (+) meter lead to a Red regulator lead (the thick Red lead only on V6 motors) and the negative (-) meter lead to either Yellow regulator lead (some models may have 1 lead that is Yellow/Black). For all models, resistance should be 100-400 ohms.

■ **When using a digital meter set for diode testing on 25 hp and smaller motors, 30/40 hp 2-cylinder motors, 135-200 hp (2.5L) carbureted motors, as well as 2001 EFI, V6 Models the manufacturer says the reading should be 0.4-0.8 volts.**

2. With an analog ohmmeter set to the R x 1k scale, the next diode test is made by connecting the negative (-) meter lead to a Red regulator lead (the thick Red lead only on V6 motors) and connecting the positive (+) meter lead to either Yellow regulator lead. Resistance should be as follows:
- 25 hp and smaller models: with an analog meter, the reading should be 40,000 ohm to infinite resistance or with a digital meter set to diode testing, the reading should be infinite, **OUCH** or **OL** (depending upon the meter).
- 30/40 hp 2-cyl models with an analog meter: the reading should be

20,000 ohm to infinite resistance or with a digital meter set to diode testing, the reading should be infinite, **OUCH** or **OL** (depending upon the meter).
- 40/40 Jet/50/55/60 hp 3-cyl models: the meter should read 40,000 ohm to infinite resistance.
- 135-200 hp (2.5L) carbureted, as well as 2001 EFI, V6 models with an analog meter: the reading should be 20,000 ohm to infinite resistance or with a digital meter set to diode testing, the reading should be infinite, **OUCH** or **OL** (depending upon the meter).

■ **On 30/40 hp motors and 40/40 Jet/50/55/60 hp motors, Mercury states to test the 1 Yellow lead first and then the other. On the first you may get a reading or it may be infinite, on the second it should remain infinite (no needle movement).**

3. With the meter still set to the R x 1k scale, test the SCR by connecting the negative (-) meter lead to either Yellow lead and the positive (+) meter lead to the regulator case ground resistance should be as follows:
- 15 hp and smaller motors with an analog meter: the reading should be 10,000 ohms (10k ohms) minimum, or with a digital meter set to resistance should read at least 900,000 ohms (900k ohms). Reverse the leads and set a digital meter to diode testing, the reading should be 1.5 V to infinite, **OUCH** or **OL** (again, depending upon the meter).
- 25 hp and smaller motors with an analog meter: the reading should be 9,000-(9k ohms) minimum, or with a digital meter set to resistance should read at least 900,000 ohms (900k ohms). Reverse the leads and set a digital meter to diode testing, the reading should be 1.5 V to infinite, **OUCH** or **OL** (again, depending upon the meter).
- 30/40 hp 2-cyl models with an analog meter: the reading should be 8,000-15,000 ohms (8k-15k ohms) or with a digital meter set to diode testing, the reading should be 1.5 V to infinite, **OUCH** or **OL** (again, depending upon the meter).
- 40/40 Jet/50/55/60 hp 3-cyl models: the meter should read 15,000 ohms to infinite resistance.
- 135-200 hp (2.5L) carbureted, as well as 2001 EFI, V6 models with an analog meter: the reading should be 8,000-15,000 ohms (8k-15k ohms) or with a digital meter set to diode testing, the reading should be 1.5 V to infinite, **OUCH** or **OL** (again, depending upon the meter).

4. With the meter still set to the R x 1k scale, test the SCR by connecting the negative (-) meter lead to either Yellow lead and the positive (+) meter lead to the regulator case ground resistance should be as follows:
- 15 hp and smaller motors with an analog meter: the reading should be 10,000 ohms (10k ohms) minimum, or with a digital meter set to diode testing, the reading should be 1.5 V to infinite, **OUCH** or **OL** (again, depending upon the meter).
- 25 hp and smaller motors with an analog meter: the reading should be 9,000-(9k ohms) minimum, or with a digital meter set to diode testing, the reading should be 1.5 V to infinite, **OUCH** or **OL** (again, depending upon the meter).

5. If you have a digital meter and are testing 25 hp and smaller motors, perform these additional circuit checks:
- With the meter set to read resistance, connect the positive (+) meter lead to the Black regulator lead and the negative (-) meter lead to the regulator case, the reading should be 0.0 ohms.

• Next move the positive (+) meter lead to the Red regulator lead, leaving the negative (-) meter lead on the regulator case, the reading should be 4.28k ohms (4,280 ohms) for 15 hp and smaller motors, or 3k-4.28k ohms (3,000-4,280 ohms) for 20/25 hp moors.

• Finally, set the meter to diode reading and connect the positive (+) meter lead to the regulator case and the negative (-) meter lead to the Red regulator case, the reading should be 0.4-0.7 volts.

6. Finally, on 30 hp and larger models: test the tachometer circuit using the meter still set to the R x 1k scale, by connecting the positive (+) meter lead to the Grey lead and the negative (-) meter lead to the regulator case ground Resistance should be as follows:

• 30/40 hp 2-cyl models with an analog meter: the reading should be 10,000-15,000 ohms (10k-15k ohms) and Mercury advises it is not testable with a digital meter.

• 40/40 Jet/50/55/60 hp 3-cyl models: the meter should read 10,000-50,000 ohms (10k-50k ohms).

• 135-200 hp (2.5L) carbureted, as well as 2001 EFI, V6 models: the meter should read 10,000-50,000 ohms (10k-50k ohms) and Mercury advises it is not testable with a digital meter.

REMOVAL & INSTALLATION

External Regulators (Flywheel Stator Models)

◆ See Figures 96, 97 and 98

In most cases the regulator (or regulators on V6 models) is easily removed as it is generally mounted to an electrical component bracket on the side of the powerhead. If you have any trouble locating, check the wiring diagrams and follow the wires. Generally speaking there are 2 Yellow wires running from the stator assembly (under the flywheel) to the regulator, as well as 2 Red wires which go from the regulator to provide the powerhead with power.

Besides disconnecting the negative battery cable for safety (to avoid accidentally grounding a hot lead), removal is a simple matter of tagging and disconnecting the wiring and then loosening the mounting bolts (usually 2). During installation, take your time to make sure all wiring is connected at tagged earlier and that it is tucked/routed out of harm's way.

Internal Regulators (Alternator Models)

◆ See Figures 93 and 94

1. Remove the alternator from the powerhead for access.

2. Loosen and remove the 3 screws and the nut which secure the end cover, then remove the end cover and insulator assembly from the back of the alternator for access to the regulator and the rectifier.

3. Loosen the 4 screws securing the regulator and brush assembly, then remove them from the back of the alternator.

▣ **Brushes are replaced as an assembly. Inspect the assembly for stuck brushes or excessive brush wear. Normal exposed brush length is 0.158 in. (4.0mm), and a wear limit of 0.059 in. (1.5mm) should be used as a guide for determining their serviceability.**

4. Installation is essentially the reverse of removal. As usual, take care to make sure fasteners are tightened, but not so tight as to crack the components or housing. The regulator screws may be tightened to 17 inch lbs. (1.9 Nm).

Fig. 96 Typical Mercury regulator mounting - 30/40 hp 2-cyl models

Fig. 97 Typical Mercury regulator mounting - 80 Jet/115/125 models

Fig. 98 Typical Mercury regulator mounting - V6 models

STARTER CIRCUIT

Starter Motor Circuits

DESCRIPTION & OPERATION

◆ See Figure 99

In the early days, all outboard engines were started by simply pulling on a rope wound around the flywheel. As time passed and owners were reluctant to use muscle power, it was necessary to replace the rope starter with some form of power cranking system. Today, many small engines are still started by pulling on a rope but others have a powered starter motor installed.

The system utilized to replace the rope method was an electric starter motor coupled with a mechanical gear mesh between the starter motor and the powerhead flywheel, similar to the method used to crank an automobile engine.

As the name implies, the sole purpose of the starter motor circuit is to control operation of the starter motor to crank the powerhead until the engine is operating. The circuit includes a relay or magnetic switch to connect or disconnect the motor from the battery. The operator controls the switch with a starter button or key switch which is sometimes referred to as the activation circuit. The starter motor relay or solenoid is basically a remote controlled switch which uses the activation circuit to turn on and off the starter motor circuit.

A neutral safety switch is installed into the circuit to permit operation of the starter motor only if the shift control lever is in neutral. This switch is a safety device to prevent accidental engine start when the engine is in gear.

The starter motor is a series wound electric motor which draws a heavy current from the battery. It is designed to be used only for short periods of time to crank the engine for starting. To prevent overheating the motor, cranking should not be continued for more than 30 seconds without allowing the motor to cool for at least 3 minutes. Actually, this time can be spent in making preliminary checks to determine why the engine fails to start.

Power is transmitted from the starter motor to the powerhead flywheel through a Bendix drive. This drive has a pinion gear mounted on screw threads. When the motor is operated, the pinion gear moves upward and meshes with the teeth on the flywheel ring gear.

When the powerhead starts, the pinion gear is driven faster than the shaft and as a result, it screws out of mesh with the flywheel. A rubber cushion is built into the Bendix drive to absorb the shock when the pinion meshes with the flywheel ring gear. The parts of the drive must be properly assembled for efficient operation. If the screw shaft assembly is reversed, it will strike the splines and the rubber cushion will not absorb the shock.

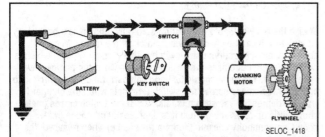

SELOC_1418

Fig. 99 A typical starting system converts electrical energy into mechanical energy to turn the engine. The components are: battery, to provide electricity to operate the starter; ignition switch (or button) to control energizing of the starter relay (solenoid); the relay which will make and break the circuit between the battery and starter motor; and finally the starter, to convert electrical energy into mechanical energy to rotate the engine

The sound of the motor during cranking is a good indication of whether the starter motor is operating properly or not. Naturally, temperature conditions will affect the speed at which the starter motor is able to crank the engine. The speed of cranking a cold engine will be much slower than when cranking a warm engine. An experienced operator will learn to recognize the favorable sounds of the powerhead cranking under various conditions.

The job of the starter motor relay is to complete the circuit between the battery and starter motor. It does this by closing the starter circuit electromagnetically, when activated by the key switch or starter button. This is a completely sealed switch, which meets SAE standards for marine applications. Do not substitute an automotive-type relay for this application. It is not sealed and gasoline fumes can be ignited upon starting the powerhead. The relay consists of a coil winding, plunger, return spring, contact disc and 4 externally mounted terminals. The relay is installed in series with the positive battery cables mounted to the 2 larger terminals. The smaller terminals connect to the neutral switch and ground.

To activate the relay, the shift lever is placed in neutral, closing the neutral switch. Electricity coming through the ignition switch goes into the relay coil winding which creates a magnetic field. The electricity then goes on to ground in the powerhead. The magnetic field surrounds the plunger in the relay, which draws the disc contact into the 2 larger terminals. Upon contact of the terminals, the heavy amperage circuit to the starter motor is closed and activates the starter motor. When the key switch is released, the magnetic field is no longer supported and the magnetic field collapses. The return spring working on the plunger opens the disc contact, opening the circuit to the starter.

When the armature plate is out of position or the shift lever is moved into forward or reverse gear, the neutral switch is placed in the open position and the starter control circuit cannot be activated. This prevents the powerhead from starting while in gear.

TROUBLESHOOTING

◆ See Figures 100 and 101

If the starter motor spins but fails to engage and crank the engine, the cause is usually a corroded or gummy Bendix drive. The drive should be removed, cleaned and given an inspection.

Before wasting too much time troubleshooting the cranking motor circuit, the following checks should be made. Many times, the problem will be corrected.

- Battery fully charged?
- Shift control lever in neutral (many models use a neutral safety switch)?
- Are there any blown fuses?
- All electrical connections clean and tight?
- Wiring in good condition, insulation not worn or frayed?

Two more potential problems may cause the powerhead to crank slowly even though the cranking motor circuit is in excellent condition: a tight or frozen powerhead and/or water in the lower unit. The following troubleshooting procedures are presented in a logical sequence, with the most common and easily corrected areas listed first in each problem area. The connection number refers to the numbered positions in the accompanying illustrations.

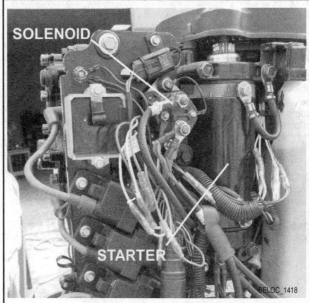

Fig. 100 Typical Mercury electrical starter components

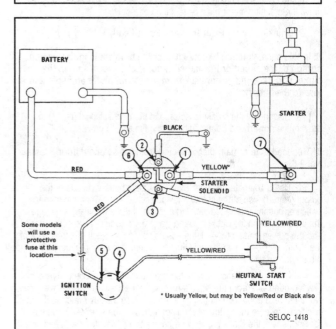

SELOC_1418

Fig. 101 Use this typical starting system diagram to help understand/troubleshoot the starter circuit. Test points correspond to circled numbers in the diagram

The following are some common starter circuit problems with lists of their potential causes.

The starter circuit may seem complicated at first, but it is actually pretty simple when it comes down to it. Power from the positive side of the battery is normally connected to one side of the starter relay/solenoid (No. 6 in the accompanying diagram). A ground cable from the battery either connects to a common ground on the powerhead or to the starter motor itself. When the starter relay/solenoid is activated (remember, a relay is essentially a remote controlled switch), it closes an internal switch connecting power from the positive battery cable to a cable (No. 6 to No. 1) which runs to the starter motor itself (No. 7).

The starter relay/solenoid is activated when power is applied to it (No. 3) from a starter button or ignition switch (No. 5 and 4) and often times through a fuse. In many cases the circuit contains a normally open neutral safety switch which closes only when the switch plunger is depressed by the shift linkage (which should only occur when the linkage is in the neutral position).

Obviously since this circuit will vary slightly depending upon the motor and rigging you should also refer to the Wiring Diagram for the particular model on which you are working to verify components and wire colors.

✳✳ SELOC WARNING

Although in some instances we've provided some resistance tests to illustrate how a component can be checked on or off the motor, DON'T perform resistance checks with the wires still connected. The meter could be damaged if you attempt to take resistance readings on a live circuit. If the component is still installed, use the voltage checks.

1. Cranking Motor Rotates Slowly:

■ In this case you KNOW that the activation circuit is working. You don't know how well the power circuit is doing (i.e. is there too much resistance somewhere in the circuit or is it a problem of too little current supply compared to the load) and you don't know if the problem is the starter motor itself or a mechanical problem with the powerhead.

   a. Battery charge is low. Charge the battery to full capacity.

   b. Electrical connections corroded or loose. Clean and tighten.

   c. Defective cranking motor. Perform an amp draw test. Lay an amp draw-gauge on the cable leading to the cranking motor. Turn the key on and attempt to crank the engine. If the gauge indicates an excessive amperage draw, the cranking motor must be replaced or rebuilt.

2. Cranking Motor Fails To Crank Powerhead:

■ In this case, you don't know whether the problem is the activation portion of the circuit or the starter motor itself? Pick a spot in the circuit and trace the power back from the starter motor and solenoid toward the battery.

   a. Disconnect the cranking motor lead from the solenoid (No. 1) to prevent the powerhead from starting during the testing process.

■ This lead is to remain disconnected from the solenoid during tests No. 2-6.

   b. Use a voltmeter to check for approximate battery voltage at the output (starter) side of the starter relay/solenoid (No. 1) when the starter switch or button is turned to the start position. If there is sufficient voltage, the problem lies in the cable or starter (or ground circuit between the starter and the powerhead/battery). If there is no voltage, check the battery side of the solenoid (No. 6), just to verify that the solenoid is getting battery power to the starter circuit.

   c. Check the solenoid side of the ground circuit by disconnecting the Black ground wire from the No. 2 terminal. Connect a voltmeter between the No. 2 terminal and a common engine ground. Turn the key switch to the start position. Observe the voltmeter reading. If there is the slightest amount of reading, check the Black ground wire connection or check for an open circuit.

3. Test Cranking Motor Relay/Solenoid:

■ The solenoid can easily be tested while it is on or off the motor. The basis for the test is to apply 12 volts across the activation circuit (No. 3 and No. 2 in the illustration) and then to check and see if there is continuity across the starter motor power supply circuit (No. 6 and No. 1). If there is continuity without power applied to the activation circuit the switch is stuck closed and faulty, or alternately, if there is no continuity with power applied the switch is stuck open and faulty.

   a. Check to see if the activation circuit (neutral safety switch and/or ignition switch/button and/or fuse) is supplying voltage to the starter solenoid. Connect a voltmeter between the engine common ground and the No. 3 terminal. Turn the ignition key switch to the start position. Observe the voltmeter reading. If the meter gives a significant voltage reading (something more than at least 0.3 volt), the solenoid is defective and must be replaced (this is true if you've already verified the ground and battery voltage to power the circuit). If however there is little or no voltage, trace the circuit back toward the ignition start button/switch (through the fuse and/or neutral start switch if equipped).

4. Test Neutral Start Switch:

■ Like the relay/solenoid, the neutral start (also known as a neutral safety) switch can be checked either on or off the motor. When equipped, most Mercurys use a switch with a plunger or lever which is normally open, but closes the switch contacts when the plunger/lever is depressed. The most simple test you can perform on the switch is to check for resistance across the switch terminals when the plunger (or lever) is depressed vs. when it is released. If the switch closes (meter shows little/no resistance) when it is depressed and opens (meter shows no continuity, meaning infinite resistance) when released, the switch is operating properly.

   a. Connect a voltmeter between the common engine ground and the ignition switch at test point No. 4. Turn the ignition key switch to the start position. Observe the voltmeter. If there is any indication of a reading at No 4, but NOT at No. 3, the neutral start switch is open or the lead is open between the No. 3 and No. 4. Repair the switch or lead, as applicable. If there was no power at No. 4, it is time to check the switch/button and power supply to the switch.

5. Test for power to the ignition switch/button:

   a. Connect a voltmeter between the common engine ground and test point No. 5.

   b. The voltmeter should indicate approximate battery voltage (about 12-volts). If the meter needle flickers (fails to hold steady), check the circuit between No. 5 and common engine ground. If meter fails to indicate voltage, replace the positive battery cable.

   c. If power is good to the button/switch, but not on the other (No 4, neutral switch) side, the button or switch is suspect.

6. Test large power supply cables:

   a. Connect the starter/solenoid cable (usually Yellow or Yellow/Red on these motors, but it can also be a black cable on some models) to the cranking motor solenoid.

   b. Connect the voltmeter between the engine common ground and No. 7.

   c. Turn the ignition key switch to the start position, or depress the start button.

   d. Observe the voltmeter. If there is no reading, check the cable for a poor connection or an open circuit. If there is any indication of a reading, and the cranking motor does not rotate, make sure the starter motor ground is good, otherwise the cranking motor must serviced or replaced.

Starter Relay/Solenoid

DESCRIPTION & OPERATION

◆ See Figures 102 and 103

All electric start motors are equipped with a remote mounted starter relay, often called the starter solenoid. All Optimax motors as well as some late-model EFI motors are equipped with a solenoid-driven Bendix starter (as opposed to a centrifugal-driven Bendix used on all other models). As such there are actually 2 components which are called solenoids. The first, a remote mounted "slave solenoid" which performs the normal relay functions and a second, starter mounted drive solenoid which mechanically engages the Bendix drive to the flywheel when power is applied by the slave solenoid.

In all cases, the starter solenoids all function in the same basic manner. The starter motor solenoid or relay is actually a remote controlled switch located in the wiring between the battery and the powerhead. Although it can be tested, it cannot be repaired, therefore, if troubleshooting indicates the switch to be faulty, it must be replaced.

Before beginning any work on the relay, disconnect the positive (+) and negative (-) leads from the battery terminal for safety. Keep in mind that the positive lead, where it connects to the solenoid, should always be hot and it is too easy to accidentally ground the tool you are using to the powerhead. This not only will give you quite a shock (we've actually seen it all but weld a wrench in place), but the resultant sparks could create a dangerous condition.

✳✳ SELOC WARNING

Disconnecting the battery leads is most important when testing or servicing the solenoid because the cranking motor lead will often be disconnected and allowed to hang free. The other end of this lead is connected to the battery. If the leads are not disconnected from the battery and the free relay end should happen to come in contact with any metal part on the powerhead, sparks would fly and the end of the lead would be burned.

As detailed in the Starter Motor Circuits section, the starter relay/solenoid has the all important function of responding to the activation circuit in order to physically close the switch contacts that will apply battery power to the starter. Or in this case of the solenoid-driven starters, the slave solenoid does that in order to apply battery power to the drive solenoid which in turn closes the switch contacts to provide battery power to the Starter itself.

The starter motor power circuit contains heavy gauge leads capable of large amp loads in order to power the motor. The activation circuit contains smaller gauge leads, since it only need carry sufficient amperage to activate the relay itself. The activation circuit applies battery power to the solenoid through a neutral safety switch (when equipped) and a starter button or switch.

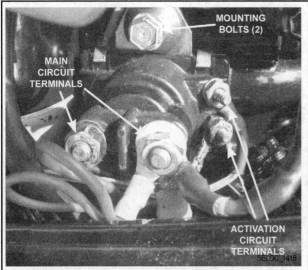

Fig. 102 Typical small motor Mercury solenoid

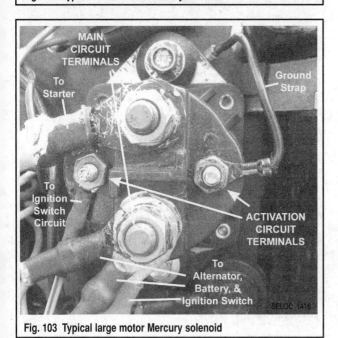

Fig. 103 Typical large motor Mercury solenoid

TESTING

◆ See Figures 102 and 103

There are typically 2 types of solenoids used on these models. The first is a slightly taller, more box-shaped solenoid where both the large terminals (main battery-to-starter power circuit terminals) are on top of the unit, while the activation circuit terminals (smaller, keyswitch or stop button wiring terminals) are on the side of the unit. Since this type of relay is generally found on inline motors (except OptiMax models) and NOT on V6 engines, we'll refer to it as the small motor relay.

The other common relay used on Mercury outboards is a small round unit where all 4 of the relay terminals are on top. On these relays the larger main circuit terminals are directly across from each other (180° apart) while the smaller activation circuit terminals are also directly across from each other (180° apart, placing them 90° from each of the main circuit terminals). Basically as you work your way around the relay it goes main terminal, activation terminal, main terminal, activation terminal. Since this type of relay is usually found on V6 motors (and Optimax inline motors) we'll refer to it as the large motor relay.

Regardless of design, function is identical. There should be no continuity across the main terminals when 12 volts of DC current is not applied to the activation terminals.

Testing can take various forms. You can manually apply voltage across the activation terminals and check for continuity using an ohmmeter across the main terminals (if the power cable has been disconnected) or check for voltage on the starter side of the circuit using voltmeter or test light (if the power cable has NOT been disconnected). A quick test is to apply voltage to the activation terminals and to listen for an audible click indicating the internal switch was thrown (though it's still a good idea to follow this up with a volt or ohm test).

Details on testing a solenoid with an ohmmeter follow:

✳✳ SELOC WARNING

The following tests must be conducted with the relay wiring disconnected (to prevent accidental starting and to prevent potential damage to the meter).

■ The wire colors provided in this procedure are true for most years and models, but if yours differs, check the Wiring Diagrams in this section. In some cases wires may not agree with either the procedure or the diagram as all or part of a harness may have been replaced previously on older outboards, or on motors rigged with an electrical starter after the factory.

The test is relatively simple. Since the relay is simply a remote controlled electric switch, first you check that the remote control (activation) circuit has continuity and the main switch (starter power) circuit does NOT. Then you apply 12 volts to that remote control circuit and make sure the main switch contacts close only when power is applied.

1. Obtain an ohmmeter or DVOM set to read resistance. Check for continuity across the terminals for the small relay leads (the remote control/activation circuit). In most cases these are the terminals for the small black and yellow/red leads (black for ground and yellow/red from the neutral or ignition switch). The meter should indicate continuity. If the meter registers no continuity, the relay is defective and must be replaced. No service or adjustment is possible.

2. Connect 1 test lead of the ohmmeter to each of the large relay terminals. There should be NO continuity.

3. Now, use a 12-volt battery connected to the remote control (activation) circuit of the relay in order to activate the relay while continuing to check for continuity across the large relay terminals. Connect the positive (+) lead from a fully charged 12-volt battery to the small terminal for the yellow/red lead and momentarily make contact with the ground lead from the battery to the small terminal for the Black lead. If a loud click sound is heard, and the ohmmeter indicates continuity, the solenoid is in serviceable condition. If, however, a click sound is not heard, and/or the ohmmeter does not indicate continuity, the solenoid is defective and must be replaced with a suitable marine type solenoid.

■ The drive solenoid mounted directly to the starter on all OptiMax and many of the EFI motors can also be tested in this way.

REMOVAL & INSTALLATION

◆ **See Figures 102 thru 106**

■ **The wire colors provided in this procedure are true for most years and models, but if yours differs, check the Wiring Diagrams in this section. In some cases wires may not agree with either the procedure or the diagram as all or part of a harness may have been replaced previously on older outboards, or on motors rigged with an electrical starter after the factory.**

Most Mercury starter relays are secured to either the side of the powerhead, or an electrical component bracket located on the side of the powerhead by 2 retaining bolts. When used, the electrical component brackets are designed to try and isolate components from the heat and vibration of the powerhead.

When servicing a solenoid, use care not to over-tighten the fasteners and crack the case. The same goes for the terminal nuts, make sure they are snug but do not over-tighten them.

On all OptiMax and most V6 motors, the solenoid function is performed by 2 related, but separate components, one a remote mounted slave solenoid and the other a drive solenoid mounted piggy-back on the starter motor assembly itself. It is usually easier to remove the starter completely from the powerhead before attempting to remove the later from the starter itself.

1. On all models, the FIRST step in solenoid removal is to DISCONNECT the battery cables for safety. Don't disregard this, cause if you do, you *will* be working with live wires.

2. Locate the solenoid on the side of the powerhead (or lower piggyback on the starter motor, as applicable). On some models, the solenoid is mounted low on the powerhead, so if necessary remove the lower cowling for additional access.

3. Push down the boots from the 2 large leads (they are usually Yellow and Red) at the relay. Take time tag and/or note each of the leads and to which terminal they are connected, then remove each of the leads from the solenoid. Disconnect the smaller leads (usually Yellow/Red and Black).

4. Unbolt and carefully remove the relay from the powerhead.

To Install:

5. Position the relay to the powerhead and secure using the retaining bolts.

6. Connect the smaller relay leads (usually Yellow/Red and Black) as tagged during removal.

7. Connect the large leads to the main relay terminals, as tagged during removal. One is from the terminal on the cranking motor, while the other is hot, coming from the positive battery terminal, and possibly the alternator, and often feeding the hot side of the ignition switch.

■ **Though Mercury does not spec it on most motors, we recommend a light coating of liquid electrical tape to seal the cable ends and terminal nuts.**

8. If equipped, position the elbow boots onto the main relay leads.

9. If removed for access, install the lower cowling assembly.

10. If the work is complete, then reconnect the battery terminals. If further work is to be carried out on the cranking system, then leave the battery cables disconnected until the work is complete.

Starter Motor

DESCRIPTION

One basic type of cranking motor is used on all powerheads covered here, however there are 2 slightly different types of drives. The majority of Mercury models, including nearly all carbureted motors and some early EFI models are equipped with a centrifugal drive Bendix starter, while ALL OptiMax motors and most late-model EFI motors (and possibly a few carbureted V6 models) are equipped with a solenoid-driven Bendix starter.

The major difference between the two is the solenoid driven model has an additional starter solenoid mounted piggyback on the starter assembly. The job of this solenoid is to move a clutch/gear assembly mechanically meshing the pinion gear with the flywheel only when the starter is engaged. On all other models, a threaded pinion spins up into position using centrifugal force when the starter motor is activated.

With the exception of the slight distinction in service necessitated by this change, most starter motor servicing is the same.

Otherwise, marine cranking motors are very similar in construction and operation to the units used in the automotive industry.

As with most starter motors, the housing is not designed to provide airflow which would be necessary for continuous or repeated operation. Therefore, never operate a cranking motor for more than 30 seconds without allowing it to cool for at least 3 minutes. Continuous operation without the cooling period can cause serious damage to or destroy the cranking motor.

REMOVAL & INSTALLATION

25 Hp & Smaller Motors

◆ **See Figure 107**

1. Before beginning any work on the starter motor, disconnect the battery cables from the battery and isolate them for safety.

2. Unlatch and remove the cowling from the powerhead for access.

3. Tag and disconnect the power cable (usually yellow) from the solenoid to the starter motor, at the terminal on the bottom of the starter.

4. These models *usually* ground through the mounting bracket, but check for a ground cable just to be sure. If present, tag and remove it.

5. Remove the mounting bracket bolts (2) threaded horizontally through the bracket toward the top of the starter motor assembly, then carefully remove the starter motor from the powerhead.

To Install:

6. Align the starter motor onto the powerhead and thread the retaining bolts, then tighten them securely. If a separate ground cable is used, be sure to position at this time.

7. Connect the power cable from the solenoid to the terminal on the bottom of the starter motor and carefully tighten the terminal nut.

Fig. 104 The main relay leads are usually protected by terminal covers (boots)

Fig. 105 . . .just gently pull them down for access

Fig. 106 View of a starter mounted solenoid

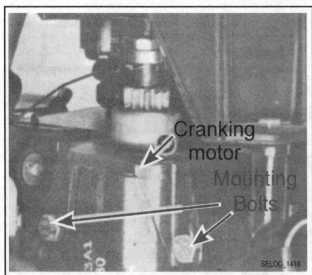

Fig. 107 Typical starter motor mounting for 25 hp and smaller motors

Fig. 108 Typical centrifugal driven starter mounting

■ Though Mercury does not spec it on most motors, we recommend a light coating of liquid electrical tape to seal the cable end and terminal nut.

8. Install the engine cowling.
9. Reconnect the battery cables and test the starter motor operation.

30 Hp & Larger Motors

◆ See Figures 108 thru 114

1. Before beginning ANY work on the starter motor, disconnect the battery cables from the battery and isolate them for safety.
2. Remove the engine top cover for access.
3. The majority of the centrifugal drive Bendix starters utilize a separate black ground cable for the starter motor, usually attached toward the top of the starter motor assembly. If equipped, tag and disconnect the ground cable either at the starter or at the powerhead (though we usually prefer to disconnect it from the starter).

■ The solenoid-driven Bendix starters usually use a ground cable that runs from the top of the starter motor body to the mounting bolt, so you will automatically disconnect it from the powerhead when you unbolt the starter later in this procedure. There are also a few centrifugal drive starters which use a ground bolt like this too.

4. Tag and disconnect the wiring from the starter as follows, depending upon the type of starter. You may have to cut one or more wire ties for access to wiring or to relocate wiring out of the way:
• On centrifugal drive starters besides the ground cable disconnected in the previous step (unless it is secured by a starter mounting bolt and therefore will be undone later) there should only be 1 main power cable running from the solenoid to a terminal towards the bottom side of the starter assembly. Tag and disconnect this lead.
• On solenoid-driven starters, besides the ground strap mentioned earlier (which will be disconnected by removing 1 of the mounting bolts later in this procedure) there are multiple leads going to the drive solenoid on the back of the starter. There should be one smaller gauge (usually Yellow/Red) activation circuit lead running from the slave solenoid to an activation terminal on the drive solenoid. Also, there should be 2 large gauge (usually Black) main power leads running either or both directly from the battery and from the battery side of the main power terminal on the slave solenoid, to a main terminal on the drive solenoid. Tag and disconnect these leads.

■ There is no need to disconnect the lead running from the other main terminal of the drive solenoid back into the body of the starter assembly.

5. Support the starter and remove the mounting bracket bolts (3 or 4 depending upon the design, though *most* 75 hp and larger models use 4 bolts). The mounting bolts are threaded horizontally through a clamp or bracket at the top and bottom of the starter. Once the bolts are removed

Fig. 109 Typical solenoid driven starter mounting

Fig. 110 Main power cable on a centrifugal driven unit

Fig. 111 Main power cable and switch wiring on a solenoid driven unit

Fig. 112 Many units have a ground strap near the top

Fig. 113 Typical upper mounting clamp. . .

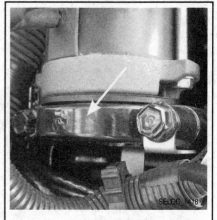

Fig. 114 . . .and lower mounting clamp

carefully remove the starter motor (with clamps, if equipped) from the powerhead.

To Install:

6. Align the starter motor (and clamps, if used) onto the powerhead and thread the retaining bolts, then tighten them securely. On starters where the ground cable is secured by a mounting bolt, be sure to position it at this time (usually under the upper right mounting bolt).

7. Reconnect the wiring as tagged during removal. If there are any questions as to the proper wiring on solenoid-driven models, refer to the information for removal, the Solenoid Testing/Removal & Installation procedures, or the wiring diagrams for more information.

■ Though Mercury does not spec it on most motors, we recommend a light coating of liquid electrical tape to seal the starter and solenoid wiring and terminal nut(s).

8. Install the engine cowling.
9. Reconnect the battery cables and test the starter motor operation.

OVERHAUL

25 Hp & Smaller Motors

◆ See Figures 115, 116 and 117

1. Remove the starter assembly as detailed earlier under Removal & Installation.

✳✳ SELOC CAUTION

Alignment marks must be scribed on the end caps and frame assembly to ensure the armature and end caps are properly installed back in their original positions. The marks will not guarantee the thru-bolts will slide through the frame assembly on the first attempt but without them aligned, it is not possible - period.

2. Scribe a matchmark on the motor frame and matching marks on the end caps (drive end and brush end). These marks will ensure the end caps are installed back to the frame assembly in the exact position from which they are removed.

3. Remove the 2 thru-bolts. Lift the end cap free of the frame assembly. It may be necessary to gently tap the end cap with a soft head mallet to help jar the cap loose.

4. Pull the armature from the frame assembly. A slight "pull" may be felt as the armature is removed, due to the magnets installed in the frame assembly. Obtain some type of strap wrench. Secure the strap wrench around the armature. Hold the armature with the strap wrench and remove the lock nut from the pinion gear end of the armature shaft. Use a quick jerk motion to "break" the nut loose.

■ The locknut should be discarded and replaced with a new one once it has been removed.

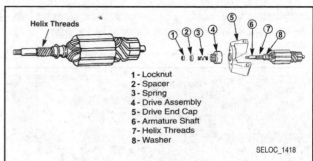

1 - Locknut
2 - Spacer
3 - Spring
4 - Drive Assembly
5 - Drive End Cap
6 - Armature Shaft
7 - Helix Threads
8 - Washer

Helix Threads

Fig. 115 Exploded view of the starter assembly - 25 hp and smaller motors

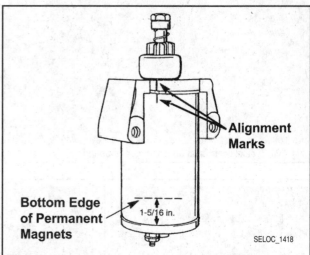

Alignment Marks

Bottom Edge of Permanent Magnets

1-5/16 in.

Fig. 116 Starter assembly, position the armature into the frame and align the matchmarks

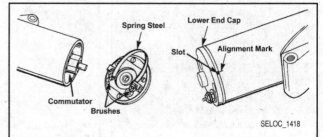

Spring Steel
Lower End Cap
Slot
Alignment Mark
Commutator
Brushes

Fig. 117 Using a piece of spring steel to retainer the brushes during assembly

5. After the locknut has been removed, disassemble and keep the parts in order as they are removed from the shaft, as shown: the spacer, spring, drive assembly (pinion gear), drive end cap and washer.

6. Refer to the information under Cleaning & Inspection to help determine serviceability of components.

To Assemble:

7. Apply a thin coating of SAE 10W oil on to the helix threads on the armature shaft and to the bushings in the drive end cap and lower end cap. BUT DO NOT over-lubricate these points.

8. Slide the washer onto the armature shaft, followed by the drive end cap.

9. Next, install the drive assembly onto the shaft, then the spring and spacer. Start a NEW locknut onto the end of the shaft.

10. Obtain a strap wrench. Prevent the armature from rotating with the strap wrench and tighten the locknut on the armature shaft.

11. Insert the assembled armature into the frame assembly (aligning the matchmarks made earlier). A slight "pull" will be felt as the armature is installed due to the magnets in the frame assembly. The armature should be into the frame so that the permanent magnets are recessed 1 5/16 in. (33.3mm) from the end cap split line.

12. If removed, install the brushes and springs to the end cap. Be sure to feed the leads into the recesses.

13. Push in on each brush and secure the brush temporarily in the retracted position with a small piece of wire or a piece of spring steel. A paper clip, bent to the proper shape, will do the job.

14. Push in on the drive end so the commutator will extend out of the starter frame, then carefully align the end cap and install it onto the frame assembly. Hold the cap tightly against the frame assembly with 1 hand and remove the wires holding the brushes in the retracted position with the other hand. Assistance from a "helper" would be most welcome, at this time.

■ **The marks made during disassembly MUST be aligned to ensure the unit is being assembled back into its original position. The marks will not guarantee the thru-bolts will slide through the frame assembly on the first attempt but without them aligned, it could be said the task is "impossible".**

15. Slide the 2 thru-bolts through the bottom end cap, through the frame assembly and into the drive end cap. The permanent magnets in the frame assembly will "attract" the thru-bolts as they are passed through. Therefore, have patience and look through the holes in the end cap to see how to move the bolts to permit them to thread into the cap. Once the bolts have been started in the threads, tighten them both securely.

16. Install the starter assembly as detailed earlier under Removal & Installation.

30 Hp & Larger Motors - Centrifugal-Driven Starter

◆ See Figures 118 thru 122

1. Remove the starter assembly as detailed earlier under Removal & Installation.

✳✳ SELOC CAUTION

Alignment marks must be scribed on the end caps and frame assembly to ensure the armature and end cap are properly installed back in their original position. The marks will not guarantee the thru-bolts will slide through the frame assembly on the first attempt but without them aligned, it is not possible - period.

2. Scribe a matchmark on the motor frame and matching marks on the end caps (drive end and brush end). These marks will ensure the end caps are installed back to the frame assembly in the exact position from which they are removed.

3. Remove the 2 thru-bolts. Lift the end cap from the bottom of the frame assembly. It may be necessary to gently tap the end cap with a soft head mallet to help jar the cap loose.

4. Pull the drive end cap and armature assembly from the frame. A slight "pull" may be felt as the armature is removed, due to the magnets installed in the frame assembly.

5. Lift upward on the pinion drive assembly in order to slip a wrench underneath to hold the armature steady by its flats. While holding the armature from turning, loosen the locknut from the top end of the armature shaft.

■ **The locknut should be discarded and replaced with a new one once it has been removed.**

6. After the locknut has been removed, disassemble and keep the parts in order as they are removed from the shaft, as shown: the spacer, spring, drive assembly (pinion gear), drive end cap and washer.

7. Refer to the information under Cleaning & Inspection to help determine serviceability of components.

To Assemble:

8. Apply a thin coating of SAE 10W oil on to the helix threads on the armature shaft and to the bushings in the drive end and commutator (brush end) cap. BUT DO NOT over-lubricate these points.

9. Slide the washer onto the armature shaft, followed by the drive end cap. Next, install the drive assembly onto the shaft, then the spring and spacer. Start a NEW locknut onto the end of the shaft.

10. Using a wrench inserted under the pinion gear to hold the armature from turning, tighten the locknut on the armature shaft.

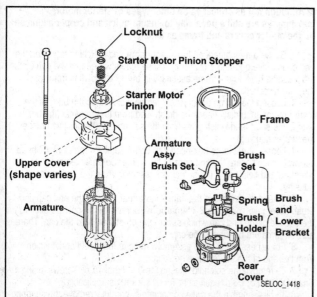

Fig. 118 Exploded view of a typical centrifugal-driven starter motor

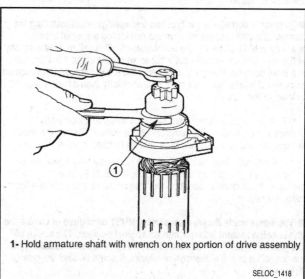

1- Hold armature shaft with wrench on hex portion of drive assembly

SELOC_1418

Fig. 119 Hold the armature with a wrench while you loosen the locknut. . .

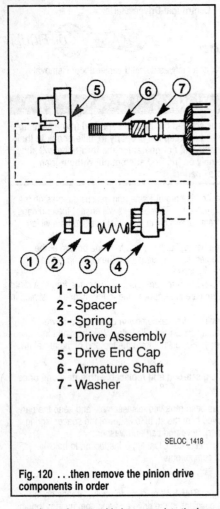

1 - Locknut
2 - Spacer
3 - Spring
4 - Drive Assembly
5 - Drive End Cap
6 - Armature Shaft
7 - Washer

SELOC_1418

Fig. 120 . . .then remove the pinion drive components in order

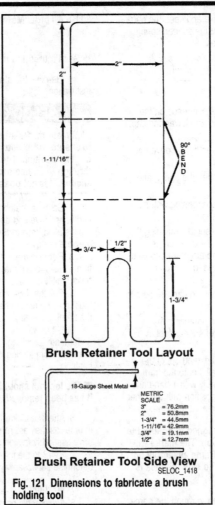

Brush Retainer Tool Layout

18-Gauge Sheet Metal

METRIC
SCALE
3" = 76.2mm
2" = 50.8mm
1-3/4" = 44.5mm
1-11/16" = 42.9mm
3/4" = 19.1mm
1/2" = 12.7mm

Brush Retainer Tool Side View
SELOC_1418

Fig. 121 Dimensions to fabricate a brush holding tool

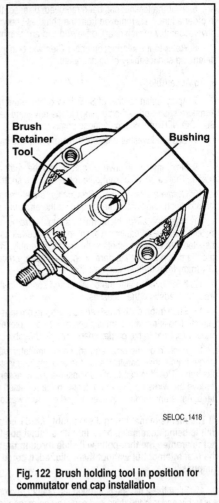

Brush Retainer Tool

Bushing

SELOC_1418

Fig. 122 Brush holding tool in position for commutator end cap installation

11. Insert the assembled armature into the frame assembly (aligning the matchmarks made earlier). A slight "pull" will be felt as the armature is installed due to the magnets in the frame assembly.

12. If removed, install the brushes and springs to the end cap. Be sure to feed the leads into the recesses. The terminal for the positive brushes is installed using an insulating washer (closest to the end cap), then a washer, split washer and finally a hex nut.

■ **To prevent damage to the brushes and springs when installing the commutator end cap, we recommend fashioning a special brush retaining tool as shown in the accompanying illustration. However, shy of this tool you can usually cut a 1/2 in. wide slot about 1 3/4 in. deep in a steel putty knife and use it to serve the same purpose (although an extra set of hands becomes more of a necessity than a luxury with this version of the tool).**

13. Slide the brush springs into the brush holder and then insert the brushes into the holder while compressing the springs. Position the special tool over the end cap and brushes to hold the brushes in place.

14. Position the lower end cap onto the frame assembly. Lower the cap as far as it will go and then remove the special tool. Now, align the mark on the cap with the mark on the frame and then install the thru-bolts and tighten them securely.

■ **The marks made during disassembly MUST be aligned to ensure the unit is being assembled back into its original position. The marks will not guarantee the thru-bolts will slide through the frame assembly on the first attempt but without them aligned, it could be said the task is "impossible".**

15. Install the starter assembly as detailed earlier under Removal & Installation.

30 Hp & Larger Motors - Solenoid-Driven Starter

◆ See Figures 124 thru 126

③ DIFFICULT

1. Remove the starter assembly as detailed earlier under Removal & Installation.

■ **Although not as critical as on other types of starter motors, matchmarks are still a good way to ensure quick and proper alignment of the starter covers and frame assembly.**

2. Scribe a matchmark on the motor frame and matching marks on the end caps (drive end and brush end). These marks will ensure the end caps are installed back to the frame assembly in the exact position from which they are removed.

3. Loosen the nut on the solenoid terminal closest to the body of the starter motor, then disconnect the brush lead from that terminal (the brush lead is the short lead which goes from the solenoid to the positive brushes in the starter housing).

4. Loosen the 2 large hex-head thru-bolts (not the internal Torx® head screws) which secure the end caps to the frame. Carefully remove the armature and field frame (brush holder) assembly from the starter drive housing.

5. If necessary remove the frame and armature from the end cap and brush assembly. There are 2 internal Torx® screws (Snap-On® E6 socket) on the outside of the end cap which secure the brush holder to the cap. There is also an armature bearing in the cap.

6. For access to the drive assembly, remove the shield and cushion from the end of the starter drive housing.

7. To remove the solenoid, loosen the 3 internal Torx® screws (using a Snap-On® E5 or E6 or equivalent Torx® socket, as applicable).

8. To disassemble the drive components, carefully remove the snapring from the stop of the starter shaft, then remove the pinion gear.

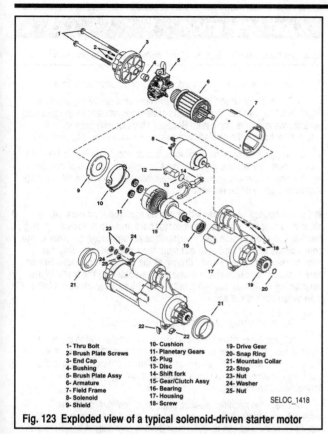

1- Thru Bolt
2- Brush Plate Screws
3- End Cap
4- Bushing
5- Brush Plate Assy
6- Armature
7- Field Frame
8- Solenoid
9- Shield
10- Cushion
11- Planetary Gears
12- Plug
13- Disc
14- Shift fork
15- Gear/Clutch Assy
16- Bearing
17- Housing
18- Screw
19- Drive Gear
20- Snap Ring
21- Mountain Collar
22- Stop
23- Nut
24- Washer
25- Nut

SELOC_1418

Fig. 123 Exploded view of a typical solenoid-driven starter motor

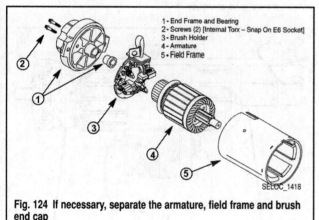

1 - End Frame and Bearing
2 - Screws (2) [Internal Torx – Snap On E6 Socket]
3 - Brush Holder
4 - Armature
5 - Field Frame

SELOC_1418

Fig. 124 If necessary, separate the armature, field frame and brush end cap

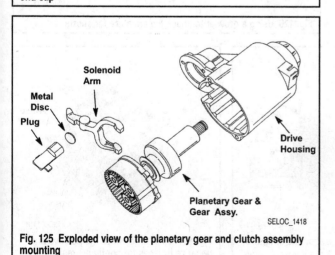

Solenoid Arm

Metal Disc

Plug

Drive Housing

Planetary Gear & Gear Assy.

SELOC_1418

Fig. 125 Exploded view of the planetary gear and clutch assembly mounting

9. Now from inside the drive assembly, remove the planetary gear and clutch assembly.

10. Remove the solenoid arm, metal disc and plug.

11. Check the drive housing needle bearing (just below where the pinion gear normally mounts) for roughness. If the bearing is worn or damaged the bearing can be removed using a suitable mandrel to drive/press the bearing from the housing.

■ **If the bearing has spun in the drive housing bore, the housing must be replaced.**

12. Refer to the information under Cleaning & Inspection to help determine serviceability of the remaining components.

To Assemble:

13. Position the solenoid arm on the planetary gear assembly, then insert the assembly into the drive housing. Install the metal disc and plug into the drive housing.

14. Install the solenoid and secure by tightening the 3 Torx® head screws to 40 inch lbs. (4.5 Nm).

Install the drive gear and secure using the snapring.

15. Install the rubber bumpers on the housing.

16. Install the custom and shield in the drive housing.

17. Place the field frame over the armature.

18. If removed, install the brushes and springs to the end cap. Be sure to feed the leads into the recesses.

19. While holding the brushes back, slide the brush plate onto the armature while at the same time aligning the brush lead grommet with the slot in the frame (this should align the first of the matchmarks made earlier).

20. Secure the end plate to the brush assembly using the 2 Torx® screws and tighten to 30 inch lbs. (3.4 Nm).

21. Install the field frame to the drive housing aligning the slot in the field frame with the plug in the drive housing (and aligning the second set of matchmarks made earlier).

22. Install the through bolts and tighten to 110 inch lbs. (12.5 Nm).

23. Reconnect the brush lead to the closest large terminal of the drive solenoid, then tighten the nut to 55 inch lbs. (6 Nm).

■ **Most models have a ground strap on the side of the motor, when equipped, Mercury now recommends coating the connection with liquid neoprene or liquid electrical tape after it is fastened, to prevent corrosion.**

24. Install the starter assembly as detailed earlier under Removal & Installation.

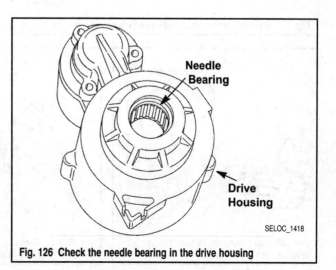

Needle Bearing

Drive Housing

SELOC_1418

Fig. 126 Check the needle bearing in the drive housing

CLEANING & INSPECTION

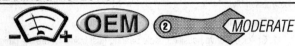

◆ **See Figures 127 thru 131**

■ **If a centrifugal drive motor does not fully engage with the flywheel the drive assembly may be binding on the helix threads of the armature shaft due to either wear or dirt. Figure out the source of binding before reassembly and clean or replace components, as necessary.**

1. Carefully clean all starter parts. Metal parts can be cleaned with a gentle parts cleaning solution, but must be thoroughly dried (compressed air is helpful) before reassembly.

2. Visually inspect the pinion teeth for signs of cracks, chips or excessive amounts of wear and replace, if necessary.

3. Visually inspect the brush holder for signs of damage or for failure to hold the brushes against the commutator.

4. Visually inspect the brushes. Any brushes which are chipped or pitted must be replaced. Make sure brush length is still acceptable. On 6-25 hp models the brushes should be at least 3/16 in. (4.8mm), while on 30 hp and larger models the brushes should be at least 1/4 in. (6.4mm).

5. Check the commutator for signs of pitting, rough or unevenly worn surfaces. Minor pitting or roughness can be removed with No. 00 grit sandpaper. However, if necessary the commutator CAN be turned on a lathe, if so keep in mind the following points:

• DO NOT turn the commutator surface excessively. Remove the minimal amount of material to return the surface to peak operating condition.

• Clean any copper particles from the slots between the commutator bars.

• Use No. 00 sandpaper to clean up the final surface and remove any burrs.

✳✳ **SELOC CAUTION**

After sanding, thoroughly clean the finished surface using solvent

6. Test the armature for unwanted shorts or grounds, as follows:

a. To check for shorts, place the armature on a growler and hold a hacksaw blade over the core while the armature is rotated. If the saw blade vibrates, the armature is shorted. Recheck after cleaning between the commutator bars. If the saw *still* vibrates, the armature must be replaced.

b. To check for unwanted grounds, use an ohmmeter (set to the R x 1 scale) to test between the armature core or shaft (center, end of armature) and the body of the armature. If the meter indicates continuity, the armature is grounded an must be replaced.

■ **Open-circuited armatures often can be repaired. The most likely place for an open is at the commutator bars and are the result of long cranking periods (which can overheat the motor, causing solder in the connections to melt and spin out from the rotating force. This causes poor connection, arcing and burning of the commutator bars). Repair bars that are not excessively burned by re-soldering the leads in the bars (using rosin flux solder) and turning the commutator on a lathe, then undercutting the mica.**

7. Test the brushes for good conductivity as follows:

a. Check the positive brushes using an ohmmeter (set to the R x 1 scale). Place 1 meter lead on each positive brush and make sure there is NO resistance. If the meter shows any resistance, check the lead to the brush and positive terminal solder connection and repair or replace the connection, as necessary.

b. Check the negative brushes using an ohmmeter (set to the R x 1 scale). Place 1 meter lead on a negative brush and the other on a bare metal part of the end cap to make sure there is continuity. A brush must be replaced if there is no continuity.

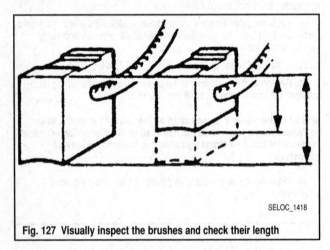

Fig. 127 Visually inspect the brushes and check their length

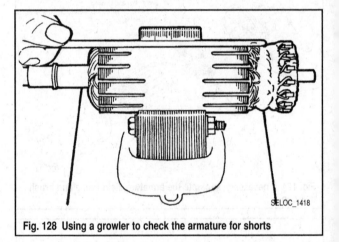

Fig. 128 Using a growler to check the armature for shorts

Fig. 129 Use an ohmmeter to check the armature for grounds. . .

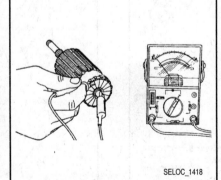

Fig. 130 . . .where you test on the armature varies slightly by design

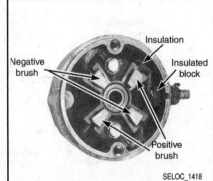

Fig. 131 Be sure to check the positive and negative brushes for continuity/resistance

ELECTRICAL SWITCH/SOLENOID SERVICE

This short section provides testing procedures for other electrical parts installed on the powerhead. If a unit fails the testing, the faulty part must be replaced. In most cases, removal and installation is through attaching hardware.

Ignition Keyswitch

TESTING

Mercury Commander 2000 Keyswitch

◆ See Figure 132

The Mercury Commander 2000 main keyswitch is normally located inside the control box or it is rigged to the dash of the boat. When installed in the control box, the box must be normally be opened to gain access to the main switch leads.

■ The wire colors on this test are only applicable to Mercury harnesses and keyswitches. Depending upon boat rigging, the craft on which you are working could deviate from these colors, at least on the switch side of the harness.

Disconnect the 5 main switch leads: the Black, Black/Yellow, Red, Yellow/Red, Purple and Yellow/Black. Obtain an ohmmeter or a DVOM set to read resistance. Check to be sure the switch is in the off position. Make contact with 1 meter lead to the Black switch lead, and the other meter lead to the Black/Yellow switch lead. The meter should indicate continuity. Keep both meter leads in place. Rotate the switch to the on position, and then to the start position. If the meter indicates continuity in either or both switch positions, the switch is defective and must be replaced.

■ Likewise, check for any continuity between the remaining 4 leads (Red, Yellow/Red, Purple and Yellow/Black). There should be no continuity with the between the remaining leads with the switch in the OFF position. Also, there should be no continuity or between them and the Black or Black/Yellow leads with the switch in ANY position.

Continue with the test according to the accompanying illustration. Move 1 meter lead to the Red wire, then check for continuity to the remaining leads as indicated for **Run**, **Start** and **Choke** positions.

Continue testing each lead for continuity to only the proper other leads as indicated by the chart for each switch position.

Once testing has been satisfactorily completed reconnect the 6 leads, tuck the leads neatly inside the control box, away from moving parts, and replace the outer cover.

Late Model 3 & 4-Position Mercury Keyswitches

◆ See Figures 133, 134 and 135

The late model Mercury keyswitch is normally located inside the control box or it is rigged to the dash of the boat. When installed in the control box, the box must be normally be opened to gain access to the main switch leads.

■ The wire colors on this test are only applicable to Mercury harnesses and keyswitches. Depending upon boat rigging, the craft on which you are working could deviate from these colors, at least on the switch side of the harness.

Unplug the wiring harness from the keyswitch. With the connector lock assembly facing downward, the terminals are identified as Pins **A** through **F**, moving counterclockwise from the top right terminal and ending at the bottom right terminal.

Using a voltmeter and the accompanying test charts, check for continuity across the appropriate pin combinations with the switch in its various positions. Continue testing each pin for continuity to only the proper other leads as indicated by the chart for each switch position.

Once testing has been satisfactorily completed reconnect the harness, tuck the wiring, away from moving parts.

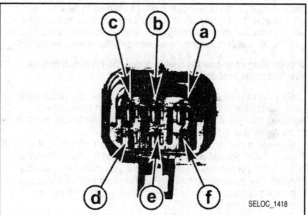

Fig. 133 Late model Mercury keyswitch wiring terminal identification

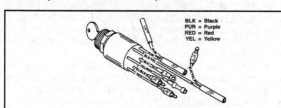

KEY POSITION	CONTINUITY SHOULD BE INDICATED AT THE FOLLOWING POINTS:					
	BLK	BLK/YEL	RED	YEL/RED	PUR	YEL/BLK
OFF	o———o					
RUN			o		o	
START			o	o		
			o		o	
CHOKE*			o		o	
				o		o

*Key switch must be positioned to "RUN" or "START" and key pushed in to actuate choke for this test.

NOTE: If meter readings are other than specified in the preceding tests, verify that switch and not wiring is faulty. If wiring checks ok, replace switch.

SELOC_1418

Fig. 132 Mercury 2000 Commander keyswitch testing

Ref. No.	Pin	Wire Color	Description
a	A	Red	12 volts
b	B	Black	Ground
c, d	C, D	Purple	Run
e	E	Black/yellow	Off
f	F	Yellow/red	Start

Meter Test Leads		Key Position	Reading (Ω)
Red	Black		
Pin B	Pin E	Off	Continuity
Pin A	Pin F	Run	Continuity
Pin A	Pin C, D		Continuity
Pin A	Pin F	Start	Continuity
Pin F	Pin C, D		Continuity
Pin A	Pin C, D		Continuity

SELOC_1418

Fig. 134 Test specs for late model Mercury 3-position keyswitch

Ref. No.	Pin	Wire Color	Description
a	A	Red	12 volts
b	B	Black	Ground
c	C	Purple/white	Accessory
d	D	Purple	Run
e	E	Black/yellow	Off
f	F	Yellow/red	Start

Meter Test Leads		Key Position	Reading (Ω)
Red	Black		
Pin B	Pin E	Off	Continuity
Pin A	Pin C	Accessories	Continuity
Pin A	Pin F	Run	Continuity
Pin A	Pin C		Continuity
Pin A	Pin F	Start	Continuity
Pin F	Pin D		Continuity
Pin A	Pin D		Continuity
Pin A	Pin C		Continuity

SELOC_1418

Fig. 135 Test specs for late model Mercury 3-position keyswitch

Kill Switch

TESTING

◆ See Figures 136 and 137

Multiple types of kill switches are used by Mercury, depending upon the model. One style, the lanyard type is usually mounted to the front panel of any powerhead not equipped with a control box. On powerheads equipped with a control box, the kill switch is often mounted on the forward side of the box. The Kill switch must have the emergency tether in place before testing.

■ **Depending upon how the boat is rigged, the kill switch may be attached to the boat dash itself.**

Other types of kill switches, including a push button stop switch is also used, predominantly by tiller control and manual start models. When equipped, this type of switch is usually found in the tiller handle, at the center of the throttle grip.

Regardless their location or design, all kill switches tend to be of a simple 2 wire, 2 switch position design. They are designed to allow continuity under only 1 switch position and to break the circuit under the other. Generally speaking they are in the no continuity position when the powerhead can run and ground some part of the ignition circuit when in the **OFF** position (meaning they have continuity).

To test a kill switch, locate and disconnect the wiring (usually Black and Black/Yellow), then connect an ohmmeter. Test first with the lanyard toggle switch in the **RUN** position or the push button no-depressed and you should find NO CONTINUITY. Next depress the push button or switch the toggle to

the **OFF** position and recheck, you should now have continuity (1 ohm or less resistance). If you have continuity or no-continuity in both positions, check the switch wiring and/or replace the switch.

Start Button

TESTING

Electric start models without an ignition keyswitch (generally meaning tiller control units) are equipped with a simple start button which us used to actuate the starter solenoid activation circuit. The switch is normally open, closing the circuit only while the button is depressed and held.

Trace the start button harness containing 2 wires from the switch to their nearest quick disconnect fitting. The colors may vary for different models, but it shouldn't make a difference, because it is only a 2 wire switch and your tests will be with 1 meter lead attached to each wire. Generally the wires are Yellow and Yellow/Red. For more details, refer to the Wiring Diagrams, in this section.

Disengage the 2 leads and connect an ohmmeter across the disconnected leads. Depress the start button. The meter should register continuity. Release the button and the meter should now register NO continuity. Both tests must be successful. If the tests are not successful, the start button must be replaced. The start button is a 1 piece sealed unit and cannot be serviced.

Neutral Safety Switch

TESTING

◆ See Figure 138

Electric start powerheads are normally equipped with a neutral safety switch in the starter circuit in order to prevent accidentally starting the motor in gear. The switch is normally either located on the powerhead where it can be directly in contact with shift linkage (generally this is on tiller control models) or it is located in the remote control housing.

In either case, function is simple, it is a normally OPEN switch, which prevents the starter activation circuit (key switch or start button) from functioning. When the shift linkage is in Neutral, the linkage will depress and hold the switch plunger, CLOSING switch contacts and allowing the circuit to be completed by the key switch or start button. The switch may be located inline in the activation circuit, between the key switch/starter button and the starter solenoid OR it may be on the ground side of the circuit, between the solenoid and ground.

To test the switch, trace the neutral safety switch leads from the switch to their nearest quick disconnect fitting. The colors of these leads will vary with the model, though they are often Yellow/tracer (such as Yellow/Red or Yellow/Black) and/or just plain Black. Of course, it shouldn't make a

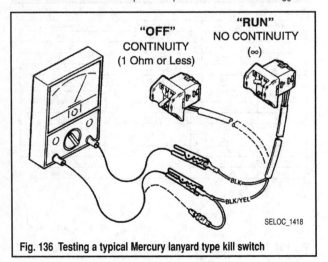

Fig. 136 Testing a typical Mercury lanyard type kill switch

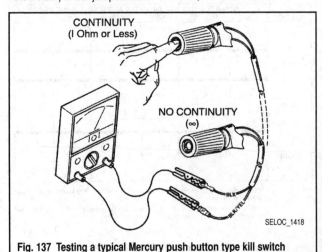

Fig. 137 Testing a typical Mercury push button type kill switch

difference, because it is only a 2 wire switch and your tests will be with 1 meter lead attached to each wire. For more details, refer to the Wiring Diagrams, in this section.

Disengage the 2 leads, then connect an ohmmeter or a DVOM set to read resistance across the 2 disconnected leads. When the shift lever is in the neutral position (usually this means the plunger is depressed/held downward into the switch body), the meter should indicate continuity.

When the lower unit is shifted into either forward or reverse gear (the plunger is normally released), the meter should indicate no continuity.

■ **Don't be alarmed if a particular switch/linkage is designed to operate in the opposite manner regarding plunger position and continuity (i.e. held downward in gear to break switch contacts). The important thing is that the switch must ONLY show continuity in the proper position for Neutral.**

The switch must pass all 3 tests to verify the safety aspect of the switch is functioning properly. If the switch fails any one or more of the tests, the switch must be adjusted (if the physical positioning of the switch allows this) or replaced (more likely on most Mercury/ models).

Remember this is a safety switch. A faulty switch may allow the powerhead to be started with the lower unit in gear - a potentially dangerous situation for the boat, crew, and passengers.

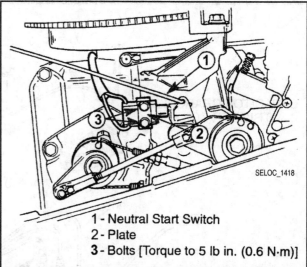

1 - Neutral Start Switch
2 - Plate
3 - Bolts [Torque to 5 lb in. (0.6 N·m)]

Fig. 138 Typical Mercury powerhead-mounted neutral safety switch

WIRING DIAGRAMS

■ The following diagrams represent the most popular models with the most popular optional equipment. For additional information on Remote Control wiring, please refer to the Remote Control section.

INDEX

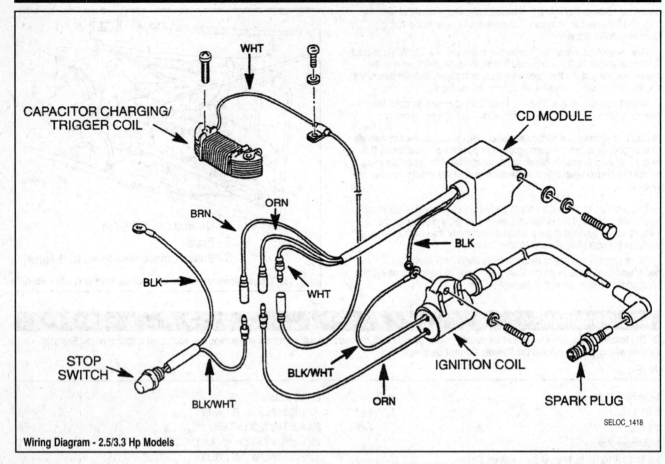

CAPACITOR CHARGING/TRIGGER COIL

WHT

CD MODULE

BRN

ORN

BLK

WHT

BLK

STOP SWITCH

BLK/WHT

BLK/WHT

BLK/WHT

ORN

IGNITION COIL

SPARK PLUG

SELOC_1418

Wiring Diagram - 2.5/3.3 Hp Models

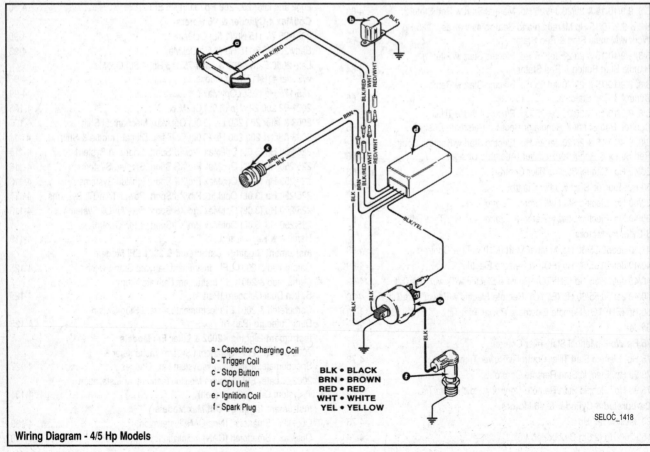

a - Capacitor Charging Coil
b - Trigger Coil
c - Stop Button
d - CDI Unit
e - Ignition Coil
f - Spark Plug

BLK • BLACK
BRN • BROWN
RED • RED
WHT • WHITE
YEL • YELLOW

SELOC_1418

Wiring Diagram - 4/5 Hp Models

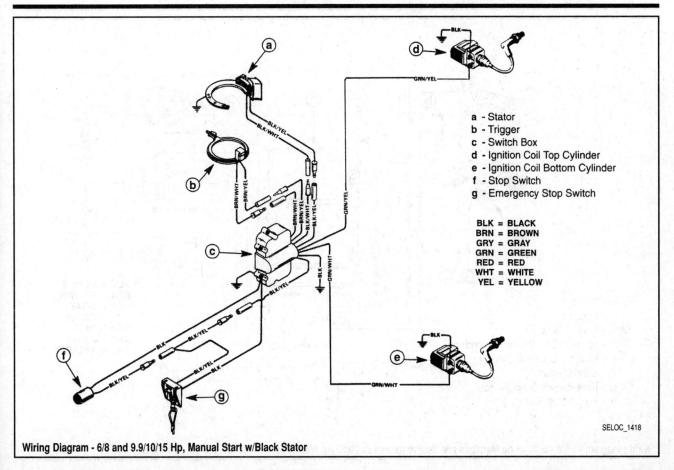

a - Stator
b - Trigger
c - Switch Box
d - Ignition Coil Top Cylinder
e - Ignition Coil Bottom Cylinder
f - Stop Switch
g - Emergency Stop Switch

BLK = BLACK
BRN = BROWN
GRY = GRAY
GRN = GREEN
RED = RED
WHT = WHITE
YEL = YELLOW

SELOC_1418

Wiring Diagram - 6/8 and 9.9/10/15 Hp, Manual Start w/Black Stator

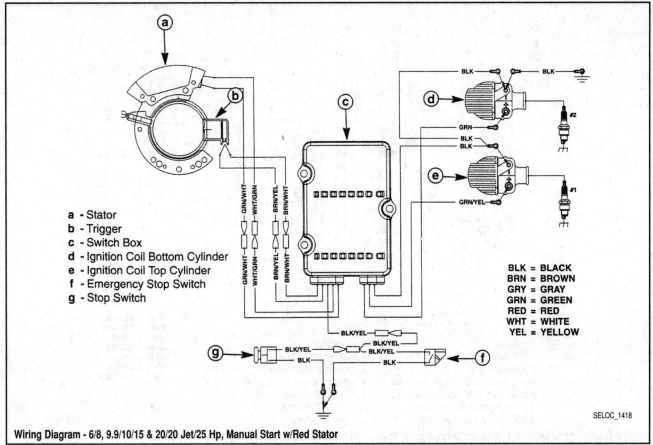

a - Stator
b - Trigger
c - Switch Box
d - Ignition Coil Bottom Cylinder
e - Ignition Coil Top Cylinder
f - Emergency Stop Switch
g - Stop Switch

BLK = BLACK
BRN = BROWN
GRY = GRAY
GRN = GREEN
RED = RED
WHT = WHITE
YEL = YELLOW

SELOC_1418

Wiring Diagram - 6/8, 9.9/10/15 & 20/20 Jet/25 Hp, Manual Start w/Red Stator

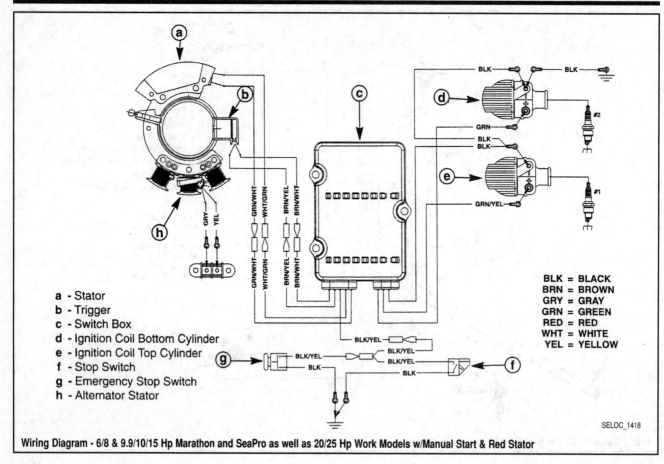

a - Stator
b - Trigger
c - Switch Box
d - Ignition Coil Bottom Cylinder
e - Ignition Coil Top Cylinder
f - Stop Switch
g - Emergency Stop Switch
h - Alternator Stator

BLK = BLACK
BRN = BROWN
GRY = GRAY
GRN = GREEN
RED = RED
WHT = WHITE
YEL = YELLOW

SELOC_1418

Wiring Diagram - 6/8 & 9.9/10/15 Hp Marathon and SeaPro as well as 20/25 Hp Work Models w/Manual Start & Red Stator

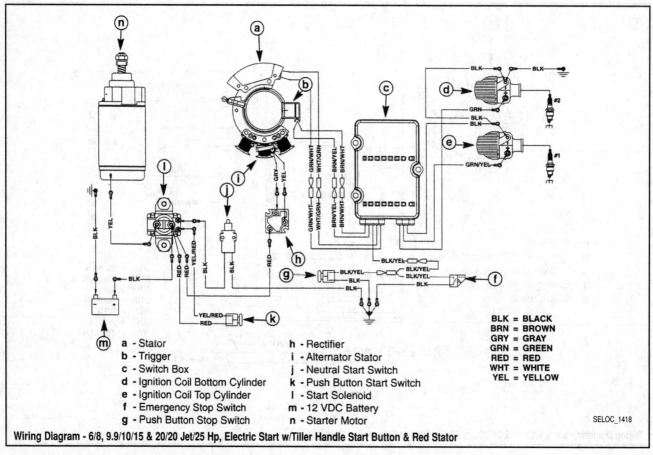

a - Stator
b - Trigger
c - Switch Box
d - Ignition Coil Bottom Cylinder
e - Ignition Coil Top Cylinder
f - Emergency Stop Switch
g - Push Button Stop Switch

h - Rectifier
i - Alternator Stator
j - Neutral Start Switch
k - Push Button Start Switch
l - Start Solenoid
m - 12 VDC Battery
n - Starter Motor

BLK = BLACK
BRN = BROWN
GRY = GRAY
GRN = GREEN
RED = RED
WHT = WHITE
YEL = YELLOW

SELOC_1418

Wiring Diagram - 6/8, 9.9/10/15 & 20/20 Jet/25 Hp, Electric Start w/Tiller Handle Start Button & Red Stator

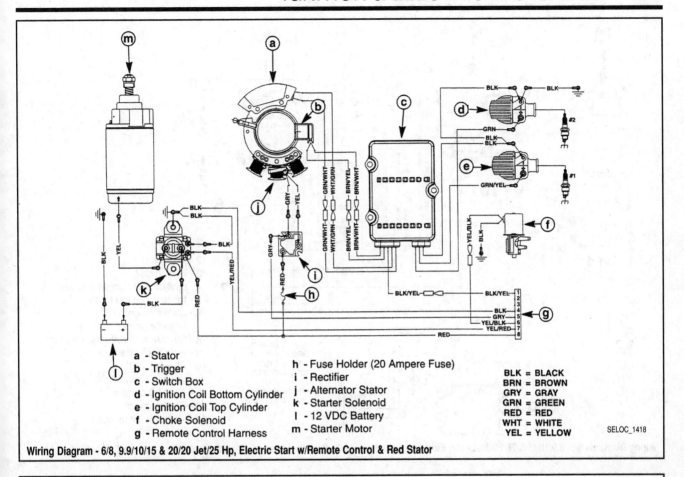

a - Stator
b - Trigger
c - Switch Box
d - Ignition Coil Bottom Cylinder
e - Ignition Coil Top Cylinder
f - Choke Solenoid
g - Remote Control Harness

h - Fuse Holder (20 Ampere Fuse)
i - Rectifier
j - Alternator Stator
k - Starter Solenoid
l - 12 VDC Battery
m - Starter Motor

BLK = BLACK
BRN = BROWN
GRY = GRAY
GRN = GREEN
RED = RED
WHT = WHITE
YEL = YELLOW

SELOC_1418

Wiring Diagram - 6/8, 9.9/10/15 & 20/20 Jet/25 Hp, Electric Start w/Remote Control & Red Stator

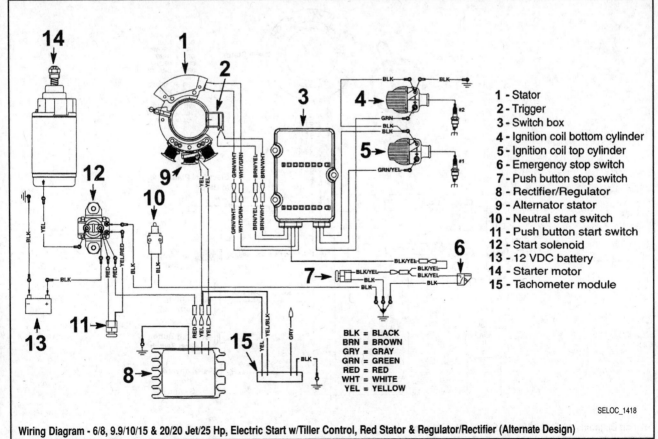

1 - Stator
2 - Trigger
3 - Switch box
4 - Ignition coil bottom cylinder
5 - Ignition coil top cylinder
6 - Emergency stop switch
7 - Push button stop switch
8 - Rectifier/Regulator
9 - Alternator stator
10 - Neutral start switch
11 - Push button start switch
12 - Start solenoid
13 - 12 VDC battery
14 - Starter motor
15 - Tachometer module

BLK = BLACK
BRN = BROWN
GRY = GRAY
GRN = GREEN
RED = RED
WHT = WHITE
YEL = YELLOW

SELOC_1418

Wiring Diagram - 6/8, 9.9/10/15 & 20/20 Jet/25 Hp, Electric Start w/Tiller Control, Red Stator & Regulator/Rectifier (Alternate Design)

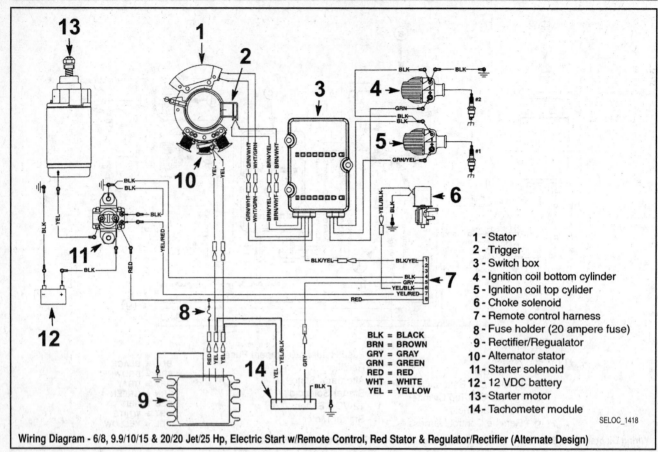

1 - Stator
2 - Trigger
3 - Switch box
4 - Ignition coil bottom cylinder
5 - Ignition coil top cylider
6 - Choke solenoid
7 - Remote control harness
8 - Fuse holder (20 ampere fuse)
9 - Rectifier/Regualator
10 - Alternator stator
11 - Starter solenoid
12 - 12 VDC battery
13 - Starter motor
14 - Tachometer module

BLK = BLACK
BRN = BROWN
GRY = GRAY
GRN = GREEN
RED = RED
WHT = WHITE
YEL = YELLOW

SELOC_1418

Wiring Diagram - 6/8, 9.9/10/15 & 20/20 Jet/25 Hp, Electric Start w/Remote Control, Red Stator & Regulator/Rectifier (Alternate Design)

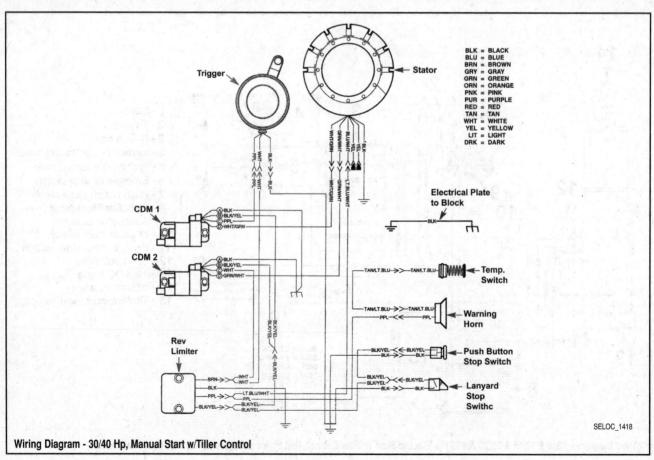

BLK = BLACK
BLU = BLUE
BRN = BROWN
GRY = GRAY
GRN = GREEN
ORN = ORANGE
PNK = PINK
PUR = PURPLE
RED = RED
TAN = TAN
WHT = WHITE
YEL = YELLOW
LIT = LIGHT
DRK = DARK

SELOC_1418

Wiring Diagram - 30/40 Hp, Manual Start w/Tiller Control

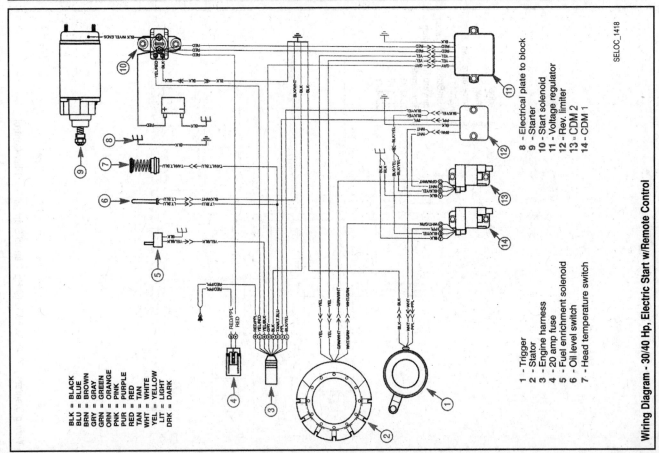

SELOC_1418

Wiring Diagram - 30/40 Hp, Electric Start w/Remote Control

BLK = BLACK
BLU = BLUE
BRN = BROWN
GRY = GRAY
GRN = GREEN
ORN = ORANGE
PNK = PINK
PUR = PURPLE
RED = RED
TAN = TAN
WHT = WHITE
YEL = YELLOW
LT = LIGHT
DRK = DARK

1 - Trigger
2 - Stator
3 - Engine harness
4 - 20 amp fuse
5 - Fuel enrichment solenoid
6 - Oil level switch
7 - Head temperature switch
8 - Electrical plate to block
9 - Starter
10 - Start solenoid
11 - Voltage regulator
12 - Rev. limiter
13 - CDM 2
14 - CDM 1

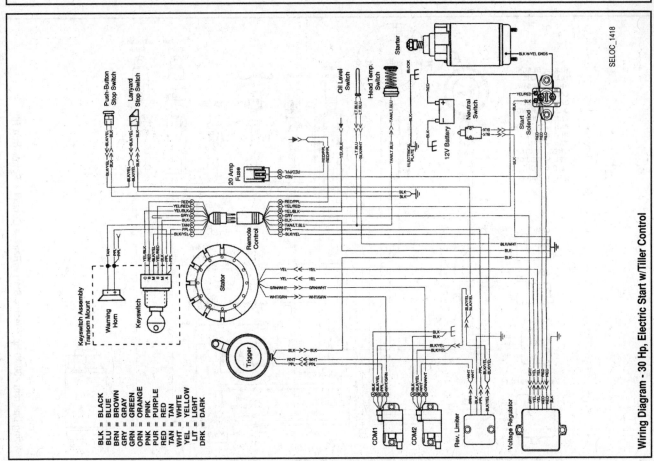

SELOC_1418

Wiring Diagram - 30 Hp, Electric Start w/Tiller Control

BLK = BLACK
BLU = BLUE
BRN = BROWN
GRY = GRAY
GRN = GREEN
ORN = ORANGE
PNK = PINK
PUR = PURPLE
RED = RED
TAN = TAN
WHT = WHITE
YEL = YELLOW
LT = LIGHT
DRK = DARK

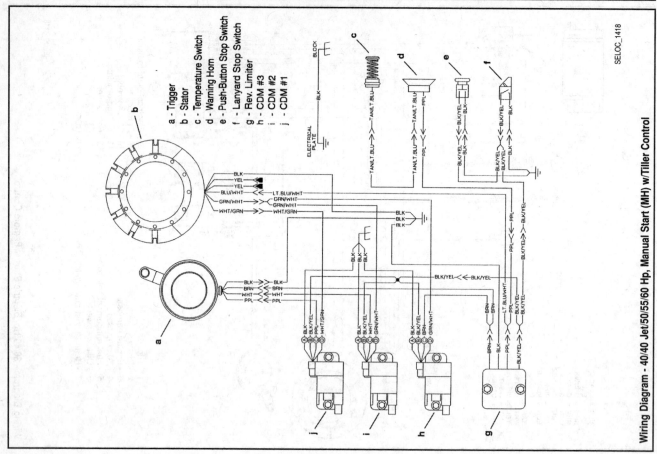

Wiring Diagram - 40/40 Jet/50/55/60 Hp, Manual Start (MH) w/Tiller Control

SELOC_1418

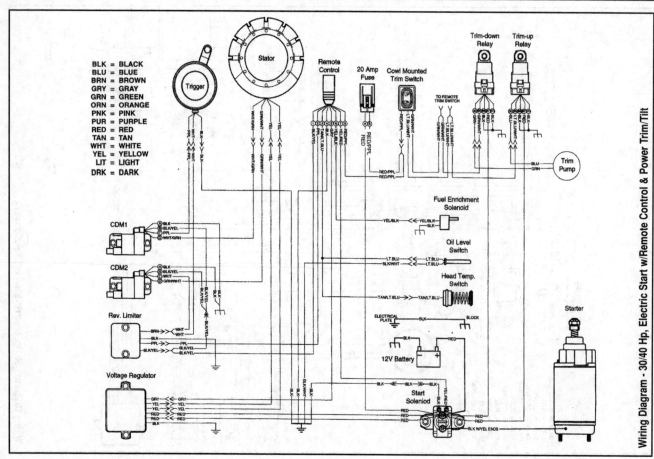

Wiring Diagram - 30/40 Hp, Electric Start w/Remote Control & Power Trim/Tilt

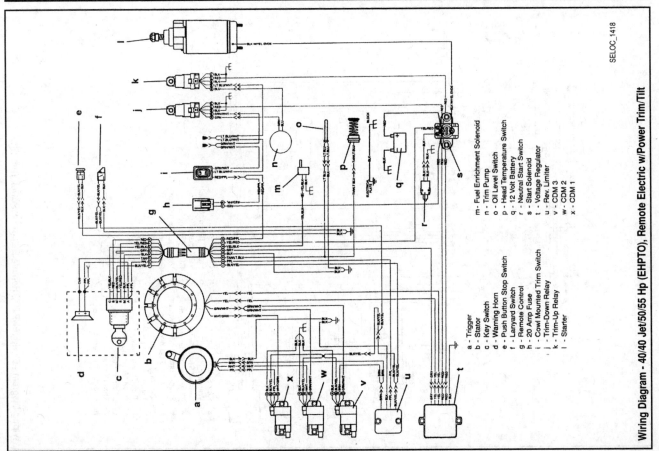

SELOC_1418

Wiring Diagram - 40/40 Jet/50/55 Hp (EHPTO), Remote Electric w/Power Trim/Tilt

a - Trigger
b - Stator
c - Key Switch
d - Warning Horn
e - Push Button Stop Switch
f - Lanyard Switch
g - Remote Control
h - 20 Amp Fuse
i - Cowl Mounted Trim Switch
j - Trim-Down Relay
k - Trim-Up Relay
l - Starter

m - Fuel Enrichment Solenoid
n - Trim Pump
o - Oil Level Switch
p - Head Temperature Switch
q - 12 Volt Battery
r - Neutral Start Switch
s - Start Solenoid
t - Voltage Regulator
u - Rev. Limiter
v - CDM 3
w - CDM 2
x - CDM 1

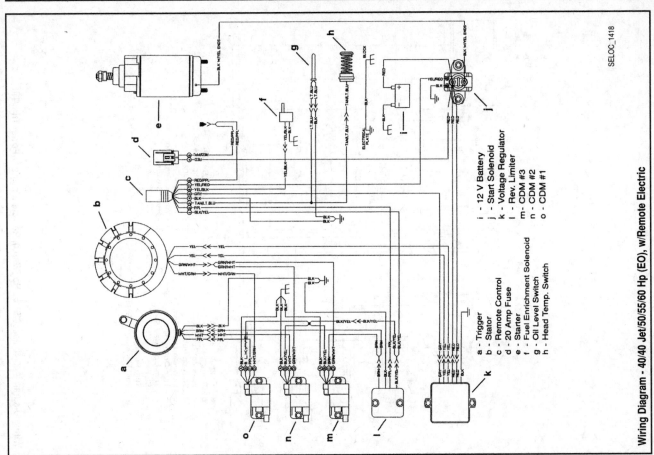

SELOC_1418

Wiring Diagram - 40/40 Jet/50/55/60 Hp (EO), w/Remote Electric

a - Trigger
b - Stator
c - Remote Control
d - 20 Amp Fuse
e - Starter
f - Fuel Enrichment Solenoid
g - Oil Level Switch
h - Head Temp. Switch

i - 12 V Battery
j - Start Solenoid
k - Voltage Regulator
l - Rev. Limiter
m - CDM #3
n - CDM #2
o - CDM #1

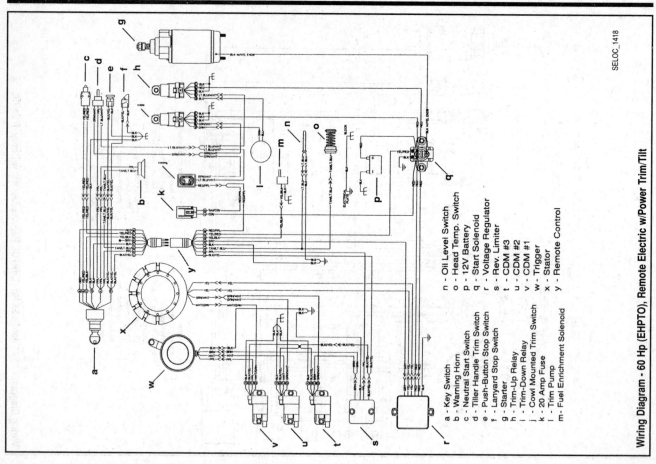

SELOC_1418

Wiring Diagram - 60 Hp (EHPTO), Remote Electric w/Power Trim/Tilt

a - Key Switch
b - Warning Horn
c - Neutral Start Switch
d - Tiller Handle Trim Switch
e - Push-Button Stop Switch
f - Lanyard Stop Switch
g - Starter
h - Trim-Up Relay
i - Trim-Down Relay
j - Cowl Mounted Trim Switch
k - 20 Amp Fuse
l - Trim Pump
m - Fuel Enrichment Solenoid

n - Oil Level Switch
o - Head Temp. Switch
p - 12V Battery
q - Start Solenoid
r - Voltage Regulator
s - Rev. Limiter
t - CDM #3
u - CDM #2
v - CDM #1
w - Trigger
x - Stator
y - Remote Control

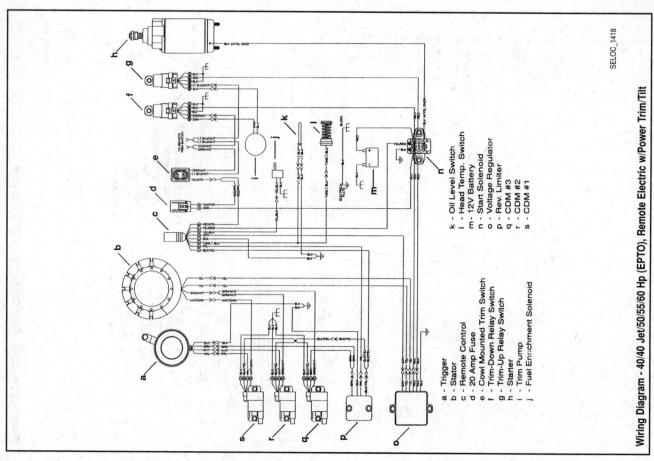

SELOC_1418

Wiring Diagram - 40/40 Jet/50/55/60 Hp (EPTO), Remote Electric w/Power Trim/Tilt

a - Trigger
b - Stator
c - Remote Control
d - 20 Amp Fuse
e - Cowl Mounted Trim Switch
f - Trim-Down Relay Switch
g - Trim-Up Relay Switch
h - Starter
i - Trim Pump
j - Fuel Enrichment Solenoid

k - Oil Level Switch
l - Head Temp. Switch
m - 12V Battery
n - Start Solenoid
o - Voltage Regulator
p - Rev. Limiter
q - CDM #3
r - CDM #2
s - CDM #1

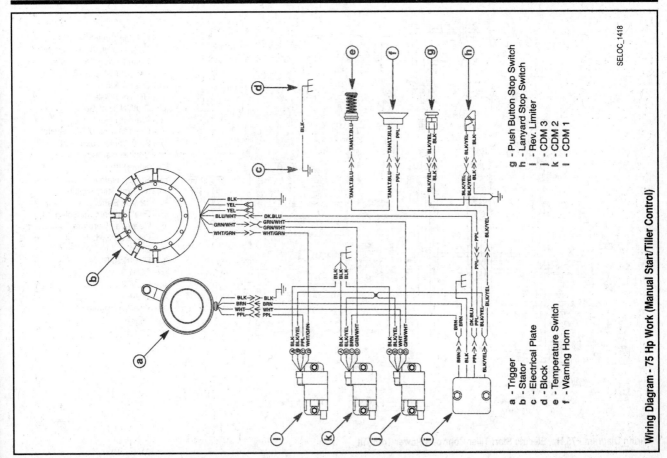

Wiring Diagram - 75 Hp Work (Manual Start/Tiller Control)

g - Push Button Stop Switch
h - Lanyard Stop Switch
i - Rev. Limiter
j - CDM 3
k - CDM 2
l - CDM 1

a - Trigger
b - Stator
c - Electrical Plate
d - Block
e - Temperature Switch
f - Warning Horn

SELOC_1418

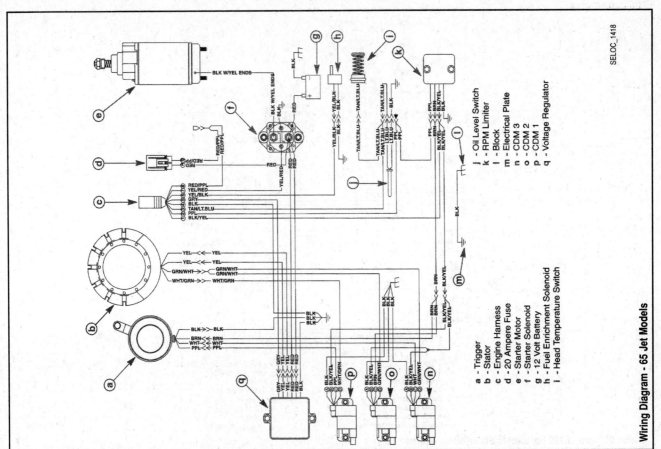

Wiring Diagram - 65 Jet Models

j - Oil Level Switch
k - RPM Limiter
l - Block
m - Electrical Plate
n - CDM 3
o - CDM 2
p - CDM 1
q - Voltage Regulator

a - Trigger
b - Stator
c - Engine Harness
d - 20 Ampere Fuse
e - Starter Motor
f - Starter Solenoid
g - 12 Volt Battery
h - Fuel Enrichment Solenoid
i - Head Temperature Switch

SELOC_1418

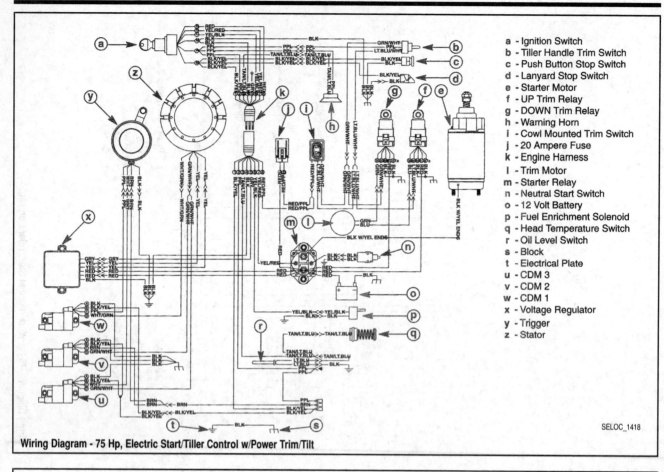

a - Ignition Switch
b - Tiller Handle Trim Switch
c - Push Button Stop Switch
d - Lanyard Stop Switch
e - Starter Motor
f - UP Trim Relay
g - DOWN Trim Relay
h - Warning Horn
i - Cowl Mounted Trim Switch
j - 20 Ampere Fuse
k - Engine Harness
l - Trim Motor
m - Starter Relay
n - Neutral Start Switch
o - 12 Volt Battery
p - Fuel Enrichment Solenoid
q - Head Temperature Switch
r - Oil Level Switch
s - Block
t - Electrical Plate
u - CDM 3
v - CDM 2
w - CDM 1
x - Voltage Regulator
y - Trigger
z - Stator

SELOC_1418

Wiring Diagram - 75 Hp, Electric Start/Tiller Control w/Power Trim/Tilt

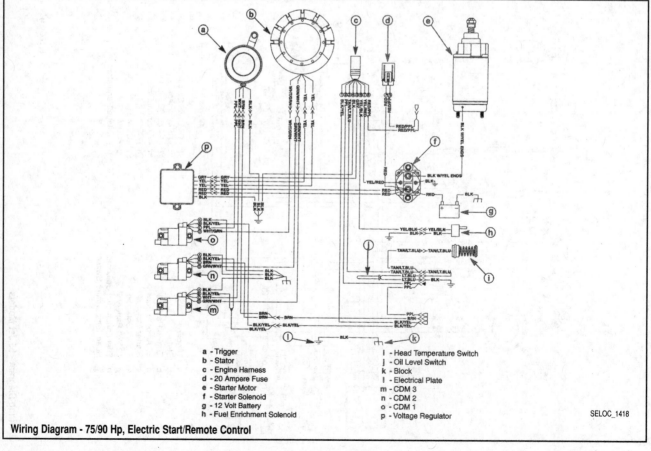

a - Trigger
b - Stator
c - Engine Harness
d - 20 Ampere Fuse
e - Starter Motor
f - Starter Solenoid
g - 12 Volt Battery
h - Fuel Enrichment Solenoid

i - Head Temperature Switch
j - Oil Level Switch
k - Block
l - Electrical Plate
m - CDM 3
n - CDM 2
o - CDM 1
p - Voltage Regulator

SELOC_1418

Wiring Diagram - 75/90 Hp, Electric Start/Remote Control

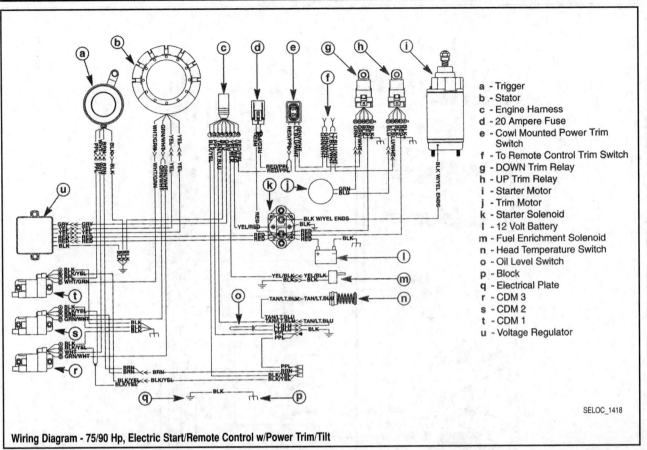

Wiring Diagram - 75/90 Hp, Electric Start/Remote Control w/Power Trim/Tilt

a - Trigger
b - Stator
c - Engine Harness
d - 20 Ampere Fuse
e - Cowl Mounted Power Trim Switch
f - To Remote Control Trim Switch
g - DOWN Trim Relay
h - UP Trim Relay
i - Starter Motor
j - Trim Motor
k - Starter Solenoid
l - 12 Volt Battery
m - Fuel Enrichment Solenoid
n - Head Temperature Switch
o - Oil Level Switch
p - Block
q - Electrical Plate
r - CDM 3
s - CDM 2
t - CDM 1
u - Voltage Regulator

SELOC_1418

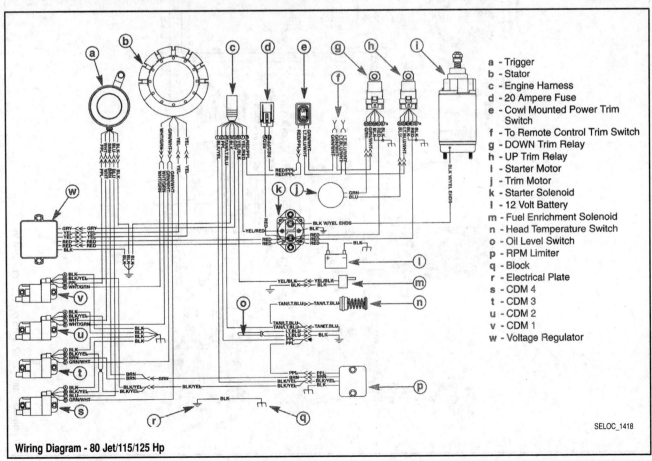

Wiring Diagram - 80 Jet/115/125 Hp

a - Trigger
b - Stator
c - Engine Harness
d - 20 Ampere Fuse
e - Cowl Mounted Power Trim Switch
f - To Remote Control Trim Switch
g - DOWN Trim Relay
h - UP Trim Relay
i - Starter Motor
j - Trim Motor
k - Starter Solenoid
l - 12 Volt Battery
m - Fuel Enrichment Solenoid
n - Head Temperature Switch
o - Oil Level Switch
p - RPM Limiter
q - Block
r - Electrical Plate
s - CDM 4
t - CDM 3
u - CDM 2
v - CDM 1
w - Voltage Regulator

SELOC_1418

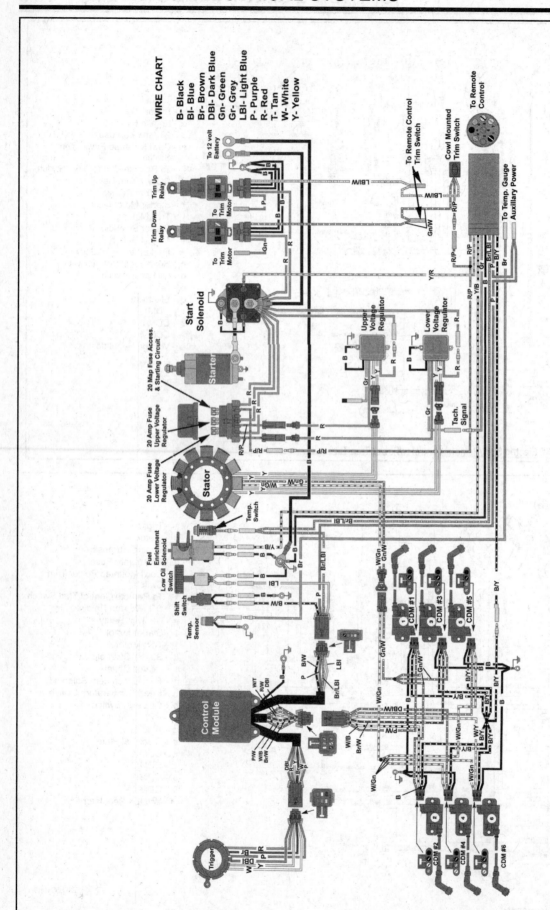

Wiring Diagram - 150/200 Hp (2.5L) Carbureted V6

SELOC_1418

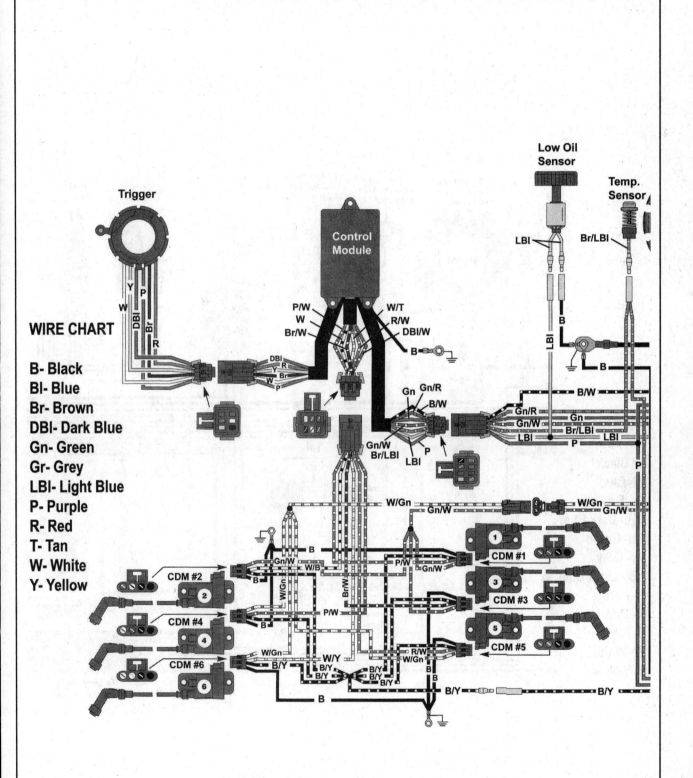

WIRE CHART

B- Black
Bl- Blue
Br- Brown
DBl- Dark Blue
Gn- Green
Gr- Grey
LBl- Light Blue
P- Purple
R- Red
T- Tan
W- White
Y- Yellow

SELOC_1418

Wiring Diagram - 2001 150/175 Hp (2.5L) V6 EFI Models (Left)

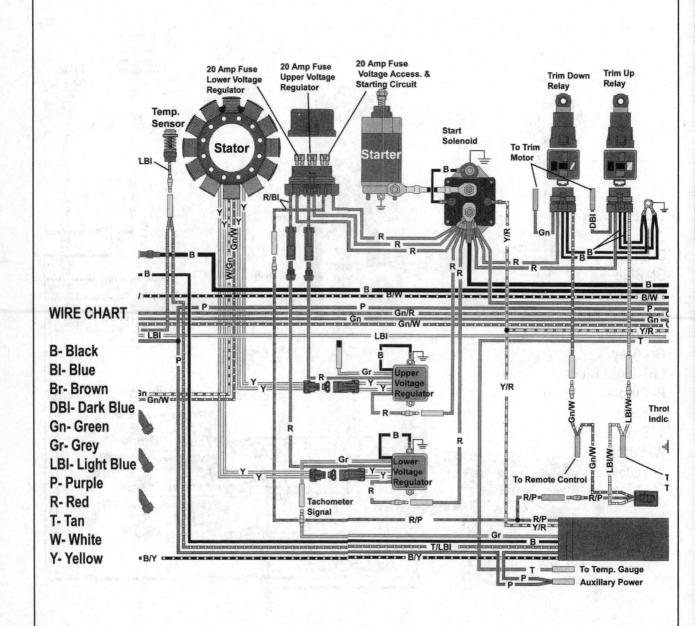

WIRE CHART

B- Black
Bl- Blue
Br- Brown
DBl- Dark Blue
Gn- Green
Gr- Grey
LBl- Light Blue
P- Purple
R- Red
T- Tan
W- White
Y- Yellow

SELOC_1418

Wiring Diagram - 2001 150/175 Hp (2.5L) V6 EFI Models (Center)

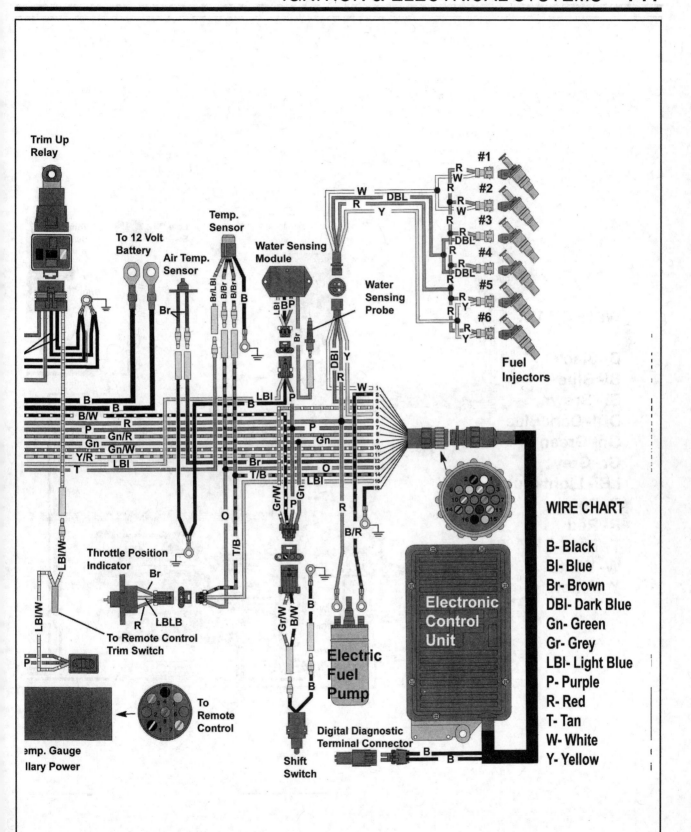

WIRE CHART

B- Black
Bl- Blue
Br- Brown
DBl- Dark Blue
Gn- Green
Gr- Grey
LBl- Light Blue
P- Purple
R- Red
T- Tan
W- White
Y- Yellow

Trim Up Relay

To 12 Volt Battery

Air Temp. Sensor

Temp. Sensor

Water Sensing Module

Water Sensing Probe

Fuel Injectors

#1 #2 #3 #4 #5 #6

Throttle Position Indicator

To Remote Control Trim Switch

To Remote Control

Temp. Gauge
llary Power

Electric Fuel Pump

Shift Switch

Digital Diagnostic Terminal Connector

Electronic Control Unit

SELOC_1418

Wiring Diagram - 2001 150/175 Hp (2.5L) V6 EFI Models (Right)

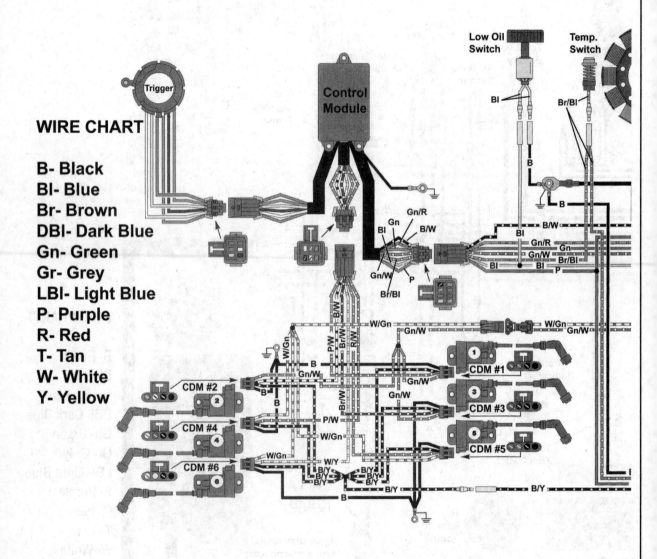

WIRE CHART

B- Black
Bl- Blue
Br- Brown
DBl- Dark Blue
Gn- Green
Gr- Grey
LBl- Light Blue
P- Purple
R- Red
T- Tan
W- White
Y- Yellow

SELOC_1418

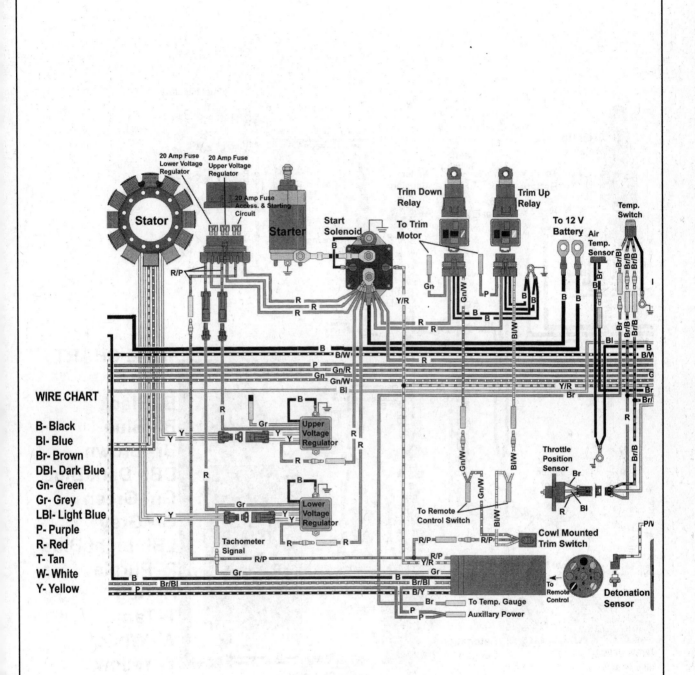

WIRE CHART

B- Black
Bl- Blue
Br- Brown
DBl- Dark Blue
Gn- Green
Gr- Grey
LBl- Light Blue
P- Purple
R- Red
T- Tan
W- White
Y- Yellow

SELOC_1418

Wiring Diagram - 2001 200 Hp (2.5L) V6 EFI Models (Center)

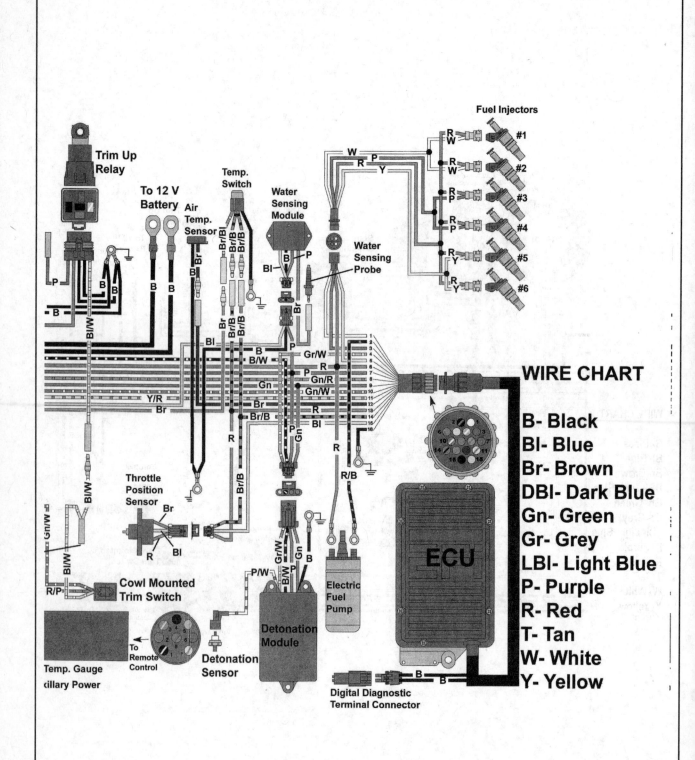

Fuel Injectors

WIRE CHART

B- Black
Bl- Blue
Br- Brown
DBl- Dark Blue
Gn- Green
Gr- Grey
LBl- Light Blue
P- Purple
R- Red
T- Tan
W- White
Y- Yellow

Trim Up Relay

To 12 V Battery

Air Temp. Sensor

Temp. Switch

Water Sensing Module

Water Sensing Probe

Throttle Position Sensor

Cowl Mounted Trim Switch

Temp. Gauge
Auxillary Power

To Remote Control

Detonation Sensor

Detonation Module

Electric Fuel Pump

ECU

Digital Diagnostic Terminal Connector

SELOC_1418

Wiring Diagram - 2001 200 Hp (2.5L) V6 EFI Models (Right)

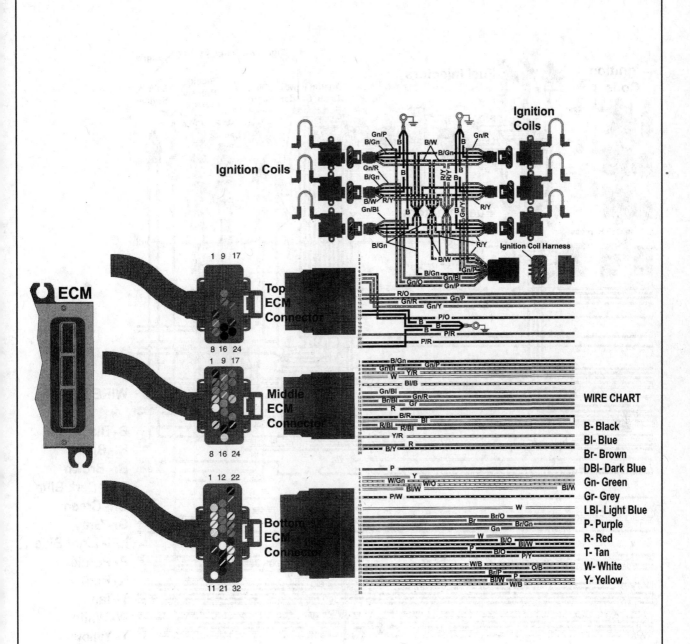

Wiring Diagram - 2002 or later 150-200 Hp (2.5L) V6 EFI Models (Left)

SELOC_1418

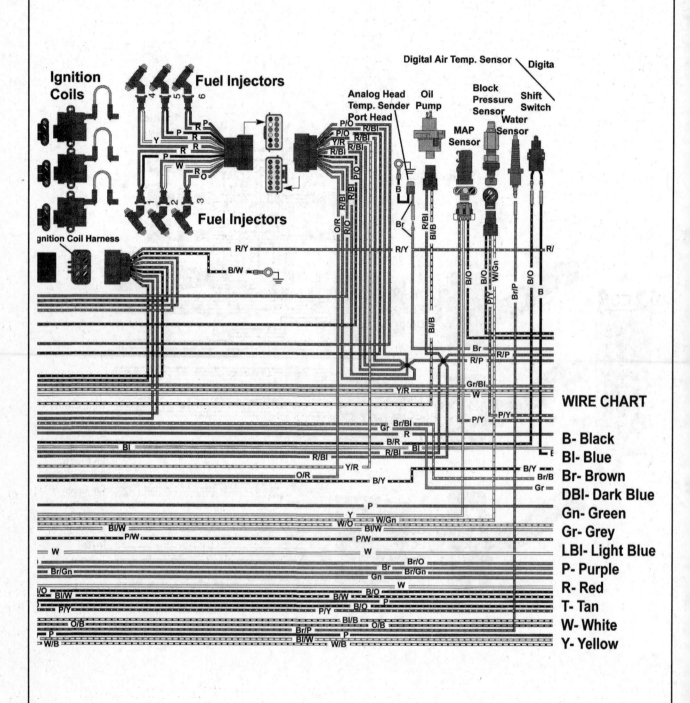

WIRE CHART

B- Black
Bl- Blue
Br- Brown
DBl- Dark Blue
Gn- Green
Gr- Grey
LBl- Light Blue
P- Purple
R- Red
T- Tan
W- White
Y- Yellow

SELOC_1418

Wiring Diagram - 2002 or later 150-200 Hp (2.5L) V6 EFI Models (Center, Left)

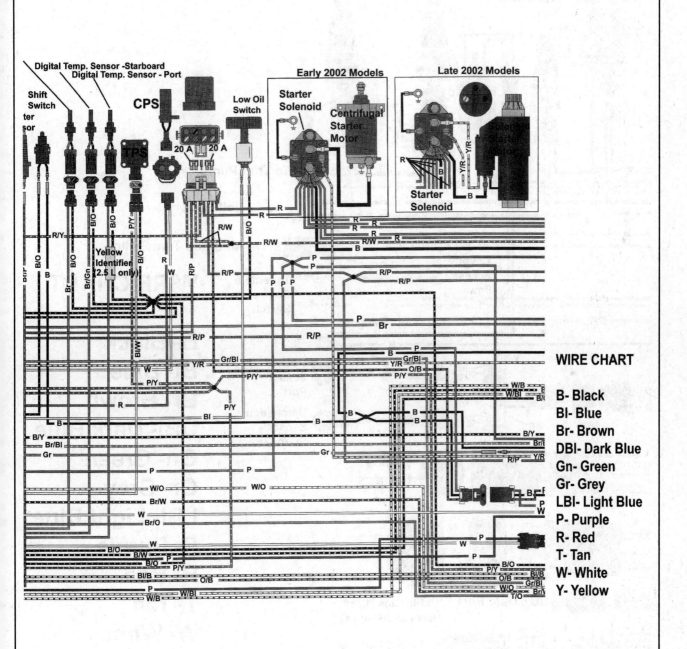

Digital Temp. Sensor -Starboard
Digital Temp. Sensor - Port

Shift
Switch
ter
sor

CPS

Low Oil
Switch

TPS

20 A   20 A

Yellow
Identifier
(2.5 L only)

Early 2002 Models

Starter
Solenoid

Centrifugal
Starter
Motor

Late 2002 Models

Starter
Solenoid

WIRE CHART

B- Black
Bl- Blue
Br- Brown
DBl- Dark Blue
Gn- Green
Gr- Grey
LBl- Light Blue
P- Purple
R- Red
T- Tan
W- White
Y- Yellow

SELOC_1418

Wiring Diagram - 2002 or later 150-200 Hp (2.5L) V6 EFI Models (Center, Right)

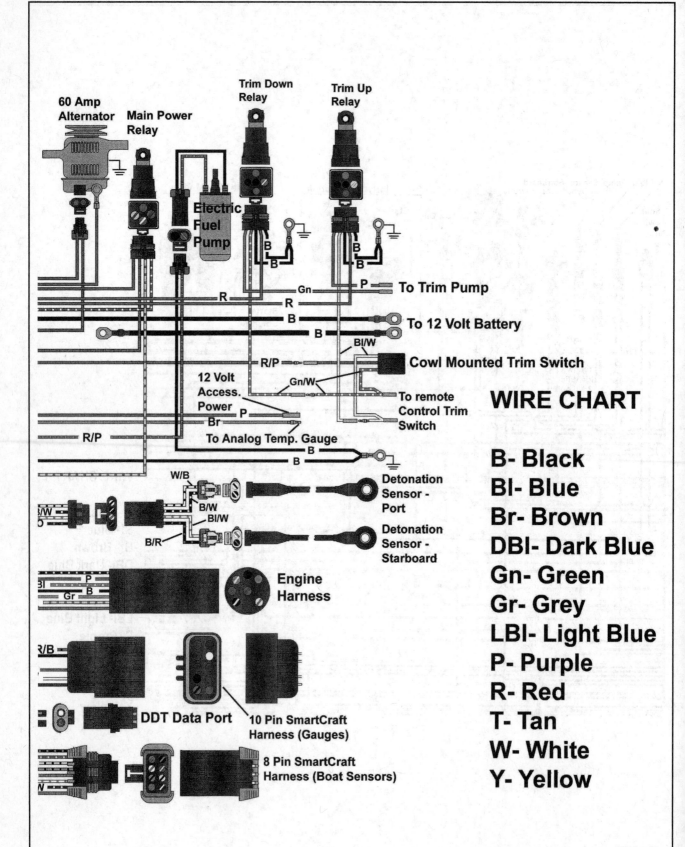

60 Amp Alternator

Main Power Relay

Trim Down Relay

Trim Up Relay

Electric Fuel Pump

B
B

B
B

P — To Trim Pump

Gn

R

R

B — To 12 Volt Battery

B

Bl/W

R/P — Cowl Mounted Trim Switch

12 Volt Access. Power

Gn/W

To remote Control Trim Switch

P

Br

R/P

To Analog Temp. Gauge

B

B

W/B

B/W

B/W

Bl/W

B/R

Detonation Sensor - Port

Detonation Sensor - Starboard

P

B

Gr

Engine Harness

R/B

DDT Data Port

10 Pin SmartCraft Harness (Gauges)

8 Pin SmartCraft Harness (Boat Sensors)

WIRE CHART

B- Black

Bl- Blue

Br- Brown

DBl- Dark Blue

Gn- Green

Gr- Grey

LBl- Light Blue

P- Purple

R- Red

T- Tan

W- White

Y- Yellow

SELOC_1418

Wiring Diagram - 2002 or later 150-200 Hp (2.5L) V6 EFI Models (Right)

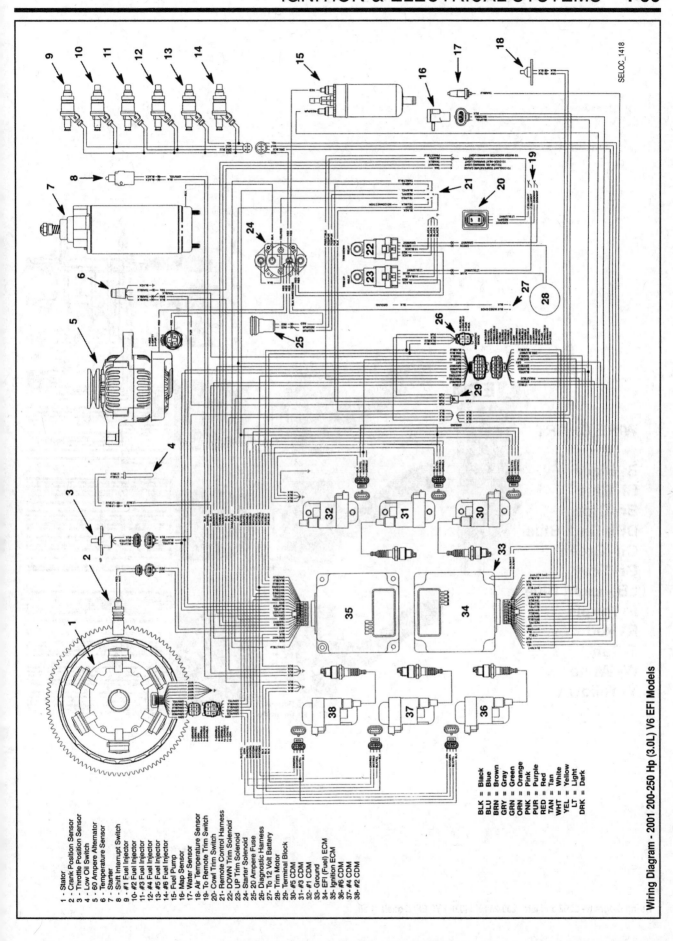

SELOC_1418

1 - Stator
2 - Crank Position Sensor
3 - Throttle Position Sensor
4 - Low Oil Switch
5 - 60 Ampere Alternator
6 - Temperature Sensor
7 - Starter
8 - Shift Interrupt Switch
9 - #1 Fuel Injector
10 - #2 Fuel Injector
11 - #3 Fuel Injector
12 - #4 Fuel Injector
13 - #5 Fuel Injector
14 - #6 Fuel Injector
15 - Fuel Pump
16 - Map Sensor
17 - Water Sensor
18 - Air Temperature Sensor
19 - To Remote Trim Switch
20 - Cowl Trim Switch
21 - Remote Control Harness
22 - DOWN Trim Solenoid
23 - UP Trim Solenoid
24 - Starter Solenoid
25 - 20 Ampere Fuse
26 - Diagnostic Harness
27 - To 12 Volt Battery
28 - Trim Motor
29 - Terminal Block
30 - #5 CDM
31 - #3 CDM
32 - #1 CDM
33 - Ground
34 - EFI (Fuel) ECM
35 - Ignition ECM
36 - #6 CDM
37 - #4 CDM
38 - #2 CDM

BLK = Black
BLU = Blue
BRN = Brown
GRY = Gray
GRN = Green
ORN = Orange
PNK = Pink
PUR = Purple
RED = Red
TAN = Tan
WHT = White
YEL = Yellow
LT = Light
DRK = Dark

Wiring Diagram - 2001 200-250 Hp (3.0L) V6 EFI Models

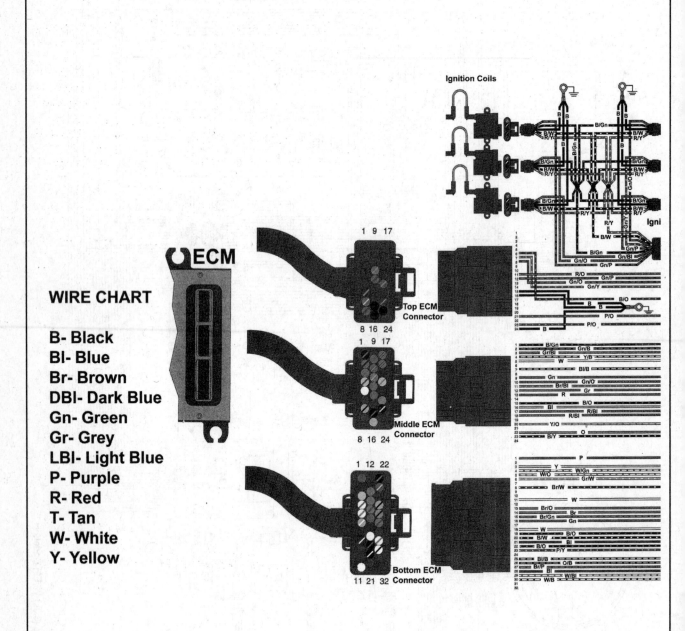

WIRE CHART

B- Black
Bl- Blue
Br- Brown
DBl- Dark Blue
Gn- Green
Gr- Grey
LBl- Light Blue
P- Purple
R- Red
T- Tan
W- White
Y- Yellow

SELOC_1418

Wiring Diagram - 2002 and later 200-250 Hp (3.0L) V6 EFI Models (Left)

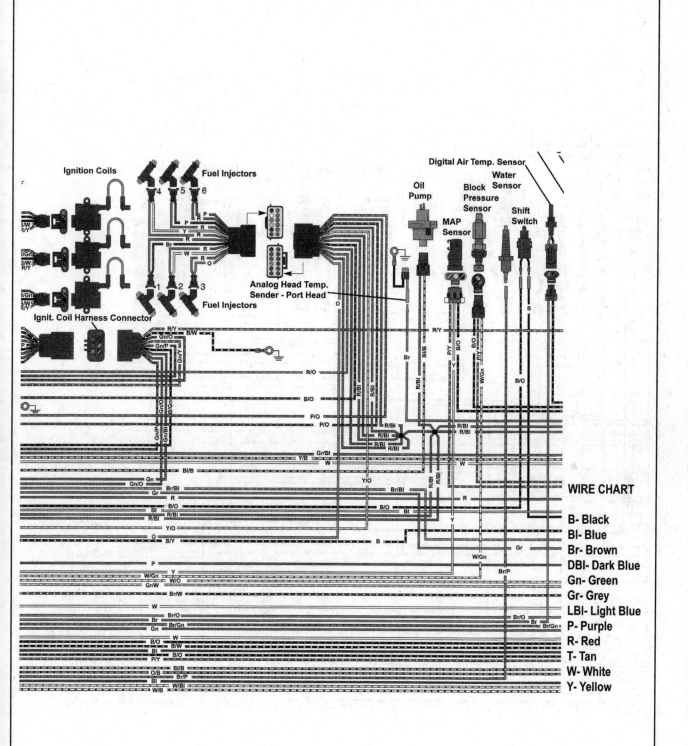

WIRE CHART

B- Black
Bl- Blue
Br- Brown
DBl- Dark Blue
Gn- Green
Gr- Grey
LBl- Light Blue
P- Purple
R- Red
T- Tan
W- White
Y- Yellow

SELOC_1418

Wiring Diagram - 2002 and later 200-250 Hp (3.0L) V6 EFI Models (Center, Left)

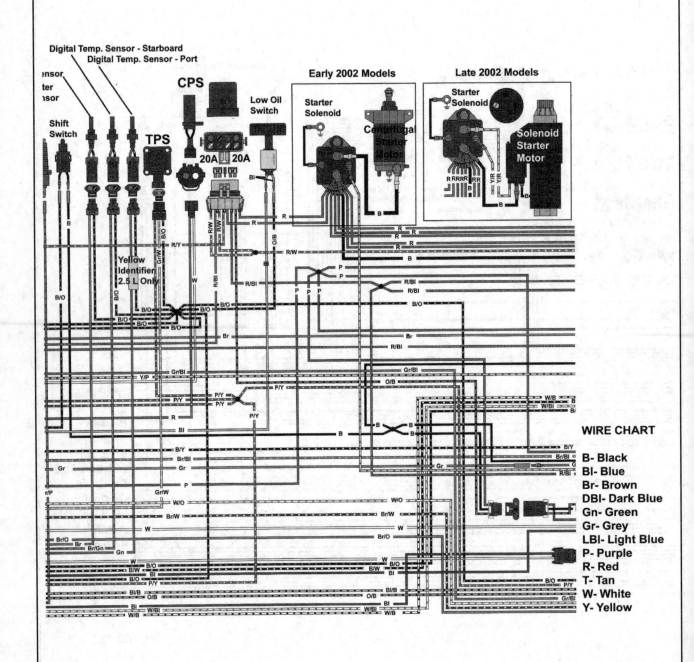

WIRE CHART

B- Black
Bl- Blue
Br- Brown
DBl- Dark Blue
Gn- Green
Gr- Grey
LBl- Light Blue
P- Purple
R- Red
T- Tan
W- White
Y- Yellow

SELOC_1418

Wiring Diagram - 2002 and later 200-250 Hp (3.0L) V6 EFI Models (Center, Right)

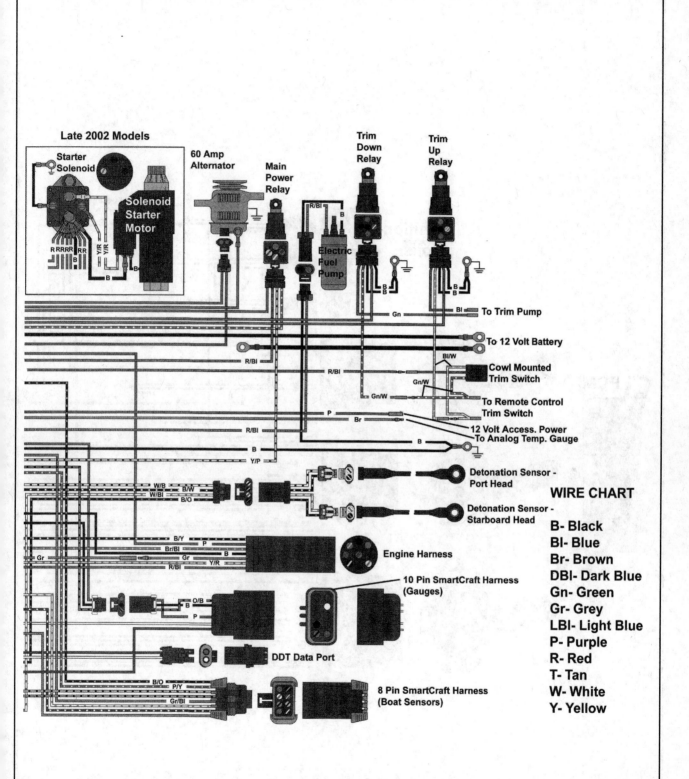

Late 2002 Models

Starter Solenoid

Solenoid Starter Motor

R R R R R R   Y/R   Y/R

B   B

60 Amp Alternator

Main Power Relay

Trim Down Relay

Trim Up Relay

R/Bl   B

Electric Fuel Pump

B   B   B   B

Gn   Bl   To Trim Pump

To 12 Volt Battery

R/Bl

R/Bl

Bl/W

Cowl Mounted Trim Switch

Gn/W

Gn/W

To Remote Control Trim Switch

P   Br   12 Volt Access. Power

R/Bl   To Analog Temp. Gauge

B   B

Y/P

Detonation Sensor - Port Head

W/B   B/W

W/Bl   B/O

Detonation Sensor - Starboard Head

B/Y   P

Br/Bl   B

Gr   Gr   Y/R

R/Bl

Engine Harness

10 Pin SmartCraft Harness (Gauges)

O/B

P

DDT Data Port

B/O

P/Y

Gr/Bl

8 Pin SmartCraft Harness (Boat Sensors)

WIRE CHART

B- Black
Bl- Blue
Br- Brown
DBl- Dark Blue
Gn- Green
Gr- Grey
LBl- Light Blue
P- Purple
R- Red
T- Tan
W- White
Y- Yellow

SELOC_1418

Wiring Diagram - 2002 and later 200-250 Hp (3.0L) V6 EFI Models (Right)

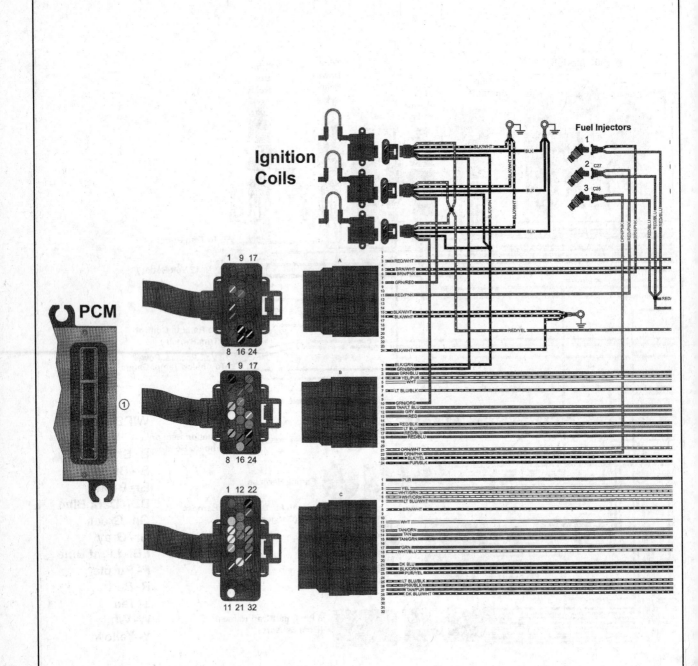

SELOC_1418

Wiring Diagram - 2001-05 75-115 Hp (1.5L) OptiMax (Left)

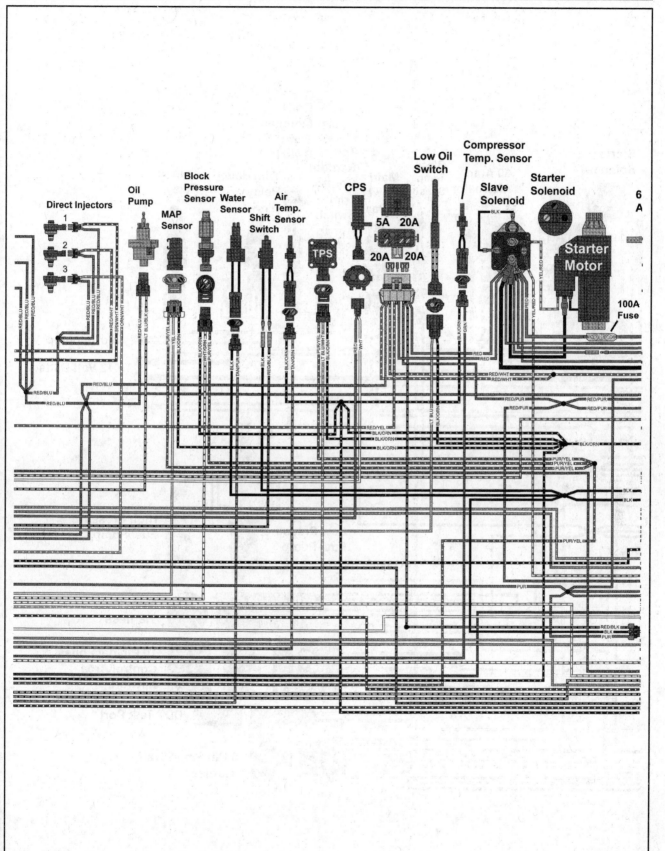

Wiring Diagram - 2001-05 75-115 Hp (1.5L) OptiMax (Center)

SELOC_1418

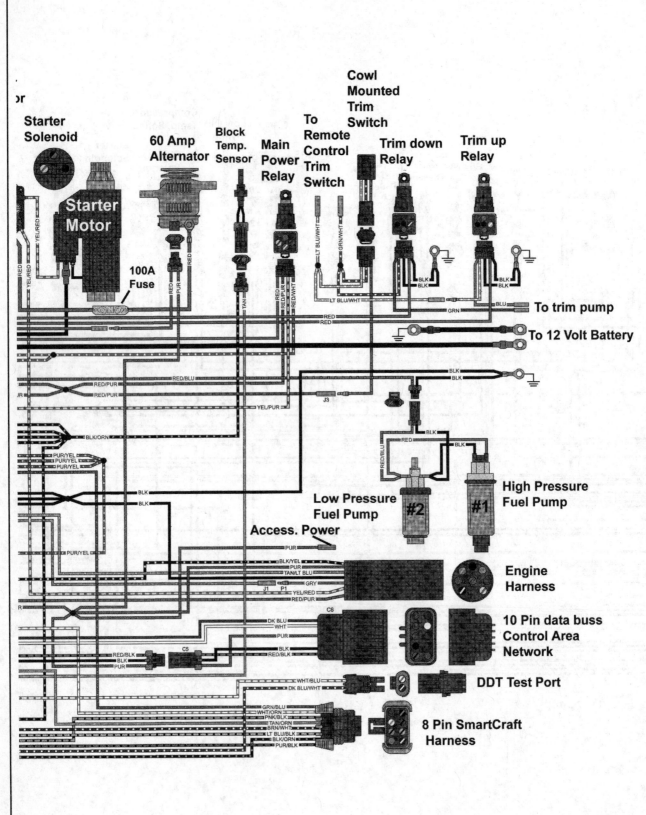

Wiring Diagram - 2001-05 75-115 Hp (1.5L) OptiMax (Right)

SELOC_1418

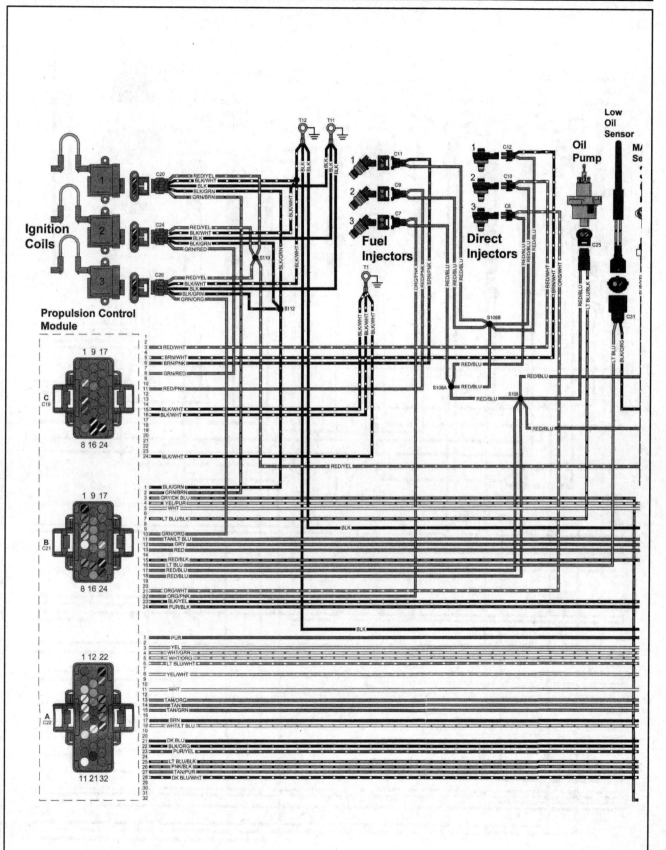

Wiring Diagram - Early 2006 75-115 Hp (1.5L) OptiMax (Left)

SELOC_1418

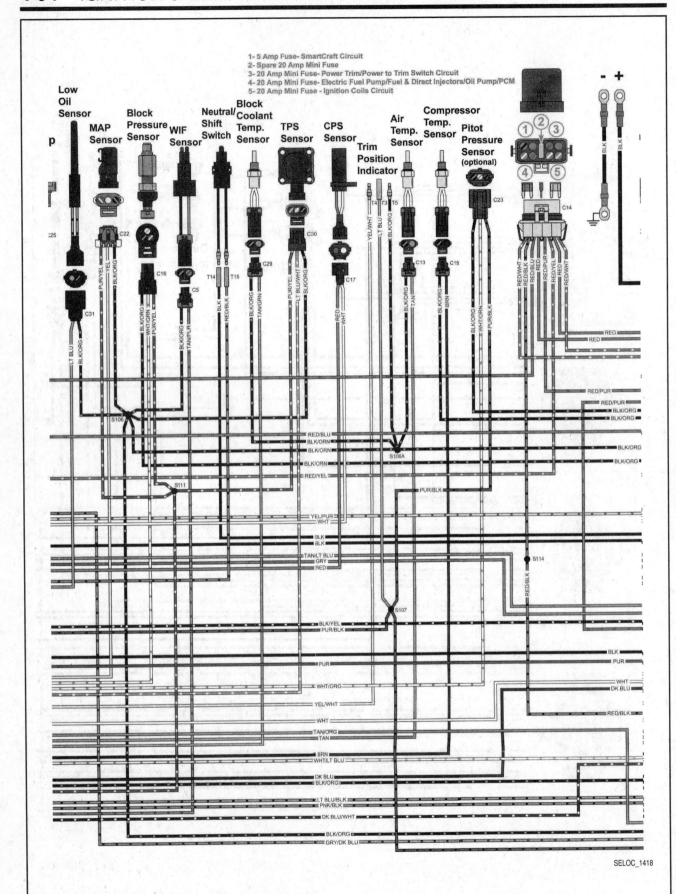

1- 5 Amp Fuse- SmartCraft Circuit
2- Spare 20 Amp Mini Fuse
3- 20 Amp Mini Fuse- Power Trim/Power to Trim Switch Circuit
4- 20 Amp Mini Fuse- Electric Fuel Pump/Fuel & Direct Injectors/Oil Pump/PCM
5- 20 Amp Mini Fuse - Ignition Coils Circuit

Wiring Diagram - Early 2006 75-115 Hp (1.5L) OptiMax (Center)

SELOC_1418

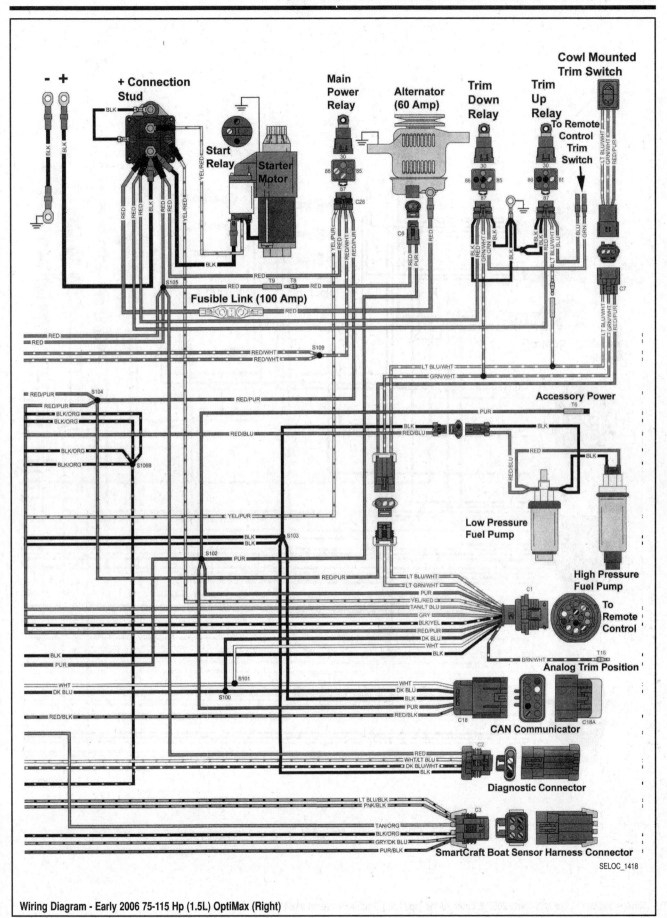

Wiring Diagram - Early 2006 75-115 Hp (1.5L) OptiMax (Right)

SELOC_1418

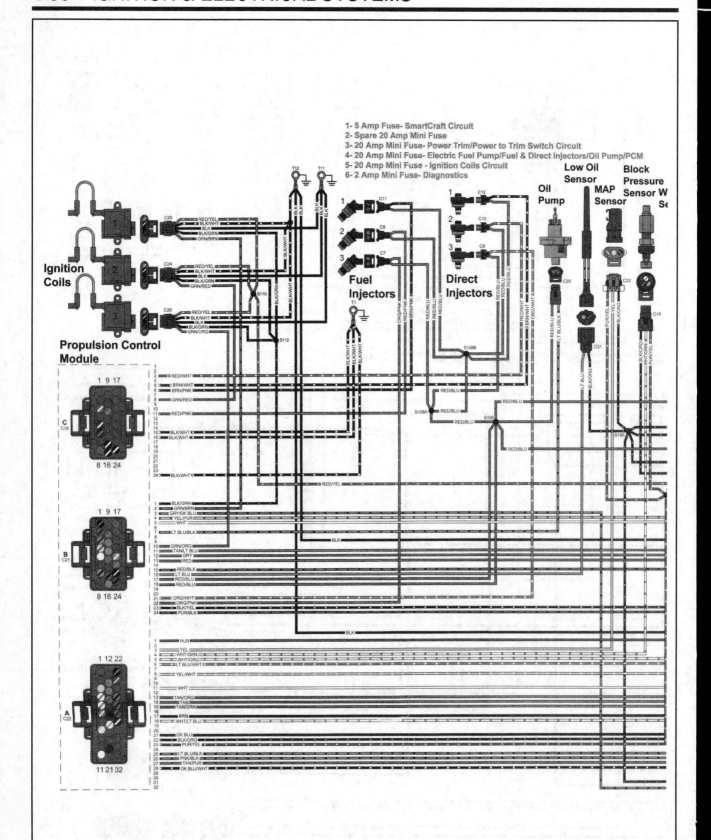

1- 5 Amp Fuse- SmartCraft Circuit
2- Spare 20 Amp Mini Fuse
3- 20 Amp Mini Fuse- Power Trim/Power to Trim Switch Circuit
4- 20 Amp Mini Fuse- Electric Fuel Pump/Fuel & Direct Injectors/Oil Pump/PCM
5- 20 Amp Mini Fuse - Ignition Coils Circuit
6- 2 Amp Mini Fuse- Diagnostics

Wiring Diagram - Late 2006/Early 2007 & Later 75-115 Hp (1.5L) OptiMax w/Serial #1B417702 & Above (Left)

SELOC_1418

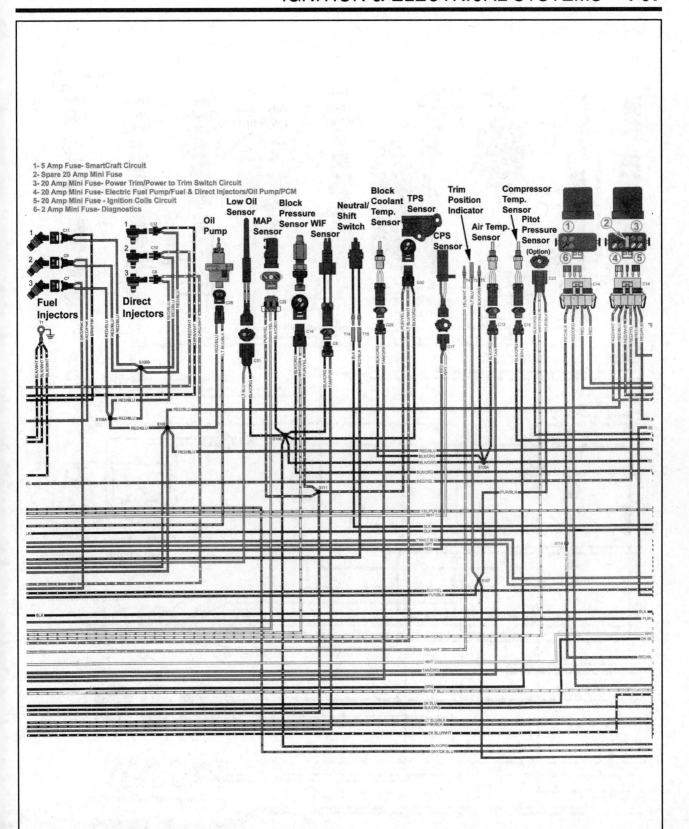

1- 5 Amp Fuse- SmartCraft Circuit
2- Spare 20 Amp Fuse
3- 20 Amp Mini Fuse- Power Trim/Power to Trim Switch Circuit
4- 20 Amp Mini Fuse- Electric Fuel Pump/Fuel & Direct Injectors/Oil Pump/PCM
5- 20 Amp Mini Fuse - Ignition Coils Circuit
6- 2 Amp Mini Fuse- Diagnostics

Wiring Diagram - Late 2006/Early 2007 & Later 75-115 Hp (1.5L) OptiMax w/Serial #1B417702 & Above (Center)

SELOC_1418

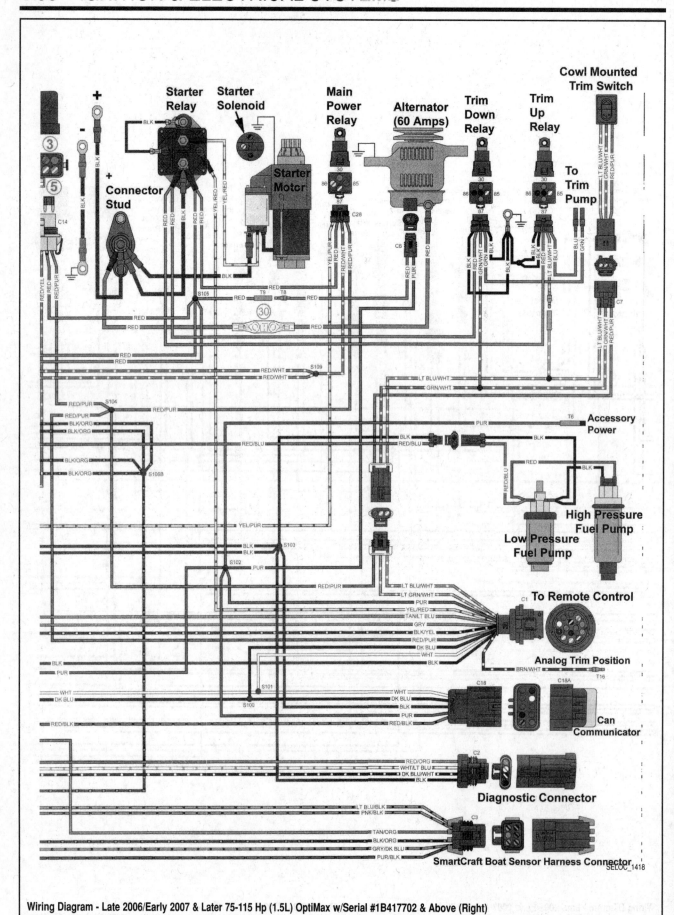

Wiring Diagram - Late 2006/Early 2007 & Later 75-115 Hp (1.5L) OptiMax w/Serial #1B417702 & Above (Right)

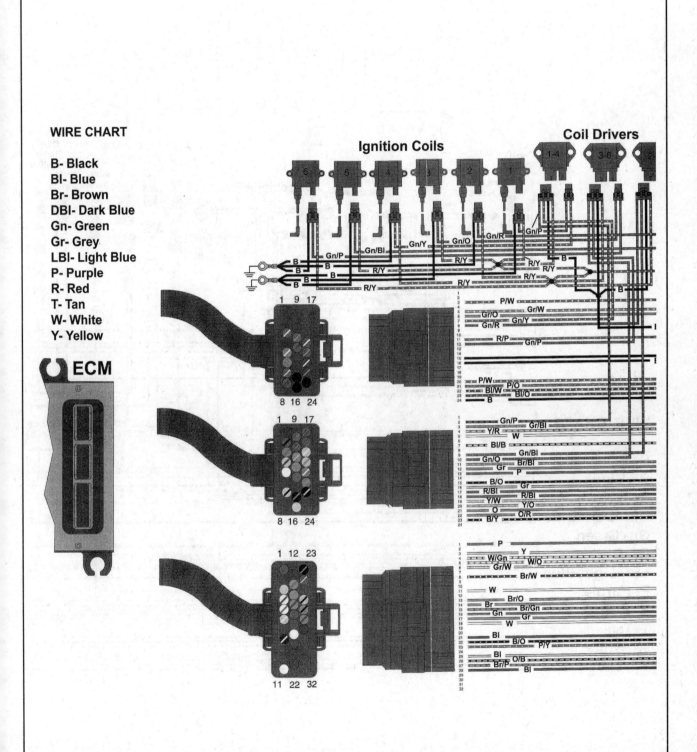

WIRE CHART

B- Black
Bl- Blue
Br- Brown
DBl- Dark Blue
Gn- Green
Gr- Grey
LBl- Light Blue
P- Purple
R- Red
T- Tan
W- White
Y- Yellow

ECM

Ignition Coils

Coil Drivers

SELOC_1418

Wiring Diagram - 135-175 Hp (2.5L) OptiMax (Left)

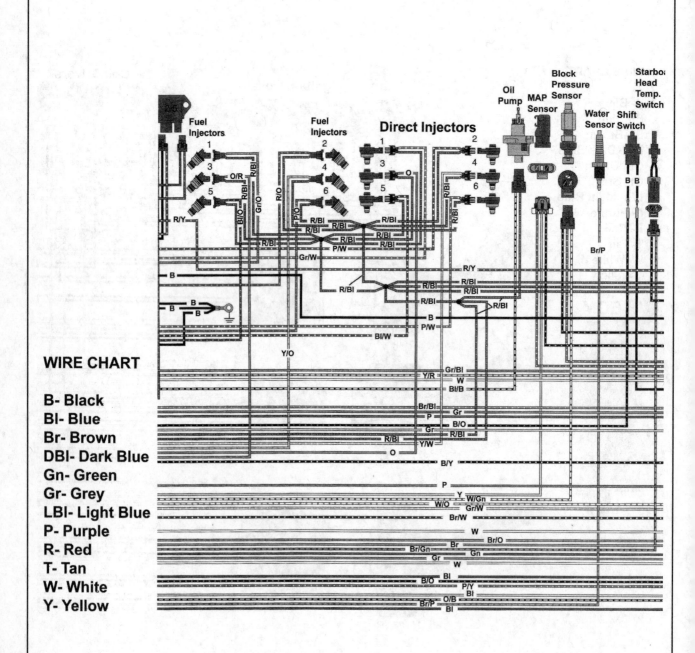

WIRE CHART

B- Black
Bl- Blue
Br- Brown
DBl- Dark Blue
Gn- Green
Gr- Grey
LBl- Light Blue
P- Purple
R- Red
T- Tan
W- White
Y- Yellow

SELOC_1418

Wiring Diagram - 135-175 Hp (2.5L) OptiMax (Center, Left)

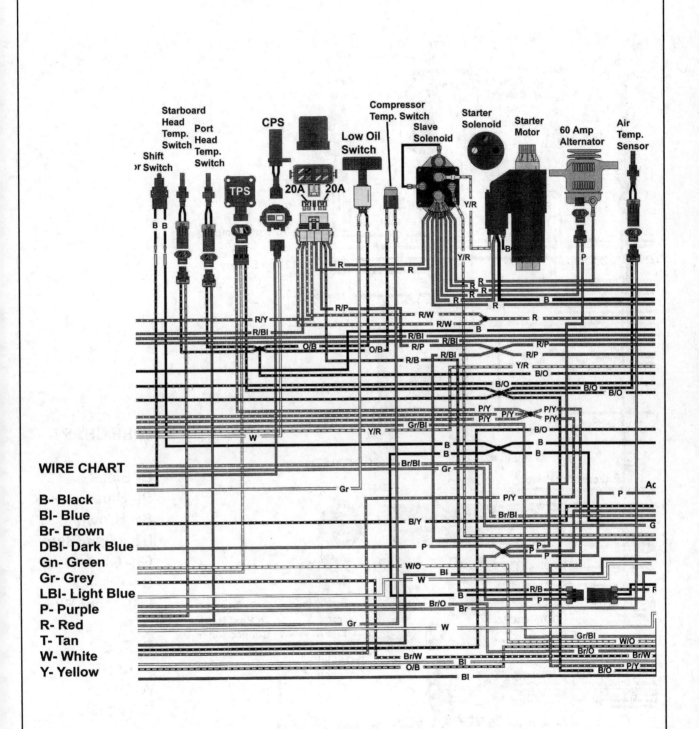

WIRE CHART

B- Black
Bl- Blue
Br- Brown
DBl- Dark Blue
Gn- Green
Gr- Grey
LBl- Light Blue
P- Purple
R- Red
T- Tan
W- White
Y- Yellow

SELOC_1418

Wiring Diagram - 135-175 Hp (2.5L) OptiMax (Center, Right)

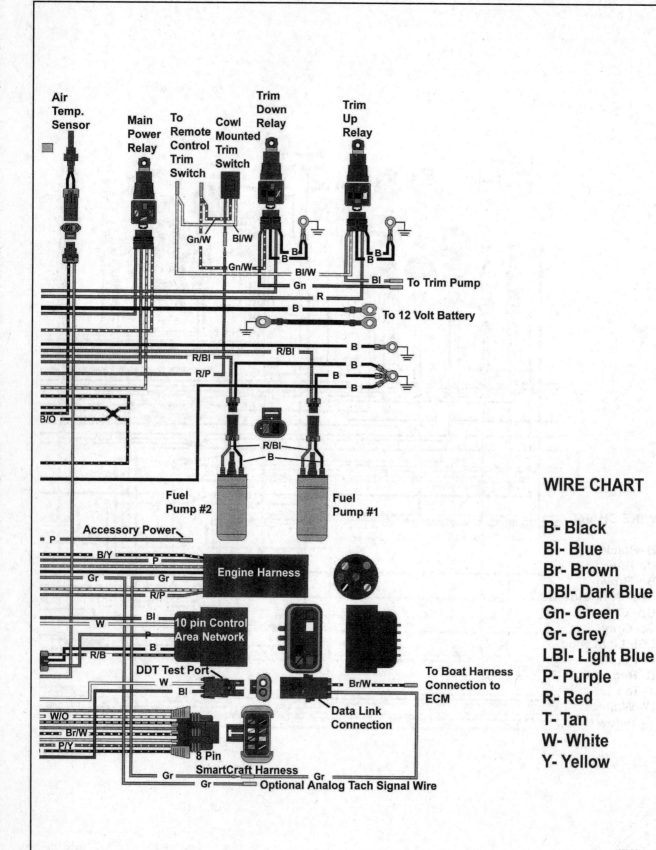

WIRE CHART

B- Black
Bl- Blue
Br- Brown
DBl- Dark Blue
Gn- Green
Gr- Grey
LBl- Light Blue
P- Purple
R- Red
T- Tan
W- White
Y- Yellow

SELOC_1418

Wiring Diagram - 135-175 Hp (2.5L) OptiMax (Right)

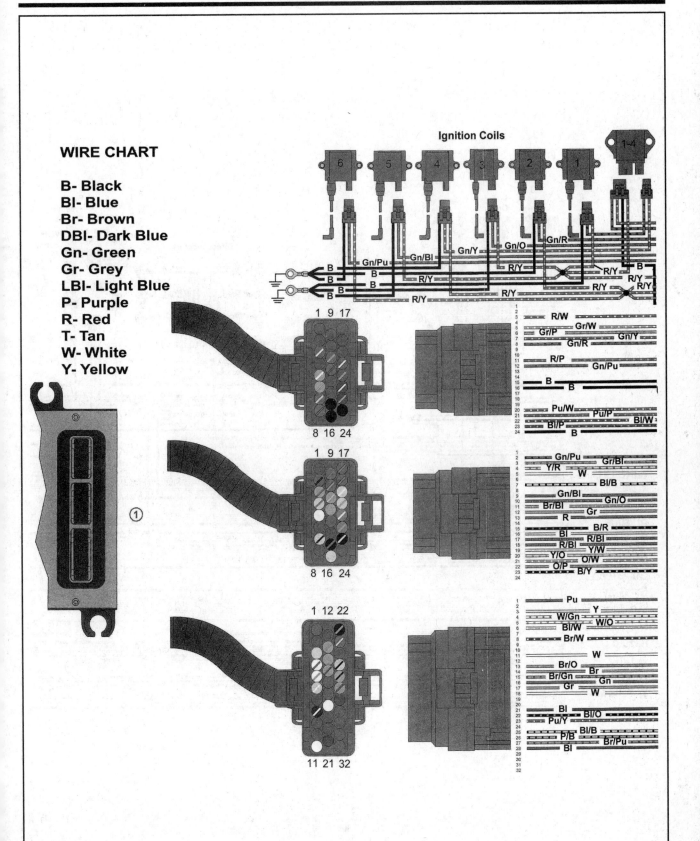

WIRE CHART

B- Black
Bl- Blue
Br- Brown
DBl- Dark Blue
Gn- Green
Gr- Grey
LBl- Light Blue
P- Purple
R- Red
T- Tan
W- White
Y- Yellow

Ignition Coils

SELOC_1418

Wiring Diagram - 2001-02 200/225 Hp (3.0L) OptiMax (Left)

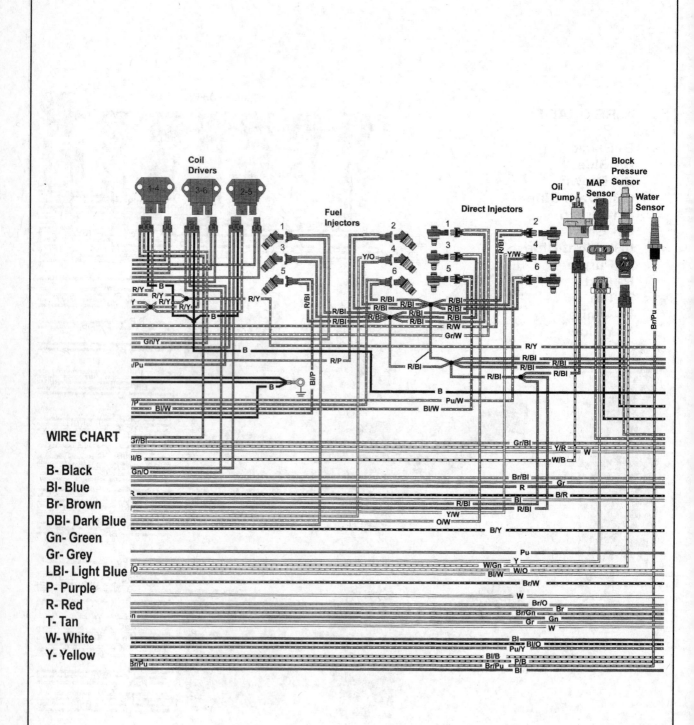

WIRE CHART

B- Black
Bl- Blue
Br- Brown
DBl- Dark Blue
Gn- Green
Gr- Grey
LBl- Light Blue
P- Purple
R- Red
T- Tan
W- White
Y- Yellow

SELOC_1418

Wiring Diagram - 2001-02 200/225 Hp (3.0L) OptiMax (Center, Left)

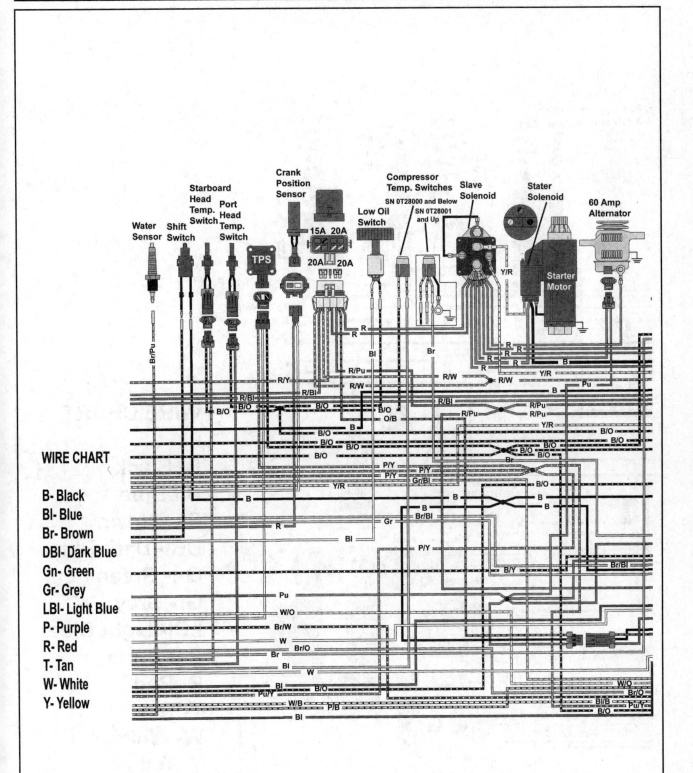

WIRE CHART

B- Black
Bl- Blue
Br- Brown
DBl- Dark Blue
Gn- Green
Gr- Grey
LBl- Light Blue
P- Purple
R- Red
T- Tan
W- White
Y- Yellow

SELOC_1418

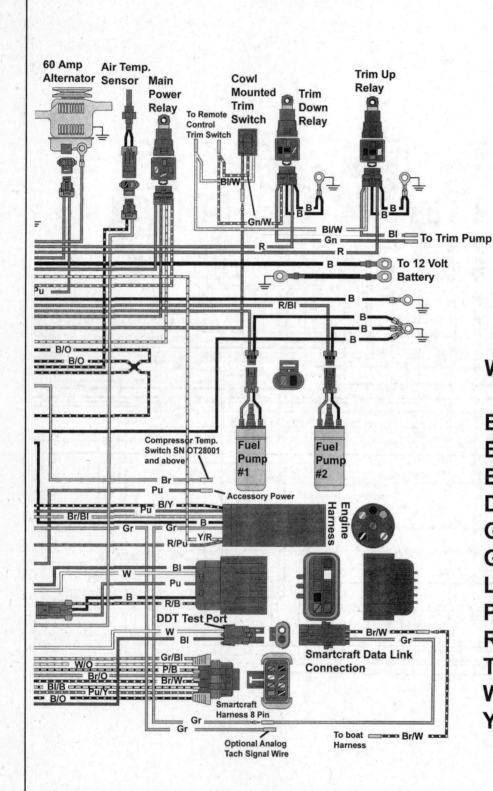

WIRE CHART

B- Black
Bl- Blue
Br- Brown
DBl- Dark Blue
Gn- Green
Gr- Grey
LBl- Light Blue
P- Purple
R- Red
T- Tan
W- White
Y- Yellow

SELOC_1418

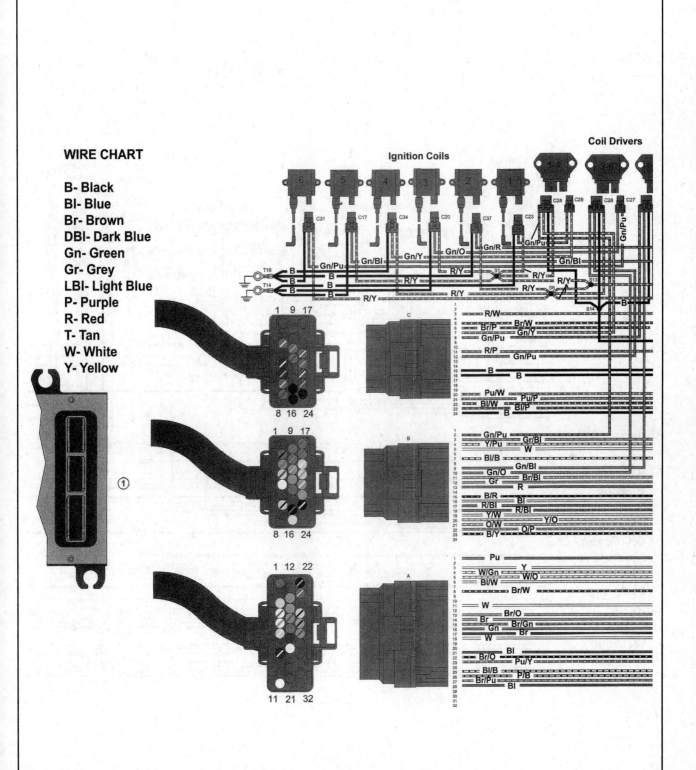

WIRE CHART

- **B-** Black
- **Bl-** Blue
- **Br-** Brown
- **DBl-** Dark Blue
- **Gn-** Green
- **Gr-** Grey
- **LBl-** Light Blue
- **P-** Purple
- **R-** Red
- **T-** Tan
- **W-** White
- **Y-** Yellow

Ignition Coils

Coil Drivers

SELOC_1418

Wiring Diagram - 2003 & later 200-250 Hp (3.0L) OptiMax, Mechanical Shift (Left)

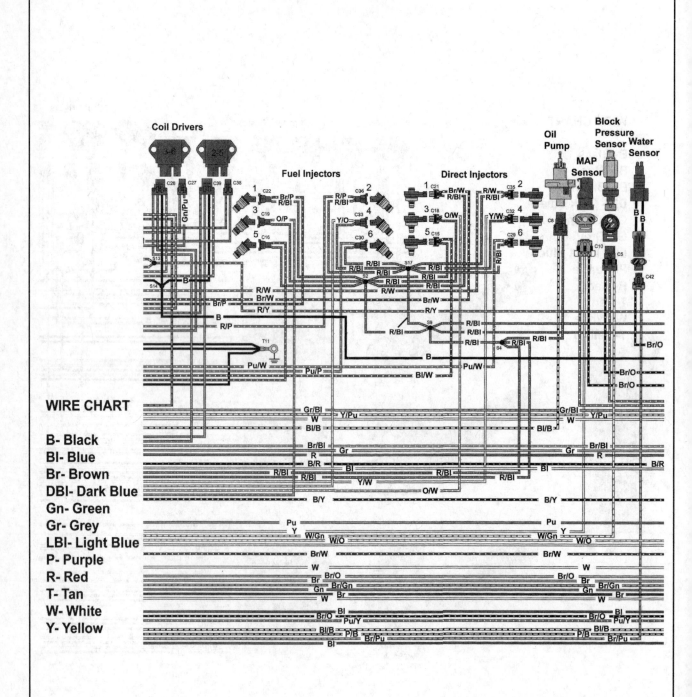

WIRE CHART

B- Black
Bl- Blue
Br- Brown
DBl- Dark Blue
Gn- Green
Gr- Grey
LBl- Light Blue
P- Purple
R- Red
T- Tan
W- White
Y- Yellow

SELOC_1418

Wiring Diagram - 2003 & later 200-250 Hp (3.0L) OptiMax, Mechanical Shift (Center, Left)

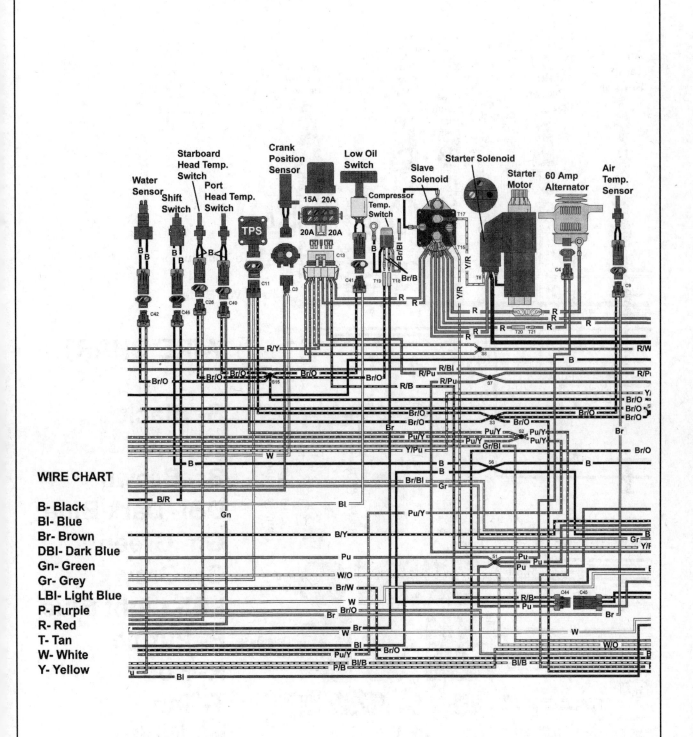

WIRE CHART

B- Black
Bl- Blue
Br- Brown
DBl- Dark Blue
Gn- Green
Gr- Grey
LBl- Light Blue
P- Purple
R- Red
T- Tan
W- White
Y- Yellow

SELOC_1418

Wiring Diagram - 2003 & later 200-250 Hp (3.0L) OptiMax, Mechanical Shift (Center, Right)

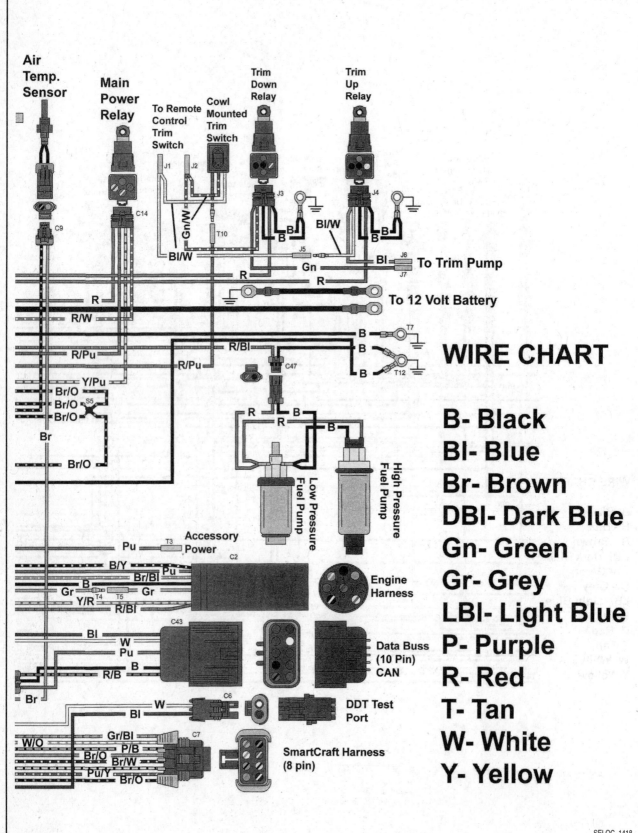

WIRE CHART

B- Black

Bl- Blue

Br- Brown

DBl- Dark Blue

Gn- Green

Gr- Grey

LBl- Light Blue

P- Purple

R- Red

T- Tan

W- White

Y- Yellow

SELOC_1418

Wiring Diagram - 2003 & later 200-250 Hp (3.0L) OptiMax, Mechanical Shift (Right)

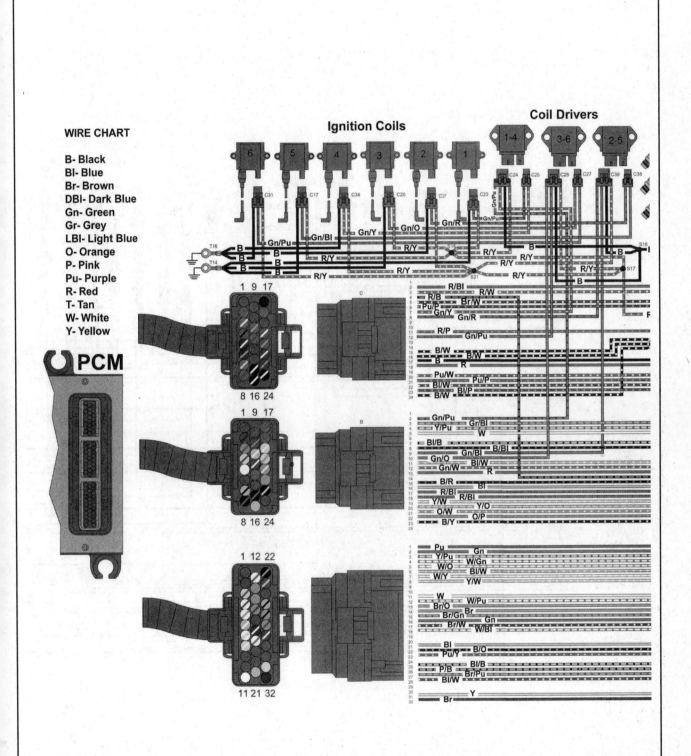

Wiring Diagram - 2003 & later 200-250 Hp (3.0L) OptiMax, Digital Throttle & Shift (Left)

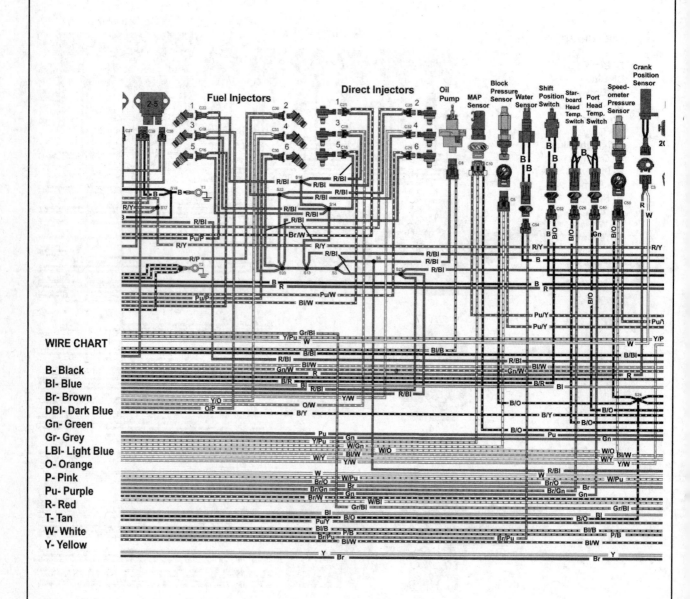

WIRE CHART

B- Black
Bl- Blue
Br- Brown
DBl- Dark Blue
Gn- Green
Gr- Grey
LBl- Light Blue
O- Orange
P- Pink
Pu- Purple
R- Red
T- Tan
W- White
Y- Yellow

SELOC_1418

Wiring Diagram - 2003 & later 200-250 Hp (3.0L) OptiMax, Digital Throttle & Shift (Center, Left)

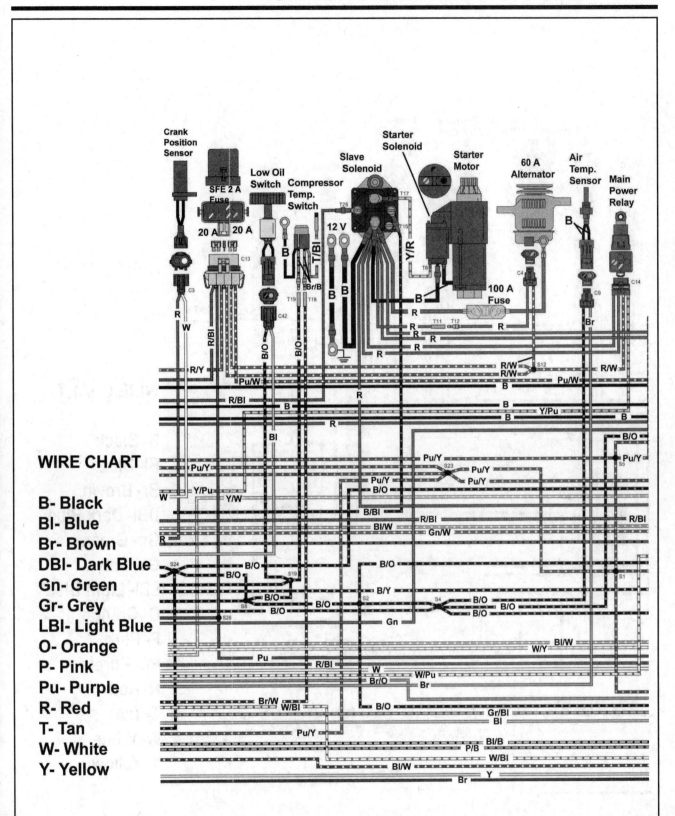

Crank
Position
Sensor

SFE 2 A
Fuse

Low Oil
Switch

Compressor
Temp.
Switch

Slave
Solenoid

Starter
Solenoid

Starter
Motor

60 A
Alternator

Air
Temp.
Sensor

Main
Power
Relay

20 A   20 A

12 V

100 A
Fuse

WIRE CHART

B- Black

Bl- Blue

Br- Brown

DBl- Dark Blue

Gn- Green

Gr- Grey

LBl- Light Blue

O- Orange

P- Pink

Pu- Purple

R- Red

T- Tan

W- White

Y- Yellow

SELOC_1418

Wiring Diagram - 2003 & later 200-250 Hp (3.0L) OptiMax, Digital Throttle & Shift (Center, Right)

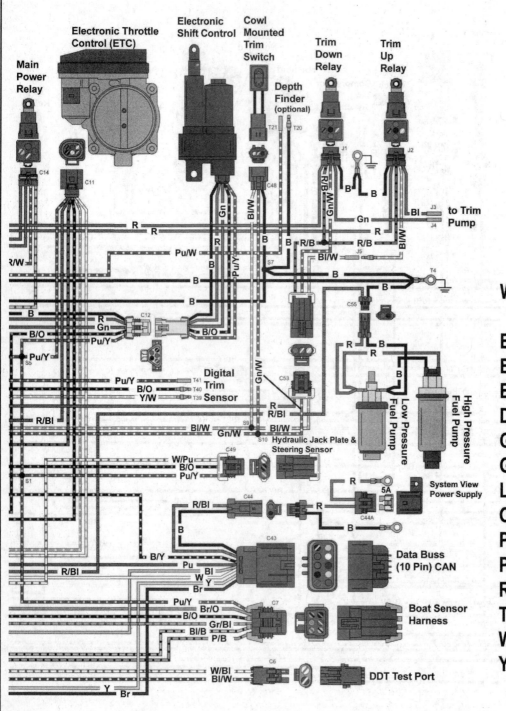

WIRE CHART

B- Black

Bl- Blue

Br- Brown

DBl- Dark Blue

Gn- Green

Gr- Grey

LBl- Light Blue

O- Orange

P- Pink

Pu- Purple

R- Red

T- Tan

W- White

Y- Yellow

SELOC_1418

Wiring Diagram - 2003 & later 200-250 Hp (3.0L) OptiMax, Digital Throttle & Shift (Right)

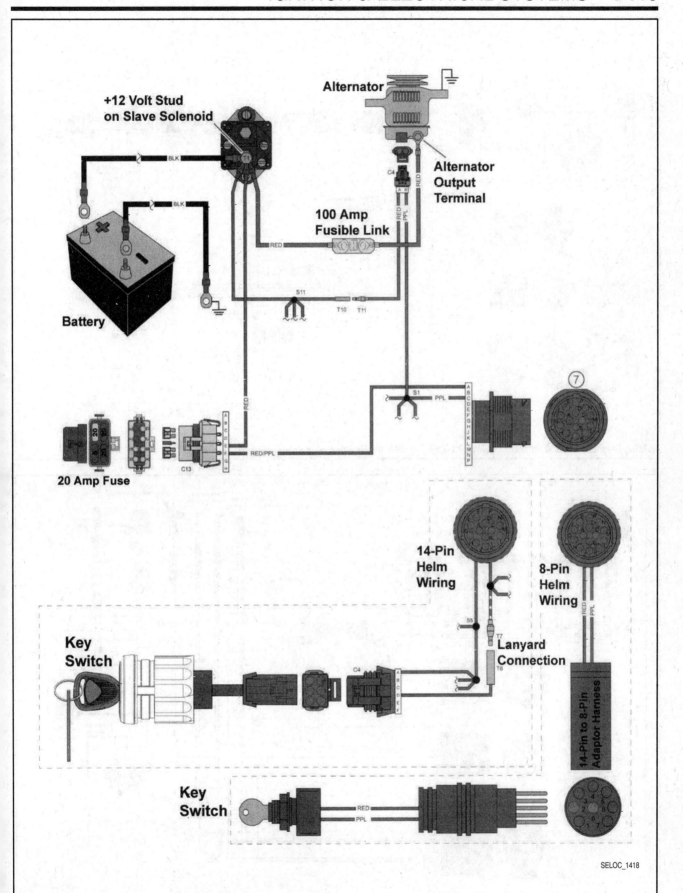

Wiring Diagram - 225/250 Hp (3.0L) OptiMax Pro/XS/Sport, Charging Systems

SELOC_1418

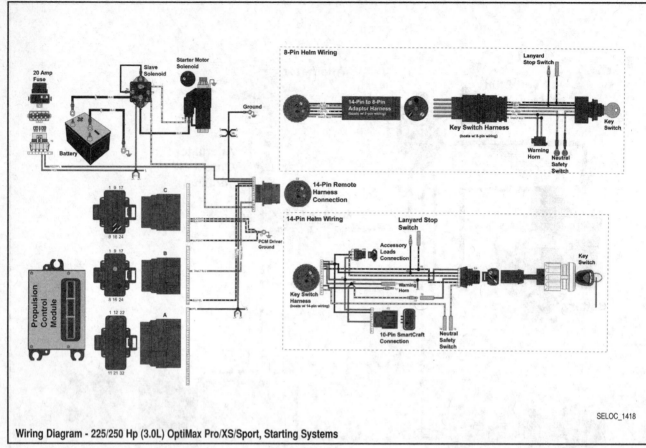

Wiring Diagram - 225/250 Hp (3.0L) OptiMax Pro/XS/Sport, Starting Systems

SELOC_1418

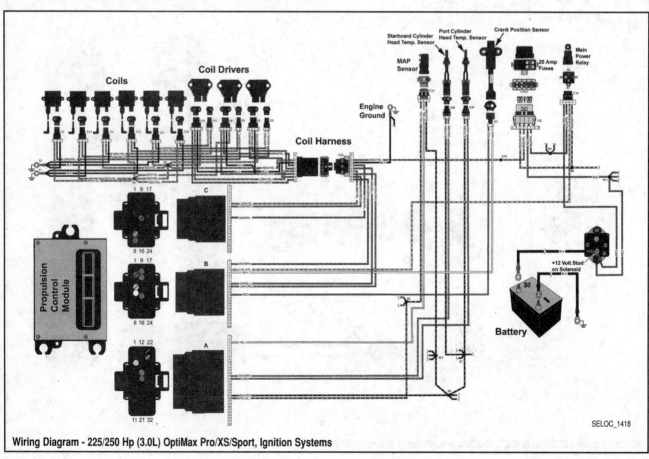

Wiring Diagram - 225/250 Hp (3.0L) OptiMax Pro/XS/Sport, Ignition Systems

SELOC_1418

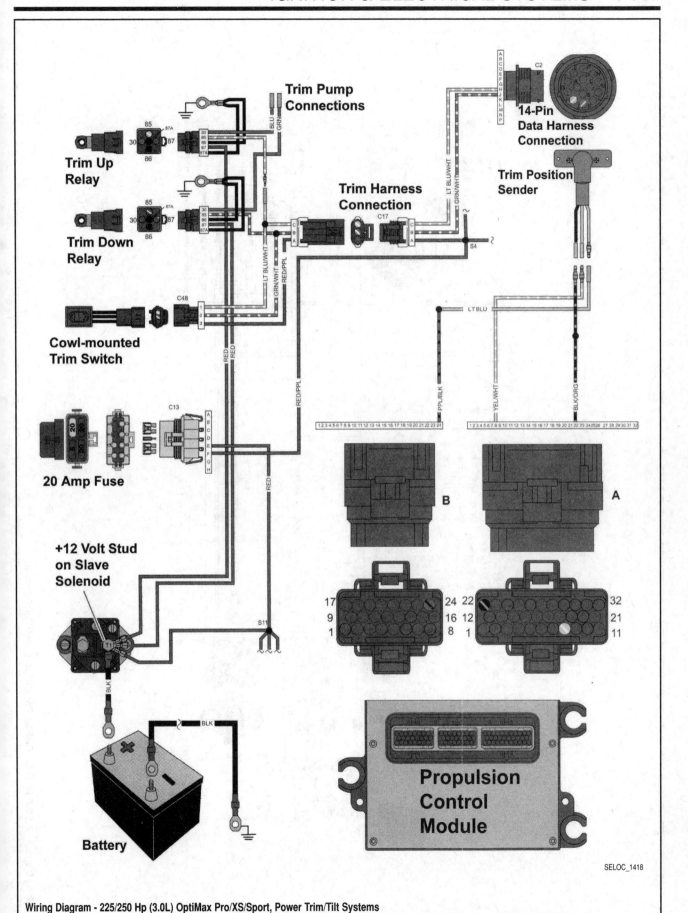

Wiring Diagram - 225/250 Hp (3.0L) OptiMax Pro/XS/Sport, Power Trim/Tilt Systems

SELOC_1418

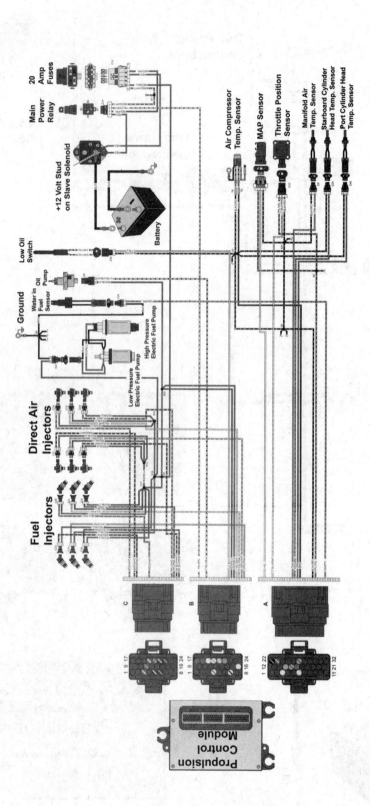

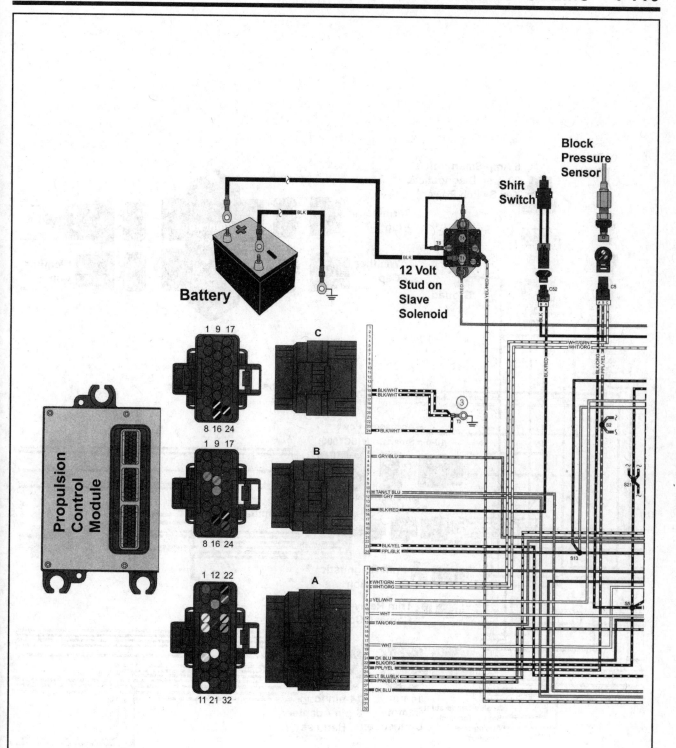

Wiring Diagram - 225/250 Hp (3.0L) OptiMax Pro/XS/Sport, 8-Pin Control Harness & Keyswitch (Left)

SELOC_1418

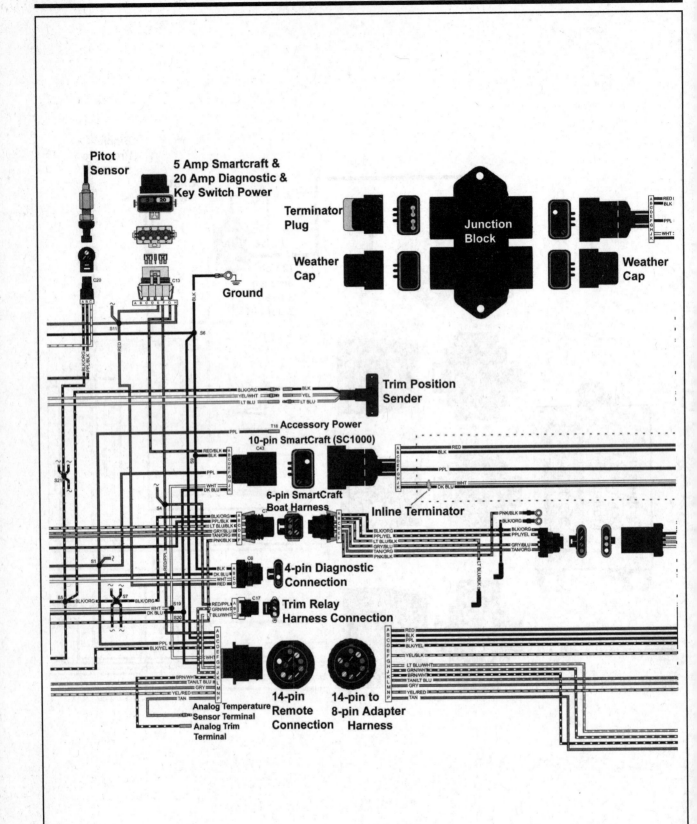

Wiring Diagram - 225/250 Hp (3.0L) OptiMax Pro/XS/Sport, 8-Pin Control Harness & Keyswitch (Center)

SELOC_1418

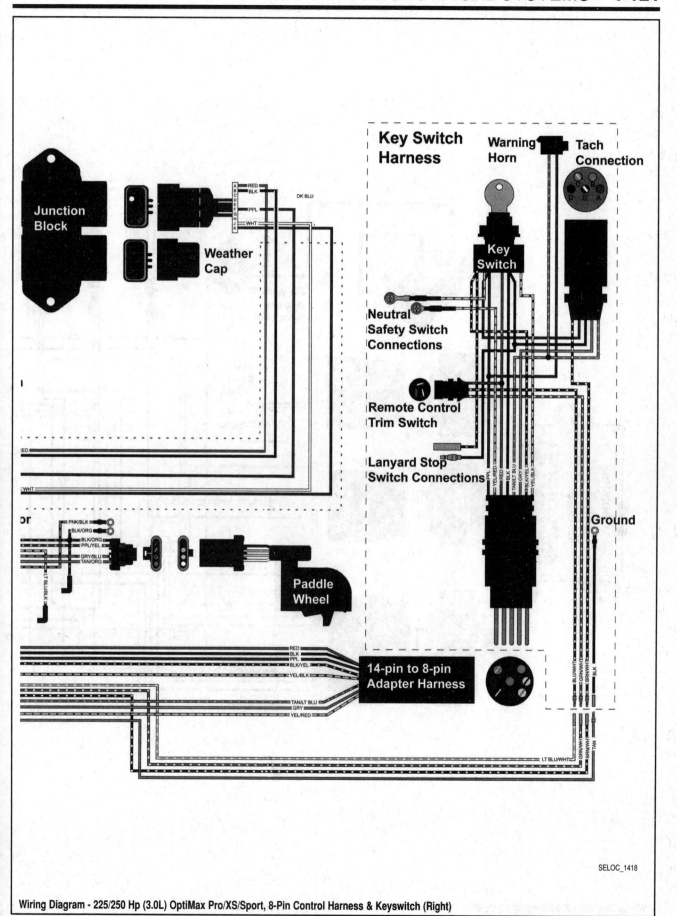

Wiring Diagram - 225/250 Hp (3.0L) OptiMax Pro/XS/Sport, 8-Pin Control Harness & Keyswitch (Right)

SELOC_1418

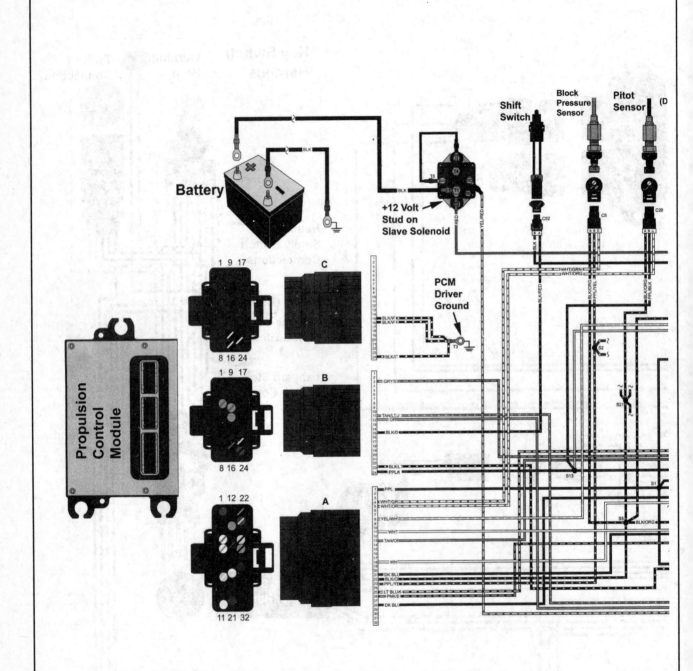

Wiring Diagram - 225/250 Hp (3.0L) OptiMax Pro/XS/Sport, 8-Pin Control Harness & Keyswitch (Left)

SELOC_1418

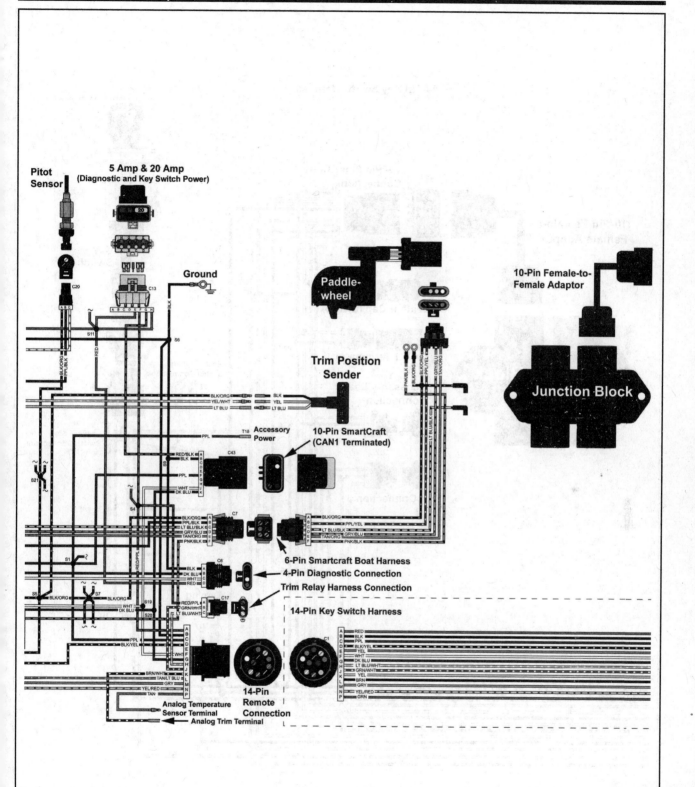

Pitot Sensor

5 Amp & 20 Amp (Diagnostic and Key Switch Power)

Ground

Paddle-wheel

10-Pin Female-to-Female Adaptor

Junction Block

Trim Position Sender

Accessory Power

10-Pin SmartCraft (CAN1 Terminated)

6-Pin Smartcraft Boat Harness
4-Pin Diagnostic Connection
Trim Relay Harness Connection

14-Pin Key Switch Harness

Analog Temperature Sensor Terminal
Analog Trim Terminal

14-Pin Remote Connection

SELOC_1418

Wiring Diagram - 225/250 Hp (3.0L) OptiMax Pro/XS/Sport, 8-Pin Control Harness & Keyswitch (Center)

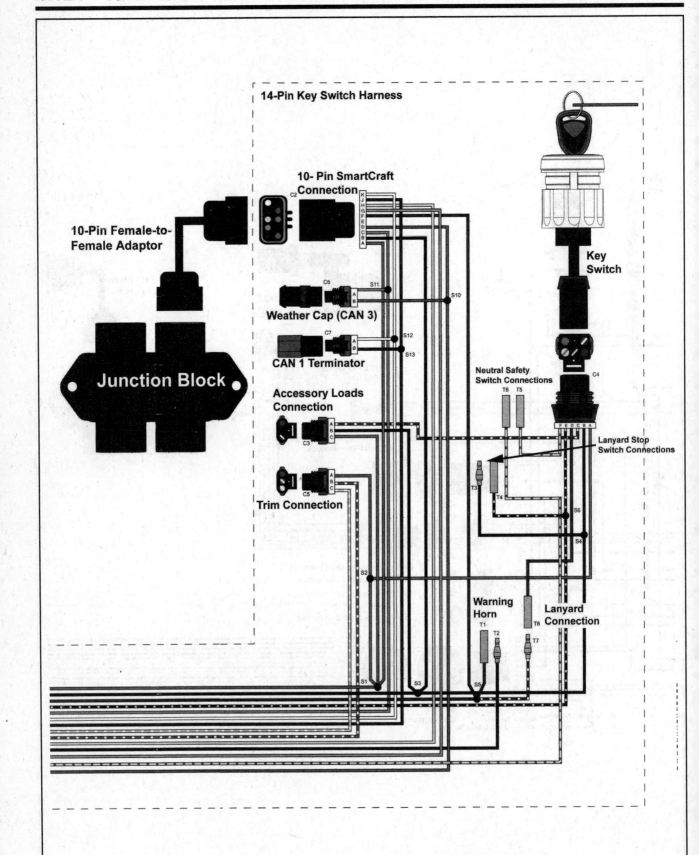

SELOC_1418

Wiring Diagram - 225/250 Hp (3.0L) OptiMax Pro/XS/Sport, 8-Pin Control Harness & Keyswitch (Right)

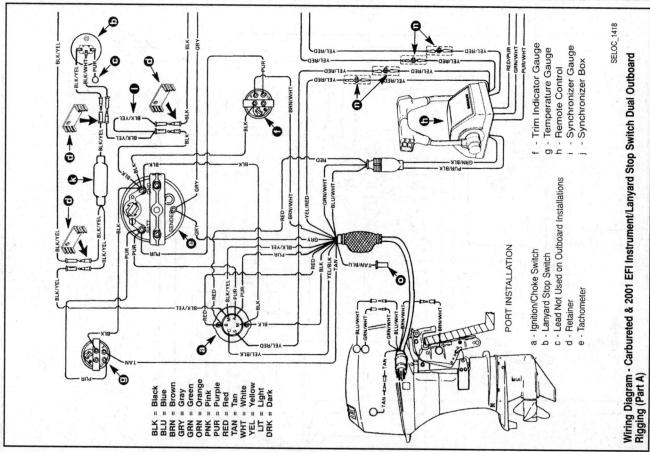

BLK = Black
BLU = Blue
BRN = Brown
GRY = Gray
GRN = Green
ORN = Orange
PNK = Pink
PUR = Purple
RED = Red
TAN = Tan
WHT = White
YEL = Yellow
LIT = Light
DRK = Dark

PORT INSTALLATION

a - Ignition/Choke Switch
b - Lanyard Stop Switch
c - Lead Not Used on Outboard Installations
d - Retainer
e - Tachometer

f - Trim Indicator Gauge
g - Temperature Gauge
h - Remote Control
i - Synchronizer Gauge
j - Synchronizer Box

SELOC_1418

Wiring Diagram - Carbureted & 2001 EFI Instrument/Lanyard Stop Switch Dual Outboard Rigging (Part A)

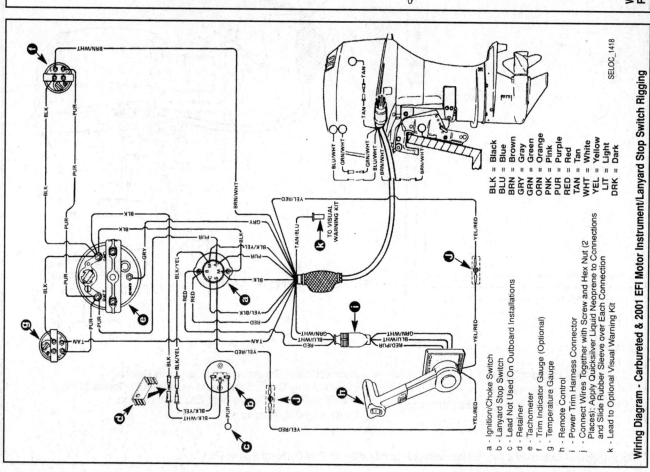

BLK = Black
BLU = Blue
BRN = Brown
GRY = Gray
GRN = Green
ORN = Orange
PNK = Pink
PUR = Purple
RED = Red
TAN = Tan
WHT = White
YEL = Yellow
LIT = Light
DRK = Dark

a - Ignition/Choke Switch
b - Lanyard Stop Switch
c - Lead Not Used On Outboard Installations
d - Retainer
e - Tachometer
f - Trim Indicator Gauge (Optional)
g - Temperature Gauge
h - Remote Control
i - Power Trim Harness Connector
j - Connect Wires Together with Screw and Hex Nut (2 Places); Apply Quicksilver Liquid Neoprene to Connections and Slide Rubber Sleeve over Each Connection
k - Lead to Optional Visual Warning Kit

SELOC_1418

Wiring Diagram - Carbureted & 2001 EFI Motor Instrument/Lanyard Stop Switch Rigging

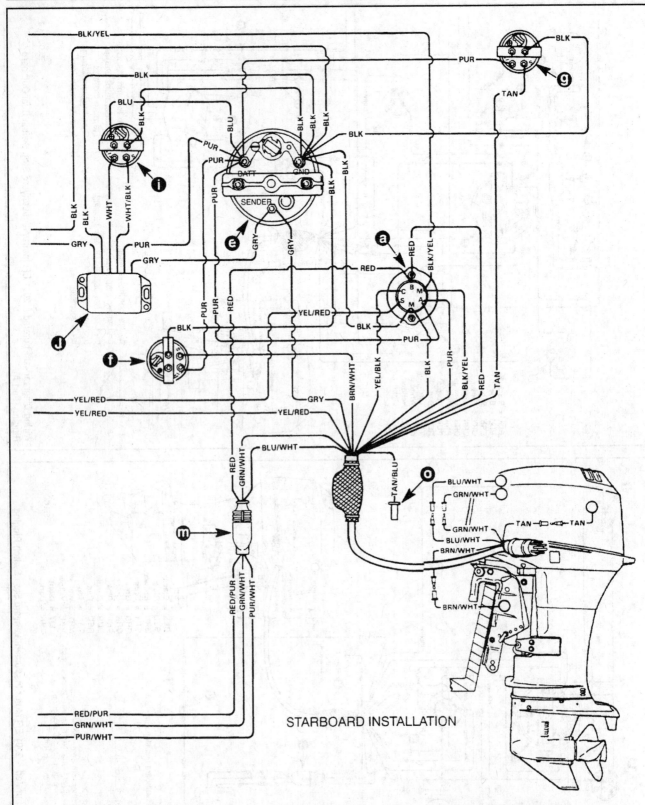

STARBOARD INSTALLATION

k - Lanyard/Diode
l - "Y" Harness
m - Power Trim Harness Connector

n - Connect Wires Together with Screw and Hex Nut (4
Places);Apply Quicksilver Liquid Neoprene to Connections
and Slide Rubber Sleeve over Each Connection.
o - Lead to Visual Warning Kit

SELOC_1418

Wiring Diagram - Carbureted & 2001 EFI Instrument/Lanyard Stop Switch Dual Outboard Rigging (Part B)

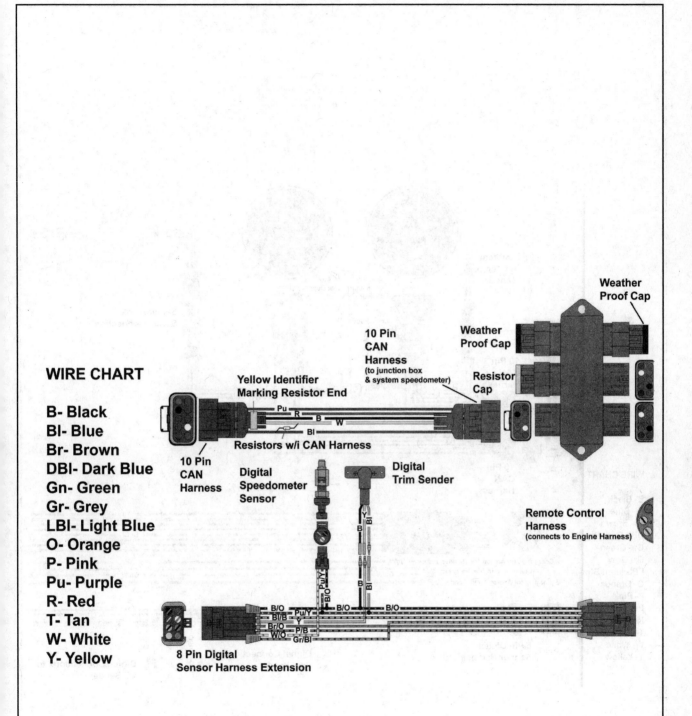

WIRE CHART

B- Black
Bl- Blue
Br- Brown
DBl- Dark Blue
Gn- Green
Gr- Grey
LBl- Light Blue
O- Orange
P- Pink
Pu- Purple
R- Red
T- Tan
W- White
Y- Yellow

Yellow Identifier
Marking Resistor End

10 Pin
CAN
Harness
(to junction box
& system speedometer)

Weather
Proof Cap

Weather
Proof Cap

Resistor
Cap

Resistors w/i CAN Harness

10 Pin
CAN
Harness

Digital
Speedometer
Sensor

Digital
Trim Sender

Remote Control
Harness
(connects to Engine Harness)

8 Pin Digital
Sensor Harness Extension

SELOC_1418

Wiring Diagram - Typical 2002 or Later EFI Motor Smartcraft System Tachometer & Speedometer Installation with System Link Gauges (Left)

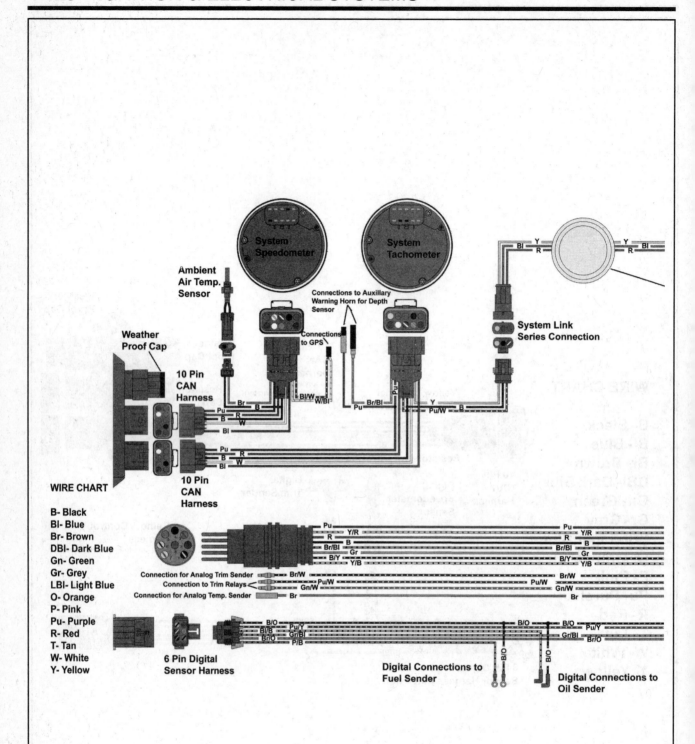

WIRE CHART

B- Black
Bl- Blue
Br- Brown
DBl- Dark Blue
Gn- Green
Gr- Grey
LBl- Light Blue
O- Orange
P- Pink
Pu- Purple
R- Red
T- Tan
W- White
Y- Yellow

SELOC_1418

Wiring Diagram - Typical 2002 or Later EFI Motor Smartcraft System Tachometer & Speedometer Installation with System Link Gauges (Center, Left)

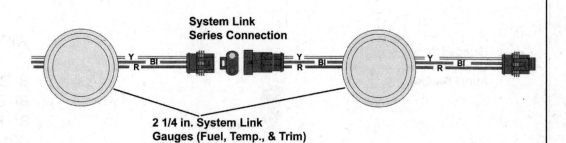

**System Link
Series Connection**

**2 1/4 in. System Link
Gauges (Fuel, Temp., & Trim)**

WIRE CHART

B- Black
Bl- Blue
Br- Brown
DBl- Dark Blue
Gn- Green
Gr- Grey
LBl- Light Blue
O- Orange
P- Pink
Pu- Purple
R- Red
T- Tan
W- White
Y- Yellow

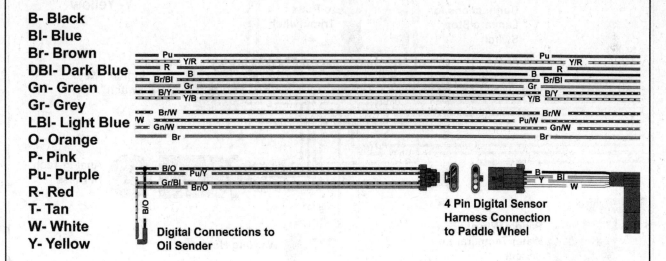

**Digital Connections to
Oil Sender**

**4 Pin Digital Sensor
Harness Connection
to Paddle Wheel**

SELOC_1418

Wiring Diagram - Typical 2002 or Later EFI Motor Smartcraft System Tachometer & Speedometer Installation with System Link Gauges (Center, Right)

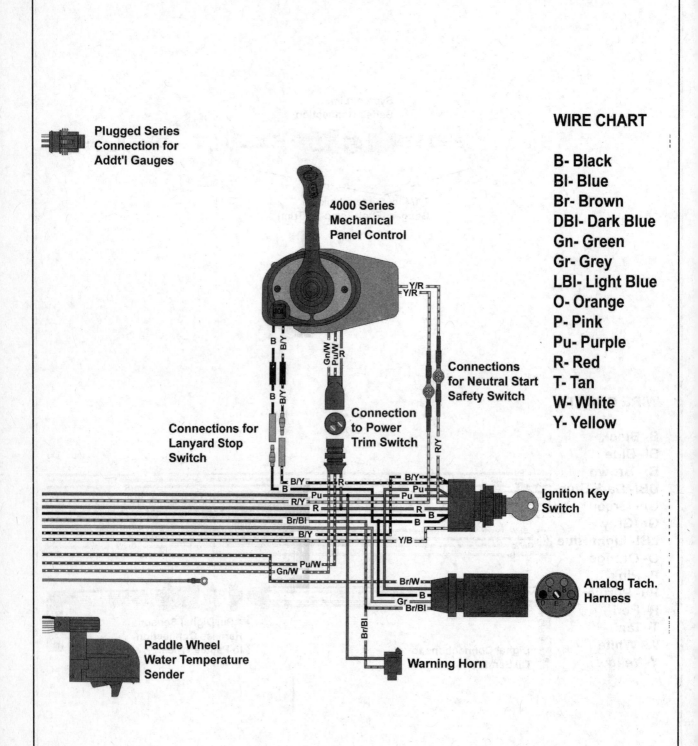

Plugged Series Connection for Addt'l Gauges

4000 Series Mechanical Panel Control

WIRE CHART

B- Black
Bl- Blue
Br- Brown
DBl- Dark Blue
Gn- Green
Gr- Grey
LBl- Light Blue
O- Orange
P- Pink
Pu- Purple
R- Red
T- Tan
W- White
Y- Yellow

Connections for Neutral Start Safety Switch

Connection to Power Trim Switch

Connections for Lanyard Stop Switch

Ignition Key Switch

Analog Tach. Harness

Paddle Wheel Water Temperature Sender

Warning Horn

SELOC_1418

Wiring Diagram - Typical 2002 or Later EFI Motor Smartcraft System Tachometer & Speedometer Installation with System Link Gauges (Right)

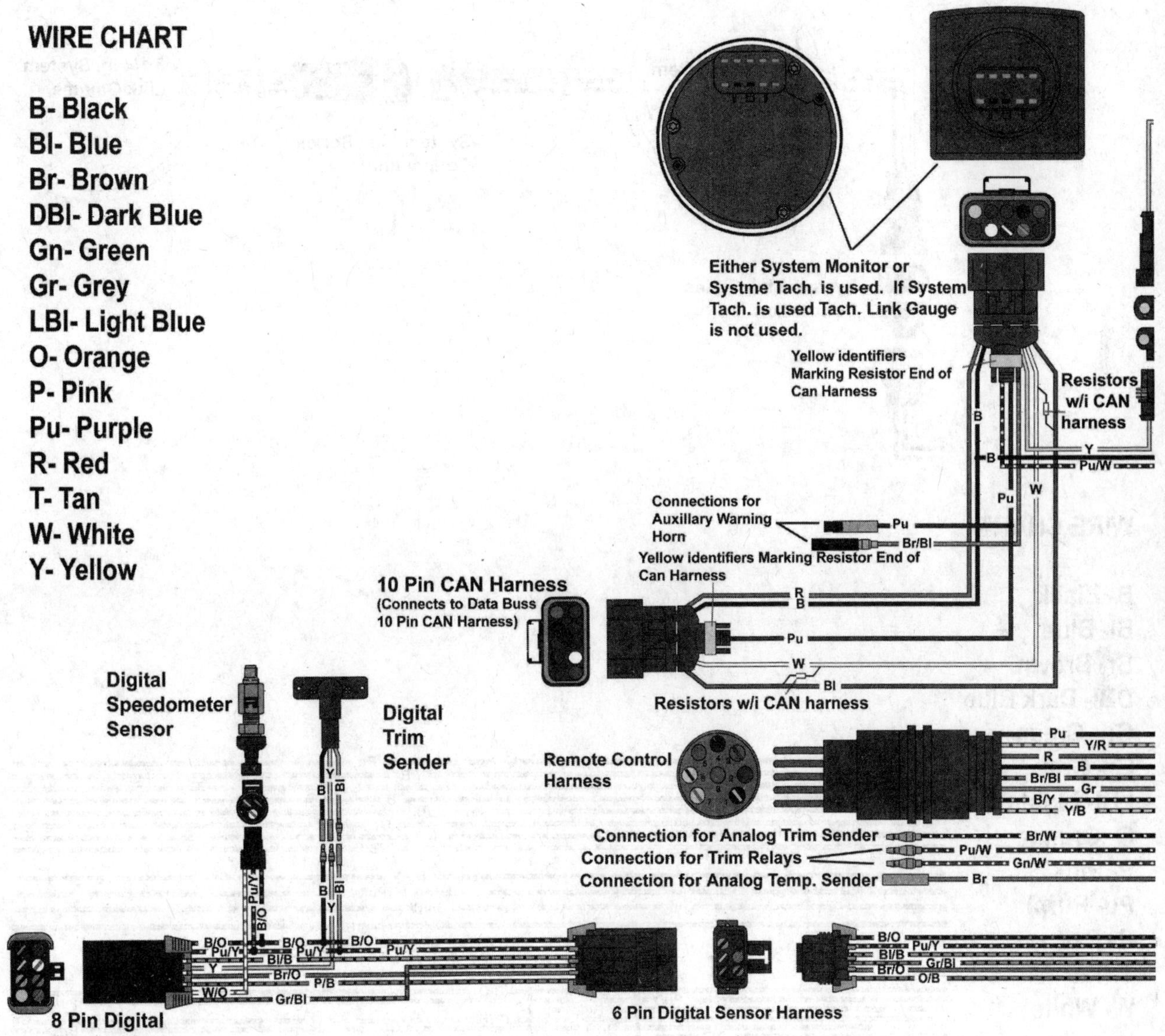

WIRE CHART

B- Black
Bl- Blue
Br- Brown
DBl- Dark Blue
Gn- Green
Gr- Grey
LBl- Light Blue
O- Orange
P- Pink
Pu- Purple
R- Red
T- Tan
W- White
Y- Yellow

Either System Monitor or Systme Tach. is used. If System Tach. is used Tach. Link Gauge is not used.

Yellow identifiers Marking Resistor End of Can Harness

Resistors w/i CAN harness

Connections for Auxillary Warning Horn

Yellow identifiers Marking Resistor End of Can Harness

Resistors w/i CAN harness

10 Pin CAN Harness
(Connects to Data Buss 10 Pin CAN Harness)

Digital Speedometer Sensor

Digital Trim Sender

Remote Control Harness

Connection for Analog Trim Sender
Connection for Trim Relays
Connection for Analog Temp. Sender

8 Pin Digital Sensor Harness Extension

6 Pin Digital Sensor Harness

SELOC_1418

Wiring Diagram - Typical 2002 or Later EFI Motor System Monitor/Tachometer Installation with System Link Gauges (Left)

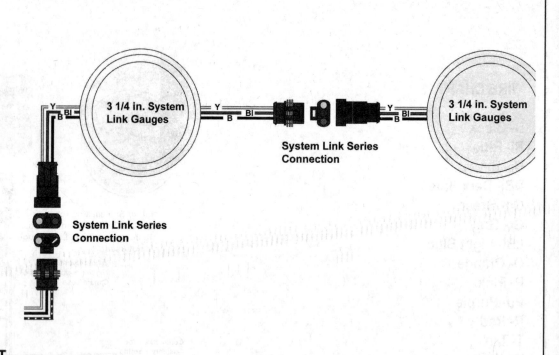

WIRE CHART

B- Black
Bl- Blue
Br- Brown
DBl- Dark Blue
Gn- Green
Gr- Grey
LBl- Light Blue
O- Orange
P- Pink
Pu- Purple
R- Red
T- Tan
W- White
Y- Yellow

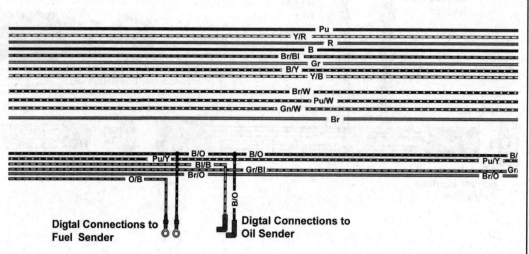

SELOC_1418

Wiring Diagram - Typical 2002 or Later EFI Motor System Monitor/Tachometer Installation with System Link Gauges (Center, Left)

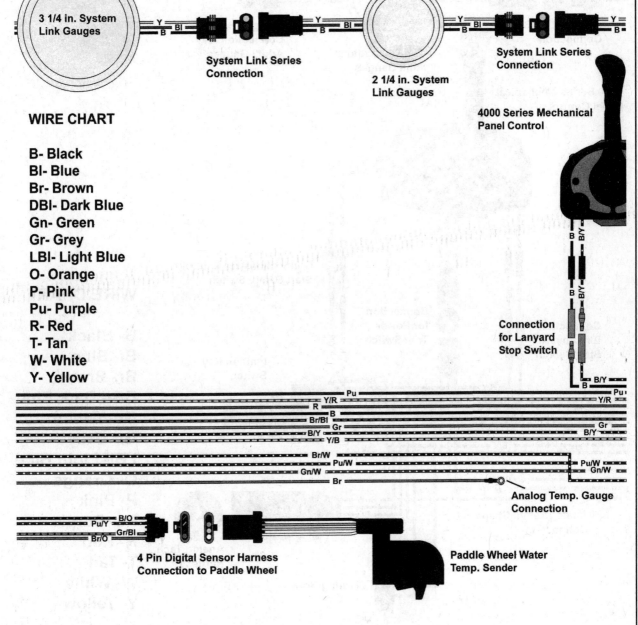

WIRE CHART

B- Black
Bl- Blue
Br- Brown
DBl- Dark Blue
Gn- Green
Gr- Grey
LBl- Light Blue
O- Orange
P- Pink
Pu- Purple
R- Red
T- Tan
W- White
Y- Yellow

3 1/4 in. System Link Gauges

System Link Series Connection

2 1/4 in. System Link Gauges

System Link Series Connection

4000 Series Mechanical Panel Control

Connection for Lanyard Stop Switch

Analog Temp. Gauge Connection

4 Pin Digital Sensor Harness Connection to Paddle Wheel

Paddle Wheel Water Temp. Sender

SELOC_1418

Wiring Diagram - Typical 2002 or Later EFI Motor System Monitor/Tachometer Installation with System Link Gauges (Center, Right)

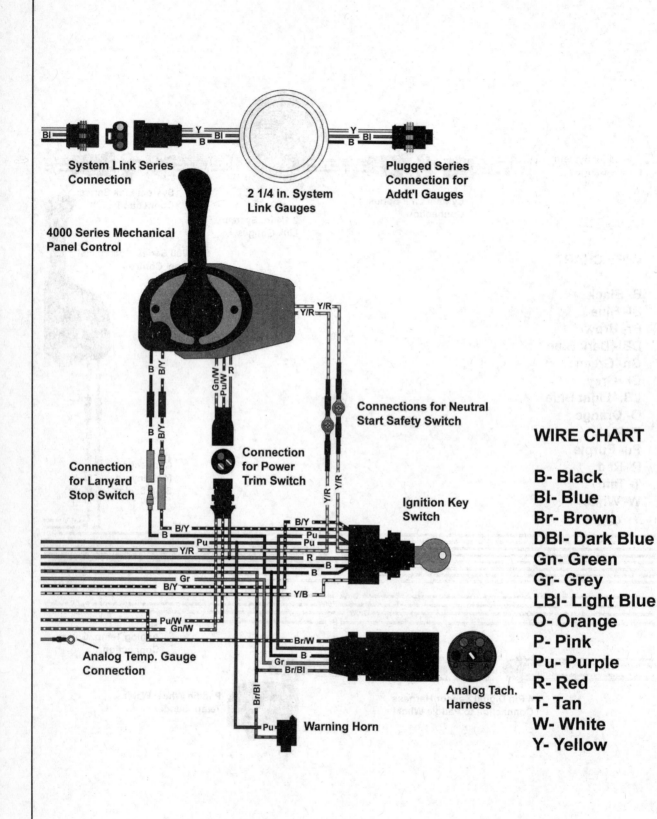

System Link Series Connection

Plugged Series Connection for Addt'l Gauges

2 1/4 in. System Link Gauges

4000 Series Mechanical Panel Control

Connections for Neutral Start Safety Switch

Connection for Lanyard Stop Switch

Connection for Power Trim Switch

Ignition Key Switch

Analog Temp. Gauge Connection

Analog Tach. Harness

Warning Horn

WIRE CHART

B- Black
Bl- Blue
Br- Brown
DBl- Dark Blue
Gn- Green
Gr- Grey
LBl- Light Blue
O- Orange
P- Pink
Pu- Purple
R- Red
T- Tan
W- White
Y- Yellow

SELOC_1418

Wiring Diagram - Typical 2002 or Later EFI Motor System Monitor/Tachometer Installation with System Link Gauges (Right)

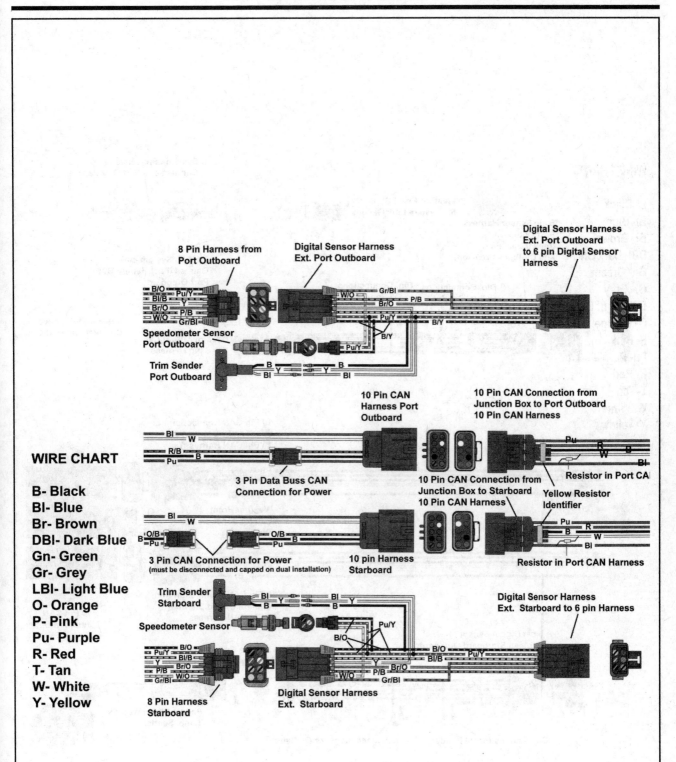

WIRE CHART

B- Black
Bl- Blue
Br- Brown
DBl- Dark Blue
Gn- Green
Gr- Grey
LBl- Light Blue
O- Orange
P- Pink
Pu- Purple
R- Red
T- Tan
W- White
Y- Yellow

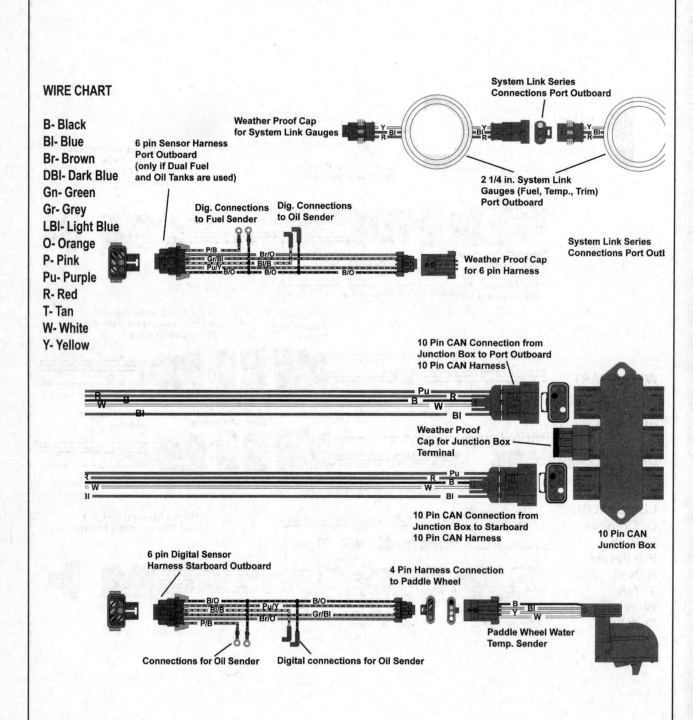

WIRE CHART

B- Black
Bl- Blue
Br- Brown
DBl- Dark Blue
Gn- Green
Gr- Grey
LBl- Light Blue
O- Orange
P- Pink
Pu- Purple
R- Red
T- Tan
W- White
Y- Yellow

Wiring Diagram - Typical 2002 or Later EFI Dual Outboard Smartcraft Installation (Center, Left)

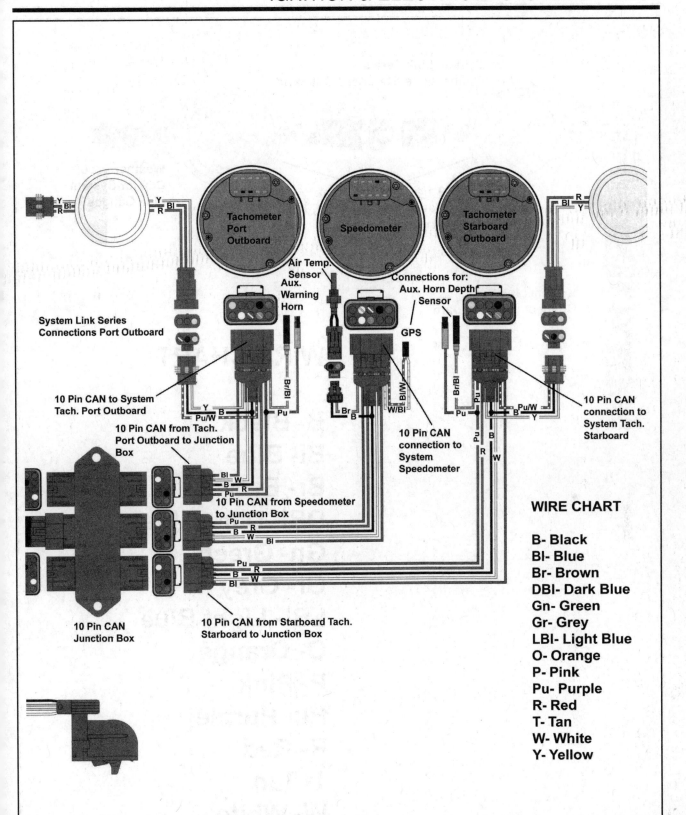

System Link Series Connections Port Outboard

Tachometer Port Outboard

Speedometer

Tachometer Starboard Outboard

Air Temp. Sensor
Aux. Warning Horn

Connections for:
Aux. Horn Depth Sensor

GPS

10 Pin CAN to System Tach. Port Outboard

10 Pin CAN from Tach. Port Outboard to Junction Box

10 Pin CAN connection to System Speedometer

10 Pin CAN connection to System Tach. Starboard

10 Pin CAN Junction Box

10 Pin CAN from Speedometer to Junction Box

10 Pin CAN from Starboard Tach. Starboard to Junction Box

Br/Bl

Y
Pu/W B Pu

Br

B

W/Bl

Bl/W

Pu
B
R

Pu/W
Y

Bl
W
R
Pu

R
B
W

Pu
B
R
W
Bl

Pu
B
R
W

WIRE CHART

B- Black
Bl- Blue
Br- Brown
DBl- Dark Blue
Gn- Green
Gr- Grey
LBl- Light Blue
O- Orange
P- Pink
Pu- Purple
R- Red
T- Tan
W- White
Y- Yellow

SELOC_1418

Wiring Diagram - Typical 2002 or Later EFI Dual Outboard Smartcraft Installation (Center, Right)

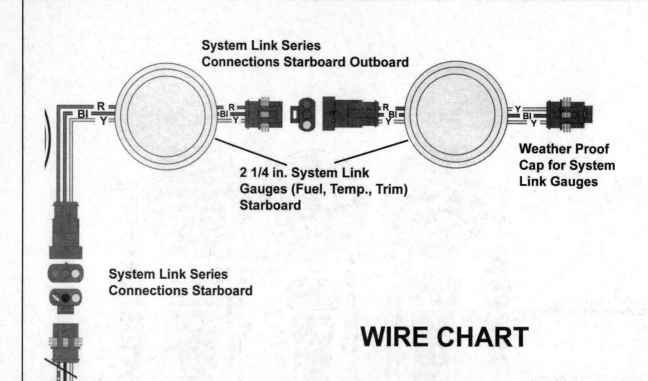

System Link Series
Connections Starboard Outboard

2 1/4 in. System Link
Gauges (Fuel, Temp., Trim)
Starboard

Weather Proof
Cap for System
Link Gauges

System Link Series
Connections Starboard

WIRE CHART

B- Black
Bl- Blue
Br- Brown
DBl- Dark Blue
Gn- Green
Gr- Grey
LBl- Light Blue
O- Orange
P- Pink
Pu- Purple
R- Red
T- Tan
W- White
Y- Yellow

SELOC_1418

Wiring Diagram - Typical 2002 or Later EFI Dual Outboard Smartcraft Installation (Right)

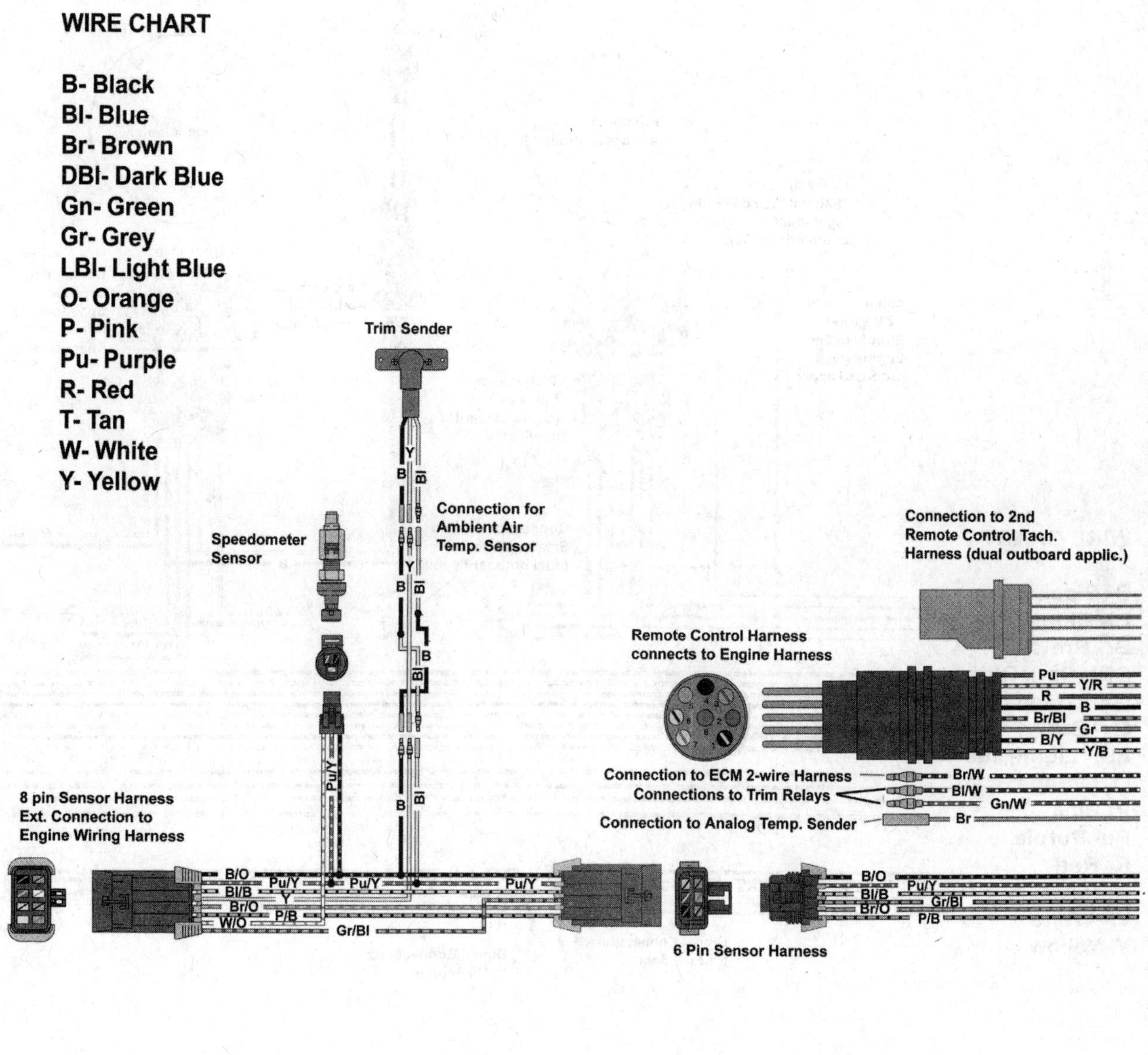

WIRE CHART

B- Black
Bl- Blue
Br- Brown
DBl- Dark Blue
Gn- Green
Gr- Grey
LBl- Light Blue
O- Orange
P- Pink
Pu- Purple
R- Red
T- Tan
W- White
Y- Yellow

Trim Sender

Speedometer
Sensor

Connection for
Ambient Air
Temp. Sensor

Y
B Bl

Y
B Bl

B

Bl
Bl

B Bl

Connection to 2nd
Remote Control Tach.
Harness (dual outboard applic.)

Remote Control Harness
connects to Engine Harness

Pu   Y/R
R   B
Br/Bl
Gr
B/Y
Y/B

Connection to ECM 2-wire Harness   Br/W
Connections to Trim Relays   Bl/W
                                         Gn/W
Connection to Analog Temp. Sender   Br

8 pin Sensor Harness
Ext. Connection to
Engine Wiring Harness

B/O
Pu/Y   Pu/Y   Pu/Y
Bl/B   Y
Br/O   P/B
W/O
Gr/Bl

Pu/Y

6 Pin Sensor Harness

B/O
Pu/Y
Bl/B   Gr/Bl
Br/O   P/B

SELOC_1418

Wiring Diagram - Typical OptiMax Motor Smartcraft (Non-CAN) Installation (Left)

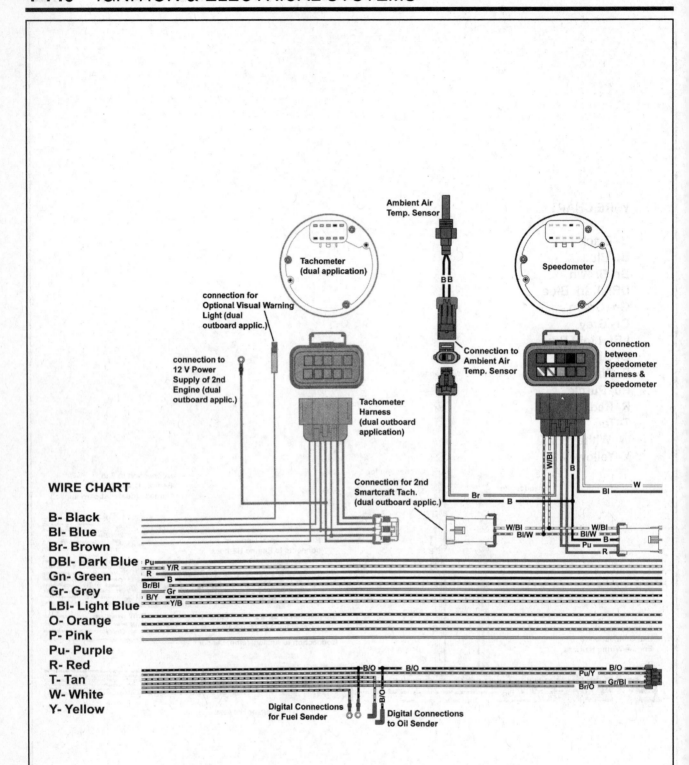

WIRE CHART

B- Black
BI- Blue
Br- Brown
DBI- Dark Blue
Gn- Green
Gr- Grey
LBI- Light Blue
O- Orange
P- Pink
Pu- Purple
R- Red
T- Tan
W- White
Y- Yellow

SELOC_1418

Wiring Diagram - Typical OptiMax Motor Smartcraft (Non-CAN) Installation (Center, Left)

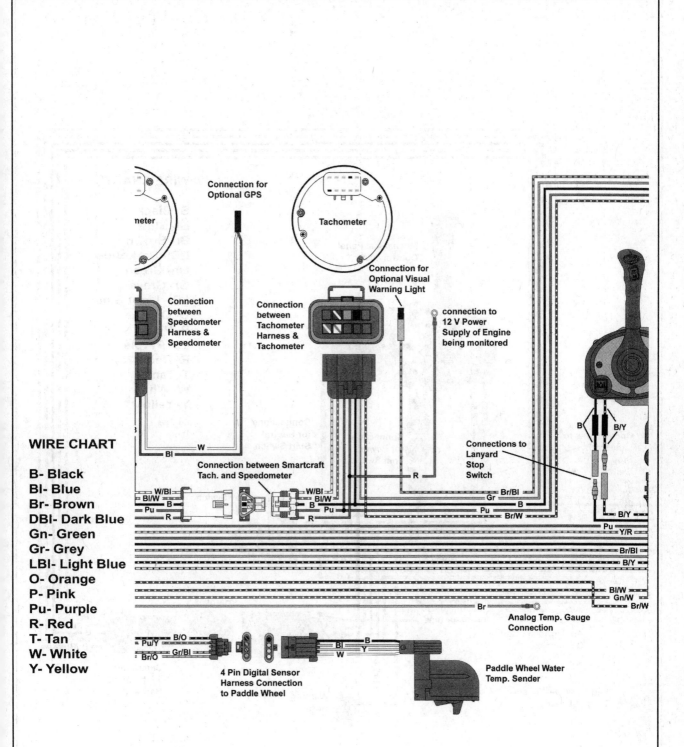

WIRE CHART

B- Black
Bl- Blue
Br- Brown
DBl- Dark Blue
Gn- Green
Gr- Grey
LBl- Light Blue
O- Orange
P- Pink
Pu- Purple
R- Red
T- Tan
W- White
Y- Yellow

Connection for Optional GPS

Tachometer

Connection for Optional Visual Warning Light

connection to 12 V Power Supply of Engine being monitored

Connection between Speedometer Harness & Speedometer

Connection between Tachometer Harness & Tachometer

Connections to Lanyard Stop Switch

Connection between Smartcraft Tach. and Speedometer

Analog Temp. Gauge Connection

4 Pin Digital Sensor Harness Connection to Paddle Wheel

Paddle Wheel Water Temp. Sender

SELOC_1418

Wiring Diagram - Typical OptiMax Motor Smartcraft (Non-CAN) Installation (Center, Right)

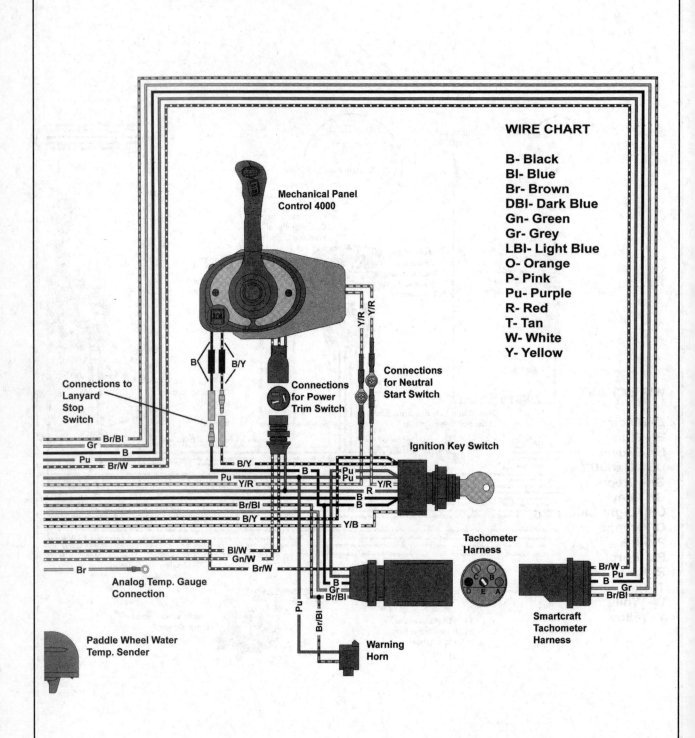

WIRE CHART

B- Black
Bl- Blue
Br- Brown
DBl- Dark Blue
Gn- Green
Gr- Grey
LBl- Light Blue
O- Orange
P- Pink
Pu- Purple
R- Red
T- Tan
W- White
Y- Yellow

Mechanical Panel Control 4000

Connections for Power Trim Switch

Connections for Neutral Start Switch

Connections to Lanyard Stop Switch

Ignition Key Switch

Tachometer Harness

Analog Temp. Gauge Connection

Smartcraft Tachometer Harness

Paddle Wheel Water Temp. Sender

Warning Horn

SELOC_1418

Wiring Diagram - Typical OptiMax Motor Smartcraft (Non-CAN) Installation (Right)

WIRE CHART

B- Black
Bl- Blue
Br- Brown
DBl- Dark Blue
Gn- Green
Gr- Grey
LBl- Light Blue
O- Orange
P- Pink
Pu- Purple
R- Red
T- Tan
W- White
Y- Yellow

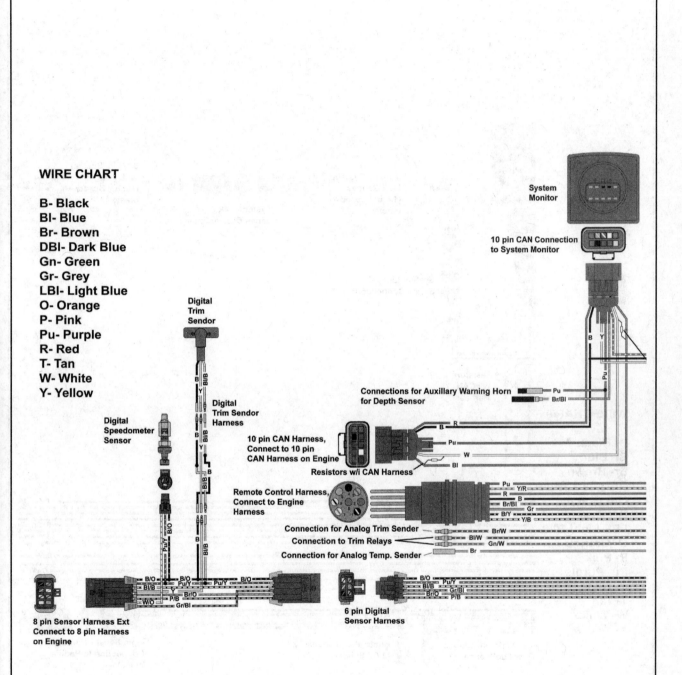

Wiring Diagram - Typical OptiMax Motor Smartcraft (CAN) Installation (Left)

SELOC_1418

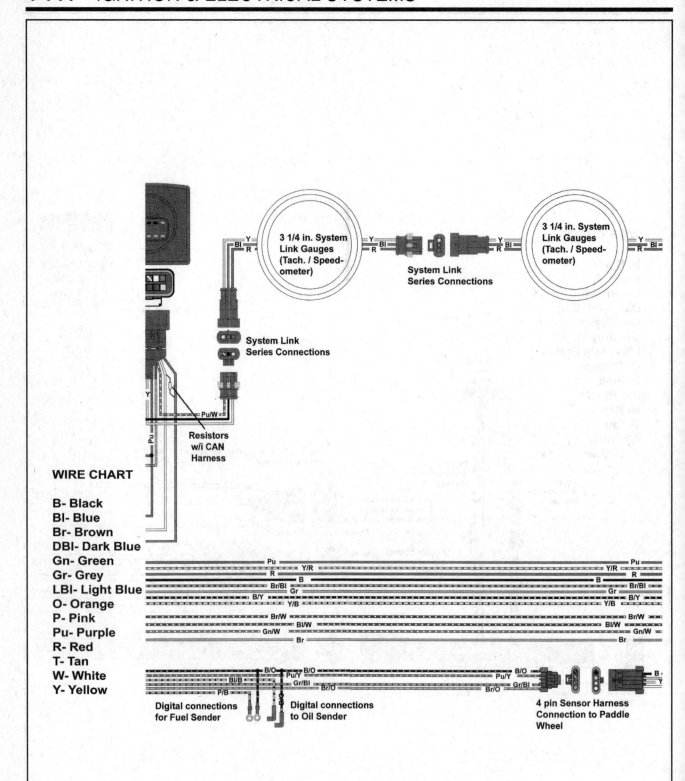

WIRE CHART

B- Black
Bl- Blue
Br- Brown
DBl- Dark Blue
Gn- Green
Gr- Grey
LBl- Light Blue
O- Orange
P- Pink
Pu- Purple
R- Red
T- Tan
W- White
Y- Yellow

3 1/4 in. System Link Gauges (Tach. / Speedometer)

System Link Series Connections

Resistors w/i CAN Harness

Digital connections for Fuel Sender

Digital connections to Oil Sender

4 pin Sensor Harness Connection to Paddle Wheel

SELOC_1418

Wiring Diagram - Typical OptiMax Motor Smartcraft (CAN) Installation (Center, Left)

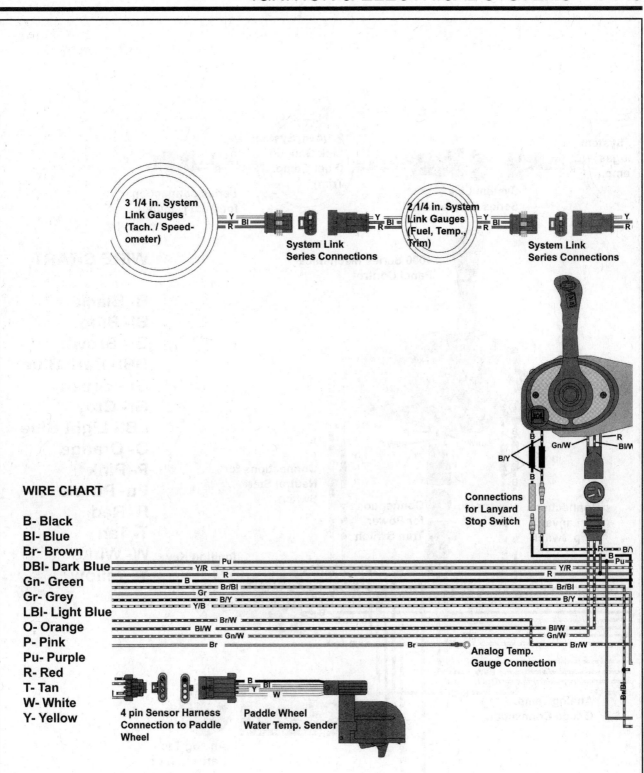

WIRE CHART

B- Black
Bl- Blue
Br- Brown
DBl- Dark Blue
Gn- Green
Gr- Grey
LBl- Light Blue
O- Orange
P- Pink
Pu- Purple
R- Red
T- Tan
W- White
Y- Yellow

3 1/4 in. System Link Gauges (Tach. / Speed-ometer)

System Link Series Connections

2 1/4 in. System Link Gauges (Fuel, Temp., Trim)

System Link Series Connections

Connections for Lanyard Stop Switch

Analog Temp. Gauge Connection

4 pin Sensor Harness Connection to Paddle Wheel

Paddle Wheel Water Temp. Sender

Wiring Diagram - Typical OptiMax Motor Smartcraft (CAN) Installation (Center, Right)

SELOC_1418

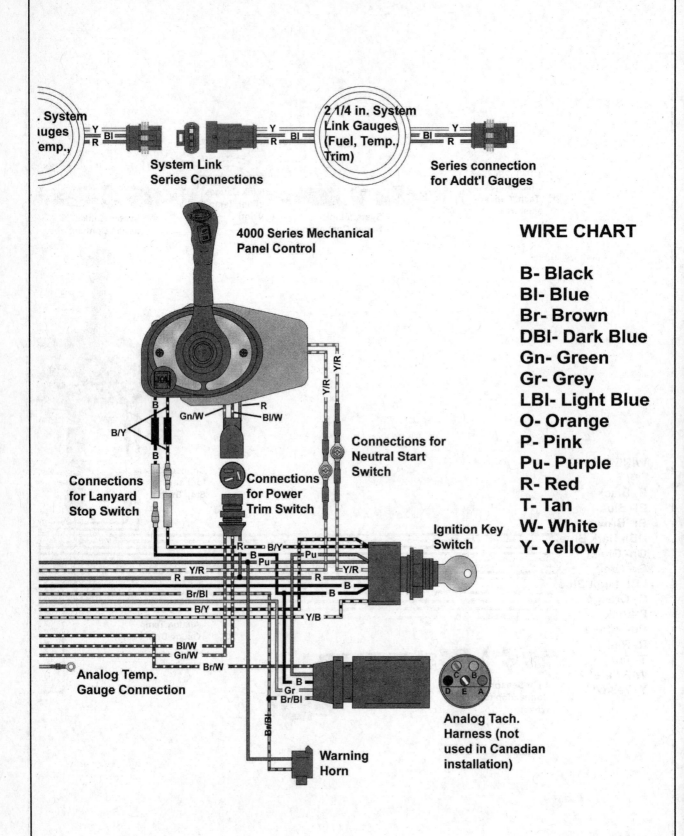

WIRE CHART

B- Black
Bl- Blue
Br- Brown
DBl- Dark Blue
Gn- Green
Gr- Grey
LBl- Light Blue
O- Orange
P- Pink
Pu- Purple
R- Red
T- Tan
W- White
Y- Yellow

Wiring Diagram - Typical OptiMax Motor Smartcraft (CAN) Installation (Right)

SELOC_1418

5

COOLING & LUBRICATION

OIL INJECTION SYSTEM

General Information

Over the years, manufacturers have tried various ways of mixing oil with the fuel on 2-stroke outboard engines. This was not only to make refueling easier, but to also provide more accurate fuel/oil ratios over the entire engine operational range.

An outboard's requirement for a fuel/oil mix does not stay constant. Mixture requirement varies depending on engine speed and load. A mixture that is too rich in oil results in excess exhaust smoke, will foul spark plugs and lead to excessive carbon deposits and pinging.

Pre-mixing the engine lubrication oil with gasoline before use is by far the simplest method. Unfortunately, it's also the messiest and the greatest polluter of the environment. Nothing complicated here, just add the correct amount of oil to a certain volume of fuel in your tank, and the job is done.

One important consideration often overlooked is the grade of 2-stroke oil used. Current standards call for TCW-3, but check with your manufacturer for their recommendation. TCW-3 has far superior additives that reduce many common outboard-engine problems such as sticking rings and carbon build-up in the combustion chamber.

It's also important to know what ratio of fuel to oil your engine manufacturer recommends. In recent years, many manufacturers have reduced the ratio from 50:1 to 100:1. This means 100 parts of gasoline to 1 part of oil or half as much oil as before.

The order in which you do things counts, too. Remember to put the correct amount of oil in the tank before adding the gasoline or at the same time as you add the gasoline, so the fuel will mix with the new oil as you fill the tank.

The second popular type of lubrication for 2-strokes is a mechanical oil pump systems. Some outboard manufacturers use a mechanically driven oil pump mounted on the engine block. The oil pump is connected to the throttle by way of a linkage arm. The theory of operation here is that the crankshaft drives a gear in the pump, creating oil pressure. As the throttle lever is advanced to increase engine speed, the linkage arm also moves, opening a valve that allows more oil to flow into the oil pump.

Many modern (usually fuel injected) outboards utilize an electronic oil pump to meter oil flow and, in such a way, can even more precisely control the amount of oil which is introduced at any engine rpm.

There are 2 schools of thought on the matter of where to inject the oil and this is where the significant difference lies between oil-injection systems. Some manufacturers inject the oil into a port in the carburetor-intake system behind the carburetor throttle plate (or throttle body on fuel injected motors). This means that only pure fuel passes through the carburetor. The oil blends with the incoming fuel/air mixture immediately after it leaves the carburetor throat, just before the mix enters the crankcase through the reed-valve plates.

Other manufacturers inject the oil from the pump into a port in the fuel supply line (in the line itself, into the engine mounted fuel pump, or the EFI vapor separator tank), where it is mixed with the fuel before it enters the carburetor.

Most outboard engines since the late 60's are equipped with a system that's designed to recycle excess oil accumulating in the crankcase. Oil that's injected into the engine is not completely consumed as it enters the combustion chamber with the air/fuel mix. This oil puddles inside the crankcase.

To remove this excess oil, manufacturers have devised a simple system of small fittings screwed directly into the side of the powerhead block assembly at key points. These fittings have hoses connected to the carburetor intake system, so the excess oil from the crankcase can be mixed with the incoming air/fuel mixture.

MERCURY OIL INJECTION (SPVR, ESP & EMP)

There are basically 3 slightly different types of oil injection found on Mercury 2-stroke motors. The first, known as Single Point Variable Ratio (SPVR) is a purely mechanical system that is found on all Carbureted models, as well as some 2001 or earlier EFI motors. The second, called Electronic Single Point (ESP) is basically an electronically operated version of the SPVR and is normally found on most EFI models (though there is some indication that early units may have used SPVR instead. The last, called Electronic Multi-Point (EMP) is found only on OptiMax motors, and is the most sophisticated of the systems.

If in doubt as to what system you are working, refer to the following system descriptions. It should be very easy to figure them out simply based on a visual inspection of the components.

Single Point Variable Ratio (SPVR) System

◆ See Figures 1 and 2

As stated earlier, all carbureted motors (plus a few early 2001 EFI motors) are equipped with a mechanical (driven off the crankshaft) oil pump which utilizes linkage attached to the throttle linkage assembly to vary oil output. On these motors oil drawn from the engine mounted reservoir is drawn into the mechanical pump, and then varied output from the pump is then introduced to the main fuel line.

On carbureted motors the oil is introduced right before the powerhead pulse-driven fuel pump. On EFI motors the oil is introduced directly into the fuel/vapor separator tank. As such, a mixture of fuel and oil is provided to the carburetors or EFI vapor separator tank (VST). Though the system works well, it obviously takes some time to respond to changes in throttle, starting 3/4 or WOT runs with a leaner oil ratio and leaving a richer ratio in the float bowls or EFI VST when you drop down to idle off plane.

Fig. 1 Powerhead mounted reservoir for oil injection system

Fig. 2 An SPVR system uses a mechanical oil pump

Smaller motors tend to use only the engine mounted oil reservoir which should be checked and filled periodically. Larger motors usually utilize a remote, boat mounted reservoir from which oil is drawn (by crankcase vacuum) to top off the engine mounted oil tank.

During powerhead break-in, pump output is normally combined with pre-mix in order to ensure a more generous supply of oil for the duration of the break-in period.

Visually, this system is easily identified by a mechanical oil injection pump which is bolted to the lower side of the powerhead. The pump has an irregular shaped cover, is normally painted black to match the powerhead, contains 2 oil lines (one inlet/one outlet) and utilizes a control arm which attaches to throttle linkage.

Electronic Single Point (ESP) System

◆ See Figures 3 and 4

Most EFI systems (generally all 2002 or later models) are equipped with an electronic oil pump. Like the SPVR equipped motors this pump still receives oil in a single line from a powerhead mounted reservoir (and usually a remote reservoir to keep that full), then pumps oil out in a single line to the fuel vapor separator tank (VST).

Fig. 3 Most ESP systems use a motor mounted and remote reservoir

Fig. 4 Oil pump from an ESP system (note only 2 oil lines)

The only real difference between this system and the SPVR is that the oil pump is electronic. So where the mechanical pump on the SPVR is driven off the crankshaft and output varied by linkage, this pump is controlled by the ECM which directly controls pump output. As such it has much finer control of pump output than the purely mechanical system.

During powerhead break-in, pump output is normally combined with pre-mix in order to ensure a more generous supply of oil for the duration of the break-in period.

This system is easily identified by the mounting of the electronic pump on the lower side of the powerhead. Though this pump also only has 2 oil lines (one inlet and 1 outlet) it has a distinctly different shape than the mechanical pump. The pump is normally a silver, cylindrical unit, mounted to a small, triangular shaped bracket. Unlike the mechanical pump there is no control arm or throttle linkage attached, but instead there is and electrical connector for pump control.

Electronic Multi-Point (EMP) Systems

◆ See Figures 5 and 6

OptiMax motors are equipped with the most advanced form of oil injection found on Mercury outboards. These are the only motors on which oil is NOT mixed with the fuel before entering the combustion chamber.

Like most other larger oil injected motors, oil is stored inside the remote oil tank in the boat. Crankcase pressure forces the oil from the remote oil into the engine mounted oil reservoir. The engine oil reservoir then feeds oil to the oil pump.

Like the ESP system, the oil pump is ECM driven and all oil distribution is controlled electronically. But *this* is the point where the ESP system differs from other Mercury electronic oil injection systems. The ECM controls oil distribution directly to the crankcase and air compressor. The oil pump has 1 oil discharge port and line per cylinder (meaning 4 or 6 depending upon the motor) that lead directly to the crankcase, as well as 1 additional discharge port and oil line which supplies oil to the air compressor. Unused oil from the air compressor returns to the plenum and is ingested through the crankcase. So these oil pump have either 5 or 7 oil discharge lines, along with an oil supply line.

The ECM is programmed to automatically increase the oil supply to the engine during the initial engine break-in period. The oil ratio is doubled during the first 120 minutes of operation whenever the engine speed exceeds 2500 rpm and is under load; below 2500 rpm the oil pump provides oil at the normal ratio.

After the engine break-in period, the oil ratio will return to normal: 300-400:1 at idle and 60:1 at wide open throttle (depending on throttle load).

■ **On some light boat applications after break-in is completed and the engine is being run at cruising speed (between 4000-5000 rpm) the fuel to oil ratio may be as high as 40:1. This results from a reduced throttle opening with a corresponding reduction in fuel consumption.**

Fig. 5 Oil pump from an EMP system (note multiple oil lines)

SELOC_1418

Fig. 6 Like other large motors systems, EMP powerheads will generally use both a remote and motor mounted oil reservoir

This system is easily identified by the mounting of the electronic pump on the lower side of the powerhead. Like the pump used on ESP systems, the pump is normally a silver, cylindrical unit, mounted to a small, triangular shaped bracket at the lower corner of the powerhead. Unlike the other electronic pump, there are multiple pump outlet lines (again either 5 or 7) in addition to the pump inlet line. Of course, there are not control levers, and only a wiring connector to provide pump control.

Troubleshooting

OIL INJECTION SYSTEM

 MODERATE

Like other systems on a 2-stroke engine, the oil injection system emphasizes simplicity. On pre-mix engines, when enough oil accumulated in the crankcase it passed into the combustion chamber where it burned along with the fuel. Since oil is a fuel just like gasoline, it burns as well. Unlike gasoline, oil doesn't burn as efficiently and may produce blue smoke that is seen coming out of the exhaust. If the engine was cold, such as during initial start-up, oil has an even harder time igniting resulting in excessive amounts of smoke.

Other problems, such oil remaining in the combustion chamber, tend to foul spark plugs and cause the engine to misfire at idle. Most of these problems have been alleviated by the introduction of automatic oil injection systems.

■ **One of the most common problems with oil injection systems is the use of poor quality injection oil. Poor quality oil tends to gel in the system, clogging lines and filters.**

It is normal for a 2-stroke engine to emit some blue smoke from the exhaust, though more efficient motors like EFI and OptiMax motors will minimize this. The blue color of the smoke comes from the burning 2-stroke oil. An excessive amount of blue smoke indicates too much oil being injected into the engine. On most engines, this is usually caused by an incorrectly adjusted injection control rod or a malfunction in the electronically controlled system (as applicable).

If the exhaust smoke is white, this is a sign of water entering the combustion chamber. Water may enter as condensation or more seriously may enter through a defective head gasket or cracked head. Usually white smoke from condensation will disappear quickly as the engine warms.

If the exhaust smoke is black, this is a sign of an excessively rich fuel mixture or incorrect spark plugs. The black color of the smoke comes from the fuel burning.

WARNING SYSTEM CIRCUIT

 EASY

All oil injected motors use some form of a low oil level sensor for the powerhead mounted oil reservoir. Generally speaking, the switch should show NO continuity when the tank is at least half full or more. Sometime between half full and empty the switch should show continuity. If the switch does not open/close as these conditions change, first check to make sure the float (usually a magnetic type located inside the tank) is not stuck. If the float is not stuck, yet the switch refuses to open/close as oil level changes, then it is defective and must be replaced.

■ **If a sensor tests bad while installed, but *not* when removed from the tank, then there is a GOOD chance that the magnetic float within the tank is stuck.**

The warning system on most outboards is designed to give a short beep upon initial keyswitch activation, in order to ensure the operator the system is still working. If however the warning horn fails to operate, the system MUST be fixed before continuing powerhead operation. Failure to do so might allow an undiagnosed problem with the Engine Overheat or Low Oil circuits which could allow the powerhead to become significantly damaged or destroyed should a lubrication or cooling problem occur.

If the **warning horn does not sound** when the key is first turned on check the following possible causes:

• Check for an open on the warning circuit tan or tan/blue (depending upon model) wire between the horn and powerhead. Disconnect this lead from the harness at the terminal on the engine block and ground it to the powerhead (you may need a jumper on some models). With the key turned on the horn should now sound. If not, check for an open circuit and/or check the horn.

• If not problems are found with the circuit or horn, suspect the warning module/ignition module/ECM could be the fault.

■ **For more details on the troubleshooting the warning system or on warning system activation, please refer to WARNING SYSTEM, later in this section.**

Oil Tank

REMOVAL & INSTALLATION

Oil tank removal and installation is not generally a tricky or difficult operation. However, it *can* be messy, so don't even think about starting it without a couple of drain pans, potentially a container to drain the tank into, and plugs, hose pinchers, caps or some other way to seal the lines to both prevent excessive oil leakage and/or system contamination.

Though oil hose routing tends to be straight forward (an inlet line from the remote tank to the top of the powerhead mounted tank, when equipped, and an outlet line from the bottom of the powerhead tank to the oil pump), it is still a good idea to mark or tag connections, as well as to note the line routing in order to prevent any possible mix-ups.

※※ SELOC CAUTION

Proper oil line routing and connections are essential for correct oil injection system operation. The line connections to the powerhead and oil pump look the same but may contain check valves of differing calibrations. Oil lines must be installed between the pump and powerhead correctly and connected to the proper fittings on the intake manifold in order for the system to operate properly.

30-60 Hp Models

 MODERATE

◆ **See Figures 7 thru 11**

1. On electric start models, disconnect the negative battery cable for safety.
2. Remove the engine cowling for access.

3. Remove the fasteners securing the tank. On most 30-50 hp models this means 2 nuts at the top of the tank (which secures the tank cover also on 2-cylinder models, for motors so equipped, reposition the cover)

4. On 55 hp and larger motors, remove the starter/oil bracket at the top rear of the tank, where it meets the electric starter assembly.

5. Disconnect the oil outlet hose. You've got a couple of choices here. You can disconnect it from the pump and either direct it into a container to drain the tank OR you can use a hose pincher or plug to cap it off keeping the oil in the tank. Lastly, you could disconnect it from the tank itself, but that tends to be the messiest option.

6. Tag and disconnect the wiring for the low oil level sensor (trace the sensor wires to the bullet connectors).

7. Remove the oil tank assembly from the powerhead mounting grommets.

8. If necessary for sensor replacement, remove the low oil sensor from the oil tank.

✳✳ SELOC CAUTION

The sensor is fragile. Always handle it with care.

To Install:

9. If removed, install the oil sensor into the tank and tighten the screw securely.

10. Carefully position the tank to the powerhead mounting grommets.

11. Connect the oil level sensor wires as tagged at the bullet connectors.

12. Reconnect the oil line.

✳✳ SELOC CAUTION

Any time the oil tank hose is disconnected, the oil injection pump must be purged (bled) of any trapped air. Failure to bleed the system could lead to powerhead seizure due to lack of adequate lubrication. For details, please refer to the Bleeding The Oil Injection System in this section.

13. If used, reposition the oil tank cover.

14. If used, reposition the starter/oil bracket to the powerhead.

15. Tighten the oil tank retaining bolts securely to the powerhead (but don't over-tighten and damage the tank!)

16. Install the engine cowling.

17. On electric start models, reconnect the negative battery cable.

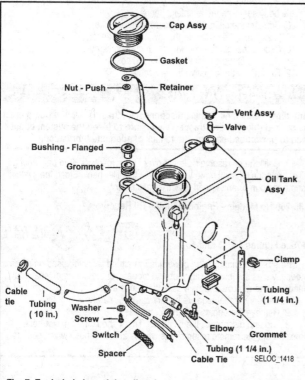

Fig. 7 Exploded view of the oil tank assembly - 30-50 Hp models (2-cylinder shown, 3-cylinder similar)

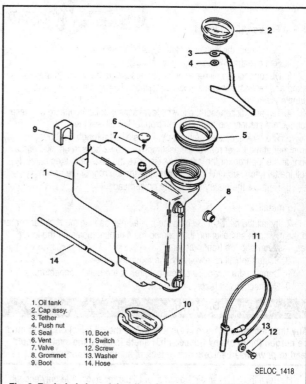

Fig. 8 Exploded view of the oil tank assembly - 55/60 Hp 3-cylinder models

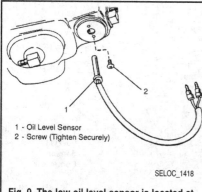

Fig. 9 The low oil level sensor is located at the bottom of the oil tank on most models

Fig. 10 On 30/40 hp twins, the tank fasteners also secure the top cover

Fig. 11 Oil tank on a 30/40 hp twin

Fig. 12 Typical oil tank mounting for 65-125 hp motors

Fig. 13 Unbolt the tank and the bottom and top (shown) brackets. . .

Fig. 14 . . .then carefully pull it from the top rear grommet

75/65 Jet/90 & 80 Jet/115/125 Hp Models

◆ See Figures 12, 13 and 14

MODERATE

1. Disconnect the negative battery cable for safety.
2. Remove the engine cowling for access.
3. Disconnect the oil reservoir outlet hose from the fitting on the oil pump and redirect it into a suitable container, allowing the reservoir to completely drain.
4. Tag and disconnect the low oil level sensor bullet connectors. These connectors are located at the bottom front of the oil reservoir.
5. The tank is usually secured by a single bolt through a rubber grommet at the lower rear of the assembly and by 2 bolts, a bracket and a boot at the top front of the assembly. Remove the fasteners, then carefully pull the tank forward out of the grommet at the top rear of the assembly.
6. Remove the reservoir from the powerhead.

To Install:

7. Install the oil reservoir onto the powerhead seating the top rear corner in the grommet and aligning the aligning the mounting holes for the lower rear fastener and top front bracket.
8. Secure the oil reservoir with the mounting screws.
9. Connect the low oil level sensor wires at the bullet connectors.
10. Connect the oil reservoir outlet hose.

✳✳ SELOC WARNING

Any time the oil tank hose is disconnected, the oil injection pump must be purged (bled) of any trapped air. Failure to bleed the system could lead to powerhead seizure due to lack of adequate lubrication.

11. Refill the oil reservoir. Do not over fill the reservoir. Add only enough oil to bring the oil level up to the bottom of the filler neck. Bleed the system and top-off again, as necessary.

75/90/115 Hp Optimax Models

◆ See Figure 15

MODERATE

1. Disconnect the negative battery cable for safety.
2. Remove the engine cowling for access.
3. Disconnect the air compressor air intake hose from the filter housing on the top back of the tank assembly.
4. Disconnect the oil outlet hose at the base of the tank, either from the oil filter or from the tank itself. Be ready to cap or plug the opening to prevent the tank from draining completely (unless you'd prefer to direct the hose into a suitable container in order to drain the tank).
5. Tag and disconnect the wiring for the low oil level sensor.
6. Remove the 2 tank mounting bolts (threaded vertically downward into the powerhead mounting bracket at the top rear of the tank, on either side of the front of the flywheel), then remove the tank assembly from the powerhead.

To Install:

7. Position the oil tank assembly onto the powerhead, aligning the throttle body seal with the air intake flange.
8. Connect the oil hose and secure using wire ties.
9. Reconnect the wiring for the low oil level sensor.
10. Secure the oil reservoir with the mounting screws.
11. Reconnect the air compressor intake hose.
12. Reconnect the negative battery cable.

✳✳ SELOC WARNING

Any time the oil tank hose is disconnected, the oil injection pump must be purged (bled) of any trapped air. Failure to bleed the system could lead to powerhead seizure due to lack of adequate lubrication.

13. Refill the oil reservoir. Do not over fill the reservoir. Add only enough oil to bring the oil level up to the bottom of the filler neck. Bleed the system (prime the pump) and top-off again, as necessary.

135-250 Hp Models - Engine Mounted Oil Reservoir

◆ See Figures 16, 17 and 18

MODERATE

The size and shape of the powerhead mounted reservoir used on these models varies somewhat with year and model, but the basics of what you need to do to remove it remain essentially the same.

1. Disconnect the negative battery cable for safety.
2. Remove the engine cowling for access.
3. Tag and disconnect the oil hoses. There should be an inlet hose coming from the remote reservoir, as well as an outlet hose which goes to

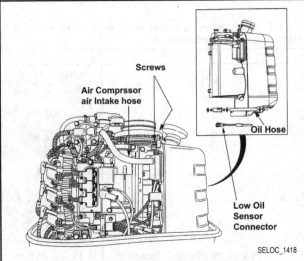

Fig. 15 Oil tank mounting - 75/90/115 Hp Optimax Models

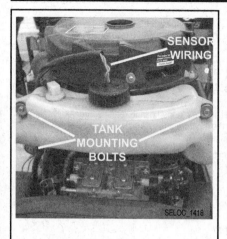

Fig. 16 Typical oil tank for a carb Mercury V6

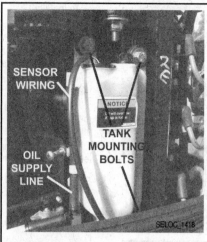

Fig. 17 Typical oil tank for an EFI Mercury V6

Fig. 18 Typical oil tank for an OptiMax Mercury V6

the oil pump. In most cases it is easiest to disconnect the inlet hose at the tank, but to disconnect the outlet hose from the oil pump itself. In the later case, keeping the outlet hose attached to the tank should allow you to drain the tank through the hose, to plug or pinch the hose to prevent leakage, or even to physically secure the end of the hose higher than the tank itself, also to prevent leakage. The choice is yours.

4. Tag and disconnect the wiring for the low oil level sensor. Again, in most cases there are 2 bullet connectors running to the sensor mounted in the tank. These models usually have the sensor mounting in the TOP of the tank for easy access/service.

5. Remove the bolts (usually 3) securing the oil reservoir to the powerhead and then remove the oil reservoir.

To Install:

6. On carbureted and EFI 2.5L motors (2001), apply a threadlocking compound, such as Loctite ® 222, or equivalent, to the threads of the mounting bolts and then secure the oil reservoir to the powerhead. Tighten the bolts to 25 inch lbs. (2.8 Nm).

7. On all other models, position the reservoir to the powerhead and secure using the retaining bolts. No threadlocker is necessary.

8. Reconnect the oil hoses and secure using new wire ties.

9. Reconnect the oil level sensor wiring.

10. Reconnect the negative battery cable.

✳✳ SELOC WARNING

Any time the oil tank hose is disconnected, the oil injection pump must be purged (bled) of any trapped air. Failure to bleed the system could lead to powerhead seizure due to lack of adequate lubrication.

11. Refill the oil reservoir. Do not over fill the reservoir. Add only enough oil to bring the oil level up to the bottom of the filler neck. Bleed the system (prime the pump) and top-off again, as necessary.

135-250 Hp Models - Remote Oil Tank

◆ **See Figures 19, 20 and 21**

The remote oil tank should be installed in an area in the boat where there is access for refilling the tank. The tank should be restrained to keep it from moving around, causing possible damage.

An acceptable means of restraining the tank would be the use of eye bolts and an elastic restraining strap across the center of the tank. Taking care that any metal hooks do not puncture the tank.

Keep in mind, when installing the tank in tight areas, that this tank will be under pressure while the engine is operating and will expand slightly. Therefore, the restraints used with the tank should be left with a small amount of slack to enable the tank expand.

Oil hoses routed through the engine well must be able to reach the hose fittings on the engine through its full range of movement and not bind, stretch or kink. Any lose of injector oil could cause serious damage to the powerhead.

CLEANING & INSPECTION

One of the most common problems with oil injection systems is the use of poor quality injection oil. This poor quality oil tends to gel in the system, clogging oil lines and filters. If this is found to be the case with your system, or if the powerhead has been sitting in storage for a length of time, it is wise to remove the oil tank and clean it with solvent.

Fig. 19 Remote oil tank ready for installation

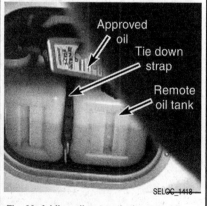

Fig. 20 Adding oil to a typical remote oil tank in the stern of a boat

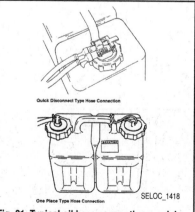

Fig. 21 Typical oil hose connections, quick-disconnect type and 1-piece type

While it is removed, take the opportunity to inspect it for damage and replace it as necessary. The oil tank is the only source of oil for the powerhead. If it should leak, the powerhead will eventually run out of injection oil, with catastrophic and very costly results. Remember, there are no parts stores when you are miles out at sea.

Oil Pump

REMOVAL & INSTALLATION

✳✳ SELOC WARNING

Proper oil line routing and connections are essential for correct oil injection system operation. The line connections to the powerhead and oil pump look the same but may contain check valves of differing calibrations. Oil lines must be installed between the pump and powerhead correctly and connected to the proper fittings on the intake manifold in order for the system to operate properly.

The only purpose for disassembling an oil injection pump is to locate a problem in oil delivery. For example, if the pump is frozen due to debris or rust, the pump can be disassembled and cleaned.

Mechanical Pumps

◆ See Figures 22 thru 25

One of the oil lines to the pump is a gravity fed oil inlet line which runs from the powerhead mounted oil tank. Once you disconnect it from the pump, the tank will want to empty through the open hose. You've got multiple solutions to this problem. You can plug, cap or pinch the line still attached to the tank to keep it from draining, or you can redirect the flow into a suitable drain pan or recovery container for re-use in filling the tank once the line is reconnected.

1. Locate the oil inlet line from the tank to the pump. Disconnect the line, then either cap/plug the line to prevent leakage or drain the tank into a suitable container using the line.

2. Tag and disconnect the oil outlet line from the pump to the fuel system. Plug or cap all open lines to prevent system contamination.

3. CAREFULLY pry the oil injection link rod free of the ball joint on the injection pump lever. Take care not to alter the length of this rod.

4. Remove the 2 bolts securing the pump to the powerhead or powerhead adapter (as applicable) and lift the pump clear. In most cases the driven gear in the powerhead should remain behind. Note how the pump shaft or worm gear tab can align with the slot in the driven gear. It is not necessary to match the exact position of the current alignment during assembly, it is only necessary to make sure the 2 properly mesh and are not forced together.

To Install:

5. Check to be sure the shaft of the oil pump will index into the slot at the center of the crankshaft driven shaft. If the 2 are no longer aligned, rotate the pump shaft to match the slot in the driven shaft. Install the oil pump with the pump shaft indexed into the slot on the driven shaft.

6. Secure the pump to the powerhead with the 2 attaching bolts. Tighten the bolts securely.

7. Reconnect the oil lines as tagged during removal (generally that means the lower line is the input from the oil tank and the upper line is the output to the fuel system).

8. Snap the oil injection link rod back onto the ball joint on the pump lever. Fill and bleed the oil injection system.

Electronic Pumps

◆ See Figures 26 and 27

The manufacturer has made no provisions for rebuilding this pump. Spare parts are not available. If any part is found to be defective and no longer fit for service the pump must be replaced.

Save the O-rings, even if they are defective. The old ring will be essential when purchasing a new ring to ensure the proper type and size is obtained.

Mercury electric oil pumps are normally rubber-mounted to a flange on the lower corner of the powerhead.

1. Disconnect the negative battery cable.

2. Tag and disconnect the wiring harness from the pump.

3. Tag and disconnect the hoses for the oil pump. Just like mechanical pumps, you'll have to make a decision whether or not to cap or plug most hoses, including the oil tank outlet hose (which can also be used to drain the tank if desired).

4. Remove the bolts (usually 3) securing the pump to the powerhead through rubber mounts, then remove the pump assembly.

To Install:

5. Position the pump assembly onto the powerhead tighten the bolts.

6. Reconnect the oil hoses in the correct locations, as tagged during removal.

7. Reconnect the wiring harness.

8. Refill and prime the oil system. Refer to the Bleeding The Oil Injection System procedures in this section.

OIL PUMP CONTROL ROD ADJUSTMENT

See the Timing & Synchronization procedures in the Maintenance & Tune-Up section for mechanical oil pump linkage adjustment.

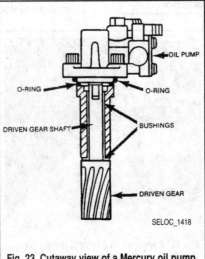

Fig. 22 Typical Mercury oil pump mounting on the powerhead

Fig. 23 Cutaway view of a Mercury oil pump

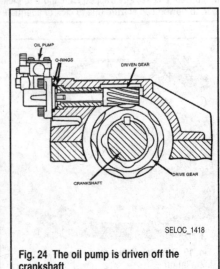

Fig. 24 The oil pump is driven off the crankshaft

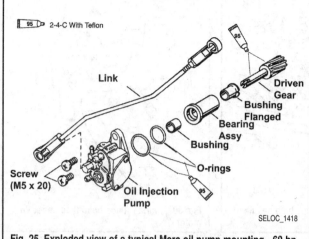

Fig. 25 Exploded view of a typical Merc oil pump mounting - 60 hp and smaller motors (larger motors similar but often with a worm bushing surrounded by O-rings and no pictured bearing assembly)

BLEEDING THE OIL INJECTION SYSTEM

Mechanical Pumps (SPVR Systems)

◆ See Figures 28, 29 and 30

Anytime the oil pump is disconnected, removed or replaced the system must be thoroughly bled to ensure the powerhead is not threatened by oil starvation. The following procedure will ensure a proper oil supply for the powerhead during initial operation. It's not a bad idea to visually spot check the clear oil lines from time to time afterwards in order to ensure there are no air bubbles entering the system.

1. Start by making sure the powerhead mounted oil reservoir is topped off.

2. With the engine shut off, hold an absorbent cloth below the oil injection pump and be prepared to catch the oil as it oozes from the bleed screw.

3. Locate the bleed screw (either a hex head or Philips head screw found just below or in the pump top cover). Loosen the bleed screw 3, maybe 4 full turns, allowing oil to exit the screw until there are no longer any air bubbles in the INLET hose (the hose coming from the oil reservoir).

■ This step filled the inlet oil line AND the pump with oil.

4. When a steady stream of oil is observed with no sign of air, tighten the bleed screw securely.

5. Next, it is necessary to run the motor in order to purge air from the pump OUTLET hose (the hose which runs *from* the oil pump to the fuel system). Hook up a source of water for the cooling system and proceed as follows (depending upon the model):

• On all inline motors, purge air from the outlet hose by running the engine on a 50:1 pre-mix fuel/oil mixture (from a portable tank). Start and run the motor at idle speed until no more air bubbles are present in the outlet hose, then shut the motor down and remove the pre-mix fuel source.

• On all V6 models Mercury says that you can just start and run the motor to purge air. If any air bubbles persist, they can be purged out of the hose by disconnecting the oil pump link rod and rotating the pump arm fully clockwise (to the pump WOT position) while operating the engine at 1000-1500 rpm. If necessary, gently pinch the fuel line between the remote fuel line connector and the oil injection pump T-fitting. This will cause the fuel pump to provide a partial vacuum that will aid in the removal of any trapped air in the system. Once the outlet hose is fully purged of air, shut the motor off and reconnect the oil pump link rod.

■ Honestly, though Mercury doesn't state it is always necessary to move the pump arm to WOT on V6 models, we'd rather finish purging the system faster and recommend ALWAYS doing this as the final step. Why run the engine waiting for the hose to self-purge at a slower rate?

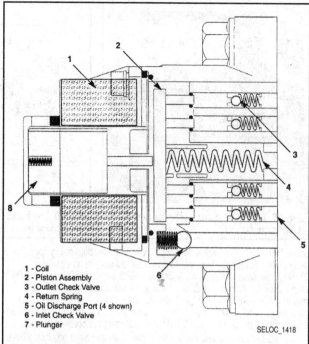

1 - Coil
2 - Piston Assembly
3 - Outlet Check Valve
4 - Return Spring
5 - Oil Discharge Port (4 shown)
6 - Inlet Check Valve
7 - Plunger

SELOC_1418

Fig. 26 Cutaway view of an electrically operated OptiMax oil injection pump assembly (EFI similar, but fewer discharge ports/passages)

Fig. 27 Typical Mercury electronic oil pump (EFI shown, OptiMax similar)

Electric Pumps (ESP and EMP Systems)

OEM ② ⚙ MODERATE

■ Anytime the oil pump is disconnected, removed or replaced the system must be thoroughly bled to ensure the powerhead is not threatened by oil starvation. The following procedure will ensure a proper oil supply for the powerhead during initial operation.

Electronically controlled oil pumps can self-prime, if given the command from the ECM. Various methods are available for this, the easiest being to

Fig. 28 Typical oil pump assembly and bleed screw

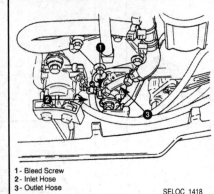

1 - Bleed Screw
2 - Inlet Hose
3 - Outlet Hose

SELOC_1418

Fig. 29 The bleed screw may be a hex head type (most inline motors). . .

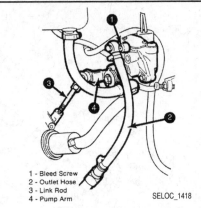

1 - Bleed Screw
2 - Outlet Hose
3 - Link Rod
4 - Pump Arm

SELOC_1418

Fig. 30 . . .or a Philips head type (most V6 motors)

use a Mercury DDT or equivalent scan tool. However, depending upon the setting you choose you may wind up with more than just a self prime. Through the scan tool you can select a Reset Break-In function which will both cause the auto prime to run, but also clear the run/fault history and restart a 120 minute powerhead break-in cycle. Alternately a scan tool can be used to simply select the Oil Pump Prime function which will run the auto-prime without touching run/fault history or restarting the break-in cycle.

If you do NOT have access to a scan tool, then you have no choice but to use the Shift Switch Activation method, which will act like the scan tool Reset Break-In function, except that it does not clear run/fault history. So using the Shift Switch Activation method *will* prime the system and start a new 120 minute break-in cycle. To use the Shift Switch Activation method, start by making sure the primer bulb from the remote oil tank is firm, then:

- Turn the ignition switch to the **ON** position without starting the motor.
- Within 10 seconds of turning the switch **ON**, move the remote shift handle from Neutral to Forward gear 3-5 times. This will start the auto-prime process AND reset the break-in function.

On V6 models, Mercury recommends using a gearcase leakage tester (#FT-8950), which is basically a hand-held vacuum or pressure pump. Ironically, although the systems are essentially the same for the 1.5L OptiMax motor, Mercury does NOT mention using this pump. This means that either something is different on the 1.5L or, more likely, that the pump is not completely necessary on V6 models. Since the pump is used to keep the oil tank pressurized (to help ensure proper oil flow) you *should* be able to cheat the system using primer bulb in the supply line from the remote oil reservoir to keep the powerhead tank full and pressurized.

1. On V6 models, connect the leakage tester/hand held pressure pump to the inlet T-fitting on the onboard oil reservoir (on the line, toward the bottom of the tank). While clamping off the inlet hose (to prevent the

pressure from back-flushing the supply line and escaping into the remote tank), manually pressurize the reservoir to 10 psi.

2. Using the DDT or Shift Select Switch, activate the oil pump prime sequence. Maintain the 10 psi pressure throughout the auto prime sequence. when the auto prime is completed, remove the leakage tester and refill the onboard oil reservoir.

TESTING OIL PUMP DISCHARGE RATE

◆ **See Figure 31**

The following procedure is designed to monitor the output of a mechanical the oil pump when it is held in the wide open throttle (WOT) position as the motor is run at a specified speed (usually 1500 rpm) while running on a tank of pre-mix.

■ **Running the motor on pre-mix is in theory an option for all EFI motors, however it could not work for OptiMax motors since the oil is not mixed with the fuel prior to the combustion chamber.**

Mercury does not specify a procedure for checking oil output of an electric oil pump, however, they DO provide a specification of the amount of oil which should be expelled in total during the auto-prime procedure. If desired you could disconnect the oil discharge line (EFI) or tag and disconnect all of the oil discharge lines (OptiMax) and direct them into a graduated cylinder in order to measure oil output, then use a Scan Tool or the Shift Switch Activation to start the auto-prime sequence (for more details, please refer to Bleeding The Oil Injection System in this section). One downside to this

Oil Tank and Pump Capacities

Model	Oil Tank Capacity Approx. Running Time	Reserve Capcity Approx. Running Time	Oil Output W/ Engine @ 1500 rpm & Pump at @ WOT
30/40 (2-cyl)	50.5 oz. (1.5L); 4.7 hours @ 5250 rpm	30 minutes @ 5250 rpm	8.5cc of oil in 10 Minutes w/ motor @ 900 rpm
40 (3-cyl)	3.0 qts. (2.8L); 7 hours @ WOT	14.5 oz. (0.43L); 30 minutes @ WOT	12-18cc of oil in 10 minutes
40 Jet/50/55/60 (3-cyl)	3.0 qts. (2.8L); 7 hours @ WOT	14.5 oz. (0.43L); 30 minutes @ WOT	19-25cc of oil in 10 minutes
65 Jet/ 75/90	3.2 qts. (3.0L); 6 hours @ WOT	1 qt. (0.95L); 1 hour @ WOT	27-32cc of oil in 10 minutes
80 Jet/115/125	5.13 qts. (4.9L); 5 hours @ WOT	1 qt. (0.95L); 50 min @ WOT	43-51cc of oil in 10 minutes
75/80 Jet/90/115 Optimax	n/a - ECM varies output	n/a - ECM varies output	102-118cc output with autoprime
110 Jet/135/150/175 Optimax	n/a - ECM varies output	n/a - ECM varies output	102-118cc output with autoprime
150 (XR6,Classic) & 200	3 ga. (11.4L); 6.6 hours @ WOT	0.94 qt. (0.89L); 30-35 min @ WOT	15cc of oil in 3 Minutes w/ motor @ 1000 rpm
150/175/200 (2.5L) EFI			
2001	3 ga. (11.4L); 6.6 hours @ WOT	0.94 qt. (0.89L); 30-35 min @ WOT	15cc of oil in 3 Minutes w/ motor @ 1000 rpm
2002-05	3 ga. (11.4L); 6.6 hours @ WOT	0.74 qt. (0.70L); 20-25 min @ WOT	26cc output with autoprime
200/225/250 EFI (3.0L)			
2001	3 ga. (11.4L); 6.0 hours @ WOT	1.50 qt. (1.45L); 30-35 min @ WOT	31.5cc of oil in 3 Minutes
2002-05	3 ga. (11.4L); 6.0 hours @ WOT	1.50 qt. (1.45L); 30-35 min @ WOT	26cc output with autoprime
200/225/250 Optimax	n/a - ECM varies output	n/a - ECM varies output	102-118cc output with autoprime ①

① Specification is for all but Pro/XS/Sport models where output during autoprime should be 80-92cc

SELOC_1418

Fig. 31 Oil Tank & Pump Capacities

procedure is that if you do not have a scan tool, the Shift Switch Activation method will reset the ECM to Break-In oil ratios for the next 120 minutes of operation.

To check the oil pump output of any mechanical pump (SPVR) proceed as follows:

1. Connect a remote fuel source to the powerhead with a 50:1 pre-mix fuel/oil mixture.

2. Install a flush device to the lower unit or place the outboard in a test tank.

3. Remove the cowling and disconnect the oil pump output line (clear hose) from the fuel line T-fitting (carbureted models) or the EFI vapor separator tank (VST). Plug the open fitting to prevent any fuel leakage while the outboard is operating.

4. Disconnect the link rod from the oil pump lever. Set the oil pump lever to the WOT position, which is normally FULLY CLOCKWISE, except on the following motors on which it is normally FULLY COUNTERCLOCKWISE:

- 40/40 Jet/50/55/60 Hp 3-Cyl Models
- 75/65 Jet/90 Hp Models
- 80 Jet/115/125 Hp Models

5. Place the oil pump output hose into a graduated container, 0-250cc or equivalent. Start the powerhead and operate it at the rpm as specified in the accompanying chart and for the length of time specified (usually either 10 minutes for inline motors or 3 minutes for V6 motors).

If the oil injection pump output is less than specified, the pump will need to be replaced.

Oil Lines

CAUTIONS

- Do not bend or twist the oil lines when installing.
- When installing clips, position the tabs toward the inside and make sure they are not in contact with other parts.
- Check the oil lines, when installed in position, do not come in contact with rods and levers during engine operation.
- Secure all valves and sensors using their original fasteners.
- Install hose protectors in their original positions.
- Extreme caution should be taken not to scratch or damage oil lines.
- Do not excessively compress an oil line when installing clamps.
- Always use factory type clamps when installing fuel lines. Never use screw type clamps.
- When installing the oil tank, ensure oil lines will not be pinched between the tank and the powerhead.

COOLING SYSTEM

◆ See Figure 32

All Mercury outboard engines are equipped with a raw water cooling system, meaning that sea, lake or river water is drawn through a water intake in the gearcase lower unit and pumped through the powerhead by a water pump impeller. The exact mounting and location of the pump/impeller varies only slightly from model-to-model, but on nearly all Mercury motors it is mounted to the lower unit along the gearcase-to-intermediate section split line. We say *nearly* all, because on certain versions of the single-cylinder 2.5 hp motor (specifically non-shifting versions) the water pump is instead mounted under a plate immediately in front of the propeller.

Poor operating habits can play havoc with the cooling system. For instance, running the engine with the water pick-up out of water can destroy the water pump impeller in a matter of seconds. Running in shallow water, kicking up debris that is drawn through the pump, can not only damage the pump itself, but send the debris throughout the entire system, causing water restrictions that create overheating.

For many boaters, annual replacement of the water pump impeller is considered cheap insurance for a trouble-free boating season. This is probably a bit too conservative for most people, but after a number of trouble-free seasons, an impeller doesn't owe you anything and you should consider taking the time and a little bit of money necessary to replace it. Remember that should an impeller fail you'll be stranded. Not to mention that a worn impeller will simply supply less cooling water than required by specification, allowing the powerhead to run hot placing unnecessary stress on components and best or risking overheating the powerhead at worst. And, to make matters even worse, should an impeller disintegrate the broken bits will likely travel up through the system, potentially forming blocks/clogs in cooling passages.

■ While impeller replacement is no longer usually an annual task, most people we talk to still recommend the impeller be replaced every 2nd or 3rd season to ensure trouble-free operation.

All but the smallest motors (the single cylinder models) are equipped with a thermostat that restricts the amount of cooling water allowed into the powerhead until the powerhead reaches normal operating temperature. The purpose of the thermostat is to increase engine performance and reduce emissions by making sure the engine warms as quickly as possible to operating temperature and remains there during use under all conditions. Running a motor without a thermostat may prevent it from fully warming, not only increasing emissions and reducing fuel economy, but it will likely lead to carbon fouling, stumbling and poor performance in general. It can even damage the motor, especially if the motor is then run under load (such as full-throttle operation) without allowing it to thoroughly warm).

On the other hand, a restricted thermostat can promote engine overheating. However the good news is that should you be caught on the water with a restricted thermostat, you should be able to easily remove it and get back to shore, just make sure you replace it before the next outing.

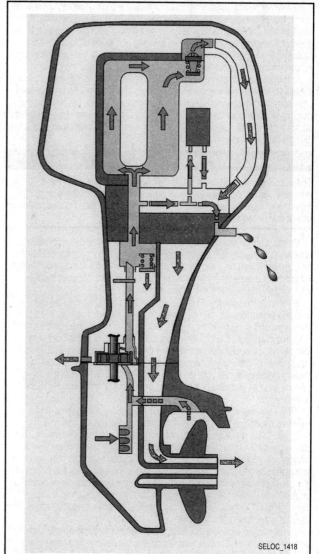

SELOC_1418

Fig. 32 Cut-away view of a typical outboard cooling system showing water flow

■ **Most V6 motors are equipped with a separate thermostat for each cylinder bank.**

In addition to a thermostat to control water flow, most mid-to-large range powerheads (generally 65 hp and larger motors) also contain a spring-loaded poppet valve designed to prevent over-pressurization of the system. The valve is either located under an extension on the thermostat housing cover (most inline motors) or is located at the base of the powerhead under its own cover (most V6 motors).

The water intake grate and cooling passages throughout the powerhead and gearcase comprise the balance of the cooling system. Both components require the most simple, but most frequent maintenance to ensure proper cooling system operation. The water intake grate should be inspected before and after each outing to make sure it is not clogged or damaged. A damaged grate could allow debris into the motor that could clog passages or damage the water pump impeller (both conditions could lead to overheating the powerhead). Cooling passages have the tendency to become clogged gradually over time by debris and corrosion. The best way to prevent this is to flush the cooling system **after each use** regardless of where you boat (salt or freshwater). But obviously, this form of maintenance is even more important on vessels used in salt, brackish or polluted waters that will promote internal corrosion of the cooling passages.

Description & Operation

◆ See Figures 33 thru 37

The water pumps used on most outboards are a displacement type water pump. Water pressure is increased by the change in volume between the impeller and the pump case. The water pump uses an impeller driven by the driveshaft, sealing between an offset housing and lower plate to create a flexing of the impeller blades. The rubber impeller inside the pump maintains an equal volume of water flow at most operating speeds.

At low speeds the pump acts like a full displacement pump with the longer impeller blades following the contour of the pump housing. As pump speed increases, and because of resistance to the flow of water, the impeller vanes bend back away from the pump housing and the pump acts like a centrifugal pump. If the impeller blades are short, they remain in contact throughout the full rpm range, supplying full pressure.

✳✳ SELOC WARNING

The outboard should never be run without water, not even for a moment. As the dry impeller tips come in contact with the pump housing or insert, the impeller will be damaged. In most cases, damage will occur to the impeller in seconds.

On most powerheads (30 hp and larger), if the powerhead overheats, a warning circuit is triggered by a temperature switch to signal the operator of an overheat condition. This should happen before major damage can occur. Reasons for overheating can be as simple as a plastic bag over the water inlet, or as serious as a leaking head gasket.

Whenever the powerhead is started and the cooling system begins pumping water through the powerhead, a water indicator stream will appear from a cooling system indicator in the engine cover. The water stream fitting may become blocked with debris (especially when lazy operators fail to flush the system after each use, yes we said LAZY, does this mean YOU?) and cease flowing. Should this occur, it might lead one to suspect a cooling system malfunction. Be sure to check and clean the opening in the fitting using a stiff piece of wire before testing or inspecting other cooling system components.

Whenever water is pumped through the powerhead it absorbs and removes excessive heat. This means that anytime a motor begins to overheat, there is not enough (or no) water flowing to the powerhead. This can happen for various reasons, including a damaged or worn impeller, clogged intake or water passages or a stuck closed/restricted thermostat. A sometimes overlooked cause of overheating is the inability of the linings of the cooling passage to conduct heat. Over time, large amounts of corrosion deposits will form, especially on engines that have not received sufficient maintenance. Corrosion deposits can insulate the powerhead passages from the raw water flowing through them.

As mentioned earlier, a thermostat is used to control the flow of engine water, to provide fast engine warm-up and to regulate water temperatures. An element in the thermostat expands when heated and contracts when cooled. The element is connected through a piston to a valve. When the element is heated, pressure is exerted against a rubber diaphragm, which forces the valve to open. As the element is cooled, the contraction allows a spring to close the valve. Thus, the valve remains closed while the water is cold, limiting circulation of water.

As the engine warms, the element expands and the thermostat valve opens, permitting water to flow through the powerhead. This opening and closing of the thermostat permits enough water to enter the powerhead to keep the engine within operating limits.

In addition to the normal flow of cooling water on carbureted and EFI motors, OptiMax motors also need to direct water flow to a fuel cooler (found in the port fuel rail on V6 models) and to the air compressor assembly. OptiMax motors have the most complicated water flow of all the Mercury 2-stroke outboards.

Troubleshooting the Cooling System

◆ See Figures 33 thru 37

Poor operating habits can play havoc with the cooling system. For instance, running the engine with the water pick-up out of water can destroy the water pump impeller in a matter of seconds. Running in shallow water, kicking up debris that is drawn through the pump, can not only damage the pump itself, but send the debris throughout the entire system, causing water restrictions that create overheating.

Symptoms of overheating are numerous and include:
• A "pinging" noise coming from the engine, commonly known as detonation
• Loss of power
• A burning smell coming from the engine
• Paint discoloration on the powerhead in the area of the spark plugs and cylinder heads

■ **If these symptoms occur, immediately seek and correct the cause (and CHECK the function of the overheat warning system)! If the engine has overheated to the point where paint has discolored, it may be too late to save the powerhead. Powerheads in this state usually require at least partial overhaul.**

So what are major causes of overheating? Well the most prevalent cause is lack of maintenance. Other causes which are directly attributable to lack of maintenance or poor operating habits are:
• Fuel system problems causing lean mixture
• Incorrect oil mixture in fuel or a problem with the oil injection system
• Spark plugs of incorrect heat range
• Faulty thermostat
• Restricted water flow through the powerhead due to sand or silt buildup
• Faulty water pump impeller
• Sticking thermostat

■ **When troubleshooting the cooling system, especially for overheat conditions, the motor should be run in a test tank or on a launched vessel (to simulate normal running conditions.). Running the motor on a flushing device may provide both a higher volume of water than the system would deliver (and often times with water that is much colder than would be found in your local lake or bay).**

A water-cooled powerhead has a lot of potential problems to consider when talking about overheating. The most overlooked tends to be the simplest, clogged cooling passages or water intake grate. Although a visual inspection of the intake grate will go a long way, the cooling passage condition can really only be checked by operating the motor or disassembling it to observe the passages.

Damaged or worn cooling system components tend to cause most other problems. And since there are relatively few components, they are easy to discuss. The most obvious is a thermostat that is damaged or corroded, which will often cause the motor to run hot or cold (depending on the position in which the thermostat is stuck, closed or open). The other most obvious component is the water pump impeller is really the heart of the cooling system and it is easy to check, easy that is once it is accessed.

■ **A thermostat or poppet valve (on larger motors) which is stuck open will prevent the motor from fully warming. Running cold the motor will likely stumble or hesitate or otherwise run rough at idle.**

Periodic inspection and replacement of the water pump impeller is a mainstay for many mariners. There are those that wouldn't really consider launching their vessel at the beginning of a season without performing this check. If the water pump is removed for inspection, check the impeller and

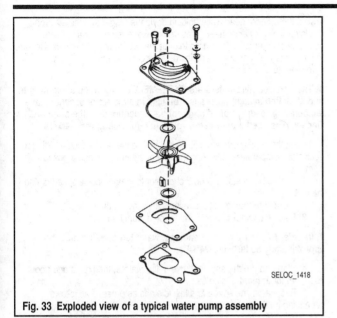

Fig. 33 Exploded view of a typical water pump assembly

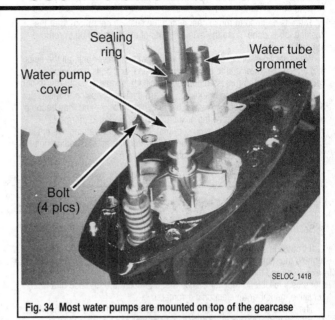

Fig. 34 Most water pumps are mounted on top of the gearcase

Fig. 35 Cutaway view of a water pump showing the relationship between the housing, impeller and driveshaft

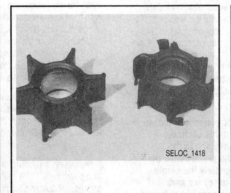

Fig. 36 New impeller (left) alongside an impeller no longer fit for service (right)

Fig. 37 Check the water intake for debris before/after each use

housing for wear, grooves or scoring that might prevent proper sealing. Check for grooves in the driveshaft where the seal rides. Any damage in these areas may cause air or exhaust gases to be drawn into the pump, putting bubbles into the water. In this case, air does not aid in cooling. When inspecting the pump, consider the following:

• Is the pump inlet clear and clean of foreign material or marine growth? Check that the inlet screen is totally open. How about the impeller?

• Try and separate the impeller hub from the rubber. If it shows signs of loosening or cracking away from the hub, replace the impeller.

• Has the impeller taken a set, and are the blade tips worn down or do they look burned? Are the side sealing rings on the impeller worn away? If so, replace the impeller.

Remember that the life of the powerhead depends on this pump, so don't reuse any parts that look damaged. Are any parts of the impeller missing! If so, they must be found. Broken pieces will migrate up the water tube into the water jacket passages and cause a restriction that could block a water passage. It can be expensive or time consuming to locate the broken pieces in the water passages, but they must be found, or major damage could occur.

■ Impellers are made of highly durable materials these days and although some people still replace them annually, it is much more common to get 2 or 3 seasons out of an impeller before replacing it. And, even after 2-3 seasons, impellers are more often replaced out of preventive maintenance than for necessary repairs.

The best insurance against breaking the impeller is to replace it at the beginning of each boating season (or after a couple of seasons, are you sensing a pattern here?), and to NEVER run it out of the water. If installing a metallic body pump housing, coat all screws with non-hardening sealing compound to retard galvanic corrosion. A water tube normally carries the water from the pump to the powerhead. Grommets seal the water tube to the water pump and exhaust housing at each end of the tube, and can deteriorate. Also, the water tube(s) should be checked for holes through the side of the tube, for restrictions, dents, or kinks.

Overheating at high rpm, but not under light load, may indicate a leaking head gasket. If a head gasket it leaking, water can go into the cylinder, or hot exhaust gases may go into the water jacket, creating exhaust bubbles and excessive heat. Remember that aluminum heads have a tendency to warp, and need to be checked (if not re-surfaced) each time they are removed. If necessary, they can be resurfaced by using emery paper and a surface block moving in a figure-eight motion. If this is done, you should also inspect the cylinders and pistons for damage. Other areas to consider are the exhaust cover gaskets and plate. Look for corrosion pin holes. This is rare, but if the outboard has been operated in salt water over the years, there may just be a problem.

If the outboard is mounted too high on the transom, air may be drawn into the water inlet or sufficient water may not be available at the water inlet. When underway the outboard anti-ventilation plate should be running at or near the bottom of the boat and parallel to the surface of the water. This will allow undisturbed water to come to the lower unit, and the water pick-up should be able to draw sufficient water for proper cooling.

Whenever the outboard has been run in polluted, brackish or saltwater, the cooling system should be flushed. Follow the instructions provided under Flushing the Cooling System in the Engine Maintenance section for more details. But in most cases, the outboard should be flushed for at least 5 minutes. This will wash the salt from the castings and reduce internal corrosion. If the outboard is small and there is no flushing tool that will fit, run the outboard in a tank, drum, or large sturdy bucket.

There is no need to run in gear during the flushing operation. After the flushing job is done, rinse the external parts of the outboard off to remove the salt spray.

When service work is done on the water pump or lower unit, all the bolts that attach the lower unit to the exhaust housing, and bolts that hold the water pump housing (unless otherwise specified), should be coated with non-hardening gasket sealing compound to guard against corrosion. If this is not done, the bolts may become seized by galvanic corrosion and may become extremely difficult to remove the next time service work is performed.

Last but not least, check to be sure that the overheat warning system is working properly. On most motors, grounding the wire at the sending unit will cause the horn to sound and/or a light should turn on.

TESTING COOLING SYSTEM EFFICIENCY

◆ See Figure 38

If trouble is suspected, cooling system efficiency can be checked by running the motor in a test tank or on a launched boat (while an assistant navigates) and monitoring cylinder head temperatures. There are 2 common methods available to monitor cylinder head temperature, the use of a heat sensitive marker or an electronic pyrometer.

The Stevens Instrument company markets a product known as the Markal Thermomelt Stik®. This is a physical marker that can be purchased to check different heat ranges. The marker is designed to leave a chalky mark behind on a part of the motor that will remain solid and chalky until it is warmed to a specific temperature, at which point the mark will melt appearing liquid and glossy. When using a Thermomelt Stik or equivalent marker, markers of 2 different heat specifications are necessary for this test. One, the low temperature range marker, should be rated just equal to or very slightly less than the range of the Thermostat Opening rating. The second, the high temperature range marker should be just equal to or slightly less than the Overheat Thermoswitch On rating.

■ **To determine normal operating temperatures for you motor, please refer to the Cooling System Specifications chart in this section. Generally speaking, engines should run at or slightly above the thermostat opening temperature or anywhere in the range below the temperature sensor continuity heat range. Unfortunately the single cylinder models have no thermostat or temperature sensor and therefore no specs, but you can probably make a decent guess based on other models in their general hp and size range that a normal operating temperature would be around 120°F (49°C) give or take 10% or so.**

Alternately, an electronic pyrometer may be used. Many DVOMs are available with thermosensor adapters that can be touched to the cylinder head in order to get a reading. Also, some instrument companies are now producing relatively inexpensive infra-red pyrometers (such as the Raytek® MiniTemp®) of a point-and-shoot design. These units are simply pointed

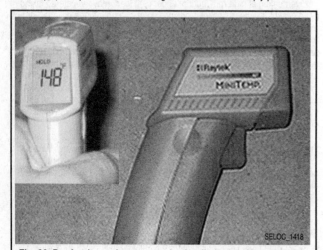

SELOC_1418

Fig. 38 Bar far the easiest way to check cylinder head temperature is with a hand-held pyrometer like the MiniTemp® from Raytek® pictured here

toward the cylinder head while holding down the trigger and the electronic display will give cylinder head temperature. For ease of use and relative accuracy of information, it is hard to beat these infra-red pyrometers. Be sure to follow the tool manufacturer's instructions closely when using any pyrometer to ensure accurate readings.

■ **The infra-red pyrometers are EXTREMELY useful when attempting to find a cooling system blockage. You can, in a matter of seconds, take temperature readings of all different parts/locations on the outboard. Once you've used a pyrometer, you won't go back to the markers.**

To test the cooling system efficiency, obtain either a Thermomelt Stik (or equivalent temperature indicating marker) or a pyrometer and proceed as follows:

1. If available, install a shop tachometer to gauge engine speed during the test.
2. Make sure the proper propeller is installed on the motor.
3. Place the motor in a test tank or on a launched craft.

■ **In order to ensure proper readings, water temperature must be approximately 60-80°F (18-24°C).**

4. Start and run the engine at idle until it warms, then run it and about 3000 rpm for **at least** 5 minutes.
5. Reduce engine speed to a low idle and as proceed as follows depending on the test equipment:
   • If using Thermomelt Stiks, make 2 marks on the powerhead/cylinder head (at the thermostat housing), one with the low-range marker and one with the high-range marker. Continue to operate the motor at idle. The low-range mark must turn liquid and glossy or the engine is being overcooled (if equipped, check the thermostat for a stuck open condition). The high-range mark must remain chalky, or the motor is overheating (if equipped, check the thermostat for a stuck closed condition and then check the cooling system passages and the water pump impeller).

■ **For most models, temperature readings should be taken on the cylinder head or powerhead (at or near the thermostat housing). For details on thermostat opening temperature specifications please refer to the Cooling System Specifications chart in this section.**

   • If using a pyrometer, take temperature readings on the cylinder head and powerhead. The meter should indicate temperatures between the thermostat opening temp and the overheat sensor off specification (when available). On smaller motors which do not use a thermostat and/or an overheat sensor temperatures should usually be in approximately 125-155°F (53-67°C) range, otherwise it is likely that the engine is being over/under cooled. If equipped, check the thermostat first for either condition and then suspect the cooling water passages and/or the impeller.
6. Increase engine speed to wide open throttle (WOT) rpm and continue to watch the markers or the reading on the pyrometer. The engine must not overheat at this speed either or the system components must be examined further.

■ **When checking the engine at speed expect temperatures to vary slightly from the idle test. Some models will run slightly hotter and some slightly cooler due to the differences in volume of water delivered by the cooling system when compared with engine load. This is normal. In all cases (when equipped), temperatures should not exceed the overheat warning sensor on temperature.**

TESTING THE THERMOSTAT

◆ See Figures 39, 40 and 41

Most 6 hp and larger motors are equipped with a thermostat that restricts the amount of cooling water allowed into the powerhead until the powerhead reaches normal operating temperature. The purpose of the thermostat is to prevent the water from starting to cool the powerhead until the powerhead has warmed to normal operating temperature. In doing this the thermostat will increase engine performance and reduce emissions.

However, this means that the thermostat is vitally important to proper cooling system operation. A thermostat can fail by seizing in either the open or closed positions, or it can, due to wear or deterioration, open or close at the wrong time. All failures would potentially affect engine operation.

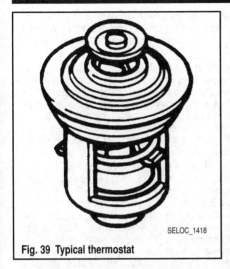

Fig. 39 Typical thermostat

SELOC_1418

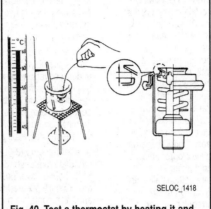

SELOC_1418

Fig. 40 Test a thermostat by heating it and observing at what temperatures it opens/closes

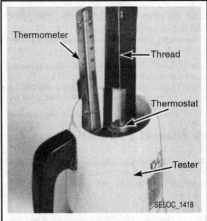

Thermometer

Thread

Thermostat

Tester

SELOC_1418

Fig. 41 Testing a t-stat using a thread between the valve and body

A thermostat that is stuck open or will not fully close, may prevent a powerhead from ever fully warming, this could lead to carbon fouling, stumbling, hesitation and all around poor performance. Although these symptoms could occur at any speed, they are more likely to affect most motors at idle when high water flow through the open orifice will allow for more cooling than the lower production of heat in the powerhead requires.

A thermostat that is stuck closed will usually reveal itself right away as the engine will not only come up to temperature quickly, but the powerhead will begin to overheat and the temperature warning circuit (if so equipped) should be triggered shortly thereafter. However, a thermostat that is stuck partially closed may be harder to notice. Cooling water may reach the powerhead and keep it within a normal operating temperature range at various engine rpm, but allow heat to build up at other rpm. Generally speaking, engines suffering from this type of thermostat failure will show symptoms at part or full throttle, but problems can occur at idle as well. Symptoms, besides overheating, may include hesitation, stumbling, increased noise and smoke from the motor and, general, poor performance.

Testing a thermostat is a relatively easy proposition. Simply remove the thermostat from the powerhead and suspend it in a container of water, then heat the water watching for the thermostat element to move (open) and noting at what temperature it accomplishes this. If you suspect a faulty or inoperable thermostat and cannot seem to verify proper opening/closing temperatures, it may be a good idea (especially since you've already gone through the trouble of removing the thermostat) to simply replace it (it's a relatively low cost part, that performs an important function). Doing so should remove it from suspicion for at least a couple of seasons.

1. Locate and remove the thermostat from the powerhead, as detailed in this section.

2. Suspend the thermostat and a thermometer in a container of water. For most accurate test results, it is best to hang the thermostat and a thermometer using lengths of string so that they are not touching the bottoms or sides of the container (this ensures that both components remain at the same temperature as the water and not the container).

■ One great trick to help tell the INSTANT the thermostat starts to open is to gently open the thermostat valve by hand and insert a length of thread through the t-stat body allowing the valve to close against the thread. Suspend the thermostat by THIS thread, and check the temperature the instant it falls off.

3. Slowly heat the water while observing the thermostat for movement. The moment you observe movement, check the thermometer and note the temperature. If the water begins to boil (reaches about 212°F/100°C at normal atmospheric pressure) and NO movement has occurred, discontinue the test and throw the piece of junk thermostat away (if you are *sure* there was no movement).

4. Remove the source of heat and allow the water to cool (you can speed this up a little by adding some cool water to the container, but if you're using a glass container, don't add too much or you'll risk breaking the container). Observe the thermostat again for movement as the water cools. When movement occurs, check the thermometer and record the temperature.

5. Compare the opening temperature with the specifications provided in the Cooling System Specifications chart in this section (or with the stamping on top of the thermostat itself).

6. Specifications for closing temperatures are not provided by the manufacturer, but typically a thermostat must close at a temperature near, but below the temperature for the opening specification. A slight modulation (repeated opening and closing) of the thermostat will normally occur at borderline temperatures during engine operation keeping the powerhead in the proper operating range.

7. Replace the thermostat if it does not operate as described, or if you are unsure of the test results and would like to eliminate the thermostat as a possible problem. Refer to the removal and installation procedure for Thermostat in this section for more details.

Thermostat

Most 6 hp and larger motors are equipped with a thermostat that restricts the amount of cooling water allowed into the powerhead until the powerhead reaches normal operating temperature. The purpose of the thermostat is to prevent the water from starting to cool the powerhead until the powerhead has warmed to normal operating temperature. In doing this the thermostat will increase engine performance and reduce emissions.

On all models so equipped, the thermostat components are mounted somewhere on the powerhead, in a cooling passage, under an access cover that is sealed using a gasket or an O-ring. On some Mercury models they are on the side of the powerhead itself (exhaust cover/block), however on MOST Mercury models they are mounted to the top of the cylinder head/cover.

REMOVAL & INSTALLATION

◆ **See Figures 42 thru 49**

② *MODERATE*

■ **For additional art showing the thermostat and cover mounting, please refer to the Powerhead section.**

On all models, the thermostat assembly is mounted under a cover on the powerhead. The size, shape and location of this cover, including the number of components and seals found underneath varies slightly by model. However, even with this said the service procedures are virtually identical for all Mercury outboards.

■ **On a few motors, the cover can be installed facing different directions. For these models, matchmark the cover to the mating surface or otherwise make a note of cover orientation to ensure installation facing the proper direction. This is especially important for covers to which hoses attach (that have been removed).**

1. Disconnect the negative battery cable (if equipped) and/or remove the spark plug wire(s) from the plug(s) and ground them on the powerhead for safety.

2. Remove the engine top cover for access.

3. Locate the thermostat housing on the powerhead and remove any interfering components (flywheel or side cover), then remove the cover and thermostat as follows:

- 6/8, 9.9/10/15 and 20/20 Jet/25 hp models: the thermostat is located under a small cover secured by 2 bolts to the top of the cylinder block cover (with a hose for the cooling stream indicator attached to a fitting at the center of the cover). The cover uses an irregularly shaped gasket which matches the cover profile, PLUS the thermostat uses a large round seal. All indications are both should be installed dry. However, should the hose fitting be removed from the cover for any reason, be sure to apply a light coating of Loctite® PST Pipe Sealant or equivalent to the threads before installation.

- 30/40 hp 2-cylinder models: the thermostat and the spring-loaded pressure poppet valve are both located under a small irregular shaped cover on the top, side of the powerhead. The cover itself is normally secured by 4 bolts and sealed with a gasket. In addition, under the smaller round portion of the cover resides a thermostat assembly and thermostat seal. Under the larger rounded portion of the cover there is (in this order working inward from the cover) a spring, screw, cup-shaped washer, diaphragm and poppet. All indications are that the cover gasket and thermostat seal are both installed dry.

- 40/40 Jet/50/55/60 hp 3-cylinder models: the thermostat on these motors (like the 30/40 hp twin) is located under a small irregular shaped cover on the top, side of the powerhead. The cover itself is normally secured by 4 bolts and sealed with a gasket. In addition, the thermostat assembly is either sealed using an O-ring located just under the thermostat assembly OR a seal which mounts on top of the assembly (this varies by year and model, but generally speaking the 40/40 Jet//50 hp models use the seal on top, while the 55/60 hp models are more likely to have the O-ring underneath). All models mount the thermostat assembly in a carrier insert which is positioned in the powerhead. Like most other Mercury motors, all indications are that the both the O-ring or seal and cover gasket are installed dry.

- 75/65 Jet/90 and 80 Jet/115/125 hp models: the thermostat and the spring-loaded pressure poppet valve are both located under a small irregular shaped cover mounted vertically on the upper end of the cylinder cover. The thermostat cover itself is normally secured by 5 bolts and sealed with a gasket. In addition, under the higher, smaller round portion of the cover resides a thermostat assembly and thermostat seal (some models, mostly 3-cyl. motors usually have a washer as well). Under the larger rounded portion of the cover there is (in this order working inward from the cover) a spring, screw, cup-shaped washer, diaphragm, poppet, carrier (not on all models) and grommet. All indications are that the cover gasket and thermostat seal are both installed dry.

- 75/90/115 hp Optimax models: the thermostat and the spring-loaded pressure poppet valve are both located under a small irregular shaped cover mounted vertically (at something of an angle) on the upper end of the cylinder head. The cover itself is normally secured by 4 bolts and sealed with a gasket. Under the higher, smaller round portion of the cover resides a thermostat assembly and thermostat seal. Under the lower, larger rounded portion of the cover there is (in this order working inward from the cover) a spring, screw, cup-shaped washer, diaphragm and poppet (and some models may use a spacer under the poppet). All indications are that the cover gasket and thermostat seal are both installed dry.

- 135-200 hp (2.5L) V6 models: A thermostat is located under a small round (carb/EFI) or roughly oval (OptiMax) shaped cover positioned at the top of each cylinder head. The covers are normally secured by 2 bolts. Because they are usually both symmetrical AND contain a hose outlet, you should either matchmark them or note their orientation before removal. Under each cover is a seal and a thermostat assembly.

In addition to the thermostats, these motors use a water pressure poppet valve assembly (often referred to as a relief valve). The assembly is mounted under a cover on the lower starboard side of the powerhead (base of the block). Under the cover reside the following components; in this order from the cover toward the block: a gasket, screw, washer, diaphragm, water deflector, plate, gasket, spring, washer (carb or 2001 EFI only), poppet, grommet and carrier. All indications are that the poppet assembly gaskets and thermostat seals are all installed dry.

- 200-250 hp (3.0L) V6 models: A thermostat is located under a small rounded/square-ish (OptiMax and 2002 or later EFI) or roughly oval (2001 EFI) shaped cover positioned at the top of each cylinder head. The covers are normally secured by 2 bolts. Because they are sometimes symmetrical and often contain a hose outlet, you should either matchmark them or note their orientation before removal. What is under each cover varies slightly by model/year. On 2001 EFI motors the roughly oval shaped cover is over a seal and thermostat. However the more rounded/square shaped covers used by 2002 or later EFI motors, as well as all OptiMax motors use a gasket for sealing, though it is true that both the OptiMax as well as 2002 EFI units (but usually not the 2003 or later EFI) will contain a thermostat seal as well. One thing that all of these motors have in common is that the threads of the cover retaining bolts must be lightly oiled (using clean 2-stroke oil) before installation and tightening.

In addition to the thermostats, these motors use a water pressure poppet valve assembly (often referred to as a relief valve). The assembly is mounted under a cover on the lower side of the powerhead exhaust adapter plate (just under the mating surface to which the block attaches). Under the cover reside the following components, in this order from the cover toward the exhaust adapter, a gasket, screw, washer, diaphragm, plate, gasket, spring, poppet, grommet and carrier. All indications are that the poppet assembly gaskets and thermostat gaskets/seals are all installed dry.

4. Loosen and remove the bolts (as noted in the previous step) securing the cover, then carefully pull the cover from the powerhead. If necessary, tap around the outside of the cover using a rubber or plastic mallet to help loosen the seal.

■ On V6 OptiMax motors, a line from the air/fuel rail assembly is sometimes in partially in the way of one or more of the bolts and makes removing the thermostat housing just a little more difficult. But in most cases there is access, using a wrench or thin socket/extension.

5. Check if the seal or gasket was removed with the cover. When a composite gasket and/or sealant was used, make sure all traces of gasket and sealant material are removed from the cover and the powerhead mounting surface. For installation purposes, take note of the direction in which the thermostat is facing before removal.

Fig. 42 Thermostat and poppet valve cover - 30/40 hp 2-cyl models

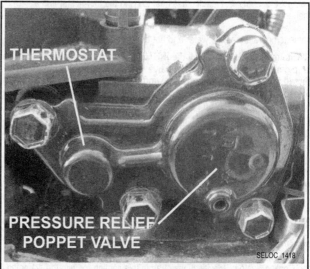

Fig. 43 Component locations under cover - 30/40 hp 2-cyl models

Fig. 44 Typical thermostat and poppet valve cover for most inline 3-cyl and 4-cyl motors

Fig. 45 Like the 30/40 hp, note component locations in relation to shape of cover

Fig. 46 Thermostat housings are found on top of the cylinder head for V6 motors

Fig. 47 This style was used on nearly all 2.5L motors and 2001 3.0L EFI models

Fig. 48 The more square shaped thermostat cover is used on most 3.0L motors. . .

Fig. 49 . . .on OptiMax models, air/fuel rail hoses may partially block access

■ None of the factory instructions we could find for these motors mentioned the use of any sealant so we suspect most SHOULD NOT use sealant on the thermostat housing gasket (or seal, or poppet/pressure relief valve gasket/seal). Also, the thermostats on most Mercury motors are installed with the flange side facing the cover.

6. Visually inspect the thermostat for obvious damage including corrosion, cracks/breaks or severe discoloration from overheating. The cause of a malfunctioning thermostat is often foreign matter stuck to the valve seat. Inspect the thermostat to make sure that it is clean and free of foreign matter. Make sure any springs have not lost tension. If necessary, refer to the Testing the Thermostat in this section for details concerning using heat to test thermostat function.

To install:

7. Install each of the thermostat (and/or poppet/pressure relief valve) components in the reverse order of removal (it's pretty obvious). Replace any gaskets, seals and/or O-rings. Pay close attention to the direction each component is installed.

8. Install the thermostat housing cover using a new gasket. Tighten the bolts until snug. Tighten the bolts alternately and evenly (for a torque specification, please refer to the Torque Specifications chart found in the Powerhead section).

9. Connect the negative battery cable and/or spark plug lead(s), then verify proper cooling system operation.

THERMOSTAT MOUNTING EXPLODED VIEWS

◆ See Figures 50 thru 59

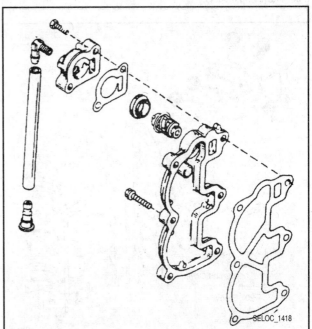

Fig. 50 Exploded view of the thermostat assembly - 6/8 and 9.9/10/15 hp models (note that 20/20 Jet/25 hp models mount in the exact same way, but t-stat and cylinder cover shape varies slightly)

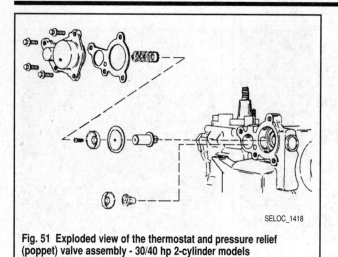

SELOC_1418

Fig. 51 Exploded view of the thermostat and pressure relief (poppet) valve assembly - 30/40 hp 2-cylinder models

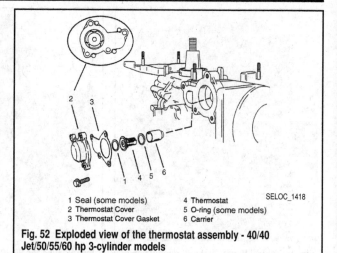

SELOC_1418

1 Seal (some models)   4 Thermostat
2 Thermostat Cover   5 O-ring (some models)
3 Thermostat Cover Gasket   6 Carrier

Fig. 52 Exploded view of the thermostat assembly - 40/40 Jet/50/55/60 hp 3-cylinder models

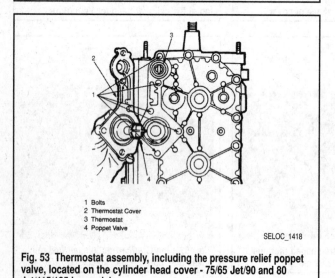

1 Bolts
2 Thermostat Cover
3 Thermostat
4 Poppet Valve

SELOC_1418

Fig. 53 Thermostat assembly, including the pressure relief poppet valve, located on the cylinder head cover - 75/65 Jet/90 and 80 Jet/115/125 hp models

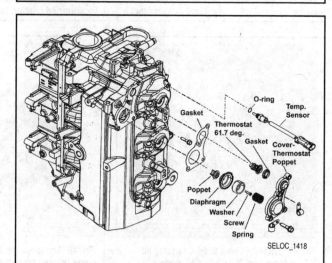

SELOC_1418

Fig. 54 Exploded view of the thermostat and pressure relief (poppet) valve assembly - 75/90/115 hp Optimax models

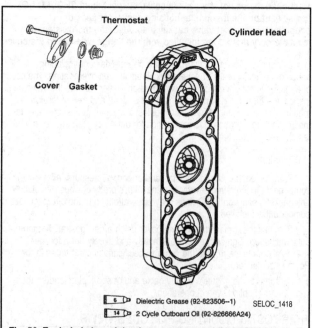

1 - Thermostat
2 - Gasket
3 - Thermostat Cover
4 - Bolt
5 - 90° Elbow
6 - Hose
7 - Sta-Strap

SELOC_1418

Fig. 55 Exploded view of the thermostat assembly - 150-200 hp (2.5L) carbureted and EFI V6 models

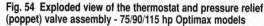

6   Dielectric Grease (92-823506--1)   SELOC_1418
14 2 Cycle Outboard Oil (92-826666A24)

Fig. 56 Exploded view of the thermostat assembly - 135-200 Hp (2.5L) OptiMax and 2001 200-250 hp (3.0L) EFI V6 models

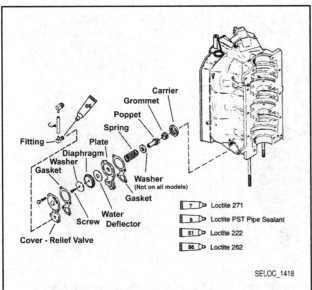

Fig. 57 Exploded view of the pressure relief valve assembly - 135-200 hp (2.5L) V6 models

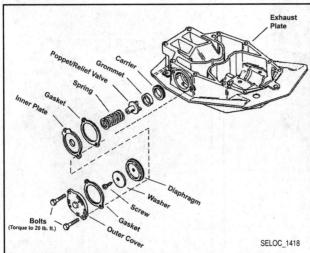

Fig. 59 Exploded view of the pressure relief valve assembly - 200-250 hp (3.0L) V6 models

Water Pump

REMOVAL & INSTALLATION

✳✳ SELOC WARNING

Since proper water pump operation is critical to outboard operation, all seals and gaskets should be replaced whenever the water pump is removed. Also, installation of a new impeller each time the water pump is disassembled is good insurance against overheating.

✳✳ SELOC CAUTION

Never turn a used impeller over and reuse it. The impeller rotates with the driveshaft and the vanes take a set in a clockwise direction. Turning the impeller over will cause the vanes to move in the opposite and result in premature impeller failure.

For additional exploded views and information on the gearcase, please refer to Gearcase in the Lower Unit section.

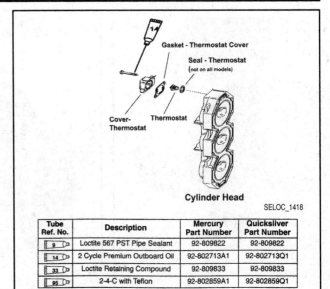

Fig. 58 Exploded view of the thermostat assembly - 200-250 hp (3.0L) OptiMax and 2002 or later EFI V6 models

Tube Ref. No.	Description	Mercury Part Number	Quicksilver Part Number
9	Loctite 567 PST Pipe Sealant	92-809822	92-809822
14	2 Cycle Premium Outboard Oil	92-802713A1	92-802713Q1
33	Loctite Retaining Compound	92-809833	92-809833
95	2-4-C with Teflon	92-802859A1	92-802859Q1

2.5/3.3 Hp (Non-Shifting) Models

◆ **See Figures 60 thru 67**

One of the features making this lower unit unique from all others covered in this service is the location of the water pump. Instead of the pump being installed on the driveshaft, as on most units, the water pump is installed on the propeller shaft.

The pump impeller may be replaced without removing or disassembling the lower unit. In fact the propeller shaft does not have to be disassembled. The only work required is to remove the propeller and a couple other simple tasks to replace an impeller.

1. Tilt the motor upward and lock it in this position for access.
2. Pull the cotter pin from the propeller shaft. A spare cotter pin is normally provided to a new owner and may be found in the spark plug access cover, courtesy of the manufacturer.
3. Slide the propeller rearward and free of the propeller shaft. Remove the shear pin. A spare shear pin may also normally be found in the spark plug access cover. Again, compliments of the manufacturer.
4. Remove the 2 bolts and washers securing the water pump cover to the lower unit.
5. Jar the water pump cover free of the lower unit using a soft head mallet. Remove the cover from the propeller shaft. Check the inside surface of the cover for signs of wear indicating foreign particles had entered the water pump.
6. Carefully pry the water pump impeller from the lower unit recess (noting the direction the impeller vanes are facing before removal). Remove and inspect the impeller drive pin.
7. Inspect the inside of the water pump housing for signs of wear or damage (or if there is a suspected problem with the bearing) it must be removed and serviced, as follows:

   a. Pull the housing from the shaft (remember there is gearcase oil behind it).

   b. Visually and physically inspect the bearing to make sure that is rolls freely and is not rusted. If any roughness can be felt and/or the bearing is obviously badly rusted, pull the bearing from the housing using a suitable slide-hammer with expander rod (such as Snap-On #CG 40-4) and 7/16 in. Collet (like Snap-On #CG 40-9).

   c. Remove and discard the old O-ring from the groove in the water pump housing flange.

   d. Using a small driver from the water pump side of the housing, carefully tap the seal out from the gearcase side of the housing.

To Install:

8. If the water pump housing was removed for service, assemble and install it as follows:

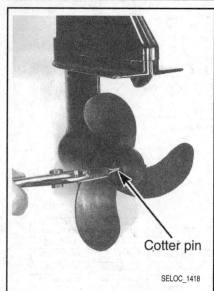

Fig. 60 Remove the propeller cotter pin. . .

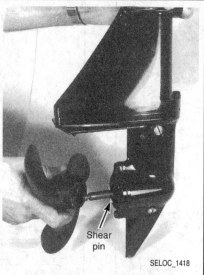

Fig. 61 . . . then remove the propeller and shear pin

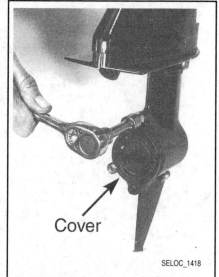

Fig. 62 Unbolt the water pump housing cover. . .

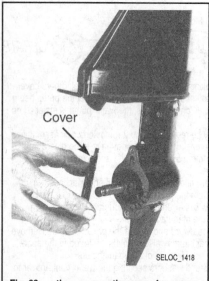

Fig. 63 . . .then remove the cover for access

Fig. 64 Carefully pry the impeller free of the housing. . .

Fig. 65 . . .then remove the impeller and drive pin

a. Coat the new housing seal lightly using 2-4-C with Teflon (or a suitable marine grade grease), then using a suitable driver or smooth socket, carefully tap the new seal into the housing. Position the seal so the spring is facing back toward the gearcase once the housing is installed.

b. Apply a light coating of gearcase lube to the new bearing assembly, then use a 13/16 in. socket positioned against the lettered side of the bearing to carefully press the bearing into the housing until it seats.

c. Apply a light coating of 2-4-C with Teflon (or a suitable marine grade grease) to the NEW housing O-ring, then install the O-ring on the housing flange.

d. Position the housing over the propeller shaft, taking care not to damage the seal or O-ring.

9. Rotate the propeller shaft until the hole in the shaft is facing the largest part of the water pump cavity in the gear housing (this opening is usually not perfectly round.) Insert the drive pin through the hole in the propeller shaft.

10. Slide a new water pump impeller onto the propeller shaft. As the impeller begins to enter the pump housing, rotate the impeller clockwise to permit the impeller vanes to curl in the proper direction. Continue to move the impeller into the housing with the cutout in the impeller indexed over the drive pin. As the name implies, the drive pin "drives" or "rotates" the impeller.

11. Install the water pump cover onto the water pump housing.

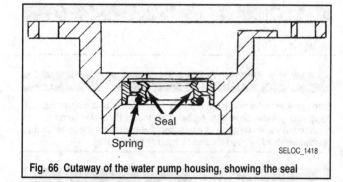

Fig. 66 Cutaway of the water pump housing, showing the seal

12. Apply Quicksilver Perfect Seal, or equivalent to the threads of the pump cover attaching bolts, then secure the cover in place with the 2 bolts (and a washer on each bolt). Tighten the bolts alternately and evenly to 50 inch lbs. (5.6 Nm).

13. Coat the propeller shaft lightly using Special Lubricant 101, 2-4-C with Teflon, Perfect Seal or a suitable marine grade grease.

14. Install the shear pin, then install the propeller using a new cotter pin.

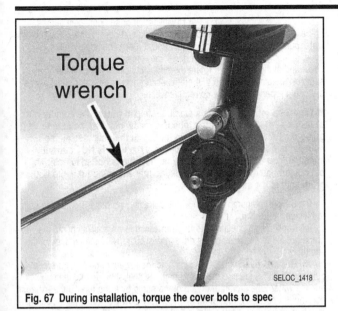

Fig. 67 During installation, torque the cover bolts to spec

2.5/3.3 Hp (Shifting) and 4/5 Hp Models

◆ See Figures 68 thru 72

The driveshaft is one continuous shaft extending from the powerhead crankshaft into the torpedo bore of the lower unit. On its path of travel from the crankshaft downward into the gearcase, the driveshaft passes through the water pump housing, so that it can rotate an impeller mounted over the shaft. Therefore, the lower unit must be removed from the exhaust housing in order to access the water pump assembly, which is mounted on top of the gearcase, along the gearcase-to-exhaust housing split line.

1. Disconnect and ground the spark plug wire for safety.

2. Rotate and swing the lower end unit in the full UP tilt position for access.

3. Remove the rubber access plug on the starboard side of the upper gear housing. If servicing a 2.5/3.3 hp unit place the gear shift in the Neutral position. If servicing a model 4/5 hp unit place the gear shift in the Reverse gear position.

4. Loosen, **but do not remove** the bolt for the shift rod clamp. Loosen the bolt only enough to allow the shift shaft to slide free of the clamp when the lower unit is removed from the exhaust housing.

5. Remove the 2 attaching bolts and washers securing the lower unit to the exhaust housing (one is located at the very front center of the housing, the other near the rear center). Pull down on the lower unit gear housing to carefully separate the lower unit from the exhaust housing. Guide the shift rod and driveshaft out of the exhaust housing as the lower unit is removed.

6. Remove the 4 bolts, flat washers and lockwashers securing the pump housing to the pump base assembly.

7. Pull up on the pump housing (on 4/5 hp motors, at the same time disengage the water tube from the lower unit housing). Slide the pump housing and gasket up and free of the driveshaft. Remove and discard the gasket.

8. Lift up and slide the water pump impeller off the driveshaft (the cartridge *usually* remains in the water pump housing, but if not, slide that off the impeller and driveshaft first).

9. Remove the drive pin from the recess in the driveshaft. Place the small drive pin into a safe place for later use.

10. If necessary, remove the guide plate and gasket.

11. Inspect the water pump housing insert for signs of wear or damage, and if necessary remove it.

12. Water tube grommets can usually be reused, but be sure to inspect them for dryness, cracking or damage and remove for replacement, as necessary.

To Install:

13. If the base plate was removed, slide the base plate gasket down the driveshaft and into place on the water pump base plate. In some cases the bolt holes are offset and/or there are a pair of alignment pins protruding from the pump base plate (meaning the gasket holes will only align properly one way), so make sure the bolt holes are properly aligned.

14. Again, only if the base plate was removed, slide the plate down the driveshaft and into place.

15. Apply just a "dab" of petroleum jelly or marine grade grease to the Woodruff key, and then place it in the driveshaft keyway. The jelly/grease will hold the key in place.

16. Slide the water pump gasket down over the driveshaft.

17. Slide the water pump impeller down the driveshaft with the keyway in the pump aligned with the Woodruff key in the driveshaft.

■ **If an old impeller is used, the impeller must be installed in the same position from which it was removed. Never turn the impeller over, thinking it will extend its life. On the contrary, the blades will probably crack and break after just a short time of operation.**

18. If removed, install the water pump cartridge into the water pump cover. There is normally a locating tab on the cartridge (insert), which must index with a slot in the cover.

19. If removed, install the water tube seal(s) to the water pump housing and/or the top of the gearcase (indexing the tabs, when present, with the holes in the mounting point). For the 1 seal used on top of the housing on 2.5/3.3 hp motors, Mercury recommends using a light coating of Bellows Adhesive on the outside of the seal (where it mates with the housing). On all water tube seals, including the 3 used on 4/5 hp motors (one on top of the gearcase, 1 on the side of the water pump and 1 on the top of the water pump) be sure to coat the inner diameter of the seal using 2-4-C with Teflon or an equivalent marine grade grease.

■ **Make sure the pump cover (housing) gasket is still in position with the bolt holes properly aligned once the cover is seated.**

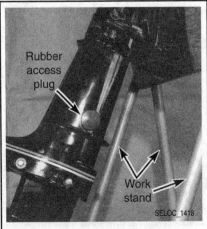

Fig. 68 Tilt the outboard for access to the gearcase

Fig. 69 Remove the rubber plug and loosen the shift coupler

Fig. 70 Remove the 2 bolts securing the gearcase

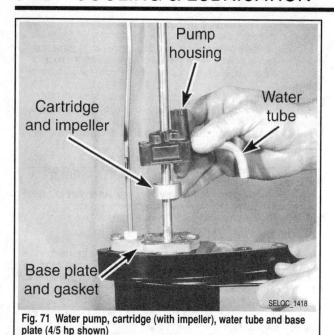

Fig. 71 Water pump, cartridge (with impeller), water tube and base plate (4/5 hp shown)

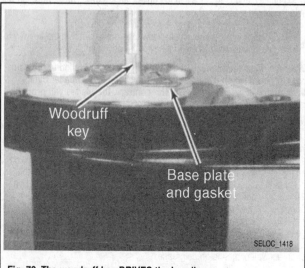

Fig. 72 The woodruff key DRIVES the impeller

20. Apply a light coating of 2-4-C with Teflon or an equivalent marine grade grease to the insert, inside the water pump housing, then slide the water pump cover down the driveshaft until it makes contact with the impeller. Apply a small amount of downward pressure on the water pump cover and at the same time rotate the driveshaft clockwise until the cover seats on the plate. Rotation of the driveshaft ensures all water pump vanes are bent properly and in the correct direction.

■ On 4/5 hp motors there is a small, L-shaped water tube coming out of the front of the pump. If the tube is already installed, be sure to guide the tube into the grommet in the top of the gearcase as the housing is positioned over the impeller. If the tube is NOT currently installed, you CAN wait until the pump housing is bolted in position, but sometimes it can be a little tricky to maneuver the tube into both grommets at the same time.

21. Secure the pump in place with the 4 attaching bolts. Tighten the bolts to 70 inch lbs. (8 Nm).

22. Apply a light coating of 2-4-C with Teflon or an equivalent marine grade grease to the sides of the driveshaft splines (NOT the top face of the shaft!).

23. Reinstall the gearcase assembly to the exhaust housing, *carefully* guiding the driveshaft into the crankshaft, the shift shaft up into the coupling and the water tube grommet on top of the pump over the tube in the exhaust housing. If necessary, rotate the flywheel *very* slowly by hand (in a clockwise direction when viewed from above) to help align the splines. On 4/5 hp motors, the gearcase should be in Reverse, so you could alternately spin the propeller slowly in the normal direction of rotation for reverse (usually counterclockwise) in order to slowly spin the splines.

24. Install the 2 bolts and washers securing the gearcase to the hosing and tighten to 50 inch lbs. (5.6 Nm) on 2.5/3.3 hp motors, or to 70 inch lbs. (8 Nm) on 4/5 hp motors.

25. Tighten the shift shaft coupling bolt securely.

26. Verify proper shifter operation, then install the rubber plug to the side of the exhaust housing.

27. Reconnect the spark plug wire.

6/8 & 9.9/10/15 Hp Models

◆ See Figures 73 thru 85

No mention is made of a cartridge or insert for the water pump housing on these models, which likely means that the housing itself serves this purpose. If this is the case and damage or excessive wear is found inside the housing during inspection, the housing assembly will have to be replaced.

1. Remove the gearcase from the exhaust housing as detailed in the Lower Unit section. Place the gearcase in a holder or a vise so you can access the water pump on the top of the gearcase-to-exhaust housing split line.

2. If necessary, pull upward and separate the water tube guide from the water pump outlet grommet.

Fig. 73 If necessary, remove the water tube

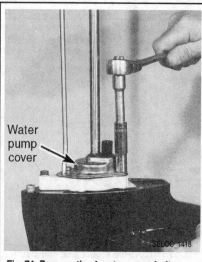

Fig. 74 Remove the 4 water pump bolts. . .

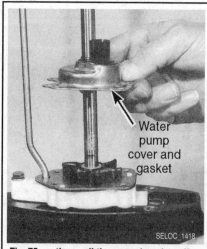

Fig. 75 . . .then pull the pump housing off the gearcase

3. Loosen and remove the 4 bolts and washers securing the water pump cover to the lower unit.

4. Slide the water pump cover and gasket up and free of the driveshaft. Discard the old gasket.

5. Slide the impeller and fiber washer up and free of the driveshaft.

6. Remove the Woodruff key from the driveshaft cutout. Place the Woodruff key in a safe place for later use.

7. Slide the lower (second) fiber washer up and free from the driveshaft.

8. Visually inspect the face plate for damage, warping or excessive wear. If necessary, slide the face plate and gasket up and free of the driveshaft. Discard the old gasket.

9. Generally speaking, a typical water pump service would be considered complete at this stage. However, if there is a suspected problem with the water pick-up tube or seal, the base plate and pick-up tube can be removed for inspection/service as follows:

a. Remove the retaining bolt from the base of the water pump. Place a screwdriver under the pump base and break the seal between the pump base and the lower unit.

b. Slide the pump base, water tube, gasket and shift shaft up and free of the driveshaft. The water tube is retained in the bottom of the pump base by a retainer and screw. The water tube will pull out of the grommet in the lower unit housing.

c. Using a pair of needlenose pliers, remove the E-clip from the shift shaft. Place the E-clip in a safe place for later use.

d. Unscrew the shift cam from the end of the shift shaft.

e. Slide the shift shaft out of the water pump base. Remove the O-rings and seal from the water pump base. Discard the seals and O-rings.

f. Turn the base plate over and remove the screw retainer securing the water pump pickup tube to the base.

g. Pull the water tube out of the base plate and remove the rubber seal on the end of the tube.

h. The water pump base plate contains dual back-to-back oil seals for the driveshaft which may be pushed or tapped out of the housing using a suitable driver, should replacement become necessary.

To Install:

10. If the water pump base and water pick-up tube were removed, install them as follows:

a. If the driveshaft seals were removed, start by installing 2 new seals. Obtain mandrel (#91-13655 and the appropriate driver handle or suitable substitutes). Place the water pump plate on the workbench as it is to be installed in the lower unit. The seals are installed back-to-back. The upper seal prevents water from entering the lower unit and the lower seal prevents lubricant from escaping. Coat both seal lips with a good grade of water resistant lubricant and coat the outer diameter of both seals using Loctite® 271.

b. Install the first oil seal using the driver with the oil seal lip facing down. Install the second oil seal with the oil seal lip facing up.

c. Install a new seal onto the end of the water tube. Turn the base plate over and insert the water tube into the bottom of the base plate. Secure the water tube and seal with the retaining ring and Phillips head screw.

d. Coat the surface of a new O-ring with a good grade of water resistant lubricant. Install the O-ring into the recess in the base plate.

e. Slide the end of the shift shaft through the O-ring and pump base plate.

f. Install the E-clip into the slotted ring on the shift shaft. This clip must be installed after the shift shaft has been installed through the water pump base plate.

g. Screw the threaded shift cam onto the end of the shift shaft. The tip of the shaft should be visible through the hole in the shift cam.

h. Place a new base plate gasket over the driveshaft and onto the lower unit housing. Be sure the small bypass hole is aligned with the lower unit housing.

i. Slide the assembled shift shaft and base plate onto the lower unit with the cam end going in first. Be sure the cam tapered surface is pointing towards the forward gear. Lower the pump base down onto the lower unit housing guiding the water tube into the grommet at the rear of the housing.

j. Apply Loctite® 271 or equivalent to the threads of the bolt and secure the water pump base plate to the lower unit housing. Tighten the bolt to 50 inch lbs. (5.6 Nm).

11. Slide a new gasket and the face plate down the driveshaft and into place on the water pump base plate. On most models, the screw holes are offset. Therefore, the gasket holes and plate will only line up properly one way.

12. Slide the lower fiber washer down the driveshaft and place it against the face plate.

13. Apply a "dab" of petroleum jelly or marine grade grease onto the Woodruff key, and then place it in the driveshaft keyway. The petroleum jelly/grease will hold the Woodruff key in place while installing the impeller.

14. Slide the water pump impeller down the driveshaft with the keyway in the impeller aligned with the Woodruff key in the driveshaft. Press the impeller into place with the keyway aligned.

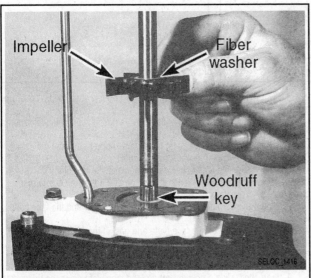

Fig. 76 Remove the upper washer and impeller. . .

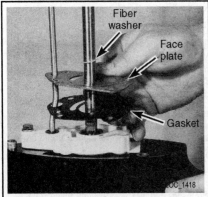

Fig. 77 . . .then the lower washer, and if necessary, the face plate

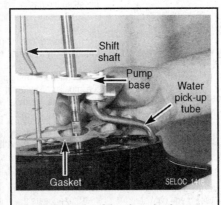

Fig. 78 The water pickup is under the water pump base plate

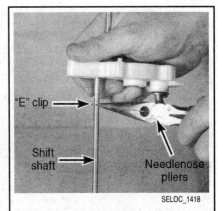

Fig. 79 If necessary remove the E-clip. . .

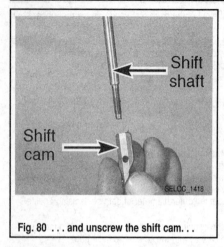

Fig. 80 . . . and unscrew the shift cam. . .

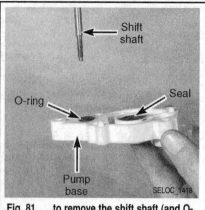

Fig. 81 . . .to remove the shift shaft (and O-ring)

Fig. 82 To remove the water pickup tube, loosen the bolt. . .

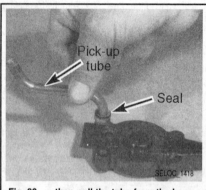

Fig. 83 . . .then pull the tube from the base plate

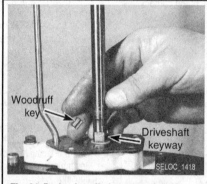

Fig. 84 During installation use a dab of grease to hold the drive key. . .

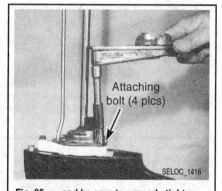

Fig. 85 . . .and be sure to properly tighten the pump bolts

■ If an old impeller is installed be sure the impeller is installed in the same manner from which it was removed. Never turn the impeller over thinking it will extend its life. On the contrary, the blades would crack and break after just a short time of operation.

15. Slide the upper fiber washer down the driveshaft and position it on top of the impeller.

16. Install a new O-ring to the bottom of the pump cover, then lubricate the O-ring and the inner pump surface with a light coating of 2-4-C with Teflon, or an equivalent marine grade grease.

17. Slide the pump cover down the driveshaft until it makes contact with the impeller. Apply a small amount of downward pressure on the water pump cover, and at the same time, rotate the driveshaft clockwise until the cover seats on the plate. Rotation of the driveshaft ensures all water pump vanes are bent in the proper direction.

■ Make sure the pump cover (housing) gasket is still in position with the bolt holes properly aligned once the cover is seated.

18. Apply Loctite® 271 or equivalent to the threads of the water pump bolts, then secure the pump in place with the 4 attaching bolts. Tighten the bolts to 50 inch lbs. (5.6 Nm).

19. If removed, lubricate the water tube seal using a light coating of 2-4-C with Teflon, or an equivalent marine grade grease, then push a new water tube seal into the water pump cover recess. Insert the water tube guide over the new seal.

20. Install the Gearcase to the exhaust housing as detailed in the Lower Unit section.

20/20 Jet/25 Hp Models

◆ See Figures 86, 87 and 88

1. Remove the gearcase from the exhaust housing as detailed in the Lower Unit section. Place the gearcase in a holder or a vise so you can access the water pump on the top of the gearcase-to-exhaust housing split line.

2. Slide the rubber ring (centrifugal slinger) on top of the pump cover up and free of the driveshaft.

3. Loosen and remove the 4 bolts and washers securing the water pump cover to the lower unit.

4. Lift and slide the water pump cover and O-ring free of the driveshaft. Discard the O-ring inside the water pump cover.

5. Lift and slide the upper fiber washer and impeller free of the driveshaft.

6. Remove the impeller Woodruff key from the driveshaft cutout and place the key in a safe place for later use.

7. Slide the lower fiber washer between the impeller and the face plate up and free of the driveshaft.

8. If necessary, lift off the face plate and gasket from the lower unit housing. Discard the old gasket.

To Install:

9. If removed, slide a new gasket and the face plate down the driveshaft and into place on the lower unit. On some models, the screw holes are offset, so the gasket holes and plate will only line up properly one way.

10. Slide the lower fiber washer down the driveshaft and into place on the face plate.

11. Apply a "dab" of petroleum jelly or marine grade grease onto the Woodruff key, and then place it in the driveshaft keyway. The petroleum jelly/grease will hold the Woodruff key in place while installing the impeller.

12. Slide the water pump impeller down the driveshaft with the keyway in the impeller aligned with the Woodruff key in the driveshaft.

13. Press the impeller into place with the keyway aligned.

■ If an old impeller is installed be sure the impeller is installed in the same manner from which it was removed. Never turn the impeller over thinking it will extend its life. On the contrary, the blades would crack and break after just a short time of operation.

14. Slide the upper fiber washer down the driveshaft and into place on the top of the impeller.

15. It appears that these models normally are equipped with a water pump housing that utilizes a replaceable cartridge insert. If necessary, replace the insert before installation.

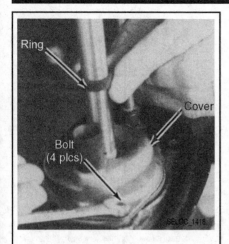

Fig. 86 Unbolt the water pump housing/cover. . .

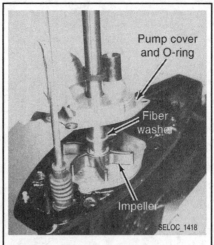

Fig. 87 . . .then lift it up the driveshaft to reach the impeller

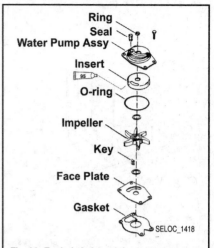

Fig. 88 Exploded view of the water pump assembly - 20/20 Jet/25 hp models

16. Install a new O-ring to the bottom of the pump cover, then lubricate the O-ring and the inner pump surface with a light coating of 2-4-C with Teflon, or an equivalent marine grade grease.

17. Slide the pump cover down the driveshaft until it makes contact with the impeller. Apply a small amount of downward pressure on the water pump cover, and at the same time, rotate the driveshaft clockwise until the cover seats on the plate. Rotation of the driveshaft ensures all water pump vanes are bent in the proper direction.

18. Apply Loctite®271 or equivalent to the threads of the water pump bolts, then secure the pump in place with the 4 attaching bolts. Tighten the bolts to 60 inch lbs. (6.8 Nm).

19. Slide a new rubber ring (centrifugal slinger) down the driveshaft and onto the top of the water pump cover.

20. Install the Gearcase to the exhaust housing as detailed in the Lower Unit section.

30/40 Hp 2-Cylinder & 40/40 Jet/50 Hp 3-Cylinder Models

◆ See Figures 89 thru 93  *MODERATE*

No mention is made of a cartridge or insert for the water pump housing on these models, which likely means that the housing itself serves this purpose. If this is the case and damage or excessive wear is found inside the housing during inspection, the housing assembly will have to be replaced.

1. Remove the Gearcase from the exhaust housing as detailed in the Lower Unit section. Place the gearcase in a holder or a vise so you can access the water pump on the top of the gearcase-to-exhaust housing split line.

■ **If the water tube seal remained in the drive shaft housing, remove the seal from the housing and reinstall on the water pump cover. Secure the seal to the cover with Loctite® 405, or equivalent.**

2. Loosen and remove the 4 bolts and washers securing the water pump cover to the lower unit.

3. Lift and slide the water pump cover and gasket free of the driveshaft. Discard the gasket.

■ **There is no insert in the water pump cover/housing on these gearcases. After removal, check the thickness of the steel at the discharge slots. If it is 0.060 in. (1.524mm) or LESS, OR if the cover has grooves in it (other than the impeller sealing groove) which are 0.030 in. (0.762mm) deep or more, the cover should be replaced.**

4. Lift and slide the upper fiber washer and impeller free of the driveshaft.

5. Remove the impeller Woodruff key from the driveshaft cutout and place the key in a safe place for later use.

6. Slide the lower fiber washer between the impeller and the base plate up and free of the driveshaft.

7. If necessary, lift off the base plate and gasket from the lower unit housing. Discard the old gasket.

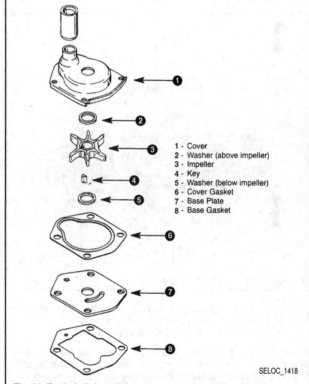

1 - Cover
2 - Washer (above impeller)
3 - Impeller
4 - Key
5 - Washer (below impeller)
6 - Cover Gasket
7 - Base Plate
8 - Base Gasket

SELOC_1418

Fig. 89 Exploded view of the water pump assembly - 30-50 hp models

■ **If the base plate is removed, be sure to check the condition of the exhaust deflector located in the top of the gearcase right behind and below the base plate mounting point. If the deflector is warped or damaged, replace it during assembly.**

To Install:

8. If removed, slide a new gasket and the base plate down the driveshaft and into place on the lower unit.

9. Position a new pump cover gasket, WITH THE NEOPRENE STRIP FACING UPWARD on top of the base plate.

10. Slide the lower fiber washer down the driveshaft and into place on the face plate.

11. Apply a "dab" of petroleum jelly or marine grade grease onto the Woodruff key, and then place it in the driveshaft keyway. The petroleum jelly/grease will hold the Woodruff key in place while installing the impeller.

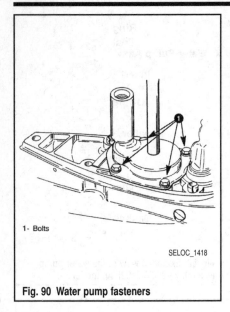

1- Bolts

SELOC_1418

Fig. 90 Water pump fasteners

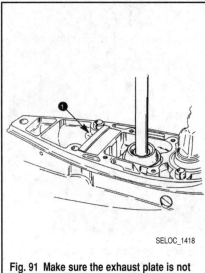

SELOC_1418

Fig. 91 Make sure the exhaust plate is not warped

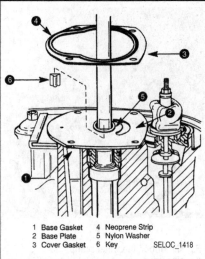

1 Base Gasket	4 Neoprene Strip
2 Base Plate	5 Nylon Washer
3 Cover Gasket	6 Key

SELOC_1418

Fig. 92 Position the new pump gasket with the neoprene seal upward. . .

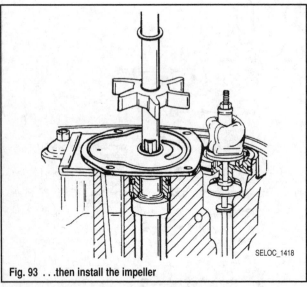

SELOC_1418

Fig. 93 . . .then install the impeller

12. Slide the water pump impeller down the driveshaft with the keyway in the impeller aligned with the Woodruff key in the driveshaft.

13. Press the impeller into place with the keyway aligned.

■ **If an old impeller is installed be sure the impeller is installed in the same manner from which it was removed. Never turn the impeller over thinking it will extend its life. On the contrary, the blades would crack and break after just a short time of operation.**

14. Slide the upper fiber washer down the driveshaft and into place on the top of the impeller.

15. Lubricate the inner pump surface with a light coating of 2-4-C with Teflon, or an equivalent marine grade grease.

16. Slide the pump cover down the driveshaft until it makes contact with the impeller. Apply a small amount of downward pressure on the water pump cover, and at the same time, rotate the driveshaft clockwise until the cover seats on the plate. Rotation of the driveshaft ensures all water pump vanes are bent in the proper direction.

■ **Make sure the pump cover (housing) gasket is still in position with the bolt holes properly aligned once the cover is seated.**

17. Apply Loctite® 271 or equivalent to the threads of the water pump bolts, then secure the pump in place with the 4 attaching bolts. Tighten the bolts to 60 inch lbs. (6.8 Nm).

18. Install the gearcase to the exhaust housing as detailed in the Lower Unit section.

55/60 Hp 3-Cylinder Models

 MODERATE

◆ **See Figures 94, 95 and 96**

No mention is made of a cartridge or insert for the water pump housing on these models, which likely means that the housing itself serves this purpose. If this is the case and damage or excessive wear is found inside the housing during inspection, the housing assembly will have to be replaced.

1. Remove the gearcase from the exhaust housing as detailed in the Lower Unit section. Place the gearcase in a holder or a vise so you can access the water pump on the top of the gearcase-to-exhaust housing split line.

■ **If the water tube seal remained in the drive shaft housing, remove the seal from the housing and reinstall on the water pump cover. Replace the tube seal if it was damaged.**

2. Loosen and remove the bolts (6 for non-Bigfoot models or 4 on Bigfoot gearcases) and washers securing the water pump cover to the lower unit. The extra 2 bolts on the non-Bigfoot models don't actually secure the water pump housing itself, but they do secure the face plate.

3. Lift and slide the water pump cover and gasket free of the driveshaft. Discard the gasket.

■ **There is no insert in the water pump cover/housing on these gearcases. After removal, check the thickness of the steel at the discharge slots. If it is 0.060 in. (1.524mm) or LESS, OR if the cover has grooves in it (other than the circular impeller sealing groove) which are 0.030 in. (0.762mm) deep or more, the cover should be replaced.**

4. Lift and slide the impeller free of the driveshaft.

5. Remove the impeller Woodruff key from the driveshaft cutout and place the key in a safe place for later use.

6. If necessary, lift off the face plate and gasket from the lower unit housing. Discard the old gasket.

■ **Like the water pump cover, the face plate should be checked for excessive grooving. If there are grooves (other than the circular impeller sealing groove) which are 0.030 in. (0.762mm) deep or more, the face plate should be replaced.**

To Install:

7. If removed, slide a new gasket and the face plate down the driveshaft and into place on the lower unit.

8. Apply a "dab" of petroleum jelly or marine grade grease onto the Woodruff key, and then place it in the driveshaft keyway. The petroleum jelly/grease will hold the Woodruff key in place while installing the impeller.

9. Slide the water pump impeller down the driveshaft with the keyway in the impeller aligned with the Woodruff key in the driveshaft.

10. Press the impeller into place with the keyway aligned.

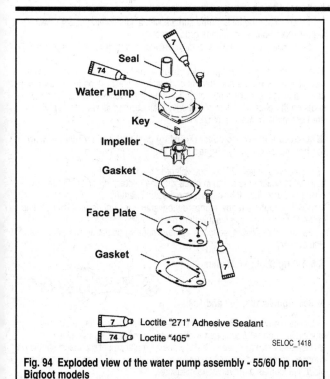

Fig. 94 Exploded view of the water pump assembly - 55/60 hp non-Bigfoot models

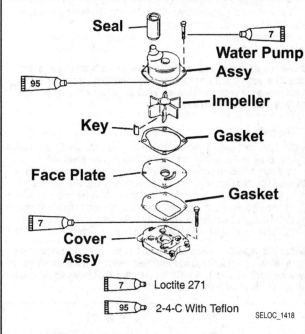

Fig. 95 Exploded view of the water pump assembly - 60 hp non-Bigfoot models

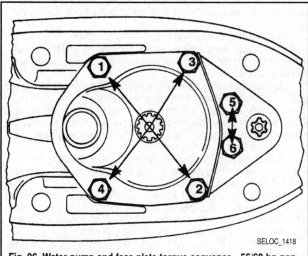

Fig. 96 Water pump and face plate torque sequence - 55/60 hp non-Bigfoot models

■ If an old impeller is installed be sure the impeller is installed in the same manner from which it was removed. Never turn the impeller over thinking it will extend its life. On the contrary, the blades would crack and break after just a short time of operation.

11. Lubricate the inner pump surface with a light coating of 2-4-C with Teflon, or an equivalent marine grade grease.

■ The Mercury instructions for non-Bigfoot motors are somewhat confusing. The SAY to lubricate the "inner diameter of the cover with Needle Bearing Assembly Lubricant" but then show an illustration of that lubricant applied to the gasket AND go on to say to "install gasket with bead toward cover." The later seems to make sense, as the assembly lube would hold the gasket in place on the cover as it is installed over the impeller.

12. Position the gasket on the cover itself (especially if it is a non-Bigfoot model with assembly lube designed to hold it in place to the cover) as opposed to the top of the gearcase which is probably more commonly done.

13. Slide the pump cover down the driveshaft until it makes contact with the impeller. Apply a small amount of downward pressure on the water pump cover, and at the same time, rotate the driveshaft clockwise until the cover seats on the plate. Rotation of the driveshaft ensures all water pump vanes are bent in the proper direction.

14. Apply Loctite® 271 or equivalent to the threads of the water pump bolts, then secure the pump in place with the 4 or 6 attaching bolts (as applicable). Tighten the bolts to 60 inch lbs. (6.8 Nm) using a crossing pattern. For non-Bigfoot models, start with the 4 pump housing bolts, then move to the 2 face plate bolts located between the housing and the shift shaft.

15. If not done already, lubricate the water tube seal using a light coating of 2-4-C with Teflon or an equivalent marine grade grease.

16. Install the gearcase to the exhaust housing as detailed in the Lower Unit section.

65 Jet-125 Hp Models

◆ See Figures 97 thru 100

No mention is made of a cartridge or insert for the water pump housing on these models, which likely means that the housing itself serves this purpose. If this is the case and damage or excessive wear is found inside the housing during inspection, the housing assembly will have to be replaced.

■ Newer models will normally NOT have isolators installed on the water pump.

1. Remove the gearcase from the exhaust housing as detailed in the Lower Unit section. Place the gearcase in a holder or a vise so you can access the water pump on the top of the gearcase-to-exhaust housing split line.

■ If the water tube seal remained in the drive shaft housing, remove the seal from the housing and reinstall on the water pump cover. Replace the tube seal if it was damaged.

2. Loosen and remove the 4 pump mounting bolts and washers.
3. Lift and slide the water pump cover and gasket free of the driveshaft. Discard the gasket.

■ There is no insert in the water pump cover/housing on these gearcases. After removal, check the thickness of the steel at the discharge slots. If it is 0.060 in. (1.524mm) or LESS, OR if the cover has grooves in it (other than the circular impeller sealing groove) which are 0.030 in. (0.762mm) deep or more, the cover should be replaced.

4. Lift and slide the impeller free of the driveshaft.

5. Remove the impeller Woodruff key from the driveshaft cutout and place the key in a safe place for later use.

6. If necessary, lift off the face plate and gasket from the water pump base which is bolted to the lower unit housing. Discard the old gasket.

■ Like the water pump cover, the face plate should be checked for excessive grooving. If there are grooves (other than the circular impeller sealing groove) which are 0.030 in. (0.762mm) deep or more, the face plate should be replaced.

To Install:

7. If removed, slide a new gasket and the face plate down the driveshaft and into place on the water pump base at the top of the lower unit.

8. Apply a "dab" of petroleum jelly or marine grade grease onto the Woodruff key, and then place it in the driveshaft keyway. The petroleum jelly/grease will hold the Woodruff key in place while installing the impeller.

9. Slide the water pump impeller down the driveshaft with the keyway in the impeller aligned with the Woodruff key in the driveshaft.

10. Press the impeller into place with the keyway aligned.

■ If an old impeller is installed be sure the impeller is installed in the same manner from which it was removed. Never turn the impeller over thinking it will extend its life. On the contrary, the blades would crack and break after just a short time of operation.

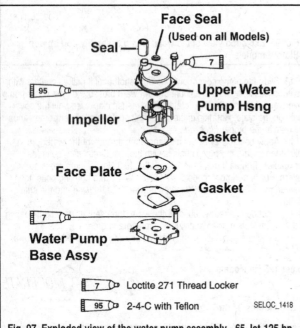

Fig. 97 Exploded view of the water pump assembly - 65 Jet-125 hp models

11. Lubricate the inner pump surface with a light coating of 2-4-C with Teflon, or an equivalent marine grade grease.

12. Position the gasket over the driveshaft and impeller, onto the face plate.

13. Slide the pump cover down the driveshaft until it makes contact with the impeller. Apply a small amount of downward pressure on the water pump cover, and at the same time, rotate the driveshaft clockwise until the cover seats on the plate. Rotation of the driveshaft ensures all water pump vanes are bent in the proper direction.

■ Make sure the pump cover (housing) gasket is still in position with the bolt holes properly aligned once the cover is seated.

14. Apply Loctite® 271 or equivalent to the threads of the water pump bolts, then secure the pump in place with the 4 attaching bolts. Tighten the bolts to 60 inch lbs. (6.8 Nm) using a crossing pattern.

15. If not done already, lubricate the water tube seal using a light coating of 2-4-C with Teflon or an equivalent marine grade grease.

16. Install the gearcase to the exhaust housing as detailed in the Lower Unit section.

135-200 Hp (2.5L) V6 Models

◆ See Figures 101, 102 and 103

Unlike many of Mercury's larger motors the gearcases found on these models are normally equipped with a water pump housing that uses a replaceable cartridge insert. During impeller replacement the insert should be checked carefully for wear or damage and replaced, as necessary (or just plain replaced as a preventative measure).

1. Remove the gearcase from the exhaust housing as detailed in the Lower Unit section. Place the gearcase in a holder or a vise so you can access the water pump on the top of the gearcase-to-exhaust housing split line.

2. Slide the centrifugal slinger up and off the driveshaft.

3. If necessary, remove the water tube guide and the seal from the top of the water pump cover. If removed, keep the guide for reuse, but the seal must be discarded.

4. Loosen and remove the 4 water pump fasteners (typically there are 3 nuts, of 2 different sizes, and a bolt along with washers).

5. Use 2 small prybars, 1 at the front of the pump housing and the other at the rear to gently break the housing free of the gasket mating surface. Lift and slide the water pump cover and gasket free of the driveshaft. Discard the gasket and clean the mating surfaces.

6. Inspect the water pump cover closely for cracks, distortion (from overheating), excessive grooves or rough surfaces and replace if necessary, using 1 of the 2 methods as follows:

• Always try this method first. Place the pump cover in a bench vise, then use a punch and hammer to carefully drive the insert down and out of the water pump cover from the topside of the assembly. If the insert won't come free, don't risk damage to the housing, proceed with method 2.

• If you can't free the insert with a punch, drill two 3/16 in. (4.8mm) holes down through the top of the water pump housing BUT NOT THROUGH THE INSERT! Place the holes to the port and starboard side of the driveshaft

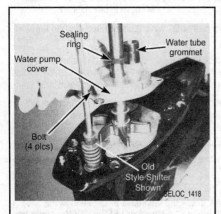

Fig. 98 Unbolt and lift the water pump housing. . .

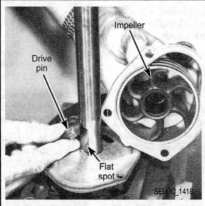

Fig. 99 . . .sometimes the impeller remains in the housing

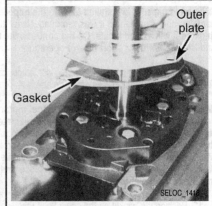

Fig. 100 If necessary, remove the face (outer/lower) plate and gasket

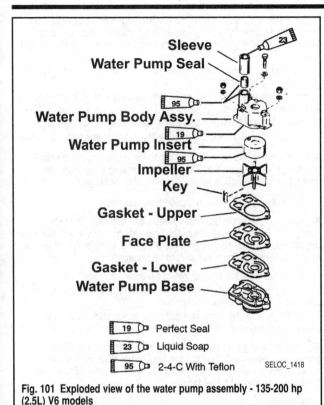

Fig. 101 Exploded view of the water pump assembly - 135-200 hp (2.5L) V6 models

19	Perfect Seal
23	Liquid Soap
95	2-4-C With Teflon

SELOC_1418

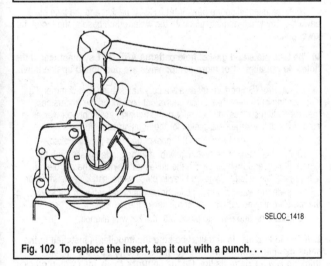

SELOC_1418

Fig. 102 To replace the insert, tap it out with a punch. . .

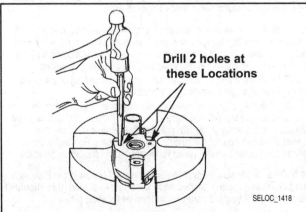

Drill 2 holes at these Locations

SELOC_1418

Fig. 103 . . .but if necessary drill 2 small holes in the cover for access

opening in the top of the cover, far enough from any edges to preserve the strength of the cover itself. Insert the tip of a punch through the hole and use a hammer to try and drive the insert out. Alternate back and forth between the 2 holes.

※※ **SELOC CAUTION**

In some extreme cases it may be necessary to split the insert open using a hammer and chisel. But proceed with caution not to damage the water pump housing.

7. Lift and slide the impeller free of the driveshaft.
8. Remove the impeller Woodruff key from the driveshaft cutout and place the key in a safe place for later use.
9. If necessary, lift off the face plate and gasket from the water pump base which is bolted to the lower unit housing. Discard the old gasket.

■ **Like the water pump cover, the face plate should be checked for excessive wear, warpage, grooving or rough surfaces, and replaced if damaged.**

To Install:

10. If removed, slide a new gasket and the face plate down the driveshaft and into place on the water pump base at the top of the lower unit. Be sure to index the gasket and plate with the dowel pin(s).
11. Slide the new face plate-to-water pump cover gasket down the driveshaft, indexing it with the dowel pin(s) as well.
12. Apply a "dab" of petroleum jelly or marine grade grease onto the Woodruff key, and then place it in the driveshaft keyway. The petroleum jelly/grease will hold the Woodruff key in place while installing the impeller.
13. Slide the water pump impeller down the driveshaft with the keyway in the impeller aligned with the Woodruff key in the driveshaft.
14. Press the impeller into place with the keyway aligned.

■ **If an old impeller is installed be sure the impeller is installed in the same manner from which it was removed. Never turn the impeller over thinking it will extend its life. On the contrary, the blades would crack and break after just a short time of operation.**

15. If removed, install a new water pump insert (cartridge) into the pump cover by first applying a light coating of Perfect Seal (or equivalent sealant) to the outer diameter of the insert, then inserting it into the cover while aligning the insert tab with the slot in the pump cover. Be sure to wipe off any excess sealant before installing the cover.
16. Lubricate the inner pump surface with a light coating of 2-4-C with Teflon, or an equivalent marine grade grease.
17. Slide the pump cover down the driveshaft until it makes contact with the impeller. Apply a small amount of downward pressure on the water pump cover, and at the same time, rotate the driveshaft clockwise until the cover seats on the plate. Rotation of the driveshaft ensures all water pump vanes are bent in the proper direction.

■ **Make sure the pump cover (housing) gasket is still in position with the bolt holes properly aligned and dowel pin(s) indexed once the cover is seated.**

18. Secure the pump in place with the 4 fasteners. Tighten the bolt to 35 inch lbs. (4 Nm) and the nuts to 50 inch lbs. (5.5 Nm) using a crossing pattern.
19. If not done already, lubricate the inside of the water tube seal using a light coating of 2-4-C with Teflon or an equivalent marine grade grease. If the seal was removed, install it so the plastic side goes INTO the pump cover.
20. If removed, install the water tube guide onto the pump cover.
21. Slide the centrifugal slinger down the driveshaft so it sits against the pump cover.
22. Install the gearcase to the exhaust housing as detailed in the Lower Unit section.

200-250 Hp (3.0L) V6 Models

OEM ② MODERATE

◆ **See Figures 104, 105 and 106**

No mention is made of a cartridge or insert for the water pump housing on these models, which likely means that the housing itself serves this purpose. If this is the case and damage or excessive wear is found inside the housing during inspection, the housing assembly will have to be replaced.

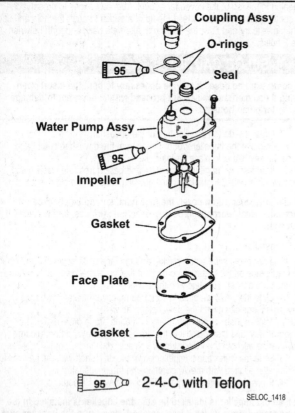

2-4-C with Teflon

SELOC_1418

Fig. 104 Exploded view of the water pump assembly - 200-250 hp (3.0L) V6 models

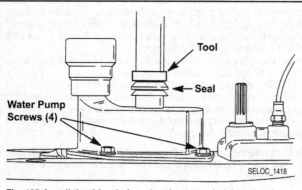

SELOC_1418

Fig. 105 Install the driveshaft seal to the proper height using the special tool. . .

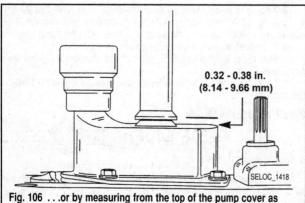

0.32 - 0.38 in.
(8.14 - 9.66 mm)

SELOC_1418

Fig. 106 . . .or by measuring from the top of the pump cover as shown

1. Remove the gearcase from the exhaust housing as detailed in the Lower Unit section. Place the gearcase in a holder or a vise so you can access the water pump on the top of the gearcase-to-exhaust housing split line.

2. Slide the driveshaft seal off the top of the water pump housing, up and free of the driveshaft.

3. If necessary remove the water tube coupling assembly, along with its 2 O-rings.

4. Loosen and remove the 4 pump mounting bolts and washers.

5. Lift and slide the water pump cover and gasket free of the driveshaft.

■ **If the water pump cover is sticking to the gearcase, use 2 small prybars, 1 on either side of the gearcase/housing to carefully pry upward on the mounting flange, breaking the gasket seal.**

6. Remove and discard the gasket from face plate.

■ **There is no insert in the water pump cover/housing on these gearcases. After removal, check the cover for grooves in it (other than the circular impeller sealing groove) which are 0.030 in. (0.762mm) deep or more, if present the cover should be replaced.**

7. Lift and slide the impeller free of the driveshaft.

8. Remove the impeller Woodruff key from the driveshaft cutout and place the key in a safe place for later use.

9. If necessary, lift off the face plate and gasket from the water pump base which is bolted to the lower unit housing. Discard the old gasket.

■ **Like the water pump cover, the face plate should be checked for excessive grooving. If there are grooves (other than the circular impeller sealing groove) which are 0.030 in. (0.762mm) deep or more, the face plate should be replaced.**

To Install:

10. If removed, slide the new SMALL HOLE gasket and the face plate down the driveshaft and into place on the water pump base at the top of the lower unit.

■ **The face plate and gasket hole patterns ARE NOT symmetrical. If the holes do not align, 1 or more components are positioned upside-down.**

11. On Pro/XS/Sport V6 models Mercury has started to recommend a light coating of Perfect Seal or an equivalent sealant on the underside (gasket mating surface) of the water pump housing assembly, as well as to both sides of the large hole gasket for the housing assembly.

12. Install the new LARGE HOLE gasket on top of the face plate.

13. Apply a "dab" of petroleum jelly or marine grade grease onto the Woodruff key, and then place it in the driveshaft keyway. The petroleum jelly/grease will hold the Woodruff key in place while installing the impeller.

14. Slide the water pump impeller down the driveshaft with the keyway in the impeller aligned with the Woodruff key in the driveshaft.

15. Press the impeller into place with the keyway aligned.

■ **If an old impeller is installed be sure the impeller is installed in the same manner from which it was removed. Never turn the impeller over thinking it will extend its life. On the contrary, the blades would crack and break after just a short time of operation.**

16. Install 2 water pump locating pins (sometimes included with the kit, otherwise easily fabricated from 2 studs the same size/thread as the water pump retaining bolts or two slightly smaller dowels). The locating pins are used to keep the gaskets in alignment as the water pump is seated.

17. Lubricate the inner pump surface with a light coating of 2-4-C with Teflon, or an equivalent marine grade grease.

18. Slide the pump cover down the driveshaft and over the locating pins until it makes contact with the impeller. Apply a small amount of downward pressure on the water pump cover, and at the same time, rotate the driveshaft clockwise until the cover seats on the plate. Rotation of the driveshaft ensures all water pump vanes are bent in the proper direction.

■ **Make sure the pump cover (housing) gasket is still in position with the bolt holes properly aligned once the cover is seated (this shouldn't be a problem if you are using some form of locating pin).**

19. Start the water pump fasteners by hand, then slowly and evenly tighten them using a crossing pattern to 60 inch lbs. (6.8 Nm).

20. If removed, lightly lubricate the water tube coupling O-rings using a light coating of 2-4-C with Teflon or an equivalent marine grade grease. Carefully install the coupling to the pump housing making sure the O-rings are not damaged.

✳✳ SELOC CAUTION

The driveshaft seal must be installed above the pump housing at the proper height OR air will be drawn into the pump potentially overheating and damaging the powerhead. Also, DO NOT grease the driveshaft during seal installation.

21. Install the driveshaft seal using 1 of the 2 following methods:

• A tool such as #26-816575A2 is normally provided in the seal kit or tool number #817275A3 is normally provided in the water pump kit. If so, use the tool to gently push the seal down into position until the tool seats against the pump housing.

• If the tool is not available, gently press the seal against the housing until it is about 0.35 in. (8.9mm) above the top of the pump housing. A tolerance of 0.32-0.38 in. (8.14-9.66mm) above the top of the housing is permissible.

22. Install the gearcase to the exhaust housing as detailed in the Lower Unit section.

WARNING SYSTEMS

Oil & Temperature Warning Systems

◆ See Figure 107

Most 30 hp and larger Mercury motors are normally equipped with a warning horn/buzzer system which is used to alert the operator of conditions that could potentially damage the powerhead.

Generally speaking a warning circuit is activated when the keyswitch is first turned on, with 1 short beep to let the operator know the circuit is working. If at anytime during engine operation the temperature or oil level sensor (the later is obviously only used on oil injected models) closes the warning circuit, the horn or buzzer will sound alerting the operator that either the powerhead is in danger of overheating or of damage from lack of sufficient oil.

Even the most basic of Mercury warning systems tend to involve a warning or timing control module or even the ignition module or ECM. In many cases, a warning circuit actuation will also have an effect on the ignition system (limiting timing to help limit engine speed) or the EFI/DFI fuel delivery system (limiting fuel delivery, also to limit engine speed and prevent damage).

The major components of the warning systems are really therefore the sensors.

Temperature sensors are mounted directly to the powerhead, normally somewhere on the cylinder head itself. Oil sensors, for what we hope are obvious reasons, are found in the powerhead mounted oil tank, either mounted in the top of the tank, inserted in oil, OR mounted in the bottom of the tank, usually in a insulated pocket that protrudes upward into the path of travel for a magnetic float mounted inside the tank itself.

CHECKING THE WARNING SYSTEM CIRCUIT

As stated earlier, most 30 hp and larger Mercury motors are normally equipped with a warning horn/buzzer system which is used to alert the operator of conditions that could potentially damage the powerhead.

Two types of temperature sensors are used. Most mid-range motors are equipped with an on/off type switch, which is activated by temperature. Most larger motors are equipped with a variable sensor or thermistor, whose internal resistance varies with temperature.

All oil injected motors use some form of a low oil level sensor for the powerhead mounted oil reservoir. Generally speaking, the switch should show NO continuity when the tank is at least half full or more. Sometime between half full and empty the switch should show continuity. If the switch does not open/close as these conditions change, first check to make sure the float (usually a magnetic assembly located inside the tank) is not stuck. If the float is not stuck, yet the switch refuses to open/close as oil level changes, then it is defective and must be replaced.

CLEANING & INSPECTION

◆ See Figure 36

Clean all water pump parts with solvent, and then dry them with compressed air.

Inspect the water pump cover and base for cracks and distortion. On some models there is a minimum wear thickness or a maximum groove depth which should be checked. Where a specification is available we've provided it in the Removal & Installation procedure earlier in this section.

If possible, always install a new water pump impeller while the lower unit is disassembled. A new impeller will ensure extended satisfactory service and give "peace of mind" to the owner. If the old impeller must be returned to service, **never** install it in reverse to the original direction of rotation. Installation in reverse will cause premature impeller failure.

Inspect the ends of the impeller blades for cracks, tears, and wear.

Check for a glazed or melted appearance, caused from operating without sufficient water. If any question exists, as previously stated, install a new impeller if at all possible.

■ We *really* can't emphasize this enough, once you've gone so far to disassemble the water pump WHAT POSSIBLE reason do you have to say a couple of bucks to reuse the old impeller. Replace it for peace of mind.

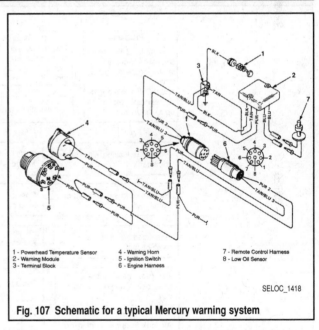

1 - Powerhead Temperature Sensor
2 - Warning Module
3 - Terminal Block
4 - Warning Horn
5 - Ignition Switch
6 - Engine Harness
7 - Remote Control Harness
8 - Low Oil Sensor

SELOC_1418

Fig. 107 Schematic for a typical Mercury warning system

■ It is not uncommon for the magnetic float in the tank to become stuck, making a sensor test bad when installed, but test good when removed from the tank.

The warning system on most outboards is designed to give a short beep upon initial keyswitch activation, in order to ensure the operator the system is still working. If however the warning horn fails to operate, the system MUST be fixed before continuing powerhead operation. Failure to do so might allow an undiagnosed problem with the Engine Overheat or Low Oil circuits which could allow the powerhead to become significantly damaged or destroyed should a lubrication or cooling problem occur.

If the **warning horn does not sound** when the key is first turned on check the following possible causes:

• Check for an open on the warning circuit tan or tan/blue (depending upon model) wire between the horn and powerhead. Disconnect this lead from the harness at the terminal on the engine block and ground it to the powerhead (you may need a jumper on some models). With the key turned on the horn should now sound. If not, check for an open circuit and/or check the horn.

• If not problems are found with the circuit or horn, suspect the warning module/ignition module/ECM could be the fault.

TROUBLESHOOTING A WARNING SYSTEM ACTIVATION

Should the warning system actuate during powerhead operation (either intermittently or constantly), the powerhead must be shut down IMMEDIATELY and the cause determined before continued engine operation. Failure to do so could allow an engine overheat or lack of oil condition to damage/destroy the powerhead.

On all OptiMax motors, plus any carbureted or V6 EFI motors, refer to the troubleshooting charts in the next section to help determine possible causes of warning system activation.

On inline carbureted motors, use the following symptoms to help determine cause of warning system activation:

If the **warning horn stays on when key is turned on, but engine is cold** suspect the following:

• OIL LEVEL TOO LOW: check the oil level and top off as necessary.

• FAULTY OIL LEVEL SWITCH: check the oil level switch to make sure continuity changes as the tank goes from empty to full. IF not, and the float is not stuck, the switch is likely faulty and must be replaced.

• FAULTY ENGINE OVERHEAT SENSOR: to check, disconnect the engine temperature sensor lead, generally a tan or tan/blue wire at the bullet connector, then turn the key to the **ON** position. If the horn stops, the sensor is likely the fault. If the horn *still* sounds, suspect the module.

If the **warning horn stays comes on when engine is running** suspect the following:

• OIL LEVEL TOO LOW: check the oil level and top off as necessary.

• ENGINE TEMPERATURE TOO EXCESSIVE: check temperature of powerhead.

• FAULTY OIL LEVEL SWITCH: locate the leads from the oil level switch (usually light blue) and disconnect them. If the horn stops sounding and the oil level was full (and switch float was not stuck), the switch is defective. If the horn continues to sound check the leads for a short. Also, you can check the oil level switch to make sure continuity changes as the tank goes from empty to full. If not, and the float is not stuck, the switch is likely faulty and must be replaced.

• FAULTY TEMPERATURE SENSOR: to check, disconnect the engine temperature sensor lead generally a tan or tan/blue wire at the bullet connector, then turn the key to the **ON** position. If the horn stops, the sensor is likely the fault. If the horn still sounds, suspect the module.

OPTIMAX & CARUBURETED/EFI V6 MOTOR WARNING SYSTEM TROUBLESHOOTING CHARTS

◆ See Figures 108 thru 119

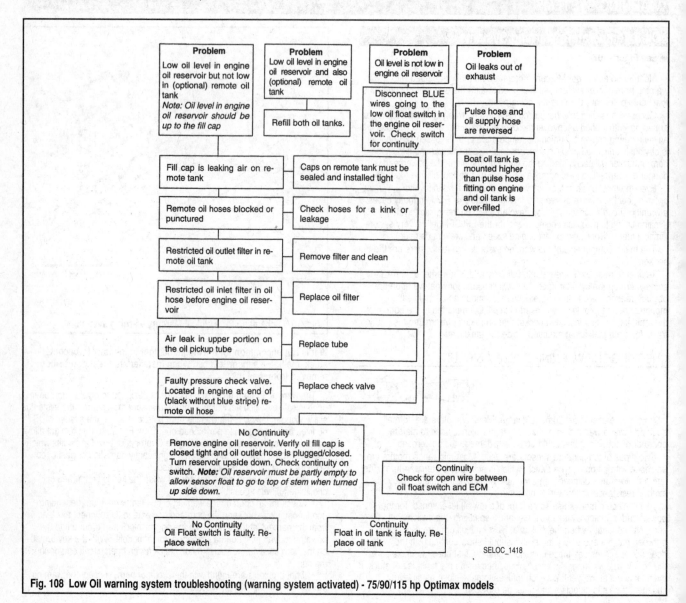

Fig. 108 Low Oil warning system troubleshooting (warning system activated) - 75/90/115 hp Optimax models

Guardian System Operation with Gauges

SmartCraft Gauge	System will sound warning horn and display the warning message.

Guardian System Activation

Condition	Result
Engine Overheat	Engine power level can be reduced to any percentage down to a fast idle speed, if overheat condition persists.
Air Compressor Overheat	Warning horn is activated but there is no power reduction.
Block Water Pressure Low	Engine power level can be reduced to any percentage down to a fast idle, if condition persists.
Throttle Position Sensor Failure	If the throttle position sensor fails or becomes disconnected, power will be limited to a maximum of approximately 4500 RPM. When the TPS is in the fail mode, the ECM will use the MAP sensor for a reference to determine fuel calibration.
Temperature Sensor (cylinder head and air compressor) Failure	If a temperature sensor should fail or become disconnected, power will be limited to a maximum of approximately 4500 RPM.
Battery Voltage (too high or too low)	Battery voltage greater than 16 volts or less than 10 volts will result in engine output power being reduced. The higher or lower the voltage is outside of these parameters, the greater the percentage of power reduction. In an extreme case, power could be reduced to a fast idle speed.
Oil Pump Failure	If the oil pump fails or an open circuit occurs between the pump and the PCM, engine power will be reduced to a fast idle.

SELOC_1418

Fig. 110 Guardian system activation - 75/90/115 hp Optimax models

Warning Horn

Function	Sound	Description
Start Up	One Beep	Normal System Test
Low Oil	Four Beeps every 2 Minutes	Oil level is low in the oil tank.
Water In Fuel	Four Beeps every 2 Minutes	Water in the fuel filter chamber reached the full level.
Cooling System Problem	Continuous	Engine Guardian System is activated. Power limit will very with level of overheat.
Oil Level is Critically Low	Continuous	Engine Guardian System is activated. Power limit will allow a fast idle.
Oil Pump Failure	Continuous	Engine Guardian System is activated. Power limit will allow a fast idle.
Engine Overspeed	Continuous	The warning horn is activated any time engine speed exceeds the maximum allowable RPM. The system will limit the engine speed to within the allowable range.
Sensor out of Range	Continuous	Engine Guardian System is activated. Power limit may activate at full throttle speed.
	Intermittent Beep	Engine Guardian System is activated. Power limit may restrict engine speed to idle.
Engine Running Cold at Slow Speed	One Beep	Engine is not reaching correct temperature while operating below 1000 RPM.

SELOC_1418

Fig. 109 Warning horn signals - 75/90/115 hp Optimax models

Problem: Oil Level in Engine Oil Reservoir Tank is Low But Not Low in Remote Oil Tank.

Possible Cause	Corrective Action
Quick disconnect on remote oil tank is not fully connected	Reconnect
Remote oil hose (blue stripe) is blocked.	Check length of hose for a kink.
Remote pulse hose (second hose) is blocked or punctured.	Check length of hose for a kink.
Remote pulse hose check valve is faulty (this valve is located at the engine end of the hose).	Replace check valve.
A restricted oil outlet filter in the remote tank.	Remove filter and clean.
Leak at upper end of remote oil tank pick-up tube.	Check tube for cracks or leaks.
Oil and Pulse hoses reversed.	Check hose connections.
Low crankcase pressure.	Check pressure from pulse hose check valve (2 psi minimum).

Problem: Warning Horn Does Not Sound When Ignition Key is Turned to "ON" Position.

Possible Cause	Corrective Action
Horn malfunction or open (TAN/BLUE) wire between horn and engine.	Use a jumper wire to ground TAN/BLUE lead (at engine bullet connector starboard temperature sensor) to engine ground. Warning horn should sound. If not, check TAN/BLUE wire between horn and engine for open circuit and check horn.
Faulty Ignition Control Module	Check if all Control Module leads are connected to harness leads. If so, module may be faulty.
Using incorrect side mount remote control or ignition/choke assembly.	See info on remote control

Problem: Warning Horn Stays on When Ignition Key is Turned to "ON" Position.

Engine overheat sensor/Control Module	If horn sounds a continuous signal, the engine overheat sensor starboard head may be faulty. Disconnect overheat sensor and turn ignition key to "ON" position. If horn still sounds a continuous signal, the Control Module is faulty. Replace module and re-test. If signal does not sound, then engine overheat sensor is faulty. Replace and re-test.

Problem: Warning Horn sounds when Engine is Running and Oil Level in Engine Reservoir is Full.

Possible Cause	Corrective Action
Defective low oil sensor (located in fill cap of engine oil reservoir).	Do not remove cap from oil reservoir. Disconnect both low oil sensor leads from terminal connectors. Connect an ohmmeter between leads. There should be no continuity through sensor. If continuity exists, sensor is faulty. Replace cap assembly.

If all of the checks are positive, the Control Module is faulty. Replace Module and re-test.

SELOC_1418

Fig. 111 Warning system troubleshooting - 135-200 hp (2.5L) V6 carbureted models and 2001 EFI models

Problem	Horn	Monitor Display	Guardian Activated	Engine Speed Reduction Activated
Power Up/System Check	Single Beep	Yes	N/A	—
Low Oil	4 Beep...2 Minutes Off	Yes	No	No
Oil Pump Electrical Failure		Yes	Yes	Yes (See Guardian System)
Over Heat	Continuous Beep	Yes	Yes	Yes (See Guardian System)
Water In Fuel	4 Beep...2 Minutes Off	Yes	No	No
Over Speed	Continuous Beep	Yes	Yes	Yes (See Guardian System)
Coolant Sensor Failure	No	Yes	No	No
MAP Sensor Failure	No	Yes	No	No
Air Temperature Sensor Failure	No	Yes	No	No
Ignition Coil Failure	No	Yes	No	No
Injector Failure	No	Yes	No	No
Horn Failure	N/A	Yes	No	No
Battery Voltage too high (16V) or too low (11V) or very low (9.5V)	No	Yes	Yes	Yes (See Guardian System)
Throttle Sensor Failure	Continuous Intermittant Beeping	Yes	Yes	Yes (See Guardian System)
Block Water Pressure	Yes	Yes	Yes	Yes (See Guardian System)
Calculated Oil Level Critical	Yes	Yes	Yes	Yes (See Guardian System)

SELOC_1418

Fig. 113 Warning horn signals - 2002 or later 150-250 hp (2.5L & 3.0L) EFI models

Problem: Oil Level in Engine Oil Reservoir Tank is Low But Not Low in Remote Oil Tank.

Possible Cause	Corrective Action
Fill cap is leaking air on the remote tank.	Make sure O-rings or gaskets are in place and caps are tight.
Quick disconnect on remote oil tank is not fully connected.	Re-connect
Remote oil hose (blue stripe) is blocked.	Check length of hose for a kink.
Remote pulse hose (second hose) is blocked or punctured.	Check length of hose for a kink or leakage.
Remote pulse hose check valve is faulty (this valve is located at the engine end of the hose).	Replace check valve.
A restricted oil outlet filter in the remote tank.	Remove filter and clean.
Air leak in upper portion of oil pickup tube.	Replace tube.

Problem: Warning Horn Does Not Sound When Ignition Key is Turned to "ON" Position.

Possible Cause	Corrective Action
Horn malfunction or open (TAN) wire between horn and engine.	Use a jumper wire to ground TAN lead (at engine terminal block) to engine ground. Warning horn should sound. If not, check TAN wire between horn and engine for open circuit and check horn.
Electronic Control Module (ECM)	Check if all ECM leads are connected to harness leads. If so, ECM may be faulty.
Using incorrect side mount remote control or ignition/choke assembly.	
Open circuit on PURPLE wire going to (+) terminal of horn.	Check for battery voltage at (+) terminal of horn when ignition key is turned on.
Engine overheat sensor	If horn sounds a continuous signal, the engine overheat sensor may be faulty. Disconnect overheat sensor and turn ignition key to "ON" position. If horn still sounds a continuous signal, the ECM is faulty. Replace ECM and re-test. If signal does not sound, then engine overheat sensor is faulty. Replace and re-test.
Electronic Control Module (ECM)	Check connections – replace ECM.

Problem: Warning Horn sounds when Engine is Running and Oil Level in Engine Reservoir is Full.

Possible Cause	Corrective Action
Defective low oil sensor	Disconnect both low oil sensor leads from terminal connectors, connect an ohmmeter between leads. There should be NO continuity through sensor. If continuity exists, sensor is faulty.

SELOC_1418

Fig. 112 Warning system troubleshooting - 2001 EFI 200-250 hp (3.0L) V6 models

Guardian System Operation with Gauges

Smartcraft Gauge/Monitor	System will sound warning horn and display the warning message.

Guardian System Activation

Function	Warning Horn		Description
	Sound		
Cooling System Problem	Continuous		Engine Guardian System is activated. Power limit will very with level of overheat. Shift outboard into neutral and check for a steady stream of water coming out of the water pump indicator hole. If no water is coming out of the water pump indicator hole or flow is intermittent, stop engine and check water intake holes for obstruction. The Guardian system must be RESET before engine will operate at higher speeds. Moving throttle lever back to idle resets the system.
Oil Level Is Critically Low	Continuous		Engine Guardian System is activated. Power limit will limit engine speed. The oil level is critically low in the engine mounted oil reservoir tank. Refill the engine mounted oil reservoir tank along with the remote oil tank.
Oil Pump Failure	Continuous		Engine Guardian System is activated. Power limit will limit engine speed. The warning horn is activated if the oil pump should ever stop functioning electrically. No lubricating oil is being supplied to the engine.
Engine Overspeed	Continuous		The warning horn is activated any time engine speed exceeds the maximum allowable RPM. The system will limit the engine speed to within the allowable range. If the overspeed condition continues, the Engine Guardian System will place the engine in power reduction. The Guardian system must be RESET before engine can resume full power. Moving throttle lever back to idle resets the system. Engine overspeed indicates a condition that should be corrected. Overspeed could be caused by incorrect propeller pitch, engine height, trim angle, etc.
Sensor out of Range	Continuous		Engine Guardian System is activated. Power limit may activate at full throttle speed.
	Intermittent Beep		Engine Guardian System is activated. Power limit may restrict engine speed to idle.

SELOC_1418

Fig. 114 Guardian system activation - 2002 or later 150-250 hp (2.5L & 3.0L) EFI models

Problem	Horn	Monitor Display	Guardian Activated	Engine Speed Reduction Activated
Power Up/System Check	Single Beep	Yes	N/A	No
Low Oil	4 Beep...2 Minutes Off	Yes	No	No
Oil Pump Electrical Failure		Yes	Yes	Yes (See Guardian System)
Over Heat	Continuous Beep	Yes	Yes	Yes (See Guardian System)
Water In Fuel	4 Beep...2 Minutes Off	Yes	No	
Over Speed	Continuous Beep	Yes	Yes	Yes
Coolant Sensor Failure	No	Yes	No	No
MAP Sensor Failure	No	Yes	No	No
Air Temperature Sensor Failure	No	Yes	No	No
Ignition Coil Failure	No	Yes	No	No
Injector Failure	No	Yes	No	No
Horn Failure	N/A	Yes	No	No
Battery Voltage too high (16V) or too low (11V) or very low (9.5V)	No	Yes	Yes	Yes (See Guardian System)
Over Heat Cyl. Head/Compressor	Continuous Beep	Yes	No	Yes (See Guardian System)
Throttle Sensor Failure	Continuous Intermittant Beeping	Yes	Yes	Yes (See Guardian System)
Block Water Pressure	Yes	Yes	Yes	Yes (See Guardian System)
Calculated Oil Level Critical	Yes	Yes	Yes	Yes (See Guardian System)

SELOC_1418

Fig. 115 Warning horn signals - 135-175 hp (2.5L) Optimax models

Guardian System Operation with Gauges

4 Function Gauge (2000 Models Only)	System will sound warning horn and illuminate appropriate light on gauge.
Smartcraft Gauge/Monitor	System will sound warning horn and display the warning message.

Guardian System Activation

Condition	Result
Engine Overheat	Engine power level can be reduced to any percentage down to an idle speed, if overheat condition persists.
Air Compressor Overheat	2000 Model – engine power level can be reduced to any percentage down to an idle speed, if overheat condition persists. 2001 Model – no power reduced.
Block Water Pressure Low	Engine power level can be reduced to any percentage down to a fast idle, if condition persists.
Throttle Position Sensor Failure	If the throttle position sensor fails or becomes disconnected, power will be limited to a maximum of approximately 4500 rpm. When the TPS is in the fail mode, the ECM will use the MAP sensor for a reference to determine fuel calibration.
Temperature Sensor (cylinder head) Failure	If a temperature sensor should fail or become disconnected, power will be reduced by 25%.
Temperature Sensor Air Compressor Failure	2000 Model – power reduced by 25%. 2001 Model – no power reduced.
Battery Voltage (too high or too low)	Battery voltage greater than 16.5 volts or less than 10.5 volts will result in engine output power being reduced. The higher or lower the voltage is outside of these parameters, the greater the percentage of power reduction. In an extreme case, power could be reduced to idle speed.
Oil Pump Failure	If the oil pump fails or an open circuit occurs between the pump and the ECM, engine power will be reduced to idle.
Calculated Oil Level Critical	After low condition occurs, ECM calculates when critical level will occur, then reduces power to idle.

SELOC_1418

Fig. 117 Guardian system activation - 135-225 hp (2.5L & 3.0L) Optimax models

Problem	Horn	Check Engine Light	Low Oil Light	Over Heat Light	Water In Fuel Light	Engine Speed Reduction Activated
Power Up/System Check	Single Beep	Yes	Yes	Yes	Yes	No
Low Oil	4 Beep...2 Minutes Off		Yes			No
Oil Pump Electrical Failure	Continuous Beep	Yes	Yes			Yes
Over Heat	Continuous Beep			Yes		Yes
Water In Fuel	4 Beep...2 Minutes Off				Yes	No
Over Speed	Continuous Beep					Yes (Activated at 5800 RPM)
Coolant Sensor Failure	No	Yes				No
MAP Sensor Failure	No	Yes				No
Air Temperature Sensor Failure	No	Yes				No
Ignition Coil Failure	No	Yes				No
Injector Failure	No	Yes				No
Horn Failure	N/A	Yes				No
Battery Voltage too high (16V) or too low (11V) or very low (9.5V)	No	Yes				Yes – If battery voltage is less than 10.4 V – RPM is reduced.
Over Heat Cyl. Head/Compressor	Continuous Beep			Yes		Yes
Throttle Sensor Failure	Continuous Intermittant Beeping	Yes				RPM reduced, ECM then refers to MAP Sensor for throttle position
Block Water Pressure	Continuous Beep			Yes		Yes
Calculated Oil Level Critical	Continuous Beep	Yes	Yes			Yes

SELOC_1418

Fig. 116 Warning horn signals - 200/225 hp (3.0L) Optimax models

Warning Horn

Function	Sound	Description
Start Up	One Beep	Normal System Test
Low Oil Reserve	Four Beeps every 2 Minutes	Oil level is low in the engine mounted oil reservoir tank. Refill the engine mounted oil reservoir tank along with the remote oil tank. Refer to Fuel & Oil Section.
Oil Level is Critically Low	Continuous	Engine Guardian System is activated. Power limit will allow a fast idle. The oil level is critically low in the engine mounted oil reservoir tank. Refill the engine mounted oil reservoir tank along with the remote oil tank. Refer to Fuel and Oil Section.
Oil Pump Failure	Continuous	Engine Guardian System is activated. Power limit will allow a fast idle. The warning horn is activated if the oil pump should ever stop functioning electrically. No lubricating oil is being supplied to the engine.

NOTE: As an option, Mercury Monitor or SmartCraft Gauges may be used to provide low oil information or oil pump operation information.

SELOC_1418

Fig. 118 Low oil warning system troubleshooting - common warning horn signals all 2.5L & 3.0L OptiMax models and 2002 or later EFI motors

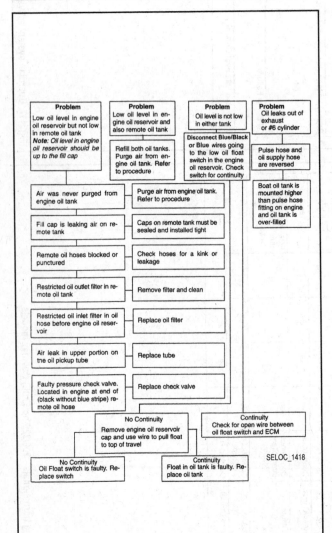

SELOC_1418

Fig. 119 Low oil warning system troubleshooting (warning system activated) - all 2.5L & 3.0L OptiMax models and 2002 or later EFI motors

Temperature Switch (Warning System)

DESCRIPTION & OPERATION

◆ See Figures 120, 121 and 122

Most 30 hp or larger Mercury outboards are equipped with either a temperature switch (and on/off switch which is activated by temperature) and/or a thermo-sensor (a variable resistor whose values change with changes in temperatures). Basically, almost all inline motors (except the OptiMax models) are equipped with a switch, where all of the V6 motors are normally equipped with a sensor.

The information under this heading is for temperature *switches*. For information on testing or servicing a temperature *sensor*, please refer to the appropriate heading (EFI or OptiMax) in the Fuel System section. On carbureted V6 motors which are equipped with a sensor instead of a switch, refer to the data on the 2001 2.5L EFI motor.

A temperature switch is generally a bi-metallic type on/off switch. Typically a temperature switch will consist of 2 different type metals attached together in a small strip. Because each metal has a different coefficient of expansion - each will expand a different amount within a given temperature range, and the strip will bend rather than extend. Depending upon whether the switch is designed to be normally closed or normally open in a given temperature range will determine how the metal switch and contacts are arranged.

Mercury motors *usually* (but it is possible that they do not always) use a normally open switch assembly that provides no continuity across the switch contacts until a given temperature is reached. The internals of the switch are designed to bend in such a manner that the switch contacts will only close at or above a certain temperature. In this way, only once the switch physically reaches a given temperature will the switch contacts close, activating the warning circuit.

The temperature switch therefore only sends an ON or OFF signal, not a variable voltage as in the thermo-sensor signal. If the powerhead temperature exceeds a specified level, the temperature switch will close a warning circuit activating a buzzer. Generally, the buzzer will continue to sound until the temperature drops to an acceptable level.

The temperature switches used by these motors normally have 2 calibrations. Each has an overheat warning ON temperature setting (the temperature at which the switch contacts will close), but they also have an overheat warning OFF temperature (which is usually significantly lower than the ON temperature). The switches are designed not to close until a certain point has been reached, but once reached, they remain closed until the powerhead (switch) has cooled well below the activation point. For details on temperature switch operational parameters, please refer to the Cooling System Specifications charts in this section.

OPERATIONAL TEMPERATURE SWITCH CHECK

◆ See Figures 120, 121 and 122

■ **This section deals with the temperature switches only (on/off temperature activated switches). For information on the variable resistance temperature sensors such as the ETS used on fuel injected motors and carbureted V6 motors, please refer to the Fuel System section.**

1. Identify the leads (often Tan, Tan with a tracer and/or Black, but check the Wiring Diagrams in the Ignition & Electrical section for the motor on which you are working to be sure) from the temperature switch.
2. Disconnect the lead or leads at the quick disconnect fittings. This won't have an effect on most motors because the circuit is broken by the switch during normal operation.
3. Start and run the motor (because you'll be running at above idle, it is really best to use a test tank or launch the craft to prevent the possibility of engine overspeed). Increase engine speed to just above 2500 rpm.
4. Touch the single lead to the powerhead (grounding the circuit) or, if a switch is equipped with 2 leads, touch them together). The warning buzzer should sound continuously and, on most models, the powerhead rpm should drop to 2000 rpm.
5. Reconnect the wiring to the switch leads, back the throttle down to idle rpm and the buzzer should be silent.
6. If the temperature switch does not operate as specified, either the switch or its circuit are faulty.

Fig. 120 A temperature switch or sensor is mounted to most Mercury cylinder heads...

Fig. 121 ...V6 motors usually have at least 1 per head (sometimes more on EFI/OptiMax motors)

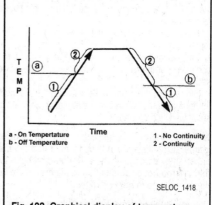

a - On Tempertature
b - Off Temperature

Time

1 - No Continuity
2 - Continuity

SELOC_1418

Fig. 122 Graphical display of temperature switch operation showing switch position (on/off , i.e. continuity/discontinuity) in relation to temperature and time

CHECKING TEMPERATURE SWITCH CONTINUITY

◆ See Figures 120 thru 124

■ This section deals with the temperature switches only (on/off temperature activated switches). For information on the variable resistance temperature sensors such as the ETS used on fuel injected motors and carbureted V6 motors, please refer to the Fuel System section.

Locate and remove the temperature switch from the powerhead. The temperature switch usually resembles a condenser, often with Tan, Tan with a tracer or Black leads (but check the Wiring Diagrams from the Ignition & Electrical System section for the motor on which you are working).

The temperature switch can normally be removed from the powerhead easily, by removing the screw holding the locktab (when equipped) and then carefully pulling it out of its recess.

✳✳ SELOC CAUTION

The temperature switch is a very fragile sensor and should never be dropped or handled in a rough manner.

Remove the temperature switch from the block and disconnect the wiring the quick disconnect fitting(s). Since the switch should have no continuity until it is heated past a point which could damage the powerhead, there are no resistance tests to be performed on this switch while it is installed.

■ Ok, that last statement was too much of a generalization. If the warning system is activated and remains activated on a cold motor, you can check the temperature switch for continuity between the lead and ground (or across the contacts should it have 2 wires, but that type of switch is not normally used on these Mercury motors). If there is continuity even with a cold motor, then the switch is bad, get rid of it!

Testing is a simple matter of watching for continuity across the switch contact and ground (using an ohmmeter or DVOM set to read resistance across the switch wiring and switch body) as the switch is slowly heated to activation temperature. The best method for this is to suspend the switch in water which is slowly heated to temperature while you watch the meter. Obtain a thermometer, an ohmmeter, and a suitable pan in which to boil water.

Place the container of water, at room temperature, on a stove. Secure the thermometer in the water in such a manner to prevent the thermometer bulb from contacting the bottom or sides of the pan (this ensures the thermometer will read the temperature of the water and NOT the pan, which could be considerably higher).

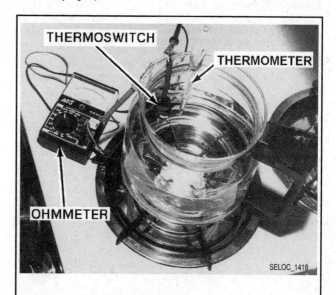

THERMOSWITCH

THERMOMETER

OHMMETER

SELOC_1418

Fig. 124 Heating a container of water to check the performance of the temperature switch

Fig. 123 Performing a resistance check on a temperature switch

Immerse the sensing end of the temperature switch into the water up to the shoulder and suspend it, also NOT touching the pan for the same reason as the thermometer.

Make contact with the Red meter lead to the switch lead, and the Black meter lead to the switch body (or other switch lead, if used). The meter should indicate no continuity, as long as the water temperature is below the continuity parameter listed in the Cooling System Specifications chart in this section.

Slowly heat the water, while watching the thermometer and meter. When the temperature rises above the activation point listed in the chart, the meter should indicate continuity.

■ A slight variation is allowable above or below the stated temperatures listed. But keep in mind that too great a variation could either allow damage to the powerhead by operating it under conditions that are too hot or could allow activation of the warning system when the powerhead is not in danger.

Remove the pan from heat and allow the sensor to slowly cool (you can add a little cool water or, if using a metal pan, immerse the pan itself in a second pan of cooler water to help speed the cooling process). Watch the meter and the thermometer as the water cools. The switch contacts must open (showing no continuity) at or near the No Continuity temperature listed in the chart.

If the meter reading is acceptable, the temperature switch is functioning correctly. If the meter readings are not acceptable, the temperature switch can be easily replaced with a new unit.

Oil Level Sensor (Switch)

DESCRIPTION & TESTING

 MODERATE

◆ See Figures 125 and 126

■ For more images of oil level sensor mounting, please refer to the Oil Tank procedures in the Oil Systems section.

As noted earlier, all Mercury oil injection systems used a powerhead mounted oil reservoir which contains an oil level switch. Though the basic design of the switch varies somewhat from model-to-model, the function does not. In all cases, the switch is designed to show no continuity as long as sufficient oil is in the tank.

The switch may be mounted physically in the oil (usually from the top of the tank), in which case it normally contains a built-in float of some type) or the switch can be mounted to the tank, but isolated from the oil (in which case it is normally a magnetic sensor which operates in conjunction with a float mounted inside the tank itself).

In either case, testing is a simple matter of making sure the switch has no continuity when oil level is sufficient, and continuity when the tank is empty. One of the most common ways to mis-diagnose the sensor/system is to

pronounce a sensor bad when the FLOAT IS STUCK! So always check/clean the float assembly before condemning a sensor.

You might notice that we've used both the term sensor and switch, but truthfully it is normally only designed to be a on/off SWITCH and not a variable sensor, even though either term is often applied.

Fig. 125 Many Mercury motors use an oil level sensor. . .

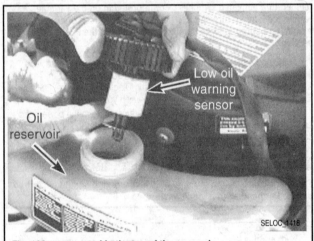
Fig. 126 . . .mounted in the top of the reservoir

COOLING SYSTEM FLOW SCHEMATICS

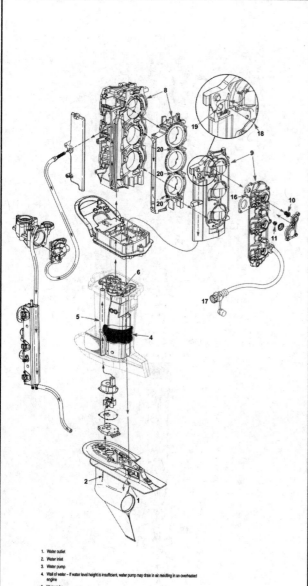

1. Water outlet
2. Water inlet
3. Water pump
4. Wall of water – If water level height is insufficient, water pump may draw in air resulting in an overheated engine
5. Water tube
6. Water dump down exhaust pipe
7. Exhaust adaptor plate
8. Cylinder block – halves separated for illustration
9. Cylinder block combustion chamber – halves separated for illustration
10. Thermostat 61° C (142° F) – If stuck closed, engine will overheat at idle. If stuck open, engine will run cold at idle.
11. Poppet valve – controls water flow at high RPM. If poppet valve is stuck open at low RPM, the engine will not reach proper operating temperature (run cold) and will run rough at idle
12. Exhaust cover – separated for illustration
13. Strainer screen for air compressor water supply – If restricted, compressor will overheat and tell-tale will be weak
14. Air compressor – water outlet from air compressor connects to fuel rail coolant inlet
15. Fuel rail – fuel cooler is built from fuel rail
16. Tell-tale outlet on bottom cowl
17. Powerhead flush
18. Thermostat water dump
19. Poppet valve opening
20. Rubber water deflectors (3) – If missing may result in uneven cooling of cylinders and scuffed pistons.

SELOC_1418

Cooling system schematic - 75-115 hp (1.5L) Optimax models

1. Cylinder Head Cover – Removed from head for illustration, normally part of head casting.
2. Thermostat (2) 143° F (61.7° C) – If stuck closed, engine will over heat at low speed. If stuck open, engine will not warm up at idle speed.
3. Check Valve for powerhead flush.
4. Poppet Valve – Controls water flow at high RPM.
   NOTE: *If poppet valve is stuck open at low RPM, the engine will not reach proper operating temperature (run cold) and will run rough at idle.*
5. Primary Water Discharge into Driveshaft Housing.
6. Water passing through thermostats dump into the adaptor plate, then discharges down the exhaust.
7. Water Dump Holes for Exhaust Cooling (2 each) 1/8 in. (3.175 mm) – if holes are plugged, tuner pipe will melt and bearing carrier prop shaft seals will be damaged.
8. Tell-Tale Outlet
9. Water Tube
10. Wall of Water – If water level height is insufficient, water pump may draw in air resulting in an overheated engine.
11. Excess water from wall of water around exhaust bucket exits around anodes.
12. Water Inlet
13. Water exiting exhaust discharge.

SELOC_1418

Cooling system schematic - 150-200 hp (2.5L) carbureted models and 2001 EFI models

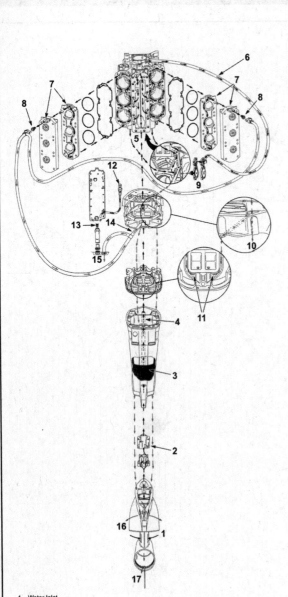

1. Water Inlet
2. Water Pump
3. Wall of Water - If water level height is insufficient, water pump may draw in air resulting in an overheated engine
4. Water Tube
5. Main Water Feed to Powerhead from Water Tube
6. Water by-pass prevents air/steam pockets at top of powerhead
7. Cylinder Head Cover - Removed from head for illustration, normally part of head casting
8. Thermostats (2) 143° F (61.7° C) - If stuck closed, engine will overheat at low speed. If stuck open, engine will not warm up at idle speed.
9. Poppet Valve - Controls water flow at high RPM. If poppet valve is stuck open at low RPM, the engine will not reach proper operating temperature (run cold) and will run rough at idle
10. Primary water discharge into driveshaft housing.
11. Water Dump Holes for exhaust cooling (2 each) 1/8 in. (3.175 mm) - if holes are plugged, tuner pipe will melt and bearing carrier prop shaft seals will be damaged.
12. Water Pressure Sensor
13. Check Valve for powerhead flush
14. Water passing through thermostats dump into adaptor plate, then discharges down the exhaust.
15. Tell-Tale Outlet
16. Excess water from wall of water around exhaust bucket exits around anodes.
17. Water Exits with Exhaust Discharge

SELOC_1418

Cooling system schematic - 2002 or later 150-200 hp (2.5L) EFI models

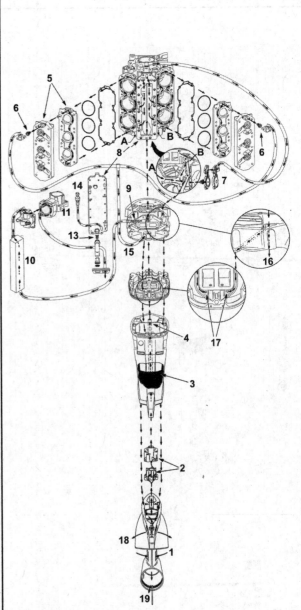

1. Water Inlet
2. Water Pump
3. Wall of Water - If water level height is insufficient, water pump may draw in air resulting in an overheated engine
4. Water Tube
5. Cylinder Head Cover - Removed from head for illustration, normally part of head casting
6. Thermostats (2) 143° F (61.7° C) - If stuck closed, engine will overheat at idle
7. Poppet Valve - Controls water flow at high RPM. If poppet valve is stuck open at low RPM, the engine will not reach proper operating temperature (run cold) and will run rough at idle
8. Exhaust Divider Plate - Separated for illustration
9. Strainer Screen for air compressor and fuel rail water supply - If restricted, compressor will overheat and tell-tale will be weak
10. Port Fuel Rail - Fuel Cooler is built into Port Fuel Rail
11. Air Compressor
12. Water Outlet from Air Compressor - Connects to tell-tale outlet on bottom cowl
13. Check Valve for powerhead flush.
14. Block Water Pressure Sensor
15. Water passing through thermostats dump into the adaptor plate, then discharges down the exhaust
16. Primary Water Discharge into Driveshaft Housing
17. Water Dump Holes Exhaust Cooling (2 each) 1/8 in. (3.175 mm) - If holes are plugged, tuner pipe will melt and bearing carrier prop shaft seals will be damaged
18. Excess water from wall of water around exhaust bucket exits around anodes
19. Water Exits with Exhaust Discharge

SELOC_1418

Cooling system schematic - 135-175 hp (2.5L) OptiMax models

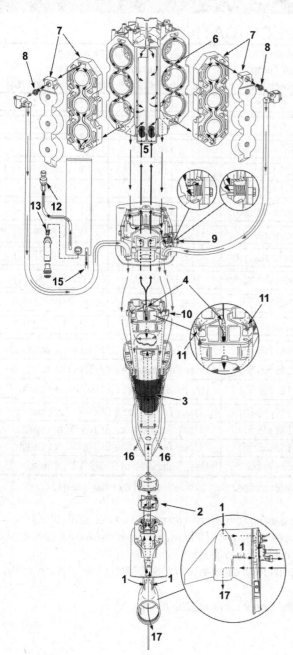

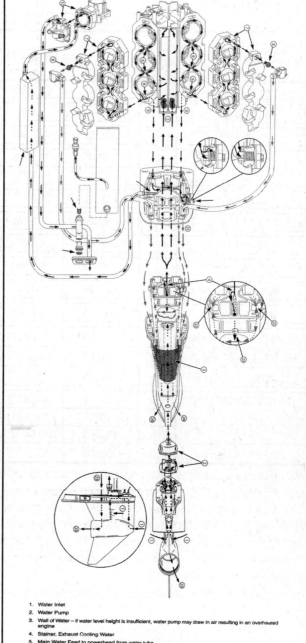

1. Water Inlet
2. Water Pump
3. Wall of Water - If water level height is insufficient, water pump may draw in air resulting in an overheated engine
4. Strainer - Exhaust cooling water
5. Main Water Feed to Powerhead from Water Tube
6. Rubber Water Dams - If missing may result in uneven cooling of cylinders and scuffed pistons
7. Cylinder Head Cover - Removed from head for illustration, normally part of head casting
8. Thermostats (2) 143° F (61.7° C) - If stuck closed, engine will overheat at low speed. If stuck open, engine will not warm up at idle speed.
9. Poppet Valve - Controls water flow at high RPM. If poppet valve is stuck open at low RPM, the engine will not reach proper operating temperature (run cold) and will run rough at idle
10. Primary water discharge into driveshaft housing.
11. Water discharge into exhaust relief passage
12. Water Pressure Sensor
13. Check Valve for powerhead flush
14. Water passing through thermostats dump into adaptor plate, then discharges down the exhaust.
15. Tell-Tale Outlet
16. Excess water from wall of of water around exhaust bucket exits around anodes.
17. Water Exits with Exhaust Discharge

SELOC_1418

Cooling system schematic - 200-250 hp (3.0L) EFI models (2002 or later shown, earlier models similar)

1. Water Inlet
2. Water Pump
3. Wall of Water – If water level height is insufficient, water pump may draw in air resulting in an overheated engine
4. Strainer, Exhaust Cooling Water
5. Main Water Feed to powerhead from water tube
6. Rubber Water Dams (6) – If missing may result in uneven cooling of cylinders and scuffed pistons
7. Cylinder Head Cover – removed from head for illustration, normally part of head casting
8. Thermostats (2) 61.7° C (143° F) – if stuck closed, engine will overheat at idle. If stuck open, engine will run cold at idle.
9. Main Water Dump from Powerhead – fills chamber in adaptor plate which feeds port fuel rail, air compressor, and poppet valve.
10. Strainer Screen for Port Fuel Rail and Air Compressor Water Supply – If restricted, compressor will overheat and tell-tale will be weak
11. Port Fuel Rail – fuel cooler is built into port fuel rail
12. Air Compressor – water outlet from air compressor connects to tell-tale outlet on bottom cowl
13. Tell-Tale outlet on bottom cowl
14. Powerhead Flush Inlet Plug
15. Check Valve for powerhead flush
16. Block Water Pressure Sensor
17. Rear Exhaust Divider Plate – separated for illustration. normally part of engine block
18. Poppet Valve – controls water flow at high RPM. If poppet valve is stuck open at low rpm, the engine will not reach proper operating temperature (run cold) and will run rough at idle
19. Poppet Valve in Closed Position
20. Poppet Valve in Open Position
21. Main Water Dump from Thermostats and Poppet Valve
22. Poppet Valve, when open, fills chamber in exhaust tube and slots provide water spray to keep exhaust relief holes clear
23. Some Water from Water Tube Strainer is diverted through exhaust tube and propeller hub with exhaust discharge
24. Excess Water from wall of water around exhaust tube exits around anodes

SELOC_1418

Cooling system schematic - 200-250 hp (3.0L) OptiMax models

COOLING SYSTEM SPECIFICATIONS

Model (HP)	No of Cyl	Year	Displace cu.in. (cc)	Thermostat Open Temp. Degrees F(C)	Temperature Switch Operational Parameters	
					Continuity F (C)	No Continuity F (C)
2.5/3.3	1	2001-05	4.6 (74.6)	n/a	n/a	n/a
4/5	1	2001-05	6.2 (102)	n/a	n/a	n/a
6/8	2	2001-05	13 (210)	120 (49) / optional	n/a	n/a
9.9/15	2	2001-05	16 (262)	120 (49)	n/a	n/a
10	2	2001	16 (262)	120 (49) / optional	n/a	n/a
20/20 Jet/25	2	2001-05	24 (400)	5 (3) above stamping	n/a	n/a
30/40	2	2001-03	39 (644)	5 (3) above stamping	182-198 (83-92)	162-178 (72-81)
40/40 Jet/50/55/60	3	2001-09	59 (967)	5 (3) above stamping	182-198 (83-92)	162-178 (72-81)
65 Jet/75/90	3	2001-09	85 (1386)	5 (3) above stamping	182-198 (83-92)	162-178 (72-81)
75/80 Jet/90/115 Optimax	3	2004-09	93 (1526)	140-145 (60-63)	Uses Temperatures SENSOR not switch	
80 Jet/115/125	4	2001-05	113 (1848)	5 (3) above stamping	182-198 (83-92)	162-178 (72-81)
110 Jet/135/150/175 Optimax	V6	2001-09	153 (2507)	140-145 (60-63)	Uses Temperature SENSOR not switch ①	
150 (XR6,Classic) and 200	V6	2001-05	153 (2507)	140-145 (60-63)	Uses Temperature SENSOR not switch ②	
150/175/200 (2.5L) EFI	V6	2001-09	153 (2507)	140-145 (60-63)	Uses Temperature SENSOR not switch ②	
200/225/250 (3.0L) EFI	V6	2001	185 (3032)	140-145 (60-63)	Uses Temperature SENSOR not switch	
	V6	2002	185 (3032)	118-123 (48-51)	Uses Temperature SENSOR not switch	
	V6	2003-05	185 (3032)	128-133 (53-56)	Uses Temperature SENSOR not switch	
200/225/250 Optimax	V6	2001-09	185 (3032)	118-122 (48-50) ④	Uses Temperature SENSOR not switch ③	

Above Stamping - Most Mercury Thermostats are stamped on the bottom with an operating temperature. They must begin to open approx 5 degrees F (3 degrees C) ABOVE stamped temperature

① Sensor in Starboard bank (and Port on 2001 or later models) activates warning horn when temperature exceeds 221F (105C)

② Sensor in Starboard bank activates warning horn when temperature exceeds 232-248F (102.3-128.9C) and deactivates warning system when temperature drops below 195-225F (89.5C). Port sensor for optional temp gauge

③ Sensors in Starboard and Port bank activates warning horn when temperature of either exceeds 180 F (82 C), then activates speed reduction system by 185 F (85 C)

④ Specification is for regular Optimax models, for Pro/XS/Sport models thermostat opens at 130F (54C)

6

POWERHEAD

POWERHEAD MECHANICAL

General Information

You can compare the major components of an outboard with the engine and drivetrain of your car or truck. In doing so, the powerhead is the equivalent of the engine and the gearcase is the equivalent of your drivetrain (the transmission/transaxle). The powerhead is the assembly that produces the power necessary to move the vehicle, while the gearcase is the assembly that transmits that power via gears, shafts and a propeller (instead of tires).

Speaking in this manner, the powerhead is the "engine" or "motor" portion of your outboard. It is an assembly of long-life components that are protected through proper maintenance. Lubrication, the use of high-quality 2-stroke oils, is probably the most important way to preserve powerhead condition (that and frequent use/service). Similarly, proper tune-ups that help maintain proper air/fuel mixture ratios and prevent pinging, knocking or other potentially damaging operating conditions are the next best way to preserve your motor. But, even given the best of conditions, components in a motor begin wearing the first time the motor is started and will continue to do so over the life of the powerhead.

Eventually, all powerheads will require some repair. The particular broken or worn component, plus the age and overall condition of the motor may help dictate whether a small repair or major overhaul is warranted. As much as you can generalize about mechanical work, the complexity of the job will vary with 2 major factors:

• The age of the motor (the older OR less well maintained the motor is) the more difficult the repair

• The larger and more complex the motor, the more difficult the repair.

Again, these are generalizations and, working carefully, a skilled do-it-yourself boater can disassemble and repair a 250 hp, fuel injected V6 powerhead, as well as a seasoned professional. But both DIYers and professionals must know their limits. These days, many professionals will leave portions of machine work (from cylinder block and piston disassembly, cleaning and inspection to honing and assembly up to a machinist). This is not because they are not capable of the task, but because that's what a machinist does day in and day out. A machinist is naturally going to be more experienced with the procedures.

If a complete powerhead overhaul is necessary on your outboard, we recommend that you find a local machine shop that has both an excellent reputation and that specializes in marine work. This is just as important and handy a resource to the professional as a DIYer. If possible, consult with the machine shop before disassembly to make sure you follow procedures or mark components, as they would desire. Some machine shops would prefer to perform the disassembly themselves. In these cases, you can usually remove the powerhead from the gearcase and deliver the entire unit to the shop for disassembly, inspection, machining and assembly.

■ There are a number of major outboard powerhead re-manufacturers throughout the U.S. and you may be able to quite cost effectively purchase a replacement powerhead for installation making it an attractive option. At least it is an option you should consider before undertaking a major overhaul.

If you decide to perform the entire overhaul yourself, proceed slowly, taking care to follow instructions closely. Consider using a digital camera (if available) to help document component mounting during the removal and disassembly procedures. This can be especially helpful if the overhaul or rebuild is going to take place over an extended amount of time. If this is your first overhaul, don't even THINK about trying to get it done in one weekend, YOU WON'T. It is better to proceed slowly, asking for help when necessary from your trusted parts counterman or a tech with experience on these motors.

Keep in mind that anytime pistons, rings and bearings have been replaced, the powerhead must be broken-in again, as if it were a brand-new motor. Once a major overhaul is completed, refer to the section on Powerhead Break-In for details on how to ensure the rings set properly without damage or scoring to the new cylinder wall or the piston surfaces. Careful break-in of a properly overhauled motor will ensure many years of service for the trusty powerhead.

IMPORTANT POINTS FOR POWERHEAD OVERHAUL

Repair Procedures

Service and repair procedures will vary slightly between individual models, but the basic instructions are quite similar. Special tools may be called out in certain instances. These tools may be purchased from the local marine dealer.

✳✳ SELOC CAUTION

When removing the powerhead, always secure the gearcase in a suitable holding fixture to prevent injury or damage if the powerhead releases from the gearcase suddenly (which often occurs on outboards that have been in service for some time or that are extensively corroded).

On most of the single cylinder motors, and a FEW of the twins, the powerhead is actually light enough to be lifted by hand. BUT, it is really a better idea to use an engine hoist when lifting most powerhead assemblies. This not only makes sure that the powerhead is secured at all times, but helps prevent injuries, which could occur if the powerhead releases suddenly. Also, keep in mind the components such as the driveshaft could be damaged if the powerhead is removed or installed at an angle other than perfectly perpendicular to the gearcase. It is a lot easier to align the powerhead when it is supported, than when you are holding the powerhead and trying to raise it from or lower it into position.

✳✳ SELOC WARNING

If powerhead removal is difficult, first check to make sure there are no missed fasteners between the gearcase (midsection/exhaust housing/intermediate housing, as applicable) and the powerhead itself. If none are found, apply suitable penetrating oil, like WD-40® or our new favorite, PB Blaster® to the mating surfaces. Give a few minutes (or a few hours) for the oil to work, then carefully pry the powerhead free while lifting it from the gearcase. Be careful not to damage the mating surfaces by using any sharp-tipped prybars or other by prying on thin/weak gearcase or powerhead bosses/surfaces.

When working on a powerhead, use either a very sturdy workbench or an engine stand to hold the assembly. Keep in mind that larger HP motors utilize powerheads that can weigh several hundred pounds.

Although removal and installation is relatively straightforward on most models, the overhaul procedures can be quite involved. Whatever portion you decide to tackle, always, always, ALWAYS take good notes and tag as many parts/hoses/connections as you can during the removal process. As they are removed, arrange components along the worksurface in the same orientation to each other as they are when installed.

Torque Sequence and Values

A torque sequence means that when installing that component you tighten bolt No. 1 a couple of turns, then No. 2, then 3, and so on. Once the bolts start to snug, you can precede one of two ways. If you don't have a torque value, you tighten the bolts securely, a 1/4 turn or so at a time still following that sequence until you feel they are securely fastened. However, if you have do have a torque value, you tighten the bolts just to the point when they start to snug, then set a torque wrench to the proper value and continue to tighten each bolt in 1-2 passes of that sequence until the proper amount of torque is placed on each of the bolts.

■ Mercury normally advises 3 passes of a torque sequence (using an increasing torque value for each pass) in order to reach a given torque specification.

When no specific torque sequence is given, you would do well to use either a criss-crossing or spiraling pattern (working from the inside out or outside in) to ensure even clamp loads and to prevent the possibility of damage in case you simply missed a torque sequence. In any case, using such patterns will never CAUSE harm, so you've got nothing to loose by proactively using them.

POWERHEAD COMPONENTS

Service procedures for the carburetors, fuel pumps, starter, and other powerhead components are given in their respective Sections of this technical guide.

CLEANLINESS

◆ **See Figure 1**

Make a determined effort to keep parts and the work area as clean as possible. Parts must be cleaned and thoroughly inspected before they are assembled, installed, or adjusted. Use proper lubricants, or their equivalent, whenever they are recommended.

2.5/3.3 Hp Powerheads

REMOVAL & INSTALLATION

◆ **See Figure 2**  MODERATE

These powerheads are small enough that you CAN remove them from the intermediate housing without removing most fuel and electrical components. We advise that you consider why the powerhead is being removed. If you are planning on disassembling the powerhead for inspection or overhaul, then you'll have to remove the fuel and electrical components anyway. If this is the case, removing them before powerhead removal is probably a good idea, if only to protect them from potential damage when lifting and moving the powerhead itself. However, the choice remains yours whether or not you follow all of the initial steps for stripping the powerhead of these components.

1. Remove the cowling from the powerhead for access.
2. For safety, disconnect the secondary ignition lead from the spark plug.
3. If desired, remove the hand-rewind starter assembly from the powerhead.
4. If desired, remove the flywheel from the powerhead.
5. If desired, remove the ignition electrical components from the powerhead.
6. Drain and remove the fuel tank from the powerhead. Remove the fuel shut-off valve (on some models this may have been mounted directly to and removed with the fuel tank assembly) and carburetor from the powerhead.
7. From the underside of the powerhead-to-intermediate housing mounting flange, loosen and remove the 6 bolts securing the powerhead.
8. Tap the side of the powerhead with a soft head mallet to shock the gasket seal, and then lift the powerhead from the intermediate housing.
9. Be sure the gasket mating surfaces are clean and free of any traces of old gasket.

To install:
10. Place the new gasket in position on the intermediate housing.

■ **Make sure the gasket is positioned properly and not obscuring any passage.**

11. Apply a light coating of marine grade grease to the driveshaft-to-crankshaft splines
12. Carefully install the powerhead onto the intermediate housing while aligning the surfaces. If necessary, slowly rotate the flywheel/crankshaft (clockwise) to align the crankshaft and driveshaft splines. Alternately the propeller shaft may be rotated (if a shifting style gearcase is in gear), but again, only in the normal direction of rotation to prevent potential damage to the water pump impeller.
13. Once the powerhead is properly seated (the crankshaft is properly splined to the driveshaft) it is time to install the retaining bolts. Apply a light coating of Perfect Seal (or an equivalent sealant or threadlocker) to the threads of the bolts that secure the powerhead to the intermediate housing. Tighten the bolts in two stages starting with the center two bolts and working outwards. Tighten the bolts alternately to 50 inch lbs. (5.6 Nm).
14. If removed, install the carburetor, fuel shut-off valve and fuel tank to the powerhead.
15. If removed, install the ignition components to the powerhead.
16. If removed, install the flywheel.
17. If removed, install the hand-rewind starter assembly.

■ **Be sure to run the motor without the cowling installed in order to check for potential fuel or oil leaks. Remedy any leaks before proceeding.**

18. Reconnect the spark plug wire, hook up a source of cooling water and test run the motor.
19. Install the cowling to the powerhead.

■ **If the powerhead was rebuilt or replaced with a remanufactured unit, don't forget to follow the proper Break-In procedures.**

DISSASSEMBLY & ASSEMBLY

 OEM MODERATE

◆ **See Figure 3**

1. If not done during powerhead removal, strip the powerhead of the hand-rewind starter, flywheel and the fuel and electrical components. For details, please refer to the appropriate sections of this guide.

SELOC_1418

Fig. 1 The exterior and interior of the powerhead must be kept clean, well lubricated and properly tuned for the owner to receive maximum enjoyment

Mounting bolt (6 plcs)

SELOC_1418

Fig. 2 Powerhead mounting bolts

2. Loosen the 2 bolts securing the end cap to the bottom of the powerhead assembly, then remove the end gap and discard the old gasket.

3. Using a crossing pattern, loosen and remove the 4 cylinder head (cover) bolts, then remove the head and discard the old gasket.

4. Using a crossing or spiraling pattern, loosen and remove the 6 crankcase cover bolts (noting the 2 different lengths and locations), then carefully separate the crankcase halves.

5. If necessary, loosen the 2 screws then remove the reed stop and reed.

6. If necessary remove the 2 oil seals from the lower end cap (oil seal housing). Mercury recommends using an internal jawed, threaded seal puller, but often a lever type seal puller will work if you're really careful not to damage the housing a sealing surfaces. Note the direction the seal lips are facing (normally both down on these models).

7. Remove and discard the oil seals from the crankshaft.

8. Carefully lift the crankshaft assembly (and the piston) from the cylinder block.

✳✳ SELOC CAUTION

DO NOT remove the ball bearings from the crankshaft unless they are being replaced. Removal will damage/destroy the bearings.

9. Inspect the crankshaft bearings and, if necessary for replacement, carefully press them from the crankshaft using a bearing separator. Be sure to protect the ends of the crankshaft during bearing removal.

✳✳ SELOC WARNING

ALWAYS use eye protection when attempting to remove the piston pin lockrings as they are prone to launching out of position unpredictably.

10. If necessary, use a piston ring expander (such as #91-24697) to remove the ring from the piston.

11. Inspect the piston assembly and, if necessary remove it from the crankshaft for cleaning or replacement. To remove the piston, first remove the G-lockring from both ends of the piston pin using a pair of needle nose pliers (note that these lockrings are often distorted slightly during removal and it is usually best to replace them).

12. Slide out the piston pin, and then the piston may be separated from the connecting rod. Push out the caged roller bearings from the small end of the connecting rod.

To assemble:

13. If removed, press a new set of ball bearings onto the crankshaft, making sure the alignment pins will be positioned toward the BOTTOM of the crankshaft. Be sure only to press on the INNER bearing race (the race that is in contact with the shaft) to prevent damaging the bearings.

14. If removed, install the piston to the crankshaft assembly, making sure to face the arrow on the dome of the piston DOWNWARD toward the exhaust port. In order to install the piston, position the caged roller bearing and the piston, then carefully insert the piston pin. Use 2 NEW G-lockrings to secure the piston pin.

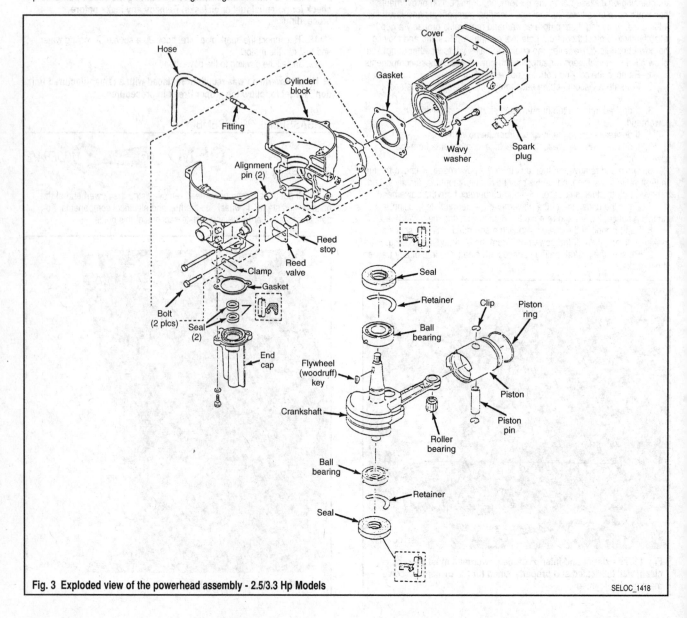

Fig. 3 Exploded view of the powerhead assembly - 2.5/3.3 Hp Models

SELOC_1418

15. Install a piston ring with the gap over the locating pin in the ring groove using an expander tool.

16. Apply a light coating of 2-4-C with Teflon or an equivalent marine grade grease to the lips of the oil seals, then position the seals onto the crankshaft with the reinforced portions of the seals facing inward toward the bearings.

17. Install the bearing retainers into the cylinder block and prepare to install the crankshaft by lubricating the piston ring, piston and cylinder wall lightly with oil.

18. No ring compressor is necessary, keep the crankshaft horizontal and carefully push the piston into the cylinder, then gently push the crankshaft down into position. Rotate the bearings as necessary so the alignment pins are positioned in the notches. Finally position the oil seals against the retainers.

19. Using a small screwdriver inserted through the exhaust port, gently press on the ring to make sure there is tension on the ring. IF the ring fails to return to position (there is no tension) then it is likely that the ring was broken during installation and the crankshaft assembly will have to be removed again to replace the ring.

20. Apply a light coating of Loctite® 271 or equivalent threadlocking compound to the threads of the reed valve retaining screws, then install the reed valve and stop making sure the stop curves upward away from the valve body.

21. Carefully clean the mating surface of the crankcase halves making sure it is completely free of grease or oil, then apply a continuous bead of Loctite Master Gasket on the cylinder block mating surface, JUST INBOARD of the bolts holes on either side of the crankshaft bore.

■ **Mercury is serious about the continuous beads of Loctite. They go so far as to instruct their technicians that IF there is a void while placing the bead to either wipe away the whole bead and start over OR to apply an additional bead, parallel to the void and overlapping the previously applied bead.**

22. Assemble the crankcase halves, then install and tighten the 6 bolts to 50 inch lbs. (5.6 Nm) using multiple passes of a spiraling pattern.

23. Install a new head gasket without any sealing substance. Position the head in place over the gasket and secure using the 4 bolts. Tighten the bolts to 85 inch lbs. (9.6 Nm) using a crossing pattern.

24. If removed, install 2 new oil seals to the lower end cap with the lips facing in the same direction as noted during removal (usually downward). Lubricate the seal lips with a light coating of 2-4-C with Teflon or an equivalent marine grade grease, then gently tap them into position using a suitable driver or smooth socket.

25. Install the lower end cap using a new gasket, then tighten the 2 bolts alternately and evenly to 50 inch lbs. (5.6 Nm).

26. Rotate the crankshaft a few times to make sure it moves freely and there is NO binding.

27. Install the Powerhead, as detailed in this section.

28. Once the powerhead is fully assembled, if cylinder block components were replaced, especially the pistons and/or rings, be sure to operate the engine as directed for new component break-in. For details, please refer to Powerhead Break-In, in this section.

CLEANING & INSPECTION

◆ **See Figure 3**

Cleaning and inspecting the components is virtually the same for any two-stroke outboard and varies mostly by specifications (which are listed in the Engine Specifications charts) or by component type. A section detailing the proper procedures, sorted mostly by component, can be found under Powerhead Refinishing.

4/5 Hp Powerheads

REMOVAL & INSTALLATION

◆ **See Figure 2**

These powerheads are small enough that you CAN remove them from the intermediate housing without removing most fuel and electrical components. We advise that you consider why the powerhead is being removed. If you are planning on disassembling the powerhead for inspection or overhaul, then you'll have to remove the fuel and electrical components anyway. If this

is the case, removing them before powerhead removal is probably a good idea, if only to protect them from potential damage when lifting and moving the powerhead itself. However, the choice remains yours whether or not you follow all of the initial steps for stripping the powerhead of these components.

1. Remove the cowling from the powerhead for access.

2. For safety, disconnect the secondary ignition lead from the spark plug.

3. If desired, remove the hand-rewind starter assembly from the powerhead.

4. If desired, remove the flywheel from the powerhead.

5. If desired, remove the ignition electrical components from the powerhead.

6. Drain and remove the fuel tank from the powerhead.

7. Remove the carburetor from the powerhead and disconnect the throttle linkage.

8. From the underside of the powerhead-to-intermediate housing mounting flange, loosen and remove the 6 bolts securing the powerhead.

9. Tap the side of the powerhead with a soft head mallet to shock and break the gasket seal or gently rock the powerhead back and forth, and then lift the powerhead from the intermediate housing.

10. Be sure the gasket mating surfaces are clean and free of any traces of old gasket.

To install:

11. Place the new gasket in position on the intermediate housing.

■ **Make sure the gasket is positioned properly and not obscuring any passage.**

12. Apply a light coating of marine grade grease to the driveshaft-to-crankshaft splines

13. Carefully install the powerhead onto the intermediate housing while aligning the surfaces. If necessary, slowly rotate the flywheel/crankshaft (clockwise) to align the crankshaft and driveshaft splines. Alternately the propeller shaft may be rotated (if the gearcase is in gear), but again, only in the normal direction of rotation to prevent potential damage to the water pump impeller.

14. Once the powerhead is properly seated (the crankshaft is properly splined to the driveshaft) it is time to install the retaining bolts. Tighten the bolts in two stages starting with the center two bolts and working outwards. Tighten the bolts alternately to 70 inch lbs. (8 Nm).

15. If removed, install the carburetor and reconnect the throttle linkage.

16. Install the fuel tank.

17. If removed, install the ignition components to the powerhead.

18. If removed, install the flywheel.

19. If removed, install the hand-rewind starter assembly.

■ **Be sure to run the motor without the cowling installed in order to check for potential fuel or oil leaks. Remedy any leaks before proceeding.**

20. Reconnect the spark plug wire, hook up a source of cooling water and test run the motor.

21. Install the cowling to the powerhead.

■ **If the powerhead was rebuilt or replaced with a remanufactured unit, don't forget to follow the proper Break-In procedures.**

DISASSEMBLY & ASSEMBLY

◆ **See Figures 4, 5 and 6**

1. If not done during powerhead removal, strip the powerhead of the hand-rewind starter, flywheel and the fuel and electrical components. For details, please refer to the appropriate sections of this guide.

2. Loosen the 2 bolts securing the end cap to the bottom of the powerhead assembly, then remove the end gap and discard the old gasket.

3. Tag and disconnect the vent hose running from the check valve to the fitting, both on the powerhead.

4. Using the reverse of the torque sequence (a crossing star pattern), loosen and remove the 5 cylinder head bolts, then remove the head and discard the old gasket.

5. Using the reverse of the torque sequence (which will basically result in a counter-clockwise spiraling pattern that starts at the top bolts, goes through the bottom bolts and ends at the middle bolts) loosen and remove the 6 crankcase cover bolts (noting the 2 different lengths and locations), then carefully separate the crankcase halves.

■ Two small prytabs are located on either side of the crankcase, along the split line, just a bit below the flywheel mounting area. If you have difficulty breaking the crankcase gasket seal, use two small prybars to gently pry at the tabs and separate the crankcase halves.

6. If necessary remove the oil seal(s) from the lower end cap (oil seal housing). Mercury recommends using an internal jawed, threaded seal puller like #91-83165M or equivalent, but often a lever type seal puller will work if you're really careful not to damage the housing a sealing surfaces.

■ Most of these models should be equipped with one lower end cap seal, but there have been versions which use two seals, so double check once you've removed the seal just to make sure another isn't still in the housing.

7. If the end cap (oil seal housing) is equipped with a lower bushing AND that bushing must be replaced, use a long punch to push the bushing out the bottom of the housing.

8. If necessary, loosen the 2 screws then remove the reed stop and reed.

9. Remove and discard the oil seals from the crankshaft.

10. Remove the locating washers from either end of the crankshaft.

11. Carefully lift the crankshaft assembly (and the piston) from the cylinder block.

✱✱ SELOC CAUTION

DO NOT remove the ball bearings from the crankshaft unless they are being replaced. Removal will damage/destroy the bearings.

12. Inspect the crankshaft bearings and, if necessary for replacement, carefully press them from the crankshaft using a bearing separator. Be sure to protect the ends of the crankshaft during bearing removal.

✱✱ SELOC WARNING

ALWAYS use eye protection when attempting to remove the piston pin lockrings as they are prone to launching out of position unpredictably.

13. If necessary, use a piston ring expander (such as #91-24697) to remove the rings from the piston.

14. Inspect the piston assembly and, if necessary remove it from the crankshaft for cleaning or replacement. To remove the piston, first remove the lockring from both ends of the piston pin. There is normally a small cutout in the side of the piston, adjacent to the piston pin bore, a small tool can usually be inserted through the cutout and under the lockring to gently pry the ring free (note that these lockrings are often distorted slightly during removal and it is usually best to replace them).

15. Slide out the piston pin, and then the piston may be separated from the connecting rod. Push out the needle roller bearing assembly from the small end of the connecting rod.

16. Check the locating pins in the crankcase halves (there are two pins total) for wear or damage and replace, as necessary.

To assemble:

17. Make sure all parts have been thoroughly cleaned and inspected. Be sure to oil all bearings and bearing surfaces lightly using a coating of fresh 2-stroke oil.

18. If removed, press a new set of ball bearings onto the top and bottom of the crankshaft. Position the bearings so the numbered side faces AWAY from the crankshaft (toward the driver). Be sure only to press on the INNER bearing race (the race that is in contact with the shaft) to prevent damaging the bearings.

19. If removed, install the piston to the crankshaft assembly, making sure the side marked UP on the dome of the piston faces UPWARD toward the flywheel end of the shaft. In order to install the piston, oil the roller bearing and position it inside the connection rod, then place the piston over the con rod and carefully insert the piston pin. Center the pin inside the piston, then secure it using 2 lockrings.

20. Using an expander tool, install the piston rings with the gaps over the locating pin in each ring groove.

21. Make sure the bearing surfaces of the crankcase halves are clean and free of any oil or debris. Make sure the locating pins are installed in the crankcase halves.

22. Prepare to install the crankshaft by lubricating the piston rings, piston and cylinder wall lightly with oil.

23. No ring compressor is necessary, keep the crankshaft horizontal and carefully push the piston into the cylinder, then gently push the crankshaft down into position. Rotate the bearings as necessary so the alignment pins are positioned in the crankcase notches.

24. Install the locating washers, pressing the large diameter half of the rings into the grooves in the crankcase block.

25. Apply a light coating of 2-4-C with Teflon or an equivalent marine grade grease to the lips of the oil seals, then position the seals onto the crankshaft with the lips facing inward toward the bearings.

26. Using a small screwdriver inserted through the exhaust port, gently press on each of the piston rings to make sure there is tension on each ring. IF a ring fails to return to position (there is no tension) then it is likely that the ring was broken during installation and the crankshaft assembly will have to be removed again to replace the ring.

27. Apply a light coating of Loctite® 271 or equivalent threadlocking compound to the threads of the reed valve retaining screws, then install the reed valve and stop making sure the stop curves upward away from the valve body. Tighten the reed valve retaining screws to 9 inch lbs. (1 Nm).

28. Again, make sure the mating surface of the crankcase halves is still completely free of grease or oil, then apply a continuous bead of Loctite Master Gasket on the cylinder block mating surface, JUST INBOARD of the bolts holes on either side of the crankshaft bore.

■ Mercury is serious about the continuous beads of Loctite. They often go so far as to instruct their technicians that IF there is a void while placing the bead to either wipe away the whole bead and start over OR to apply an additional bead, parallel to the void and overlapping the previously applied bead.

29. Assemble the crankcase halves, then install and tighten the 6 bolts to 91 inch lbs. (10.3 Nm) using at least 3 passes of the torque sequence (a clockwise spiraling pattern that starts at the inner left bolt and works its way outward to the top right bolt).

30. Install a new head gasket without any sealing substance. Position the head in place over the gasket and secure using the 5 bolts. Tighten the bolts to 215 inch lbs. (24.4 Nm) using at least 3 passes of a star-crossing pattern.

31. Install the breather/check valve hose to the powerhead.

32. If removed, install the new oil seal(s) to the lower end cap with the lips facing in the same direction as noted during removal (usually downward). On models that utilize 2 seals there is normally a spacer between the seals. Lubricate the seal lips with a light coating of 2-4-C with Teflon or an equivalent marine grade grease, then gently tap the seal(s) into position using a suitable driver or smooth socket.

33. If equipped and if removed, install a new bushing to the bottom of the end cap (oil seal housing). Like the oil seal(s), use a suitable driver or socket to gently tap the bushing into position.

34. Install the lower end cap using a new gasket, then tighten the 2 bolts alternately and evenly to 70 inch lbs. (8 Nm).

35. Rotate the crankshaft a few times to make sure it moves freely and there is NO binding.

36. Install the Powerhead, as detailed in this section.

37. Once the powerhead is fully assembled, if cylinder block components were replaced, especially the pistons and/or rings, be sure to operate the engine as directed for new component break-in. For details, please refer to Powerhead Break-In, in this section.

CLEANING & INSPECTION

◆ See Figure 4

Cleaning and inspecting the components is virtually the same for any two-stroke outboard and varies mostly by specifications (which are listed in the Engine Specifications charts) or by component type. A section detailing the proper procedures, sorted mostly by component, can be found under Powerhead Refinishing.

One somewhat unique component to this powerhead is the breather hose and check valve attached to the powerhead. Make sure that air will only pass one way through this hose/valve assembly. If necessary, the assembly can be replaced by pushing a new valve into a length of same diameter/length/compound hose (or reusing the old hose if you've pushed the old valve out). Install the check valve 1 in. (25.4mm) into the end of a hose using a 5/32 in (3.9mm) punch to press against the ONE HOLE SIDE of valve.

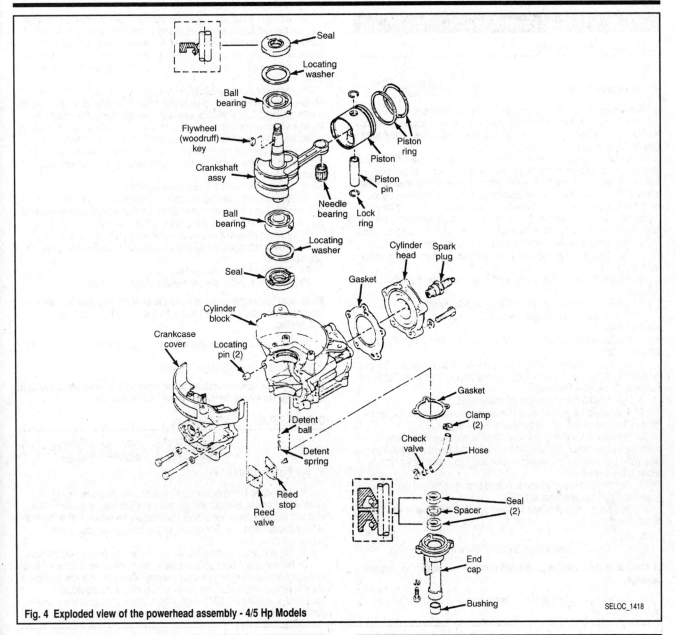

Fig. 4 Exploded view of the powerhead assembly - 4/5 Hp Models

SELOC_1418

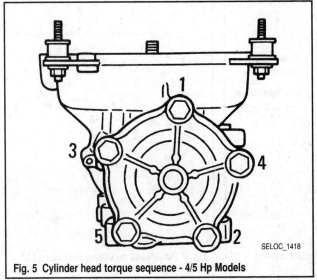

Fig. 5 Cylinder head torque sequence - 4/5 Hp Models

SELOC_1418

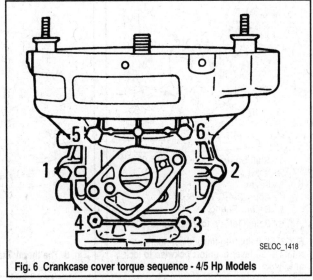

Fig. 6 Crankcase cover torque sequence - 4/5 Hp Models

SELOC_1418

6/8 and 9.9/10/15 Hp Powerheads

REMOVAL & INSTALLATION

◆ See Figures 7, 8 and 9    MODERATE

These powerheads are small enough that you CAN remove them from the intermediate housing without removing most fuel and electrical components. However, they're starting to get large enough that the chances of damaging those components starts to go up.

We advise that you consider why the powerhead is being removed. If you are planning on disassembling the powerhead for inspection or overhaul, then you'll have to remove the fuel and electrical components anyway. If this is the case, removing them before powerhead removal is probably a good idea, if only to protect them from potential damage when lifting and moving the powerhead itself. However, the choice remains yours whether or not you follow all of the initial steps for stripping the powerhead of these components.

1. Remove the cowling from the powerhead for access.
2. For safety, tag and disconnect the secondary ignition leads from the spark plugs.
3. If desired, remove the hand-rewind starter assembly from the powerhead.
4. If desired, remove the flywheel from the powerhead.
5. If desired, remove the ignition/electrical components from the powerhead.
6. If equipped and desired, remove the electric starter assembly from the powerhead.
7. Disconnect the throttle shift cables or linkage, as necessary. On most models the shift shaft is locked in place on the yoke using a lever that pivots clockwise off the yoke to release.
8. Remove the carburetor from the powerhead.
9. Disconnect the cooling system indicator hose from either the telltale fitting in the cowling or from the thermostat housing.
10. From the underside of the powerhead-to-intermediate housing mounting flange, loosen and remove the 4 bolts and 2 nuts (on each lower side of the powerhead there should be one nut, surrounded by 2 bolts) securing the powerhead.
11. Gently rock the powerhead back and forth, and then lift the powerhead from the intermediate housing.
12. Be sure the gasket mating surfaces are clean and free of any traces of old gasket.

To install:
13. Place the new gasket in position on the intermediate housing.

■ **Make sure the gasket is positioned properly and not obscuring any passage.**

14. Apply a light coating of marine grade grease to the driveshaft-to-crankshaft splines

15. Carefully install the powerhead onto the intermediate housing while aligning the surfaces. If necessary, slowly rotate the flywheel/crankshaft (clockwise) to align the crankshaft and driveshaft splines. Alternately the propeller shaft may be rotated (if the gearcase is in gear), but again, only in the normal direction of rotation to prevent potential damage to the water pump impeller.
16. Once the powerhead is properly seated (the crankshaft is properly splined to the driveshaft) it is time to install the retaining bolts and nuts. Apply a light coating of Loctite #680 or equivalent threadlocker to the threads, then install and tighten the fasteners in at least two stages starting with the nuts at center and working outward to the bolts to 100 inch lbs. (11.3 Nm).
17. Reconnect the cooling system indicator hose to the telltale fitting in the cowling or to the thermostat housing, depending on which end was disconnected.
18. If removed, install the carburetor assembly.
19. Reconnect the throttle and shift cables/linkage, as necessary. On most models there is a lever on the shift shaft that rotates counterclockwise over the yoke to lock the shaft in contact with the linkage.
20. If equipped (and if removed), install the electric starter assembly.
21. If removed, install the ignition and electrical components to the powerhead.
22. If removed, install the flywheel.
23. If removed, install the hand-rewind starter assembly.

■ **Be sure to run the motor without the cowling installed in order to check for potential fuel or oil leaks. Remedy any leaks before proceeding.**

24. Reconnect the spark plug wires, hook up a source of cooling water and test run the motor.
25. Install the cowling to the powerhead.

■ **If the powerhead was rebuilt or replaced with a remanufactured unit, don't forget to follow the proper Break-In procedures.**

DISASSEMBLY & ASSEMBLY

 OEM ③ DIFFICULT

◆ See Figures 10 thru 14

1. If not done during powerhead removal, strip the powerhead of the hand-rewind starter, flywheel and the fuel and electrical components. The shift and throttle linkage assemblies must usually be removed from the side of the powerhead as well. For details, please refer to the appropriate sections of this guide.
2. Tag and disconnect the bleed hoses to ensure proper reinstallation.
3. Remove the 3 bolts securing the intake manifold/reed block assembly to the powerhead, then remove the assembly. Though earlier models used a block assembly which could be disassembled to replace individual components, starting in 1997 the reed block assemblies used were non-serviceable. (However the rubber seal and O-ring MAY be available).

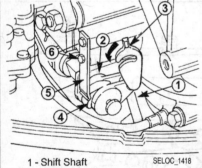

1 - Shift Shaft
2 - Yoke
3 - Lever
4 - Apply 2-4-C w/Teflon
5 - Detent Spring
6 - Bolts [Torque to 25 lb. in. (2.8 N·m)]

SELOC_1418

Fig. 7 Shift shaft-to-linkage connection, rotate shaft lever counterclockwise to lock onto yoke

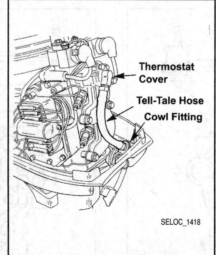

Thermostat Cover
Tell-Tale Hose
Cowl Fitting

SELOC_1418

Fig. 8 The thermostat-to-telltale cooling system hose

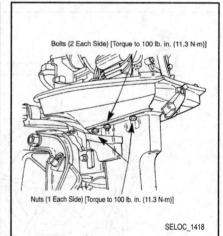

Bolts (2 Each Side) [Torque to 100 lb. in. (11.3 N·m)]

Nuts (1 Each Side) [Torque to 100 lb. in. (11.3 N·m)]

SELOC_1418

Fig. 9 Powerhead mounting fasteners (3 to each side)

4. Remove the 2 bolts and separate the thermostat cover (and thermostat, if equipped) from the top of the cylinder head cover.

5. Using the reverse of the torque sequence (meaning start with the outer bolts and work inward), loosen and remove the 6 cylinder head bolts, then remove the head and discard the old gasket.

✳✳ SELOC CAUTION

Exhaust cover bolts are some of the most likely bolts to become frozen from corrosion on outboard motors. Before removal it's USUALLY a good idea to hit them all with penetrating oil and then to place the socket and driver over each bolt and tap on them lightly using a mallet to help loosen the corrosion on the bolt threads before you even try to loosen them. A hand impact driver can also work wonders here, but avoid air powered impact tools, as they are much more likely to shear the bolt threads.

6. Using the reverse of the torque sequence (meaning start with the outer bolts and work inward), loosen and remove the 11 exhaust cover bolts, then remove the cover, followed by the cover gasket, manifold (also sometimes called the plate) and the manifold gasket. As usual, discard the old gaskets.

■ **If you suspect the powerhead has been overheated at some point and/or the spark plugs are grayish colored (which is a possible sign of water intrusion), inspect the exhaust manifold for warpage and/or signs that the gaskets were not properly aligned.**

7. Remove the 2 bolts securing the long, vertical intake cover to the side of the powerhead. Remove the cover and gasket (6/8 hp models) or O-ring (9.9-15 hp models).

8. Using the reverse of the torque sequence (meaning start with the outer bolts and work inward) loosen and remove the 6 crankcase cover-to-cylinder block bolts. Use a soft mallet to tap around the edge of the cover-to-block mating surface, then carefully separate the cover from the block.

9. Carefully lift the crankshaft assembly (and the pistons) from the cylinder block.

10. If equipped remove the center main bearing halves and sleeve halves from the crankshaft and/or cylinder block.

■ **On Some 9.9-15 hp motors you must first remove a retaining ring from the groove at the center of the main bearing race, then remove the races and bearing halves.**

11. If equipped, remove the sealing ring from the groove just inboard of the crankshaft throw.

12. Remove the upper crankshaft seal and bearing.

13. Remove the lower crankshaft seal, stuffer washer and retaining ring.

14. Remove and discard the old coupling seal from the end of the crankshaft. The best method to accomplish this is to use an Expanding Rod and Collet (Snap-On® numbers CG40-4 and CG40-15 or equivalents) along with a small slide-hammer.

15. Before going any further, LABEL the pistons and all matching components (connecting rod, rod cap etc) to make sure the proper components are all removed, stored and reinstalled together. Also, take some time to make matchmarks between things like the connecting rod caps and the rods or the piston and the connecting rod to make sure all components are installed facing the same directions as they were before they were disassembled. It is true that many of these components have factory alignment or identification marks, but it is never a bad idea to double-check them or mark them yourself before proceeding.

16. If the pistons are to be removed from the connecting rods, FIRST unbolt and remove the rods from the crankshaft. Alternately and evenly loosen the 2 connecting rod cap bolts, going back and forth with no more than 1 full turn on each bolt without moving back to the other bolt. Proceed with this pattern until the connecting rod cap can be removed. Support the piston as you remove the bolts so the piston and rod do not fall free and become damaged. Keep track of the bearings as the cap and connecting rod for each piston assembly is removed from the crankshaft.

17. Disassemble each piston for further inspection and component replacement, as follows:

✳✳ SELOC WARNING

ALWAYS use eye protection when attempting to remove the piston pin lockrings as they are prone to launching out of position unpredictably.

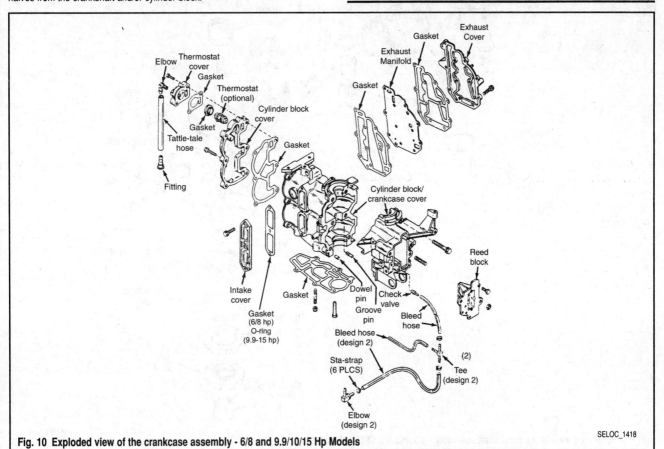

Fig. 10 Exploded view of the crankcase assembly - 6/8 and 9.9/10/15 Hp Models

SELOC_1418

a. To remove the connecting rod from each piston, first remove the lockring from both ends of the piston pin. Use a pair of needlenose pliers to grasp and remove the piston locking. Keep in mind that these lockrings are often distorted slightly during removal and so they must be discarded are replaced with new lockrings during installation.

■ If necessary, a piston cradle can be fabricated from a large block of wood. The important design features of the cradle are that it is semi-circular on top to hold (or cradle, get it?) the piston and that there is a bore at the bottom center into which the wrist pin can tapped or pressed. The final important feature of the cradle is that it is sturdy enough to withstand the force of the shop press, BUT, keep in mind that the fit is loose on these pistons and significant force should not be necessary. It is NOT uncommon for someone to simply support the piston in their lap when working on these motors.

b. While carefully supporting the piston to prevent damage use a soft-faced mallet and driver (such as Piston Pin Tool 91-13663 or equivalent) to gently tap the pin from the bore in the piston. As the pin is removed, KEEP CLOSE TRACK of the locating washers and the loose needle bearings (there should be 24 needles per piston).

■ As the wrist pin is removed from the piston you must retain the 24 loose needle bearings and the 2 locating washers from the small end of the connecting rod. If you can, stop tapping or pressing once the wrist pin JUST clears the bottom of the connecting rod bore. Then carefully withdraw the rod along with the needle bearings (place a thin bladed tool like a putty knife or your finger under the rod hole as it is pulled free of the piston bore/wrist pin). Inspect the bearings, if any of the needles are lost or worn, replace the entire bearing assembly.

c. Use a Piston Ring Expander (such as #91-24697) to remove the rings from the pistons. The manufacturer does not recommend reusing the rings. Retain them in sets for inspection purposes (marking their original locations), but discard them before installation.

■ We've said it before. We know there will be some circumstances that will prompt people to reuse rings, but for the most part, if you've come THIS far, it is probably a good idea to bite the bullet and replace the rings to ensure long-lasting performance from the powerhead.

d. Repeat for the other piston.

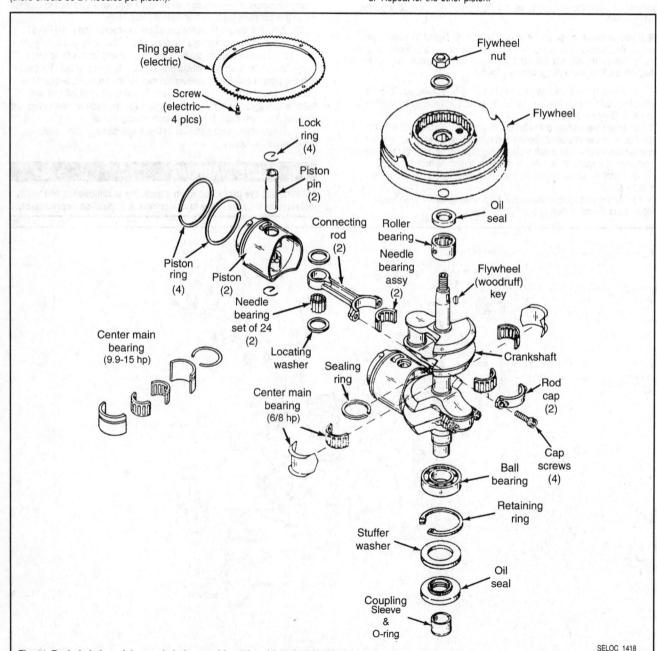

Fig. 11 Exploded view of the crankshaft assembly - 6/8 and 9.9/10/15 Hp Models

SELOC_1418

※※ SELOC CAUTION

DO NOT remove the lower ball bearing from the crankshaft unless it is being replaced. Removal will damage/destroy the bearing assembly.

18. Inspect the lower crankshaft bearing and, if necessary for replacement, carefully press it from the crankshaft using a bearing separator. Be sure to protect the ends of the crankshaft during bearing removal.

■ **Mercury and Seloc recommend that ALL seals be replaced during a rebuilding procedure of this magnitude.**

To assemble:

19. Make sure all parts have been thoroughly cleaned and inspected. Be sure to oil all bearings and bearing surfaces lightly using a coating of fresh 2-stroke oil. Loose needle bearings, once cleaned, should be spread out on a clean worksurface, like a sheet of paper and coated lightly using Needle Bearing Assembly Lubricant.

20. If removed, press a new ball bearing assembly onto the bottom of the crankshaft. Position the bearing with the open side of ball retainer facing the crank. Be sure only to press on the INNER bearing race (the race that is in contact with the shaft) to prevent damaging the bearings. Press the bearing into position until the inner race is seated firmly on the shoulder of the crankshaft.

21. Lubricate the O-ring of the new coupling seal using a light coating of 2-4-C with Teflon or equivalent marine grade grease, then use a driver to gently push the seal into position on the crank and bearing assembly.

22. Install a new sealing ring to the crankshaft, just inboard of the lower throw for the upper piston.

23. If disassembled for inspection and/or component replacement, assemble each piston as follows:

a. Apply a light coating of Needle Bearing Assembly grease (or 2-4-C with Teflon or an equivalent gasoline soluble marine grade grease) to the wrist pin bearing needles and the locating washers, then hold the lower locating washer in place while you insert the 24 needles around the inner circumference of the connecting rod wrist pin bore. Install the Piston Pin Tool # 91-13663A1 or equivalent into the bore to align the needles and hold them in place during assembly.

■ **The wrist pin bearing tool is, essentially, a short round pin, roughly the same outer diameter of the wrist pin, but very short, just long enough to insert it through the connecting rod wrist pin bore and install the locating washers on either side. A similarly sized dowel pin can be substituted, or an old connecting rod wrist pin from the same size piston could be used. If substituting another rod or drift pin, make sure the surface is smooth and clean/free of all corrosion. Also, make sure the outer diameter of the tool is smaller than the wrist pin and bore to prevent the possibility of binding or damage.**

b. Once all the needles are in position, inspect to make sure one is not missing. Generally speaking, if the tip of an awl can be inserted between the needles, one or more needles is/are missing.

※※ SELOC CAUTION

NEVER replace just one needle. If one or more needles are damaged or missing, replace the entire set.

c. With the bearings and the wrist pin tool in position, install the upper locating washer. Again, a dab of grease should hold the thrust washer in position.

d. Apply a light coating of clean engine oil or assembly lube to the wrist pin and the pin bore in the piston.

e. Support the piston (ideally in a piston cradle for ease of assembly but the wrist pin fit should be loose on both sides so this is not absolutely necessary). Insert the wrist pin into the bore and gently press it through the bore until it just starts to appear inside the piston skirt. Stop pressing, then position the connecting rod with the needle bearings, locating washers and wrist pin bearing installation tool inside the piston skirt. Continue to press the wrist pin into the position, through the thrust washers and bearings, pushing the wrist pin bearing installer through the other side and out of the piston.

f. Once the wrist pin is properly positioned and centered in the piston, install new lockrings into the grooves in both ends of the piston bore.

g. Use a ring expander to install the piston rings on each piston, making sure to install the RECTANGULAR ring first in the bottom groove and the HALF-KEYSTONE (or tapered ring) in the top groove. Each piston ring groove should contain a dowel pin, around which the ring gap must be

situated. If not the rings will likely break during installation (or if by they don't by some miracle, during the first start-up). Be sure to install the tapered ring in the top groove.

■ **When using the ring expander, spread each ring JUST enough to slip over the piston.**

h. Repeat for the other piston.

24. Install each piston and connecting rod assembly to the crankshaft as follows:

i. Apply a light coating of 2-4-C with Teflon or a suitable marine grade grease to the connecting rod big-end roller bearing halves.

j. Position both halves of the big-end roller bearing onto the crank pin.

k. Clean the connecting rod bolts using solvent and dry with compressed air, then apply a light coating of 2-stroke engine oil to the bolt threads.

l. Position the piston/connecting rod assembly to the crankshaft (facing in the same direction as noted and marked during removal, then place the connecting rod end cap over the crank pin (also aligning the matchmarks made during removal) and start the bolts by hand.

■ **These motors use fractured end caps, meaning the end cap and connecting rod were originally one piece. As a result the caps will only fit properly in one direction and when they fit the fracture lines will mate PERFECTLY.**

m. Alternately and evenly tighten the end cap retaining bolts, no more than one turn before moving to the other bolt. Once they start to tighten, use multiple passes to tighten them to 100 inch lbs. (11.3 Nm).

n. Repeat for the other piston/connecting rod assembly.

25. Install the retaining ring (with chamfered side facing AWAY from the lower crankshaft ball bearing), stuffer washer and lower crankshaft oil seal to the bottom of the crankshaft assembly.

26. Install the upper crankshaft roller bearing and oil seal.

※※ SELOC CAUTION

Like with all bearings, when replacing a roller bearing half, ALWAYS replace both halves together.

27. Apply a light coating of 2-4-C with Teflon or a suitable marine grade grease to the center main bearing surface of the crankshaft, then install the main bearing roller halves. On 9.9-15 hp models so equipped, position the race halves with the retaining ring groove TOWARDS the flywheel end of the crankshaft while making sure the facture lines of the race halves are correctly mated. Secure the center main bearing race halves together with the retaining ring, then verify the ring bridges both fracture lines of the race.

28. Make sure the bearing surfaces of the crankcase halves are clean and free of any oil or debris.

29. Prepare to install the crankshaft by lubricating the piston rings, piston and cylinder wall lightly with oil.

30. No ring compressor is necessary, keep the crankshaft horizontal and carefully push the pistons into the cylinder. On the smaller models the center main bearing half will not be installed yet, so don't FULLY seat the crankshaft into the block until you do that. Install the center main bearing sleeve and bearing into the cylinder block, then the second half of the sleeve and bearing. On larger models where the bearing was installed earlier, align the hole in the center of the main bearing race with the pin in the block.

31. Position the alignment boss of the upper crankshaft roller bearing into the notch in the cylinder block.

32. Press the retaining ring (bevel facing down) FIRMLY into the groove in the cylinder block and then gently seat the crankshaft into position.

33. Carefully push the seals inward to seat.

34. Using a small screwdriver inserted through the exhaust ports, gently press on each of the piston rings to make sure there is tension on each ring. IF a ring fails to return to position (there is no tension) then it is likely that the ring was broken during installation and the crankshaft assembly will have to be removed again to replace the ring.

35. Again, make sure the mating surface of the crankcase halves (cover and block) is still completely free of grease or oil, then apply a continuous bead of Loctite Master Gasket on the cylinder block mating surface, JUST INBOARD of the bolts holes on either side of the crankshaft bore.

■ **Mercury is serious about the continuous beads of Loctite. They often go so far as to instruct their technicians that IF there is a void while placing the bead to either wipe away the whole bead and start over OR to apply an additional bead, parallel to the void and overlapping the previously applied bead.**

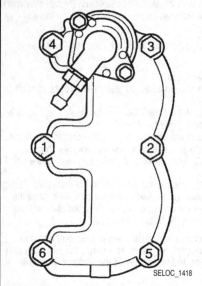

Fig. 12 Cylinder head (block cover) torque sequence - 6/8 and 9.9/10/15 Hp Models

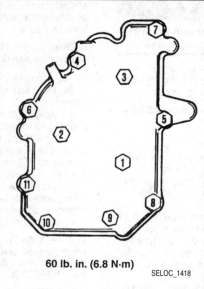

Fig. 13 Exhaust cover torque sequence - 6/8 and 9.9/10/15 Hp Models

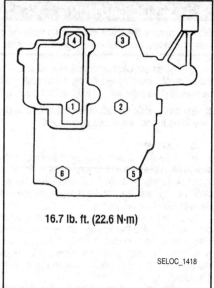

Fig. 14 Crankcase cover torque sequence - 6/8 and 9.9/10/15 Hp Models

36. Assemble the crankcase cover to the block, then install and tighten the 6 bolts to 16.7 ft. lbs. (22.7 Nm) using at least 3 passes of the torque sequence (a pattern that starts at the inner left bolt and works its way outward to the bottom left bolt).

37. Rotate the crankshaft a few times to make sure it moves freely and there is NO binding.

38. Install the reed valve assembly (making sure the O-ring and seal are in position) and secure using the retaining bolts. Tighten the bolts to 60 inch lbs. (6.8 Nm).

39. Install the long, vertical intake cover to the side of the powerhead using a new gasket or an O-ring that is in good condition (as applicable). Tighten the cover bolts to 60 inch lbs. (6.8 Nm).

40. Install the exhaust manifold and cover using new gaskets and the 11 retaining bolts. Tighten the bolts in at least 3 passes of the torque sequence (which works outward from the center bolts) to 60 inch lbs. (6.8 Nm).

41. Install a new cylinder head/block cover gasket without any sealing substance. Position the head in place over the gasket and secure using the 6 bolts. Tighten the bolts to 60 inch lbs. (6.8 Nm) using at least 3 passes of the torque sequence which starts at the center and works outward.

42. If equipped, install the thermostat, then install the thermostat cover and tighten the bolts to, yup, you guessed it, 60 inch lbs. (6.8 Nm).

43. Reconnect the bleed hoses as tagged during removal.

44. Install the Powerhead, as detailed in this section.

45. Once the powerhead is fully assembled, if cylinder block components were replaced, especially the pistons and/or rings, be sure to operate the engine as directed for new component break-in. For details, please refer to Powerhead Break-In, in this section.

CLEANING & INSPECTION

◆ See Figures 10 and 11

Cleaning and inspecting the components is virtually the same for any two-stroke outboard and varies mostly by specifications (which are listed in the Engine Specifications charts) or by component type. A section detailing the proper procedures, sorted mostly by component, can be found under Powerhead Refinishing.

20/20 Jet/25 Hp Powerheads

REMOVAL & INSTALLATION

◆ See Figures 15 and 16

These powerheads are small enough that you CAN remove them from the intermediate housing without removing most fuel and electrical components.

However, they're starting to get large enough that the chances of damaging those components starts to go up.

We advise that you consider why the powerhead is being removed. If you are planning on disassembling the powerhead for inspection or overhaul, then you'll have to remove the fuel and electrical components anyway. If this is the case, removing them before powerhead removal is probably a good idea, if only to protect them from potential damage when lifting and moving the powerhead itself. However, the choice remains yours whether or not you follow all of the initial steps for stripping the powerhead of these components.

1. Remove the cowling from the powerhead for access.

2. For safety, tag and disconnect the secondary ignition leads from the spark plugs.

3. If desired, remove the starter assembly from the powerhead (meaning the electrical and/or hand rewind starter, as equipped).

4. If desired, remove the flywheel from the powerhead.

5. If desired, remove the ignition/electrical components from the powerhead.

6. Disconnect the throttle shift cables or linkage, as necessary.

7. Remove the carburetor from the powerhead.

8. Disconnect the cooling system indicator hose from either the telltale fitting in the cowling or from the thermostat housing.

9. From the underside of the powerhead-to-intermediate housing mounting flange, loosen and remove the 4 bolts (2 on each side, or one per corner) securing the access panel, then remove the panel to reach the powerhead mounting fasteners.

10. Loosen and remove the 6 bolts (3 on each side) securing the powerhead to the intermediate housing.

11. Gently rock the powerhead back and forth, and then lift the powerhead from the intermediate housing. Place the housing on a suitable stand or worksurface.

12. Be sure the gasket mating surfaces are clean and free of any traces of old gasket.

To install:

13. Place the new gasket in position on the intermediate housing.

■ **Make sure the gasket is positioned properly and not obscuring any passage.**

14. Apply a light coating of marine grade grease to the driveshaft-to-crankshaft splines

15. Carefully install the powerhead onto the intermediate housing while aligning the surfaces. If necessary, slowly rotate the flywheel/crankshaft (clockwise) to align the crankshaft and driveshaft splines. Alternately the propeller shaft may be rotated (if the gearcase is in gear), but again, only in the normal direction of rotation to prevent potential damage to the water pump impeller.

16. Once the powerhead is properly seated (the crankshaft is properly splined to the driveshaft) install and tighten the 6 retaining bolts to 20 ft. lbs.

(27.1 Nm). Tighten the bolts using at least 2 passes of a crossing pattern that starts and the center and works outward.

17. Install the access/trim cover to the bottom of the cowling and tighten the 4 retaining bolts to 85 inch lbs. (9.6 Nm).

18. Reconnect the cooling system indicator hose to the telltale fitting in the cowling or to the thermostat housing, depending on which end was disconnected.

19. If removed, install the carburetor assembly.

20. Reconnect the throttle and shift cables/linkage, as necessary.

21. If removed, install the ignition and electrical components to the powerhead.

22. If removed, install the flywheel.

23. If removed, install the electric starter and/or hand-rewind starter assembly.

■ **Be sure to run the motor without the cowling installed in order to check for potential fuel or oil leaks. Remedy any leaks before proceeding.**

24. Reconnect the spark plug wires, hook up a source of cooling water and test run the motor.

25. Install the cowling to the powerhead.

■ **If the powerhead was rebuilt or replaced with a remanufactured unit, don't forget to follow the proper Break-In procedures.**

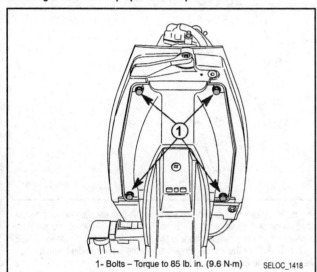

1- Bolts – Torque to 85 lb. in. (9.6 N·m)   SELOC_1418

Fig. 15 Access/trim cover fasteners (must be removed to access the powerhead mounting bolts)

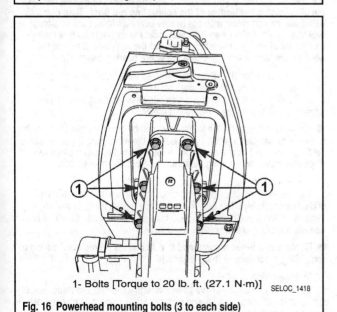

1- Bolts [Torque to 20 lb. ft. (27.1 N·m)]   SELOC_1418

Fig. 16 Powerhead mounting bolts (3 to each side)

DISSASSEMBLY & ASSEMBLY

◆ **See Figures 17 thru 21**

1. If not done during powerhead removal, strip the powerhead of the hand-rewind starter, flywheel and the fuel and electrical components. For details, please refer to the appropriate sections of this guide. The shift and throttle linkage assemblies must often be removed from the side of the powerhead as well.

2. Remove the 3 bolts securing the intake manifold/reed block assembly to the powerhead, then remove the assembly. Though earlier models used a block assembly which could be disassembled to replace individual components, Mercury states that these reed block assemblies are non-serviceable. Check the intake manifold/reed block seal for signs of excessive wear, cuts or abrasions. IF the seal is damaged, the ENTIRE REED BLOCK ASSEMBLY must be replaced.

3. Remove the 2 bolts and separate the thermostat cover and thermostat, if equipped) from the top of the cylinder head cover.

4. Using a pattern that is roughly the reverse of the torque sequence (meaning start with the outer bolts and work inward), loosen and remove the balance of the 7 cylinder head/block cover bolts (you removed 2 of them in the last step when you removed the thermostat), then remove the head and discard the old gasket.

✳✳ SELOC CAUTION

Exhaust cover bolts are some of the most likely bolts to become frozen from corrosion on outboard motors. Before removal it's USUALLY a good idea to hit them all with penetrating oil and then to place the socket and driver over each bolt and tap on them lightly using a mallet to help loosen the corrosion on the bolt threads before you even try to loosen them. A hand impact driver can also work wonders here, but avoid air powered impact tools, as they are much more likely to shear the bolt threads.

5. Using the reverse of the torque sequence (meaning start with the outer bolts and work inward), loosen and remove the 9 exhaust cover bolts, then remove the cover, followed by the cover gasket, manifold (also sometimes called the plate) and the manifold gasket. As usual, discard the old gaskets.

■ **If you suspect the powerhead has been overheated at some point and/or the spark plugs are grayish colored (which is a possible sign of water intrusion), inspect the exhaust manifold for warpage and/or signs that the gaskets were not properly aligned.**

6. On the port side of the powerhead, remove the 4 bolts securing the 2 transfer port covers (2 bolts/cover), then remove the covers and inspect the O-rings.

7. On the starboard side of the powerhead, remove the 10 bolts securing the 4 transfer port covers (2 bolts each for 2 of the covers, 3 bolts each for the other 2 covers), then remove the covers and inspect the O-rings.

8. Using the reverse of the torque sequence (meaning start with the outer bolts and work inward) loosen and remove the 6 crankcase cover-to-cylinder block bolts. Use a soft mallet to tap around the edge of the cover-to-block mating surface, then carefully separate the cover from the block.

9. Carefully lift the crankshaft assembly (and the pistons) from the cylinder block.

10. Slide the upper and lower roller bearings and seals off the crankshaft ends. Discard the old seals.

11. Remove the retaining ring from the groove in the center main bearing race, then remove the races and bearing halves along with the thrust washers. MARK and/or NOTE the location of each piece to ensure it is returned to the same position in relation to the other pieces of the center main bearing assembly.

12. Before going any further, LABEL the pistons and all matching components (connecting rod, rod cap etc) to make sure the proper components are all removed, stored and reinstalled together. Also, take some time to make matchmarks between things like the connecting rod caps and the rods or the piston and the connecting rod to make sure all components are installed facing the same directions as they were before they were disassembled. It is true that many of these components have factory alignment or identification marks, but it is never a bad idea to double-check them or mark them yourself before proceeding.

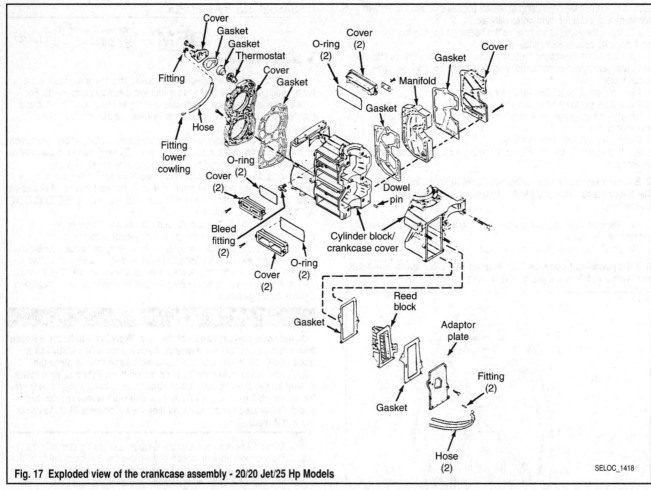

Fig. 17 Exploded view of the crankcase assembly - 20/20 Jet/25 Hp Models

SELOC_1418

13. If the pistons are to be removed from the connecting rods, FIRST unbolt and remove the rods from the crankshaft. Alternately and evenly loosen the 2 connecting rod cap bolts, going back and forth with no more than 1 full turn on each bolt without moving back to the other bolt. Proceed with this pattern until the connecting rod cap can be removed, then DISCARD the old connecting rod cap bolts. Support the piston as you remove the bolts so the piston and rod do not fall free and become damaged. Also, be sure to keep track of the bearings and bearing cages as the cap and connecting rod for each piston assembly is removed from the crankshaft. It appears that some models may use loose needles on the connecting rod big end bearing. If so all needles must be recovered and reused or the full set must be replaced.

14. Disassemble each piston for further inspection and component replacement, as follows:

✱✱ SELOC WARNING

ALWAYS use eye protection when attempting to remove the piston pin lockrings as they are prone to launching out of position unpredictably.

a. To remove the connecting rod from each piston, first remove the lockring from both ends of the piston pin. Use a pair of needlenose pliers to grasp and remove the piston locking. Keep in mind that these lockrings are often distorted slightly during removal and so they must be discarded are replaced with new lockrings during installation.

■ **If necessary, a piston cradle can be fabricated from a large block of wood. The important design features of the cradle are that it is semi-circular on top to hold (or cradle, get it?) the piston and that there is a bore at the bottom center into which the wrist pin can be tapped or pressed. The final important feature of the cradle is that it is sturdy enough to withstand the force of the shop press, BUT, keep in mind that the fit is loose on these pistons and significant force should not be necessary. It is NOT uncommon for someone to simply support the piston in their lap when working on these motors.**

b. While carefully supporting the piston to prevent damage use a soft-faced mallet and driver (such as Piston Pin Tool #91-76160A2 or equivalent) to gently tap the pin from the bore in the piston. As the pin is removed, KEEP CLOSE TRACK of the locating washers and the loose needle bearings (there should be 27 needles per piston).

■ **As the wrist pin is removed from the piston you must retain the 27 loose needle bearings and the 2 locating washers from the small end of the connecting rod. If you can, stop tapping or pressing once the wrist pin JUST clears the bottom of the connecting rod bore. Then carefully withdraw the rod along with the needle bearings (place a thin bladed tool like a putty knife or your finger under the rod hole as it is pulled free of the piston bore/wrist pin). Inspect the bearings, if any of the needles are lost or worn, replace the entire bearing assembly.**

c. Use a Piston Ring Expander (such as #91-24697) to remove the rings from the pistons. The manufacturer does not recommend reusing the rings. Retain them in sets for inspection purposes (marking their original locations), but discard them before installation.

■ **We've said it before. We know there will be some circumstances that will prompt people to reuse rings, but for the most part, if you've come THIS far, it is probably a good idea to bite the bullet and replace the rings to ensure long-lasting performance from the powerhead.**

d. Repeat for the other piston.

15. Remove and discard the old coupling seal (wear sleeve) from the end of the crankshaft. The best method to accomplish this is to use an Expanding Rod and Collet (Snap-On® numbers CG40-4 and CG40-15 or equivalents) along with a small slide-hammer.

■ **Mercury and Seloc recommend that ALL seals be replaced during a rebuilding procedure of this magnitude.**

To assemble:

16. Make sure all parts have been thoroughly cleaned and inspected. Be sure to oil all bearings and bearing surfaces lightly using a coating of fresh 2-stroke oil. Loose needle bearings, once cleaned, should be spread out on a

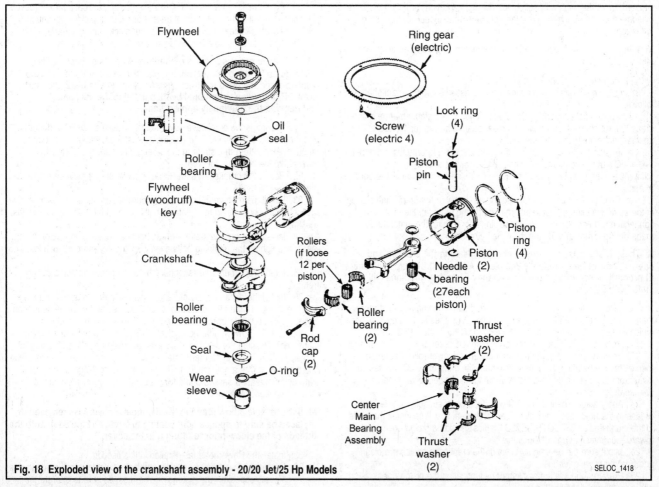

Fig. 18 Exploded view of the crankshaft assembly - 20/20 Jet/25 Hp Models

SELOC_1418

clean worksurface, like a sheet of paper and coated lightly using Needle Bearing Assembly Lubricant.

17. Apply a light coating of Loctite® 271 to the carrier surface on the crankshaft, then square the carrier to the shaft using a 7/8 in. DEEP socket and carefully tap or press the carrier (wear sleeve) onto the crankshaft).

18. Apply a light coating of 2-4-C with Teflon or equivalent marine grade grease, to the crankshaft center main roller bearings and thrust washers (the grease will help hold them in place on the crankshaft). Position the roller bearings and thrust washers onto the crankshaft, then install the bearing race halves so that the retainer ring groove faces FORWARD toward the flywheel end of the crankshaft. Secure the halves using the retainer ring, making sure the ring bridges both fracture lines of the bearing race.

19. Verify the center main bearing assembly is properly installed by pushing on the thrust washers towards either end of the crankshaft, then us a feeler gauge to check clearance between the crankshaft throws and the thrust washers. If clearance exceeds 0.030 in. (0.762mm) the thrust washers must be replaced.

20. If disassembled for inspection and/or component replacement, assemble each piston as follows:

a. Apply a light coating of Needle Bearing Assembly grease (or 2-4-C with Teflon or an equivalent gasoline soluble marine grade grease) to the wrist pin bearing needles and the locating washers, then hold the lower locating washer in place while you insert the 27 needles around the inner circumference of the connecting rod wrist pin bore. Install the Piston Pin Tool # 91-76160A2 or equivalent into the bore to align the needles and hold them in place during assembly.

■ The wrist pin bearing tool is, essentially, a short round pin, roughly the same outer diameter of the wrist pin, but very short, just long enough to insert it through the connecting rod wrist pin bore and install the locating washers on either side. A similarly sized dowel pin can be substituted, or an old connecting rod wrist pin from the same size piston could be used. If substituting another rod or drift pin, make sure the surface is smooth and clean/free of all corrosion. Also, make sure the outer diameter of the tool is smaller than the wrist pin and bore to prevent the possibility of binding or damage.

b. Once all the needles are in position, inspect to make sure one is not missing. Generally speaking, if the tip of an awl can be inserted between the needles, one or more needles is/are missing.

✳✳ SELOC CAUTION

NEVER replace just one needle. If one or more needles are damaged or missing, replace the entire set.

c. With the bearings and the wrist pin tool in position, install the upper locating washer. Again, a dab of grease should hold the thrust washer in position.

d. Apply a light coating of clean engine oil or assembly lube to the wrist pin and the pin bore in the piston.

■ Remember to position all components in the same orientation from which they were removed. For instance, the domes of most pistons are marked with a stamped UP for reference as to which end faces the flywheel. Likewise, there is often a raised rib on the wrist pin boss (looking under the piston skirt) which also corresponds to the side which faces the flywheel.

e. Support the piston (ideally in a piston cradle for ease of assembly but the wrist pin fit should be loose on both sides so this is not absolutely necessary). Insert the wrist pin into the bore and gently press it through the bore until it just starts to appear inside the piston skirt. Stop pressing, then position the connecting rod with the needle bearings, locating washers and wrist pin bearing installation tool inside the piston skirt. Continue to press the wrist pin into the position, through the thrust washers and bearings, pushing the wrist pin bearing installer through the other side and out of the piston.

f. Once the wrist pin is properly positioned and centered in the piston, install new lockrings into the grooves in both ends of the piston bore.

g. Use a ring expander to install the piston rings on each piston, making sure to install the RECTANGULAR ring first in the bottom groove and the HALF-KEYSTONE (or tapered ring) in the top groove. Each piston ring groove should contain a dowel pin, around which the ring gap must be situated. If not the rings will likely break during installation (or if by they don't by some miracle, during the first start-up). Be sure to install the tapered ring in the top groove.

■ **When using the ring expander, spread each ring JUST enough to slip over the piston.**

h. Repeat for the other piston.

21. Install each piston and connecting rod assembly to the crankshaft as follows:

a. Apply a light coating of 2-4-C with Teflon or a suitable marine grade grease to the connecting rod big-end roller bearing halves (or 12 loose needles, depending upon the design).

b. Position both halves of the big-end roller bearing onto the crank pin.

c. Clean the connecting rod bolts using solvent and dry with compressed air, then apply a light coating of 2-stroke engine oil to the bolt threads.

d. Position the piston/connecting rod assembly to the crankshaft (facing in the same direction as noted and marked during removal, then place the connecting rod end cap over the crank pin (also aligning the matchmarks made during removal) and start the bolts by hand.

■ **These motors use fractured end caps, meaning the end cap and connecting rod were originally one piece. As a result the caps will only fit properly in one direction and when they fit the fracture lines will mate PERFECTLY.**

e. Alternately and evenly tighten the end cap retaining bolts, no more than one turn before moving to the other bolt. Once they start to tighten, use multiple passes to tighten them to 150 inch lbs./12.5 ft. lbs. (16.8 Nm).

f. Repeat for the other piston/connecting rod assembly.

22. Apply a light coating of oil to the upper and lower roller bearings, then install the bearings onto the crankshaft with the numbered side facing away from the pistons (toward the flywheel for the upper, toward the driveshaft for the lower bearing).

23. Apply a light coating of 2-4-C with Teflon or a suitable marine grade grease to the new upper and lower crankshaft seal lips, then install the seals to the crankshaft. FOR BOTH SEALS the lips should be facing downward toward the driveshaft end of the crank.

24. Make sure the bearing surfaces of the crankcase halves are clean and free of any oil or debris.

25. Prepare to install the crankshaft by lubricating the piston rings, piston and cylinder wall lightly with oil.

26. No ring compressor is necessary, keep the crankshaft horizontal and carefully push the pistons into the cylinder. Align the hole in the center of the main bearing race with the pin in the block.

27. Position the alignment boss of the upper crankshaft roller bearing into the notch in the cylinder block.

28. Carefully push the seals inward to seat.

29. Using a small screwdriver inserted through the exhaust ports, gently press on each of the piston rings to make sure there is tension on each ring. IF a ring fails to return to position (there is no tension) then it is likely that the ring was broken during installation and the crankshaft assembly will have to be removed again to replace the ring.

30. Again, make sure the mating surface of the crankcase halves (cover and block) is still completely free of grease or oil, then apply a continuous bead of Loctite Master Gasket on the cylinder block mating surface, JUST INBOARD of the bolts holes on either side of the crankshaft bore.

■ **Mercury is serious about the continuous beads of Loctite. They often go so far as to instruct their technicians that IF there is a void while placing the bead to either wipe away the whole bead and start over OR to apply an additional bead, parallel to the void and overlapping the previously applied bead.**

31. Assemble the crankcase cover to the block, then install and tighten the 6 bolts to 30 ft. lbs. (40.7 Nm) using at least 3 passes of the torque sequence (a pattern that starts at the inner left bolt and works its way outward to the bottom right bolt).

32. Rotate the crankshaft a few times to make sure it moves freely and there is NO binding.

33. Install the reed valve assembly/carburetor adapter plate (using a new gasket) and secure using the retaining bolts. Tighten the bolts to 80 inch lbs. (9 Nm).

34. Install the port and starboard transfer port covers using NEW O-ring seals and tighten the retaining bolts evenly on each cover to 40 inch lbs. (4.5 Nm).

35. If removed, route and reconnect the bleed hoses as noted during removal.

36. Install the exhaust manifold and cover using new gaskets and the 9 retaining bolts. Tighten the bolts in at least 3 passes of the torque sequence (which works outward from the center bolts) to 140 inch lbs. (15.8 Nm).

37. Install a new cylinder head/block cover gasket without any sealing substance. Position the head in place over the gasket and secure using the 7 bolts (that includes the 2 thermostat housing bolts, so be sure to reposition the thermostat with a new gasket). Tighten the bolts to 140 inch lbs. (15.8 Nm) using at least 3 passes of the torque sequence which starts at the center and works outward.

■ **If the tell-tale elbow fitting in the thermostat cover was removed or replaced be sure to apply a light coating of Loctite® Pipe Sealant to the threads of the elbow prior to fitting it to the cover.**

38. Install the Powerhead, as detailed in this section.

39. Once the powerhead is fully assembled, if cylinder block components were replaced, especially the pistons and/or rings, be sure to operate the engine as directed for new component break-in. For details, please refer to Powerhead Break-In, in this section.

CLEANING & INSPECTION

◆ **See Figures 17 and 18**

Cleaning and inspecting the components is virtually the same for any two-stroke outboard and varies mostly by specifications (which are listed in the Engine Specifications charts) or by component type. A section detailing the proper procedures, sorted mostly by component, can be found under Powerhead Refinishing.

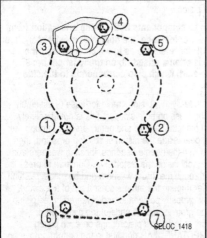

Fig. 19 Cylinder head (block cover) torque sequence - 20/20 Jet/25 Hp Models

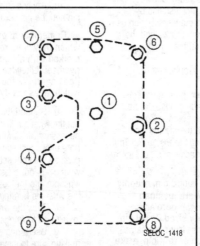

Fig. 20 Exhaust cover torque sequence - 20/20 Jet/25 Hp Models

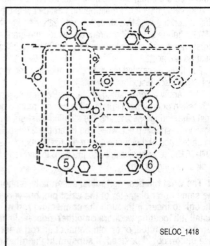

Fig. 21 Crankcase cover torque sequence - 20/20 Jet/25 Hp Models

30/40 Hp 2-Cylinder Powerheads

REMOVAL & INSTALLATION

◆ See Figures 22 thru 27

These powerheads are starting to be large enough that we really recommend you both strip all fuel and ignition components before removal AND be sure to use an engine hoist or other lifting device for safety and to prevent damage.

Besides. consider why the powerhead is being removed. If you are planning on disassembling the powerhead for inspection or overhaul, then you'll have to remove the fuel and electrical components anyway. Removing them before powerhead removal is a good idea and will help to protect them from potential damage when lifting and moving the powerhead itself.

1. Remove the cowling from the powerhead for access.

2. For safety, either disconnect the battery cables (electric start models) and/or tag and disconnect the secondary ignition leads from the spark plugs (manual start models).

3. On remote control models, disconnect the remote control harness at the main wiring harness connector. Disconnect any additional remote control harness leads from the bullet connectors.

4. On tiller control models, disconnect the safety lanyard switch and stop button assembly bullet connectors.

5. Disconnect the fuel supply line from the lower cowl, then disconnect the fuel line either at the engine tray or fuel filter.

6. Disconnect the throttle and shift cables.

7. Disconnect the black ground lead between the engine tray and the powerhead.

8. Disconnect the tell-tale water discharge hose from the fitting on the powerhead.

9. On manual start models, disconnect the anti-start in gear cable.

10. Disconnect the neutral safety switch leads.

11. On the manual start models, tag and disconnect the primer bulb lines from the intake manifold and carburetor fittings.

12. If equipped, remove the hand rewind starter assembly.

13. On electric start models, remove the flywheel cover.

14. Remove the bolt, nut and flat washer securing the trim cover (upper driveshaft housing cover) to the driveshaft housing. Remove the trim cover for access to the powerhead mounting bolts.

15. Remove the 6 bolts (3 on each side) securing the powerhead to the driveshaft housing.

16. Remove the plastic cap from the center of the flywheel and install the lifting eye (#91-90455T) into the flywheel a minimum of 5 turns.

■ At this point double check for any cables, hoses, wires or linkages that will interfere with the removal of the powerhead.

17. Gently rock the powerhead back and forth just enough to break the gasket seal between the powerhead and driveshaft housing. When the seal breaks loose, lift the powerhead off the driveshaft and driveshaft housing using a suitable lifting hoist.

18. Move the powerhead to a clean work area and placed in a suitable holding fixture.

To install:

19. Thoroughly clean all old gasket material from the driveshaft housing and powerhead mating surfaces.

20. Lubricate the SIDES of the splines on the driveshaft with Quicksilver Special Lubricant No. 101 or equivalent. Wipe off any excess lubricant from the top of the driveshaft as any lubricant left on top of the driveshaft could prevent it from seating in the crankshaft.

21. Place a new gasket onto the driveshaft housing.

22. Install the powerhead on the driveshaft housing. You may have to rotate the crankshaft (or conversely the propshaft to rotate the driveshaft if the motor is in gear) in order to index the crankshaft and driveshaft splines.

23. Install and tighten the powerhead mounting bolts evenly to 29 ft. lbs. (39.3 Nm) using a couple of passes of a pattern that starts at the center and works outward.

24. Install the trim cover over the driveshaft housing and secure using the bolt, washer and nut. Tighten the fasteners securely.

Fig. 22 Remove the cowling. . .

Fig. 23 . . .for access to the powerhead

Fig. 24 Remove the manual starter or flywheel cover. . .

Fig. 25 . . .and strip the powerhead of lines/connections

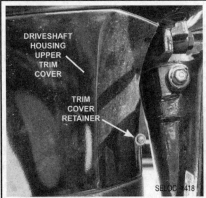

Fig. 26 Remove the trim cover for access to powerhead mounting bolts

Fig. 27 Most models have a plastic cover in the flywheel which must be removed to install the lifting eye

25. Remove the lifting eye and install the plastic cover over the flywheel retaining nut.

26. On electric start models, install the flywheel cover.

27. If equipped, install the hand rewind starter assembly.

28. On the manual start models, reconnect the primer bulb lines to the intake manifold and carburetor fittings, as tagged during removal.

29. Reconnect the neutral safety switch leads.

30. On manual start models, reconnect the anti-start in gear cable.

31. Reconnect the tell-tale water discharge hose to the fitting on the powerhead.

32. Reconnect the black ground lead between the engine tray and the powerhead.

33. Reconnect the throttle and shift cables.

34. Reconnect the fuel supply line to the lower cowl and reconnect the fuel line either at the engine tray or fuel filter.

35. On tiller control models, reconnect the safety lanyard switch and stop button assembly bullet connectors.

36. On remote control models, reconnect the remote control harness at the main wiring harness connector. Reconnect any additional remote control harness leads from the bullet connectors which were tagged during removal.

■ **Be sure to run the motor without the cowling installed in order to check for potential fuel or oil leaks. Remedy any leaks before proceeding.**

37. On electric start equipped models, reconnect the battery cables.

38. Reconnect the spark plug wires, hook up a source of cooling water and test run the motor.

39. Install the cowling to the powerhead.

■ **If the powerhead was rebuilt or replaced with a remanufactured unit, don't forget to follow the proper Break-In procedures.**

DISSASSEMBLY & ASSEMBLY

◆ **See Figures 28 thru 31**

1. If not done during powerhead removal, strip the powerhead of the remaining fuel and electric components. For details, where applicable, please refer to the appropriate sections of this guide.

- Electric starter motor
- Ignition Module (switch box)
- Ignition coil
- Starter solenoid
- Voltage regulator/rectifier
- Flywheel
- Stator coil
- Trigger coil
- Air intake silencer
- Carburetor and linkage
- Fuel pump
- Fuel enrichment valve
- Shift cable latch assembly
- Control cable anchor bracket
- Warning module
- Oil pump

2. Tag and disconnect the bleed lines. There are normally two lines on these models, a shorter one on the side of the powerhead which forms a small "U" shape between 2 fittings and a longer one on the end of the powerhead which snakes its way up from the bottom of corner of the powerhead up to the top of the intake/reed block manifold.

3. Remove the thermostat and poppet valve assembly from the top side of the powerhead.

4. Using a few passes in the reverse of the torque sequence (which results in a clockwise spiraling pattern that starts at the lower right bolt and works inward, ending with the 2nd bolt from the top on the right), remove the 8 bolts securing the intake manifold/reed block assembly to the powerhead, then remove the assembly (manifold, reed plate and gasket).

5. Lay the powerhead down on its side (the block side) for access to the lower end cap bolts.

6. Remove the 3 bolts securing the lower end cap (seal holder) assembly to the crankcase, then remove the assembly. The assembly contains 2 seals, a smaller one mounted with the lips facing away from the crankshaft and a larger one with the lips facing toward the crank. The

assembly is also sealed using an O-ring. Both the seals and the O-ring should be replaced to ensure an oil tight powerhead.

7. Using a few passes in the reverse of the torque sequence loosen and remove the 10 crankcase cover-to-cylinder block bolts. Use a soft mallet to tap around the edge of the cover-to-block mating surface, then carefully separate the cover from the block.

■ **On most Mercury twin cylinder motors you can usually remove the crankshaft and pistons as an assembly. However, as the motors get larger, it becomes easier to unbolt the connecting rods first and lift out the crankshaft separately from the pistons. These motors are likely the breaking point for the two methods and you can probably do either, but we've recommended the later here (removing them separately).**

8. Take a second to matchmark the connecting rod caps to the connecting rods. They MUST not only be reinstalled on the same rods, but facing the same direction or bearing failure will occur.

9. Unbolt the connecting rod caps to free the crankshaft. Alternately and evenly loosen the 2 connecting rod cap bolts, going back and forth with no more than 1 full turn on each bolt without moving back to the other bolt. Proceed with this pattern until the connecting rod cap can be removed. MAKE sure you keep track of the rod cap and bearing cage as they must be returned to the original connecting rod and crankshaft journal if they are to be reused. It's best to label them AND to reinstall them to the connecting rod as soon as the crankshaft is out of the way.

10. Carefully lift the crankshaft assembly from the connecting rods and cylinder block.

11. If necessary, disassemble the crankshaft and bearings as follows:

a. Slide the upper seal off the top of the crankshaft.

b. Mercury doesn't say whether or not the top bearing is pressed in place. Generally speaking, if the bearing is pressed onto the shaft it cannot be removed without a bearing separator and SHOULD NOT be removed unless it is being replaced, as a bearing separator will normally damage the bearing during removal. However, on some models in this size/hp range Mercury has used a loose fit bearing that can be slid on or off the crank without damage.

c. Remove the retaining ring from the groove in the center main bearing race, then remove the races and bearing halves, but proceed slowly as most models will use 14 free rollers which must be kept together as a set (and the whole set must be replaced if even one roller is lost or damaged).

d. While wearing a pair of safety glasses, carefully remove the ring seal from the edge of the lower crankshaft throw that faces the center main bearing.

e. The lower bearing IS pressed onto the shaft and removal will destroy it, so ONLY remove it if necessary to replace the bearing or for access to the oil pump drive gear (on oil injected models), as necessary. To remove the bearing first remove the snapring, then use a shop press and bearing separator to free the bearing from the crankshaft.

f. If necessary, remove the oil pump drive gear from the bottom of the crankshaft.

12. Before going any further, LABEL the pistons and all matching components (connecting rod, rod cap etc) to make sure the proper components are all removed, stored and reinstalled together. You should have started earlier when you matchmarked the connecting rod caps, but go a step further now to make sure you also mark the direction the remaining components are facing. Remember when reusing these components (caps, rods, pistons) you must make sure all components are installed in the same locations and facing the same directions as they were before they were disassembled. It is true that many of these components have factory alignment or identification marks, but it is never a bad idea to double-check them or mark them yourself before proceeding.

13. Carefully pull each piston out of the crankshaft bore for inspection and disassembly, as necessary. If it is necessary to disassemble them for further inspection and component replacement, as follows:

■ **We've said it before. We know there will be some circumstances that will prompt people to reuse rings, but for the most part, if you've come THIS far, it is probably a good idea to bite the bullet and replace the rings to ensure long-lasting performance from the powerhead.**

a. Use a Piston Ring Expander (such as #91-24697) to remove the rings from the pistons. The manufacturer does not recommend reusing the rings. Retain them in sets for inspection purposes (marking their original locations), but discard them before installation.

✳✳ SELOC WARNING

ALWAYS use eye protection when attempting to remove the piston pin lockrings as they are prone to launching out of position unpredictably.

b. To remove the connecting rod from each piston, first remove the lockring from both ends of the piston pin. Use an awl inserted just under the lip of the ring to gently lever it out of the piston. Keep in mind that these lockrings are often distorted slightly during removal and so they must be discarded are replaced with new lockrings during installation.

■ If necessary, a piston cradle can be fabricated from a large block of wood. The important design features of the cradle are that it is semi-circular on top to hold (or cradle, get it?) the piston and that there is a bore at the bottom center into which the wrist pin can tapped or pressed. The final important feature of the cradle is that it is sturdy enough to withstand the force of the shop press, BUT, keep in mind that the fit is loose on these pistons and significant force should not be necessary. It is NOT uncommon for someone to simply support the piston in their lap when working on these motors.

c. While carefully supporting the piston to prevent damage use a soft-faced mallet and driver (such as Piston Pin Tool #91-74607A2 or equivalent) to gently tap the pin from the bore in the piston. As the pin is removed, KEEP CLOSE TRACK of the locating/thrust washers and the loose needle bearings (there should be 29 needles per piston).

■ As the wrist pin is removed from the piston you must retain the 29 loose needle bearings and the 2 locating/thrust washers from the small end of the connecting rod. If you can, stop tapping or pressing once the wrist pin JUST clears the bottom of the connecting rod bore. Then carefully withdraw the rod along with the needle bearings (place a thin bladed tool like a putty knife or your finger under the rod hole as it is pulled free of the piston bore/wrist pin). Inspect the bearings, if any of the needles are lost or worn, replace the entire bearing assembly.

d. Repeat for the other piston.

■ Mercury and Seloc recommend that ALL seals be replaced during a rebuilding procedure of this magnitude.

To assemble:

14. Prepare the following components for installation later, as follows:

a. If the reed valves were removed from the plate, position the reeds over the locating pins on the plate, then place the stops over them and secure using the bolt and washer on the front with the nut on the other side of the plate. Be sure to tighten the bolt and nut to 60 inch lbs. (6.8 Nm).

b. If the end cap seals were removed, use a suitable driver to gently press or tap the replacement seals into the end cap. Remember the smaller seal should face downward AWAY from the crankshaft, while the larger seal faces upward TOWARD the crankshaft. Lubricate the seal lips as well as the replacement housing O-ring using a light coating of 2-4-C with Teflon or an equivalent marine grade grease. Also, RIGHT before installation, apply a light coating of Perfect Seal or an equivalent sealant to the end cap flange.

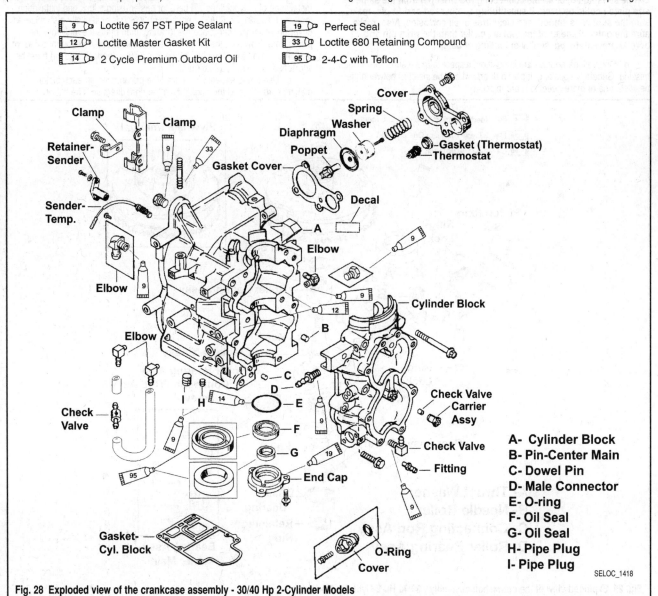

9 ⫍ Loctite 567 PST Pipe Sealant
12 ⫍ Loctite Master Gasket Kit
14 ⫍ 2 Cycle Premium Outboard Oil
19 ⫍ Perfect Seal
33 ⫍ Loctite 680 Retaining Compound
95 ⫍ 2-4-C with Teflon

A- Cylinder Block
B- Pin-Center Main
C- Dowel Pin
D- Male Connector
E- O-ring
F- Oil Seal
G- Oil Seal
H- Pipe Plug
I- Pipe Plug

SELOC_1418

Fig. 28 Exploded view of the crankcase assembly - 30/40 Hp 2-Cylinder Models

Any grease used inside the powerhead for assembly MUST be gasoline soluble. The use of a non-soluble grease may cause severe powerhead damage.

15. Make sure all parts have been thoroughly cleaned and inspected. Be sure to oil all bearings and bearing surfaces lightly using a coating of fresh 2-stroke oil. Loose needle bearings, once cleaned, should be spread out on a clean worksurface, like a sheet of paper and coated lightly using Needle Bearing Assembly Lubricant.

16. If disassembled for inspection and/or component replacement, assemble each piston as follows:

a. Apply a light coating of Needle Bearing Assembly grease (or 2-4-C with Teflon or an equivalent gasoline soluble marine grade grease) to the wrist pin bearing needles and the locating/thrust washers, then hold the lower locating washer in place while you insert the 29 needles around the inner circumference of the connecting rod wrist pin bore. Install the Piston Pin Tool # 91-74607A2 or equivalent into the bore to align the needles and hold them in place during assembly.

■ **The wrist pin bearing tool is, essentially, a short round pin, roughly the same outer diameter of the wrist pin, but very short, just long enough to insert it through the connecting rod wrist pin bore and install the locating washers on either side. A similarly sized dowel pin can be substituted, or an old connecting rod wrist pin from the same size piston could be used. If substituting another rod or drift pin, make sure the surface is smooth and clean/free of all corrosion. Also, make sure the outer diameter of the tool is smaller than the wrist pin and bore to prevent the possibility of binding or damage.**

b. Once all the needles are in position, inspect to make sure one is not missing. Generally speaking, if the tip of an awl can be inserted between the needles, one or more needles is/are missing.

NEVER replace just one needle. If one or more needles are damaged or missing, replace the entire set.

c. With the bearings and the wrist pin tool in position, install the upper locating/thrust washer. Again, a dab of grease should hold the thrust washer in position.

d. Apply a light coating of clean engine oil or assembly lube to the wrist pin and the pin bore in the piston.

■ **Remember to position all components in the same orientation from which they were removed. For instance, the domes of most pistons are marked with a stamped UP for reference as to which end faces the flywheel.**

e. Support the piston (ideally in a piston cradle for ease of assembly but the wrist pin fit should be loose on both sides so this is not absolutely necessary). Insert the wrist pin into the bore and gently press it through the bore until it just starts to appear inside the piston skirt. Stop pressing, then position the connecting rod with the needle bearings, locating washers and wrist pin bearing installation tool inside the piston skirt. Continue to press the wrist pin into the position, through the thrust washers and bearings, pushing the wrist pin bearing installer through the other side and out of the piston.

f. Once the wrist pin is properly positioned and centered in the piston, install new lockrings into the grooves in both ends of the piston bore. Installation of the lockrings will be a lot easier using a lock ring installation tool (such as #91-77109A3 or equivalent) which is essentially a shouldered driver on which you can place the lockring, then press it evenly into the piston bore. However, the stock tool referenced above must be modified for use on these motors, the shaft of the tool (where the lockrings sit) must be shortened to 1.050 in. (26.67mm).

g. Use a ring expander to install the piston rings on each piston, making sure to install the rings facing the right direction. The flat or

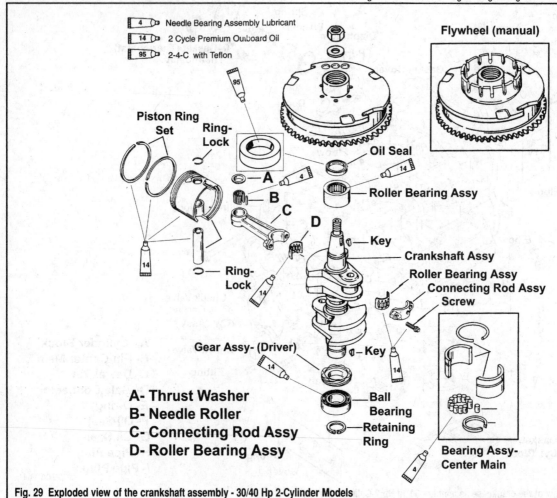

4	Needle Bearing Assembly Lubricant
14	2 Cycle Premium Outboard Oil
95	2-4-C with Teflon

A- Thrust Washer
B- Needle Roller
C- Connecting Rod Assy
D- Roller Bearing Assy

Fig. 29 Exploded view of the crankshaft assembly - 30/40 Hp 2-Cylinder Models

SELOC_1418

RECTANGULAR side must be facing downward, while the tapered or HALF-KEYSTONE side should be facing upward. In most cases, replacement rings will be marked with a letter or dot on the side which faces up. Also, each piston ring groove should contain a dowel pin, around which the ring gap must be situated. If not the rings will likely break during installation (or if by they don't by some miracle, during the first start-up). Be sure to install the tapered ring in the top groove.

■ **When using the ring expander, spread each ring JUST enough to slip over the piston.**

   h. Repeat for the other piston.

17. Prepare to install the pistons by lubricating the piston rings, pistons and cylinder walls lightly with oil.

18. No ring compressor is necessary, just verify that the piston is facing the proper direction (the UP mark on the dome is towards the flywheel end of the powerhead), then gently insert the piston into the cylinder bore. Repeat for the other piston, positioning the pistons as best you can to facilitate crankshaft installation later.

19. If the crankshaft was disassembled, install the oil pump drive gear and/or bearings to prepare it for installation as follows:

   a. If removed, slide the oil pump gear into position on the crankshaft while aligning the keyway and key. Next oil and install the bearing using a press and suitable driver. Once the bearing is in position tight against the gear install the retaining ring.

   b. Wearing a pair of safety glasses to protect your eyes, carefully install the seal ring to the shoulder of lower crankshaft throw which faces the center main bearing.

   c. Apply a light coating of Needle Bearing Assembly Lube or equivalent marine grade grease, to the crankshaft center journal in order to hold the 14 main roller bearings. Position the roller bearings followed by the bearing race halves so that the retainer ring groove faces FORWARD toward the flywheel end of the crankshaft. Secure the halves using the retainer ring, making sure the ring bridges both fracture lines of the bearing race.

   d. If removed, oil and install the upper main bearing making sure the race is positioned with the alignment hole facing toward the lower gear end of the crankshaft.

20. Place the pistons fully upward in their bores (i.e. away from the crankshaft) so as not to interfere with crankshaft installation (this way you can insert the bearing cage and pull them up to the crankshaft journal AFTER the crankshaft is seated). Carefully lower the crankshaft into position, aligning the holes in the center main and upper bearings with the alignment pins in the block.

21. Install each piston and connecting rod assembly to the crankshaft as follows:

   a. Apply a light coating of 2-stroke engine oil to the bearing cages, then install the lower half in the connecting rod and pull the rod upward into position over the crank pin.

   b. Position the other bearing half over the crank pin.

   c. Clean the connecting rod bolts using solvent and dry with compressed air, then apply a light coating of 2-stroke engine oil to the bolt threads.

   d. Position the piston/connecting rod assembly to the crankshaft (facing in the same direction as noted and marked during removal, then place the connecting rod end cap over the crank pin (also aligning the matchmarks made during removal) and start the bolts by hand.

■ **These motors use fractured end caps, meaning the end cap and connecting rod were originally one piece. As a result the caps will only fit properly in one direction and when they fit the fracture lines will mate PERFECTLY.**

   e. Alternately and evenly tighten the end cap retaining bolts, no more than one turn before moving to the other bolt. Once they start to tighten, use multiple passes to tighten them to first to 15 inch lbs. (1.7 Nm), next to 120 inch lbs (13.6 Nm) and then finally an additional 90 degrees (1/4 turn).

   f. Repeat for the other piston/connecting rod assembly.

22. Apply a light coating of 2-4-C with Teflon or a suitable marine grade grease to the new upper crankshaft seal lips, then slide the seal into position with the lips facing downward toward the crankshaft.

23. Make sure the bearing surfaces of the crankcase halves are clean and free of any oil or debris, then apply a continuous bead of Loctite Master Gasket on the cylinder block mating surface, JUST INBOARD of the bolts holes on either side of the crankshaft bore. Make sure the sealer is extended right to the edge at each center main journal to prevent blow-by between the cylinders.

■ **Mercury is serious about the continuous beads of Loctite. They often go so far as to instruct their technicians that IF there is a void while placing the bead to either wipe away the whole bead and start over OR to apply an additional bead, parallel to the void and overlapping the previously applied bead.**

24. Apply a light coating of Perfect Seal to the assembled end cap, then position it over the bottom of the crank.

25. Assemble the crankcase cover to the block, then install and finger-tighten the 10 retaining bolts followed by the 3 end cap cover retaining bolts.

26. Using at least 3 passes of the torque sequence, tighten the crankcase cover bolts to 16.5 ft. lbs. (22.4 Nm).

27. Tighten the end cap bolts to 16.5 ft. lbs. (22.4 Nm) using at least 3 passes of the following sequence, starting with the bolt at the lower corner of the end cap (toward the edge of the powerhead in the approximately 7 o'clock position when looking at the bottom of the assembly and working clockwise to the other corner bolt at about 10 o'clock position and finally to the inboard bolt at 3 o'clock position.

28. Rotate the crankshaft a few times to make sure it moves freely and there is NO binding.

29. Install the intake manifold, reed valve block plate and a new gasket, then secure using the retaining bolts. Tighten the bolts using at least 3 passes of the torque sequence to 16.5 ft. lbs. (22.4 Nm).

30. Install the thermostat and poppet valve assembly using a new thermostat housing gasket to the side of the powerhead. Tighten the bolts to 16.5 ft. lbs. (22.4 Nm).

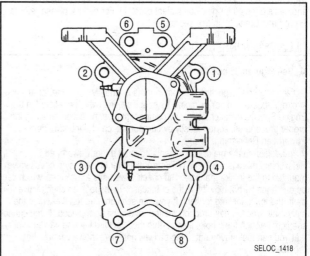

SELOC_1418

Fig. 30 Intake manifold torque sequence - 30/40 Hp 2-Cylinder Models

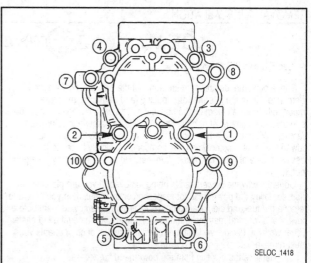

SELOC_1418

Fig. 31 Crankcase cover torque sequence - 30/40 Hp 2-Cylinder Models

31. If removed, route and reconnect the bleed hoses as noted during removal.

32. The following fuel and electric components CAN be installed to the powerhead before the powerhead is mounted to the intermediate housing. For details, where applicable, please refer to the appropriate sections of this guide.

- Oil pump
- Warning module
- Control cable anchor bracket
- Shift cable latch assembly
- Fuel enrichment valve
- Fuel pump
- Carburetor and linkage
- Air intake silencer
- Trigger coil
- Stator coil
- Flywheel
- Voltage regulator/rectifier
- Starter solenoid
- Ignition coil
- Ignition Module (switch box)
- Electric starter motor

33. Install the Powerhead, as detailed in this section.

34. Once the powerhead is fully assembled, if cylinder block components were replaced, especially the pistons and/or rings, be sure to operate the engine as directed for new component break-in. For details, please refer to Powerhead Break-In, in this section.

CLEANING & INSPECTION

◆ **See Figures 28 and 29**

Cleaning and inspecting the components is virtually the same for any two-stroke outboard and varies mostly by specifications (which are listed in the Engine Specifications charts) or by component type. A section detailing the proper procedures, sorted mostly by component, can be found under Powerhead Refinishing.

In addition to the major components covered under Powerhead Refinishing, there is a 3/16 in. (7.76mm) diameter brass casing check valve mounted in the intake manifold. The check valve uses a nylon ball which can be damaged by hot blow-by in the crankcase. To inspect the check valve peer through it, looking for light. If you can see light through the whole the nylon ball is bad, it has probably melted, and must be replaced. If you cannot see light through the whole, insert a thin piece of wire into the valve to verify that you can feel movement. If the ball will not move the valve must be replaced.

40/40 Jet/50/55/60 Hp 3-Cylinder Powerheads

REMOVAL & INSTALLATION

◆ **See Figures 27, 32 and 33**

These powerheads are large enough that the only practical way to lift them safely from the intermediate housing (and perhaps even more importantly to install them on the housing mating with the driveshaft splines) is to use an engine hoist or other lifting device for safety and to prevent damage. When using a hoist you don't HAVE to completely strip the powerhead of all electrical and fuel components, simply because there is less of chance of them becoming damaged. However, it still may be a good idea.

Consider why the powerhead is being removed. If you are planning on disassembling the powerhead for inspection or overhaul, then you'll have to remove the fuel and electrical components anyway. Removing them before powerhead removal may make the rest of powerhead removal a tad easier. Either way you decide, we'll list the minimum amount of components you must remove before lifting the powerhead free.

1. Remove the cowling from the powerhead for access.

2. For safety, either disconnect the battery cables (electric start models) and/or tag and disconnect the secondary ignition leads from the spark plugs (manual start models).

3. On remote control models, disconnect the remote control harness at the main wiring harness connector. Disconnect any additional remote control harness leads from the bullet connectors.

4. On tiller control models, disconnect the safety lanyard switch and stop button/horn wiring bullet connectors.

5. Disconnect the trim switch and trim relay wires.

6. Disconnect the fuel supply line from the lower cowl, then disconnect the fuel line either at the engine tray or fuel filter.

7. Disconnect the throttle and shift cables.

8. If desired, remove the oil tank assembly.

9. Disconnect the blue oil warning module wires.

10. Disconnect the black ground lead between the engine tray and the powerhead.

11. Disconnect the tell-tale water discharge hose from the fitting on the powerhead.

12. On manual start models, disconnect the anti-start in gear cable.

13. Disconnect the neutral safety switch leads.

14. On the manual start models, tag and disconnect the primer bulb lines from the intake manifold and carburetor fittings.

15. If equipped, remove the hand rewind starter assembly.

16. On electric start models, remove the flywheel cover.

17. Remove the bolt, nut and flat washer securing the trim cover (upper driveshaft housing cover) to the driveshaft housing. Remove the trim cover for access to the powerhead mounting bolts.

18. Install the lifting eye (#91-90455 or equivalent) into the flywheel a minimum of 5 turns.

19. Remove the 6 bolts (3 on each side) securing the powerhead to the driveshaft housing.

■ **At this point double check for any cables, hoses, wires or linkages that will interfere with the removal of the powerhead.**

20. Gently rock the powerhead back and forth just enough to break the gasket seal between the powerhead and driveshaft housing. When the seal breaks loose, lift the powerhead off the driveshaft and driveshaft housing using a suitable lifting hoist.

21. Move the powerhead to a clean work area and placed in a suitable holding fixture.

To install:

22. Thoroughly clean all old gasket material from the driveshaft housing and powerhead mating surfaces.

23. Lubricate the SIDES of the splines on the driveshaft with Quicksilver Special Lubricant No. 101 or an equivalent marine grade grease. Wipe off any excess lubricant from the top of the driveshaft as any lubricant left on top of the driveshaft could prevent it from seating in the crankshaft.

24. Place a new gasket onto the driveshaft housing. It is CRITICAL that you use the correct gasket. There are 2 different style gaskets available and the NEWER style (#27-828553) replaces the old style (#27-812865). The old style was good only for 50/55/60 hp models, while the new style is good for all 40/50/55/60 hp models. IF you were to install the old style (for 50/55/60 hp motors only gasket) on a 40/50/55/60 hp model motor severe powerhead damage could occur. The new style gasket is visually identified by a tab on one side of the gasket, right between 2 bolt holes, near the rounded end of the gasket. In all cases, if you install the NEW part no OR a gasket which is the same as what was removed, you'll be fine.

25. Install the powerhead on the driveshaft housing. You may have to rotate the crankshaft (or conversely the propshaft to rotate the driveshaft if the motor is in gear) in order to index the crankshaft and driveshaft splines.

26. Install and tighten the powerhead mounting bolts evenly to 28 ft. lbs. (38 Nm) using a couple of passes of a pattern that starts at the center and works outward. Remove the lifting eye and lifting device.

27. Install the trim cover over the driveshaft housing and secure using the bolt, washer and nut. Tighten the fasteners to 80 inch lbs. (9 Nm).

28. On electric start models, install the flywheel cover.

29. If equipped, install the hand rewind starter assembly.

30. On the manual start models, reconnect the primer bulb lines to the intake manifold and carburetor fittings, as tagged during removal.

31. Reconnect the neutral safety switch leads.

32. On manual start models, reconnect the anti-start in gear cable.

33. Reconnect the tell-tale water discharge hose to the fitting on the powerhead.

34. Reconnect the black ground lead between the engine tray and the powerhead.

35. Reconnect the blue oil warning module wires.

36. If removed, install the oil tank assembly.

37. Reconnect the throttle and shift cables.

38. Reconnect the fuel supply line to the lower cowl and reconnect the fuel line either at the engine tray or fuel filter.

39. Reconnect the trim switch and trim relay wires.

40. On tiller control models, reconnect the safety lanyard switch and stop button assembly bullet connectors.

41. On remote control models, reconnect the remote control harness at the main wiring harness connector. Reconnect any additional remote control harness leads from the bullet connectors which were tagged during removal.

■ **Be sure to run the motor without the cowling installed in order to check for potential fuel or oil leaks. Remedy any leaks before proceeding.**

42. If applicable, reconnect the battery cables.

43. Reconnect the spark plug wires, hook up a source of cooling water and test run the motor.

44. Install the cowling to the powerhead.

■ **If the powerhead was rebuilt or replaced with a remanufactured unit, don't forget to follow the proper Break-In procedures.**

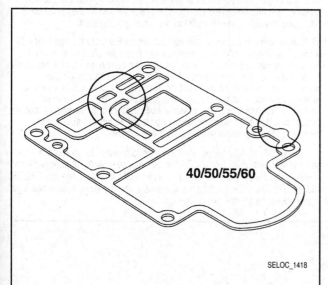

Fig. 32 New style gasket (# 27-828553) good for all 40/50/55/60 models

SELOC_1418

Fig. 33 Old style gasket (# 27-812865) ONLY good on 50/55/60 models (NOT for use on 40/50/55/60 models)

SELOC_1418

DISSASSEMBLY & ASSEMBLY

OEM   ③   DIFFICULT

◆ **See Figures 34 thru 37**

1. If not done during powerhead removal, strip the powerhead of the remaining fuel and electric components. For details, where applicable, please refer to the appropriate sections of this guide.

- Ignition Module (CDM)
- Flywheel
- Stator coil
- Trigger coil
- Electric starter motor
- Starter solenoid
- Voltage regulator/rectifier
- Fuel pump
- Air intake silencer
- Carburetor and linkage
- Fuel enrichment valve
- Oil pump
- Shift cable latch assembly
- Control cable anchor bracket

2. If necessary, tag and disconnect the bleed line(s).

3. Remove the thermostat valve assembly from the top side of the powerhead.

4. Using a few passes in the reverse of the torque sequence, remove the 14 bolts securing the intake manifold/reed block assembly to the powerhead, then remove the assembly (manifold, manifold gasket, reed plate and reed plate gasket). Discard the old gaskets.

5. On the crankcase cover, just behind the reed plate mounting surface there are 2 check valves/check valve holders. You can see them at the top center of each of the lower 2 of the 3 openings for the reed valves. It's a good idea to unscrew and remove these holders in order to inspect or replace the valves.

6. Lay the powerhead down on its side (the block side) for access to the lower end cap bolts.

7. Remove the 3 bolts securing the lower end cap (seal holder) assembly to the crankcase, however, don't remove the end cap assembly yet, as it is easier once the crankcase cover is separated.

8. Using a few passes in the reverse of the torque sequence loosen and remove the 14 crankcase cover-to-cylinder block bolts. Use a soft mallet to tap around the edge of the cover-to-block mating surface, then carefully separate the cover from the block.

9. Remove the lower end cap assembly. The assembly contains 2 seals, a smaller one mounted with the lips facing away from the crankshaft and a larger one with either the lips facing TOWARD the crank on 40/50 models or AWAY from the crank on 55/60 models (so pay attention to how they are facing when you remove them). The assembly is also sealed using an O-ring. Both the seals and the O-ring should be replaced to ensure an oil tight powerhead.

■ **On most Mercury inline motors you can usually remove the crankshaft and pistons as an assembly. However, as the motors get larger, it becomes easier to unbolt the connecting rods first and lift out the crankshaft separately from the pistons. Though you can probably do either we've recommended the later here (removing them separately) to better protect the crank and pistons.**

10. Take a second to matchmark the connecting rod caps to the connecting rods. They MUST not only be reinstalled on the same rods, but facing the same direction or bearing failure will occur.

11. Unbolt the connecting rod caps to free the crankshaft. Alternately and evenly loosen the 2 connecting rod cap bolts, going back and forth with no more than 1 full turn on each bolt without moving back to the other bolt. Proceed with this pattern until the connecting rod cap can be removed. MAKE sure you keep track of the rod cap and bearing cage as they must be returned to the original connecting rod and crankshaft journal if they are to be reused. It's best to label them AND to reinstall them to the connecting rod as soon as the crankshaft is out of the way.

■ **The connecting rod bolts used on these motors are both angle torqued and normally of the stretch-type meaning they are distorted slightly when tightened. For this reason they MUST be replaced with new bolts during assembly.**

12. Carefully lift the crankshaft assembly from the connecting rods and cylinder block.

13. If necessary, disassemble the crankshaft and bearings as follows:

a. Slide the upper seal off the top of the crankshaft.

b. Mercury doesn't say whether or not the top bearing is pressed in place. Generally speaking, if the bearing is pressed onto the shaft it cannot be removed without a bearing separator and SHOULD NOT be removed unless it is being replaced, as a bearing separator will normally damage the bearing during removal. However, on some models in this size/hp range Mercury has used a loose fit bearing that can be slid on or off the crank without damage.

c. Remove the retaining ring from the groove in each of the center main bearing races, then remove the races and bearing halves, but proceed slowly as there should be 28 free rollers which must be kept together as a set (and the whole set must be replaced if even one roller is lost or damaged) under each set of bearing races.

d. While wearing a pair of safety glasses, carefully remove the ring seal from the edge of the crankshaft throw for each of the center main bearings.

e. The lower bearing IS pressed onto the shaft and removal will destroy it, so ONLY remove it if necessary to replace the bearing or for access to the oil pump drive gear, as necessary. To remove the bearing first remove the snapring, then use a shop press and bearing separator to free the bearing from the crankshaft.

f. If necessary, remove the oil pump drive gear from the bottom of the crankshaft.

14. Before going any further, LABEL the pistons and all matching components (connecting rod, rod cap etc) to make sure the proper components are all removed, stored and reinstalled together. You should have started earlier when you matchmarked the connecting rod caps, but go a step further now to make sure you also mark the direction the remaining components are facing. Remember when reusing these components (caps, rods, pistons) you must make sure all components are installed in the same locations and facing the same directions as they were before they were disassembled. It is true that many of these components have factory alignment or identification marks, but it is never a bad idea to double-check them or mark them yourself before proceeding.

15. Carefully pull each piston out of the crankshaft bore for inspection and disassembly, as necessary. If it is necessary to disassemble them for further inspection and component replacement, as follows:

■ We've said it before. We know there will be some circumstances that will prompt people to reuse rings, but for the most part, if you've come THIS far, it is probably a good idea to bite the bullet and replace the rings to ensure long-lasting performance from the powerhead.

a. Use a Piston Ring Expander (such as #91-24697) to remove the rings from the pistons. The manufacturer does not recommend reusing the rings. Retain them in sets for inspection purposes (marking their original locations), but discard them before installation.

✳✳ SELOC WARNING

ALWAYS use eye protection when attempting to remove the piston pin lockrings as they are prone to launching out of position unpredictably.

b. To remove the connecting rod from each piston, first remove the lockring from both ends of the piston pin. Use an awl inserted just under the lip of the ring to gently lever it out of the piston. Keep in mind that these lockrings are often distorted slightly during removal and so they must be discarded are replaced with new lockrings during installation.

■ If necessary, a piston cradle can be fabricated from a large block of wood. The important design features of the cradle are that it is semi-circular on top to hold (or cradle, get it?) the piston and that there is a bore at the bottom center into which the wrist pin can tapped or pressed. The final important feature of the cradle is that it is sturdy enough to withstand the force of the shop press, BUT, keep in mind that the fit is loose on these pistons and significant force should not be necessary. It is NOT uncommon for someone to simply support the piston in their lap when working on these motors.

c. While carefully supporting the piston to prevent damage use a soft-faced mallet and driver (such as Piston Pin Tool #91-74607A3 or equivalent) to gently tap the pin from the bore in the piston. As the pin is removed, KEEP CLOSE TRACK of the locating/thrust washers and the loose needle bearings (there should be 29 needles per piston).

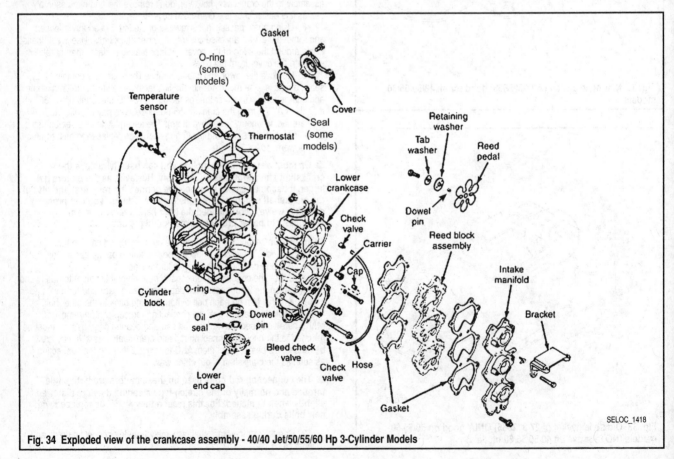

SELOC_1418

Fig. 34 Exploded view of the crankcase assembly - 40/40 Jet/50/55/60 Hp 3-Cylinder Models

■ As the wrist pin is removed from the piston you must retain the 29 loose needle bearings and the 2 locating/thrust washers from the small end of the connecting rod. If you can, stop tapping or pressing once the wrist pin JUST clears the bottom of the connecting rod bore. Then carefully withdraw the rod along with the needle bearings (place a thin bladed tool like a putty knife or your finger under the rod hole as it is pulled free of the piston bore/wrist pin). Inspect the bearings, if any of the needles are lost or worn, replace the entire bearing assembly.

    d. Repeat for the other pistons.

■ Mercury and Seloc recommend that ALL seals be replaced during a rebuilding procedure of this magnitude.

To assemble:

16. Prepare the following components for installation later, as follows:

    a. If the reed valves were removed from the plate, position the reeds over the locating pins on the plate, then place the stops or retaining washers over them and secure using the bolt and washer. Be sure to apply a light coating of Loctite® 271 or an equivalent threadlocker to the threads of the retaining bolt, then tighten to 60 inch lbs. (6.8 Nm).

    b. If the end cap seals were removed, use a suitable driver to gently press or tap the replacement seals into the end cap. Remember the smaller seal should face downward AWAY from the crankshaft, while the larger seal faces either TOWARD the crank on 40/50 models or AWAY from the crank on 55/60 models.

Lubricate the seal lips as well as the replacement housing O-ring using a light coating of clean 2-stroke oil. Also, RIGHT before installation, apply a light coating of Perfect Seal or an equivalent sealant to the end cap flange.

✱✱ **SELOC CAUTION**

Any grease used inside the powerhead for assembly MUST be gasoline soluble. The use of a non-soluble grease may cause severe powerhead damage.

17. Make sure all parts have been thoroughly cleaned and inspected. Be sure to oil all bearings and bearing surfaces lightly using a coating of fresh 2-stroke oil. Loose needle bearings, once cleaned, should be spread out on a clean worksurface, like a sheet of paper and coated lightly using Needle Bearing Assembly Lubricant.

18. If disassembled for inspection and/or component replacement, assemble each piston as follows:

    a. Apply a light coating of Needle Bearing Assembly grease (or 2-4-C with Teflon or an equivalent gasoline soluble marine grade grease) to the wrist pin bearing needles and the locating/thrust washers, then hold the lower locating washer in place while you insert the 29 needles around the inner circumference of the connecting rod wrist pin bore. Install the Piston Pin Tool #91-74607A3 or equivalent into the bore to align the needles and hold them in place during assembly.

■ The wrist pin bearing tool is, essentially, a short round pin, roughly the same outer diameter of the wrist pin, but very short, just long enough to insert it through the connecting rod wrist pin bore and install the locating washers on either side. A similarly sized dowel pin can be substituted, or an old connecting rod wrist pin from the same size piston could be used. If substituting another rod or drift pin, make sure the surface is smooth and clean/free of all corrosion. Also, make sure the outer diameter of the tool is smaller than the wrist pin and bore to prevent the possibility of binding or damage.

    b. Once all the needles are in position, inspect to make sure one is not missing. Generally speaking, if the tip of an awl can be inserted between the needles, one or more needles is/are missing.

✱✱ **SELOC CAUTION**

NEVER replace just one needle. If one or more needles are damaged or missing, replace the entire set.

    c. With the bearings and the wrist pin tool in position, install the upper locating/thrust washer. Again, a dab of grease should hold the thrust washer in position.

    d. Apply a light coating of clean engine oil or assembly lube to the wrist pin and the pin bore in the piston.

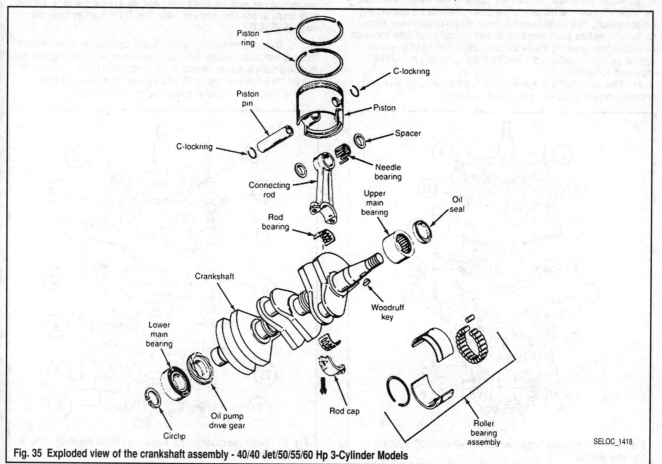

Fig. 35 Exploded view of the crankshaft assembly - 40/40 Jet/50/55/60 Hp 3-Cylinder Models

SELOC_1418

■ Remember to position all components in the same orientation from which they were removed. For instance, the domes of most pistons are marked with a stamped UP for reference as to which end faces the flywheel.

e. Support the piston (ideally in a piston cradle for ease of assembly but the wrist pin fit should be loose on both sides so this is not absolutely necessary). Insert the wrist pin into the bore and gently press it through the bore until it just starts to appear inside the piston skirt. Stop pressing, then position the connecting rod with the needle bearings, locating washers and wrist pin bearing installation tool inside the piston skirt. Continue to press the wrist pin into the position, through the thrust washers and bearings, pushing the wrist pin bearing installer through the other side and out of the piston.

f. Once the wrist pin is properly positioned and centered in the piston, install new lockrings into the grooves in both ends of the piston bore. Installation of the lockrings will be a lot easier using a lock ring installation tool (such as #91-77109A3 or equivalent) which is essentially a shouldered driver on which you can place the lockring, then press it evenly into the piston bore. However, the stock tool referenced above must be modified for use on these motors, the shaft of the tool (where the lockrings sit) must be shortened to 1.050 in. (26.67mm).

g. Use a ring expander to install the piston rings on each piston (the top ring is chromed, while the bottom should be steel), making sure to install the rings facing the right direction. The flat or RECTANGULAR side must be facing downward, while the tapered or HALF-KEYSTONE side should be facing upward. In most cases, replacement rings will be marked with a letter or dot on the side which faces up. Also, each piston ring groove should contain a dowel pin, around which the ring gap must be situated. If not the rings will likely break during installation (or if they don't by some miracle, during the first start-up). Be sure to install the tapered ring in the top groove.

■ When using the ring expander, spread each ring JUST enough to slip over the piston.

h. Repeat for the other pistons.
19. Prepare to install the pistons by lubricating the piston rings, pistons and cylinder walls lightly with oil.
20. No ring compressor is necessary, just verify that the piston is facing the proper direction (the UP mark on the dome is towards the flywheel end of the powerhead), then gently insert the piston into the cylinder bore. Repeat for the other pistons, positioning them as best you can to facilitate crankshaft installation later (generally this means at the top of their travel so you can position the bearing half and pull them up to the connecting rod AFTER the crankshaft is in place).
21. If the crankshaft was disassembled, install the oil pump drive gear and/or bearings to prepare it for installation as follows:

a. If removed, slide the oil pump gear into position on the crankshaft while aligning the keyway and key. Next oil and install the bearing using a press and suitable driver. Once the bearing is in position tight against the gear install the retaining ring.

b. Wearing a pair of safety glasses to protect your eyes, carefully install the seal rings to the shoulder of the crankshaft throws for the center main bearings.

c. Apply a light coating of Needle Bearing Assembly Lube or equivalent marine grade grease, to the crankshaft center journal in order to hold the 28 main roller bearings for EACH center main bearing. Position the roller bearings followed by the bearing race halves so that the retainer ring grooves face FORWARD toward the flywheel end of the crankshaft. Secure each pair of halves using the retainer ring, making sure the ring bridges both fracture lines on each set of bearing races.

d. If removed, oil and install the upper main bearing.
22. Place the pistons fully upward in their bores (i.e. away from the crankshaft) so as not to interfere with crankshaft installation (this way you can insert the bearing cage and pull them up to the crankshaft journal AFTER the crankshaft is seated). Carefully lower the crankshaft into position, aligning the larger holes in the center main and upper bearings with the alignment pins in the block.
23. Install each piston and connecting rod assembly to the crankshaft as follows:

a. Apply a light coating of 2-stroke engine oil to the bearing cages, then install the lower half in the connecting rod and pull the rod upward into position over the crank pin.

b. Position the other bearing half over the crank pin.

c. Clean the connecting rod bolts holes using solvent and dry with compressed air, then apply a light coating of 2-stroke engine oil to the threads of the NEW connecting rod cap bolts.

d. Position the piston/connecting rod assembly to the crankshaft (facing in the same direction as noted and marked during removal, then place the connecting rod end cap over the crank pin (also aligning the matchmarks made during removal) and start the bolts by hand.

■ These motors use fractured end caps, meaning the end cap and connecting rod were originally one piece. As a result the caps will only fit properly in one direction and when they fit the fracture lines will mate PERFECTLY.

e. Alternately and evenly tighten the end cap retaining bolts, no more than one turn before moving to the other bolt. Once they start to tighten, use multiple passes to tighten them to first to 15 inch lbs. (1.7 Nm), next to 120 inch lbs (13.6 Nm) and then finally an additional 90 degrees (1/4 turn).

f. Repeat for the other piston/connecting rod assemblies.

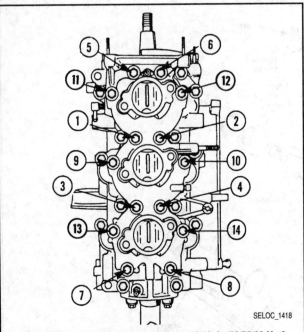

Fig. 36 Intake manifold torque sequence - 40/40 Jet/50/55/60 Hp 3-Cylinder Models

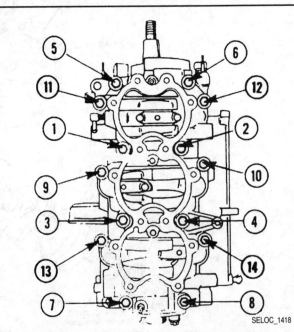

Fig. 37 Crankcase cover torque sequence - 40/40 Jet/50/55/60 Hp 3-Cylinder Models

24. Apply a light coating of clean 2-stroke engine oil to the new upper crankshaft seal lips, then slide the seal into position with the lips facing downward toward the crankshaft.

25. Make sure the bearing surfaces of the crankcase halves are clean and free of any oil or debris, then apply a continuous bead of Loctite Master Gasket on the mating surfaces of the crankcase cover and cylinder block, JUST INBOARD of the bolts holes on either side of the crankshaft bore. Make sure the sealer is extended right to the edge at each center main journal to prevent blow-by between the cylinders.

■ Mercury is serious about the continuous beads of Loctite. They often go so far as to instruct their technicians that IF there is a void while placing the bead to either wipe away the whole bead and start over OR to apply an additional bead, parallel to the void and overlapping the previously applied bead.

26. Apply a light coating of Perfect Seal to the crankcase mating flange of the assembled end cap, then position it over the bottom of the crank.

27. Assemble the crankcase cover to the block, then install and finger-tighten the 14 retaining bolts (note there are 2 sizes) followed by the 3 end cap cover retaining bolts.

28. Using at least 3 passes of the torque sequence, tighten the crankcase cover bolts to 18 ft. lbs. (24 Nm).

29. Tighten the end cap bolts to 18 ft. lbs. (24 Nm) using at least 3 passes of the following sequence, starting with the bolt at the upper corner of the end cap (toward the edge of the powerhead in the approximately 10 o'clock position when looking at the bottom of the assembly and working clockwise to the inboard bolt at about 3'clock position and finally back out to the other corner bolt at 7 o'clock position.

30. Rotate the crankshaft a few times to make sure it moves freely and there is NO binding.

31. If removed, install the check valve/holders to the crankcase cover.

32. Install the reed valve block plate and a new gasket, followed by the intake manifold with a new gasket, then secure using the retaining bolts. Tighten the bolts using at least 3 passes of the torque sequence to 18 ft. lbs. (24 Nm).

33. Install the thermostat valve assembly using a new thermostat housing gasket to the side of the powerhead. Tighten the bolts to 18 ft. lbs. (24 Nm).

34. If removed, route and reconnect the bleed hose(s) as noted during removal.

35. The following fuel and electric components CAN be installed to the powerhead before the powerhead is mounted to the intermediate housing. For details, where applicable, please refer to the appropriate sections of this guide.

- Control cable anchor bracket
- Shift cable latch assembly
- Oil pump
- Fuel enrichment valve
- Carburetor and linkage
- Air intake silencer
- Fuel pump
- Voltage regulator/rectifier

- Starter solenoid
- Electric starter motor
- Trigger coil
- Stator coil
- Flywheel
- Ignition Module (CDM)

36. Install the Powerhead, as detailed in this section.

37. Once the powerhead is fully assembled, if cylinder block components were replaced, especially the pistons and/or rings, be sure to operate the engine as directed for new component break-in. For details, please refer to Powerhead Break-In, in this section.

CLEANING & INSPECTION

◆ See Figures 34 and 35

Cleaning and inspecting the components is virtually the same for any two-stroke outboard and varies mostly by specifications (which are listed in the Engine Specifications charts) or by component type. A section detailing the proper procedures, sorted mostly by component, can be found under Powerhead Refinishing.

In addition to the major components covered under Powerhead Refinishing, there are two 3/16 in. (7.76mm) diameter brass casing check valves mounted in the intake manifold. The check valve uses a nylon ball which can be damaged by hot blow-by in the crankcase. To inspect each check valve peer through it, looking for light. If you can see light through the whole the nylon ball is bad, it has probably melted, and must be replaced. If you cannot see light through the whole, insert a thin piece of wire into the valve to verify that you can feel movement. If the ball will not move the valve must be replaced.

75/65 Jet/90 Hp and 80 Jet/115/125 Hp Powerheads

REMOVAL & INSTALLATION

◆ See Figures 27 and 38 thru 41

These powerheads are large enough that the only practical way to lift them safely from the intermediate housing (and perhaps even more importantly to install them on the housing mating with the driveshaft splines) is to use an engine hoist or other lifting device for safety and to prevent damage. When using a hoist you don't HAVE to completely strip the powerhead of all electrical and fuel components, simply because there is less of chance of them becoming damaged. However, it still may be a good idea.

Consider why the powerhead is being removed. If you are planning on disassembling the powerhead for inspection or overhaul, then you'll have to remove the fuel and electrical components anyway. Removing them before powerhead removal may make the rest of powerhead removal a tad easier. Either way you decide, we'll list the minimum amount of components you must remove before lifting the powerhead free.

Fig. 38 Remove the top cowling. . .

Fig. 39 . . .and the lower STARBOARD and. . .

Fig. 40 PORT cowls for access

1. Remove the cowling from the powerhead for access.
2. For safety, disconnect the battery cables.
3. Tag and disconnect the bullet connectors for the cowl mounted trim switch harness.
4. Disconnect the tell-tale water discharge hose from the fitting on the exhaust cover.
5. Remove the 2 bolts securing the aft end of the lower cowling (on the inside of the cowling just below the latch assembly), then move to the front, exterior of the cowling on the port side and unthread the bolt securing the forward portion of the lower cowling. Now moving inside the cowling again, loosen the PORT side nut securing the front latch to the bottom cowl, then remove the STARBOARD and PORT bottom cowls for additional access.
6. Locate the shift arm linkage on the PORT side of the powerhead, just a little below and forward from the exhaust cover (right below the throttle linkage). Disconnect the shift arm linkage at the stud (it is usually secured with a nut, a couple of washers and a bushing).
7. On the Starboard side of the motor, right below the electric starter, locate, tag and disconnect the wiring from the Up and Down power trim relays.
8. Remove the electric starter.
9. Tag and disconnect the wiring running to the electrical plate, then unbolt and remove the plate assembly complete with all plate mounted ignition and electrical components.
10. Tag and disconnect the fuel supply hose.
11. Install the lifting eye (#91-90455 or equivalent) into the flywheel a minimum of 5 turns.
12. Remove the 8 fasteners (3 on each side and 2 on the backside), usually nuts and washers on studs) securing the powerhead to the driveshaft housing.

■ **At this point double check for any cables, hoses, wires or linkages that will interfere with the removal of the powerhead.**

13. Using a suitable engine hoist, carefully lift the powerhead straight up and off the driveshaft/driveshaft housing while breaking the gasket seal.
14. Move the powerhead to a clean work area and placed in a suitable holding fixture.

To install:
15. Thoroughly clean all old gasket material from the driveshaft housing and powerhead mating surfaces.
16. Place a new gasket over the locating pins on the driveshaft housing.
17. Lubricate the SIDES of the splines on the driveshaft using 2-4-C with Teflon or an equivalent marine grade grease. Wipe off any excess lubricant

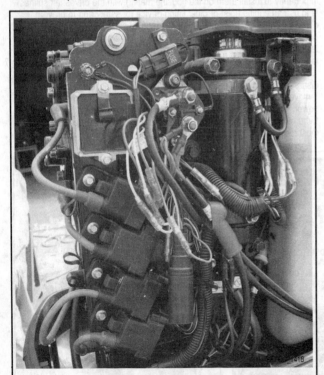

Fig. 41 Tag and disconnect all wiring on the electrical component plate

from the top of the driveshaft as any lubricant left on top of the driveshaft could prevent it from seating in the crankshaft.
18. Double-check that the tip-seal in the crankshaft seal carrier is in place.
19. SLOWLY lower the powerhead onto the driveshaft housing. You may have to rotate the crankshaft (or conversely the propshaft to rotate the driveshaft if the motor is in gear) in order to index the crankshaft and driveshaft splines. Continue lowering the powerhead until is fully seated on the housing and alignment pins.
20. Clean the threads of the powerhead mounting fasteners (studs, nuts, bolts) using Loctite® 7649 Primer Dry, then apply a light coating of Loctite® 271 or equivalent threadlocking compound to the threads. Install and tighten the powerhead mounting bolts evenly to 45 ft. lbs. (61 Nm) using a couple of passes of a pattern that starts at the center and works outward. Remove the lifting eye and lifting device.
21. Reconnect the fuel supply hose.
22. Install the electrical component plate along with the ignition and electrical components that are attached. Reconnect the wiring as tagged during removal.
23. Install the electric starter assembly.
24. Reconnect the wiring for the UP and DOWN trim relays, found just underneath the starter.
25. Reconnect the shift linkage on the stud, secure using the bushing, washers and retaining nut.
26. Reconnect the bullet connectors to the cowl mounted trim switch.
27. Reconnect the telltale hose to the fitting on the exhaust cover.
28. Align the Starboard and Port lower cowling plates, then secure using them using the fasteners. First install the 2 bolts threaded horizontally just under the rear latch (inside the cover) and tighten them to 65 inch lbs. (7.5 Nm). Next, move to the bolt on the outside front of the cowling, also tightening it to 65 inch lbs. (7.5 Nm). Lastly, inside the front of the cover, securely tighten the PORT nut that secures the front latch to the bottom cowl.

■ **Be sure to run the motor without the cowling installed in order to check for potential fuel or oil leaks. Remedy any leaks before proceeding.**

29. Reconnect the battery cables.
30. Hook up a source of cooling water and test run the motor.
31. Install the cowling to the powerhead.

■ **If the powerhead was rebuilt or replaced with a remanufactured unit, don't forget to follow the proper Break-In procedures.**

DISSASSEMBLY & ASSEMBLY

◆ **See Figures 42 thru 53**

1. Loosen the 6 (3-cylinder) or 8 (4-cylinder) retaining screws, then remove the air box cover from the air intake silencer assembly. The screws are normally Philips head and are spread around the top and sides of the cover perimeter.
2. Remove the air intake silencer assembly after removing the following air box screws:
• On the starboard side of the engine, at the base of the air box, remove the lower oil tank support screw.
• At the top of the air box, on the starboard side, remove the upper oil tank support screw (threaded vertically downward).
• Some early-models may have a voltage regulator bolted to the top port side of the cover, if so, disconnect the wiring and remove the regulator retaining bolts, then remove the regulator.
• Remove the 4 (3-cyl) or 6 (4-cyl) nuts and the 2 screws from the interior of the air box (on either side of each carburetor throttle opening).
• Once the fasteners have all been removed, carefully pull the air box off the carburetor studs.
3. Tag and disconnect the oil lines, then remove the oil reservoir.
4. Disconnect the oil pump linkage.
5. Tag and disconnect the main fuel hose and the fuel enrichment hose, then lift the carburetors off the powerhead as an assembled unit (remember they were positioned on the studs and held in place by the air box and retaining nuts removed earlier.

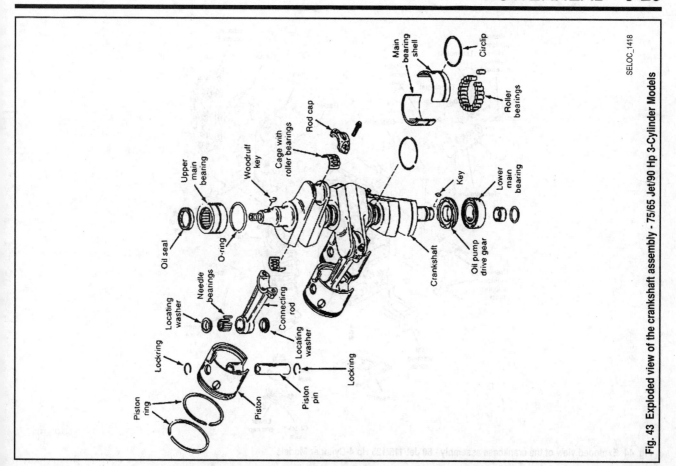

SELOC_1418

Fig. 43 Exploded view of the crankshaft assembly - 75/65 Jet/90 Hp 3-Cylinder Models

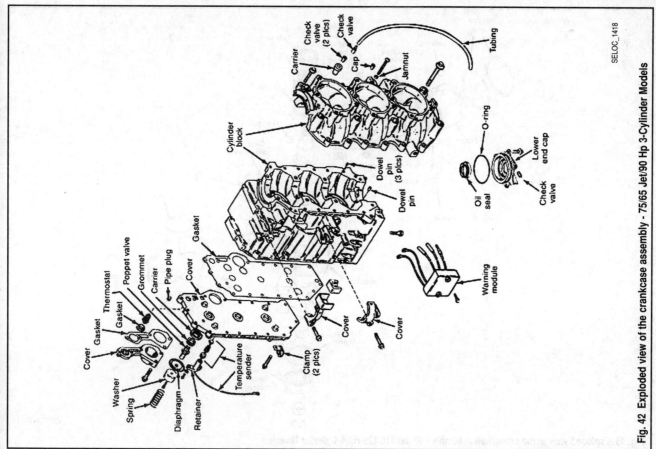

SELOC_1418

Fig. 42 Exploded view of the crankcase assembly - 75/65 Jet/90 Hp 3-Cylinder Models

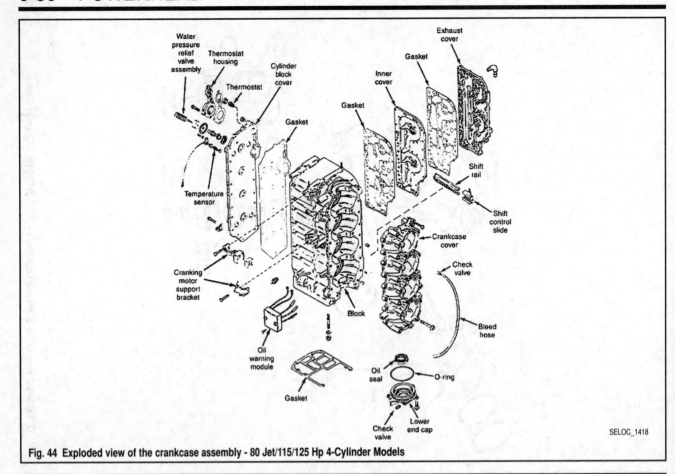

Fig. 44 Exploded view of the crankcase assembly - 80 Jet/115/125 Hp 4-Cylinder Models

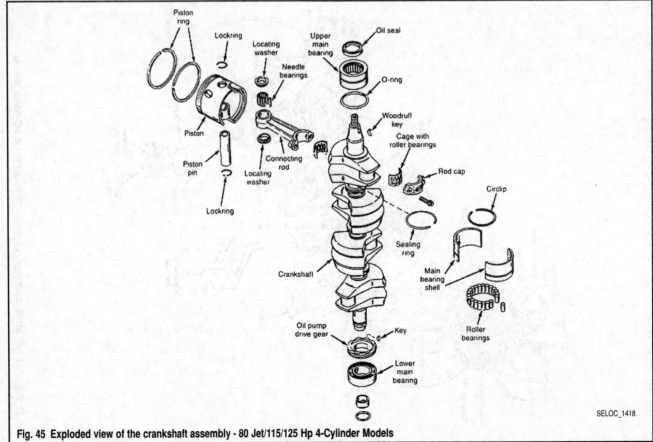

Fig. 45 Exploded view of the crankshaft assembly - 80 Jet/115/125 Hp 4-Cylinder Models

6. Tag and disconnect all wiring for the ignition/electrical component plate, then remove the plate and all attached components from the side of the powerhead.

7. As equipped, tag and disconnect the wiring, then remove the warning/rev limiter module from the side of the powerhead (normally just below the ignition/electrical component plate).

8. Using a few passes in the reverse of the torque sequence, remove the 15 (3-cyl) or 18 (4-cyl) bolts securing the intake manifolds/reed block assemblies to the powerhead, then remove the assemblies and discard the old gaskets. The 3-cylinder motors utilize an individual manifold per cylinder, but all should be unbolted and removed as an assembly as they are connected by hoses. The 4-cylinder motors use 2 manifolds (one per 2 cylinders) each of which also contains a cover and gasket on the front of the manifold assembly. In all cases a single reed valve block can be found for each cylinder, which is also secured by the manifold retaining bolts sealed on both sides using gaskets (sealed to the crankcase cover on one side and the intake manifold on the other). Discard all of the old gaskets.

■ **When you remove the intake manifolds the reed blocks will normally be left behind, held by the gasket seal. Use a small prytool to free up each of the reed blocks and remove them from the crankcase cover.**

9. Remove the Electric Starter Motor, as detailed in the Ignition and Electrical section.

10. Remove the Flywheel, followed by the Stator and Trigger Coils. Refer to the Ignition and Electrical section for details, as necessary.

11. Remove the Fuel Pump, as detailed in the Fuel System section (though at this point, it should only involve removing the 2 mounting bolts, which is simple if you don't confuse them with the pump cover assembly bolts).

12. Remove the Oil Pump, as detailed in the Lubrication and Cooling System section (though, like the fuel pump, at this point all the should really be necessary is to remove the 2 retaining bolts.

13. Remove the oil pump shaft and housing.

14. If necessary, tag and disconnect the bleed lines. On 4-cylinder models, at the very least, it is necessary to disconnect the accelerator pump hoses from the #3 and #4 cylinder check valve fittings (on the lower side of the block).

15. For all models it is a good idea to remove the check valves from the block and check their condition as follows:

   a. Spray a small amount of WD-40 into the barded end of the check valve, then blow out excess with high pressure air (wearing safety goggles, duh).

   b. Apply suction to the barbed end using a small hand-held air vacuum pump. Continue to apply suction until the valve opens at about 22-28 in. hg. (74.5-94.8kPa) of vacuum.

   c. Within 30 seconds after stopping suction vacuum should not fall below 5 in. hg. (16.9kPa).

✳✳ SELOC CAUTION

The accelerator check valves should NOT be installed in the block while honing operations are being performed. Also, they should NOT be reinstalled until ALL block cleaning procedures have been completed. Failure to heed this will likely lead to check valve damage.

16. Using a few passes in the reverse of the torque sequence loosen and remove the cylinder block cover-to-cylinder block bolts. The bolts are also used to secure the thermostat housing, so once removed, carefully break the thermostat housing gasket seal and remove the t-stat components. Next, use a soft mallet to tap around the edge of the cover-to-block mating surface, then carefully separate the cover from the block. If necessary, use 2 small prytools in the cast pry-slots, towards the bottom of the cover to carefully break the gasket seal and remove it from the cylinder block.

17. Using a few passes in the reverse of the torque sequence loosen and remove the exhaust cover bolts. Next, use a soft mallet to tap around the edge of the cover-to-block mating surface, then carefully separate the cover from the block. If necessary, use a small prytools at the cast pry-slot(s) at the bottom corner(s) of the cover to carefully break the gasket seal and remove it from the cylinder block.

18. Alternately and evenly loosen and remove the 3 bolts securing the lower end cap to the crankcase, however, don't remove the end cap assembly yet, as it is easier once the crankcase cover is separated.

19. Using a few passes in the reverse of the 2 torque sequences (starting with the smaller outer bolts and then working on the larger inner bolts) loosen and remove the 2 sets of crankcase cover-to-cylinder block

bolts. Use a soft mallet to tap around the edge of the cover-to-block mating surface, then carefully separate the cover from the block.

20. Remove the lower end cap assembly. The assembly uses a single seal which normally faces upward toward the crankshaft (but double-check this before you remove the seal from the housing). The assembly is also sealed using an O-ring. Both the seal and the O-ring should be replaced to ensure an oil tight powerhead.

■ **On most Mercury inline motors you can usually remove the crankshaft and pistons as an assembly. However, on these larger motors, it becomes easier to unbolt the connecting rods first and lift out the crankshaft separately from the pistons. Though you can probably do either we've recommended the later here (removing them separately) to better protect the crank and pistons.**

21. Take a second to matchmark the connecting rod caps to the connecting rods. They MUST not only be reinstalled on the same rods, but facing the same direction or bearing failure will occur.

■ **The connecting rod bolt caps can be loosened or installed using a 3/8 in. 12-point socket.**

22. Unbolt the connecting rod caps to free the crankshaft. Alternately and evenly loosen the 2 connecting rod cap bolts, going back and forth with no more than 1 full turn on each bolt without moving back to the other bolt. Proceed with this pattern until the connecting rod cap can be removed. MAKE sure you keep track of the rod cap and bearing cage as they must be returned to the original connecting rod and crankshaft journal if they are to be reused. It's best to label them AND to reinstall them to the connecting rod as soon as the crankshaft is out of the way.

23. Carefully lift the crankshaft assembly from the connecting rods and cylinder block.

24. If necessary, disassemble the crankshaft and bearings as follows:

   a. Slide the upper main bearing, seal and O-ring assembly off the top of the crankshaft.

   b. Slide the lower wear seal and O-ring off the bottom of the crankshaft. Pop out the old O-ring using an awl or pick and discard.

   c. Remove the retaining ring from the groove in each of the center main bearing races, then remove the races and bearing halves, but proceed slowly as there should be 32 free rollers which must be kept together as a set (and the whole set must be replaced if even one roller is lost or damaged) under each set of bearing races.

   d. While wearing a pair of safety glasses, carefully remove the 2 sealing rings from each center main journal.

   e. The lower bearing IS pressed onto the shaft and removal will destroy it, so ONLY remove it if necessary to replace the bearing or for access to the oil pump drive gear, as necessary. To remove the bearing use a shop press and bearing separator to free the bearing from the crankshaft.

   f. If necessary, remove the oil pump drive gear from the bottom of the crankshaft.

25. Before going any further, LABEL the pistons and all matching components (connecting rod, rod cap etc) to make sure the proper components are all removed, stored and reinstalled together. You should have started earlier when you matchmarked the connecting rod caps, but go a step further now to make sure you also mark the direction the remaining components are facing. Remember when reusing these components (caps, rods, pistons) you must make sure all components are installed in the same locations and facing the same directions as they were before they were disassembled. It is true that many of these components have factory alignment or identification marks, but it is never a bad idea to double-check them or mark them yourself before proceeding.

26. Carefully pull each piston out of the crankshaft bore for inspection and disassembly, as necessary. If it is necessary to disassemble them for further inspection and component replacement, as follows:

■ **We've said it before. We know there will be some circumstances that will prompt people to reuse rings, but for the most part, if you've come THIS far, it is probably a good idea to bite the bullet and replace the rings to ensure long-lasting performance from the powerhead.**

   a. Use a Piston Ring Expander (such as #91-24697) to remove the rings from the pistons. The manufacturer does not recommend reusing the rings. Retain them in sets for inspection purposes (marking their original locations), but discard them before installation.

✳✳ SELOC WARNING

ALWAYS use eye protection when attempting to remove the piston pin lockrings as they are prone to launching out of position unpredictably.

b. To remove the connecting rod from each piston, first remove the lockring from both ends of the piston pin. Use an awl inserted just under the lip of the ring to gently lever it out of the piston. Keep in mind that these lockrings are often distorted slightly during removal and so they must be discarded are replaced with new lockrings during installation.

■ If necessary, a piston cradle can be fabricated from a large block of wood. The important design features of the cradle are that it is semi-circular on top to hold (or cradle, get it?) the piston and that there is a bore at the bottom center into which the wrist pin can tapped or pressed. The final important feature of the cradle is that it is sturdy enough to withstand the force of the shop press, BUT, keep in mind that the fit is loose on these pistons and significant force should not be necessary. It is NOT uncommon for someone to simply support the piston in their lap when working on these motors.

c. While carefully supporting the piston to prevent damage use a soft-faced mallet and driver (such as Piston Pin Tool #91-74607A3 or equivalent) to gently tap the pin from the bore in the piston. As the pin is removed, KEEP CLOSE TRACK of the locating/thrust washers and the loose needle bearings (there should be 29 needles per piston).

■ As the wrist pin is removed from the piston you must retain the 29 loose needle bearings and the 2 locating/thrust washers from the small end of the connecting rod. If you can, stop tapping or pressing once the wrist pin JUST clears the bottom of the connecting rod bore. Then carefully withdraw the rod along with the needle bearings (place a thin bladed tool like a putty knife or your finger under the rod hole as it is pulled free of the piston bore/wrist pin). Inspect the bearings, if any of the needles are lost or worn, replace the entire bearing assembly.

d. Repeat for the other pistons.

■ Mercury and Seloc recommend that ALL seals be replaced during a rebuilding procedure of this magnitude.

To assemble:
27. Prepare the following components for installation later, as follows:
a. If the reed valves were removed from the plate, position the reeds over the locating pins on the plate, then place the retaining washer and a new tab washer over them and secure using the bolt. Be sure to apply a light coating of Loctite® 271 or an equivalent threadlocker to the threads of the retaining bolt, then tighten to 80 inch lbs. (9 Nm).

✳✳ SELOC CAUTION

Under NO circumstances should the reed bolt torque exceed 100 inch lbs. (11.5 Nm).

b. If the lower end cap seal was removed, use a suitable driver to gently press or tap the replacement seal into the end cap. Remember the seal should face upward (TOWARD the crank). Before installation be sure to apply a light coating of Loctite® 271 or equivalent threadlocking compound on the outer face of the new seal, then after installation wipe away any traces of excess Loctite. Once installed lubricate the seal lips and the new housing O-ring with a light coat of 2-4-C with Teflon or an equivalent marine grade grease. Also, RIGHT before installation, apply a light coating of Perfect Seal or an equivalent sealant to the end cap flange.

✳✳ SELOC CAUTION

Any grease used inside the powerhead for assembly MUST be gasoline soluble. The use of a non-soluble grease may cause severe powerhead damage.

28. Make sure all parts have been thoroughly cleaned and inspected. Be sure to oil all bearings and bearing surfaces lightly using a coating of fresh 2-stroke oil. Loose needle bearings, once cleaned, should be spread out on a clean worksurface, like a sheet of paper and coated lightly using Needle Bearing Assembly Lubricant.
29. These motors use 2 (3-cyl) or 3 (4-cyl) check valves installed in the crankcase cover along the intake manifold/reed valve block mating surfaces (at the center of the flange between cylinders). The assembly consists of a valve and a carrier. If the valve is stuck or burnt it can be pushed out of the carrier, a replacement inserted and the assembly installed. Make sure the single hole of the valve is installed facing the tapered end of the carrier. When the carrier is positioned in the crankcase cover align the tab on the carrier with the slot in the cover.
30. If disassembled for inspection and/or component replacement, assemble each piston as follows:

a. Apply a light coating of Needle Bearing Assembly grease (or 2-4-C with Teflon or an equivalent gasoline soluble marine grade grease) to the wrist pin bearing needles and the locating/thrust washers, then hold the lower locating washer in place while you insert the 29 needles around the inner circumference of the connecting rod wrist pin bore. Install the Piston Pin Tool #91-74607A3 or equivalent into the bore to align the needles and hold them in place during assembly.

■ The wrist pin bearing tool is, essentially, a short round pin, roughly the same outer diameter of the wrist pin, but very short, just long enough to insert it through the connecting rod wrist pin bore and install the locating washers on either side. A similarly sized dowel pin can be substituted, or an old connecting rod wrist pin from the same size piston could be used. If substituting another rod or drift pin, make sure the surface is smooth and clean/free of all corrosion. Also, make sure the outer diameter of the tool is smaller than the wrist pin and bore to prevent the possibility of binding or damage.

b. Once all the needles are in position, inspect to make sure one is not missing. Generally speaking, if the tip of an awl can be inserted between the needles, one or more needles is/are missing.

✳✳ SELOC CAUTION

NEVER replace just one needle. If one or more needles are damaged or missing, replace the entire set.

c. With the bearings and the wrist pin tool in position, install the upper locating/thrust washer. Again, a dab of grease should hold the thrust washer in position.
d. Apply a light coating of clean engine oil or assembly lube to the wrist pin and the pin bore in the piston.

■ Remember to position all components in the same orientation from which they were removed. For instance, the domes of most pistons are marked with a stamped UP for reference as to which end faces the flywheel.

e. Support the piston (ideally in a piston cradle for ease of assembly but the wrist pin fit should be loose on both sides so this is not absolutely necessary). Insert the wrist pin into the bore and gently press it through the bore until it just starts to appear inside the piston skirt. Stop pressing, then position the connecting rod with the needle bearings, locating washers and wrist pin bearing installation tool inside the piston skirt. Continue to press the wrist pin into the position, through the thrust washers and bearings, pushing the wrist pin bearing installer through the other side and out of the piston.
f. Once the wrist pin is properly positioned and centered in the piston, install new lockrings into the grooves in both ends of the piston bore. Installation of the lockrings will be a lot easier using a lock ring installation tool (such as #91-77109A2 or equivalent) which is essentially a shouldered driver on which you can place the lockring, then press it evenly into the piston bore.
g. Use a ring expander to install the piston rings on each, making sure to install the rings facing the right direction. The flat or RECTANGULAR side must be facing downward, while the tapered or HALF-KEYSTONE side should be facing upward. In most cases, replacement rings will be marked with a letter or dot on the side which faces up. Also, each piston ring groove should contain a dowel pin, around which the ring gap must be situated. If not the rings will likely break during installation (or if by they don't by some miracle, during the first start-up). Be sure to install the tapered ring in the top groove.

■ When using the ring expander, spread each ring JUST enough to slip over the piston.

h. Repeat for the other pistons.
31. Prepare to install the pistons by lubricating the piston rings, pistons and cylinder walls lightly with oil.
32. No ring compressor is necessary, just verify that the piston is facing the proper direction (the UP mark on the dome is towards the flywheel end of the powerhead), then gently insert the piston into the cylinder bore. Repeat for the other pistons, positioning them as best you can to facilitate crankshaft installation later (generally this means at the top of their travel so you can position the bearing half and pull them up to the connecting rod AFTER the crankshaft is in place).

■ After each piston is installed, check the ring tension through the exhaust port using a screwdriver. Press gently on each ring making sure it springs back outward. If it does not, the ring was likely broken during installation and the piston must be removed for cleaning, inspection and new rings.

33. If the crankshaft was disassembled, install the oil pump drive gear and/or bearings to prepare it for installation as follows:

a. If removed, slide the oil pump gear into position on the crankshaft while aligning the keyway and key. Next oil and install the bearing using a press and suitable driver. Once the bearing is in position tight against the gear install the wear sleeve with its new O-ring. To install the wear sleeve first apply a light coating of Loctite® 271 or equivalent to the crankshaft, then use a block of wood to tap the wear sleeve squarely onto the crankshaft until is seats lightly.

✳✳ SELOC CAUTION

The wear sleeve is a lightweight 31-gauge stainless steel. Be careful not to collapse the end contour and distort the shape otherwise it will not mate/seal properly with the lower end cap and seal assembly.

b. Carefully install the seal rings to the center main bearing journals. There should be 2 rings per journal on these models. Be sure to install the 2 rings into the slots with their gaps positioned 180 degrees apart.

c. Apply a light coating of Needle Bearing Assembly Lube, 2-4-C with Teflon or equivalent gasoline soluble marine grade grease, to the crankshaft center journal in order to hold the 32 main roller bearings for EACH center main bearing. Position the roller bearings followed by the bearing race halves so that the holes in the races are positioned toward the LOWER gear end of the crankshaft. Secure each pair of halves using the retainer ring.

d. If removed, oil and install the upper main bearing, seal and O-ring assembly. The seal should be positioned with the lips facing downward toward the crankshaft and the holes in the bearing race should also be toward the lower end of the crankshaft.

■ **Mercury recommends placing all of the pistons fully downward in their bores (upward toward you and the crank if the block is facing downward) with all of the connecting rods facing toward one side of the block. This is fine if you've got a helping hand to reposition them as necessary as the crank is placed in position. Otherwise, use our method as noted below.**

34. Place the pistons fully upward in their bores (i.e. away from the crankshaft) so as not to interfere with crankshaft installation (this way you can insert the bearing cage and pull them up to the crankshaft journal AFTER the crankshaft is seated). Carefully lower the crankshaft into position, aligning the holes in the center main and upper bearings with the alignment pins in the block.

35. Install each piston and connecting rod assembly to the crankshaft as follows:

a. Apply a light coating of 2-stroke engine oil to the bearing cages, then install the lower half in the connecting rod and pull the rod upward into position over the crank pin.

b. Position the other bearing half over the crank pin.

c. Clean the connecting rod bolts holes using solvent and dry with compressed air, then apply a light coating of 2-stroke engine oil to the threads of the connecting rod cap bolts.

d. Position the piston/connecting rod assembly to the crankshaft (facing in the same direction as noted and marked during removal, then place the connecting rod end cap over the crank pin (also aligning the matchmarks made during removal) and start the bolts by hand.

■ **These motors use fractured end caps, meaning the end cap and connecting rod were originally one piece. As a result the caps will only fit properly in one direction and when they fit the fracture lines will mate PERFECTLY.**

e. Alternately and evenly tighten the end cap retaining bolts, no more than one turn before moving to the other bolt. Once they start to tighten, use multiple passes to tighten them to first to 15 inch lbs. (1.7 Nm), next to 30 ft. lbs. (40.5 Nm) and then finally an additional 90 degrees (1/4 turn).

f. Repeat for the other piston/connecting rod assemblies.

36. Apply a light coating of Perfect Seal to the crankcase mating flange of the assembled lower end cap, then position it over the bottom of the crank. Install 2 end cap bolts finger-tight to hold it in place as the crankcase cover is installed.

37. Make sure the bearing surfaces of the crankcase halves are clean and free of any oil or debris, then apply a continuous bead of Loctite Master Gasket on the mating surfaces of the crankcase cover and cylinder block, JUST INBOARD of the bolts holes on either side of the crankshaft bore. Make sure the sealer is extended right to the edge at each center main journal to prevent blow-by between the cylinders.

■ **Mercury is serious about the continuous beads of Loctite. They often go so far as to instruct their technicians that IF there is a void while placing the bead to either wipe away the whole bead and start over OR to apply an additional bead, parallel to the void and overlapping the previously applied bead.**

38. Assemble the crankcase cover to the block, then install and finger-tighten the 2 sets of retaining bolts (note there are 2 sizes, larger ones are installed inboard) followed by the last of the lower end cap cover retaining bolts.

39. Using at least 3 passes of the 2 crankcase cover torque sequences (starting with the larger, inner bolts, then moving to the smaller outer bolts), tighten the crankcase cover bolts. Tighten the larger bolts to 25 ft. lbs. (34 Nm) and the smaller bolts to 18 ft. lbs. (24.5 Nm).

40. Tighten the lower end cap bolts using multiple passes in a clockwise pattern around the housing to 18 ft. lbs. (24.5 Nm).

41. Rotate the crankshaft a few times to make sure it moves freely and there is NO binding.

■ **Major cover replacement gaskets from Mercury are normally impregnated with a coating that causes the gasket to retain torque. It is NOT necessary to use sealant on these style gaskets or Loctite on the bolts.**

42. Install the cylinder head cover along with a new gasket, then install and finger-tighten all of the retaining bolts, EXCEPT those used for the thermostat housing.

■ **Remember to reposition the clips under bolts number 2 and 10 in the torque sequence on 3-cylinder powerheads. Keep the clips positioned parallel to the edge of the cover while the bolts are tightened.**

43. Install the thermostat and poppet assembly to the cylinder head cover, then depress the cover over the assemblies and install/finger-tighten the retaining bolts.

44. Tighten the cylinder head cover bolts using at least 3 passes of the torque sequence to 18 ft. lbs. (24 Nm) for the normal cover bolts and to 25 ft. lbs. (34 Nm) for the bolts that also secure the thermostat housing.

45. Install the exhaust plate (with a new gasket) followed by the exhaust cover (with a new gasket) to the side of the cylinder block, then secure using the retaining bolts. Tighten the bolts using at least 3 passes of the torque sequence to 18 ft. lbs. (24 Nm).

46. Install the oil injection pump shaft and housing, but be sure to check/replace the housing O-rings, as necessary.

✳✳ SELOC CAUTION

MAKE SURE that the oil pump drove gears engage properly, as failure to do this could result in TOTAL POWERHEAD FAILURE. To be sure, once the oil pump shaft and housing are installed, rotate the crankshaft and watch to make sure the oil pump shaft turns!

47. Install the oil injection pump and tighten the retaining bolts to 60 inch lbs. (7 Nm).

48. Install the fuel pump and tighten the retaining bolts to 40 inch lbs. (4.5 Nm).

49. Reconnect the fuel/oil lines and secure using new wire ties.

50. Position the reed block and intake manifold assemblies to the intake manifold using new gaskets, then secure using the retaining bolts. Tighten the bolts using at least 3 passes of the torque sequences to 18 ft. lbs. (24 Nm).

51. On 4-cylinder models, install the accelerator pump check valves and hoses to the side of the powerhead.

52. If equipped, install the accelerator pump to the side end of the powerhead and tighten the bolts to 130 inch lbs. (14.5 Nm).

53. Reconnect the bleed line hoses as tagged and noted during removal.

54. As equipped, install the warning/rev limiter module to the side of the powerhead and reconnect the wiring as tagged during removal.

55. Install the ignition/electrical component plate, along with all attached components, to the side of the powerhead, then reconnect the wiring as tagged during removal.

56. Install the Trigger Coil, Stator Coil and Flywheel, as detailed in the Ignition and Electrical section.

57. Install the Electric Starter Motor, as detailed in the Ignition and Electrical section.

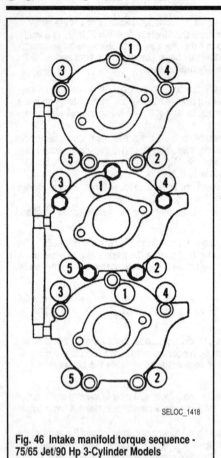

Fig. 46 Intake manifold torque sequence - 75/65 Jet/90 Hp 3-Cylinder Models

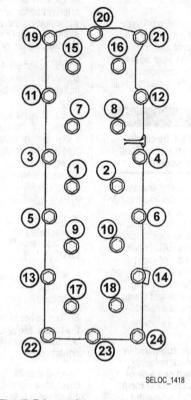

Fig. 47 Exhaust plate cover torque sequence - 75/65 Jet/90 Hp 3-Cylinder Models

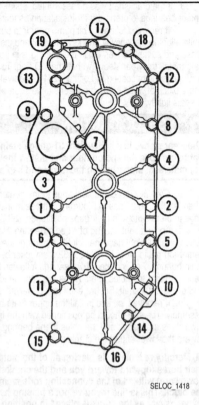

Fig. 48 Cylinder block cover torque sequence - 75/65 Jet/90 Hp 3-Cylinder Models

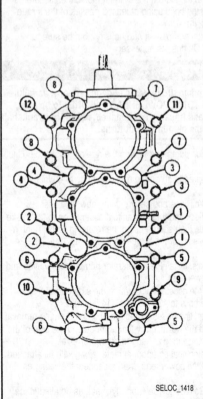

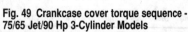

Fig. 49 Crankcase cover torque sequence - 75/65 Jet/90 Hp 3-Cylinder Models

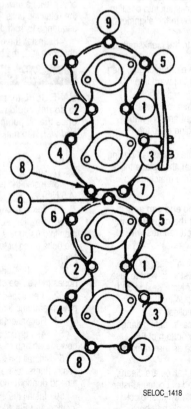

Fig. 50 Intake manifold torque sequence - 80 Jet/115/125 Hp 4-Cylinder Models

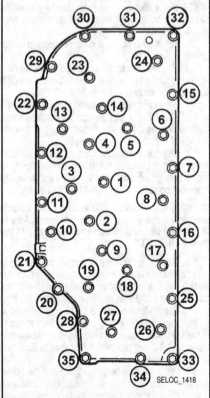

Fig. 51 Exhaust plate cover torque sequence - 80 Jet/115/125 Hp 4-Cylinder Models

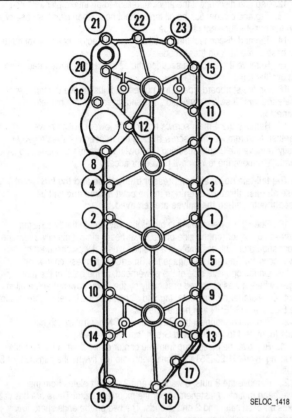

Fig. 52 Cylinder block cover torque sequence - 80 Jet/115/125 Hp 4-Cylinder Models

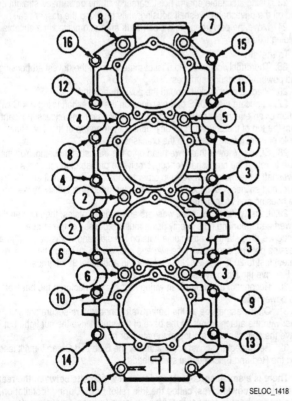

Fig. 53 Crankcase cover torque sequence - 80 Jet/115/125 Hp 4-Cylinder Models

58. Position new carburetor gaskets over the carburetor mounting studs, then install the carburetor assembly. Be sure to reconnect the manifold hose as tagged during removal. Connect the input fuel hose to the top carburetor, the input fuel hose to the fuel enrichment valve AND the input hose to the accelerator pump (if equipped). Secure the hoses with new wire ties. Reconnect the throttle linkage.

59. Install the oil tank assembly.

60. Install the air box/intake silencer assembly and secure using the retaining bolts/nuts. Tighten the nuts and screws on the front center of the cover (which also secure the carburetors) to 45 ft. lbs. (61 Nm).

61. Install the air box cover and secure using the retaining screws.

62. Install the Powerhead, as detailed in this section.

63. Once the powerhead is fully assembled, if cylinder block components were replaced, especially the pistons and/or rings, be sure to operate the engine as directed for new component break-in. For details, please refer to Powerhead Break-In, in this section.

CLEANING & INSPECTION

◆ **See Figures 42 thru 45**

Cleaning and inspecting the components is virtually the same for any two-stroke outboard and varies mostly by specifications (which are listed in the Engine Specifications charts) or by component type. A section detailing the proper procedures, sorted mostly by component, can be found under Powerhead Refinishing.

'75/90/115 Hp OptiMax Powerheads

REMOVAL & INSTALLATION

◆ **See Figures 27 and 54 thru 56**

These powerheads are large (and expensive) enough that the only practical way to lift them safely from the intermediate housing (and perhaps even more importantly to install them on the housing mating with the driveshaft splines) is to use an engine hoist or other lifting device for safety and to prevent damage. When using a hoist you don't HAVE to completely strip the powerhead of all electrical and fuel components, simply because there is less of chance of them becoming damaged. However, it still may be a good idea.

Consider why the powerhead is being removed. If you are planning on disassembling the powerhead for inspection or overhaul, then you'll have to remove the fuel and electrical components anyway. Removing them before powerhead removal may make the rest of powerhead removal a tad easier. Either way you decide, we'll list the minimum amount of components you must remove before lifting the powerhead free.

1. Remove the cowling from the powerhead for access.

2. For safety, disconnect the battery cables.

3. Disconnect the remote fuel tank hose from the outboard.

4. On the front/port side of the motor locate the 2 small bolts which are threaded downward into the lower engine cowling to secure the remote control harness clamp. Loosen the bolts and remove the clamp.

5. Make sure the outboard shift lever is in the NEUTRAL position.

6. On the side of the powerhead locate where the throttle/shift cable barrels are locked into the bracket. There is a pivoting barrel retainer latch which secures the cables. Pivot the latch clockwise, off the barrels to free them. Then follow the cables to the ends there they attach to the powerhead linkage, remove the retaining nuts and washers, and disconnect the cable ends from the linkage.

7. On the bottom of the shift arm (the vertical stud to which the lower of the two cables in the bracket attach to) there is a shift arm attaching nut, loosen and remove the nut.

8. Remove the clamp and disconnect the input fuel line.

9. If equipped with the optional remote oil tank, tag and disconnect the tank hose connections.

10. Tag and disconnect the remote control harness. It is normally a large barrel-shaped connecter found at the base of the powerhead (just under the Vapor Separator Tank assembly). The connector is secured to the lower engine cowling in two small brackets and secured by a wire loom.

11. Remove the flywheel cover from the top of the powerhead.

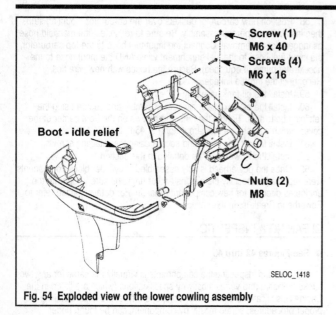

Fig. 54 Exploded view of the lower cowling assembly

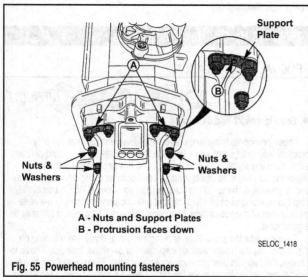

Fig. 55 Powerhead mounting fasteners

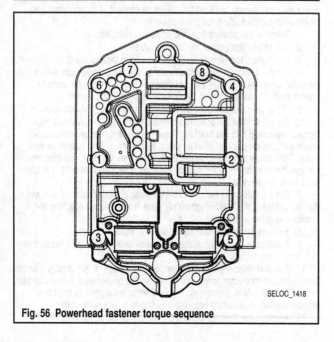

Fig. 56 Powerhead fastener torque sequence

12. Tag and disconnect the air compressor-to-air intake plenum hose.

13. Tag and disconnect the oil tank output hose, then plug the hose or cap the outlet to prevent oil spillage.

14. Tag and disconnect the low oil sensor connector (normally found at the base of the tank on these models).

15. Remove the 2 screws securing the oil tank to the powerhead, then remove the tank.

16. On the starboard side of the powerhead, just above the trim switch locate the trim/tilt switch connector. Tag and disengage the harness connector.

17. Remove the 4 main screws holding the lower cowling halves together, then remove the 1 screw holding the front cowl halves together. Lastly, remove the 2 nuts attaching the cowl haves to the driveshaft housing. Carefully remove the lower cowl halves for access.

■ **The tell-tale hose, air compressor dump hose and the IVST vent hose all pass through openings in the cowl halves and will be disconnected when the halves are removed.**

18. Once the lower cowling is out of the way, locate the SmartCraft harness connector (on the side of the powerhead, almost directly under the vapor separator tank assembly in a position where it was obscured by the lower cowling). Tag and disengage the SmartCraft harness connector.

19. On the opposite side of the powerhead, at the base of the motor (again where it was obscured previously by the lower cowling) and a bit in front of the starter motor locate, tag and disconnect the bullet connectors for the BLUE and GREEN trim harness leads.

20. Tag and disconnect the RED and BLACK neutral switch harness leads (right next to the Blue and Green trim switch leads).

21. Remove the plastic cap from the center of the flywheel, then install the lifting eye (#91-90455T1 or equivalent) into the flywheel a minimum of 5 turns.

22. Remove the 8 nuts, 6 washers and 2 support plates from the powerhead mounting fasteners under the mounting base. There are basically 3 nuts on each side, and 2 on the ends. The two on the ends share the support plates with the adjacent 2 on either side.

■ **At this point double check for any cables, hoses, wires or linkages that will interfere with the removal of the powerhead.**

23. Using a suitable engine hoist, carefully lift the powerhead straight up and off the driveshaft/driveshaft housing while breaking the gasket seal.

24. Move the powerhead to a clean work area and placed in a suitable holding fixture.

To install:

25. Thoroughly clean all old gasket material from the driveshaft housing and powerhead mating surfaces.

26. Place a new gasket around the powerhead studs.

27. Lubricate the SIDES of the splines on the driveshaft using 2-4-C with Teflon or an equivalent marine grade grease. Wipe off any excess lubricant from the top of the driveshaft as any lubricant left on top of the driveshaft could prevent it from seating in the crankshaft.

28. SLOWLY lower the powerhead onto the driveshaft housing. You may have to rotate the crankshaft (or conversely the propshaft to rotate the driveshaft if the motor is in gear) in order to index the crankshaft and driveshaft splines. Continue lowering the powerhead until is fully seated on the housing and alignment pins.

29. Loosely install the nuts, washers and support plates, then tighten the powerhead mounting nuts evenly using multiple passes of the torque sequence (that starts at the middle nuts on the side and works outward in a crossing pattern, ending with the two end nuts on the inside of the support plates) first to 20 ft. lbs. (27 Nm) and then an additional 90 degrees. Remove the lifting eye and lifting device.

30. Reconnect the SmartCraft wiring harness connector at the base of the powerhead.

31. On the same side as the SmartCraft connector, re-secure the shift lever with the attaching nut on the base of the stud. Drive the nut tight, but leave the shift cable off the top of the stud for now.

32. Reconnect the bullet connectors for the Blue and Green trim/tilt leads and the Red and Black neutral switch leads.

■ **There is a small, sorta-rectangular boot which sits between the rear of the lower cowl halves, called the idle relief boot. During installation, be sure the boot is in position. Also, be sure to keep an eye on all of the lower wiring and hoses to make sure none are pinched or damaged during installation of the lower cowling halves.**

33. Carefully position the lower cowl halves, then secure using the bolts and nuts. Be sure to install the screw which holds the front cowl halves together.

34. Position the oil tank assembly onto the engine aligning the throttle body seal with the air intake flange.

35. Reconnect the oil hose and secure using new wire ties.

36. Reconnect the low oil sensor harness.

37. Secure the oil tank assembly with the 2 screws at the top of the tank, threaded downward into the top of the powerhead.

■ **The shift cable should be the first cable to move as the remote control handle is shifted out of NEUTRAL. The shift cable barrel fits into the lower of the two holes in the barrel retainer.**

38. Install the shift cable to the barrel retainer bracket and the top of the shift lever stud. If the powerhead or the shift cable was replaced, properly adjust the cable and barrel as follows:

a. Place both the remote control shifter and the powerhead shift lever in NEUTRAL.

b. Measure the distance between the center of the barrel pocket and the shift lever pin (the top of the shift lever stud, where the end of the cable will be attached). Push inward on the end of the shift cable until you can feel resistance, then adjust the position of the barrel cable to obtain the proper distance (as measured) between the center of the cable end and the center of the barrel. Thread the barrel inward or outward as necessary.

c. Insert the cable barrel into the retainer pocket, then position the cable end over the pin and secure using the washer and locknut. Tighten the locknut securely, then back off 1/4 of a turn to prevent binding.

d. Check the shift cable by shifting the remote into FORWARD and verifying that the propeller is locked in gear (if not, readjust the barrel closer to the cable end).

e. Next shift the remote into NEUTRAL and verify that the propeller shaft now turns freely (if not, readjust the barrel further away from the cable end).

f. While slowly turning the propeller by hand, shift the remote control into REVERSE. The propeller shaft should be locked into reverse gear (if not, adjust the barrel away from the cable end).

g. Shift the remote control back to NEUTRAL. The propeller must turn freely, without drag (if not, adjust the barrel closer to the cable end).

■ **If at anytime during these checks the cable barrel is readjusted, go back and start the shift checks over with Neutral again, then FORWARD, REVERSE and finish be rechecking the shift from REVERSE back to NEUTRAL.**

39. Install the throttle cable to the top hole in the barrel retainer bracket and to the pin on the bottom of the throttle lever. If the powerhead or the throttle cable was replaced, properly adjust the cable and barrel as follows:

a. Place the throttle cable end over the pin on the bottom of the throttle arm, then secure using the washer and locknut.

b. Adjust the cable barrel so the installed throttle cable will hold the throttle arm against the idle stop, then place the barrel into the top hole in the barrel retainer.

c. Rotate the latch on the barrel retainer counterclockwise and up/over the cable barrels, locking the barrels and latch in place.

40. Position the fuel hose, wiring and cables through the opening in the lower engine cowl front cover. Place the neoprene wrap around the wiring, hoses and control cables, then position the cover and cowl seal.

41. Reconnect the remote control wiring harness connector, then push the wire retainer over the ends of the connector (the retainer will hold the connector firmly together). Position the harness connection and retainer into the cable holder molded into the side of the lower engine cowl. The arms of the wire retainer should fit over the outer edges of the cowling molded retainer.

42. Install the cover and cowl seal, tightening the retaining screws to 65 inch lbs. (7.3 Nm).

43. Install the flywheel cover.

44. Reconnect the battery cables.

45. Refill and properly prime the oil system.

■ **Be sure to run the motor without the cowling installed in order to check for potential fuel or oil leaks. Remedy any leaks before proceeding.**

46. Hook up a source of cooling water and test run the motor.

47. Install the cowling to the powerhead.

■ **If the powerhead was rebuilt or replaced with a remanufactured unit, don't forget to follow the proper Break-In procedures.**

DISSASSEMBLY & ASSEMBLY

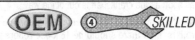

◆ **See Figures 57 thru 61**

■ **All ignition and electrical components should remain attached to the electrical component plate, in this way the plate along with those components can be removed as an assembly.**

1. If not done during powerhead removal, strip the powerhead of the remaining fuel and electric components. For details, where applicable, please refer to the appropriate sections of this guide.
   - Electric starter motor
   - Electrical component plate (along with the attached PCM, ignition coil and starter solenoid)
   - Alternator
   - Flywheel
   - Direct fuel injection components (including air/fuel rail and injectors, fuel pump/vapor separator tank assembly). This includes removing the air plenum assembly. Be sure to loosen the bolts using multiple passes in the reverse of the torque sequence.
   - On-board oil reservoir tank
   - Oil pump and bleed hoses. Be sure to tag all of the hoses and note routing before removal
   - Throttle/shift control cables and linkage
   - Thermostat and poppet valve assembly

2. Using a few passes in the reverse of the torque sequence, remove the 15 bolts securing the reed block adaptor plate to the crankcase half. A single reed valve block assembly is used for each cylinder and is secured to the backside of the adaptor plate by 2 screws each. In addition each reed valve block is sealed to the plate with an individual gasket, while the entire adaptor plate and reed valve assembly is then sealed to the crankcase with one large gasket. Once the adaptor plate bolts are removed, separate the plate and reed blocks from the crankcase, then remove and discard the large gasket.

■ **DO NOT remove the reed blocks from the adaptor plate unless replacement is necessary. If removed, a new gasket must be used between each reed block and the adaptor plate.**

3. Alternately and evenly loosen and remove the 3 bolts securing the lower end cap to the crankcase, however, don't remove the end cap assembly yet, as it is easier once the crankcase cover is separated.

4. Using a few passes in the reverse of the 2 torque sequences (starting with the smaller outer bolts and then working on the larger inner bolts) loosen and remove the 2 sets of crankcase cover-to-cylinder block bolts. Use a soft mallet to tap around the edge of the cover-to-block mating surface, then carefully separate the cover from the block.

5. Remove the lower end cap assembly. The assembly may be equipped with either a single seal which normally faces upward toward the crankshaft or dual seals, at least the inner of which normally faces upward toward the crankshaft (double-check seal positioning before removal from the housing). The assembly is also sealed using an O-ring. Both the seal(s) and the O-ring should be replaced to ensure an oil tight powerhead. If necessary, use a punch and hammer to drive out the old seal(s).

■ **On most Mercury inline motors you can usually remove the crankshaft and pistons as an assembly (and this one is no exception). However, on these larger motors, it becomes easier to unbolt the connecting rods first and lift out the crankshaft separately from the pistons. Though you can probably do either we've recommended the later here (removing them separately) to better protect the crank and pistons.**

6. Take a second to matchmark the connecting rod caps to the connecting rods. They MUST not only be reinstalled on the same rods, but facing the same direction or bearing failure will occur.

■ **The connecting rod bolt caps can be loosened or installed using a 3/8 in. 12-point socket.**

7. Unbolt the connecting rod caps to free the crankshaft. Alternately and evenly loosen the 2 connecting rod cap bolts, going back and forth with no more than 1 full turn on each bolt without moving back to the other bolt. Proceed with this pattern until the connecting rod cap can be removed. MAKE sure you keep track of the rod cap and bearing cage as they must be returned to the original connecting rod and crankshaft journal if they are to

be reused. It's best to label them AND to reinstall them to the connecting rod as soon as the crankshaft is out of the way.

■ **The connecting rod bearings on some motors may utilize loose rollers. If so there should be 16 loose rollers per bearing set. Be sure to keep track of the rollers, because if even just ONE is lost, the entire bearing set must be replaced.**

8. Carefully lift the crankshaft assembly from the connecting rods and cylinder block.

9. If necessary, disassemble the crankshaft and bearings as follows:

a. Slide the upper main bearing and seal off the top of the crankshaft. The Inspect the bearing for rust, pits, discoloration or roughness and replace, as necessary, but ALWAYS replace the seal.

b. Slide the lower wear seal and O-ring off the bottom of the crankshaft. Pop out the old O-ring using an awl or pick and discard.

c. Remove the retaining ring from the groove in each of the center main bearing races, then remove the races and bearing halves (which normally DO NOT use free rollers on these models, but proceed slowly JUST IN CASE).

■ **ONLY remove the crankshaft sealing rings IF replacement is necessary. Honestly, the sealing rings normally do not require replacement UNLESS they are broken.**

d. If replacement is necessary, while wearing a pair of safety glasses, carefully remove the sealing ring from the shoulder on each center main journal.

e. The lower bearing IS pressed onto the shaft and removal will destroy it, so ONLY remove it if necessary to replace the bearing. To remove the bearing first remove the snapring using a pair of snapring pliers, then use a shop press and bearing separator to free the bearing from the crankshaft.

10. Before going any further, LABEL the pistons and all matching components (connecting rod, rod cap etc) to make sure the proper components are all removed, stored and reinstalled together. You should have started earlier when you matchmarked the connecting rod caps, but go a step further now to make sure you also mark the direction the remaining components are facing. Remember when reusing these components (caps, rods, pistons) you must make sure all components are installed in the same locations and facing the same directions as they were before they were disassembled. It is true that many of these components have factory alignment or identification marks, but it is never a bad idea to double-check them or mark them yourself before proceeding.

11. Carefully pull each piston out of the crankshaft bore for inspection and disassembly, as necessary. If it is necessary to disassemble them for further inspection and component replacement, as follows:

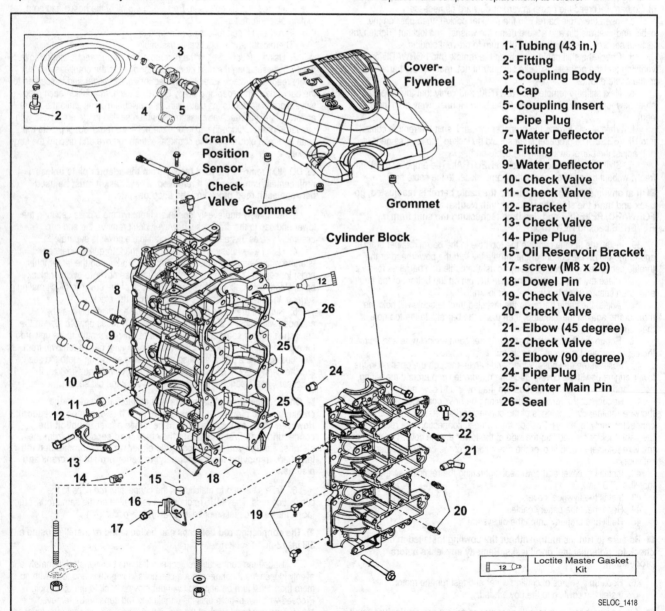

Fig. 57 Exploded view of the crankcase assembly - 75/90/115 Hp OptiMax Models

■ We've said it before. We know there will be some circumstances that will prompt people to reuse rings, but for the most part, if you've come THIS far, it is probably a good idea to bite the bullet and replace the rings to ensure long-lasting performance from the powerhead.

a. Use a Piston Ring Expander (such as #91-24697) to remove the rings from the pistons. The manufacturer does not recommend reusing the rings. Retain them in sets for inspection purposes (marking their original locations), but discard them before installation.

✳✳ SELOC WARNING

ALWAYS use eye protection when attempting to remove the piston pin lockrings as they are prone to launching out of position unpredictably.

b. To remove the connecting rod from each piston, first remove the lockring from both ends of the piston pin. Use an awl inserted just under the lip of the ring to gently lever it out of the piston. Keep in mind that these lockrings are often distorted slightly during removal and so they must be discarded are replaced with new lockrings during installation.

■ If necessary, a piston cradle can be fabricated from a large block of wood. The important design features of the cradle are that it is semi-circular on top to hold (or cradle, get it?) the piston and that there is a bore at the bottom center into which the wrist pin can tapped or pressed. The final important feature of the cradle is that it is sturdy enough to withstand the force of the shop press, BUT, keep in mind that the fit is loose on these pistons and significant force should not be necessary. It is NOT uncommon for someone to simply support the piston in their lap when working on these motors.

■ GREAT TIP - Use a heat lamp to gently warm the piston prior to piston pin disassembly, it will make the pin slide out MUCH easier!

c. While carefully supporting the piston to prevent damage use a soft-faced mallet and driver (such as Piston Pin Tool #91-92973A1 or equivalent) to gently tap the pin from the bore in the piston. As the pin is removed, KEEP CLOSE TRACK of the locating/thrust washers and the loose needle bearings (there should be 34 needles per piston).

■ As the wrist pin is removed from the piston you must retain the 34 loose needle bearings (IF you plan on reusing them) and the 2 locating/thrust washers from the small end of the connecting rod. If you can, stop tapping or pressing once the wrist pin JUST clears the bottom of the connecting rod bore. Then carefully withdraw the rod along with the needle bearings (place a thin bladed tool like a putty knife or your finger under the rod hole as it is pulled free of the piston bore/wrist pin). Inspect the bearings, if any of the needles are lost or worn, replace the entire bearing assembly. HONESTLY though, Mercury recommends replacing the needles at this time so it is probably best NOT to reuse them on this motor.

d. Repeat for the other pistons.

12. There are 3 Allen screws in the top of the cylinder block which cover water dam plugs. The screws should be removed so the plugs can be inspected for damage from high engine temperatures which might occur due to lack of cooling water. Replace the plugs, as necessary, keeping in mind that the plugs are necessary to maintain proper coolant flow around the cylinder bores. Missing or damaged plugs could cause powerhead overheating.

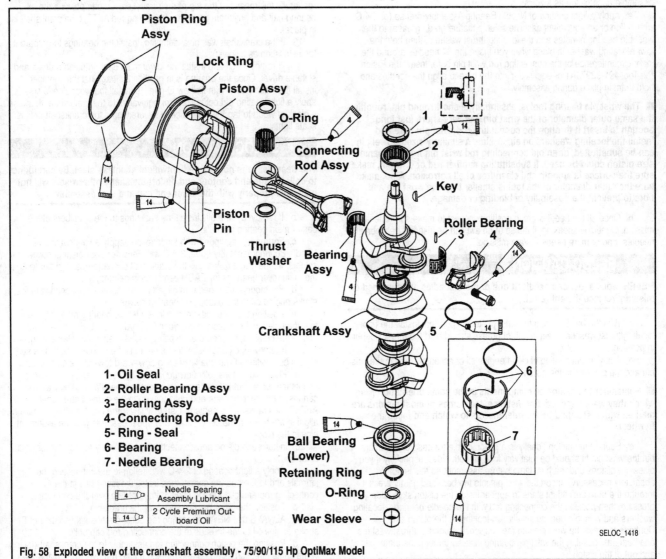

1- Oil Seal
2- Roller Bearing Assy
3- Bearing Assy
4- Connecting Rod Assy
5- Ring - Seal
6- Bearing
7- Needle Bearing

	Needle Bearing Assembly Lubricant
14	2 Cycle Premium Outboard Oil

SELOC_1418

Fig. 58 Exploded view of the crankshaft assembly - 75/90/115 Hp OptiMax Model

■ **Mercury and Seloc recommend that ALL seals be replaced during a rebuilding procedure of this magnitude (that is all except any crankshaft sealing rings that aren't broken).**

To assemble:

13. Prepare the following components for installation later, as follows:

a. If the reed valves blocks were removed from the adaptor plate, position the reeds blocks to the plate using new gaskets. Tighten the block retaining screws (2 per block) to 90 inch lbs. (10.2 Nm).

b. If the lower end cap seal was removed, use a suitable driver to gently press or tap the replacement seal(s) into the end cap. Remember the single seal (or at least the inner if 2 are used) should face upward (TOWARD the crank). After installation lubricate the seal lips with a light coat of 2-4-C with Teflon or an equivalent marine grade grease. Lubricate the new O-ring using a light coat of 2-stroke engine oil.

✳✳ SELOC CAUTION

Any grease used inside the powerhead for assembly MUST be gasoline soluble. The use of a non-soluble grease may cause severe powerhead damage.

14. Make sure all parts have been thoroughly cleaned and inspected. Be sure to oil all bearings and bearing surfaces lightly using a coating of fresh 2-stroke oil. Loose needle bearings, once cleaned, should be spread out on a clean worksurface, like a sheet of paper and coated lightly using Needle Bearing Assembly Lubricant.

15. If disassembled for inspection and/or component replacement, assemble each piston as follows:

a. Apply a light coating of Needle Bearing Assembly grease (or 2-4-C with Teflon or an equivalent gasoline soluble marine grade grease) to the wrist pin bearing needles and the locating/thrust washers, then hold the lower locating washer in place while you insert the 34 needles around the inner circumference of the connecting rod wrist pin bore. Install the Piston Pin Tool #91-92973A1 or equivalent into the bore to align the needles and hold them in place during assembly.

■ **The wrist pin bearing tool is, essentially, a short round pin, roughly the same outer diameter of the wrist pin, but very short, just long enough to insert it through the connecting rod wrist pin bore and install the locating washers on either side. A similarly sized dowel pin can be substituted, or an old connecting rod wrist pin from the same size piston could be used. If substituting another rod or drift pin, make sure the surface is smooth and clean/free of all corrosion. Also, make sure the outer diameter of the tool is smaller than the wrist pin and bore to prevent the possibility of binding or damage.**

b. Once all the needles are in position, inspect to make sure one is not missing. Generally speaking, if the tip of an awl can be inserted between the needles, one or more needles is/are missing.

✳✳ SELOC CAUTION

NEVER replace just one needle. If one or more needles are damaged or missing, replace the entire set.

c. With the bearings and the wrist pin tool in position, install the upper locating/thrust washer. Again, a dab of grease should hold the thrust washer in position.

d. Apply a light coating of clean engine oil or assembly lube to the wrist pin and the pin bore in the piston.

■ **Remember to position all components in the same orientation from which they were removed. For instance, the domes of most pistons are marked with a stamped UP for reference as to which end faces the flywheel.**

e. Support the piston (ideally in a piston cradle for ease of assembly but the wrist pin fit should be relatively lose on both sides, especially if you place the pistons under a heat lamp before assembly, so this is not absolutely necessary). Insert the wrist pin into the bore and gently press it through the bore until it just starts to appear inside the piston skirt. Stop pressing, then position the connecting rod with the needle bearings, locating washers and wrist pin bearing installation tool inside the piston skirt. Continue to press the wrist pin into the position, through the thrust washers and bearings, pushing wrist pin bearing installer through the other side and out of the piston.

f. Once the wrist pin is properly positioned and centered in the piston, install new lockrings into the grooves in both ends of the piston bore. Installation of the lockrings will be a lot easier using a lock ring installation tool (such as #91-93004A2 or equivalent) which is essentially a shouldered driver on which you can place the lockring, then press it evenly into the piston bore.

g. Use a ring expander to install the piston rings on each piston making sure to install the rings facing the right direction. The flat or RECTANGULAR side must be facing downward, while the tapered or HALF-KEYSTONE side should be facing upward. In most cases, replacement rings will be marked with a letter or dot on the side which faces up (rings for this motor are often labeled **1T** the "t" standing for top or facing upward). Also, each piston ring groove should contain a dowel pin, around which the ring gap must be situated. If not the rings will likely break during installation (or if by they don't by some miracle, during the first start-up). Be sure to install the tapered ring in the top groove.

■ **When using the ring expander, spread each ring JUST enough to slip over the piston.**

h. Repeat for the other pistons.

16. Prepare to install the pistons by lubricating the piston rings, pistons and cylinder walls lightly with oil.

17. No ring compressor is necessary, just verify that the piston is facing the proper direction (the UP mark on the dome is towards the flywheel end of the powerhead), then gently insert the piston into the cylinder bore. If necessary, use a small flat-bladed tool to gently compress each piston ring as it starts into the bottom of the bore. Repeat for the other pistons, positioning them as best you can to facilitate crankshaft installation later (generally this means at the top of their travel so you can position the bearing half and pull them up to the connecting rod AFTER the crankshaft is in place).

18. If the crankshaft was disassembled, install the bearings to prepare it for installation as follows:

a. If removed, oil and install the lower main bearing using a press and suitable driver. Once the bearing is in position tight against the shoulder install the wear sleeve with its new O-ring. To install the wear sleeve first apply a light coating of Loctite® 271 or equivalent to the crankshaft, then use a block of wood to tap the wear sleeve squarely onto the crankshaft until is seats lightly.

✳✳ SELOC CAUTION

The wear sleeve is normally a lightweight stainless steel. Be careful not to collapse the end contour and distort the shape otherwise it will not mate/seal properly with the lower end cap and seal assembly.

b. If removed, carefully install the seal rings to the shoulders of the center main bearing journals.

c. For models equipped with loose rollers, apply a light coating of Needle Bearing Assembly Lube, 2-4-C with Teflon or equivalent gasoline soluble marine grade grease, to the crankshaft center journal in order to hold the main roller bearings for EACH center main bearing.

d. For models equipped with caged rollers, apply a light coating of 2-stroke engine oil to the cages and bearing halves.

e. Position the roller bearings followed by the bearing race halves so that the holes in the races are positioned toward the LOWER (driveshaft) end of the crankshaft. Secure each pair of halves using the retainer ring.

f. If removed, oil and install the upper main bearing and seal. The seal should be positioned with the lips facing downward toward the crankshaft.

19. Place the pistons fully upward in their bores (i.e. away from the crankshaft) so as not to interfere with crankshaft installation (this way you can insert the bearing cage and pull them up to the crankshaft journal AFTER the crankshaft is seated). Carefully lower the crankshaft into position, aligning the holes in the center main and upper bearings with the alignment pins in the block.

20. Install each piston and connecting rod assembly to the crankshaft as follows:

a. Apply a light coating of 2-4-C with Teflon to the connecting rod journals and to the bearing cages, then install the lower half in the connecting rod and pull the rod upward into position over the crank pin.

b. Position the other bearing half over the crank pin.

c. ALWAYS use NEW connecting rod bolts. Apply a light coating of 2-stroke engine oil to the threads of the connecting rod cap bolts.

d. Position the piston/connecting rod assembly to the crankshaft (facing in the same direction as noted and marked during removal, then place the

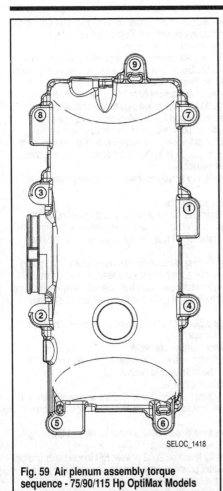

Fig. 59 Air plenum assembly torque sequence - 75/90/115 Hp OptiMax Models

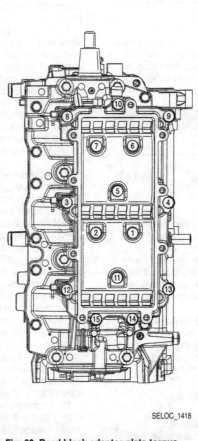
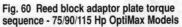

Fig. 60 Reed block adaptor plate torque sequence - 75/90/115 Hp OptiMax Models

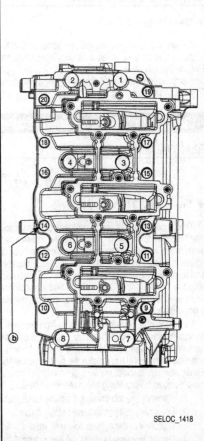

Fig. 61 Crankcase cover torque sequence - 75/90/115 Hp OptiMax Models

connecting rod end cap over the crank pin (also aligning the matchmarks made during removal) and start the bolts by hand.

■ **These motors use fractured end caps, meaning the end cap and connecting rod were originally one piece. As a result the caps will only fit properly in one direction and when they fit the fracture lines will mate PERFECTLY. Before the bolts are fully tightened (and stretched) run a pencil over the fracture lines, if the pencil catches, loosen the screws, retighten and check again).**

e. Alternately and evenly tighten the end cap retaining bolts, no more than one turn before moving to the other bolt. Once they start to tighten, use multiple passes to tighten them to first to 15 inch lbs. (1.7 Nm), next to 30 ft. lbs. (40.6 Nm) and then finally an additional 90 degrees (1/4 turn).

f. Once tightened, insert an awl or punch into the opening in the bottom of the connecting rod cap. Push gently on the bearing races. The races MUST rotate smoothly. If the races do not the connecting rod cap is not properly tightened, it must be loosened and retightened again (although at this point that would mean the bolts would have to be replaced again). If proper alignment cannot be achieved, you are going to have to discard the connecting rod.

g. Repeat for the other piston/connecting rod assemblies.

21. Make sure the seal area is lubricated lightly with 2-stroke engine oil, then position the assembled lower end cap over the bottom of the crank. Loosely install 2 end cap bolts to hold it in place as the crankcase cover is installed.

22. Make sure the bearing surfaces of the crankcase halves are clean and free of any oil or debris. Clean the surface using Loctite® Primer 203 or equivalent (usually included in the Master Gasket kit). Next apply a continuous bead of Loctite Master Gasket on the mating surfaces of the crankcase cover and cylinder block, JUST INBOARD of the bolts holes on either side of the crankshaft bore.

23. Position and new seal on the crankcase block.

24. Assemble the crankcase cover to the block, then install and finger-tighten the 8 large center main bolts. Tighten the bolts a little at a time to

compress the crankshaft seal rings and draw the cover down onto the block. Then use at least 3 passes of the torque sequence to tighten the bolts to 28 ft. lbs. (37.9 Nm).

25. Next, install the smaller crankcase cover flange bolts and finger-tighten. Again use at least 3 passes of the torque sequence to tighten the smaller flange bolts to 18 ft. lbs. (24.4 Nm).

26. Tighten the lower end cap bolts using multiple passes in a clockwise pattern around the housing to 18 ft. lbs. (24.4 Nm).

27. Rotate the crankshaft a few times to make sure it moves freely and there is NO binding.

28. Position a new reed block adaptor plate assembly gasket to the crankcase cover with the BEADED side FACING the air plenum, then install the adaptor plate and reed block assembly. Tighten the bolts using at least 3 passes of the torque sequences to 18 ft. lbs. (24.4 Nm).

29. Install the air plenum assembly over the reed block adaptor plate. Tighten the bolts to 18 ft. lbs. (24.4 Nm) using at least 3 passes of the torque sequence.

30. Either now or after the powerhead reinstalled (some prefer to wait so as to protect these components) install the remaining fuel and electric components to the powerhead. For details, where applicable, please refer to the appropriate sections of this guide.

• Thermostat and poppet valve assembly
• Throttle/shift control cables and linkage
• Oil pump and bleed hoses. Be sure to connect all of the hoses and route them as noted before removal.
• On-board oil reservoir tank
• Direct fuel injection components (including air/fuel rail and injectors, fuel pump/vapor separator tank assembly).
• Flywheel
• Alternator
• Electrical component plate (along with the attached PCM, ignition coil and starter solenoid)
• Electric starter motor

CLEANING & INSPECTION

◆ **See Figures 57 and 58**

Cleaning and inspecting the components is virtually the same for any two-stroke outboard and varies mostly by specifications (which are listed in the Engine Specifications charts) or by component type. A section detailing the proper procedures, sorted mostly by component, can be found under Powerhead Refinishing.

135-200 Hp (2.5L) V6 Powerheads

REMOVAL & INSTALLATION

◆ **See Figures 27 and 62 thru 66**

These powerheads are large (and relatively expensive) that the only practical way to lift them safely from the intermediate housing (and perhaps even more importantly to install them on the housing mating with the driveshaft splines) is to use an engine hoist or other lifting device for safety and to prevent damage. When using a hoist you don't HAVE to completely strip the powerhead of all electrical and fuel components, simply because there is less of chance of them becoming damaged. However, it still may be a good idea.

Consider why the powerhead is being removed. If you are planning on disassembling the powerhead for inspection or overhaul, then you'll have to remove the fuel and electrical components anyway. Removing them before powerhead removal may make the rest of powerhead removal a tad easier. Either way you decide, we'll list the minimum amount of components you must remove before lifting the powerhead free.

1. Remove the top cowling from the powerhead for access.
2. For safety, disconnect the battery cables.
3. Disconnect the remote fuel tank hose from the outboard.
4. On the front/port side of the motor locate the 2 small screws which are threaded downward into the lower engine cowling to secure the remote control harness clamp (the bridge which comprises the top half of the small circular hose/cable opening in the lower cowl). Loosen the screw and remove the clamp.
5. For OptiMax models, it is recommended that you cut the sta-strap that secures the tell-tale hose to the fitting on the lower cowl, then disconnect the hose from the fitting now instead of later. As the hose connects to a removable insert in the cowling on most models we've seen, we don't know why this would be necessary now, as it would be easier after the next step since the lower cowling being out of the way would provide more clearance/access, so we're just passing on the recommendation (perhaps in case there are some cowlings where the other end of the hose is attached to the cowling itself and not an insert).
6. For all models, remove the 4 bolts securing the lower cowl halves, then carefully remove the 2 pieces of the lower cowling. Normally 3 of the bolts are located on the inside of the lower cowl, one as the front of the powerhead, and two at the rear. The 4th bolt is normally found on the outside of the housing, toward the front of the powerhead, just above the swivel bracket. USUALLY, all fasteners are threaded from the port side of the motor.

■ **Remove the cowling slowly, checking for hoses or wiring connections which might still be attached. Tag and disconnect anything which would interfere with complete removal.**

7. For EFI and carbureted models, tag and disconnect the following hoses and wires:
   • The remote oil tank pressure hose and supply hose to reservoir. The pressure hose connects to the powerhead just below the electric starter.
   • The remote control harness. This connector is a large round wiring connector that is normally secured in a small bracket vertically on the starboard side of the powerhead. On carbureted models and 2001 or earlier EFI models it is found toward the rear of the powerhead, just in front of the starboard cylinder head. On 2002 or later EFI models it is found toward the front of the powerhead, a bit above the oil pump, near the rear edge of the air plenum.
   • The 2 power trim bullet connectors, which are found toward the base of the powerhead on the starboard side.

• For carbureted models or 2001 and earlier EFI models, also unplug the 2 connectors from the trim relays bolted to the starboard side of the powerhead, just below the remote control harness.
• For 2002 or later EFI models, disconnect the battery cables from the powerhead (negative from the block right below the starter and positive from the starter solenoid). Also on these models locate and disconnect the SmartCraft harness connector at the front base of the motor, on the starboard side. Lastly on 2002 or later models go to the back of the powerhead and locate the water pressure/speedometer pressure sensors at the top left of the electrical component bracket. Tag and disconnect the hoses for the sensors either at the sensors themselves, or by tracing them back to a connector. The gray hose is for the water pressure sensor, black hose for the speedometer sensor.

8. For OptiMax models, tag and disconnect the following hoses and wires:
   • The remote oil tank pressure hose.
   • The remote control harness. This connector is a large round wiring connector that is normally secured in a small bracket vertically on the front, starboard side of the powerhead (a bit above the oil pump, near the rear edge of the air plenum).
   • The warning gauge harness, also on the starboard side of the motor, toward the front of the powerhead, a bit below the remote control harness.
   • The Blue and Green trim harness leads from the trim relays (which are mounted to the base of the electrical component bracket on the starboard side of the powerhead).
   • System Monitor/Digital Sensor harness connectors toward the front, starboard side base of the motor.
   • Water pressure sensor harness connector.

9. Make sure the outboard shift lever is in the NEUTRAL position, if necessary manually slide the shift lever to Neutral.
10. Disconnect the throttle cable.
11. Remove the locknut that secures the shift cable latch assembly, the latch, flat washer, nylon wear plate, spring and shift cable from the control cable anchor bracket.
12. Loosen the clamp, then disconnect the input fuel line at the fitting where it passes through the lower cowling.
13. If equipped (normally found on 2002 or later EFI models with analog gauges) disconnect the water pressure hose and the speedometer hose where they pass through the lower cowling.
14. For OptiMax models, disconnect the port fuel rail inlet water hose from the fitting on the exhaust adaptor plate, then remove the excess air hose (which runs from the port fuel rail) from the fitting on the plate. Lastly, remove the thermostat outlet hose from the plate.
15. For EFI and carbureted models, remove the water hose from the fitting on the exhaust adaptor plate.
16. If equipped, remove the flywheel cover from the top of the powerhead. Although some flywheel covers have access holes to mount the powerhead lifting device, it is safer to remove the cover first.
17. Remove the plastic cap from the center of the flywheel, then install the lifting eye (#91-90455 or equivalent) into the flywheel a minimum of 5 turns.
18. Using a few passes in the reverse of the torque sequence, which results in a crossing pattern that starts at the outer fasteners and works inward, remove the 10 nuts (5 per side) that secure the powerhead base.

■ **At this point double check for any cables, hoses, wires or linkages that will interfere with the removal of the powerhead.**

19. Using a suitable engine hoist, carefully lift the powerhead straight up and off the driveshaft/driveshaft housing while breaking the gasket seal.
20. Move the powerhead to a clean work area and placed in a suitable holding fixture.

To install:
21. Thoroughly clean all old gasket material from the driveshaft housing and powerhead mating surfaces.
22. Place a new gasket around the powerhead studs.
23. Lubricate the SIDES of the splines on the driveshaft using 2-4-C with Teflon or an equivalent marine grade grease. Wipe off any excess lubricant from the top of the driveshaft as any lubricant left on top of the driveshaft could prevent it from seating in the crankshaft.
24. SLOWLY lower the powerhead onto the driveshaft housing. You may have to rotate the crankshaft (or conversely the propshaft to rotate the driveshaft if the motor is in gear) in order to index the crankshaft and driveshaft splines. Continue lowering the powerhead until is fully seated on the housing and alignment pins.

Fig. 62 The lower cowling is secured from the port side

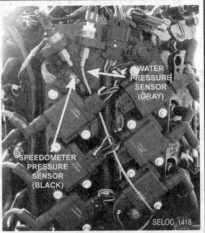

Fig. 63 Water pressure and speedometer sensors used on some EFI models

Fig. 64 Be sure to place the shift lever in Neutral. . .

Fig. 65 . . .and remove the flywheel cover

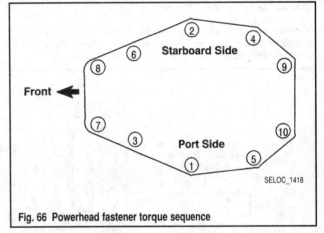

Fig. 66 Powerhead fastener torque sequence

25. Loosely install the nuts and washers, then tighten the powerhead mounting nuts evenly using at least 3 passes of the torque sequence (that starts at the middle nuts on the side and works outward in a crossing pattern, ending with the outer nuts on the rear of the powerhead) to 20 ft. lbs. (27 Nm).

26. Remove the lifting eye and lifting device and install flywheel cover and/or plastic cap, as applicable.

27. For OptiMax motors, reconnect the fuel cooler inlet and outlet water hoses, thermostat hose and the air hose from the port fuel rail to the fittings on the exhaust adaptor plate. Secure all of the hoses using new wire ties.

28. For EFI and carbureted models, reconnect the following hoses or harness connectors at this point:

• Reconnect the water hose to the fitting on the exhaust adaptor plate and secure using a new wire tie.

• Reconnect the oil tank pressure hose and supply hose to the reservoir.

• Reconnect the remote control harness, as well as the power trim wires. On carbureted and early EFI models this includes the trim relay connectors.

• On 2002 or later EFI models, reconnect the battery cables to the block and engage the SmartCraft harness connector. Reconnect the water hoses to the water pressure sensor (gray hose) and the speedometer pressure sensor (black hose) on the top left of the electrical component bracket, above the ignition coils.

29. Install the lower cowl halves and secure using the 4 retaining bolts.

30. Reconnect the input fuel line.

31. For 2002 or later EFI models, reconnect the water pressure and speedometer pressure hoses.

32. For OptiMax models, reconnect the following wiring harnesses:

• The Blue and Green trim harness leads to the solenoids.

• The remote control harness connector.
• The warning gauge harness.
• The system monitor/digital sensor harness connectors.
• The water pressure sensor harness
• Make sure there is sufficient slack in the engine wiring harness, battery cables, fuel hose and oil hoses that are all routed between the control clamp and the engine attachment point to prevent them from becoming stressed or pinched, then carefully place the neoprene wrap over the wiring, hoses and control cables.

33. Install the harness retainer bridge over the wiring/hose outlet at the front port side of the lower cowling then secure using the 2 retaining screws.

34. If not done already, slide the outboard shift lever into the neutral position.

35. Install the shift cable and secure using the spring, nylon wear plate, flat washer, latch and locknut.

36. Install the throttle cable. On models secured with a locknut, tighten the nut and then back off 1/4 turn.

■ If necessary refer to the Timing and Synchronization procedures to check powerhead set up.

37. Reconnect the battery cables.

38. If drained or disconnected, refill and properly prime the oil system, respectively.

■ Be sure to run the motor without the cowling installed in order to check for potential fuel or oil leaks. Remedy any leaks before proceeding.

39. Hook up a source of cooling water and test run the motor.
40. Install the cowling to the powerhead.

■ If the powerhead was rebuilt or replaced with a remanufactured unit, don't forget to follow the proper Break-In procedures.

DISSASSEMBLY & ASSEMBLY

◆ See Figures 67 thru 72

■ All ignition and electrical components should remain attached to the electrical component plate, in this way the plate along with those components can be removed as an assembly.

1. If not done during powerhead removal, strip the powerhead of the remaining fuel (carburetor, EFI or OptiMax as applicable) and electric components. For details, where applicable, please refer to the appropriate sections of this guide:
- Electric starter motor
- Electrical component plate (along with the attached PCM, ignition coil and starter solenoid for fuel injected models, and trim relays on 2002 or later EFI models, or ignition modules, starter solenoid and voltage regulator on carbureted models).
- Trim solenoids (carbureted or 2001 and earlier EFI only)
- Control module (carbureted or 2001 and earlier EFI only)
- Detonation module (2001 and earlier 200 hp EFI only)
- Flywheel
- Alternator (OptiMax or 2002 and later EFI) or stator and trigger assembly (Carbureted and 2001 or earlier EFI)
- On-board oil reservoir tank
- Oil pump (on models with an electronic oil pump the assembly is normally mounted to the air plenum, and will be removed with the assembly so all you need to do is tag and disconnect the lines at this point).

- OptiMax Direct fuel injection components (including air/fuel rail and injectors, fuel pump/vapor separator tank assembly). This includes removing the air plenum assembly. Be sure to loosen the bolts using multiple passes in the reverse of the torque sequence. Also be sure to unbolt and remove the air compressor assembly (tag and disconnect the hoses, then unbolt and remove the assembly).

■ Be careful during air plenum removal as on most models the air plenum on OptiMax models secures the reed block/adaptor plate.

- EFI fuel injection components (including air plenum, fuel rail with injectors, vapor separator tank assembly and low pressure pulse fuel pump)
- Carburetors, linkage and air intake silencer assembly.
- Bleed hoses. Be sure to tag all of the hoses and note routing before removal
- Throttle/shift control cables and linkage
- Thermostat assemblies
2. Remove the water pressure (poppet) valve assembly from the lower side of the powerhead.
3. Using a few passes in the reverse of the torque sequence, remove the bolts securing the cylinder heads to the engine block, then carefully remove the heads and gaskets.
4. Using a few passes in the reverse of the torque sequence, remove the bolts securing the exhaust plate to the engine block, then carefully remove the plate, gasket and seal.
5. If not done earlier, remove the reed block housing from the cylinder block. On carbureted and EFI motors the reed block housing is part of the intake manifold assembly and is secured to the block by up to 8 retaining bolts along the flange. However, keep in mind that there should also be 12

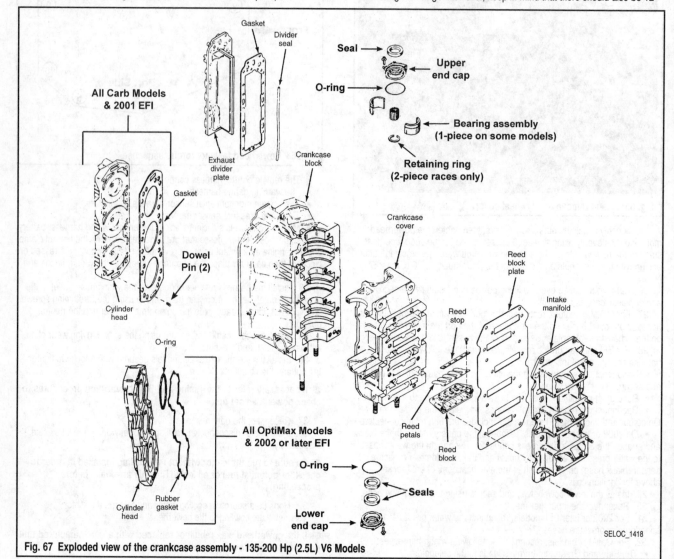

Fig. 67 Exploded view of the crankcase assembly - 135-200 Hp (2.5L) V6 Models

SELOC_1418

reed block retaining bolts which should NOT be removed unless component replacement is required. On OptiMax motors, the adaptor plate is secured to the crankcase cover by the air plenum bolts which were removed earlier.

■ **DO NOT remove the reed blocks from the adaptor plate unless replacement is necessary.**

6. Alternately and evenly loosen and remove the bolts from both the upper and lower end caps, however, don't remove the end caps yet, as it is easier once the crankcase cover is separated.

■ **There is only a torque sequence for the larger inner crankcase cover bolts, however it is a good idea to mimic the basic concept of that sequence for the smaller cover flange bolts.**

7. Using a few passes in the reverse of the torque sequences (starting with the smaller outer bolts and then working on the larger inner bolts) loosen and remove the 2 sets of crankcase cover-to-cylinder block bolts. Use a soft mallet to tap around the edge of the cover-to-block mating surface, then carefully separate the cover from the block. The casting for most crankcase covers include 2 or more prytabs on the sides/ends of the cover, so if necessary, use 2 prybars to help separate it from the engine block.

■ **Unlike Mercury inline motors (where you can usually remove the crankshaft and pistons as an assembly) the V6 motors really require that you remove all of the pistons separately from the crankshaft.**

8. Before going any further, LABEL (number and matchmark) the pistons and all matching components (connecting rod, rod cap etc) to make sure the proper components are all removed, stored and reinstalled together. Be SURE to matchmark the connecting rod caps to the connecting rods. They MUST not only be reinstalled on the same rods, but facing the same direction or bearing failure will occur.

■ **Remember when reusing these components (caps, rods, pistons and bearings) you must make sure all components are installed in the same locations and facing the same directions as they were before they were disassembled. It is true that many of these components have factory alignment or identification marks, but it is never a bad idea to double-check them or mark them yourself before proceeding.**

■ **The connecting rod bolt caps can be loosened or installed using a 5/16 in. 12-point socket.**

9. Unbolt the connecting rod caps to free each piston from the crankshaft. Alternately and evenly loosen the 2 connecting rod cap bolts, going back and forth with no more than 1 full turn on each bolt without moving back to the other bolt. Proceed with this pattern until the connecting rod cap can be removed. MAKE sure you keep track of the rod cap and bearing cage as they must be returned to the original connecting rod and crankshaft journal if they are to be reused. It's best to label them AND to reinstall them to the connecting rod as soon as the piston is removed.

10. As each connecting rod cap and bearing assembly is removed to free the piston, carefully push the piston out of the top of the block, then reassemble the connecting rod and bearings to it for cleaning, inspection and storage. Repeat for the remaining pistons. If it is necessary to disassemble the pistons for further inspection and component replacement, as follows:

■ **We've said it before. We know there will be some circumstances that will prompt people to reuse rings, but for the most part, if you've come THIS far, it is probably a good idea to bite the bullet and replace the rings to ensure long-lasting performance from the powerhead.**

a. Use a Piston Ring Expander (such as #91-24697) to remove the rings from the pistons. The manufacturer does not recommend reusing the rings. Retain them in sets for inspection purposes (marking their original locations), but discard them before installation.

✳✳ SELOC WARNING

ALWAYS use eye protection when attempting to remove the piston pin lockrings as they are prone to launching out of position unpredictably.

b. To remove the connecting rod from each piston, first remove the lockring from both ends of the piston pin. Use an awl inserted just under the lip of the ring to gently lever it out of the piston. Keep in mind that these lockrings are often distorted slightly during removal and so they must be discarded are replaced with new lockrings during installation.

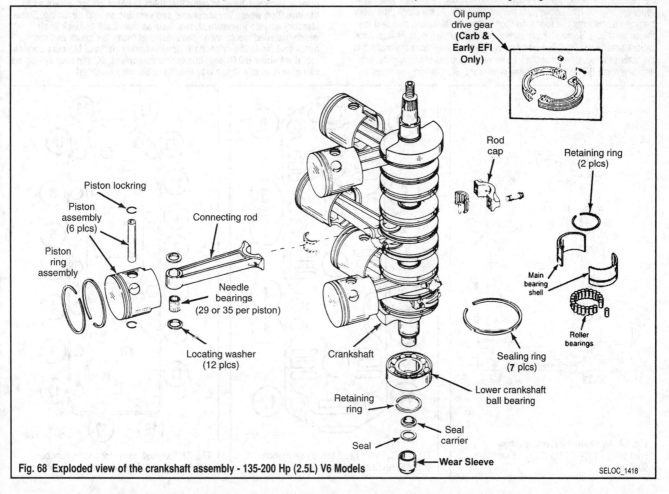

Fig. 68 Exploded view of the crankshaft assembly - 135-200 Hp (2.5L) V6 Models

SELOC_1418

■ If necessary, a piston cradle can be fabricated from a large block of wood. The important design features of the cradle are that it is semi-circular on top to hold (or cradle, get it?) the piston and that there is a bore at the bottom center into which the wrist pin can tapped or pressed. The final important feature of the cradle is that it is sturdy enough to withstand the force of the shop press, BUT, keep in mind that the fit is loose on these pistons and significant force should not be necessary. It is NOT uncommon for someone to simply support the piston in their lap when working on these motors.

■ GREAT TIP - Use a heat lamp to gently warm the piston prior to piston pin disassembly, it will make the pin slide out MUCH easier!

    c. While carefully supporting the piston to prevent damage use a soft-faced mallet and driver (such as Piston Pin Tool #91-76159A1 or equivalent for carbureted and EFI models or #91-92973A1 for OptiMax models) to gently tap the pin from the bore in the piston. As the pin is removed, KEEP CLOSE TRACK of the locating/thrust washers and the loose needle bearings (there should be 29 needles per piston on most models such as carbureted or EFI, but there should be 35 needles per piston on OptiMax motors).

■ As the wrist pin is removed from the piston you must retain the loose needle bearings (IF you plan on reusing them) and the 2 locating/thrust washers from the small end of the connecting rod. If you can, stop tapping or pressing once the wrist pin JUST clears the bottom of the connecting rod bore. Then carefully withdraw the rod along with the needle bearings (place a thin bladed tool like a putty knife or your finger under the rod hole as it is pulled free of the piston bore/wrist pin). Inspect the bearings, if any of the needles are lost or worn, replace the entire bearing assembly. HONESTLY though, Mercury and Seloc both recommend replacing the needles at this time so it is probably best NOT to reuse them on this motor.

    d. Repeat for the other pistons.

  11. Remove the upper and lower end cap assemblies from the crankshaft and block. Remove and discard the O-rings. Note the direction the seal lips are facing before removal, there is normally a single seal in the upper caps which faces downward toward the crank. At the same time there is some disagreement as to the direction where the lower end cap seal lips should be facing. Some Mercury service literature shows them as dual back-to-back seals which installed would mean the inner seal faces the crank and the outer seal faces the driveshaft. However other parts of Mercury service literature show BOTH seal lips facing downward. SO it is very important to

note the way it was sealed previously and reused that method during assembly. Use a punch and hammer to drive out the old seals. Be sure to inspect the roller bearing contained in the upper end cap. If the bearing is damaged, replace the end cap and roller bearing as an assembly.

  12. Carefully lift the crankshaft assembly from the cylinder block.

  13. If necessary, disassemble the crankshaft and bearings as follows:

    a. Remove the retaining ring from the groove in each of the center main bearing races, then remove the races and bearings (which may consist of either loose needles or caged bearing halves, though the later is far more common). Like all bearings and races they MUST be kept together and returned to the same journal if they are to be reused.

■ ONLY remove the crankshaft sealing rings IF replacement is necessary. Honestly, the sealing rings normally do not require replacement UNLESS they are broken.

    b. If replacement is necessary, while wearing a pair of safety glasses, carefully remove the sealing ring from the grooves in the shoulders on the crankshaft. The best way to remove the sealing rings is with a universal piston ring expander like #91-24697 or equivalent.

    c. The lower bearing IS pressed onto the shaft and removal will destroy it, so ONLY remove it if necessary to replace the bearing or for additional access to the oil pump drive gear (carbureted models or 2001 and earlier EFI models only), as necessary. To remove the bearing first remove the snapring using a pair of snapring pliers, then use a shop press and bearing separator to free the bearing from the crankshaft.

    d. On carbureted models or 2001 and earlier EFI models only, if necessary loosen and remove the 2 screws securing the halves of the oil pump drive gear together, then remove the gear.

■ Mercury and Seloc recommend that ALL seals be replaced during a rebuilding procedure of this magnitude (that is all except any crankshaft sealing rings that aren't broken).

  To assemble:

■ Remember the direction the lower end cap seal lips were facing prior to removal. Keep in mind that there is some disagreement as to the direction where the lower end cap seal lips should be facing. Some Mercury service literature shows them as dual back-to-back seals which installed would mean the inner seal faces the crank and the outer seal faces the driveshaft. However other parts of Mercury service literature show BOTH seal lips facing downward. SO the best advice we can give is to mimic the way the old seals were installed.

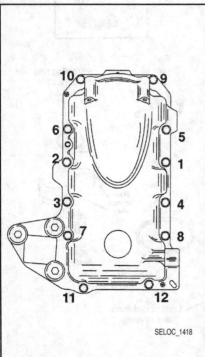

Fig. 69 Air plenum assembly torque sequence - 135-175 Hp (2.5L) V6 OptiMax Models

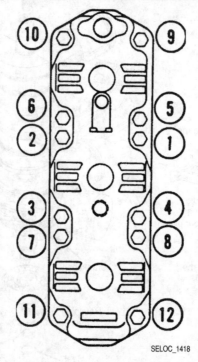

Fig. 70 Cylinder head torque sequence - 135-200 Hp (2.5L) V6 Models

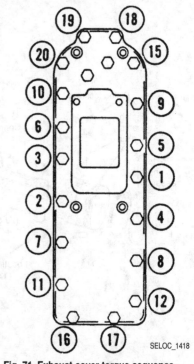

Fig. 71 Exhaust cover torque sequence - 135-200 Hp (2.5L) V6 Models

14. Install new seals to the upper and lower end caps. Before installing each seal, apply a light coating of Loctite® 271 or equivalent to the outer body of the seal (which contacts the housing when installed), then carefully tap the seal squarely into place using a suitable driver. After installation lubricate the seal lips and the NEW housing O-rings with a light coat of 2-4-C with Teflon or an equivalent marine grade grease.

❄❄ SELOC CAUTION

Any grease used inside the powerhead for assembly MUST be gasoline soluble. The use of a non-soluble grease may cause severe powerhead damage.

15. Make sure all parts have been thoroughly cleaned and inspected. Be sure to oil all bearings and bearing surfaces lightly using a coating of fresh 2-stroke oil. Loose needle bearings, once cleaned, should be spread out on a clean worksurface, like a sheet of paper and coated lightly using Needle Bearing Assembly Lubricant.

16. If the crankshaft was disassembled, install the bearings to prepare it for installation as follows:

a. On Carbureted models or 2001 and earlier EFI models, if removed install the oil pump drive gear. The gear halves are secured using 2 screws which must be cleaned with Loctite® 7649 Primer and then coated lightly using Loctite® 271 threadlocking compound (or their equivalents) before installing and securely tightening them.

b. If removed, oil and install the lower main bearing using a press and suitable driver. Once the bearing is in position tight against the shoulder install the wear sleeve with the seal carrier and new seal.

c. If removed, carefully install the seal rings to the grooves in the crankshaft. The best way to install the sealing rings is with a universal piston ring expander like #91-24697 or equivalent.

d. For models equipped with caged rollers, apply a light coating of 2-stroke engine oil to the cages and bearing halves.

■ **Although most models use caged rollers, IF you should come across one with loose rollers, apply a light coating of Needle Bearing Assembly Lube, 2-4-C with Teflon or equivalent gasoline soluble marine grade grease, to the crankshaft center journal in order to hold the main roller bearings for EACH center main bearing.**

e. Position the roller bearings followed by the bearing race halves so that the larger of the 3 holes in the races are positioned toward the LOWER (driveshaft) end of the crankshaft. Secure each pair of halves using the retainer ring (making sure the ring bridges the fracture lines of the race halves).

17. It's time to test fit the crankshaft to the block. For starters, check the mating surfaces in the block and crankcase cover for the crankshaft sealing rings. IF there are wear grooves present the rings on the new crankshaft will have to fit into the grooves without binding the crank. Before installation, remove any burrs that might exist on the groove edges, then lubricate the seals with light oil and position the crank along with the upper and lower end caps. Place the crankcase cover over the crankshaft, then rotate the shaft several times to check if the sealing rings are binding. IF there is any evidence of binding recheck the grooves for burrs. If the binding cannot be eliminated the block will have to be replaced.

18. Lubricate the crankshaft sealing rings and the upper lower crankshaft ends (oil seal mating areas) using a light coat of 2-stroke engine oil.

19. Position the crankshaft into the block with all of the crankshaft seal rings gaps facing straight up and with the holes in all of the bearing races facing the dowels. Gently push the shaft down into position to seat the races on the dowel pins.

20. Position the upper and lower end caps to the crankshaft and cylinder block. Loosely install the cap-to-block retaining bolts but DO NOT tighten until later, after the crankcase cover is installed.

21. If disassembled for inspection and/or component replacement, assemble each piston as follows:

a. Apply a light coating of Needle Bearing Assembly grease (or 2-4-C with Teflon or an equivalent gasoline soluble marine grade grease) to the wrist pin bearing needles and the locating/thrust washers, then hold the lower locating washer in place while you insert the 29 (Carb and EFI) or 35 (OptiMax) needles around the inner circumference of the connecting rod wrist pin bore. Install the Piston Pin Tool either #91-74607A3 (Carb and EFI) or #91-92973A1 (OptiMax) or equivalent into the bore to align the needles and hold them in place during assembly.

■ **The wrist pin bearing tool is, essentially, a short round pin, roughly the same outer diameter of the wrist pin, but very short, just long enough to insert it through the connecting rod wrist pin bore and install the locating washers on either side. A similarly sized dowel pin can be substituted, or an old connecting rod wrist pin from the same size piston could be used. If substituting another rod or drift pin, make sure the surface is smooth and clean/free of all corrosion. Also, make sure the outer diameter of the tool is smaller than the wrist pin and bore to prevent the possibility of binding or damage.**

b. Once all the needles are in position, inspect to make sure one is not missing. Generally speaking, if the tip of an awl can be inserted between the needles, one or more needles is/are missing.

❄❄ SELOC CAUTION

NEVER replace just one needle. If one or more needles are damaged or missing, replace the entire set.

c. With the bearings and the wrist pin tool in position, install the upper locating/thrust washer. Again, a dab of grease should hold the thrust washer in position.

d. Apply a light coating of clean engine oil or assembly lube to the wrist pin and the pin bore in the piston.

■ **Remember to position all components in the same orientation from which they were removed. For instance, the domes of most pistons are marked with a stamped UP for reference as to which end faces the flywheel.**

e. Support the piston (ideally in a piston cradle for ease of assembly but the wrist pin fit should be relatively lose on both sides, especially if you place the pistons under a heat lamp before assembly, so this is not absolutely necessary). Insert the wrist pin into the bore and gently press it through the bore until it just starts to appear inside the piston skirt. Stop pressing, then position the connecting rod with the needle bearings, locating washers and wrist pin bearing installation tool inside the piston skirt. Continue to press the wrist pin into the position, through the thrust washers and bearings, pushing the wrist pin bearing installer through the other side and out of the piston.

f. Once the wrist pin is properly positioned and centered in the piston, install new lockrings into the grooves in both ends of the piston bore. Installation of the lockrings will be a lot easier using a lock ring installation tool (such as #91-77109A3 for Carb/EFI or #91-93004A2 for OptiMax or equivalent) which is essentially a shouldered driver on which you can place the lockring, then press it evenly into the piston bore.

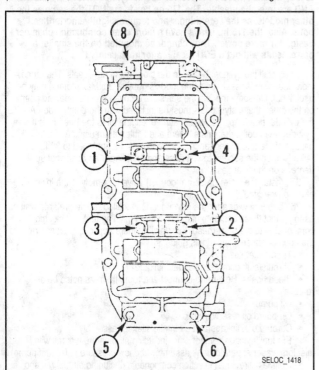

SELOC_1418

Fig. 72 Crankcase cover torque sequence - 135-200 Hp (2.5L) V6 Models

■ **Most carbureted 200 hp models and some 2001 or earlier 200 hp motors use a fully tapered (Keystone) ring for the top ring, while most of the rest of the motors use a half-tapered (half-keystone) ring everywhere.**

g. Use a ring expander to install the piston rings on each piston making sure to install the rings facing the right direction. Unless equipped with a full Keystone ring, the flat or RECTANGULAR side must be facing downward, while the tapered or HALF-KEYSTONE side should be facing upward. In most cases, replacement rings will be marked with a letter or dot on the side which faces up. Also, each piston ring groove should contain a dowel pin, around which the ring gap must be situated. If not the rings will likely break during installation (or if by they don't by some miracle, during the first start-up). Be sure to install the tapered ring in the top groove.

■ **When using the ring expander, spread each ring JUST enough to slip over the piston.**

h. Repeat for the other pistons.

■ **No ring compressor is necessary, just verify that the piston is facing the proper direction (the UP mark on the dome is towards the flywheel end of the powerhead). Additionally, pistons marked with a P on the dome (under the UP stamping) must be installed in the Port bank, while pistons marked with an S on the dome must be installed in the Starboard bank.**

22. Install each piston and connecting rod assembly to the crankshaft as follows:

a. Remove the connecting rod cap and bearings.

b. Prepare to install the piston by lubricating the journal surfaces and bearings using a light coating of 2-4-C with Teflon or an equivalent marine grade grease. Lubricate the piston rings, piston and cylinder wall lightly with 2-stroke engine oil.

c. Gently insert the piston into the cylinder bore. If necessary, use a small flat-bladed tool to gently compress each piston ring as it starts into the bore. Work slowly, watching to make sure the connecting rod doesn't contact and damage the cylinder wall or the crankshaft journal.

d. Install the lower half of the bearing cage assembly into the connecting rod, then pull the rod upward into position over the crank pin.

e. Position the other bearing half over the crank pin.

f. Apply a light coating of 2-stroke engine oil to the threads of the connecting rod cap bolts.

g. Position the piston/connecting rod assembly to the crankshaft (facing in the same direction as noted and marked during removal, then place the connecting rod end cap over the crank pin (also aligning the matchmarks made during removal) and start the bolts by hand.

■ **These motors use fractured end caps, meaning the end cap and connecting rod were originally one piece. As a result the caps will only fit properly in one direction and when they fit the fracture lines will mate PERFECTLY. Before the bolts are fully tightened run a pencil over the fracture lines, if the pencil catches, loosen the screws, retighten and check again).**

h. Alternately and evenly tighten the end cap retaining bolts (again using a 5/16 in. 12-point socket), no more than one turn before moving to the other bolt. Once they start to tighten, use multiple passes to tighten them to first to 15 inch lbs. (1.7 Nm), next to 30 ft. lbs. (40.6 Nm) and then finally an additional 90 degrees (1/4 turn).

i. Insert a small awl or screwdriver through the exhaust port to check the rings for tension. Press gently on each ring and make sure it springs back outward. If a ring has not tension, then it was likely broken during installation and the piston must be removed in order to check for damage and to replace the ring.

j. After each piston is installed, rotate the crankshaft a few times to make sure there is no binding or catching.

k. Repeat for the other piston/connecting rod assemblies.

23. Make sure the bearing surfaces of the crankcase halves are COMPLETELY clean and free of any oil or debris. Clean the surface using Loctite® 7649 Primer or equivalent.

24. On Carbureted and EFI models install new gasket strips into he grooves on either side of the crankcase cover, trimming the end of each gasket strip flush with the cover edge.

25. Next apply a thin, even, continuous bead of Loctite® Master Gasket 203 or equivalent on the mating surfaces of the crankcase cover and cylinder block, JUST INBOARD of the bolts holes on either side of the crankshaft bore. Continue a thin strip of the bead inward on each block/cover webbing toward the edge of each crankshaft sealing ring mating surface, but be sure

to leave some room for expansion between the edge of the bead and the sealing ring.

26. Assemble the crankcase cover to the block, then install and finger-tighten the 8 large center main bolts (after the thread and heads have been lightly oiled). Tighten the bolts a little at a time to compress the crankshaft seal/gasket material and draw the cover down onto the block. Then use at least 3 passes of the torque sequence to tighten the bolts to 38 ft. lbs. (51.5 Nm).

27. Next, install the smaller crankcase cover flange bolts and finger-tighten. Again use at least 3 passes of a torque sequence that mimics the larger bolt sequence (starts at the center and works outward in a criss-crossing pattern) to tighten the smaller flange bolts to 180 inch lbs. (20.3 Nm).

28. Tighten the upper and lower cap bolts using multiple passes in a clockwise pattern around the housing to 150 inch lbs. (17 Nm) for the upper cap and 80 inch lbs. (9 Nm) for the lower cap.

29. Again, rotate the crankshaft a few final times to make sure it moves freely and there is NO binding.

30. If the reed valve blocks were removed from the adaptor plate (and integral intake manifold on carbureted and EFI models), install them to the plate using a new gasket. Tighten the retaining bolts to 80 inch lbs. (9 Nm) for Carbureted and EFI models or to 105 inch lbs. (12 Nm) for OptiMax models.

31. On carbureted or EFI models, install the intake manifold/reed block adaptor plate assembly to the crankcase cover and secure using the retaining bolts (usually 8). Tighten the bolts to 105 inch lbs. (12 Nm) using at least 3 passes of a criss-crossing sequence that starts at the center and ends at the upper and lower bolts.

32. For OptiMax motors, install the air plenum assembly along with the reed block adaptor plate to the crankcase cover and secure using the 12 retaining bolts. Tighten the bolts to 175 inch lbs. (19.8 Nm) using at least 3 passes of the torque sequence (a counterclockwise spiraling pattern that begins with the right side center bolt and works outward toward the bolt on the bottom right).

33. Install the exhaust divider plate assembly. First position a new seal into the slot in the block and apply a light coating of Loctite® 271 or equivalent threadlock to the threads of the retaining bolts. Next, place the gasket and manifold/plate to the block and secure using the bolts. Tighten the bolts to 180 inch lbs./15 ft. lbs. (20 Nm) for carbureted and EFI models or to 200 inch lbs. (22.5 Nm) for OptiMax motors, in all cases using at least 3 passes of the torque sequence.

■ **For OptiMax motors the 115/135/150 hp cylinder heads are the same AND are interchangeable. The 175 hp heads are NOT the same as the other models, as they are thinner and require slightly shorter head bolts. Also, the 175 hp heads have a more open combustion chamber design for more power. They cannot be mounted on the smaller hp powerheads without a ECU recalibration/replacement.**

34. Install the cylinder heads to the block using new seals. The round individual cylinder seals on some models (mostly OptiMax motors) may be directional. On some models the seals are grooved on one side and pointy on the other, these style seals must be installed with the pointy side FACING the cylinder head. Apply a light coating of engine oil to the threads of the cylinder head bolts, then install them and tighten them using AT LEAST 3 passes of the torque sequence. The bolts must be tightened to 30 ft. lbs. (40.5 Nm) on the second to last pass and then finally an additional 90 degrees on the final pass.

35. Install the water pressure (poppet) valve assembly to the lower side of the powerhead.

36. Either now or after the powerhead reinstalled (some prefer to wait so as to protect these components) install the remaining fuel and electric components to the powerhead. For details, where applicable, please refer to the appropriate sections of this guide.
• Thermostat assemblies
• Throttle/shift control cables and linkage
• Bleed hoses. Be sure to connect and route them as noted before removal
• Oil pump
• On-board oil reservoir tank
• Carburetors, linkage and air intake silencer assembly.
• EFI fuel injection components (including air plenum, fuel rail with injectors, vapor separator tank assembly and low pressure pulse fuel pump)
• OptiMax Direct fuel injection components (including air/fuel rail and injectors, fuel pump/vapor separator tank assembly). This includes installing the air compressor assembly (reconnecting the hoses and tagged during removal).

• Alternator (OptiMax or 2002 and later EFI) or stator and trigger assembly (carbureted and 2001 EFI)
• Flywheel
• Detonation module (2001 200 hp EFI only)
• Control module (carbureted or 2001 EFI only)
• Trim solenoids (carbureted or 2001 EFI only)
• Electrical component plate (along with the attached PCM, ignition coil and starter solenoid for fuel injected models, and trim relays on 2002 or later EFI models, or ignition modules, starter solenoid and voltage regulator on carbureted models).
• Electric starter motor

CLEANING & INSPECTION

◆ **See Figures 67 and 68**

Cleaning and inspecting the components is virtually the same for any 2-stroke outboard and varies mostly by specifications (which are listed in the Engine Specifications charts) or by component type. A section detailing the proper procedures, sorted mostly by component, can be found under Powerhead Refinishing.

200-250 Hp (3.0L) V6 Powerheads

REMOVAL & INSTALLATION

◆ **See Figures 27, 66 and 73 thru 78**

These powerheads are as large (and relatively expensive) as they get, so the only practical way to lift them safely from the intermediate housing (and perhaps even more importantly to install them on the housing mating with the driveshaft splines) is to use an engine hoist or other lifting device for safety and to prevent damage. When using a hoist you don't have to completely strip the powerhead of all electrical and fuel components, simply because there is less of chance of them becoming damaged. However, it still may be a good idea.

Consider why the powerhead is being removed. If you are planning on disassembling the powerhead for inspection or overhaul, then you'll have to remove the fuel and electrical components anyway. Removing them before powerhead removal may make the rest of powerhead removal a tad easier. Either way you decide, we'll list the minimum amount of components you must remove before lifting the powerhead free.

1. Remove the top cowling from the powerhead for access.
2. For safety, disconnect the battery cables.
3. Disconnect the remote fuel tank hose from the outboard.
4. On the front/port side of the motor locate the 2 small screws which are threaded downward into the lower engine cowling to secure the remote control harness clamp (the bridge which comprises the top half of the small circular hose/cable opening in the lower cowl). Loosen the screw and remove the clamp.
5. Remove the 4 bolts securing the lower cowl halves, then carefully remove the 2 pieces of the lower cowling. On conventional models, normally 3 of the bolts are located on the inside of the lower cowl, one as the front of the powerhead, and 2 at the rear. There is normally an access cover in the lower cowling to help with access to the lower of the rear screws. The 4th bolt is normally found on the outside of the housing, toward the front of the powerhead, just above the swivel bracket. On Pro/XS/Sport models, normally the upper 2 fasteners are threaded inside the covers, while the lower 2 are normally threaded from outside the covers (though the rear lower is often hidden behind a plug). Usually, the fasteners on all models are threaded from the port side of the motor.

■ **Remove the cowling slowly, checking for hoses or wiring connections which might still be attached. Tag and disconnect anything which would interfere with complete removal. On some models a wire tie is securing the tell-tale hose to a fitting on the cowl that must be removed. Also, if necessary, unplug the wiring from the cowl mounted trim switch.**

6. On EFI models, tag and disconnect the following hoses and wires:
• On 2001 models, trace the remove oil tank pulse hose to the fitting on the powerhead, then tag and disconnect it.
• On 2002 and later models, disconnect the oil tank pressure hose from the back of the electric oil pump.
• The remote control harness. This connector is a large round wiring connector that is normally secured in a small bracket vertically on the

starboard side of the powerhead. On 2001 EFI models it is found toward the rear of the powerhead, between the electrical component cover and the starter. On 2002 or later EFI models it is also found toward the rear of the powerhead, along the cylinder head-to-block mating line, just behind the ECM.
• If not done earlier, the cowl mounted tilt switch harness. On 2002 or later models, there are 2 power trim bullet connectors, which are found toward the base of the powerhead on the starboard side.
• 2001 EFI models are either equipped with 2 trim solenoids or 2 trim relays. If equipped with the round bodied solenoids, tag and remove the Blue, Green and Black harness leads from them. If equipped with the square bodied relays, unbolt and remove the ground harness from the electrical plate mounting bolt, then remove the relay positive (red) leads from the battery side of the starter solenoid. On relay models, also tag and disconnect the trim relay harness from each relay, the Blue/White and Green/White leads from the lower cowl trim switch harness (if not done already) AND the Blue and Green power trim bullet leads from the trim motor harness.
• On 2002 or later EFI models, disconnect the battery cables from the powerhead (negative from the block right below the starter and positive from the starter solenoid). Also on these models, locate and disconnect the SmartCraft harness connector at the front base of the motor, on the starboard side.
• On 2002 or later EFI models, move around port side of the motor and disconnect the oil supply hose for the powerhead mounted tank. Next, just a little below that unbolt the shift lever bracket assembly. Then move to the base of the port cylinder head and disconnect the gray water pressure hose.
7. On OptiMax models, tag and disconnect the following hoses and wires:
• The remote oil tank hose connector.
• The remote control harness. This connector is a large round wiring connector that is normally secured in a small bracket vertically on the rear, starboard side of the powerhead (just behind the ECM, along the cylinder head-to-block mating line).
• The 2 multi-pin SmartCraft harness connectors at the lower front, starboard side of the powerhead.
• If not done already, the Blue and Green trim harness bullet connector leads (just below the trim relays which are mounted to the base of the electrical component bracket on the starboard side of the powerhead).
8. Make sure the outboard shift lever is in the Neutral position (for EFI models) or Forward position (for OptiMax models), if necessary manually slide the shift lever into position.
9. Remove the locknuts and flat washers that secure both the throttle (upper) cable and the shift (lower) cable.
10. Pivot the cable anchor latch about a 1/4 turn or more counterclockwise to release the cable barrels. Disconnect the cables from the powerhead.
11. Loosen the clamp, then disconnect the input fuel line at the fitting where it passes through the lower cowling.
12. If equipped (normally found on 2002 or later EFI models with analog gauges) disconnect the water pressure hose (gray) and the speedometer hose (black) where they pass through the lower cowling.
13. Tag and disconnect the various cooling system hoses. It is usually easiest (and safest) to cut the wire-ties and disconnect the hoses from the fittings on the powerhead:
• The tell-tale hose which runs to the fitting on the lower cowl insert.
• If equipped, the flush hose which also runs to the fitting on the lower cowl insert.
• The port and starboard thermostat bypass/discharge hoses. On Optimax and 2001 EFI models it is sometimes easier to disconnect them at the thermostat housing fittings, but on other models, like the 2002 or later EFI, it is easiest to disconnect them from the fittings at the base of the powerhead/adaptor assembly.
• On OptiMax models, disconnect the fuel tail inlet water hose from the fitting on the powerhead base/adaptor assembly.
14. If equipped, remove the flywheel cover from the top of the powerhead. Although some flywheel covers have access holes to mount the powerhead lifting device, it is safer to remove the cover first.
15. Remove the plastic cap from the center of the flywheel, then install the lifting eye (#91-90455 or equivalent) into the flywheel a minimum of 5 turns.
16. Using a few passes in the reverse of the torque sequence, which on most models results roughly in a crossing pattern that starts at the outer fasteners and works inward, remove the 10 nuts (5 per side) that secure the powerhead base.

■ **At this point double check for any cables, hoses, wires or linkages that will interfere with the removal of the powerhead.**

Fig. 73 Remove the top cowling for access

Fig. 74 ...to start disconnecting the necessary hoses...

Fig. 75 ...and wiring, such as the cowl trim switch

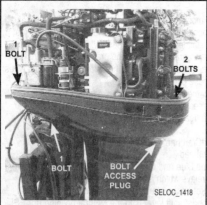

Fig. 76 Unbolt and remove the lower cowling for better access

Fig. 77 Disconnect the shift cable from the shifter bracket (shown)

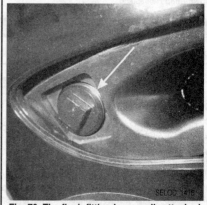

Fig. 78 The flush fitting is normally attached to an lower cowling insert bracket

17. Using a suitable engine hoist, carefully lift the powerhead straight up and off the driveshaft/driveshaft housing while breaking the gasket seal.

18. Move the powerhead to a clean work area and placed in a suitable holding fixture.

To install:

19. Thoroughly clean all old gasket material from the driveshaft housing and powerhead mating surfaces.

20. Place a new gasket around the powerhead studs.

21. Lubricate the sides of the splines on the driveshaft using 2-4-C with Teflon or an equivalent marine grade grease. Wipe off any excess lubricant from the top of the driveshaft as any lubricant left on top of the driveshaft could prevent it from seating in the crankshaft.

22. Slowly lower the powerhead onto the driveshaft housing. You may have to rotate the crankshaft (or conversely the propshaft to rotate the driveshaft if the motor is in gear) in order to index the crankshaft and driveshaft splines. Continue lowering the powerhead until is fully seated on the housing and alignment pins.

23. Loosely install the nuts and washers (locknuts on some models, like the Pro/XS/Sport), then tighten the powerhead mounting nuts evenly using at least 3 passes of the torque sequence (that roughly starts at the middle nuts on the side and works outward in a crossing pattern, ending with the outer nuts on the rear of the powerhead, but check the accompanying sequences, the Pro/XS/Sport models are a little different.). On all models, on the first pass just finger-tighten them. With conventional models on the next 2 passes tighten them to 20 ft. lbs. (27 Nm) and then to a final torque of 50 ft. lbs. (68 Nm). On Pro/XS/Sport models on the next 2 passes tighten them to 25 ft. lbs. (34 Nm) and then on the final pass tighten only nuts 1-8 in the sequence to a final torque of 40 ft. lbs. (54 Nm).

24. Remove the lifting eye and lifting device and install flywheel cover and/or plastic cap, as applicable.

25. Reconnect the various cooling system hoses as tagged during removal:

• On OptiMax models, reconnect the fuel tail inlet water hose to the fitting on the powerhead base/adaptor assembly.

• The port and starboard thermostat bypass/discharge hoses. On OptiMax and 2001 EFI models it is sometimes easier to disconnect them at the thermostat housing fittings, but on other models, like the 2002 or later EFI, it is easiest to disconnect them from the fittings at the base of the powerhead/adaptor assembly.

• If equipped, the flush hose which also runs to the fitting on the lower cowl insert.

• The tell-tale hose which runs to the fitting on the lower cowl insert.

■ Positioning the lower cowling is helpful, especially when routing and connecting remote hoses or wiring. However, the lower cowling can make reconnection of certain wiring, hoses or components near the base of the powerhead a little more difficult, so skip ahead as necessary to reconnect any wiring or components which may be harder to reconnect later. For instance, on 2002 and later EFI motors, if the shift lever bracket assembly was disconnected, it is usually easier to get at it now.

26. Reconnect any components which might be more/too difficult to reach once the lower cowling is put in position, such as on 2002 or later EFI motors, if removed, position and secure the shift lever bracket assembly. Also on these motors, reconnect the gray water pressure sensor hose near the base of the port cylinder head.

27. Reposition and install the lower cowl halves taking care not to pinch any hoses or wires, then secure the lower cowling using the 4 retaining bolts. Install the access plug for the lower rear bolt to the cowling.

28. Reconnect the input fuel line and secure using the retaining clamp or a new wire tie, as applicable. If equipped, reconnect the gray water pressure hose or the black speedometer hose (normally found on late-model EFI motors with analog gauges).

29. On EFI models, reconnect the following hoses or harness connectors (if they were not connected before cowling installation):

• On 2001 models, reconnect the oil tank pulse hose to the fitting on the powerhead.

• On 2002 and later models, reconnect the oil tank pressure hose to the back of the electric oil pump.

• The remote control harness.

• If not done earlier, the cowl mounted tilt switch harness. On 2002 or later models, there are 2 power trim bullet connectors, which are found toward the base of the powerhead on the starboard side.

• 2001 and earlier EFI models are either equipped with 2 trim solenoids or 2 trim relays. If equipped with the round bodied solenoids, reconnect the Blue, Green and Black harness leads to them as tagged during removal. If equipped with the square bodied relays, reconnect the ground harness to the electrical plate mounting bolt, then reconnect the relay positive (red) leads to the battery side of the starter solenoid. On relay models also reconnect the trim relay

harness to each relay, the BLUE/WHITE and GREEN/WHITE leads to the lower cowl trim switch harness (if not done already) AND the BLUE and GREEN power trim bullet leads to the trim motor harness.

• For 2002 or later EFI models, reconnect the battery cables to the powerhead (negative to the block right below the starter and positive to the starter solenoid). Also on these models reconnect the SmartCraft harness connector at the front base of the motor, on the starboard side.

• For 2002 or later EFI models, move around port side of the motor and reconnect the oil supply hose for the powerhead mounted tank.

30. For OptiMax models, reconnect the following wiring harnesses (if not done earlier before the cowling was installed):

• The Blue and Green trim harness bullet leads.

• The 2 multi-pin SmartCraft harness connectors at the lower front, starboard side of the powerhead.

• The remote control harness connector.

• The remote oil tank hose connector.

■ **The shift cable should be the first cable to move as the remote control handle is shifted out of NEUTRAL. The shift cable barrel fits into the lower of the two holes in the barrel retainer.**

31. If components were replaced, properly adjust and set-up the shifter linkage as follows:

a. Before installing the shift cable, first use a pencil to determine the full forward and full reverse positions, so a middle position can be marked for neutral. Move the shift lever all the way forward and draw a mark where on the cable where it aligns with the cable end guide. Next move the shift lever all the way to reverse and make a second mark. Now, make a mark that is centered between the 2 and make sure the cable guide is aligned with that center mark when it is attached to the motor.

b. Manually slide the shift cable linkage on the powerhead fully forward until resistance is felt, then fully rearward until resistance is felt. Center the cable anchor pin between these two points.

c. Align the shift cable end guide with the center mark made on the cable earlier and adjust the cable barrel until the barrel fits in the retainer and the cable end fits on the linkage anchor pin.

d. Install the shift cable to the barrel retainer bracket and the top of the shift lever stud and secure using the washer and locknut. Tighten the locknut securely, then back off 1/4 of a turn to prevent binding.

e. Check the shift cable by shifting the remote into FORWARD and verifying that the propeller is locked in gear (if not, readjust the barrel closer to the cable end).

f. Next shift the remote into NEUTRAL and verify that the propeller shaft now turns freely (if not, readjust the barrel further away from the cable end).

g. While slowly turning the propeller by hand, shift the remote control into REVERSE. The propeller shaft should be locked into reverse gear (if not, adjust the barrel away from the cable end).

h. Shift the remote control back to NEUTRAL. The propeller must turn freely, without drag (if not, adjust the barrel closer to the cable end).

■ **If at anytime during these checks the cable barrel is readjusted, go back and start the shift checks over with Neutral again, then FORWARD, REVERSE and finish be rechecking the shift from REVERSE back to NEUTRAL.**

32. Install the throttle cable to the top hole in the barrel retainer bracket and to the pin on the bottom of the throttle lever. If the powerhead or the throttle cable was replaced, properly adjust the cable and barrel as follows:

a. Place the throttle cable end over the pin on the bottom of the throttle arm, then secure using the washer and locknut (tighten the locknut, then back off 1/4 turn).

b. Adjust the cable barrel so the installed throttle cable will hold the throttle arm against the idle stop, then place the barrel into the top hole in the barrel retainer.

c. Rotate the latch on the barrel retainer counterclockwise and up/over the cable barrels, locking the barrels and latch in place.

d. Check the throttle adjustment by shifting the outboard into and out of gear (both Forward and Reverse) a few times to activate the throttle linkage. Be sure to rotate the propeller shaft when shifting into Reverse. Then return the remote to the Neutral position and place a thin piece of paper between the idle adjustment screw and idle stop. Adjustment is correct when the paper can be removed without tearing BUT still has some drag on it. If necessary, readjust the cable barrel.

33. Make sure there is sufficient slack in the engine wiring harness, battery cables, fuel hose and oil hoses that are all routed between the control clamp and the engine attachment point to prevent them from becoming stressed or pinched, then carefully place the neoprene wrap over the wiring, hoses and control cables.

34. Install the harness retainer bridge over the wiring/hose outlet at the front port side of the lower cowling then secure using the 2 retaining screws.

■ **If necessary refer to the Timing and Synchronization procedures to check powerhead set up.**

35. Reconnect the battery cables.

36. If drained or disconnected, refill and properly prime the oil system, respectively.

■ **Be sure to run the motor without the cowling installed in order to check for potential fuel or oil leaks. Remedy any leaks before proceeding.**

37. Hook up a source of cooling water and test run the motor.

38. Install the cowling to the powerhead.

■ **If the powerhead was rebuilt or replaced with a remanufactured unit, don't forget to follow the proper Break-In procedures.**

DISSASSEMBLY & ASSEMBLY

◆ **See Figures 79 thru 85**

■ **All ignition and electrical components should remain attached to the electrical component plate, in this way the plate along with those components can be removed as an assembly.**

1. If not done during powerhead removal, strip the powerhead of the remaining fuel (EFI or OptiMax as applicable) and electric components. For details, where applicable, please refer to the appropriate sections of this guide:

• Electric starter motor

• Electrical component plate (along with the attached components such as the ECM, ignition coils/CDMs, and starter solenoid and, for 2002 or later EFI models, trim relays).

• If necessary (to prevent damage during cleaning) remove powerhead mounted sensors such as CPS, TPS and temperature sensors.

• Alternator

• Flywheel

• Stator (for 2001 or earlier EFI models)

• Electronic shift/throttle assembly (OptiMax models, so equipped)

• On-board oil reservoir tank

• Oil pump (the pump is normally mounted to the air plenum on these motors, and will be removed with the assembly so all you need to do is tag and disconnect the lines at this point).

• OptiMax Direct fuel injection components (including air/fuel rail and injectors, fuel pump/vapor separator tank assembly). This includes removing the air plenum assembly. Be sure to loosen the bolts using multiple passes in the reverse of the torque sequence (which results in a counterclockwise spiraling pattern that starts with the right lower bolt and works inward toward the right center bolt). Also be sure to unbolt and remove the air compressor assembly (tag and disconnect the hoses, then unbolt and remove the assembly).

■ **Be careful during air plenum removal as on the air plenum on OptiMax models secures the reed block/adaptor plate.**

• EFI fuel injection components (including air plenum, fuel rail with injectors, vapor separator tank assembly and low pressure pulse fuel pump). Like on OptiMax motors, when removing the air plenum it is a good idea to loosen the bolts using multiple passes in the reverse of the torque sequence (which results in a counterclockwise spiraling pattern that starts with the right lower bolt and works inward toward the right center bolt).

• Bleed hoses. Be sure to tag all of the hoses and note routing before removal

• Throttle/shift control cables and linkage (except electronic shift models)
• Thermostat assemblies

2. Remove the water pressure (poppet) valve assembly from the lower side of the powerhead adaptor plate.

3. Using a few passes in the reverse of the torque sequence, remove the bolts securing the cylinder heads to the engine block, then carefully remove the heads and gaskets.

4. If not done earlier, remove the reed block housing from the cylinder block. On most motors the adaptor plate is secured to the crankcase cover by the air plenum bolts which were removed earlier.

■ **DO NOT remove the reed blocks from the adaptor plate unless replacement is necessary.**

5. Alternately and evenly loosen and remove the 4 bolts from lower end cap, however, don't remove the end cap yet, as it is easier once the crankcase cover is separated.

6. Using a few passes in the reverse of the 2 torque sequences (starting with the smaller outer flange bolts and then working on the larger inner bolts, in both cases working from the top/bottom bolts towards the center bolts) loosen and remove the 2 sets of crankcase cover-to-cylinder block bolts. Use a soft mallet to tap around the edge of the cover-to-block mating surface, then carefully separate the cover from the block.

■ **Unlike Mercury inline motors (where you can usually remove the crankshaft and pistons as an assembly) the V6 motors really require that you remove all of the pistons separately from the crankshaft.**

7. Before going any further, LABEL (number and matchmark) the pistons and all matching components (connecting rod, rod cap etc) to make sure the proper components are all removed, stored and reinstalled together. Be SURE to matchmark the connecting rod caps to the connecting rods. They must not only be reinstalled on the same rods, but facing the same direction or bearing failure will occur.

■ **Remember when reusing these components (caps, rods, pistons and bearings) you must make sure all components are installed in the same locations and facing the same directions as they were before they were disassembled. It is true that many of these components have factory alignment or identification marks, but it is never a bad idea to double-check them or mark them yourself before proceeding.**

■ The connecting rod bolt caps can be loosened or installed using a 3/8 in. 12-point socket. Although we recommend you using the rod bolts to temporarily reassemble the caps to the rods once the pistons are removed, we'd like to warn you that the bolts are normally stretched and weakened when they are torqued, so we recommend replacing them with new bolts during final assembly.

8. Unbolt the connecting rod caps to free each piston from the crankshaft. Alternately and evenly loosen the 2 connecting rod cap bolts, going back and forth with no more than 1 full turn on each bolt without moving back to the other bolt. Proceed with this pattern until the connecting rod cap can be removed. Make sure you keep track of the rod cap and bearing cage as they must be returned to the original connecting rod and crankshaft journal if they are to be reused. It's best to label them AND to reinstall them to the connecting rod as soon as the piston is removed.

■ On some models the rollers may come loose from the connecting rod bearing cages. If so, be sure to keep track of them and make sure they are returned to the same cage. There are usually 16 rollers per rod.

9. As each connecting rod cap and bearing assembly is removed to free the piston, carefully push the piston out of the top of the block, then reassemble the connecting rod and bearings to it for cleaning, inspection and storage. Repeat for the remaining pistons. If it is necessary to disassemble the pistons for further inspection and component replacement, as follows:

■ We've said it before. We know there will be some circumstances that will prompt people to reuse rings, but for the most part, if you've come this far, it is probably a good idea to bite the bullet and replace the rings to ensure long-lasting performance from the powerhead.

a. Use a Piston Ring Expander (such as #91-24697) to remove the rings from the pistons. The manufacturer does not recommend reusing the rings. Retain them in sets for inspection purposes (marking their original locations), but discard them before installation.

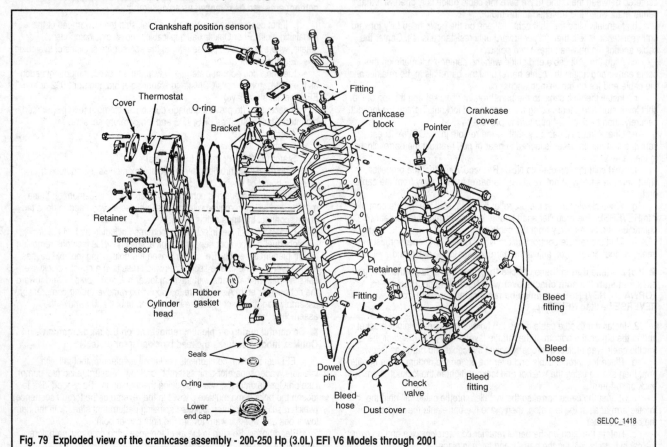

Fig. 79 Exploded view of the crankcase assembly - 200-250 Hp (3.0L) EFI V6 Models through 2001

SELOC_1418

✳✳ SELOC WARNING

ALWAYS use eye protection when attempting to remove the piston pin lockrings as they are prone to launching out of position unpredictably.

b. To remove the connecting rod from each piston, first remove the lockring from both ends of the piston pin. Use an awl inserted just under the lip of the ring to gently lever it out of the piston. Keep in mind that these lockrings are often distorted slightly during removal and so they must be discarded are replaced with new lockrings during installation.

■ If necessary, a piston cradle can be fabricated from a large block of wood. The important design features of the cradle are that it is semi-circular on top to hold (or cradle, get it?) the piston and that there is a bore at the bottom center into which the wrist pin can tapped or pressed. The final important feature of the cradle is that it is sturdy enough to withstand the force of the shop press, BUT, keep in mind that the fit is loose on these pistons and significant force should not be necessary. It is NOT uncommon for someone to simply support the piston in their lap when working on these motors.

■ GREAT TIP - Use a heat lamp to gently warm the piston prior to piston pin disassembly, it will make the pin slide out MUCH easier!

c. While carefully supporting the piston to prevent damage use a soft-faced mallet and driver (such as Piston Pin Tool #91-92973A1) to gently tap the pin from the bore in the piston. As the pin is removed, KEEP CLOSE TRACK of the locating/thrust washers and the loose needle bearings (there should be 34 needles per piston).

■ As the wrist pin is removed from the piston you must retain the loose needle bearings (IF you plan on reusing them) and the 2 locating/thrust washers from the small end of the connecting rod. If you can, stop tapping or pressing once the wrist pin JUST clears the bottom of the connecting rod bore. Then carefully withdraw the rod along with the needle bearings (place a thin bladed tool like a putty knife or your finger under the rod hole as it is pulled free of the piston bore/wrist pin). Inspect the bearings, if any of the needles are lost or worn, replace the entire bearing assembly. HONESTLY though, Mercury and Seloc both recommend replacing the needles at this time so it is probably best NOT to reuse them on this motor.

d. Repeat for the other pistons.

10. Remove the lower end cap assembly from the crankshaft and block. Remove and discard the O-ring. Note the direction the seal lips are facing before removal as it may vary slightly with model. It appears that one early (2001 and earlier) EFI models there were two seals installed back to back, a smaller diameter seal whose lips would face down toward the driveshaft, and a larger diameter seal whose lips would face up toward the crank. However, it also looks all OptiMax motors, as well as late-model EFI powerheads (2002 or later) use 2 similar or identical sized seals which are both installed with the lips facing downward toward the driveshaft. Either way, it is a good idea to note the direction the seals were facing previously before removal and reuse that method during assembly. Use a punch and hammer to drive out the old seals.

11. Carefully lift the crankshaft assembly from the cylinder block.

12. If necessary, disassemble the crankshaft and bearings as follows:

a. Inspect the upper roller bearing and replace the bearing/carrier assembly, as necessary. If the upper roller bearing is to be reused, it is still a

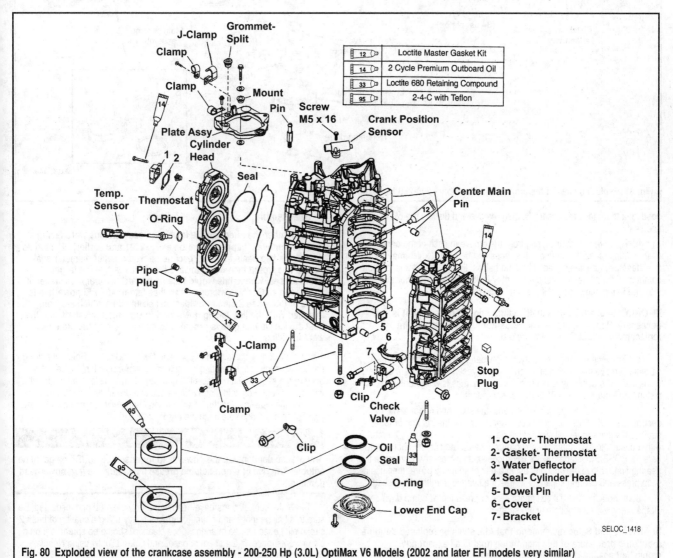

12	Loctite Master Gasket Kit	
14	2 Cycle Premium Outboard Oil	
33	Loctite 680 Retaining Compound	
95	2-4-C with Teflon	

1- Cover- Thermostat
2- Gasket- Thermostat
3- Water Deflector
4- Seal- Cylinder Head
5- Dowel Pin
6- Cover
7- Bracket

SELOC_1418

Fig. 80 Exploded view of the crankcase assembly - 200-250 Hp (3.0L) OptiMax V6 Models (2002 and later EFI models very similar)

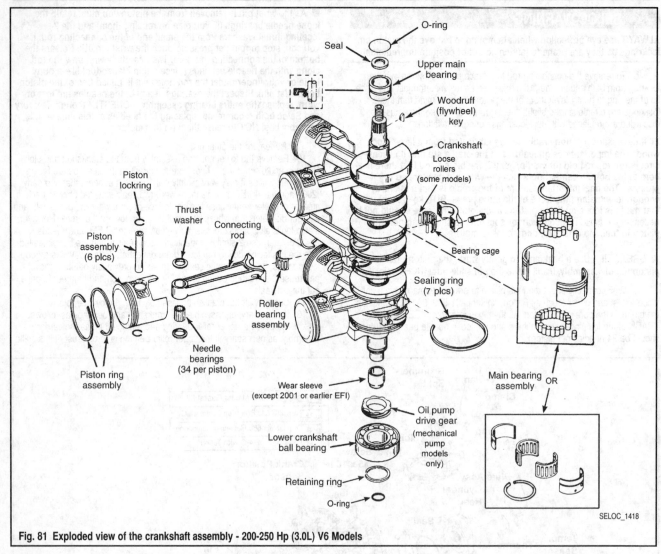

Fig. 81 Exploded view of the crankshaft assembly - 200-250 Hp (3.0L) V6 Models

good idea to remove the assembly, then remove and discard the old O-ring and seal.

b. Remove the retaining ring from the groove in each of the center main bearing races, then remove the races and bearings (which may consist of either loose needles or caged bearing halves, though the later is far more common). Like all bearings and races they MUST be kept together and returned to the same journal if they are to be reused.

■ ONLY remove the crankshaft sealing rings IF replacement is necessary. Honestly, the sealing rings normally do not require replacement UNLESS they are broken.

c. If replacement is necessary, while wearing a pair of safety glasses, carefully remove the sealing ring from the grooves in the shoulders on the crankshaft. The best way to remove the sealing rings is with a universal piston ring expander like #91-24697 or equivalent.

d. The lower bearing IS pressed onto the shaft and removal will destroy it, so ONLY remove it if necessary to replace the bearing or for additional access to the oil pump drive gear (which was used only on models with a mechanical oil pump, meaning mostly carbureted models not covered here, or POSSIBLY some EARLY EFI models), as necessary. To remove the bearing first remove the snapring using a pair of snapring pliers, then use a shop press and bearing separator to free the bearing from the crankshaft.

e. IF equipped, and if equipped, slide or press the oil pump drive gear off the keyway on the bottom of the shaft.

■ Mercury and Seloc recommend that ALL seals be replaced during a rebuilding procedure of this magnitude (that is all except any crankshaft sealing rings that aren't broken).

To assemble:

■ Remember the direction the lower end cap seal lips were facing prior to removal. It appears that on early (2001 and earlier) EFI models there were two seals installed back to back, a smaller diameter seal whose lips would face down toward the driveshaft, and a larger diameter seal whose lips would face up toward the crank. However, it also looks all OptiMax motors, as well as late-model EFI powerheads (2002 or later) use 2 similar or identical sized seals which are both installed with the lips facing downward toward the driveshaft. But the best advice we can give is probably to mimic the way the old seals were installed.

13. Install new seals to the upper and lower end caps. Before installing each seal, apply a light coating of Loctite® 271 or equivalent to the outer body of the seal (which contacts the housing when installed), then carefully tap the seal squarely into place using a suitable driver. After installation lubricate the seal lips and the NEW housing O-ring with a light coat of 2-4-C with Teflon or an equivalent marine grade grease.

✳✳ SELOC CAUTION

Any grease used inside the powerhead for assembly MUST be gasoline soluble. The use of a non-soluble grease may cause severe powerhead damage.

14. Make sure all parts have been thoroughly cleaned and inspected. Be sure to oil all bearings and bearing surfaces lightly using a coating of fresh 2-stroke oil. Loose needle bearings, once cleaned, should be spread out on a clean worksurface, like a sheet of paper and coated lightly using Needle Bearing Assembly Lubricant.

15. If the crankshaft was disassembled, install the bearings to prepare it for installation as follows:

a. If equipped, slide the oil pump drive gear (with the flange side facing downward toward the driveshaft) onto the rear of the crankshaft. Align the slot in the gear with the keyway on the crankshaft and gently seat it against the shoulder of the crank.

b. If removed, oil and install the lower main bearing using a press and suitable driver. Once the bearing is in position tight against the shoulder install the wear sleeve with the seal carrier and new seal.

c. If removed, carefully install the seal rings to the grooves in the crankshaft. The best way to install the sealing rings is with a universal piston ring expander like #91-24697 or equivalent.

d. For models equipped with caged rollers, apply a light coating of 2-stroke engine oil to the cages and bearing halves.

■ Although many models use caged rollers, if you should come across one with loose rollers, apply a light coating of Needle Bearing Assembly Lube, 2-4-C with Teflon or equivalent gasoline soluble marine grade grease, to the crankshaft center journal in order to hold the main roller bearings for EACH center main bearing.

e. Position the roller bearings followed by the bearing race halves so that holes in each race is positioned toward the lower (driveshaft) end of the crankshaft. Secure each pair of halves using the retainer ring (making sure the ring bridges the fracture lines of the race halves).

f. If removed (and we recommended it, so I hope you did it!), lubricate a new upper bearing oil seal and O-ring using a light coating of 2-4-C with Teflon, then install them to the bearing housing. Be sure to coat both the seal lips AND the outer diameter of the seal body with the grease, then position the seal with the lips facing downward (toward the crankshaft) and press it into position using a suitable driver. Oil and install the upper bearing assembly over the crankshaft with the hole in the bearing casing facing downward, toward the crankshaft.

16. It's time to test fit the crankshaft to the block. For starters, check the mating surfaces in the block and crankcase cover for the crankshaft sealing rings. If there are wear grooves present the rings on the new crankshaft will have to fit into the grooves without binding the crank. Before installation, remove any burrs that might exist on the groove edges. Also, check to make sure the bearing retaining dowel pins are in position in the block. Finally lubricate the seals with light oil and position the crank to the block along with the upper and lower end caps. Place the crankcase cover over the crankshaft, then rotate the shaft several times to check if the sealing rings are binding. If there is any evidence of binding, recheck the grooves for burrs. If the binding cannot be eliminated the block will have to be replaced.

17. Lubricate the crankshaft sealing rings and the upper lower crankshaft ends (oil seal mating areas) using a light coat of 2-stroke engine oil.

18. Position the crankshaft into the block with all of the crankshaft seal rings gaps facing straight up and with the holes in all of the bearing races facing the dowels. Gently push the shaft down into position to seat the races on the dowel pins.

19. Position the lower end cap to the crankshaft and cylinder block. Loosely install the cap-to-block retaining bolts but DO NOT tighten them until later, after the crankcase cover is installed (and the end cap-to-crankcase cover bolts are also installed and loosely tightened).

20. If disassembled for inspection and/or component replacement, assemble each piston as follows:

a. Apply a light coating of Needle Bearing Assembly grease (or 2-4-C with Teflon or an equivalent gasoline soluble marine grade grease) to the wrist pin bearing needles and the locating/thrust washers, then hold the lower locating washer in place while you insert the 34 needles around the inner circumference of the connecting rod wrist pin bore. Install the Piston Pin Tool # 91-92973A1 or equivalent into the bore to align the needles and hold them in place during assembly.

■ The wrist pin bearing tool is, essentially, a short round pin, roughly the same outer diameter of the wrist pin, but very short, just long enough to insert it through the connecting rod wrist pin bore and install the locating washers on either side. A similarly sized dowel pin can be substituted, or an old connecting rod wrist pin from the same size piston could be used. If substituting another rod or drift pin, make sure the surface is smooth and clean/free of all corrosion. Also, make sure the outer diameter of the tool is smaller than the wrist pin and bore to prevent the possibility of binding or damage.

b. Once all the needles are in position, inspect to make sure one is not missing. Generally speaking, if the tip of an awl can be inserted between the needles, one or more needles is/are missing.

NEVER replace just 1 needle. If one or more needles are damaged or missing, replace the entire set.

c. With the bearings and the wrist pin tool in position, install the upper locating/thrust washer. Again, a dab of grease should hold the thrust washer in position.

d. Apply a light coating of clean engine oil or assembly lube to the wrist pin and the pin bore in the piston.

■ Remember to position all components in the same orientation from which they were removed. For instance, the domes of most pistons are marked with a stamped UP for reference as to which end faces the flywheel.

e. Support the piston (ideally in a piston cradle for ease of assembly but the wrist pin fit should be relatively lose on both sides, especially if you place the pistons under a heat lamp before assembly, so this is not absolutely necessary). Insert the wrist pin into the bore and gently press it through the bore until it just starts to appear inside the piston skirt. Stop pressing, then position the connecting rod with the needle bearings, locating washers and wrist pin bearing installation tool inside the piston skirt. Continue to press the wrist pin into the position, through the thrust washers and bearings, pushing the wrist pin bearing installer through the other side and out of the piston.

f. Once the wrist pin is properly positioned and centered in the piston, install new lock-rings into the grooves in both ends of the piston bore. Installation of the lock-rings will be a lot easier using a lock ring installation tool (such as #91-93004A2 for EFI and for Pro/XS/Sport OptiMax models or #91-93004A3 for conventional OptiMax models or equivalent) which is essentially a shouldered driver on which you can place the lockring, then press it evenly into the piston bore.

■ Beginning around 2001, OE piston rings for these motors should be stamped with both an alpha code "T" for TOP of ring and numeric code either "1" for first/ top ring or "2" for second/bottom ring.

g. Use a ring expander to install the piston rings on each piston making sure to install the rings facing the right direction. The flat or rectangular side must be facing downward, while the tapered or half-keystone side should be facing upward. In most cases, replacement rings will be marked with a letter or dot on the side which faces up. Also, each piston ring groove should contain a dowel pin, around which the ring gap must be situated. If not the rings will likely break during installation (or if by they don't by some miracle, during the first start-up). Be sure to install the tapered ring in the top groove.

■ When using the ring expander, spread each ring just enough to slip over the piston.

h. Repeat for the other pistons.

■ Although you can get away without a ring compressor when assembling MOST Mercury powerheads, these are large enough with heavy duty enough rings that a compressor is recommended. Verify that the piston is facing the proper direction (the UP mark on the dome is towards the flywheel end of the powerhead). Additionally, pistons marked with a 'P' or a single dot on the dome (under the UP stamping) must be installed in the Port bank, while pistons marked with an 'S' or 2 dots on the dome must be installed in the Starboard bank.

21. Install each piston and connecting rod assembly to the crankshaft as follows:

a. Remove the connecting rod cap and bearings.

b. Prepare to install the piston by lubricating the journal surfaces and bearings using a light coating of 2-4-C with Teflon or an equivalent marine grade grease. Lubricate the piston rings, piston, ring compressor and cylinder wall lightly with 2-stroke engine oil.

c. Place the piston in the ring compressor and tighten carefully, just enough to compress the rings slightly.

d. Gently insert the piston into the cylinder bore. Work slowly, watching to make sure the connecting rod doesn't contact and damage the cylinder wall or the crankshaft journal.

e. Install the lower half of the bearing cage assembly into the connecting rod, then pull the rod upward into position over the crank pin.

f. Position the other bearing half over the crank pin.

g. Apply a light coating of 2-stroke engine oil to the threads of the NEW connecting rod cap bolts (remember, the old bolts are weakened from the torque process and should not be reused).

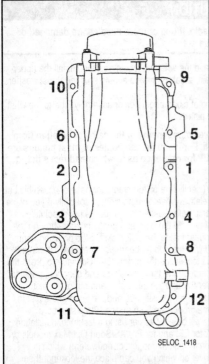

Fig. 82 Air plenum assembly torque sequence - 200-250 Hp (3.0L) V6 Models

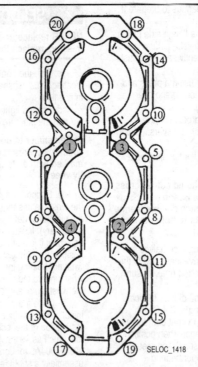

Fig. 83 Cylinder head torque sequence - 200-250 Hp (3.0L) V6 Models

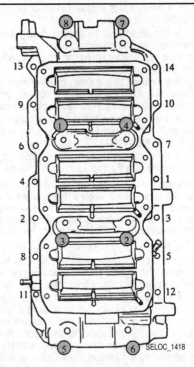

Fig. 84 Crankcase cover torque sequence - 200-250 Hp (3.0L) V6 Models

h. Position the piston/connecting rod assembly to the crankshaft (facing in the same direction as noted and marked during removal, then place the connecting rod end cap over the crank pin (also aligning the matchmarks made during removal). Lightly oil the threads and then start the bolts by hand.

■ **These motors use fractured end caps, meaning the end cap and connecting rod were originally 1 piece. As a result, the caps will only fit properly in one direction and when they fit the fracture lines will mate PERFECTLY. Before the bolts are fully tightened run a pencil over the fracture lines, if the pencil catches, loosen the screws, retighten and check again).**

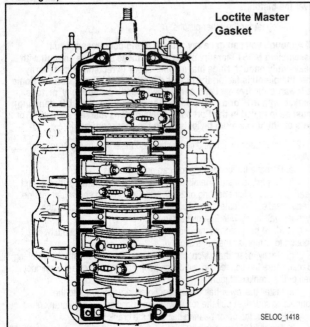

Fig. 85 Sealant application to the cylinder block prior to crankcase cover installation

i. Alternately and evenly tighten the end cap retaining bolts (again using a 3/8 in. 12-point socket), no more than 1 turn before moving to the other bolt. Once they start to tighten, use multiple passes to tighten them to specification. On most models, tighten the bolts first to 15 inch lbs. (1.7 Nm), next to 30 ft. lbs. (40.6 Nm) and then finally an additional 90° (1/4 turn). However, for the top 2 bolts on 2001 EFI models, tighten them to 20 ft. lbs. (27 Nm), then the additional 90° (1/4 turn).

j. After each piston is installed, rotate the crankshaft a few times to make sure there is no binding or catching.

k. Repeat for the other piston/connecting rod assemblies.

22. Make sure the bearing surfaces of the crankcase halves are COMPLETELY clean and free of any oil or debris. Clean the surface using Loctite® 7649 Primer, Loctite® Primer 203, or equivalent.

23. Next apply a thin, even, continuous bead of Loctite® Master Gasket 203 or equivalent on the mating surface of the block, JUST INBOARD of the main flange bolts holes on either side of the block (though make sure you surround both of the top most bolt holes/one on either side of the crank, as well as the 3 bottom most bolt holes/two on one side, one on the other side of the crank. Also, continue a thin strip of the bead inward on each block/cover webbing toward the edge of each crankshaft sealing ring mating surface or on either side of the inner bolt holes, so there is 2 strips of sealant toward each center main bearing, but be sure to leave some room for expansion between the edge of the bead and the sealing ring.

24. Assemble the crankcase cover to the block, then install and finger-tighten the 8 large center main bolts (after the threads and heads are lightly oiled). Tighten the bolts a little at a time to compress the crankshaft seal/gasket material and draw the cover down onto the block. Use at least 3 passes of the torque sequence to tighten the bolts to 30 ft. lbs. (40.5 Nm), then on the FINAL pass, tighten each of the bolts an additional 90° (1/4 turn).

25. Next, install the smaller crankcase cover flange bolts and finger-tighten. Again use at least 3 passes the torque sequence (starts at the center and works outward in a criss-crossing pattern) to tighten the smaller flange bolts to 21 ft. lbs. (28.5 Nm).

26. Tighten the lower end cap bolts using multiple passes in a clockwise pattern around the housing to 85 inch lbs. (9.5 Nm).

27. Again, rotate the crankshaft a few final times to make sure it moves freely and there is NO binding.

28. If the reed valve blocks were removed from the adaptor plate, install them to the plate using a new gasket. Tighten the retaining bolts to 90 inch lbs. (10 Nm).

29. Install the air plenum assembly along with the reed block adaptor plate to the crankcase cover and secure using the 12 retaining bolts. Tighten the bolts to 100 inch lbs. (11.5 Nm) for most models or 125 inch lbs. (14 Nm) for Pro/XS/Sport models. In either case use at least 3 passes of the torque sequence (a counterclockwise spiraling pattern that begins with the right side center bolt and works outward toward the bolt on the bottom right).

■ **On EFI motors, the 3.0L Work powerhead utilizes a low compression cylinder head which can be identified from other 225/250 hp heads by a difference in the casting numbers. The 3.0L Work motors uses the number 812851-C4 or 812851C-13, whereas the other 225/250 hp models use 812851-C2 or 812851C15. In either case, the heads are NOT interchangeable with other models.**

30. Install the cylinder heads to the block using new seals (making sure the thermostats are facing upward and normally the spark plugs are facing inward). The round individual cylinder seals on some models (mostly OptiMax motors) may be directional. On some models the seals are grooved on one side and pointy on the other, these style seals must be installed with the pointy side FACING the cylinder head. Apply a light coating of engine oil to the threads of the cylinder head bolts, then install them and tighten them using at least 3 passes of the torque sequence. The bolts must be tightened to 20 ft. lbs. (27 Nm) on the second to last pass and then finally an additional 90° on the final pass.

■ **On OptiMax motors there are a number of differences between different model years you must be aware off. First off, models through 2002 used a different diameter O-ring (4.089 in/103.86mm) than the 2003 and later models (3.950 in./100.33mm). Even though the earlier O-rings will fit the 2003 or later models, they must NOT be used or leakage may occur. Also, the 2003 and later O-ring must be stretched slightly to fit the 2003 and later cylinder heads. AND, the 2003 and later O-ring CAN be retrofitted to 2001-02 models. Next, the design of the cylinder head used in 2001-02 is different from both earlier and later models, as it was redesigned to increase hp output by about 3-4 hp and the ECM was recalibrated accordingly. So don't try to interchange heads out of those year spans.**

31. Install the water pressure (poppet) valve assembly to the lower side of the powerhead.

32. Either now or after the powerhead reinstalled (some prefer to wait so as to protect these components) install the remaining fuel and electric components to the powerhead. For details, where applicable, please refer to the appropriate sections of this guide.
- Thermostat assemblies
- Throttle/shift control cables and linkage (except electronic shift models)
- Bleed hoses. Be sure to connect and route them as tagged and noted during removal.
- EFI fuel injection components (fuel rail with injectors, vapor separator tank assembly and low pressure pulse fuel pump).
- OptiMax Direct fuel injection components (including air/fuel rail and injectors, fuel pump/vapor separator tank assembly). Also be sure to install the air compressor assembly (connecting the hoses as tagged during removal).
- Oil pump (probably installed along with the air plenum earlier, now is the time to reconnect the hoses, as tagged during removal).
- On-board oil reservoir tank
- Electronic shift/throttle assembly (OptiMax models, so equipped)
- Stator (for 2001 EFI models)
- Flywheel
- Alternator
- If removed (to prevent damage during cleaning) install the powerhead mounted sensors such as CPS, TPS and temperature sensors.
- Electrical component plate (along with the attached components such as the ECM, ignition coils/CDMs, and starter solenoid and, for 2002 or later EFI models, trim relays).
- Electric starter motor

CLEANING & INSPECTION

◆ **See Figures 79, 80 and 81**

Cleaning and inspecting the components is virtually the same for any 2-stroke outboard and varies mostly by specifications (which are listed in the Engine Specifications charts) or by component type. A section detailing the proper procedures, sorted mostly by component, can be found under Powerhead Refinishing.

POWERHEAD RECONDITIONING

Determining Powerhead Condition

Anything that generates heat and/or friction will eventually burn or wear out (for example, a light bulb generates heat, therefore its life span is limited). With this in mind, a running powerhead generates tremendous amounts of both; friction is encountered by the moving and rotating parts inside the powerhead and heat is created by friction and combustion of the fuel. However, the powerhead has systems designed to help reduce the effects of heat and friction and provide added longevity. The oil injection system combines oil with the fuel to reduce the amount of friction encountered by the moving parts inside the powerhead, while the cooling system reduces heat created by friction and combustion. If either system is not maintained, a break-down will be inevitable. Therefore, you can see how regular maintenance can affect the service life of your powerhead.

There are a number of methods for evaluating the condition of your powerhead. A Tune-Up Compression Test or Leak Test (as detained in the Engine Maintenance and Tune-Up section) can reveal the condition of your pistons, piston rings, cylinder bores and head gasket(s). Leak tests can also determine the condition of all engine seals and gaskets. Because the 2-stroke powerhead is a pump, the crankcase must be sealed against pressure created on the down stroke of the piston and vacuum created when the piston moves toward top dead center. If there are air leaks into the crankcase, insufficient fuel will be brought into the crankcase and into the cylinder for normal combustion.

Buy or Rebuild?

◆ **See Figures 86 and 87**

Once you've determined that a powerhead is worn out, you must make some decisions. The question of whether or not a powerhead is worth

rebuilding is largely a subjective matter and one of personal worth. Is the powerhead a popular one, or is it an obsolete model? Are parts available? Is the outboard it's being put into worth keeping? Would it be less expensive to buy a new powerhead, have your powerhead rebuilt by a machine shop, rebuild it yourself or buy a used powerhead? Or would it be simpler and less expensive to buy another outboard? If you have considered all these matters and more and have still decided to rebuild the powerhead, then it is time to decide how you will rebuild it.

■ **The editors at Seloc feel that most powerhead machining should be performed by a professional machine shop. Don't think of it as wasting money, rather, as insurance that the job has been done right the first time. There are many expensive and specialized tools required to perform such tasks as boring and honing a powerhead. Even inspecting the parts requires expensive micrometers and gauges to properly measure wear and clearances. Also, a machine shop can deliver to you clean and ready to assemble parts, saving you time and aggravation. Your maximum savings will come from performing the removal, disassembly, assembly and installation of the powerhead and purchasing or renting only the tools required to perform the above tasks. Depending on the particular circumstances, you may save 40 to 60 percent of the cost doing these yourself.**

A complete rebuild or overhaul of a powerhead involves replacing or reconditioning all of the moving parts (pistons, rods, crankshaft, etc.) with new or remanufactured ones and machining the non-moving wearing surfaces of the block and heads. Unfortunately, this may not be cost effective. For instance, your crankshaft may have been damaged or worn, but it can be machined for a minimal fee.

So, as you can see, you can replace everything inside the powerhead, but, it is wiser to replace only those parts which are really needed and, if possible, repair the more expensive ones.

Powerhead Overhaul Tips

Most powerhead internal overhaul procedures are fairly standard. In addition to the specific Disassembly & Assembly procedures provided earlier, this section should serve as a guide to acceptable rebuilding procedures. Examples of standard rebuilding practice are given and should be used along with specific details concerning your particular powerhead.

Competent and accurate machine shop services will ensure maximum performance, reliability and powerhead life. In most instances it really does make more sense to remove, clean and inspect the component, buy the necessary parts and deliver these to a shop for actual machine work.

Much of the assembly work (crankshaft, bearings, pistons, connecting rods and other components) is well within the scope of most people's tools and abilities. You will have to decide for yourself the depth of involvement you desire in a powerhead repair or rebuild.

CAUTIONS

Aluminum is extremely popular for use in powerheads, due to its low weight. Observe the following precautions when handling aluminum parts:
• Never hot tank aluminum parts, the caustic hot tank solution will eat the aluminum
• Remove all aluminum parts (identification tag, etc.) from powerhead parts prior to hot tanking
• Always coat threads lightly with oil or anti-seize compounds (or threadlocking compounds when specified) before installation, to prevent seizure
• Never overtighten bolts or spark plugs especially in aluminum threads

When assembling the powerhead, any parts that will be exposed to frictional contact must be prelubed to provide lubrication at initial start-up. Any product specifically formulated for this purpose can be used, but there are times when Mercury specifies 2-stroke oil or 2-4-C with Teflon or some other grease. Unfortunately, Mercury isn't very consistent with this, often specifying one type of lubricant for a bearing or seal on one motor, and a different type on another motor. We suspect that might be a sign that most of the time it really doesn't matter, but we still think it is probably safer to stick with the specific recommendations whenever possible.

When semi-permanent (locked, but removable) installation of bolts or nuts is desired, threads should be cleaned and coated with Loctite® or another similar, commercial non-hardening sealant.

CLEANING

Before the powerhead and its components are inspected, they must be thoroughly cleaned. You will need to remove any varnish, oil sludge and/or carbon deposits from all of the components to insure an accurate inspection. A crack in the block or cylinder head can easily become overlooked if hidden by a layer of sludge or carbon.

Most of the cleaning process can be carried out with common hand tools and readily available solvents or solutions. Carbon deposits can be chipped away using a hammer and a hard wooden chisel. Old gasket material and varnish or sludge can usually be removed using a scraper and/or cleaning

solvent. Extremely stubborn deposits may require the use of a power drill with a wire brush. Always follow any safety recommendations given by the manufacturer of the tool and/or solvent. You should always wear eye protection during any cleaning process involving scraping, chipping or spraying of solvents.

■ **If using a wire brush, use extreme care around any critical machined surfaces (such as the gasket surfaces, bearing saddles, cylinder bores, etc.). Use of a wire brush is not recommended on any aluminum components.**

An alternative to the mess and hassle of cleaning the parts yourself is to drop them off at a local machine shop. They will, more than likely, have the necessary equipment to properly clean all of the parts for a nominal fee.

✳✳ SELOC CAUTION

Always wear eye protection during any cleaning process involving scraping, chipping or spraying of solvents.

Remove any plugs, seals or other components and carefully wash and degrease all of the powerhead components including the fasteners and bolts. Small parts should be placed in a metal basket and allowed to soak. Use pipe cleaner type brushes and clean all passageways in the components.

Use a ring expander to remove the rings from the pistons. Clean the piston ring grooves with a ring groove cleaner or a piece of broken ring. Scrape the carbon off of the top of the piston. You should never use a wire brush on the pistons as pieces of the brush may be left behind, which can become hot spots damaging the piston when it is returned to service. After preparing all of the piston assemblies in this manner, wash and degrease them again.

Powerhead Preparation

◆ See Figure 88

To properly rebuild a powerhead, you must first remove it from the outboard, then disassemble and inspect it. Ideally you should place your powerhead on a stand. This affords you the best access to the components. Follow the manufacturer's directions for using the stand with your particular powerhead.

Now that you have the powerhead on a stand, it's time to strip it of all but the necessary components. Before you start disassembling the powerhead, you may want to take a moment to draw some pictures, fabricate some labels or get some containers to mark and hold the various components and the bolts and/or studs which fasten them. Modern day powerheads use a lot of little brackets and clips which hold wiring harnesses and such and these holders are often mounted on studs and/or bolts that can be easily mixed up. The manufacturer spent a lot of time and money designing your outboard and they wouldn't have wasted any of it by haphazardly placing brackets, clips or fasteners. If it's present when you disassemble it, put it back when you assemble it, you will regret not remembering that little bracket which holds a wire harness out of the path of a rotating part.

Fig. 86 The question of whether or not a powerhead is worth rebuilding is largely a subjective matter and one of personal worth. This powerhead is not worth much in its present condition

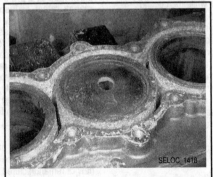

Fig. 87 A burned piston like this one will be replaced during an overhaul. The condition which caused the hole in the top of the piston must be identified and corrected or the same thing will happen again

Fig. 88 Larger powerheads will require the use of a lifting eye and engine hoist

■ Digital cameras can be a God send when it comes to powerhead overhaul. If you've got access to one, take the time to photograph each side and the top of the assembled powerhead. Those pictures can become invaluable when it comes time to reassemble and you failed to properly tag something or to note the proper routing.

You should begin by unbolting any accessories attached to the powerhead. Remove any covers remaining on the powerhead. The idea is to reduce the powerhead to the bare necessities (cylinder head(s), cylinder block, crankshaft, pistons and connecting rods), plus any other 'in block' components.

Cylinder Block and Head

GENERAL INFORMATION

◆ See Figures 89 thru 99

The cylinder block is made of aluminum and may have cast-in iron cylinder liners. It is the major part of the powerhead and care must be given to this part when service work is performed. Mishandling or improper service procedures performed on this assembly may make scrap out of an otherwise good casting. The cylinder assembly casting and other major castings on the outboard are expensive and need to be cared for accordingly.

There are generally three parts to the cylinder assembly, the cylinder block, the cylinder head and the crankcase half. We say generally, because there appears to be at least one model on which the block does not have a removable head, however the vast majority of motors in general and these Mercury powerheads in particular do. The cylinder block and crankcase half are married together and line bored to receive the crankshaft bearings, reed blocks and on some powerheads sealing rings. After this operation they are treated as one casting.

■ Remember that anything done to the mating surfaces during service work will change the inner bore diameter for the main bearings, reed blocks and sealing rings and possibly prevent the block and crankcase mating surfaces from sealing.

The only service work allowed on the mating surface is a lapping operation to remove nicks from the service. Carefully guard this surface when other service work is being performed. The different sealing materials used to seal the mating surfaces are sealing strips, sealing compound and Loctite®.

Since the 2-stroke powerhead operates like a pump with one inlet and one outlet for each cylinder, special sealing features must be designed into

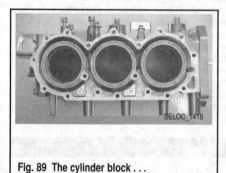
Fig. 89 The cylinder block . . .

Fig. 90 . . . crankcase half . . .

Fig. 91 . . . and cylinder head make up the major components of the cylinder assembly

Fig. 92 An exhaust port can be seen above the lip of the piston. The inlet port is on the opposite side of the cylinder wall

Fig. 93 The block and crankcase are machined to fit perfectly. They cradle the spinning crankshaft

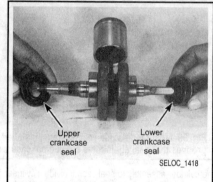
Fig. 94 On smaller motors, seals are installed to the ends of the crankshaft

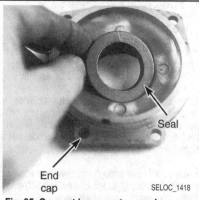
Fig. 95 On most larger motors, end caps are used to secure the crankshaft seals

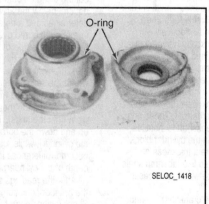
Fig. 96 End caps are sealed to the crankcase/block with O-rings. . .

Fig. 97 . . .and may be installed at both the upper. . .

Lower end cap

SELOC_1418

Fig. 98 . . .and lower ends of the block

Sealing ring

SELOC_1418

Fig. 99 Most models use sealing rings, installed in crankshaft grooves that seal against the block/crankcase webbing

the cylinder assembly to seal each individual cylinder in a multi-Cylinder powerhead. Each inlet manifold must be completely sealed both for vacuum and pressure. One way of doing this internally is with a labyrinth seal, which is located between two adjacent cylinders next to the crankshaft. It may be of aluminum or brass, formed in the assembly and machined with small circular grooves running very close to a machined area on the crankshaft. The tolerance is so close that fuel residue puddling in the seal effectively completes the seal between the cylinder block and crankcase halves against the crankshaft. Crankcase pressures are therefore retained to each individual cylinder. No repair of the labyrinth seal is made. If damage has occurred to the seal, the main bearings have allowed the crankshaft to run out and rub.

Another method of internal sealing between the crankcases is with seal rings (the most popular method on these Mercury powerheads). These rings are installed in grooves in the crankshaft. When the crankshaft is installed, the sealing rings mate up to and seal against the web in the cylinder block crankcase halves and crankshaft. Sealing rings of different thickness are available for service work. The side tolerance is close, so like labyrinth seals, puddled fuel residue will effectively complete the seal between crankcases and crankshaft.

To seal the ends of the cylinder assembly around the crankshaft, O-rings are installed around the end caps (when equipped) and/or neoprene seals are installed inside the cap and seal against the crankshaft.

INSPECTION

◆ See Figures 100, 101 and 102

Everytime the cylinder head is removed, the cylinder head and cylinder block deck should be checked for warping. Do this with a straight edge or a surface block. If the cylinder head or cylinder block deck are warped, the surface should be machined flat by a competent machine shop. Minor warpage may be cured by using emery paper in a figure eight motion on a surface block until the surface is true.

Inspect the cylinder head and cylinder block for cracks and damage to the bolt holes caused by galvanic corrosion. On models which do not use a cylinder head, check the cylinder dome for holes or cracks caused by overheating and pre-ignition. The spark plug threads may also be damaged by overtorquing the spark plug.

Quite often the small bolts around the cylinder block sealing area are seized by corrosion. If white powder is evident around the bolts, stop. Galvanic corrosion is probably seizing the shank of the bolt and possibly the threads as well. Putting a wrench on them may just twist the head off, creating one big mess. Know the strength of the bolt and stop before it breaks. If it does break, don't reach for an easy out, it won't work.

A good way to service these seized bolts is with localized heat (from a heat gun, not a torch) and a good penetrating oil. Heat the aluminum casting, not the bolt. This releases the bolt from the corrosive grip by creating clearance between the bolt, the corrosion and the aluminum casting. Be careful because too much heat will melt the casting. Many bolts can be released in this way, preventing drilling out the total bolt and Heli-Coiling the hole or tapping the hole for an oversize bolt.

To help prevent bolts from seizing due to corrosion, coat threads with a good anti-seize compound.

Cylinder Bores

GENERAL INFORMATION

The purpose of the cylinder bore is to help lock in combustion gases, provide a guided path for the piston to travel within, provide a lubricated surface for the piston rings to seal against and transfer heat to the cooling system. These functions are carried out through all engine speeds. To function properly the cylinder has to have a true machined surface and must have the proper finish installed on it to retain lubricant.

INSPECTION

◆ See Figures 103 and 104

The roundness of the cylinder diameter and the straightness of the cylinder wall should be inspected carefully. Micrometer readings should be taken at several points to determine the cylinder condition. Start at the bottom using an outside micrometer or dial bore gauge. By starting at the bottom, below the area of ring travel, cylinder bore diameter can be determined and a determination can be made if the powerhead is standard or has been bored oversize. Take the second measurement straight up from the first in the area of the ports and note that the cylinder is larger here. This is the area where the rings ride and it has worn slightly. Take the third measurement within a half inch of the top of the cylinder, straight up from where the second measurement was taken. These three measurements should be repeated with the measuring instrument turned 90° clockwise.

Comparing the measurements taken will tell you how much the cylinder is worn from the rings. Taper tells you how much that wear varies as you move up and down the bore (difference between the readings higher and lower in the cylinder), while roundness tells you how much elliptical the cylinder has become (differences in readings taken at 90 degree angles along the same depth of the cylinder).

After the readings are taken, you will have enough information to access the cylinder condition. Compare these readings to the Engine Rebuilding Specifications chart for the motor in question. Look at whatever specifications are supplied for Cylinder Bore, Bore Wear Limit, Cylinder Out of Round and/or Cylinder Taper, as applicable. This will tell you if the rings

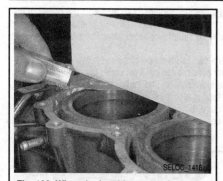

Fig. 100 When the head is removed, check the head and block deck for warping using a straight edge and feeler gauge

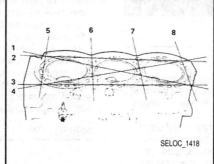

Fig. 101 When inspecting components for warping, check in multiple directions

Fig. 102 To help prevent bolts from seizing due to corrosion, coat threads with a good antiseize compound.

can simply be replaced or if the cylinder will need to be overbored. While measuring the cylinder, you should also be noting if there is a cross-hatched pattern on the cylinder walls. Also note any scuffing or deep scratches.

REFINISHING

If the cylinder is out of round, worn beyond specification, scored or deeply scratched, reboring will be necessary. If the cylinder is within specification, it can be deglazed with a flex hone and new rings installed.

■ **Some cylinders are chrome plated and require special service procedures. Consult a qualified machine shop when dealing with chrome plated cylinders.**

Almost all engine block refinishing must be performed by a machine shop. If the cylinders are not to be rebored, then the cylinder glaze can be removed with a ball hone. When removing cylinder glaze with a ball hone, use a light or penetrating type oil to lubricate the hone. Do not allow the hone to run dry as this may cause excessive scoring of the cylinder bores and wear on the hone. If new pistons are required, be sure to save the old parts to use as a reference for positioning.

When deglazing, it is important to retain the factory surface of the cylinder wall. The cross-hatched pattern on the cylinder wall is used to retain oil and seal the rings. As the piston rings move up and down the wall, a glaze develops. The hone is used to remove this glaze and reestablish the basket weave pattern. The pattern and the finish is has a satin look and makes an excellent surface for good retention of 2-stroke oil on the cylinder wall.

There is nothing magic about the crosshatch angle but there should be one similar to what the factory used. (approximately 20-40°). Too steep an angle or too flat a pattern is not acceptable and as it is not good for ring

seating. Since the hone reverses as it is being pushed down and pulled up the cylinder wall, many different angles are created. Multiple criss-Crossing angles are the secret for longevity of the cylinder and the rings. The pattern allows 2-stroke oil to flow under the piston ring bearing surface and prevents a metal-to-metal contact between the cylinder wall and piston ring. The satin finish is necessary to prevent early break-in scuffing and to seat the ring correctly.

After the cylinder hone operation has been completed, one very important job remains. The grit that was developed in the machining process must be thoroughly cleaned up. Grit left in the powerhead will find its way into the bearings and piston rings and become embedded into the piston skirts, effectively grinding away at these precision parts. Relate this to emery paper applied to a piece of steel or steel against a grinding stone. The effect is removal of material from the steel. Grit left in the powerhead will damage internal components in a very short time.

Wiping down the cylinder bores with an oil or solvent soaked rag does not remove grit. Cleaning must be thorough so that all abrasive grit material has been removed from the cylinders. It is important to use a scrub brush and plenty of soapy water. Remember that aluminum is not safe with all cleaning compounds, so use a mild dish washing detergent that is designed to remove grease. After the cylinder is thought to be clean, use a white paper towel to test the cylinder. Rub the paper towel up and down on the cylinder and look for the presence of gray color on the towel. The gray color is grit. Re-scrub the cylinder until it is perfectly clean and passes the paper towel test. When the cylinder passes the test, immediately coat it with 2-stroke oil to prevent rust from forming.

■ **Rust forms very quickly on clean, oil free metal. Immediately coat all clean metal with 2-stroke oil to prevent the formation of rust.**

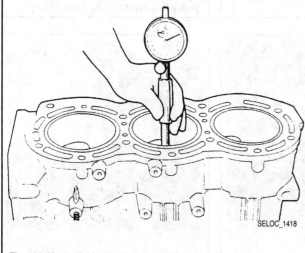

Fig. 103 Use a dial bore gauge to measure roundness and straightness of the cylinder wall

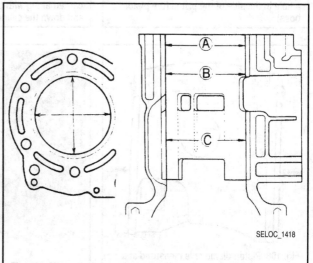

Fig. 104 Measure at several points to determine the cylinder condition

Pistons

GENERAL INFORMATION

◆ See Figures 105 thru 108

A Piston is the moveable end of a cylinder. The cylinder bore provides a guided path for the piston allowing a small clearance between the piston skirt and cylinder wall. This clearance allows for piston expansion (which occurs with heat) and controls piston rock within the cylinder.

One common modern piston design is such that the head of the piston directs incoming fuel toward the top of the cylinder and outgoing exhaust to the exhaust port in the cylinder wall. This design is called a deflector type piston head. The deflector dome deflects the incoming fuel upward to the spark plug end of the cylinder, partially cooling the cylinder and spark plug tip. It also purges the spent gases from the cylinder. In essence, the incoming fuel charge is chasing out the exhaust gases from the cylinder.

Not all piston designs are of the deflector head type. Other pistons have a small convex crown on the piston head. In this case, port design aids in directing the incoming fuel upward. The piston head bears the brunt of the combustion force and heat. Most of the heat is transferred from the piston head through the rings to the cylinder wall and then on to the cooling system.

The piston design can be round, cam ground or barrel shaped. The cam ground design allows for expansion of the piston in a controlled manner. As the piston heats up, expansion take place and the piston moves out along the piston pin becoming more round as it warms up. Barrel shaped pistons rock very slightly in the bore which helps to keep the rings free.

The piston has machined ring grooves in which the rings are installed. They are carried along with the piston as it travels up and down the cylinder wall. There is one small pin in each ring groove to prevent the ring from rotating. The piston skirt is the bearing area for thrust and rides on the cylinder wall oil film. The side thrust of the piston is dependent upon piston pin location. If the pin is in the center of the piston, then there will be more

thrust. If the pin is offset a few thousandths of an inch from the center of the piston, there will be less thrust. A used piston will have one side of the piston skirt show more signs of wear than the opposite side. The side showing wear is the major thrust side.

Thrust is caused by the pendulum action of the rod following the crankshaft rotation, which pulls the rod out from under the piston. The combustion pressure therefore pushes and thrusts the piston skirt against the cylinder wall. Some heat is also transferred at this point. The other skirt receives only minor pressure. Some pistons have small grooves circling the skirts to retain oil in the critical area between the skirt and the cylinder wall.

INSPECTION

◆ See Figures 109 and 110

The piston needs to be inspected for damage. Check the head for erosion caused by excessive heat, lean mixtures and out of specification timing/synchronization. Examine the ring land area to see if it is flat and not rounded over. Also look for burned through areas caused by pre-ignition. Check the skirt for scoring caused by a break through of the oil film, excessive cylinder wall temperatures, incorrect timing/synchronization or inadequate lubrication.

To measure the piston diameter, place an outside micrometer on the piston skirt at the specified location. Normally, all pistons in a given powerhead should read the same. Check the specifications for placement of the micrometer when measuring pistons. Generally on all but the smallest motors, there is a specific place on the piston. This is especially true of barrel shaped pistons that are larger in the middle than they are at the top and bottom.

If the piston looks reasonably good after cleaning, take a close look at the ring lands. Wear may develop on the bottom of the ring lands. This wear is usually uneven, causing the ring to push on the higher areas and loads the

Fig. 105 The piston pin is inserted through a hole in the side of the piston (the piston boss)

Fig. 106 A piston has machined grooves in which the rings are installed. They are carried along with the piston as it travels up and down the cylinder wall

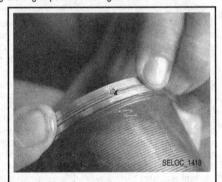

Fig. 107 There is one small pin in each ring groove to prevent the ring from rotating

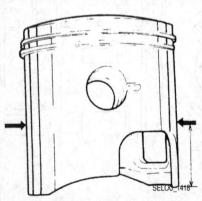

Fig. 108 Piston diameter is measured at a specific point which the manufacturer will normally specify

Fig. 109 This piston is severely scored from lack of lubrication and should not be reused

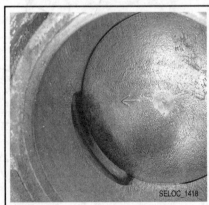

Fig. 110 Pistons should be installed with marks facing as noted during disassembly

ring unevenly when inertia is the greatest. Such uneven support of the ring will cause ring breakage and the piston will need to be replaced.

When installing a new ring in the groove, measure the ring side clearance against specification. Also check the see if the ring pins are there and that they have not loosened. Measure the skirt to see if the piston is collapsed.

Pistons must be oriented in a particular manner with regards to the cylinder for proper operation. Most Mercury pistons are marked with a stamping such as the word **UP** which denotes which side of the piston should be facing the flywheel. Some piston manufacturers will use other markings, such as an arrow which must be faced in a certain direction (such as toward the exhaust port or toward the flywheel etc). Always note markings before the pistons are removed, and, if pistons are to be replaced, compare the new markings and shape of the piston to determine proper orientation.

Piston Pins

GENERAL INFORMATION

◆ See Figures 111, 112 and 113

A hole placed in the side of the piston, commonly referred to as the piston boss, is used to mount the piston to the piston pin. The combustion pressure is transferred to the piston pin and connecting rod, then on to the crankshaft where it is converted to rotary motion.

The pin is fitted to the piston bosses. The piston pin is the inner bearing race for the bearing mounted in the small end of the connecting rod. This transfers the combustion pressures into the connecting rod and allows the rod to swing with a pendulum-like action.

Piston pins are secured into both piston bosses. All have retainers and in addition some types (usually not Mercurys) use a press fit to secure the pin. There are some models which use a slip fit. These may require special installation techniques.

Another type of pin fitting is loose on one side and tight on the other. This type aids in removal of the pin without collapsing the piston. With this design, always press on the pin from the loose boss side. The piston is marked on the inside of the piston skirt with the word "loose" to identify the loose boss. Always press with the loose side up and press the pin all the way through and out. When installing, press with the loose side up.

In all pressing operations, set the piston in a cradle block to support the piston. Some pistons require heating to expand the piston bosses so the pin can be pressed out without collapsing the piston. Other pistons just have a slip fit.

■ **Though the fitting is loose on most Mercury pistons, warming them first with a heat lamp is usually a good idea to ensure the pin will slide freely from the piston pin boss.**

INSPECTION

◆ See Figures 114 thru 119

Check the piston pin retainer grooves for evidence of the retainers moving as they may have been distorted. Always replace the retainers once they

have been removed. If there is evidence of wear in any of these areas, the piston should be replaced.

Inspect piston pin for wear in the bearing area. Rust marks caused by water will leave a needle bearing imprint. Chatter marks on the pin indicate that the piston pin should be replaced. If these marks are not too heavy, they may possibly be cleaned with emery paper for loose needle bearings or crocus cloth for caged bearings.

If the piston pin checks out visually, measure its outside diameter and compare that measurement with the inside diameter of the piston pin bore. Proper clearance is vital to providing enough lubrication.

MOST Mercury models however, do not provide a piston pin-to-bore clearance specification.

Piston Rings

GENERAL INFORMATION

◆ See Figure 120

The piston rings seal the piston to the cylinder bore, just as other seals are used on the crankshaft and lower unit. To perform correctly, the rings must conform to the cylinder wall and maintain adequate pressure to insure their sealing action at required operating speeds and temperatures. There are different designs used throughout the outboard industry. A given manufacturer will select a ring design that meets the operating requirements of the powerhead. This may be a standard ring, a pressure back (Keystone) ring or a combination of rings.

■ **Most Mercury rings are half-Keystone (tapered on top and rectangular on the bottom).**

The functions of the piston ring include sealing the combustion gases so they cannot pass between the piston and the cylinder wall into the crankcase (which would upset the pulse while allowing power to escape) and maintaining an oil film in conjunction with the cylinder wall finish throughout the ring travel area. The rings also transfer heat picked up by the piston during combustion. This heat is transferred into the cylinder wall and thus to the cooling system. There are either two or three rings per piston, which perform these functions.

■ **An oil control ring is not used on 2-stroke engines.**

All piston rings used on these motors are of the compression type. This means that they are for sealing the clearance between piston and cylinder wall. They are not allowed to rotate on the piston as automotive piston rings do. Instead, they are prevented from rotating by a pin in the piston ring groove. If the ring was allowed to turn, a ring end could snap into the cylinder port and become broken.

The ring ends are specially machined to compensate for the pin. As the rings warm up in a running powerhead they expand, thereby requiring a specific end gap between the ring ends for expansion. This ring gap decreases upon warm-up, effectively limiting blowing gases (from the combustion process) from going into the crankcase. The rings ride in a piston ring groove with minimal side clearance, which gives them support as they move up and down the cylinder wall. With this support, combustion gas

Fig. 111 The holes in the bottom of this piston pin bore provide oiling to the piston pin . . .

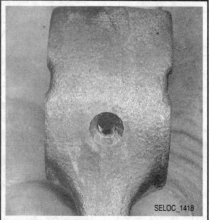

Fig. 112 A similar hole in the connecting rod also oils the pin

Fig. 113 Most Mercury motors use a floating pin design (which slides out easily once the retainers are removed)

Fig. 114 Measuring the piston pin bore inside diameter. You can compare this with the pin outside diameter to determine pin-to-bore clearance

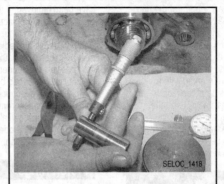

Fig. 115 Measuring the piston pin outside diameter at the point where the pin aligns with the piston pin bore . . .

Fig. 116 . . . and also at the point where the pin aligns with the connecting rod bore

Fig. 117 Typical caged bearing design used to support piston pins. . .

Fig. 118 . . .or a loose needle bearing design

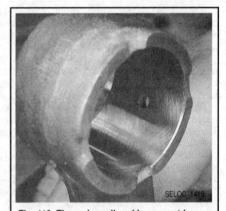

Fig. 119 The rod small end bore must be perfectly round to prevent bearing trouble

Fig. 120 The piston rings seal the piston to the cylinder bore

pressure and oil effectively seal the piston ring against the ring land and the cylinder wall. As long as the oil mix is correct and temperatures remain where they should, the rings will provide service for many hours of operation.

INSPECTION

◆ See Figures 121 thru 125

One of the first indications of ring trouble is the loss of compression and performance. When compression has been lost or lowered because of the ring not sealing, the ring is either broken or stuck with carbon, gum or varnish. Improper oil mixing and stale gasoline provide the carbon, gum and varnish which cause the rings to stick. Low octane fuel that allows pre-

ignition/detonation, improperly adjusted timing/synchronization and lean fuel mixtures can damage the ring land, causing the ring to stick or break.

■ **Running the outboard out of the water for even a few seconds can have damaging effects on the rings, pistons, cylinder walls and water pump.**

To determine if the rings fit the cylinder and piston, two measurements are taken; ring gap and ring side clearance. To determine these measurements, the ring is pushed into the cylinder bore using the piston skirt, so it will be square. Position each ring, one at a time, at the bottom of the cylinder (the smallest diameter) and using a feeler gauge, measure the expansion space between the ring ends. This is known as the ring gap measurement. Compare this measurement against the Ring Gap specifications (where provided). If the measurement is too small, the ring must be filed to increase the gap. If it is too large, either the bore is too large or the ring is not correct for the powerhead.

After ring end gap has been determined, position each ring in the piston ring groove and using a feeler gauge, measure between the ring and the piston ring land. Compare this measurement against the Ring Side Clearance specifications (where provided). If the measurement is too small, the ring groove may be compressed. Inspect the ring groove and ring land condition. If it is too large the ring may not correct for the powerhead.

Connecting Rods

GENERAL INFORMATION

The connecting rod transfers the combustion pressure from the piston pin to the crankshaft, changing the vertical motion into rotary motion. In doing so, the connecting rod swings back and forth on the piston pin like a pendulum while it is traveling up and down. It goes down by combustion pressure and goes up by flywheel momentum and/or other power strokes on a multi-Cylinder powerhead. The connecting rod can be of aluminum on smaller horsepower fishing outboards or more often of steel on larger horsepower models.

Fig. 121 Use a feeler gauge to measure ring gap with the ring installed in the cylinder

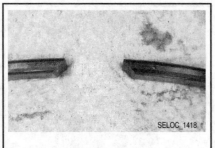

Fig. 122 Some rings are square while other rings have a notched shape

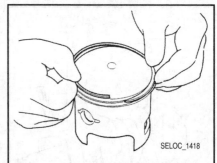

Fig. 123 It is possible to install rings by placing the ring in the groove and carefully working it around the piston. . .

Fig. 124 . . . however the best method is to use a ring expander

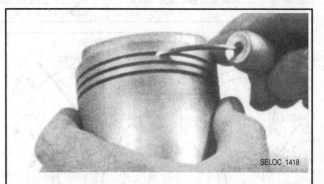

Fig. 125 Clean the piston ring grooves using a special tool or piece of a broken ring

Most connecting rod designs use a steel liner with needle bearings in the large end and needle bearings in the small end (often loose needles on these models).

The steel rod is a bearing race at both the large and small ends of the rod. It is hardened to withstand the rolling pressures applied from the loose or caged needle bearings. Like most connecting rod designs, these rods use two piece, fractured end-caps which must be bolted into position in only the one proper alignment (never 180 degrees out).

The connecting rods are generally mist lubricated. Some of the rods have a trough design in the shank area. Oil holes may be drilled into the bearing area at both ends of this trough. Oil mist that falls out of the fuel will settle into the rod trough and collect. As the rod moves in and out, the oil is sloshed back and forth in the trough and out the oil holes into the rod or piston pin bearings. This provides sufficient lubrication for these bearings. When the rod is equipped with oil holes, the oil holes have to be placed in the upward position toward the tapered end of the crankshaft when reassembled.

INSPECTION

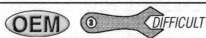

◆ **See Figures 126 thru 129**

Damage to the connecting rod can be caused by lack of lubrication and will result in galling of the bearing and eventual seizing to the crankshaft. Over speeding of the powerhead may also cause the upper shank area of the rod to stretch and break near the piston pin.

Steel rods are inspected in the bearing areas, much like you would inspect a roller bearing. Look for scoring, pit marks, chatter marks, rust and color change. A blue color indicates overheating of the bearing surface. Minor rust marks or scoring may be cleaned up using crocus cloth for caged needle bearings or emery paper for loose needle bearings. A piece of round stock, cut with a slot in one end to accept a small piece of emery paper and mounted in a drill motor, can be used to clean up the rod ends.

The rod also needs to be checked to see if it is bent or has a twist in it. To do this, remove the piston and place the rod on a surface plate or a piece of flat glass (automotive widow). Using a flash light behind the rod and looking

from in front of the rod, check for any light which can be seen under the rod ends. If light can be seen shining under the rod ends, the rod is bent and it must be replaced. You can also use a 0.002 in. (0.05mm) feeler gauge. See if it will start under the machined area of the rod. If it will, the rod is bent.

Examine the rod bolts and studs for damage and replace the nuts where used. Keep in mind that Mercury recommends that about half their connecting rod bolts should only be used once. This is normally recommended when, during torquing, the bolts are stretched and therefore weakened (so they may break if reused). Ironically on bolts of this type which are stretched when tightened NORMALLY use an angle torque as the final step in the tightening sequence (as does nearly all the Mercury connecting rods). So we do wonder if ALL of the connecting rod bolts need to be replaced on these motors, regardless of the fact that Mercury doesn't always specifically recommend it. Certainly, we suspect it is a worthwhile expense, if only for piece of mind.

Always reinstall the rod back on the same journal from which it was removed. The needle bearings, rod bearing surface and crankshaft journal are all mated to each other once the powerhead has been run.

■ **When installing the connecting rod, typically the long sloping side must be installed toward the exhaust side of the cylinder assembly and if there is a hole in the connecting rod, position the oil hole upward. Many piston designs are marked with the word "UP". This side should be placed toward the tapered end of the crankshaft.**

Crankshaft

GENERAL INFORMATION

◆ **See Figures 130 and 131**

The crankshaft is used to convert vertical motion received from the piston and connecting rod into rotary motion, which turns the driveshaft. It mounts the flywheel, which imparts a momentum to smooth out pulses between power strokes. It also provides sealing surfaces for the upper and lower seals and provides a surface for the labyrinth seal or sealing rings to hold and/or a groove in which sealing rings are installed to seal pressures into each crankcase.

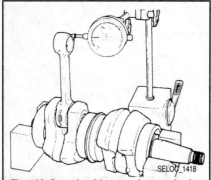

Fig. 126 On rods with not end caps, check side clearance using a dial gauge (rod deflection). On rods with end caps, use a feeler gauge

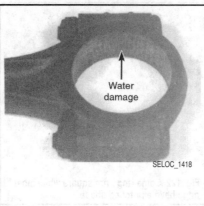

Fig. 127 Lack of lubrication will result in galling the bearing

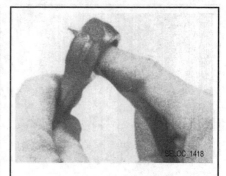

Fig. 128 Minor rust or scoring can be cleaned using crocus cloth (caged needles) or emery paper (loose needles)

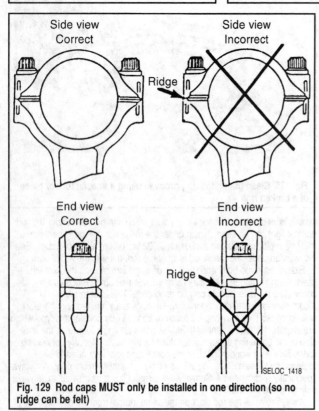

Fig. 129 Rod caps MUST only be installed in one direction (so no ridge can be felt)

Mounted main bearings control the axial movement of the crankshaft as it accomplishes its functions. The crankshaft bearing journals are case hardened to be able to withstand the stresses applied by the floating needle bearings used for connecting rod and main bearings. In essence, the crankshaft journals are the inner bearing races for the needle bearings.

INSPECTION

◆ See Figures 132 and 133

Pressure from the power stoke applied to the crankshaft rod journal by the needle bearings has a tendency to wear the journal on one side. During crankshaft inspection, the journals should also be measured with a micrometer to determine if they are round and straight. They should also be inspected for scoring, pitting, rust marks, chatter marks and discoloration caused by heat.

Check the sealing surfaces for grooves worn in by the upper and lower crankshaft seals. Take a look at the splined area which receives the driveshaft. Inspect the side of the splines for wear. This wear can be caused

by lack of lubrication or improper lubricant applied during a seasonal service. An exhaust housing/lower unit that has received a sudden impact can be warped and this can also cause spline damage in the crankshaft.

The crankshaft cannot be repaired because of the case hardening and the possibility of changing the metallurgical properties of the material during the welding and machine operation. Also, there are normally no oversized bearings available. Repairs are limited to cleaning up the journal surface with 320 emery paper when loose needle bearings are run on the journal. Where caged roller bearings are run, the journal may be polished with crocus cloth.

The tapered end of the crankshaft has a spline or a keyway and key which times the flywheel to the crankshaft. Inspect the spline or key and keyway for damage. The crankshaft taper should be clean and free of scoring, rust and lubrication. The taper must match the flywheel hub. If someone has hit the flywheel with a heavy hammer or has used an improper puller to remove the flywheel, the flywheel and hub may be warped. Place the flywheel on the tapered end of the crankshaft and check the fit. If there is any rocking indication a distorted hub, replace the flywheel. Always use a puller which pulls from the bolt pattern or threaded inner hub of the flywheel. Never use a puller on the outside of the flywheel.

The taper is used to lock the flywheel hub to the crankshaft. When mounting the flywheel to the crankshaft the taper on the crankshaft and in the flywheel hub must be cleaned with a fast evaporating solvent. No lubrication should be applied on the crankshaft taper or flywheel hub. The flywheel nut must be torqued to specification to obtain a press fit between the flywheel hub and the crankshaft taper. If the nut is not brought to specifications, the flywheel may spin on the crankshaft causing major damage.

The flywheel key is for alignment purposes and sets the flywheel's relative position to the crankshaft. Check the key for partial shearing on the side. If there is any indication of shearing, replace the key. Also check the keyway in the flywheel and crankshaft for damage. If there is damage which will allow incorrect positioning of the flywheel, the powerhead timing will be off.

Bearings

GENERAL INFORMATION

◆ See Figure 134

Needle bearings are used to carry the load which is applied to the piston and rod. This load is developed in the combustion process and the bearings reduce the friction between the crankshaft and the connecting rod. They roll with little effort and at times have been referred to as anti-friction bearings, as they reduce friction by reducing the surface area that is in contact with the crankshaft and the connecting rod.

These needle bearings are of two types, loose and caged. When loose bearings are used, there can be upwards of 35 loose bearings floating between the rod journal of the crankshaft and the connecting rod. These bearings are aided in rolling by the movement of the crank pin journal and the connecting rod pendulum action. The surface installed on the journal and rod encourages needle rotation because of its relative roughness. If the journal and rod surface was polished with crocus cloth, the loose needle bearings would have the tendency to scoot, wearing both surfaces. So, journals and rods which uses the loose needle bearings are cleaned up sing 320 grit emery paper.

Fig. 130 Typical crankshaft assembly with connecting rods that use end caps

Fig. 131 Typical pressed together crankshaft assembly with a one piece connecting rod

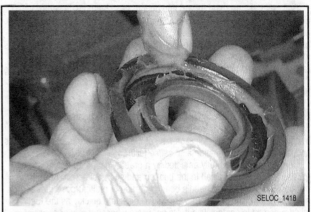

Fig. 132 Crankshaft seals should always be lubricated prior to installation

Fig. 133 Most crankshafts use two types of seals, an O-ring and sealing ring

Fig. 134 Typical caged bearing assembly

Caged needle bearings used a reduced number of needles and the needles are kept separated and are encouraged to roll by the cage. The cage also controls end movement of the bearings. Because of the cage, the journal and rod surfaces can be smoother, so these surfaces are polished with crocus cloth.

Main bearings are used to mount and control the axial movement of the crankshaft. They are either ball, needle or split race needle bearings. The split race needle bearings are held together with a ring and are sandwiched between the crankcase and cylinder assembly. The split race bearings are commonly used as center main bearings, as this is the only type of bearing that can be easily installed in this location. The ball bearings may be mounted as top or bottom mains on the crankshaft.

The bearing is made up of three parts - the inner race, needle and the outer race. In most industrial applications, the outer or inner race of a needle bearing assembly is held in a fixed position by a housing or shaft. The connecting rod needle bearings in the outboard powerhead have the same basic parts, but differ in that both inner and outer races are in motion. The outer race-the connecting rod-is swinging like a clock pendulum. The inner race-the crankshaft-is rotating and the needle bearing is floating between the two races.

INSPECTION

◆ See Figure 134

When the powerhead is disassembled and inspection of the parts is made, then by necessity along with examining the needle bearings, the crankshaft main bearing journal, rod journal and connecting rod bearing surfaces are also examined. The surfaces of all three of these parts can give a tremendous amount of information and the examination will determine if the parts are reusable.

Surfaces should be examined for scoring, pitting, chatter marks, rust marks, spalling and discoloration from overheating of the bearing surfaces. Minor scoring or pitting and rust marks may be cleaned up and the surfaces brought back to a satisfactory condition. This is done using crocus cloth for caged needle bearings and 320 grit emery paper for loose needle bearings. This is not a metal removing process, rather just a clean up of the surfaces.

Needle bearings are used as main bearings and are inspected for the same conditions as listed above. There are no oversized bearings available for rod or main bearings. Because of the hardness of the crankshaft (a bearing race), it should not be turned or welded up in order to bring it back to standard size. The welding process may stress the metallurgical properties of the crankshaft, developing cracks.

The caged rod bearings and split race main bearings are inspected for the same condition as loose needle bearings, plus the cage is examined for wear, cracks and breaks.

Ball bearings are used for top and bottom main bearings in some powerheads. These may be pressed onto the crankshaft or pressed into the end cap. To examine these bearings, wash, dry, oil and check them on the crankshaft or in the bearing cap. Turn the bearing by hand and feel if there is any roughness or catching. Try to wobble the bearing by grasping the outer race, (inner race) checking for looseness of the bearing. Replace the bearing if any of these conditions are found. If the bearing is pressed off (out) the bearing will probably be damaged and should be replaced.

If new or used bearings are contaminated with grit or dirt particles at the time of installation, abrasion will naturally follow. Many bearing failures are due to the introduction of foreign material into the internal parts of the bearing during assembly. Misalignment of the rod cap, torque of the rod bolts and lack of proper lubrication also cause failures.

Bearing failure is usually detected by a gradual rise in operating noise, excessive looseness (axial) in the bearing and shaft deflection. Keep the work area clean and use needle bearing grease or multipurpose grease to hold the bearings in place. This grease will dissipate quickly as the fuel mixture comes in contact with it. Do not use a wheel bearing or chassis grease as this will cause damage to the bearings. Oil the ball bearings with 2-stroke oil upon installation. Remember to keep them clean.

Reed Valves

GENERAL INFORMATION

◆ See Figure 135

All Mercury/Mariner 2-Stroke outboard motors are equipped with one set of reed valves per cylinder. The reed valves allow the air/fuel mixture from the carburetor to enter the crankcase, but not exit. They are in essence, one way check valves.

Reed valves are essentially maintenance free and cause very few problems. However, if a reed valve does not seal, the air/fuel mixture will escape the crankcase and not be transported to the combustion chamber. Some slight spitting of fuel back out of the carburetor throat at idle can be considered normal, but any substantial discharge of fuel from the carburetor throat indicates reed valve failure.

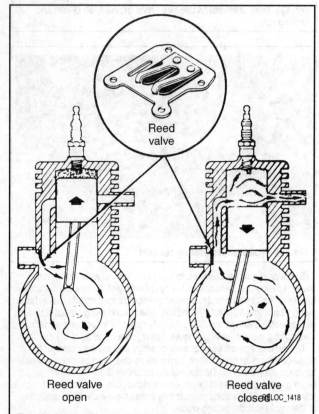

Fig. 135 Reed valves allow the air/fuel mixture from the carburetor to enter the crankcase, but not exit. They are in essence, one way check valves

INSPECTION

◆ See Figures 136 and 137

On most models, the reeds can be inspected with the carburetor or air plenum/throttle body assembly removed. A small flashlight and dental mirror can be used to inspect for broken, cracked or chipped reeds. If reed damage is discovered, it is important to attempt to locate the missing pieces of the reed petal. The reed petals are made of stainless steel and will cause internal engine damage if allowed to pass through the crankcase and combustion chamber.

Anytime the reed valves are removed from the powerhead, the reeds should be inspected for excessive stand open. When available, reed valve specifications are listed in the Engine Rebuilding Specifications chart for each powerhead. On models with reed stop specifications, the reed stop opening should also be measured. The easiest way to measure the reed stop opening is to use a drill bit or feeler gauge of the specified diameter. The shank (smooth portion) of the drill bit/gauge should just fit between the top surface of the closed reed petal and the inner edge of the reed stop. Nearly all Modern Mercury reed stops are NOT adjustable.

Reed petals must never be turned over and reinstalled. This can lead to a preloaded condition. Preloaded reeds require a higher crankcase vacuum level to open. This causes acceleration and carburetor calibration problems. The reed petals should be flush to nearly flush along the entire length of the reed block mating surface with no preload. Most larger engines have a maximum stand open specification.

Many new models use rubber coated reed blocks. These reed blocks cushion the impact as the reed closes and improves sealing. Reed block assembly part numbers automatically supersede where applicable. Rubber coated reed blocks are serviced as assemblies only.

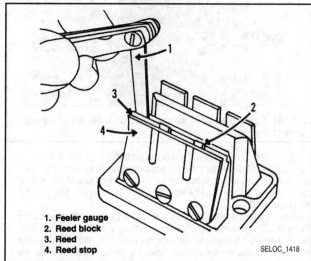

1. Feeler gauge
2. Reed block
3. Reed
4. Reed stop

Fig. 136 Measuring the reed opening on a typical Mercury reed block

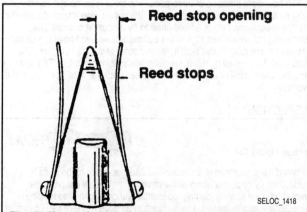

Reed stop opening

Reed stops

Fig. 137 Measure between the reed stop and the top of the closed reed

POWERHEAD BREAK-IN

Break-in Procedures

Anytime a new or rebuilt powerhead is installed (this includes a powerhead whose wear components such as pistons and rings or main bearings are replaced) the motor must undergo proper break-in.

By following break-in procedures largely consisting of specific engine operating limitations during the first 1-5 hours or so of operation, you will help you will help ensure a long and trouble-free life. Failure to follow these recommendations may allow components to seat improperly, causing accelerated wear and premature powerhead failure.

On all motors, special attention is required to the engine oil during initial break-in. Pay close attention to the special fuel/oil mixture requirements. Especially during break-in, pay close attention to all pre and post operation checks. This goes double when checking for fuel, oil or water leaks. At each start-up and frequently during operation, check for presence of the cooling indicator stream.

At the completion of break-in, double-check the tightness of all exposed engine fasteners.

Even though Mercury technically ends the break-in period of a motor after the first 1-5 hours of operation, it is a good idea to continue top pay closer attention to the motor until the completion of the entire first 10 hours of engine operation (and completion of a break-in/10 hour service).

During break-in, one of the most important things you can do is to **vary** the engine speed. This allows parts to wear in under conditions throughout the powerband, not just at idle or mid-throttle.

■ **During break-in, check your hourmeter or a watch frequently and be sure to change the engine speed at least every 15 minutes (that means between every 2-3 tenths on the hourmeter). However, on some motors, such as all Optimax and EFI motors, Mercury recommends that you change your engine speed about every TWO minutes!**

Be sure to **always** allow the engine to reach operating temperature before setting the throttle anywhere above idle. This means you should always start and run the motor for at least 60 seconds before advancing the throttle. Of course, it is just as bad to allow a motor to idle excessively (especially in cold water) during break-in, so don't allow it to idle for more than 10 minutes.

✳✳ SELOC WARNING

NEVER run the engine out of the water, unless a flush fitting is used to provide a source of cooling. Remember that the water pump can be destroyed in less than a minute just from a lack of water. The powerhead will suffer damage in very little time as well, but even if it is not overheated out of water, reduced cooling from a damaged water pump impeller could destroy it later. Don't risk it.

■ **Although a flush fitting can be used to run an engine, it CANNOT be used to break-in a motor, as you cannot run an engine much above idle on a flush fitting without risking damage from a runaway powerhead (the motor running overspeed).**

During engine break-in MOST motors must use a higher ratio of oil in the fuel, this is true whether or not they are equipped with an oil injection system.

For pre-mix carbureted motors, oil of the proper break-in ratio must be mixed in the main fuel supply and run for the duration of break-in for that model. With the exception of the single cylinder motors, as well as the 20-25 hp twins, this means a break-in fuel:oil ratio of 25:1. On the singles and the 20-25 hp twins, no special ratio is used and Mercury recommends breaking them in with the normal fuel:oil ratio of 50:1.

On carbureted or EFI motors that are equipped with a MECHANICAL oil injection system, a combination of the injection system output and pre-mix will yield the proper break-in fuel:oil ration. On all of these motors the desired END result is a ratio of 25:1 and since the pump output will basically equal a ratio of 50:1, all you need to do is run tanks of 50:1 pre-mix in the main fuel supply at the same time and the end result will be 25:1.

■ **Don't run the higher fuel:oil ratio any longer than absolutely necessary to clear the fuel tank after the 1-5 hours (depending upon the motor) is over. Doing so could lead to excessive carbon fouling of the spark plugs and excessive carbon deposits in the combustion chambers.**

On EFI or OptiMax motors equipped with an electronic oil injection system NO PRE-MIX should be used. Instead the oil injection system itself must be placed in break-in mode and it will automatically increase oiling as necessary for the duration of powerhead break-in. There are basically 2 ways to initiate a break-in sequence on these motors. The first is to use a scan tool. The second, if no scan tool is available, is to use the Shift Switch Activation method, which will act like the scan tool Reset Break-In function, except that it does not clear run/fault history. Using this method will both prime the oil system and start a new 120 minute break-in cycle. To use the Shift Switch Activation method, start by making sure the primer bulb from the remote oil tank is firm, then:

- Turn the Ignition Keyswitch to ON without starting the motor.
- Within 10 seconds of turning the keyswitch ON, move the remote shift handle from Neutral to Forward Gear 3-5 times. This will start the Auto-Prime process AND reset the break-in function.

For more details please refer to Bleeding the Oil Injection System, in the Cooling and Lubrication System section.

On ALL oil injection systems (both mechanical and electronic) top off the oil tank and place a piece of tape or make a mark on the side of the tank housing to note the oil level. Watch this level and make sure it is dropping slowly throughout the break-in period. This will ensure you the system is working. Also, it is a good habit to get into as it will help you gauge system and engine operation long after engine break-in. Keep in mind that on models with a remote oil tank in addition to the powerhead mounted tank, you'll want to mark the REMOTE tank, because the powerhead mounted one will periodically refill itself.

✳✳ SELOC WARNING

Do not operate the engine at full throttle except for very short periods, until after break-in.

As we've said, proper break-in varies slightly from 1 to 5 hours, depending upon the model as follows:
- 2.5/3.3 hp, 6/8 hp, 9.9/10/15 hp, 20/20 Jet/25 hp and 30/40 hp motors - 1 hour total
- 4/5 hp motors - 2 hours total
- 40/40 Jet/50/55/60 hp and larger motors - 5 hours total (with the greatest restrictions on the first 2 hours)

To properly break-in a powerhead, proceed as follows:

1. Each time you start the motor, IMMEDIATELY check to be sure the water pump is operating. If the water pump is operating, a fine stream will be discharged from the exhaust relief hole at the rear of the driveshaft housing. Also, each time you start the powerhead, allow it to idle until it reaches normal operating temperature (for at least 60 seconds, but NO MORE than 5-10 minutes) before increasing speed above idle.

■ **If the cooling water indicator stream is missing, shut the powerhead down and figure out why before proceeding.**

2. The VERY first time you start the motor, allow it to idle for **5-10** minutes. Make sure the motor runs at the slowest possible speed (a fast idle in neutral is best).

3. For the remainder of the first hour operate the motor in gear at various speeds, but DO NOT exceed 1/2 throttle or so. One exception however is made on larger motors, for 65 hp and larger motors SHORT BURSTS of full throttle operation are acceptable, but only for periods up to about 10 seconds. IF this is enough to plane the boat, go ahead and do that sometimes, as long as you can reduce throttle to about 1/2 or so and stay on plane.

■ **After the first hour, you are technically done with operating restrictions on 2.5/3.3 hp, 6/8 hp, 9.9/10/15 hp, 20/20 Jet/25 hp and 30/40 hp motors, HOWEVER, we strongly suggest you consider continuing break-in procedures through at least a second hour of operation.**

4. For the second hour of operation continue to vary speeds. If you have an easily planning boat, accelerate at 3/4 throttle or slightly more JUST long enough to get the boat up onto plane, then immediately reduce speed, but try to keep the boat on plane (as that lessons the amount of load on the motor). Continue to vary the engine speed and operate the motor under different load conditions. For 65 hp and larger motors, at least a couple of times during the hour, run the engine AT FULL THROTTLE for ONE MINUTE, then allow at least 5-10 minutes of operation at 3/4 throttle or less to allow the motor to cool.

■ Build good habits now. It is NEVER good for the motor to operate at full throttle for any length of time then suddenly drop off to idle. You should always decelerate the motor gradually, allowing the pistons to cool as the speed is reduced.

■ After the second hour, the break-in period is officially over for 4/5 hp motors

5. For the remainder of the first 5 hours (for hours 3-5) continue the ritual of warming the motor and varying engine speed. Full throttle operation should NOT exceed 5 minutes allowing sufficient time to slowly cool the pistons and operate under other conditions before the next full throttle run.

6. After the 5th hour of operation, run the engine as normal and on a normal fuel:oil ratio. Before removing the pre-mix from the fuel tank on mechanical models verify that the system has been working by making a visual check of oil level.

■ While the engine is operating during the initial period, check the fuel, exhaust, and water systems for leaks. Tilt the motor up out of the water at the end of each day to closely inspect the drive unit and look for signs of damage or leakage (of course, if you're trailering, this should be done to protect the skeg before pulling it out on the trailer anyway, right?).

SPECIFICATION CHARTS

Engine Rebuilding Specifications — 2.5/3.3 HP

Component	Maximum	
	Standard (in.)	Metric (mm)
Crankshaft runout	0.001	0.05
Connecting rod deflection	0.022-0.056	0.6-1.5
Cylinder bore ①	1.85	47.05
Cylinder bore wear limit ②	1.852	47.04
Cylinder out of round	0.002	0.05
Cylinder taper	0.002	0.05
Piston to cylinder clearance	0.002-0.005	0.06-0.15
Piston ring end gap	0.006-0.012	0.18-0.33
Piston ring side clearance	0.0003-0.0010	0.01-0.05
Maximum reed stop opening	0.236-0.244	6.0-6.2

① 5 mm oversize: 1.869 in. (47.55 mm)

② Oversize bore: 1.871 in. (47.52 mm)

SELOC_1418

Engine Rebuilding Specifications — 4/5 HP

Component	Maximum	
	Standard (in.)	Metric (mm)
Crankshaft runout	0.002	0.05
Connecting rod side clearance	0.005-0.015	0.13-0.37
Cylinder bore ①	2.165	54.991
Cylinder out of round	0.003	0.08
Cylinder taper	0.003	0.08
Piston diameter ②	2.164	54.965
Piston to cylinder clearance	0.0012-0.0024	0.03-0.06
Piston ring end gap	0.008-0.016	0.2-0.4
Piston ring side clearance (TOP)	0.0012-0.0023	0.03-0.07
Piston ring side clearance (2nd)	0.0008-0.0024	0.02-0.06
Maximum reed stop opening	0.24-0.248	6.0-6.2
Cylinder head warpage	0.002	0.05

NOTE measure piston diameter at a point 90 degrees to the piston pin centerline and approx. 5/8 in. (16mm) from the bottom of the piston skirt

①   Oversize: 2.185 in. (55.499 mm)

②   Oversize: 2.184 in. (55.473 mm)

SELOC_1418

Engine Rebuilding Specifications — 6/8 HP and 9.9/10/15 HP

Component	Standard (in.)	Maximum Metric (mm)
Crankshaft runout	0.003	0.076
Top main bearing jounal	0.7517	19.093
Center main bearing journal	0.8108	20.594
Bottom ball bearing journal	0.7880	20.015
Connecting rod journal	0.8125	20.638
Connecting rod small end I.D.	0.8195	20.815
Connecting rod big end I.D.	1.0635	27.0129
Cylinder bore		
6/8 hp	2.125	53.975
9.9/15 hp	2.375	60.325
Cylinder out of round	0.004	0.1016
Cylinder taper	0.004	0.1016
Bore type	Cast iron	
Piston diameter ①		
6/8 hp	2.123	53.92
9.9/15 hp	2.373	60.27
Piston to cylinder clearance	0.002-0.005	0.05-0.13
Piston ring end gap	0.010-0.018	0.25-0.46
Maximum reed opening	0.007	0.178
Reed stop opening	19/64 (0.296)	7.54

① Measured 90 degrees to piston pin and 0.10 in. (2.54mm) from bottom edge of skirt

SELOC_1418

Engine Rebuilding Specifications — 20/20 Jet/25 HP

Component	Standard (in.)	Maximum Metric (mm)
Crankshaft runout	0.003	0.076
Top main bearing jounal	1.251	31.77
Center main bearing journal	1.000	25.40
Bottom ball bearing journal	1.125	28.58
Connecting rod journal	0.883	22.43
Crankshaft endplay	0.004-0.019	0.10-0.64
Connecting rod small end I.D.	0.897	22.78
Connecting rod big end I.D.	1.196	30.38
Cylinder bore	2.562	65.01
Cylinder out of round	0.003	0.076
Bore type:		
0G202749 and below	Chrome (cannot be rebored or efficiently honed)	
0G202750 and abovel	Mercosil (can be bored 0.030 in. oversize)	
Cylinder taper	0.003	0.08
Piston diameter ①		
Standard	2.5583-2.5593	64.98-65.00
Acceptable	2.5578-2.5588	64.97-64.99
Piston to cylinder clearance	0.003-0.004	0.076-0.101
Piston ring end gap	0.011-0.025	0.28-0.64
Maximum reed opening	0.007	0.178

① Measured 90 degrees to piston pin and 0.50 in. (12.7mm) from bottom edge of skirt
  Also note that piston diameter will be 0.001-0.0015 less if coating is worn off piston (it is used)

SELOC_1418

Engine Rebuilding Specifications — 30/40 HP 2-Cylinder Motors

Component	Standard (in.)	Maximum Metric (mm)
Crankshaft runout	0.003	0.076
Top main bearing jounal	1.375	34.925
Center main bearing journal	1.216	30.886
Bottom ball bearing journal	1.385	35.179
Connecting rod journal	1.181	29.997
Connecting rod small end I.D.	0.957	24.307
Connecting rod big end I.D.	1.499	38.074
Cylinder bore		
Standard	2.993	76.022
0.015 oversize	3.007	76.377
Cylinder out of round	0.003	0.076
Bore type	Cast iron	
Cylinder taper	0.003	0.076
Cylinder finish hone I.D.	2.993	76.022
Piston diameter ①		
Standard	2.987-2.989	75.870-75.920
0.015 oversize	3.002-3.004	76.251-76.301
Piston ring end gap	0.010-0.018	0.254-0.457
Maximum reed opening	0.02	0.508
Reed stop		
30 hp	0.09	2.286
40 hp	Non-Adjustable	

① Measured 90 degrees to piston pin and 0.50 in. (12.7mm) from bottom edge of skirt

SELOC_1418

Engine Rebuilding Specifications — 40/40 Jet/50/55/60 HP
3-Cylinder Motors

Component	Standard (in.)	Maximum Metric (mm)
Cylinder bore		
Standard	2.988	75.895
0.015 oversize	3.003	76.276
0.030 oversize	3.018	76.657
Cylinder out of round	0.003	0.08
Bore type	Cast Iron	
Cylinder taper	0.003	0.08
Piston diameter ①		
Standard	2.950	74.93
0.015 oversize	2.965	75.31
0.030 oversize	2.980	75.69
Reed stop (standard opening)		
40 hp	0.09	2.286
40 Jet, 50, 55, 60 hp	Non-Adjustable	
Maximum reed opening	0.02	0.5
Reed thickness	0.01	0.254

① Measured 90 degrees to piston pin and 0.50 in. (12.7mm) from bottom edge of skirt   SELOC_1418

Engine Rebuilding Specifications — 65 Jet/75/90/80 Jet/100/115/125 HP
Carbureted 1386cc and 1848cc Models

Component	Standard (in.)	Maximum Metric (mm)
Crankshaft runout	0.006	0.152
Cylinder bore	3.5	
Standard	3.501	88.93
0.015 oversize	3.516	89.31
0.030 oversize	3.531	89.69
Bore type	Cast Iron	
Cylinder out of round	0.003	0.076
Cylinder taper	0.003	0.076
Piston diameter ①		
Standard	3.495	88.77
0.015 oversize	3.510	89.15
0.030 oversize	3.525	89.54
Maximum reed opening	0.020	0.508

① Measured 90 degrees to piston pin and 0.50 in. (12.7mm) from bottom edge of skirt   SELOC_1418

Engine Rebuilding Specifications
75/80 Jet/90/115 HP OptiMax 1526cc Models

Component	Standard (in.)	Standard Metric (mm)
Crankshaft		
Runout	0.002	0.0508
Main bearing journal	1.6250-1.6255	41.2750-41.2877
Bottom main bearing journal	1.3783-1.3788	35.0088-35.0215
Crankshaft pin journal	1.4168-1.4173	35.9867-35.9994
Connecting rod		
Wrist pin bore diameter	1.1054-1.1059	28.0771-28.0898
Crankshaft pin diameter	1.7341-1.7346	44.0461-44.0588
Cylinder compression	90-110 psi	620-758 kPa
Cylinder bore		
Standard	3.6265	92.1131
0.015 oversize	3.6415	92.4941
0.030 oversize	3.6565	92.8750
Cylinder bore type	Cast Iron	
Cylinder out of round	0.003	0.076
Cylinder taper	0.003	0.076
Piston diameter ①		
Standard	3.6185-3.6195	91.9099-91.9353
0.015 oversize	3.6335-3.6345	92.2909-92.3163
0.030 oversize	3.6485-3.6495	92.6719-92.6973
Piston		
Ring end gap	0.010-0.018	0.25-0.46
Wrist pin bore diameter	0.9180-0.9183	23.3174-23.3260
Wrist pin diameter	0.9176-0.9179	23.3090-23.3166
Reed stand open	0.020	0.50

① Measured 90 degrees to piston pin and 0.945 in. (24mm) from bottom edge of skirt   SELOC_1418

Engine Rebuilding Specifications
110 Jet/135/150/175/200 HP
2507cc Models

Component	Maximum Standard (in.)	Metric (mm)
Crankshaft runout	0.006	0.152
Cylinder head warpage (max)	0.004	0.1
Cylinder bore		
Standard	3.501	88.925
0.015 oversize	3.516	89.306
Bore type	Cast Iron	
Cylinder out of round	0.003	0.076
Cylinder taper	0.003	0.076
Piston diameter ①		
Except Optimax		
Standard	3.493-3.495	88.723-88.773
0.015 oversize	3.508-3.510	89.104-89.154
Optimax		
Standard	3.492-3.493	88.6968-88.7222
0.015 oversize	3.507-3.508	89.0778-89.1032
Piston ring gap		
Except Optimax	0.018-0.025	0.45-0.64
Optimax	0.010-0.018	0.25-0.45
Maximum reed opening	0.02	0.50

① Measured 90 degrees to piston pin and 0.50 in. (12.7mm) from bottom edge of skirt

SELOC_1418

Engine Rebuilding Specifications — 200/225/250 HP
Optimax and EFI 3032cc Models

Component	Standard (in.)	Metric (mm)
Crankshaft runout	0.002	0.0508
Cylinder bore – Except Pro/XS/Sport		
Standard	3.6265	92.1131
.015 in. (0.381 mm) oversize	3.6415	92.4941
.030 in. (0.762 mm) oversize	3.6565	92.8751
Cylinder bore – Pro/XS/Sport		
Standard	3.6250	92.0750
Cylinder bore type	Cast Iron	
Cylinder distortion/taper/wear	0.003	0.076
Piston diameter ①		
All except 2003 & later Optimax		
Standard	3.6205-3.6215	91.9607-91.9861
.015 in. (0.381 mm) oversize	3.6355-3.6365	92.3413-92.3667
.030 in. (0.762 mm) oversize	3.6505-3.6515	92.7223-92.7477
2003 & later Optimax		
Standard	3.6185-3.6195	91.9099-91.9353
.015 in. (0.381 mm) oversize	3.6335-3.6345	92.2909-92.3163
.030 in. (0.762 mm) oversize	3.6485-3.6495	92.6719-92.6973
Additional piston dimension Pro/XS/Sport		
Ring end gap (top or bottom)	0.010-0.018	0.254-0.457
Wrist pin bore diameter	0.92691-0.94175	23.543-23.920
Wrist pin diameter	0.91768-0.91798	23.309-23.316
Reed stand open	0.020	0.50
Cylinder compression		
Except Pro/XS/Sport	90-110 psi	621-758 kPa
Pro/XS/Sport	100-110 psi	689-758 kPa

Note: we're aware that some standard to metric conversions are questionable, but those are the figures that Mercury provides, use your best judgment

① Measured 90 degrees to piston pin and 0.50 in. (12.7mm) from bottom edge of skirt for models through 2002, or measured 0.945 in. (24.0mm) from the bottom edge for 2003 and later models

SELOC_1418

Engine Torque Specifications

Component/Model	Standard (ft. lbs.)	Metric (Nm)
Conventional bolt/nut		
8mm (M5)	36 in. lbs. / 3.0 ft. lbs.	4
10mm (M6)	70 in. lbs. / 6.0 ft. lbs.	8
12mm (M8)	156 in. lbs. / 13 ft. lbs.	18
14mm (M10)	26	36
17mm (M12)	31	42
Cylinder block cover (cylinder head and/or cover)		
2.5/3.3 hp	85 in. lbs.	9.6
4/5 hp	215 in. lbs.	24.4
6/8 hp and 9.9/10/15 hp	60 in. lbs.	6.8
20/20 Jet/25 hp	140 in. lbs.	15.8
65 Jet-125 hp (as applicable)	18	24.5
135-225 hp (2.5L) - oil the threads	30 ft. lbs. + 90 degrees	40.5 Nm + 90 degrees
200-250 hp (3.0L) - oil the threads		
All except top 2 bolts on 2001 EFI	20 ft. lbs. + 90 degrees	27 Nm + 90 degrees
Top 2 bolts on 2001 EFI	30 ft. lbs. + 90 degrees	40.5 Nm + 90 degrees
Connecting rod bolt		
6/8 hp and 9.9/10/15 hp	100 in. lbs.	11.3
20/20 Jet/25 hp	150 in. lbs.	12.5
30/40 hp (2-cyl)	15 in. lbs., then 120 in. lbs. + 1/4 turn	1.7, then 13.6 + 1/4 turn
40/40 Jet/50/55/60 hp (3-cyl)	120 in. lbs. + 1/4 turn	13.56 + 1/4 turn
65 Jet-125 hp	15 in. lbs., then 30 ft. lbs. + 1/4 turn	1.7, then 40.6 + 1/4 turn
135-200 hp (2.5L) - oil the threads	15 in. lbs., then 30 ft. lbs. + 1/4 turn	1.7, then 40.6 + 1/4 turn
200-250 hp (3.0L) - oil the threads		
Except 2001 EFI	15 in. lbs., then 30 ft. lbs. + 1/4 turn	1.7, then 40.6 + 1/4 turn
2001 EFI		
All except top 2 bolts	20 ft. lbs. + 1/4 turn	27 + 1/4 turn
Top 2 bolts	30 ft. lbs. + 1/4 turn	40.5 + 1/4 turn
Crankcase to cylinder block		
2.5/3.3 hp	50 in. lbs.	5.6
4/5 hp	90 in. lbs.	10.3
6/8 hp and 9.9/10/15 hp	16.5	22.6
20/25 hp	30	40.7
30/40 hp (2-cyl)	16.5	22.4
40/40 Jet/50/55/60 hp (3-cyl)	18.4	24.9
65 Jet-125 hp		
Large bolts (M10)		
Except Optimax	25	34
Optimax	28	37.9
65 Jet-125 hp (small bolts, M8)		
Except Optimax	18.5	24.5
Optimax	18.0	24.4
135-200 hp (2.5L) - oil the threads		
3/8 in. bolts	38	51.5
5/16 in. bolts	15	20.3
200-250 hp (3.0L) - oil the threads		
M8 x 1.25 x 35mm bolts	21	28.5
M10 x 1.5 x 80mm bolts	30 ft. lbs. + 1/4 turn	40.5 + 1/4 turn

Engine Torque Specifications

Component/Model	Standard (ft. lbs.)	Metric (Nm)
Crankcase upper end cap (seal holder)		
135-200 hp (2.5L)	150 in. lbs.	16.9
Crankcase lower end cap (seal holder)		
2.5/3.3 hp	50 in. lbs.	6.0
135-200 hp (2.5L)	80 in. lbs.	6.8
200-250 hp (3.0L)	85 in. lbs.	9.6
Exhaust cover		
6/8 hp and 9.9/10/15 hp	60 in. lbs.	6.8
20/20 Jet/25 hp	12	15.8
65 Jet-115 hp (as applicable)	18	24.5
135-200 hp (2.5L) - exhaust divider plate		
Except Optimax - apply Loctite 271	180 in. lbs. / 15 ft. lbs.	20
Optimax - apply Loctite 271	200 in. lbs.	22.5
Flywheel nut (or bolt)		
2.5/3.3 hp	30	4.1
4/5 hp	40	52.2
6/8 hp and 9.9/10/15 hp	50	67.8
20/20 Jet/25 hp (bolt)	57.5	78.0
30/40 hp (2-cyl)	95	129
40/40 Jet/50/55/60 hp (3-cyl)	100	135.6
65 Jet-125 hp	120	163
135-200 hp (2.5L)		
Except Optimax	120	163
Optimax	125	169.5
200-250 hp (3.0L)	125	169.5
Gearcase (not JET drive) to mid-section		
2.5/3.3 hp	50 in. lbs.	5.6
4/5 hp	70 in. lbs. / 6.0 ft. lbs.	8
6/8 hp and 9.9/10/15 hp	15	20.3
20-125 hp	40	54
135-200 hp (2.5L)	50	68
200-250 hp (3.0L)		
Nuts	55	74.6
Bolt in trim tab recess ②	45	61
Hand rewind starter mount		
4-15 hp	70 in. lbs. / 6.0 ft. lbs.	7.9
20/20 Jet/25 hp	90 in. lbs.	10
Intake cover (transfer port cover)		
6/8 hp and 9.9/10/15 hp	60 in. lbs.	6.8
Intake manifold (or reed block adapter plate)		
20/20 Jet /25 hp (carb adpt plate bolts/nuts)	80 in. lbs.	9.0
30/40 hp (2-cyl)	16.5	22.4
40/40 Jet/50/55/60 hp (3-cyl)	18.9	24.9
65 Jet-125 hp	18	24.5
135-200 hp (2.5L)	105 in. lbs.	12
200-250 hp (3.0L) - air plenum		
Except Pro/XS/Sport	100 in. lbs.	11.5
Pro/XS/Sport apply Loctite 271	125 in. lbs.	14

Engine Torque Specifications

Component/Model	Standard (ft. lbs.)	Metric (Nm)
Spark plugs		
2.5/3.3 hp	240 inch lbs / 20 ft. lbs	27.1
4/5 hp	168 inch. lbs. / 14 ft. lbs	19.7
6-250 hp	240 inch lbs. / 20 ft. lbs. (or finger tight + 1/4 turn)	27.1 (or finger tight + 1/4 turn)
Starter motor to crankcase		
20/20 Jet/25 hp	18.0	24.8
30/40 hp (2-cyl)	16.6	22.5
40/40 Jet/50/55/60 hp (3-cyl)	16.5	22.3
65 Jet-200 hp		
Except Optimax	210 in. lbs / 17.5 ft. lbs	23.5
Optimax		
75-115 hp	18	24.4
135-200 hp	23 (top bolts) / 21 (bottom bolts)	31 (top bolts) / 28.5 (bottom bolts)
200-250 hp (3.0L)	23 (top bolts) / 21 (bottom bolts)	31 (top bolts) / 28.5 (bottom bolts)
Water pump housing/cover		
2.5/3.3 hp (pump behind propeller)	50 in. lbs.	5.6
2.5/3.3 hp and 4/5 hp (along gearcase split)	70 in. lbs. / 6.0 ft. lbs.	8
6/8 hp and 9.9/10/15 hp	50 in. lbs.	5.6
20 Jet, 65 Jet & 80 Jet	70 in. lbs.	7.9
20/25 hp	60 in. lbs.	6.8
30-125 hp (Except 65 Jet & 80 Jet)	60 in. lbs.	6.8
135-200 hp (2.5L)	35 in. lbs.	3.9
200-250 hp (3.0L)	60 in. lbs.	6.8
Water pump base		
4/5 hp	70 in. lbs. / 6.0 ft. lbs.	8
6/8 hp and 9.9/10/15 hp	50 in. lbs.	5.6
55/60 hp (3-cyl)	60 in. lbs.	6.8
75-125 hp	60 in. lbs.	6.8
Thermostat housing		
6/8 hp and 9.9/10/15 hp	60 in. lbs.	6.8
20/20 Jet/25 hp	140 in. lbs.	15.8
30/40 hp (2-cyl)	16.5	22.4
40/40 Jet/50/55/60 hp (3-cyl)	18.4	24.9
75-125 hp		
Except Optimax	25	34
Optimax	18	24.4
135-200 hp (2.5L)	200 in. lbs. / 16.5 ft. lbs.	22.5
200-250 hp (3.0L)		
EFI		
2001	100 in. lbs.	11.3
2002-05	see cylinder head bolts - top 2	
Optimax		
Except Pro/XS/Sport	140 in. lbs.	16
Pro/XS/Sport	100 in. lbs.	11.5

SELOC_1418

Engine Torque Specifications

Component/Model	Standard (ft. lbs.)	Metric (Nm)
Pinion gear fastener		
20/20 Jet/25 hp	15	20.3
30-60 hp (exc 60 hp Bigfoot)	50	67.8
60 hp Bigfoot and 75-125 hp	70	94.9
135-200 hp (2.5L)	75	101.7
200-250 hp (3.0L)	100	136
Except 2001 EFI apply Loctite 271	70	94.9
2001 EFI apply Loctite 271	75	101
Powerhead to mid-section		
2.5/3.3 hp	50 in. lbs.	5.6
4/5 hp	70 in. lbs. / 6.0 ft. lbs.	8
6/8 hp and 9.9/10/15 hp	100 in. lbs.	11.3
20/20 Jet/25 hp	20	27.1
30/40 hp (2-cyl)	29	39
40/40 Jet/50/55/60 hp (3-cyl)	28	38
65 Jet-125 hp		
Except Optimax	45	61
Optimax	20 ft. lbs. + 90 degrees	27 Nm + 90 degrees
135-200 hp (2.5L)		
Except EFI	20	27
EFI	25	33.8
200-250 hp (3.0L)	50	67.8
Except Pro/XS/Sport		
Pro/XS/Sport		
Stage I	Snug in sequence	Snug in sequence
Stage II	25 ①	34
Stage III ①	40 ①	54 ①
Propeller nut		
4/5 hp	150 in. lbs. / 12.5 ft. lbs.	17
6/8 hp and 9.9/10/15 hp	70 in. lbs. / 6.0 ft. lbs.	7.9
20/25 hp	16.7	22.6
30-250 hp	55	74.6
Reed mounting screw		
20/20 Jet /25 hp (carb adpt plate bolts/nuts)	80 in. lbs.	9.0
30/40 hp and 40/50/55/60 hp	60 in. lbs.	6.8
65 Jet-125 hp (except Optimax)	80 in. lbs.	9.0
Reed block bolts		
75/90/115 hp Optimax	90 in. lbs.	10.2
135-200 hp (2.5L)	105 in. lbs.	12
200-250 hp (3.0L)	90 in. lbs.	10

SELOC_1418

① There are 10 nuts in a crossing sequence, the last two are port side, middle and one back from middle, those last two are not tightened to 40 ft. lbs. in the final sequence, leave them at 25 ft. lbs.

② If equipped

7

LOWER UNIT

LOWER UNIT

General Information

Generally speaking, the lower unit is considered to be the part of the outboard below the exhaust housing. The unit contains the propeller shaft, the driven and pinion gears, the driveshaft from the powerhead and the water pump. Torque is transferred from the powerhead's crankshaft to the gearcase by a driveshaft. A pinion gear at the bottom of the driveshaft meshes with a drive gear in the gearcase to change the vertical power flow into a horizontal flow through the propeller shaft. The power head driveshaft rotates clockwise continuously when the engine is running, but propeller rotation is controlled by the gear train shifting mechanism.

The lower units normally found on all but a few of the smallest motors are equipped with shifting capabilities. The forward and, if equipped, reverse gears together with the clutch, shift assembly and related linkage are all housed within the lower unit.

On outboards with a reverse gear, a sliding clutch engages the appropriate gear in the gearcase when the shift mechanism is placed in forward or reverse respectively. This creates a direct coupling that transfers the power flow from the pinion to the propeller shaft. A similar clutch mechanism is used on a handful of small hp motors that contain both a forward and neutral shift position, but no reverse gear.

Generally lower units are available in 1 or 2 major types, depending upon the size of the motor. Many of the mid-range units are available with standard or Bigfoot (heavy duty/high thrust) models. The Bigfoot units are simply beefier units with higher gear ratios, designed for more push and less speed. The larger motors are generally available in both conventional and counter-rotating models.

On counter-rotating models, the propeller rotates in the opposite direction than on regular models and is used when dual engines are mounted in the boat to equalize the propeller's directional churning force on the water. This allows the operator to maintain a true course instead of being pulled off toward one side while underway.

Another type of lower unit available on some mid-range units is a jet drive propulsion system. Water is drawn in from the forward edge of the lower unit and forced out under pressure to propel the boat forward.

In all cases, the lower unit is easily removed without removing the entire outboard from the boat.

Shifting Principles

STANDARD ROTATING UNIT

Non-Reverse Type

◆ See Figure 1

There are 2 types of non-reversing type lower units normally used only on the smallest of outboards (2.5/3.3 hp units). The first is a direct drive unit - the pinion gear on the lower end of the driveshaft is in constant mesh with the forward gear. The second is a Forward-Neutral shifting unit, in which shift assembly allows the drive gear on the propshaft to move in and out of contact with the pinion gear on the lower end of the driveshaft.

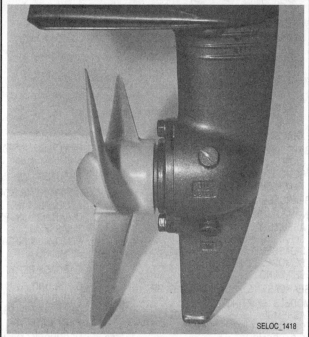

Fig. 1 Typical non-reversing gearcase

In both cases, reverse action of the propeller is accomplished by the operator swinging the engine with the tiller handle 180° and holding it in this position while the boat moves sternward. When the operator is ready to move forward again, they simply swing the tiller handle back to the normal forward position.

Reverse Type

◆ See Figures 2, 3 and 4

The most common lower unit design (whether it be standard, Bigfoot, or counter-rotating) is equipped with a clutch dog permitting operation in neutral, forward and reverse. Although some of the smaller units use a step shifter mechanism, many mid-range and larger units use a rotating shifter cam. On these mid-range and larger units unit when the outboard is shifted from neutral to reverse, the shift rod is rotated. This action moves the plunger and clutch dog toward the reverse gear. When the unit is shifted from reverse to neutral and then to forward gear, the rotation of the shift rod is reversed and the clutch dog is moved toward the forward gear.

These shift mechanisms basically work as follows. The pinion gear on the driveshaft is in constant mesh with both the forward and reverse gears, which when unlocked from the propeller shaft, can spin freely on their bearings. However, there is a clutch dog which is splined to the propeller shaft and can slide back and for the between the gears (the direction and amount of travel is controlled by the shifter assembly). When the clutch dog

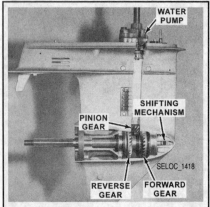

Fig. 2 Cutaway view of a lower unit showing the shift assembly and water pump

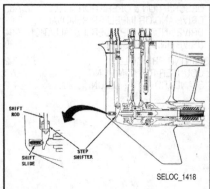

Fig. 3 Cross section of a typical lower unit showing a step shifter mechanism

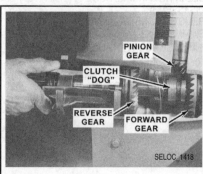

Fig. 4 Cut away view of a standard rotation lower unit with major parts identified

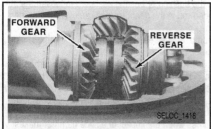

Fig. 5 Cutaway of a typical counter-rotating unit with the clutch dog in NEUTRAL

Fig. 6 Cutaway of a typical counter-rotating unit with the clutch dog in FORWARD

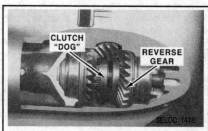

Fig. 7 Cutaway of typical counter-rotating unit with the clutch dog in REVERSE

is pushed toward the forward or reverse gears, the ramped ears of the dog spline with an inside face of the gear, locking the gear into contact with the propeller shaft, causing the shaft to turn in the same direction and at the same rate as that gear. Once the shift mechanism pulls the clutch dog away, the propeller shaft will again freewheel, slowing to an eventual stop unless the clutch dog is engaged again with one of the gears.

From this explanation, an understanding of wear characteristics can be appreciated. The pinion gear and the clutch dog receive the most wear, followed by the forward gear, with the reverse gear receiving the least wear. All three gears, the forward, reverse and pinion, are spiral bevel type gears.

Generally a mixture of ball bearings, tapered roller bearings and caged or loose needle bearings may be found in each unit. The type bearing used is normally indicated in the procedures.

COUNTER-ROTATING UNIT

◆ See Figures 5, 6 and 7

■ Counter-rotational shifting is usually accomplished without modification to the shift cable at the shift box. In fact, the normal setup is essential for correct shifting. The only real special equipment the counter-rotating unit requires is the installation of a left-hand propeller.

The same shifting mechanisms used on standard units may also be found on their counter-rotating counterparts. There are usually a few minor differences between most standard and counter-rotating units. The biggest differences are in the shifting mechanism (in the gearcase itself). The counter-rotating shift mechanism is a mirror image of the standard shift mechanism. Mirror image shifting mechanisms produce counter-rotation of the propeller shaft.

Part of this difference is that normally the forward and reverse gears are reversed in position when compared to standard units (well, there's no actual difference in the gears, but the gears that perform the functions are reversed): what would be the forward gear on a standard unit becomes the reverse gear on a counter-rotating unit and what would normally be the reverse gear on a standard unit, becomes the forward gear on a counter-rotating unit. The pinion gear remains the same and driveshaft rotation remains the same as on a standard lower unit.

Exhaust Gases

◆ See Figure 8

At low powerhead speed, the exhaust gases from the powerhead typically escape from an idle hole in the intermediate housing on many motors. As powerhead rpm increases to normal cruising rpm or high-speed rpm, these gases are forced down through the intermediate housing and lower unit, then out with cycled water through the propeller.

Water for cooling is pumped to the powerhead by the water pump and is expelled with the exhaust gases. The water pump impeller is installed on the driveshaft for most motors (and is installed on the propshaft on the few that are not). Therefore, the water output of the pump is directly proportional to powerhead rpm.

Troubleshooting

Troubleshooting (other than disassembly and inspection) must be done before the unit is removed from the powerhead to permit isolating the problem to one area. Always attempt to proceed with troubleshooting in an orderly manner. The shot-in-the-dark approach will only result in wasted time, incorrect diagnosis, frustration and unnecessary replacement of parts.

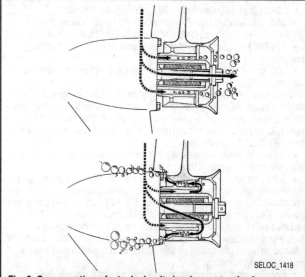

Fig. 8 Cross section of a typical unit showing route of exhaust gases with the unit in forward (top) or reverse (bottom) gear

The following steps are presented in a logical sequence with the most prevalent, easiest and less costly items to be checked listed first.

1. Check the propeller and the rubber hub. See if the hub is shredded or loosen and slipping. If the propeller has been subjected to many strikes against underwater objects, it could slip on its hub. If the hub appears to be damaged, replace it with a new hub. Replacement of the hub must be done by a propeller rebuilding shop equipped with the proper tools and experience for such work.

2. Verify the ignition switch is **OFF**, to prevent possible personal injury, should the engine start. Shift the unit into reverse gear and at the same time have an assistant turn the propeller shaft to ensure the clutch is fully engaged.

3. If the shift handle is hard to move, the trouble may be in the lower unit shift rod or in the cable itself, requiring an adjustment or a repair, or in the shift box.

4. Disconnect the remote control cable at the engine and then remove the remote control shift cable. Operate the shift lever. If shifting is still hard, the problem is in the shift cable or control box.

5. If the shifting feels normal with the remote control cable disconnected, the problem must be in the lower unit. To verify the problem is in the lower unit, have an assistant turn the propeller and at the same time move the shift cable back and forth. Determine if the clutch engages properly.

Gearcase Repair and Overhaul Tips

All Mercury gearcases use the same basic inner design in that the pinion gear is splined to the bottom of a vertical driveshaft and secured by a retaining nut (or a clip on the absolute smallest of gearcases). The driveshaft uses the pinion gear to transfer the rotational power from the crankshaft to the propeller shaft. Both the driveshaft and propeller shaft ride in replaceable bearings and a bath of gear oil. Gear oil is kept inside the case itself (and water is kept out) by propeller shaft and driveshaft housing seals at either end (bottom rear and top center) of the gearcase.

The size and type of the shafts, bearings, gears and seals will vary slightly from model-to-model. Also, all but the smallest gearcases contain Neutral and Reverse shifting capabilities. Generally speaking one of 2 major designs are employed, a vertical push-pull shifter retained by a pinned, bolted or jam nut/adjuster nut coupler assembly) is used by mostly by smaller gearcases, but can be found on models up to 50 hp, while a horizontally ratcheting shifter which connected through splines to the shift lever in the intermediate housing is used on most larger outboards.

We've divided coverage from removal and installation to complete overhaul procedures into a number of major categories based on gearcase types:
- 2.5/3.3 Hp (1-Cylinder) motors equipped with non-shifting gearcases
- 2.5/3.3 Hp and 4/5/6 Hp (1-Cylinder) with shifting (either F/N or F/N/R) gearcases
- 6-25 Hp (2-Cylinder) models
- 30/40 Hp (2-Cylinder) and 40/50 hp (3-Cylinder) models
- 55/60 (3-Cylinder) Non-Bigfoot, 60 hp (3-Cylinder) Bigfoot, 75/65 Jet/90 Hp, 75/90/115 Hp Optimax and 80 Jet/115/125 Hp Models
- 135-200 Hp (2.5L) V6 Models (both standard and counter-rotating versions)
- 200-250 Hp (3.0L) V6 Models (both standard and counter-rotating versions)

We've included diagrams and photographs from as many gearcases on which we could realistically get our hands, however keep in mind that subtle differences will occur from model-to-model or year-to-year. To ensure a successful overhaul a few general notes should be followed throughout the overhaul procedure, as follows:
- As with all repair or overhaul procedures, make sure the work area is clean and free of excessive dirt or moisture.
- Take your time when removing and disassembling components taking a moment with each to either matchmark its positioning or note its orientation in relation to other components. This goes ESPECIALLY for the placement of any shims which might be used (keep in mind that all models are NOT equipped with shims). Absent specific instructions for checking and re-shimming a gearcase (which occurs when bearings, gears and/or shafts are replaced on some gearcases) all shims must be returned to their original positions during assembly.
- Seal placement and orientation will vary on some models. Before removing any seal be sure to first note the direction in which the seal lips are facing and be sure to install the replacement seal (and you ALWAYS replace seals when they are removed) facing the same direction.
- Unless otherwise directed seal lips are packed with a light coating of Quicksilver 2-4-C W/ Teflon or an equivalent water resistant lubricant. When used, O-rings are also coated with 2-4-C or some form of marine grease during installation.
- Needle bearings which are pressed in place are generally destroyed by the removal process and should be replaced when separated from a gear, shaft or bore. Ball Bearings can usually be freed without damage, but use common sense. Once the bearing is free, clean it thoroughly, then rotate the inner/outer cage feeling for rough spots. Replace any bearing that is not in top shape.
- The correct size driver must be used for bearing or seal installation. Generally speaking a driver must be smooth (especially for seals, to prevent it from cutting the sealing surface) and should be sized to spread as much of the force over the seal or bearing surface as possible. When a bearing is installed ON a shaft, the driver MUST contact the inner race (never push on only the outer race in this case or the bearing will be damaged). When a bearing is installed IN a bore, the driver MUST contact the outer race (in this circumstance, do NOT push only on the inner race or the bearing will be damaged).
- Driveshaft bearings or bushings are often pressed straight into position from above and/or pulled into position from below. Although Mercury often has a special tool available to help draw a bushing or bearing into position in the lower portion of the gearcase, all you REALLY need to accomplish this is a sufficiently long threaded bolt with a nut and a washer that is the size of the driver you'd want to contact that bushing or bearing. You'll also need a nut and a plate to place on top of the gearcase. Place the threaded rod through one nut and plate and then down into the gearcase through the bushing/bearing (so the plate is resting against the top of the gearcase, the upper nut is resting against the plate and the threads of the rod are just protruding down into the pinion mounting area of the gearcase.). Next install the washer and lower nut over the bottom threads of the rod to support the bearings or busing. Slowly draw the bearing or bushing into position in the gearcase by holding the threaded rod from turning while at the same time turning the upper nut to draw the entire bolt, bushing/bearing, washer and

nut assembly upward until the bushing or bearing seats. Then loosen the nut on the bottom of the threaded rod in the gearcase and remove the threaded rod, plate, nut and washer.

In all cases the water pump, driveshaft oil seal or propeller shaft oil seal may be replaced without complete gearcase disassembly.

Although Mercury calls for various special tools for gearcase overhaul MOST of them come down to the appropriate sized driver (see the bullet about drivers earlier in this section), OR a universal puller or a slide hammer generally with inside jaws. Whenever possible, we prefer the use of a puller over a slide hammer, just because the force is applied more linearly, however there are simply times when the impact of a slide hammer makes the job easier, so use common sense when determining which tool you'd prefer to use for bearing, race or potentially even shaft removal.

Lastly, we've tried to provide the best procedures possible for gearcase overhaul, and in the most logical order. However, in most cases procedures for some of the earlier procedures can be performed alone, without additional procedures. For instance water pump or propeller shaft bearing carrier/seal housing procedures can normally be performed completely independently of each other. However, since the driveshaft oil seal housing is found UNDER the water pump, you'd obviously have to remove the pump first for access.

Gearcase - 2.5/3.3 Hp Models (W/Non-Shifting Gearcases)

REMOVAL & INSTALLATION

◆ See Figures 9, 10 and 11

■ On this ONE model, the water pump is mounted in the lower portion of the gearcase, between the propeller and the propeller shaft bearing carrier. So on this ONE model, there is NO need to remove the gearcase in order to service the water pump.

1. Disconnect and ground the spark plug lead to the powerhead to prevent the accidental starting of the powerhead.
2. Close the fuel supply shut off valve.
3. If necessary for gearcase overhaul, position a suitable container under the lower unit, and then remove the OIL LEVEL and OIL screw and the screws. Allow the gear lubricant to drain into the container. As the lubricant drains, catch some with your fingers from time to time, and rub it between your thumb and finger to determine if any metal particles are present. If metal is detected in the lubricant, the unit must be completely disassembled, inspected, and the damaged parts replaced. Check the color of the lubricant as it drains. A whitish or creamy color indicates the presence of water in the lubricant. Check the drain pan for signs of water separation from the lubricant. The presence of any water in the gear lubricant is bad news. The unit must be completely disassembled, inspected, the cause of the problem determined, and then corrected.
4. If necessary, remove the Propeller, as detailed in the Maintenance and Tune-Up section.
5. Position the outboard in the full up position.
6. Remove the two bolts securing the lower unit gear housing to the mid-section. One bolt is located on the front side of the mid-section (threaded downward into the top of the gear housing) and the other goes into the mid-section from the underneath side of the lower gear housing.
7. Separate the lower gear housing from the mid-section. It may be necessary to gently tap the seam around the lower gear housing with a soft-head mallet to jar it loose. The upper rectangular driveshaft and sleeve, together with the water tube, will usually remain in the mid-section when the two units are separated. The lower driveshaft will remain in the lower gear housing.

To install:
8. Apply a coating of multi-purpose lubricant (like 2-4-C with Teflon) onto the outside diameter of the long tube's end. Carefully insert the long tube into the mid-section housing, through the lower crankshaft seal.
9. Apply a coating of multi-purpose lubricant (like 2-4-C with Teflon) to the outside diameter of the rectangular driveshaft at BOTH ends. Insert the driveshaft into the tube and push it onto the end of the crankshaft.
10. Apply a thin coating of multi-purpose lubricant (like 2-4-C with Teflon) to the inside diameter of the driveshaft and water tube seals in the lower gear housing.

Fig. 9 If necessary, drain the gearcase oil

Fig. 10 Unbolt the gearcase. . .

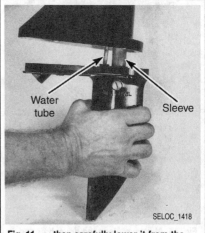

Fig. 11 . . .then carefully lower it from the housing

11. Begin to bring the midsection housing and the lower gear housing together. As the two units come closer, rotate the propeller shaft slightly to index the upper end of the lower driveshaft with the lower end of the upper rectangular driveshaft. At the same time, feed the water tube into the water tube seal. Push the lower gear housing and the mid-section together.

12. Secure the lower gear housing and the mid-section together with the two bolts (using a lockwasher and regular washer for each). The lockwasher should be installed between the bolt head and the regular washer.

13. Tighten the bolts alternately and evenly to 50 inch lbs. (5.6 Nm).

14. If not done already, properly refill the gearcase lubricant and check for leaks.

15. If removed, install the Propeller, as detailed in the Maintenance and Tune-Up section.

16. Reconnect the spark plug lead.

DISASSEMBLY

◆ See Figures 12 thru 18

1. As detailed earlier in this section, remove the gearcase from the outboard.

2. If note done during removal, drain the gearcase oil and remove the propeller.

3. Remove the water pump and hosing assembly, as detailed in the Lubrication and Cooling System section.

4. To remove the pinion gear (which is securing the propeller shaft and driveshaft), use a small prytool or STURDY screwdriver to "pop" the circlip out of the groove in the lower end of the lower driveshaft. This clip holds the pinion gear onto the driveshaft. The clip may not come free on the first try, but have patience and it will come free.

5. With one hand, pull the lower drive shaft straight up and out of the lower gear housing and at the same time catch the pinion gear with the other hand.

6. Pull the propeller shaft out of the lower gear housing. The drive gear and the shim material on the shaft will come with it (well the shim material SHOULD come out with it). Save the shim material.

7. Check the condition of the forward propeller shaft bearing. This bearing should not need to be removed unless it appears to be unfit for further service. If the bearing does not rotate freely or is rusted, it should be replaced. Remove the bearing using a suitable bearing puller (Mercury recommends using a slide hammer along with an Expander Rod and 3/8 in. Collet such as Snap-On tools #CG-41-11 and #CG-41-13 or equivalents). If a suitable puller is not available, use the alternate bearing removal method detailed later in this procedure.

8. Remove and discard the water tube seal and the driveshaft oil seal from the upper portion of the lower housing.

9. If the driveshaft bearing is no longer fit for service, use a pair of snap-ring pliers and remove the snap-ring securing the driveshaft bearing in place. Use a suitable bearing puller to remove the bearing (Mercury recommends using a slide hammer along with an Expander Rod and 7/16 in. Collet such as Snap-On tools #CG-40-4 and #CG-40-9 or equivalents).

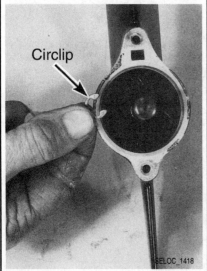

Fig. 12 Remove the circlip to free the pinion gear. . .

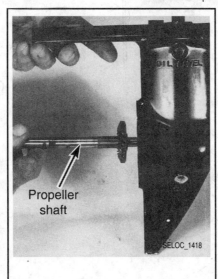

Fig. 13 . . .then remove the propeller shaft

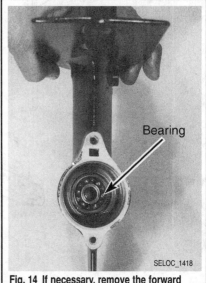

Fig. 14 If necessary, remove the forward gear bearing. . .

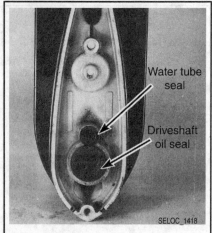

Fig. 15 Always replace the driveshaft and water tube seals

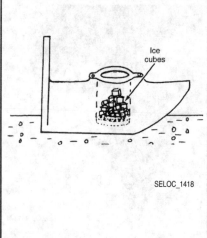

Fig. 16 Alternate bearing removal method...

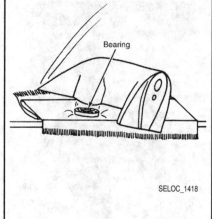

Fig. 17 ...be careful not to crack/damage the housing

10. If a suitable bearing puller is not available remove the driveshaft upper and/or the forward gear bearings as follows:

■ **The following are to be performed ONLY if the bearings must be removed and a suitable bearing puller is not available. The procedure will change the temperature between the bearing retainer and the housing substantially, hopefully about 80°F (50°C). This change will contract one metal - the bearing retainer, and expand the other metal - the housing, giving perhaps as much as 0.003 in. (0.08mm) clearance to allow the bearing to fall free. Read the complete steps before commencing the work because three things are necessary: a freezer, refrigerator, or ice chest; some ice cubes or crushed ice; and a container large enough to immerse about 1-1/2 in. (3mm) of the forward part of the lower gear housing in boiling water.**

a. After all parts, including all seals, grommets, etc., have been removed from the lower gear housing, place the gear housing in a freezer, preferably overnight. If a freezer is not available try an electric refrigerator or ice chest.

b. The next morning, obtain a container of suitable size to hold about 1-1/2 in. (3mm) of the forward part of the lower gear housing. Fill the container with water and bring to a rapid boil. While the water is coming to a boil, place a folded towel on a flat surface for padding.

c. After the water is boiling, remove the lower gear housing from its cold storage area. Fill the propeller shaft cavity with ice cubes or crushed ice. Hold the lower gear housing by the trim tab end and the lower end of the housing. Now, immerse the lower unit in the boiling water for about 20 or 30 seconds.

d. Quickly remove the housing from the boiling water; dump the ice; and with the open end of the housing facing downward, slam the housing onto the padded surface. presto, the bearing should fall out.

e. If the bearing fails to come free, try the complete procedure a second time. Failure on the second attempt will require the use of a bearing puller.

11. Once the driveshaft bearing is removed from the housing, lift the sleeve out as well.

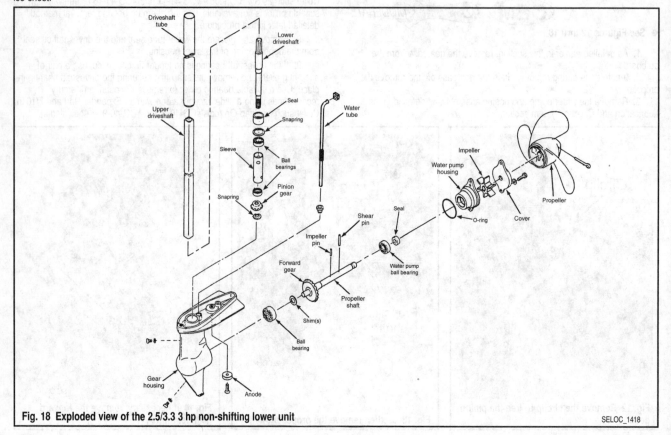

Fig. 18 Exploded view of the 2.5/3.3 3 hp non-shifting lower unit

12. If the lower driveshaft bearing must be replaced, use a 3/4 in. socket and a driver rod to tap it down into the propeller shaft cavity so it can be removed from the gearcase housing.

CLEANING & INSPECTION

◆ See Figure 18

1. Clean all metallic water pump and gearcase parts with solvent, and then dry them with compressed air. Inspect the water pump cover and base for cracks and distortion. If possible, always install a new water pump impeller while the lower unit is disassembled. A new impeller will ensure extended satisfactory service and give "peace of mind" to the owner. If the old impeller must be returned to service, never install it in reverse to the original direction of rotation. Installation in reverse will cause premature impeller failure.

2. Inspect the ends of the impeller blades for cracks, tears, and wear. Check for a glazed or melted appearance, caused from operating without sufficient water. If any question exists, as previously stated, install a new impeller if at all possible.

3. Inspect the seal surface of the driveshaft and replace IF it is deeply grooved so as to allow water into the housing. Also, check the shaft splines and replace the shaft if they are twisted or worn.

4. Clean the pinion gear and the propeller shaft with solvent. Dry the cleaned parts with compressed air.

5. Check the pinion gear and the drive gear for abnormal wear. Make sure that no gear teeth are chipped or worn.

6. Inspect the bearing surface of the propeller shaft. Check the shaft surface for pitting, scoring, grooving, imbedded particles, uneven wear and discoloration.

7. Replace the propeller shaft if the seal surface is deeply grooved, as to allow water to enter the gear housing.

8. Make sure the shear pin and impeller drive pin holes in the propeller shaft are not worn.

9. Check the straightness of the propeller shaft with a set of V-blocks. Rotate the propeller on the blocks.

10. Good shop practice dictates installation of new O-rings and oil seals regardless of their appearance.

ASSEMBLY

■ It is a good idea to temporarily install the bearing carrier into the housing before starting to tap on the prop shaft to seat the bearing. In this way the bearing carrier will act as a guide, keeping the propeller shaft straight.

1. If removed, apply a light coating of oil to the outer diameter of a NEW forward gear/propeller shaft bearing. Place the bearing into the housing with the numbered side of the bearing facing outward toward the propeller shaft. Place a washer over the end of the propeller shaft (between the gear and the bearing) to protect the shaft and bearing then use a soft-faced mallet to tap on the end of the propeller shaft (using it like a driver) to seat the bearing into the gearcase.

2. If the lower driveshaft bearing was removed, you'll need a way to pull it up into position. Use a 3/8 in. (9.5mm) threaded rod which is taller than the gearcase housing, along with a couple of nuts, and some LARGE washers (one on top that will go over the whole gearcase, and one on the bottom that is about 1 in. (25.4mm) in outer diameter (with just a 3/8 in./9.5mm hole in the middle). Proceed as follows:

a. Place the threaded rod down through the driveshaft opening in the gearcase.

b. Apply a light coating of oil to the outer diameter of the replacement bearing, then using a nut and the 1 in. (25.4mm) washer, place it onto the threaded rod with the numbered side of the bearing facing upward into the gearcase.

c. Place a very large washer or metal plate over the top of the threaded rod and position it down against the gearcase, then thread another nut all the way down the rod until it contacts the plate.

d. Making sure the rod is perfectly straight and centered in the gearcase, tighten the upper nut to draw the rod (and washer and bearing) upward until the bearing is seated in the bore, flush with the housing.

e. Remove the threaded rod, nuts and washers.

3. Position the driveshaft sleeve into the housing, on top of the lower driveshaft bearing.

4. If the upper driveshaft bearing was removed, oil the outer diameter of the replacement and position it at the top of the gearcase opening with the numbered side facing upward. Use a 13/16 in. socket and a driver to carefully tap the bearing down into position until it bottoms.

5. Install the snap ring to secure the upper driveshaft bearing in place.

6. Apply a light coating of Quicksilver Bellows Adhesive to the outer diameter of a new driveshaft oil seal and water tube seal. Install the seals, then coat the inner diameter of the seals lightly using a suitable marine grade lubricant (like 2-4-C with Teflon).

7. Place the shim material saved during disassembly onto the propeller shaft ahead of the drive gear. The shim material should give the same backlash between the pinion and drive gear as before disassembly. Insert the propeller shaft into the lower gear housing. Push the propeller shaft into the forward propeller shaft bearing as far as possible.

8. Slide the lower driveshaft down through the lower gear housing with the splined end going in first - the square end up.

9. Hold the driveshaft with one hand and slide the pinion gear up onto the lower end of the driveshaft with the other hand. The splines of the pinion gear will index with the splines of the driveshaft if the gear is rotated slightly as it is moved upward.

10. Now, comes the hard part. Snap the Circlip in the groove on the end of the driveshaft to secure the pinion gear in place. If the first attempt is not successful, try again. Take a break, have a cup of coffee, tea, (A BEER!) whatever, then give it another go.

11. Check the gearcase backlash by pulling up on the driveshaft and holding it so it will not rotate, then pushing in on the propeller shaft and attempting to rotate the shaft back and forth. There should be some play between the gears (you should hear/feel a slight clicking sound as the propeller shaft if rotated back and forth). Mercury states that the AMOUNT of play between these gears is not critical, HOWEVER NO LASH is NOT acceptable. If there is no play or way too much play, remove the driveshaft and propeller shaft again and change the thickness of the propeller shaft/forward gear shim(s).

12. Install the Water Pump and housing assembly, as detailed under Lubrication and Cooling.

13. Refill the gearcase with oil and check for leaks.

14. Install the Propeller, as detailed in the Maintenance and Tune-Up section.

15. Install the Gearcase assembly, as detailed earlier in this section.

SHIMMING

◆ See Figure 19

Backlash between two mating gears is the amount of movement one gear will make before turning its mating gear when one gear is rotated back and forth.

1. Pull upward on the lower driveshaft with one hand and at the same time push inward and rotate the propeller shaft left and right with the other hand. The manufacturer recommends there should be a small amount of

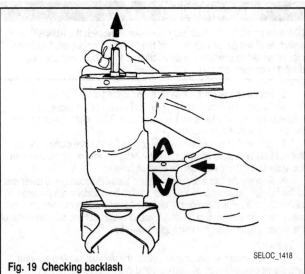
SELOC_1418
Fig. 19 Checking backlash

backlash "play" between the two gears, but does not specify exactly how much. Mercury states that it is unacceptable to have NO backlash.

■ **If there is no backlash between the pinion gear and the drive gear, either one or both of two conditions may exist. The propeller shaft may not be indexed completely into the forward propeller shaft bearing or there is too much shim material ahead of the drive gear. Therefore, first try moving the propeller shaft further into the bearing. Let's hope the "no backlash" condition is corrected. If not, the driveshaft and propeller shaft will have to be removed, shim material removed, and the parts assembled, then the backlash checked again.**

Gearcase - 2.5/3.3 Hp and 4/5/6 Hp Models (W/Shifting Gearcases)

REMOVAL & INSTALLATION

 MODERATE

◆ **See Figures 20 thru 23**

This section covers the both the 2.5/3.3 hp motors which are equipped with shifting F/N gearcases, as well as the 4/5 hp motors which are all equipped with shifting F/N/R gearcases.

1. For safety, disconnect the high tension spark plug lead and remove the spark plug before working on the lower unit.
2. Shift the gearcase into Neutral for 2.5/3.3 hp models or Reverse for 4/5 hp models.
3. If necessary for access or for overhaul purposes, remove the Propeller, as detailed in the Maintenance and Tune-Up section.

■ **It's actually a good idea to at least loosen the OIL LEVEL/VENT screw first, to make sure it will come out without a problem. You can't refill the gearcase, if you can't get the vent screw loose.**

4. If necessary for gearcase overhaul purposes, position a suitable clean container under the lower unit. Remove the **FILL** screw on the bottom of the lower unit, and then the **OIL LEVEL/VENT** screw. The vent screw must be removed to allow air to enter the lower unit behind the lubricant. Allow the gear lubricant to drain into the container.

As the lubricant drains, catch some with your fingers from time-to-time and rub it between your thumb and finger to determine if there are any metal particles present. Examine the fill plug. A small magnet imbedded in the end of the plug will pickup any metal particles. If metal is detected in the lubricant, the unit must be completely disassembled, inspected, and the damaged parts replaced.

Check the color of the lubricant as it drains. A whitish or creamy color indicates the presence of water in the lubricant. Check the container for signs of water separation from the lubricant. The presence of any water in the lubricant is bad news. The unit must be completely disassembled; inspected; the cause of the problem determined; and then corrected.

✳✳ SELOC WARNING

Check to be sure the spark plug has been removed; the RUN/STOP switch is in the STOP position and the battery, if equipped, has been disconnected to prevent any possibility of the powerhead starting during propeller removal.

5. Rotate and swing the lower end unit in the full UP tilt position.
6. Remove the rubber access plug on the side of the upper gear housing (it is normally found on the Port side of 2.5/3.3 hp models and the Starboard side of 4/5 hp models).
7. Loosen, but do not remove the bolt for the shift rod clamp. Loosen the bolt only enough to allow the shift shaft to slide free of the clamp when the lower unit is removed from the exhaust housing.
8. Remove the two attaching bolts and washers securing the lower unit to the exhaust housing (both threaded upward from underneath the anti-cavitation plate). Pull down on the lower unit gear housing and separate the lower unit from the exhaust housing. Guide the shift rod and driveshaft out of the exhaust housing as the lower unit is lowered.

To install:
9. If not already in that position, swing the exhaust housing outward until the tilt lock lever can be actuated, and then engage the tilt lock.

10. Both the lower unit and the shift lever must be in matching positions when assembling the lower unit. If necessary, move the shift lever to NEUTRAL for 2.5/3.3 hp motors or to REVERSE for 4/5 hp motors. This will enable the upper shift shaft to align with the clamp on the lower shift shaft.

■ **If necessary to shift the lower unit into reverse on 4/5 hp motors, rotate the propeller shaft while pushing down on the shift shaft to shift the lower unit. When in reverse, the propeller shaft will not rotate more than a few degrees in either direction.**

11. Coat the driveshaft splines with a light coating of Quicksilver 2-4-C or a suitable multi-purpose lubricant. Take care not to use an excessive amount of lubricant on the driveshaft splines on any Mercury Outboard.

■ **An excessive amount of lubricant on top of the driveshaft to crankshaft splines will be trapped in the clearance space. This trapped lubricant will not allow the driveshaft to fully engage with the crankshaft.**

12. Insert the driveshaft into the exhaust housing and at the same time align the water tube with the water pump cover outlet.
13. Feed the lower shift rod into the coupler on the upper shift rod. Maintain the lower unit mating surface parallel with the exhaust housing mating surface.
14. Push the lower unit toward the exhaust housing and at the same time rotate the flywheel to permit the crankshaft splines to index with the driveshaft splines.
15. Secure the lower unit in place with the two attaching bolts and washers. Tighten the bolts alternately to 50 inch lbs. (5.6 Nm) for 2.5/3.3 hp motors or to 70 inch lbs. (8 Nm) for 4/5 hp motors.
16. Tighten the shift shaft coupler bolt securely, then verify that the outboard is shifting properly.
17. Install the plug into the intermediate housing to cover the shift coupler bolt.
18. If the gearcase was drained refill it with fresh gear oil.
19. If the propeller was removed, install the propeller.
20. Reconnect the spark plug wire.

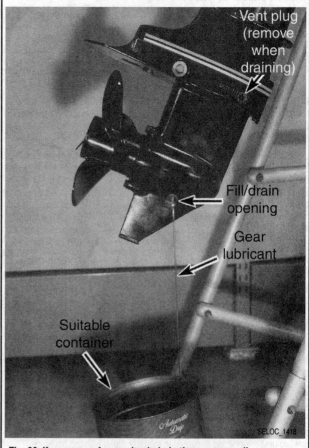

Fig. 20 If necessary for overhaul, drain the gearcase oil

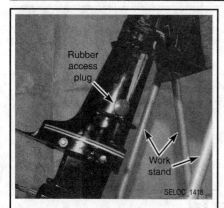

Fig. 21 Remove the rubber plug from the midsection. . .

Fig. 22 . . . for access to the shift linkage

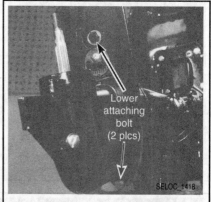

Fig. 23 Remove the gearcase attaching bolts

DISASSEMBLY

OEM ③ DIFFICULT

Bearing Carrier & Propeller Shaft

◆ See Figures 24 thru 31

■ For gearcase exploded views, please refer to Cleaning & Inspection in this section.

■ Most of the accompanying photos show a F/N/R gearcase, so the propeller shaft will differ slightly from F/N models.

1. Loosen and remove the two bolts that secure the propeller shaft bearing carrier to the lower rear of the gearcase housing.

2. There are a few different methods that are generally use to free the bearing carrier and propeller shaft from the housing, use the one that you prefer:

• Method 1 - Secure the lower unit in a vise equipped with soft jaws, in a holding fixture, or clamp it to something solid with a wood clamp. Grasp the propeller shaft and pull the shaft and bearing carrier assembly out of the lower unit.

• Method 2 - using a rubber mallet, carefully tap the ears of the bearing carrier assembly so they rotate the bearing carrier cap 90° (1/4 turn), which will break the gasket seal between the carrier and housing. Now, GENTLY tap the bearing carrier cap with a soft headed mallet. Tap evenly on both sides of the cap to jar the bearing carrier free.

• Method 3 - Clamp the propeller shaft in a vise equipped with soft jaws. Obtain a soft headed mallet and strike the lower unit just above the bearing carrier to jar the bearing carrier free.

3. With the bearing carrier removed from the lower unit housing, pull the cam follower out the end of the propeller shaft.

4. For 4/5 hp models, tap the end of the propeller shaft on the table and remove the small round guide from inside the propeller shaft.

5. Slide the bearing carrier off the end of the propeller shaft.

6. For 4/5 hp models, Slide the reverse gear and washer off the propeller shaft. If a small shim is on the back side of the reverse gear, be sure to save this shim.

7. For 2.5/3.3 hp models, use a small awl or screwdriver to compress the spring (located deeper in the shaft than the clutch), then remove the clutch assembly and spring from the propeller shaft. Replace the clutch if teeth are rounded chipped, or if wear is evident from the cam follower or spring.

■ Rounded jaws indicate improper shift cable adjustment, idle speed too high while shifting OR shifting too slowly.

8. For 4/5 hp models, disassemble the clutch/shifting mechanism as follows:

a. Place the cam follower back into the end of the propeller shaft and rotate the propeller shaft in a vertical position.

b. Press down on the propeller shaft to release the spring tension against the sliding clutch. Using a 1/8 in. pin punch, push the cross pin out through the opposite side of the sliding clutch.

c. Remove the punch and withdraw the cross pin from the sliding clutch. When the cross pin is removed, the sliding clutch should fall free of the propeller shaft.

9. Remove the spring from inside the propeller shaft.

10. Remove the O-ring from the bearing carrier and discard the O-ring.

■ In most cases, the propeller shaft bearing will be damaged during the removal procedure. Examine the bearing carefully, perform the following work only if the propeller bearing is damaged and is no longer fit for further service OR if the propeller seal must be replaced (as the seal cannot be replaced without bearing removal).

11. Grasp the inner race of the propeller shaft ball bearing. Attempt to work the race back-and-forth. The attempt should result in very little, if any "play". A very slight amount of side "play" is acceptable, because there is a little clearance in the bearing.

12. Lubricate the ball bearing with light oil. Check the action of the bearing by rotating the outer bearing race. The bearing should have a

Fig. 24 Remove the bearing carrier retaining bolts. . .

Fig. 25 Gently tap the bearing carrier free of the gearcase

Fig. 26 Remove the cam follower from the propeller shaft. . .

Fig. 27 . . .but don't loose the guide

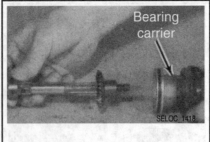
Fig. 28 Separate the propeller shaft from the bearing carrier

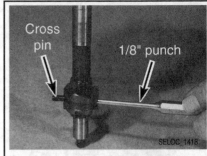
Fig. 29 Drive out the cross-pin using a punch. . .

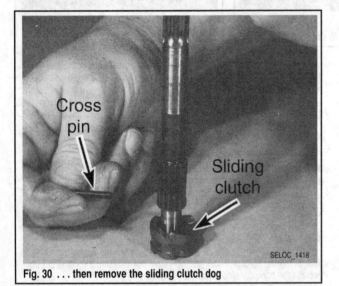
Fig. 30 . . . then remove the sliding clutch dog

Fig. 31 Remove the clutch spring from inside the propeller shaft

smooth action and no rust stains. If the ball bearing sounds or feels rough or catches, the bearing should be removed and discarded.

13. Removal of the bearing will distort it and therefore, it cannot be installed a second time. Be absolutely sure a new part is available before removing the bearing. Unfortunately, the bearing, good or bad, must be removed before the propeller shaft seal can be driven out, and of course they must be replaced with new ones.

14. Note the position of the bearing in relation to the carrier. Look for any embossed numbers or letters and observe how one shoulder of the bearing is flush with the carrier.

15. Remove the bearing using either a slide hammer with suitable collet and expanding rod (such as Snap-On #CG 41-13 and #CG 41-11) for 2.5/3.3 hp motors or a universal internal jawed puller (#91-83165M) for 4/5 hp motors.

16. Check the positioning of the seal in the propeller shaft bearing carrier (2.5/3.3 hp motors usually use a single lipped seal positioned so the lips are facing back toward the bearing, while 4/5 hp motors usually use a dual-lipped seal, but either way, note the positioning for installation purposes), then drive the old seal out of the housing from the propeller side using a small punch or driver.

Driveshaft & Water Pump Base

◆ See Figures 32 thru 35

■ Most of the accompanying photos show a F/N/R gearcase, so the shifter assembly will differ slightly from F/N models.

1. Remove the Water Pump assembly as detailed in the Lubrication and Cooling System section.

2. For 4/5 hp motors, remove the bolt securing the shift shaft assembly to the top of the driveshaft and water pump base. Carefully pull upward to remove the shift shaft and bushing. Remove the shaft O-ring from the bore in the base.

■ For 2.5/3.3 hp motors, it is usually easier to remove the shift shaft after the water pump base plate has been removed. Also on these motors, it may be easier to pry the water pump base free from the gearcase housing at the reinforced point on the side of the base, adjacent to the driveshaft.

3. Grasp the driveshaft, then lift up and remove the driveshaft and water pump base as an assembly from the lower unit housing. Remove and discard the gasket.

4. Reach inside the lower unit bearing carrier cavity and remove the pinion gear which was splined to the driveshaft and should have fallen into the case when the shaft was withdrawn.

■ Be alert for any shims behind the pinion gear. Save these shims, because they will be critical during the assembling work. Also be aware of any shims which may be installed on top of the driveshaft upper bearing.

5. Slide the water pump base off the driveshaft bearing assembly.

6. Remove the seals, O-rings and gasket materials from the water pump base assembly and/or from the top of the gearcase (there should be a water pickup tube seal in the top of the gearcase on 4/5 hp models). This base is normally equipped with a dual-lipped seal that is removed or installed from underneath the housing. Use an internal jawed seal puller to remove it from the bore in the underside of the base.

7. The upper driveshaft bearing is pressed onto the shaft itself. On 2.5/3.3 hp motors, inspect the bearing and, if replacement is required, use a bearing separator to carefully push it off the shaft. On 4/5 hp motors the upper driveshaft bearing, as well as the sleeve and bushing are all NOT serviceable, if replacement of any one part is required, the entire driveshaft assembly must be replaced.

■ If possible use Mercury 91-17351 for removal and installation of the driveshaft lower needle bearing on 4/5 hp motors. However, a suitable sized driver MAY be substituted. On 2.5/3.3 hp motors a 3/4 in. socket can be used as a driver for the lower driveshaft bearing.

8. On 2.5/3.3 hp motors, remove the driveshaft sleeve/spacer, if necessary for access.

9. If the driveshaft lower needle bearing must be removed, double-check the installed height of the current bearing before disturbing it (as the replacement is driven in from the top to a certain depth using the specified tool, OR if the special tool is not available, using a suitable driver and the measurement). If the bearing is to be replaced, use a driver to tap it out of its bore and into the gearcase. Reach in and remove the bearing.

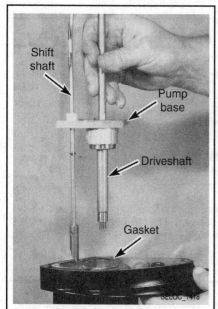

Fig. 32 View of the driveshaft, water pump and shift shaft base

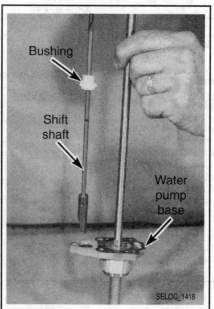

Fig. 33 Remove the shift shaft and bushing from the assembly

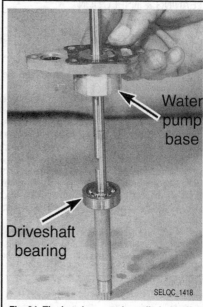

Fig. 34 The housing must be pulled up, off the driveshaft. . .

Fig. 35 . . .for access to the oil seal in the underside of the housing

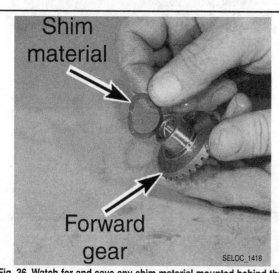

Fig. 36 Watch for and save any shim material mounted behind the forward gear

Forward Gear and Bearing

◆ See Figures 36 and 37

1. Tilt the lower unit. The forward gear will fall into your hand. If the gear fails to fall free, strike the open end of the lower unit on a block of wood to jar the gear free.

2. Although Mercury claims these gearcases, when installed on these powerheads, are NOT adjustable when it comes to lash, watch for and save any shim material (if any is found behind the forward gear). If present, this shim material is critical to obtaining the correct backlash during installation. The shim material will be located between the forward gear and the bearing.

3. The forward gear bearing is a one piece ball bearing and will not - or should not - just fall out of the lower unit cavity. The bearing must be "pulled" from the lower unit ONLY if it is unfit for further service.

4. Obtain puller (#91-27780) or a slide hammer with a jaw expander attachment. Hook the jaws inside the inner bearing race and use the slide hammer to jar the race free.

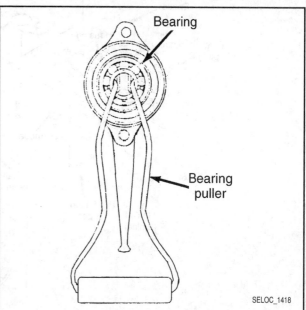

Fig. 37 If replacement is necessary, use a puller to remove the forward bearing

CLEANING & INSPECTION

◆ **See Figures 39 and 40**

1. Clean all water pump parts with solvent, and then dry them using compressed air.

2. Inspect the water pump cover and base for cracks and distortion. If possible always install a new water pump impeller while the lower unit is disassembled. A new impeller will insure extended satisfactory service. If the old impeller must be used, never install it in reverse to the original direction of rotation. Reverse installation could lead to premature impeller failure and subsequent powerhead damage.

3. Inspect the bearing surface of the propeller shaft.

4. Check the shaft surface of propeller shaft for pitting, scoring, grooving, imbedded particles, uneven wear and discoloration.

5. Check the straightness of the propeller shaft with a set of machinist V-blocks.

6. Clean the pinion gear and the propeller shaft with solvent and dry the components with compressed air.

7. Check the pinion gear and drive gear for abnormal wear.

ASSEMBLY

Sliding Clutch Assembly

2.5/3.3 Hp Models

◆ **See Figures 39 and 41**

1. Slide the spring down into the propeller shaft, then insert a narrow screwdriver or awl into the slot from the side to compress the spring sufficiently to insert the clutch into the shaft.

2. Apply a dab of Needle Bearing Assembly grease or some other suitable marine lubricant to the end of the cam follower (as this will help hold the follower in place), then insert the follower into the end of the shaft.

3. Place the propeller shaft aside for installation later.

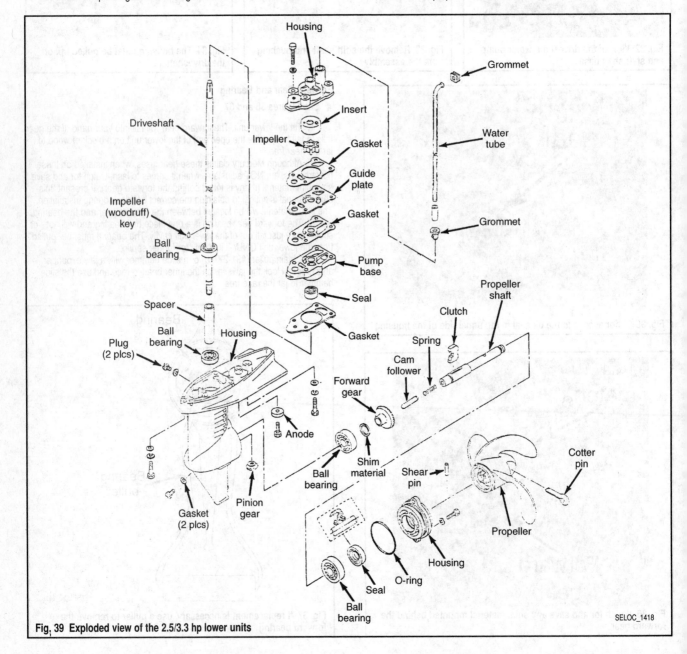

Fig. 39 Exploded view of the 2.5/3.3 hp lower units

SELOC_1418

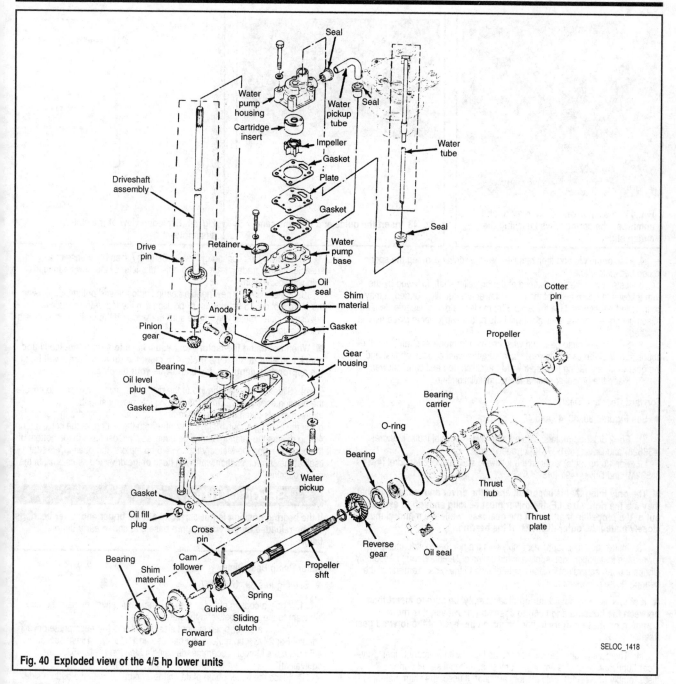

Fig. 40 Exploded view of the 4/5 hp lower units

SELOC_1418

4/5 Hp Models

◆ See Figures 40 and 42 thru 44

1. Guide the sliding clutch over the splines of the propeller shaft with the hole in the clutch aligned with the hole in the propeller shaft. The clutch may be installed either way, but it's best to install it so the side with the least amount of wear faces the forward gear.

■ If you don't have 3 arms, the easiest way to proceed is to secure a small screwdriver or awl in the bench vise so you can hold the propeller shaft over it, compressing the spring, as you use your remaining hand to install the cross pin.

2. Slide the spring down into the propeller shaft. Insert a narrow screwdriver into the slot (or hold the shaft up to the vise mounted screwdriver) and compress the spring until approximately 1/4 in. (12mm) is obtained between the top of the slot and the screwdriver.

3. Insert the cross pin into the sliding clutch and through the space held open by the screwdriver.

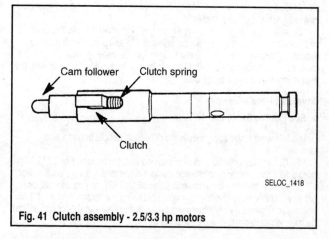

SELOC_1418

Fig. 41 Clutch assembly - 2.5/3.3 hp motors

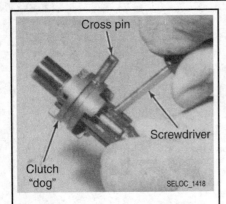

Fig. 42 Use a narrow screwdriver to and compress the spring while installing the clutch slider

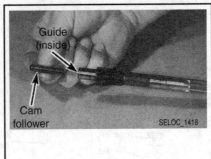

Fig. 43 Insert the guide and then the cam follower

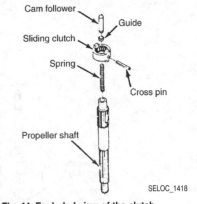

Fig. 44 Exploded view of the clutch assembly

4. Center the pin and then remove the screwdriver allowing the spring to pop back into place.

5. Install the guide into the end of the propeller shaft, followed by the cam follower. As both ends of the cam follower are equally rounded, it may be installed either way. Like the 2.5/3.3 hp models, a dab of Needle Bearing Assembly grease or the equivalent will help hold the follower in place as you proceed.

6. Slide the reverse gear and washer onto the propeller shaft (from the opposite end of the cam follower). Make sure the washer goes on first and the gear teeth are facing forward toward the clutch dog and cam follower.

7. Place the propeller shaft aside for installation later.

Forward Gear and Bearing

◆ See Figures 39, 40, 45 and 46

1. For 2.5/3.3 hp motors, obtain the following special tools: Mandrel (#38628) and Driver (#91-37312) or equivalents.

2. For 4/5 hp motors, obtain the following special tools: Mandrel (#91-84532M) and Driver (#91-84529M).

■ The only thing that is important about the driver and mandrel is that they are the right size. I.E. the driver must be long enough to extend out of the propeller shaft end of the gearcase, while the mandrel must closely match the outer-diameter of the bearing.

3. Install the ball bearing assembly with the numbered side facing toward the installation tool. Apply a light coating of Needle Bearing Assembly lubricant or an equivalent marine grade grease to the outer diameter of the bearing (to help in assembly).

■ If shims were removed during disassembly, be sure to insert them between the housing and bearing assembly. However, it is more common on these motors to find shims on the back of the forward gear itself.

4. Thoroughly lubricate the bearing and gear with Mercury Super Duty Gear Lubricant, or equivalent. Insert any shim material saved during disassembly (if present), onto the forward gear. If used, the shim material should give the same amount of backlash between the pinion gear and the forward gear as before disassembly.

5. Insert the lubricated forward gear and shim material into the lower unit with the teeth of the gear facing outward. On some models it may be necessary to press the gear inward to seat it in the bearing, when this is the case, use the propeller shaft and bearing carrier as a guide to make sure the gear is inserted square to the bearing.

Driveshaft Needle Bearing

◆ See Figures 39, 40 and 47

The following steps apply only if the driveshaft needle bearing was removed.

1. Obtain a needle bearing removal and installation tool. For 2.5/3.3 hp motors Mercury simply recommends using a driver and a 3/4 in. socket. For 4/5 hp motors, Mercury recommends using #91-17351, or you can still use just any suitably sized driver (however, if not using the factory tool on 4/5 hp motors, you'll need the measurement taken during disassembly of how deep the needle bearing sits in the gearcase).

2. Apply a light coating of Needle Bearing Assembly grease or a suitable marine grade grease to the outer diameter of the new bearing to help with installation.

3. Either slide the new needle bearing onto the end of the tool (if using the Merc tool on 4/5 hp motors) or position it in the top of the gearcase driveshaft bore. Either way, make sure the side with the embossed numbers is facing upward.

■ When not using the factory tool you'll have to stop periodically and check installed depth. One trick is to mark the driver where it will be at the appropriate depth so you'll know as your tapping.

4. Slide the tool into the top of the driveshaft cavity. Take care to ensure the bearing does not tilt. Drive the bearing into place using the measurements taken during disassembly (for 2.5/3.3 hp motors this should place the bearing about 1/8 in. / 3.17mm) from the bottom of the bore, but on 4/5 hp motors the bearing should be even less, almost flush to the bottom of the bore). If using the Mercury tool on 4/5 hp motors, the bearing will seat itself at the correct location when the head of the drive rod seats against the guide bushing.

✳✳ SELOC CAUTION

If the bearing position is not correct, both the upper and lower bearings will be reloaded by the water pump base and cause early bearing failure.

Water Pump Base Plate

◆ See Figures 39, 40 and 48

1. Apply a coating of 2-4-C with Teflon or an equivalent marine grade lubricant to the cavity between the new seal lips.

2. Obtain a driver or equivalent size socket and the appropriate driver handle. For 2.5/3.3 hp motors you can use an 11/16 in or 16mm socket. For 4/5 hp motors Mercury recommends using a Mandrel (#91-84530M or equivalent).

3. Place the water pump plate on the workbench with the bottom side facing up.

4. Install the oil seal using the driver with the oil seal (spring side), lip facing up. Tap or drive the seal slowly down into the pump base until the seal bottoms in the bore.

5. On 2.5/3.3 hp motors, lubricate and install a new shift shaft O-ring into the pump base assembly.

Driveshaft, Pinion Gear and Water Pump Base

◆ See Figures 39, 40 and 49 thru 51

1. If removed on 2.5/3.3 hp motors, press a new lower driveshaft bearing into position on the shaft. The safest way to do this is to fabricate a driver from a metal tube. The driver should be 3 5/8 in. (92.07mm) long, with an outer-diameter of 3/4 in. (19.05mm) and in inner diameter of 13/32 in. (10.31mm). Position the driveshaft so the crankshaft end is resting on a padded surface, then use the fabricated tool to carefully tap the bearing down until the tool is flush with the pinion end of the shaft. Also, at this point, be sure to position the driveshaft sleeve into the gearcase housing, above the lower needle bearing.

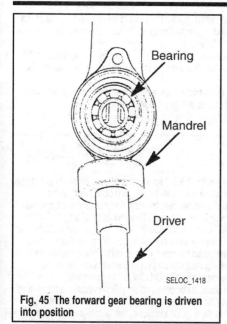

Fig. 45 The forward gear bearing is driven into position

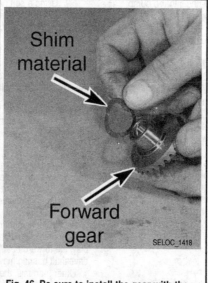

Fig. 46 Be sure to install the gear with the shims saved during disassembly

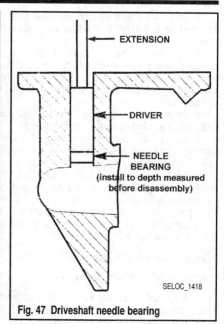

Fig. 47 Driveshaft needle bearing

Fig. 48 Install the seal from the underside of the water pump base plate

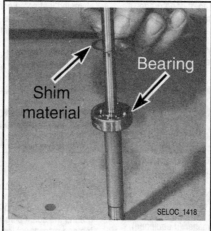

Fig. 49 If used, place the shim(s) over the driveshaft and bearing

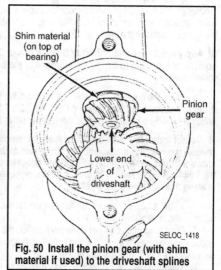

Fig. 50 Install the pinion gear (with shim material if used) to the driveshaft splines

2. Slide any shim material saved during disassembly (more likely on 4/5 hp motors only) over the driveshaft and down onto the bearing. If present, the shim material should give the same amount of backlash between the pinion gear and the other two gears as before disassembling.

■ On 4/5 hp motors, mercury normally directs technicians to assemble the shifter shaft components AFTER the driveshaft and water pump base assembly has been installed to the gearcase, but from our experience we've found you can position the shifter before the assembly is installed (and you usually need to on 2.5/3.3 hp models anyway). However, on 4/5 hp motors, the choice is really yours.

3. Insert the pinion gear into the lower unit. Now, with the other hand, lower the driveshaft, shift shaft and pump base down into the lower unit with the shift cam going in first. Slowly rotate the driveshaft until the driveshaft splines index with the splines of the pinion gear.

4. Check to make sure there is at least SOME backlash/freeplay between the forward gear and pinion gear at this point.

■ On F/N shifting gearcases for 2.5/3.3 hp motors, Mercury does suggest that during assembly you can check backlash by pushing downward on the driveshaft and upward on the pinion gear (since it floats on the shaft splines) then rocker the forward gear back and forth. You should hear a light clicking sound, resulting from 0.002-0.006 in. (0.050-0.152mm) of free-play. If however an excessive amount of free-play, about 0.012 in. (0.30mm) or more, exists components should be inspected for wear. If there is NO or insufficient free-play, the forward bear/bearing may not be properly seated (or the lower driveshaft bearing may be positioned TOO LOW in the gearcase).

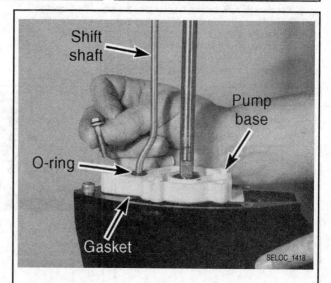

Fig. 51 Install the shift shaft and secure using the retaining bolt on 4/5 hp models

5. If not done already, lubricate and install a new O-ring into the pump base opening for the shift shaft. If applicable, slide a new shift shaft bushing down onto the shift shaft.

6. For 2.5/3.3 hp motors, install the shift shaft in the top of the gearcase, making sure the tapered side of the shift cam faces toward the pinion shaft. It is a good idea to lightly lubricate the shaft with 2-4-C with Teflon before installation (it helps to lubricate the shaft O-ring and/or bushing, as applicable).

7. Position a new water pump base plate gasket on the gearcase. Be sure the pins and bolt holes are correctly aligned.

8. Lubricate the seal in the pump base lightly with a good grade of water resistant lubricant. Slide the water pump base down the driveshaft and over the upper bearing.

9. For 4/5 hp motors, if not done already, install the shift shaft in the top of the water pump base plate and into the gearcase, making sure the tapered side of the shift cam faces toward the pinion shaft. Install the shift shaft retainer plate and tighten the retaining bolt to 70 inch lbs. (8 Nm).

10. Install the Water Pump assembly as detailed in the Lubrication and Cooling System section.

Bearing Carrier and Propeller Shaft

◆ **See Figures 39, 40 and 52 thru 54**

1. If the seal and bearing were removed, place the bearing carrier on the workbench with the propeller end facing down.

2. Obtain driver/installation tool (either a 13/16 in. socket for 2.5/3.3 hp models or #91-83174M for 4/5 hp models). Apply a coating of good grade water resistant lubricant to the lip of the seal, and then install the seal with the flat side down (if the sides are different).

■ **Apply a light coating of Needle Bearing Assembly grease or an equivalent lubricant to the outside surface of the bearing to ease installation.**

3. Place the bearing, numbered side up, into the bearing carrier. Obtain driver and suitable mandrel (either #37312 for 2.5/3.3 hp models or #91-84536M for 4/5 hp models) and tap the bearing down into the bearing carrier.

■ **Do not forget to install any shim material removed from behind the reverse gear during disassembling. This shim material, and in some cases a thrust washer, must be installed between the reverse gear and the ball bearing assembly in order to ensure proper mesh between the reverse gear and the pinion gear.**

4. Install a new O-ring around the bearing carrier. Apply a coating of water resistant lubricant to the O-ring.

5. On 4/5 hp models, if not done earlier, slide the washer (if equipped), and reverse gear onto the end of the propeller shaft with the teeth of the gear facing away from the bearing carrier.

6. Slide the bearing carrier over the propeller shaft.

7. Check to be sure the cam follower is on the end of the propeller shaft. This is a loose piece and easily forgotten. Again, if not done earlier, apply a liberal amount of water resistant lubricant on the cam follower and insert the follower into the end of the propeller shaft.

8. Insert the propeller shaft and bearing carrier into the lower unit until the end of the shaft indexes into the forward gear.

9. Using a soft head mallet, lightly tap around the circumference of the carrier until it is fully seated in place. Check to be sure the two bolt holes are aligned with the lower unit holes.

10. Install the two washers and bolts. Tighten the bolts alternately to 50 inch lbs. (5.6 Nm) for 2.5/3.3 hp motors or to 70 inch lbs. (8 Nm) for 4/5 hp motors.

11. Refill the gearcase and check for leaks, then install the gearcase to the outboard as detailed earlier in this section.

SHIMMING

On F/N shifting gearcases for 2.5/3.3 hp motors, Mercury does suggest that during assembly you can check backlash by pushing downward on the driveshaft and upward on the pinion gear (since it floats on the shaft splines) then rocker the forward gear back and forth. You should hear a light clicking sound, resulting from 0.002-0.006 in. (0.050-0.152mm) of free-play. If however an excessive amount of free-play, about 0.012 in. (0.30mm) or more, exists components should be inspected for wear. If there is NO or insufficient free-play, the forward bear/bearing may not be properly seated (or the lower driveshaft bearing may be positioned TOO LOW in the gearcase).

Other than this, the manufacturer gives no specific instructions for setting up the backlash on these units. However, IF any shim material found during disassembly was placed in its original locations, the backlash should be acceptable (and if any shim material was found, somebody MUST have shimmed it at some point right?).

On these gearcases, generally the manufacturer states: "The amount of play between the gears is not critical, but no play is unacceptable." Therefore, if after the assembly work is complete, the gears are indeed locked, the unit must be disassembled and shim material removed from behind the forward or reverse gears, using a "trial and error" method. If the lower unit is allowed to operate without "some" backlash, very heavy wear on the three gears will take place almost immediately.

Gearcase - 6-25 Hp Models

REMOVAL & INSTALLATION

◆ **See Figures 55 thru 58**

1. For safety and to prevent accidental engine start, disconnect and ground the spark plug leads to the powerhead or for electric start models, tag and disconnect the battery cables.

2. If necessary for overhaul purposes, remove the Propeller, as detailed in the Maintenance and Tune-Up section.

■ **It's actually a good idea to at least loosen the OIL LEVEL/VENT screw first, to make sure it will come out without a problem. You can't refill the gearcase, if you can't get the vent screw loose.**

3. If necessary for gearcase overhaul purposes, position a suitable clean container under the lower unit. Remove the FILL screw on the bottom of the lower unit, and then the OIL LEVEL/VENT screw. The vent screw must

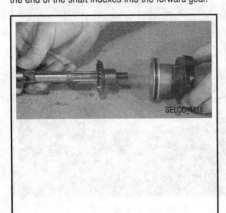

Fig. 52 Install the propeller shaft to the bearing carrier...

Fig. 53 ...then, if not done earlier, install the cam follower

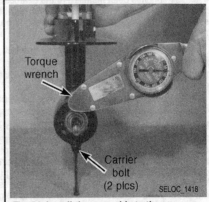

Fig. 54 Install the assembly to the gearcase and secure

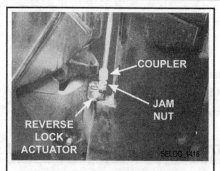

Fig. 55 For two piece shift shafts, loosen the jam nut and coupler on the shift rod. . .

Fig. 56 . . .otherwise, unlatch and remove the retainer at the top of the shift shaft

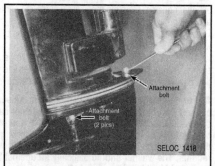

Fig. 57 Then unbolt the gearcase (6-15 hp models shown)

be removed to allow air to enter the lower unit behind the lubricant. Allow the gear lubricant to drain into the container.

As the lubricant drains, catch some with your fingers from time-to-time and rub it between your thumb and finger to determine if there are any metal particles present. Examine the fill plug. A small magnet imbedded in the end of the plug will pickup any metal particles. If metal is detected in the lubricant, the unit must be completely disassembled, inspected, and the damaged parts replaced.

Check the color of the lubricant as it drains. A whitish or creamy color indicates the presence of water in the lubricant. Check the container for signs of water separation from the lubricant. The presence of any water in the lubricant is bad news. The unit must be completely disassembled; inspected; the cause of the problem determined; and then corrected.

4. Because of the difference in shifter couplings between most 6-15 hp motors and most 20/25 hp motors, on the 6-15 hp models, shift the outboard into Forward (it should make access to the reverse lock actuator easier). We suggest leaving the 20/25 hp motors in Neutral.

5. If not done already, rotate the outboard unit to the full UP position and engage the tilt lock pin.

■ **On most 20/25 hp electric start models, you will have to remove the starter for access to the shift shaft retainer for the 1-piece shaft.**

6. Locate and detach the shift shaft coupling depending upon the style of shift shaft as follows. Most models have a 1-piece shift shaft which is secured by a latch assembly at the base of the powerhead, while a few might contain a 2-piece unit which is secured by an adjuster and locknut assembly at the front of the intermediate housing/swivel assembly. The quickest way to tell the difference is to follow the shift shaft up from the gearcase nose cone (it should first be visible part of the way up the intermediate housing where it will be exposed on most models for the reverse lock actuator). If the shaft is NOT exposed there or if you cannot find a threaded adjuster (coupler) and Jam nut (as shown, in an accompanying photo) then you are likely dealing with a 1-piece shift shaft. Once you've determined what type of shift shaft you have, then proceed as follows depending upon the type:

• 1-piece shift shafts - look under the engine cowling at the front base of the powerhead (roughly in a straight line above the gearcase nose cone) and locate the top of the shift shaft. Unlatch the retainer by pivoting it clockwise about 1/4 turn off the shaft retainer. Then remove the retainer to free the shaft.

• 2-piece shift shafts - mark the threads to show the exact position of the top of the coupler and the bottom of the jam nut, then loosen the jam nut and unthread the adjuster (coupler) until the upper and lower shafts are free.

■ **By measuring or marking the jam nut and coupler positions on the threads before removal you may save yourself some trouble when adjusting the shifter during installation.**

7. For 6-15 hp motors, although the shift shaft coupling designs vary slightly, all should have a small reverse lock actuator, visible on the shift rod at the front of the motor. Scribe a matchmark onto the shaft to make sure the actuator will be returned to the same position, then loosen the retaining screw and remove the actuator.

8. For 20/25 hp motors, there should be a small flat washer, near the top o the shaft (retained by a tang on the shaft and a wire loop for the reverse lock rod on the outboard). The washer will likely pull off the shaft as the gearcase is lowered, but be sure to retain it, as it is necessary for reassembly.

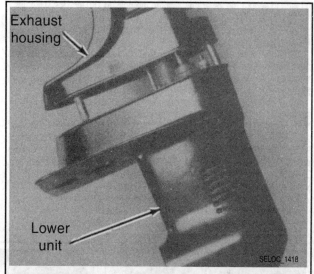

Fig. 58 Upon installation, guide the 2 shafts up into the housing as the lower unit is lifted into place

9. Remove the bolts and washers securing the lower unit to the exhaust housing. For 6-15 hp models there should be 3 bolts, one threaded upward on each side and one threaded downward at the front of the gearcase. On 20/25 hp models there should be 2 bolts threaded upward on either side of the gearcase.

10. Pull down on the lower unit gear housing and separate the lower unit from the exhaust housing.

11. Guide the shift rod and driveshaft out of the exhaust housing as the lower unit is removed.

To install:

12. For 6-15 hp motors, make sure the motor linkage is in the FORWARD gear position, but temporarily place the gearcase in NEUTRAL, then check the length of any equipped with a 1-piece shift rod. As measured from the top of the water pump base to the center of the hole for the retainer on top of the rod, the length should be as follows. If necessary, rotate shift shaft CLOCKWISE to decrease distance or COUNTERCLOCKWISE to increase distance:

• Short shaft models - 16 1/2 in. (419mm)
• Long shaft models - 22 in. (559mm)
• Short shaft models - 27 1/2 in. (698mm)

13. Once checked or adjusted, place the shift rod on 6-15 hp gearcases into the FORWARD position. Also on these models, place a 1/4 in. 6.4mm diameter bead of RTV silicone sealer on the water pump base (as the rear of the base, where it meets the gearcase).

14. If not done already, swing the exhaust housing outward until the tilt lock lever can be actuated, and then engage the tilt lock.

15. Apply a liberal coating of 2-4-C Marine Lubricant With Teflon to the driveshaft splines (but NOT to the top of the shaft as it may prevent the shaft from fully seating in the crankshaft).

16. Position the driveshaft and shift shaft into the exhaust housing. Guide the two shafts up into the housing as the lower unit is lifted to meet the

exhaust housing. Maintain the lower unit mating surface parallel with the exhaust housing mating surface.

17. As the lower unit approaches closer to the exhaust housing, align the water tube with the water pump tube guide.

18. If necessary rotate the flywheel slightly to permit the crankshaft splines to index with the driveshaft splines (alternately you COULD put the gearcase in gear and slowly turn the propeller shaft in the proper direction).

■ **For models with a 1-piece shift shaft, make sure the shaft is roughly aligned with the latch assembly so the retainer can be installed and secured. On 20/25 hp motors, MAKE SURE you place the flat washer over top of the shift shaft before it passes through the wire loom for the reverse lock rod located inside the engine cowl, just under the retainer. The washer should basically sit against the underside of the loom, held by the tab on the shaft.**

19. Apply a light coating of Loctite® 271 or an equivalent threadlocking compound on the threads of the gearcase retaining bolts, then thread them in place to secure the lower unit. Tighten the bolts to 180 inch lbs./15 ft. lbs. (20 Nm) for 6-15 hp models or to 40 ft. lbs. (54 Nm) for 20/25 hp models.

20. For models with a 1-piece shift shaft, install the retainer then pivot the latch locking it into place.

21. For models with a 2-piece shift shaft, bring the two halves of the shift shaft together and thread the coupler back into position to join them. Either match the old marks/measurement or adjust the shifter so that the gearcase properly engages and releases Forward and Reverse gears.

22. For 6-15 hp motors, align the matchmarks made earlier, then install the reverse lock actuator loosely on the shift shaft and check adjustment of the lock rod. Basically the rod can be positioned in one of two variances, so the outboard can ONLY be tilted up in Forward gear OR so it can be tiled in Forward AND Neutral. For the later, the actuator must be positioned so there is 0.25 in. (6.4mm) of clearance between the reverse lock hook and actuator. Once positioned properly, tighten the retaining screw to 45 inch lbs. (5.1 Nm).

23. If the gearcase was drained refill it with fresh gear oil.

24. If the propeller was removed, install the propeller.

25. Reconnect the spark plug wires and/or the battery cables.

DISASSEMBLY

Propeller Shaft and Bearing Carrier

◆ **See Figures 59 thru 65**

■ **For gearcase exploded views, please refer to Cleaning & Inspection in this section.**

✻✻ SELOC CAUTION

The threads on the bearing carrier are LEFT HAND. Be sure to turn the bearing carrier tool in CLOCKWISE for removal and in the COUNTERCLOCKWISE for assembly.

1. For 20/25 hp models, loosen and remove the 3 screws that secure the O-ring retainer plate and O-ring to the bearing carrier. Then, remove the plate and O-ring from the gear housing.

2. Obtain the appropriate bearing carrier tool (basically a special spanner-like socket), either #91-13664 for 6-15 hp models or #91-93843—1 for 20/25 hp models. Unscrew the bearing carrier by turning CLOCKWISE, then grasp the propeller shaft and pull to remove the bearing carrier and propeller shaft assembly from the lower unit gear housing.

■ **At this point the cam follower on the front (inner) end of the propeller shaft is free to slide out of the shaft. Recover the follower if it falls into the gearcase during removal.**

3. Begin disassembly of the propeller shaft by separating the bearing carrier, propeller shaft and reverse gear. Although Mercury claims most of these gearcases, when installed on these powerheads, are NOT adjustable when it comes to lash, watch for and save any shim material from the back side of the reverse gear. If present, the shim material is critical in obtaining the correct backlash during assembling.

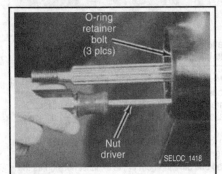

Fig. 59 On Bigfoot models, remove the bolts securing the O-ring retainer plate

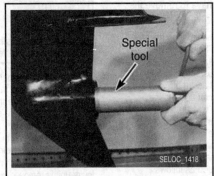

Fig. 60 Unscrew the LEFT-HAND THREAD bearing carrier by turning CLOCKWISE

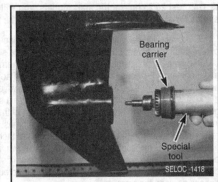

Fig. 61 Remove the bearing carrier and propeller shaft assembly

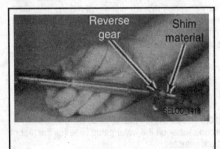

Fig. 62 Slide the reverse gear from the shaft and retain any bearing material

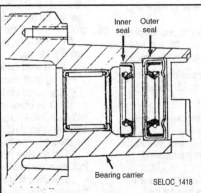

Fig. 63 Use a prybar, seal remover or even a punch from the opposite side to remove the seals

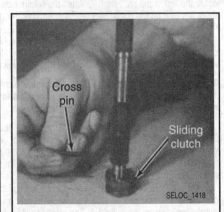

Fig. 64 Remove the cross pin, then remove the sliding clutch. . .

4. For 6-15 hp models, use a punch to carefully tap the cross-pin out of the sliding clutch on the propeller shaft. Be sure to tap on the end of the cross-pin that is NOT grooved. Remove and discard the pin (which must be replaced during assembly).

5. On 20/25 hp models, use a screwdriver or punch to carefully unwind the cross-pin retainer spring from the sliding clutch. To remove the cross-pin on thee models, insert the cam follower back into the end of the propeller shaft (with the flat end facing inward toward the cross-pin). Gently push the propeller shaft and cam follower against a flat surface then use a punch to push the cross-pin out of the sliding clutch.

6. When the cross pin is removed the sliding clutch should fall free of the propeller shaft. Examine the clutch on both ends for broken, chipped or rounded off teeth. Rounded jaws may be caused by improper shift cable adjustment, excessively high engine idle speed and/or operator shifting too slowly.

7. Check the cam follower for worn, flat or blunt end.

8. Remove the cam follower spring from inside the propeller shaft. Check the spring to be sure it is not broken or distorted.

9. Now, it's time to turn your attention to the bearing carrier. Remove and discard the O-ring on the bearing carrier.

✳✳ SELOC WARNING

Perform the following work only if the seal(s) have been damaged and are no longer fit for service. Removal of the seal(s) destroys their sealing qualities. Therefore, the seal/s cannot be installed a second time. Be absolutely sure new seal/s are available before removing the old ones. Of course, on the other hand, we think it is rare that you would want to open up a gearcase and not replace the seals. Keep in mind, they're the first line of defense for protecting the gearcase and all of it's internal components.

10. Inspect the condition of the seals in the bearing carrier. If the seals appear to be damaged and replacement is required there are multiple methods available to remove them. In most cases, you can either use a punch to drive the 2 seals from the rear (gearcase) side of the carrier (this is recommended on the 6-15 hp models), or in other cases a seal remover or pry tool can be used to carefully extract them (this is recommended on 20/25 hp models). If necessary a slide hammer and expanding jaw attachment will do the trick. However in all cases, it is usually easiest to proceed with the carrier carefully mounted in a soft-jawed vise for stability when removing the seal. Either way, be sure to check the position of the seal lips before removal (for installation purposes). In most cases, there should be dual back-to-back seals mounted in these bearing carriers.

■ **Perform the following work only if the bushing (6-15 hp) or needle bearing (20/25 hp) is damaged and is no longer fit for further service. Removal of the bushing or bearing will distort it. Therefore, bearings with prior service cannot be installed a second time. Be absolutely sure a new part is available before removing the bearing. Unfortunately, the oil seals, good or bad, must be removed before the bearing can be driven out, and of course they must be replaced with new seals. But we've already covered why they should be replaced anyway.**

11. Note the position of the bushing or bearing in relation to the carrier. Look for any embossed numbers or letters and which shoulder of the bearing is flush in the carrier. Remove the bushing or bearing using a punch and a hammer, or obtain a suitable sized socket and driver, and then drive the bushing/bearing from the carrier.

Driveshaft and Pinion Gear Removal

◆ **See Figures 66 and 67**

If not done already, remove the water pump assembly, as detailed in the Lubrication and Cooling section. On 6-15 hp motors, instructions contained there include removal of the water pump base plate and shift shaft assembly, as well as replacing the dual back-to-back seals found in the underside of the plate.

For 6-15 hp models, pull upward on the driveshaft with one hand and with the other hand, reach into the lower unit cavity and grasp the pinion gear. As the driveshaft is removed the pinion gear and thrust washer will fall free in your hand. When removing the driveshaft watch for and save any shim material found on top of the pinion gear. Although Mercury claims most of these gearcases, when installed on these powerheads, are NOT adjustable when it comes to lash, if present, this shim material is critical to obtain the correct backlash during installation.

On 20/25 hp models, if not done already, you can pull the shift shaft assembly from the gear housing at this time. Be sure to remove and replace the outer O-ring from the shift shaft retainer. Next, clamp the driveshaft in a vise equipped with soft jaws and remove the either nut or bolt on the end of the driveshaft securing the pinion gear to the shaft. Pull the driveshaft out of the lower unit housing. Remove the pinion gear and tapered roller bearing.

Forward Gear and Bearing

◆ **See Figure 69**

1. Tilt the lower unit. With the pinion removed, the forward gear and tapered roller bearing should fall into your hand. If the gear and bearing fail to fall free, strike the open end of the lower unit on a block of wood to jar the gear and bearing free.

2. Examine the tapered bearing for rust, pitted or discolored (bluish), rollers. If the bearing is worn or damaged than it must be replaced.

3. If the bearing must be replaced, use a Universal Puller Plate/Bearing Separator (#91-37241 or equivalent), and press the forward gear off the tapered roller bearing. Remove and save any shim material.

4. If the bearing is replaced or if the race mounted in the gearcase nose cone otherwise requires replacement use a suitable expanding (internal) jawed puller and either puller plate or a slide hammer to remove the race from the gearcase.

Driveshaft Seals, Bushings and/or Needle Bearings

◆ **See Figures 70 and 71**

The configuration of the 20/25 hp and 6-15 hp gearcases differ slightly when it comes to the configuration of the driveshaft seals, bushing (if applicable) and/or bearings.

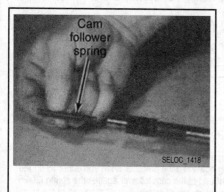

Fig. 65 . . . and the cam follower spring from the shaft

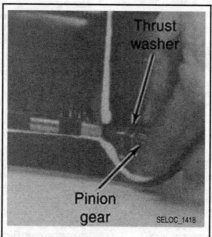

Fig. 66 On 6-15 hp models the pinion nut is just splined in place

Fig. 67 On 20/25 hp models, use a suitable wrench to loosen the pinion nut or bolt

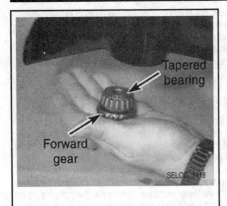

Fig. 69 The forward gear and tapered bearing should fall free of the gearcase once the pinion is removed

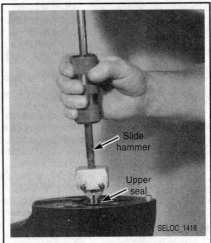

Fig. 70 Removing seals from the top of the 20/25 hp gearcase

Fig. 71 Use a threaded rod to draw the bearing cup puller into the bottom of the 20/25 hp gearcase

The 6-15 hp models utilize dual back-to-back seals mounted in the top of the water pump base plate. Under the plate an upper bushing is installed in the top of the gear housing with a lubrication sleeve underneath it. At the bottom of the housing, just beneath the sleeve and right above the pinion gear is a needle bearing. Be sure to measure the installed height of the needle bearing before removal to ensure proper installation.

20/25 hp models utilize dual back-to-back mounted upper seal assembly that is installed directly into the top of the gearcase housing. Underneath the seals you'll find a roller bearing and a sleeve. The sleeve should not be removed from the gearcase, and if you're not using the special tools for bearing removal/installation, you should check the installed height of the bearing before removal.

■ Perform the following work only if the seals, bushing and/or needle bearing(s) have been damaged and are no longer fit for service. Removal of the seals and bearings destroy the sealing qualities of the seal, and the bearing cages are distorted. Therefore, the seals and bearings cannot be installed a second time. Be absolutely sure a new seal or bearing is available before removing the old component.

1. Examine the upper seal for deterioration, hardness and wear. Only one seal is visible for inspection, but if the seal appears to be damaged or excessively worn, both the seals should be replaced. The seals are installed back-to-back. The upper seal prevents water from entering the lower unit and the other seal prevents lubricant from escaping.

■ For 6-15 hp motors, more details on replacing the seals which are found in the underside of the water pump base plate, can be found under Water Pump in the Lubrication and Cooling System section.

2. Remove the driveshaft seals, bushing and/or needle bearings and forward gear bearing race. Use a punch or seal removal tool to remove the seals from the water pump base plate on or 6-15 hp models. Use a Slide Hammer tool (#91-34569A1) or a universal slide hammer with a jaw expander attachment to remove the seals from the top of the gearcase on 20/25 hp models.

■ Reminder, before removing the driveshaft bushings or bearings, ALWAYS measure the installed height for use during installation when special tools are not available.

3. Inspect the driveshaft bushing and bearing surface to determine the condition of the bushing and bearing. If worn or damaged, replace the shaft, bushing AND bearing. HOWEVER, IF either appears to be spinning it their bores, the gear housing itself must be replaced.

4. If the bushing and/or bearing requires replacement on 6-15 hp models, proceed as follows:

a. Using an expanding rod (such as Snap-On® CG 40-4) and collet (such as Snap-On® CG 40A6) and a suitable slide hammer, remove the upper bushing from the top of the gearcase.

b. Check the top of the driveshaft bore for a burr that is sometimes formed when the upper bushing is installed. If present it may prevent the sleeve from being removed, so carefully remove the burr with a knife.

c. Remove the lubrication sleeve for access to the lower bearing.

d. Use a suitable driver to tap the bearing out into the bottom of the gearcase.

5. If the upper bearing requires replacement on 20/25 hp models, proceed as follows:

a. Using a suitable driver carefully tap the bearing THROUGH the sleeve and out into the bottom of the gearcase.

b. To remove the lower bearing cup, use a threaded rod to draw the bearing cup puller (#91-44385, which is essentially a split bushing which will compress slightly to enter the bearing cup, then will expand as the rod through the center is tightened, thereby gripping the bearing cup) up into the cup from within the gearcase (by tightening the nut). Then, tap on the top of the threaded rod to drive the cup out into the bottom of the gearcase. Loosen and remove the threaded rod, then remove the cup and puller tool.

CLEANING & INSPECTION

OEM ③ DIFFICULT

◆ See Figures 72 thru 75

1. Clean all water pump parts with solvent, and then dry them using compressed air.

2. Inspect the water pump cover and base for cracks and distortion. If possible always install a new water pump impeller while the lower unit is disassembled. A new impeller will insure extended satisfactory service. If the old impeller must be used, never install it in reverse to the original direction of rotation. Reverse installation could lead to premature impeller failure and subsequent powerhead damage.

3. Inspect the bearing surface of the propeller shaft.

4. Check the shaft surface of propeller shaft for pitting, scoring, grooving, imbedded particles, uneven wear and discoloration.

5. Check the straightness of the propeller shaft with a set of machinist V-blocks.

6. Clean the pinion gear and the propeller shaft with solvent and dry the components with compressed air.

7. Check the pinion gear and drive gear for abnormal wear.

ASSEMBLY

OEM ③ DIFFICULT

Sliding Clutch

◆ See Figures 72 thru 77

1. Apply a small amount of lower unit lubricant onto the spring and cam follower. Slide the spring down into the propeller shaft and either insert the cam follower with the flat end of the cam follower against the spring (20/25 hp models) or insert a small punch or screwdriver into the end of the shaft against the spring (6-15 hp models).

2. Slip the sliding clutch into position over the shaft. On 20/25 hp models, the clutch must be installed with the longer end of the clutch towards the reverse gear. Guide the sliding clutch over the splines of the propeller shaft with the hole in the sliding clutch aligned with the slot in the propeller shaft.

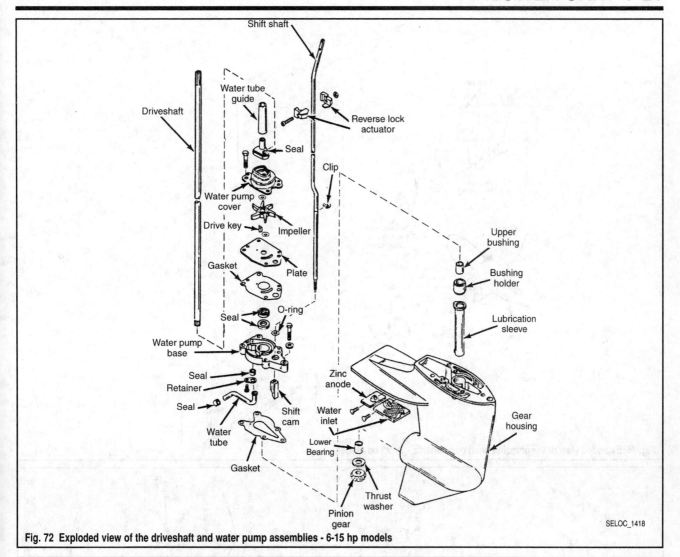

Fig. 72 Exploded view of the driveshaft and water pump assemblies - 6-15 hp models

SELOC_1418

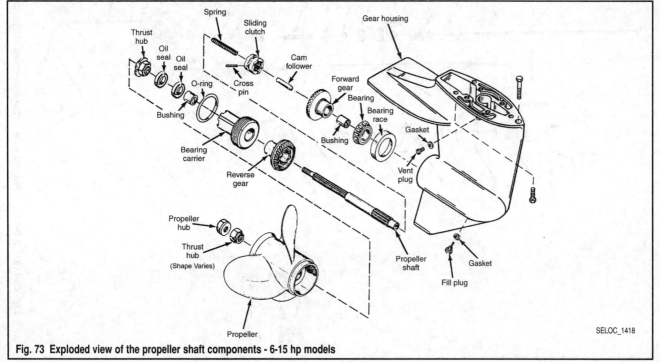

Fig. 73 Exploded view of the propeller shaft components - 6-15 hp models

SELOC_1418

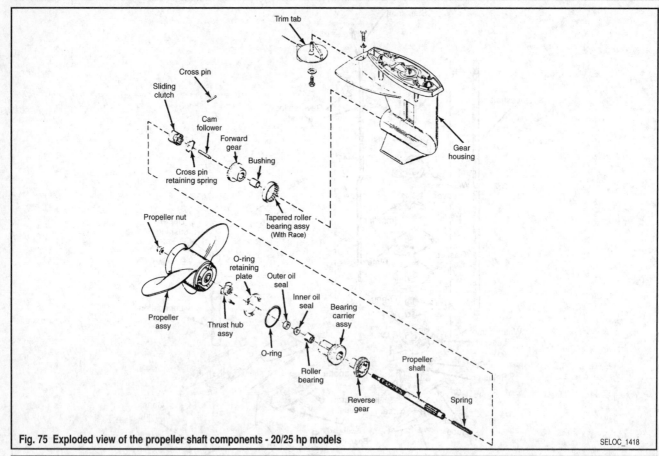

Fig. 75 Exploded view of the propeller shaft components - 20/25 hp models

SELOC_1418

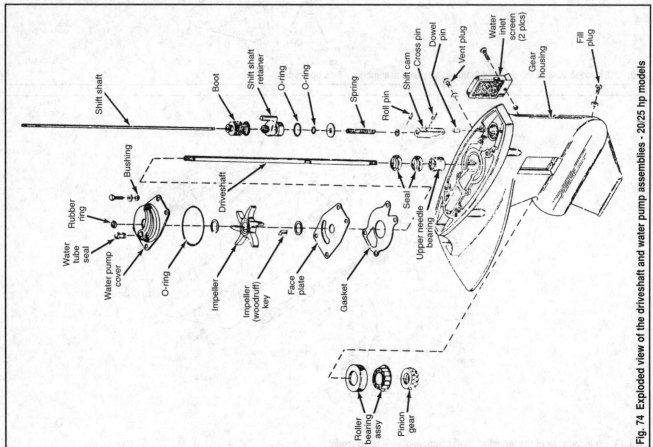

Fig. 74 Exploded view of the driveshaft and water pump assemblies - 20/25 hp models

3. Rotate the propeller shaft vertically and compress the spring by pushing the shaft and the cam follower (20/25 hp models) or punch/screwdriver (6-15 hp models) against a solid object. For 6-15 hp models it is easier to put the other end of the shaft or punch in a vise so you can pull the shaft back of the tool with the vise holding the tool still.

4. While holding the shaft against spring tension install the cross-pin as follows:

a. For 6-15 hp models, insert the NON-grooved end of a NEW pin in through the sliding clutch and shaft until it contacts the punch or screwdriver. Continue to push the pin into place as you slowly and carefully pull the shaft back from the vise, withdrawing the punch or screwdriver.

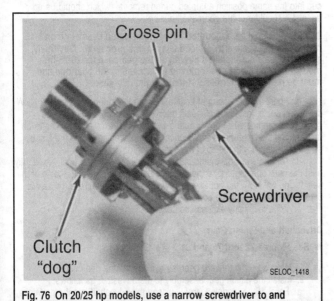

Fig. 76 On 20/25 hp models, use a narrow screwdriver to and compress the spring while installing the clutch slider

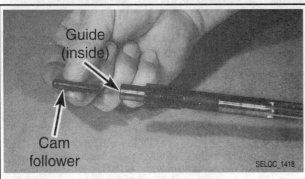

Fig. 77 Insert the guide and then the cam follower

b. For 20/25 hp models, hold the spring compressed, and at the same time, insert the cross pin into the sliding clutch pushing the screwdriver or punch out the other side. Center the cross-pin in the sliding clutch. Install the cross-pin retainer spring into the recess on the sliding clutch. Pull out the cam follower.

5. Place a dab of multi-purpose lubricant onto both ends of the cam follower. Insert the cam follower with the flat end of the cam follower against the spring and the rounded end protruding. This position will allow smooth operation between the shift cam and the follower.

6. Slide the reverse gear and washer onto the propeller shaft. Do not forget to install any shim material removed from behind the reverse gear during disassembling. This shim material and, in some cases a thrust washer, must be installed between the reverse gear and the ball bearing assembly in order to ensure proper mesh between the reverse gear and the pinion gear.

7. Place the propeller shaft aside for installation later, once the propeller shaft bearing carrier is ready.

Forward Gear and Bearing

◆ **See Figures 72 thru 75, 78 and 79**

1. Insert any shim material saved during disassembly into the lower unit bearing race cavity. If used, the shim material should give the same amount of backlash between the pinion gear and the forward gear as before disassembling. Apply a small amount of multipurpose lubricant to the bore for the forward bearing race.

2. Insert the bearing race squarely into the lower unit housing. Using mandrel #91-13658 (6-15 hp models) or #91-36571 (20/25 hp models) and a suitable driver, press the bearing race down into the bore recess.

■ **Mercury states that you can use the bearing carrier tool as a driver on 6-15 hp models, or even the propeller shaft (if a suitable nut is threaded onto the end in order to take the tapping or pressing force) on 20/25 hp models.**

3. Position the forward gear tapered bearing over the forward gear. Use a suitable mandrel and press the bearing flush against the shoulder of the gear. Always press on the inner race (the portion of the bearing that contacts the gear), never on the cage or the rollers.

4. Thoroughly lubricate the tapered roller bearing and gear with Mercury Super Duty Gear Lubricant, or equivalent. Insert the lubricated forward gear and bearing into the lower unit with the teeth of the gear facing outward.

Driveshaft Needle Bearing(s)

◆ **See Figures 72 thru 75 and 80**

The lower bearing or bearing cup on these models is drawn up into position in the gearcase from underneath (near the pinion positioning). Although Mercury sells special tools for this, all you REALLY need to accomplish this is a sufficiently long threaded bolt with a nut and a washer that is the size of the driver you'd want to contact that bushing or bearing. You'll also need a nut and a plate to place on top of the gearcase.

Place the threaded rod through one nut and plate and then down into the gearcase through the bushing/bearing (so the plate is resting against the top of the gearcase, the upper nut is resting against the plate and the threads of the rod are just protruding down into the pinion mounting area of the

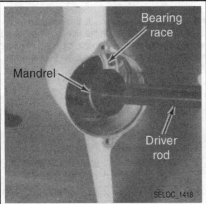

Fig. 78 If removed, install a new forward bearing race using a suitable driver

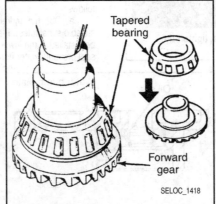

Fig. 79 If separated, press a new bearing onto the forward gear

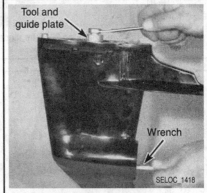

Fig. 80 Use a special tool or threaded rod and washers to draw the lower bearing or cup into position (as applicable)

gearcase.). Next install the washer and lower nut over the bottom threads of the rod to support the bearings or bushing. Slowly draw the bearing or bushing into position in the gearcase by holding the threaded rod from turning while at the same time turning the upper nut to draw the entire bolt, bushing/bearing, washer and nut assembly upward until the bushing or bearing seats. Then loosen the nut on the bottom of the threaded rod in the gearcase and remove the threaded rod, plate, nut and washer.

■ **One potential problem with using something other than the special tool to install the bearing or bearing cup is that IF there is no shoulder against which the bearing or cup is mounting you'll need to measure to ensure it is installed to the correct depth (height), we recommended that you take these measurements before removal.**

1. For 6-15 hp models, proceed as follows:
a. The lower needle bearing has to be pulled up into the housing from the housing gear cavity. Therefore, a special tool (#91-824790A1) is the best way to do this. However, that or threaded rod type tool is about the only way this bearing can be installed.
b. Using a special tool (such as the ones mentioned earlier), position the plate on top of the lower unit. Lower the long threaded bushing installer into the lower unit from the top until the threads protrude into the lower unit cavity.
c. Now, insert the bearing onto the mandrel (so the numbered/lettered side will be facing down toward the bottom of the gearcase) and install this over the threaded rod in the gear cavity. Insert and hold the threaded nut holder onto the threaded rod. With a wrench, turn the rod nut clockwise and draw the bushing up into the lower housing. Install it to the same height as measured before removal. It appears from the illustrations that this is somewhere above the flush line on the bottom of the gearcase.
d. Remove the special tool and insert the lubrication sleeve down into the housing.
e. Using the same installation tool install, but with the mandrel position reversed (on top), install the upper bushing down into the top of the gearcase. Alternately, a driver CAN be used to CAREFULLY tap the bushing into position.

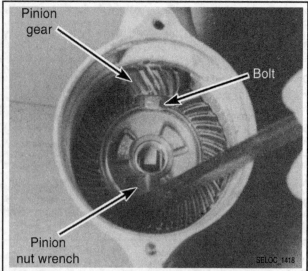

Fig. 81 Install the pinion and, on 20/25 hp models, tighten the pinion fastener

2. For 20/25 hp models, proceed as follows:
a. Place the upper bearing over the bearing retainer bore with the lettered side facing upward, then using a suitable driver and mandrel carefully press or tap the bearing into the gearcase. Continue to press or tap the bearing into the housing sleeve until the bearing is JUST below the oil hole, which should place it a total depth of about 1 in. (25.4mm) below the top of the gearcase.
b. Next, install the lower bearing cup into the driveshaft bore using either the special tool set from Mercury (#91-44385, #91-29310, #11-24156, #91-44385 and #12-34961) or a suitable set threaded rod and set of washers/nuts as described earlier. Using the tools, carefully draw the bearing cup into the same position as noted during removal (which should be up against a shoulder of the gearcase housing).

■ **Since the seals are mounted in the water pump base plate on 6-15 hp models the seals are serviced later in this procedure. Obviously, since the seals are installed directly in the gearcase for 20/25 hp models they needed to be removed for access to the bearings and must be replaced regardless of their original condition.**

c. Apply a light coating of Loctite® 274 to the outer diameter of 2 NEW driveshaft seals. Install the oil seals back-to-back using a suitable size socket and hammer. The upper seal prevents water from entering the lower unit and the lower seal prevents lubricant from escaping. Place the first oil seal into the bore with the lip of the seal facing down.
d. Tap the seals down into position. Install the first seal until it is slightly past the bore opening. Place the second seal into the bore with the lip of the seal facing up. Now, tap both seals down into the bore until the top seal is 3/16 in. (4.7mm) below the bore.
e. Wipe away any excess Loctite®.

Driveshaft and Pinion Gear

◆ **See Figures 72 thru 75 and 81**

1. Make sure the forward gear is positioned in the gearcase.
2. Install the pinion gear assembly to the driveshaft as follows:
a. For 6-15 hp models, carefully insert the driveshaft into the gearcase, then lift the driveshaft slightly as you position the thrust washer (with the grooved side facing downward toward the pinion gear). Next, position the pinion gear over the splines. Rock or rotate the driveshaft slightly as necessary to align the splines.
b. For 20/25 hp models, insert the lower roller bearing into the cup, then position the pinion gear into the housing with the teeth meshed with the forward gear. While holding the gear in position carefully insert the driveshaft, rotating it slightly back and forth until the splines align. Apply a light coating of Loctite® 271 to the threads of the pinion nut (or bolt, as equipped), then install and tighten the fastener to 159 inch lbs. (18 Nm).

Water Pump Plate

◆ **See Figures 72 thru 75 and 82 thru 84**

The manufacturer has produced a new style water pump base plate and gasket for the 6-15 hp models. The older style plate and gasket could fail under certain conditions causing excessive water leakage. A newer style plate and silicon impregnated gasket has replaced the previous model which is no longer available. Replacements will always be the new style.

1. On the 6-15 hp models, prepare the water pump base plate as follows:
a. Obtain mandrel (#91-13655 and the appropriate driver handle or suitable substitutes). Place the water pump plate on the workbench as it is to be installed in the lower unit. The seals are installed back-to-back. The upper seal prevents water from entering the lower unit and the lower seal prevents

Fig. 82 Old and new style water pump base plates - 6-15 hp models

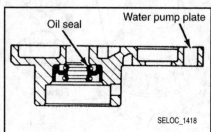

Fig. 83 Seal installation to the 6-15 hp water pump base plates

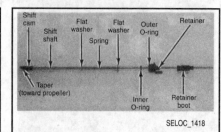

Fig. 84 Shift shaft component arrangement - 20/25 hp models

lubricant from escaping. Coat both seal lips with a good grade of water resistant lubricant and coat the outer diameter of both seals using Loctite® 271.

b. Install the first oil seal using the driver with the oil seal lip facing down. Install the second oil seal with the oil seal lip facing up.

c. Install a new seal onto the end of the water tube. Turn the base plate over and insert the water tube into the bottom of the base plate. Secure the water tube and seal with the retaining ring and Phillips head screw.

d. Coat the surface of a new O-ring with a good grade of water resistant lubricant. Install the O-ring into the recess in the base plate.

e. Slide the end of the shift shaft through the O-ring and pump base plate.

f. Install the E-clip into the slotted ring on the shift shaft. This clip must be installed after the shift shaft has been installed through the water pump base plate.

g. Screw the threaded shift cam onto the end of the shift shaft. The tip of the shaft should be visible through the hole in the shift cam.

h. Place a new base plate gasket over the driveshaft and onto the lower unit housing. Be sure the small bypass hole is aligned with the lower unit housing.

i. Slide the assembled shift shaft and base plate onto the lower unit with the cam end going in first. Be sure the cam tapered surface is pointing towards the forward gear. Lower the pump base down onto the lower unit housing guiding the water tube into the grommet at the rear of the housing.

j. Apply Loctite® 271 or equivalent to the threads of the bolt and secure the water pump base plate to the lower unit housing. Tighten the bolt to 50 inch lbs. (5.6 Nm).

2. For 20/25 hp models, proceed as follows:

a. Assemble the shift shaft components as shown in the accompanying illustration. Be sure new inner and outer O-rings are installed. Lightly coat the O-rings with multipurpose lubricant to assist in assembly.

b. Carefully examine the retainer boot, if there is any evidence of deterioration or excessive wear, replace the boot.

c. Slide the assembled shift shaft into the lower unit with the cam end going in first. Be sure the tapered end of the cam is facing towards the propeller.

3. Install the Water Pump assembly along with a new base plate gasket as detailed in the Lubrication and Cooling System section.

Bearing Carrier

◆ See Figures 72 thru 75 and 85 thru 89

■ The bearing carrier used on these models utilize two seals installed back-to-back. One seal prevents lubricant in the lower unit from escaping and the other prevents water from entering. Be sure to install the seals correctly with the lips of the seals facing in the proper direction.

Prepare the bearing carrier for installation with the propeller shaft.

1. For 6-15 hp models, proceed as follows:

a. If the bushing was removed, position the carrier with the threaded side (gearcase side) facing upward. Apply a light coating of gear lube to the bushing, then press or carefully tap the bushing down into the housing using a suitable mandrel (like #91-824785A1).

b. If the seals were removed, invert the carrier with the threaded (gearcase) side facing downward. Apply a light coating of Loctite® 271 or equivalent to the OUTER diameter of the 2 new seals. Install the first (inner) seal with the lip facing inward toward the bushing. Use the same mandrel as recommended for bushing installation to carefully push or tap the seal down into position. Next install the outer (fish line cutter) seal with the lip facing upward toward the mandrel.

c. Apply a light coating of 2-4-C with Teflon between the lips of the seals, to the threads of the bearing carrier and pilot diameter and to a NEW bearing carrier O-ring.

d. Install the O-ring onto the bearing carrier.

2. For 20/25 hp models, proceed as follows:

a. If the bearing was removed, place the bearing carrier onto a hard surface with the threaded end down. Apply a light coating of oil to the bearing bore in the carrier. Insert the needle roller bearing into the bearing bore with the lettered end of the bearing facing up. Using a suitable size socket and driver, press or tap the needle bearing down into the bore until the lower edge of the bearing is 1.1406 in. (28.9mm) from the threaded end (currently the bottom) of the bearing carrier housing.

b. Apply a light coating of Loctite® 271 or equivalent to the OUTER diameter of the 2 new seals. Position the inner seal into the seal bore with the lip of the seal facing down. Using a suitable size socket and driver, press or tap the seal down into the bore until the top of the seal is at a depth of 0.875 in. (22.2mm) from the end (currently the top) of the bearing carrier housing.

c. Apply a light coating of 2-4-C with Teflon between the seals.

d. Position the outer seal into the seal bore with the lip of the seal facing up. Using a suitable size socket and driver, press or tap the seal down into the bore until the top of the seal is at a depth of 0.875 in. (22.2mm) from the end (currently the top) of the bearing carrier housing.

e. Apply a light coating of 2-4-C with Teflon to the lips of the seals. Also apply this lubricant to the bearing carrier threads, to the O-ring groove and to the pilot diameter.

3. Insert the propeller shaft with the reverse gear, through the bearing carrier. Check the cam follower on the end of the propeller shaft. This is a loose piece and is easily forgotten.

4. Using the same method as was used in the disassembly process, insert the bearing carrier into the lower unit.

5. Screw the bearing carrier into the lower unit housing COUNTERCLOCKWISE (since it has left-hand thread). Using the carrier tool referenced during disassembly, tighten the bearing carrier to 85 ft. lbs. (115 Nm) for 6-15 hp models or to 80 ft. lbs. (109 Nm) for 20/25 hp models.

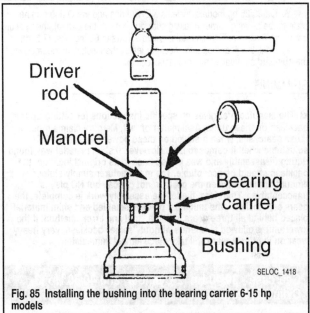

Driver rod

Mandrel

Bearing carrier

Bushing

SELOC_1418

Fig. 85 Installing the bushing into the bearing carrier 6-15 hp models

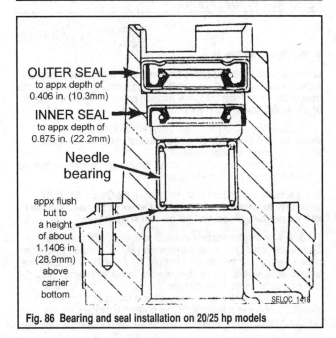

OUTER SEAL
to appx depth of
0.406 in. (10.3mm)

INNER SEAL
to appx depth of
0.875 in. (22.2mm)

Needle bearing

appx flush but to a height of about 1.1406 in. (28.9mm) above carrier bottom

SELOC_1418

Fig. 86 Bearing and seal installation on 20/25 hp models

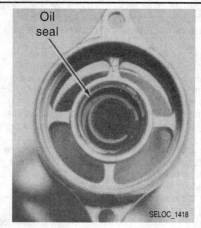

Fig. 87 Dual back-to-back seals are installed at the rear of the carrier

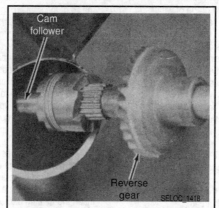

Fig. 88 Sometimes it is easier to install the propeller shaft separately from the bearing carrier (be sure the cam follower stays in position)

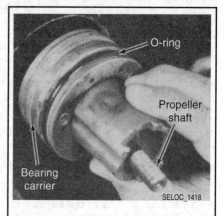

Fig. 89 Bearing carrier installation (20/25 hp shown, others similar)

6. On 20/25 hp models, install a NEW O-ring and the O-ring retainer plate to the bearing carrier (making sure not to pinch the O-ring), then secure it with the 3 screws. Tighten retainer plate screws to 65 inch lbs. (7 Nm).

7. Refill the gearcase and check for leaks, then install the gearcase to the outboard as detailed earlier in this section.

SHIMMING

■ The manufacturer gives no specific instructions for setting up the backlash on these units. As a matter of fact, Mercury claims most of these gearcases, when installed on these powerheads, are NOT adjustable when it comes to lash. However, if shim material was found during disassembly and was placed back in its original location, the backlash should be acceptable. The manufacturer simply states: "The amount of play between the gears is not critical, but NO play is unacceptable." Therefore, if after the assembly work is complete, the gears are "locked", the unit must be disassembled and shim material placed behind all three gears, using a "trial and error" method. If the lower unit is allowed to operate without "some" backlash, very heavy wear on the three gears will take place almost immediately.

Gearcase - 30/40 Hp (2-Cylinder) and 40/50 Hp (3-Cylinder) Models

REMOVAL & INSTALLATION

 MODERATE

◆ See Figures 90 thru 94

■ Unlike the 4-stroke models in the same size range, these 2-stroke midrange motors use a 2-piece shift-shaft assembly joined at the center front of the motor (near the swivel housing cover/lower motor mount) by a threaded coupler and lock nut.

1. For safety and to prevent accidental engine start, disconnect and ground the spark plug leads to the powerhead or for electric start models, tag and disconnect the battery cables.

2. If necessary for overhaul purposes, remove the Propeller, as detailed in the Maintenance and Tune-Up section.

■ It's actually a good idea to at least loosen the OIL LEVEL/VENT screw first, to make sure it will come out without a problem. You can't refill the gearcase, if you can't get the vent screw loose.

3. If necessary for gearcase overhaul purposes, position a suitable clean container under the lower unit. Remove the FILL screw on the bottom of the lower unit, and then the OIL LEVEL/VENT screw. The vent screw must be removed to allow air to enter the lower unit behind the lubricant. Allow the gear lubricant to drain into the container.

As the lubricant drains, catch some with your fingers from time-to-time and rub it between your thumb and finger to determine if there are any metal particles present. Examine the fill plug. A small magnet imbedded in the end

of the plug will pickup any metal particles. If metal is detected in the lubricant, the unit must be completely disassembled, inspected, and the damaged parts replaced.

Check the color of the lubricant as it drains. A whitish or creamy color indicates the presence of water in the lubricant. Check the container for signs of water separation from the lubricant. The presence of any water in the lubricant is bad news. The unit must be completely disassembled; inspected; the cause of the problem determined; and then corrected.

4. If not done already, rotate the outboard unit to the full UP position and engage the tilt lock pin.

5. Shift the gearcase into Neutral.

6. Locate the shift shaft coupling at the front of the motor about halfway between the bottom of the lower engine cowling and the top of the gearcase, just above a shelf and just below the cover for the lower motor mount. There is a coupler (adjuster) nut threaded on top of a jam nut. Loosen the jam nut JUST enough to unthread the coupler. By not moving the jam nut too far you can use the position during installation to keep from having to adjust the coupler. However, if the nut interferes with gearcase removal, simply mark the shaft at that point before removing the nut to accomplish the same goal.

7. Matchmark the trim tab to the bottom of the gearcase housing (to preserve alignment during installation), then remove the screw and washer securing the trim tab to the case. Remove the trim tab for access.

8. Remove the gearcase retaining nut and washer that is mounted high in the trim tab cavity.

9. Remove the 4 bolts and washers securing the lower unit to the exhaust housing.

10. Pull down on the lower unit gear housing and separate the lower unit from the exhaust housing.

11. Guide the shift rod and driveshaft out of the exhaust housing as the lower unit is removed.

To install:

12. If not done already, swing the exhaust housing outward until the tilt lock lever can be actuated, and then engage the tilt lock.

13. Apply a liberal coating of 2-4-C Marine Lubricant With Teflon to the driveshaft splines (but NOT to the top of the shaft as it may prevent the shaft from fully seating in the crankshaft).

14. Position the driveshaft and shift shaft into the exhaust housing. Guide the two shafts up into the housing as the lower unit is lifted to meet the exhaust housing. Maintain the lower unit mating surface parallel with the exhaust housing mating surface.

15. As the lower unit approaches closer to the exhaust housing, align the water tube with the water pump tube guide.

16. If necessary rotate the flywheel slightly to permit the crankshaft splines to index with the driveshaft splines (alternately you COULD put the gearcase in gear and slowly turn the propeller shaft in the proper direction).

17. Secure the lower unit in place with the 4 attaching bolts. Tighten the bolts to 40 ft. lbs. (54 Nm).

18. Reconnect the shift shaft by threading the coupler back into position, then tightening the jam nut securely against it.

■ Once the shift shaft is reconnected and the jam nut is tightened, the BOTTOM of the Jam nut should be about flush with the top of the spray plate on the housing just below the nut.

Fig. 90 Locate the shift shaft coupling at the front of the powerhead. . .

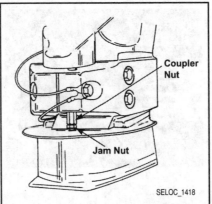

Fig. 91 . . .loosen the jam nut and unthread the coupler

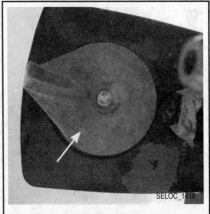

Fig. 92 Next, remove the trim tab. . .

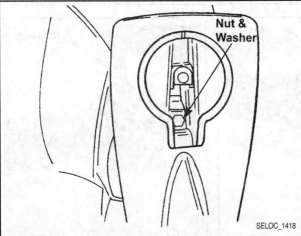

Fig. 93 . . .for access to the gearcase nut and washer located underneath

Fig. 94 In addition to the nut, these gearcases use 4 retaining bolts (2 on either side)

19. Install the gearcase retaining nut and washer located in the trim tab cavity. Tighten the nut to 40 ft. lbs. (54 Nm).

20. Align the matchmark made earlier, then install the trim tab and secure using the bolt and washer. Tighten the retaining bolt to 16 ft. lbs. (22 Nm).

■ **Some Mercury service literature says that the torque spec should be 23.3 ft. lbs. (31.6 Nm) for 30/40 hp motors, but we don't think the size of the bolt varies between the powerheads using this gearcase, so we'd probably use the lower torque value on them as well, just to be safe. If you're worried about it, you might consider using a small amount a threadlocking compound to make sure it doesn't loosen in service and/or to check it after the first couple of outings to make sure it is secure with the lower torque.**

21. If the gearcase was drained refill it with fresh gear oil and check for leaks.

22. If the propeller was removed, install the propeller.

23. Reconnect the spark plug wires and/or the battery cables.

DISASSEMBLY

OEM ③ DIFFICULT

Bearing Carrier and Propeller Shaft

◆ See Figures 95 thru 101

■ **For gearcase exploded views, please refer to Cleaning & Inspection in this section.**

1. If not done already, remove the gearcase from the motor, then clamp the skeg in a vice equipped with soft jaws or between two pieces of soft wood.

2. The bearing carrier on these gearcases are retained by 2 bolts. Sometimes the bolts are also secured by lock tabs. If so equipped, bend the lock tabs away from the bolts with a punch.

3. Loosen and remove the 2 bearing carrier retaining bolts.

4. Two methods are available to remove the bearing carrier from these lower units. The first and safest method is to use an internal jawed puller or slide hammer with internal jaws (such as #91-27780). The second and less desirable method is to tap the carrier so the bolt ears are no longer aligned at 12 and 6 o'clock and to use a mallet to actually drive the lower unit off the carrier. In both methods, heat carefully applied to the outside of the lower unit, can assist in removing the bearing carrier.

5. Remove the propeller shaft and bearing carrier gearcase. When pulling the propeller shaft free, be aware that the cam follower will be free to fall from the front of the shaft (and often does), so locate and keep track of it. Set the assembly aside for disassembly later.

6. Remove and discard the bearing carrier O-ring.

7. To replace bearings or seals in the bearing carrier, proceed as follows:

    a. Clamp the bearing carrier in a vise equipped with soft jaws. Use a slide hammer and internal jawed puller to remove the reverse gear. If the reverse gear ball bearing remained in the bearing carrier, remove the bearing using a slide hammer.

■ **You've got 2 choices on seal removal. IF the needle bearing inside the carrier needs to be replaced you can drive it out the rear (propeller side) of the carrier and, in doing so, it will take the seals out with it. However, if ONLY seal replacement is necessary, use small internal jawed puller or seal remover (or if absolutely necessary a small pry tool) to remove the seals from the propeller end of the carrier.**

    b. If the needle bearing requires replacement, place the carrier so the oil seal (propeller) end is facing downward, then insert a driver through the reverse gear side of the carrier. Drive the bearing, along with the oil seals out the back of the carrier.

Fig. 95 Bearing carrier retaining bolts

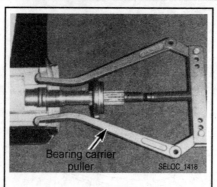

Fig. 96 Bearing carrier removal with a threaded, jawed puller

Fig. 97 An internal jawed puller and slide hammer can separate the reverse gear and bearing from the carrier

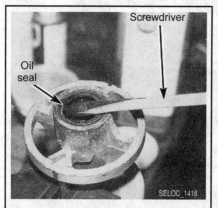

Fig. 98 If only the seals are being replaced CAREFULLY pry or pull them from the housing

Fig. 99 Remove the cross pin retaining spring. . .

Fig. 100 . . .then remove the pin and sliding clutch

Fig. 101 Finally, remove the spring from the end of the propeller shaft

c. If ONLY the oil seals need to be removed for replacement, remove then by carefully pulling or prying them from the bore in the carrier. Take note of seal orientation, most models use two seals mounted back-to-back.

8. To disassemble the propeller shaft for cleaning, inspection and/or parts replacement, proceed as follows:

a. Position the propeller shaft cam follower against a solid object. Insert a thin blade screwdriver or an awl under the first coil of the cross-pin retainer spring and rotate the propeller shaft to unwind the spring from the sliding clutch. Take care not to over-stretch the spring.

b. Push against the cam follower and at the same time push the cross-pin out of the sliding clutch with a punch.

c. Release the pressure on the cam follower and slide the clutch forward off the propeller shaft. Now, tip the propeller shaft and allow the cam follower and spring to slide out of the propeller shaft. Set the unit aside for cleaning and inspecting.

Driveshaft, Shift Shaft and Bearings

◆ See Figures 102 and 103

1. Remove the Water Pump assembly as detailed in the Lubrication and Cooling System section.

2. Obtain a Driveshaft Holding Tool (#91-825196), which is essentially a large nut with internal splines that match the splines on top of the driveshaft. Use a box end wrench through the bearing carrier cavity on the pinion gear nut and the special tool on the end of the driveshaft with a breaker bar. Hold the pinion nut with the box end wrench and rotate the driveshaft counterclockwise until the nut is released from the driveshaft.

3. Rotate the driveshaft back and forth slightly while pulling upward on it to free the pinion gear and pinion bearing, while starting to withdraw the shaft itself from the gearcase.

4. Remove the pinion gear and the tapered roller bearing from the lower unit.

5. Withdraw the driveshaft from the lower unit. The ball bearing which is pressed on the shaft should come out along with the round water pump base/seal carrier. Lift the water pump base and seal carrier off the shaft and place it aside.

6. Remove the forward gear and bearing assembly from the race in the gearcase nose cone.

7. If necessary, pull upward and remove the shift shaft assembly from the gearcase.

8. If the driveshaft bearing MUST be replaced it must be pushed off the shaft (which will destroy the bearing, so don't do it UNLESS it must be replaced). Believe it or not, the recommended method from Mercury is to place the driveshaft in a vise with the jaws closed just tight enough to support the bearing (but not contact the shaft), then tap on the top of the driveshaft to separate the shaft and bearing (supporting the shaft from underneath so it doesn't fall and become damaged when the bearing gives

way). Alternately, a Universal Puller Plate (#C-91-37241) can be positioned on a press to free the shaft from the bearing.

9. If the pinion bearing requires replacement you'll have to push the lower bearing race out of the gearcase. Mercury has a special tool, 91-825200A1, which looks a little like a funny spring-loaded piece of exercise equipment designed to strengthen your grip when you squeeze. The tool is put in position through the lower portion of the gearcase, squeezed and inserted up into the race. A driver (#91-13779) is then inserted down from the top of the gearcase and into the middle of the spring loaded tool. Between the size of the driver and the spring loaded halves of the tool, the assembly is wedged into the race, then the driver can be tapped to push the race down into the bottom of the gearcase.

■ **The water pump base seals are normally installed back-to-back, but double-check the old orientation before removal, just to be sure.**

10. The round water pump base/seal carrier should be serviced to ensure proper sealing. Using a suitable driver, push or tap the seals out top of the base assembly from underneath. Also, remove and discard the old base O-ring.

Forward Gear and Bearing

◆ **See Figures 104 and 105**

1. After the pinion gear is removed, the forward gear and bearing can be lifted out of the lower unit. The tapered bearing race will remain within the lower unit.

2. If bearing and/or race replacement is necessary, remove the forward bearing race and shim material using a slide hammer and an internal jawed puller (such as #91-27780).

■ **The forward gear does not have to be removed in order to perform an adequate job of cleaning and inspecting. Therefore, it is not necessary to remove the gear unless it is unfit for further service. And, keep in mind that IF the gear and bearing are separated, you'll have to replace the bearing.**

3. If the needle bearing (which is pressed INTO the gear) requires replacement, place the gear and bearing assembly on a press with the gear facing upward, then use a suitable mandrel to press the bearing out of the center of the assembly.

4. If the tapered outer bearing requires replacement, position a universal puller plate (#91-37241) between the forward gear and the tapered bearing. Place the puller plate and gear on a press with the gear on the bottom.

5. Press the gear out of the bearing with a suitable mandrel. If the bearing cannot be removed, clamp the forward gear in a vise equipped with soft jaws. Use a punch and hammer to drive the roller bearing out of the forward gear.

■ **Once the bearing has been removed it cannot be used a second time because the roller cage will be damaged during the removal operation.**

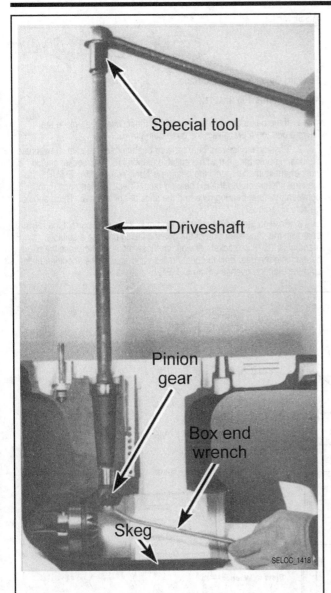

Fig. 102 Use a special driveshaft tool to turn the shaft, loosening the pinion nut

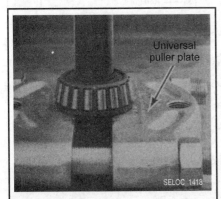

Fig. 103 A bearing separator can be used to remove or install the upper driveshaft bearing (tapered bearing shown, but most utilize a roller ball bearing these days)

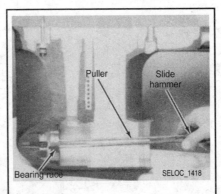

Fig. 104 If the forward gear bearing is replaced, use a slide-hammer and internal jaw puller to remove the race from the nose cone

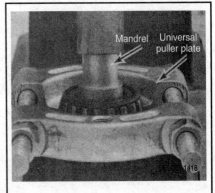

Fig. 105 If necessary, use a puller plate to separate the forward gear and bearing

CLEANING & INSPECTION

◆ See Figures 106 and 107

1. Clean all water pump parts with solvent, and then dry them using compressed air.

2. Inspect the water pump cover and base for cracks and distortion. If possible always install a new water pump impeller while the lower unit is disassembled. A new impeller will insure extended satisfactory service. If the old impeller must be used, never install it in reverse to the original direction of rotation. Reverse installation could lead to premature impeller failure and subsequent powerhead damage.

3. Inspect the bearing surface of the propeller shaft.

4. Check the shaft surface of propeller shaft for pitting, scoring, grooving, imbedded particles, uneven wear and discoloration.

5. Check the straightness of the propeller shaft with a set of machinist V-blocks.

6. Clean the pinion gear and the propeller shaft with solvent and dry the components with compressed air.

7. Check the pinion gear and drive gear for abnormal wear.

ASSEMBLY

Forward Gear, Bearing and Race

◆ See Figures 106 thru 112

1. If the bearings were removed from the forward gear, place the forward gear on a press with the gear teeth down.

2. Position the forward gear tapered bearing over the gear. Now, press the bearing onto the gear with a suitable mandrel (a 1 1/8 socket should work on these models) until the bearing is firmly seated. MAKE SURE the mandrel pushes on the INNER bearing race. Check for clearance (gap) between the inner bearing race and the shoulder of the gear. There should be zero clearance.

3. Position the needle roller bearing over the center bore of the forward gear with the numbered side of the bearing facing up. Use a suitable mandrel (a 15/16 in. socket will work on these models) and press the roller bearing into the gear until the bearing is seated against the shoulder. Make sure the mandrel pushes on the OUTER bearing race.

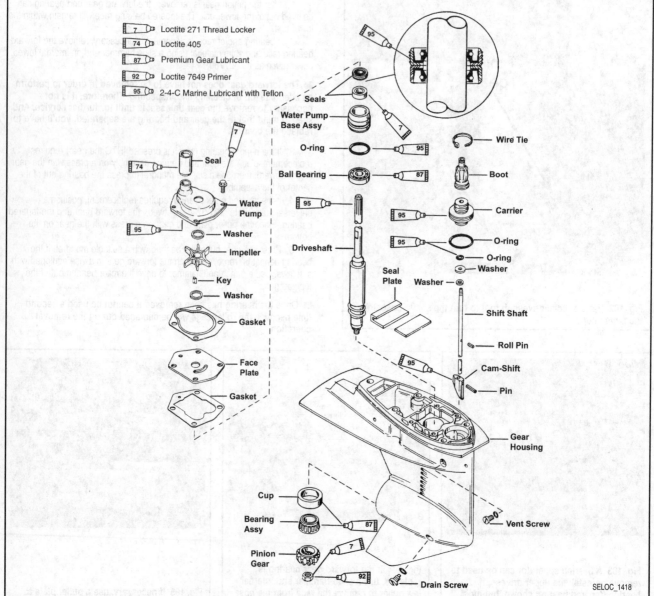

Fig. 106 Exploded view of the driveshaft, shifter and water pump assemblies - 30/40 Hp (2-Cylinder) and 40/50 hp (3-Cylinder) models

SELOC_1418

■ The bearing carrier is used as a pilot while installing the forward gear bearing cup in the next step. Therefore, the bearing carrier must have been assembled to include at least the propeller shaft roller bearing.

■ Earlier versions of these gearcases may be equipped with shims behind the forward bearing race. Mercury tells us that the shims should have been eliminated on these models, however, if you came across any during disassembly, be sure to reinstall them behind the new race during assembly.

4. Position the tapered bearing race squarely over the bearing bore in the front portion of the lower unit.

5. Obtain a Bearing Driver Mandrel (#91-36571). Place the tool over the tapered bearing race.

6. Insert the propeller shaft into the hole in the center of the bearing cup. Lower the bearing carrier assembly down over the propeller shaft, and then lower it into the lower unit. The bearing carrier will serve as a pilot to ensure proper bearing race alignment.

7. Use a brass mallet (so as to protect the prop shaft) to drive the propeller shaft against the bearing driver cup until the tapered bearing race is seated.

8. Withdraw the propeller shaft and bearing carrier, then lift out the driver cup.

9. Position the forward gear assembly into the forward bearing race.

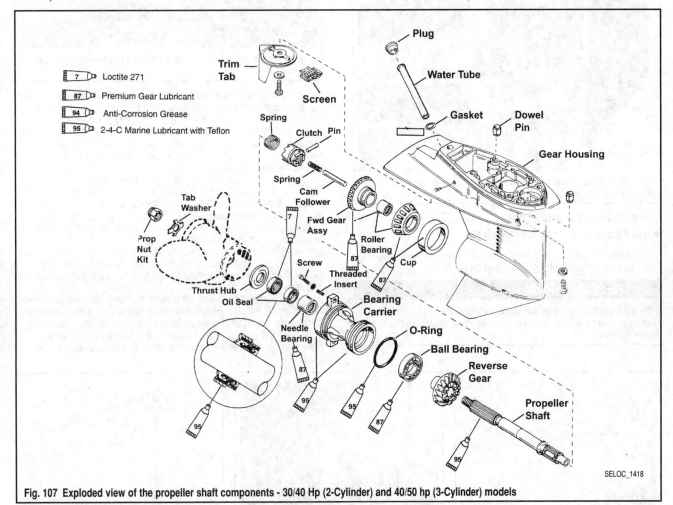

Fig. 107 Exploded view of the propeller shaft components - 30/40 Hp (2-Cylinder) and 40/50 hp (3-Cylinder) models

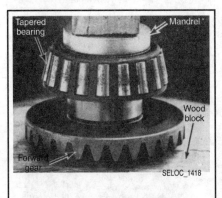

Fig. 108 If replacement is necessary press a new tapered bearing onto the forward gear

Fig. 109 If the needle bearing was removed, install a new bearing into the forward gear

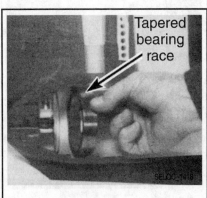

Fig. 110 The forward bearing race is positioned by hand then seated with a driver

Fig. 111 Use the bearing carrier and prop shaft to center and install the forward bearing race

Driveshaft, Shift Shaft and Bearings

◆ See Figures 106 and 107, and 113 thru 115

1. Apply a coating of 2-4-C Marine Lubricant With Teflon to the NEW O-rings, then install them on the shift shaft and carrier.

✳✳ SELOC WARNING

If the shift shaft is bottomed out in the gearcase the cross pin in the clutch dog will be bent by the cam follower when tightening the bearing carrier screws.

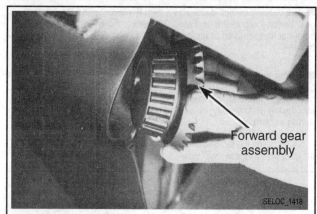

Fig. 112 Once the race is installed, position the forward gear and bearing

2. Position the carrier (with boot) on the shaft, then install the shaft and carrier assembly into the gear housing. Don't bottom the shaft out in the housing, pull upward on the shaft until the boot is NOT deformed.

3. Secure the boot using a new wire tie, then position the shift shaft so the ramp (tapered portion on the bottom) faces toward the propeller shaft. If necessary, look in through the propeller shaft opening as you twist the shift rod into the proper position.

4. If the pinion bearing race was removed, you'll need to draw the replacement up into position in the gearcase using a long threaded rod (#91-31229), a support plate (#91-293100), some washers (#12-34961), a nut (#11-24156), a thrust bearing (#31-85560), a pilot (#91-825199) to keep everything centered) AND finally, a bearing cup sized mandrel (#91-825198) or equivalents. The right combination of a long threaded rod with nuts and washers SHOULD do the trick, but pick the sizes carefully. Position the race into the housing with the numbers facing UP and the tapered side facing DOWN, then assemble the tools and draw the face up into the gearcase until it is seated.

5. If the driveshaft upper roller bearing was removed, position the new bearing over the shaft and support the inner race with a bearing separator or, if necessary, a soft-jawed vise (just making sure the vice doesn't touch the shaft). Thread the old pinion nut 3/4 of the way onto the shaft in order to protect the splines, then use a press to carefully seat the bearing on the shaft.

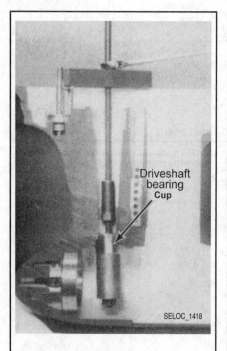

Fig. 113 If removed, draw a new pinion bearing race into position

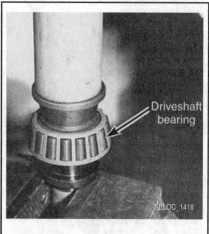

Fig. 114 Pressing a new bearing onto the driveshaft (old style tapered bearing shown)

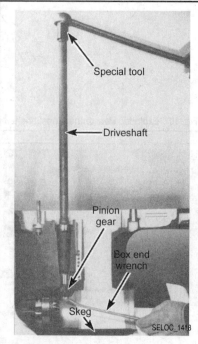

Fig. 115 Tighten the pinion using the driveshaft spline adaptor

6. Remove the driveshaft and bearing assembly from the press, then remove the pinion nut. Clean the threads of A NEW pinion nut and the driveshaft with Loctite® 7649 Primer.

7. Carefully insert the driveshaft into the gearcase while holding the pinion gear and bearing in position, engage the splines of pinion gear to the driveshaft.

8. Apply a coating of Loctite® 271 or equivalent threadlocking compound to the threads of the new pinion nut, then thread the nut onto the bottom of the driveshaft.

9. Using the splined driveshaft holding tool (#91-825196) with a large enough socket and torque wrench on top and a pinion nut wrench on the bottom, tighten the pinion nut to 50 ft. lbs. (68 Nm).

10. If the water pump base seals were removed, apply a light coating of Loctite® 271 to the OUTER DIAMETER of the metal cased seal. Position it with the lips facing downward, then press it in through the top of the pump base until it bottoms. Next, with the lips of the neoprene outer diameter seal facing upward, press it into the case until it seats against the first seal.

11. Apply a light coating of 2-4-C Marine Lubricant With Teflon to the lips of both seals in the carrier and to a NEW water pump base O-ring, then install the O-ring to the groove in the carrier.

12. CAREFULLY position the carrier over the driveshaft and slide it down into position in the gearcase.

13. Install the Water Pump assembly as detailed in the Lubrication and Cooling System section.

Bearing Carrier Assembly

◆ See Figures 106, 107 and 116 thru 120

1. Position the propeller shaft needle roller bearing into the propeller end of the bearing carrier and apply a light coating of Special Lubricant 101 (or 2-4-C Marine Lubricant With Teflon or an equivalent) to the outer diameter of the bearing.

2. Press the bearing into the bearing carrier with an Oil Seal and Bearing Driver (#91-817011) or a suitable mandrel.

■ If the driver is not available, press the bearing into the carrier to a depth of 0.82 in. (20mm) below the propeller shaft end of the carrier, then install the first seal to a depth of 0.44 in. (11mm) and the second seal to a depth of 0.04 in. (1mm) below the end of the carrier.

3. Coat the outer diameter of both propeller shaft oil seals with Loctite® 271 or equivalent threadlocking compound.

4. Obtain an Oil Seal Driver (#91-817007) or an equivalent mandrel. Place one seal on the longer shoulder side of the driver tool with the lip of the seal away from the shoulder (facing inward toward the bearing and the carrier).

5. Press the seal into the bearing carrier until the seal driver bottoms against the bearing carrier. Place the second seal on the short shoulder side of the seal driver with the lip of the seal toward the shoulder.

6. Press the seal into the bearing carrier until the seal driver bottoms against the bearing carrier. Clean excess Loctite® from the seals.

7. Position the reverse gear on a press with the gear teeth facing down and apply a light coating of gear lube to the bearing mating surface. Place the ball bearing over the gear and press the bearing onto the gear with a suitable mandrel (1 1/4 in. socket).

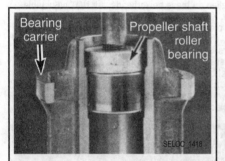

Fig. 116 If removed, press the new needle roller bearing in through the propeller side of the carrier

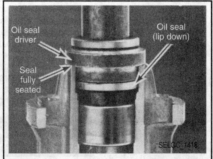

Fig. 117 Install the FIRST oil seal with the lips facing inward toward the carrier

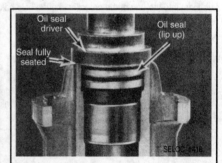

Fig. 118 Then install the SECOND oil seal with the lips facing up toward the propeller

Fig. 119 Press the bearing onto the reverse gear (thrust washer not always used)...

Fig. 120 ...then press the reverse gear assembly into the carrier

8. Place a new O-ring into the bearing carrier groove (it can be easier to install the O-ring without damage if it is positioned before the reverse gear and bearing is pressed into the carrier.

9. Coat the O-ring and oil seals with multipurpose lubricant, or equivalent.

10. Place the bearing carrier over the gear and bearing assembly. Press the bearing carrier onto the bearing using a suitable mandrel. If you position the gear facing upward in the press, you can use a 3/4 in. socket to press against the gear itself. We actually recommend the later method.

Propeller Shaft Assembly

◆ See Figures 106, 107, 121 and 122

1. Insert the spring into the end of the propeller shaft.

2. Insert the flat end of the cam follower into the front end of the propeller shaft.

3. Slip the sliding clutch over the clutch splines with the cross-pin hole aligned with the cross-pin slot in the propeller shaft. Also, make sure the SHORT SHOULDER of the clutch is facing the FORWARD GEAR.

4. Position the cam follower against a solid object and push against the cam follower to compress the spring. Hold the propeller shaft in this position and at the same time, insert the cross-pin. Release pressure on the spring.

■ **If you have trouble compressing the follower spring sufficiently using the follower, remove it and substitute a 3/16 in. Allen wrench or similar sized tool.**

5. Install the cross-pin retainer spring over the sliding clutch. Take care not to over-stretch the spring.

6. Remove the cam follower and insert just a little Multipurpose Lubricant into the end of the propeller shaft to help hold the follower in position during assembly. Install the cam follower.

Propeller Shaft and Bearing Carrier Installation

◆ See Figures 106, 107 and 123

1. Insert the propeller shaft into the center of the forward gear assembly, making sure the cam follower does not dislodge.

2. Slide the bearing carrier into the lower unit. Take care not to damage the propeller shaft oil seals.

Fig. 121 Insert the spring, then slide the clutch dog in position

Fig. 122 Secure the clutch dog with the cross pin, then install the retaining spring

3. Push the bearing carrier into the lower unit and at the same time slowly rotate the driveshaft to allow the pinion gear teeth to engage with the reverse gear teeth.

■ **Before discarding the original TAB washers and bolts, measure them to be certain they weren't already replaced at some point.**

4. DISCARD THE ORIGINAL TAB WASHERS or the thin 0.063 in. (1.60mm) flat washers and original 25MM LONG SCREWS. Instead, install thicker flat washers and longer screws. Specifically obtain two 1.18 in. (30mm) long screws (10-855940-30) and two 0.090 in. (2.29mm) thick washers.

5. Apply a light coating of Loctite® 271 or equivalent threadlocking compound to the threads of the new screws. Install the screws and washers, then tighten the screws to 19. ft. lbs. (25 Nm).

6. Refill the gearcase and check for leaks, then install the gearcase to the outboard as detailed earlier in this section.

■ **Mercury recommends using a pressure gauge installed to the vent plug in order to pressurize and check the gearcase before it is returned to service. Apply 10-15 psi (68.9-103.4 kPa) of pressure and watch the gauge for 5 minutes. Then rotate the driveshaft and propshaft and move the shift rod making sure the pressure does not drop. IF pressure drops, submerge the gearcase (with pressure applied) and watch for bubbles. Bottom line, the gearcase MUST hold pressure for at least 5 minutes.**

SHIMMING

■ **The manufacturer gives no specific instructions for setting up the backlash on these units. As a matter of fact, Mercury claims these gearcases, when installed on these powerheads, are NOT adjustable when it comes to lash. However, if shim material was found during disassembly and was placed back in its original location, the backlash should be acceptable. The manufacturer simply states: "The amount of play between the gears is not critical, but NO play is unacceptable." Therefore, if after the assembly work is complete, the gears are "locked", the unit must be disassembled and shim material placed behind all three gears, using a "trial and error" method. If the lower unit is allowed to operate without "some" backlash, very heavy wear on the three gears will take place almost immediately.**

Gearcase - 55/60 (3-Cylinder) Non-Bigfoot, 60 hp Bigfoot and 75-125 Hp Models

There are basically 3 similar design gearcases covered in this section. The first is the gearcase used on 55/60 3-Cylinder Non-Bigfoot models. The second is found on 60 hp Bigfoot models. The third is found on all non-jet 75-125 hp models. In all cases, there are far more similarities than differences.

Similarities between the 3 gearcase designs include:

• Water pump, base and seal assemblies (nearly identical from the impeller up, but the size and shapes of the base plates vary, plus the 55/60 hp non-Bigfoot models use a seal carrier which is separate from the base plate)

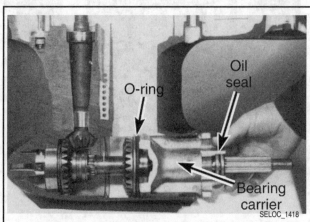

Fig. 123 With the propeller shaft in position, carefully insert the bearing carrier assembly

• Rotating/ratcheting shift assemblies (nearly identical on all but the 55/60 hp non-Bigfoot models, which are still very similar)

• Propeller shaft bearing carrier (along with seals and bearings, again, identical except the 55/60 hp non-Bigfoot models, which are still very similar).

Most differences between the gearcases come in the form of the driveshaft and bearing design and/or the propshaft shifter components. Other differences are minor, limited to a single bearing, shim or washer. All differences are clearly stated throughout the service and overhaul procedures.

■ **For gearcase exploded views, please refer to Cleaning & Inspection in this section.**

REMOVAL & INSTALLATION

◆ **See Figures 124, 125 and 126**

✳✳ SELOC CAUTION

Although the design of the gearcase used on 75-115 hp carbureted 2-strokes is the same as that used on 75-115 hp OptiMax motors, the two gearcases ARE NOT interchangeable. The driveshaft on gearcases meant for OptiMax models are 0.625 in. (15.87mm) LONGER than the one used on gearcases for the carbureted 2-strokes. As a result, installation of the carbureted model gearcase on an OptiMax motor would result in crankshaft failure.

Unlike the smaller Mercury gearcases, the rotating/ratcheting shift mechanism used on these models utilizes a splined shift shaft connection, making it un-necessary to locate and disconnect shift linkage before gearcase removal.

1. For safety and to prevent accidental engine start, disconnect and ground the spark plug leads to the powerhead or tag and disconnect the battery cables.

2. If necessary for overhaul purposes, remove the Propeller, as detailed in the Maintenance and Tune-Up section.

■ **It's actually a good idea to at least loosen the OIL LEVEL and/or VENT screw(s) first, to make sure it will come out without a problem. You can't refill the gearcase, if you can't get the vent screw loose.**

3. If necessary for gearcase overhaul purposes, position a suitable clean container under the lower unit. Remove the FILL screw on the bottom of the lower unit, and then the OIL LEVEL and/or VENT screw(s). The vent screw must be removed to allow air to enter the lower unit behind the lubricant. Allow the gear lubricant to drain into the container.

As the lubricant drains, catch some with your fingers from time-to-time and rub it between your thumb and finger to determine if there are any metal particles present. Examine the fill plug. A small magnet imbedded in the end of the plug will pickup any metal particles. If metal is detected in the lubricant, the unit must be completely disassembled, inspected, and the damaged parts replaced.

Check the color of the lubricant as it drains. A whitish or creamy color indicates the presence of water in the lubricant. Check the container for signs of water separation from the lubricant. The presence of any water in the lubricant is bad news. The unit must be completely disassembled; inspected; the cause of the problem determined; and then corrected.

4. If not done already, rotate the outboard unit to the full UP position and engage the tilt lock lever.

5. Shift the gearcase into FORWARD.

6. All models utilize 4 gearcase fasteners toward the front and middle of the housing (2 on either side), in addition to a locknut and washer underneath the anti-cavitation plate. For some models (including the 40/50/60 hp 996cc motors) the locknut and washer obscured by the trim tab, if so scribe a line between the trim tab and the anti-cavitation plate. This mark will ensure the trim tab will be installed back at the original angle, then unbolt and remove the trim tab for access.

7. Remove the gearcase fastening locknut and washer from the center rear of the anti-cavitation plate (For most models it is found directly in front of the trim tab, however on some, like the 55/60 hp non-Bigfoot models, it is in the recess above the tab so you'll first have to matchmark and remove the tab for access).

8. Remove the 4 bolts (or nuts, depending upon the model) and washers securing the lower unit (2 on either side) to the exhaust housing.

9. Pull down on the lower unit gear housing and CAREFULLY separate the lower unit guiding it straight out of the exhaust housing. Guide the shift rod and driveshaft out of the exhaust housing as the lower unit is removed.

10. Place the lower unit in a suitable holding fixture or work area.

To install:

■ **Always fill the lower unit with lubricant and check for leaks before installing the unit to the driveshaft housing (could save you the trouble of removing it again right away).**

11. If a pressure tester is available, thread the adapter into the vent plug bore, then slowly pressurize the gearcase housing to 10-12 psi (69-83 kPa) pressure. Watch the gauge, as pressure must hold for at least 5 minutes. Slowly rotate the driveshaft and propshaft while watching the gauge. If the pressure drops, submerge the unit in water and recheck, watching for air bubbles to determine the pressure leak. Fix the leak before proceeding.

12. Take time to remove any old gasket material from the fill and vent recesses and from the screws.

13. Place the lower unit in an upright vertical position. Fill the lower unit with Super-Duty Lubricant, or equivalent, through the fill opening at the bottom of the unit. Never add lubricant to the lower unit without first removing the vent screw and having the unit in its normal operating position - vertical. Failure to remove the vent screw will result in air becoming trapped within the lower unit. Trapped air will not allow the proper amount of lubricant to be added.

14. Continue filling slowly until the lubricant begins to escape from the vent or level opening (as applicable) with no air bubbles visible.

15. Use a new gasket and install the vent and/or level screw(s). Slide a new gasket onto the fill screw. Remove the lubricant tube and quickly install the fill screw.

16. If no pressure tester was available, visibly check the lower unit for leaks. Recheck for leaks after the first time the unit is warmed to normal operating temperatures from use.

17. If not done already, swing the exhaust housing outward until the tilt lock lever can be actuated, and then engage the tilt lock. Also, if the powerhead shift linkage was moved from the Forward position, be sure to reposition it now. On most models that means the front of the shift block must extend about 1/8 in. (3.2mm) past the front of the rail.

■ **On 55/60 non-BF models, if the water tube seal was removed from the top of the water pump or otherwise replaced, be sure to seal it to the water pump housing using Loctite® 405 or an equivalent sealant.**

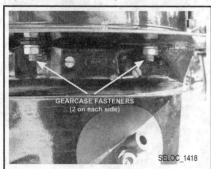

Fig. 124 These gearcases are secured by 4 fasteners on the side. . .

Fig. 125 . . .and one locknut underneath the anti-cavitation plate

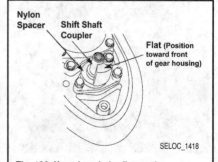

Fig. 126 If equipped, the flat on the mechanical lock out shifter tab must face forward during installation

18. Apply a coating of 2-4-C Marine Lubricant with Teflon to the inner diameter of the water tube seal, to the shift shaft coupler and to the driveshaft splines. The water tube seal used on 75 hp and larger models should be installed with the labyrinth end facing the water tube in the driveshaft housing and the tapered end facing the water pump.

■ **DO NOT put lubricant on the ENDS of the shafts, as it could prevent the shaft splines from fully seating in the shift coupler and the crankshaft.**

19. Finish preparing the gearcase for installation, depending upon the model/gearcase, proceed as follows (not all gearcases require additional steps):
   • For all except 55/60 non-BF models, apply a light bead of RTV 587 or equivalent silicone sealer to the top of the gearcase housing, along the seal area just behind the rear edge of the water pump housing.
   • Some models are equipped with a mechanical reverse lock on the top of the shifter mechanism. When so equipped instead of a plain rounded shift shaft coupling on top of the gearcase, there will be a coupling that contains a small, semi-circular, almost triangular tab on the front of the spline. When so equipped, make sure the flat is positioned toward the front of the gear housing.

20. For all gearcases, apply a light coating of Loctite® 271 or equivalent threadlocking compound to the threads of the 4 gearcase fasteners (not on the locknut which goes under the anti-cavitation plate).

21. Carefully bring the gearcase into alignment with the exhaust housing and slowly insert it STRAIGHT UP and into position, while aligning the driveshaft, shift shaft and water tube seal.

■ **It may be necessary to move the shift block (located under the engine cowl) slightly to help align the upper shift shaft splines with the shift shaft coupler splines during gearcase installation. Likewise, it may be helpful to slowly turn the propeller in the normal direction of rotation to align the driveshaft-to-crankshaft splines.**

22. Thread the gearcase fasteners until they are all hand-tight, then alternately and evenly tighten the fasteners to 40 ft. lbs. (54 Nm).

23. If removed for access to the gearcase locknut and washer found underneath, reinstall the trim tab. Be sure to align the matchmarks made earlier to preserve trim tab adjustment, then tighten the tab retaining bolt to 22 ft. lbs. (30 Nm).

24. If the propeller was removed, install the propeller.

25. Reconnect the spark plug wires and/or the battery cables.

DISASSEMBLY

OEM ③ *DIFFICULT*

Unlike most of the smaller Mercury gearcases, the lower unit found on these models contains various shims for adjusting pinion height and forward/reverse gear backlash. Always keep track of the shims during removal, since if the gearcase or the shafts/bearings are being reused the shims should be returned to the same positions. Also, if the gearcase and/or shafts/bearings are replaced, the shims can be sometimes be used as a good starting point for the shimming procedures.

■ **For gearcase exploded views, please refer to Cleaning & Inspection in this section.**

Bearing Carrier Removal & Disassembly

◆ See Figures 127, 128 and 129

✳✳ SELOC WARNING

If you don't have access to the special Mercury mandrels designed to install bearings to the proper depth inside the bearing carrier, measure the installed height of all bearings before removal (except of course the reverse gear bearing, on 55/60 non-BF models where it is pressed directly onto the gear). Use the measured depths at installation time to ensure the bearings are positioned properly.

1. Remove the two bolts or nuts and washers securing the bearing carrier in the lower unit.

■ **There are multiple ways to remove the bearing carrier on these models. The method recommended by Mercury (and therefore the preferred method) involves the use of a threaded puller with long internal jaws that will gently grab and pull the carrier off while pushing against the propshaft. We've listed a number of other possibilities, depending upon the tools at hand, however keep in mind that all of the alternatives involve some measure of risk to the gearcase and/or carrier assembly.**

2. Free the bearing carrier assembly from the gearcase using one of the following methods:
   a. Install an internal jawed puller (#91-46086A1) and a suitable drive bolt. Position the propeller thrust hub to maintain an outward pressure on the puller jaws. Tighten the puller bolt to break the seal and free the bearing carrier.
   b. If the puller is having a problem freeing a frozen carrier, remove the puller bolt and connect a slide hammer (such as #91-34569A1 or equivalent) to the Puller Jaws (#91-46086A1). Use a few sharp quick blows with the slide hammer to hopefully break the gasket seal and free the bearing carrier.
   c. If no puller jaws are available, CAREFULLY use a soft head mallet and tap the ears of the bearing carrier to offset the carrier from the housing. Now, tap opposite "ears" alternately and evenly on the back side to remove the carrier from the lower unit housing.
   d. The final, and least desirable method is using a mallet to free a frozen carrier. Clamp the propeller shaft in a vise equipped with soft jaws in a horizontal position. Use a mallet and strike the lower unit with quick sharp blows midway between the anti-cavitation plate and the propeller shaft. This action will drive the lower unit off the bearing carrier. Take care not to drop the lower unit when the unit finally comes free of the carrier.

■ **If the lower unit refuses to move, it may be necessary to carefully apply heat to the lower unit in the area of the carrier and at the same time attempt to move it off the carrier.**

3. Once the seal is broken, remove the propeller shaft and bearing carrier from the gearcase. Be careful not to loose the cam follower from the shifter end of the propeller shaft (and the 3 check balls found in the shaft behind it for all by 55/60 non-BF models). Separate the propeller shaft and bearing carrier assembly, placing each aside for disassembly.

4. For all but the 55/60 hp non-BF models, the reverse gear should pull free of the carrier, so lift it from the carrier along with the thrust bearing and

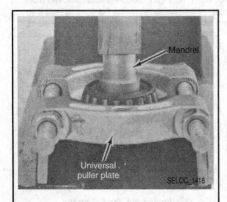

Fig. 127 Start by removing the 2 bearing carrier fasteners

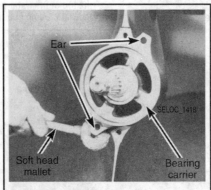

Fig. 128 If a puller is not available, use a soft mallet to turn the carrier slightly, then gently tap it out of the gearcase

Fig. 129 Remove and discard the old carrier O-ring

thrust washer. Replace any damaged components. The bearing on these models is pressed into the carrier. If the bearing does not roll freely or shows any sign of corrosion, clamp the carrier in a vise equipped with soft jaws. Obtain and use Slide Hammer (91-34569A1) or an equivalent slide hammer with internal jawed puller to free the bearing from the carrier.

5. For 55/60 hp non-BF models, the reverse gear is pressed into the carrier. Check for damage. If the bearing or gear is damaged and requires replacement use a slide hammer with internal jaws or a suitable bearing puller (#91-27780). On these models, if the bearing requires replacement you'll need to use a press, universal bearing separator (like #91-37241) and a suitable driver (#91-37312) to separate the bearing from the gear.

6. Inspect the two seals and the condition of the bearing at the rear end of the carrier. If the seals have failed and have allowed water to enter the lower unit, the bearings are no longer fit for further service.

7. If only the seals are being replaced, use a small pry tool or seal extractor to carefully remove both seals. On most models these seals are installed back-to-back but double-check seal orientation before removal.

8. If both the bearing and seals are being removed you can drive or press them all out together. Obtain a Bearing Removal and Installation Kit (#91-31229A-7). Obtain a Driver Rod (#91-37323) and either Mandrel (#91-36569) for all except the 55/60 hp non-BF models or Mandrel (#91-37312) for the 55/60 hp non-BF models or another suitable substitute mandrel. Insert the removal tools into the forward end of the carrier and press or drive the needle bearing and seals together out the rear end of the carrier.

Propeller Shaft Disassembly

◆ **See Figures 130 thru 134**

To disassemble the propeller shaft for cleaning, inspection and/or parts replacement, proceed as follows:

1. Insert a thin blade screwdriver or an awl under the first coil of the cross-pin retainer spring and rotate the propeller shaft to unwind the spring from the sliding clutch. Take care not to over-stretch the spring.

2. Position the propeller shaft cam follower against a solid object. Push against the cam follower to depress the clutch spring and hold against the spring pressure during the next step.

3. Push the cross-pin out of the sliding clutch with a punch, then slide the clutch forward off the propeller shaft. Pull the shaft slowly back away from the solid object to gently release spring pressure.

4. Now, tip the propeller shaft and allow the cam follower and spring components to slide out of the propeller shaft. The components vary slightly by gearcase model as follows:
- For 55/60 hp non-BF models, remove the cam follower, then remove the guide block and the spring.
- For 60 hp BF model, as well as all 75 hp and larger models, remove the cam follower, then remove the 3 metal balls, the guide block and finally the spring.

5. For all models, check the propeller shaft for straightness using a dial-gauge and a pair of v-blocks. Set the shaft in the blocks so it is resting on the bearing surfaces, then set and zero the dial gauge to check run-out right behind (toward the gearcase side) of the propeller splines. Shaft run-out must not exceed 0.006 in. (0.152mm) for 55/60 hp non-BF models or 0.009 in. (0.228mm) for all other models.

6. Inspect the oil seal surface for grooving or pitting. Mercury only provides specs for the surface of the shaft on 55/60 hp non-BF models, on which they say grooving must not exceed 0.005 in. (0.12mm).

Driveshaft and Bearings

◆ **See Figures 135 thru 142**

1. If not done already, remove the Water Pump assembly as detailed in the Lubrication and Cooling System section.

2. Remove the water pump base, as follows:
- For 55/60 hp non-BF models, use a small pr tool to carefully pry under the ears or in the slot(s) provided on either side of the housing. Once free, lift the base carefully off the driveshaft, then remove and discard the base O-ring. Check the condition of the seal and plate on the gearcase and, if necessary, remove them for replacement.
- For all except the 55/60 hp non-BF models, loosen and remove the 6 water pump base retaining bolts, then CAREFULLY pry at the tabs provided (one at the front and one at the rear of the base) to free the base from the gearcase. Lift the base carefully from the driveshaft, then remove and discard the base gasket.

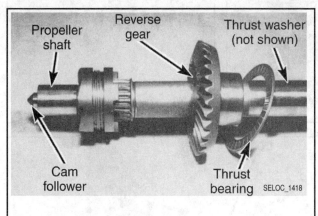

Fig. 130 Typical propeller shaft assembly

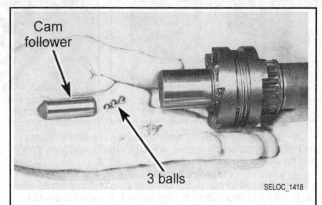

Fig. 131 On all Except 55/60 hp non-BF models there are 3 balls behind the cam follower

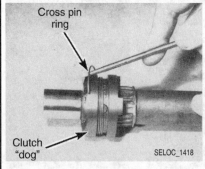

Fig. 132 To remove the sliding clutch, first remove the pin retaining spring. . .

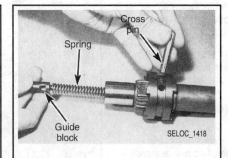

Fig. 133 . . .then remove the cross pin, guide block (if equipped) and spring. . .

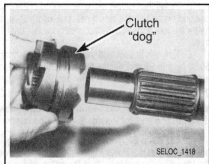

Fig. 134 . . .and slide the clutch dog from the shaft

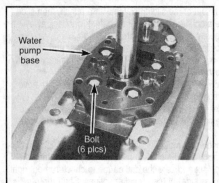

Fig. 135 Remove the water pump base plate (except 55/60 non-BF shown). . .

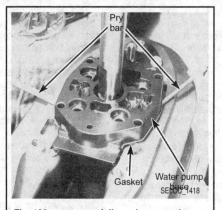

Fig. 136 . . .pry carefully and remove the baseplate. . .

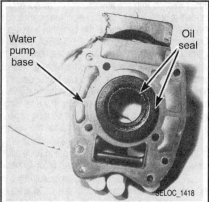

Fig. 137 . . .for access to the driveshaft oil seals

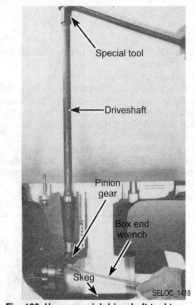

Fig. 138 Use a special driveshaft tool to turn the shaft, loosening the pinion nut

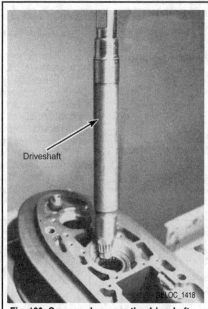

Fig. 139 Grasp and remove the driveshaft from the gearcase. . .

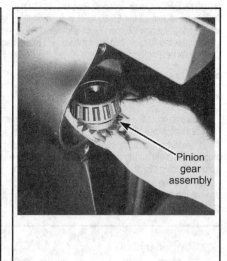

Fig. 140 . . .then remove the pinion gear (and bearing on most models)

3. For installation purposes, note the directions in which the oil seals are facing (they are usually installed back-to-back) then carefully pry them from the water pump base.

4. Select a box end wrench the same size as the pinion nut. This tool will be used to prevent the nut from turning as the driveshaft is rotated. Obtain the proper driveshaft holding tool for the model gearcase being serviced as follows:

- For all 55/60 hp (BF and non-BF) models use #91-817070
- For all 75 hp and larger models use #91-56775

5. Install the holding tool onto the end of the driveshaft. Now, with the box end wrench on the pinion gear nut, use an appropriate wrench and rotate the tool and driveshaft counterclockwise until the pinion gear nut is free.

6. Remove the nut and the pinion gear assembly. On all except the 55/60 hp non-BF models the pinion bearing should come free with the assembly. On 55/60 hp non-BF models the bearing is pressed up into the gearcase.

■ **The forward gear and bearing assembly is now free, remove it from the gearcase and set it aside for attention later.**

7. For 60 hp BF models and all 75 hp and larger models, obtain and position a Universal Puller Plate (#C-91-37241) between the pinion gear and the tapered roller bearing. Place the puller plate and gear, with the gear on the bottom, in an arbor press. Use a suitable mandrel and press the gear free of the bearing. The mandrel must contact the gear collar, but clear the bearing cage.

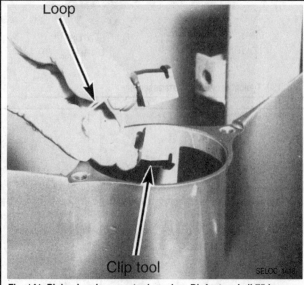

Fig. 141 Pinion bearing race tool used on Bigfoot and all 75 hp or larger motors

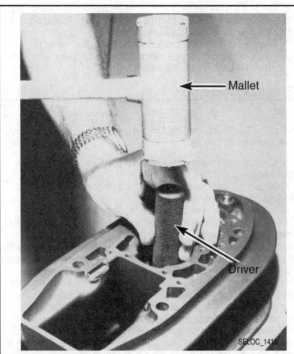

Fig. 142 Removing the pinion bearing (55/60 hp non-BF models) or race (all others)

■ **Once the bearing has been removed, it cannot be used a second time. The roller cage will be distorted during the removal process.**

8. Pull the driveshaft up and out of the lower unit housing. For 55/60 hp non-BF models you will first need the use of an upper driveshaft bearing retainer tool (#91-43506 or equivalent) to loosen the round bearing retainer nut.

■ **When removing the driveshaft on 55/60 hp non-BF motors, watch for and retain the shims used to properly position the pinion gear, they are located under the upper driveshaft bearing. On all other models, the shims used to position the pinion gear are mounted above the pinion bearing race.**

9. For 55/60 hp non-BF models, proceed as follows to service the driveshaft bearings:
    a. The upper bearing is pressed onto the driveshaft. If replacement is necessary, use a universal puller plate to separate the bearing from the shaft.
    b. The lower bearing is pressed into the gearcase from the top. If replacement is necessary, carefully drive the bearing down the rest of the way and out of the driveshaft bore into the propshaft bore of the gearcase using a driver such as #91-817058A1. It is usually a good idea to measure the installed depth of the bearing before removal, to use as a reference

during installation. And if you don't have the required Mercury driver for installation, this measurement is a necessity.

10. For all 60 hp Bigfoot models, as well as all 75 and larger models, service the driveshaft and bearings as follows:
    a. Inspect the condition of the wear sleeve at the lower end of the driveshaft. If the sleeve is worn or distorted it will allow water to enter the lower unit. If replacement is necessary, support the driveshaft in a universal bearing separator tool, resting over an open vice. Carefully, using a soft head mallet, tap the upper splined driveshaft end to force the wear sleeve up and free of the driveshaft.
    b. Remove and discard the rubber sealing ring around the driveshaft.
    c. If the upper driveshaft bearing requires replacement, use a suitable internal jawed puller assembly and puller bridge or support (#91-83165M or equivalent) to carefully pull the bearing from the gearcase bore. It is usually a good idea to measure the installed depth of the bearing before removal, to use as a reference during installation.
    d. If the oil sleeve requires replacement use the same puller assembly that was used to remove the upper bearing to free it from the gearcase. Again, checking the installed height of the sleeve before removal is usually a good idea.

■ **The upper driveshaft bearing and sleeve do NOT need to be removed in order to service the pinion bearing race, but be careful not to damage them if they remain in position. ALSO, after removal, remember to retain any shims installed above the race.**

    e. Lastly, if the pinion bearing and/or race requires replacement you'll have to push the lower bearing race out of the gearcase. Mercury has a number of special tools each of which look a little like a funny spring-loaded piece of exercise equipment designed to strengthen your grip when you squeeze. The tool is put in position through the lower portion of the gearcase, squeezed and inserted up into the race. A driver is then inserted down from the top of the gearcase and into the middle of the spring loaded tool. Between the size of the driver and the spring loaded halves of the tool, the assembly is wedged into the race, then the driver can be tapped to push the race down into the bottom of the gearcase. The actual tool number varies with the model gearcase, gear ratio and the number of teeth on the pinion gear, but on all of these models it should be either #91-14308T01 or #91-14308T02.

Forward Gear, Bearing, and Race

◆ **See Figures 143, 144 and 145**

The forward gear and bearing assembly is inserted in a race in the nose cone and basically held in position by the pinion gear. A tapered bearing which rolls in the nose cone race is pressed onto the outside of the gear and a needle bearing is pressed inside the gear. The needle bearing supports the forward end of the propeller shaft.

As usual, removal of the bearings will normally render them unfit for service, so make sure they are shot and replacements are available before bearings are separated from the gear.

After the pinion gear and driveshaft have been removed, the forward gear assembly and associated bearing can be lifted out of the housing.

■ **The forward gear tapered roller bearing is pressed onto the short shaft of the gear assembly.**

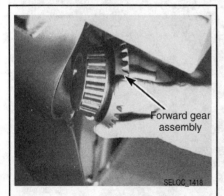

Fig. 143 Remove the forward gear and bearing assembly

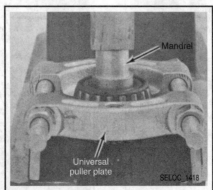

Fig. 144 Remove the tapered roller bearing from the forward gear using a bearing separator

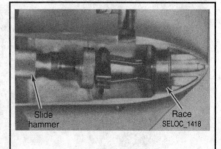

Fig. 145 Removing the forward gear bearing race using a slide-hammer

1. Most, but not all (usually 75 hp and larger) forward gears utilize a small snapring at the bearing end (not toothed end) of the gear. The snapring is used to secure the internal needle bearing. If the bearing must be removed, first check for the presence of this snapring and remove it, if present.

2. If the needle bearing is no longer fit for further service, it can be removed by tapping, with a blunt punch and hammer, around the bearing cage from the aft end - the gear end - of the bearing. This action will destroy the bearing cage. Therefore, be sure a replacement bearing is on hand, before attempting to remove the defective bearing.

3. If the forward gear tapered roller bearing must be replaced, use a universal bearing separator and a press to push the bearing off the gear.

■ Remember that the bearing and race are a set, so if the bearing is replaced you'll also need to replace the nose cone race. A wear pattern will have been worn in the race by the old bearing. Therefore, if the race is not replaced, the new bearing will be quickly worn by the old worn race.

4. Remove the forward gear tapered roller bearing race using a slide hammer.

5. SAVE any shim material from behind the race after the race is removed. The same amount of shim material will probably be used during assembly.

Shift Shaft Removal

◆ See Figures 146 and 147

For 55/60 hp non-BF motors the shift shaft and seal assembly is inserted into the housing and is secured by a slight interference fit only, however on all other models the assembly is secured in a housing that is bolted to the top of the gearcase.

1. Remove the shift shaft coupler and spacer from the top of the shift shaft splines.

2. For 55/60 hp non-BF models, to remove the shaft, grasp it just below or toward the bottom of the splines using a pair of pliers (take care to prevent damaging the shaft/splines). Pull upward slightly to start to free the shaft and seal housing from the gearcase.

3. For all 60 hp BF models, as well as all 75 hp and larger motors, remove the 2 bolts securing the shaft housing, then use 2 small pry tools to carefully pry the housing (and shaft) upward to just start to free the shaft assembly from the gearcase.

4. With one hand reaching into the gearcase to hold the shift cam in position and to help protect the orientation of the cam, pull upward to remove free the shaft and seal housing from the shift coupler and from the gearcase.

5. Separate the seal housing from the shaft. Remove the old clip from the shaft (located just below the seal housing).

6. Remove and discard the seal housing O-ring, then use a small pry tool to carefully remove the seal from the top of the shaft (if damaged). Note the direction the seal is facing before removal.

7. Reach into the housing and pull out the shift cam. Attempt to preserve the orientation of the cam (which side is facing up and which side faces down) for reference come installation.

CLEANING & INSPECTION

◆ See Figures 148 thru 160

1. Clean all water pump parts with solvent, and then dry them with compressed air.

2. Inspect the water pump cover and base for cracks and distortion, possibly caused from overheating.

3. Inspect the face plate and water pump insert for grooves and/or rough surfaces. If possible, always install a new water pump impeller while the lower unit is disassembled. A new impeller will ensure extended satisfactory service and give "peace of mind" to the owner. If the old impeller must be returned to service, never install it in reverse to the original direction of rotation. Installation in reverse will cause premature impeller failure.

4. Inspect the impeller side seal surfaces and the ends of the impeller blades for cracks, tears, and wear. Check for a glazed or melted appearance, caused from operating without sufficient water. If any question exists, and as previously stated, install a new impeller if at all possible.

5. Clean all bearings with solvent, dry them with compressed air, and inspect them carefully. Be sure there is no water in the air line. Direct the air stream through the bearing. Never spin a bearing with compressed air. Such action is highly dangerous and may cause the bearing to score from lack of lubrication. After the bearings are clean and dry, lubricate them with Quicksilver Formula 50-D lubricant or equivalent. Do not lubricate tapered bearing cups until after they have been inspected.

6. Inspect all ball bearings for roughness, catches, and bearing race side wear. Hold the outer race, and work the inner bearing race in-and-out, to check for side wear.

7. Determine the condition of tapered bearing rollers and inner bearing race, by inspecting the bearing cup for pitting, scoring, grooves, uneven wear, imbedded particles, and discoloration caused from overheating. Always replace tapered roller bearings as a set.

8. Inspect all bearing surface of the shaft roller bearing support. Check the shaft surface for pitting, scoring, grooving, imbedded particles, uneven wear and discoloration caused from overheating. The shaft and bearing must be replaced as a set if either is unfit for continued service.

■ Remember to check shaft surfaces at oil seal contact areas, as grooves or wear in that area will allow leakage.

Inspect the sliding clutch of the propeller shaft. Check the reverse gear side clutch "dogs". If the "dogs" are rounded one of three causes may be to blame

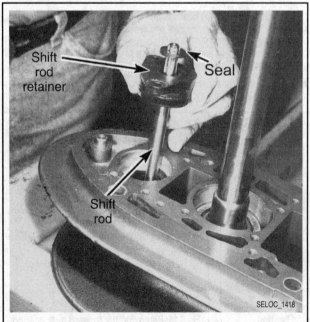

Fig. 146 Removing the shift shaft assembly (bolted-on style shown)

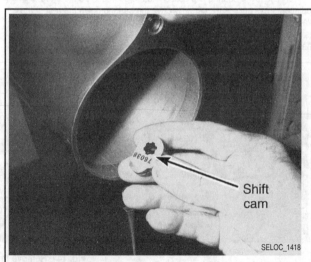

Fig. 147 Remove the shift cam, noting the orientation of the part number (facing up or down)

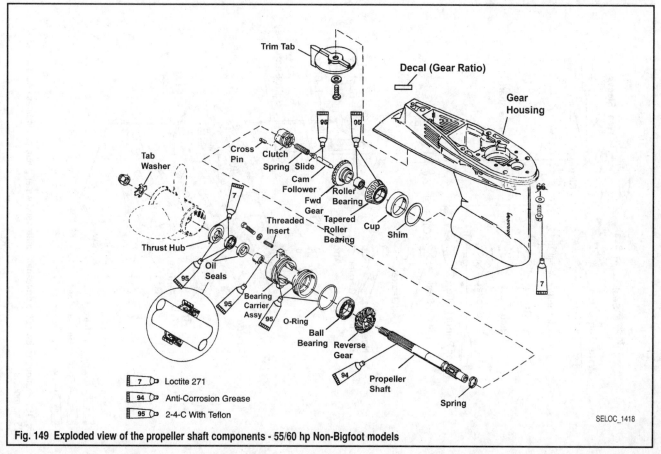

7 — Loctite 271
94 — Anti-Corrosion Grease
95 — 2-4-C With Teflon

SELOC_1418

Fig. 149 Exploded view of the propeller shaft components - 55/60 hp Non-Bigfoot models

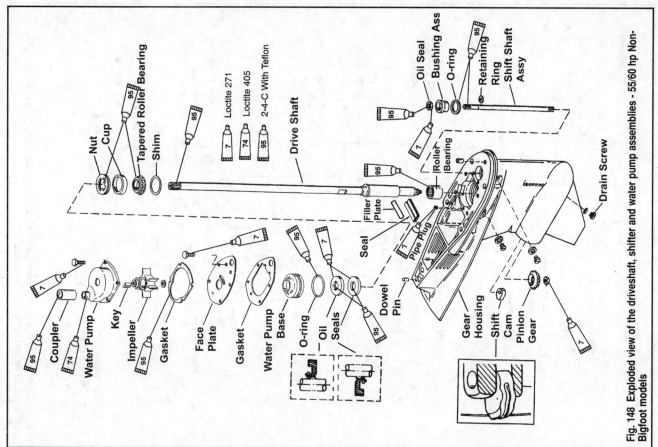

Fig. 148 Exploded view of the driveshaft, shifter and water pump assemblies - 55/60 hp Non-Bigfoot models

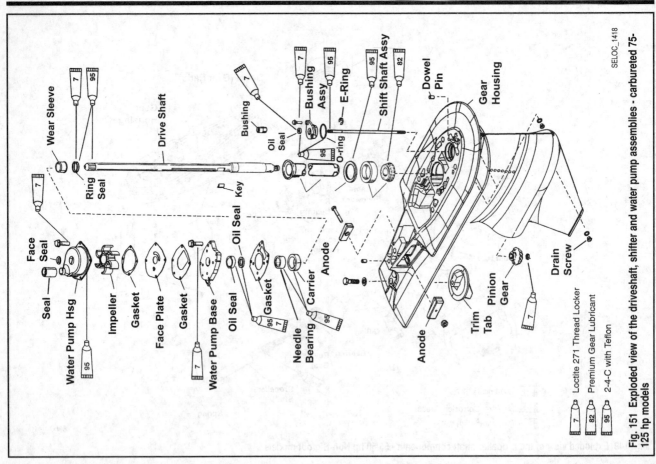

SELOC_1418

Fig. 151 Exploded view of the driveshaft, shifter and water pump assemblies - carbureted 75-125 hp models

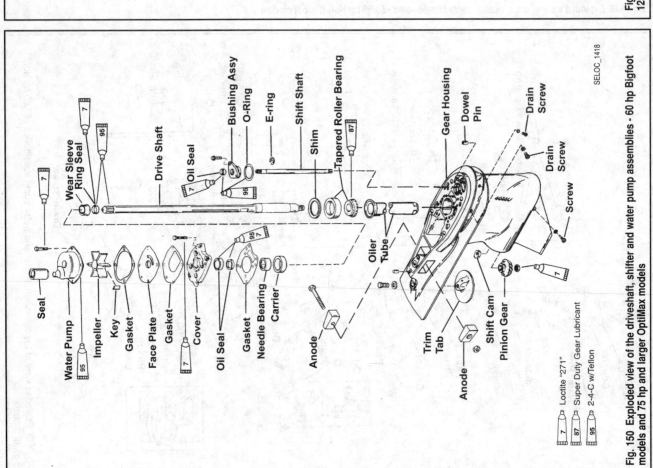

SELOC_1418

Fig. 150 Exploded view of the driveshaft, shifter and water pump assemblies - 60 hp Bigfoot models and 75 hp and larger OptiMax models

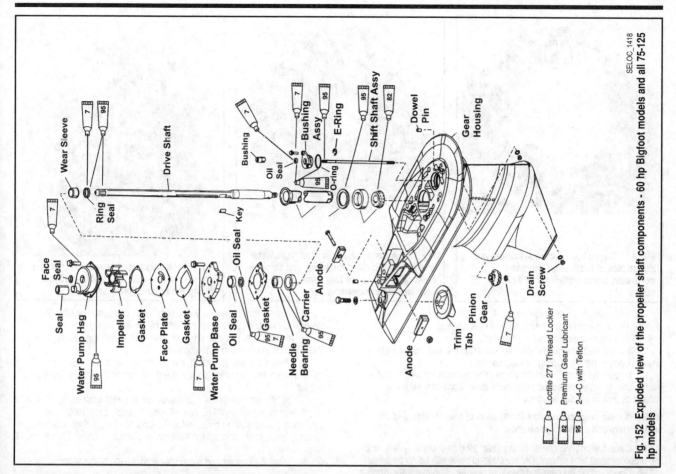

Fig. 152 Exploded view of the propeller shaft components - 60 hp Bigfoot models and all 75-125 hp models

Fig. 153 The clutch "dog" and matching splines on the propeller shaft should be closely inspected

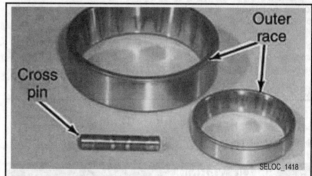

Fig. 154 Grooves on the cross pin or marked wear patterns on the outer roller bearing races are evidence of premature failure of these or associated parts

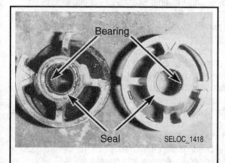

Fig. 155 Comparison of a worn bearing carrier (left) with a new one (right)

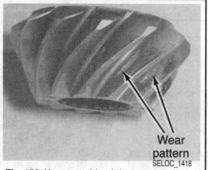

Fig. 156 Unacceptable pinion gear wear pattern, probably caused by inadequate lubrication in the lower unit

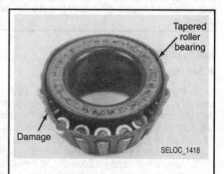

Fig. 157 Distorted tapered roller bearing unfit for further service

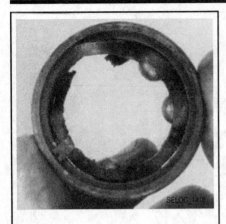

Fig. 158 Caged ball bearing set destroyed due to lack of lubrication, vibration, corrosion, metal particles, or all of the above

Fig. 159 A rusted and corroded gear. Water was allowed to enter the lower unit through a bad seal and cause this damage to the gear and other expensive parts

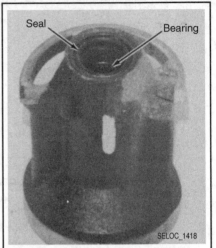

Fig. 160 This damaged bearing carrier was "frozen" in the lower unit and destroyed during the removal process

a. Improper shift cable adjustment.

b. Running engine at too high an rpm while shifting.

c. Shifting from neutral to reverse gear too quickly.

9. Inspect the cam follower and replace it and the shift cam if there is any evidence of pitting, scoring, or rough surfaces.

10. Inspect the propeller shaft roller bearing surfaces for pitting, rust marks, uneven wear, imbedded metal particles or signs of overheating caused by lack of adequate lubrication.

■ **Good shop practice requires installation of new O-rings and oil seals regardless of their appearance.**

11. Clean the bearing carrier, pinion gear, drive gear clutch spring, and the propeller shaft with solvent. Dry the cleaned parts with compressed air.

12. Check the pinion gear and the drive gear for abnormal wear. Apply a coating of light-weight oil to the roller bearing. Rotate the bearing and check for cracks or catches.

13. Inspect the propeller shaft oil seal surface to be sure it is not pitted, grooved, or scratched. Inspect the roller bearing contact surface on the propeller shaft for pitting, grooves, scoring, uneven wear, imbedded metal particles, and discoloration caused from overheating.

ASSEMBLY

◆ **See Figures 148 thru 152**

✳✳ SELOC WARNING

Before beginning the installation work, count the number of teeth on the pinion gear and on the reverse gear. Not necessary to count the forward gear. Knowing the number of teeth on the pinion and reverse gear will permit obtaining the correct tool setup for the shimming procedure.

Shift Shaft Installation

◆ **See Figures 148 thru 152 and 161**

1. Place the cam into the forward portion of the lower unit in the slot provided in the cast webbing while aligning the hole in the cam with the shift shaft pilot bore in the gearcase. The stamped (numbered) face of the shift cam must the same direction as noted during removal, which is normally with the numbers on the cam facing UPWARD for these models.

2. Install a new O-ring around the shift rod seal housing.

3. Apply a light coating of Loctite® 271, Perfect Seal, or an equivalent to the outer diameter of a new shift shaft seal then position it into the top of the shift shaft seal retainer (usually with the lips facing upward, but orient it the same way as noted during removal). Tap the seal gently downward into position.

4. Apply a light coating of 2-4-C with Teflon or equivalent lubricant to the new O-ring and to the inner diameter of the new seal.

5. Install a new clip onto the shift shaft, then install the seal retainer CAREFULLY over the top of the splines.

6. Place the lower shift shaft, the short spline end, into the shift shaft cavity. Rotate the shift shaft to allow the shaft splines to index with the cam splines.

7. Tap the retainer gently and evenly into the shift shaft cavity.

8. For all except 55/60 hp non-BF models, apply Loctite® 271 or equivalent threadlocking compound to the threads of both shaft retainer securing bolts. Install and tighten the bolts to a torque value of 60 inch lbs. (7 Nm).

9. Install the coupler and nylon spacer to the top of the shaft.

Forward Gear, Bearing and Race

◆ **See Figures 148 thru 152, 162 and 163**

✳✳ SELOC WARNING

The bearing carrier is used as a pilot while installing the forward gear bearing race in the next step. Therefore, if disassembled, the bearing carrier must be temporarily assembled to include at least the propeller shaft roller bearing.

1. Place the same amount of shim material saved during disassembly into the lower unit. If the shim material was lost, or if a new lower unit is being used, begin with approximately 0.010 in. (0.25mm) material. Coat the forward bearing race bore with Quicksilver 2-4-C with Teflon or equivalent lubricant.

2. Position the tapered bearing race squarely over the bearing bore in the front portion of the lower unit. Obtain a Bearing Driver cup tool (#91-31106 for all except 55/60 hp non-BF models), or driver cup tool (#91-817009 for the 55/60 hp non-BF models). Place the tool over the tapered bearing race.

3. Insert the propeller shaft into the hole in the center of the bearing race. Lower the bearing carrier assembly down over the propeller shaft, and then lower it into the lower unit. The bearing carrier will serve as a pilot to ensure proper bearing race alignment.

4. Use a soft-faced mallet (to protect the propeller shaft) and drive the propeller shaft against the bearing driver cup until the tapered bearing race is seated against the shim material. Withdraw the propeller shaft and bearing carrier, then lift out the driver cup.

5. If the forward gear bearings were removed, place the forward gear on a press with the gear teeth down. Apply a light coating of 2-4-C with Teflon or an equivalent lubricant to the bearing and cage surfaces.

6. Position the forward gear tapered bearing over the gear, then press the bearing onto the gear with a suitable mandrel until the bearing is firmly seated.

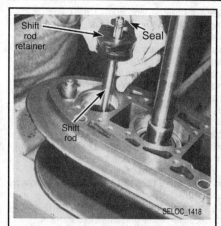

Fig. 161 Install the shift shaft, carefully aligning the splines on the shaft with those on the shift cam

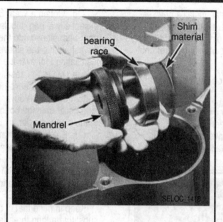

Fig. 162 Image showing the positioning of the driver (mandrel), race and shim

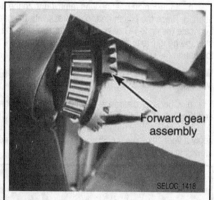

Fig. 163 Once the race is driven in position, insert the forward gear and bearing assembly

■ Press on the inner bearing race only for the outer tapered roller bearing assembly. Pressing on the outer bearing cage will distort the bearing. Because the gear shoulder is longer than the bearing on most models be sure to use a tube type mandrel to be sure the bearing is fully seated.

7. Check for clearance (gap) between the inner bearing race and the shoulder of the gear. There should be zero clearance.

8. On all except 55/60 hp non-BF models check the reverse gear side of the sliding clutch, it will have either 3 or 6 jaws. On models with a 3-jaw clutch, the needle bearing should be driven into the forward gear until it is 0.155 in. (3.94mm) below the surface of the gear. On 6-jaw models, drive the bearing in until flush with the surface.

9. Position the needle roller bearing over the center bore of the forward gear with the numbered side of the bearing facing up. Use a suitable mandrel and press the roller bearing into position either flush with the surface or slightly below (as noted earlier) or until seated on non-Bigfoot models.

10. If equipped, install a retaining ring into the groove on the inside of the forward gear to secure the needle bearing.

11. Insert the forward gear assembly into the forward gear bearing race.

Driveshaft and Bearings

◆ See Figures 148 thru 152 and 164 thru 168

1. If the pinion gear and/or upper bearings were removed on 55/60 hp non-BF models install the replacements, as follows:

a. Use a driver (#91-817058A1 or equivalent) to tap the new pinion bearing (which has been lubricated with 2-4-C, Needle Bearing Assembly lubricant or equivalent and positioned with the numbered side facing upward) down into the gearcase from the top of the bore. Continue to gently tap the bearing until the top of the bearing is 7.05-7.07 in. (179.0-179.5mm) below the upper surface of the gearcase housing.

b. Lubricate a new upper driveshaft bearing with 2-4-C, Needle Bearing Assembly lubricant or equivalent, then use a suitable mandrel to push the bearing onto the driveshaft until it seats against the driveshaft's shoulder.

c. Upon installation of the driveshaft and pinion gear, also be sure to install the race and shim or shims (saved from disassembly, or if lost start with a 0.15 in./0.361mm shim). Tighten the upper driveshaft bearing retainer using the special tool (#91-43506 or equivalent high-strength spanner) to 75 ft. lbs. (102 Nm).

2. For 60 hp Bigfoot models and all 75-125 hp models, install the driveshaft bearings as follows:

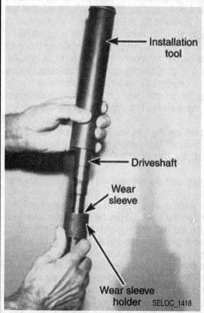

Fig. 164 You'll need a pipe-like special tool to install the wear sleeve (all except 55/60 hp non-BF models)

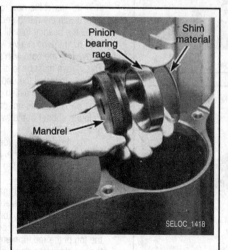

Fig. 165 Like the forward gear race, the pinion race and shims are installed using a suitably sized mandrel (all except 55/60 hp non-BF models)

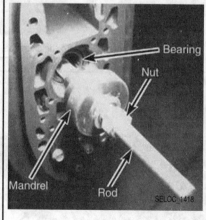

Fig. 166 Upper driveshaft bearing installation (all except 55/60 hp non-BF models)

a. If the Pinion bearing was removed from the gear, install a new bearing. Place the pinion gear on a press with the gear teeth down. Lubricate (using 2-4-C or equivalent) and position the pinion tapered bearing over the gear. Press the bearing onto the gear with a suitable mandrel until the bearing is firmly seated.

■ **Press on the inner bearing race only. Pressing on the bearing cage will distort the bearing.**

b. Check for clearance (gap) between the pinion inner bearing race and the shoulder of the gear. There should be zero clearance.

c. Install a new rubber sealing ring around the driveshaft, and then apply a light coating of Loctite® 271 or equivalent threadlocking compound around the ring.

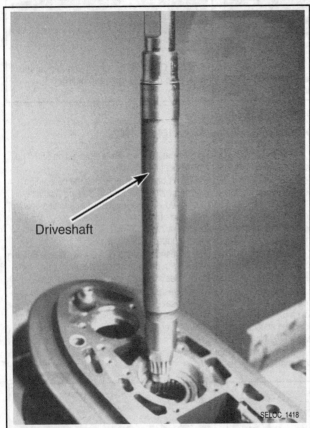

Driveshaft

Fig. 167 Insert the driveshaft into the gearcase...

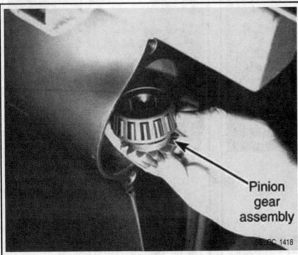

Pinion gear assembly

Fig. 168 ...then install the pinion gear (and bearing on most models)

d. Obtain a Wear Sleeve Installation Tool (#91-14310A1), which is essentially a long, tubular driver which will fit over the driveshaft. Insert the new wear sleeve into the sleeve holder. Slide the bottom end of the driveshaft into the sleeve and holder. Slide the long collar portion of the installation tool over the top end of the driveshaft.

e. Move the assembled tool and driveshaft to an arbor press. Position the base of the sleeve holder onto a suitable support which will allow the bottom end of the driveshaft to pass through. Press on the top surface of the long collar until the bottom surface of the collar seats against the top surface of the sleeve holder.

f. Set the driveshaft aside for later installation.

g. If the pinion gear bearing was replaced, use the Bearing Installation Tool Kit (#91-14309T01 or #T02), including the long threaded rod (#91-31229), nut (#11-24156), upper bore mandrel (#91-13781) and the appropriate pinion bearing race mandrel (#91-13781 for 75-125 hp motors or #91-13780 for 60 hp BF models).

Keep in mind that in the absence of these special tools you can usually substitute a long threaded rod, some nuts and suitably sized washers.

h. Using the threaded rod and drivers, set the pinion gear bearing race onto the shorter mandrel with the taper toward the mandrel. Place the same amount of shim material, removed during disassembly, on top of the race.

■ **If the shim material was misplaced or not recorded, begin with material totaling 0.025 in. (0.635mm). Remember, shim material must be measured individually and the thickness of each added to arrive at the total thickness. Never use a micrometer to measure the total thickness of several pieces of shim material. Such a procedure results in a very inaccurate measurement and will lead to incorrect shimming of the lower unit.**

i. Insert the threaded rod thru the shim material, the bearing race, and into the lower mandrel (or through the lower washer and secure using two nuts tightened against each other to keep them from loosening).

■ **Lubricate the outer diameter of the bearing race using 2-4-C with Teflon or an equivalent lubricant.**

j. At the top of the bore install the upper mandrel or a large washer, then secure the assembly with a nut. Tighten the nut to draw the threaded rod (and bearing race assembly) upward pulling the race into the bore until seated.

k. Apply a light coating of 2-4-C with Teflon or an equivalent lubricant to the inner and outer surfaces of the bearing sleeve.

l. Identify the end of the new roller bearing with the embossed numbers. Identify the end of the bearing sleeve with a slight taper. Place the tapered end down on the arbor press support plate.

m. Insert the new roller bearing down into the larger end of the bearing sleeve with the numbered end facing upward. Obtain a suitable mandrel and press the bearing in until it is flush with the sleeve.

n. If the oil sleeve was removed from the driveshaft bore, install the sleeve into the bore with the tab on the upper portion of the sleeve facing aft. This sleeve must be in place before the upper driveshaft bearing is installed.

o. To install the upper driveshaft bearing, obtain the same special tools used to install the pinion gear bearing race earlier in this procedure.

p. Thread the nut onto the rod until the nut is about 2/3 of the way up. Hold the shorter of the two mandrels inside the lower unit under the driveshaft bore.

q. Insert the threaded rod into the bore and thread the mandrel onto the rod until the mandrel is secure. Place the assembled roller bearing and sleeve over the threaded rod, into the cavity, and with the flush/numbered side facing upward.

r. Slide the other mandrel over the threaded rod with the shoulder side down to fit inside the sleeve.

s. Thread the nut on the rod down against the upper mandrel until all slack (clearance) is removed from the setup. With the proper size wrench on the top of the rod, prevent the rod from rotating, and at the same time rotate the nut clockwise down against the upper mandrel. As the nut is rotated the two mandrels will come closer together. This action will seat the bearing and sleeve assembly. When the shoulder of the upper mandrel makes contact with the upper face of the lower unit, the bearing/sleeve assembly is correctly positioned in the driveshaft bore.

t. Remove the special tools.

3. For all models, it's time to install the driveshaft itself. With one hand, lower the driveshaft into the top of the lower unit. With the other hand, hold the pinion gear up in the lower unit below the driveshaft cavity and with the teeth of the gear indexed (meshed) with the teeth of the forward gear.

4. Rotate and insert the driveshaft until the driveshaft splines align and engage with the splines of the pinion gear. Continue to insert the driveshaft into the pinion gear until the tapered bearing is against the bearing race.

■ Mercury sells new pinion nuts with a drylock patch on the threads. If you've got a new nut for installation don't install it until after backlash has been checked and possibly adjusted. Instead, use the old pinion nut for this step, and save the new nut for final assembly. If you don't have a new nut, no biggie, Mercury just advises the use of Loctite® 271, or an equivalent threadlocking compound during final assembly.

5. Install the old pinion gear nut onto the driveshaft. Hold the pinion nut with a socket wrench. Pad the area where the socket wrench flex handle will contact the lower unit while the pinion nut is being tightened.

■ On some models there is a recessed portion of the pinion nut and, when present, it should face TOWARD the pinion gear.

6. Obtain the proper driveshaft holding tool for the model gearcase being serviced as follows:
• For all 55/60 hp (BF and non-BF) models use #91-817070
• For all 75 hp and larger models use #91-56775
7. Place the special wrench over the upper end of the driveshaft. Use a torque wrench and socket to tighten the nut to 50 ft. lbs. (67 Nm) for 55/60 hp non-BF models or to 70 ft. lbs. (95 Nm) for 60 hp BF models and for all 75-125 hp models.

✳✳ SELOC WARNING

After the pinion gear depth and the forward gear backlash have been set, the old nut must either be replaced with a new pinion gear nut OR the nut must be removed and the threads coated with Loctite® 271 or an equivalent threadlocking compound.

8. Now, if you've replaced the gearcase housing, or if you've replaced one or more of the bearings SKIP ahead to the Shimming procedures in order to check backlash and determine if some of the shims must be replaced. However, if only seals have been replaced, then you can proceed with the assembly (assuming you used a new nut or coated the old pinion nut with Loctite during assembly).

Propeller Shaft Assembly

◆ See Figures 148 thru 152 and 169 thru 172

1. Align the hole through the clutch "dog" with the slot in the propeller shaft, and then slide the "dog" onto the clutch with the "square" teeth facing the propeller end of the shaft (the reverse gear). The teeth with the "slanted" ramps (the ratcheting clutch teeth) face forward.
2. Insert the spring into the end of the propeller shaft. If equipped, insert the guide block, stepped end, into the front end of the propeller shaft with the cross-pin hole aligned with the cross-pin hole In the sliding clutch.
3. For 60 hp Bigfoot and all 75-125 hp models, install the 3 metal balls. Use a light coating of 2-4-C with Teflon or an equivalent lubricant to hold the balls in position.
4. For all models, apply a light coating of 2-4-C with Teflon to the cam follower, then insert the follower into the front end of the propeller shaft.
5. Position the cam follower against a solid object and push against the cam follower to compress the spring.
6. Hold the propeller shaft in this position and at the same time, use a punch to align the guide block cross-pin opening with the sliding clutch cross-pin hole.
7. Remove the punch and insert the cross-pin.
8. Install the cross-pin ring over the sliding clutch. Take care not to over stretch the spring.

■ When handling the assembled propeller shaft, Take care to keep the forward end tilted slightly upward to keep the small balls or other loose parts in place inside the shaft.

Bearing Carrier Assembly and Installation

◆ See Figures 148 thru 152, 173 and 174

Be sure to lubricate the inner and outer diameters of the bearings, as well as the inner diameter only of the oil seals using 2-4-C with Teflon or an equivalent lubricant prior to installation. The outer diameter of the oil seals should be coated using Loctite® 271 or an equivalent threadlocking compound. After installing the seals, wipe away any excess Loctite before proceeding.

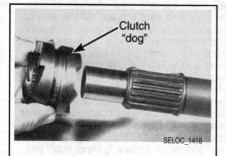

Fig. 169 Slide the clutch dog into position on the propeller shaft. . .

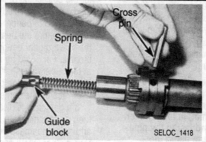

Fig. 170 . . . then insert the spring and any remaining components before securing the assembly with the cross pin

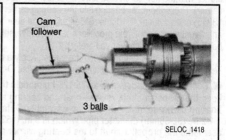

Fig. 171 On some models that includes 3 metal balls which are inserted right before the cam follower

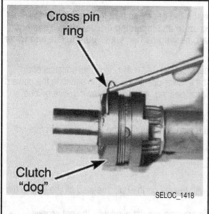

Fig. 172 Install the retaining spring to secure the cross pin

Fig. 173 No matter what, ALWAYS at LEAST replace the bearing carrier O-ring

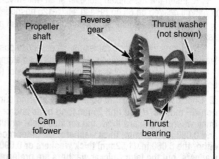

Fig. 174 On most models the propeller shaft utilizes a reverse gear thrust bearing and washer (except 55/60 hp non-BF models)

1. Position the propeller shaft needle roller bearing into the aft (propeller) end of the bearing carrier with the numbered side toward the aft end (toward the driver). Press the needle bearing into the bearing carrier with a suitable mandrel. For 55/60 hp non-BF models mandrel #91-817011 will give you the correct depth when the upper edge of the tool seats against the carrier. For 60 hp BF models, as well as all 75-125 hp models, position mandrel #91-13945 underneath the carrier (at the reverse gear side) and then use mandrel #91-15755 to push the bearing down until it seats.

■ **If the Mercury tools are not available, position the bearing to the depth measured prior to removal.**

2. Obtain a Oil Seal Driver (#91-31108 for all except the 55/60 hp non-BF models which use #91-817007). Place one seal on the longer shoulder side of the driver tool with the lip of the seal away from the shoulder

3. Press the seal into the bearing carrier until the seal driver bottoms against the bearing carrier for all except 55/60 hp non-BF models where the seal should be installed to the proper depth as noted.

■ **On 55/60 hp non-BF models if the Mercury mandrel is not available, make sure the first (inner) seal is installed to a depth of 0.82 in. (20mm) below the end of the bearing carrier and the second (outer) seal to a depth of 0.04 in. (1mm) below the end of the carrier.**

4. Place the second seal on the short shoulder side of the seal driver with the lip of the seal toward the shoulder. Press the seal into the bearing carrier until the seal driver bottoms against the bearing carrier for all except 55/60 hp non-BF models where the seal should be installed to the proper depth as noted. Clean excess Loctite from the seals.

5. Install the reverse gear and bearing as follows, depending upon the model:

a. For 55/60 hp non-BF gearcases, use a suitable mandrel to press the roller bearing onto the back of the reverse gear until it bottoms. As usual, make sure the mandrel only presses on the portion of the bearing in contact with the gear (inner). Then use a mandrel to press the assembled gear and bearing into the forward (gearcase) end of the bearing housing until it bottoms out.

b. For 60 hp Bigfoot models, as well as all 75-125 hp models, position the larger propeller shaft roller bearing into the forward end of the bearing carrier with the numbered side toward the tool (forward end, away from the carrier). Press the roller bearing into the bearing carrier using mandrel (91-13945) or equivalent substitute. Slide the reverse gear and thrust bearing and thrust washer onto the propeller shaft (or conversely position them to the bearing housing, whichever you prefer).

6. Install a new O-ring around the carrier. Apply a coating of 2-4-C with Teflon or and equivalent Lubricant onto the installed O-ring.

■ **For 60 hp BF models and all 75-125 hp models there is a trick to keep the propeller shaft, reverse gear and bearing all in position in the bearing carrier to install them as an assembly. Obtain a length of 1 1/4-1 1/2 in. (32-38mm) diameter pipe which is 6 in. (152mm) in length, then assemble the propeller shaft to the bearing carrier and fasten the pipe to the outer end (over the propeller splines) and secure using the propeller nut and tabbed washer. This will hold everything securely in position so you can install the bearing carrier and propeller shaft as an assembly.**

7. Insert the assembled propeller shaft (with reverse gear and thrust bearing if desired on 60 hp BF and all 75-125 hp models) into the lower unit or use the tip listed earlier so you can install the shaft and carrier as an assembly.

8. Rotate the bearing carrier, as necessary, to position the word **TOP** facing upward after installation (if the carrier is so stamped).

9. Following final installation, coat the threads of the attaching bolts with Loctite ® 271.

■ **Check the washers and bolts for the bearing carrier. For 55/60 hp non-BF motors, if you find 25mm long bolts and/or 0.063 in. (1.6mm) thick washers discard them and replace them with 30mm long bolts and 0.090 in. (2.29mm) thick washers. On other models you may use either the 0.060 in. (1.52mm) thick washers or 0.090 in. (2.29mm) thick washers, but the later (thicker washers are preferred, and keep in mind the two different washers would require different torque specs).**

10. For 55/60 hp non-BF motors, install and tighten the bolts (or nuts on models equipped with studs) to 225 inch lbs./18.8 ft. lbs. (19 Nm).

11. For 60 hp BF models, as well as all 75-125 hp motors, install and tighten the bolts (or nuts on models equipped with studs) to either 22 ft. lbs. (30 Nm) if the thicker washers are used or to 25 ft. lbs. (34 Nm) if the thinner washers are used.

12. If not done earlier during driveshaft installation (assuming no preload check needed to be performed) install the water pump base, followed by the water pump assembly, keeping the following points in mind:

• The dual seals should be coated on the outer-diameter with Loctite® 271 or equivalent and installed back-to-back so the inner or lower seal faces the gearcase and the upper or outer seal faces the water pump.

• A NEW O-ring should be used, and coated (along with the inner lips of the seals) with 2-4-C with Teflon or an equivalent lubricant.

• For 55/60 hp non-BF motors the seals should be installed so the upper seal is 0.04 in. (1.02mm) from the top edge of the water pump base.

• For 60 hp BF models, as well as all 75-125 hp motors apply a light coating of the same threadlocking material used on the seals to the threads of the water pump base retaining bolts, then install the bolts and tighten to 60 inch lbs. (7 Nm).

13. Refill the gearcase and check for leaks, then install the gearcase to the outboard as detailed earlier in this section.

■ **Mercury recommends using a pressure gauge installed to the vent plug in order to pressurize and check the gearcase before it is returned to service. Apply 10-15 psi (68.9-103.4 kPa) of pressure and watch the gauge for 5 minutes. Then rotate the driveshaft and propshaft and move the shift rod making sure the pressure does not drop. IF pressure drops, submerge the gearcase (with pressure applied) and watch for bubbles. Bottom line, the gearcase MUST hold pressure for at least 5 minutes.**

SHIMMING

Backlash

◆ **See Figure 175**

For any gearcase to operate properly the pinion gear on the driveshaft must mesh properly with the forward and reverse gears. And, remember that these gears remain in contact with the pinion gear at all times, regardless of whether or not the sliding clutch has locked one of them to the propeller shaft. If these gears are not properly meshed, there is a greater risk of breaking teeth or, more commonly, rapid wear leading to gearcase failure.

On some of the smaller Mercury gearcases there is little or no adjustment for the positioning of these gears. However, the gearcases used on these medium to large sized Mercury motors are all equipped with shim material that is used to perform minute adjustments on the positioning of the bearings (and therefore the bearings and gears). Shim material can be found between the forward gear bearing and race assembly and the nose cone on all models. For 55/60 hp non-BF models pinion adjusting shim material is found at the upper driveshaft bearing. For 60 hp BF models as well as all 75-125 hp models the pinion gear shim material is found directly above the pinion gear race (in the same general fashion as the nose cone gear).

If a gearcase is disassembled and reassembled replacing only the seals, then no shimming should be necessary. However, if shafts, gears/bearings and/or the gearcase housing itself is replaced it is necessary to check the Pinion Gear Depth and the Forward Gear Backlash to ensure proper gear positioning.

Pinion Gear Depth

◆ **See Figures 176 thru 179**

■ **This procedure starts with the gearcase mostly reassembled (following our procedures up to and including driveshaft reassembly, but with the propeller shaft, bearing carrier, water pump plate and water pump still removed).**

Obtain a Bearing Preload Tool (#91-14311A2). This tool consists of a spring, which seats against the lower unit and is used to simulate the upward driving force on the pinion gear bearing. Install the following, one by one, over the driveshaft.

a. Plate (used on 55/60 hp non-BF models only)

b. Adaptor, with the flat face down against the lower unit (or plate on 55/60 hp non-BF models)

c. Thrust bearing

d. Thrust washer

e. Spring

f. Special large bolt, with the nut on the bolt threaded all the way to the top of the threads.

g. Sleeve (the holes in the sleeve must align with the set screws, typically you'll use the 3/4 in./19mm ID sleeve, but others are usually included in the kit)

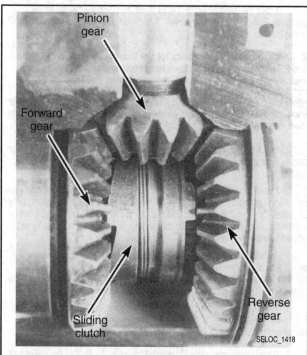

Fig. 175 Backlash is the amount of play allowed by the position of the forward and reverse gears in relation to the pinion gear

1. Align the two holes in the sleeve with the Allen set screws in the special large bolt head. These set screws grip the driveshaft and prevent the tool from slipping on the driveshaft. Tighten the two set screws securely with the Allen key provided with the tool kit.

2. Measure the distance between the bottom surface of the bolt head and the top surface of the nut. Record this measurement.

3. Use a wrench to hold the bolt head steady and another wrench to tighten the nut down against the spring. Continue tightening the nut until the distance between the bottom surface of the bolt head and the top surface of the nut equals 1.00 in. (25.4mm) plus the distance recorded when the tool was installed.

4. The driveshaft has now been forced upwards placing an upward preload on the pinion gear bearing.

5. Obtain the correct Pinion Gear Locating Tool (#91-817008A2 for 55/60 hp non-BF models or #91-12349A2 for 60 hp BF models or for all 75-125 hp models). For the later tool you'll need to assemble it by placing

sliding the gauging block and the split collar onto the tool (with the gauging block numbers facing away from the split collar and toward the cross-hatched section of the tool shaft, on the opposite end of the tool from the snapring). Do NOT tighten the collar retaining screw at this time, however, use the 2 screws to secure the gauging block to the split collar.

6. Insert the tool into the installed forward gear assembly. For all except the 55/60 hp non-BF models, if positioned correctly, one of the flats on the gauging block should be positioned directly under the pinion gear teeth. On the 55/60 hp non-BF models the gauging surface is already part of the tool and it should be located under the near edge of the pinion gear (where the reverse gear would contact).

■ Notice the gauging block on the tool (#91-12349A2) has a series of flat sides, identified by the numbers 1 thru 8. The eight different numbered faces are needed for use with different combinations of gear ratios. Only a few gear ratios are used on units covered here. At the beginning of this section, we asked you to count and record the number of teeth on the pinion and reverse gears, now it's time to use that data (yes it's true that there may be a sticker on the gearcase which lists the ratio, but if gears were replaced, well, you never know right).

7. For 60 hp BF models, and for all 75-125 hp models, the gauging block accepts various gauging flats. After inserting it once and making sure the collar/block is under the near edge of the reverse gear, remove the tool and tighten the collar retaining bolt, THEN select and install the appropriate numbered gauge flat and locating disc depending upon the model (and more specifically the gear ratio or number of pinion teeth vs. number of reverse gear teeth, as follows):

• For 60 hp BF motors with 2.31:1 (13/30 gear teeth) use flat No. 8 and Locating disc No. 3.

• For 75-90 hp motors with 2.31:1 (13/30 gear teeth) or 2.33:1 (12/28 gear teeth) use flat No. 8 and Locating disc No. 3.

• For 115-125 hp motors with 2.07:1 (14/29 gear teeth) use flat No. 2 and Locating disc No. 3.

■ No other gear ratio combinations SHOULD BE possible for units covered in this manual. If another "odd-ball" gear ratio is encountered, somewhere, sometime, someone has purposely changed the gear ratio to suit their own requirements, possibly for operation of the outboard at different altitudes. Therefore, these factory specifications would no longer apply.

8. Let's assume everything is as it should be and the appropriate tool face to use is determined. Position this face (either No. 8 or No. 2 or the single face on the tool for 55/60 hp non-BF models) under the pinion gear teeth.

9. On 60 hp BF models and all 75-125 hp non-Bigfoot models, install the locating disc over the tool shank with the access hole facing upward. Push the disc all the way into the lower unit until it rests against a machined shoulder of the bore. This disc supports the tool shank in the lower unit bore.

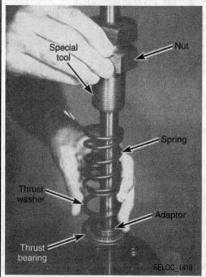

Fig. 176 Installing the pinion gear pre-loading tool

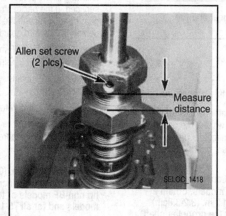

Fig. 177 Set the nut to the proper distance to set pre-load

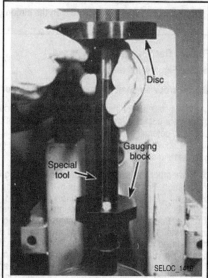

Fig. 178 Installing the proper flat/locating disc (all except 55/60 hp non-BF models)

10. The pinion depth is the distance between the bottom of the pinion gear teeth and the flat surface of the gauging block.

11. Measure this distance by inserting a long feeler gauge into the access hole and between the pinion gear teeth and gauging block flat face.

12. For all units covered in this section, the correct clearance is 0.025 in. (0.64mm). If the clearance is correct, remove only the Pinion Gear Locating Tools from the inside of the gearcase. Leave the Pinion Gear Bearing Preload Tool in place on the driveshaft in order to check Forward Gear Backlash.

13. If the clearance is not correct, add or subtract shim material from behind the pinion gear bearing race or upper driveshaft bearing (as applicable) to lower or raise the gear.
  • If the pinion gear depth was found to be greater than 0.025 in. (0.64mm), then that amount of shim material must be added to lower the gear.
  • If the pinion gear depth was found to be less than 0.025 in. (0.64mm), then that amount of shim material must be removed to raise the gear.

■ The difference between 0.025 in. (0.64mm) and the actual clearance measured is the amount of shim material needed to correct the pinion gear depth.

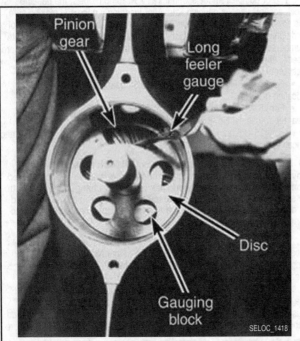

Fig. 179 Checking the gap between the pinion gear and the gauging surface

Forward Gear Backlash

◆ See Figures 180 and 181

■ The propeller shaft is temporarily installed for this procedure.

Final installation of the shaft is usually performed later because an old pinion gear nut is normally installed on the driveshaft without the use of Loctite to determine the pinion gear depth (in case it needs to be disassembled again to make changes to the pinion gear bearing or forward gear bearing shims).

■ Remember the trick we mentioned in Bearing Carrier Assembly and Installation. For 60 hp BF models or all 75-125 hp models, obtain a piece of PVC pipe, 6 in. (15.2cm) long by 1 1/4 -1 1/2 in. (3.2-3.8cm) diameter and a washer large enough to slide over the propeller shaft threads. Before installing the assembled bearing carrier into the lower unit, install the PVC pipe over the propeller end of the shaft, followed by the large washer and propeller nut to hold it all together. Hand tighten the nut. The pipe will exert a pressure on the bearing carrier, thrust washer and reverse gear and will allow each component to be held in alignment while the shaft is installed. Once the shaft is well seated inside the lower unit, the propeller nut, washer and PVC pipe is removed.

1. Insert the assembled propeller shaft and bearing carrier to the gearcase. Rotate the bearing carrier, if necessary, to position the word **TOP** facing upward after installation.

■ Check the washers and bolts for the bearing carrier. For 55/60 hp non-BF motors, if you find 25mm long bolts and/or 0.063 in. (1.6mm) thick washers discard them and replace them with 30mm long bolts and 0.090 in. (2.29mm) thick washers. On other models you may use either the 0.060 in. (1.52mm) thick washers or 0.090 in. (2.29mm) thick washers, but the later (thicker washers are preferred, but if they're not used, remember the torque specs are different).

2. For 55/60 hp non-BF motors, install and tighten the bolts (or nuts on models equipped with studs) to a torque value of 225 inch lbs./18.8 ft. lbs. (19 Nm).

3. For 60 hp BF models, as well as all 75-125 hp non-Bigfoot motors, install and tighten the bolts (or nuts on models equipped with studs) to either 22 ft. lbs. (30 Nm) if the thicker washers are used or to 25 ft. lbs. (34 Nm) if the thinner washers are used. On these models, remove the propeller nut, tabbed washer and length of PVC pipe at this point.

4. Install a bearing carrier puller with the arms of the puller on the carrier and the center bolt on the end of the propeller shaft. Tighten the puller center bolt to 45 inch lbs. (5 Nm). This action places a preload on the forward gear, pushing the forward gear into the forward gear bearing.

5. Rotate the driveshaft about five to ten full revolutions to properly seat the forward gear tapered roller bearing. Then recheck the torque value on the center puller bolt.

6. Attach the appropriate Backlash Dial Indicator Rod to the driveshaft and position a dial indicator (using a magnetic base or a threaded rod with nuts and plates to position it alongside the Indicator Rod) and THEN point the dial indicator to the appropriate number on the indicator rod tool, depending upon the model as follows:
  • 55/60 hp non-BF models - use backlash indicator tool #91-19660-1 and point the tool to the No. 3 mark.
  • 60 hp BF models - use backlash indicator tool #91-78473 and point the tool to the No. 4 mark on the indicator rod tool.
  • 75-90 hp carbureted models - use backlash indicator tool #91-78473 and point the tool to the No. 4 mark on the indicator rod tool.
  • 75-90 hp OptiMax models - use backlash indicator tool #91-19660-1 and point the tool to the No. 4 mark on the indicator rod tool.
  • 115-125 hp motors - use backlash indicator tool #91-19660-1 and point the tool to the No. 1 mark on the indicator rod tool.

7. Rotate the driveshaft back-and-forth and observe movement of the dial indicator. Total movement is the forward gear backlash. Check the following listing proper amount of backlash permissible for the unit being serviced. If the backlash is too great, add shim material behind the forward gear bearing race. If the backlash is too small, remove shim material from behind the forward gear bearing race.
  • 55/60 hp non-BF models, as well as 75-115 hp Optimax models - there should be 0.013-0.019 in. (0.33-0.48mm) of backlash.
  • 60 hp BF models - there should be 0.012-0.019 in. (0.30-0.48mm) of backlash.
  • 75-90 hp carbureted models - there should be 0.012-0.019 in. (0.30-0.48mm) of backlash.
  • 115-125 hp carbureted models - there should be 0.015-0.022 in. (0.38-0.55mm) of backlash.

■ For each 0.001 in. (0.025mm) of shim which is added or removed expect backlash to increase or decrease 0.00125 in. (0.032mm) for 55/60 hp non-BF models or to change 0.001 in. (0.025mm) for 60 hp BF models and for all 75-125 hp models.

8. Remove all special tools. Remove the bearing carrier and propeller shaft.

9. Remove the pinion nut and either replace it with a new nut which has threadlocking compound already applied OR apply a light coating of Loctite® 271 or equivalent threadlocking compound to the threads. Proceed with final assembly from the pinion nut tightening through installation of the Bearing Carrier and Propeller Shaft assembly.

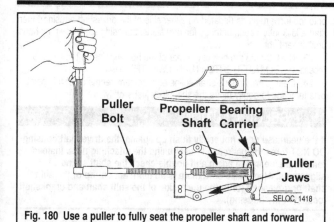

Fig. 180 Use a puller to fully seat the propeller shaft and forward gear. . .

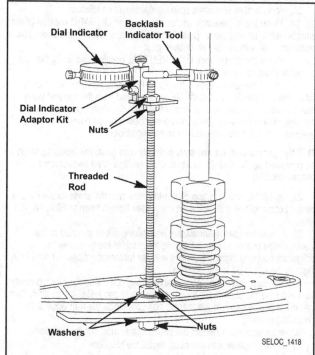

Fig. 181 . . .then check forward gear backlash using the appropriate indicator tool

Gearcase - 135-200 Hp (2.5L) V6 Models

There are basically 2 similar (almost identical) design gearcases covered in this section. The 4.75 in (120.65mm) torpedo diameter standard rotation gearcase AND the 4.75 in (120.65mm) torpedo diameter counter rotation gearcase. The basic design of the gearcase, water pump, driveshaft and shift shaft are all almost the same between the 2 units.

Differences occur in the placement of gears as well as the type and location of bearings between the 2 units. These differences are necessitated by the fact that on standard rotation units the FORWARD bearing, located in the nose cone, takes the majority of thrust through use (as most use is in Forward gear on the average outboard). On counter-rotation units the FORWARD bearing (which also takes the majority of the thrust in use) is located on the propshaft, up against the bearing carrier instead of in the nose cone. Small differences in the bearing assemblies help to strengthen the assemblies to perform their jobs as such.

■ **For gearcase exploded views, please refer to Cleaning & Inspection in this section.**

REMOVAL & INSTALLATION

◆ **See Figures 182, 183 and 184**

Unlike most of the smaller Mercury gearcases, the rotating/ratcheting shift mechanism used on these models utilizes a splined shift shaft connection, making it un-necessary to locate and disconnect shift linkage before gearcase removal.

1. For safety disconnect the battery cables.
2. If necessary for overhaul purposes, remove the Propeller, as detailed in the Maintenance and Tune-Up section.

■ **It's actually a good idea to at least loosen the OIL LEVEL and/or VENT screw(s) first, to make sure it will come out without a problem. You can't refill the gearcase, if you can't get the vent screw loose.**

3. If necessary for gearcase overhaul purposes, position a suitable clean container under the lower unit. Remove the FILL screw on the bottom of the lower unit, and then the OIL LEVEL and/or VENT screw(s). The vent screw must be removed to allow air to enter the lower unit behind the lubricant. Allow the gear lubricant to drain into the container.

As the lubricant drains, catch some with your fingers from time-to-time and rub it between your thumb and finger to determine if there are any metal particles present. Examine the fill plug. A small magnet imbedded in the end of the plug will pickup any metal particles. If metal is detected in the lubricant, the unit must be completely disassembled, inspected, and the damaged parts replaced.

Check the color of the lubricant as it drains. A whitish or creamy color indicates the presence of water in the lubricant. Check the container for signs of water separation from the lubricant. The presence of any water in

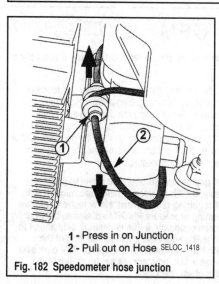

1 - Press in on Junction
2 - Pull out on Hose SELOC_1418

Fig. 182 Speedometer hose junction

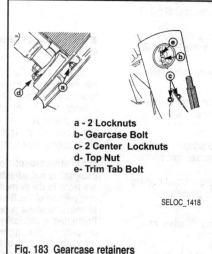

a - 2 Locknuts
b - Gearcase Bolt
c - 2 Center Locknuts
d - Top Nut
e - Trim Tab Bolt

SELOC_1418

Fig. 183 Gearcase retainers

Fig. 184 View of the 2 forward locknuts, plus the top nut

the lubricant is bad news. The unit must be completely disassembled; inspected; the cause of the problem determined; and then corrected.

4. If not done already, rotate the outboard unit to the full UP position and engage the tilt lock lever.

5. Shift the gearcase into NEUTRAL.

6. Matchmark the trim tab to the anti-cavitation plate to preserve alignment, then loosen the bolt which secures the tab (the trim tab is normally secured from above by a bolt that is threaded vertically downward through an access panel in the top of the driveshaft housing, under a plastic cover). Remove the trim tab from the cavitation plate for access to the gearcase retaining bolt located inside the trim tab cavity. Loosen and remove the previously hidden gearcase retaining bolt.

7. Still working on the underside of the anti-cavitation plate, move forward to about the middle of the plate, the point where the gearcase itself starts to extend downward from the plate. On either side of the housing there is a gearcase retaining center locknut tucked into a slight recess in the plate. Loosen and remove these 2 gearcase retaining locknuts.

8. At the front lower motor mount/pivot point, locate the speedometer hose junction. While pressing in on the speedometer hose junction, carefully pull out on the hose itself to disconnect it.

9. Also at the front lower motor mount/pivot point, locate the top locknut for the front gearcase mounting stud (it's right there threaded vertically down into contact with the gearcase/mount assembly. Loosen and remove the locknut.

10. Alternately and evenly loosen the 2 forward gearcase mounting locknuts (but DON'T try to remove the one nut before the opposite side is loosened or the driveshaft housing could be damaged). Once they are sufficiently loosened, pull down on the lower unit gear housing and CAREFULLY separate the lower unit guiding it straight out of the exhaust housing (removing these last 2 nuts as this occurs). Guide the shift rod and driveshaft out of the exhaust housing as the lower unit is removed.

11. Place the lower unit in a suitable holding fixture or work area.

To install:

■ **Always fill the lower unit with lubricant and check for leaks before installing the unit to the driveshaft housing (could save you the trouble of removing it again right away).**

12. Take time to remove any old gasket material from the fill and vent recesses and from the screws.

13. Place the lower unit in an upright vertical position. Fill the lower unit with Super-Duty Lubricant, or equivalent, through the fill opening at the bottom of the unit. Never add lubricant to the lower unit without first removing the vent screw and having the unit in its normal operating position - vertical. Failure to remove the vent screw will result in air becoming trapped within the lower unit. Trapped air will not allow the proper amount of lubricant to be added.

14. Continue filling slowly until the lubricant begins to escape from the vent or level opening (as applicable) with no air bubbles visible.

15. Use a new gasket and install the vent and/or level screw(s). Slide a new gasket onto the fill screw. Remove the lubricant tube and quickly install the fill screw.

16. Visibly check the lower unit for leaks. Recheck for leaks after the first time the unit is warmed to normal operating temperatures from use.

17. If not done already, swing the exhaust housing outward until the tilt lock lever can be actuated, and then engage the tilt lock.

18. Place the gearcase in the FORWARD gear position, also, be sure to most the guide block anchor pin up near the powerhead, to the FORWARD position also.

19. Apply a coating of 2-4-C Marine Lubricant with Teflon to the splines of the shift shaft coupler and the driveshaft.

■ **DO NOT put lubricant on the ENDS of the shafts, as it could prevent the shaft splines from fully seating in the shift coupler and the crankshaft.**

20. Apply a light bead of G.E. Silicone Sealer or an equivalent sealant, to the top of the divider block located on the top of the gearcase, right behind the water pump assembly.

21. Carefully bring the gearcase into alignment with the exhaust housing and slowly insert it STRAIGHT UP and into position, while aligning the driveshaft, shift shaft and water tube seal.

■ **If necessary, slowly turn the propeller in the normal direction of rotation to align the driveshaft-to-crankshaft splines.**

22. While still holding the gearcase in position, place the flat washers

onto the forward studs (located on either side of the driveshaft housing), then start a locknut on each of these forward studs, threading them down by hand until finger-tight.

23. Next, thread the bolt at the rear of the housing, inside the trim tab recess, but DO NOT TIGHTEN IT FULLY AT THIS TIME.

24. Make sure the shift shaft spline engagement seems ok at this point by attempting to gently move the shift block just slightly.

✴✴ SELOC CAUTION

If the gearcase does not easily flush up against the driveshaft housing DO NOT force the gearcase to seat using the attaching nuts. Instead remove the gearcase again and double-check the shaft spline alignments. Also, make sure there is no lubricant on top of the shift shaft or driveshaft (or in the underside of the shift shaft and crankshaft located in the housing).

25. Alternately and evenly tighten the 2 forward locknuts which were installed earlier (on either side of the lower motor mount/swivel attachment point) to 50 ft. lbs. (68 Nm).

26. Verify proper shift shaft spline engagement as follows:

a. Place the guide block anchor pin into the FORWARD position (if not already) while turning the prop shaft in the normal direction of rotation. The propeller shaft should not turn once in gear.

b. Place the guide block in the NEUTRAL position and verify that the propeller shaft now turns freely.

c. Place the guide block in the REVERSE gear position, now slowly rotate the FLYWHEEL CLOCKWISE when viewed from the top and verify that the propeller shaft now turns in the normal direction of reverse gear for the gearcase in question (meaning counterclockwise on normal rotation gearcases or clockwise on counter rotation gearcases).

■ **If the gearcase does not shift properly into gear, the housing must be removed again and the shaft splines realigned as necessary for proper shifting.**

27. Install the washers and center locknuts onto the studs located at the center bottom of the anti-cavitation plate, then tighten them to 50 ft. lbs. (68 Nm).

28. Install the special flat washer and the top nut to the stud at the leading edge of the driveshaft housing (front top of the gearcase-to-driveshaft coupling, right in front of the lower motor mount/swivel point). Tighten the nut to 50 ft. lbs. (68 Nm).

29. Securely tighten the gearcase retaining bolt inside the trim tab cavity, then align the matchmarks made during removal and install the trim tab itself. Install the plastic cap into the trim tab bolt opening at the rear edge of the driveshaft housing.

30. If applicable, reconnect the speedometer tube into the junction.

31. If the propeller was removed, install the propeller.

32. Reconnect the battery cables.

DISASSEMBLY

■ **For gearcase exploded views, please refer to Cleaning & Inspection in this section.**

Unlike most of the smaller Mercury gearcases, the lower unit found on these models contains various shims for adjusting pinion height and forward/reverse gear backlash. Always keep track of the shims during removal, since if the gearcase or the shafts/bearings are being reused the shims should be returned to the same positions. Also, if the gearcase and/or shafts/bearings are replaced, the shims can be sometimes be used as a good starting point for the shimming procedures.

■ **On standard rotation gearcases, although the Reverse gear backlash is not adjustable, it CAN be used to help determine if parts are worn to the point of requiring replacement. Many techs will follow that portion of the Shimming procedure PRIOR to disassembly in order to help determine gearcase condition. If this is desired, skip ahead to Shimming, specifically the Reverse Gear backlash portion of the procedures, starting with the installation of a length of PCV pipe over the propeller shaft and use a dial indicator (as set-up earlier in the Shimming procedures) to check backlash.**

Bearing Carrier Removal & Disassembly

◆ See Figures 185 and 186

■ Because the shift shaft actually splines with a cam that sits in the propeller shaft actuator rod, you CANNOT remove the propeller shaft until AFTER you've removed the shift shaft.

✳✳ SELOC WARNING

If you don't have access to the special Mercury mandrels designed to install bearings to the proper depth inside the bearing carrier, measure the installed height of all bearings and/or races before removal. Use the measured depths at installation time to ensure the bearings are positioned properly (especially in positions where there are no shoulders to drive the bearings or races against).

1. Remove the gearcase from the mid-section, as detailed earlier in this section. Mount the case in a holding fixture or a soft-jawed vise, positioning the propeller shaft horizontally.

2. If necessary for access or for further disassembly, remove the Water Pump assembly, as detailed in the Lubrication and Cooling System section.

3. Using a long spanner wrench, like a Shift Shaft Bushing Tool (#91-31107), unthread the shift shaft bushing (BUT DO NOT REMOVE IT FROM THE SHAFT yet).

4. Make sure the gearcase is in NEUTRAL.

5. Now, moving down to the propeller shaft bearing carrier and locate any locktab(s) that is(are) bent over to secure the large cover nut. Using a punch, carefully unbend the locktab(s) so the cover nut can be spun free.

6. Using a large spanner like a Cover Nut Tool (#91-61069) carefully loosen and unthread the cover nut. Once the cover nut is removed, pull the locktab washer from the housing.

✳✳ SELOC CAUTION

With the cover nut removed you need to exorcise EXTREME seloc caution around the propeller shaft. DO NOT APPLY ANY sideways force on the shaft or you might snap the neck of the clutch actuator rod!!!

7. Install an internal jawed puller such as Long Puller Jaws (#91-46086A1) and a suitable drive bolt. Position the propeller thrust hub to maintain an outward pressure on the puller jaws. Tighten the puller to preload the propeller shaft.

8. Now, tighten the puller bolt to break the seal and free the bearing carrier from the gearcase, the carrier alignment key should come free with it. DO NOT attempt to remove the propeller shaft yet.

■ If the bearing carrier refuses to move, it may be necessary to carefully apply heat to the lower unit in the area of the carrier and at the same time attempt to move it off the carrier.

9. With the gearcase in Neutral, lift the shift shaft free of the splines of the cam sitting in the propeller shaft actuator rod. If necessary, protect the top of the shaft with a thin piece of aluminum and use a pair of pliers to get a good grip. BUT BE CAREFUL not to score or damage the shaft and to make sure that you DO NOT rotate the shaft as it is pulled upward. Remove the shaft and set it aside for possible disassembly later.

10. At this point the propeller shaft (along with the cam follower and shift cam) should simply pull free of the gearcase, HOWEVER IF it is hung up, DO NOT lever it or try to pull it with a slide hammer or lever, you'll just break the end of the propshaft that way. Instead, if it hangs up, try to free it with one of these methods:

• It's possible that the shift cam is not in the Neutral position. Try pushing the propeller shaft back into the gearcase against the nose cone gear, then shine a flashlight down through the shift shaft bore in the top of the gearcase. If the splined hole in the shift cam is visible, reinstall the shift shaft and rotate it to neutral, then remove the shift shaft, followed by the propeller shaft.

• Another solution is to push the propeller shaft back into the nose gear, then slide the bearing carrier back into the gear housing in order to support the shaft. Remove the gear housing from the support fixture and place it on its left side (as viewed from the propeller end). Gently strike the upper leading end of the gearcase using a rubber mallet and this will hopefully dislodge the cam follower into a clearance pocket on the left side of the gearcase housing. Remove the bearing carrier, then pull the propeller shaft out of the gear housing and recover the shift cam (though this last part might have to wait until the nose cone gear is removed for clearance).

11. Remove and discard the old O-ring from the bearing carrier.

12. On standard rotation gearcases, the propshaft gear (Reverse on these models) is pressed into the housing, if it or the bearing must be replaced, position the carrier in a soft-jawed vise and use a jawed slide hammer to remove the gear. If the bearing stays behind, use the same assembly to remove it from the carrier. If the bearing stays on the gear, use a bearing separator to push it off the gear.

13. On counter rotation gearcases, the propshaft gear (Forward on these models) is installed ON the shaft in a bearing adapter which will be serviced later under Propeller shaft disassembly. For now, if the thrust bearing, thrust washer or thrust bearing surface on the propeller shaft shows any sign of rust, pitting or bluing from lack of lubricant they should be removed from the rear of the carrier for examination or replacement.

■ The dual back-to-back oil seals in the bearing carrier can be pried out separately (if the needle bearing is not being replaced) OR can be driven out from behind while removing the needle bearing.

14. If the needle bearing inside the carrier is NOT going to be replaced, invert the bearing carrier so the propeller end is facing upward and use a small prytool to carefully remove the dual back-to-back oil seals.

15. If the needle bearing inside the carrier IS going to be replaced, position the carrier in a suitable shop press with the front side (gear side) facing upward. Use a suitable driver and mandrel to press the needle bearing and seals out the propeller end of the carrier.

Fig. 185 The bearing carrier is secured by a large cover nut and tabbed washer

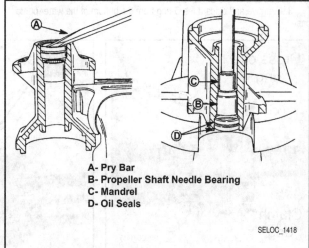

A- Pry Bar
B- Propeller Shaft Needle Bearing
C- Mandrel
D- Oil Seals

SELOC_1418

Fig. 186 Removing the oil seals only (left) or needle bearing AND oil seals (right)

Shift Shaft Disassembly

There is not a whole lot to shift shaft service on these motors. Start by removing the shift shaft from the gearcase as detailed under Bearing Carrier Removal & Disassembly. Then it's just a matter of cleaning and inspecting the shaft/bushing, and perhaps replacing the oil seal and O-ring

1. Carefully slide the busing from the shift shaft.

2. Remove and discard both the O-ring and the oil seal (noting the direction the lips are facing before removal). The oil seal can either be pried out or removed using a driver and hammer.

3. Thoroughly clean the bushing and shaft with solvent and dry with compressed air. Make sure the splines are good on both ends and that there are no grooves, corrosion or excessive wear on the bushing sealing surface.

Propeller Shaft Disassembly

◆ See Figures 187, 188 and 189

To disassemble the propeller shaft for cleaning, inspection and/or parts replacement, proceed as follows:

1. Before disassembly use a pair of V-blocks and a dial gauge to check the shaft run-out. Dial indicator movement must be 0.006 in. (0.152mm) or less, otherwise the shaft must be replaced.

2. Remove the shift cam from the follower (if it was removed with the propeller shaft, if not, remember to recover it from the gearcase later).

3. Insert a thin blade screwdriver or an awl under the first coil of the cross-pin retainer spring and rotate the propeller shaft to unwind the spring from the sliding clutch. Take care not to over-stretch the spring.

4. Remove the tiny detent pin from the clutch, then push the cross-pin out of the sliding clutch with a punch. At this point the sliding clutch and clutch actuator/cam follower should be free to slide from the propeller shaft. Note, once the clutch actuator/cam follower assembly is removed from the propeller shaft, the rod can separate from the actuator.

5. Four counter rotation units, remove the forward gear and bearing adaptor assembly, followed by the 2 thrust races and 2 keepers all from the prop shaft. If necessary, disassemble the gear and bearing adaptor as follows:

   a. Remove the forward gear from the bearing adaptor.

   b. Remove the thrust bearing and spacer shim from the adaptor.

   c. Remove the thrust washer and O-ring from the adaptor.

   d. Remove the thrust bearing and thrust washer from the bearing adaptor.

   e. Check the bearing in the adaptor. If replacement is necessary use a hammer and chisel to break the bearing loose from the adaptor (but be careful not to damage the adaptor).

Driveshaft and Bearings

◆ See Figures 190 and 191

1. If not done already, remove the Water Pump assembly as detailed in the Lubrication and Cooling System section.

2. If not done already, remove the water pump base by using 2 small prytools (one at either end) and some padding to protect the gearcase. Lift the base up and off the driveshaft.

3. Remove and discard the O-ring from the bottom of the water pump base.

4. For installation purposes, note the directions in which the oil seals are facing (they are usually installed back-to-back) then carefully pry them from the water pump base.

5. Obtain a long/large spanner such as the Bearing Retainer tool (#91-43506) and use it to loosen and remove the driveshaft bearing retainer in the top of the gearcase.

■ **You can use a socket and flex-head driver or a large wrench to hold the pinion nut. In either case, it's a good idea to pad the gearcase housing where the tool might contact it in order to prevent potential damage to the housing.**

6. Select a box end wrench the same size as the pinion nut. This tool will be used to prevent the nut from turning as the driveshaft is rotated. Obtain the proper driveshaft holding tool (#91-34377A1) and install it onto the end of the driveshaft. Now, with the box end wrench on the pinion gear nut, use an appropriate wrench and rotate the tool and driveshaft counterclockwise until the pinion gear nut is free. Remove the pinion nut and the holding tool.

7. Now position and SECURE the gearcase so the driveshaft is horizontal and the propshaft bore is vertical with the nose cone pointing downward. Position a large block of wood over the top (water pump mounting surface) of the gearcase and hold the driveshaft firmly while tapping the wood with a mallet to free the driveshaft.

✳✳ SELOC CAUTION

DO NOT let the gearcase fall as severe damage may occur.

8. Once the driveshaft is free, slide it out of the gearcase, but be sure to retain any shims which may have been placed under the driveshaft's upper tapered bearing.

9. Reach into the propeller shaft bore and retrieve the pinion gear, as well as the nose cone gear assembly.

10. If the upper tapered bearing requires replacement, use a bearing separator (or even the jaws of a vise) to hold the bearing while you push or gently tap the shaft free.

11. Remove the 18 loose needles from the outer race of the driveshaft lower needle bearing. IF inspection of the driveshaft needle bearing surface shows that bearing replacement is required, the 18 loose needle bearings must be reinstalled in the bearing race to provide a surface for the mandrel to drive against.

■ **The Nose cone gear MUST be removed prior to needle bearing removal.**

12. If replacement is necessary, use a suitable mandrel (#91-37263), pilot (#91-36571) and driver rod (#91-37323) or the equivalents to drive the lower needle bearing assembly down into the gearcase propshaft bore.

Nose Cone Gear, Bearing, and Race

◆ See Figures 192, 193 and 194

The nose cone (Forward on standard rotation or Reverse for counter rotation units) gear and bearing assembly is inserted in a race in the nose cone and basically held in position by the pinion gear.

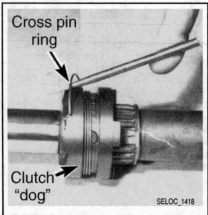

Fig. 187 To remove the sliding clutch, first remove the pin retaining spring. . .

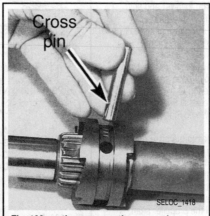

Fig. 188 . . .then remove the cross pin, followed by the actuator (not shown)

Fig. 189 . . .and slide the clutch dog from the shaft

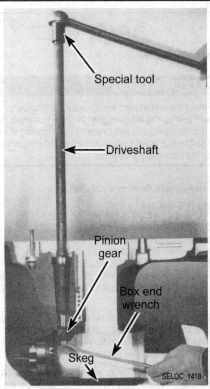

Fig. 190 Use a special driveshaft tool to turn the shaft, loosening the pinion nut

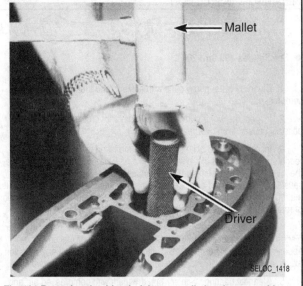

Fig. 191 Removing the driveshaft lower needle bearing assembly

■ Remember that the bearing and race are a set, so if the bearing is replaced you'll also need to replace the nose cone race. A wear pattern will have been worn in the race by the old bearing. Therefore, if the race is not replaced, the new bearing will be quickly worn by the old worn race.

4. Remove the forward gear tapered roller bearing race using a slide hammer.

5. SAVE any shim material from behind the race after the race is removed. The same amount of shim material will probably be used during assembly.

6. For counter rotation units:

a. Remove the thrust bearing and thrust washer from the gear bearing cup.

b. Using an internal jawed puller assembly (preferably with a puller plate over the end of the gearcase and not a slide hammer) free the bearing adaptor from the gearcase. Retain shims for reuse, HOWEVER inspect them first, as they may be damaged during removal. IF they were damaged, measure them and obtain replacements of the same size for use during installation.

■ A mandrel can also be used to press the needle bearing out from the nose cone gear. On counter rotation units, Mercury notes to use #91-63569.

7. For all units, if the needle bearing inside the gear is no longer fit for further service, it can be removed by tapping, with a blunt punch and hammer, around the bearing cage from the aft end - the gear end - of the bearing. This action will destroy the bearing cage. Therefore, be sure a replacement bearing is on hand, before attempting to remove the defective bearing.

On standard rotation units, a tapered bearing which rolls in the nose cone race is pressed onto the outside of the gear and a needle bearing is pressed inside the gear. The needle bearing supports the forward end of the propeller shaft.

On counter rotation units, the nose cone gear is installed in a bearing adaptor assembly (not unlike the one used for the propshaft mounted gear and bearing on these models) and includes a roller bearing which is pressed into the adaptor. The gear on these models also has a needle bearing pressed into the center.

As usual, removal of the bearings will normally render them unfit for service, so make sure they are shot and replacements are available before bearings are separated from the gear.

After the pinion gear and driveshaft have been removed, the nose cone gear can be lifted out of the housing. To disassemble them and their components, proceed as follows depending upon the model.

1. If not done already, lift the gear out of the nose cone.
2. For standard rotation units:
3. If the forward gear tapered roller bearing must be replaced, use a universal bearing separator and a press to push the bearing off the gear.

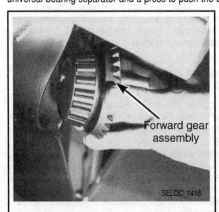

Fig. 192 Remove the nose cone gear (and bearing on standard rotation units)

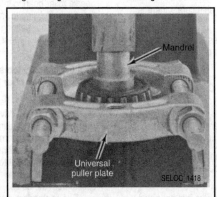

Fig. 193 On standard rotation units, remove the tapered roller bearing using a bearing separator

Fig. 194 Removing the gear race using a slide-hammer (standard, shown) or the adaptor using a puller (counter rotation)

CLEANING & INSPECTION

◆ **See Figures 195 thru 206**

1. Clean all water pump parts with solvent, and then dry them with compressed air.

2. Inspect the water pump cover and base for cracks and distortion, possibly caused from overheating.

3. Inspect the face plate and water pump insert for grooves and/or rough surfaces. If possible, always install a new water pump impeller while the lower unit is disassembled. A new impeller will ensure extended satisfactory service and give "peace of mind" to the owner. If the old impeller must be returned to service, never install it in reverse to the original direction of rotation. Installation in reverse will cause premature impeller failure.

4. Inspect the impeller side seal surfaces and the ends of the impeller blades for cracks, tears, and wear. Check for a glazed or melted appearance, caused from operating without sufficient water. If any question exists, and as previously stated, install a new impeller if at all possible.

5. Clean all bearings with solvent, dry them with compressed air, and inspect them carefully. Be sure there is no water in the air line. Direct the air stream through the bearing. Never spin a bearing with compressed air. Such action is highly dangerous and may cause the bearing to score from lack of lubrication. After the bearings are clean and dry, lubricate them with Quicksilver Formula 50-D lubricant or equivalent. Do not lubricate tapered bearing cups until after they have been inspected.

6. Inspect all ball bearings for roughness, catches, and bearing race side wear. Hold the outer race, and work the inner bearing race in-and-out, to check for side wear.

7. Determine the condition of tapered bearing rollers and inner bearing race, by inspecting the bearing cup for pitting, scoring, grooves, uneven wear, imbedded particles, and discoloration caused from overheating. Always replace tapered roller bearings as a set.

8. Inspect the bearing surface of the shaft roller bearing support. Check the shaft surface for pitting, scoring, grooving, imbedded particles, uneven wear and discoloration caused from overheating. The shaft and bearing must be replaced as a set if either is unfit for continued service.

■ **Remember to check shaft surfaces at oil seal contact areas, as grooves or wear in that area will allow leakage.**

Inspect the sliding clutch of the propeller shaft. Check the reverse gear side clutch "dogs". If the "dogs" are rounded one of three causes may be to blame

   a. Improper shift cable adjustment.

   b. Running engine at too high an rpm while shifting.

   c. Shifting from neutral to reverse gear too quickly.

9. Inspect the cam follower and replace it and the shift cam if there is any evidence of pitting, scoring, or rough surfaces.

10. Inspect the propeller shaft roller bearing surfaces for pitting, rust marks, uneven wear, imbedded metal particles or signs of overheating caused by lack of adequate lubrication.

■ **Good shop practice requires installation of new O-rings and oil seals regardless of their appearance.**

11. Clean the bearing carrier, pinion gear, drive gear clutch and the propeller shaft with solvent. Dry the cleaned parts with compressed air.

12. Check the pinion gear and the drive gear for abnormal wear. Apply a coating of light-weight oil to the roller bearing. Rotate the bearing and check for cracks or catches.

13. Inspect the propeller shaft oil seal surface to be sure it is not pitted, grooved, or scratched. Inspect the roller bearing contact surface on the propeller shaft for pitting, grooves, scoring, uneven wear, imbedded metal particles, and discoloration caused from overheating.

ASSEMBLY

◆ **See Figures 195 thru 198**

✳✳ SELOC WARNING

Before beginning the installation work, count the number of teeth on the pinion gear and on the forward gear in order to determine gear ratio for backlash checking/adjustment procedures. Knowing the number of teeth on the pinion and nose cone is necessary to obtain the correct tool setup for the shimming procedure.

Driveshaft Lower Needle Bearing Installation

◆ **See Figures 195 and 197**

If removed from the gearcase, the first step in assembly should be to pull the new needle bearing assembly into the housing. This must be done BEFORE the installation of the nose cone gear.

Although special tools are noted for use to accomplish this, if the are not available, you can usually get away with using a long bolt, a series of appropriately sized washers, some locknut and possibly a large plate through which to thread the bolt.

■ **IF a needle bearing failure has occurred AND, the original bearing race has turned in the gear housing, the housing itself must be replaced. Loose fitting needle bearing will move out of position and cause repeated failures to other components.**

1. Apply a thin coat of 2-4-C with Teflon or an equivalent marine grade grease to the driveshaft needle bearing bore in the gear housing.

2. Reach in through the propeller shaft bore and place the needle bearing in the driveshaft bore, with the numbers facing UPWARD toward the top of the gearcase.

3. Assemble an installation tool to draw the bearing up into the gearcase. If using the Mercury recommended tools, use a Puller Rod (#91-31229), Nut (#11-24156), Pilot (#91-36571), Top Plate (#91-29310) and Mandrel (#91-92788). Using the installation tool, draw the bearing up until it bottoms into the gearcase shoulder (using care not to exert excessive force one the bearing at this point).

Bearing Carrier Assembly

◆ **See Figures 196 and 198**

Because the propshaft gear and bearing components vary somewhat between standard rotation and counter rotation units, some of the procedures in this section will differ depending upon the type of unit.

Be sure to lubricate the inner and outer diameters of the bearings, as well as the inner diameter of the oil seals using 2-4-C with Teflon or an equivalent lubricant prior to installation. The outer diameter of the oil seals should be coated using Loctite® 271 or an equivalent threadlocking compound. After installing the seals, wipe away any excess Loctite before proceeding.

1. If removed from the bearing carrier, position the propeller shaft needle roller bearing into the aft (propeller) end of the bearing carrier with the numbered side toward the aft end (toward the driver). Press the needle bearing into the bearing carrier with a suitable mandrel (#91-15755) which will give you the correct depth when the upper edge of the tool seats against the carrier.

■ **If the Mercury tools are not available, position the bearing to the depth measured prior to removal.**

2. Obtain an Oil Seal Driver (#91-31108). Place one seal on the longer shoulder side of the driver tool with the lip of the seal away from the shoulder

3. Press the seal into the bearing carrier until the seal driver bottoms against the bearing carrier

4. Place the second seal on the short shoulder side of the seal driver with the lip of the seal toward the shoulder. Press the seal into the bearing carrier until the seal driver bottoms against the bearing carrier. Clean excess Loctite from the seals.

5. For standard rotation units, proceed as follows to install the reverse gear and bearing:

   a. Position the thrust washer over the reverse gear (the washer is tapered at the outer diameter, so make sure the LARGER diameter is facing

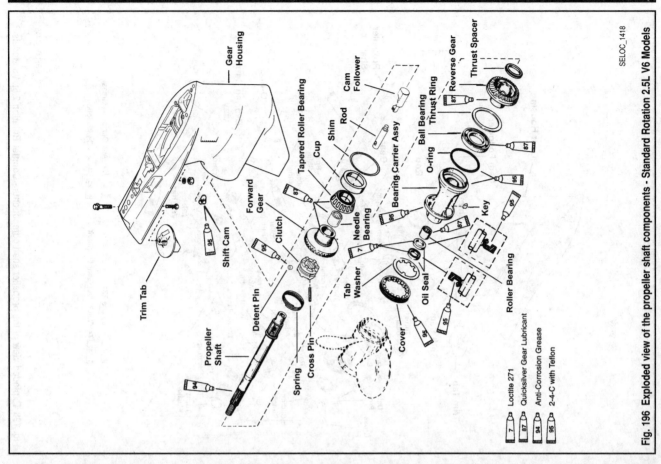

SELOC_1418

Gear Housing

Tapered Roller Bearing

Cup

Shim

Rod

Cam Follower

Bearing Carrier Assy

Forward Gear

Clutch

Needle Bearing

Ball Bearing

Thrust Ring

O-ring

Reverse Gear

Thrust Spacer

87

87

95

95

95

Key

87

Shift Cam

95

Trim Tab

95

Detent Pin

Tab Washer

7

Roller Bearing

Propeller Shaft

Spring

Cross Pin

Oil Seal

Cover

95

95

94

Loctite 271

Quicksilver Gear Lubricant

Anti-Corrosion Grease

2-4-C with Teflon

7   87   94   95

Fig. 196 Exploded view of the propeller shaft components - Standard Rotation 2.5L V6 Models

SELOC_1418

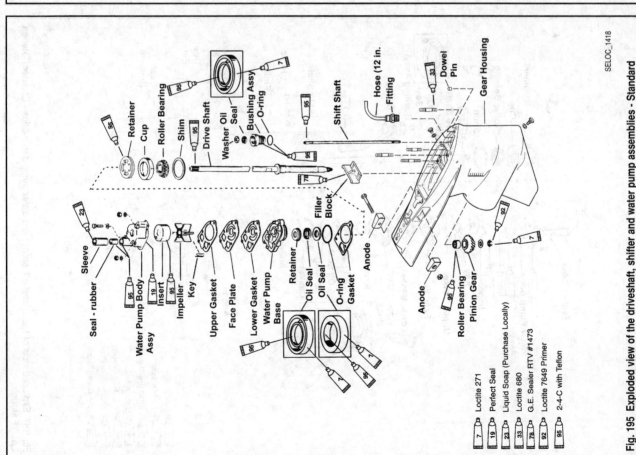

Retainer

Cup

Roller Bearing

Shim

Drive Shaft

Washer Oil Seal

Bushing Assy

O-ring

Shift Shaft

Hose (12 in.)

Fitting

Dowel Pin

Gear Housing

95

95

95

95

33

78

92

7

Sleeve

Seal - rubber

Water Pump Body Assy

Insert

Impeller

Key

Upper Gasket

Face Plate

Lower Gasket

Water Pump Base

Retainer

Oil Seal

Oil Seal

O-ring

Gasket

Filler Block

Anode

Anode

Roller Bearing

Pinion Gear

23

95

19

95

95

95

7

95

Loctite 271

Perfect Seal

Liquid Soap (Purchase Locally)

Loctite 680

G.E. Sealer RTV #1473

Loctite 7649 Primer

2-4-C with Teflon

7   19   23   33   78   92   95

Fig. 195 Exploded view of the driveshaft, shifter and water pump assemblies - Standard Rotation 2.5L V6 Models

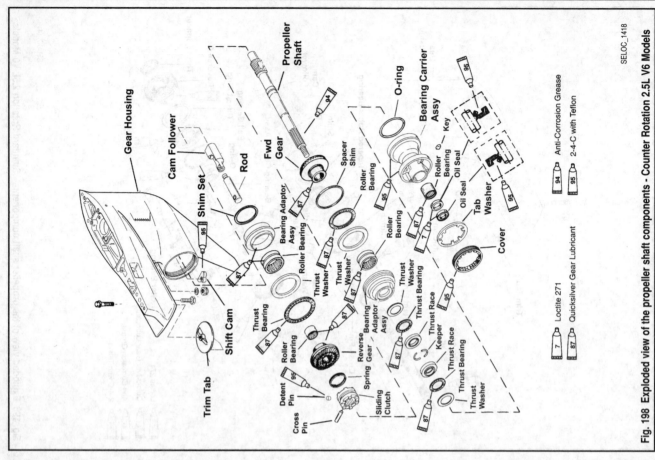

Fig. 198 Exploded view of the propeller shaft components - Counter Rotation 2.5L V6 Models

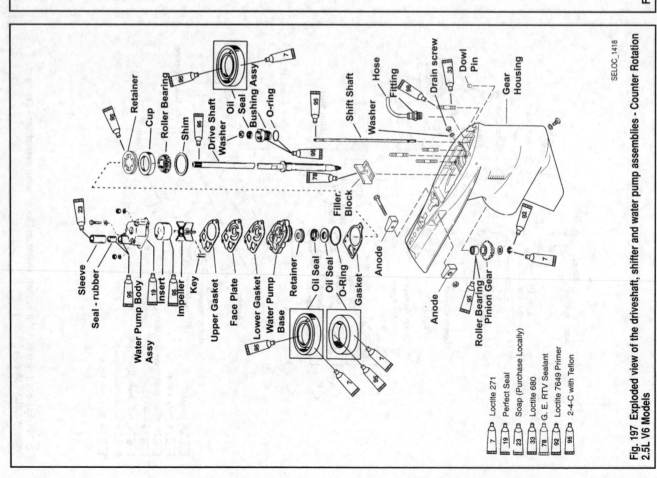

Fig. 197 Exploded view of the driveshaft, shifter and water pump assemblies - Counter Rotation 2.5L V6 Models

the gear). Then, use a suitable mandrel to press the roller bearing onto the back of the reverse gear until it bottoms. As usual, make sure the mandrel only presses on the portion of the bearing in contact with the gear (inner). Then use a mandrel to press the assembled gear and bearing into the forward (gearcase) end of the bearing housing until it bottoms out.

    b. Install a new O-ring over the bearing carrier and position it between the carrier and thrust washer. Apply a coating of 2-4-C with Teflon or and equivalent Lubricant onto the installed O-ring.

    6. For counter rotation units, use a suitable mandrel to press the forward gear bearing INTO the bearing adaptor until the bearing is just flush with the lip of the adaptor.

Nose Cone Gear, Bearing and Race

◆ See Figures 196, 198 and 207 thru 208

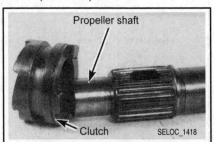

Fig. 199 The clutch "dog" and matching splines on the propeller shaft should be closely inspected

Fig. 200 Grooves on the cross pin or marked wear patterns on the outer roller bearing races are evidence of premature failure of these or associated parts

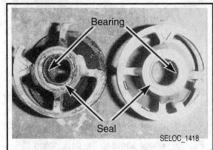

Fig. 201 Comparison of a worn bearing carrier (left) with a new one (right)

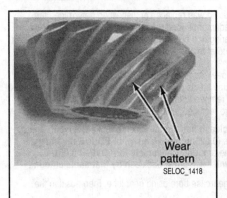

Fig. 202 Unacceptable pinion gear wear pattern, probably caused by inadequate lubrication in the lower unit

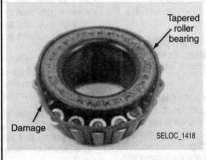

Fig. 203 Distorted tapered roller bearing unfit for further service

Fig. 204 Caged ball bearing set destroyed due to lack of lubrication, vibration, corrosion, metal particles, or all of the above

Fig. 205 A rusted and corroded gear. Water was allowed to enter the lower unit through a bad seal and cause this damage to the gear and other expensive parts

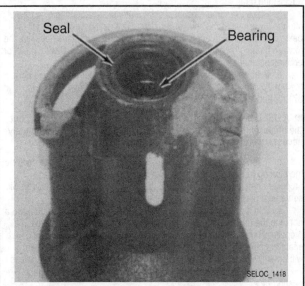

Fig. 206 This damaged bearing carrier was "frozen" in the lower unit and destroyed during the removal process

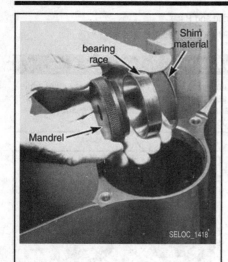

Fig. 207 Image showing the positioning of the driver (mandrel), race and shim (standard rotation)

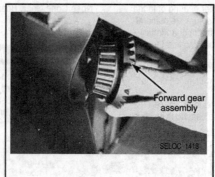

Fig. 208 Once the race is driven in position, insert the gear and bearing assembly

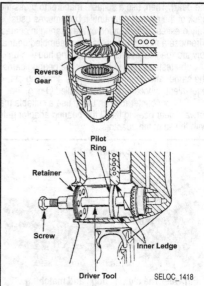

Fig. 209 Seating the reverse gear adaptor assembly (counter rotation units)

Because the nose cone gear components vary somewhat between standard rotation and counter rotation units, some of the procedures in this section will differ depending upon the type of unit.

■ On counter rotation units don't confuse the Forward (propeller shaft) and Reverse (nose cone) gears. The Reverse gear has both A SHORTER HUB and a SMALLER INNER DIAMETER needle bearing bore.

1. If the needle bearing was removed from the nose cone gear, install the replacement at this time. Position the gear with the teeth facing downward on a press, then use a suitable mandrel (#91-86943) to press against the numbered side of the bearing until the tool bottoms against the gear. For standard rotation units there are 2 needle bearings in this gear, press the first one in with the LONG shoulder of the tool and the second with the SHORT shoulder. But for counter rotation units there should be only one needle bearing, and for this reason, press with the SHORT shoulder of the tool only.

2. For standard rotation units, install the shims, race, gear and bearing assembly as follows:
   a. First, install the tapered roller bearing to the gear itself. Use a suitable mandrel to push on the bearing inner race until it is seated firmly against the gear.
   b. Place the same amount of shim material saved during disassembly into the lower unit. If the shim material was lost, or if a new lower unit is being used, begin with approximately 0.010 in. (0.25mm) material. Coat the forward bearing race bore with gear lube or Quicksilver 2-4-C with Teflon or equivalent lubricant.
   c. Position the tapered bearing race squarely over the bearing bore in the front portion of the lower unit. Obtain a Bearing Driver Cup tool (#91-87120) and place the tool over the tapered bearing race.

■ A USED propeller shaft is REALLY preferred for the next step, but the disassembled shaft from this gearcase CAN be used if absolutely necessary.

   d. Insert the disassembled propeller shaft into the hole in the center of the bearing race. Lower the bearing carrier assembly down over the propeller shaft, and then lower it into the lower unit. The bearing carrier will serve as a pilot to ensure proper bearing race alignment.
   e. Use a soft-faced mallet (to protect the propeller shaft) and thread a nut onto the end of the shaft until flush with the shaft itself, then drive the propeller shaft against the bearing driver cup until the tapered bearing race is seated against the shim material. Remove the nut from the shaft, then withdraw the propeller shaft and bearing carrier, then lift out the driver cup.
   f. Oil and install the forward gear assembly into the forward gear bearing race.

3. For counter rotation units, install the reverse gear and bearing cup adaptor assembly, along with the necessary shims:

   a. If the gear housing was replaced or inspection determined the reverse gear bearing adaptor needed to be replaced, position the new reverse gear roller bearing in a press with the numbered/lettered side facing up. Use a suitable mandrel to press the bearing cup adaptor down over the bearing.
   b. If the reverse gear, adaptor, large thrust bearing and race were not replaced, install the same shims (or suitable replacements of the same thickness) which were removed during disassembly. HOWEVER, if any of these components were replaced, start with 0.008 in. (0.51mm) of shims.

■ If reverse gear backlash has already been checked, keep in mind that ADDING 0.001 in. (0.025mm) of shims will REDUCE reverse gear backlash by that same amount. Conversely REMOVING 0.001 in. (0.025mm) of shims will INCREASE backlash by that same amount.

   c. Lubricate the gearcase bore using gear lube, then position the shim(s) into the bore.
   d. Next, install the bearing adaptor over top of the shims in the gearcase bore.
   e. Now, position the gear WITHOUT the thrust race or bearing into the adaptor.
   f. Install a Pilot Ring (#91-18603) over a Driver Tool (#91-18605) and seat the pilot ring against the inner ledge of the gearcase (not quite as deep at the driveshaft bore). Install a Screw (#10-18602) into the retainer and gently tighten the screw against the driver tool while holding the retainer securely. Continue to apply pressure against the driver rod until the reverse gear/gearing cup adaptor JUST SEATS into the gearcase. DO NOT force the assembly down excessively or the reverse gear bearing will likely be damaged. You will know when the adaptor begins to seat because the force necessary to turn the screw will increase dramatically.
   g. Once the adaptor is seated, remove the screw, retainer, driver tool, pilot ring and reverse gear. Apply a light coating of gear lube to the thrust bearing, then install the thrust race and bearing onto the bearing adaptor.
   h. Install the reverse gear into the adaptor.

Driveshaft and Pinion Gear

◆ See Figures 195 and 197

1. If the tapered bearing was removed from the driveshaft, apply a light coat of gear oil to the bearing inner diameter. Install a used pinion nut onto the end of the driveshaft, leaving about 1/16 in. (1.6mm) of nut threads exposed (i.e. making sure the driveshaft threads are NOT extending past the nut to protect them from damage). Place the bearing over the driveshaft, then position the driveshaft/bearing over a bearing separator and suitable mandrel (an old shaft inner race will do the job) so that the force of pressing on the pinion nut will go right to the inner race of the new bearing. In this way, carefully press the bearing onto the shaft, then remove the nut.

2. Position the pinion gear in the gear housing, right below the driveshaft bore, with the teeth meshed with the teeth of the nose gear.

3. Insert the driveshaft carefully down into the gearcase, rotating it slightly as necessary to engage the shaft splines with the splines in the pinion gear. Continue to insert the driveshaft until the tapered bearing is against the bearing race.

■ **Mercury sells new pinion nuts (sometimes with a drylock patch already on the threads). If you've got a new nut for installation don't install it until after backlash has been checked and possibly adjusted. Instead, use the old pinion nut for this step, and save the new nut for final assembly. If you don't have a new nut, no biggie, Mercury just advises the use of Loctite® 7649 Primer to clean the threads and then the use Loctite®271, or an equivalent threadlocking compound during final assembly.**

4. Install the old pinion gear nut onto the driveshaft and hand tighten for now.

5. Position the old driveshaft shims saved from disassembly, OR, if the shims were lost or the gearcase replaced, start with about 0.010 in. (0.254mm) of shims.

6. Install the upper bearing race and retainer (with the word **OFF** visible), and use a Bearing Retainer Tool (#91-43506) to tighten the bearing retainer to 100 ft. lbs. (135.5 Nm).

7. Next, use a socket and flex handle or a large wrench to hold the pinion nut from turning, then use the Driveshaft Holding Tool (#91-34377A1 or #91-90094) over the top splines of the driveshaft to turn the shaft and tighten the nut to 75 ft. lbs. (101.5 Nm).

✳✳ SELOC WARNING

After the pinion gear depth and the forward gear backlash have been set, the old nut must either be replaced with a new pinion gear nut OR the nut must be removed and the threads clean with Loctite® 7649 Primer and finally coated with Loctite® 271 or an equivalent threadlocking compound.

8. Now, if you've replaced the gearcase housing, or if you've replaced one or more of the bearings SKIP ahead to the Shimming procedures in order to check backlash and determine if some of the shims must be replaced. However, if only seals have been replaced, then you can proceed with the assembly (assuming you used a new nut or coated the old pinion nut with Loctite during assembly).

Shift Shaft Assembly

◆ **See Figures 195 thru 197**

1. Place the shift shaft bushing on a press with the threaded side down.

2. Apply a light coating of Loctite® 271 or equivalent to the outside diameter of a new oil seal, then press the seal into the shift shaft bushing with the lip of the seal facing upward. Wipe away any excess Loctite from the seal and bushing.

3. Install the rubber washer against the oil seal, then install the O-ring over the threads, up against the bushing shoulder.

4. Apply a light coating of 2-4-C with Teflon to the O-ring, oil seal and to the threads of the shift shaft bushing.

5. Place the bushing on the shaft for installation later, once the propshaft is in position.

Propeller Shaft Assembly

◆ **See Figures 196, 198 and 210 thru 212**

1. Slide the clutch dog onto the propeller shaft splines, aligning the cross pin hole and the detent hole with the appropriate slot and detent in the shaft. On standard rotation gearcases, the grooved rings on the sliding clutch must be positioned facing toward the propeller end of the shaft, however on counter rotation gearcases the grooved rings may face either end.

2. If separated, place a small dab of 2-4-C with Teflon or an equivalent marine grade grease on the connecting point between the actuator rod and cam follower, then connect the rod to the follower.

3. Insert the assembled actuator rod and cam follower into the propeller shaft, aligning the hole in the rod with the hole in the clutch dog. If necessary, use a small tapered punch to help align then, then install the cross pin (using it to push the punch out if used).

4. Place a small dab of 2-4-C with Teflon or an equivalent marine grade grease on the rounded end of the detent pin, then install the pin into the hole of the sliding clutch with the rounded end facing toward the propeller shaft.

5. Install the cross-pin ring over the sliding clutch. Take care not to over stretch the spring. Spirally wrap the spring into the groove on the sliding clutch, making sure the straight end of the spring is against the side of the groove.

6. Place a small dab of 2-4-C with Teflon or an equivalent marine grade grease on the shift cam, then place the cam into the pocket on the end of the cam follower with the numbered or lettered side of the cam facing upward (it may even be stamped with the word **UP**. The flat side of the cam must be parallel to the shaft and outer edge of the pocket.

Bearing Carrier, Propeller Shaft and Shift Shaft Installation

◆ **See Figures 195 thru 198**

Basically, because of the shift cam which resides in the follower pocket, the propeller shaft must be positioned in the gearcase so that the shift shaft can be installed, then and remaining components such as the propeller shaft gear and/or bearing carrier can then be installed into the gearcase and overtop of the shaft.

The only real difference in the procedures comes from the slight difference in the propeller shaft assembly between the standard rotation units (where the prop shaft gear, or REVERSE in this case, is mounted in the bearing carrier) and the counter rotation units (where the prop shaft gear, FORWARD in this case, and bearing adaptor are installed on the prop shaft itself). The assembly procedures take into account that you have to install the forward gear and bearing assembly piece-by-piece onto the already positioned propeller shaft for counter rotation units.

As always, be sure to lubricate bearings or thrust surfaces with gear oil and oil seals using 2-4-C with Teflon or an equivalent lubricant prior to installation.

1. Place the gearcase a suitable support fixture so the propeller shaft bore is horizontal.

2. If not done earlier, apply a light coating of 2-4-C with Teflon or equivalent marine grade grease to the shift shaft splines, bushing O-ring, and the threads of the cam bushing. It's also a good idea to

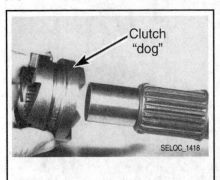

Fig. 210 Slide the clutch dog into position on the propeller shaft. . .

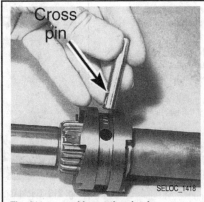

Fig. 211 . . . and insert the clutch actuator/follower assembly before securing the assembly with the cross pin

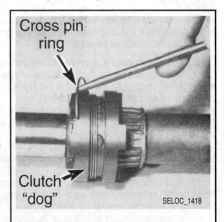

Fig. 212 Install the retaining spring to secure the detent and cross pins

3. Make sure the shift cam is sitting properly in the cam follower pocket, then grab the propeller shaft with one hand and the shift shaft assembly with the other (or better yet, have an assistant grab the shift shaft).

4. While peering down (flashlight helps) the shift shaft bore in the top of the case housing, insert the propeller shaft into the gearcase until you can see the splines on the shift cam. Now holding the propeller in this position, insert the shift shaft down through the bore and mesh the splines on the shaft with those in the hole of the shift cam. It may be necessary to rotate the cam back and forth VERY slightly.

5. Thread the shift shaft bushing down into the gearcase to hold everything in place, but DO NOT tighten fully at this time.

6. For counter rotation units, it's time to install the rest of the propeller shaft/gear/bearing components, as follows:

a. Turn the gearcase so that the propeller shaft bore opening is now facing upward, this will make it easier to install and seat the components.

b. Install the appropriate spacer shim into the gear housing.

c. Apply a light coating of gear lube to the thrust bearing, then position the bearing onto the Forward gear bearing adaptor. Next place the Forward gear over the thrust bearing and the bearing adaptor.

d. Insert the forward gear installation tool (#91-815850) or a suitable driver into the back of the bearing adaptor and gear assembly (with the gear facing downward away from the tool), then using the tool place the assembly over the propeller shaft and into the gear housing. Using the tool, apply downward pressure to slide the bearing adaptor into position then remove the tool from the assembly and from the gearcase housing.

e. Next, install the thrust race on top of the bearing adaptor.

f. Apply a light coating of gear lube to the small thrust bearing, then install it onto the thrust race.

g. Install the thrust collar with the STEPPED SIDE FACING DOWNWARD (toward the thrust bearing).

h. Pull up very slightly on the propeller shaft JUST enough to gain access to the groove in the shaft, then install the 2 keepers into the groove and allow the propeller shaft to seat again.

i. Install the second thrust collar with the STEPPED SIDE FACING UPWARD (away from the first thrust bearing).

j. Apply a light coating of gear lube to the second thrust bearing, then install it onto the second thrust collar.

k. Apply a light coating of gear lube to the second thrust bearing race, then install it into the surface inside of the BEARING CARRIER.

l. Apply a light coating of 2-4-C with Teflon or equivalent marine grade grease to the bearing carrier O-ring and install the O-ring to the bearing adaptor.

7. If not done already, make sure to apply a light coating of gear lube to the bearing carrier needle bearing and pack the space between the oil seals with a coating of 2-4-C with Teflon, or an equivalent marine grade grease. Use the same grease to apply a light coating to the outer diameter of the bearing carrier, where it contacts the gearcase.

8. Carefully install the bearing carrier into the gearcase. On standard rotation units, you should rotate the driveshaft very slowly in a clockwise direction (viewed from above) to help mesh the reverse gear in the bearing carrier with the pinion gear as the bearing carrier is seated. Also, during installation, make sure the carrier keyway is aligned with the gear housing keyway, and position the key.

9. Once the carrier is seated, place the large tab washer against the bearing carrier.

■ **The cover nut torque is extremely high. Make sure the gearcase is either mounted in a strong stand or bolted to a driveshaft housing to avoid possible gearcase damage.**

10. Apply a light coating of marine grade grease to the threads of the cover nut, then install the nut into the housing, making sure the word **OFF** and arrow are visible. Use the cover nut tool (#91-61069) or equivalent to tighten the cover nut to 210 ft. lbs. (285 Nm).

11. Bend one of the lock tabs of the lock washer into the cover nut (normally only ONE will align). Bend the remaining tabs of the washer toward the front of the gear housing.

12. Using the shift shaft bushing tool (#91-31107) tighten the shift shaft bushing to 50 ft. lbs. (68 Nm) for standard rotation units or to 30 ft. lbs. (41 Nm) for counter rotation units.

13. If not done earlier during driveshaft installation (assuming no preload check needed to be performed) install the water pump base, followed by the water pump assembly, keeping the following points in mind:

• The dual seals should be coated on the outer-diameter with Loctite® 271 or equivalent and installed back-to-back so the inner or lower seal faces the gearcase and the upper or outer seal faces the water pump.

• A NEW O-ring should be used, and coated (along with the inner lips of the seals) with 2-4-C with Teflon or an equivalent lubricant.

• If the divider block was removed from the top of the gearcase use RTV sealer to seal the seams between the divider block and the gear housing, then install the water pump base using a new base gasket.

14. Refill the gearcase and check for leaks, then install the gearcase to the outboard as detailed earlier in this section.

SHIMMING

Backlash

◆ **See Figure 213**

For any gearcase to operate properly the pinion gear on the driveshaft must mesh properly with the forward and reverse gears. And, remember that these gears remain in contact with the pinion gear at all times, regardless of whether or not the sliding clutch has locked one of them to the propeller shaft. If these gears are not properly meshed, there is a greater risk of breaking teeth or, more commonly, rapid wear leading to gearcase failure.

On some of the smaller Mercury gearcases there is little or no adjustment for the positioning of these gears. However, the gearcases used on these large Mercury motors are all equipped with shim material that is used to perform minute adjustments on the positioning of the bearings (and therefore the bearings and gears). Shim material can be found in various positions, depending upon whether it is a standard or counter rotation unit.

For standard rotation units shim(s) can be found between the forward gear bearing/race assembly and the nose cone, as well as underneath the upper driveshaft bearing.

For counter rotation units, shim(s) can be found between the forward gear and thrust bearing on the prop shaft, between the reverse gear bearing adaptor and nose cone of the gearcase, and, like the standard rotation units, underneath the upper driveshaft bearing.

If a gearcase is disassembled and reassembled replacing only the seals, then no shimming should be necessary. However, if shafts, gears/bearings and/or the gearcase housing itself is replaced it is necessary to check the Pinion Gear Depth along with the Forward and Reverse Gear Backlash to ensure proper gear positioning.

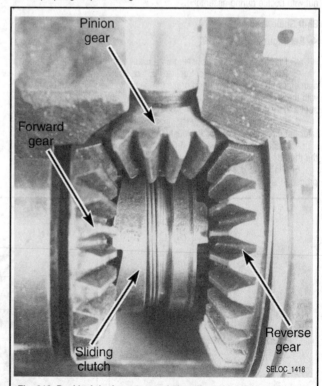

Fig. 213 Backlash is the amount of play allowed by the position of the forward and reverse gears in relation to the pinion gear (standard rotation shown, gear positions are reversed in counter rotation units)

Pinion Gear Depth

◆ See Figures 214, 215 and 216

■ This procedure starts with the gearcase mostly reassembled (following our procedures up to and including driveshaft reassembly, but with the shift shaft, propeller shaft, bearing carrier, water pump plate and water pump still removed).

Obtain a Bearing Preload Tool (#91-14311A1). This tool consists of a spring, which seats against the lower unit and is used to simulate the upward driving force on the pinion gear bearing. Install the following, one by one, over the driveshaft.

    a. Adaptor, with the flat face down against the lower unit

    b. Thrust bearing

    c. Thrust washer

    d. Spring

    e. Special large bolt, with the nut on the bolt threaded all the way to the top of the threads

    f. Sleeve (the holes in the sleeve must align with the set screws

1. Align the two holes in the sleeve with the Allen set screws in the special large bolt head. These set screws grip the driveshaft and prevent the tool from slipping on the driveshaft. Tighten the two set screws securely with the Allen key provided with the tool kit.

2. Measure the distance between the bottom surface of the bolt head and the top surface of the nut. Record this measurement.

3. Use a wrench to hold the bolt head steady and another wrench to tighten the nut down against the spring. Continue tightening the nut until the distance between the bottom surface of the bolt head and the top surface of the nut equals 1.00 in. (25.4mm) plus the distance recorded when the tool was installed.

4. The driveshaft has now been forced upwards placing an upward preload on the pinion gear bearing. Turn the driveshaft 2 or more turns to make sure the bearings are seated.

5. Obtain the correct Pinion Gear Locating Tool (#91-74776) along with Flat # 7 and Disc # 2 from the tool.

6. Insert the locating tool (with the appropriate flat and disc in place), making sure the tool bottoms on the bearing carrier shoulder in the housing.

7. The pinion depth is the distance between the bottom of the pinion gear teeth and the flat surface of the gauging block.

8. Measure this distance by inserting a long feeler gauge into the access hole and between the pinion gear teeth and gauging block flat face.

9. The correct clearance is 0.025 in. (0.64mm). If the clearance is correct, remove only the Pinion Gear Locating Tools from the inside of the gearcase. Leave the Pinion Gear Bearing Preload Tool in place on the driveshaft in order to check Gear Backlash.

10. If the clearance is not correct, add or subtract shim material from below the upper driveshaft bearing to lower or raise the gear.

• If the pinion gear depth was found to be greater than 0.025 in. (0.64mm), then that amount of shim material must be added to lower the gear.

• If the pinion gear depth was found to be less than 0.025 in. (0.64mm), then that amount of shim material must be removed to raise the gear.

■ The difference between 0.025 in. (0.64mm) and the actual clearance measured is the amount of shim material needed to correct the pinion gear depth.

11. Once Pinion Gear Backlash is set, be sure to either install and torque a new pre-coated pinion nut or coat the pinion nut with Loctite® 271 and retorque.

Forward and Reverse Gear Backlash - Standard Rotation Units

◆ See Figures 217 and 218

■ The propeller shaft is temporarily installed for this procedure.

Final installation of the shaft is usually performed later both because an old pinion gear nut is normally installed on the driveshaft without the use of Loctite to determine the pinion gear depth (in case it needs to be disassembled again to make changes to the pinion gear bearing or forward gear bearing shims) and because the shift cam is not installed during this procedure.

1. First, check and adjust the Pinion Gear Depth, as detailed earlier in this section. Leave the pinion bearing preload tool installed on top of the gearcase.

2. Next, insert the assembled propeller shaft MINUS the shift cam, then install the bearing carrier into the housing and thread the cover nut tighten against the carrier (but there is not need to torque the cover nut to spec or anything that tight).

3. Install a bearing carrier puller with the arms of the puller on the carrier and the center bolt on the end of the propeller shaft. Tighten the puller center bolt to 45 inch lbs. (5 Nm). This action places a preload on the forward gear, pushing the forward gear into the forward gear bearing.

4. Attach the a Backlash Dial Indicator Rod (#91-78473) to the driveshaft. Position a dial indicator (using a magnetic base or a threaded rod with nuts and plates to position it alongside the Indicator Rod) and point the dial indicator to the appropriate number on the indicator rod tool, depending upon the model as follows:

• For gearcases with a ratio of 1.87:1 (15 teeth on the pinion gear) point the tool to the No. 1 mark.

• For gearcases with a ratio of 2.0:1 (14 teeth on the pinion gear) point the tool to the No. 2 mark.

• For gearcases with a ratio of 2.3:1 (13 teeth on the pinion gear) point the tool to the No. 4 mark.

5. Recheck the puller torque, then rotate the driveshaft back-and-forth and observe movement of the dial indicator. Total movement is the forward gear backlash. Check the following listing to determine the proper amount of

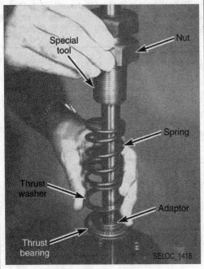

Fig. 214 Installing the pinion gear pre-loading tool

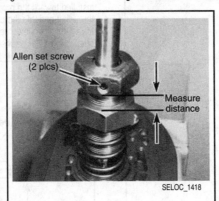

Fig. 215 Set the nut to the proper distance to set pre-load

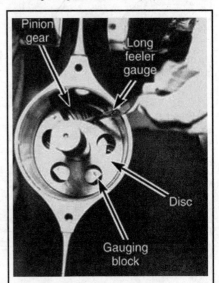

Fig. 216 Checking the gap between the pinion gear and the gauging surface

backlash permissible for the unit being serviced. If the backlash is too great, add shim material behind the forward gear bearing race. If the backlash is too small, remove shim material from behind the forward gear bearing race.

• For gearcases with a ratio of 1.87:1 (15 teeth on the pinion gear) there should be 0.018-0.027 in. (0.46-0.69mm) of backlash.

• For gearcases with a ratio of 2.0:1 (14 teeth on the pinion gear) there should be 0.015-0.022 in. (0.38-0.56mm) of backlash.

• For gearcases with a ratio of 2.3:1 (13 teeth on the pinion gear) there should be 0.018-0.023 in. (0.46-0.58mm) of backlash.

■ **For each 0.001 in. (0.025mm) of shim which is added or removed expect backlash to increase or decrease about 0.001 in. (0.025mm).**

6. Although the reverse gear backlash is not adjustable on these models, it can still be checked as follows:

a. Remove the propeller shaft and install the shift cam, then reinstall it along with the shift shaft and the bearing carrier

b. Shift the gearcase into Reverse.

c. Obtain a piece of PVC pipe, 5 1/2 in. (14cm) long by 1 1/2 in. (3.8cm) diameter and install it over the propeller shaft using the tab washer and propeller nut. Tighten the nut to 45 inch lbs. (5 Nm).

d. Gently rock the driveshaft back and forth while watching the dial indicator. Backlash should be about 0.030-0.050 in. (0.762-1.27mm).

■ **If the Reverse gear backlash is not in spec either the gearcase is improperly assembled OR the parts are excessively worn and must be replaced.**

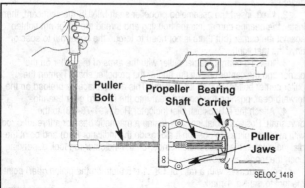

Fig. 217 Use a puller to fully seat the propeller shaft and forward gear. . .

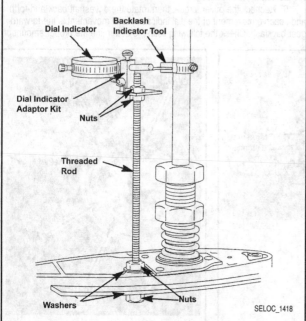

Fig. 218 . . .then check gear backlash using the appropriate indicator tool

Reverse and Forward Gear Backlash - Counter Rotation Units

◆ **See Figures 218 and 219**

Because of differences in the propeller shaft and gear/bearing assemblies BOTH the Reverse AND Forward gear lashes are adjustable on counter rotation units.

■ **The propeller shaft is temporarily installed for the second half of this procedure. The first half is performed using a suitable driver tool to center and preload the nose cone gear/bearing assembly.**

1. First, check and adjust the Pinion Gear Depth, as detailed earlier in this section. Leave the pinion bearing preload tool installed on top of the gearcase.

2. Next, insert the following tools to center and preload the Reverse gear in the nose cone of the gearcase. Insert a Driver Tool (#91-18605) and slide the Pilot Ring (#91-18603) over the driver until it rests against the ledge in the gearcase which is right before the pinion gear. Next install a Thread Retainer (#91-18604) into the gearcase bearing carrier cover nut threads, then tighten the Screw (#91-18602) to 45 inch lbs. (5 Nm) to preload the gear/bearing assembly.

3. If not done already for Pinion Gear Depth measurement, thread the stud adaptor from the Bearing Preload Tool (#91-14311A1) all of the way onto the stud then install the Backlash Indicator Tool (#91-78473) and a suitable dial indicator. Position the indicator on the line marked 1 for models with 15 teeth on the pinion gear (a ratio of 1.87:1) or on the line marked 2 for models with 14 teeth on the pinion gear (a ratio of 2.0:1).

4. Turn the driveshaft back and forth very lightly and watch the dial indicator to determine Reverse gear backlash. Lash should be 0.030-0.050 in. (0.76-1.27mm). If the Reverse gear backlash is out of spec, remove the Reverse gear assembly and add or subtract shim material to bring it in spec. If the backlash is too great, add shim material. If the backlash is too small, remove shim material. In either case, aim for the middle of the spec and keep in mind that lash should change by about the same amount as the thickness of the shims you add or subtract. Once the adjustment is made, reassemble and recheck. On the second assembly, if you're confident, you can use Loctite on the pinion nut and treat it as part of the final assembly.

5. Next it's time to check the Forward Gear (propeller shaft gear) backlash. Prepare by installing a Load Washer (#12-37429) over a #44-93003 propeller shaft.

6. Now, assemble the bearing carrier, bearing adaptor, thrust washer, thrust bearing and forward gear all onto the propeller shaft. Position the starting shim (whatever was removed) against the shoulder inside the gearcase (right before the pinion nut) and install the propeller shaft and bearing carrier assembly.

7. Install the tab washer and the cover nut, then tighten the cover nut to 100 ft. lbs. (135.5 Nm) in order to seat the Forward gear assembly in the gearcase.

8. Obtain a piece of PVC pipe, 5 1/2 in. (14cm) long by 1 1/2 in. (3.8cm) diameter and drill a small 3/8 (22.2mm) hole through the side which will be installed toward the propeller nut, this way a screwdriver can be inserted through it and into the propeller shaft splines in order to prevent the splines from turning while tighten in the retaining nut. THEN install it over the propeller shaft using a flat washer and the propeller nut. Tighten the nut to 45 inch lbs.

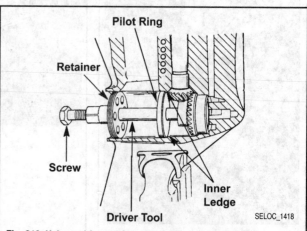

Fig. 219 Using a driver tool to seat the Reverse gear when checking backlash

(5 Nm) in order to seat the Forward gear against the thrust bearing and help keep the propeller shaft from moving while checking the lash.

9. If not done already for Pinion Gear Depth measurement, thread the stud adaptor from the Bearing Preload Tool (#91-14311A1) all of the way onto the stud then install the Backlash Indicator Tool (#91-78473) and a suitable dial indicator. Position the indicator on the line marked 1 for models with 15 teeth on the pinion gear (a ratio of 1.87:1) or on the line marked 2 for models with 14 teeth on the pinion gear (a ratio of 2.0:1).

10. Turn the driveshaft back and forth very lightly and watch the dial indicator to determine Forward gear backlash. Lash should be 0.018-0.027 in. (0.46-0.69mm) for models with a 1.87:1 gear ratio (15 pinion gear teeth) or 0.015-0.022 in. (0.38-0.56mm) for models with a 2.00:1 gear ratio (14 pinion gear teeth). If the Forward gear backlash is out of spec add or subtract shim material to bring it in spec. If the backlash is too great, remove shim material. If the backlash is too small, add shim material. In either case, aim for the middle of the spec and keep in mind that lash should change by about the same amount as the thickness of the shims you add or subtract. Once the adjustment is made, reassemble and recheck.

11. Whether or not the backlash is within specification you'll have to remove the test propeller shaft assembly so the propeller shaft can be properly assembled and installed (without the load washer or test shaft).

12. Proceed with final assembly from the pinion nut tightening (if not done already) through installation of the Bearing Carrier and Propeller Shaft assembly.

Gearcase - Conventional 200-250 Hp (3.0L) V6 Models

■ **For gearcase exploded views, please refer to the Cleaning & Inspection section.**

There are basically 4 similar (2 of which are almost identical) design gearcases found on 200-250 hp V6 models, 2 of which are covered in this section; the other 2 of which can be found in the Gearcase - 225/250 Hp (3.0L) V6 Pro/Sport XS Models section.

The 4.75 in (120.65mm) torpedo diameter standard rotation gearcase AND the 4.75 in (120.65mm) torpedo diameter counter-rotation gearcases are nearly identical and are found on most models, including all EFI and conventional OptiMax models. The basic design of the gearcase, water pump, driveshaft and shift shaft are all almost the same between the 2 units.

Most differences occur in the placement of gears as well as the type and location of bearings between the 2 units. These differences are necessitated by the fact that on standard rotation units the Forward bearing, located in the nose cone, takes the majority of thrust through use (as most use is in Forward gear on the average outboard). On counter-rotation units the Forward bearing (which also takes the majority of the thrust in use) is located on the propshaft, up against the bearing carrier instead of in the nose cone. Small differences in the bearing assemblies help to strengthen the assemblies to perform their jobs as such.

■ **One caveat to the 2 gearcases mentioned above is that there is some evidence that EFI models through 2001 MAY have used a slightly different version of these gearcases. Differences would be very few, but might include one less tapered roller bearing on the upper driveshaft, a slightly different shaped shifter housing and a lack of a retaining ring for the nose cone gear/bearing assembly on standard rotation units. As usual, should the differences make any difference in repair procedures, it will be noted in the appropriate steps.**

REMOVAL & INSTALLATION

◆ **See Figures 182 and 220**

Like all of the other large Mercury gearcases, the rotating/ratcheting shift mechanism used on these models utilizes a splined shift shaft connection, making it un-necessary to locate and disconnect shift linkage before gearcase removal.

1. For safety disconnect the battery cables.

2. If necessary for overhaul purposes, remove the Propeller, as detailed in the Maintenance and Tune-Up section.

■ **It's actually a good idea to at least loosen the OIL LEVEL and/or VENT screw(s) first, to make sure it will come out without a problem. You can't refill the gearcase, if you can't get the vent screw loose.**

3. If necessary for gearcase overhaul purposes, position a suitable clean container under the lower unit. Remove the FILL screw on the bottom

of the lower unit, and then the OIL LEVEL and/or VENT screw(s). The vent screw must be removed to allow air to enter the lower unit behind the lubricant. Allow the gear lubricant to drain into the container.

As the lubricant drains, catch some with your fingers from time-to-time and rub it between your thumb and finger to determine if there are any metal particles present. Examine the fill plug. A small magnet imbedded in the end of the plug will pickup any metal particles. If metal is detected in the lubricant, the unit must be completely disassembled, inspected, and the damaged parts replaced.

Check the color of the lubricant as it drains. A whitish or creamy color indicates the presence of water in the lubricant. Check the container for signs of water separation from the lubricant. The presence of any water in the lubricant is bad news. The unit must be completely disassembled; inspected; the cause of the problem determined; and then corrected.

4. If not done already, rotate the outboard unit to the full UP position and engage the tilt lock lever.

5. Shift the gearcase into NEUTRAL.

6. Matchmark the trim tab to the anti-cavitation plate to preserve alignment, then loosen the bolt which secures the tab (the trim tab is secured from above by a bolt that is threaded vertically downward through an access panel in the top of the driveshaft housing, usually under a plastic or rubber cover. Remove the trim tab from the cavitation plate for access to the gearcase retaining bolt located inside the trim tab cavity. Loosen and remove the previously hidden gearcase retaining bolt.

7. If applicable, at the front lower motor mount/pivot point, locate the speedometer hose junction. While pressing in on the speedometer hose junction, carefully pull out on the hose itself to disconnect it.

■ **Some early-model EFI units may be equipped with a speedometer hose but NOT the junction assembly described here. On these models the hose is often secured by a wire tie which must be cut to detach the hose from the fitting.**

8. Alternately and evenly loosen the 4 gearcase mounting locknuts (2 on each side), but DON'T try to remove any one nut before the opposite side is loosened or the driveshaft housing could be damaged. Once they are sufficiently loosened, pull down on the lower unit gear housing and CAREFULLY separate the lower unit guiding it straight out of the exhaust housing. Guide the shift rod and driveshaft out of the exhaust housing as the lower unit is removed.

9. Place the lower unit in a suitable holding fixture or work area.

To install:

■ **Always fill the lower unit with lubricant and check for leaks before installing the unit to the driveshaft housing (could save you the trouble of removing it again right away).**

10. Take time to remove any old gasket material from the fill and vent recesses and from the screws.

11. Place the lower unit in an upright vertical position. Fill the lower unit with Super-Duty Lubricant, or equivalent, through the fill opening at the bottom of the unit. Never add lubricant to the lower unit without first removing

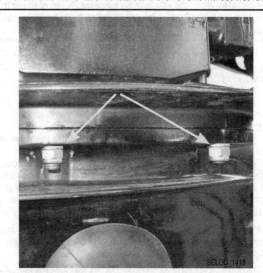

Fig. 220 Gearcase retainers

the vent screw and having the unit in its normal operating position - vertical. Failure to remove the vent screw will result in air becoming trapped within the lower unit. Trapped air will not allow the proper amount of lubricant to be added.

12. Continue filling slowly until the lubricant begins to escape from the vent or level opening (as applicable) with no air bubbles visible.

13. Use a new gasket and install the vent and/or level screw(s). Slide a new gasket onto the fill screw. Remove the lubricant tube and quickly install the fill screw.

14. Visibly check the lower unit for leaks. Recheck for leaks after the first time the unit is warmed to normal operating temperatures from use.

15. If not done already, swing the exhaust housing outward until the tilt lock lever can be actuated, and then engage the tilt lock.

16. Apply a coating of 2-4-C Marine Lubricant with Teflon to the splines of the shift shaft coupler and the driveshaft.

■ **DO NOT put lubricant on the ENDS of the shafts, as it could prevent the shaft splines from fully seating in the shift coupler and the crankshaft.**

17. Apply a light bead of G.E. Silicone Sealer or an equivalent sealant, to the top of the divider block located on the top of the gearcase, right behind the water pump assembly.

■ **On early-model EFI gearcases without the speedometer junction described earlier, you may have to feed the speedometer tube through the opening in the driveshaft housing as the gearcase is raised into position.**

18. Carefully bring the gearcase into alignment with the exhaust housing and slowly insert it STRAIGHT UP and into position, while aligning the driveshaft, shift shaft and water tube seal.

■ **If necessary, slowly turn the crankshaft in the normal direction of rotation (clockwise when viewed from above) using the flywheel to align the driveshaft-to-crankshaft splines. Alternately, you can shift both the gearcase and the powerhead linkage into Forward and turn the propeller shaft to align the driveshaft splines, but that may make shift shaft alignment slightly more difficult.**

19. While still holding the gearcase in position, place the flat washers onto the studs (located on either side of the driveshaft housing), then start a locknut on each of these studs, threading them down by hand until finger-tight.

20. Next, thread the bolt at the rear of the housing, inside the trim tab recess, but DO NOT TIGHTEN IT FULLY AT THIS TIME.

21. Make sure the shift shaft spline engagement seems ok at this point by attempting to gently move the shift block just slightly.

✱✱ SELOC CAUTION

If the gearcase does not easily flush up against the driveshaft housing DO NOT force the gearcase to seat using the attaching nuts. Instead remove the gearcase again and double-check the shaft spline alignments. Also, make sure there is no lubricant on top of the shift shaft or driveshaft (or in the underside of the shift shaft and crankshaft located in the housing).

22. Alternately and evenly tighten 2 of the locknuts which were installed earlier (one on either side of the lower motor mount/swivel attachment point) to 55 ft. lbs. (75 Nm).

23. Verify proper shift shaft spline engagement as follows:

a. Place the guide block anchor pin into the FORWARD position (if not already) while turning the prop shaft in the normal direction of rotation. The propeller shaft should not turn once in gear.

b. Place the guide block in the NEUTRAL position and verify that the propeller shaft now turns freely.

c. Place the guide block in the REVERSE gear position, now slowly rotate the FLYWHEEL CLOCKWISE when viewed from the top and verify that the propeller shaft now turns in the normal direction of reverse gear for the gearcase in question (meaning counterclockwise on normal rotation gearcases or clockwise on counter rotation gearcases).

■ **If the gearcase does not shift properly into gear, the housing must be removed again and the shaft splines realigned as necessary for proper shifting.**

24. If not done already, place washers and locknuts onto the remaining studs on either side of the gearcase, then tighten them to 55 ft. lbs. (75 Nm).

25. Tighten the gearcase retaining bolt inside the trim tab cavity to 45 ft. lbs. (61 Nm), then align the matchmarks made during removal and install the trim tab itself. Install the plastic cap into the trim tab bolt opening at the rear edge of the driveshaft housing.

■ **On some early-EFI motors Mercury suggested replacing the trim tab bolt using a new bolt with a locking "patch" each time the tab was installed. Alternately, we'd recommend using a light coating of a threadlocking compound whenever you have removed the trim tab. Either way tighten the bolt to 40 ft. lbs. (54 Nm).**

26. If applicable, reconnect the speedometer tube into the junction.

27. If the propeller was removed, install the propeller.

28. Reconnect the battery cables.

DISASSEMBLY

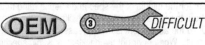

■ **For gearcase exploded views, please refer to Cleaning & Inspection in this section.**

Unlike most of the smaller Mercury gearcases, the lower unit found on these models contains various shims for adjusting pinion height and forward/reverse gear backlash. Always keep track of the shims during removal, since if the gearcase or the shafts/bearings are being reused the shims should be returned to the same positions. Also, if the gearcase and/or shafts/bearings are replaced, the shims can be sometimes be used as a good starting point for the shimming procedures.

■ **On standard rotation gearcases, although the Reverse gear backlash is not adjustable, it CAN be used to help determine if parts are worn to the point of requiring replacement. Many techs will follow that portion of the Shimming procedure PRIOR to disassembly in order to help determine gearcase condition. If this is desired, skip ahead to Shimming to check backlash.**

Also unique to Mercury gearcases is the shifter set-up. The design has the shaft extending into splines on a C-shaped shift crank that cups the end of the propeller shaft. The design is such that the propeller shaft is not removed until AFTER the driveshaft and pinion assembly has been unbolted and removed giving the propeller shaft sufficient clearance and play to be freed from this assembly. The result is that it is difficult-to-impossible to get around the special tool needed to secure the pinion nut (Pinion Nut Wrench/MR slot #91-61067A3)

Bearing Carrier Removal & Disassembly

◆ **See Figures 186, 221 and 222**

■ **Because the shift shaft actually splines with a C-shaped cup that holds the front of the propeller shaft, the bearing carrier is removed independently of the propeller shaft (which can only be removed after the driveshaft later in the Disassembly procedure).**

✱✱ SELOC WARNING

If you don't have access to the special Mercury mandrels designed to install bearings to the proper depth inside the bearing carrier, measure the installed height of all bearings and/or races before removal. Use the measured depths at installation time to ensure the bearings are positioned properly (especially in positions where there are no shoulders to drive the bearings or races against).

1. Remove the gearcase from the mid-section, as detailed earlier in this section. Mount the case in a holding fixture or a soft-jawed vise, so you have access to the propeller shaft bore.

2. If necessary for access or for further disassembly, remove the Water Pump assembly, as detailed in the Lubrication and Cooling System section.

3. Locate the one tab on the tabbed washer which is bend down over the bearing carrier retainer (cover nut) to keep it from loosening. We say one tab because normally only one of them should be in position at a time, but it doesn't hurt to check the rest. Using a small punch, carefully bend the tab back so it no longer interferes.

4. Using a large spanner like a Cover Nut Tool (#91-61069) carefully loosen and unthread the cover nut. Once the cover nut is removed, pull the locktab washer from the housing. IF the cover nut is instead corroded into

place and cannot be turned, drill holes in it, roughly positioned at 120 degree intervals around the face of the retainer (drill two holes almost next to each other, then two more holes, each at 120 degree intervals). Once the holes are drilled use a hammer and chisel to fracture and remove the remainder, BUT USE EXTREME SELOC CAUTION not to damage the bearing carrier.

5. Install an internal jawed puller such as Long Puller Jaws (#91-46086A1) and a suitable drive bolt. Position the puller jaws on the outer ring of the carrier, close to the bosses. Tighten the puller to preload the propeller shaft.

6. Now, tighten the puller bolt to break the seal and free the bearing carrier from the gearcase. DO NOT attempt to remove the propeller shaft yet.

■ **If the bearing carrier refuses to move, it may be necessary to carefully apply heat to the lower unit in the area of the carrier and at the same time attempt to move it off the carrier.**

7. Check the condition of the bearing carrier itself, make sure the portion of the carrier which contacts the gearcase is not too corroded. If it shows signs of excessive corrosion, replace it to preserve proper gearcase sealing and to minimize the chance of it seizing in the housing.

8. On standard rotation gearcases, remove and discard the O-ring from between the carrier and thrust washer. On these gearcases the propshaft gear (Reverse on these models) is pressed into the housing, if it or the bearing must be replaced, position the carrier in a soft-jawed vise and use a jawed slide hammer to remove the gear. If the bearing stays behind, use the

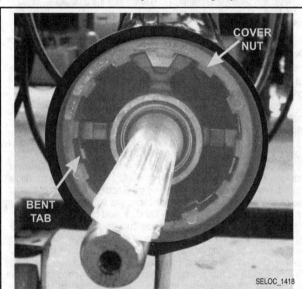

Fig. 221 The bearing carrier is secured by a large cover nut and tabbed washer

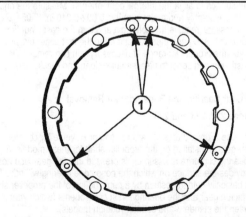

1- Drilled Holes   SELOC_1418

Fig. 222 If the carrier nut is frozen drill it out as shown, then remove it with a chisel

same assembly to remove it from the carrier. If the bearing stays on the gear, use a bearing separator to push it off the gear.

9. On counter rotation gearcases, the propshaft gear (Forward on these models) is installed ON the shaft along with a bearing adapter and series of thrust bearing/washer components. Remove them from the propshaft in the gearcase, one at a time in the following order, noting both their orientation and positioning as each is removed (it's a good idea to install them over a dowel or shaft to preserve this as they are removed). Reach into the gearcase and remove the following components:

a. Start with the thrust washer (it may be in the gearcase on the propshaft, OR it may be stuck to the inside surface of the bearing carrier).

b. Remove the aft thrust bearing.

c. Remove the aft thrust collar, then lift up on the propeller shaft while pushing down on the forward thrust collar so you can remove the 2 keepers. Once the keepers are out of the way, remove the forward thrust collar.

d. Remove the forward thrust bearing.

e. Use a length of 1/8 in. (3mm) diameter wire to form a hooking tool that can be inserted into the gearcase and under the forward gear bearing adaptor. Use the tool to gently lift the bearing adaptor and remove it from the gearcase/propshaft.

f. Shift the gearcase into Forward, then remove the O-ring from the inside of the gearcase.

g. Use the a Forward Gear Installation Tool (#91-815850) to remove the thrust race, bearing and the forward gear together. HOWEVER, keep in mind that the thrust race has a tight fit in the bore and IF this doesn't work, make a small hook on the end of the stiff piece of wire and use it to remove the thrust race/gear while carefully applying heat to the outside of the gearcase.

h. Remove and retain the forward gear shim from the gearcase.

■ **The dual back-to-back oil seals in the bearing carrier can be pried out separately (if the needle bearing is not being replaced) OR can be driven out from behind while removing the needle bearing.**

10. If the needle bearing inside the carrier is NOT going to be replaced, invert the bearing carrier so the propeller end is facing upward and use a small prytool to carefully remove the dual back-to-back oil seals.

11. If the needle bearing inside the carrier IS going to be replaced, position the carrier in a suitable shop press with the front side (gear side) facing upward. Use a suitable driver and mandrel to press the needle bearing and seals out the propeller end of the carrier.

Driveshaft and Bearings Removal/Service

◆ See Figure 223

1. If not done already, remove the Water Pump assembly as detailed in the Lubrication and Cooling System section. Be sure to remove the water face plate and gasket as well for access to the oil seal carrier and driveshaft components.

2. If not done already, remove the upper driveshaft oil seal carrier by using 2 small prytools (one in the slot provided on either side). Lift the base up and off the driveshaft.

3. Remove and discard the O-ring from the bottom of the seal carrier.

4. For installation purposes, note the directions in which the oil seals are facing (they are usually installed back-to-back) then carefully pry them from the seal carrier OR use a punch to drive them out the flared bottom of the carrier.

■ **OK, here comes the pain in the butt part of this gearcase, trying to loosen the pinion nut with the propeller shaft still installed.**

5. Obtain the following special tools, Drive Shaft Nut Wrench (#91-56775), Drive Shaft Bearing Retainer Wrench (#91-43506) and Pinion Nut Wrench/MR Slot (#91-61067A3), then proceed as follows to unbolt the pinion nut:

a. Position the bearing retainer wrench down over the driveshaft for use after the pinion nut is JUST loosened.

b. Place the drive shaft nut wrench onto the top of the driveshaft.

c. Insert the pinion nut wrench into the gearcase with the slot facing the pinion gear. If necessary, lift and rotate the driveshaft slightly to help align the pinion gear nut into the slot in the nut wrench.

d. Install the bearing carrier back into the gear housing BACKWARDS to help support the proper shaft and keep the pinion nut wrench aligned.

e. Using the driveshaft nut wrench just LOOSEN, but DO NOT REMOVE the pinion nut by rotating the driveshaft counterclockwise. IF the driveshaft is broken (or breaks during this procedure) remove the bearing carrier and place the gearcase in Forward gear, then use a Propshaft Nut

Wrench (#91-61077) to turn the propshaft counterclockwise, which will turn the gears and loosen the pinion nut.

f. Now, completely unscrew the driveshaft bearing retainer.

g. Finish unscrewing the pinion nut by rotating the driveshaft or propeller shaft counterclockwise.

h. Remove all of the specialty tools.

6. Once the driveshaft is free, slide it out of the gearcase, but be sure to retain any shims which may have been placed between the driveshaft's tapered bearing and shoulder in the gearcase bore.

7. CAREFULLY move the propshaft down toward the port side of the gearcase, then reaching inside and retrieve the pinion gear, washer and nut from inside the housing.

■ A few of the early-model EFI motors covered here MAY use a gearcase with only ONE upper tapered roller bearing, however MOST use 2. Other than buying parts, the difference in inconsequential. The same method can be used to remove both upper bearings at the same time from the later type of gearcase.

8. If the tapered bearing(s) requires replacement, use a bearing separator (or even the jaws of a vise) to hold the bearing(s) while you push or gently tap the shaft free. Also, if the bearings are replaced, you'll need to remove the bearing cup (and shims) which are mounted in the upper portion of the gearcase. Use a slide hammer with a pair of internal puller jaws to grab on the inner lip of the bearing cup and pull it from the gearcase.

9. Inspect the driveshaft lower bearing surface (right above the pinion gear splines) where the needles roll. If the driveshaft is pitted, grooved/scored, worn unevenly, discolored from overheating, has embedded particles or you otherwise suspect the bearing, the needles and race should be replaced as a set. However, you'll need to remove the propshaft and nose cone gear first (for access and to prevent damage), then come back to this procedure in order to proceed.

10. Once the propshaft has been removed, make sure all 18 loose needles are inside the lower race, then install a Bearing Driver (#91-36569) inside the bearing, along with the Pilot Washer (#91-36571) and the Driver Rod (#91-37323). Use a mallet to drive the rod, washer and driver downward, driving the race and needle bearing assembly down into the propshaft bore of the gearcase. Remove and discard the old race and needles.

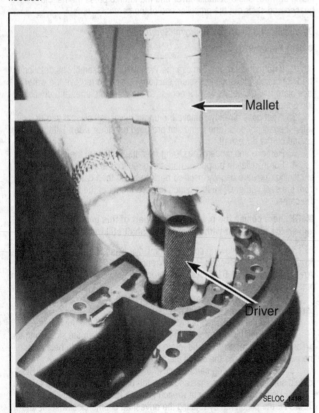

Fig. 223 Removing the driveshaft lower needle bearing assembly

Propeller Shaft Removal and Disassembly

◆ See Figures 224, 225 and 226

Once the driveshaft and pinion gear assembly are removed you finally have the free-play necessary to remove and service the propeller shaft.

1. For standard rotation models, position the gearcase so the propeller shaft is sticking straight up, then tilt the shaft toward the port side of the gearcase (to allow the shift spool to disengage from the shift shaft crank) and remove the shaft assembly pulling straight up and out of the case. The nose cone gear (Forward on these models) should come out of the bearing cup and gearcase along with the shaft.

2. For counter rotation models Mercury recommends positioning the gearcase so the propshaft is pointing straight downward. Remove the shaft by tilting it to the port side of the housing (to allow the shift spool to disengage from the shift shaft crank) and then lower the shaft out of the housing. During removal, the rollers of the nose cone gear (Reverse on these models) may become dislodged. If this happens, check the bearing cage to make sure it has not been damaged and, if so, just snap the rollers back into position. HOWEVER, if the cage was damaged, you'll have to replace the assembly. Either way, be sure to locate and retain the thrust race and thrust bearing which should be on top of the Reverse gear, but may also stick inside the bearing adaptor.

To disassemble the propeller shaft assembly for cleaning, inspection and/or parts replacement, proceed as follows:

3. Insert a thin blade screwdriver or an awl under the first coil of the cross-pin and detent pin retainer spring and rotate the propeller shaft to unwind the spring from the sliding clutch. Take care not to over-stretch the spring.

■ IF after removal the spring does not coil back to its normal position, the spring has been over-stretched and must be replaced.

4. Remove the tiny detent pin from the clutch, then push the cross-pin out of the sliding clutch with a punch. At this point the sliding clutch, nose cone gear and shift spool can all be removed from the shaft.

5. In addition to the usual regiments of cleaning and inspection of the shaft and bearing surfaces (detailed later in this section under Cleaning & Inspection), be sure to check the following measurements:

a. Use a pair of v-blocks and a dial gauge to check the shaft run-out at a point on the smooth portion of the shaft JUST inboard of the propeller splines. Dial indicator movement should generally be 0.009 in. (0.23mm) or less (with the exception of some early-model EFI gearcases with the single upper driveshaft bearing, as a very, very small additional amount of run-out was ok on them, as their recommended spec was 0.010 in. (0.25mm). If run-out exceeds the proper specification, the shaft must be replaced.

b. For standard rotation models, measure the propeller shaft forward-to-reverse shoulder length (from shoulder-to-shoulder across the splines for the clutch assembly) and replace the shaft if the measurement is under 2.040 in. (51.82mm). Also on these models, measure the thickness of the REVERSE thrust washer to check for wear or a taper. Replace the Reverse thrust washer is the thickness is less than 0.240 in. (6.096mm).

c. On all models, you should inspect the shift spool for wear, especially where it comes in contact with the crank on the shift shaft. Make sure the spool spins freely (one trick to help ensure this is to tap the castle nut end against a firm surface to align the internal components). Measure the end-play of the spool assembly and make sure it is 0.002-0.010 in. (0.05-0.25mm). If necessary, remove the cotter pin and turn the castle nut first clockwise down until snug and then backing off to the first cotter pin slot which gives you the proper clearance. If necessary you can remove the cotter pin, castle nut and spool to disassemble, clean and reassemble components.

Nose Cone Gear Bearing and Race/Adaptor Removal

◆ See Figures 227, 228 and 229

The nose cone (Forward on standard rotation or Reverse for counter rotation units) gear is installed on the propeller shaft for these models. A nose cone bearing cup or race assembly is pressed into the gearcase nose cone and provides the surface on which the nose cone bearing will ride. The gear itself is basically held in position by a combination of the propeller shaft splines, the pinion gear and the bearing/thrust components (which vary slightly between the standard and counter rotation models).

On standard rotation units, a tapered bearing which rolls in the nose cone race is pressed onto the outside of the gear and a needle bearing is pressed inside the gear. The needle bearing supports the forward end of the propeller shaft.

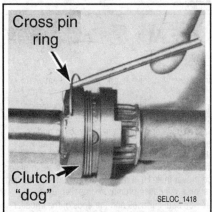

Fig. 224 To remove the sliding clutch, gear and spool, first remove the pin retaining spring. . .

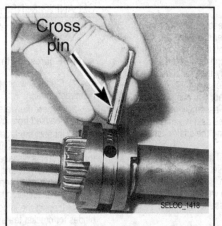

Fig. 225 . . .then remove detent and cross pins

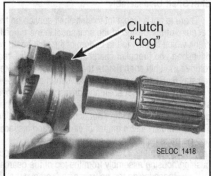

Fig. 226 . . .at this point the clutch dog (as well as the gear and spool not shown) will slide from the shaft

On counter rotation units, when installed the nose cone gear sits in a bearing adaptor assembly (not unlike the one used for the propshaft mounted gear and bearing on these models) and includes a roller bearing which is pressed into the adaptor. The gear on these models also has a needle bearing pressed into the center.

As usual, removal of the bearings will normally render them unfit for service, so make sure they are shot and replacements are available before bearings are separated from the gear.

After the pinion gear, driveshaft and propeller shaft have all been removed, the nose cone race or race/adaptor assembly can be lifted removed from the housing for service. Likewise, once the propeller shaft is disassembled, the nose cone gear can be serviced as well. To disassemble the nose cone gear and bearings, proceed as follows depending upon the model.

1. If necessary to replace the internal needle bearing (or the external tapered roller bearing on standard rotation units), remove the nose cone gear from the propeller shaft.

2. For standard rotation units:

3. If the forward gear tapered roller bearing must be replaced, use a universal bearing separator and a press to push the bearing off the gear.

■ Remember that the bearing and race are a set, so if the bearing is replaced you'll also need to replace the nose cone race. A wear pattern will have been worn in the race by the old bearing. Therefore, if the race is not replaced, the new bearing will be quickly worn by the old worn race.

4. Remove the forward gear tapered roller bearing race from the gearcase either using an internal jawed puller and puller plate adaptor (preferred method) or using a slide hammer (if necessary, but use seloc caution not to damage the gearcase).

5. SAVE any shim material from behind the race after the race is removed. The same amount of shim material will probably be used during assembly.

6. For counter rotation units:

a. Using an internal jawed puller assembly (preferably with a puller plate over the end of the gearcase and not a slide hammer) free the bearing adaptor from the gearcase. Retain shims for reuse, HOWEVER inspect them first, as they may be damaged during removal. IF they were damaged, measure them and obtain replacements of the same size for use during installation.

b. If necessary, use a suitable mandrel to press the roller bearing out of the bearing adaptor, then discard the old bearing. If disassembled, lubricate the adaptor bore using 2-4-C with Teflon or suitable marine grade grease, then press the new roller bearing into the adaptor with the numbered side of the bearing facing the driver shoulder. Drive the new bearing into the adaptor until the bearing is flush with the adaptor face.

c. On all except some early-model EFI gearcases (with the single upper driveshaft bearing) there is normally a snapring inside the back of the nose cone gear which helps secure the internal needle bearing. If the bearing is to be replaced, first check for and, if present, remove the snapring.

■ If removal is necessary, a mandrel or a hammer and punch may be used to press the needle bearing out from the nose cone gear.

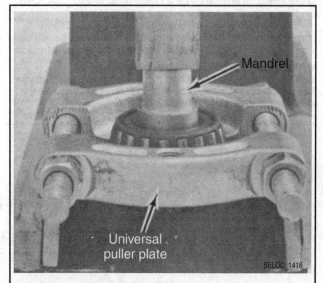

Fig. 227 On standard rotation units, remove the tapered roller bearing using a bearing separator

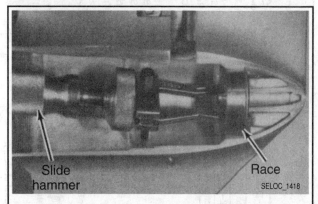

Fig. 228 Removing the gear race using a slide-hammer (standard, shown) or the adaptor using a puller (counter rotation)

7. For all units, if the needle bearing inside the gear is no longer fit for further service, it can be removed by pressing with a mandrel or tapping, with a blunt punch and hammer, around the bearing cage from the aft end - the gear end - of the bearing. This action will destroy the bearing cage. Therefore, be sure a replacement bearing is on hand, before attempting to remove the defective bearing.

Shift Shaft Disassembly

There is not a whole lot to shift shaft service on these motors, but because of the design of the shift spool in the propeller shaft and the shift crank which splines to the bottom of the shift shaft, you can't remove it until driveshaft and propeller shafts are removed. Well, technically you CAN remove the shift shaft before that point, but it's unlikely that you'd be able to reinstall it without removing the driveshaft and propeller shaft.

Once the propeller shaft and driveshaft are removed, you can unbolt the shift shaft bushing/housing (2 bolts), then remove the assembly from the gearcase, pulling the shaft out of the shift crank splines as it is withdrawn. Then it's mostly just a matter of cleaning and inspecting the shaft/bushing, and perhaps replacing the oil seal and O-ring

1. Loosen the 2 retaining bolts, then carefully pull the shift shaft and bushing/housing assembly from the top of the gearcase.

2. Reach inside the gearcase and remove the shift crank. Clean it with solvent and dry it, then inspect the splined portion, as well as the diameter of the spool that goes over the locating pin in the gearcase for damage. Replace, as necessary.

3. Carefully remove the shift shaft and rubber washer from the bushing/housing assembly.

4. Remove and discard the bushing/housing gasket, the O-ring and the oil seal (noting the direction the lips are facing before removal). The oil seal can either be pried out or removed using a driver and hammer.

5. Thoroughly clean the bushing and shaft with solvent and dry with compressed air. Make sure the splines are good on both ends and that there are no grooves, corrosion or excessive wear on the bushing sealing surface.

■ **The speedometer connector can be removed from the bushing/housing assembly for cleaning, inspecting and/or replacement. If so the threads of the connector must be coated with Perfect Seal or an equivalent sealant during installation.**

CLEANING & INSPECTION

◆ **See Figures 229 thru 239**

1. Clean all water pump parts with solvent, and then dry them with compressed air.

2. Inspect the water pump cover and base for cracks and distortion, possibly caused from overheating.

3. Inspect the face plate and water pump insert for grooves and/or rough surfaces. If possible, always install a new water pump impeller while the lower unit is disassembled. A new impeller will ensure extended satisfactory service and give "peace of mind" to the owner. If the old impeller must be returned to service, never install it in reverse to the original direction of rotation. Installation in reverse will cause premature impeller failure.

4. Inspect the impeller side seal surfaces and the ends of the impeller blades for cracks, tears, and wear. Check for a glazed or melted appearance, caused from operating without sufficient water. If any question exists, and as previously stated, install a new impeller if at all possible.

5. Clean all bearings with solvent, dry them with compressed air, and inspect them carefully. Be sure there is no water in the air line. Direct the air stream through the bearing. Never spin a bearing with compressed air. Such action is highly dangerous and may cause the bearing to score from lack of lubrication. After the bearings are clean and dry, lubricate them with Quicksilver Formula 50-D lubricant or equivalent. Do not lubricate tapered bearing cups until after they have been inspected.

6. Inspect all ball bearings for roughness, catches, and bearing race side wear. Hold the outer race, and work the inner bearing race in-and-out, to check for side wear.

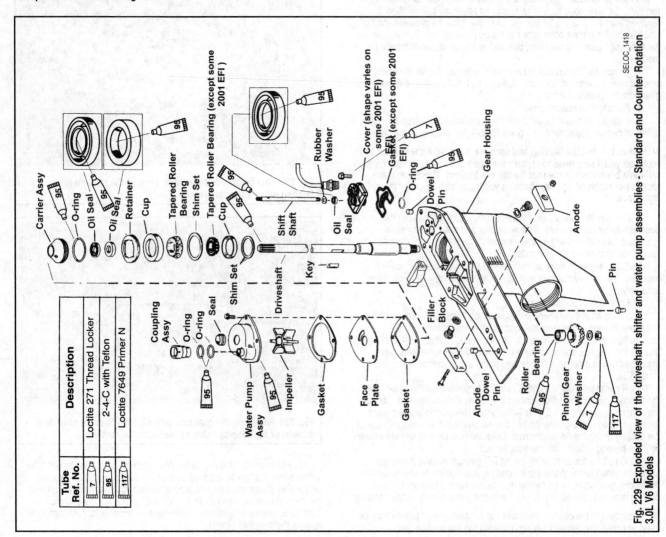

Fig. 229 Exploded view of the driveshaft, shifter and water pump assemblies - Standard and Counter Rotation 3.0L V6 Models

SELOC_1418

Tube Ref. No.	Description
7	Loctite 271 Thread Locker
95	2-4-C with Teflon
117	Loctite 7649 Primer N

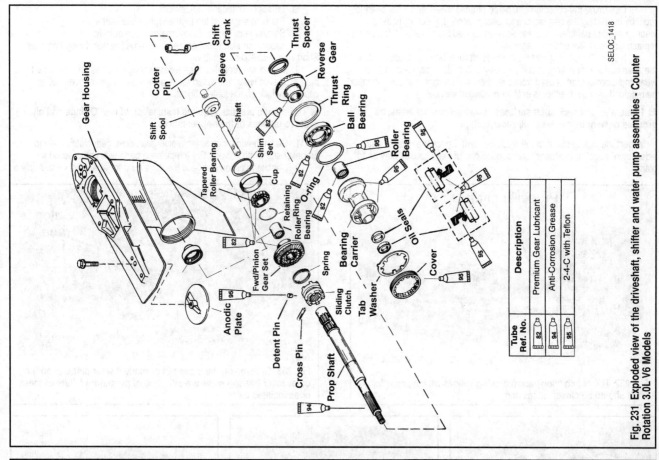

Fig. 231 Exploded view of the driveshaft, shifter and water pump assemblies - Counter Rotation 3.0L V6 Models

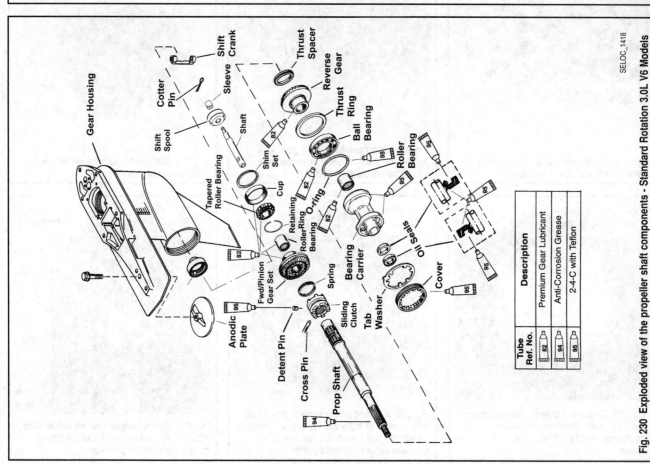

Fig. 230 Exploded view of the propeller shaft components - Standard Rotation 3.0L V6 Models

7. Determine the condition of tapered bearing rollers and inner bearing race, by inspecting the bearing cup for pitting, scoring, grooves, uneven wear, imbedded particles, and discoloration caused from overheating. Always replace tapered roller bearings as a set.

8. Inspect the bearing surface of the shaft roller bearing support. Check the shaft surface for pitting, scoring, grooving, imbedded particles, uneven wear and discoloration caused from overheating. The shaft and bearing must be replaced as a set if either is unfit for continued service.

■ **Remember to check shaft surfaces at oil seal contact areas, as grooves or wear in that area will allow leakage.**

Inspect the sliding clutch of the propeller shaft. Check the reverse gear side clutch "dogs". If the "dogs" are rounded one of three causes may be to blame

a. Improper shift cable adjustment.
b. Running engine at too high an rpm while shifting.
c. Shifting from neutral to reverse gear too quickly.

9. Inspect the shift spool assembly as noted earlier during Propeller Shaft Removal & Disassembly.

10. Inspect the propeller shaft roller bearing surfaces for pitting, rust marks, uneven wear, imbedded metal particles or signs of overheating caused by lack of adequate lubrication.

■ **Good shop practice requires installation of new O-rings and oil seals regardless of their appearance.**

11. Clean the bearing carrier, pinion gear, drive gear clutch and the propeller shaft with solvent. Dry the cleaned parts with compressed air.

12. Check the pinion gear and the drive gear for abnormal wear. Apply a

Fig. 232 The clutch "dog" and matching splines on the propeller shaft should be closely inspected

Fig. 233 Grooves on the cross pin or marked wear patterns on the outer roller bearing races are evidence of premature failure of these or associated parts

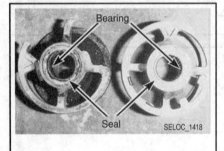

Fig. 234 Comparison of a worn bearing carrier (left) with a new one (right)

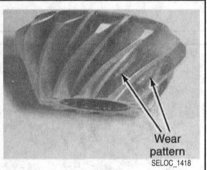

Fig. 235 Unacceptable pinion gear wear pattern, probably caused by inadequate lubrication in the lower unit

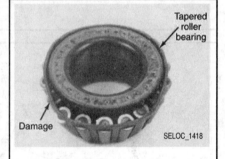

Fig. 236 Distorted tapered roller bearing unfit for further service

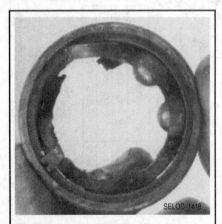

Fig. 237 Caged ball bearing set destroyed due to lack of lubrication, vibration, corrosion, metal particles, or all of the above

Fig. 238 A rusted and corroded gear. Water was allowed to enter the lower unit through a bad seal and cause this damage to the gear and other expensive parts

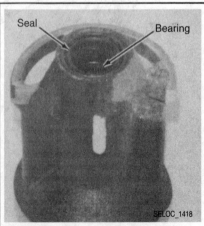

Fig. 239 This damaged bearing carrier was "frozen" in the lower unit and destroyed during the removal process

coating of light-weight oil to the roller bearing. Rotate the bearing and check for cracks or catches.

13. Inspect the propeller shaft oil seal surface to be sure it is not pitted, grooved, or scratched. Inspect the roller bearing contact surface on the propeller shaft for pitting, grooves, scoring, uneven wear, imbedded metal particles, and discoloration caused from overheating.

ASSEMBLY

◆ See Figures 229, 230 and 231

✱✱ SELOC WARNING

Before beginning the installation work, count the number of teeth on the pinion gear and on the forward gear in order to determine gear ratio for backlash checking/adjustment procedures. Knowing the number of teeth on the pinion and nose cone is necessary to obtain the correct tool setup for the shimming procedure.

Driveshaft Pinion Needle Bearing Installation

◆ See Figures 229

If removed from the gearcase, the first step in assembly should be to pull the new needle bearing assembly into the housing. This must be done BEFORE the installation of the nose cone gear.

Although special tools are noted for use to accomplish this, if the are not available, you can usually get away with using a long bolt, a series of appropriately sized washers, some locknut and possibly a large plate through which to thread the bolt.

1. Apply a thin coat of gear lube to the driveshaft needle bearing bore in the gear housing.

2. Place the NEW needle bearing (complete with the cardboard shipping tube still in place) on top of a suitable driver (such as #91-38628) with the numbers on the bearing facing away from the driver.

3. Reach in through the propeller shaft bore and place the needle bearing in the driveshaft bore, with the numbers facing UPWARD toward the top of the gearcase.

4. Assemble an installation tool to draw the bearing up into the gearcase. If using the Mercury recommended tools use a Puller Rod (#91-31229), Nut (#11-24156), Pilot (#91-36571), Top Plate (#91-29310), Washer (#12-34961) on top of the plate and as already noted, a Driver (#91-38628). Using the installation tool, draw the bearing up until it bottoms into the gearcase shoulder (using care not to exert excessive force one the bearing at this point).

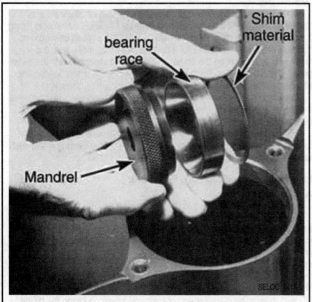

Fig. 240 Image showing the positioning of the driver (mandrel), race and shim (standard rotation)

Nose Cone Gear Bearing Race or Adaptor

◆ See Figures 230, 231, 240 and 241

✱✱ SELOC WARNING

On these motors a special bearing cup installation tool is normally used that contains centering rings so no special pilot is needed to keep the race/adaptor centered during installation. However, on many other gearcases Mercury recommends using an assembled bearing carrier as the pilot, which may be feasible on these models IF the special tools are not available.

Because the nose cone gear components vary somewhat between standard rotation and counter rotation units, some of the procedures in this section will differ depending upon the type of unit.

■ On counter rotation units don't confuse the Forward (propeller shaft) and Reverse (nose cone) gears. The FORWARD gear has both A SHORTER HUB and a SMALLER INNER DIAMETER needle bearing bore.

1. If the needle bearing was removed from the nose cone gear, install the replacement at this time. Position the gear with the teeth facing downward on a press, then use a suitable mandrel (#91-877321A1 for standard rotation or #91-86943 for counter rotation units) to press against the numbered side of the bearing until the tool bottoms against the gear. On most models (except some early EFI units) install the snapring to the back of the gear once the bearing is seated inside.

2. For standard rotation units, finish preparing the gear/bearing and install the shims along with the race as follows:

a. First, install the tapered roller bearing to the gear itself. Use a suitable mandrel to push on the bearing inner race until it is seated firmly against the gear. Position the gear aside for installation on the propeller shaft later.

b. Place the same amount of shim material saved during disassembly into the lower unit. If the shim material was lost, or if a new lower unit is being used, begin with approximately 0.020 in. (0.51mm) material. Coat the forward bearing race bore with gear lube.

c. Position the tapered bearing race squarely over the bearing bore in the front portion of the lower unit. Obtain a Bearing Cup Installation Tool (#91-18605A1) and place the tool over the tapered bearing race along with a Driver Cup (#91-31106).

d. Use the installation tool (by turning the hex head screw threaded into the tool base) until the tapered bearing race is seated against the shim material. Remove the installation tool assembly.

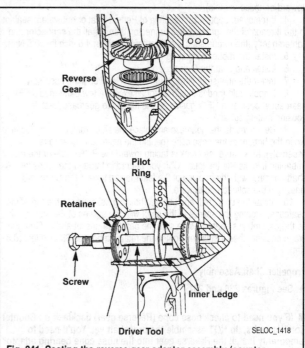

Fig. 241 Seating the reverse gear adaptor assembly (counter rotation units)

3. For counter rotation units, install the reverse gear bearing cup adaptor assembly, along with the necessary shims as follows:

a. If the gear housing was replaced or inspection determined the reverse gear bearing adaptor needed to be replaced (and you haven't pressed the new bearing into position yet), lubricate the adaptor bore using 2-4-C with Teflon or suitable marine grade grease, then press the new roller bearing into the adaptor with the numbered side of the bearing facing the driver shoulder. Drive the new bearing into the adaptor until the bearing is flush with the adaptor face.

b. If the reverse gear, adaptor, large thrust bearing and race were not replaced, install the same shims (or suitable replacements of the same thickness) which were removed during disassembly. HOWEVER, if any of these components were replaced, start with 0.008 in. (0.51mm) of shims.

■ **If reverse gear backlash has already been checked, keep in mind that ADDING 0.001 in. (0.025mm) of shims will REDUCE reverse gear backlash by that same amount. Conversely REMOVING 0.001 in. (0.025mm) of shims will INCREASE backlash by that same amount.**

c. Lubricate the gearcase bore using 2-4-C with Teflon, then position the shim(s) into the bore.

d. Next, install the bearing adaptor over top of the shims in the gearcase bore.

e. Now, position the gear WITHOUT the thrust race or thrust bearing into the adaptor.

f. Install the Bearing Adaptor Installation tool (#91-18605A1) into the bore as straight as possible. Lubricate the threads of the tool with 2-4-C with Teflon or equivalent, then turn the hex-head screw of the tool until the reverse gear/gearing cup adaptor JUST SEATS into the gearcase. DO NOT force the assembly down excessively or the reverse gear bearing will likely be damaged. You will know when the adaptor begins to seat because the force necessary to turn the screw will increase dramatically.

g. Once the adaptor is seated, remove the installation tool AND the reverse gear (since it needs to be out of the way to install the shift shaft assembly).

Shift Shaft Assembly

◆ **See Figures 229**

1. If disassembled, place the shift shaft bushing/housing on a work surface with the top of the housing facing upward.

2. Apply a light coating of 2-4-C with Teflon to the new O-ring and to the inner and outer surfaces of the new seal.

3. Using a suitable driver, carefully seal the new seal in the housing with the lips facing the same way as noted during removal (usually facing upward on these models).

4. If removed, apply a light coating of Perfect Seal or equivalent sealant to the threads of the speedometer connector, then install the connector and tighten gently (the torque spec is extremely low, about 4.5 inch lbs./0.5 Nm).

5. Install the rubber washer against the oil seal.

6. Install and seat the O-ring.

7. Insert the shift shaft through the washer/seal/bushing assembly.

8. If applicable (and it should be for all except a few early-model EFI gearcases) position a NEW gasket on the top of the gearcase-to-shift housing mating surface.

9. Reach inside the gearcase and place the shift crank over the locating pin in the bottom of the nose cone (behind the nose cone gear race assembly). Make sure the crank is facing toward the PORT side of the motor (meaning it looks like the letter "C" if you peered in through the propeller shaft opening, with the opening of the letter facing toward the Starboard side) with the splined end upward.

10. Install the shift shaft and bushing/housing assembly to the top of the gearcase, aligning the splines of the shaft with the splines of the shift crank as the assembly is seated. Make sure the gearcase gasket and/or O-ring(s) are in position, then install and tighten the retaining bolts to 60 inch lbs. (6.8 Nm).

Propeller Shaft Assembly

◆ **See Figures 230 and 231**

■ **IF you need to check nose cone (Reverse gear) backlash on Counter Rotation units, do NOT assemble the propshaft yet. You'll need to temporarily install the Reverse gear into the nose cone bearing adaptor without the propshaft in order to accomplish this. If necessary, skip ahead to Shimming to find out what you'll need to do.**

1. Slide the clutch dog onto the propeller shaft splines, aligning the cross pin hole and the detent hole with the appropriate slot and detent in the shaft.

2. Slide the nose cone gear assembly onto the propshaft, then install the shift spool, aligning the holes in the spool with the slot in the prop shaft and the hole for the clutch cross-pin.

3. Insert the cross-pin through the clutch, shaft and shift spool.

4. Place a small dab of 2-4-C with Teflon or an equivalent marine grade grease on the end of the detent pin, then install the pin into the hole of the sliding clutch.

5. Install the cross-pin ring over the sliding clutch. Take care not to over stretch the spring. Spirally wrap the spring into the groove on the sliding clutch, making sure the spring is wound straight onto the clutch, not overlapping itself. The spring must sit tight/flat against the clutch or it has been overstretched and must be replaced.

6. Position the propeller shaft aside for installation later.

Bearing Carrier Assembly

◆ **See Figures 230 and 231**

Because the propshaft gear and bearing components vary somewhat between standard rotation and counter rotation units, some of the procedures in this section will differ depending upon the type of unit.

Be sure to lubricate the inner and outer diameters of the bearings with gear lubricant prior to installation. Also, make sure to coat the inner diameter of the oil seals using 2-4-C with Teflon or an equivalent lubricant. The outer diameter of the oil seals should be coated using Loctite® 271 or an equivalent threadlocking compound. After installing the seals, wipe away any excess Loctite before proceeding.

1. If removed from the bearing carrier, oil and position the propeller shaft needle roller bearing into the aft (propeller) end of the bearing carrier with the numbered side toward the aft end (toward the driver). Press the needle bearing into the bearing carrier with a suitable mandrel such as 91-15755 which will give you the correct depth when the upper edge of the tool seats against the carrier.

■ **If the Mercury tools are not available, position the bearing to the depth measured prior to removal.**

2. Obtain an Oil Seal Driver (#91-31108). Place one seal on the longer shoulder side of the driver tool with the lip of the seal away from the shoulder

3. Press the seal into the bearing carrier until the seal driver bottoms against the bearing carrier

4. Place the second seal on the short shoulder side of the seal driver with the lip of the seal toward the shoulder. Press the seal into the bearing carrier until the seal driver bottoms against the bearing carrier. Clean excess Loctite from the seals.

5. For standard rotation units, proceed as follows to install the reverse gear and bearing:

a. Position the thrust washer over the reverse gear (the washer is tapered at the outer diameter, so make sure the LARGER diameter is facing the gear). Then, use a suitable mandrel to press a NEW ball bearing onto the back of the reverse gear until it bottoms. As usual, make sure the mandrel only presses on the portion of the bearing in contact with the gear (inner). Then use a mandrel to press the assembled gear and bearing into the forward (gearcase) end of the bearing housing until it bottoms out.

b. Install a new O-ring over the bearing carrier and position it between the carrier and thrust washer. Apply a coating of 2-4-C with Teflon and equivalent Lubricant onto the installed O-ring.

6. For counter rotation units, use a suitable mandrel to press the forward gear bearing INTO the bearing adaptor until the bearing is just flush with the lip of the adaptor.

Propeller Shaft, Driveshaft and Pinion Gear

◆ **See Figure 229**

1. If the tapered bearings were removed (there is only one on some early-model EFI gearcases) from the driveshaft, apply a light coat of gear oil to the bearing inner diameter(s). Install a used pinion nut onto the end of the driveshaft, leaving about 1/16 in. (1.6mm) of nut threads exposed (i.e. making sure the driveshaft threads are NOT extending past the nut to protect them from damage). Place the bearing(s) over the driveshaft (with the larger diameter end facing the pinion gear on single-bearing driveshaft assemblies OR with the smaller diameter end of the smaller bearing facing the pinion and the larger diameter end of the larger bearing facing both the smaller bearing and the pinion on most models which use 2 bearings on the

driveshaft). Then position the driveshaft/bearing over a bearing separator and suitable mandrel (an old shaft inner race will do the job) so that the force of pressing on the pinion nut will go right to the inner race of the new bearing(s). In this way, carefully press the bearing(s) onto the shaft, then remove the old nut.

2. If the pinion bearing needle bearings have fallen out, place all 18 needles into the needle bearing outer race using 2-4-C with Teflon or an equivalent marine grade grease to hold the needles in position.

3. For counter rotation units, lubricate the large reverse gear thrust bearing and thrust race with gear lube, then position the thrust race, followed by the thrust bearing into the bearing adaptor in the nose cone of the gear housing.

■ On Counter-Rotation models do NOT install the propeller shaft yet if you need to check gearcase Shimming and backlash. Instead, temporarily install the Reverse gear (nose cone gear) WITHOUT the propeller shaft and continue with driveshaft installation, follow the instructions in Shimming, then once you're certain all shims are properly selected finish installation as detailed here.

4. For standard rotation motors, prepare to install the propeller shaft by tilting the propeller end of the propeller shaft toward the Port side of the gearcase and rotating the shift shaft from reverse to neutral as you insert the propshaft and shift spool into the shift crank.

5. For counter rotation motors, prepare to install the propeller shaft, as follows:

a. Rotate the shift crank toward the aft end of the gear housing until it touches against the bearing adaptor (hold it in this position).

✳✳ SELOC CAUTION

When inserting the propshaft into the gear housing, be VERY careful as the needle bearings in the reverse gear bearing adaptor CAN become dislodged. If necessary, remove the propshaft again and inspect the bearing assembly to make sure nothing has been damaged or knocked out of position. IF a needle has become loose but not damaged it can normally be snapped back into place.

b. Tilt the propeller end of the propshaft toward the port side of the gearcase and begin to slowly lower it into the gear housing.

c. Continue to slowly insert the tilted shaft until the reverse gear hub comes into contact with the bearing adaptor and the shaft is fully inserted into the gear. At this point SLOWLY and GENTLY move the shaft toward the center of the housing and lower the reverse gear into the bearing adaptor. Hopefully the shift spool will engage with the shift crank as the propshaft is centered. If not, pull back and repeat.

6. Operate the shift shaft to ensure the crank is sitting properly on the spool. The sliding clutch should move forward when the shift shaft is turned clockwise or aft when the shaft is turned counterclockwise.

7. If not done already, slide the rubber washer down the top of the shift shaft so that it JUST touches the oil seal in the bushing/housing assembly.

8. Place the shims for the upper driveshaft tapered bearing(s) into the driveshaft housing bore. If the original shims were not retained or if the pinion gear, driveshaft, driveshaft bearing(s), tapered bearing cap or gearcase was replaced, start with 0.038 in. (0.96mm) of shim material. IF you don't need to check Pinion Gear Depth and Upper Driveshaft Bearing Clearance on models with 2 upper driveshaft bearings, install the lower bearing cup, HOWEVER, if these clearances need to be checked, leave the lower bearing cup out of the assembly temporarily, it will be reinstalled after the Pinion Gear Depth has been checked.

9. To ease installation glue the pinion gear nut washer to the bottom of the pinion gear using 3M Adhesive or Bellows Adhesive.

10. If the backlash has not changed, apply a light coating of Loctite® 271 or equivalent threadlocking material to the threads of a NEW PINION NUT. However, if you're just installing the driveshaft to check gearcase Shimming, then use the old nut and leave off the threadlocking compound at this time. Place the pinion nut into the MR slot in the pinion nut holding tool (#91-61067A3) with the flat side facing away from the pinion gear.

11. Position the pinion gear in the gear housing, right below the driveshaft bore, with the teeth meshed with the teeth of the nose gear.

12. Insert the driveshaft carefully down into the gearcase, rotating it slightly as necessary to engage the shaft splines with the splines in the pinion gear. Continue to insert the driveshaft until the tapered bearing is against the bearing race.

■ It may be necessary to lift up on the driveshaft JUST slightly in order to install the pinion nut tool.

13. Holding the pinion nut in position with the pinion nut tool, carefully turn the driveshaft to thread the nut.

14. Apply a light coating of 2-4-C with Teflon to the threads of the bearing retainer. Insert the upper tapered bearing cut and the retainer over the driveshaft and into the gearcase. Start to thread the retainer.

15. Install the bearing carrier into the gear housing BACKWARDS to hold the propeller shaft and the pinion nut wrench in position.

16. Tighten the pinion nut using a driveshaft holding tool (#91-56775) to turn the driveshaft clockwise. Tighten to 70 ft. lbs. (95 Nm).

17. Remove the bearing carrier, pinion nut adapter and the driveshaft holding tool, THEN install the driveshaft bearing retainer wrench (#91-43506) and reinstall the driveshaft holding tool. Use the retainer wrench to tighten the driveshaft bearing retainer to 100 ft. lbs. (135 Nm).

18. Now, if you've replaced the gearcase housing, or if you've replaced one or more of the bearings SKIP ahead to the Shimming procedures in order to check backlash and determine if some of the shims must be replaced. However, if only seals have been replaced, then you can proceed with the assembly.

Bearing Carrier Installation

◆ See Figures 229, 230 and 231

The only real difference in the procedures comes from the slight difference in the propeller shaft assembly between the standard rotation units (where the prop shaft gear, or REVERSE in this case, is mounted in the bearing carrier) and the counter rotation units (where the prop shaft gear, FORWARD in this case, and bearing adaptor are installed on the prop shaft itself). The assembly procedures take into account that you have to install the forward gear and bearing assembly piece-by-piece onto the already positioned propeller shaft for counter rotation units.

As always, be sure to lubricate bearings or thrust surfaces with gear oil and oil seals using 2-4-C with Teflon or an equivalent lubricant prior to installation.

1. Place the gearcase a suitable support fixture. For standard rotation units the propshaft bore can be vertical OR horizontal, but for counter rotation units it must be vertical.

2. For counter rotation units, it's time to install the rest of the propeller shaft/gear/bearing components, as follows:

a. Install the appropriate spacer shim into the gear housing.

b. Apply a light coating of gear lube to the thrust bearing, then position the bearing onto the Forward gear bearing adaptor. Next place the Forward gear over the thrust bearing and the bearing adaptor.

c. Insert the forward gear installation tool (#91-815850) or a suitable driver into the back of the bearing adaptor and gear assembly (with the gear facing downward away from the tool), then using the tool place the assembly over the propeller shaft and into the gear housing. Using the tool, apply downward pressure to slide the bearing adaptor into position then remove the tool from the assembly and from the gearcase housing.

d. Next, install the thrust race on top of the bearing adaptor.

e. Apply a light coating of gear lube to the small thrust bearing, then install it onto the thrust race.

f. Install the thrust collar with the STEPPED SIDE FACING DOWNWARD (toward the thrust bearing).

g. Pull up very slightly on the propeller shaft JUST enough to gain access to the groove in the shaft, then install the 2 keepers into the groove and allow the propeller shaft to seat again.

h. Install the second thrust collar with the STEPPED SIDE FACING UPWARD (away from the first thrust bearing).

i. Apply a light coating of gear lube to the second thrust bearing, then install it onto the second thrust collar.

j. Apply a light coating of gear lube to the second thrust bearing race, then install it into the surface inside of the BEARING CARRIER.

k. Apply a light coating of 2-4-C with Teflon or equivalent marine grade grease to the bearing carrier O-ring and install the O-ring to the bearing adaptor.

3. If not done already, make sure to apply a light coating of gear lube to the bearing carrier needle bearing and pack the space between the oil seals with a coating of 2-4-C with Teflon, or an equivalent marine grade grease. Use the same grease to apply a light coating to the outer diameter of the bearing carrier, where it contacts the gearcase.

4. Carefully install the bearing carrier into the gearcase. On standard rotation units, you should rotate the driveshaft very slowly in a clockwise direction (viewed from above) to help mesh the reverse gear in the bearing carrier with the pinion gear as the bearing carrier is seated.

■ During installation be sure to align the V-shaped notch in the top of the carrier with the alignment hole in the gear housing bore.

5. Once the carrier is seated, place the large tab washer against the bearing carrier. If you properly aligned the V-shaped notch in the top of the carrier with the alignment hole in the gear housing bore, the tab washer should install with the external tab inserted into the hole in the gear housing. At this point the V-shaped tab and the V-notch in the carrier should be in alignment.

■ The cover nut torque is extremely high. Make sure the gearcase is either mounted in a strong stand or bolted to a driveshaft housing to avoid possible gearcase damage.

6. Apply a light coating of 2-4-C with Teflon or an equivalent marine grade grease to the threads of the cover nut and threads in the gearcase housing, then install the nut into the housing, making sure the word **OFF** and arrow (if applicable) are visible. Use the cover nut tool (#91-61069) or equivalent to tighten the cover nut to 210 ft. lbs. (285 Nm).

7. Bend one of the lock tabs of the lock washer outward, into the cover nut (normally only ONE will align). Bend the remaining tabs of the washer inward toward the front of the gear housing.

■ If one tab of the lock washer DOES NOT align between 2 notches so it can be bent down locking the retainer continue to tighten the retainer until one does. DO NOT loosen the retainer from spec in order to align a tab!

8. If not done earlier during driveshaft installation (assuming no preload check needed to be performed) install the oil seal carrier, followed by the water pump assembly, keeping the following points in mind:

• The dual seals in the carrier should be installed back-to-back so the inner or lower seal faces the gearcase and the upper or outer seal faces the water pump.

• A NEW O-ring should be used, and coated (along with the inner lips of the seals) with 2-4-C with Teflon or an equivalent lubricant.

• If the divider block was removed from the top of the gearcase it must be properly positioned, then install the water pump base using a new base gasket.

9. Refill the gearcase and check for leaks, then install the gearcase to the outboard as detailed earlier in this section.

SHIMMING

Backlash

◆ See Figure 242

For any gearcase to operate properly the pinion gear on the driveshaft must mesh properly with the forward and reverse gears. And, remember that these gears remain in contact with the pinion gear at all times, regardless of whether or not the sliding clutch has locked one of them to the propeller shaft. If these gears are not properly meshed, there is a greater risk of breaking teeth or, more commonly, rapid wear leading to gearcase failure.

On some of the smaller Mercury gearcases there is little or no adjustment for the positioning of these gears. However, the gearcases used on these large Mercury motors are all equipped with shim material that is used to perform minute adjustments on the positioning of the bearings (and therefore the bearings and gears). Shim material can be found in various positions, depending upon whether it is a standard or counter rotation unit.

For standard rotation units shim(s) can be found between the forward gear bearing/race assembly and the nose cone, as well as underneath the upper driveshaft bearing.

For counter rotation units, shim(s) can be found between the forward gear and thrust bearing on the prop shaft, between the reverse gear bearing adaptor and nose cone of the gearcase, and, like the standard rotation units, underneath the upper driveshaft bearing.

If a gearcase is disassembled and reassembled replacing only the seals, then no shimming should be necessary. However, if shafts, gears/bearings and/or the gearcase housing itself is replaced it is necessary to check the Pinion Gear Depth and the Forward and Reverse Gear Backlash to ensure proper gear positioning.

Pinion Gear Depth

◆ See Figures 243, 244 and 245

■ This procedure starts with the gearcase mostly reassembled (following our procedures up to and including driveshaft reassembly, but with propshaft gear not installed on the shaft itself for counter rotation units, and for all units the propshaft bearing carrier, driveshaft upper seal carrier, water pump plate and water pump still removed).

There are two possible tools you can use to determine Pinion Gear Depth. The first, #91-12349A2 must be used WITHOUT the propeller shaft installed, which means the nose cone gear must be removed from the propeller shaft. This tool is usually better served on counter rotation units where we warned you earlier to proceed this way. However, on all units you can ALSO use #91-56048T and the advantage of this second tool is that it can be used WITH the propeller shaft and nose cone gear installed.

Obtain a Bearing Preload Tool (#91-14311A1). This tool consists of a spring, which seats against the temporarily installed water pump base plate from your lower unit and is used to simulate the upward driving force on the pinion gear bearing. Install the following, one by one, over the driveshaft.

a. The water pump base plate from your gearcase.
b. Collared thrust washer
c. Flat thrust washer
d. Thrust bearing
e. Flat thrust washer
f. Spring
g. Special large bolt, with the nut on the bolt threaded all the way to the top of the threads

1. Pull up on the driveshaft and tighten the 2 Allen set screws in the special large bolt head. These set screws grip the driveshaft and prevent the tool from slipping on the driveshaft. Tighten the two set screws securely.

2. Measure the distance between the bottom surface of the bolt head and the top surface of the nut. Record this measurement.

3. Use a wrench to hold the bolt head steady and another wrench to tighten the nut down against the spring. Continue tightening the nut until the distance between the bottom surface of the bolt head and the top surface of the nut equals 1.00 in. (25.4mm) plus the distance recorded when the tool was installed.

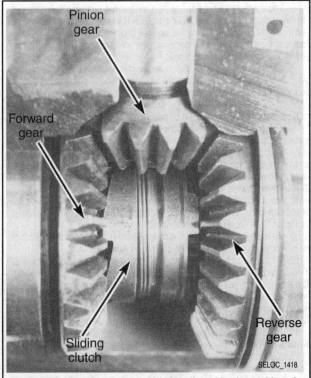

Fig. 242 Backlash is the amount of play allowed by the position of the forward and reverse gears in relation to the pinion gear (standard rotation shown, gear positions are reversed in counter rotation units)

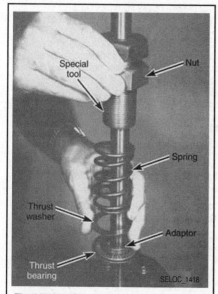

Fig. 243 Installing the pinion gear pre-loading tool

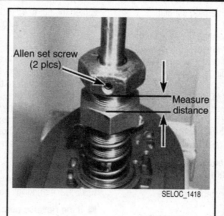

Fig. 244 Set the nut to the proper distance to set pre-load

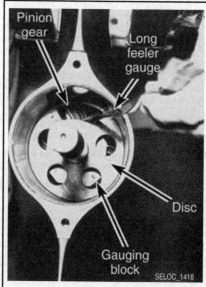

Fig. 245 Checking the gap between the pinion gear and the gauging surface

4. The driveshaft has now been forced upwards placing an upward preload on the pinion gear bearing. Turn the driveshaft 2 or more turns to make sure the bearings are seated.

5. Obtain the correct Pinion Gear Locating Tool (either #91-12349A2 along with Flat # 4 and Disc # 2 from the tool if the propshaft is not installed or #91-56048T if the propshaft IS installed).

6. Insert the locating tool (with the appropriate flat and disc in place), making sure the tool bottoms on the bearing carrier shoulder in the housing.

7. The pinion depth is the distance between the bottom of the pinion gear teeth and the flat surface of the gauging block.

8. Measure this distance by inserting a long feeler gauge into the access hole and between the pinion gear teeth and gauging block flat face. Be sure to take measurements at 3 different locations, rotating the driveshaft 120 degrees each time. Take the average of the 3 readings (add the three reading together and divide the total by 3).

9. The correct clearance is 0.025 in. (0.64mm). If the average clearance is correct, skip the next step, but leave the tools in place.

10. If the clearance is not correct, add or subtract shim material from below the upper driveshaft bearing to lower or raise the gear.

• If the pinion gear depth was found to be greater than 0.025 in. (0.64mm), then that amount of shim material must be added to lower the gear.

• If the pinion gear depth was found to be less than 0.025 in. (0.64mm), then that amount of shim material must be removed to raise the gear.

■ **The difference between 0.025 in. (0.64mm) and the actual clearance measured is the amount of shim material needed to correct the pinion gear depth.**

11. For early-model EFI units with a single upper driveshaft bearing, skip to the last step in this procedure. For models with 2 upper driveshaft bearings continue as follows: once the correct pinion gear height is achieved, remove the Pinion Gear Depth preload tool and loosen the pinion nut, then remove the upper retainer and driveshaft again for access. Oil and install the LOWER bearing cup (which was left out during assembly on purpose) and reinstall the pinion gear and driveshaft to the gearcase.

12. Reinstall the upper driveshaft bearing cut and the upper bearing retainer. Tighten the upper bearing retainer once more to 100 ft. lbs. (135 Nm) and the pinion nut once again to 70 ft. lbs. (95 Nm).

13. Now, push down on the driveshaft and check clearance between the pinion gear and the shoulder of the Pinion Gear Locating Tool. Clearance (for these 2 upper driveshaft bearing gearcases) must be 0.020-0.024 in. (0.51-0.61mm). If clearance is NOT as specified change shims under the LOWER BEARING CUP ONLY. Change the lower bearing cup shims as necessary to obtain the proper clearance without changing the pinion gear depth.

✳✳ SELOC CAUTION

DO NOT change shims under the upper bearing cup in order to change the clearance measured in the last step, it won't do that, you'll wind up changing Pinion Gear Depth instead!

14. Remove only the Pinion Gear Locating Tools from the inside of the gearcase. If you didn't have to check the Upper Driveshaft Bearing Clearance leave the Pinion Gear Bearing Preload Tool in place on the driveshaft in order to check Gear Backlash. If you DID have to check it, reinstall the Pinion Gear Bearing Preload Tool and check the Forward and Reverse Gear Backlash as detailed later in this section.

Forward and Reverse Gear Backlash - Standard Rotation Units

◆ **See Figures 246 and 247**

■ **The bearing carrier is temporarily installed for this procedure.**

Final installation of the bearing carrier (and possibly propeller shaft if the forward bearing requires any shimming) is usually performed later because an old pinion gear nut is normally installed on the driveshaft without the use of Loctite to determine the pinion gear depth (in case it needs to be disassembled again to make changes to the pinion gear bearing or forward gear bearing shims).

1. First, check and adjust the Pinion Gear Depth, as detailed earlier in this section. Leave the pinion bearing preload tool installed on top of the gearcase.

2. Next, if not done earlier, install the propeller shaft and nose cone gear assembly.

3. Install the bearing carrier, if necessary turn the driveshaft slightly to align the pinion gear teeth to the reverse gear. Align the V-shaped notch in the bearing carrier and the V-shaped tab on the tabbed washer with the alignment hole in the housing (as detailed earlier under Bearing Carrier Installation), then lubricate and install the bearing carrier retainer nut. Tighten the retainer down fully by hand, but do NOT torque at this time.

4. Install a bearing carrier puller with the arms of the puller on the carrier and the center bolt on the end of the propeller shaft. Tighten the puller center bolt to a torque value of 45 inch lbs. (5 Nm). This action places a preload on the forward gear, pushing the forward gear into the forward gear bearing.

5. Attach the proper Backlash Dial Indicator Rod (either #91-78473 for 1.87:1 ratio gearcases or #91-53459 for 1.75:1 ratio gearcases, as calculated using the number of forward gear teeth divided by the number of pinion gear teeth) to the driveshaft. Position a dial indicator (using a magnetic base or a threaded rod with nuts and plates to position it alongside the Indicator Rod) and point the dial indicator to the I mark on the indicator tool. Tighten the indicator tool onto the driveshaft and rotate the shaft so the needle in the dial

makes at least one full revolution then comes back to zero on the dial indicator scale.

6. Recheck the puller torque, then rotate the driveshaft back-and-forth and observe movement of the dial indicator. Total movement is the forward gear backlash, this is the first reading. Loosen the indicator tool and move it 90 degrees (1/4 turn) then repeat and record the reading. Do this for a total of 4 readings (each at 1/4 turn from the previous).

7. Add all 4 readings together and divide by four to get the average and THAT is your measured forward gear backlash, it should be 0.017-0.028 in. (0.431-0.711mm) for all gearcase ratios. If the backlash is too great, add shim material behind the forward gear bearing race. If the backlash is too small, remove shim material from behind the forward gear bearing race.

■ **For each 0.001 in. (0.025mm) of shim which is added or removed expect backlash to increase or decrease about 0.001 in. (0.025mm).**

8. Although the reverse gear backlash is not adjustable on these models, it should still be checked as follows:

a. Remove the bearing carrier puller installed earlier in this procedure and install the Pinion Nut Holder (#91-61067A3) used to thread the pinion nut earlier over top of the propeller shaft (large end toward the bearing carrier). Install the washer (#12-54048) and the propeller retaining nut, then

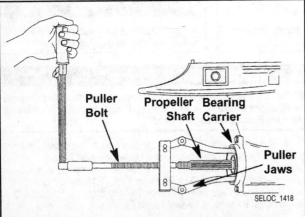

Fig. 246 Use a puller to fully seat the propeller shaft and forward gear. . .

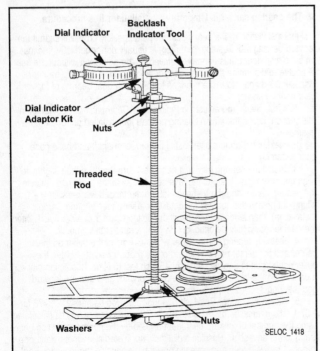

Fig. 247 . . .then check gear backlash using the appropriate indicator tool

tighten the nut to 45 inch lbs. (5 Nm). Rotate the driveshaft 3 full turns clockwise and then retorque the propeller nut once again to 45 inch lbs. (5 Nm).

b. If removed, reinstall the backlash indicator tool used for Forward Gear Backlash and the dial indicator. Once again point the dial indicator to the I mark on the indicator tool. Tighten the indicator tool onto the driveshaft and rotate the shaft so the needle in the dial makes at least one full revolution then comes back to zero on the dial indicator scale.

c. Gently rock the driveshaft back and forth while watching the dial indicator. Total movement is the reverse gear backlash, this is the first reading. Loosen the indicator tool and move it 90 degrees (1/4 turn) then repeat and record the reading. Do this for a total of 4 readings (each at 1/4 turn from the previous).

9. Add all 4 readings together and divide by four to get the average and THAT is your measured forward gear backlash, it should be 0.030-0.050 in. (0.76-1.27mm) for all gearcase ratios.

■ **If the Reverse gear backlash is not in spec either the gearcase is improperly assembled OR the parts are excessively worn and must be replaced.**

10. To remove the bearing preload tool, first remove the dial and indicator tools, then screw the bottom nut of the bearing preload tool all the way up against the bottom of the bolt head. Loosen the two Allen screws, then carefully remove all remaining components of the tool.

11. Finish assembly of the gearcase using Loctite on the pinion nut and the correct shims (if any need to be changed).

Reverse and Forward Gear Backlash - Counter Rotation Units

◆ **See Figures 218 and 248**

Because of differences in the propeller shaft and gear/bearing assemblies BOTH the Reverse AND Forward gear lashes are adjustable on counter rotation units.

■ **To make these backlash checks, first the nose cone (Reverse) gear is temporarily installed without the propeller shaft in order to check gear backlash. Then, propeller shaft is temporarily installed (not fully assembled) for the second half of this procedure to check the prop shaft (Forward) gear lash. The first half is performed using a suitable driver tool to center and preload the nose cone gear/bearing assembly.**

1. First, check and adjust the Pinion Gear Depth and, if applicable the Upper Driveshaft Bearing Clearance, as detailed earlier in this section. Either leave installed or reinstall the pinion bearing preload tool on top of the gearcase.

2. Next, insert the Bearing Adaptor Installation Tool (#91-18605A1) to center and preload the Reverse gear in the nose cone of the gearcase. Insert driver and slide the pilot ring over the driver until it rests against the ledge in the gearcase which is right before the pinion gear. Next install thread retainer into the gearcase bearing carrier cover nut threads, then tighten screw to 45 inch lbs. (5 Nm) to preload the gear/bearing assembly.

3. Attach the proper Backlash Dial Indicator Rod (either #91-78473 for 1.87:1 ratio gearcases or #91-53459 for 1.75:1 ratio gearcases, as calculated using the number of forward gear teeth divided by the number of pinion gear teeth) to the driveshaft. Position a dial indicator (using a magnetic base or a threaded rod with nuts and plates to position it alongside the Indicator Rod) and point the dial indicator to the I mark on the indicator tool. Tighten the indicator tool onto the driveshaft and rotate the shaft so the needle in the dial makes at least one full revolution then comes back to zero on the dial indicator scale.

4. Recheck the torque on the bearing adaptor tool, then rotate the driveshaft back-and-forth and observe movement of the dial indicator. Total movement is the reverse gear backlash, this is the first reading. Loosen the indicator tool and move it 90 degrees (1/4 turn) then repeat and record the reading. Do this for a total of 4 readings (each at 1/4 turn from the previous).

5. Add all 4 readings together and divide by four to get the average and THAT is your measured reverse gear backlash, it should be 0.040-0.060 in. (1.0-1.5mm) for all gearcase ratios. If the backlash is too great, add shim material behind the reverse gear bearing adaptor. If the backlash is too small, remove shim material from behind the reverse gear bearing adaptor.

■ **For each 0.001 in. (0.025mm) of shim which is added or removed expect backlash to increase or decrease about 0.001 in. (0.025mm).**

6. Remove the bearing adaptor installation tool.

7. Next it's time to check the Forward Gear (propeller shaft gear) backlash. Prepare by installing some of the forward gear and propeller shaft

components (just like normal shaft and gear installation). Start by placing a load washer on the propeller shaft and temporarily installing it into the gearcase (this is a disassembled shaft without the shift spool or clutch dog, it just has the load washer).

8. Installing the forward gear spacer shim into the gear housing.

9. If not done already, oil and install the thrust washer, thrust bearing and forward gear into the bearing adaptor. Use the Forward Gear Installation Tool (#91-815850) inserted through the back of the adaptor to hold and lower the assembly against the shim in the gearcase. Install the tool with the bearing adaptor assembly over the propeller shaft and push downward to seat the adaptor, then remove the tool leaving the adaptor assembly behind.

10. Now, install the bearing carrier over the propeller shaft, pushing down until it is fully seated. Align the V-shaped notch in the bearing carrier and the V-shaped tab on the tabbed washer with the alignment hole in the housing (as detailed earlier under Bearing Carrier Installation), then lubricate and install the bearing carrier retainer nut. Tighten the retainer down fully to 210 ft. lbs. (285 Nm) using the Bearing Carrier Cover Nut tool (#91-61069).

11. Obtain a piece of PVC pipe, 5 in. (12.7cm) long by 2 in. (5.08cm) diameter and drill a small 3/8 (22.2mm) hole through the side which will be installed toward the propeller nut, this way a screwdriver can be inserted through it and into the propeller shaft splines in order to prevent the splines from turning while tighten in the retaining nut. THEN install it over the propeller shaft using a flat washer and the propeller nut. Tighten the nut to 45 inch lbs. (5 Nm) in order to seat the Forward gear against the thrust bearing and help keep the propeller shaft from moving while checking the lash.

12. If not done already for Pinion Gear Depth measurement or Reverse gear backlash measurement install the pinion bearing preload tool on top of the gearcase.

13. If not done already for Reverse gear backlash measurement attach the proper Backlash Dial Indicator Rod (either #91-78473 for 1.87:1 ratio gearcases or #91-53459 for 1.75:1 ratio gearcases, as calculated using the number of forward gear teeth divided by the number of pinion gear teeth) to the driveshaft. Position a dial indicator (using a magnetic base or a threaded rod with nuts and plates to position it alongside the Indicator Rod) and point the dial indicator to the I mark on the indicator tool. Tighten the indicator tool onto the driveshaft and rotate the shaft so the needle in the dial makes at least one full revolution then comes back to zero on the dial indicator scale.

14. Recheck the torque on the bearing adaptor tool, then rotate the driveshaft back-and-forth and observe movement of the dial indicator. Total movement is the Forward gear backlash, this is the first reading. Loosen the indicator tool and move it 90 degrees (1/4 turn) then repeat and record the reading. Do this for a total of 4 readings (each at 1/4 turn from the previous).

15. Add all 4 readings together and divide by four to get the average and THAT is your measured forward gear backlash, it should be 0.017-0.028 in. (0.431-0.711mm) for all gearcase ratios. If the backlash is too great, remove shim material from in front of the forward gear bearing adaptor. If the backlash is too small, add shim material in front of the forward gear bearing adaptor.

■ **For each 0.001 in. (0.025mm) of shim which is added or removed expect backlash to increase or decrease about 0.001 in. (0.025mm).**

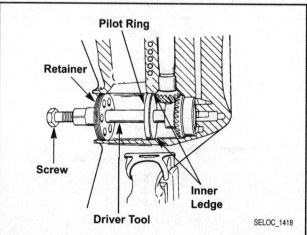

Fig. 248 Using a bearing adaptor tool to seat the Reverse gear when checking backlash

Labels in figure: Pilot Ring, Retainer, Screw, Driver Tool, Inner Ledge

SELOC_1418

16. To remove the bearing preload tool, first remove the dial and indicator tools, then screw the bottom nut of the bearing preload tool all the way up against the bottom of the bolt head. Loosen the two Allen screws, then carefully remove all remaining components of the tool.

17. Finish assembly of the gearcase using Loctite on the pinion nut and the correct shims (if any need to be changed).

Gearcase - 225/250 Hp (3.0L) V6 Pro/Sport XS Models

◆ **See Figures 250 and 251**

There are basically 2 similar design gearcases covered in this section. The Pro XS and the Sport XS gearcases. The basic design of the gearcase, water pump, driveshaft and propeller shaft are all almost the same between the 2 units (and as such, are VERY similar to the conventional rotation 200-250 hp gearcases covered earlier in this section).

Slight differences between these gearcases and the conventional gearcases covered earlier can be found in the driveshaft seal carrier, as well as the propeller shaft bearing carriers. The Sport XS models also contain additional differences, specifically in the more streamlined shape of the gearcase housing itself, as well as differences in the shifter assembly and an additional water passage plate that is found under the water pump assembly.

REMOVAL & INSTALLATION

◆ **See Figures 182 and 220**

MODERATE

Like all of the other large Mercury gearcases, the rotating/ratcheting shift mechanism used on these models utilizes a splined shift shaft connection, making it un-necessary to remove shift linkage before gearcase removal.

1. For safety disconnect the battery cables.

2. If necessary for overhaul purposes (it's necessary to drain the gear oil on most of these models) or for safety/access, remove the propeller as detailed in the Maintenance & Tune-Up section.

■ **It's actually a good idea to at least loosen the Oil Level and/or Vent screw(s) first, to make sure it will come out without a problem. You can't refill the gearcase, if you can't get the vent screw loose.**

3. If necessary for gearcase overhaul purposes, position a suitable clean container under the lower unit. Remove the Fill screw on the bottom of the lower unit, and then the Oil Level and/or Vent screw(s). The vent screw must be removed to allow air to enter the lower unit behind the lubricant. Allow the gear lubricant to drain into the container.

As the lubricant drains, catch some with your fingers from time-to-time and rub it between your thumb and finger to determine if there are any metal particles present. Examine the fill plug. A small magnet embedded in the end of the plug will pickup any metal particles. If metal is detected in the lubricant, the unit must be completely disassembled, inspected, and the damaged parts replaced.

Check the color of the lubricant as it drains. A whitish or creamy color indicates the presence of water in the lubricant. Check the container for signs of water separation from the lubricant. The presence of any water in the lubricant is bad news. The unit must be completely disassembled; inspected; the cause of the problem determined; and then corrected.

4. If not done already, rotate the outboard unit to the full UP position and engage the tilt lock lever.

5. Shift the gearcase into Neutral.

6. These models are normally equipped with a round anode where you would find a trim tab on a conventional model. This anode must be removed to access the gearcase retaining bolt which is found hidden above it in the gearcase cavity. Remove the rubber plug on the rear edge of the gearcase housing (above the anti-cavitation plate), then loosen the bolt in the recess which secures the anode plate from above. Remove the anode plate from the anti-cavitation plate for access to the gearcase retaining bolt. Loosen and remove the previously hidden gearcase retaining bolt.

■ **Some units may be equipped with a speedometer hose that is often secured by a wire tie which must be cut to detach the hose from the fitting. You may need to drop the housing first for access to the fitting.**

7. Alternately and evenly loosen the 4 gearcase mounting locknuts (2 on each side), but DON'T try to remove any 1 nut before the opposite side is loosened or the driveshaft housing could be damaged. Once they are sufficiently loosened, pull down on the lower unit gear housing and

CAREFULLY separate the lower unit guiding it straight out of the exhaust housing. Guide the shift rod and driveshaft out of the exhaust housing as the lower unit is removed.

8. Place the lower unit in a suitable holding fixture or work area.

To install:

■ **Always fill the lower unit with lubricant and check for leaks before installing the unit to the driveshaft housing (could save you the trouble of removing it again right away).**

9. Take time to remove any old gasket material from the fill and vent recesses and from the screws.

10. Place the lower unit in an upright vertical position. Fill the lower unit with Super-Duty Lubricant, or equivalent, through the fill opening at the bottom of the unit. Never add lubricant to the lower unit without first removing the vent screw and having the unit in its normal operating position - vertical. Failure to remove the vent screw will result in air becoming trapped within the lower unit. Trapped air will not allow the proper amount of lubricant to be added.

11. Continue filling slowly until the lubricant begins to escape from the vent or level opening (as applicable) with no air bubbles visible.

12. Use a new gasket and install the vent and/or level screw(s). Slide a new gasket onto the fill screw. Remove the lubricant tube and quickly install the fill screw.

13. Visibly check the lower unit for leaks. Recheck for leaks after the first time the unit is warmed to normal operating temperatures from use.

14. If not done already, swing the exhaust housing outward until the tilt lock lever can be actuated, and then engage the tilt lock.

15. Apply a coating of 2-4-C Marine Lubricant with Teflon to the splines of the shift shaft coupler and the driveshaft.

■ **DO NOT put lubricant on the ENDS of the shafts, as it could prevent the shaft splines from fully seating in the shift coupler and the crankshaft.**

■ **On some models you may have to feed the speedometer tube through the opening in the driveshaft housing as the gearcase is raised into position.**

16. Make sure the powerhead gear shifter assembly is in the Neutral position.

17. Carefully bring the gearcase into alignment with the exhaust housing and slowly insert it STRAIGHT UP and into position, while aligning the driveshaft, shift shaft and water tube seal.

■ **If necessary, slowly turn the crankshaft in the normal direction of rotation (clockwise when viewed from above) using the flywheel to align the driveshaft-to-crankshaft splines. Alternately, you can shift both the gearcase and the powerhead linkage into Forward and turn the propeller shaft to align the driveshaft splines, but that may make shift shaft alignment slightly more difficult.**

18. While still holding the gearcase in position, place the flat washers onto the studs (located on either side of the driveshaft housing), then start a locknut on each of these studs, threading them down by hand until finger-tight.

■ **Mercury recommends installing a NEW gearcase retaining bolt in the anode plate access, one that is equipped with a locking patch. Although, alternately you might be able to reuse the old bolt if it is coated with a suitable medium strength threadlocking compound.**

19. Make sure the shift shaft spline engagement seems ok at this point by attempting to gently move the shift block just slightly.

20. Next, thread the bolt at the rear of the housing, inside the anode plate recess, but DO NOT TIGHTEN IT FULLY AT THIS TIME.

✳✳ SELOC CAUTION

If the gearcase does not easily flush up against the driveshaft housing DO NOT force the gearcase to seat using the attaching nuts. Instead remove the gearcase again and double-check the shaft spline alignments. Also, make sure there is no lubricant on top of the shift shaft or driveshaft (or in the underside of the shift shaft and crankshaft located in the housing).

21. Alternately and evenly tighten 2 of the locknuts which were installed earlier (one on either side of the lower motor mount/swivel attachment point) to 55 ft. lbs. (75 Nm).

22. Verify proper shift shaft spline engagement as follows:
a. Place the guide block anchor pin into the Forward position (if not already) while turning the prop shaft in the normal direction of rotation. The propeller shaft should not turn once in gear.
b. Place the guide block in the Neutral position and verify that the propeller shaft now turns freely.
c. Place the guide block in the Reverse gear position, now slowly rotate the FLYWHEEL CLOCKWISE when viewed from the top and verify that the propeller shaft now turns in the normal direction of reverse gear for the gearcase in question (meaning counterclockwise on normal rotation gearcases or clockwise on counter-rotation gearcases).

■ **If the gearcase does not shift properly into gear, the housing must be removed again and the shaft splines realigned as necessary for proper shifting.**

23. If not done already, place washers and locknuts onto the remaining studs on either side of the gearcase, then tighten them to 55 ft. lbs. (75 Nm).
24. Tighten the gearcase retaining bolt securely inside the anode plate cavity, then install the anode plate. Install the plastic cap into the trim tab bolt opening at the rear edge of the driveshaft housing.
25. If applicable, reconnect the speedometer tube into the junction.
26. If the propeller was removed, install the propeller.
27. Reconnect the battery cables.

DISASSEMBLY

◆ **See Figures 250 and 251**

Unlike most of the smaller Mercury gearcases, the lower unit found on these models contains various shims for adjusting pinion height and forward/reverse gear backlash. Always keep track of the shims during removal, since if the gearcase or the shafts/bearings are being reused the shims should be returned to the same positions. Also, if the gearcase and/or shafts/bearings are replaced, the shims can be sometimes be used as a good starting point for the shimming procedures.

Also unique to the gearcases on these largest V6 Mercurys is the shifter set-up. The design has the shaft extending into splines on a C-shaped shift crank that cups the end of the propeller shaft. The design is such that the propeller shaft is not removed until AFTER the driveshaft and pinion assembly has been unbolted and removed giving the propeller shaft sufficient clearance and play to be freed from this assembly. The result is that it is difficult-to-impossible to get around the special tool needed to secure the pinion nut (Pinion Nut Wrench/MR slot #91-61067003)

Pre-Disassembly Inspection

◆ **See Figures 249, 250 and 251**

Prior to disassembling the gearcase on Pro/Sport XS models Mercury advises to use a dial gauge to check propeller shaft run-out and deflection in order to determine the condition of certain internal components (including the shaft itself, the reverse gear and the thrust washer).

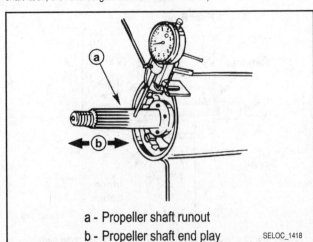

a - Propeller shaft runout
b - Propeller shaft end play

SELOC_1418

Fig. 249 Use a dial gauge to perform a pre-disassembly inspection

1. Position a dial indicator so that it can read movement side-to-side on the propeller shaft. Push the shaft to one side and zero the indicator, then push the shaft to the other side and record the amount of movement. A deflection of 0.003 in. (0.08mm) or more means the propeller shaft bearing is worn and should be replaced.

2. Next rotate the propeller shaft while watching the dial indicator for movement. A total movement of 0.009 in. (0.229mm) or more means the shaft is bent and should be replaced.

3. Lastly, position the dial gauge on the tip (rear end of the shaft). Push inward and zero the gauge, then pull outward and record total end play. If movement is more than 0.045 in. (1.143mm) then you should inspect the reverse shoulder on the propeller shaft, the reverse gear and the thrust washer for excessive wear and replace as necessary.

Bearing Carrier Removal & Disassembly

◆ See Figures 186, 221, 222, 250 and 251

■ Because the shift shaft actually splines with a C-shaped cup that holds the front of the propeller shaft, the bearing carrier is removed independently of the propeller shaft (which can only be removed after the driveshaft later in the Disassembly procedure).

✳✳ SELOC WARNING

If you don't have access to the special Mercury mandrels designed to install bearings to the proper depth inside the bearing carrier, measure the installed height of all bearings and/or races before removal. Use the measured depths at installation time to ensure the bearings are positioned properly (especially in positions where there are no shoulders to drive the bearings or races against).

1. Remove the gearcase from the mid-section, as detailed earlier in this section. Mount the case in a holding fixture or a soft-jawed vise, so you have access to the propeller shaft bore.

2. If necessary for access or for further disassembly, remove the water pump assembly, as detailed in the Lubrication & Cooling System section.

3. Locate the 1 tab on the tabbed washer which is bent down over the bearing carrier retainer (cover nut) to keep it from loosening. We say 1 tab because normally only 1 of them should be in position at a time, but it doesn't hurt to check the rest (but they are normally pressed inward). Using a small punch, carefully bend the tab back so it no longer interferes.

4. Using a large spanner like Carrier Retainer Nut Wrench #91-840393 carefully loosen and unthread the cover nut. Once the cover nut is removed, pull the locktab washer from the housing. If the cover nut is instead corroded into place and cannot be turned, drill holes in it, roughly positioned at 120° intervals around the face of the retainer (drill 2 holes almost next to each other at the top, then 2 more holes, each at 120° intervals). Once the holes are drilled use a hammer and chisel to fracture and remove the remainder, BUT USE EXTREME SELOC CAUTION not to damage the bearing carrier.

5. Install an internal jawed puller such as Long Puller Jaws (#91-46086A1) and a suitable drive bolt. Position the puller jaws on the outer ring of the carrier, close to the bosses. Tighten the puller to preload the propeller shaft.

6. Now, tighten the puller bolt to break the seal and free the bearing carrier from the gearcase. DO NOT attempt to remove the propeller shaft yet.

■ If the bearing carrier refuses to move, it may be necessary to carefully apply heat to the lower unit in the area of the carrier and at the same time attempt to move it off the carrier.

7. Check the condition of the bearing carrier itself, make sure the portion of the carrier which contacts the gearcase is not too corroded. If it shows signs of excessive corrosion, replace it to preserve proper gearcase sealing and to minimize the chance of it seizing in the housing.

8. Remove and discard the carrier O-ring. On these gearcases the propshaft gear (Reverse on these models) is pressed into the housing, if it or the bearing must be replaced, position the carrier in a soft-jawed vise and use a jawed slide hammer to remove the gear. If the bearing stays behind, use the same assembly to remove it from the carrier. If the bearing stays on the gear, use a bearing separator to push it off the gear.

■ The dual back-to-back oil seals in the bearing carrier must be pried out separately if they are to be replaced, or if access is necessary to the needle bearing and bearing snapring.

9. If necessary for seal replacement or for access to the needle bearing/snapring, invert the bearing carrier so the propeller end is facing

upward and use a small prytool to carefully remove each of the dual back-to-back oil seals.

10. If the needle bearing inside the carrier requires replacement, first remove the snapring with a suitable pair of snapring pliers, then position the carrier in a suitable shop press with the front side (gear side) facing upward. Use a suitable driver and mandrel to press the needle bearing out the propeller end of the carrier.

Driveshaft & Bearings Removal/Service

◆ See Figures 223, 250 and 251

■ On Sport XS models there is an additional plate (water passage plate) mounted on top of the gearcase. Though it shouldn't interfere with removal of the driveshaft, should the passage plate need to come off you'll have to remove the shift shaft assembly first (since it is mounted on top of the front plate's front end).

1. If not done already, remove the water pump assembly as detailed in the Lubrication & Cooling System section. Be sure to remove the water face plate and gasket as well for access to the oil seal carrier and driveshaft components.

2. If not done already, remove the Bearing Carrier and reverse gear assembly as detailed earlier in this section.

3. If not done already, remove the upper driveshaft oil seal carrier by first removing the 2 retaining bolts, then using 2 small pry-tools (in the slots provided on either side). Lift the base up and off the driveshaft.

4. Remove and discard the O-ring from the bottom of the seal carrier.

5. For installation purposes, note the directions in which the oil seals are facing (they are usually installed back-to-back) then carefully pry them from the seal carrier OR use a punch to drive them out the flared bottom of the carrier.

6. Slide a Driveshaft Bearing Retainer Wrench (#91-43506T) down over the driveshaft and use it to loosen the upper driveshaft bearing retainer. Unthread the retainer and remove the tool.

■ OK, here comes the pain in the butt part of this gearcase, trying to loosen the pinion nut with the propeller shaft still installed.

7. Obtain the following special tools, Bearing Carrier Installation Tool (#91-840388), Driveshaft holder tool (#91-56775T) and Pinion Nut Wrench (#91-61067003) and then proceed as follows to unbolt the pinion nut:

a. Place the driveshaft holder (nut wrench) onto the top of the driveshaft.

b. Insert the pinion nut wrench into the gearcase with the slot facing the pinion gear. If necessary, lift and rotate the driveshaft slightly to help align the pinion gear nut into the slot in the nut wrench.

c. Install the bearing carrier back into the gear housing BACKWARDS to help support the proper shaft and keep the pinion nut wrench aligned.

d. Using the driveshaft holding tool (nut wrench) and suitable socket with a breaker bar to loosen the pinion nut.

e. Remove all of the specialty tools.

8. Once the driveshaft is free, slide it out of the gearcase, but be sure to retain any shims which may have been placed between the driveshaft's tapered bearing and shoulder in the gearcase bore.

■ If the pinion gear is seized on the driveshaft position the driveshaft in a soft jawed vise (clamped as close as possible to the water pump studs), then use a block of wood to protect the gearcase as you tap on the wood (and therefore the case) to drive it off the shaft. BE SURE TO HAVE AN ASSISTANT hold the case so that it doesn't become damaged should it fall away suddenly when the pinion releases.

9. CAREFULLY move the propshaft down toward the port side of the gearcase, then reaching inside and retrieve the pinion gear, washer and nut from inside the housing.

10. If the tapered bearings require replacement, use a bearing separator (or even the jaws of a vise) to hold the bearing(s) while you push or gently tap the shaft free. Also, if the bearings are replaced, you'll need to remove the bearing cup (and shims) which are mounted in the upper portion of the gearcase. Use a slide hammer with a pair of internal puller jaws to grab on the inner lip of the bearing cup and pull it from the gearcase.

11. Inspect the driveshaft lower bearing surface (right above the pinion gear splines) where the needles roll. If the driveshaft is pitted, grooved/scored, worn unevenly, discolored from overheating, has embedded particles or you otherwise suspect the bearing, the needles and race should be replaced as a set. However, you'll need to remove the propshaft and forward gear first (for access and to prevent damage), then come back to this procedure in order to proceed.

12. Once the propshaft and forward gear has been removed, make sure all 18 loose needles are inside the lower race, then install a suitable bearing driver (such as from the Bearing Removal and Installation Kit #91-31229A7). Use a mallet to drive the bearing assembly down into the propshaft bore of the gearcase. Remove and discard the old race and needles.

Propeller Shaft Removal & Disassembly

◆ See Figures 224 thru 226, 250 and 251

Once the driveshaft and pinion gear assembly are removed you finally have the free-play necessary to remove and service the propeller shaft (along with the forward gear).

1. Tilt the shaft toward the port side of the gearcase (to allow the shift spool to disengage from the shift shaft crank) and remove the shaft assembly pulling up and out of the case. The forward gear and the shift spool should come out of the bearing cup and gearcase along with the shaft, but the shift crank will be left behind. Mercury seems to state that you can usually free the shift crank from the gear housing at this point though so use care if you do not intend to dislodge it.

To disassemble the propeller shaft assembly for cleaning, inspection and/or parts replacement, proceed as follows:

2. Insert a thin blade screwdriver or an awl under the first coil of the cross-pin and detent pin retainer spring and rotate the propeller shaft to unwind the spring from the sliding clutch. Take care not to over-stretch the spring.

■ IF after removal the spring does not coil back to its normal position, the spring has been over-stretched and must be replaced.

3. Remove the 2 tiny detent pins from the clutch, then use a cross-pin T-handle tool (such as #91-86642-2) to unthread and remove the cross-pin from the sliding clutch with a punch. At this point the sliding clutch, forward gear and shift spool can all be removed from the shaft.

4. In addition to the usual regiments of cleaning and inspection of the shaft and bearing surfaces (detailed later in this section under Cleaning & Inspection), be sure to check the following measurements:

a. Use a pair of V-blocks and a dial gauge to check the shaft run-out at a point on the smooth portion of the shaft just inboard of the propeller splines. Dial indicator movement should generally be 0.009 in. (0.23mm) or less. If run-out exceeds the proper specification, the shaft must be replaced.

b. Measure the propeller shaft forward-to-reverse shoulder length (from shoulder-to-shoulder across the splines for the clutch assembly) and replace the shaft if the measurement is under 2.040 in. (51.82mm). Also on these models, measure the thickness of the Reverse thrust washer to check for wear or a taper. Replace the Reverse thrust washer is the thickness is less than 0.240 in. (6.096mm).

c. On all models, you should inspect the shift spool for wear, especially where it comes in contact with the crank on the shift shaft. Make sure the spool spins freely (one trick to help ensure this is to tap the castle nut end against a firm surface to align the internal components). Measure the end-play of the spool assembly and make sure it is 0.002-0.010 in. (0.05-0.25mm). If necessary, remove the cotter pin and turn the castle nut first clockwise down until snug and then backing off to the first cotter pin slot which gives you the proper clearance. If necessary you can remove the cotter pin, castle nut and spool to disassemble, clean and reassemble components.

Nose Cone Gear Bearing & Race/Adapter Removal

◆ See Figures 227, 228, 250 and 251

The forward gear is installed on the propeller shaft for these models. A forward gear bearing cup is pressed into the gearcase nose cone and provides the surface on which the bearing will ride. The gear itself is basically held in position by a combination of the propeller shaft splines, the pinion gear and the bearing/thrust components.

A tapered bearing which rolls in the nose cone race is pressed onto the outside of the gear and a needle bearing is pressed inside the gear. The needle bearing supports the forward end of the propeller shaft.

As usual, removal of the bearings will normally render them unfit for service, so make sure they are shot and replacements are available before bearings are separated from the gear.

After the pinion gear, driveshaft and propeller shaft have all been removed, the forward gear race can be lifted removed from the housing for service. Likewise, once the propeller shaft is disassembled, the forward gear can be serviced as well. To disassemble the gear and bearings, proceed as follows:

1. If necessary to replace the internal needle bearing or the external tapered roller bearing, remove the forward gear from the propeller shaft.

2. If the forward gear tapered roller bearing must be replaced, use a universal bearing separator and a press to push the bearing off the gear.

■ Remember that the bearing and race are a set, so if the bearing is replaced you'll also need to replace the nose cone race. A wear pattern will have been worn in the race by the old bearing. Therefore, if the race is not replaced, the new bearing will be quickly worn by the old worn race.

3. Remove the forward gear tapered roller bearing race from the gearcase either using an internal jawed puller and puller plate adapter (preferred method) or using a slide hammer (if necessary, but use seloc caution not to damage the gearcase).

4. SAVE any shim material from behind the race after the race is removed. The same amount of shim material will probably be used during assembly.

5. If the needle bearing inside the gear is no longer fit for further service, it can be removed by pressing with a mandrel or tapping, with a blunt punch and hammer, around the bearing cage from the aft end - the gear end - of the bearing. This action will destroy the bearing cage. Therefore, be sure a replacement bearing is on hand, before attempting to remove the defective bearing.

Shift Shaft Disassembly

◆ See Figures 250 and 251

There is not a whole lot to shift shaft service on these motors, as a matter of fact Mercury specifically states that with the exceptions of the shift crank that is found inside the gearcase you CAN remove the shift shaft and housing without disassembling any of the other components. We assume that is because the shift crank will remain in position as long as the rest of the gearcase is not disturbed.

If necessary to remove the shift shaft and housing/bushing assembly you can unbolt the shift shaft bushing/housing (2 bolts), then remove the assembly from the gearcase, pulling the shaft out of the shift crank splines as it is withdrawn. Then it's mostly just a matter of cleaning and inspecting the shaft/bushing, and perhaps replacing the oil seal and O-ring

1. Loosen the retaining bolts (normally only 2 bolts on these models, however the Sport XS also has 2 bolts that secure the water passage plate which you may wish to remove as well), then carefully pull the shift shaft and bushing/housing assembly from the top of the gearcase.

2. On Sport XS models, if necessary, remove the water passage plate and discard the old gasket.

3. If the gearcase is disassembled (and not done already), reach inside the gearcase and remove the shift crank. Clean it with solvent and dry it, then inspect the splined portion, as well as the diameter of the spool that goes over the locating pin in the gearcase for damage. Replace, as necessary.

4. Carefully remove the shift shaft and rubber washer from the bushing/housing assembly.

5. Remove and discard the bushing/housing gasket, the O-ring and the oil seal (noting the direction the lips are facing before removal). The oil seal can either be pried out or removed using a driver and hammer.

6. Thoroughly clean the bushing and shaft with solvent and dry with compressed air. Make sure the splines are good on both ends and that there are no grooves, corrosion or excessive wear on the bushing sealing surface.

■ On Pro XS models the speedometer connector can be removed from the bushing/housing assembly for cleaning, inspecting and/or replacement. If so coat the threads of the connector with Perfect Seal or an equivalent sealant during installation.

CLEANING & INSPECTION

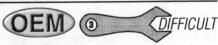

◆ See Figures 232 thru 239, 250 and 251

1. Clean all water pump parts with solvent, and then dry them with compressed air.

2. Inspect the water pump cover and base for cracks and distortion, possibly caused from overheating.

3. Inspect the face plate and water pump insert for grooves and/or rough surfaces. If possible, always install a new water pump impeller while the

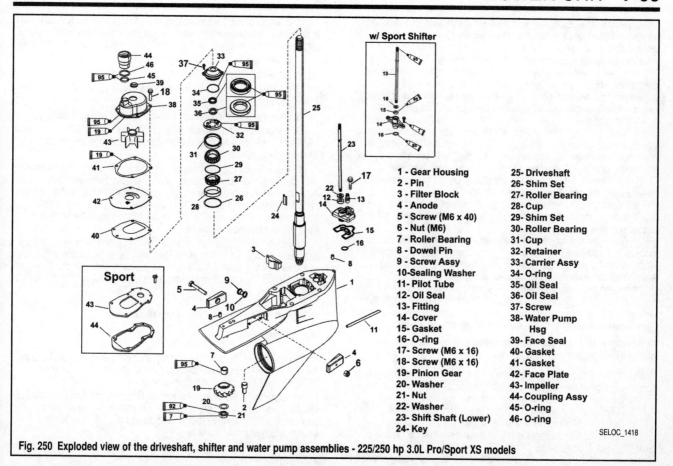

1 - Gear Housing	25- Driveshaft
2 - Pin	26- Shim Set
3 - Filter Block	27- Roller Bearing
4 - Anode	28- Cup
5 - Screw (M6 x 40)	29- Shim Set
6 - Nut (M6)	30- Roller Bearing
7 - Roller Bearing	31- Cup
8 - Dowel Pin	32- Retainer
9 - Screw Assy	33- Carrier Assy
10-Sealing Washer	34- O-ring
11- Pilot Tube	35- Oil Seal
12- Oil Seal	36- Oil Seal
13- Fitting	37- Screw
14- Cover	38- Water Pump
15- Gasket	Hsg
16- O-ring	39- Face Seal
17- Screw (M6 x 16)	40- Gasket
18- Screw (M6 x 16)	41- Gasket
19- Pinion Gear	42- Face Plate
20- Washer	43- Impeller
21- Nut	44- Coupling Assy
22- Washer	45- O-ring
23- Shift Shaft (Lower)	46- O-ring
24- Key	

SELOC_1418

Fig. 250 Exploded view of the driveshaft, shifter and water pump assemblies - 225/250 hp 3.0L Pro/Sport XS models

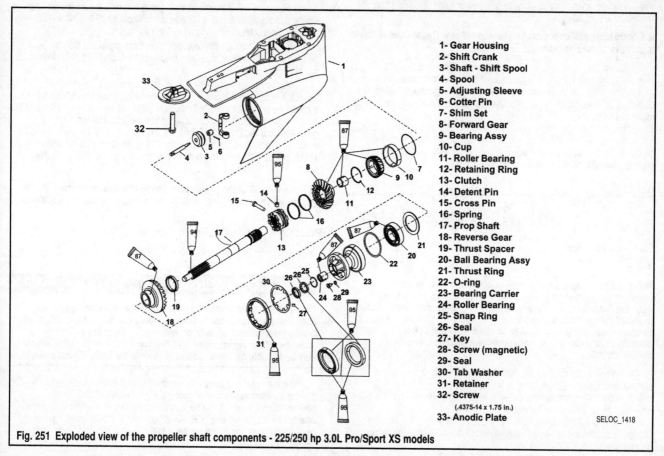

1- Gear Housing
2- Shift Crank
3- Shaft - Shift Spool
4- Spool
5- Adjusting Sleeve
6- Cotter Pin
7- Shim Set
8- Forward Gear
9- Bearing Assy
10- Cup
11- Roller Bearing
12- Retaining Ring
13- Clutch
14- Detent Pin
15- Cross Pin
16- Spring
17- Prop Shaft
18- Reverse Gear
19- Thrust Spacer
20- Ball Bearing Assy
21- Thrust Ring
22- O-ring
23- Bearing Carrier
24- Roller Bearing
25- Snap Ring
26- Seal
27- Key
28- Screw (magnetic)
29- Seal
30- Tab Washer
31- Retainer
32- Screw
(.4375-14 x 1.75 in.)
33- Anodic Plate

SELOC_1418

Fig. 251 Exploded view of the propeller shaft components - 225/250 hp 3.0L Pro/Sport XS models

lower unit is disassembled. A new impeller will ensure extended satisfactory service and give "peace of mind" to the owner. If the old impeller must be returned to service, never install it in reverse to the original direction of rotation. Installation in reverse will cause premature impeller failure.

4. Inspect the impeller side seal surfaces and the ends of the impeller blades for cracks, tears, and wear. Check for a glazed or melted appearance, caused from operating without sufficient water. If any question exists, and as previously stated, install a new impeller if at all possible.

5. Clean all bearings with solvent, dry them with compressed air, and inspect them carefully. Be sure there is no water in the air line. Direct the air stream through the bearing. Never spin a bearing with compressed air. Such action is highly dangerous and may cause the bearing to score from lack of lubrication. After the bearings are clean and dry, lubricate them with Quicksilver Formula 50-D lubricant or equivalent. Do not lubricate tapered bearing cups until after they have been inspected.

6. Inspect all ball bearings for roughness, catches, and bearing race side wear. Hold the outer race, and work the inner bearing race in-and-out, to check for side wear.

7. Determine the condition of tapered bearing rollers and inner bearing race, by inspecting the bearing cup for pitting, scoring, grooves, uneven wear, embedded particles, and discoloration caused from overheating. Always replace tapered roller bearings as a set.

8. Inspect the bearing surface of the shaft roller bearing support. Check the shaft surface for pitting, scoring, grooving, embedded particles, uneven wear and discoloration caused from overheating. The shaft and bearing must be replaced as a set if either is unfit for continued service.

■ **Remember to check shaft surfaces at oil seal contact areas, as grooves or wear in that area will allow leakage.**

Inspect the sliding clutch of the propeller shaft. Check the reverse gear side clutch dogs. If the dogs are rounded 1 of 3 causes may be to blame:
  a. Improper shift cable adjustment.
  b. Running engine at too high an rpm while shifting.
  c. Shifting from neutral to reverse gear too quickly.

9. Inspect the shift spool assembly as noted earlier during Propeller Shaft Removal & Disassembly.

10. Inspect the propeller shaft roller bearing surfaces for pitting, rust marks, uneven wear, embedded metal particles or signs of overheating caused by lack of adequate lubrication.

■ **Good shop practice requires installation of new O-rings and oil seals regardless of their appearance.**

11. Clean the bearing carrier, pinion gear, drive gear clutch and the propeller shaft with solvent. Dry the cleaned parts with compressed air.

12. Check the pinion gear and the drive gear for abnormal wear. Apply a coating of light-weight oil to the roller bearing. Rotate the bearing and check for cracks or catches.

13. Inspect the propeller shaft oil seal surface to be sure it is not pitted, grooved, or scratched. Inspect the roller bearing contact surface on the propeller shaft for pitting, grooves, scoring, uneven wear, embedded metal particles, and discoloration caused from overheating.

ASSEMBLY

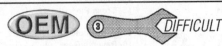

Shift Shaft Assembly

◆ See Figure 250

■ **On Pro XS models the factory service literature suggests fully installing the shift shaft assembly at this point in the procedure, HOWEVER on Sport XS models the literature specifically states ONLY to install the shaft itself at this time, to temporarily hold the coupler in position inside the gearcase, then to come back and finish shaft and housing installation only after the driveshaft and its components have been installed.**

1. If disassembled, place the shift shaft bushing/housing on a work surface with the top of the housing facing upward.

2. Apply a light coating of 2-4-C with Teflon to the new O-ring and to the inner and outer surfaces of the new seal.

3. Using a suitable driver, carefully seal the new seal in the housing with the lips facing the same way as noted during removal (usually facing upward on these models).

4. If removed, apply a light coating of Perfect Seal or equivalent sealant to the threads of the speedometer connector, then install the connector and tighten gently (the torque spec is extremely low, about 4.5 inch lbs./0.5 Nm).

5. Install the rubber washer against the oil seal.

6. Install and seat the O-ring.

7. On Sport XS models, if removed install the water passage plate using a new gasket.

8. Insert the shift shaft through the washer/seal/bushing assembly.

9. If applicable (and it should be for all Pro XS models, but not for Sport XS models) position a NEW gasket on the top of the gearcase-to-shift housing mating surface.

10. Reach inside the gearcase and place the shift crank over the locating pin in the bottom of the nose cone (behind the nose cone gear race assembly). Make sure the crank is facing toward the Port side of the motor (meaning it looks like the letter "C" if you peered in through the propeller shaft opening, with the opening of the letter facing toward the Starboard side) with the splined end upward.

11. Install the shift shaft and bushing/housing assembly to the top of the gearcase, aligning the splines of the shaft with the splines of the shift crank as the assembly is seated. Make sure the gearcase gasket and/or O-ring(s) are in position, then install and tighten the retaining bolts to 60 inch lbs. (6.8 Nm).

Driveshaft Pinion Needle Bearing Installation

◆ See Figure 250

If removed from the gearcase, the first step in assembly should be to pull the new needle bearing assembly into the housing. This must be done BEFORE the installation of the nose cone gear.

Although special tools are noted for use to accomplish this, if the are not available, you can usually get away with using a long bolt, a series of appropriately sized washers, some locknuts and possibly a large plate through which to thread the bolt.

1. Apply a thin coat of gear lube to the driveshaft needle bearing bore in the gear housing.

2. Place the NEW needle bearing (complete with the cardboard shipping tube still in place) on top of a suitable driver (from kit #91-31229A7) with the numbers on the bearing facing away from the driver.

3. Reach in through the propeller shaft bore and place the needle bearing in the driveshaft bore, with the numbers facing UPWARD toward the top of the gearcase.

4. Assemble an installation tool to draw the bearing up into the gearcase. If using the Mercury recommended tools use the Puller Rod through the bore, Nut threaded downward against the Top Plate on top of the housing, the Pilot on the upper tapered bearing shoulder inside the gearcase, and the Driver (with bearing) on the bottom of the Rod. Using the installation tool, draw the bearing up until it presses against the gearcase shoulder (using care not to exert excessive force on the bearing at this point).

■ **Make sure all of the needles are still in place in the bearing. If necessary use some 2-4-C with Teflon to help hold them in position.**

Nose Cone Gear Bearing Race or Adapter

◆ See Figures 240, 241 and 250

✳✳ SELOC WARNING

On these motors a special bearing cup installation tool is normally used that contains centering rings to keep the race/adapter centered during installation. However, on many other gearcases Mercury recommends using an assembled bearing carrier as the pilot, which may be feasible on these models IF the special tools are not available.

1. If the needle bearing was removed from the nose cone gear, install the replacement at this time. Position the gear with the teeth facing downward on a press, then use a suitable mandrel (#91-33491) to press against the numbered side of the bearing until the tool bottoms against the gear. If equipped, install the snapring to the back of the gear once the bearing is seated inside.

2. Next, position the thrust washer on to the forward gear, then install the tapered roller bearing to the gear itself. Use a suitable mandrel to push on the bearing inner race until it is seated firmly against the gear. Position the gear aside for installation on the propeller shaft later.

3. Place the same amount of shim material saved during disassembly into the lower unit. If the shim material was lost, or if a new lower unit is being used, begin with approximately 0.020 in. (0.51mm) material. Coat the forward bearing race bore with gear lube.

4. Position the tapered bearing race squarely over the bearing bore in the front portion of the lower unit. Obtain a Bearing Cup Installation Tool (#91-18605A2) and place the tool over the tapered bearing race along with Driver Cup (#91-87120T).

5. Use the installation tool (by turning the hex head screw threaded into the tool base) until the tapered bearing race is seated against the shim material. Remove the installation tool assembly.

Propeller Shaft Assembly

◆ See Figures 230 and 231

1. Slide the clutch dog onto the propeller shaft splines, aligning the cross pin holes and the detent hole with the slot in the shaft (making sure the grooved end of the clutch is positioned facing the direction the shift spool will be installed).

2. Slide the forward gear and bearing assembly onto the propshaft, then install the shift spool, aligning the holes in the spool with the slot in the prop shaft and the hole for the clutch cross-pin.

3. Insert the cross-pin through the clutch, shaft and shift spool. Use the cross-pin T-handle tool (#91-86642-2) to thread and tighten the cross pin. Make sure the flat tang on the cross-pin is installed so that it runs parallel to the spring groove.

4. Place a small dab of 2-4-C with Teflon or an equivalent marine grade grease on the end of the 2 detent pins, then install the pins into the holes of the sliding clutch.

5. Install the cross-pin spring over the sliding clutch. Take care not to over stretch the spring. Spirally wrap the spring into the groove on the sliding clutch, making sure the spring is wound straight onto the clutch, not overlapping itself. The spring must sit tight/flat against the clutch or it has been overstretched and must be replaced.

6. Position the propeller shaft aside for installation later.

Bearing Carrier Assembly

◆ See Figure 251

Be sure to lubricate the inner and outer diameters of the bearings with gear lubricant prior to installation. Also, make sure to coat the inner diameter of the oil seals using 2-4-C with Teflon or an equivalent lubricant. The outer diameter of the oil seals should be coated using Loctite(r) 680 or an equivalent adhesive compound. After installing the seals, wipe away any excess Loctite before proceeding.

1. If removed from the bearing carrier, oil and position the propeller shaft needle roller bearing into the aft (propeller) end of the bearing carrier with the numbered side toward the aft end (toward the driver). Press the needle bearing into the bearing carrier with a suitable mandrel such as #91-816292 which will give you the correct depth when the upper edge of the tool seats against the carrier.

■ **If the Mercury tools are not available, position the bearing to the depth measured prior to removal.**

2. Once the needle bearing is positioned, install the snapring with the sharp edge of the ring facing the seal end of the carrier.

3. Clean the seal bore of any grease or oil.

4. Obtain an Oil Seal Driver (#91-817569T). Place 1 seal on the longer shoulder side of the driver tool with the lip of the seal away from the shoulder.

5. Press the seal into the bearing carrier until the seal driver bottoms against the bearing carrier, then clean away any excess adhesive.

6. Place the second seal on the short shoulder side of the seal driver with the lip of the seal toward the shoulder. Press the seal into the bearing carrier until the seal driver bottoms against the bearing carrier. Clean excess Loctite from the seals and lubricate both seal lips with 2-4-C with Teflon or an equivalent grease.

7. Next prepare and install the reverse gear assembly. Position the thrust washer over the reverse gear. Then, use a suitable mandrel to press a NEW ball bearing onto the back of the reverse gear until it bottoms. As usual, make sure the mandrel only presses on the portion of the bearing in contact with the gear (inner). Then use a mandrel to press the assembled

gear and bearing into the forward (gearcase) end of the bearing housing until it bottoms out (protect the carrier by using the pilot washer on the aft end of the carrier to take the direct force from the press).

8. Install a new O-ring over the bearing carrier and into the groove. Apply a coating of 2-4-C with Teflon or and equivalent Lubricant onto the installed O-ring.

Propeller Shaft, Driveshaft & Pinion Gear

◆ See Figure 251

1. If the tapered bearings were removed from the driveshaft, apply a light coat of gear oil to the bearing inner diameter(s). Install a used pinion nut onto the end of the driveshaft, leaving about 1/16 in. (1.6mm) of nut threads exposed (i.e. making sure the driveshaft threads are NOT extending past the nut to protect them from damage). Place the bearings over the driveshaft (with the smaller diameter end of the smaller bearing facing the pinion and the larger diameter end of the larger bearing facing both the smaller bearing and the pinion). Position the driveshaft/bearing assembly over a bearing separator and suitable mandrel (an old shaft inner race will do the job) so that the force of pressing on the pinion nut will go right to the inner race of the new bearing(s). In this way, carefully press the bearing(s) onto the shaft, then remove the old nut.

2. Double-check to make sure the needles of the pinion needle bearing assembly are in position still. If the pinion bearing needle bearings have fallen out, place all 18 needles into the needle bearing outer race using 2-4-C with Teflon or an equivalent marine grade grease to hold the needles in position.

3. Prepare to install the propeller shaft by tilting the propeller end of the propeller shaft toward the Port side of the gearcase and rotating the shift shaft from reverse to neutral as you insert the propshaft and shift spool into the shift crank.

4. Operate the shift shaft to ensure the crank is sitting properly on the spool. The sliding clutch should move forward when the shift shaft is turned clockwise or aft when the shaft is turned counterclockwise.

5. If not done already, slide the rubber washer down the top of the shift shaft so that it JUST touches the oil seal in the bushing/housing assembly.

6. Install the lower tapered roller bearing (race) into the gearcase housing bore.

7. Place the shims for the upper driveshaft tapered bearing(s) into the driveshaft housing bore. If the original shims were not retained or if the pinion gear, driveshaft, driveshaft bearing(s), tapered bearing cap or gearcase was replaced, start with 0.020 in. (0.508mm) of shim material. IF you don't need to check Pinion Gear Depth and Upper Driveshaft Bearing Clearance, install the lower bearing cup, However, if these clearances need to be checked, leave the lower bearing cup out of the assembly temporarily, it will be reinstalled after the Pinion Gear Depth has been checked.

8. To ease installation glue the pinion gear nut washer to the bottom of the pinion gear using 3M Adhesive or Bellows Adhesive.

9. If this is final assembly (components affecting backlash have not changed or you've already checked/adjusted the shims), apply a light coating of Loctite(r) 271 or equivalent threadlocking material to the threads of the pinion nut (for many years Mercury suggested always using a new nut during assembly). However, if you're just installing the driveshaft to check gearcase Shimming, then leave off the threadlocking compound at this time. Place the pinion nut into the MR slot in the pinion nut holding tool (#91-61067003) with the flat side facing away from the pinion gear.

10. Position the pinion gear in the gear housing, right below the driveshaft bore, with the teeth meshed with the teeth of the nose gear.

11. Insert the driveshaft carefully down into the gearcase, rotating it slightly as necessary to engage the shaft splines with the splines in the pinion gear. Continue to insert the driveshaft until the tapered bearing is against the bearing race.

■ **It may be necessary to lift up on the driveshaft slightly in order to install the pinion nut tool.**

12. Holding the pinion nut in position with the pinion nut tool, carefully turn the driveshaft to thread the nut.

13. Install the shims and upper driveshaft tapered roller bearing cup.

14. Apply a light coating of 2-4-C with Teflon to the threads of the bearing retainer. Insert the retainer over driveshaft and into the gearcase (over top of the tapered bearing cup). Start to thread the retainer.

15. Install the bearing carrier into the gear housing BACKWARDS to hold the propeller shaft and the pinion nut wrench in position.

16. Tighten the pinion nut using a Driveshaft Holding Tool (#91-56775T) to turn the driveshaft clockwise. Tighten to 75-80 ft. lbs. (102-109 Nm).

17. Remove the bearing carrier, pinion nut adapter and the driveshaft holding tool, THEN install the Driveshaft Bearing Retainer Wrench (#91-43506T). Use the retainer wrench to tighten the driveshaft bearing retainer to 100 ft. lbs. (135.5 Nm).

18. Now, if you've replaced the gearcase housing, or if you've replaced 1 or more of the bearings SKIP ahead to the Shimming procedures in order to check backlash and determine if some of the shims must be replaced. However, if only seals have been replaced, then you can proceed with the assembly.

Bearing Carrier Installation

◆ See Figures 250 and 251

As always, be sure to lubricate bearings or thrust surfaces with gear oil and oil seals using 2-4-C with Teflon or an equivalent lubricant prior to installation.

1. If not done already, make sure to apply a light coating of gear lube to the bearing carrier needle bearing and pack the space between the oil seals with a coating of 2-4-C with Teflon, or an equivalent marine grade grease. Use the same grease to apply a light coating to the outer diameter of the bearing carrier, where it contacts the gearcase.

2. Carefully install the bearing carrier into the gearcase. Rotate the driveshaft very slowly in a clockwise direction (viewed from above) to help mesh the reverse gear in the bearing carrier with the pinion gear as the bearing carrier is seated.

■ **During installation be sure to align the notch in the bottom of the carrier with the notch in the gear housing bore.**

3. Once the carrier is seated, place the large tab washer against the bearing carrier. If you properly aligned the notch in the bottom of the carrier with the notch in the gear housing bore, the tab washer should install with the triangular-shaped tab aligned with the triangular-shaped notch in the carrier (the notch in the carrier should be right behind the tab).

■ **The cover nut torque is extremely high. Make sure the gearcase is either mounted in a strong stand or bolted to a driveshaft housing to avoid possible gearcase damage.**

4. Apply a light coating of 2-4-C with Teflon or an equivalent marine grade grease to the threads of the cover nut and threads in the gearcase housing, then install the nut into the housing. Use the cover nut tool (#91-840393) or equivalent to tighten the cover nut to 210 ft. lbs. (285 Nm).

5. Bend 1 of the lock tabs of the lock washer outward, into the cover nut (normally only ONE will align). Bend the remaining tabs of the washer inward toward the front of the gear housing.

■ **If 1 tab of the lock washer DOES NOT align between 2 notches so it can be bent down locking the retainer continue to tighten the retainer until one does. DO NOT loosen the retainer from spec in order to align a tab!**

6. If not done earlier during driveshaft installation (assuming no preload check needed to be performed) install the oil seal carrier, followed by the water pump assembly, keeping the following points in mind:

• The dual seals in the carrier should be installed back-to-back so the inner or lower seal faces the gearcase and the upper or outer seal faces the water pump. The outside of the seals should be coated with Loctite(r) 380 or an equivalent adhesive compound prior to installation.

• A NEW O-ring should be used, and coated (along with the inner lips of the seals) with 2-4-C with Teflon or an equivalent lubricant.

• If the divider block was removed from the top of the gearcase it must be properly positioned, then install the water pump base using a new base gasket.

• Once the carrier is installed, tighten the 2 retaining bolts to 60 inch lbs. (7 Nm).

7. Refill the gearcase and check for leaks, then install the gearcase to the outboard as detailed earlier in this section.

SHIMMING

Backlash

◆ See Figure 242

For any gearcase to operate properly the pinion gear on the driveshaft must mesh properly with the forward and reverse gears. And, remember that these gears remain in contact with the pinion gear at all times, regardless of whether or not the sliding clutch has locked 1 of them to the propeller shaft.

If these gears are not properly meshed, there is a greater risk of breaking teeth or, more commonly, rapid wear leading to gearcase failure.

On some of the smaller Mercury gearcases there is little or no adjustment for the positioning of these gears. However, the gearcases used on these large Mercury motors are all equipped with shim material that is used to perform minute adjustments on the positioning of the bearings (and therefore the bearings and gears). Shim material can be found in various positions.

On these standard rotation units shim(s) can be found between the forward gear bearing/race assembly and the nose cone, as well as underneath and between the upper driveshaft bearings.

If a gearcase is disassembled and reassembled replacing only the seals, then no shimming should be necessary. However, if shafts, gears/bearings and/or the gearcase housing itself is replaced it is necessary to check the Pinion Gear Depth and the Forward and Reverse Gear Backlash to ensure proper gear positioning.

Pinion Gear Depth

◆ See Figures 243 thru 245

■ **This procedure starts with the gearcase mostly reassembled (following our procedures up to and including driveshaft reassembly, but with the propshaft bearing carrier, driveshaft upper seal carrier, water pump plate and water pump still removed).**

There are 2 possible tools you can use to determine Pinion Gear Depth. The first, #91-12349A05 must be used WITHOUT the propeller shaft installed, which means the nose cone gear must be removed from the propeller shaft. However, you can ALSO use #91-56048001 and the advantage of this second tool is that it can be used WITH the propeller shaft and forward gear installed.

Obtain a Bearing Preload Tool (#91-14311A04). This tool consists of a spring, which seats against thrust washer, thrust bearing and bearing race all positioned on top of the driveshaft bearing retainer. It is used to simulate the upward driving force on the pinion gear bearing. Install the following, 1 by 1, over the driveshaft.

    a. Collared bearing race

    b. Thrust bearing

    c. Flat thrust washer

    d. Spring

    e. Special large stud, with the nuts on the bolt threaded all the way to the top of the threads

1. Pull up on the driveshaft and tighten the 2 Allen set screws in the top nut. These set screws grip the driveshaft and prevent the tool from slipping on the shaft. Tighten the 2 set screws securely.

2. Measure the distance between the bottom surface of the top nut and the top surface of the bottom nut. Record this measurement.

3. Use a wrench to hold the top nut steady and another wrench to tighten the bottom nut down against the spring. Continue tightening the nut until the distance between the bottom surface of the top nut and the top surface of the bottom nut equals 1.00 in. (25.4mm) plus the distance recorded when the tool was installed.

4. The driveshaft has now been forced upwards placing an upward preload on the pinion gear bearing. Turn the driveshaft 2 or more turns (clockwise when viewed from above) to make sure the bearings are seated.

5. Obtain the correct Pinion Gear Locating Tool either #91-12349A05 along with Flat # 4 and Disc # 2 from the tool (if the propshaft is not installed) or #91-56048001 if the propshaft IS installed.

6. Insert the locating tool (with the appropriate flat and disc in place for the first tool), making sure the tool bottoms on the bearing carrier shoulder in the housing.

7. The pinion depth is the distance between the bottom of the pinion gear teeth and the flat surface of the gaging block.

8. Measure this distance by inserting a long feeler gauge into the access hole and between the pinion gear teeth and gaging block flat face. Be sure to take measurements at 3 different locations, rotating the driveshaft 120° each time. Take the average of the 3 readings (add the 3 readings together and divide the total by 3).

9. The correct clearance is 0.024-0.026 in. (0.610-0.660mm). If the average clearance is correct, skip the next step, but leave the tools in place.

10. If the clearance is not correct, add or subtract shim material from below the upper driveshaft tapered roller bearing cup to lower or raise the

gear. You should be able to remove the cup by wiggling the driveshaft back and forth or by turning the gearcase housing and shaking it (obviously, only after the tools have been removed).

• If the pinion gear depth was found to be greater than 0.026 in. (0.660mm), then at least that amount of shim material must be added to lower the gear.

• If the pinion gear depth was found to be less than 0.024 in. (0.610mm), then at least that amount of shim material must be removed to raise the gear.

■ **Try, if you can, to center the final result in the acceptable range, so use the difference between 0.025 in. (0.64mm) and the actual clearance measured to determine the ideal amount of shim material needed to correct the pinion gear depth.**

11. Once the correct pinion gear height is achieved, remove the Pinion Gear Depth preload tool and loosen the pinion nut, then remove the upper retainer and driveshaft again for access. Oil and install the LOWER bearing cup (which was left out during assembly on purpose) and reinstall the pinion gear and driveshaft to the gearcase.

■ **At this point the lower bearing cup, pinion shims and driveshaft with the upper and lower tapered roller bearings will be installed in the driveshaft bore.**

12. Reinstall the driveshaft preload tool but this time, place a large washer between the upper and lower nuts. This washer must protrude outward between the nuts, and provide a surface for a dial gauge to read shaft end play (vertical movement of the shaft). Tighten the set-screws on the upper nut, then position a dial indicator to read vertical shaft movement on the flat surface of the washer.

■ **There should be no preload on the bearing at this point and the shaft should rotate freely.**

13. Now push down or pull up on the driveshaft and zero the dial gauge, then move the shaft fully in the opposite direction and record total movement (play). Whatever measurement you get at this point (the amount of play) is equal to the amount of shims that must be added beneath the LOWER BEARING CUP.

✳✳ SELOC WARNING

DO NOT change the shims under the upper bearing cup or your will change the pinion height!

14. Once the appropriate shims are installed, position the lower bearing cup, pinion shims and the driveshaft assembly with the tapered roller bearings. Install and tighten the upper bearing retainer to 100 ft. lbs. (135 Nm) and the pinion gear nut to 75 ft. lbs. (102 Nm).

■ **The word OFF on the upper bearing retainer must be visible (facing upward) when installed.**

15. Move on and check the Forward & Reverse Gear Backlash.

Forward & Reverse Gear Backlash

◆ **See Figures 246 and 247**

■ **The bearing carrier is temporarily installed for this procedure.**

Final installation of the bearing carrier (and possibly propeller shaft if the forward bearing requires any shimming) is usually performed later because the pinion gear nut has to be removed and coated with Loctite.

1. First, check and adjust the Pinion Gear Depth, as detailed earlier in this section. Leave the pinion bearing preload tool installed on top of the gearcase.

2. Next, if not done earlier, install the propeller shaft and nose cone gear assembly.

3. Install the bearing carrier, if necessary turn the driveshaft slightly to align the pinion gear teeth to the reverse gear. Align the notch in the bearing carrier and the notch in the housing (as detailed earlier under Bearing Carrier Installation), then lubricate and install the bearing carrier retainer nut. Tighten the retainer down fully by hand, then torque to specification as detailed earlier in this section (but there is no need to secure the lock-tabs).

4. Install a bearing carrier puller with the arms of the puller on the carrier and the center bolt on the end of the propeller shaft. Tighten the puller center bolt to a torque value of 45 inch lbs. (5 Nm). This action places a preload on propeller shaft and the forward gear, pushing the forward gear into the forward gear bearing.

5. Attach the proper Backlash Dial Indicator Rod (#91-53459 for either the Sport XS 1.62:1 ratio gearcase or the Pro XS 1.75:1 ratio gearcase, you can double-check this ratio by calculating using the number of forward gear teeth divided by the number of pinion gear teeth) to the driveshaft. Position a dial indicator (using a magnetic base or a threaded rod with nuts and plates to position it alongside the Indicator Rod) and point the dial indicator to the **1** mark on the indicator tool. Tighten the indicator tool onto the driveshaft and rotate the shaft so the needle in the dial makes at least 1 full revolution then comes back to zero on the dial indicator scale.

6. Recheck the puller torque, then rotate the driveshaft back-and-forth and observe movement of the dial indicator. Total movement is the forward gear backlash, this is the first reading. Loosen the indicator tool and move it 90° (1/4 turn) then repeat and record the reading. Do this for a total of 4 readings (each at 1/4 turn from the previous).

7. Add all 4 readings together and divide by 4 to get the average and THAT is your measured forward gear backlash, it should be 0.020-0.025 in. (0.508-0.635mm) for all gearcase ratios. If the backlash is too great, add shim material behind the forward gear bearing race. If the backlash is too small, remove shim material from behind the forward gear bearing race.

■ **For each 0.001 in. (0.025mm) of shim which is added or removed expect backlash to increase or decrease about 0.001 in. (0.025mm).**

8. Although the reverse gear backlash is not adjustable on these models, it should still be checked as follows:

a. Remove the bearing carrier puller installed earlier in this procedure and install the Pinion Nut Holder (#91-61067003) used to thread the pinion nut earlier over top of the propeller shaft (large end toward the bearing carrier). Install the Belleville washer and the propeller retaining nut, then tighten the nut to 45 inch lbs. (5 Nm). Rotate the driveshaft 3 full turns clockwise and then re-torque the propeller nut once again to 45 inch lbs. (5 Nm).

b. If removed, reinstall the backlash indicator tool used for Forward Gear Backlash and the dial indicator. Once again point the dial indicator to the **1** mark on the indicator tool. Tighten the indicator tool onto the driveshaft and rotate the shaft so the needle in the dial makes at least 1 full revolution then comes back to zero on the dial indicator scale.

c. Gently rock the driveshaft back and forth while watching the dial indicator. Total movement is the reverse gear backlash, this is the first reading. Loosen the indicator tool and move it 90° (1/4 turn) then repeat and record the reading. Do this for a total of 4 readings (each at 1/4 turn from the previous).

■ **During this measurement it should not feel as is the pinion gear is turning the forward gear.**

9. Add all 4 readings together and divide by 4 to get the average and THAT is your measured reverse gear backlash, it should be 0.035-0.060 in. (0.89-1.52mm) for all gearcase ratios.

■ **If the Reverse gear backlash is not in spec either the gearcase is improperly assembled OR the parts are excessively worn and must be replaced.**

10. To remove the bearing preload tool, first remove the dial and indicator tools, then screw the bottom nut of the bearing preload tool all the way up against the bottom of the bolt head. Loosen the 2 Allen screws, then carefully remove all remaining components of the tool.

11. Finish assembly of the gearcase using Loctite on the pinion nut and the correct shims (if any need to be changed).

JET DRIVE

Jet Drive Assembly

DESCRIPTION & OPERATION

◆ See Figure 252

The jet drive unit is designed to permit boating in areas that would be impossible for a boat equipped with a conventional propeller drive system. The housing of the jet drive barely extends below the hull of the boat allowing passage in ankle deep water, white water rapids and over sand bars or in shoal water which would foul a propeller drive unit.

A jet drive provides reliable propulsion with a minimum of moving parts. Simply stated, water is drawn into the unit through an intake grille by an impeller driven by a driveshaft off the crankshaft of the powerhead. The water is immediately expelled under pressure through an outlet nozzle directed away from the stern of the boat.

As the speed of the boat increases and reaches planing speed, the jet drive discharges water freely into the air and only the intake grille makes contact with the water.

The jet drive is provided with a gate arrangement and linkage to permit the boat to be operated in reverse. When the gate is moved downward over the exhaust nozzle, the pressure stream is reversed by the gate (it bounces off or is turned around the thrust the opposite direction by the gate) and the boat moves sternward.

Conventional controls are used for powerhead speed, movement of the boat, shifting and power trim and tilt.

JET DRIVE AND/OR IMPELLER REMOVAL

◆ See Figures 253 thru 261  *MODERATE*

■ **Refer to Cleaning & Inspection for an exploded view of the jet drive assembly.**

1. For safety, either disconnect the negative battery cable and/or tag and disconnect the spark plug leads, grounding them to the head.
2. Shift the drive into Neutral, then disconnect the shift linkage.
3. Remove the six bolts securing the intake grille to the jet casing.
4. Ease the intake grille from the jet drive housing.
5. Pry the tab or tabs of the tabbed washer away from the impeller retaining nut to allow the nut to be removed.
6. Loosen and then remove the impeller retaining nut.
7. Remove the tabbed washer and spacers. Make a careful count of the spacers behind the washer. If the unit is relatively new, there could be as many as eight spacers stacked together. If the unit has been in service for any length of time it will not be unusual to find some of the spacers behind the impeller, which is removed in the following step.
8. Remove the jet impeller from the shaft. If the impeller is frozen to the shaft, obtain a block of wood and a hammer. Carefully tap the impeller in a clockwise direction to release the shear key.

Fig. 252 Typical jet drive outboard

9. Slide the nylon sleeve and shear key free of the driveshaft and any spacers found behind the impeller. Make a note of the number of spacers at both locations - behind the impeller and on top of the impeller, under the nut and tabbed washer.

■ **If all you are doing is adjusting drive impeller clearance, stop here and skip ahead to Jet Drive And/Or Impeller Installation.**

10. Typically 1 external bolt (threaded downward at the rear top of the housing) and 4 internal bolts (threaded upward through the outer diameter of the bearing carrier) are used to secure the jet drive to the intermediate housing. The external bolt is located at the aft end of the anti-cavitation plate.
11. The four internal bolts are located inside the jet drive housing, as indicated in the accompanying illustration. Remove the five attaching bolts.

■ **DO NOT mistake the 4 bearing carrier bolts for the 4 jet drive assembly retaining bolts. The jet drive bolts are found just outboard of the bearing carrier.**

12. Lower the jet drive from the intermediate housing. If used, remove the locating pin(s) from the front and rear of the upper jet housing.

■ **There are usually 2 locating pins on these models, but on other jet drives we've seen as many as 6 so check just to be sure. Make careful note of the size and location of each when they are removed, to assist during assembly.**

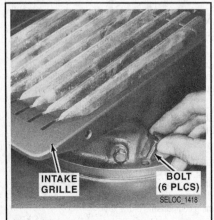

Fig. 253 Unbolt and remove the water intake grille for access to the impeller

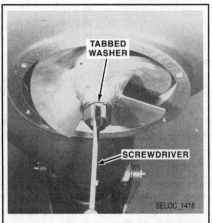

Fig. 254 To remove the impeller, bend back the locktab. . .

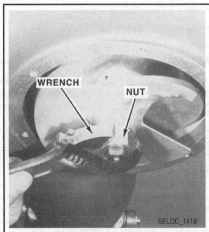

Fig. 255 . . .then loosen and remove the nut

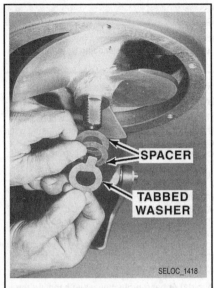

Fig. 256 Note the location of the tabbed washer and shims. . .

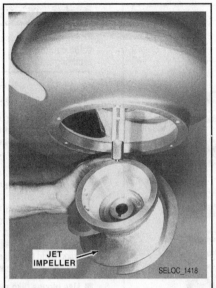

Fig. 257 . . .then lower the impeller from the drive

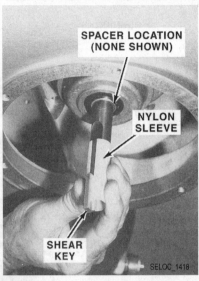

Fig. 258 Remove the nylon sleeve, shear key and any spacers which were

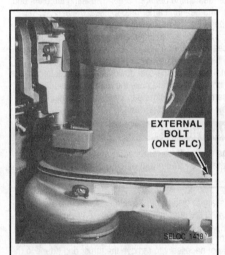

Fig. 259 To unbolt the drive remove any external bolts (usually 1). . .

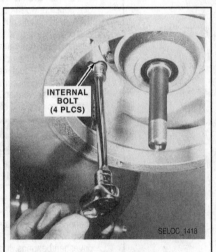

Fig. 260 . . .and any internal bolts (usually 4 just outside the bearing carrier)

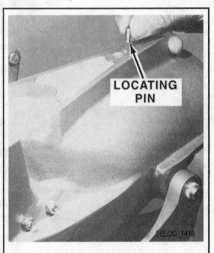

Fig. 261 When you lower the jet drive, check for the dowel (locating) pin(s)

DRIVE UNIT DISASSEMBLY

■ Refer to Cleaning & Inspection for an exploded view of the jet drive assembly.

1. Remove the jet drive assembly from the outboard as detailed earlier in this section.

2. Unbolt and remove the Water Pump/Impeller assembly, as detailed in the Cooling and Lubrication section. Basically, it means, removing the four bolts and washers from the water pump housing, then removing the water pump components: the water pump housing, the inner cartridge (if removable), the water pump impeller (with upper and lower fiber washers on some models), the Woodruff/Impeller drive key from its recess in the driveshaft, the housing gasket, base plate and usually base plate gasket.

3. Remove the 4 bolts securing the bearing carrier assembly to the jet drive housing, then remove the driveshaft and bearing assembly from the housing.

■ For further disassembly of the driveshaft and bearing carrier, please refer to Bearing Carrier Overhaul, later in this section.

BEARING CARRIER OVERHAUL

◆ See Figures 262 thru 265

■ Refer to Cleaning & Inspection for an exploded view of the jet drive assembly.

A slightly involved procedure must be followed to dismantle the bearing carrier including "torching" off the bearing housing. Naturally, excessive heat will destroy the seals and possibly damage the bearings. Therefore, this procedure should be avoided unless you are planning (or at least open to the possibility) of a complete overhaul to the Jet Drive. Otherwise, use great care or leave this part of the service work to someone who is more familiar with Jet Drives.

However, if you have the experience and the shop tooling required to perform the torch heating and bearing press tasks, use the following procedures and illustrations to perform this work. Refer to the exploded view provided under Cleaning & Inspection in this section to help identify internal components, but remember that assemblies may vary slightly. A photographic exploded view is also provided in this section.

The following procedures pickup the work after the bearing carrier and driveshaft assembly has been removed from the jet pump housing.

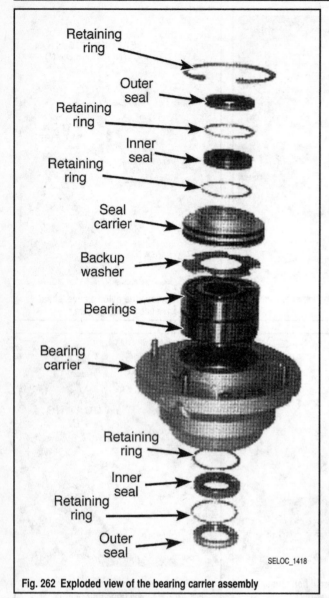

Fig. 262 Exploded view of the bearing carrier assembly

Labels in figure:
- Retaining ring
- Outer seal
- Retaining ring
- Inner seal
- Retaining ring
- Seal carrier
- Backup washer
- Bearings
- Bearing carrier
- Retaining ring
- Inner seal
- Retaining ring
- Outer seal

SELOC_1418

✳✳ SELOC CAUTION

The retaining ring is under considerable force when installed. Use seloc caution and wear eye protection when removing or installing this retaining ring.

1. Reach inside the top of the bearing carrier with a pair of snapring pliers and remove the large beveled snapring.

■ As an alternative to the safer step listed below, Mercury does note on that you can heat the housing, then holding the housing (using oven mitts of the like) you can bump the impeller/crankshaft side of the driveshaft against a wooden block to try and slide the carrier down off the bearings.

2. Apply heat evenly all around the bearing carrier using a hand held propane torch or equivalent item. Apply only enough heat to the bearing carrier so it is very warm to the touch. Now, invert the assembly in a bearing press and support the carrier between two plates as shown in the accompanying illustration. Press on the impeller end of the driveshaft and drive the shaft and bearings out the bottom of the carrier. Have an assistant standing by to catch the shaft as it falls free of the bearing carrier, otherwise the splined end of the shaft could be damaged if it strikes a concrete floor.

■ Use gloves (like oven mitts) or hand protection when handling the components such as the shaft, as should still be a bit hot.

3. Remove the outer seal in the bearing carrier by placing the blunt end of a punch down through the bearing carrier onto the seal metal case. Tap the seal down and out the bottom of the bearing carrier. Next, grasp the tab end of the spiral retaining ring below the inner seal. Unwind the spiral retaining ring from the bearing carrier.

4. Again, place the blunt end of a punch down through the bearing carrier onto the seal metal case. Tap the seal down and out the bottom of the bearing carrier. Discard both seals and remove the remaining spiral retaining ring inside the bearing carrier. Thoroughly clean the bearing carrier in cleaning solvent and blow dry with compressed air.

✳✳ SELOC WARNING

The driveshaft bearings need not be removed, unless they are unfit for further service. Once they are removed, they cannot be installed and used a second time.

5. Now if it is necessary to disassemble the bearings from the driveshaft, start by removing the snapring (if so equipped) using a pair of snapring pliers and safety glasses. Then support the shaft in a shop press using a bearing separator and press the bearings off the shaft. Discard both driveshaft bearings. It is not necessary to remove the thrust ring (found in the driveshaft groove above the upper ball bearing) unless it is damaged.

Fig. 263 After heating, use a shop press to separate the driveshaft from the carrier

(Labels: Arbor or bearing press; Bearing carrier; SELOC_1418)

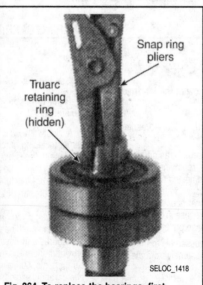

Fig. 264 To replace the bearings, first remove the snapring (if equipped). . .

(Labels: Snap ring pliers; Truarc retaining ring (hidden); SELOC_1418)

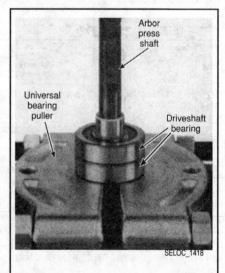

Fig. 265 . . . then use a shop press the free the bearings from the shaft

(Labels: Arbor press shaft; Universal bearing puller; Driveshaft bearing; SELOC_1418)

6. If necessary, remove and discard the retaining ring in the shaft itself.

■ **After cleaning and inspection, reassemble the bearing and driveshaft unit using new bearings and seals. Be sure to coat the inside bore of the seal surface, as well as the bearing carrier housing with marine grade grease prior to assembly. Also, be sure to fill the seal cavities and lips with grease. Install new O-rings in all locations.**

7. When installing the driveshaft, heat the bearing carrier until it JUST feels warm to the touch (DO NOT heat it excessively!), then position the bearing carrier over the driveshaft (squaring up the inner bore with the ball bearing) and gently press the carrier down until it bottoms against the bearing. DO NOT use a shop press, as only a light pressing motion should be necessary. Though, if trouble is encountered you should be able to safely tap the bearing carrier down into position using a rubber mallet.

CLEANING & INSPECTION

◆ **See Figures 266, 267 and 268**

Wash all parts, except the driveshaft assembly, in solvent and blow them dry with compressed air. Rotate the bearing assembly on the driveshaft to inspect the bearings for rough spots, binding and signs of corrosion or damage.

Saturate a shop towel with solvent and wipe both extensions of the driveshaft.

Replace the bearing lubricant as follows:

Lightly wipe the exterior of the bearing assembly with the same shop towel. Do not allow solvent to enter the three lubricant passages of the bearing assembly. The best way to clean these passages is not with solvent - because any solvent remaining in the assembly after installation will

continue to dissolve good useful lubricant and leave bearings and seals dry. This condition will cause bearings to fail through friction and seals to dry up and shrink - losing their sealing qualities.

The only way to clean and lubricate the bearing assembly is after installation to the jet drive - via the exterior lubrication fitting.

If the old lubricant emerging from the hose coupling is a dark, dirty, gray color, the seals have already broken down and water is attacking the bearings. If such is the case, it is recommended the entire driveshaft bearing assembly be taken to the dealer for service of the bearings and seals.

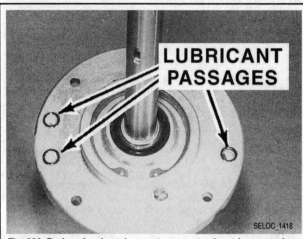

Fig. 266 During cleaning take care to prevent solvent from entering the lubrication passages

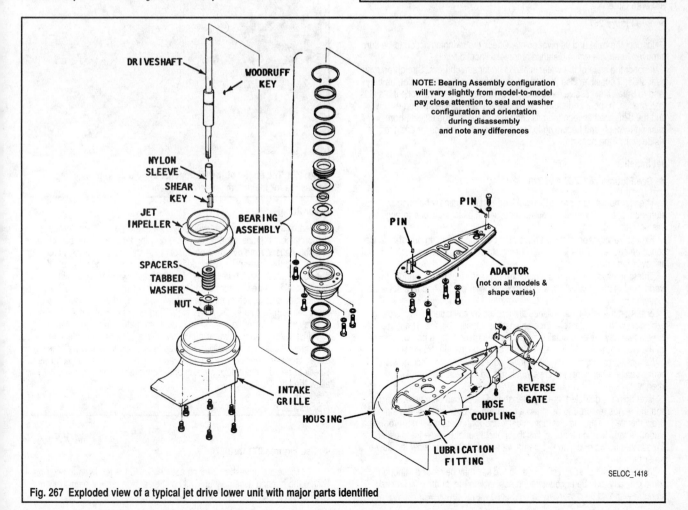

Fig. 267 Exploded view of a typical jet drive lower unit with major parts identified

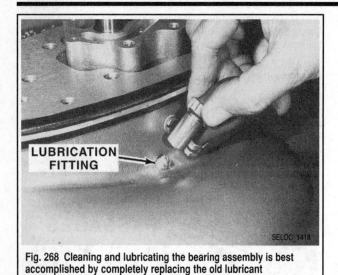

Fig. 268 Cleaning and lubricating the bearing assembly is best accomplished by completely replacing the old lubricant

Driveshaft and Associated Parts

◆ See Figure 267

Inspect the threads and splines on the driveshaft for wear, rounded edges, corrosion and damage.

Carefully check the driveshaft to verify the shaft is straight and true without any sign of damage.

Inspect the jet drive housing for nicks, dents, corrosion, or other signs of damage. Nicks may be removed with No. 120 and No. 180 emery cloth.

Reverse Gate

◆ See Figure 267

Inspect the gate and its pivot points. Check the swinging action to be sure it moves freely the entire distance of travel without binding.

Inspect the slats of the water intake grille for straightness. Straighten any bent slats, if possible. Use the utmost care when prying on any slat, as they tend to break if excessive force is applied. Replace the intake grille if a slat is lost, broken, or bent and cannot be repaired. The slats are spaced evenly and the distance between them is critical, to prevent large objects from passing through and becoming lodged between the jet impeller and the inside wall of the housing.

Jet Impeller

◆ See Figures 267, 269 and 270

The jet impeller is a precisely machined and dynamically balanced aluminum spiral. Observe the drilled recesses at exact locations to achieve this delicate balancing.

Excessive vibration of the jet drive may be attributed to an out-of-balance condition caused by the jet impeller being struck excessively by rocks, gravel or cavitation burn.

The term cavitation burn is a common expression used throughout the world among people working with pumps, impeller blades and forceful water movement.

Burns on the jet impeller blades are caused by cavitation air bubbles exploding with considerable force against the impeller blades. The edges of the blades may develop small dime size areas resembling a porous sponge, as the aluminum is actually eaten by the condition just described.

Excessive rounding of the jet impeller edges will reduce efficiency and performance. Therefore, the impeller should be inspected at regular intervals.

If rounding is detected, the impeller should be placed on a work bench and the edges restored to as sharp a condition as possible, using a file. Draw the file in only one direction. A back-and-forth motion will not produce a smooth edge. Take care not to nick the smooth surface of the jet impeller. Excessive nicking or pitting will create water turbulence and slow the flow of water through the pump.

Inspect the shear key. A slightly distorted key may be reused although some difficulty may be encountered in assembling the jet drive. A cracked shear key should be discarded and replaced with a new key.

Fig. 269 The slats of the grille must be carefully inspected and any bent slats straightened for maximum jet drive performance

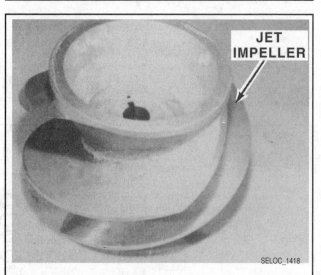

Fig. 270 The edges of the jet impeller should be kept as sharp as possible for maximum jet drive efficiency

Water Pump

Clean all water pump parts with solvent and then blow them dry with compressed air. Inspect the water pump housing for cracks and distortion, possibly caused from overheating. Inspect the steel plate, the thick aluminum spacer and the water pump cartridge for grooves and/or rough spots. If possible always install a new water pump impeller while the jet drive is disassembled. A new water pump impeller will ensure extended satisfactory service and give peace of mind to the owner. If the old water pump impeller must be returned to service, never install it in reverse of the original direction of rotation. Installation in reverse will cause premature impeller failure.

If installation of a new water pump impeller is not possible, check the sealing surfaces and be satisfied they are in good condition. Check the upper, lower and ends of the impeller vanes for grooves, cracking and wear. Check to be sure the indexing notch of the impeller hub is intact and will not allow the impeller to slip.

ASSEMBLY

◆ See Figures 271 thru 274

1. Place the driveshaft bearing assembly into the jet drive housing. Rotate the bearing assembly until all bolt holes align. There is normally only one correct position.

2. Install and tighten the bearing carrier retaining bolts, depending upon the model, as follows:

• For 20 Jet models, coat the threads of the screws lightly with Loctite® 271 or an equivalent threadlocker, then install and tighten them to 30 inch lbs. (3.4 Nm).

• For 40 Jet models, install and tighten the bolts to 70 inch lbs. (7.9 Nm). No threadlock is specifically mentioned for these units, but we fail to see what harm it would be to use if desired.

• For 65/80 Jet models, torque will vary with the size of the bolts. For 0.250-20 x 0.875 bolts tighten to 70 inch lbs. (8 Nm), but for 0.312-18 x 1.00 bolts (which should be more common on these later models) tighten the bolts to 155 inch lbs. (17.5 Nm).

3. Install the Water Pump/Impeller assembly, as detailed in the Lubrication and Cooling system section.

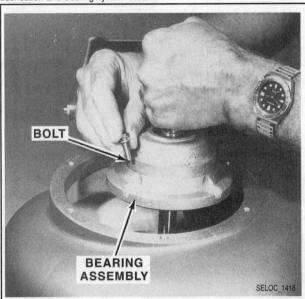

Fig. 271 If removed, install the bearing assembly into the jet drive housing

JET DRIVE AND/OR IMPELLER INSTALLATION

◆ **See Figures 275 thru 283**

1. If removed, install the locating pin(s).

■ **Mercury specifically recommends coating the threads of the jet drive retaining bolts using Loctite® 271 for 20 Jet models, but NOT for larger models. We fail to see the harm in using the Loctite for other models, in fact, we'd prefer to use a threadlocking compound both to help prevent the possibility of corrosion and to make sure the bolts cannot loosen in service.**

2. Raise the jet drive unit up and align it with the intermediate housing, with the small pins indexed into matching holes in the adaptor plate. Install the internal bolts (normally 4).

3. Install the one external bolt at the aft end of the anti-cavitation plate.

4. Tighten the 4 internal bolts to 25 ft. lbs. (34 Nm) and the one external bolt to 22 ft. lbs. (30 Nm).

5. Place the required number of spacers up against the bearing housing. Slide the nylon sleeve over the driveshaft and insert the shear key into the slot of the nylon sleeve with the key resting against the flattened portion of the driveshaft.

■ **If installing a NEW impeller, place all or most of the spacers under the impeller to hold it higher up in the wider tapered portion of the intake. As the impeller wears, shims will be transferred to the top of the impeller in order to hold it further down into the tapered intake to maintain the proper amount of clearance. For more details, please refer to CHECKING IMPELLER CLEARANCE, under Jet Drive Impeller, in the Maintenance section.**

6. Slide the jet impeller up onto the driveshaft, with the groove in the impeller collar indexing over the shear key.

7. Place the remaining spacers over the driveshaft.

8. Tighten the nut firmly (Mercury does not publish a torque and simply says to "drive tight" which means to us less than a full grunt, basically snug and a teeny bit more) regardless, the tabbed washer is really there to keep it from loosening. If neither of the two tabs on the tabbed washer aligns with the sides of the nut, remove the nut and washer. Invert the tabbed washer. Turning the washer over will change the tabs by approximately 15°. Install and tighten the nut. The tabbed washer is designed to align with the nut in one of the two positions described.

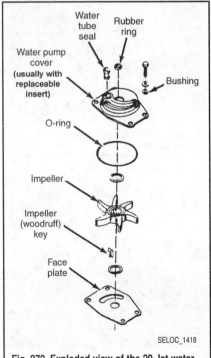

Fig. 272 Exploded view of the 20 Jet water pump

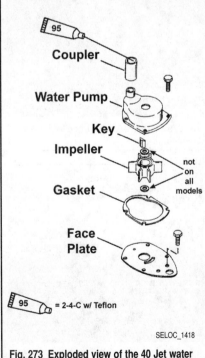

Fig. 273 Exploded view of the 40 Jet water pump

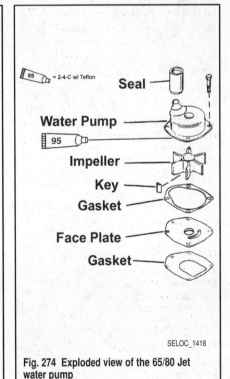

Fig. 274 Exploded view of the 65/80 Jet water pump

9. Bend the tabs up against the nut to prevent the nut from backing off and becoming loose.

10. Install the intake grille onto the jet drive housing with the slots facing aft. Install and tighten the six securing bolts. Tighten the intake grate retaining bolts to 100 inch lbs. (11.5 Nm) for the 20 Jet and 40 Jet or to 155 inch lbs. (17.5 Nm) for the 65/80 Jet.

11. Reconnect the shift linkage.

12. Reconnect the battery cables and/or the spark plug wires.

Jet Drive Adjustments

GATE LINKAGE ADJUSTMENTS

Gate Adjustment

◆ See Figures 284, 285 and 286

To adjust the reverse gate on the jet drive housing, proceed as follows:
1. Set the shifter handle to the **neutral** detent.

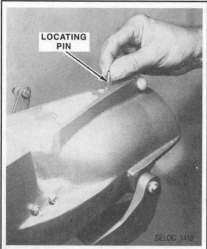

Fig. 275 Install the locating pins and raise the drive into position

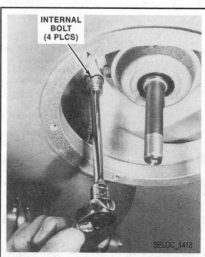

Fig. 276 Install and tighten the 4 internal drive mounting bolts. . .

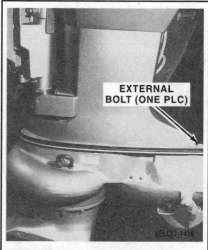

Fig. 277 . . .along with any external bolts (there is usually just 1)

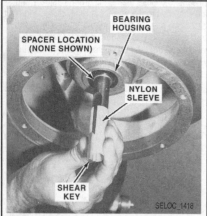

Fig. 278 Install any shims, the nylon sleeve and shear key. . .

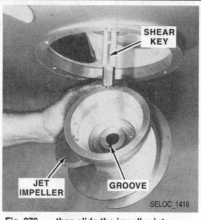

Fig. 279 . . .then slide the impeller into position. . .

Fig. 280 . . .followed by the shims and tabbed washer

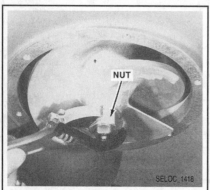

Fig. 281 Install and tighten the impeller retaining nut. . .

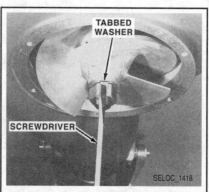

Fig. 282 . . .then bend the washer tab into place to secure the nut

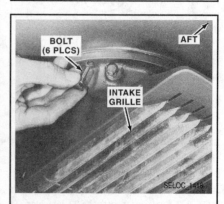

Fig. 283 Install the intake grille

2. Hold the reverse gate up and check to make sure there is about a 15/32 in. (12mm) gap between the top of the gate and the water flow passage.

3. If adjustment is necessary, have an assistant continue to hold the reverse gate up and check the clearance, while you loosen the cam screw and rotate the eccentric nut until the clearance is correct.

4. Tighten the cam screw, then recheck the adjustment.

5. Move the shifter handle to the **forward** detent.

6. With the gate now in the **forward** detent try to lift up on the reverse gate by hand (moving it back toward **neutral**). The gate should NOT move. If necessary, adjust the eccentric nut again, so the **neutral** gap is correct, but the gate will not move upward in the **forward** position.

Remote Cable Adjustment

◆ **See Figures 284 thru 288**

Proper cable adjustment is necessary in order to keep water pressure (caused by boat movement) from moving the reverse gate upward blocking normal thrust.

✳✳ SELOC CAUTION

Should be reverse gate shift suddenly and unexpectedly over the thrust nozzle, the boat will stop violently, shifting and possibly ejecting passengers and cargo. Proper adjustment is critical.

This procedure starts with the cable disconnected, for replacement, re-rigging or installation of a jet drive assembly.

1. Pull upward on the shift cam (1) until the roller is in the far end of the **forward** range (2).

2. Shift the remote control to the **forward** position. Temporarily, push the insert the cable guide onto the shift cam stud, then pull firmly on the cable casing to remove all free-play. Next, adjust the cable trunnion until it aligns with the anchor bracket, then remove the cable from the cam stud.

3. Connect the cable trunnion to the anchor bracket, then turn 90° to lock the trunnion in position.

4. Push the cable guide back onto the cam stud, then install the washer and finger-tighten the locknut.

5. Shift the remote control to the **neutral** position (3). The cam roller must snap into the **neutral** detent when you pull upward on the reverse gate with moderate pressure. If the gate does not shift properly into **neutral**, lengthen the cable slightly and recheck the adjustment.

■ **Remember, it is CRITICAL that the reverse gate remains locked into the FORWARD position when the remote is shifted into FORWARD. You should NOT be able to move the gate out of this position by hand.**

6. Tighten the cam locknut securely, then loosen it 1/8-1/4 of a turn in order to allow free movement of the cam.

7. Secure the cable loosely to the steering cable using a wire tie, but be sure NOT to restrict shift cable movement.

■ **With the engine running in NEUTRAL and the warm-up lever raised, the engine will run rough due to exhaust restriction from the reverse gate. This is normal.**

Tiller Shift Rod Adjustment

◆ **See Figures 284 thru 288**

Proper shift rod adjustment is necessary in order to keep water pressure (from boat movement) from moving the reverse gate upward blocking normal thrust.

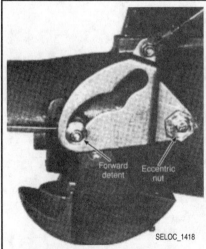

Fig. 284 Reverse gate in the FORWARD position with the cam roller at the lower end of the cutout in the bracket

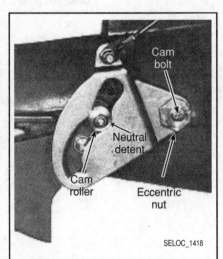

Fig. 285 Reverse gate in the NEUTRAL position, with the cam roller indexed into the neutral detent

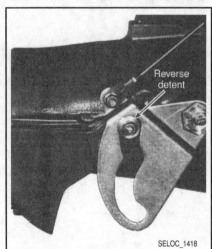

Fig. 286 Reverse gate in the REVERSE position, with the cam roller at the upper end of the cutout in the bracket

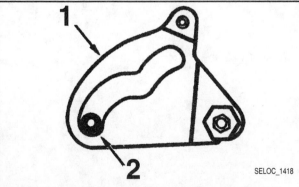

Fig. 287 When adjusting the remote cable/shift rod first move the cam (1) to the FORWARD position (2) . . .

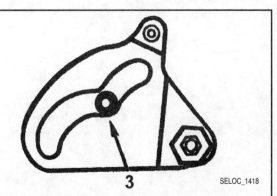

Fig. 288 . . . next, check that the cam snaps into the NEUTRAL position (3)

✳✳ SELOC CAUTION

Should be reverse gate shift suddenly and unexpectedly over the thrust nozzle, the boat will stop violently, shifting and possibly ejecting passengers and cargo. Proper adjustment is critical.

This procedure starts with the shift rod (and connector pin) disconnected, for replacement, re-rigging or installation of a jet drive assembly.

1. Pull upward on the shift cam (1) until the roller is in the far end of the **forward** range (2).

2. Move the shift handle to the **forward** position, then thread the connector onto the shift rod until it is aligned with the hole in the shift cam.

3. Install the connector pin, then secure using the cotter pin and washer.

4. Move the shifter to the **neutral** position (3). The cam roller must snap into the **neutral** detent when you pull upward on the reverse gate with moderate pressure. If the gate does not shift properly into **neutral**, remove the connector pin again and rotate the connector in order to lengthen the shift rod VERY slightly, then recheck the adjustment.

■ Remember, it is CRITICAL that the reverse gate remains locked into the FORWARD position when the handle is shifted into FORWARD. You should NOT be able to move the gate out of this position by hand.

5. Tighten the adjuster locknut securely.

■ With the engine running in NEUTRAL and the twist grip in the START position or higher, the engine will run rough due to exhaust restriction from the reverse gate. This is normal.

TRIM ADJUSTMENT

◆ See Figure 289

During operation, if the boat tends to pull to port or starboard, the flow fins may be adjusted to correct the condition. These fins are located at the top and bottom of the exhaust tube.

If the boat tends to pull to starboard (which is common with these drives), bend the trailing edge of each fin approximately 1/16 in. (1.5mm) toward the starboard side of the jet drive. Naturally, it follows logically that if the boat tends to pull to port, you should bend the fins toward the port side (but this is far less common).

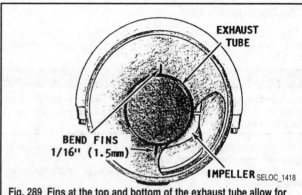

Fig. 289 Fins at the top and bottom of the exhaust tube allow for minor port/starboard trim thrust adjustment

8

TRIM AND TILT

MECHANICAL TILT (UNASSISTED)

Introduction

◆ See Figures 1 and 2

All outboard installations are equipped with some means of raising or lowering (pivoting), the complete unit for efficient operation under various load, boat design, water conditions, and for trailering to and from the water. The correct trim angle ensures maximum performance and fuel economy as well as a more comfortable ride for the crew and passengers.

The most simple form of tilt is a Mechanical Tilt adjustment consisting of a series of holes in the transom mounting bracket through which an adjustment pin passes to secure the outboard unit at the desired angle.

Such a mechanical arrangement works quite well for the smallest of units, but the larger (and heavier) outboard become, the more they require some form of assist or power system. Pretty much the majority of Mercury 2-strokes (except a few of the smallest the single and twin cylinder models), utilize either a manual gas assist or a power trim/tilt system. In some cases, simply the trim or option level of a given HP range motor could be the difference between being equipped with 1 system or the other.

Two versions of the PTT system are used, a single ram unit which is found on inline models or a 3 ram (two trim, 1 tilt) unit which is found only on V6 models.

Description & Operation

◆ See Figures 1 and 2

The system employed by all of the 1-cylinder and many of the 2-cylinder Mercury 2-strokes consists of a simple mechanical tilt pin arrangement.

On these motors, a change in the tilt angle of the outboard unit is accomplished by inserting the tilt adjustment pin through one of a series of holes in the transom mounting bracket. These holes allow the operator to obtain the desired boat trim under various speeds and loading conditions.

The tilt angle of a lower unit is properly set when the anti-cavitation plate is approximately parallel with the bottom of the boat. The boat trim is corrected by stopping the boat, removing the adjustment pin, tilting the outboard upward or downward, as desired and then installing the pin through the new hole exposed in the transom mounting bracket.

To raise the bow of the boat, the outboard is raised 1 hole at-a-time until the operator is satisfied with the boat's performance. If the bow is to be lowered, the lower unit is lowered 1 hole at-a-time.

Performance will generally be improved if the bow is lowered during operation in rough water. The boat should never be operated with the lower unit set at an excessive raised position. Such a tilt angle will cause the boat to "porpoise", which is very dangerous in rough water. Under such conditions the helmsperson does not have complete control at all times.

Instead of making extreme changes in the lower unit angle, it is far better to shift passengers and/or the load to obtain proper performance.

In order to obtain maximum efficiency and safety from the boat and outboard unit, the tilt pin must be installed in the proper position. The wide range of boat designs with their various transom angles, requires a determination be made for each outboard installation.

Actually, the tilt pin is only required if the boat handles improperly in the full trimmed "IN" position at WOT (wide open throttle). Usually this occurs when the transom angle is too large.

1. Move the outboard unit inward or outward until the anti-cavitation plate is parallel to the boat bottom. With the outboard in this position, notice the position of the swivel bracket in relation to the clamp bracket tilt pin holes. Now, install the tilt pin into the first full pin hole closer to the transom.

2. Lock the tilt pin in position by pushing in on the pin compressing the lock spring. Close the clevis on the end of the pin and release the tilt pin. The tilt pin is secured by the clevis on the end of the tilt pin.

Some earlier models utilize a cotter pin and washer on the end of the tilt pin. After installing the tilt pin through the desired clamp bracket hole, slide the washer over the end of the tilt pin and insert the cotter pin through the hole at the end of the tilt pin. Open the ends of the cotter pin to secure the tilt pin in place.

The angle of the lower unit is properly set when the boat is operating to give maximum performance, including comfort and safety.

Servicing

◆ See Figures 3 thru 8

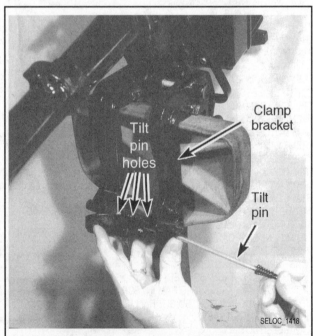

Fig. 2 With the outboard in the proper position, insert the tilt pin into the tilt pin hole

Fig. 1 Typical manual tilt pin arrangement on a smaller outboard unit

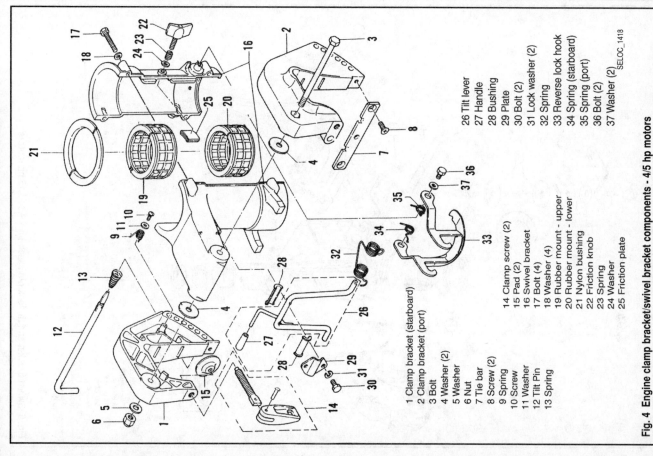

Fig. 4 Engine clamp bracket/swivel bracket components - 4/5 hp motors

1 - Clamp bracket (starboard)
2 - Clamp bracket (port)
3 - Bolt
4 - Washer (2)
5 - Washer
6 - Nut
7 - Tie bar
8 - Screw (2)
9 - Spring
10 - Screw
11 - Washer
12 - Tilt Pin
13 - Spring
14 - Clamp screw (2)
15 - Pad (2)
16 - Swivel bracket
17 - Bolt (4)
18 - Washer (4)
19 - Rubber mount - upper
20 - Rubber mount - lower
21 - Nylon bushing
22 - Friction knob
23 - Spring
24 - Washer
25 - Friction plate
26 - Tilt lever
27 - Handle
28 - Bushing
29 - Plate
30 - Bolt (2)
31 - Lock washer (2)
32 - Spring
33 - Reverse lock hook
34 - Spring (starboard)
35 - Spring (port)
36 - Bolt (2)
37 - Washer (2)

SELOC_1418

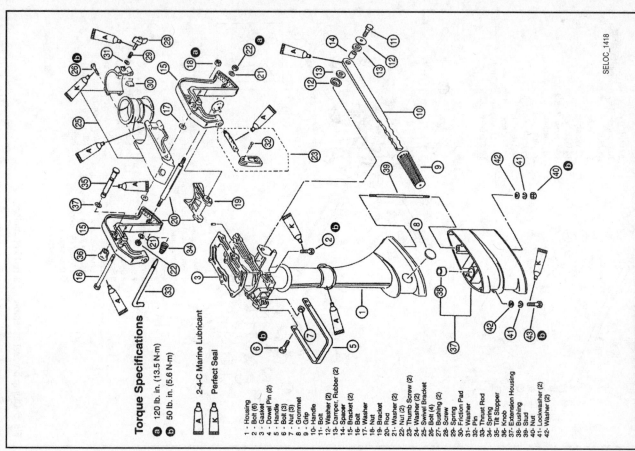

Fig. 3 Engine clamp bracket/swivel bracket components - 2.5/3.3 hp motors

SELOC_1418

Torque Specifications

ⓐ 120 lb. in. (13.5 N·m)
ⓑ 50 lb. in. (5.6 N·m)

Ⓐ 2-4-C Marine Lubricant
Ⓚ Perfect Seal

1 - Housing
2 - Bolt (6)
3 - Gasket
4 - Dowel Pin (2)
5 - Handle
6 - Bolt (3)
7 - Nut (3)
8 - Grommet
9 - Grip
10 - Handle
11 - Bolt
12 - Washer (2)
13 - Damper, Rubber (2)
14 - Spacer
15 - Bracket (2)
16 - Bolt
17 - Washer
18 - Nut
19 - Bracket
20 - Rod
21 - Washer (2)
22 - Nut (2)
23 - Thumb Screw
24 - Spring
25 - Swivel Bracket
26 - Bolt (4)
27 - Bushing (2)
28 - Screw
29 - Spring
30 - Friction Pad
31 - Washer (2)
32 - Pin
33 - Thrust Rod
34 - Spring
35 - Tilt Stopper
36 - Knob
37 - Extension Housing
38 - Bushing
39 - Stud
40 - Nut
41 - Lockwasher (2)
42 - Washer (2)

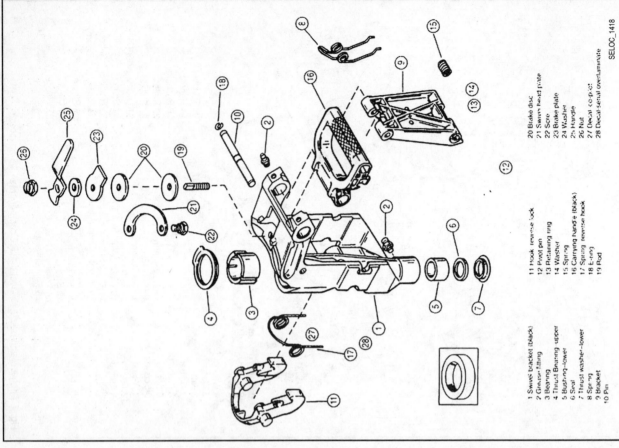

1 Swivel bracket (black)
2 Grease fitting
3 Bearing
4 Thrust bushing-upper
5 Bushing-lower
6 Seal
7 Thrust washer-lower
8 Spring
9 Bracket
10 Pin
11 Hook-reverse lock
12 Pivot pin
13 Retaining ring
14 Washer
15 Spring
16 Carrying handle (black)
17 Spring-reverse hook
18 E-ring
19 Rod
20 Brake disc
21 Swivel head plate
22 Screw
23 Brake plate
24 Washer
25 Handle
26 Nut
27 Decal-co-pilot
28 Decal-serial overturn note

SELOC_1418

Fig. 6 Swivel bracket and tilt lock pin assembly - 9.9/10/15 hp motors

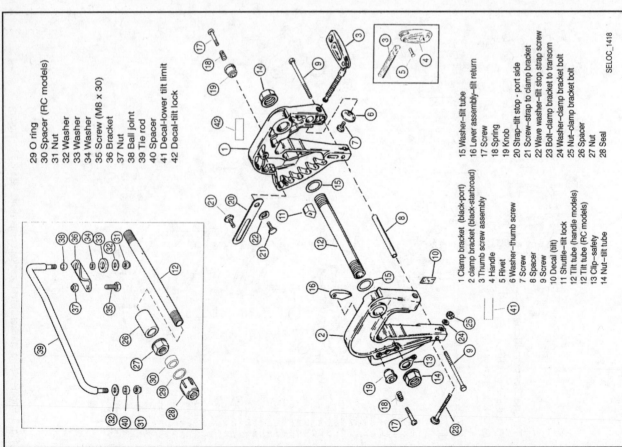

29 O ring
30 Spacer (RC models)
31 Nut
32 Washer
33 Washer
34 Washer
35 Screw (M8 x 30)
36 Bracket
37 Nut
38 Ball joint
39 Tie rod
40 Spacer
41 Decal-lower tilt limit
42 Decal-tilt lock

1 Clamp bracket (black-port)
2 clamp bracket (black-starbroad)
3 Thumb screw assembly
4 Handle
5 Rivet
6 Washer–thumb screw
7 Screw
8 Spacer
9 Screw
10 Decal (tilt)
11 Shuttle-tilt lock
12 Tilt tube (handle models)
12 Tilt tube (RC models)
13 Clip-safety
14 Nut-tilt tube

15 Washer-tilt tube
16 Lever assembly–tilt return
17 Screw
18 Spring
19 Knob
20 Strap-tilt stop - port side
21 Screw–strap to clamp bracket
22 Wave washer–tilt stop strap screw
23 Bolt–clamp bracket to transom
24 Washer–clamp bracket bolt
25 Nut-clamp bracket bolt
26 Spacer
27 Nut
28 Seal

SELOC_1418

Fig. 5 Clamp bracket assembly - 9.9/10/15 hp motors

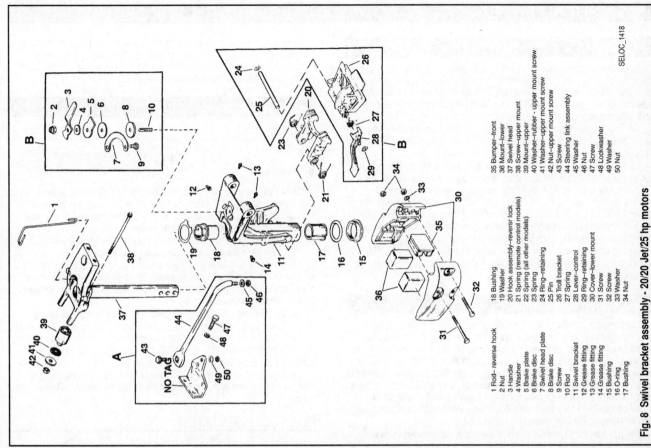

SELOC_1418

1 Rod– reverse hook
2 Nut
3 Handle
4 Washer
5 Brake plate
6 Brake disc
7 Swivel head plate
8 Brake disc
9 Screw
10 Rod
11 Swivel bracket
12 Grease fitting
13 Grease fitting
14 Grease fitting
15 Bushing
16 O-ring
17 Bushing

18 Bushing
19 Washer
20 Hook assembly–reversr lock
21 Spring (remote control models)
22 Spring (all other models)
23 Spring
24 Ring–retaining
25 Pin
26 Troll bracket
27 Spring
28 Lever–control
29 Ring–retaining
30 Cover–lower mount
31 Screw
32 Screw
33 Washer
34 Nut

35 Bumper–front
36 Mount–lower
37 Swivel head
38 Screw–upper mount
39 Mount–upper
40 Washer–rubber - upper mount screw
41 Washer–upper mount screw
42 Nut–upper mount screw
43 Screw
44 Steering link assembly
45 Washer
46 Nut
47 Screw
48 Lockwasher
49 Washer
50 Nut

Fig. 8 Swivel bracket assembly - 20/20 Jet/25 hp motors

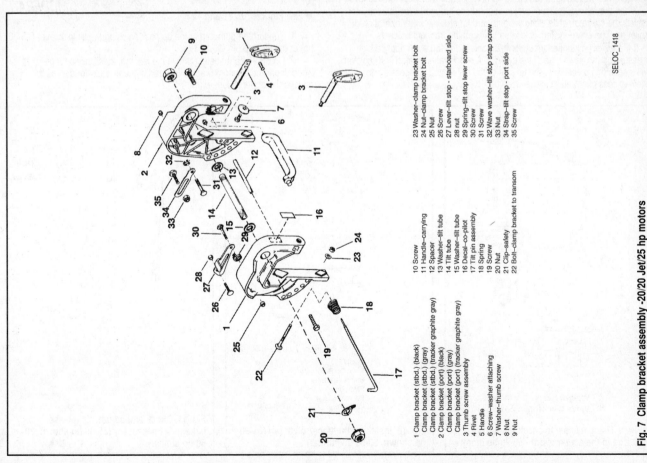

SELOC_1418

1 Clamp bracket (stbd.) (black)
Clamp bracket (stbd.) (gray)
Clamp bracket (stbd.) (tracker graphite gray)
2 Clamp bracket (port) (black)
Clamp bracket (port) (gray)
Clamp bracket (port) (tracker graphite gray)
3 Thumb screw assembly
4 Rivet
5 Handle
6 Screw–washer attaching
7 Washer–thumb screw
8 Nut
9 Nut

10 Screw
11 Handle–carrying
12 Spacer
13 Washer–tilt tube
14 Tilt tube
15 Washer–tilt tube
16 Decal–co-pilot
17 Tilt pin assembly
18 Spring
19 Screw
20 Nut
21 Clip–safety
22 Bolt–clamp bracket to transom

23 Washer–clamp bracket bolt
24 Nut–clamp bracket bolt
25 Nut
26 Screw
27 Lever–tilt stop - starboard side
28 nut
29 Spring–tilt stop lever screw
30 Screw
31 Screw
32 Wave washer–tilt stop strap screw
33 Nut
34 Strap–tilt stop - port side
35 Screw

Fig. 7 Clamp bracket assembly -20/20 Jet/25 hp motors

GAS ASSIST TILT SYSTEM

Introduction

All outboard installations are equipped with some means of raising or lowering (pivoting), the complete unit for efficient operation under various load, boat design, water conditions, and for trailering to and from the water. The correct trim angle ensures maximum performance and fuel economy as well as a more comfortable ride for the crew and passengers.

The most simple form of tilt is a mechanical tilt adjustment consisting of a series of holes in the transom mounting bracket through which an adjustment pin passes to secure the outboard unit at the desired angle.

Such a mechanical arrangement works quite well for the smallest of units, but the larger (and heavier) outboard become, the more they require some form of assist or power system. Pretty much the majority of Mercury 2-strokes (except a few of the smallest single and twin cylinder models), utilize either a Manual Gas Assist or a power trim/tilt system. In some cases, simply the trim or option level of a given HP range motor could be the difference between being equipped with one system or the other.

Two versions of the PTT system are used, a single ram unit which is found on inline models or a 3 ram (2 trim, 1 tilt) unit which is found only on V6 models.

Description & Operation

The budget method of tilting a mid-range outboard (found most often on tiller control models) is the manual gas assist tilt system. It can normally found on certain Mercury 2-stroke models from 30-90 hp. The system found on the 30-50 hp models is a unique assembly to those motors but operates in *basically* the same fashion (even though the internals are different and it is a non-serviceable unit). The only service possible on the 30-50 hp models is removal and installation for replacement of the entire unit. The systems found on the 55-90 hp motors are mechanically identical to each other and can be disassembled for overhaul/repair service.

The gas assist tilt system consists of the shock rod cylinder, valve block, nitrogen filled accumulator (on 55-90 hp models) and a manual release lever.

When the manual release lever is opened, the nitrogen filled accumulator assists the operator while tilting the outboard for shallow water operation or trailering. The entire system is located between the transom brackets.

If the lower unit strikes an underwater object, the force will apply a high pressure on the down side of the cylinder. A shock valve located in the piston will relieve this pressure to the up side of the piston and if necessary, to allow excess oil to return to the reservoir.

If the outboard has trouble staying in the selected tilt position, check for external fluid leaking from the accumulator or shock rod cylinder. Make repairs as necessary. Make sure that the manual release lever and control rod operate freely and make any adjustments if needed.

Gas Assist Damper

TESTING

◆ See Figure 9

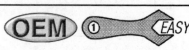

On all models, if the unit does not appear to be functioning properly, start with a visual inspection. If debris or leaking is found the assembly must be replaced on 50 hp and smaller unit; or, on 55-90 hp motors, the unit *may* also be disassembled for inspection or repairs

To check function of the gas assist cylinder proceed as follows:

1. Raise the outboard and secure using the tilt lock lever.
2. Check the adjustment on the manual release cam (the linkage just to the side of the cylinder and accumulator). The cam must open and close freely. If necessary, adjust the length of the linkage as necessary for the system to function properly.
3. On 55-90 hp motors, check a discharged accumulator as follows:
   a. Using a spring-loaded scale, tilt the outboard by applying pressure to the top of the engine cowling (while the tilt valve lever is in the open position).
   b. It must take 35-50 ft. lbs. (47-68 Nm) of pulling force to tilt the outboard from a full Down position to a full Up position.
   c. If it requires more than 50 ft. lbs. (68 Nm) of torque, the nitrogen charge in the accumulator is discharged and the accumulator will need to be replaced.

REMOVAL & INSTALLATION

◆ See Figures 10, 11 and 12

1. Support the outboard in the up position using the tilt lock lever (55-90 hp) or the tilt stop pin (30-50 hp).
2. On 55-90 hp models, remove the link rod, then position a piece of wood under the transom bracket (instead of the tilt lock) for access to remove the upper tri-lobe pin.

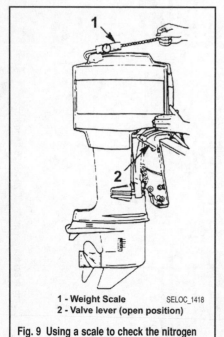

1 - Weight Scale
2 - Valve lever (open position)
SELOC_1418

Fig. 9 Using a scale to check the nitrogen charge in the accumulator - 55-90 hp motors

95 2-4-C with Teflon

1- Pivot Pin
2- Pivot Pin Bore
3- Shock Rod Bore
4- Pivot Pin
5- Swivel Bracket
6- Shock Rod
SELOC_1418

Fig. 10 Installing the upper pivot pin (55-90 hp shown, 30-50 hp similar)

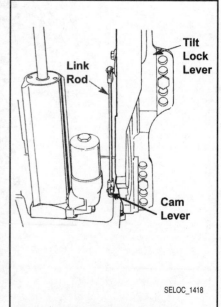

Link Rod

Tilt Lock Lever

Cam Lever

SELOC_1418

Fig. 11 Check and adjust the release linkage, as necessary (55-90 hp shown, 30-50 hp similar)

3. Remove the upper tri-lobe pin (which is used to secure the upper pivot pin). On 30-50 hp motors, you can usually grab the pin with a set of needle-nose pliers and pull downward to free it. On 55-90 hp motors, it is usually best to use a suitable punch to remove (drive down) the upper dowel pin.

■ **After removing the upper tri-lobe pin on 55-90 hp motors, reposition the tilt lock and remove the piece of wood.**

1. Use a suitable punch to drive out the upper pivot pin freeing the top of the gas cylinder.

2. On 55-90 hp motors, proceed as follows:

a. If necessary for access, remove the anode.

b. Using a punch drive out the lower dowel pin from the lower pivot pin (in the same manner as the upper tri-lobe pin was removed). Generally for 55/60 hp motors the pin must be driven UPWARD, while on 65 Jet and larger motors the pin will either be driven UPWARD (like the 55/60) or

DOWNWARD, depending on how it was inserted. In either case, retain the pin for installation.

3. On 30-50 hp motors, proceed as follows:

a. Disconnect the link rod from the cam lever.

b. Remove the tilt lock pin.

c. Remove the nuts and washers securing the lower pivot pin.

d. Remove the anode screw in order to free the ground strap.

4. Use a punch to carefully drive/tap out the lower pivot pin. On 30-50 hp motors, retain the anchor pin bushing from the clamp bracket and trim unit.

5. Tilt the shock absorber assembly (top first) out from the clamp bracket and remove the assembly.

■ **On 30-50 hp motors, stop here, no further disassembly is possible. If the unit is damaged, it must be replaced. On 55-90 hp motors, disassembly and component replacement is possible - refer to the Overhaul procedure later in this section.**

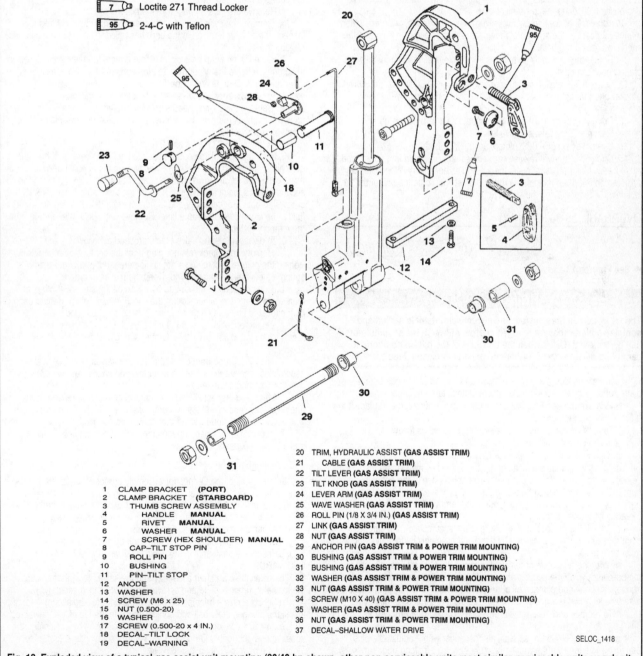

| 7 | Loctite 271 Thread Locker |
| 95 | 2-4-C with Teflon |

1	CLAMP BRACKET **(PORT)**
2	CLAMP BRACKET **(STARBOARD)**
3	THUMB SCREW ASSEMBLY
4	HANDLE **MANUAL**
5	RIVET **MANUAL**
6	WASHER **MANUAL**
7	SCREW (HEX SHOULDER) **MANUAL**
8	CAP–TILT STOP PIN
9	ROLL PIN
10	BUSHING
11	PIN–TILT STOP
12	ANODE
13	WASHER
14	SCREW (M6 x 25)
15	NUT (0.500-20)
16	WASHER
17	SCREW (0.500-20 x 4 IN.)
18	DECAL–TILT LOCK
19	DECAL–WARNING

20	TRIM, HYDRAULIC ASSIST **(GAS ASSIST TRIM)**
21	CABLE **(GAS ASSIST TRIM)**
22	TILT LEVER **(GAS ASSIST TRIM)**
23	TILT KNOB **(GAS ASSIST TRIM)**
24	LEVER ARM **(GAS ASSIST TRIM)**
25	WAVE WASHER **(GAS ASSIST TRIM)**
26	ROLL PIN (1/8 X 3/4 IN.) **(GAS ASSIST TRIM)**
27	LINK **(GAS ASSIST TRIM)**
28	NUT **(GAS ASSIST TRIM)**
29	ANCHOR PIN **(GAS ASSIST TRIM & POWER TRIM MOUNTING)**
30	BUSHING **(GAS ASSIST TRIM & POWER TRIM MOUNTING)**
31	BUSHING **(GAS ASSIST TRIM & POWER TRIM MOUNTING)**
32	WASHER **(GAS ASSIST TRIM & POWER TRIM MOUNTING)**
33	NUT **(GAS ASSIST TRIM & POWER TRIM MOUNTING)**
34	SCREW (M10 X 40) **(GAS ASSIST TRIM & POWER TRIM MOUNTING)**
35	WASHER **(GAS ASSIST TRIM & POWER TRIM MOUNTING)**
36	NUT **(GAS ASSIST TRIM & POWER TRIM MOUNTING)**
37	DECAL–SHALLOW WATER DRIVE

SELOC_1418

Fig. 12 Exploded view of a typical gas assist unit mounting (30/40 hp shown, other non-serviceable units most similar, serviceable units use don't use nuts to secure the lower pivot pin in place)

To Install:

6. Apply a light coating of Quicksilver 2-4-C with Teflon, or an equivalent lubricant to the lower pivot pin hole and pivot pin surface (and on the 30-50 hp motors, to the bushings).

7. Start the lower pivot pin (and bushings on the 30-50 hp motors) into the pivot pin hole. At the same time, position the gas assist cylinder assembly as you guide the pivot pin fully into position.

■ **On 55/60 hp motors (or any 65 Jet or larger models where the lower dowel was installed from the top), since the dowel pin for the lower pivot pin installs from the top, you can actually start it into the hole even before the lower pivot pin is fully seated (before the trim assembly is installed). On all 55-90 hp motors, use a small punch to gently drive the lower pivot pin into position until it is flush with the outer surface.**

8. On 30-50 hp motors, install the nuts and washers to anchor the lower pivot pin, then tighten them securely. Install the ground strap and anode screw. Install the tilt lock pin.

9. On 55-90 hp motors, use a small punch to drive the lower dowel pin into position securing the lower pivot pin.

10. Apply a light coating of Quicksilver 2-4-C with Teflon, or an equivalent marine lubricant to the surface of the upper pivot pin, pivot pin hole and the shock rod (gas tube) hole.

11. Rotate the assembly so the shock rod hole is aligned for installation of the upper pivot pin.

12. Using a soft mallet, carefully drive the upper pivot pin into the swivel bracket and through the shock rod until the pivot pin is flush with the swivel bracket.

13. Drive the upper tri-lobe (dowel) pin into its hole until seated.

14. Reconnect the release valve linkage.

15. If removed on 55-90 hp motors, install the anode.

16. Check the manual release cam adjustment. The cam must open and close freely. Make adjustments to the link rod as necessary. When properly adjusted on 55-90 hp motors, the rod end must snap onto the ball of the tilt lock lever *without* moving the lock lever or cam lever.

OVERHAUL (55-90 HP ONLY)

◆ **See Figures 13 and 14**

✳✳ SELOC WARNING

The tilt system is pressurized. The accumulator itself is sealed and contains a high-pressure nitrogen charge. Although the accumulator can be removed (ONLY when the tilt rod is in the full Up position), it cannot be disassembled further and must be replaced if faulty.

As with all rebuilds, be sure to remove and discard all O-rings during disassembly. Make sure the replacement O-rings are the proper size.

1. Remove the gas tilt assist assembly from the outboard and place it in a soft-jawed vise for access.

2. To remove the accumulator assembly, proceed as follows:
   a. Make sure the tilt rod is in the full UP position.
   b. Open the cam shaft valve (DOWN position).
   c. Position a small drain pan under the assembly, then locate the velocity valve (the slotted screw-head valve at the base of the assembly, right under the accumulator). Loosen the velocity valve until it just starts to drip fluid. Once the dripping stops loosen and remove the accumulator (it threads into the valve block).
   d. Check the plunger on the bottom of the accumulator. If it can be compressed into the accumulator by hand, the assembly is defective and must be replaced.
   e. With the accumulator removed, remove the O-ring (it usually still in the bore into which the accumulator was threaded), then remove the conical spring, steel ball and plunger. All of these components should be available in a service kit and it is a good idea to replace them regardless of condition.

3. To remove and disassemble the shock rod (tilt rod), proceed as follows:
   a. Using #91-74951 or an equivalent spanner with 2 adjustable pins that insert into the rod (the wrench has 1/4 in. x 5/16 in. long pegs) unscrew and remove the tilt rod end cap assembly.
   b. Carefully lift the rod and end cap from the cylinder.

■ **The only serviceable items on the shock rod assembly are the O-rings, wiper ring and the piston. If the shock requires ANY other repair, replace it as an assembly.**

   c. Place the shock rod in a soft-jawed vise, then loosen and remove the 3 screws on the lower plate.
   d. Carefully lift the plate from the rod, keeping CLOSE track of the 5 balls, 5 seats and 5 springs retained under the plate.
   e. Remove and discard the O-ring from the lower plate.
   f. With the shock rod still (or back) in the vise, use a heat gun to *gently* apply heat to loosen the piston (still at the bottom of the rod).
   g. While the piston is hot, use the spanner wrench that you used on the cap to now loosen the piston, but do not remove it completely. Allow the piston to cool and remove it from the shock rod.
   h. Inspect the check valve in the piston for debris and clean, as necessary. If the check valve cannot be cleaned, replace the piston as an assembly.
   i. Clean the shock and components with compressed air (but *don't* lose the springs, seats and balls).
   j. Remove the inner O-ring from the piston bore.
   k. Remove the cylinder end cap assembly from the shock rod and inspect. If the wiper (located in the cap) has failed to keep the rod cleaned, replace the wiper.
   l. Place the end cap on a clean work surface, then use a small prytool to carefully remove the wiper from the top center of the cap. Remove the inner and outer O-rings and discard.

4. To remove/service the valve block, proceed as follows:
   a. Remove the 2 Allen head screws (threaded horizontally) from the base of the shock rod cylinder in order to separate the valve block. Carefully remove the valve block.
   b. Remove the 2 O-rings and 2 dowels.

5. If the memory piston requires service, you can remove it from the bore either using a pair of lockring pliers to grab it and pull it upward OR by using compressed air blown into the center O-ring hole. When using the second method, point the cylinder bore facing downward toward a shop rag or towel to both prevent injury *and* to help prevent damage. Expect some fluid to be expelled at the same time. Remove and discard the O-ring from the memory piston.

6. To disassemble the valve block, proceed as follows:
   a. Remove the check retainer plug from the center side of the body. Behind the plug expect to find a small diameter spring, a larger diameter spring a steel ball and a small rod. Keep the pieces in order.
   b. Remove each of the slotted hydraulic oil transfer valve plugs and their components (on either side of the check retainer plug). Behind each plug expect to find an O-ring, a spring, a large steel ball and a rod.
   c. On the opposite side of the valve block from the transfer valves, remove the slotted velocity valve, O-ring, spring and spool (surge valve) assembly.
   d. Ninety degrees from either the transfer valves or velocity valve assemblies, remove the screw, retainer clip and spacer retainer clip, along with the cam and shaft seal.

7. Clean and inspect all components which are to be reused using engine cleaner and low-pressure compressed air. DO NOT use cloth rags which may leave residue behind that could clog a valve.

8. Inspect all machined surfaces for burrs or scoring and repair or replace, as necessary.

To Assemble:

9. Identify all of the O-rings for assembly purposes. Lubricate the O-rings using Quicksilver Power Trim and Steering Fluid (or, if not available, using automatic transmission fluid), that is all *except* the cam shaft O-ring which should be lubricated with 2-4-C with Teflon or equivalent.

10. Assemble the valve block, as follows:
   a. Install the lubricated O-ring and back-up seal to the cam.
   b. Install the shaft seal in the valve block with the LIPS FACING OUTWARD.
   c. Install the cam shaft assembly into the valve block, then secure using the insulator, retainer plate and screw. Tighten the screw securely.
   d. Install a new lubricated O-ring into the bore for the accumulator, then install the plunger, steel ball and conical spring in the top of the valve block.
   e. Install the velocity valve (surge valve assembly) by positioning the spool, spring, lubricated O-ring and the velocity screw plug into the block. Tighten the plug to 75 inch lbs. (8.5 Nm).
   f. Install the check retainer assembly by positioning the rod (plunger), large diameter spring, ball, small diameter spring and finally the plug.

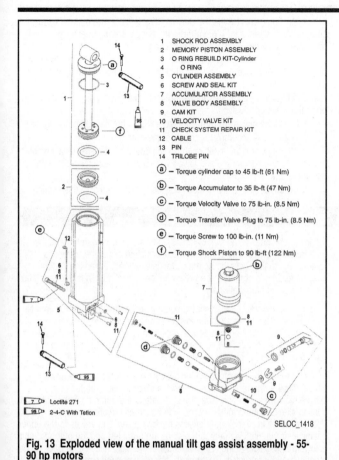

1   SHOCK ROD ASSEMBLY
2   MEMORY PISTON ASSEMBLY
3   O RING REBUILD KIT-Cylinder
4   O RING
5   CYLINDER ASSEMBLY
6   SCREW AND SEAL KIT
7   ACCUMULATOR ASSEMBLY
8   VALVE BODY ASSEMBLY
9   CAM KIT
10 VELOCITY VALVE KIT
11 CHECK SYSTEM REPAIR KIT
12 CABLE
13 PIN
14 TRILOBE PIN

ⓐ – Torque cylinder cap to 45 lb-ft (61 Nm)

ⓑ – Torque Accumulator to 35 lb-ft (47 Nm)

ⓒ – Torque Velocity Valve to 75 lb-in. (8.5 Nm)

ⓓ – Torque Transfer Valve Plug to 75 lb-in. (8.5 Nm)

ⓔ – Torque Screw to 100 lb-in. (11 Nm)

ⓕ – Torque Shock Piston to 90 lb-ft (122 Nm)

▭ 7 ▭ Loctite 271
▭ 95 ▭ 2-4-C With Teflon

SELOC_1418

Fig. 13 Exploded view of the manual tilt gas assist assembly - 55-90 hp motors

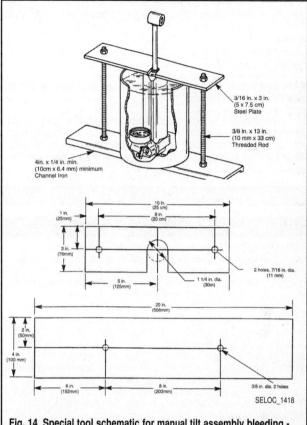

3/16 in. x 3 in.
(5 x 7.5 cm)
Steel Plate

3/8 in. x 13 in.
(10 mm x 33 cm)
Threaded Rod

4in. x 1/4 in. min.
(10cm x 6.4 cm) minimum
Channel Iron

1 in.
(25mm)

10 in.
(25 cm)

8 in.
(20 cm)

3 in.
(76mm)

5 in.
(125mm)

1 1/4 in. dia.
(30m)

2 holes, 7/16 in. dia.
(11 mm)

20 in.
(508mm)

2 in.
(50mm)

4 in.
(100 mm)

6 in.
(152mm)

8 in.
(203mm)

3/8 in. dia. 2 holes

SELOC_1418

Fig. 14 Special tool schematic for manual tilt assembly bleeding - 55-90 hp motors

g. Install each of the transfer valve assemblies by positioning the components in each bore. Install the rod (plunger), steel ball, spring and then screw plug with new, lubricated O-ring. Tighten each of the screw plugs to 75 inch lbs. (8.5 Nm).

h. Install the new lubricated O-rings and dowel pins to the mating surface of the valve block itself, then position the block to the shock rod cylinder.

i. Install the valve block retaining screws (Allen head screws) and tighten to 100 inch lbs. (11 Nm).

11. To assemble the shock (tilt) rod, proceed as follows:

a. Install the lubricated O-rings to the end cap and to the piston.

b. Install a new rod wiper to the center of the cap.

c. Clamp the shock rod in a soft-jawed vise, then position the end cap onto the rod (with the O-ring facing the bottom end of the rod and the threads facing the top end of the rod).

d. Apply a light coating of Loctite® 271 or equivalent threadlocking compound to the threads on the shock rod, then install the shock rod piston.

✳✳ SELOC CAUTION

To prevent damage to the shock rod piston, MAKE SURE the spanner wrench has 1/4 in. x 5/16 in. (6.4 x 8mm) long pegs.

e. Using the spanner, tighten the piston securely. If an adapter is available for a torque wrench, tighten the piston to 90 ft. lbs. (122 Nm).

f. Install the 5 sets of balls, seats and springs to the shock rod piston, then position the plate to secure them. Tighten the 3 plate screws to 35 inch lbs. (4.0 Nm).

g. Remove the shock assembly from the vise.

■ **There are 2 possible methods for shock rod installation and fluid filling/bleeding. The first method involves a soft-jawed vise, while the second uses a specially fabricated tool assembly.**

12. To install the shock rod with the accumulator and fill/bleed the fluid using only a soft-jawed vise, proceed as follows:

a. Position the assembly in a soft-jawed vise.

b. With the manifold cam lever in the closed (up position), slowly fill the cylinder and manifold to the top using Quicksilver Power Trim & Steering Fluid (or equivalent). Allow time for the bubbles to disperse.

c. Install the lubricated O-ring to the memory piston, then use a pair of lockring pliers (like Snap-On® SRP2 or Craftsman 4735) position the memory piston on top of the cylinder. Now, open the cam lever (down position) and carefully push the memory piston down just below the cylinder threads and CLOSE the cam lever (Up position).

d. Fill the top of the cylinder again with fluid all the way to the top, then install the shock rod assembly on top of the memory piston. Again open the cam lever (down position) as you push the shock rod down to 1/8 in. below the cylinder threads, then again CLOSE the cam lever (Up position).

e. Fill the top of the shock rod assembly with fluid to the top of the cylinder. Open the cam lever (down position) and screw the cylinder cap down into position.

f. Tighten the end cap securely using the spanner wrench. If a torquing adapter is available, tighten the end cap to 45 ft. lbs. (61 Nm).

g. Close the cam lever (Up position).

h. Open and close the cam lever, watching for bubbles coming from the accumulator check ball hole. When the bubbles stop, fill the accumulator opening to the top with fluid.

i. Grease the threads on the accumulator and opening using 2-4-C with Teflon or equivalent, then start the accumulator into the threads and OPEN the cam lever (Down position). Tighten the accumulator securely or, if an adapter is available, tighten it to 35 ft. lbs. (47 Nm).

■ **If the filling procedure was done correctly it should be difficult to turn the cylinder rod assembly by hand.**

13. To install the shock rod with the accumulator and fill/bleed the fluid using the specially fabricated tool (as detailed in the accompanying illustration), proceed as follows:

a. Install the lubricated O-ring to the memory piston and install the piston into the bore.

b. Install the shock rod assembly. Tighten the end cap securely using the spanner wrench. If a torquing adapter is available, tighten the end cap to 45 ft. lbs. (61 Nm).

c. With the shock rod in the full UP position and the manifold cam lever open (facing down), secure the tilt system to the retaining tool and container (a No. 10 can or 3 lbs. coffee can may be used to fabricate the heart of the container).

d. Fill the container nearly full using Quicksilver Power Trim & Steering Fluid (or equivalent).

■ **The fluid level MUST remain above the accumulator opening during the entire bleeding process.**

e. Bleed the unit by pushing the rod down slowly (taking 18-20 seconds *per stroke*) until the rod is stopped at the base. Hold this position until all air bubbles exit the accumulator base, then pull upward slowly on the rod 3 in. (76mm) from the base again, wait for all air bubbles to exit the accumulator base.

f. Continue to cycle the rod down and up in the same pattern, using short strokes, again 3 in. (76mm) from the base, allowing bubbles to dissipate each time. Repeat about 5-8 times.

g. Allow the unit to stand for 5 minutes, then proceed to cycle the unit another 2-3 more times using short strokes, no bubbles should appear from the port this time.

h. With the oil level *well* above the accumulator port, slowly pull the rod up to the FULL UP position.

i. Install the accumulator making sure NO air bubbles enter the system.

j. Tighten the accumulator snugly at this time.

k. With the cam lever still open (facing down) remove the tilt assembly from the oil and secure in a soft-jawed vise. Using an adapter, tighten the accumulator to 35 ft. lbs. (47 Nm).

14. Install the gas assist tilt assembly as detailed earlier in this section.

POWER TILT/TRIM - SMALL MOTOR, SINGLE RAM INTEGRAL

Introduction

All outboard installations are equipped with some means of raising or lowering (pivoting), the complete unit for efficient operation under various load, boat design, water conditions, and for trailering to and from the water. The correct trim angle ensures maximum performance and fuel economy as well as a more comfortable ride for the crew and passengers.

The most simple forms include both mechanical tilt adjustment and manual gas assist adjustment. The former consists of a series of holes in the transom mounting bracket through which an adjustment pin passes to secure the outboard unit at the desired angle. And the later of a sealed hydraulic unit which uses a nitrogen accumulator to help lift the outboard and a release valve to allow the outboard to pivot back downward when desired.

Either way, such mechanical arrangements works quite well for most smaller outboards, however the larger (and heavier) outboards become, the more it just makes sense and is convenient to use a power system. It is probably safe to say that the majority of Mercury 2-strokes utilize a power trim/tilt system, but only a handful of them are the 15 hp or smaller like those covered in this sub-section.

This is the smallest of the single-ram units found on Mercury models.

Description & Operation

Although a purely mechanical system is used on many smaller outboards, the larger (and therefore heavier) an outboard becomes, the more likely they are to utilize an electro-hydraulic Power Trim and Tilt (PTT) system. Therefore, although you should expect most 30 hp and larger Mercury 2-stroke outboards utilize some form of a PTT system, most smaller models are actually manual tilt.

Basically, this power trim/tilt systems consist of a housing with an electric motor, a gear driven hydraulic pump and hydraulic reservoir which is attached via external hydraulic lines to the single trim/tilt cylinder which is self-contained in its own housing. The cylinder performs a double function as trim/tilt cylinders and also as shock absorbers, should the lower unit strike an underwater object while the boat is underway.

The necessary valves, check valves, relief valves and hydraulic passageways are incorporated for efficient operation. A manual release valve is provided to permit the outboard unit to be raised or lowered should the battery fail to provide the necessary current to the electric motor or if a malfunction should occur in the hydraulic system.

The gear driven pump operates in much the same manner as an oil circulation pump installed on motor vehicles. The gears rotate in either direction, depending on the desired cylinder movement. One side of the pump is considered the suction side and the other the pressure side, when the gears rotate in a given direction. These sides are reversed, the suction side becomes the pressure side and the pressure side becomes the suction side when gear movement is changed to the opposite direction.

Generally 2 relays may be used for the electric motor. The relays are usually located at the bottom cowling pan, where they are fairly well protected from moisture.

■ **As a convenience, most models contain an auxiliary trim/tilt switch installed on the exterior cowling.**

When the up portion of the trim/tilt switch is depressed, the up circuit, through the relay, is closed and the electric motor rotates in a clockwise direction. Pressurized oil from the pump passes through a series of valves to the lower chamber of the trim cylinders, the pistons are extended and the outboard unit is raised. The fluid in the upper chamber of the pistons is routed back to the reservoir as the piston is extended. When the desired position for trim is obtained, the switch on the control handle is released and the outboard is held stationary.

If the trim cylinder pistons should become fully extended, such as in a tilt up situation, fluid pressure in the lower chamber of the trim cylinders increases. This increase in pressure opens an up relief valve and the fluid is routed to the reservoir. The sound of the electric motor and the pump will have a noticeable change.

When the down portion of the trim/tilt switch is depressed, the down circuit, through the relay, is closed and the electric motor rotates in a counterclockwise direction. The pressure side of the pump now becomes the suction side and the original suction side becomes the pressure side. Pressurized oil from the pump passes through a series of valves to the upper chamber of the trim cylinders, the pistons are retracted and the outboard unit is lowered. The fluid in the lower chamber of the pistons is routed back to the reservoir as the retracted is extended. When the desired position for trim is obtained, the switch on the control handle is released and the outboard is held stationary.

If the trim cylinder pistons should become fully retracted, such as in a tilt down situation, fluid pressure in the upper chamber of the trim cylinders increases. This increase in pressure opens an up relief valve and the fluid is routed to the reservoir. The sound of the electric motor and the pump will have a noticeable change.

In the event the outboard lower unit should strike an underwater object while the boat is underway, the tilt piston would be suddenly and forcibly extended, moved upward. For this reason, the lower end of the tilt piston is capped with a free piston. This free piston normally moves up and down with the tilt piston.

The free piston also moves upward but at a much slower rate than the tilt piston. The action of the tilt piston separating from the free piston causes two actions. First, the hydraulic fluid in the upper chamber above the piston is compressed and pressure builds in this area. Second, a vacuum is formed in the area between the tilt piston and the free piston.

This vacuum in the area between the 2 pistons sucks fluid from the upper chamber. The fluid fills the area slowly and the shock of the lower unit striking the object is absorbed. After the object has been passed the weight of the outboard unit tends to retract the piston. The fluid between the tilt piston and the free piston is compressed and forced through check valves to the reservoir until the free piston reaches its original neutral position.

A manual relief valve, located on the unit, allows easy manual tilt of the outboard should electric power be lost. The valve opens when the screw is turned counterclockwise, allowing fluid to flow through the manual passage. When the relief valve screw is turned fully clockwise, the manual passage is closed and the outboard is locked in position (unless the hydraulic system changes the position of the trim/tilt ram).

A thermal valve is normally used to protect the trim/tilt motor and allow it to maintain a designated trim angle. Oil in the upper chamber is pressurized when force is applied to the outboard from the rear while cruising. Oil is directed through the right side check valve and activates the thermal valve to release oil pressure and lessen the strain on the motor and pump.

Troubleshooting

GETTING STARTED (FIRST THINGS TO CHECK)

Any time a problem develops in the power trim/tilt system the first step is to determine whether it is electrical or hydraulic in nature. After the determination is made, then the appropriate steps can be taken to remedy the problem.

The first step in troubleshooting is to make sure all the connectors are properly plugged in and that all the terminals and wires are free of corrosion. The simple act of disconnecting and connecting a terminal may sometimes loosen corrosion that is preventing a proper electrical connection. Inspect each terminal carefully and coat each with dielectric grease to prevent corrosion.

The next step is to make sure the battery is fully charged and in good condition. While checking the battery, perform the same maintenance on the battery cables as you did on the electrical terminals. Disconnect the cables (negative side first), clean and coat them and then reinstall them. If the battery is past its useful life, replace it. If it only requires a charge, charge it.

Check the power trim/tilt fuse, as appropriate. Many systems will have a fuse to prevent large current draws from damaging the system. If this fuse is blown, the system will cease to function. This is a good indicator that you may have problems elsewhere in the electric system. Fuses don't blow without cause.

After inspecting the electrical side of the system, check the hydraulic fluid level and top it off as necessary. Remember to position the motor properly (full tilt up or down, as applicable) to get an accurate measurement of fluid level. A slight decrease in the level of hydraulic fluid may cause the system to act sporadically.

Finally, make sure the manual release valve is in the proper position. A slightly open manual release valve may prevent the system from working properly and mimic other more serious problems.

Just remember to check the simple things first. If these simple tests do not diagnose the cause of the problem, then it is time to investigate more deeply. Check the hydraulic system. Inspect the entire power trim/tilt electrical harness with a multi-meter, checking for excessive resistance and proper voltage.

COMPREHENSIVE TESTING

Wiring Diagram & Circuit Testing

◆ See Figures 15 thru 19

Use the accompanying wiring diagrams and troubleshooting charts to diagnose and repair electrical problems with the PTT systems on 15 hp and smaller motors.

Before starting any electrical troubleshooting, check the following:
• Check for disconnected wires
• Make sure all connections are clean (of corrosion) and tight
• Make sure the plug-in connectors are fully engaged
• Make sure the battery is fully charged

The numbers listed in the electrical troubleshooting chart corresponds to the numbered locations on the appropriate wiring diagrams.

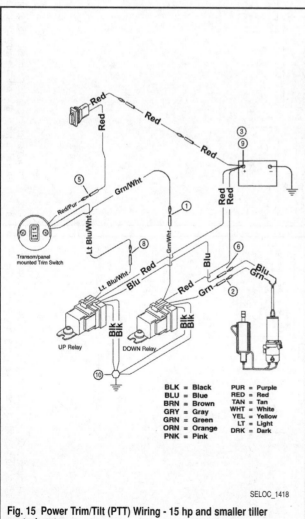

SELOC_1418

Fig. 15 Power Trim/Tilt (PTT) Wiring - 15 hp and smaller tiller control motors

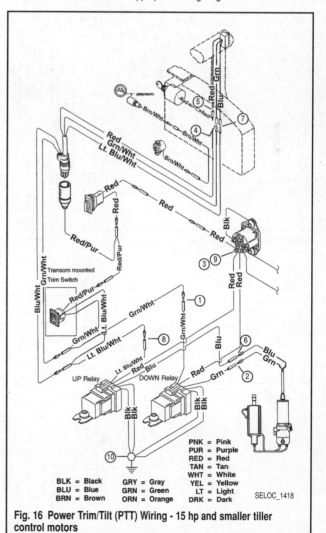

SELOC_1418

Fig. 16 Power Trim/Tilt (PTT) Wiring - 15 hp and smaller tiller control motors

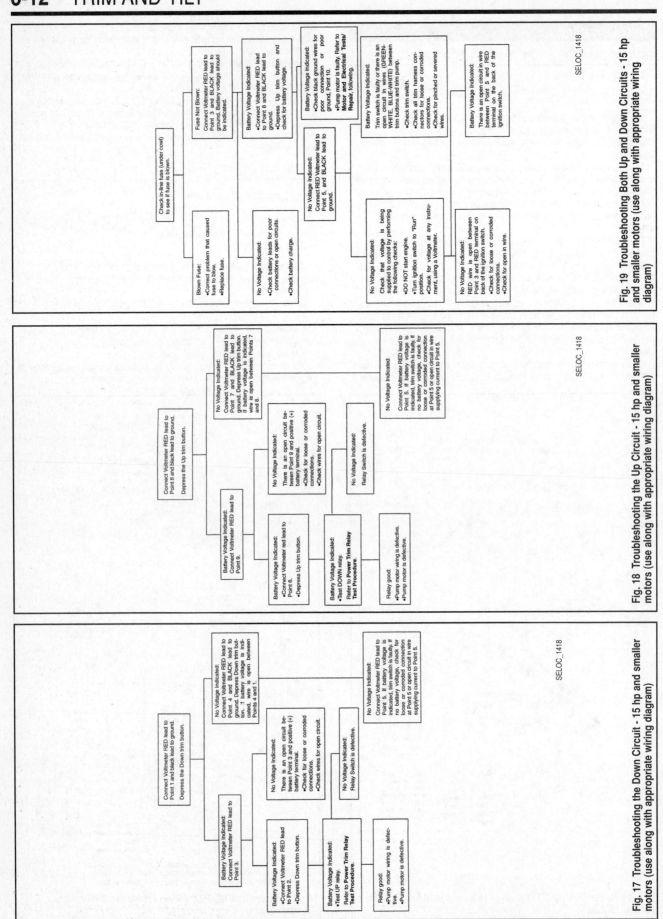

Check in-line fuse (under cowl) to see if fuse is blown.

Fuse Not Blown:
Connect Voltmeter RED lead to Point 3 and BLACK lead to ground. Battery voltage should be indicated.

Blown Fuse:
Correct problem that caused fuse to blow.
•Replace fuse.

Battery Voltage Indicated:
•Connect Voltmeter RED lead to Point 8 and BLACK lead to ground.
•Depress Up trim button and check for battery voltage.

No Voltage Indicated:
•Check battery leads for poor connections or open circuits.
•Check battery charge.

Battery Voltage Indicated:
•Check black ground wires for poor connection or poor ground, Point 10.
•Pump motor is faulty. Refer to **Motor and Electrical Tests/Repair,** following.

No Voltage Indicated:
Connect RED Voltmeter lead to Point 5, and BLACK lead to ground.

No Voltage Indicated:
Check that voltage is being supplied to control by performing the following checks:
•DO NOT start engine.
•Turn ignition switch to "Run" position.
•Check for voltage at any instrument, using a Voltmeter.

Battery Voltage Indicated:
Trim switch is faulty or there is an open circuit in wires (GREEN-WHITE, BLUE-WHITE) between trim buttons and trim pump.
•Check trim switch.
•Check all trim harness connectors for loose or corroded connections.
•Check for pinched or severed wires.

No Voltage Indicated:
RED wire is open between Point 3 and RED terminal on back of the ignition switch.
•Check for loose or corroded connections.
•Check for open in wire.

Battery Voltage Indicated:
There is an open circuit in wire between Point 5 and RED terminal on the back of the ignition switch.

SELOC_1418

Fig. 19 Troubleshooting Both Up and Down Circuits - 15 hp and smaller motors (use along with appropriate wiring diagram)

Connect Voltmeter RED lead to Point 8 and black lead to ground.

Depress the Up trim button.

No Voltage Indicated:
Connect Voltmeter RED lead to Point 7 and BLACK lead to ground. Depress Up trim button. If battery voltage is indicated, wire is open between Points 7 and 8.

Battery Voltage Indicated:
Connect Voltmeter RED lead to Point 9.

No Voltage Indicated:
Connect Voltmeter RED lead to Point 5. If battery voltage is indicated, trim switch is faulty, if no battery voltage, check for loose or corroded connection at Point 5 or open circuit in wire supplying current to Point 5.

Battery Voltage Indicated:
•Connect Voltmeter red lead to Point 6.
•Depress Up trim button.

No Voltage Indicated:
There is an open circuit between Point 9 and positive (+) battery terminal.
•Check for loose or corroded connections.
•Check wires for open circuit.

Battery Voltage Indicated:
•Test DOWN relay.
Refer to **Power Trim Relay Test Procedure.**

No Voltage Indicated:
Relay Switch is defective.

Relay good:
•Pump motor wiring is defective.
•Pump motor is defective.

SELOC_1418

Fig. 18 Troubleshooting the Up Circuit - 15 hp and smaller motors (use along with appropriate wiring diagram)

Connect Voltmeter RED lead to Point 1 and black lead to ground.

Depress the Down trim button.

No Voltage Indicated:
Connect Voltmeter RED lead to Point 4 and BLACK lead to ground. Depress Down trim button. If battery voltage is indicated, wire is open between Points 4 and 1.

Battery Voltage Indicated:
Connect Voltmeter RED lead to Point 3.

No Voltage Indicated:
Connect Voltmeter RED lead to Point 5. If battery voltage is indicated, trim switch is faulty, if no battery voltage, check for loose or corroded connection at Point 5 or open circuit in wire supplying current to Point 5.

Battery Voltage Indicated:
•Connect Voltmeter RED lead to Point 2.
•Depress Down trim button.

No Voltage Indicated:
There is an open circuit between Point 3 and positive (+) battery terminal.
•Check for loose or corroded connections.
•Check wires for open circuit.

Battery Voltage Indicated:
•Test UP relay.
Refer to **Power Trim Relay Test Procedure.**

No Voltage Indicated:
Relay Switch is defective.

Relay good:
•Pump motor wiring is defective.
•Pump motor is defective.

SELOC_1418

Fig. 17 Troubleshooting the Down Circuit - 15 hp and smaller motors (use along with appropriate wiring diagram)

PTT Hydraulic Cylinder

REMOVAL & INSTALLATION

◆ See Figures 20 and 21

1. Raise the outboard unit to the full up position. If the hydraulic system is inoperative, first rotate the manual release valve counterclockwise about 3-4 complete turns and then manually lift the unit to the full up position.

2. Using an engine hoist and a suitable strap, support the outboard in the raised position.

3. On either side of the transom bracket loosen and remove the fasteners (through bolts and nuts) securing the tilt stop straps.

4. Free the upper bracket, then remove the upper pivot pin from the bracket.

■ **Expect the hydraulic lines to leak fluid when they are loosened. Clean the area before loosening them and position a suitable drain pan. Also keep some shop towels handy.**

5. Loosen, but do not remove the hydraulic lines at the cylinder, then move to the other end of the lines and disconnect them from the pump assembly.

6. Loosen and remove the nuts on either end of the lower pivot pin, then while supporting the PTT cylinder, use a suitable driver to carefully tap out the lower pin (freeing the spacers as well). Remove the cylinder from the outboard.

7. If necessary, carefully remove the hydraulic lines from the cylinder.

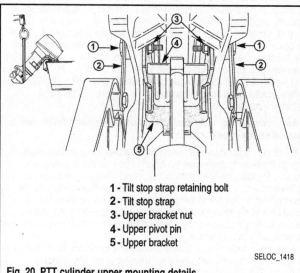

1 - Tilt stop strap retaining bolt
2 - Tilt stop strap
3 - Upper bracket nut
4 - Upper pivot pin
5 - Upper bracket

SELOC_1418

Fig. 20 PTT cylinder upper mounting details

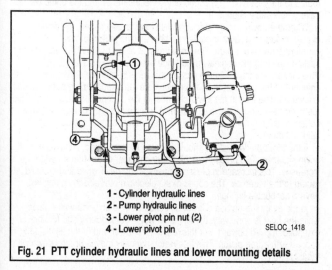

1 - Cylinder hydraulic lines
2 - Pump hydraulic lines
3 - Lower pivot pin nut (2)
4 - Lower pivot pin

SELOC_1418

Fig. 21 PTT cylinder hydraulic lines and lower mounting details

To Install:

■ **Be sure to lubricate the pivot pins and bores using 2-4-C with Teflon or an equivalent marine lubricant during assembly.**

8. If removed, install and carefully tighten (just snug at this point, they will be tightened later when both ends are connected.

9. If removed, grease and install the nylon bushing into the lower bore on the cylinder pivot.

10. Insert the lower pivot pin into the support bracket, then install the spacer and position the cylinder (aligning it with the pivot pin). Gently push the pin through the cylinder until it is flush. Insert the other spacer and push the pin into place, then install the nuts and tighten to 120 inch. lbs. (13.6 Nm).

11. Connect the hydraulic lines to the pump assembly, just finger tight at first, taking extra care not to cross thread the flare nuts.

12. Now tighten all of the hydraulic lines, being careful not to over-tighten and strip them.

13. Install the upper pivot pin into the upper bracket, then guide the pivot through the trim cylinder eye and into the upper bracket on the other side.

14. Position the upper bracket to the swivel bracket and secure using the 2 shoulder bolts (inserted through the tilt stop straps). Install a washer and Nylock nut onto each shoulder bolt, then tighten to 144 inch lbs. (16.3 Nm).

15. Remove the support and bleed the hydraulic system of any air, topping it off as necessary.

PTT Hydraulic Pump

REMOVAL & INSTALLATION

◆ See Figure 22

1. Raise the outboard unit to the full up position. If the hydraulic system is inoperative, first rotate the manual release valve counterclockwise about 3-4 complete turns and then manually lift the unit to the full up position.

2. Using an engine hoist and a suitable strap, support the outboard in the raised position.

3. Tag and disconnect the PTT motor wiring from the solenoid harness and carefully fee the wiring down to the pump noting how it is routed.

■ **Expect the hydraulic lines to leak fluid when they are loosened. Clean the area before loosening them and position a suitable drain pan. Also keep some shop towels handy.**

4. Loosen, but do not remove the hydraulic lines at the cylinder, then move to the other end of the lines and disconnect them from the pump assembly.

5. While supporting the pump, loosen and remove the 3 bolts securing the pump to the side of the support bracket, then remove the pump assembly.

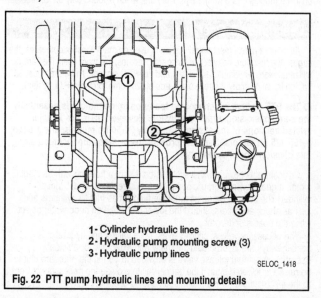

1 - Cylinder hydraulic lines
2 - Hydraulic pump mounting screw (3)
3 - Hydraulic pump lines

SELOC_1418

Fig. 22 PTT pump hydraulic lines and mounting details

To Install:

6. Position the pump to the support bracket and install the 3 mounting bolts. Tighten the bolts to 80 inch lbs. (9 Nm).

7. Carefully thread the hydraulic lines to the pump, watching to make sure you do not cross thread the flare nut, then, once you are certain they are properly threaded, tighten them securely, then move to the other end and tighten them at the cylinder as well.

8. Route and connect the PTT motor wires, as noted during removal.

9. Remove the support and bleed the hydraulic system of any air, topping it off as necessary.

HYDRAULIC SYSTEM BLEEDING

This trim/tilt system with 1 large cylinder has been designed with a "self bleeding" system. Fluid is pumped from 1 side of the cylinder to the opposite side as the piston extends or retracts. Therefore, any air trapped on 1 end of the cylinder, is pushed out back to the reservoir, when the piston bottoms out. Usually 2 or 3 cycles of the system will remove all entrapped air from the cylinder. However, if the components were replaced or the system was completely drained for some reason, it might take a *lot* of cycling. Under those circumstances it might make more sense to open the manual release valve first and fill the system.

The following short section outlines the necessary steps to perform during the cycling sequence just mentioned.

Whenever possible use Quicksilver Power Trim & Steering Fluid to fill the system. The Mercury PTT fluid is a highly refined hydraulic fluid, containing a high detergent content and additives to keep seals pliable. A high grade automatic transmission fluid, Dexron® II, may also be used if the recommended fluid is not available.

1. Install the system and open the manual release valve 3 turns.

2. Make sure the area around the fill plug is clean and free of any dirt or corrosion, then remove the fill plug and add fluid to the system until it is topped off, then reinstall the plug.

3. Manually raise and lower the trim several times, then recheck the fluid level and top off again, as necessary.

4. Close the manual release valve.

5. Cycle the system to fully raise and lower the unit 3 times.

6. Fully extend the trim rod and recheck the fluid level. Top off the fluid, as necessary. If you needed to add a significant amount of fluid, repeat cycling the unit up and down another 3 times, then recheck the fluid level. Continue this pattern until the fluid remains at the proper level without topping it off again.

OVERHAUL

Mercury provides no information regarding the possible overhaul of the PTT components on these models.

POWER TILT/TRIM - MID-RANGE MOTOR, SINGLE RAM INTEGRAL

Introduction

All outboard installations are equipped with some means of raising or lowering (pivoting), the complete unit for efficient operation under various load, boat design, water conditions, and for trailering to and from the water. The correct trim angle ensures maximum performance and fuel economy as well as a more comfortable ride for the crew and passengers.

The most simple forms include both mechanical tilt adjustment and manual gas assist adjustment. The former consists of a series of holes in the transom mounting bracket through which an adjustment pin passes to secure the outboard unit at the desired angle. And the latter of a sealed hydraulic unit which uses a nitrogen accumulator to help lift the outboard and a release valve to allow the outboard to pivot back downward when desired.

Either way, such mechanical arrangements works quite well for smaller (most 1-cyl and 2-cyl, and perhaps even a few of the more basic 3-cyl motors) outboards, however the larger (and heavier) outboards become, the more it just makes sense and is convenient to use a power system. Pretty much the majority of Mercury 2-strokes, utilize a power trim/tilt system. In some cases, simply the trim or option level of a given HP range motor could be the difference between being equipped with one system or the other.

Two versions of the PTT system are used, a single ram unit which is found on inline models or a 3 ram (2 trim, 1 tilt) unit which is found only on V6 models.

Description & Operation

Although a purely mechanical system is used on many smaller outboards, the larger (and therefore heavier) an outboard becomes, the more likely they are to utilize an electro-hydraulic power trim and tilt (PTT) system. Therefore, most 30 hp and larger Mercury 2-stroke outboards utilize some form of a PTT system.

■ **The PTT system which may be found on inline motors is essentially the same basic single ram unit design. However, there are 2 major related versions of the system, 1 used by 30-50 hp motors, and 1 used by 55-125 hp motors. Separate procedures or steps are included along this division where necessary.**

Basically, this power trim/tilt systems consist of a housing with an electric motor, a gear driven hydraulic pump, hydraulic reservoir and 1 trim/tilt cylinder. The cylinder performs a double function as trim/tilt cylinders and also as shock absorbers, should the lower unit strike an underwater object while the boat is underway.

The necessary valves, check valves, relief valves and hydraulic passageways are incorporated internally and externally for efficient operation. A manual release valve is provided to permit the outboard unit to be raised or lowered should the battery fail to provide the necessary current to the electric motor or if a malfunction should occur in the hydraulic system.

The gear driven pump operates in much the same manner as an oil circulation pump installed on motor vehicles. The gears rotate in either direction, depending on the desired cylinder movement. One side of the pump is considered the suction side and the other the pressure side, when the gears rotate in a given direction. These sides are reversed, the suction side becomes the pressure side and the pressure side becomes the suction side when gear movement is changed to the opposite direction.

Generally 2 relays may be used for the electric motor. The relays are usually located at the bottom cowling pan, where they are fairly well protected from moisture.

■ **As a convenience, most models contain an auxiliary trim/tilt switch installed on the exterior cowling.**

When the up portion of the trim/tilt switch is depressed, the up circuit, through the relay, is closed and the electric motor rotates in a clockwise direction. Pressurized oil from the pump passes through a series of valves to the lower chamber of the trim cylinders, the pistons are extended and the outboard unit is raised. The fluid in the upper chamber of the pistons is routed back to the reservoir as the piston is extended. When the desired position for trim is obtained, the switch on the control handle is released and the outboard is held stationary.

If the trim cylinder pistons should become fully extended, such as in a tilt up situation, fluid pressure in the lower chamber of the trim cylinders increases. This increase in pressure opens an up relief valve and the fluid is routed to the reservoir. The sound of the electric motor and the pump will have a noticeable change.

When the down portion of the trim/tilt switch is depressed, the down circuit, through the relay, is closed and the electric motor rotates in a counterclockwise direction. The pressure side of the pump now becomes the suction side and the original suction side becomes the pressure side. Pressurized oil from the pump passes through a series of valves to the upper chamber of the trim cylinders, the pistons are retracted and the outboard unit is lowered. The fluid in the lower chamber of the pistons is routed back to the reservoir as the retracted is extended. When the desired position for trim is obtained, the switch on the control handle is released and the outboard is held stationary.

If the trim cylinder pistons should become fully retracted, such as in a tilt down situation, fluid pressure in the upper chamber of the trim cylinders increases. This increase in pressure opens an up relief valve and the fluid is routed to the reservoir. The sound of the electric motor and the pump will have a noticeable change.

In the event the outboard lower unit should strike an underwater object while the boat is underway, the tilt piston would be suddenly and forcibly extended, moved upward. For this reason, the lower end of the tilt piston is capped with a free piston. This free piston normally moves up and down with the tilt piston.

The free piston also moves upward but at a much slower rate than the tilt piston. The action of the tilt piston separating from the free piston causes 2 actions. First, the hydraulic fluid in the upper chamber above the piston is compressed and pressure builds in this area. Second, a vacuum is formed in the area between the tilt piston and the free piston.

This vacuum in the area between the 2 pistons sucks fluid from the upper chamber. The fluid fills the area slowly and the shock of the lower unit striking the object is absorbed. After the object has been passed the weight of the outboard unit tends to retract the piston. The fluid between the tilt piston and the free piston is compressed and forced through check valves to the reservoir until the free piston reaches its original neutral position.

A manual relief valve, located on the stern bracket, allows easy manual tilt of the outboard should electric power be lost. The valve opens when the screw is turned counterclockwise, allowing fluid to flow through the manual passage. When the relief valve screw is turned fully clockwise, the manual passage is closed and the outboard is locked in position (unless the hydraulic system changes the position of the trim/tilt ram).

A thermal valve is used to protect the trim/tilt motor and allow it to maintain a designated trim angle. Oil in the upper chamber is pressurized when force is applied to the outboard from the rear while cruising. Oil is directed through the right side check valve and activates the thermal valve to release oil pressure and lessen the strain on the motor and pump.

Troubleshooting

GETTING STARTED (FIRST THINGS TO CHECK)

Any time a problem develops in the power trim/tilt system the first step is to determine whether it is electrical or hydraulic in nature. After the determination is made, then the appropriate steps can be taken to remedy the problem.

The first step in troubleshooting is to make sure all the connectors are properly plugged in and that all the terminals and wires are free of corrosion. The simple act of disconnecting and connecting a terminal may sometimes loosen corrosion that is preventing a proper electrical connection. Inspect each terminal carefully and coat each with dielectric grease to prevent corrosion.

The next step is to make sure the battery is fully charged and in good condition. While checking the battery, perform the same maintenance on the battery cables as you did on the electrical terminals. Disconnect the cables (negative side first), clean and coat them and then reinstall them. If the battery is past its useful life, replace it. If it only requires a charge, charge it.

Check the power trim/tilt fuse, as appropriate. Many systems will have a fuse to prevent large current draws from damaging the system. If this fuse is blown, the system will cease to function. This is a good indicator that you may have problems elsewhere in the electric system. Fuses don't blow without cause.

After inspecting the electrical side of the system, check the hydraulic fluid level and top it off as necessary. Remember to position the motor properly (full tilt up or down, as applicable) to get an accurate measurement of fluid level. A slight decrease in the level of hydraulic fluid may cause the system to act sporadically.

Finally, make sure the manual release valve is in the proper position. A slightly open manual release valve may prevent the system from working properly and mimic other more serious problems.

Just remember to check the simple things first. If these simple tests do not diagnose the cause of the problem, then it is time to investigate more deeply. Check the hydraulic system. Inspect the entire power trim/tilt electrical harness with a multi-meter, checking for excessive resistance and proper voltage.

COMPREHENSIVE TESTING

Symptom Diagnosis

◆ See Figures 23 and 24

Use the accompanying symptom charts to diagnose and repair problems with the PTT system used on 30-125 hp motors. Before starting, make be sure to visually inspect the system for leaks or damage. Also, make sure the oil level is correct.

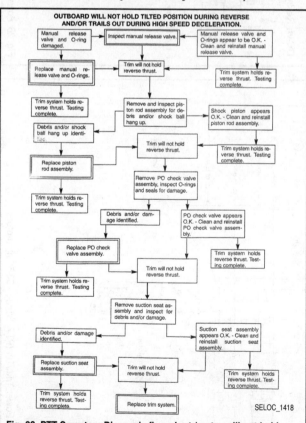

Fig. 23 PTT Symptom Diagnosis flow-chart (motor will not hold tilted position during reverse and/or trails out during high speed deceleration)

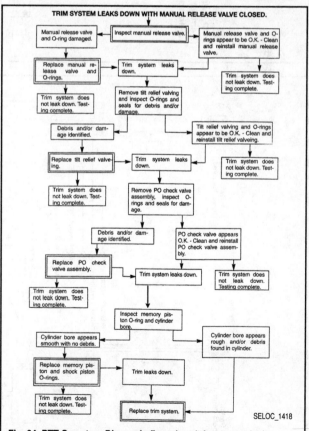

Fig. 24 PTT Symptom Diagnosis flow-chart (trim system leaks down with manual release valve closed)

Wiring Diagram & Circuit Testing

◆ See Figures 25 thru 28

Use the accompanying wiring diagrams and troubleshooting charts to diagnose and repair electrical problems with the PTT systems on 25-115 hp motors.

Before starting any electrical troubleshooting, check the following:
- Check for disconnected wires
- Make sure all connections are clean (of corrosion) and tight
- Make sure the plug-in connectors are fully engaged
- Make sure the battery is fully charged

The numbers listed in the electrical troubleshooting chart corresponds to the numbered locations on the appropriate wiring diagrams.

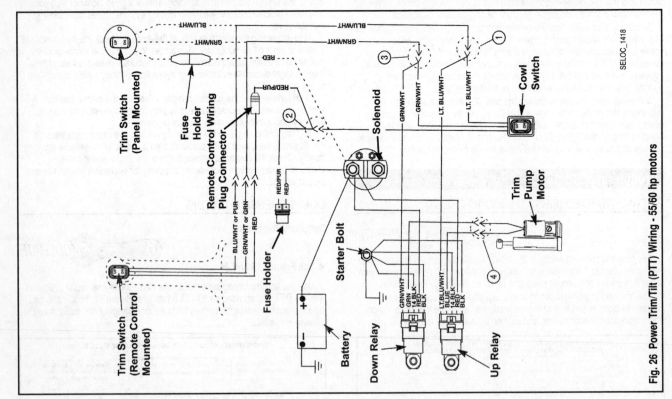

Fig. 26 Power Trim/Tilt (PTT) Wiring - 55/60 hp motors

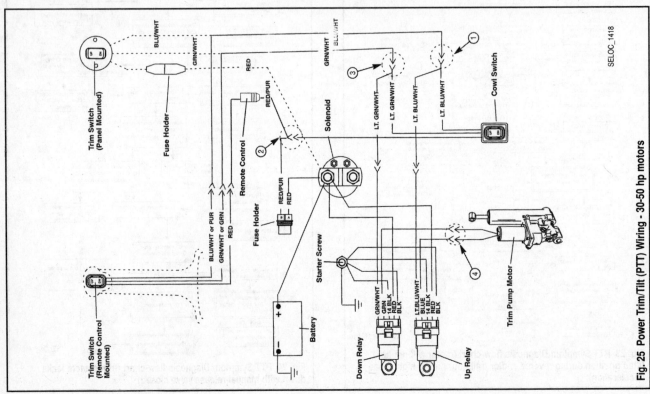

Fig. 25 Power Trim/Tilt (PTT) Wiring - 30-50 hp motors

Problem	Possible Cause	Remedy
Trim Switch "UP" is inoperative, but the Cowl Switch "UP" does operate.	1. Open wire between Wire Connection (1) and Trim Switch. 2. Faulty Trim Switch.	1. Check for an open connection or cut wire. 2. Replace.
Cowl Switch "UP" is inoperative, but the Trim Switch "UP" does operate.	1. Open wire between Wire Connection (2) and Solenoid. 2. Faulty Cowl Switch.	1. Check for an open connection or cut wire. 2. Replace.
Trim Switch "UP" and Cowl Switch "UP" are both inoperative.	1. Open wire between Wire Connection (1) and the "Up" Relay. 2. Open BLK wire between ground and "UP" Relay. 3. Open RED wire between Solenoid and "UP" Relay. 4. Faulty "UP" Relay.	1. Check for an open connection. 2. Check for an open connection. 3. Check for an open connection. 4. Replace.
Trim Switch "DOWN" is inoperative, but the Cowl Switch "DOWN" does operate.	1. Open wire between Wire Connection (3) and Trim Switch. 2. Faulty Trim Switch.	1. Check for an open connection or cut wire. 2. Replace.
Cowl Switch "DOWN" is inoperative, but the Trim Switch "DOWN" does operate.	1. Open wire between Wire Connection (2) and Solenoid. 2. Faulty Cowl Switch.	1. Check for a open connection or cut wire. 2. Replace
Trim Switch "DOWN" and Cowl Switch "DOWN" are both inoperative.	1. Open wire between Wire Connection (3) and the "Up" Relay. 2. Open BLK wire between ground and "DOWN" Relay. 3. Open RED wire between Solenoid and "DOWN" Relay. 4. 20 AMP Fuse blown.	1. Check for an open connection. 2. Check for an open connection. 3. Check for an open connection. 4. Replace fuse. Locate the cause of the blown fuse. Check electrical wiring for a shorted circuit.
Trim Switch "UP" and "DOWN" are both inoperative, but the Cowl Switch does operate.	2. Faulty trim switch. 3. Wire is open between fuse holder and solenoid. 4. Wire is open between fuse holder and trim switch.	2. Replace. 3. Check for a open connection or cut wire. 4. Check for a loose or corroded connection.
Trim Switch and Cowl Switch are both inoperative.	1. One of the Trim Pump Motor wires is open between the motor and the Relays. 2. Faulty trim pump motor.	1. Check wire connections (4) for loose or corroded condition. 2. If voltage is present at connections (4) when the appropriate trim button is pressed, then motor is faulty. Replace motor.
Trim system operates (motor runs) without pressing the switches.	1. The Trim or Cowl switch is shorted.	1. Replace.

SELOC_1418

Fig. 28 Electrical Troubleshooting PTT systems on 30-125 hp motors (use along with appropriate wiring diagram)

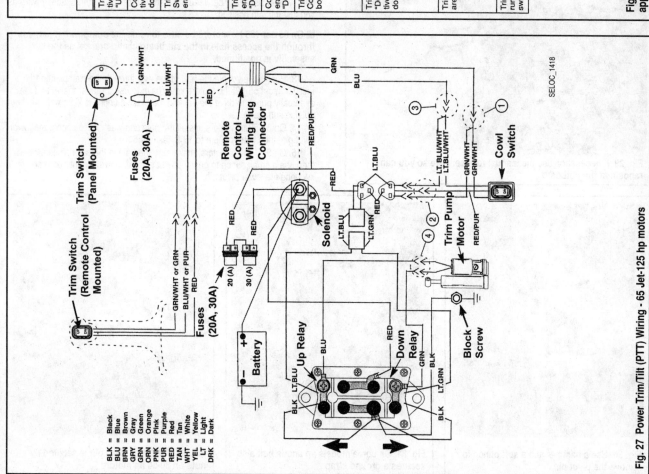

BLK = Black
BLU = Blue
BRN = Brown
GRY = Gray
GRN = Green
ORN = Orange
PNK = Pink
PUR = Purple
RED = Red
TAN = Tan
WHT = White
YEL = Yellow
LT = Light
DRK = Dark

SELOC_1418

Fig. 27 Power Trim/Tilt (PTT) Wiring - 65 Jet-125 hp motors

PTT Assembly
(Mid-Range Motor, Single Ram Assembly)

REMOVAL & INSTALLATION

 MODERATE

◆ **See Figures 29 thru 32**

If the unit is removed for overhaul it is usually a good idea to refill and bench bleed it before installation. This is easier on the motor and battery as you will wind up running it up and down a number of times during the bleeding process. For more details, please refer to Hydraulic System Bleeding later in this section.

1. Raise the outboard unit to the full Up position. If the hydraulic system is inoperative, first rotate the manual release valve counterclockwise about 3-4 complete turns and then manually lift the unit to the full Up position.

2. Set the tilt lock pin in place to support the outboard.

3. Tag and disconnect the PTT motor electrical wires (usually Blue and Green) and release any clamp(s)/J-clip(s). Be sure to note the wire routing and the position of any clamp(s)/J-clip(s) for installation purposes.

4. Remove the upper tri-lobe pin (which is used to secure the upper pivot pin, on 55 hp and larger motors, it's at the center of the pin where it passes through the trim ram upper rod). You can usually grab the pin with a set of needle-nose pliers and pull downward to free it.

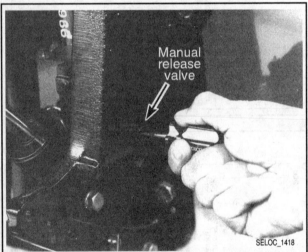

Fig. 29 If necessary, use the manual release valve so you can reposition the outboard

5. Use a suitable punch to drive out the upper pivot pin freeing the top of the trim/tilt rod.

6. Remove the anode.

7. On 30-50 hp motors, remove the nuts and washers securing the lower pivot pin.

8. On 55-125 hp motors, using a punch drive out the lower dowel pin from the lower pivot pin. The dowel pin on these motors usually contains a flanged head on one end and so can only be driven out in one direction. For some that means driving upward, for others than means driving downward. When possible, we always suggest reinstallation from the top, so that it cannot fall out if loosens in service.

9. Use a punch to carefully drive/tap out the lower pivot pin. On 30-50 hp motors, retain the pivot pin/anchor pin bushings from the clamp bracket and trim unit.

10. Tilt the shock absorber assembly (top first) out from the clamp bracket and remove the assembly.

To Install:

■ **Be sure to lubricate the pivot pins and bores using 2-4-C with Teflon or an equivalent marine lubricant during assembly.**

11. On 30-50 hp motors, lubricate and install the lower pivot pin bushings into the clamp brackets and trim unit.

■ **One trick to help position the trim unit if you're working solo (without enough extra hands) is to just start the lower pivot pin in the clamp bore BEFORE you position the unit, this way you can hold the unit with 1 hand and push the lower pin the rest of the way in with the other. On 55 hp and larger motors Mercury even recommends placing the dowel pin in the bore resting against the pivot pin (at least on models where the dowel pin is inserted from the top), that way when you slide the pivot pin into position, the dowel pin should start to fall into the pivot pin.**

12. Lubricate and start to position the lower pivot pin, then position the trim cylinder assembly between the clamp brackets. Tilt the bottom inward as you insert it into position.

■ **On 65 Jet-125 hp motors be sure to route the pump wiring harness through the access hole in the starboard clamp bracket as the assembly is positioned.**

13. Push the pivot pin the rest of the way through the clamps and trim unit. On most 55-125 hp motors you'll have to use a rubber mallet and eventually punch to drive the pin into the bracket until it is flush with the outside surface.

14. On 30-50 hp motors, install the nuts and washers to anchor the lower pivot pin, then tighten them to 18 ft. lbs. (24.5 Nm).

15. On 55-125 hp motors, make sure the hole in the pivot pin is aligned, then use a small punch to gently drive the lower dowel pin into position, securing the lower pivot pin.

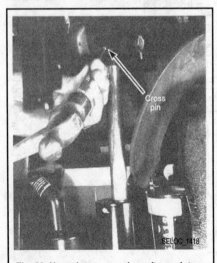

Fig. 30 Use a hammer and a soft punch to remove the pivot pin

Fig. 31 On some models an anode bolt also secures a ground strap

Fig. 32 The lower pivot pin is secured by nuts on 30-50 hp motors

16. Install the sacrificial anode. On some models 1 of the anode bolts is used to secure a ground strap, if applicable, be sure to position the strap between the anode and the bracket during installation.

17. Rotate the assembly so the shock rod hole is aligned for installation of the upper pivot pin.

■ **Before installing the upper pivot pin on 65 Jet-125 hp motors, make sure the cross hole in the tilt rod eye is positioned facing downward toward the rear of the motor. If the trim ram is reversed (so the lower portion of the cross hole faces forward, the trim sender if equipped, will not operate.**

18. Using a soft mallet, carefully drive the lubricated upper pivot pin into the swivel bracket and through the shock rod until the pivot pin is flush with the swivel bracket.

■ **On 30-50 hp motors, the grooved end of the upper pivot pin should be inserted first (from the port side of the assembly, so the grooved end would be facing the starboard side).**

19. Drive the upper tri-lobe (dowel) pin into its hole until seated.

20. Route the pump wiring back into position and secure using any clamp(s) and/or J-clip(s), as noted during removal. Reconnect the wiring as tagged during removal.

21. Check the fluid level and top off or fill and bleed the system, as necessary. For details, please refer to Hydraulic Bleeding in this section.

22. Check to be sure the battery has a full charge and then move the outboard unit through several cycles, full down to full up for trailering. Observe the system for satisfactory operation electrical and hydraulic. Check the system for possible leaks.

HYDRAULIC SYSTEM BLEEDING

◆ **See Figures 33 and 34**

This trim/tilt system with 1 large cylinder has been designed with a "self bleeding" system. Fluid is pumped from 1 side of the cylinder to the opposite side as the piston extends or retracts. Therefore, any air trapped on 1 end of the cylinder, is pushed out back to the reservoir, when the piston bottoms out. Usually 2 or 3 cycles of the system will remove all entrapped air from the cylinder. However, if the components were replaced or the system was completely drained for some reason, it might take a LOT of cycling. Under those circumstances it might make more sense to open the manual release valve first and fill the system.

The following short section outlines the necessary steps to perform during the cycling sequence just mentioned. For most models the system may be bled on or off the motor. For units which have been overhauled it might be a better idea to bleed it off the motor so that there is no load on the assembly as you run it back and forth.

Whenever possible use Quicksilver Power Trim and Steering Fluid to fill the system. The Mercury PTT fluid is a highly refined hydraulic fluid, containing a high detergent content and additives to keep seals pliable. A high grade automatic transmission fluid, Dexron® II, may also be used if the recommended fluid is not available.

1. Either install the system or secure it in a soft-jawed vise, whichever you prefer.

2. Make sure the area around the fill plug is clean and free of any dirt or corrosion, then remove the fill plug and add fluid to the system until it is topped off, then reinstall the plug.

■ **The manufacturer does not specifically recommend initially bleeding an empty system by cycling the pump with the release valve still open, however it should speed up the process on those models as well.**

3. Close the manual release valve.

4. Cycle the system to fully raise and lower the unit 3 times. If you're bleeding a unit that is NOT installed on the outboard, manually connect the wires to a 12 volt power source (battery) as follows:

• To extend the trim ram (move upward) connect the Blue lead to Positive and the Green lead to Negative.

• To retract the trim ram (move downward) connect the Green lead to Positive and the Blue lead to Negative.

5. Fully extend the trim rod and recheck the fluid level. Top off the fluid, as necessary. If you needed to add a significant amount of fluid, repeat cycling the unit up and down another 3 times, then recheck the fluid level. Continue this pattern until the fluid remains at the proper level without topping it off again.

OVERHAUL

◆ **See Figures 35 thru 45**

As with all rebuilds, be sure to remove and discard all O-rings during disassembly. Make sure the replacement O-rings are the proper size.

Removal of certain components requires special tools from Snap-On CG 41-11 and CG 41-14 with the 5/16 in. end. HOWEVER, you *can* fabricate a suitable tool out of a length of 0.060 in. (1.524mm) stainless steel rod. The final tool will have a 4 1/2 in. (11.43cm) long shaft which leads to a 90° bend and a 3 1/4 in. (8.25cm) long shaft that then ends in a 1/4 in. (6.35mm) loop which faces back towards the original shaft.

1. Remove the PTT unit assembly from the outboard as detailed earlier in this section.

Fill plug

Manual release valve

Fig. 33 Bench bleeding a PTT unit (the manual release valve may be opened and closed to expedite the process)

Fig. 34 During bleeding, be sure to frequently check and top off the reservoir

2. Remove the reservoir cap and the manual release valve. Tilt the unit slightly and drain the hydraulic fluid out the manual release valve bore and or out the reservoir opening (whichever is easier).

3. Place it in a soft-jawed vise for access.

4. To remove the trim motor, proceed as follows:

a. Loosen and remove the 4 screws (usually Allen head, but could be Philips too) securing the trim motor to the assembly. On some models, 1 of the bolts may secure a ground strap, be sure to note the location for installation purposes.

b. Remove the motor and seal from the assembly. On 55-125 hp units the motor/reservoir comes off as one unit, then the coupler is also removed. The coupler contains valves and other components and may also be disassembled.

5. To disassemble the pump, coupler and manifold assemblies on 55-125 hp motors, proceed as follows:

■ **To service the coupler, use the special tools from Snap-On CG 41-11 and CG 41-14 with the 5/16 in. end to remove the spool (or fabricate one from a 0.060 stainless steel rod.). Also, be sure to inspect the poppet assembly for debris under the valve tip, near the rubber seat, and replace if any is found.**

a. Remove the plugs and components from each side of the housing. Behind each plug there should be (in this order) a spring, a check valve with a separate poppet and a valve seat. In addition, on one side, there is a spool under the valve seat. Keep all components sorted/arranged in order for assembly purposes.

b. Remove the 3 screws that secure the pump to the coupler base, then remove the pump.

c. Remove the filter and seal from under the pump.

d. Remove the suction seat assembly from the bore in the base (seat, ring, ball and spring).

e. Turn your attention to the manifold at the base of the cylinder. Remove the 2 Allen head screws (threaded horizontally) from the base of the cylinder in order to separate the manifold. Carefully remove the manifold.

f. Remove tilt relief components from the side of the manifold, in the following order, the tilt limit spool, spool housing, poppet and spring.

6. To remove the pump and disassemble the manifold/check valve assemblies on 30-50 hp motors, proceed as follows:

■ **On 30-50 hp models, the pump itself IS NOT serviceable (although it**

can be replaced) **make no attempt to disassemble it.**

a. Remove the 2 Allen head screws securing the oil pump to the base assembly. Remove the pump, O-ring and filter. Then remove the 2 springs and metal ball(s) from the housing.

■ **Use the special tools from Snap-On CG 41-11 and CG 41-14 with the 5/16 in. end (or fabricate one from a 0.060 stainless steel rod) to remove the pilot valve in the next step.**

b. Unscrew the plug from the side of the manifold/housing assembly (near the manual release valve) and remove the tilt relief valve components, in order. Behind the plug you'll find a shim (on most, but not all models), a spring, a poppet assembly, a pilot valve (with 2 O-rings) and an actuator pin (also with an O-ring).

■ **Be sure to inspect each of the poppet valves for debris under the valve tip, near the rubber seat, and replace if any is found.**

c. At the base of the manifold assembly, facing downward, remove the plug and the suction seat assembly. Behind the plug you'll find a ball, the suction seat and a filter.

d. On the front of rear of the manifold assembly remove the plugs for the pilot check valves. Behind each plug you'll find a spring and a poppet. Once they are removed, use the 0.060 wire or removal tool to push out the spool and one seat. Then from the opposite side, use a punch to push out the remaining seat.

■ **The pilot check valve assembly components go through the manifold starting at either side, in this order, plug, spring, poppet, seat, spool, seat, poppet, spring and finally plug. In this way the assemblies are mirror images of each other with a spool in the middle.**

7. To remove and disassemble the shock (trim/tilt) rod, proceed as follows:

a. Using the special tool (#91-74951) or an equivalent spanner with 2 adjustable pins that insert into the rod (the wrench has 1/4 in. x 5/16 in. long pegs) unscrew and remove the tilt rod end cap assembly.

b. Carefully lift the rod and end cap from the cylinder.

■ **The *ONLY* serviceable items on the shock rod itself are the O-rings and wiper ring. If the shock rod requires *any* other repair, replace it as an assembly.**

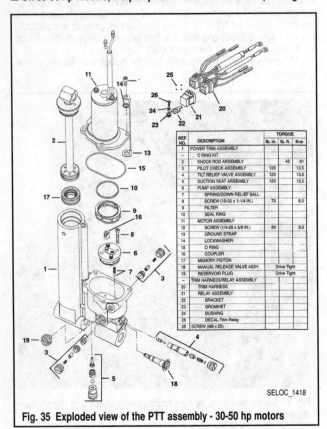

REF. NO.	DESCRIPTION	TORQUE		
		lb. in.	lb. ft.	N·m
1	POWER TRIM ASSEMBLY			
–	O RING KIT			
2	SHOCK ROD ASSEMBLY		45	61
3	PILOT CHECK ASSEMBLY	120		13.5
4	TILT RELIEF VALVE ASSEMBLY	120		13.5
5	SUCTION SEAT ASSEMBLY	120		13.5
6	PUMP ASSEMBLY			
7	SPRING/DOWN RELIEF BALL			
8	SCREW (10-32 x 1-1/4 IN.)	70		8.0
9	FILTER			
10	SEAL RING			
11	MOTOR ASSEMBLY			
12	SCREW (1/4-28 x 5/8 IN.)	80		9.0
13	GROUND STRAP			
14	LOCKWASHER			
15	O RING			
16	COUPLER			
17	MEMORY PISTON			
18	MANUAL RELEASE VALVE ASSY.	Drive Tight		
19	RESERVOIR PLUG	Drive Tight		
–	TRIM HARNESS/RELAY ASSEMBLY			
20	TRIM HARNESS			
21	RELAY ASSEMBLY			
22	BRACKET			
23	GROMMET			
24	BUSHING			
25	DECAL-Trim Relay			
26	SCREW (M6 x 25)			

SELOC_1418

Fig. 35 Exploded view of the PTT assembly - 30-50 hp motors

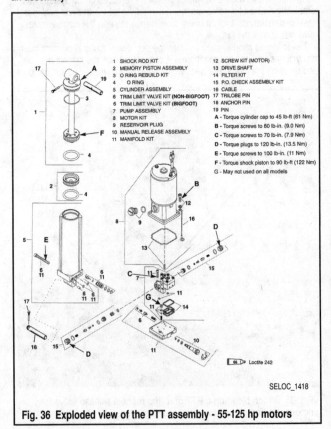

1 SHOCK ROD KIT
2 MEMORY PISTON ASSEMBLY
3 O RING REBUILD KIT
4 O RING
5 CYLINDER ASSEMBLY
6 TRIM LIMIT VALVE KIT (NON-BIGFOOT)
6 TRIM LIMIT VALVE KIT (BIGFOOT)
7 PUMP ASSEMBLY
8 MOTOR KIT
9 RESERVOIR PLUG
10 MANUAL RELEASE ASSEMBLY
11 MANIFOLD KIT
12 SCREW KIT (MOTOR)
13 DRIVE SHAFT
14 FILTER KIT
15 P.O. CHECK ASSEMBLY KIT
16 CABLE
17 TRILOBE PIN
18 ANCHOR PIN
19 PIN

A - Torque cylinder cap to 45 lb-ft (61 Nm)
B - Torque screws to 80 lb-in. (9.0 Nm)
C - Torque screws to 70 lb-in. (7.9 Nm)
D - Torque plugs to 120 lb-in. (13.5 Nm)
E - Torque screws to 100 lb-in. (11 Nm)
F - Torque shock piston to 90 lb-ft (122 Nm)
G - May not used on all models

66 ▭ Loctite 242

SELOC_1418

Fig. 36 Exploded view of the PTT assembly - 55-125 hp motors

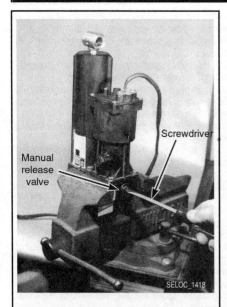

Fig. 37 Drain the hydraulic fluid through the reservoir or the release valve bore

Fig. 38 To remove the trim rod, loosen the cap with a spanner wrench. . .

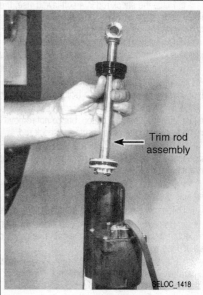

Fig. 39 . . .then lift the trim rod from the cylinder

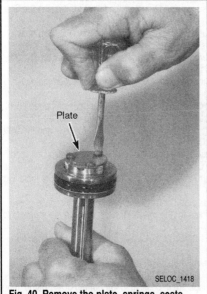

Fig. 40 Remove the plate, springs, seats and balls from the bottom of the piston

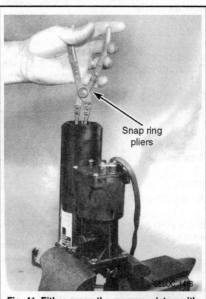

Fig. 41 Either grasp the memory piston with pliers. . .

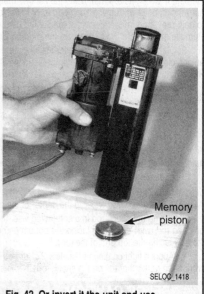

Fig. 42 Or invert it the unit and use compressed air to push it out

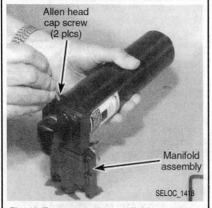

Fig. 43 To separate the manifold, remove the 2 Allen screws. . .

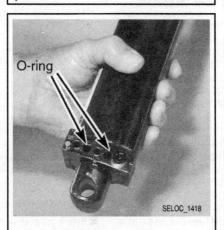

Fig. 44 Remove the old O-rings. . .

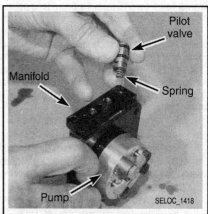

Fig. 45 . . .and all the valve components from the manifold assembly

c. Place the shock rod in a soft-jawed vise, then loosen and remove either single large bolt (30-50 hp models) or the 3 screws (55-125 hp models) on the lower plate.

d. Carefully lift the plate from the rod, keeping CLOSE track of the 5 balls, 5 seats and 5 springs retained under the plate.

e. Remove and discard the O-ring from the shock piston.

f. With the shock rod still (or back) in the vise, use a heat gun to *gently* apply heat to loosen the piston (still working at the bottom of the rod).

g. While the piston is hot, use the spanner wrench you used on the cap to just initially loosen the piston, but do not remove it completely. Allow the piston to cool and remove it from the shock rod.

h. Inspect the check valve in the piston for debris and clean, as necessary. If the check valve cannot be cleaned, replace the piston as an assembly.

i. Clean the shock and components with compressed air (but don't lose the springs, seats and balls).

j. Remove the inner O-ring from the piston bore.

k. Remove the cylinder end cap assembly from the shock rod and inspect. If the wiper (located in the cap) has failed to keep the rod cleaned, replace the wiper.

l. Place the end cap on a clean work surface, then use a small prytool to carefully remove the wiper from the top center of the cap. Remove the inner and outer O-rings and discard.

8. If the memory piston requires service, you can remove it from the bore either using a pair of lockring pliers (like Snap-On® SRP2 or Craftsman 4735) to grab it and pull it upward OR by using compressed air blown into the center O-ring hole. When using the second method, point the cylinder bore facing downward toward a shop rag or towel to both prevent injury *and* to help prevent damage. Expect some fluid to be expelled at the same time. Remove and discard the O-ring from the memory piston.

9. Clean and inspect all components which are to be reused using engine cleaner and low-pressure compressed air. DO NOT use cloth rags which may leave residue behind that could clog a valve.

10. Inspect all machined surfaces for burrs or scoring and repair or replace, as necessary.

To Assemble:

11. Identify all of the O-rings for assembly purposes. Lubricate the O-rings using Quicksilver Power Trim & Steering Fluid (or, if not available, using automatic transmission fluid).

12. To assemble the shock (trim/tilt) rod, proceed as follows:

a. Install the lubricated O-rings to the end cap and to the shock piston.

b. Install a new rod wiper to the center of the cap.

c. Clamp the shock rod in a soft-jawed vise, then position the end cap onto the rod (with the O-ring facing the bottom end of the rod and the threads facing the top end of the rod).

d. Apply a light coating of Loctite® 271 or equivalent threadlocking compound to the threads on the shock rod, then install the shock rod piston.

✳✳ SELOC CAUTION

To prevent damage to the shock rod piston, MAKE SURE the spanner wrench has 1/4 x 5/16 in. (6.4 x 8mm) long pegs.

e. Using the spanner, tighten the piston securely. If an adapter is available for a torque wrench, tighten the piston to 45 ft. lbs. (61 Nm) on 30-50 hp motors, or to 90 ft. lbs. (122 Nm) on 55-125 hp motors.

f. Install the 5 sets of balls, seats and springs to the shock rod piston, then position the plate to secure them. Tighten the large bolt on 30-50 hp motors securely. On 55-125 hp motors, tighten the 3 plate screws to 35 inch lbs. (4.0 Nm).

g. Remove the shock assembly from the vise.

13. To install the shock (trim/tilt) rod to the cylinder, proceed as follows:

a. Position the cylinder assembly in a soft-jawed vise.

b. Install the lubricated O-ring to the memory piston (if not done already) and install the piston into the bore. Push the piston all the way to the bottom of the bore.

c. Fill the cylinder to within 3 in. (76.2mm) from the top with PTT fluid.

d. Install the shock rod assembly until the PTT fluid flows through the oil blow off passage in the piston, then fill the rest of the cylinder to a point JUST below the threads for the end cap.

e. Install and tighten the end cap securely using the spanner wrench. If a torquing adapter is available, tighten the end cap to 45 ft. lbs. (61 Nm).

14. To assemble the manifold/check valves and install the pump on 30-50 hp motors, proceed as follows:

a. Starting with the tilt relief valve assembly, install the new, lubricated O-rings onto the actuator pin, the pilot valve and the plug.

b. Place the actuator pin in the pilot valve, then insert the assembly into the bore in the side of the manifold. Seat the valve using a 9/32 in. or 7mm socket in contact with the OUTER diameter of the pilot valve.

c. Next, install the poppet, spring and shim (if equipped). Install the plug and tighten to 120 inch lbs. (13.5 Nm).

d. Install the suction seat assembly to the bore which faces downward out the bottom of the manifold assembly. Start with installing the new, lubricated O-ring(s) for the seat components.

e. Next, install the filter and suction seat using a 9/32 in. or 7mm socket in contact with the OUTER diameter of the suction seat.

f. Install the ball and plug, tighten the plug to 120 inch lbs. (13.5 Nm).

g. Now install the pilot check valve assembly, as usual with new, lubricated, O-rings. Install 1 of the seats into the manifold and push it into place using a 9/32 in. or 7mm socket in contact with the OUTER diameter of the seat. Once in position install the corresponding poppet, spring and plug.

h. Move to the opposite side of the manifold assembly and position the spool and other seat, again using the socket to push the seat into place.

i. Install the remaining poppet, spring and plug. Tighten both of the pilot check valve plugs to 120 inch lbs. (13.5 Nm).

j. Finally, it is time to install the pump assembly. Position the relief ball and spring into the manifold.

k. Make sure the O-rings are in position in the bores on the bottom of the pump, then place the square cut O-ring on the filter and position the filter over the pump.

l. Install the pump to the manifold and secure using the 2 Allen head screws. Tighten the screws to 70 inch lbs. (8 Nm).

15. To assemble the pump, coupler and manifold assemblies on 55-125 hp motors, proceed as follows:

a. Starting with the trim limit assembly, install the new, lubricated O-rings on the spool housing and trim limit spool.

b. Install the proper spring into the bore. If the spring is being replaced using a kit with multiple options the heavy spring is used on 65 Jet-125 hp motors, the medium spring is used on 40-60 hp Bigfoot models, and the light spring is used on standard thrust 30-60 hp models.

c. After the spring, position the poppet, spool housing and trim limit spool.

d. Now, lubricate the O-rings and install the manual release valve. If the E-clip was removed, be sure to install the clip over the valve before installing it into the manifold.

e. Install the new lubricated O-rings and dowel pins to the base of the cylinder (at the mating surface for the manifold), then position the manifold to the cylinder.

f. Install the manifold retaining screws (Allen head screws) and tighten to 100 inch lbs. (11 Nm).

g. Now it's time to prepare the oil pump and coupler. Start by installing the spring, ball, lubricated O-ring and plastic seat to the top of the coupler.

h. Make sure the O-rings are in position in the bottom of the pump, then install the filter and filter seal UNDER the pump.

i. Position the pump and filter assembly to the coupler and tighten the 3 retaining screws to 70 inch lbs. (8 Nm).

j. Lastly, install the valve assemblies into either side of the pump housing. Install the spool, followed by the seat (with lubricated O-ring), poppet, check valve, spring and plug (also with lubricated O-ring). Repeat on the other side starting with the seat. Tighten the check valve plugs to 120 inch lbs. (13.5 Nm).

16. On 30-50 hp motors, if not done already, lubricate the O-rings and install the manual release valve. If the E-clip was removed, be sure to install the clip over the valve before installing it into the manifold.

17. If removed, install the coupler into the top of the pump.

18. Position a new seal, then install the motor, aligning the motor-to-pump coupler.

19. Install the motor retaining bolts (making sure the ground wire is in position under 1 of the bolts). Tighten the motor retaining bolts to 80 inch lbs. (9 Nm).

20. Install and bleed the PTT assembly as detailed in both procedures located earlier in this section.

■ Before installation it is usually a good idea to bench bleed a PTT assembly which has been overhauled. For details, please refer to **Hydraulic System Bleeding**, earlier in this section.

POWER TILT/TRIM SYSTEM - 3 RAM INTEGRAL

Introduction

All outboard installations are equipped with some means of raising or lowering (pivoting), the complete unit for efficient operation under various load, boat design, water conditions, and for trailering to and from the water. The correct trim angle ensures maximum performance and fuel economy as well as a more comfortable ride for the crew and passengers.

There is really only one way to deal with motors the size of the large V6 Mercury outboards, and that is through a power trim/tilt (PTT) system. The systems used on V6 models contain 3 rams (two trim, 1 tilt).

Two similar versions of the system are used, the Showa (Type I) and the Oildyne (Type II). Differences, when significant, are noted in procedures or illustrations.

Description & Operation

◆ **See Figures 46 thru 49**

There have been multiple designs of this trim/tilt system used on V6 Mercury outboards over the years. On these late-model motors however, there are basically only 2 designs that were used in production. Both

systems work identically but engineering changes and improvements have resulted in some minor component changes along with modification to the disassembly and assembly procedures.

Due to the manufacturers constant improvements from one model year to the next, some outboard models may have any either of the designs. And since Mercury is good for changing things without updating their tech literature, you might find small differences which disagree with the way we've broken them down by model or type, so just keep your wits about you when attempting overhaul. However most of the procedures are identical. Where differences do occur, these differences are clearly indicated. Therefore, before starting work, carefully examine the trim/tilt system on the outboard unit being serviced, then match the physical characteristics with the illustrations provided, to ensure the proper procedures are followed when differences are indicated. Once the model is identified, proceed with the work for that particular model. Generally speaking the following systems should be found on the following models:

• Type I (Showa) PTT system - All 2.5L carbureted and EFI models plus some of the 2.5L OptiMax models. This system is also used on some 3.0L EFI models.

• Type II (Oildyne) PTT system - Some 2.5L OptiMax models and some 2002 or later 3.0L EFI models, as well as all 3.0L OptiMax models.

In all cases, the trim/tilt system consists of an electric motor, pump, pressurized fluid reservoir, 2 small trim cylinders and 1 large tilt cylinder.

The remote control throttle lever contains switches for adjusting the trim angle, raising the outboard for shallow water operation and tilting the outboard fully up for tilting.

An electronic trim position indicator (optional equipment on some models), may be installed to provide the helmsperson with a visual reference of the outboard trim angle. This indicator system consists of a transducer mounted on one end of the tilt cylinder hinge pin and a trim gauge mounted on the control panel. As the angle of the outboard changes, the transducer senses the change altering the voltage level to the gauge on the control panel.

Operation is a relatively simple manner with the pump pushing fluid in 1 of 2 directions, depending upon which side of the motor harness to which the power is being applied.

Depressing the up button will actuate the UP solenoid and close the circuit to the electric motor. The electric motor will drive the pump forcing fluid into the UP side of the 2 trim cylinders. During upward motion hydraulic pressure will vary slightly based on model and circumstance. On Type I (Showa) systems, the maximum up pressure should be 1300 psi (8963 kPa). On Type II (Oildyne) systems, up pressure varies from 850-1150 psi (5861-7929 kPa) unloaded to approximately 3000 psi (20,684 kPa) against full engine thrust.

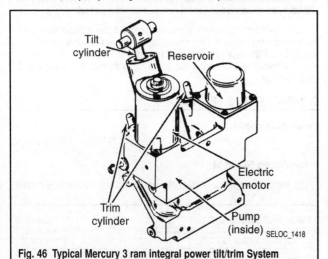

Fig. 46 Typical Mercury 3 ram integral power tilt/trim System

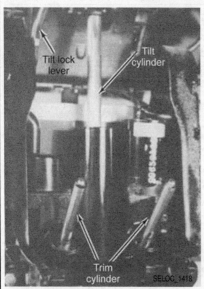

Fig. 47 The outboard unit to the full up position by the larger center tilt cylinder. The tilt lock lever has been engaged to take the weight from the cylinder

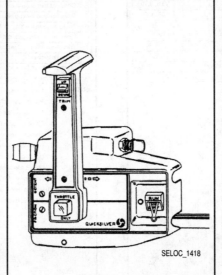

Fig. 48 Controls for this system do not use a separate tilt button. Holding the UP button will activate the center tilt cylinder to raise the outboard

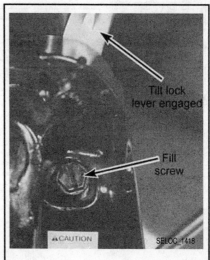

Fig. 49 The system is pressurized! Only open the fill screw when the tilt and trim rams are all fully extended (outboard in the full up position) to depressurize the system

Depressing the down button will close the DOWN circuit and actuate the DOWN solenoid. The electric motor will operate in the opposite direction (from the UP direction), forcing hydraulic fluid into the DOWN side of the tilt cylinder. This action will move the outboard unit downward. When the desired angle of trim is obtained the button is released and movement ceases. During downward motion hydraulic pressure will vary slightly based on model and circumstance, but in all cases pressure is lowered for the return to the reservoir. On Type I (Showa) systems, the maximum down pressure should be 500 psi (3447 kPa). On Type II (Oildyne) systems, up pressure varies from 640-1050 psi (4413-7239 kPa).

To tilt the outboard upward (not just trim it up), depress the up button and hold. This action will close the up circuit and activate the up solenoid. The 2 trim cylinders will extend slowly to the full trim up position. Hold the up button depressed and the tilt cylinder will continue to move the outboard unit upward to the full up position for tilting.

■ **As a safety measure to prevent accidental movement of the outboard while tilting to and from the water, it is strongly recommended a tilt bracket always be used to mechanically lock the outboard unit in the up position. Most models are equipped with such a bracket, but if not, a bracket may be purchased at modest cost from the local marine store. With the bracket in place, the unit may be tilted in confidence over rough roads without fear of the outboard being jarred suddenly to the DOWN position.**

Troubleshooting

GETTING STARTED (FIRST THINGS TO CHECK)

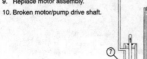

Any time a problem develops in the power trim/tilt system the first step is to determine whether it is electrical or hydraulic in nature. After the determination is made, then the appropriate steps can be taken to remedy the problem.

The first step in troubleshooting is to make sure all the connectors are properly plugged in and that all the terminals and wires are free of corrosion.

The simple act of disconnecting and connecting a terminal may sometimes loosen corrosion that is preventing a proper electrical connection. Inspect each terminal carefully and coat each with dielectric grease to prevent corrosion.

The next step is to make sure the battery is fully charged and in good condition. While checking the battery, perform the same maintenance on the battery cables as you did on the electrical terminals. Disconnect the cables (negative side first), clean and coat them and then reinstall them. If the battery is past its useful life, replace it. If it only requires a charge, charge it.

Check the power trim/tilt fuse, as appropriate. Many systems will have a fuse to prevent large current draws from damaging the system. If this fuse is blown, the system will cease to function. This is a good indicator that you may have problems elsewhere in the electric system. Fuses don't blow without cause.

After inspecting the electrical side of the system, check the hydraulic fluid level and top it off as necessary. Remember to position the motor properly (full tilt up or down, as applicable) to get an accurate measurement of fluid level. A slight decrease in the level of hydraulic fluid may cause the system to act sporadically.

Finally, make sure the manual release valve is in the proper position. A slightly open manual release valve may prevent the system from working properly and mimic other more serious problems.

■ **The hydraulic system should not allow leak down of more than 1 in. (25.4mm) as measured at the tilt ram in a 24 hour period.**

Just remember to check the simple things first. If these simple tests do not diagnose the cause of the problem, then it is time to investigate more deeply. Check the hydraulic system. Inspect the entire power trim/tilt electrical harness with a multi-meter, checking for excessive resistance and proper voltage.

COMPREHENSIVE TESTING

Symptom Diagnosis

◆ **See Figures 50 and 51**

Use the symptom charts to help narrow problems with the hydraulic system.

CONDITION OF TRIM SYSTEM	PROBLEM
A. Trim motor runs; trim system does not move up or down.	1, 2, 5, 10
B. Does not trim full down. Up trim OK.	2, 3, 4
C. Does not trim full up. Down trim OK.	1, 6
D. Partial or "Jerky" down/up.	1, 3
E. "Thump" noise when shifting.	2, 3, 6, 7
F. Does not trim under load.	5, 8, 9, 10
G. Does not hold trim position under load.	2, 5, 6
H. Trail out when backing off from high speed.	3, 4
I. Leaks down and does not hold trim.	2, 5, 7
J. Trim motor working hard and trims slow up and down.	8, 9
K. Trims up very slow.	1, 2, 8, 9
L. Starts to trim up from full down position when "IN" trim button is depressed.	3, 4
M. Trim position will not hold in reverse.	3, 4

PROBLEM
1. Low oil level.
2. Pump assembly faulty.
3. Tilt ram piston ball not seated (displaced, dirt, nickel seat).
4. Tilt ram piston O-ring leaking or cut.
5. Manual release valve leaking (check condition of O-rings) (Valve not fully closed).
6. Lower check valve not seating in port side trim ram.
7. Upper check valve not seating in port side trim ram.
8. Check condition of battery.
9. Replace motor assembly.
10. Broken motor/pump drive shaft.

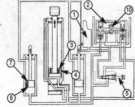

Fig. 50 Hydraulic system troubleshooting by symptom - Type I (Showa systems)

CONDITION OF TRIM SYSTEM	PROBLEM
A. Trim motor runs; trim system does not move up or down.	1, 5, 10, 11
B. Does not trim full down. Up trim OK.	3, 4, 5
C. Does not trim full up. Down trim OK.	1, 5
D. Partial or "Jerky" down/up.	1, 3
E. "Thump" noise when shifting.	3
F. Does not trim under load.	5, 8, 9
G. Does not hold trim position under load.	5, 6, 7
H. Trail out when backing off from high speed.	3, 4
I. Leaks down and does not hold trim.	5, 6, 7
J. Trim motor working hard and trims slow up and down.	8, 9
K. Trims up very slow.	1, 2, 5, 6, 8, 9
L. Starts to trim up from full down position when "IN" trim button is depressed.	3, 4
M. Trim position will not hold in reverse.	3, 4

PROBLEM 1. Low oil level.
2. Pump assembly faulty.
3. Tilt ram piston ball not seated (displaced, dirt).
4. Tilt ram piston O-ring leaking or cut.
5. Manual release valve leaking (check condition of O-rings) (Valve not fully closed).
6. Lower check valve not seating in port side trim ram.
7. Upper check valve not seating in port side trim ram.
8. Check condition of battery.
9. Replace motor assembly.
10. Broken motor/pump drive shaft.
11. Air pocket under pump.

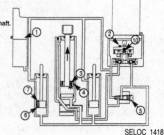

SELOC_1418

Fig. 51 Hydraulic system troubleshooting by symptom - Type II (Oildyne systems)

SELOC_1418

Wiring Diagram & Circuit Testing

MODERATE

◆ **See Figures 52 thru 56**

Use the accompanying wiring diagram and troubleshooting charts to diagnose and repair electrical problems with the PTT systems on V6 motors.

Before starting any electrical troubleshooting, check the following:
- Check for disconnected wires
- Make sure all connections are clean (of corrosion) and tight
- Make sure the plug-in connectors are fully engaged
- Make sure the battery is fully charged

The numbers listed in the electrical troubleshooting chart corresponds to the numbered locations on the wiring diagram.

Power Trim Relay Testing

The trim motor relays used in this PTT system are used to connect the trim motor to either positive or ground in order to allow the motor to run in both directions.

If the motor will not run UP it could be the UP relay is not making contact with 12 volts OR if could be that the DOWN relay is not making contact to ground.

If the motor will not run DOWN it could be the DOWN relay is not making contact with 12 volts OR if could be that the UP relay is not making contact to ground.

When the system is NOT energized, *both* relays should connect the heavy motor leads to ground.

If the troubleshooting charts point you in the direction of testing the power trim relays (because the system does not operate in 1 direction), proceed as follows to determine which relay is faulty:

1. Disconnect the heavy gauge pump wires from the trim control relay.

2. Check for continuity between the heavy leads from the trim relays to ground. There should be full continuity between the Green lead and ground, as well as the Blue lead and ground. If one relay does not contain continuity, replace THAT relay.

3. Connect a voltmeter to the heavy Blue lead and to ground, you should have 12 volts on the Blue lead when the UP switch is pushed in. Next check between the Green lead and ground, now you should have 12 volts when the DOWN switch is pushed in. Replace the relay that does *not* switch the lead to positive.

CONDITION OF TRIM SYSTEM	PROBLEM
A. Trim motor does not run when trim button is depressed.	1, 2, 4, 5, 6, 7, 8
B. Trim system trims opposite of buttons.	3
C. Cowl mounted trim buttons do not activate trim system.	2, 4, 5, 6, 7

PROBLEM

1. Battery low or discharged.
2. Open circuit in trim wiring.
3. Wiring reversed in remote control, cowl switch or trim leads.
4. Wire harness corroded through.
5. Internal motor problem (brushes, shorted armature).
6. Blown fuse(s).
7. Trim switch failure.
8. Verify relays are functioning correctly.

SELOC_1418

Fig. 52 Electrical troubleshooting by symptom

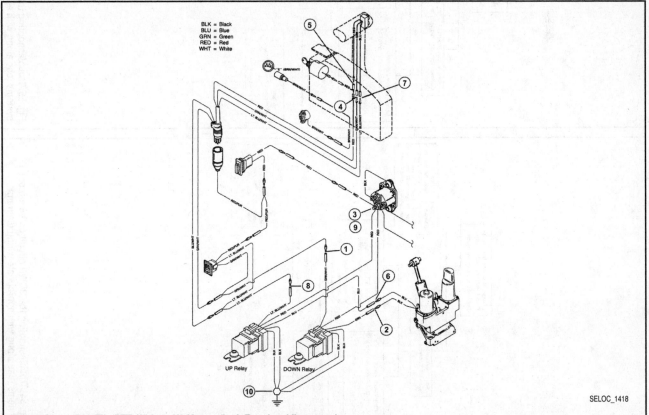

SELOC_1418

Fig. 53 Power Trim/Tilt (PTT) Wiring - V6 Motors (both Type I and II systems)

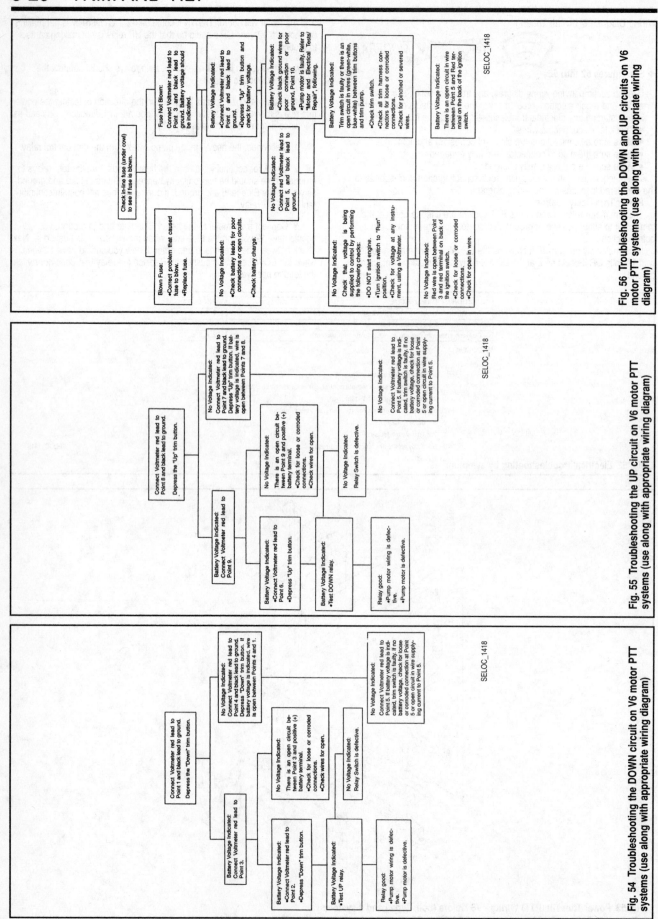

Check in-line fuse (under cowl) to see if fuse is blown.

Blown Fuse:
•Correct problem that caused fuse to blow.
•Replace fuse.

Fuse Not Blown:
Connect Voltmeter red lead to Point 3 and black lead to ground. Battery voltage should be indicated.

No Voltage Indicated:
•Check battery leads for poor connections or open circuits.
•Check battery charge.

Battery Voltage Indicated:
•Connect Voltmeter red lead to Point 8 and black lead to ground.
•Depress "Up" trim button and check for battery voltage.

No Voltage Indicated:
Connect red Voltmeter lead to Point 5, and black lead to ground.

Battery Voltage Indicated:
•Check black ground wires for poor connection or poor ground, Point 10.
•Pump motor is faulty. Refer to "Motor and Electrical Tests/Repair", following.

No Voltage Indicated:
Check that voltage is being supplied to control by performing the following checks:
•DO NOT start engine.
•Turn ignition switch to "Run" position.
•Check for voltage at any instrument, using a Voltmeter.

Battery Voltage Indicated:
Trim switch is faulty or there is an open circuit in wires (green-white, blue-white) between trim buttons and trim pump.
•Check trim switch.
•Check all trim harness connectors for loose or corroded connections.
•Check for pinched or severed wires.

No Voltage Indicated:
Red wire is open between Point 3 and red terminal on back of the ignition switch.
•Check for loose or corroded connections.
•Check for open in wire.

Battery Voltage Indicated:
There is an open circuit in wire between Point 5 and Red terminal on the back of the ignition switch.

SELOC_1418

Fig. 56 Troubleshooting the DOWN and UP circuits on V6 motor PTT systems (use along with appropriate wiring diagram)

Connect Voltmeter red lead to Point 8 and black lead to ground.

Depress the "Up" trim button.

No Voltage Indicated:
Connect Voltmeter red lead to Point 7 and black lead to ground. Depress "Up" trim button. If battery voltage is indicated, wire is open between Points 7 and 8.

Battery Voltage Indicated:
Connect Voltmeter red lead to Point 9.

No Voltage Indicated:
There is an open circuit between Point 9 and positive (+) battery terminal.
•Check for loose or corroded connections.
•Check wires for open.

No Voltage Indicated:
Connect Voltmeter red lead to Point 5. If battery voltage is indicated, trim switch is faulty. If no battery voltage, check for loose or corroded connection at Point 5 or open circuit in wire supplying current to Point 5.

Battery Voltage Indicated:
•Connect Voltmeter red lead to Point 6.
•Depress "Up" trim button.

Battery Voltage Indicated:
•Test DOWN relay.

No Voltage Indicated:
Relay Switch is defective.

Relay good:
•Pump motor wiring is defective.
•Pump motor is defective.

SELOC_1418

Fig. 55 Troubleshooting the UP circuit on V6 motor PTT systems (use along with appropriate wiring diagram)

Connect Voltmeter red lead to Point 1 and black lead to ground.

Depress the "Down" trim button.

No Voltage Indicated:
Connect Voltmeter red lead to Point 4 and black lead to ground. Depress "Down" trim button. If battery voltage is indicated, wire is open between Points 4 and 1.

Battery Voltage Indicated:
Connect Voltmeter red lead to Point 3.

Battery Voltage Indicated:
•Connect Voltmeter red lead to Point 2.
•Depress "Down" trim button.

No Voltage Indicated:
There is an open circuit between Point 3 and positive (+) battery terminal.
•Check for loose or corroded connections.
•Check wires for open.

No Voltage Indicated:
Connect Voltmeter red lead to Point 5. If battery voltage is indicated, trim switch is faulty. If no battery voltage, check for loose or corroded connection at Point 5 or open circuit in wire supplying current to Point 5.

Battery Voltage Indicated:
•Test UP relay.

No Voltage Indicated:
Relay Switch is defective.

Relay good:
•Pump motor wiring is defective.
•Pump motor is defective.

SELOC_1418

Fig. 54 Troubleshooting the DOWN circuit on V6 motor PTT systems (use along with appropriate wiring diagram)

PTT Assembly (3 Ram Assembly)

REMOVAL & INSTALLATION

OEM ② ◁ MODERATE

◆ **See Figures 57 thru 61**

If the unit is removed for overhaul it is usually a good idea to refill and bench bleed it before installation. This is easier on the motor and battery as you will wind up running it up and down a number of times during the bleeding process. For more details, please refer to Hydraulic System Bleeding later in this section.

In order to safely support the outboard while the PTT assembly is removed and installed you'll need to fabricate a support tool out of 3/8 in. (9.5mm) diameter metal rod. The tool should be just shy of 14 in. (35.56cm) in length with a 90 degree bend, 2 in. (50.8mm) long at either end. Both bends should be toward the same side of the rod, and holes should be drilled through the sides of the bent portions of the rod about 1/4 in. (6.35mm) from the each end, so that retaining clips can be installed through the rod to lock it in place. The length of the finished rod should be 14 in. (35.56cm) as measured from the center-to-center of the bend rod tips.

■ **Mercury recommends that 6 PTT unit mounting screws should not be reused on Type II systems, but instead new screws with a pre-applied patch lock should be installed. However, if necessary, you *should* be safe reusing the screws as long as you coat the threads with a suitable threadlocking compound.**

1. Remove the clamps on the transom bracket to free the PTT w...
2. Raise the outboard unit to the full up position. If the hydraulic s... is inoperative, first rotate the manual release valve counterclockwise a... 4 complete turns and then manually lift the unit to the full up position.
3. Engage the tilt lock lever.
4. If applicable, remove the trim indicator.
5. Install the support tool which was fabricated out of 3/8 in. (9.5mm) diameter rod.
6. Tag and disconnect the PTT motor electrical wires either from the solenoids (Blue, Green and Black wires for some Type I) or at the bullet connectors (Blue and Green wires for other Type I units and *all* Type II units). Be sure to note the wire routing and the position of any clamp(s)/J-clip(s) for installation purposes.
7. Open the PTT filler cap to release any remaining system pressure.
8. On models equipped with a thru-the-tilt steering system, remove the steering link arm from the end of the steering cable and remove the cable retaining nut from the side of the tilt tube.
9. Working on the starboard side of the engine bracket, loosen and remove the 2 outboard transom mounting bolts (upper and lower on this side), then loosen the tilt tube nut at the top foremost side of the starboard bracket. Loosen the tilt tube nut just until the nut is flush with the end of the tilt tube thread.
10. Now, remove the 3 screws and washers at the bottom of the starboard bracket which secure the PTT assembly to the bracket (discard these screws on Type II systems and replace with new patch lock bolts during installation). Carefully slide the starboard bracket outward, away from the motor.

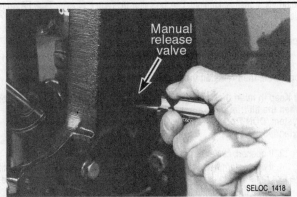

Fig. 57 If necessary, open the manual valve so you can tilt the outboard

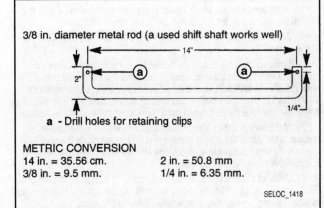

Fig. 58 Fabricate this special Support Rod tool

3/8 in. diameter metal rod (a used shift shaft works well)

14"
2"
1/4"

a - Drill holes for retaining clips

METRIC CONVERSION
14 in. = 35.56 cm.   2 in. = 50.8 mm
3/8 in. = 9.5 mm.   1/4 in. = 6.35 mm.

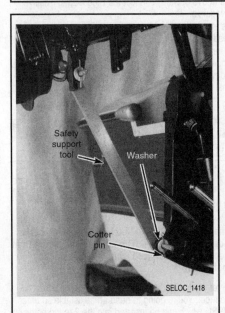

Fig. 59 Install the support rod tool

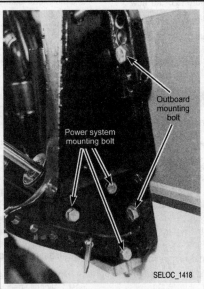

Fig. 60 Loosen the starboard transom and PTT mounting bolts

Fig. 61 Remove the cross pin, then push out the piston swivel retaining pin

11. Locate the cross pin which is used to secure the top of the tilt rod by holding the upper swivel pin in place. Using a punch, carefully drive the cross pin out (and discard), then carefully push out the upper swivel pin.

12. Finally, remove the 3 screws and washers at the bottom of the port bracket which secure the PTT assembly to the bracket (discard these screws on Type II systems and replace with new patch lock bolts during installation). Carefully remove the PTT assembly from the port bracket and outboard assembly.

13. Check the upper swivel pin bushings in the swivel bracket for wear and replace, as necessary. Also, it's a good idea to replace the trim rod shoes in the swivel bracket.

To Install:

14. Inspect the PTT assembly for signs of exposed metal. Paint any exposed metal surface on the housing (NOT the rods) to help prevent corrosion.

15. Unless you are using new patch lock bolts, apply a light coating of Loctite® 271 or an equivalent threadlocking compound to the threads of the PTT assembly retaining bolts.

16. Position the trim system to the starboard transom bracket and tilt tube nut. Tighten the bolts to 40 ft. lbs. (54 Nm) for Type I PTT units or to 45 ft. lbs. (61 Nm) for Type II PTT units. Start but do NOT secure the tilt tube nut.

■ **Remember, connecting the Blue wire to Positive and the Green wire to Negative will extend the ram. If the ram moves too far and must be retracted slightly, reverse the wires so the Green wire goes to Positive and the Blue wire goes to Negative.**

17. If not done already, use a 12-volt power source to extend the tilt tam fully so the end of the ram aligns with the upper swivel shaft hole.

18. Install the upper swivel pin with the slotted end facing the PORT side of the motor.

19. Using a screwdriver, turn the swivel pin as necessary to align the cross hole with the tilt ram hole. Keep in mind that the slot on the end of the pivot pin *should* align with the cross hole, and can be used as a reference during this alignment. Carefully insert a small tapered punch to help align the holes, then remove the punch and tap a NEW cross pin in until it is flush.

20. Make sure the starboard bracket (and PTT unit) is in the proper position, then install the 3 bolts securing the PTT assembly to the port bracket. Tighten the bolts to 40 ft. lbs. (54 Nm) on Type I PTT units or to 45 ft. lbs. (61 Nm) on Type II PTT units.

21. Connect the trim motor wires to the relays, as tagged during removal. Route the trim wires and noted during removal. If there are any questions as to proper wiring connections, please refer to Wiring Diagrams.

■ **The 2 power leads going to the trim motor should be encased in conduit tubing. If the tubing was NOT installed, be sure to order new tubing and cut it to the appropriate length.**

22. Apply marine sealant to the shanks of the transom mount bolts, then install and tighten the bolts securely. Apply marine sealer to the threads of the mounting bolts, secure with flat washers and locknuts. Be *sure* the installation is watertight.

✳✳ SELOC CAUTION

DO NOT us an air gun or impact driver to tighten the transom mount bolts.

23. Tighten the tilt tube nut securely.

24. On models with thru-the-tilt steering, tighten the steering cable retaining nut securely to the tilt tube.

25. Apply a coating of liquid neoprene to any exposed electrical connections.

26. Check the fluid level and top off or fill and bleed the system, as necessary. For details, please refer to Hydraulic Bleeding in this section.

27. Remove the tilt support rod tool that was installed during removal.

28. Check to be sure the battery has a full charge and then move the outboard unit through several cycles, full down to full up for trailering. Observe the system for satisfactory operation electrical and hydraulic. Check the system for possible leaks.

HYDRAULIC SYSTEM BLEEDING

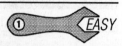

This trim/tilt system with 1 large cylinder has been designed with a "self bleeding" system. Fluid is pumped from one side of the cylinders to the opposite side as the pistons extend or retract. Therefore, any air trapped on one end of the cylinders, is pushed out back to the reservoir, when the pistons bottom out. Usually 2 or 3 cycles of the system will remove all entrapped air from the cylinder. However, if the components were replaced or the system was completely drained for some reason, it might take a LOT of cycling.

■ **When refilling an empty system run the trim motor in short bursts ONLY until the pump is primed and the trim rams start to move. If the motor is run excessively prior to priming driveshaft failure could occur.**

The following short section outlines the necessary steps to perform during the cycling sequence just mentioned. For most models the system may be bled on or off the motor. For units which have been overhauled it might be a better idea to bleed it off the motor so that there is no load on the assembly as you run it back and forth. Or conversely, you can bleed it on the motor BEFORE the tilt ram is connected and with the motor locked upward by the Support Rod Tool fabricated for the removal process. This way the motor can be run using the normal switch circuit but without having to move the load of the outboard itself.

Whenever possible, use Quicksilver Power Trim & Steering Fluid to fill the system. The Mercury PTT fluid is a highly refined hydraulic fluid, containing a high detergent content and additives to keep seals pliable. A high grade automatic transmission fluid, Dexron® II or III, Type F or even Type FA, may also be used if the recommended fluid is not available.

1. Either install the system or secure it in a soft-jawed vise, whichever you prefer.

2. Make sure the area around the fill plug is clean and free of any dirt or corrosion, then remove the fill plug and add fluid to the system until it is topped off, **then reinstall the plug**.

■ **Keep in mind that the system (and reservoir) is normally pressurized when the tilt/trim rams are retracted, so except to initially fill the system, DON'T open the reservoir cap unless the rams are all fully extended.**

3. If opened, close the manual release valve.

4. If the system was completely empty, run it with very short bursts until the trim rams start to move as an indication that the pump has been primed.

5. Cycle the system to fully lower and raise the unit 4 times. If you're bleeding a unit that is NOT installed on the outboard, manually connect the wires to a 12 volt power source (battery) as follows:

• To extend the rams (move upward) connect the Blue lead to Positive and the Green lead to Negative.

• To retract the rams (move downward) connect the Green lead to Positive and the Blue lead to Negative.

6. Fully extend the trim/tilt rams and recheck the fluid level. Top off the fluid, as necessary. If you needed to add a significant amount of fluid, repeat cycling the unit up and down another 4 times, then recheck the fluid level. Continue this pattern until the fluid remains at the proper level without topping it off again.

Trim Rod and Cap

REMOVAL & INSTALLATION

◆ See Figures 62 thru 65

■ **The trim cylinders themselves are integrated into the casting of the tilt/trim assembly and cannot be removed. The trim rod and piston from each cylinder can be removed and serviced without removing the tilt/trim assembly from the boat.**

Trim rod removal is a relatively simple matter, but requires the use of at least 1 slightly specialized tool. You're going to need an adjustable spanner wrench (such as #91-74951) which can be inserted into the 2 round holes on top of the rod end cap and used to unthread the cap.

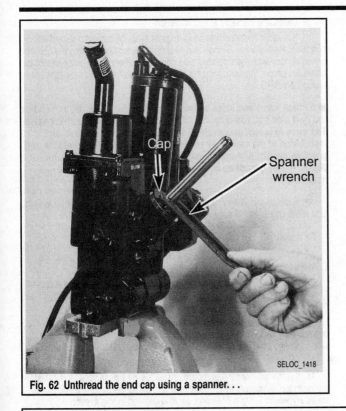

Fig. 62 Unthread the end cap using a spanner. . .

Fig. 63 . . .then remove the trim rod from the cylinder

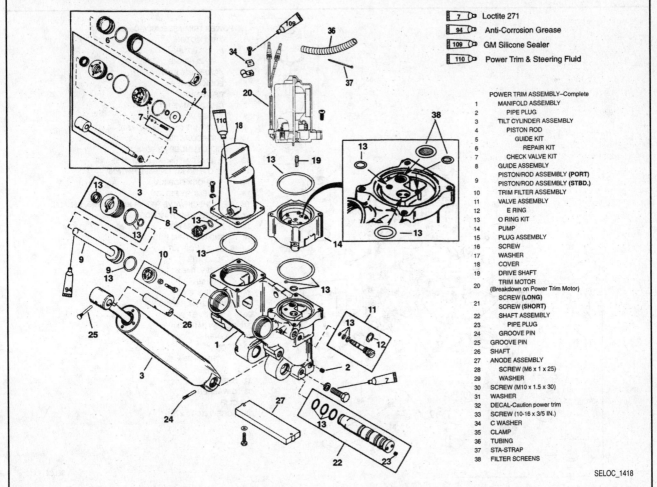

7	Loctite 271
94	Anti-Corrosion Grease
109	GM Silicone Sealer
110	Power Trim & Steering Fluid

POWER TRIM ASSEMBLY–Complete
1 MANIFOLD ASSEMBLY
2 PIPE PLUG
3 TILT CYLINDER ASSEMBLY
4 PISTON ROD
5 GUIDE KIT
6 REPAIR KIT
7 CHECK VALVE KIT
8 GUIDE ASSEMBLY
9 PISTON/ROD ASSEMBLY (PORT)
9 PISTON/ROD ASSEMBLY (STBD.)
10 TRIM FILTER ASSEMBLY
11 VALVE ASSEMBLY
12 E RING
13 O RING KIT
14 PUMP
15 PLUG ASSEMBLY
16 SCREW
17 WASHER
18 COVER
19 DRIVE SHAFT
20 TRIM MOTOR
(Breakdown on Power Trim Motor)
21 SCREW (LONG)
21 SCREW (SHORT)
22 SHAFT ASSEMBLY
23 PIPE PLUG
24 GROOVE PIN
25 GROOVE PIN
26 SHAFT
27 ANODE ASSEMBLY
28 SCREW (M6 x 1 x 25)
29 WASHER
30 SCREW (M10 x 1.5 x 30)
31 WASHER
32 DECAL-Caution power trim
33 SCREW (10-16 x 3/5 IN.)
34 C WASHER
35 CLAMP
36 TUBING
37 STA-STRAP
38 FILTER SCREENS

SELOC_1418

Fig. 64 Exploded view of the PTT assembly - Type I (Showa) systems

Once the cap is removed the rod can still be difficult to pull out of the bore considering some remaining hydraulic lock, so the use of a Trim Rod Removal tool (#91-44486A1) may be helpful. The removal tool actually clamps around the end of the rod so that you have a better pulling surface without the danger of scoring or damaging the rod itself. Remember just grabbing the rod with a pair of vise grips or something like that will likely lead to damaging the surface of the rod which will likely cause the system to leak once pressurized.

1. Prior to removal or disassembly, take note of the applicable exploded view and ensure all components are positioned in order for installation reference.

2. If desired for overhaul of other components, remove the PTT assembly from the motor as detailed earlier in this section and install it in a soft-jawed bench vise. Use a 12-volt battery source to fully extend the trim rams.

3. If the PTT assembly is to remain on the outboard, tilt the motor fully upward and engage the tilt lock lever.

4. Remove the reservoir fill plug in order to bleed off any remaining system reservoir pressure.

5. Rotate the manual release valve 3-4 turns COUNTERCLOCKWISE in order to bleed the remaining pressure from the system.

6. Place a drain pan under the trim system to catch all the fluid that's going to escape when you remove the cap and when you remove the trim rod.

7. Using a spanner wrench with the 2 tips indexed into the 2 recesses in the cylinder end cap, remove the end cap by rotating the cap in a counterclockwise direction until it is free. On some models there may be anywhere from 1-5 trim limit reducer under each cap. When equipped, each reducer will limit the amount of total trim by 2 degrees.

8. After the end cap is free, carefully withdraw the piston straight up and out of the cylinder. IF removal is difficult, install a trim rod removal tool carefully on the rod to prevent any damage, then lift and pull the rod from the cylinder.

9. Clean and inspect the rod assembly.

10. It is recommended that all O-rings or seals be replaced regardless of condition. Remove the O-rings and seal from the trim cap. Place the cap on a worksurface with the threaded side downward, then use a small prytool to carefully remove the seal. Install a replacement seal with the new seal lip facing UPWARD.

■ **A check valve and valve screen is found on the lower (piston) end of the Port side trim rod only. DO NOT attempt to remove the check valve. The valve is preset to operate at a specific pressure. Removal and installation of the valve may result in improper operating pressure and possible system damaged. Instead check the valve and screen for debris. If they cannot be cleaned, you'll have to replace the rod assembly.**

11. Replace the O-ring on the bottom of the rod (the rod piston). On Port side rods, check and, if necessary, clean the check valve and valve screen.

To Install:

12. Make sure all components are clean and free of any dirt or debris. Contaminants in the system will lead to a malfunction.

13. After new O-rings have been installed, coat the surface of the piston, O-rings and seal lip with PTT fluid or a suitable Automatic Transmission Fluid (ATF). Carefully slide the piston into the cylinder.

14. Thread the end cap onto the end of the cylinder and then tighten the cap securely with a spanner wrench. If a torque wrench adapter is available when working on a Type II system, tighten the end cap to 70 ft. lbs. (95 Nm).

15. Lubricate the top of the piston rod with anti-corrosion grease.

16. If removed from the outboard, install the PTT assembly as detailed earlier in this section AND properly refill and bleed the system as detailed under Hydraulic System Bleeding, as also detailed in this section.

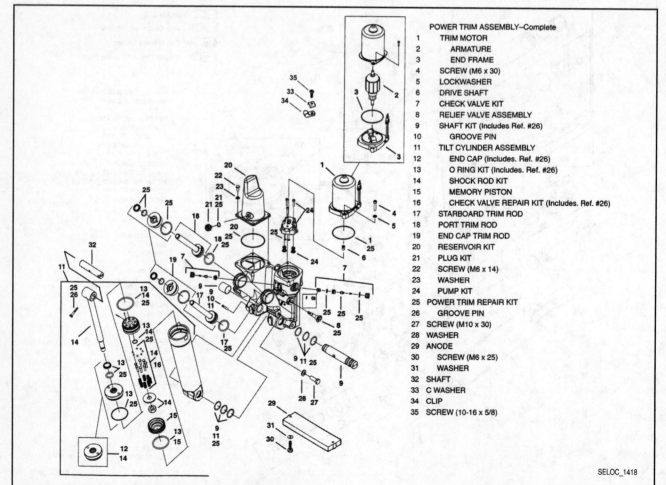

	POWER TRIM ASSEMBLY–Complete
1	TRIM MOTOR
2	ARMATURE
3	END FRAME
4	SCREW (M6 x 30)
5	LOCKWASHER
6	DRIVE SHAFT
7	CHECK VALVE KIT
8	RELIEF VALVE ASSEMBLY
9	SHAFT KIT (Includes Ref. #26)
10	GROOVE PIN
11	TILT CYLINDER ASSEMBLY
12	END CAP (Includes. Ref. #26)
13	O RING KIT (Includes. Ref. #26)
14	SHOCK ROD KIT
15	MEMORY PISTON
16	CHECK VALVE REPAIR KIT (Includes. Ref. #26)
17	STARBOARD TRIM ROD
18	PORT TRIM ROD
19	END CAP TRIM ROD
20	RESERVOIR KIT
21	PLUG KIT
22	SCREW (M6 x 14)
23	WASHER
24	PUMP KIT
25	POWER TRIM REPAIR KIT
26	GROOVE PIN
27	SCREW (M10 x 30)
28	WASHER
29	ANODE
30	SCREW (M6 x 25)
31	WASHER
32	SHAFT
33	C WASHER
34	CLIP
35	SCREW (10-16 x 5/8)

SELOC_1418

Fig. 65 Exploded view of the PTT assembly - Type II (Oildyne) systems

Tilt Rod

REMOVAL & INSTALLATION

OEM ② MODERATE

◆ See Figures 64 thru 68

■ The tilt cylinder itself is a separate piece from the PTT assembly and it can be removed (though only if the assembly is first removed from the outboard). However, if only the trim rod and piston requires service, they can be removed from the cylinder and serviced without removing the tilt/trim assembly from the boat.

Trim rod removal is a relatively simple matter, but requires the use of 1 slightly specialized tool. You're going to need an adjustable spanner wrench (#91-74951) which can be inserted into the 2 round holes on top of the rod end cap and used to unthread the cap.

There are 2 designs for tilt rod pistons used on Type I systems, Design 1 uses a bolt at the bottom end, while Design 2 uses a nut. It is interesting to note that a Design 1 rod (with the bolt at the bottom) can be used in cylinders originally equipped with Design 2 rods (with a nut at the bottom). This also applies to Type II systems, which are originally equipped only with a rod of the Design 2 (nut at the bottom) type. *However,* Design 2 pistons (with the nut at the bottom) WILL NOT fit cylinders originally equipped with Design 1 pistons. ALSO, keep in mind that memory pistons vary between the 2 designs and the original memory piston for the cylinder must be used (or the same replacement type).

1. Prior to removal or disassembly, take note of the applicable exploded view and ensure all components are positioned in order for installation reference.

2. If desired for overhaul of other components or to remove/replace the cylinder itself, remove the PTT assembly from the motor as detailed earlier in this section and install it in a soft-jawed bench vise. Use a 12-volt battery source to fully extend the trim rams. To remove the ram AND cylinder from the PTT assembly, use a small punch to carefully drive out the cross pin (which secures the lower swivel pin into the bottom of the PTT assembly and the bore on the bottom of the cylinder). Next, use the punch to carefully drive the swivel pin out the side of the assembly.

3. If the PTT assembly is to remain on the outboard, tilt the motor fully upward and engage the tilt lock lever.

4. Remove the reservoir fill plug in order to bleed off any remaining system reservoir pressure.

5. Rotate the manual release valve 3-4 turns COUNTERCLOCKWISE in order to bleed the remaining pressure from the system.

6. Place a drain pan under the trim system or cylinder to catch all the fluid that's going to escape when you remove the cap and when you remove the trim rod.

■ If you're disassembling a cylinder that has been removed from the PTT assembly, you'll have to secure it in a soft-jawed bench vise to keep it from turning when trying to loosen the end cap.

7. Using a spanner wrench with the 2 tips indexed into the 2 recesses in the cylinder end cap, remove the end cap by rotating the cap in a counterclockwise direction until it is free.

8. After the end cap is free, carefully withdraw the piston straight up and out of the cylinder.

9. Clean and inspect the rod assembly.

10. Clamp the rod in a vise equipped with soft jaws with the jaws gripping the upper end. Tighten the vise just good and snug to prevent any possible damage to the piston end. Loosen the nut or bolt, then slowly and carefully lift off the washer (as spring loaded check valves assemblies are secured under it). Remove the check valve assemblies (normally 7), then remove the O-ring and the piston.

■ There is normally a check valve held in by a roll pin. This valve cannot be removed, however it can be cleaned. If the check valve is defective, the piston assembly must be replaced

11. With the piston removed you are not free to slide the end cap up and free of the rod. Remove and discard the O-ring(s) from the piston assembly (some types use both an external AND an internal O-ring on the piston just like the end cap).

12. Disassemble the end cap, depending upon the system type as follows:

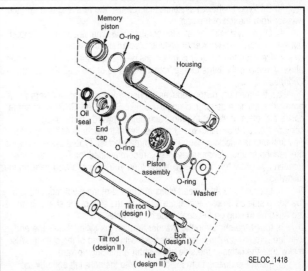

Fig. 66 Exploded view of the stock tilt rods normally found on Type I (Showa) systems

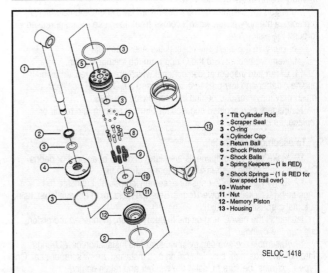

1 - Tilt Cylinder Rod
2 - Scraper Seal
3 - O-ring
4 - Cylinder Cap
5 - Return Ball
6 - Shock Piston
7 - Shock Balls
8 - Spring Keepers – (1 is RED)
9 - Shock Springs – (1 is RED for low speed trail over)
10 - Washer
11 - Nut
12 - Memory Piston
13 - Housing

Fig. 67 Exploded view of the stock tilt rod normally found on Type II (Oildyne) systems

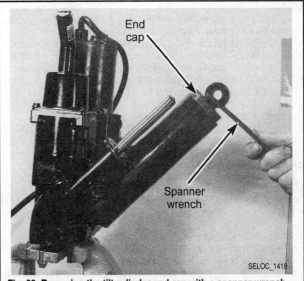

Fig. 68 Removing the tilt cylinder end cap with a spanner wrench

a. On Type I (Showa) systems, first remove the retaining ring and the washer from the top of the cap.

b. On all systems, remove the scraper seal. The seal can be popped out with a small screwdriver or prytool. Discard the old seal.

c. Remove the small O-ring from inside the cap bore. And, if not done already, remove the large O-ring from the outside of the cap. Discard the old O-rings.

13. If the memory piston requires service, you can remove it from the bore either using a pair of lockring pliers (like Snap-On® SRP4 or in some cases a Craftsman 4735) to grab it and pull it upward OR by using compressed air blown into the shaft opening at the bottom of the cylinder. The first method (locking pliers) is recommended for Type II systems, while the second method (compressed air) is recommended for Type I systems. However, in order to use the second method you'll have to also perform the following:

a. Locate the 2 Allen head plugs at the end of the pivot rod, one is on the same side as 3 holes which are found on the side of the shaft. Remove the Allen head plug on the same side as the holes.

b. Insert the shaft into the cylinder through the cylinder until the end with the remaining Allen head bolt is flush with the end of the cylinder. Also, make sure the holes in the side of the shaft are facing upward.

c. Insert the nozzle from the air hose into the bore left open by the removed Allen head bolt. In this way air will be directed through the shaft and out the hole in the shaft side, where it can push on the underside of the memory piston in the cylinder bore.

d. Invert the cylinder, facing it downward onto some shop cloths (to protect the memory piston when it comes free), then use the compressed air to push the piston free.

e. Remove the shaft and reinstall the Allen head plug.

f. Remove and discard the O-ring from the memory piston.

14. Clean and inspect all components which are to be reused using engine cleaner and low-pressure compressed air. DO NOT use cloth rags which may leave residue behind that could clog a valve.

15. Inspect all machined surfaces for burrs or scoring and repair or replace, as necessary.

To assemble and install:

16. Make sure all components are clean and free of any dirt or debris. Contaminants in the system will lead to a malfunction.

17. Identify all of the O-rings for assembly purposes. Lubricate the O-rings using Quicksilver Power Trim & Steering Fluid (or, if not available, using automatic transmission fluid).

18. Install the new O-ring on the memory piston, then insert the piston cup into the cylinder.

19. Assemble the end cap by lubricating and installation new O-rings (internal and external), then lubricating and installing a new scraper seal. On Type I systems, be sure to install the washer and retaining ring.

20. If not done already, secure the top end of the rod into a soft-jawed vise for ease of assembly.

21. Install the end cap over the rod and slide it down to the top of the rod, just above the vise jaws.

22. Install the shock piston (with new O-ring or O-rings depending upon the design), then position the check valve assemblies.

23. Carefully install the washer and bolt or locknut to secure the check valve assemblies. Tighten the fastener securely. No torque spec if given for the stock rods on Type I (Showa) systems, however for Type II (Oildyne) systems the locknut should be tightened to 95 ft. lbs. (129 Nm).

24. If the cylinder is removed from the PTT assembly, clamp it securely in a vise.

25. Install the tilt rod to the trim cylinder, sliding it slowly down into position until the end cap can be threaded. Tighten the end cap securely. No torque spec if given for the stock rods on Type I (Showa) systems, however for Type II (Oildyne) systems IF you have a suitable spanner adaptor for your torque wrench tighten the end cap to 45 ft. lbs. (61 Nm).

26. If the cylinder was removed from the PTT assembly, lubricate the pivot shaft using PTT fluid and position the cylinder to the assembly for installation. Mercury recommends using an alignment tool (#91-11230) to initially position the cylinder to the assembly, then installing the pivot shaft as the tool is withdrawn. You may have success using a wooden dowel or length of pipe of approximately the same outer diameter of the shaft. When installing the shaft, make sure the grooved end of the shaft is facing the bore for the retaining cross-pin. Once the pivot shaft is installed, carefully tap in a new cross pin (knurled end facing up toward the hammer/driver) until the pin is flush.

Trim Motor & Pump

TESTING THE MOTOR

◆ **See Figure 69**

The motors used on these models are simple 2 wire assemblies. When power is applied to one side (ground to the other) the pump will run in one direction. When power and ground are switched to the opposite contacts, the pump will run in the opposite direction.

The quickest functional test that can be performed is simply to connect the pump leads to 12 volt battery source. Connect 1 to positive and 1 to negative and listen, feel or watch for the pump to run. Reverse the connections and watch for the listen, feel or watch for the pump to run in the opposite direction. If the motor fails either one or both of these tests, it requires service or replacement.

Any further testing or inspection can only be done once the pump is removed and disassembled for access to the commutator.

Once the motor is disassembled, use a Growler (per the tool manufacturers' instructions) to check the armature for a short (in much the same way as a starter motor would be checked). Also, use a DVOM to check between the armature shaft (short thin end) and the commutator (longer, thicker end) to see if the armature is grounded. There should be NO continuity. If there is, the armature is grounded. A shorted or grounded armature must be replaced. If the commutator is worn it may be turned on an armature conditioner or a lathe and then cleaned with "00" grit sandpaper.

With the armature installed in the brushes you can check the wiring as follows:

• Check for continuity between the Blue and Green or Blue and Black motor wires (depending upon which pair is used, there should be continuity with 0 or near 0 resistance.

• Check for continuity between the Green or Black motor wire (again, it should only have one of these) and the frame (motor housing), there should be NO continuity.

• Check for continuity between the Blue motor wire and the frame (motor housing), there should be NO continuity.

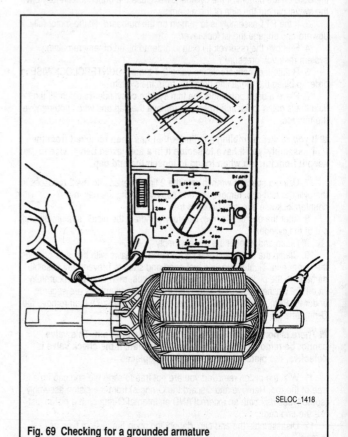

Fig. 69 Checking for a grounded armature

SELOC_1418

REMOVAL & INSTALLATION

◆ See Figures 64 and 65

■ The motor and/or pump itself is(are) mounted to the top, side of the PTT assembly casting. Both can normally be removed from the casting without first removing the assembly from the outboard, though doing so may make access, especially to some of the mount bolts, easier.

The motor is bolted to the top of the pump (Type I, Showa system) or to the PTT casting right above the pump (Type II, Oildyne systems). In either case, the motor obviously must be removed for access to the pump itself. Both motor and pump removal is a simple matter of removing a few fasteners and lifting the component away from the casting.

1. If desired for overhaul of other components, remove the PTT assembly from the motor as detailed earlier in this section and install it in a soft-jawed bench vise. Use a 12-volt battery source to fully extend the trim rams.

2. If the PTT assembly is to remain on the outboard, tilt the motor fully upward and engage the tilt lock lever.

3. Remove the reservoir fill plug in order to bleed off any remaining system reservoir pressure.

■ The pump cannot be serviced. Therefore, if troubleshooting indicates the pump has failed, it must be replaced.

4. If the PTT assembly is still installed, tag and disconnect the PTT motor wiring.

5. Remove the 2 bolts securing the pump to the top of the PTT assembly. Lift the electric motor, along with the coupler and O-ring from the assembly. Remove and discard the old O-ring.

6. On Type I (Showa systems) proceed as follows to remove the pump:

a. Remove the bolts securing the base of the pump to the PTT casting (there are normally 2, but diagrams show that there could be as many as 4, though more likely longer screws may be used on the motor for some models that would go through the pump housing and into the casting).

b. Lift upward on the pump housing to remove the assembly from the PTT casting.

c. If necessary, remove and retain the pump shaft which receives the motor coupler.

d. Remove and discard the 3 O-ring (1 large for the pump housing, and 2 smaller for fluid passages) from underneath the pump or the top of the PTT assembly.

7. On Type II (Showa systems) proceed as follows to remove the pump:

a. From the bore in the PTT assembly above which the motor was mounted, remove the upper 3 Allen head screws (in a triangular pattern, 2 on one end, 1 of the other, in all cases, the screws closest to the pump mounting screw holes).

b. Once the screws are removed, lift the pump assembly from the bore in the PTT casting. Inspect the filter and O-ring assemblies on the bottom of the pump and replace, as necessary (we always recommend AT LEAST replacing O-rings).

c. If necessary, remove the plugs from the sides of the manifold/PTT casting (just below where the pump mounts). Behind each plug is a spring, poppet and check valve assembly, with single spool in the middle separating each assembly. Check the tip of the poppets for debris around the base of the nipples and replace if found. O-rings should be checked for cuts or abrasions, but again, we prefer to just replace them.

8. To disassemble/service the motor on all systems, proceed as follows:

a. Remove the 2 screws securing the motor cover to the end cap, then carefully remove the cover and armature (respectively) from the end cap.

b. Remove and discard the cover-to-end cap O-ring.

c. Carefully inspect the power cord for cuts or tears which might allow water to enter the motor. If cuts or tears are found, Mercury notes to replace the cord on Type I systems or the cord and motor on Type II systems.

d. Inspect the brushes for pitting, chips or to make sure they are not excessively worn. If the distance between the brush pigtail and end of the brush holder slot is 1/16 in. or less, either the brush carb (Type I systems) or the complete motor end (Type II systems) must be replaced (as separate brushes are not normally available for Type II systems, but check with your local parts supplier JUST to be sure).

e. Inspect the armature as detailed earlier under Testing The Motor.

To Install:

ALL components must be CLEAN and free of any debris before assembly and installation or contamination may result in PTT system failure.

9. To assemble the motor, proceed as follows:

a. Install the armature into the end cap/brush card assembly.

b. Install new O-rings to both sides of the end cap.

c. Clamp a set of Vise-Grip® (or suitable locking pliers) onto the bottom of the armature shaft (which is protruding through the end cap). This will hold the armature in position as the cover/frame assembly is installed over top.

d. Carefully position the cover/frame assembly over top of the end cap, positioning the wiring harness retainer hole over the tab in the end cap.

e. Secure the cover/frame to the end cap and secure using the 2 screws, then remove the locking pliers.

10. On Type II (Showa systems) proceed as follows to install the pump:

a. If removed, install the spool with the check valve, poppet, spring and plug on either side in the manifold casting.

b. If removed, install the filters and O-rings into the bottom of the pump.

c. Install the pump into the manifold making sure the flat is facing toward the starboard transom bracket, then install and tighten the 3 retaining screws to 60 inch lbs. (7 Nm).

11. On Type I (Showa systems) proceed as follows to install the pump:

a. Lubricate and install 4 new O-rings (2 large O-rings, 1 for the base of the pump, 1 for the top of the pump where the motor installs, and 2 small passage O-rings which are installed bellow the pump).

b. Position the pump to the PTT assembly, making sure the flat on the side of the casting is facing the starboard transom bracket.

■ Keep track the pump driveshaft, as it is loose and may fall out of position.

12. Fill the pump with PTT fluid prior to installing the motor.

13. Carefully position the motor, aligning the motor and the pump driveshaft as they are positioned and making sure the wiring harness is properly routed.

14. On Type I systems, install the 2 bolts retaining the motor to the pump and tighten securely.

15. On Type II systems, install the 2 bolts retaining the motor to the PTT manifold/casting assembly, then tighten to 80 inch lbs. (9 Nm).

16. If the PTT assembly is already installed, properly reroute and connect the motor wiring, as tagged during removal.

17. If the PTT assembly was removed from the outboard, install it at this time.

18. Properly bleed the Hydraulic System as detailed earlier in this section.

Trim Sender & Gauge

TESTING THE TRIM SENDER

Analog Sending Units

1. Check the trim sender Black lead for proper continuity to ground.

2. Trim the outboard to the fully DOWNWARD position, then place the ignition switch to the ON position.

3. Disconnect the Brown/White trim sender wire from the trim sender harness.

4. Connect an ohmmeter (an analog one will work easier, but a DVOM will work too) set to read resistance between the Brown/White lead going back to the sending unit and a good ground, then begin trimming the motor UPWARD. The ohmmeter needle should move (or resistance should change in the DVOM display). If there is no change in resistance, the sender or wiring is defective and must be replaced.

Digital Sending Units

The digital trim sender used on some late-models requires a 5 volt DC reference signal from the ECU. You can check VERY, VERY carefully for this signal by back-probing the sender bullet connectors (using a thin probe or paper clip inserted into the rear of the connector, just DON'T ALLOW this to short!!!!).

Place the positive (red) lead from the DVOM into the back of the Blue wire connector for the sender, and the negative (black) lead from the DVOM into the back of the Black wire connector for the sender. Turn the ignition switch to the **ON/RUN** position (without starting the motor) and make sure there is 4-5 volts shown on the meter. Turn the switch to the **OFF** position and disconnect the DVOM.

Output voltage from the sender should rise or fall as the motor is trimmed up or down respectively. At full upward trim, output voltage should be about 3.5-4.5 Volts, while at full downward trim output voltage should be about 1.0-2.0 volts. To check, use a DVOM much in the same way you checked for the reference voltage, *however,* you'll need to change the connection points.

Place the positive (red) lead from the DVOM into the back of the Yellow wire connector for the sender, and the negative (black) lead from the DVOM into the back of the Black wire connector for the sender. Turn the ignition switch to the **ON/RUN** position (without starting the motor) and make sure the voltage varies as the motor is moved from FULL UP to FULL DOWN trim positions.

ADJUSTING AN ANALOG SENDER

◆ See Figure 70

■ This procedure pertains only to analog trim senders, as digital trim senders are mounted in a fixed position and are not adjustable.

1. Turn the ignition key to the **RUN** position without starting the outboard.
2. Trim the motor to the full IN position, and check the needle on the gauge. It should read fully downward. If NOT, trim the motor fully UP for access to the sender and engage the tilt lock lever. Loosen the sender screws and reposition the sender slightly as follows:

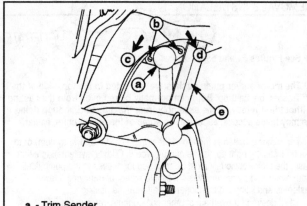

a - Trim Sender
b - Screws, Loosen to Rotate Sender
c - Turn Sender **Counterclockwise** to raise needle reading
d - Turn Sender **Clockwise** to Lower Needle Reading
e - Tilt Lock Lever

SELOC_1418

Fig. 70 Adjusting an analog trim sender gauge

• Turning the sender CLOCKWISE will LOWER the needle reading.
• Turning the sender COUNTERCLOCKWISE will RAISE the needle reading.

3. Once adjusted, retighten the sender screws and trim the unit FULLY DOWNWARD again to recheck the adjustment. Continue adjusting until the gauge reads accurately.

TRIM SENDER & GAUGE WIRING

◆ See Figure 71

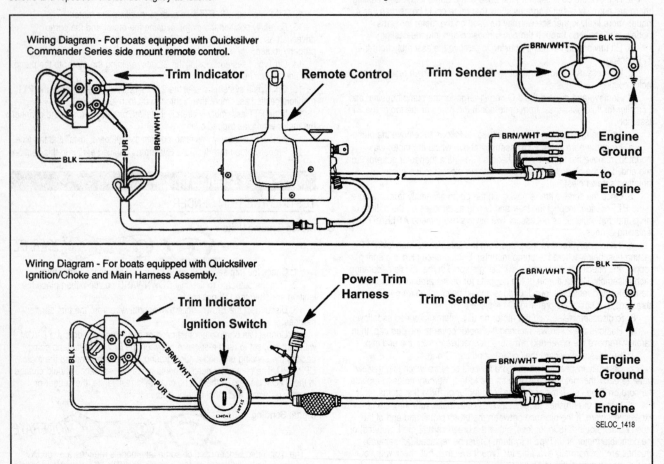

Wiring Diagram - For boats equipped with Quicksilver Commander Series side mount remote control.

Wiring Diagram - For boats equipped with Quicksilver Ignition/Choke and Main Harness Assembly.

SELOC_1418

Fig. 71 Trim Sender and Gauge Wiring

9

REMOTE CONTROLS

REMOTE CONTROLS - MECHANICAL UNITS

Commander Side Mount Remote Control Box (Mercury)

DESCRIPTION & OPERATION

■ The type of remote control mated to the outboard is determined by the boat manufacturer or the rigging dealer. Because there are such a range of remote control units that could be mated to Mercury outboards, there is no way to cover them all in detail here. This section deals with one of the most typical Mercury remote control unit. Illustrations are of the most common example of this unit. Although the procedures apply to most Mercury remotes and many aftermarket units, care and observation must be used when working on any unit as differences may exist in assemblies. When a component or mounting differs from the components covered here, use common sense, or contact the manufacturer of the unit for more information.

The remote control unit allows the helmsperson to control throttle operation and shift movements from a location other than where the outboard unit is mounted.

In most cases, the remote control box is mounted approximately halfway forward (mid-ship) on the starboard side of the boat or on the starboard side of the center console (depending upon the style/type of boat).

Several control box types are used, but most usually house a key switch, engine stop switch, choke switch, neutral safety switch, warning buzzer and the necessary wiring and cable hardware to connect the control box to the outboard unit.

TROUBLESHOOTING

One of the things taken for granted on most boats is the engine controls and cables. Depending on how they are originally routed, they will either last the life of the vessel or can be easily damaged. These cables should be routinely maintained by careful inspection for kinks or other damage and lubrication with a marine grade grease.

If the cables do not operate properly, have a helper operate the controls at the helm while you observe the cable and linkage operation at the powerhead. Make sure nothing is binding, bent or kinked. Check the hardware that secures the cables to the boat and powerhead to make sure they are tight. Inspect the clevis and cotter pins in the ends of the cables and also give some attention to the cable release hardware.

Another area of concern is the neutral start switch that is sometimes located in the control box and sometimes on the powerhead. This switch may fall out of adjustment and prevent the engine from being started. It is easily inspected using a multimeter by performing a continuity check.

REMOVAL & INSTALLATION

◆ See Figures 1 thru 6

1. Turn the ignition key to the off position. For safety and to protect from accidental attempt to start if wires are crossed, disconnect the high tension leads from the spark plugs, with a twisting motion. Others prefer to just disconnect the negative battery cable and be done with it.

2. Disconnect the remote control wiring harness plug from the outboard trim/tilt motor and pump assembly.

3. If applicable, disconnect the tachometer wiring plug from the forward end of the control housing.

4. Remove the locknuts, flat washers, and bolts (usually 3) securing the control housing to the mounting panel. One normally is located next to the run button (the ignition safety stop switch), and a second is usually beneath the control handle on the lower portion of the plastic case. The third is located behind the control handle when the handle is in the neutral position. Shift the handle into forward or reverse position to remove the bolt, then shift it back into the neutral position for the following steps.

5. Pull the remote control housing away and free of the mounting panel. Remove the plastic cover from the back of the housing. Lift off the access cover from the housing. (Some "Commander" remote control units do not have an access cover.)

6. Remove the screws (usually 2) securing the cable retainer over the throttle cable, wiring harness, and shift cable. Unscrew the Phillips-head screws (usually 2) securing the back cover to the control module, and then lift off the cover.

7. To disconnect the throttle cable, proceed as follows:

a. Loosen the cable retaining nut and raise the cable fastener enough to free the throttle cable from the pin. Lift the cable from the anchor barrel recess.

b. Remove the grommet.

8. To disconnect the shift cable, proceed as follows:

a. Shift the outboard unit into reverse gear by depressing the neutral lock bar on the control handle and moving the control handle into the reverse position.

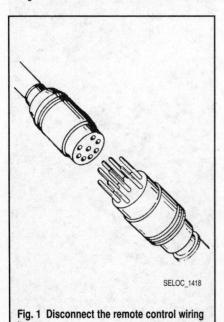

Fig. 1 Disconnect the remote control wiring harness

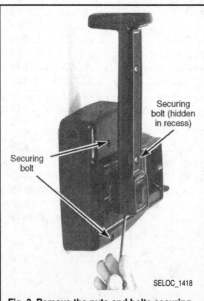

Fig. 2 Remove the nuts and bolts securing the control housing to the mounting panel

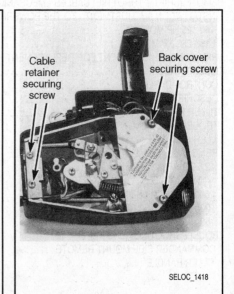

Fig. 3 You now have access to the back cover or the cable retainer

b. Loosen, but do not remove, the shift cable retainer nut (you'll usually need a 3/8 in. deep socket) as far as it will go without removing it. Raise the shift cable fastener enough to free the shift cable from the pin.

■ **Do not attempt to shift into reverse while the cable fastener is loose. An attempt to shift may cause the cable fastener to strike the neutral safety micro-switch and cause it damage.**

c. Lift the wiring harness out of the cable anchor barrel recess and remove the shift cable from the control housing.

To Install:

9. Position the control housing in place on the mounting panel and secure it with the long bolts, flat washers, and locknuts. Usually, one is

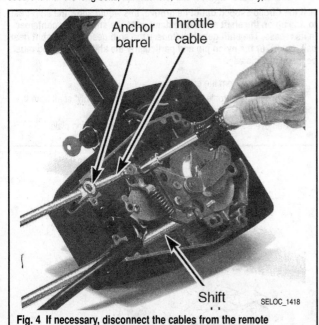

Fig. 4 If necessary, disconnect the cables from the remote

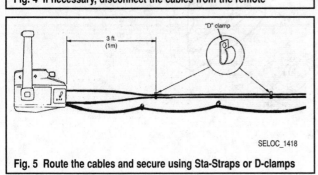

Fig. 5 Route the cables and secure using Sta-Straps or D-clamps

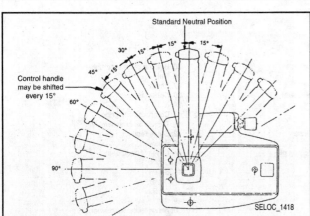

Fig. 6 The neutral position of the remote control handle may be changed to meet the owner's preference

located next to the run button (the ignition safety stop switch). The second is beneath the control handle on the power portion of the plastic case. The third bolt goes in behind the control handle when the handle is in the neutral position.

10. In order to install this bolt, shift the handle into the forward or reverse position, and then install the bolt. After the bolt is secure, shift the handle back to the neutral position for the next few steps.

11. The Commander shift box is now ready for installation.

12. Connect the tachometer wiring plug to the forward end of the control housing.

■ **Clean the prongs of the connector with crocus cloth to ensure the best connection possible. Exercise care while cleaning to prevent bending the prongs.**

13. Connect the remote control wiring harness plug from the outboard trim/tilt motor and pump assembly.

14. Install the high-tension leads to their respective spark plugs.

15. Route the wiring harness alongside the boat and fasten with the Sta-Straps. Check to be sure the wiring will not be pinched or chafe on any moving part and will not come in contact with water in the bilge. Route the shift and throttle cables the best possible way to make large bends and as few as possible. Secure the cables approximately every 3 feet (one meter).

16. The neutral position of the remote control handle may be changed to any one of a number of convenient angles to meet the owner's preference. The change is accomplished by shifting the handle 1 spline on the shaft at a time. Each spline equals 15°of arc, as shown.

DISASSEMBLY

◆ **See Figure 7**

■ **On non-power trim/tilt units, it is not necessary to remove the cover of the control handle.**

1. Depress the neutral lock bar on the control handle and shift the control handle back to the neutral position. Remove the Phillips head screws (usually 2) which secure the cover to the handle, and then lift off the cover. The push button trim switch will come free with the cover, the toggle trim switch will stay in the handle body.

2. Unsnap and then remove the wire retainer. Carefully unplug the trim wires and straighten them out from the control panel hub for ease of removal later.

3. Back-off the set screw at the base of the control handle to allow the handle to be removed from the splined control shaft.

4. Grasp the "throttle only" button and pull it off the shaft.

✳✳ SELOC CAUTION

Take care not to damage the trim wires when removing the control handle, on power trim models.

5. Remove the control handle.
6. Lift the neutral lockring from the control housing.

■ **Take care to support the weight of the control housing to avoid placing any unnecessary stress on the control shaft during the following disassembling steps.**

7. Remove the Phillips-head screws (usually 3) securing the control module to the plastic case. Normally, 2 are located on either side of the bearing plate and 1 is in the recess where the throttle cable enters the control housing.

8. Back-out the detent adjustment screw and the control handle friction screw until their heads are flush with the control module casing. This action will reduce the pre-load from the 2 springs on the detent ball for later removal.

■ **As this next step is performed, count the number of turns for each screw as they are backed-out and record the figure somewhere. This will be a tremendous aid during assembling.**

9. Remove the locknuts (usually 2) securing the neutral safety switch to the plate assembly and lift out the micro-switch from the recess in the assembly.

10. Remove the Phillips-head screw securing the retaining clip to the control module.

11. Support the module in your hand and tilt it until the shift gear spring, shift nylon pin (earlier models have a ball), shift gear pin, another ball the shift gear ball (inner), fall out from their recess. If the parts do not fall out into your hand, attach the control handle and ensure the unit is in the neutral position. The parts should come free when the handle is in the neutral position.

12. Arrange the parts from the control module recess. As the parts are removed and cleaned, keep them in order, ready for installation.

13. Remove the Phillips-head screws (usually 3) securing the bearing plate assembly to the control module housing.

14. Lift out the bearing plate assembly from the control module housing.

15. Uncoil the trim wires from the recess in the remote control module housing and lift them away with the trim harness bushing attached.

16. Remove the detent ball, the detent ball follower, and the 2 compression springs (located under the follower), from their recess in the control module housing.

17. If it is not part of the friction pad, remove the control handle friction sleeve from the recess in the control module housing.

18. Pull the throttle link assembly from the module. Remove the compression spring from the throttle lever. It is not necessary to remove this spring unless there is cause to replace it. At this point there is the least amount of tension on the compression spring. Therefore, now would be the time to replace it, if required.

19. Lift the shift pinion gear (with attached shift lever), off the pin on the bearing plate. The nylon bushing may come away with the shift lever and shift pinion gear as an assembly. Do not attempt to separate them. Both are replaced if one is worn.

20. On the non-power trim/tilt units only, remove the trim harness bushing and wiring harness retainer from the control shaft (on non-power trim/tilt units, these 2 items act as spacers.)

21. On all other units, remove the shift gear retaining ring from its groove with a pair of Circlip pliers.

■ If the Circlip slipped out of its groove, this would allow the shift gear to ride up on the shaft and cause damage to the small parts contained in its recess. The shift gear ball (inner), the shift gear pin, the shift gear ball (outer), or the nylon pin and particularly the shift gear spring must be inspected closely.

22. Lift the gear from the control shaft.

23. Remove the "throttle only" shaft pin and "throttle only" shaft from the control shaft.

24. Remove the step washer from the base of the bearing plate.

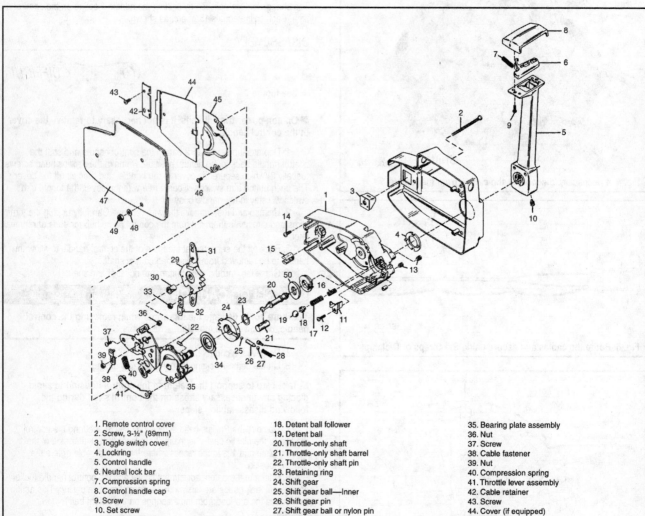

1. Remote control cover	18. Detent ball follower	35. Bearing plate assembly
2. Screw, 3-½" (89mm)	19. Detent ball	36. Nut
3. Toggle switch cover	20. Throttle-only shaft	37. Screw
4. Lockring	21. Throttle-only shaft barrel	38. Cable fastener
5. Control handle	22. Throttle-only shaft pin	39. Nut
6. Neutral lock bar	23. Retaining ring	40. Compression spring
7. Compression spring	24. Shift gear	41. Throttle lever assembly
8. Control handle cap	25. Shift gear ball—Inner	42. Cable retainer
9. Screw	26. Shift gear pin	43. Screw
10. Set screw	27. Shift gear ball or nylon pin	44. Cover (if equipped)
11. Retaining clip	28. Shift gear spring	45. Cover
12. Screw	29. Shift pinion gear	46. Screw
13. Set screw	30. Bushing	47. Cover
14. Screw	31. Shift lever	48. Washer
15. Grommet	32. Cable fastener	49. Nut
16. Compression spring	33. Nut	50. Wiring harness retainer
17. Compression spring	34. Step washer	

SELOC_1418

Fig. 7 Exploded view of a typical Mercury control box

CLEANING & INSPECTING

◆ **See Figure 7**

1. Clean all metal parts with solvent, and then blow them dry with compressed air.

■ **Never allow nylon bushings, plastic washers, nylon pins, wiring harness retainers, and the like, to remain submerged in solvent more than just a few moments. The solvent will cause these type parts to expand slightly. They are already considered a "tight fit" and even the slightest amount of expansion would make them very difficult to install. If force is used, the part is most likely to be distorted.**

2. Inspect the control housing plastic case for cracks or other damage allowing moisture to enter and cause problems with the mechanism.

3. Carefully check the teeth on the shift gear and shift lever for signs of wear. Inspect all ball bearings for nicks or grooves which would cause them to bind and fail to move freely.

4. Closely inspect the condition of all wires and their protective insulation. Look for exposed wires caused by the insulation rubbing on a moving part, cuts and nicks in the insulation and severe kinking which could cause internal breakage of the wires.

5. Inspect the surface area above the groove in which the circlip is positioned for signs of the circlip rising out of the groove. Such action would occur if the clip had lost its "spring" or if it had worn away the top surface of the groove. If the circlip slipped out of its groove, the shift gear would be able to ride up on the shaft and cause damage to the small parts contained in its recess.

6. The shift gear ball (inner), the shift gear pin, the shift gear ball (outer), or the nylon pin and particularly the shift gear spring must be inspected closely.

7. Inspect the "throttle only" shaft for wear along the ramp. In the early model units, this shaft was made of plastic. Later models have a shaft of stainless steel. Check for excessive wear or cracks on the ramp portion of the shaft. Also check the lower "stop" tab to be sure it has not broken away.

■ **Good shop practice dictates a thin coat of multipurpose lubricant be applied to all moving parts as a precaution against the enemy - moisture. Of course the lubricant will help to ensure continued satisfactory operation of the mechanism.**

ASSEMBLY

◆ **See Figure 7**

■ **The Commander control shift box, like others, has a number of small parts that must be assembled in only one order - the proper order. Therefore, the work should not be rushed or attempted if the person assembling the unit is under pressure. Work slowly, exercise patience, read ahead before performing the task, and follow the steps closely.**

1. Place the step washer over the control shaft and ensure the steps of the washer seat onto the base of the bearing plate.

2. Rotate the control shaft until the "throttle only" shaft pin hole is aligned centrally between the neutral detent notch and the control handle friction pad. Lower the "throttle only" shaft into the barrel of the control shaft. Secure the shaft in this position with the "throttle only" shaft pin.

■ **When the pin is properly installed, it should protrude slightly in line with the plastic bushing.**

3. Make an attempt to gently pull the "throttle only" shaft out of the control shaft. If the shaft and pin are properly installed, the attempt should fail.

4. Place the shift gear over the control shaft, and check to be sure the "throttle only" shaft pin clears the gear.

5. Install the retaining ring over the control shaft with a pair of Circlip pliers. Check to be sure the ring snaps into place within the groove.

6. On non-power tilt/trim units only, slide the wiring harness retainer and the trim harness bushing over the control shaft. The trim harness bushing is placed "stepped side" UP and the notched side toward the forward side of the control housing.

7. On power tilt/trim units only, insert the trim harness bushing into the recess of the remote control housing and carefully coil the wires, as shown in the illustration. Ensure the black line on the trim harness is positioned at the exact point shown for correct installation. The purpose of the coil is to allow slack in the wiring harness when the control handle is shifted through a full cycle. The bushing and wires move with the handle.

8. Position the bushing, shift pinion gear and shift lever onto the pin on the bearing plate, with the shift gear indexing with the shift pinion gear.

9. Install the 2 compression springs, the detent ball follower and the detent ball into their recess in the control module housing.

10. If the friction sleeve is not a part of the friction pad, then place the control handle friction sleeve into its recess in the control module housing.

11. Thread the detent adjustment screw and the control handle friction screw the exact number of turns as was turned out during disassembly. A fine adjustment may be necessary after the unit is completely assembled.

12. Place the compression spring (if removed), in position on the shift lever and shift pinion gear assembly against the bearing plate. Use a rubber band to secure the shift pinion gear to the bearing plate. Lower the complete bearing plate assembly into the control module housing.

13. Secure the bearing plate assembly to the control module housing with the 3 Phillips head screws, and then remove the rubber band.

14. Insert the gear shift ball (inner), into the recess of the shift gear and hole in the "throttle only" shaft barrel. Now, insert the shift gear pin into the recess with the rounded end of the pin away from the control shaft. Insert the nylon pin or shift gear ball (outer), into the same recess. Next, insert the shift gear spring.

15. Hold these small parts in place and at the same time secure them with the retaining clip and the Phillips head screw. On power trim/tilt units, this retaining clip also secures the trim wire to the control module.

16. Insert the neutral safety micro-switch into the recess of the plate assembly and secure it with the locknuts (usually 2).

17. Arrange the parts so that they are cleaned and ready for installation into the control module recess.

18. Secure the control module to the plastic control housing case with the Phillips head screws (usually 3). Normally, 2 are located on either side of the bearing plate and the third in the recess where the throttle cable enters the control housing.

19. Temporarily install the control handle onto the control shaft. Shift the unit into forward detent only, not full forward, to align the holes for installation of the throttle link. After the holes are aligned, remove the handle. Install the throttle link.

20. Again, temporarily install the control handle onto the control shaft. This time shift the unit into the neutral position, and then remove the handle.

21. Place the neutral lockring over the control shaft, with the index mark directly beneath the small boot on the front face of the cover.

22. Install the control onto the splines of the control shaft, taking care not to cut, pinch, or damage the trim wires on the power trim/tilt unit.

✳✳ SELOC WARNING

When positioning the control handle, ensure the trim wire bushing is aligned with its locating pin against the corresponding slot in the control handle. On a power trim/tilt unit: if this bushing is not installed correctly, it will not move with the control handle as it is designed to move - when shifted. This may pinch or cut the trim wires, causing serious problems. On a non-power trim/tilt unit, misplacement of this bushing (it is possible to install this bushing upside down) will not allow the control handle to seat properly against the lockring and housing. This situation will lead to the Allen screw at the base of the control handle to be incorrectly tightened to seat against the splines on the control shaft, instead of gripping the smooth portion of the shaft. Subsequently the control handle will feel "sloppy" and could cause the neutral lock to be ineffective.

✳✳ SELOC WARNING

If this handle is not seated properly, a slight pressure on the handle could throw the lower unit into gear, causing serious injury to crew, passengers, and the boat.

23. Push the "throttle only" button in place on the control shaft.

24. Ensure the control handle has seated properly, and then tighten the set screw at the base of the handle to 70 inch lbs. (7.9 Nm).

■ **Failure to tighten the set screw to the required torque value, could allow the handle to disengage with a loss of throttle and shift control.**

25. If the handle cover was removed during disassembly, slide the hooked end of the neutral lock rod into the slot in the neutral lock release. Route the trim wires in the control handle in their original locations. Connect them with the wires remaining in the handle and secure the connections with the wire retainer. Install the handle cover and tighten the Phillips head screws (usually 2).

26. Move the wiring harness clear of the barrel recess. Thread the shift cable anchor barrel to the end of the threads, away from the cable converter, and place it into the recess. Hook the pin on the end of the cable fastener through the outer hole in the shift lever.

27. Depress the neutral lock bar on the control handle and shift the handle into the reverse position. Take care to ensure the cable fastener will clear the neutral safety micro-switch. The access hole is now aligned with the locknut.

■ **Check to be sure the pin on the cable fastener is all the way through the cable end and the shift lever. A pin partially engaging the cable and the shift lever may cause the cable fastener to bend when the nut is tightened.**

28. Tighten the locknut (it normally requires a 3/8 in. deep socket) to 20-25 inch lbs. (2.26-2.82 Nm). Position the wiring harness over the installed shift cable.

29. Install the grommet into the throttle cable recess.

30. Thread the throttle cable anchor barrel to the end of the threads, away from the cable connector, and then place it into the recess over the grommet. Hook the pin on the end of the cable fastener through the outer hole in the shift lever.

■ **Check to be sure the pin on the cable fastener is all the way through the cable end and the throttle lever. A pin partially engaging the cable and throttle lever may cause the cable fastener to bend when the nut is tightened.**

31. Tighten the locknut to 20-25 inch lbs. (2.26-2.82 Nm).

32. Position the control module back cover in place and secure it with the Phillips-head screws (usually 2). Tighten the screws to 60 inch lbs. (6.78 Nm). Install the cable retainer plate over the 2 cables and secure it in place with the Phillip-head screws (also usually 2).

33. Place the plastic access cover over the control housing.

Additional Remote Units

As noted earlier the engine manufacturer has little control over the actual remote control to which any given motor is eventually connected. Mercury has over the years produced dozens of different control housings that can be used with their motors and you could potentially find almost any of them rigged to a given powerhead. The following section includes exploded views of many of the more popular units which can be used to help in disassembly or component replacement.

EXPLODED VIEWS

◆ **See Figures 8 thru 16**

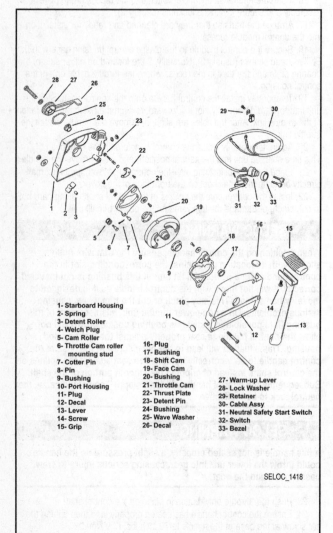

1- Starboard Housing
2- Spring
3- Detent Roller
4- Welch Plug
5- Cam Roller
6- Throttle Cam roller mounting stud
7- Cotter Pin
8- Pin
9- Bushing
10- Port Housing
11- Plug
12- Decal
13- Lever
14- Screw
15- Grip
16- Plug
17- Bushing
18- Shift Cam
19- Face Cam
20- Bushing
21- Throttle Cam
22- Thrust Plate
23- Detent Pin
24- Bushing
25- Wave Washer
26- Decal
27- Warm-up Lever
28- Lock Washer
29- Retainer
30- Cable Assy
31- Neutral Safety Start Switch
32- Switch
33- Bezel

SELOC_1418

Fig. 8 Exploded view of the standard single lever full gear shift control

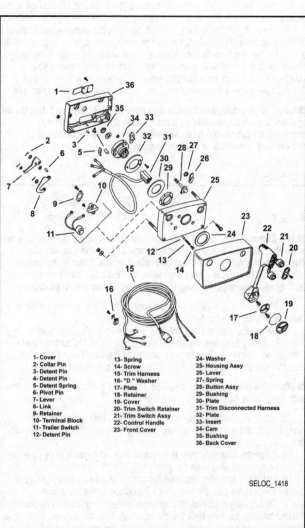

1- Cover
2- Collar Pin
3- Detent Pin
4- Detent Pin
5- Detent Spring
6- Pivot Pin
7- Lever
8- Link
9- Retainer
10- Terminal Block
11- Trailer Switch
12- Detent Pin
13- Spring
14- Screw
15- Trim Harness
16- "D" Washer
17- Plate
18- Retainer
19- Cover
20- Trim Switch Retainer
21- Trim Switch Assy
22- Control Handle
23- Front Cover
24- Washer
25- Housing Assy
26- Lever
27- Spring
28- Button Assy
29- Bushing
30- Plate
31- Trim Disconnected Harness
32- Plate
33- Insert
34- Cam
35- Bushing
36- Back Cover

SELOC_1418

Fig. 9 Exploded view of the side mount mechanical shift

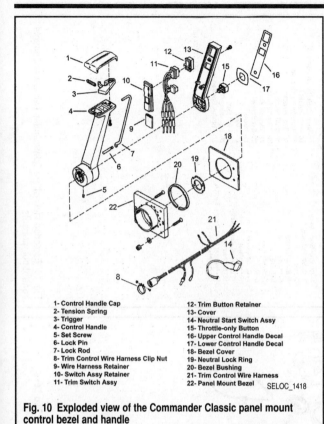

1- Control Handle Cap
2- Tension Spring
3- Trigger
4- Control Handle
5- Set Screw
6- Lock Pin
7- Lock Rod
8- Trim Control Wire Harness Clip Nut
9- Wire Harness Retainer
10- Switch Assy Retainer
11- Trim Switch Assy
12- Trim Button Retainer
13- Cover
14- Neutral Start Switch Assy
15- Throttle-only Button
16- Upper Control Handle Decal
17- Lower Control Handle Decal
18- Bezel Cover
19- Neutral Lock Ring
20- Bezel Bushing
21- Trim Control Wire Harness
22- Panel Mount Bezel

SELOC_1418

Fig. 10 Exploded view of the Commander Classic panel mount control bezel and handle

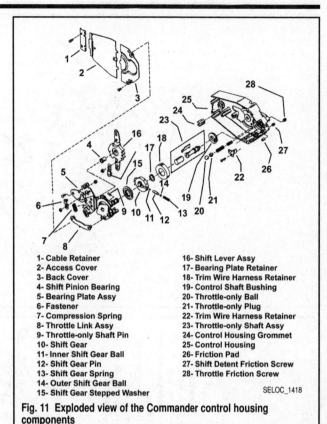

1- Cable Retainer
2- Access Cover
3- Back Cover
4- Shift Pinion Bearing
5- Bearing Plate Assy
6- Fastener
7- Compression Spring
8- Throttle Link Assy
9- Throttle-only Shaft Pin
10- Shift Gear
11- Inner Shift Gear Ball
12- Shift Gear Pin
13- Shift Gear Spring
14- Outer Shift Gear Ball
15- Shift Gear Stepped Washer
16- Shift Lever Assy
17- Bearing Plate Retainer
18- Trim Wire Harness Retainer
19- Control Shaft Bushing
20- Throttle-only Ball
21- Throttle-only Plug
22- Trim Wire Harness Retainer
23- Throttle-only Shaft Assy
24- Control Housing Grommet
25- Control Housing
26- Friction Pad
27- Shift Detent Friction Screw
28- Throttle Friction Screw

SELOC_1418

Fig. 11 Exploded view of the Commander control housing components

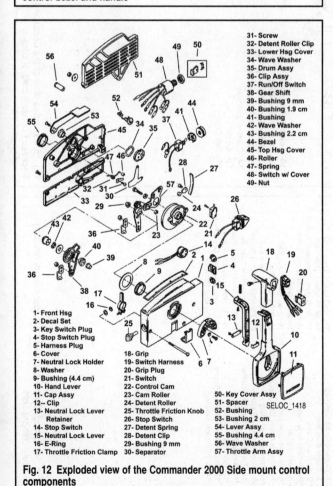

31- Screw
32- Detent Roller Clip
33- Lower Hsg Cover
34- Wave Washer
35- Drum Assy
36- Clip Assy
37- Run/Off Switch
38- Gear Shift
39- Bushing 9 mm
40- Bushing 1.9 cm
41- Bushing
42- Wave Washer
43- Bushing 2.2 cm
44- Bezel
45- Top Hsg Cover
46- Roller
47- Spring
48- Switch w/ Cover
49- Nut

1- Front Hsg
2- Decal Set
3- Key Switch Plug
4- Stop Switch Plug
5- Harness Plug
6- Cover
7- Neutral Lock Holder
8- Washer
9- Bushing (4.4 cm)
10- Hand Lever
11- Cap Assy
12- Clip
13- Neutral Lock Lever Retainer
14- Stop Switch
15- Neutral Lock Lever
16- E-Ring
17- Throttle Friction Clamp
18- Grip
19- Switch Harness
20- Grip Plug
21- Switch
22- Control Cam
23- Cam Roller
24- Detent Roller
25- Throttle Friction Knob
26- Stop Switch
27- Detent Spring
28- Detent Clip
29- Bushing 9 mm
30- Separator
50- Key Cover Assy
51- Spacer
52- Bushing
53- Bushing 2 cm
54- Lever Assy
55- Bushing 4.4 cm
56- Wave Washer
57- Throttle Arm Assy

SELOC_1418

Fig. 12 Exploded view of the Commander 2000 Side mount control components

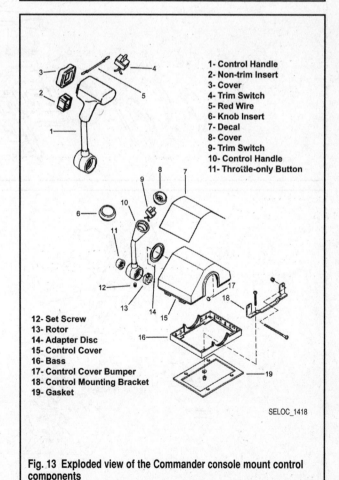

1- Control Handle
2- Non-trim Insert
3- Cover
4- Trim Switch
5- Red Wire
6- Knob Insert
7- Decal
8- Cover
9- Trim Switch
10- Control Handle
11- Throttle-only Button

12- Set Screw
13- Rotor
14- Adapter Disc
15- Control Cover
16- Bass
17- Control Cover Bumper
18- Control Mounting Bracket
19- Gasket

SELOC_1418

Fig. 13 Exploded view of the Commander console mount control components

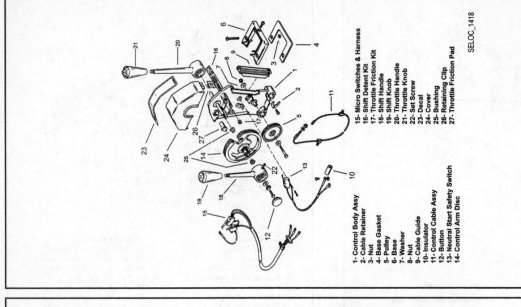

1- Control Body Assy
2- Cable Retainer
3- Nut
4- Base Gasket
5- Pulley
6- Base
7- Washer
8- Nut
9- Cable Guide
10- Insulator
11- Control Cable Assy
12- Button
13- Neutral Start Safety Switch
14- Control Arm Disc

15- Micro Switches & Harness
16- Shift Detent Kit
17- Throttle Friction Kit
18- Shift Handle
19- Shift Knob
20- Throttle Handle
21- Throttle Knob
22- Set Screw
23- Decal
24- Cover
25- Bushing
26- Retaining Clip
27- Throttle Friction Pad

Fig. 16 Exploded view of the Commander Two Lever Console control components

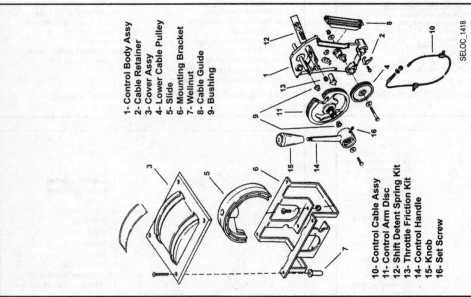

1- Control Body Assy
2- Cable Retainer
3- Cover Assy
4- Lower Cable Pulley
5- Slide
6- Mounting Bracket
7- Wellnut
8- Cable Guide
9- Bushing

10- Control Cable Assy
11- Control Arm Disc
12- Shift Detent Spring Kit
13- Throttle Friction Kit
14- Control Handle
15- Knob
16- Set Screw

Fig. 15 Exploded view of the Commander Two Lever Recessed Console control components

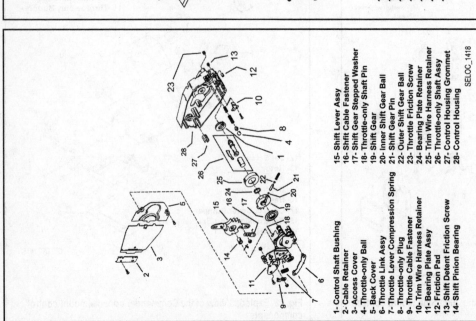

1- Control Shaft Bushing
2- Cable Retainer
3- Access Cover
4- Throttle-only Ball
5- Back Cover
6- Throttle Link Assy
7- Throttle Lever Compression Spring
8- Throttle-only Plug
9- Throttle Cable Fastener
10- Trim Wire Harness Retainer
11- Bearing Plate Assy
12- Friction Pad
13- Shift Detent Friction Screw
14- Shift Pinion Bearing

15- Shift Lever Assy
16- Shift Cable Fastener
17- Shift Gear Stepped Washer
18- Throttle-only Shaft Pin
19- Shift Gear
20- Inner Shift Gear Ball
21- Shift Gear Pin
22- Outer Shift Gear Ball
23- Throttle Friction Screw
24- Bearing Plate Retainer
25- Trim Wire Harness Retainer
26- Throttle-only Shaft Assy
27- Control Housing Grommet
28- Control Housing

Fig. 14 Exploded view of the Commander control housing module components

Remote (Shift & Throttle) Cables

RIGGING & MEASUREMENT FOR REPLACEMENT

◆ See Figure 17

The control cables should be replaced if inspection reveals any signs of damage, wear or even fraying at the exposed ends. Remember that loss of one or more cables while underway could cause loss of control at worst or, at best, strand the boat. Check cable operation frequently and inspect the cables at each service. Replace any that are hard to move or if excessive play is noted. Never replace just 1 cable (unless a freak accident caused damage to the cable), if 1 cable is worn or has failed, assume the other is in like condition and will soon follow.

Before removal, mark the cable mounting points on the remote control using a permanent marker. This will help ensure easy and proper installation of the replacement. Unless there is a reason to doubt the competence of the person who originally rigged the craft, match the cable lengths as closely as possible. Otherwise, re-measure and determine proper cable lengths as if this was a new rigging.

■ **On boats with significant freeboard drop or unusual routing of cables it may be necessary to increase the length of the cables in order to route them properly.**

When rigging an engine to a new boat, determine cable length by measuring the distance from the remote to the transom along the gunwale, and the distance from the gunwale to the each motor, then add 18 in. (457mm) to those combined measurements. Purchase a cable of the same length (or one that is **slightly** longer than that measurement). Replacement cables are normally available in 1 ft. increments.

✳✳ SELOC CAUTION

To prevent the danger of cable binding or other conditions that could cause a loss of steering control when underway, take care to route all cables with the fewest number and most gentle bends possible. A bend should NEVER have a radius of less than 8 in. (203mm).

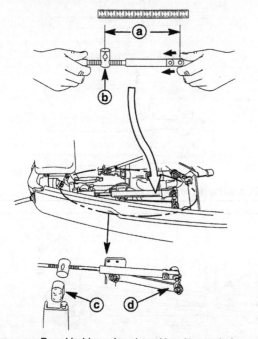

a - Move cable barrel to attain the measured distance
b - Cable barrel

c - Barrel holder– place barrel into bottom hole
d - Retainer

Fig. 18 Shift cable installation and adjustment (30-50 hp shown, others similar, but barrel bracket shape will vary)

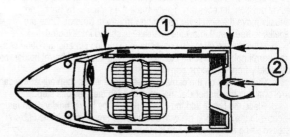

1 - Distance from remote control to transom
2 - Distance from gunwale to each engine

SELOC_1418

Fig. 17 For cable replacement or rigging, measure the distance from the remote to transom, and from the gunwale to the engine centerline then add 18 in. (457mm)

INSTALLATION & ADJUSTMENT - TO THE POWERHEAD

40-125 hp (3- and 4-Cylinder) Models

◆ See Figures 18 and 19

This procedure assumes that the shift and throttle cables are not connected to the powerhead. This procedure can also be used to disconnect and reset the remote cables in case there is a problem with the current adjustment, but it SHOULD not be a periodic adjustment. Once properly set, the cables SHOULD require no further adjustment.

■ **Keep in mind that the shift cable is normally the FIRST cable to move when the remote control handle is moved into gear. Always install the shift cable first.**

1. Install the shift cable to the powerhead as follows:

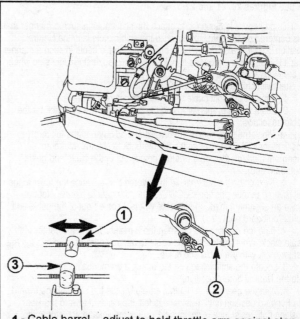

1 - Cable barrel – adjust to hold throttle arm against stop
2 - Throttle arm
3 - Barrel holder – place barrel into top hole

SELOC_1418

Fig. 19 Throttle cable installation and adjustment (30-50 hp shown, others similar)

a. Make sure both the remote control handle and gearcase are in the Neutral position. If necessary slowly rotate the propeller by hand and manually move the shift lever on the powerhead to Neutral. Once the propeller is free, you'll know the gearcase is shifted into Neutral.

b. Measure the distance between the mounting pin and the dead center of the barrel holder (on some models, the size/shaft of the barrel holder may vary slightly from the accompanying illustration).

c. Push inward on the cable until resistance is felt, then adjust the cable barrel until it is the proper distance measured in the previous step.

d. Place the barrel into the bottom hole in the barrel holder, then fasten the cable to the mounting pin with a retainer (on most models the cable is fastened using a flat washer and locknut).

e. Check shift adjustment by making sure the remote shifts the gearcase into forward, neutral and reverse (you'll likely have to slowly spin the propeller by hand while attempting to make all/some of those shifts).

2. Install the throttle cable to the powerhead (after the shift cable is already installed) as follows:

a. Make sure both the remote control handle and gearcase are in the Neutral position. If necessary slowly rotate the propeller by hand and manually move the shift lever on the powerhead to Neutral. Once the propeller is free, you'll know the gearcase is shifted into Neutral.

b. Connect the cable to the throttle lever, then secure with the retainer (again, this is normally a flat washer and locknut).

c. Adjust the cable barrel so that the installed throttle cable will hold the throttle arm AGAINST THE STOP.

d. Check the throttle adjustment by shifting the outboard into gear and out again a few times, activating the throttle linkage.

■ **Be sure to SLOWLY rotate the propeller when attempting to shift into reverse.**

e. Return the remote to Neutral, then place a thin piece of paper between the throttle arm and idle stop. Adjustment is correct if the paper can be removed with a slight drag felt, but WITHOUT tearing. Adjust the cable barrel additionally as necessary.

f. With the cable barrel inserted into the barrel holder, lock the holder using the cable latch.

135-250 hp (V6) Models

◆ **See Figures 20, 21 and 22**

This procedure is designed to attach the shift cable in such a manner that it is centered on the shift linkage while the shift handle is in the Neutral position. And it is also designed to attach the throttle cable in such a manner that it is just barely preloaded to hold the throttle against the idle stop when in the Neutral position.

Start with shift cable, as follows:

1. Connect the shift cable to the remote control handle.

2. Use a pencil or marker to find the center point on the slack for the cable as follows:

a. Place the shift handle in Forward and advance to the full throttle position, then SLOWLY return the handle back to Neutral. Make an alignment mark on the cable where it comes out of the cable end guide (outboard end).

b. Next place the shift handle in Reverse and advance the lever to the full reverse throttle position. SLOWLY return the handle back to Neutral. Make an alignment mark on the cable where it comes out of the cable end guide (outboard end).

c. Find the center point between the 2 marks. This is the center of the cable slack. When connecting the cable to the outboard be sure to align this center mark with the cable end guide.

3. Locate the shift linkage on the motor, then manually move the linkage into the Neutral detent position.

4. With the gearcase in Neutral, carefully slide the shift linkage toward the Forward position until resistance is felt, then slide it toward the Rear position until resistance is again felt. Position the linkage actuator in the middle of these two points. With the linkage actuator centered, align the cable end guide with the center point alignment mark and slip the end guide over the shift actuator stud.

5. Next, adjust the cable barrel so that it slips freely into the barrel retainer.

6. Now it is time to secure the cable end. Typically, for 2.5L models you'll use a retainer assembly, but on 3.0L models you'll use a flat washer and locknut. For models which utilize a locknut, tighten the locknut against the cable end, then back it off 1/4 of a turn.

Next, it's time to install/adjust the throttle cable, as follows:

7. Position the remote control handle in Neutral.

8. Position the engine throttle lever against the idle stop.

9. Install the throttle cable to the throttle lever. On early models secure using the anchor pin. On late models secure using the plastic washer and locknut, tighten the locknut, then back it off 1/4 of a turn.

10. Hold the throttle lever against the idle stop, then adjust the throttle cable barrel so it will slip into the retainer with a *very* light preload against the idle stop.

11. Lock the cable barrels in place using the retainer.

12. Check preload on the throttle cable by placing a thin piece of paper between the idle stop screw and the idle stop. Preload is correct if the paper can just be removed without tearing, but still has some drag on it. Readjust the cable barrel, if necessary.

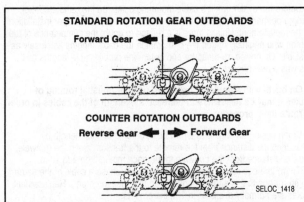

Fig. 20 Locating Neutral on the powerhead shift linkage (retainer type shown)

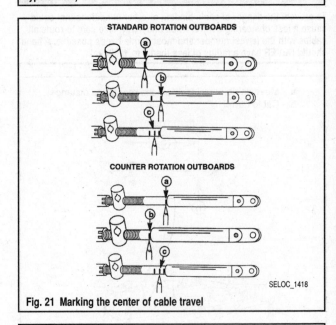

Fig. 21 Marking the center of cable travel

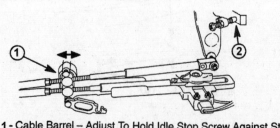

1 - Cable Barrel – Adjust To Hold Idle Stop Screw Against Stop
2 - Idle Stop Screw

SELOC_1418

Fig. 22 Throttle cable adjustment (2.5L shown, 3.0L similar)

REMOTE CONTROLS - DIGITAL THROTTLE & SHIFT (DTS) UNITS

Description & Operation

◆ **See Figure 23**

Beginning in or around 2004 Mercury began to offer the SmartCraft Digital Throttle and Shift (DTS) control system as an option to conventional remote controls on the largest of their outboards (pretty much only the 200 hp and larger OptiMax motors covered here). The initial version used a 10-pin engine connection.

Shortly thereafter a revised 14-pin version was introduced. Although most of the functions and procedures remain identical between the two, they are different systems and do not share parts. The biggest noticeable differences are the 14-pin panel mount remote control uses a shift lock on the handle; and the 14-pin console mount remote control uses a Throttle Only or Station Select button on the console.

Updated versions were introduced in multiple model years after that, including 2006 and 2007. Again, basic designs and functions are the same, but components are often not interchangeable.

DTS COMMAND MODULE

The command module is, very simply, a small computer that is meant to manage communications between the PCM and ERC. Once the key is turned to the Run position, a 12V signal is sent throughout the module harness, turning on the module. The power relay (if equipped), System View (if equipped) and any other gauges connected to the junction box will also be powered up. The voltage signal is also sent back down the data cable (via the purple wire) to the engine PCM.

CONTROL AREA NETWORK (CAN) DATA HARNESS

SmartCraft DTS utilizes 2 sets of CAN lines. CAN1 connects engine(s) and command modules to the gauges so that engine data such as temperature, pressure, depth, speed and rpm can be transmitted through the system and to the gauges at the helm.

CAN2 carries throttle and shift data only and is isolated between the engine and command module. The systems are redundant so as a back-up, if CAN2 loses connection than CAN1 will take over and transmit the throttle and shift data. When this happens, the system will move into Guardian mode and will automatically reduce power to the engine.

14-pin units utilize a DTS terminator resistor as a CAN line signal conditioner by placing a known load on the line to ensure proper communication between the command module and the PCM. There is generally a resistor at each end of the line (one near the helm and one near the engine).

PROPULSION CONTROL MODULE (PCM)

The PCM receives throttle shift, start, trim and shut off commands from the DTS command module via 2 control area network (CAN) signal and ground circuits. These commands initiate a software process which will control and perform all shift, throttle, start and stop functions normally controlled by the conventional remote control system. The PCM verifies and quantifies the shift and throttle commands from the DTS command module to ensure that the electronic shift controller (ESC) and throttle controller (ETC) are correctly positioning the throttle and shifters.

ELECTRONIC THROTTLE CONTROL (ETC)

The ETC is a single bore cylindrical valve and DC electric motor used to control the position of the throttle valve. Positive spring force is used to return the throttle valve to a preset position when there is no signal to the DC motor. Idle airflow is controlled by the throttle valve - there is no need for a conventional idle air control system.

There are 2 throttle position potentiometers (TPS), operating in opposite, that provide a positive feedback signal. As voltage increases in 1 potentiometer, it will decrease in the other. Interestingly though, the PCM processes the signals from both sensors in a manner that the digital diagnostic terminal display will show both TPS sensors increasing and decreasing in tandem.

ELECTRONIC SHIFT CONTROL (ESC)

The ESC, or shift actuator, is used as a means of shifting the drive unit into forward, neutral and reverse gears; without any conventional mechanical cables from the remote control at the helm. The actuator motor rotates a ball/screw assembly through reduction gears in the actuator. The screw shaft will then extend or retract the actuator shaft, while the gear set rotates a potentiometer in the actuator at the same time.

The PCM provides a 5 V reference signal to the potentiometer which will confirm the actual position of the actuator shaft.

ELECTRONIC REMOTE CONTROL (ERC)

Although it looks almost identical to a conventional remote control handle, the ERC handle position is what determines the throttle position and also the shift direction. The handle has a range of 78° (+/-) from the vertical, or neutral position; and is held in the neutral position by a spring detent mechanism which prevents any unintended movement.

There are 3 potentiometers in the EDC, arranged along the handle's axis of rotation, which communicate handle position to the DTS command module. The movement of the handle will vary the voltage output of the potentiometers.

Voltage from 1 potentiometer is used to set the engine speed, while the other two are used (in combination) to specify shift direction. Handle movement will determine, and activate, shift direction prior to increasing the engine speed - output from the 2 shift potentiometers is used as a redundancy check against the throttle command.

The shift potentiometer voltages are used by the DTS command module, which functions as follows:

• Reads the voltage from the 2 sensors and checks for consistency to accurately determine the position of the handle

• Creates a shift direction command

• Passes this command, with the voltages, on to the propulsion control module (PCM).

The PCM will then create its own shift command to be sent to the ESC once it verifies that the signal from the command module is valid. Once it receives the signal and command from the command module, the PCM will:

• Compares the voltage signals and command module shift command to verify consistency across the 3 signals/commands

• Creates its own shift command and then sends it back to the DTS command module, which then confirms that this new signal has the same values as that which it just sent to the PCM (and that the interpretation of them is accurate). Once done, the PCM then resend the signal back to the command module confirming the original signal

• Finally, the PCM forwards this command signal on to the ESC.

This shift command, essentially just voltage applied to the DC motor connections in the ESC/actuator, will determine the direction that the motor will rotate; thus determining the shift direction.

Throttle setting is changed via movement of the ERC handle. As mentioned, the handle position specifies both the throttle and shifter setting, which is then communicated to the DTS command module via the 3

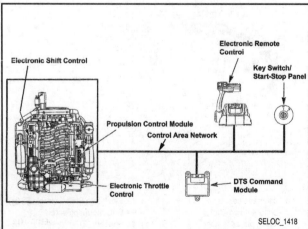

SELOC_1418

Fig. 23 SmartCraft DTS operational schematic (early-10-pin units shown)

potentiometers. The command module uses these potentiometer voltage signals to verify the throttle position as follows:

• It reads the signals from the ERC and checks for consistency by comparing the voltage from the throttle potentiometer with that from the 2 shift control potentiometers

• Upon verifying the consistency of the 3 voltage signals, it creates a command for the indicated throttle setting

• It then passes the throttle setting command (and voltages) on to the PCM.

The PCM verifies the command and then must set the commanded throttle position. It will do this as follows:

• • Compare the command module throttle command with the ERC potentiometer voltages to confirm consistency

• Creates its own throttle setting command

• Once created, it will send this command, and the potentiometer voltages back to the command module. The command module will then confirm that the new PCM command and voltages match those that it originally sent to the PCM, and, if so, send them back to the PCM

• Once this is done, the PCM will send the command and voltages through to the electronic throttle control (ETC).

This command is basically a simple torque specification of the DC throttle motor. The throttle is opened and closed by the motor acting against the throttle spring via a train of spur gears. Once this torque is applied to the throttle plate, its resultant position, determined by the 2 potentiometers in the throttle body, is sent back to the PCM. The PCM then confirms that the throttle position is equal to the command that it sent originally.

As a back up, the PCM also monitors the manifold absolute pressure (MAP) and the DC motor duty cycle (via its pulse width modulation {PWM}).

KEY SWITCH & START/STOP PANEL

The key switch, obviously, supplies power to the DTS command module, which then powers on the remainder of the SmartCraft system (including the PCM and all gauges).

You may have a 3-position switch (OFF, ON, START), a 4-position switch (OFF, ACC, ON, START), or a 2-position switch (ON, OFF) coupled with an START/STOP push button. Dual engine set-ups will have one of each for each engine. Turning the 3- or 4-position switch to the **START** position will start the engine, or turning the 2-position switch to the **ON** position and pressing the **START** button.

■ **The 4-position switch is only available on the 14-pin versions of SmartCraft.**

The signal from the switch feeds the DTS command module - if the key is ON and the module senses a switch closure at the **START** position when the engine is not running, it will initiate cranking. If the engine is already running, it will send a stop command. Pulling the lanyard or turning the key to the **OFF** position will also stop the engine.

The DTS system also incorporates SmartStart - allowing for push button starting. Rather than holding down the Start button or holding the key in the **START** position until the engine actually starts, SmartStart takes control of the process immediately. As soon as the button is pushed or the key is held over, the command module will send a signal to the engine PCM, which will then close the starter slave solenoid. The slave solenoid then closes the starter solenoid and the engine begins to crank. Upon a 3-4 second time-out expiring or the engine reaching 400 rpm, the slave solenoid circuit will be opened and the cranking will stop.

Electronic Shift Control (ESC)

SHIFT ACTUATOR REMOVAL & INSTALLATION

◆ **See Figures 24 and 25**

1. Disconnect the negative battery cable for safety.

■ **The shift actuator is located on the port side of the motor, just behind the bottom cowl.**

2. Locate the ESC shift actuator wiring harness on the pigtail coming out of one end of the actuator. Slide the connector aftward to free it from the mounting bracket, the tag and unplug the harness moving the engine harness out of the way.

3. Disconnect the hitch pin assembly (this means removing the cotter pin at the rear of the hitch pin, then pulling the pin out of the bellcrank and actuator).

4. At the other end of the actuator from the hitch pin, loosen and remove the mounting bolt, then lift the actuator off of the mounting plate.

✳✳ SELOC CAUTION

Do not rotate the piston while the actuator is removed or you will change the adjusted length..

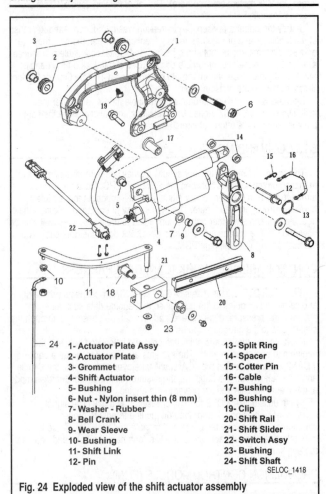

1- Actuator Plate Assy	13- Split Ring
2- Actuator Plate	14- Spacer
3- Grommet	15- Cotter Pin
4- Shift Actuator	16- Cable
5- Bushing	17- Bushing
6- Nut - Nylon insert thin (8 mm)	18- Bushing
7- Washer - Rubber	19- Clip
8- Bell Crank	20- Shift Rail
9- Wear Sleeve	21- Shift Slider
10- Bushing	22- Switch Assy
11- Shift Link	23- Bushing
12- Pin	24- Shift Shaft

SELOC_1418

Fig. 24 Exploded view of the shift actuator assembly

1 - Harness connector	4 - Electronic Shift Control (ESC)
2 - Mounting bracket	5 - ESC mounting screw
3 - Hitch pin assembly	6 - Spacers (2)

SELOC_1418

Fig. 25 Shift actuator mounting

To Install:

■ All new actuators are pre-set at the factory and equipped with plastic ties to ensure no inadvertent movement of the piston. If you are installing a new actuator, simply cut the ties and install the unit. No adjustment is necessary.

5. Position the actuator to the mounting plate while aligning the hitch pin assembly, then install the mounting bolt and tighten to 20 ft. lbs. (27 Nm).

6. Place 1 spacer on either side of the shift piston linkage, then connect the piston to the bellcrank the hitch pin assembly. Secure the hitch pin using the cotter pin.

7. Route the shift actuator wiring and reconnect it to the engine harness, then secure the connector to the mounting bracket.

8. Reconnect the negative battery cable and check the shift actuator adjustment.

ACTUATOR ADJUSTMENT

◆ See Figure 26

■ As noted earlier, all new actuators are pre-set at the factory and equipped with plastic ties to ensure no inadvertent movement of the piston. If you are installing a new actuator, simply cut the ties and install the unit. No adjustment should be necessary.

1. Turn the ignition key to the **ON** position without starting the engine and make sure that the remote control handle is in Neutral.

2. Measure from the center point of the actuator mounting bolt to the center of the hitch pin in the bellcrank.

3. If the measurement is not 7.742-7.822 in. (19.666-19.866mm), remove the unit and rotate the actuator lever piston clockwise to shorten the distance or counterclockwise to lengthen it.

■ The ESC adjustment length is monitored by the PCM and the engine will usually not start if the adjustment is out of specification.

SHIFT OVERRIDE PROCEDURE

◆ See Figure 27

If your vessel has the optional System View display installed and shows an error message reading **System Alarm** along with an active alarm in the system directory that reads **ESCERC POS DIFF** and/or the engine will not start or shift into gear, there is likely a problem with the shift actuator.

The actuator can be disengaged so that rather than stranding the vessel, the drive can be manually shifted into Neutral for starting and then into Forward gear. Engine speed will be limited to 1000-1200 rpm, but at least you'll be able to move the boat.

1. Turn the ignition switch to the **OFF** position.

2. Remove the top cowl for access.

■ The shift actuator is located on the port side of the motor, just behind the bottom cowl.

3. Tag and disconnect the wiring harness at the actuator assembly and move it out of the way.

4. Remove the cotter pin, then free the hitch pin from the bellcrank making sure you retain the 2 spacers (there was 1 on either side of the release pin).

5. Make sure that the remote control handle is in Neutral and the shift lever is Neutral, the later can be determined by making sure the shift lever bellcrank aligns with the mounting bolt for the shift plate just above the actuator.

6. You should now be able to start the engine. Once done, manually move the shift lever into a gear position as shown.

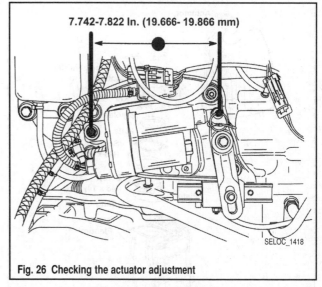

7.742-7.822 In. (19.666- 19.866 mm)

SELOC_1418

Fig. 26 Checking the actuator adjustment

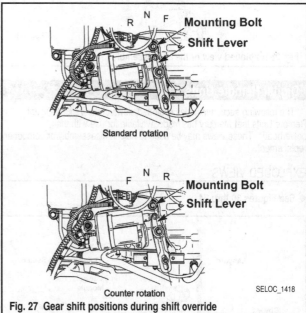

R N F — Mounting Bolt — Shift Lever

Standard rotation

F N R — Mounting Bolt — Shift Lever

Counter rotation

SELOC_1418

Fig. 27 Gear shift positions during shift override

Electronic Throttle Control (ETC)

REMOVAL & INSTALLATION

◆ See Figures 28 and 29

1. Disconnect the negative battery cable for safety.

2. Remove the air intake silencer (attenuator) assembly.

3. Tag and disconnect the electrical harness at the ETC unit.

4. Remove the 4 mounting nuts and lift the ETC unit off of the isolator. There are 2 plates that mount between the isolator and the air plenum.

5. Inspect the isolator and throttle body mating surfaces for any kind of marring or degradation that might affect the seal.

6. Inspect the O-ring on the lower isolator plate - we suggest simply replacing it.

7. For installation make sure the plates and isolator are in position, then align the ETC unit on the mounting studs and install the nuts. Tighten them, alternately, to 65 inch lbs. (7.5 Nm).

8. Reconnect the harness and install the flame arrestor.

9. Reconnect the negative battery cable.

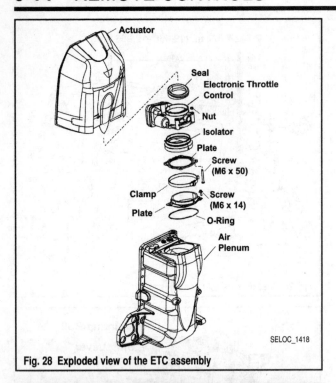

Fig. 28 Exploded view of the ETC assembly

Digital Remote Units

The following section includes exploded views of the various Digital Remote Units that Mercury has made available for use with their powerheads. These views may be used to help in disassembly or component replacement.

EXPLODED VIEWS

◆ See Figures 30 thru 34

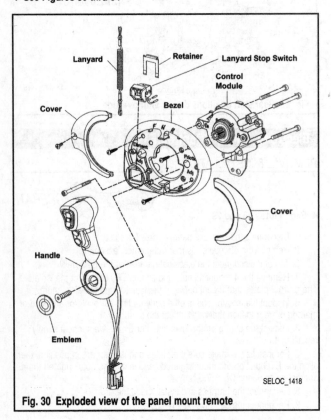

Fig. 30 Exploded view of the panel mount remote

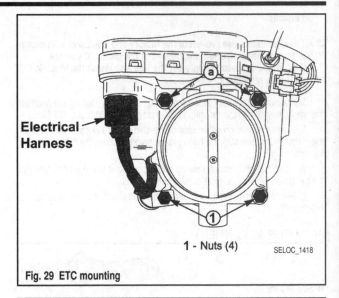

Fig. 29 ETC mounting

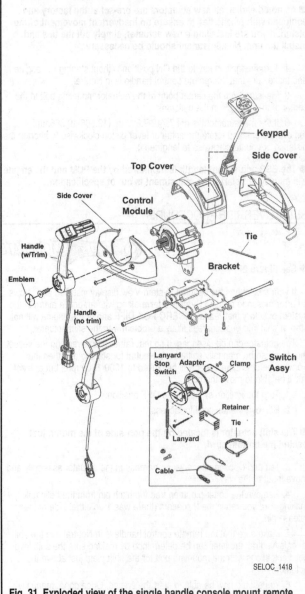

Fig. 31 Exploded view of the single handle console mount remote

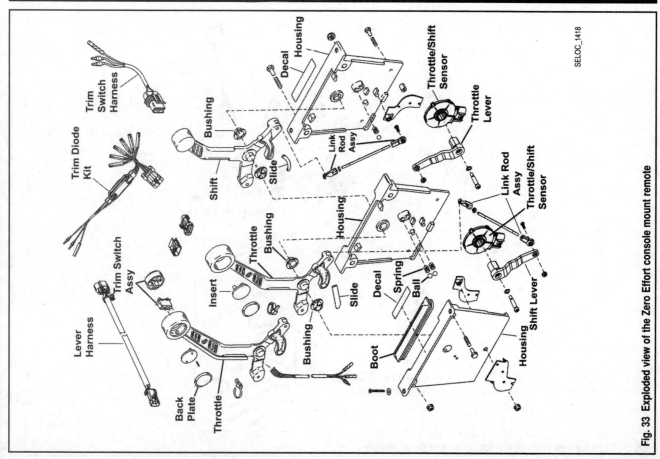

SELOC_1418

Fig. 33 Exploded view of the Zero Effort console mount remote

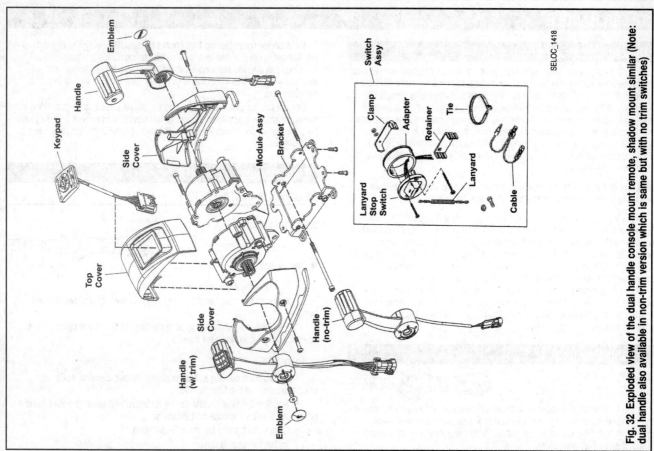

SELOC_1418

Fig. 32 Exploded view of the dual handle console mount remote, shadow mount similar (Note: dual handle also available in non-trim version which is same but with no trim switches)

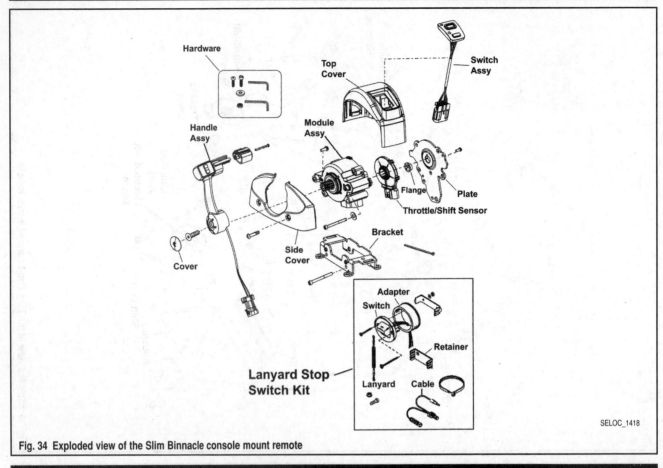

Fig. 34 Exploded view of the Slim Binnacle console mount remote

TILLER HANDLE

Description & Operation

Steering control for most outboards begins at the tiller handle and ends at the propeller. Tiller steering is the simplest form of small outboard control. All components are mounted directly to the engine and are easily serviceable.

For most (but not all models) throttle control is performed via a throttle grip mounted to the tiller arm. As the grip is rotated a cable opens and closes the throttle lever on the engine. An adjustment thumbscrew is usually located near the throttle grip to allow adjustment of the turning resistance. In this way, the operator does not have to keep constant pressure on the grip to maintain engine speed.

■ **A few of the smallest models, like the 2.5/3.3 hp singles, utilize a separate throttle lever instead of a tiller arm twist grip.**

An emergency engine stop switch is used on most outboards to prevent the engine from continuing to run without the operator in control. This switch is controlled by a small clip which keeps the switch open during normal engine operation. When the clip is removed, a spring inside the switch closes it and completes a ground connection to stop the engine. The clip is connected to a lanyard that is worn around the helmsman's wrist.

Some tiller systems utilize a throttle stopper system which limits throttle opening when the shift lever is in neutral and reverse. This prevents over revving the engine under no-load conditions and also limits the engine speed when in reverse.

Troubleshooting

If the tiller steering system seems loose, first check the engine for proper mounting. Ensure the engine is fastened to the transom securely. Next, check the tiller hinge point where it attaches to the engine and tighten the hinge pivot bolt as necessary.

Excessively tight steering that cannot be adjusted using the tension adjustment is usually due to a lack of lubrication. Once the swivel case bushings run dry, the steering shaft will get progressively tighter and eventually seize. This condition is generally caused by a lack of periodic maintenance.

Correct this condition by lubricating the swivel case bushings and working the outboard back and forth to spread the lubricant. However, this may only be a temporary fix. In severe cases, the swivel case bushings may need to be replaced.

Tiller Handle

REMOVAL & INSTALLATION

2.5/3.3 Hp Models

◆ See Figure 35

1. Remove the engine cover.
2. Remove the tiller handle pivot bolt and carefully remove the tiller handle.
3. Remove the mounting collar, mounting rubber, distance collar and special washer from the tiller handle.

To Install:

4. Install the mounting collar, mounting rubber, distance collar and special washer on the tiller handle.
5. Position the tiller handle on the outboard and install the tiller handle pivot bolt. Tighten the pivot bolt securely.
6. Check the handle for smooth operation.
7. Install the engine cover.

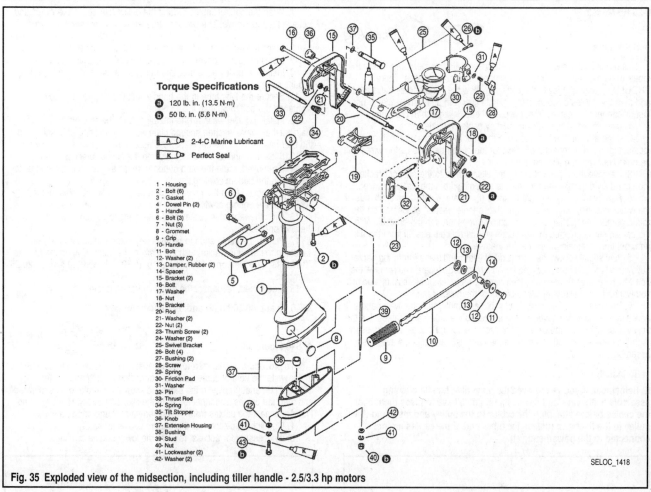

Torque Specifications

ⓐ 120 lb. in. (13.5 N·m)

ⓑ 50 lb. in. (5.6 N·m)

Ⓐ 2-4-C Marine Lubricant

Ⓚ Perfect Seal

1 - Housing
2 - Bolt (6)
3 - Gasket
4 - Dowel Pin (2)
5 - Handle
6 - Bolt (3)
7 - Nut (3)
8 - Grommet
9 - Grip
10- Handle
11- Bolt
12- Washer (2)
13- Damper, Rubber (2)
14- Spacer
15- Bracket (2)
16- Bolt
17- Washer
18- Nut
19- Bracket
20- Rod
21- Washer (2)
22- Nut (2)
23- Thumb Screw (2)
24- Washer (2)
25- Swivel Bracket
26- Bolt (4)
27- Bushing (2)
28- Screw
29- Spring
30- Friction Pad
31- Washer
32- Pin
33- Thrust Rod
34- Spring
35- Tilt Stopper
36- Knob
37- Extension Housing
38- Bushing
39- Stud
40- Nut
41- Lockwasher (2)
42- Washer (2)

SELOC_1418

Fig. 35 Exploded view of the midsection, including tiller handle - 2.5/3.3 hp motors

4/5 Hp Models

◆ **See Figure 36**

1. Loosen the throttle cable set screw and pull the throttle wire from the throttle arm of the carburetor.

2. Remove the cable jacket, bolts and lock washer.

3. Remove the plate.

4. Pull the tiller handle from the bottom cowling. Then remove the inner bushing.

To Install:

5. Install the inner bushing into the bottom cowling. Lubricate the bushing with marine grade lubricant.

6. Install the tiller handle, routing the throttle wire into the bottom cowling.

7. Install the plate over the throttle wire.

8. Secure the tiller handle to the bottom cowling using the plate, lockwashers and bolts. Tighten the bolt to 70 inch lbs. (8Nm).

9. Thread the jacket over the throttle wire and up into the recess of the plate, then route the wire to the carburetor.

10. Rotate the tiller handle twist grip to the idle position (fully clockwise), then place the throttle wire jacket (carburetor end of the wire) into the retaining bracket, threading the wire through the cable retainer.

11. With the carburetor throttle arm against the idle speed screw, carefully pull the slack from the throttle cable and secure into the retainer by tighten the screw.

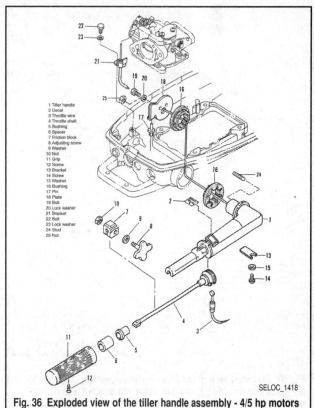

1 Tiller handle
2 Decal
3 Throttle wire
4 Throttle shaft
5 Bushing
6 Spacer
7 Friction block
8 Adjusting screw
9 Washer
10 Nut
11 Grip
12 Screw
13 Bracket
14 Screw
15 Washer
16 Bushing
17 Pin
18 Plate
19 Bolt
20 Lock washer
21 Bracket
22 Bolt
23 Lock washer
24 Stud
25 Nut

SELOC_1418

Fig. 36 Exploded view of the tiller handle assembly - 4/5 hp motors

6-25 Hp Models

◆ **See Figure 37**

This procedure is designed to carefully separate the tiller handle from the motor while the control cables are still connected at both ends. There SHOULD be enough play in the cables to allow the handle to be separated and then the cables disconnected from the tiller arm pulley. However, if not, first disconnect the cables from the powerhead linkage, then proceed.

1. If equipped, disconnect the battery cables for safety.

2. If the stop button is mounted in the tiller rod (as it is on most models), locate, tag and disconnect the bullets for the button harness so the wires can be pulled out with the tiller handle.

3. Remove the 2 bolts securing the tiller handle to the anchor bracket, then carefully separate the tiller handle assembly from the swivel bracket.

4. Remove the bushing, flat washers (there are 2, with 1 on each side of the wave washer), wave washer and tiller washer, all located on the round portion of the tiller handle which protrudes through the bracket. With the cables still connected on both ends you can slide these all from the tiller arm, but they'll remain on the cables themselves.

5. Just slightly down the underside of tiller handle shaft from the cable pulley (and the washers, bushing you just removed), locate and remove the bolt which secures the retainer, then remove the retainer itself. At this point you will be free to slide the tiller tube out of the pulley case.

6. Remove the pulley case (and cable assembly) from the tiller handle. If necessary, remove the cover bolt for access to the pulley and cables. Lift the pulley assembly from the case and replace the control cables, as necessary. With the cables free you can remove the bushing, flat washer, wave washer assembly, as necessary.

To Install:

■ Remember, if you've removed the main tiller handle bushing assembly (with wave and flat washers) you'll have to place them over the cables before attaching the cables to the pulley and installing the pulley in the tiller arm (unless the other end of the cables are not connected to the powerhead yet).

7. Wrap the control cables around the pulley. The top cable wraps and locks in the top groove. The bottom cable wraps and locks in the bottom groove.

8. Place the pulley and cable assembly into the pulley case.

9. Install the pulley cover and secure it with the retaining bolt.

10. Install the pulley assembly into the tiller handle and slide the tiller tube into the pulley.

11. Secure the tiller tube in the handle with the retainer and bolt. Tightened the bolt to 50 inch lbs. (5.6Nm).

12. Install the tiller washer (the tab aligns with the slot in the handle), plain washer, wave washer, second plain washer and flanged bushing into the tiller handle, over the cable/harness assembly.

13. Slide the tiller handle assembly into the anchor bracket.

14. If removed, route the stop button harness through the fuel connector opening in the bottom cowling.

15. If removed from the powerhead, route the control cables through the opening in the bottom cowling.

16. Align the tabs of the inner and outer flanged bushings with the slots in the anchor bracket.

17. Pull on the cable ends to remove any slack and secure the tiller handle to the anchor bracket with the plate and 2 bolts. Tighten the bolts to 70 inch lbs. (7.9 Nm) on 6-15 hp motor, or to 80 inch lbs. (9.0 Nm) on 20-25 hp motors.

18. If equipped, reconnect the negative battery cable.

30/40 hp and 40/50 hp (2- and 3-Cylinder) Models

◆ **See Figure 38**

This procedure is designed to carefully separate the tiller handle from the motor while the control cables are still connected at both ends. There SHOULD be enough play in the cables to allow the handle to be separated and then the cables disconnected from the tiller arm pulley. However, if not, first disconnect the cables from the powerhead linkage, then proceed.

1. If equipped, disconnect the battery cables for safety.

2. If necessary for access, remove the lower engine cowling.

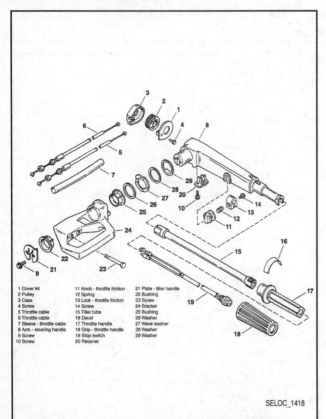

1 Cover kit
2 Pulley
3 Case
4 Screw
5 Throttle cable
6 Throttle cable
7 Sleeve - throttle cable
8 Arm - steering handle
9 Screw
10 Screw
11 Knob - throttle friction
12 Spring
13 Lock - throttle friction
14 Screw
15 Tiller tube
16 Decal
17 Throttle handle
18 Grip - throttle handle
19 Stop switch
20 Retainer
21 Plate - tiller handle
22 Bushing
23 Screw
24 Bracket
25 Bushing
26 Washer
27 Wave washer
28 Washer
29 Washer

SELOC_1418

Fig. 37 Exploded view of the tiller handle assembly - 20/20 Jet/25 hp motors shown (6-15 hp motors same, but bracket is built into swivel head)

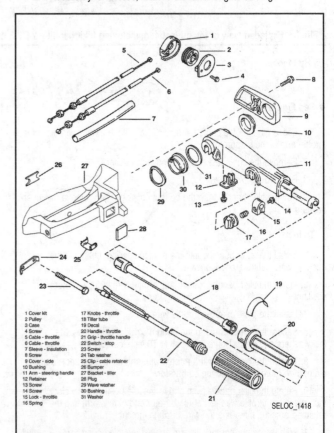

1 Cover kit
2 Pulley
3 Case
4 Screw
5 Cable - throttle
6 Cable - throttle
7 Sleeve - insulation
8 Screw
9 Cover - side
10 Bushing
11 Arm - steering handle
12 Retainer
13 Screw
14 Screw
15 Lock - throttle
16 Spring
17 Knobs - throttle
18 Tiller tube
19 Decal
20 Handle - throttle
21 Grip - throttle handle
22 Switch - stop
23 Screw
24 Tab washer
25 Clip - cable retainer
26 Bumper
27 Bracket - tiller
28 Plug
29 Wave washer
30 Bushing
31 Washer

SELOC_1418

Fig. 38 Exploded view of the tiller handle assembly - 30/40 hp and 40/50 hp (2- and 3-Cylinder) motors

3. If the stop button is mounted in the tiller rod (as it is on most models), locate, tag and disconnect the bullets for the button harness so the wires can be pulled out with the tiller handle.

4. Remove the bolt securing the tiller handle cover (which affectively holds the tiller handle in position by squeezing it between the bushings) to the anchor bracket, then carefully separate the tiller handle assembly from the swivel bracket. The outer bushing may come free with the cover, or it may remain on the tiller handle, either way, note its orientation, then remove it and place it aside.

5. Remove the wave washer, bushing and flat washer from the inside of the tiller arm (facing the steering bracket) or from the bracket itself. With the cables still connected on both ends you can slide these all from the tiller arm, but they'll remain on the cables themselves.

6. Just slightly down the underside of tiller handle shaft from the cable pulley (and the, bushings you just removed), locate and remove the bolt which secures the retainer, then remove the retainer itself. At this point you will be free to slide the tiller tube out of the pulley case.

7. Remove the pulley case (and cable assembly) from the tiller handle. If necessary, remove the cover bolt for access to the pulley and cables. Lift the pulley assembly from the case and replace the control cables, as necessary. With the cables free you can remove the bushing, flat washer, wave washer assembly, as necessary.

To Install:

■ Remember, if you've removed the main tiller handle bushing assembly (with wave and flat washers) you'll have to place them over the cables before attaching the cables to the pulley and installing the pulley in the tiller arm (unless the other end of the cables are not connected to the powerhead yet).

8. Wrap the control cables around the pulley. The top cable wraps and locks in the top groove. The bottom cable wraps and locks in the bottom groove.

9. Place the pulley and cable assembly into the pulley case.

10. Install the pulley cover and secure it with the retaining bolt. Tighten the bolt to 35 inch lbs. (4 Nm).

11. Install the pulley assembly into the tiller handle and slide the tiller tube into the pulley.

12. Secure the tiller tube in the handle with the retainer and bolt. Tighten the bolt to 35 inch lbs. (4 Nm).

13. Install the wave washer, inner bushing and flat washer into the steering bracket and/or tiller handle, over the cable/harness assembly.

14. Slide the tiller handle assembly into the anchor bracket.

15. If removed, route the stop button harness through the fuel connector opening in the bottom cowling.

16. If removed from the powerhead, route the control cables through the opening in the bottom cowling.

17. Install the outer bushing and finally the side cover to secure the bushing and handle assembly. Install and tighten the cover retaining bolt to 130 inch lbs. (14.7 Nm).

18. If removed for access, install the lower engine cowling.

19. If equipped, reconnect the negative battery cable.

55 Hp and Larger Models

◆ See Figure 39

On these models there is no removable pulley at the end of these tiller handles (as used on most 6-50 hp motors) which makes it at least feasible to remove the handle without first disconnecting the cables at the powerhead. On these motors therefore, you must start by at LEAST disconnecting the throttle cable.

This procedure is designed for the removal of both the tiller arm and shift lever, however, you may be able to skip the shift lever steps, leaving the lever in place for some models.

1. If equipped, disconnect the battery cables for safety.

2. For access, remove the lower engine cowling.

3. On 55/60 hp motors, proceed as follows:

a. Remove the nut securing the throttle cable to the engine.

b. Release the latch and remove the throttle cable with grommet from the bottom cowl.

c. Remove the cotter key from the shift rod.

4. On 65 Jet and larger motors, proceed as follows:

a. Remove the nuts securing the shift link rod and throttle cable to the engine. Release the latch and remove the shift link rod and throttle cable from the anchor bracket.

b. Remove the cotter key, washer, bushing and shift link rod from the shift lever.

5. Loosen and remove the screws (usually 4 on 55/60 hp motors or 5 on larger motors) securing the cover to the underside of the tiller handle bracket (just under the run/stop switch). Remove the cover for access to the stop switch wiring.

■ Disconnect the wiring so the tiller handle bracket can be removed for access, depending upon the model.

6. On 55/60 hp motors, disconnect the stop switch wire at the bullet connector and screw securing the ground wire.

7. On 65 Jet and larger motors, proceed as follows to disconnect all the necessary wiring:

a. Disconnect the key switch and tiller stop switch ground leads at the bullet connectors.

b. Remove the nuts securing the key switch and trim switch to the tiller handle bracket.

c. Remove the clip securing the lanyard stop switch and remove the switch from the bracket.

d. Loosen the 2 screws and remove the harness retainer from the motor.

e. Remove the grommet from the tiller bracket.

f. Unplug the connectors from the UP and DOWN PTT relays (on the side of the powerhead, just below the starter motor assembly).

g. Remove the key switch, trim switch and their wiring harnesses from the tiller bracket.

8. Remove the plug from the tiller handle bracket.

9. Remove the bolt and nut from the tiller handle bracket.

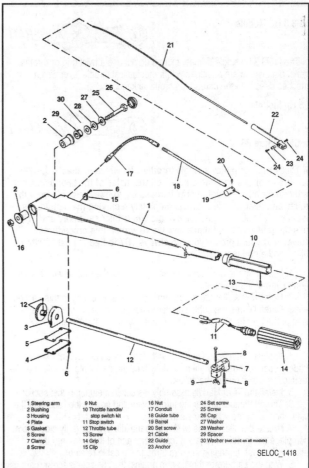

1 Steering arm	9 Nut	16 Nut	24 Set screw
2 Bushing	10 Throttle handle/	17 Conduit	25 Screw
3 Housing	stop switch kit	18 Guide tube	26 Cap
4 Plate	11 Stop switch	19 Barrel	27 Washer
5 Gasket	12 Throttle tube	20 Set screw	28 Washer
6 Screw	13 Screw	21 Cable	29 Spacer
7 Clamp	14 Grip	22 Guide	30 Washer (not used on all models)
8 Screw	15 Clip	23 Anchor	

SELOC_1418

Fig. 39 Exploded view of the tiller handle assembly - 55 hp and larger motors

10. Remove the tiller handle, nylon bushings, stainless steel spacer and flat washers from the bracket.

11. If necessary, bend the tab washer away from the bolt securing the shift lever, then remove the bolt and lever from the bracket.

To Install:

12. If removed, install the shift lever with detent assembly to the bracket. Position the lever, followed by (in this order for 55/60 hp motors) the 2 wave washers, flat washer, tabbed washer and the retaining bolt. On 65 jet and larger motors, position the tabbed washer so it aligns with the slot in the bracket. In all cases tighten the bolt securely, then on 55/60 hp motors, bend the tab in place to secure it.

13. If removed on 65 jet and larger motors, install the spring and detent pin into the tiller handle bracket.

14. Install the tiller handle with bushings to the bracket. Secure in place with the washers, bolt and nut. Tighten the bolt so that it allows tiller handle movement.

15. Tighten the nut to 40 ft. lbs. (54 Nm).

16. Install the plug.

17. Connect the various wiring which was tagged and removed during removal. If necessary, refer again to the removal procedure earlier for the wiring differences, based on model.

18. Install the tiller bracket cover and secure using the retaining screws.

19. Install the cotter key and components to the shift rod. On 65 jet and larger motors connect the shift rod to the lever using the bushing, washer and cotter key.

20. Install and, as necessary, adjust the throttle cable. On 55/60 hp motors, be sure to seat the throttle cable grommet.

21. Install the lower engine cowling.

22. If equipped, reconnect the negative battery cable.

Control Cables

REMOVAL & INSTALLATION

2.5/3.3 Hp Models

The 2.5/3.3 hp single cylinder motors do not use a twist grip or throttle cable, but instead are equipped with a mechanical throttle lever which attaches directly to the carburetor throttle linkage.

4/5 Hp Models

◆ See Figure 36

■ These models utilize a single throttle wire which is attached to the carburetor at one end and a pulley on the inside of the tiller assembly at the other end. Though it MAY be possible to free the cable from the pulley and tiller handle without completely removing the handle, it's usually a lot easier to just remove the whole tiller assembly and THEN take the pulley out and that's how we've written this procedure. However, if you'd prefer, you can always TRY to free the cable with the tiller handle still installed first.

1. Loosen the throttle cable set screw and pull the throttle wire from the throttle arm of the carburetor.

2. Remove the tiller handle assembly, as detailed earlier in this section.

3. Carefully free the cable end barrel from the pulley in the underside base of the tiller handle, then pull the cable out of the throttle shaft.

To Install:

4. Apply a light coating of 2-4-C with Teflon or an equivalent marine grade grease to the pulley, then carefully feed the cable into the tiller handle, seating the barrel end in the pulley.

5. Install the Tiller Handle assembly, as detailed earlier in this section.

6. Thread the jacket over the throttle wire and up into the recess of the plate, then route the wire to the carburetor.

7. Rotate the tiller handle twist grip to the idle position (fully clockwise), then place the throttle wire jacket (carburetor end of the wire) into the retaining bracket, threading the wire through the cable retainer.

8. With the carburetor throttle arm against the idle speed screw, carefully pull the slack from the throttle cable and secure into the retainer by tighten the screw.

6-25 Hp Models

◆ See Figure 37

■ These models utilize a dual throttle cables which are attached to the carburetor linkage at one end and a pulley on the inside of the tiller assembly at the other end. Though it MAY be possible to free the cables from the pulley and tiller handle without completely removing the handle, it's usually a lot easier to just remove the whole tiller assembly and THEN take the pulley out.

1. On all tiller handle *shift* models (meaning, except the few side shift 6-15 hp models), place the tiller handle twist grip in the Neutral position, then remove the throttle link rod (horizontally mounted rod running from the cable pulley/cam on the powerhead to the primary throttle lever) from both the throttle cam and primary throttle lever.

2. Loosen the jam nuts securing the control cables to the anchor bracket.

3. Unwrap and remove the control cables from the cable pulley/cam assembly.

■ If the cables are not being replaced, mark the top cable with a piece of tape to help aid in reassembly.

4. On 20-25 hp models, remove the nut securing the access cover, then remove the cover from the side of the powerhead.

5. Tag and disconnect the Black/Yellow and Black stop button wires.

6. IF you can get the tiller handle in a position where you can remove the retainer bolt and retainer from the underside and slide the throttle shaft forward enough that you can free the cable pulley WITHOUT removing the handle, you're golden, IF NOT, you'll have to follow the procedures found earlier in this section for removing the Tiller Handle assembly.

7. Remove the pulley case (and cable assembly) from the tiller handle. Remove the cover bolt for access to the pulley and cables. Lift the pulley assembly from the case and replace the control cables, as necessary.

To Install:

8. Wrap the control cables around the tiller handle pulley. The top cable wraps and locks in the top groove. The bottom cable wraps and locks in the bottom groove.

9. Place the tiller pulley and cable assembly into the pulley case.

10. Install the tiller pulley cover and secure it with the retaining bolt.

11. Install the tiller pulley assembly into the tiller handle and slide the tiller tube into the pulley.

12. Secure the tiller tube in the handle with the retainer and bolt. Tighten the bolt to 50 inch lbs. (5.6 Nm).

13. If removed, install the tiller handle assembly, as detailed earlier in this section.

14. Route the control cable assembly from the inside of the bottom cowl through to the linkage and bracket on the powerhead.

15. On all tiller handle shift models (meaning, except the few side shift 6-15 hp models), connect the cables to the throttle linkage as follows:

a. Place the tiller handle twist grip in the Reverse position. In this position, route the extended cable over top of the throttle pulley/cam and secure the cable into the groove in the pulley. Place the cable jacket into the top notch of the cable anchor bracket.

b. Next, twist the tiller handle grip to the Forward position. In this position, route the remaining cable (now extended) under the throttle pulley/cam and secure the cable into the groove in the pulley. Place the cable jacket into the lower notch of the cable anchor bracket.

c. On 6-15 hp models, position the tab washers so as to lock the control cables to the holes in the cable anchor bracket.

d. Rotate the tiller handle twist grip to Neutral (and on 20-25 hp motors, then rotate the handle to Slow). Adjust the jam nuts to remove slack from the throttle cables, while still allowing full travel of the linkage and carburetor throttle shutter. After adjustment, verify that Forward, Neutral and Reverse can be selected smoothly while advancing or retarding the throttle. If problems are found, check first for misrouted, kinked, pinched cables or loose attaching bolts, otherwise readjust as necessary.

e. On 20-25 hp motors, install the access cover and secure by tightening the nut and bolt to 50 inch lbs. (5.6 Nm).

16. On side shift versions of the 6-15 hp motors, connect the cables to the throttle linkage as follows:

a. Rotate the tiller handle twist grip to the **FAST** position and place the gear shift lever in the Forward position. In this position, route the extended cable over top of the throttle pulley/cam and secure the cable into the groove in the pulley. Place the cable jacket into the top notch of the cable anchor bracket.

b. Next, twist the tiller handle grip to the **SLOW** position. In this position, route the remaining cable (now extended) under the throttle pulley/cam and secure the cable into the groove in the pulley. Place the cable jacket into the lower notch of the cable anchor bracket.

c. With the tiller handle twist grip still in the **Slow** position, adjust the jam nuts to remove slack from the throttle cables, while still allowing full travel of the linkage and carburetor throttle shutter. If problems are found, check first for misrouted, kinked, pinched cables or loose attaching bolts, otherwise readjust as necessary.

30/40 hp and 40/50 hp (2- and 3-Cylinder) Models

◆ See Figures 38, 40 and 41  MODERATE

■ These models utilize a dual throttle cables which are attached to the carburetor linkage at one end and a pulley on the inside of the tiller assembly at the other end. Though it *may* be possible to free the cables from the pulley and tiller handle without completely removing the handle, it's usually a lot easier to just remove the whole tiller assembly and *then* take the pulley out.

1. Place the twist grip in the Neutral position.
2. Loosen the jam nuts securing the control cables to the anchor bracket.
3. Unwrap and remove the control cables from the cable pulley/cam assembly.

■ IF the cables are not being replaced, mark the cables to help aid in reassembly. Also, if they are not being replaced you may want to make matchmarks on the adjusters to save yourself the time of going through the whole adjustment procedure after installation.

4. IF you can get the tiller handle in a position where you can remove the retainer bolt and retainer from the underside and slide the throttle shaft forward enough that you can free the cable pulley WITHOUT removing the handle, you're golden, IF NOT, you'll have to follow the procedures found earlier in this section for removing the Tiller Handle assembly.
5. Remove the pulley case (and cable assembly) from the tiller handle. Remove the cover bolt for access to the pulley and cables. Lift the pulley assembly from the case and replace the control cables, as necessary.

To Install:

6. Wrap the control cables around the tiller pulley. The top cable wraps and locks in the top groove. The bottom cable wraps and locks in the bottom groove.
7. Place the tiller pulley and cable assembly into the pulley case.
8. Install the tiller pulley cover and secure it with the retaining bolt. Tighten the bolt to 35 inch lbs. (4 Nm).
9. Install the tiller pulley assembly into the tiller handle and slide the tiller tube into the pulley.
10. Secure the tiller tube in the handle with the retainer and bolt. Tighten the bolt to 35 inch lbs. (4 Nm).
11. If removed, install the tiller handle assembly, as detailed earlier in this section.
12. Route the control cable assembly from the inside of the bottom cowl through to the linkage and bracket on the powerhead. Place the cables on the pulley and position the adjusters in the cable brackets.
13. Now at this point if you're reinstalling cables which were properly adjusted you can simple reset them to the matchmarks made earlier, however, if no marks were made OR you have replaced the cable, proceed as follows for proper adjustment:

a. Make sure the gear shift is in the Neutral position and the motor is off, then loosen the cam follower screw (Philips head screw threaded through the follower tab just above and to the left of the pulley, to the left of the roller).
b. Next, back off the idle speed screw until the throttle shutter positioner does NOT touch the taper of the idle speed screw and the throttle plates are fully closed.
c. Now with the throttle at idle, place the cam follower roller in gentle contact with the throttle cam, centering the roller with the raised mark on the cam by adjusting the position of the throttle cable sleeves in the mounting

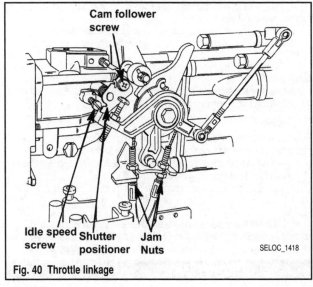

Fig. 40 Throttle linkage

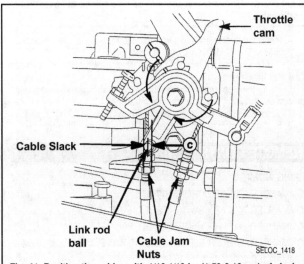

Fig. 41 Position the cables with 1/16-1/18 in. (1.59-3.18mm) of slack to prevent binding

bracket. With the cam follower resting on the cam, tighten the cam follower screw again.

■ When positioning the throttle cables, you must leave 1/16-1/8 in. (1.59-3.18mm) of slack to prevent the cables from binding. Rock the throttle cam side-to-side and measure the amount of throttle cam travel at the link rod ball.

d. Once the cables are properly positioned, tighten the cable jam nuts to hold them securely.

55 Hp & Larger Models

 MODERATE

◆ See Figure 39

■ These models utilize a single throttle cable with cable end and barrel assembly that runs through a conduit that passes through the cowling and connects into the innards of the tiller handle. The best way to service the cable is to completely remove the tiller handle assembly from the motor, then take it apart for cable replacement.

1. Remove the tiller handle assembly (with throttle cable) from the motor, as detailed earlier in this section.
2. After removing the tiller handle, use a flat tip screw driver to gently pry/push the rubber grip off the tiller handle.
3. Remove the screw from the twist grip.

4. Cut the strap securing the stop switch harness and remove the screw from the harness J-clip.

5. Remove the stop switch and twist grip from the tiller handle.

6. Remove the throttle cable anchor screws and remove the cable guide.

7. Remove the tiny Allen screw from the brass barrel and remove the barrel.

8. Unscrew the stainless conduit from the tiller handle.

9. Pull the throttle cable from the tiller handle.

To Install:

10. Insert the throttle cable (curved end facing up) into the tiller handle gear assembly while rotating the tiller arm counterclockwise.

11. Retract the throttle cable into the gear assembly until approximately 17 inches (43cm) extends from the tiller arm.

12. Slide the stainless steel conduit over the throttle cable and thread it into the tiller arm until it is lightly seated. Rotate the conduit counterclockwise 1 full turn from a lightly seated position.

13. Slide the brass barrel over the throttle cable tube and position the barrel to face towards the tiller handle. Secure the barrel to the tube with the Allen screw approximately 3.5 inches (89mm) from the stainless conduit.

✳✳ SELOC CAUTION

Do not overtighten the screw as it may crush the tube and bind the throttle cable.

14. Install the throttle cable guide onto the throttle cable with the anchor and 2 screws. The guide hold should face up (out).

15. Position the throttle arm slot to face the stop switch harness exit hole in the tiller handle.

16. Route the stop switch harness through the twist grip, into the throttle arm and out through the side of the tiller handle.

17. Secure the twist grip to the throttle arm with the attaching screw.

18. Secure the stop switch harness to the throttle arm with a wire tie.

■ **Allow enough slack in the harness (rotate the throttle grip in both directions) before securing the harness to the handle assembly.**

19. Attach the harness to the tiller arm with the J-clip, allowing enough slack in the harness for full throttle operation.

20. Install the rubber cover on the twist grip by aligning the ridges on the plastic twist grip with the grooves inside the rubber grip. Applying soapy water to the rubber grip will ease installation.

21. Install the tiller handle assembly.

22. Rotate the throttle twist grip fully clockwise to the idle position.

23. Back out the set screw from the throttle cable barrel until 2 or 3 threads of the set screw are exposed.

24. Route the cable around the powerhead and place the barrel near the receptacle guide.

25. Place the throttle cable on the peg of the throttle lever, then secure it using a washer and the locknut. Tighten the nut until snug, then back off 1/4 turn.

26. Apply a small amount of Loctite® 271 onto the exposed threads of the throttle cable barrel set screw. (Do not tighten the screw quite yet). Position the barrel into the upper hole of the barrel receptacle, facing so you can reach the Allen head screw.

✳✳ SELOC CAUTION

Do not exceed 1/8th turn on the set screw after it has bottomed out.

27. With the throttle lever held lightly against the stop and the twist grip at idle, turn the set screw of the throttle cable barrel until it bottoms out on the tube, then tighten the screw an additional 1/8 turn.

28. Secure the barrel receptacle using the barrel retainer.

29. Check the preload on the throttle cable by placing a thin piece of paper between the idle stop screw and idle stop. Preload is correctly set when the paper can be remove without tearing, but having some drag. If the paper is too loose or too tight, readjust the throttle cable barrel if necessary.

REMOTE CONTROL WIRING

WIRING INDEX

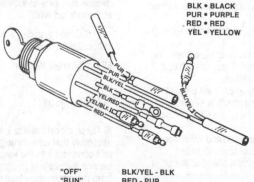

```
BLK • BLACK
PUR • PURPLE
RED • RED
YEL • YELLOW
```

"OFF"	BLK/YEL - BLK
"RUN"	RED - PUR
"START"	RED - PUR - YEL/RED
PUSH (CHOKE)*	RED - YEL/BLK

*Key switch must be positioned to "RUN" OR "START" and key pushed in to actuate choke, for this continuity test.

SELOC_1418

Wiring Diagram - Typical Mercury Commander 2000 Key/Choke Switch

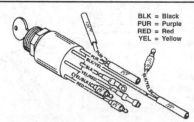

BLK = Black
PUR = Purple
RED = Red
YEL = Yellow

KEY POSITION	CONTINUITY SHOULD BE INDICATED AT THE FOLLOWING POINTS:					
	BLK	BLK/YEL	RED	YEL/RED	PUR	YEL/BLK
OFF	o—	—o				
RUN			o—	—	—o	
START			o—	—o		
				o—	—o	
			o—	—	—o	
CHOKE*			o—	—		—o
			o—	—	—o	
					o—	—o

*Key switch must be positioned to "RUN" or "START" and key pushed in to actuate choke for this test.

NOTE: If meter readings are other than specified in the preceding tests, verify that switch and not wiring is faulty. If wiring checks ok, replace switch.

SELOC_1418

Wiring Diagram and Continuity Test Chart for Commander 2000 Key Switches

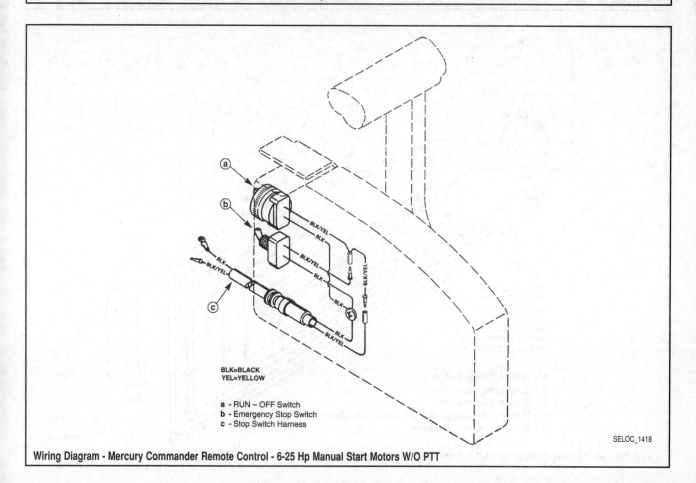

BLK=BLACK
YEL=YELLOW

a - RUN – OFF Switch
b - Emergency Stop Switch
c - Stop Switch Harness

SELOC_1418

Wiring Diagram - Mercury Commander Remote Control - 6-25 Hp Manual Start Motors W/O PTT

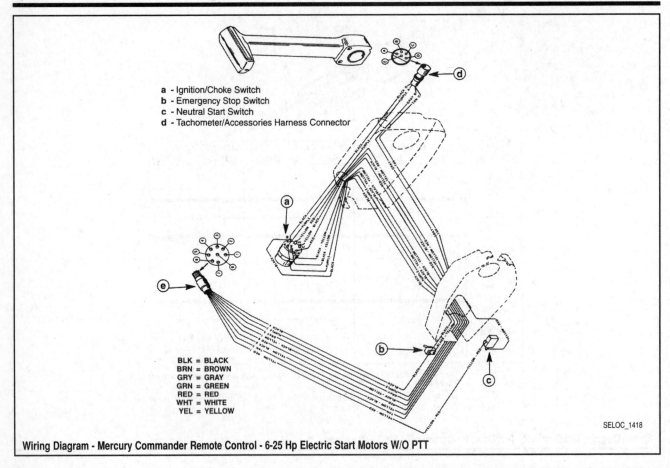

a - Ignition/Choke Switch
b - Emergency Stop Switch
c - Neutral Start Switch
d - Tachometer/Accessories Harness Connector

BLK = BLACK
BRN = BROWN
GRY = GRAY
GRN = GREEN
RED = RED
WHT = WHITE
YEL = YELLOW

SELOC_1418

Wiring Diagram - Mercury Commander Remote Control - 6-25 Hp Electric Start Motors W/O PTT

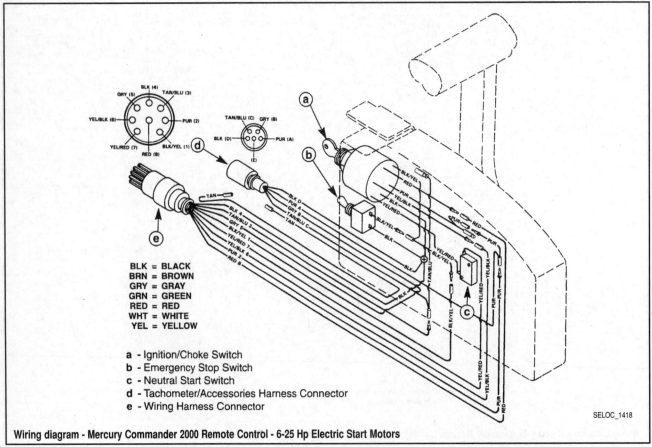

BLK = BLACK
BRN = BROWN
GRY = GRAY
GRN = GREEN
RED = RED
WHT = WHITE
YEL = YELLOW

a - Ignition/Choke Switch
b - Emergency Stop Switch
c - Neutral Start Switch
d - Tachometer/Accessories Harness Connector
e - Wiring Harness Connector

SELOC_1418

Wiring diagram - Mercury Commander 2000 Remote Control - 6-25 Hp Electric Start Motors

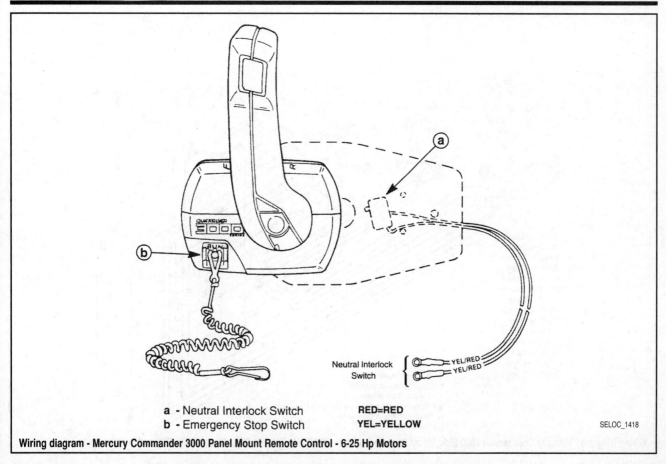

a - Neutral Interlock Switch
b - Emergency Stop Switch

Neutral Interlock Switch } YEL/RED, YEL/RED

RED=RED
YEL=YELLOW

SELOC_1418

Wiring diagram - Mercury Commander 3000 Panel Mount Remote Control - 6-25 Hp Motors

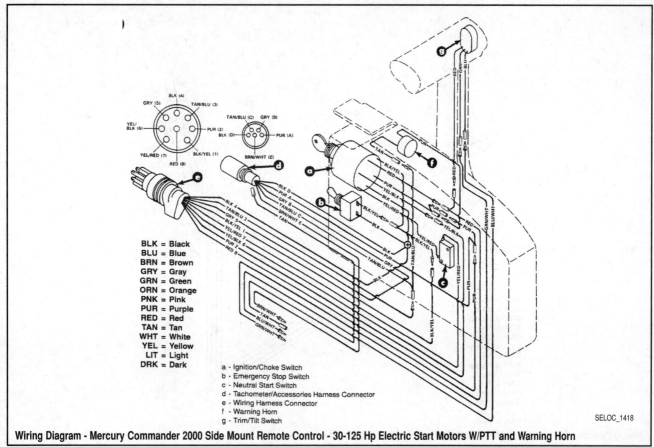

BLK = Black
BLU = Blue
BRN = Brown
GRY = Gray
GRN = Green
ORN = Orange
PNK = Pink
PUR = Purple
RED = Red
TAN = Tan
WHT = White
YEL = Yellow
LIT = Light
DRK = Dark

a - Ignition/Choke Switch
b - Emergency Stop Switch
c - Neutral Start Switch
d - Tachometer/Accessories Harness Connector
e - Wiring Harness Connector
f - Warning Horn
g - Trim/Tilt Switch

SELOC_1418

Wiring Diagram - Mercury Commander 2000 Side Mount Remote Control - 30-125 Hp Electric Start Motors W/PTT and Warning Horn

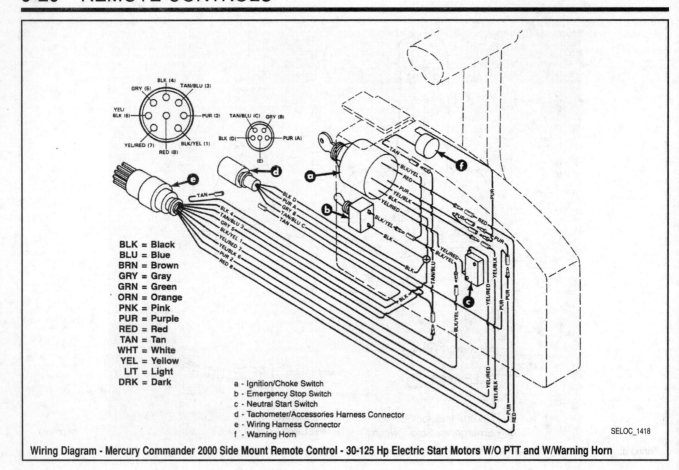

BLK = Black
BLU = Blue
BRN = Brown
GRY = Gray
GRN = Green
ORN = Orange
PNK = Pink
PUR = Purple
RED = Red
TAN = Tan
WHT = White
YEL = Yellow
LIT = Light
DRK = Dark

a - Ignition/Choke Switch
b - Emergency Stop Switch
c - Neutral Start Switch
d - Tachometer/Accessories Harness Connector
e - Wiring Harness Connector
f - Warning Horn

SELOC_1418

Wiring Diagram - Mercury Commander 2000 Side Mount Remote Control - 30-125 Hp Electric Start Motors W/O PTT and W/Warning Horn

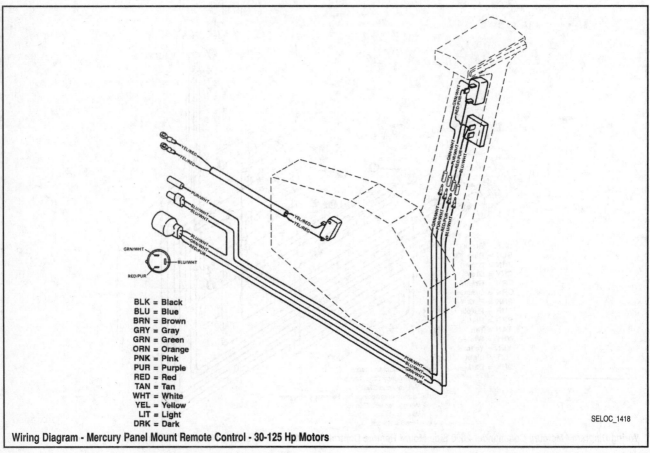

BLK = Black
BLU = Blue
BRN = Brown
GRY = Gray
GRN = Green
ORN = Orange
PNK = Pink
PUR = Purple
RED = Red
TAN = Tan
WHT = White
YEL = Yellow
LIT = Light
DRK = Dark

SELOC_1418

Wiring Diagram - Mercury Panel Mount Remote Control - 30-125 Hp Motors

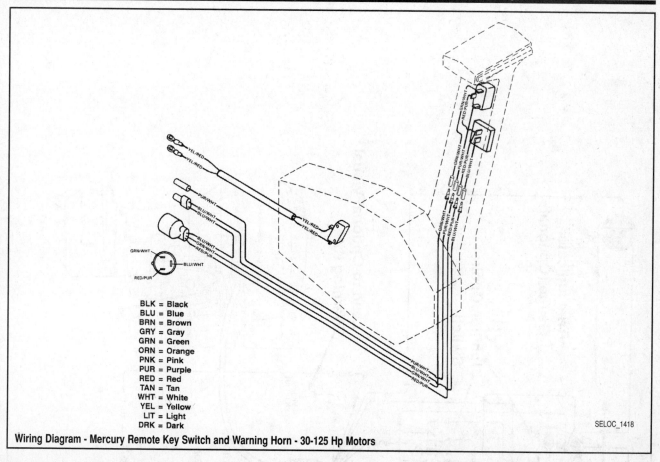

BLK = Black
BLU = Blue
BRN = Brown
GRY = Gray
GRN = Green
ORN = Orange
PNK = Pink
PUR = Purple
RED = Red
TAN = Tan
WHT = White
YEL = Yellow
LIT = Light
DRK = Dark

SELOC_1418

Wiring Diagram - Mercury Remote Key Switch and Warning Horn - 30-125 Hp Motors

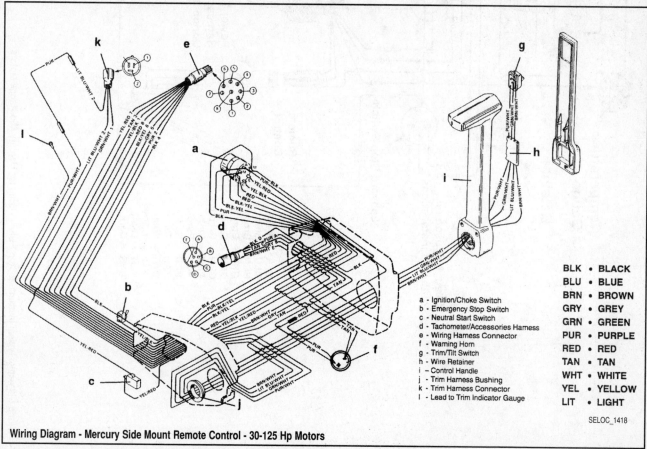

a - Ignition/Choke Switch
b - Emergency Stop Switch
c - Neutral Start Switch
d - Tachometer/Accessories Harness
e - Wiring Harness Connector
f - Warning Horn
g - Trim/Tilt Switch
h - Wire Retainer
i - Control Handle
j - Trim Harness Bushing
k - Trim Harness Connector
l - Lead to Trim Indicator Gauge

BLK • BLACK
BLU • BLUE
BRN • BROWN
GRY • GREY
GRN • GREEN
PUR • PURPLE
RED • RED
TAN • TAN
WHT • WHITE
YEL • YELLOW
LIT • LIGHT

SELOC_1418

Wiring Diagram - Mercury Side Mount Remote Control - 30-125 Hp Motors

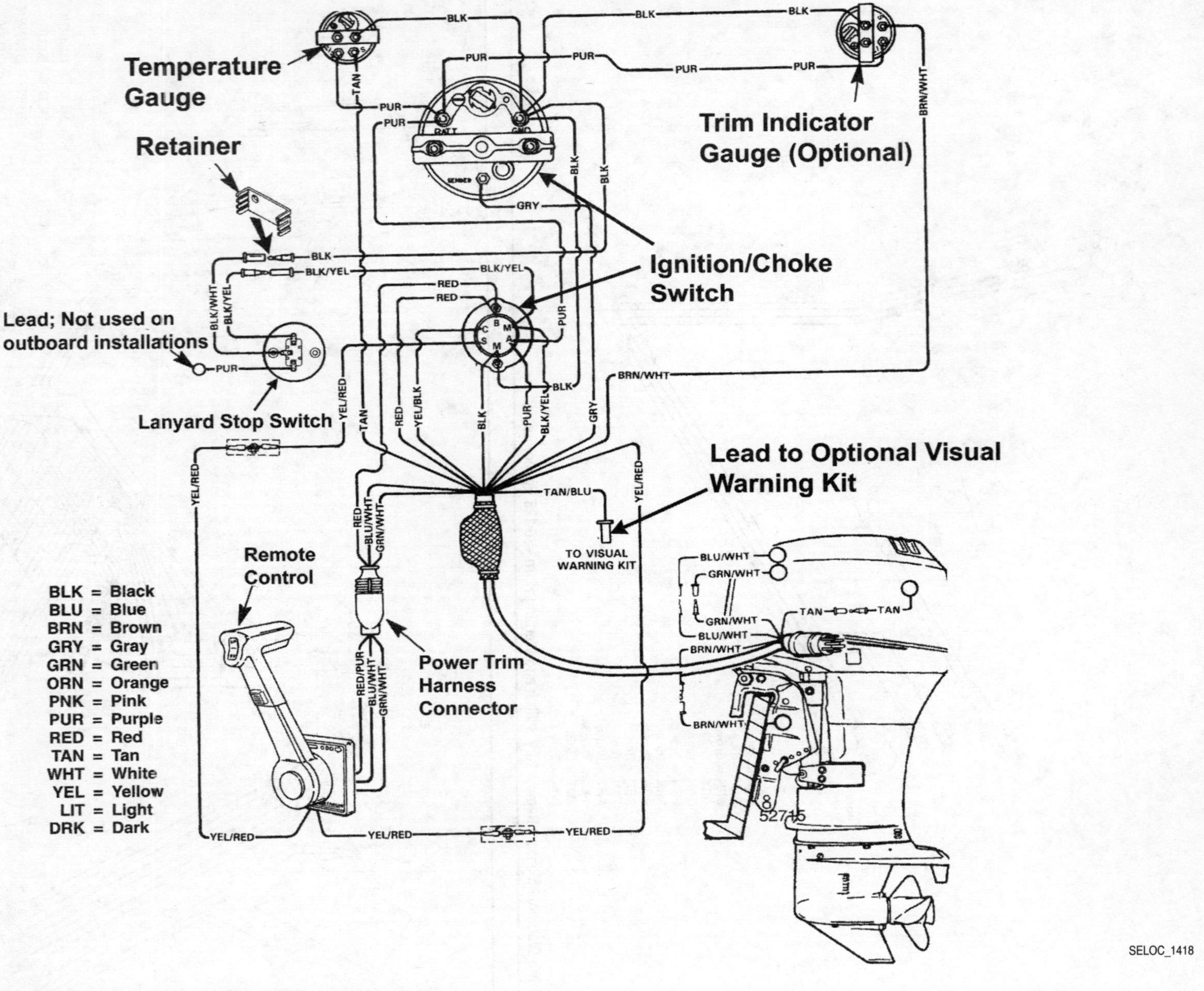

SELOC_1418

Trim Indicator Gauge (Optional)

Ignition/Choke Switch

Lead to Optional Visual Warning Kit

Temperature Gauge

Retainer

Lead; Not used on outboard installations

Lanyard Stop Switch

Remote Control

Power Trim Harness Connector

TO VISUAL WARNING KIT

BLK = Black
BLU = Blue
BRN = Brown
GRY = Gray
GRN = Green
ORN = Orange
PNK = Pink
PUR = Purple
RED = Red
TAN = Tan
WHT = White
YEL = Yellow
LIT = Light
DRK = Dark

Wiring Diagram - Typical Instrument/Stop Switch Wiring for Mercury Mid-Range Motors

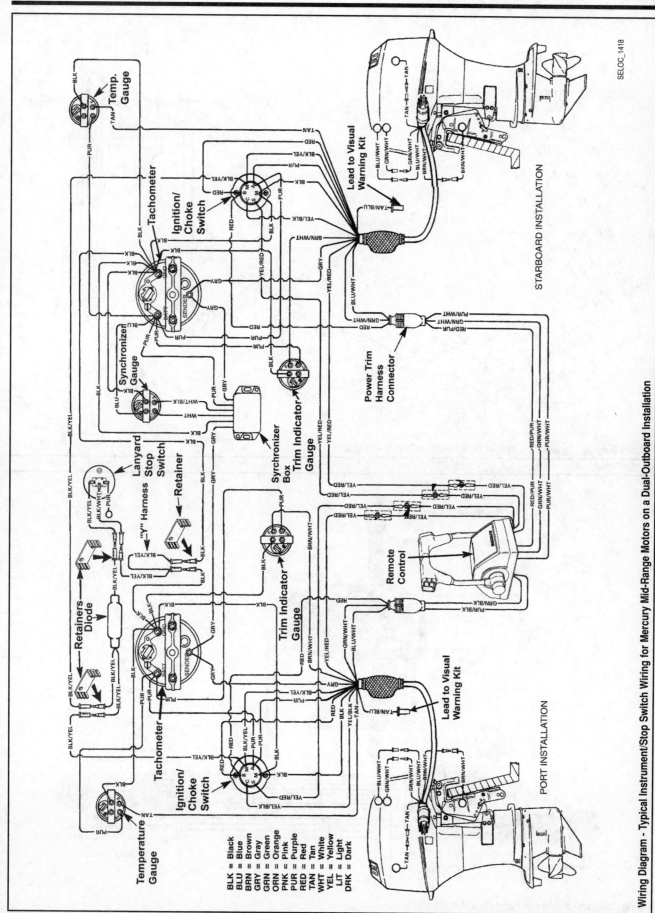

SELOC_1418

STARBOARD INSTALLATION

PORT INSTALLATION

BLK = Black
BLU = Blue
BRN = Brown
GRY = Gray
GRN = Green
ORN = Orange
PNK = Pink
PUR = Purple
RED = Red
TAN = Tan
WHT = White
YEL = Yellow
LIT = Light
DRK = Dark

Wiring Diagram - Typical Instrument/Stop Switch Wiring for Mercury Mid-Range Motors on a Dual-Outboard Installation

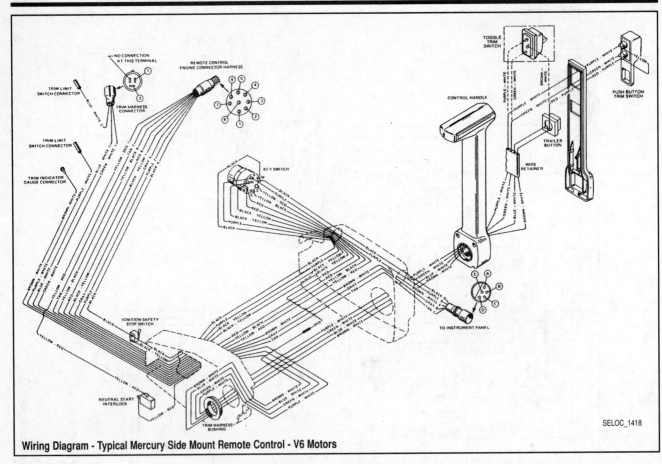

Wiring Diagram - Typical Mercury Side Mount Remote Control - V6 Motors

SELOC_1418

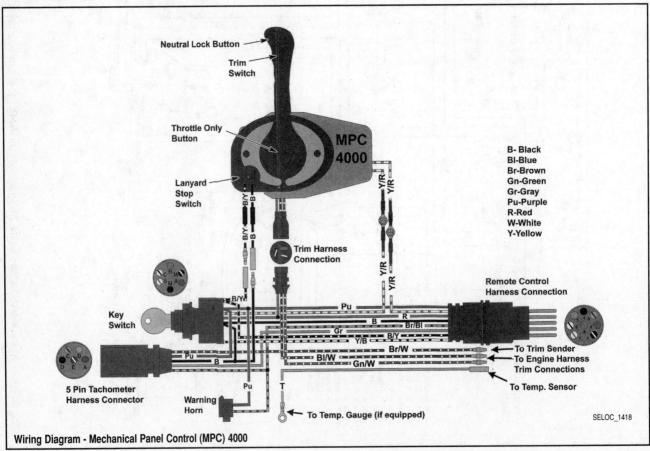

Wiring Diagram - Mechanical Panel Control (MPC) 4000

SELOC_1418

10

HAND REWIND STARTER

HAND REWIND STARTER

Description and Operation

The main components of a hand rewind starter (recoil starter) are the cover, roper, rewind spring and pawl arrangement. Pulling the rope rotates the pulley, winds the spring and activates the pawl into engagement with the starter hub at the top of the flywheel. Once the pawl engages the hub, the powerhead is spun as the rope unwinds from the pulley.

Releasing the rope on rewind starter moves the pawl out of mesh with the hub. The powerful clock-type spring recoils the pulley in the reverse direction to rewind the rope to the original position.

Newer design manual starters employ the same principles of inertia found in electric starter motors. A nylon pinion gear slides and engages the flywheel ring gear as the starter rope is pulled. At the same time, the rewind spring is tightened so there starter rope can be recoiled.

A predetermined ratio designed into the starter provides maximum cranking speed for easy starts and easy pulls. Some models have a cam follower that prevents activation of the manual starter if the shift lever is not in the neutral position. This system is commonly referred to as a Neutral Start Interlock system.

Troubleshooting The Hand Rewind Starter

Repair on hand rewind starter units is generally confined to rope, pawl, nylon pinion gear (on inertia starters) and occasionally spring replacement.

✳✳ SELOC CAUTION

When replacing the recoil starter spring extreme caution must be used. The spring is under tension and can be dangerous if not released properly.

Starters which use friction springs to assist pawl action may suffer from bent springs. This will cause the amount of friction exerted to not be correct and the pawl will not be moved into engagement.

Models equipped with a neutral start interlock system may experience a no-start condition due to a misadjusted interlock cable. The hand rewind starter should only function when the shift handle is in the **NEUTRAL** position.

2.5/3.3 Hp Models

REMOVAL & DISASSEMBLY

◆ **See Figures 1 thru 10**

1. Remove the powerhead cowling (which is normally done by releasing the two snaps, one on each side, and then lifting the cowling free).
2. Remove the three bolts securing the rewind starter assembly to the powerhead, then remove the starter assembly.

To Disassemble:

■ **The pull rope is secured in the handle recess with a figure "8" knot tied close to the end. Excess rope beyond the knot can be cut off. Burn the end of a non-fiber rope with a match to prevent it from unraveling.**

3. Turn the starter housing over with the sheave facing up. Pull on the handle to gain some slack in the rope; hold the sheave from rewinding; and then remove the handle by untying the not and carefully pulling the rope out of the handle. After the handle is removed, ease the grip on the sheave and allow the sheave to slowly, but completely rewind with the rope inside the sheave.
4. Carefully snap the circlip out of the groove in the center shaft, then lift the spacer off the friction plate and free of the center shaft.
5. Lift the friction plate slightly and snap the friction spring out of the hole in the center shaft. Remove the friction plate. The return spring may come with the plate. Remove the spring cover, and then the friction spring.
6. Remove the ratchet from the top of the sheave.

✳✳ SELOC CAUTION

The rewind spring is a potential hazard. The spring is under tremendous tension when it is wound. If the spring should accidentally be released, severe personal injury could result from being struck by the spring with force. Therefore, the following two steps must be performed with care to prevent personal injury to self and others in the area.

7. Very carefully "rock" the sheave and at the same time lift the sheave about 1/2 in. (13mm). The spring will disengage from the sheave and remain in the housing.
8. Now, IF you're replacing the spring slowly turn the housing over and gently place it on the floor with the spring facing the floor. Tap the top of the housing with a mallet and the spring will fall free of the housing and partially unwind almost instantly and with considerable force, but it will be contained within the housing. Tilt the container with the opening away from you. The spring will be released from the housing and unwind rapidly. Turn the housing over and unhook the end of the spring from the peg in the housing.
9. If you're replacing the rope, untie the knot in the end of the old starter rope and pull the rope free of the sheave.

CLEANING & INSPECTION

Wash all parts except the rope and the handle in solvent, and then blow them dry with compressed air.

Remove any trace of corrosion and wipe all metal parts with an oil dampened cloth.

Inspect the rope. Replace the rope if it appears to be weak or frayed. If the rope is frayed, check the holes through which the rope passes for rough edges or burrs. Remove the rough edges or burrs with a file and polish the surface until it is smooth.

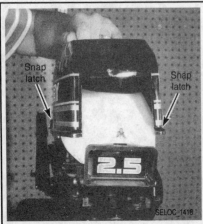
Fig. 1 Remove the cowling for access...

Fig. 2 ...then unbolt and remove the starter assembly

Fig. 3 Turn the starter over for access

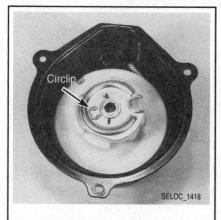

Fig. 4 Carefully remove the circlip. . .

Fig. 5 . . .then remove the spacer followed by. . .

Fig. 6 . . .the friction plate, return spring, spring cover, friction spring. . .

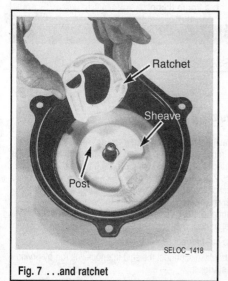

Fig. 7 . . .and ratchet

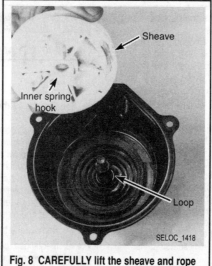

Fig. 8 CAREFULLY lift the sheave and rope

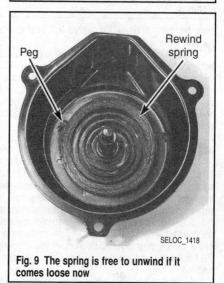

Fig. 9 The spring is free to unwind if it comes loose now

Inspect the starter spring end hooks. Replace the spring if it is weak, corroded or cracked. Inspect the tab on the spring retainer plate. This tab is inserted into the inner loop of the spring. Therefore, be sure it is straight and solid. Inspect the inside surface of the sheave rewind recess for grooves or roughness. Grooves may cause erratic rewinding of the starter rope.

ASSEMBLY & INSTALLATION

◆ See Figures 11, 12 and 13

Wear a good pair of heavy gloves while winding and installing the spring. The spring will develop tension and the edges of the spring steel are extremely sharp. The gloves will help prevent cuts to the hands and fingers.

✳✳ SELOC WARNING

It is strongly recommended a pair of safety goggles or a face shield be worn while the spring is being installed. As the work progresses over a large steel spring is wound into a *SMALL* circumference. If the spring is accidentally released, it will lash out with tremendous ferocity and very likely could cause personal injury to the installer or other persons nearby.

■ Replacement springs normally come pre-coiled and secured by tabs. Be sure to properly position and anchor these springs in the housing before releasing the tabs.

1. Apply a light coating of 2-4-C with Teflon or an equivalent marine grade grease to the inside rewind spring surface in the starter housing. Also,

Fig. 10 If necessary, replace the rope

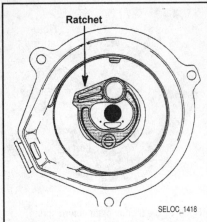

Fig. 11 Position the ratchet properly on the sheave

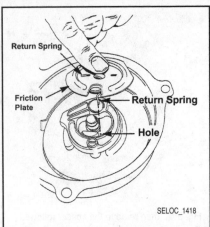

Fig. 12 Aligning the return spring into the friction plate

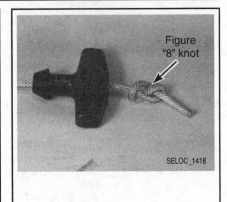

Fig. 13 Secure the rope in the handle using a figure 8 knot

once the spring is in position, apply a light coating of this grease to the outer surface of the spring itself.

2. If you find yourself with the need to hand coil a spring for installation, hold the spring in a coil in one gloved hand, then with the other hand, hook the looped end of the spring over the peg in the housing. Feed the spring clockwise into the housing. The easiest way to accomplish this task is to hold the spring in one gloved hand and with the other gloved hand, rotate the housing counterclockwise. Continue working the spring until it is all confined within the housing.

3. If you are simply reinstalling a pre-wound spring, place outer hooked end of the spring over the catch and position the spring into the housing.

4. If the rope is being replaced, feed one end of a new rope through the hole in the sheave, and then tie a figure "8" knot in the end of the rope. Pull on the rope until the knot is confined inside the sheave recess. Wind the entire pull rope counterclockwise around the sheave (when viewed from the side of the sheave which will be facing down toward the flywheel).

5. Slowly and carefully lower the sheave down over the center shaft of the housing. As the sheave goes into the housing the hook on the lower face of the sheave must index into the loop at the inner end of the spring. Once the hook is indexed into the spring, the sheave may be fully seated in the housing.

6. Position the ratchet in place on the sheave, with the flat side of the ratchet against the sheave and the round hole indexed over the post.

7. Slide the friction spring down the center shaft. This spring serves as a spacer and exerts an upward pressure on the friction plate. Install the spring cover on top of the spring.

■ Two holes are located in the sheave on the same side of the center shaft. With the holes on the side of the shaft facing you, one end of the friction spring must index into the right hole.

8. Hook one end of the return spring into the slot of the friction plate. Now, lower the friction plate onto the center shaft and index the free end of the spring into the right hole in the sheave.

9. Place the spacer over the hole on the friction plate.

10. Push down on the friction plate, and then snap the circlip into the groove on the center shaft to secure the plate and associated parts in place.

11. Place the pull rope into the notch in the sheave. With the rope in the notch, rotate the sheave three complete turns counter-clockwise. Hold tension on the rope and at the same time feed the free end of the rope through the rope guide in the starter housing. Continue to hold tension on the sheave for the next step.

12. Feed the free end of the rope through the handle and tie a figure "8" knot in the end. But, be sure to leave a reserve length of rope about 1/4-1/2 in. (6.3-12.7mm) in length.

13. Pull the rope back into the handle recess. Relax the tension on the sheave and allow the sheave to rewind until the handle is against the starter housing.

To install:

14. Install the rewind hand starter assembly onto the powerhead and secure it in place with the three attaching bolts. Tighten the bolts securely.

15. Slowly and carefully crank the powerhead with the hand rewind starter and check the completed work.

16. Install the cowling down over the powerhead and secure it in place with the two snap retainers.

4/5 Hp Models

REMOVAL & DISASSEMBLY

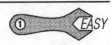

◆ See Figures 14 thru 24

There are two potential designs for the starter assemblies used on these motors. Type I (shown in most of the photos and the exploded view) contains an oval-shaped pawl attached to 3 pawl springs. Type II (shown in the exploded view) contains 2 separate ratchet tabs which operate with separate return springs and a friction plate. Some service procedures are similar between the two, which is why we've combined them here, however disassembly steps relating to those components below the sheave will usually differ and be separated in the procedures.

1. Remove the top cowl for access.

2. If equipped, carefully disconnect the shift interlock link rod by prying it free of the cam on the starter.

3. Remove the three bolts securing the starter legs to the powerhead.

4. Remove the starter and place it upside down on a suitable work surface.

To Disassemble:

5. Slowly and carefully pull out on the starter rope until it is fully unwound, then hold the sheave in this position (a firm grip should be able to accomplish this, but if necessary you may be able to use locking pliers or a C-clamp and a block of wood (to protect the housing) to accomplish this too. The next step is a heck of a lot easier if you can use two hands.

6. With the rope fully would out, carefully free the knot from one end (usually the sheave end is easiest). Untie or cut the knot, then slide the rope and handle from the starter assembly and put aside. If the rope or handle is to be replaced, free the knot from the handle end, then untie or cut the end and remove it from the handle.

7. Slowly, while you control the rotation, allow the sheave to unwind, releasing the spring tension.

8. For Type I units, disassemble the sheave and starter drive components as follows:

a. Carefully pry the circlip from the center of the pawl post using a narrow slotted screwdriver.

b. Lift the pawl, with the 3 springs attached free of the sheave.

c. Remove the nut (top side) bolt and washer (bottom side) from the center of the housing and sheave.

d. Remove the sheave bushing from the sheave.

✱✱ SELOC CAUTION

Wear a good pair of heavy gloves and safety glasses while performing the following tasks.

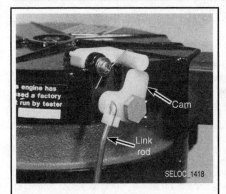

Fig. 14 If equipped, disconnect the interlock link rod

Fig. 15 Remove the 3 starter retaining bolts...

Fig. 16 ...then lift the assembly from the powerhead

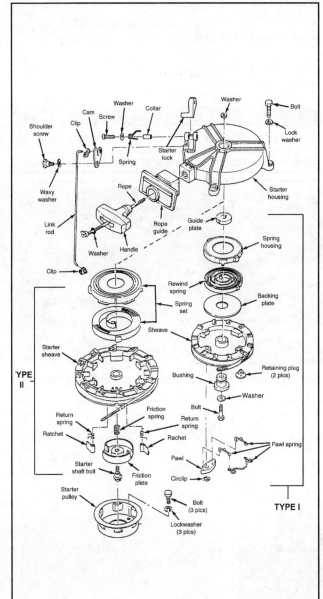

Fig. 17 Exploded view of the hand rewind starter assembly - 4/5 hp motors

SELOC_1418

✳✳ SELOC WARNING

The rewind spring is a potential hazard. The spring is under tremendous tension when it is wound. If the spring should accidentally be released, severe personal injury could result from being struck by the spring with force. Therefore, the following steps must be performed with care to prevent personal injury to self and others in the area. Do not attempt to remove the spring unless it is unfit for service and a new spring is to be installed.

e. Carefully lift out the sheave (along with the spring assembly contained between the sheave and the spring housing) from the starter housing assembly.

f. Pry out the two retainer plugs from the side of the sheave which normally faces the flywheel.

g. From the spring housing side of the sheave, lift out the tabbed guide plate from the center of the spring housing.

h. With the spring housing facing downward, rotate the sheave in a clockwise direction, while holding the spring housing stationary until the two locking tabs disengage.

9. Very carefully lift out the spring housing containing the coiled spring from the sheave and, if necessary, remove the backing plate.

10. On Type II units, disassemble the sheave and starter drive components as follows:

a. Remove the bolt from the center of the sheave, then carefully lift the sheave and spring assembly from the starter housing

b. From the flywheel side of the sheave, remove the friction plate, friction spring, 2 ratchets and 2 ratchet return springs.

✳✳ SELOC CAUTION

Wear a good pair of heavy gloves and safety glasses while performing the following tasks.

✳✳ SELOC WARNING

The rewind spring is a potential hazard. The spring is under tremendous tension when it is wound. If the spring should accidentally be released, severe personal injury could result from being struck by the spring with force. Therefore, the following steps must be performed with care to prevent personal injury to self and others in the area. Do not attempt to remove the spring unless it is unfit for service and a new spring is to be installed.

c. If necessary carefully lift the spring housing and coiled spring out of the starter sheave.

11. If the spring must be replaced, there are multiple ways to release it from the spring housing. Mercury recommends that you carefully unwind the coiled spring by hand. However, many techs prefer to jar it loose and let it unwind on the floor. To do the later, Obtain two pieces of wood, a short 2 in. x 4 in. (5cm x 10cm) will work fine. Place the two pieces of wood approximately 8 in. (20 cm) apart on the floor. Center the housing on top of the wood with the spring side facing down. Check to be sure the wood is not touching the spring. Stand behind the wood, keeping away from the

openings as the spring unwinds with considerable force. Tap the sheave with a soft mallet. The spring retainer plate will drop down releasing the spring. The spring will fall and unwind almost instantly and with force.

12. If necessary, remove the rope guide from the starter housing.

■ **If the starter interlock components require service, the large bolt is normally Loctited in place and requires a heat source to loosen.**

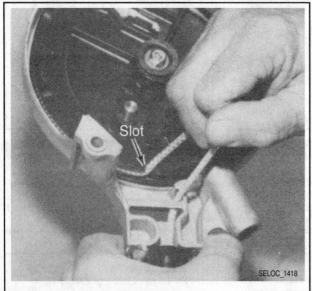

Fig. 18 You can manually unwind the rope. . .

CLEANING & INSPECTION

◆ **See Figure 17**

1. Wash all parts except the rope and the handle in solvent, and then blow them dry with compressed air.
2. Remove any trace of corrosion and wipe all metal parts with an oil-dampened cloth.
3. Inspect the rope. Replace the rope if it appears to be weak or frayed. If the rope is frayed, check the holes through which the rope passes for rough edges or burrs. Remove the rough edges or burrs with a file and polish the surface until it is smooth.
4. Inspect the starter spring end hooks. Replace the spring if it is weak, corroded or cracked.
5. Inspect the inside surface of the sheave rewind recess for grooves or roughness. Grooves may cause erratic rewinding of the starter rope.
6. Coat the entire length of the used rewind spring (a new spring will be coated with lubricant from the package), with low-temperature lubricant.

ASSEMBLY & INSTALLATION

◆ **See Figures 17, 25, 26 and 27**

1. If removed, install the starter interlock components. Be sure to apply a coating of Loctite® 271 or equivalent threadlocking compound to the bolt threads, then tighten securely.
2. If removed, install the rope guide.

✳✳ SELOC CAUTION

Wear a good pair of gloves while winding and installing the spring. The spring will develop tension and the edges of the spring steel are extremely sharp. The gloves will prevent cuts to the hands and fingers.

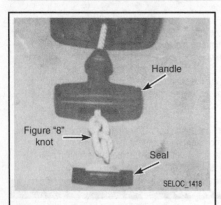

Fig. 19 . . .then release and replace it without further disassembly

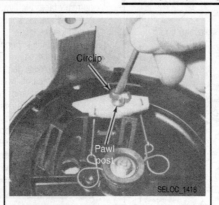

Fig. 20 To disassemble a Type I, start with the circlip. . .

Fig. 21 . . .then remove the pawl and spring assembly

Fig. 22 Remove the nut, bolt and washer. . .

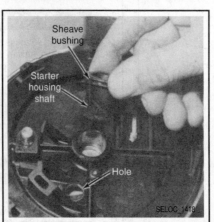

Fig. 23 . . .and sheave bushing, then the sheave is free

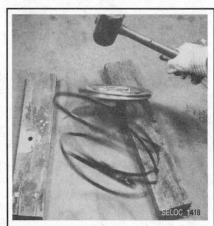

Fig. 24 One spring release method

3. When installing a new spring, apply a light coating of Quicksilver multi-purpose lubricant or equivalent anti-seize lubricant to the inside surface of both the spring and starter housings (where the spring will make contact and travel in use).

4. A new spring will normally be wound and held in a steel hoop. Hook the outer end of the new spring onto the starter housing post, and then place the spring inside the housing. Carefully remove the steel hoop. The spring should unwind slightly and seat itself in the housing.

5. When installing an old spring wind the old spring loosely in one hand in a clockwise direction. Hook the outer end of the spring onto the slot in the removable housing. Rotate the sheave clockwise and at the same time feed the spring into the housing in a counterclockwise direction. Continue working the spring into the housing until the entire length has been confined. Or conversely, you can hook the spring over the housing, then slowly wind the unwound spring clockwise, directly into the housing.

6. Assemble the spring, sheave and drive components on Type II designs as follows:

a. Install the ratchets and pawl springs into the underside of the starter sheave, positioning the ratchets so they are parallel to each other and the center bolt mounting hole, each ratchet facing inward.

b. Install the sheave and spring housing into the starter assembly housing.

c. Install the friction spring and friction plate into the starter sheave. Lubricate the friction plate with a small amount of 2-4-C with Teflon or equivalent marine lubricant.

d. Apply a light coating of Loctite® 271 or equivalent threadlocking compound to the center bolt threads, then install and tighten the bolt to 70 inch lbs.

7. Begin assembling the spring and sheave on Type I designs as follows:

a. Install the backing plate over the sheave hub. Position the spring housing over the sheave, with the two tabs on the housing aligned with the two grooves in the sheave.

b. Rotate spring housing counterclockwise to engage the two locking tabs against the grooves. Insert the guide plate into the sheave hub with the tab of the plate entering the inner loop of the spring. Insert the retainers into the slots next to the locking tabs of the spring housing.

c. Turn the sheave over and using a pair of needle nose pliers, pull the two ends of the retainers through until seated.

■ **For all models, start installation of the rope at this point (though the final method for winding the spring and rope will differ between the 2 types.**

8. Melt the tips of the rope to prevent fraying. Insert one melted end into the starter handle and the other into through the hole in the starter sheave. Tie a figure "8" knot in the each end of the rope leaving about one inch (2.5 cm) beyond the knot. Tuck the end of the rope beyond the knot into the groove, if so equipped, next to the knot.

9. Wind the rope and spring on Type II designs as follows:

a. Turn the sheave 2 1/2 turns CLOCKWISE as the rope coils into the sheave, then HOLD the rope in the slot while turning the sheave 3 turns COUNTERCLOCKWISE, then allow the sheave to slowly rewind the remainder of the loose rope.

b. Pull the starter handle slowly to check for smooth operation of the ratchet pawls, free movement of the starter rope and that the starter rope rewinds completely.

c. The Type II design is now ready for installation.

10. Wind the rope and spring on Type I designs, and install the remainder of the sheave drive components as follows:

a. While facing the flywheel side of the sheave, wind the rope in a COUNTERCLOCKWISE direction two turns around the sheave, ending and placing it in the slot/notch in the sheave.

b. Carefully lower the sheave into the starter housing. At the same time make sure the tab on the guide plate slides into the hole of the starter housing. Keep the rope in the notch while doing this.

c. Apply a light coating of 2-4-C with Teflon or equivalent lubricant to the outer diameter of the bushing, then slide the sheave bushing into the sheave and starter housing shaft.

d. First install the bold (along with the washer) through the bottom of the sheave and thread it into the housing, tightening it to 70 inch lbs. (8 Nm). Next, coat the exposed threads of the bolt with Loctite® 271 or equivalent, then install the nuts. While holding the bolt from turning, tighten the nut to 70 inch lbs. (8 Nm).

e. Holding the starter rope in the notch in the side of the sheave, wind the sheave COUNTERCLOCKWISE against spring tension. Hold the sheave from turning and pull out any remaining slack in the rope via the starter handle, then hold the handle tightly and slowly release the tension on the sheave allowing it to rewind clockwise drawing the remaining rope into the housing.

f. With the beveled end of the pawl facing inward toward the springs (and center bolt when installed) hook each end of the pawl spring into the two small holes in the pawl, from the underneath side of the pawl. The short ends of the spring will then be on the upper surface of the pawl. Move the spring up against the center of the sheave shaft, and then slide the center of the pawl onto the pawl post.

■ **Before installation it is a good idea to lubricate the pawl peg of the sheave and the groove bushing with 2-4-C with Teflon or an equivalent marine grade grease.**

g. Carefully snap the circlip into place over the pawl post to secure the pawl in place.

h. Check the action of the rewind starter before further installation work proceeds. Pull out the starter rope with the handle, then allow the spring to slowly rewind the rope. The starter should rewind smoothly and take up all the rope to lightly seat the handle against the starter housing.

To install:

11. Position the rewind starter in place on the powerhead.

■ **The no-start-in-gear protection system cannot normally be adjusted. If the system fails to prevent the sheave from rotating in any shift position except neutral, each component should be thoroughly inspected. The part should be replaced if there is any evidence of excessive wear or distortion.**

12. Secure the starter legs to the powerhead with the bolts, and tighten them to a torque value of 70 inch lbs. (8 Nm).

13. If equipped, snap the shift interlock link rod into the plastic cam on the starter housing.

■ **If the link rod adjustment at the shift lever and inside the starter housing was undisturbed, the no-start-in-gear protection system should perform satisfactorily. When the unit is not in neutral, the starter lock should drop down to block the sheave and prevent it from rotating. This means an attempt to pull on the rope with the lower unit in any gear except neutral should fail.**

Fig. 25 Tie a figure 8 knot to secure the handle...

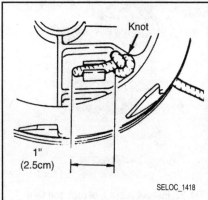

Fig. 26 ...and sheave end of the rope

Fig. 27 Installing a type I pawl and spring assembly

14. If the no-start-in-gear protection system fails to function properly, first remove the rewind hand starter from the powerhead. Inspect the blocking surface of the starter lock, the condition of the cam and the return action of the spring. Because no adjustment of the length of the link rod is possible, make sure the lower end of the rod is indeed connected to the shift mechanism below the bottom cowling and the rod itself is not binding or bent.

15. Install the cowling assembly.

6-25 Hp Models

REMOVAL & ROPE REPLACEMENT

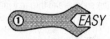

◆ **See Figures 28 thru 35**

1. Remove the top cowling for access.

✳✳ SELOC CAUTION

DO NOT turn or cock the fuel filter assembly when removing it from the starter mounting, pull STRAIGHT downward. Failure to heel this may cause the fuel line connection on the filter to break off.

2. Insert a narrow blade screwdriver between the fuel filter and the lower edge of the hand starter housing. Exert a slight downward pressure on the screwdriver and at the same time pull downward on the filter with the other hand. The fuel filter should "pop" free. Move the filter and connecting hoses to one side out of the way.

3. Unsnap the interlock link rod free of the lower lock lever.

4. Remove the three attaching bolts securing the hand rewind starter to the powerhead.

5. Lift the starter housing up and free of the powerhead.

To remove the rope:

6. Disengage the starter interlock from the rear of the starter sheave.

7. On 20-25 hp motors, remove the rope guide support.

8. Pull about 1 inch (30cm) of rope out of the starter housing and tie a loose knot in the rope to prevent the rope from rewinding back into the housing. Do not tighten the knot because it will be necessary to untie the knot with one hand later.

9. Remove the handle retainer (from the outer center of the handle); feed the rope out through the handle; untie (or cut) the knot; then pull the handle free of the rope.

10. Hold the sheave in position against spring tension and untie the knot you used earlier to secure the rope. To remove the rope completely at this stage, turn the sheave COUNTERCLOCKWISE until it stops indicating the spring is wound tight, then turn it slowly back CLOCKWISE *JUST* until the knot recess aligns with the rope hole. In this position free the rope knot from the sheave and pull out the rest of the rope.

11. If you're just replacing the rope, skip ahead to Rope Replacement and Installation. If however, you're overhauling the entire starter assembly, proceed with Disassembly. In either case, you're still holding the sheave against spring pressure, DO NOT allow it to release suddenly.

Fig. 28 Carefully pull downward to unsnap the fuel filter. . .

Fig. 29 . . .then free the interlock linkage

Fig. 30 Remove the 3 retaining bolts. . .

Fig. 31 . . .then lift the assembly from the powerhead

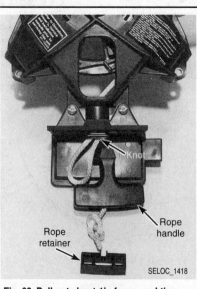

Fig. 32 Pull out about 1' of rope and tie a knot to hold it. . .

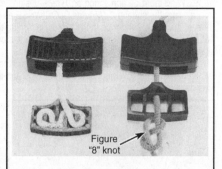

Fig. 33 . . .then free the rope from the starter handle

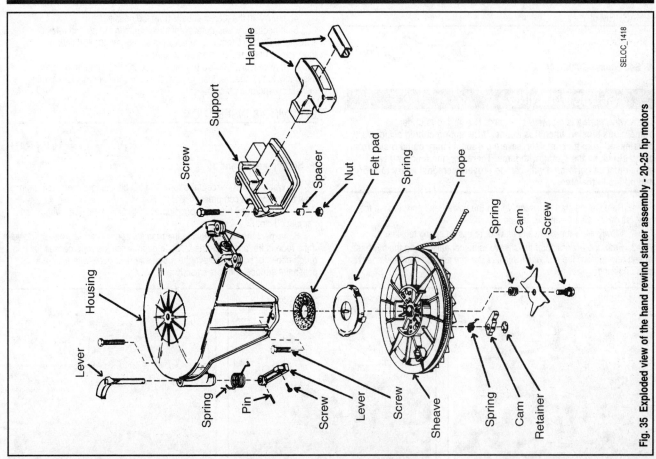

Fig. 35 Exploded view of the hand rewind starter assembly - 20-25 hp motors

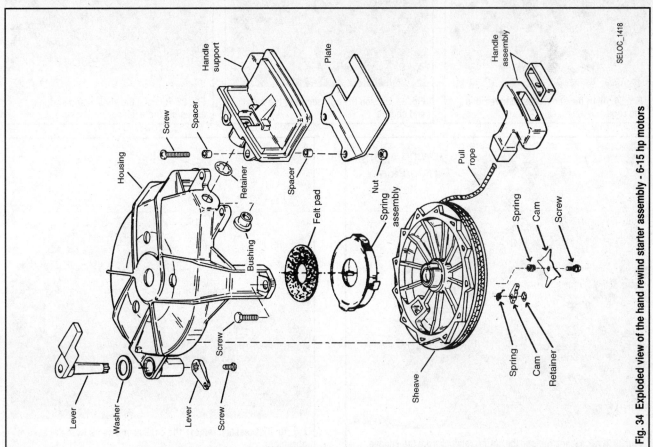

Fig. 34 Exploded view of the hand rewind starter assembly - 6-15 hp motors

DISASSEMBLY

◆ See Figures 34 thru 40

✳✳ SELOC WARNING

The rewind spring is a potential hazard. The spring is under tremendous tension when it is wound. If the spring should accidentally be released from its container, severe personal injury could result from being struck by the spring with force. Therefore, the following two steps must be performed with care to prevent personal injury to self and others in the area.

1. Remove the starter assembly from the powerhead and remove the rope, as detailed earlier in this section.
2. Slowly allow the sheave to unwind, releasing spring tension.
3. From the underside of the sheave (side normally facing the flywheel), loosen and remove the bolt which secures the star shaped cam and sheave to the housing.

4. Gently lift the starter sheave from the housing.
5. If necessary, separate the sheave from the spring and retainer housing. DO NOT attempt to remove the spring from the retainer housing, as they are normally replaced as an assembly.
6. If necessary, remove the circlip securing each cam and spring assembly to the sheave, then remove the cams and springs for cleaning, inspection and/or replacement.

CLEANING & INSPECTION

◆ See Figures 34 and 35

1. Wash all parts except the rope and the handle in solvent, and then blow them dry with compressed air.
2. Remove any trace of corrosion and wipe all metal parts with an oil-dampened cloth.
3. Inspect the rope. Replace the rope if it appears to be weak or frayed. If the rope is frayed, check the holes through which the rope passes for rough edges or burrs. Remove the rough edges or burrs with a file and polish the surface until it is smooth.

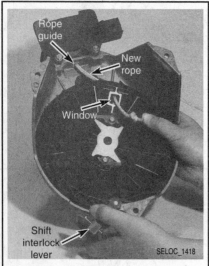

Fig. 36 Align the knot recess and remove the rope

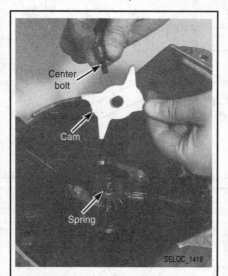

Fig. 37 Loosen the bolt securing the cam and sheave

Fig. 38 Remove the sheave and spring assembly (note felt pad)

Fig. 39 If necessary, remove the spring/retainer from the sheave

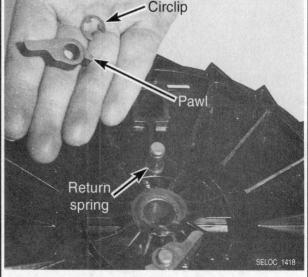

Fig. 40 If necessary, remove the circlips and cams from the sheave assembly

4. Inspect the starter spring end hooks. Replace the spring if it is weak, corroded or cracked. Inspect the tab on the spring retainer plate. This tab is inserted into the inner loop of the spring. Therefore, be sure it is straight and solid. Inspect the inside surface of the sheave rewind recess for grooves or roughness. Grooves may cause erratic rewinding of the starter rope.

■ **If starter operation was erratic or excessively noisy, check the starter clutch for damage from lack of lubrication. If necessary, replace the complete sheave assembly with a pre-lubricated starter clutch installed.**

ASSEMBLY

◆ **See Figures 34 and 35**

1. If removed, install each cam and spring assembly to the sheave using a circlip. The star-shaped inner cam will be installed along with the sheave right before the rope is installed.

2. If removed, install the felt pad to the starter housing.

3. Apply a light coating of 2-4-C with Teflon to the rewind spring assembly, as well as the cam tension spring which mounts in the center of the housing shaft.

4. Install the rewind spring assembly (and retainer housing) into the starter housing, positioning the inner end of the spring in the notch provided on the boss at the center of the starter housing.

5. Next, position the sheave into the starter housing while aligning the notches in the sheave with the appropriate notches on the outside (180 degrees apart from each other) of the recoil spring assembly.

■ **If for some reason the sheave won't seat fully, try to inspect the recoil spring assembly. On some models you should be able to peer through the lock lever access hole in the housing to help make sure the sheave and spring retainer are in proper alignment.**

6. Install the star shaped cam and secure, along with the sheave, to the housing using the retaining bolt. Tighten the bolt to 70 inch lbs. (7.9 Nm).

■ **After tightening the bolt, make sure the sheave moves freely. If not, the alignment of the sheave and recoil spring is in question.**

7. Apply a light coating of 2-4-C with Teflon or an equivalent marine grade grease to the contact surfaces to the edge of the cam that contacts both cam pawls.

8. Install the starter rope and the starter assembly.

ROPE REPLACEMENT & INSTALLATION

◆ **See Figures 34 and 35**

1. If the rope is being replaced, prepare a new length of suitable rope. On 6-15 hp motors the rope should be 66 in. (167.6 cm) in overall length. On 20-25 hp motors the rope should be 62 in. (157.5 cm) in overall length. Be sure to burn/melt the ends to keep them from unwrapping.

2. If removed from the handle, push one end of the rope through the handle and make a figure 8 knot, then position the knot in the retainer and seat the retainer back in the handle.

3. If the spring tension was released (during overhaul or because the assembly was put aside while preparing a new rope), carefully wind the sheave up by hand turning it COUNTERCLOCKWISE until it stops.

4. Now turn the sheave CLOCKWISE with spring tension AT LEAST 1 turn plus whatever is necessary to align the knot recess with the rope hole in the starter housing. Holding the sheave in this position, push the new rope through the rope holes in the starter housing and sheave.

5. Tie a figure 8 knot in the end of the rope and pull the knot back into the knot recess in the sheave.

6. Hold tension on the rope and gently ease your grasp on the sheave, allowing the sheave to rotate and slowly draw the rewound spring into the starter sheave.

7. If removed on 20-25 hp motors, install the rope guide support.

To install:

8. Slide the starter assembly down over the crankshaft and into position on the powerhead.

9. Secure the starter with the attaching hardware. Tighten the bolts alternately and evenly to 70 inch lbs. (7.9 Nm) on 6-15 hp motors or to 90 inch lbs. (10 Nm) on 20-25 hp motors.

10. Snap the interlock link rod into the lower lock lever.

11. Snap the clear plastic fuel filter into place on the bracket on the starter housing (again, making sure NOT to cock or turn the filter).

12. Check operation of the starter assembly. Install the top cowl.

30 Hp & Larger Models

REMOVAL & DISASSEMBLY

◆ **See Figures 41, 42 and 43**

On these models the rope guide support is mounted semi-independently from the starter housing assembly. As such, Mercury tends to recommend removing the handle from the rope in order to remove the starter housing while leaving the rope guide behind. We've included that procedure here, as it is handy when removing the assembly for rope replacement or overhaul. *However,* if you are not planning on replacing any part of the assembly you may wish to skip that step and instead remove the rope guide and rope along with the starter as an assembly.

1. Remove the top cowling for access.

2. Pull the starter rope out from the handle rest about 2 ft. (60 cm), then tied an overhand knot between the starter housing at the handle rest. This will hold the rope from retracting into the housing assembly once the handle is removed. Then, free the rope retainer from the end of the handle, remove the knot and untie (or cut it if the rope is being replaced). Slide the handle from the rope.

■ **If you make a matchmark between the starter interlock cable and the retainer bracket before removal, you may ease or skip adjustment during installation.**

3. Locate the interlock linkage and cable connection at the top end of the cable (on the starter housing). Remove the retaining clip, then unthread the screw which secures the cable to the housing. Disconnect the cable at this end.

4. Remove the 4 retaining bolts, then lift the starter housing assembly from the motor.

To disassemble:

5. Invert the starter housing on a suitable worksurface, then unbolt and remove the interlock cam retainer (2 or 3 bolts depending on the model). Remove the starter interlock cam lever, cam lever spring and interlock lever.

✳✳ SELOC WARNING

The rewind spring is a potential hazard. The spring is under tremendous tension when it is wound. If the spring should accidentally be released from its container, severe personal injury could result from being struck by the spring with force. Therefore, the following 2 steps must be performed with care to prevent personal injury to self and others in the area.

6. Untie the knot which is holding the starter rope, then slowly allow the sheave to unwind, releasing spring tension (and drawing the rope all the way or most of the way into the housing).

7. From the underside of the sheave (side normally facing the flywheel), loosen and remove the bolt which secures the star shaped cam and sheave to the housing.

8. Gently lift the starter sheave from the housing.

9. If necessary, separate the sheave from the spring and retainer housing. DO NOT attempt to remove the spring from the retainer housing, as they are normally replaced as an assembly.

10. If necessary, remove the circlip securing each cam pawl and spring assembly to the sheave, then remove the cams and springs for cleaning, inspection and/or replacement.

11. If you're replacing the rope, free the knot from the recess in the sheave, then unwind and pull the rope out of the assembly.

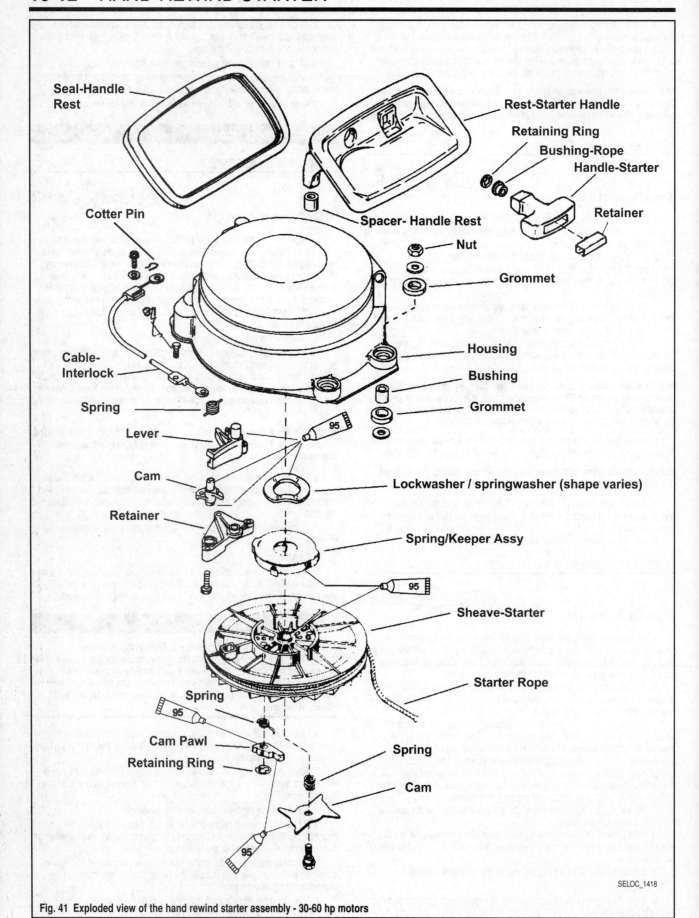

Seal-Handle Rest

Rest-Starter Handle

Retaining Ring

Bushing-Rope

Handle-Starter

Retainer

Cotter Pin

Spacer- Handle Rest

Nut

Grommet

Housing

Cable-Interlock

Bushing

Grommet

Spring

Lever

95

Cam

Lockwasher / springwasher (shape varies)

Retainer

Spring/Keeper Assy

95

Sheave-Starter

Starter Rope

Spring

95

Cam Pawl

Spring

Retaining Ring

Cam

95

SELOC_1418

Fig. 41 Exploded view of the hand rewind starter assembly - 30-60 hp motors

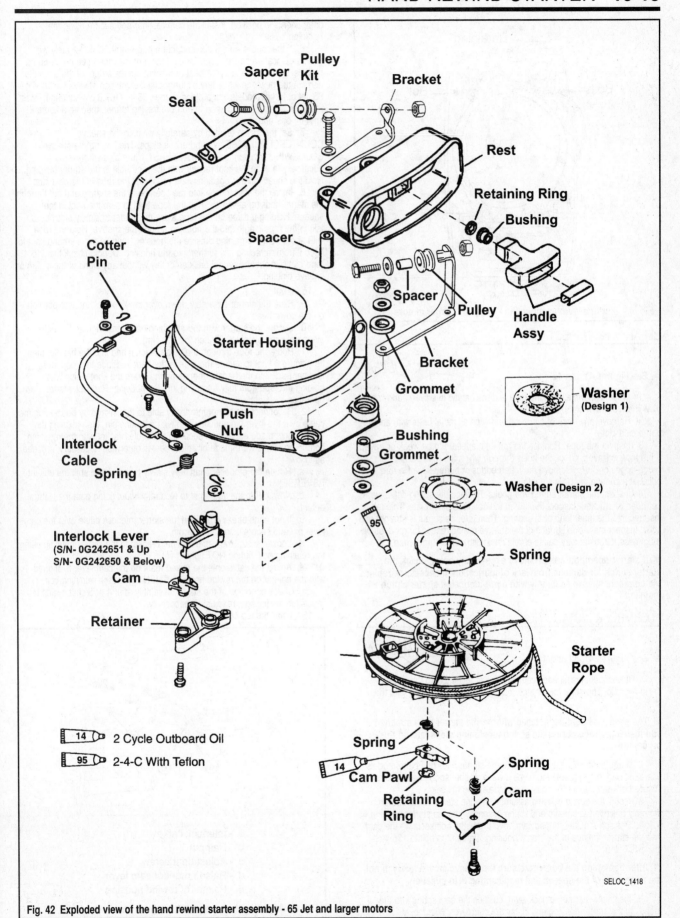

Fig. 42 Exploded view of the hand rewind starter assembly - 65 Jet and larger motors

SELOC_1418

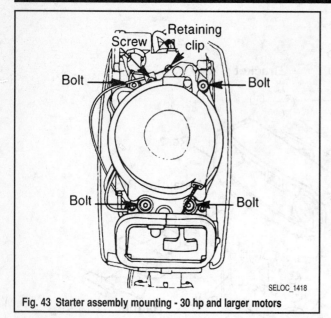

Fig. 43 Starter assembly mounting - 30 hp and larger motors

CLEANING & INSPECTION

◆ See Figures 41 and 42

MODERATE

1. Wash all parts except the rope and the handle in solvent, and then blow them dry with compressed air.

2. Remove any trace of corrosion and wipe all metal parts with an oil-dampened cloth.

3. Inspect the rope. Replace the rope if it appears to be weak or frayed. If the rope is frayed, check the holes through which the rope passes for rough edges or burrs. Remove the rough edges or burrs with a file and polish the surface until it is smooth.

4. Inspect the starter spring end hooks. Replace the spring if it is weak, corroded or cracked. Inspect the tab on the spring retainer plate. This tab is inserted into the inner loop of the spring. Therefore, be sure it is straight and solid. Inspect the inside surface of the sheave rewind recess for grooves or roughness. Grooves may cause erratic rewinding of the starter rope.

■ If starter operation was erratic or excessively noisy, check the starter clutch for damage from lack of lubrication. If necessary, replace the complete sheave assembly with a pre-lubricated starter clutch installed.

ASSEMBLY & INSTALLATION

◆ See Figures 41, 42 and 44

MODERATE

1. If removed, install each cam and spring assembly to the sheave using a circlip. The star-shaped inner cam will be installed along with the sheave.

■ The sheave should be installed without the starter rope attached, you install the rope during the spring tensioning steps later in this procedure.

2. If equipped, install the felt pad to the starter housing (some models use a felt pad, but probably more use a wave washer/spring washer that installs between the starter housing and the spring housing.

3. Install the spring housing assembly into the starter sheave, then carefully position the sheave and spring assembly into the starter housing.

4. Position the star shaped cam, then install the bolt securing the cam and the starter sheave to the starter housing. Tighten the bolt to 135 inch lbs. (15.3 Nm).

■ After tightening the bolt, make sure the sheave moves freely. If not, the alignment of the sheave and recoil spring is in question.

5. Install the starter interlock lever, position the cam spring into the recess in the starter housing and install the cam lever. With these components in position, install the cam retainer and secure using the retaining bolts.

6. If the rope is being replaced, cut a new length. Unfortunately we could not locate a specification for the length of the rope used on 30-60 hp models, so for those, do your best to reconstruct the length of the old rope and cut the new one just a tiny bit longer (since you can always trim it). On 65 Jet or larger models, cut a piece of rope 66 in. (167.6 cm) in length. Melt the ends with a flame to keep them from fraying further, then tie a figure 8 knot on one end.

7. Set the spring tension by carefully winding the sheave COUNTERCLOCKWISE by hand until it stops. Then, bring the sheave CLOCKWISE with spring tension AT LEAST 1 turn plus whatever is necessary to align the knot recess with the rope hole in the starter housing. Hold the sheave in this position while you install the rope in the next step.

8. Install the starter rope into the opening in the sheave and out through the starter housing assembly. Pull the rope tight to seat the knot in the sheave. Holding tension on the rope allow the starter to slowly unwind, taking the rope length into the housing around the sheave. Hold the rope with about 12 in. remaining outside the housing, then tie an overhand knot to keep it from retracting any further into the housing. Double-check that the rope can pull out and retract back smoothly without any signs of the spring or sheave binding.

To install:

9. Slide the starter assembly down over the flywheel and into position on the powerhead.

10. Secure the starter with the attaching hardware. Tighten the bolts alternately and evenly to 90 inch lbs. (10 Nm).

11. Route the rope through the guide bracket and into the handle, then tie a figure 8 knot to secure the rope in the handle. Place the rope and retainer back into the handle, then untie the overhand knot which was holding the rope from fully retracting into the housing. Recheck starter operation.

12. Reconnect the starter interlock cable to the linkage by positioning the cable over the lever pin and securing with a locking pin, then aligning the matchmarks made earlier and securing the cable bracket with the retaining bolt. IF no marks were made or adjustment is otherwise necessary, proceed as follows:

a. Rotate the propeller shaft very slowly by hand and shift the unit into REVERSE.

b. Now, return the gear shift to Neutral *without* going past the Neutral detent.

c. If not done already, position the starter interlock cable over the came lever pin and secure using a cotter pin.

d. Loosely install the cable retainer bracket to the starter housing using the retaining bolt, but do NOT tighten yet.

e. Adjust the cable until the raised mark on the cam lever is aligned with the pointer on the rewind housing, THEN tighten the retaining bolt.

f. Check operation of the starter assembly after 4 or 5 shift cycles to reposition the linkage, readjust IF necessary.

13. Install the top cowl.

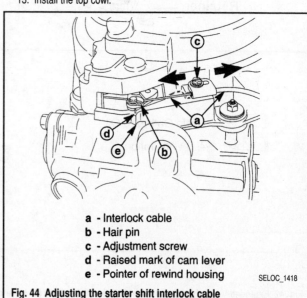

a - Interlock cable
b - Hair pin
c - Adjustment screw
d - Raised mark of cam lever
e - Pointer of rewind housing

SELOC_1418

Fig. 44 Adjusting the starter shift interlock cable